Dietary Reference Intakes (DRIs): Recommended Intakes for Individuals, Elements
Food and Nutrition Board, Institute of Medicine, National Academies

Life Stage Group	Calcium (mg/d)	Chromium (µg/d)	Copper (µg/d)	Fluoride (mg/d)	Iodine (µg/d)	Iron (mg/d)	Magnesium (mg/d)	Manganese (mg/d)	Molybdenum (µg/d)	Phosphorus (mg/d)	Selenium (µg/d)	Zinc (mg/d)	Potassium (g/d)	Sodium (g/d)	Chloride (g/d)
Infants															
0–6 mo	210*	0.2*	200*	0.01*	110*	0.27*	30*	0.003*	2*	100*	15*	2*	0.4*	0.12*	0.18*
7–12 mo	270*	5.5*	220*	0.5*	130*	11	75*	0.6*	3*	275*	20*	3	0.7*	0.37*	0.57*
Children															
1–3 y	500*	11*	340	0.7*	90	7	80	1.2*	17	460	20	3	3.0*	1.0*	1.5*
4–8 y	800*	15*	440	1*	90	10	130	1.5*	22	500	30	5	3.8*	1.2*	1.9*
Males															
9–13 y	1300*	25*	700	2*	120	8	240	1.9*	34	1250	40	8	4.5*	1.5*	2.3*
14–18 y	1300*	35*	890	3*	150	11	410	2.2*	43	1250	55	11	4.7*	1.5*	2.3*
19–30 y	1000*	35*	900	4*	150	8	400	2.3*	45	700	55	11	4.7*	1.5*	2.3*
31–50 y	1000*	35*	900	4*	150	8	420	2.3*	45	700	55	11	4.7*	1.5*	2.3*
51–70 y	1200*	30*	900	4*	150	8	420	2.3*	45	700	55	11	4.7*	1.3*	2.0*
>70 y	1200*	30*	900	4*	150	8	420	2.3*	45	700	55	11	4.7*	1.2*	1.8*
Females															
9–13 y	1300*	21*	700	2*	120	8	240	1.6*	34	1250	40	8	4.5*	1.5*	2.3*
14–18 y	1300*	24*	890	3*	150	15	360	1.6*	43	1250	55	9	4.7*	1.5*	2.3*
19–30 y	1000*	25*	900	3*	150	18	310	1.8*	45	700	55	8	4.7*	1.5*	2.3*
31–50 y	1000*	25*	900	3*	150	18	320	1.8*	45	700	55	8	4.7*	1.5*	2.3*
51–70 y	1200*	20*	900	3*	150	8	320	1.8*	45	700	55	8	4.7*	1.3*	2.0*
>70 y	1200*	20*	900	3*	150	8	320	1.8*	45	700	55	8	4.7*	1.2*	1.8*
Pregnancy															
14–18 y	1300*	29*	1000	3*	220	27	400	2.0*	50	1250	60	12	4.7*	1.5*	2.3*
19–30 y	1000*	30*	1000	3*	220	27	350	2.0*	50	700	60	11	4.7*	1.5*	2.3*
31–50 y	1000*	30*	1000	3*	220	27	360	2.0*	50	700	60	11	4.7*	1.5*	2.3*
Lactation															
14–18 y	1300*	44*	1300	3*	290	10	360	2.6*	50	1250	70	13	5.1*	1.5*	2.3*
19–30 y	1000*	45*	1300	3*	290	9	310	2.6*	50	700	70	12	5.1*	1.5*	2.3*
31–50 y	1000*	45*	1300	3*	290	9	320	2.6*	50	700	70	12	5.1*	1.5*	2.3*

NOTE: This table presents Recommended Dietary Allowances (RDAs) in **bold type** followed by an asterisk (*), RDAs and AIs may both be used as goals for individual intake. RDAs are set to meet the needs of almost all (97 to 98 percent) individuals in a group. For healthy breastfed infants, the AI is the mean intake. The AI for other life stage and gender group is believed to cover needs of all individuals in the group, but lack of data or uncertainty in the data prevent being able to specify with confidence the percentage of individuals covered by this intake.

SOURCES: *Dietary Reference Intakes for Calcium, Phosphorous, Magnesium, Vitamin D, and Fluoride* (1997); *Dietary Reference Intakes for Thiamin, Riboflavin, Niacin, Vitamin B₆, Folate, Vitamin B₁₂, Pantothenic Acid, Biotin, and Choline* (1998); *Dietary Reference Intakes for Vitamin C, Vitamin E, Selenium, and Carotenoids* (2000); *Dietary Reference Intakes for Vitamin A, Vitamin K, Arsenic, Boron, Chromium, Copper, Iodine, Iron, Manganese, Molybdenum, Nickel, Silicon, Vanadium, and Zinc* (2001); and *Dietary Reference Intake for Water, Potassium, Sodium, Chloride, and Sulfate* (2004). These reports may be accessed via http://www.nap.edu.

Reprinted with permission from the National Academies Press, Copyright © 2000, National Academy of Sciences.

Dietary Reference Intakes (DRIs): Recommended Intakes for Individuals, Macronutrients
Food and Nutrition Board, Institute of Medicine, National Academies

Life Stage Group	Total Water[a] (L/d)	Carbohydrate (g/d)	Total Fiber (g/d)	Fat (g/d)	Linoleic Acid (g/d)	α-Linoleic Acid (g/d)	Protein[b] (g/d)
Infants							
0–6 mo	0.7*	60*	ND	31*	4.4*	0.5*	9.1*
7–12 mo	0.8*	95*	ND	30*	4.6*	0.5*	11.0[c]
Children							
1–3 y	1.3*	130	19*	ND	7*	0.7*	13
4–8 y	1.7*	130	25*	ND	10*	0.9*	19
Males							
9–13 y	2.4*	130	31*	ND	12*	1.2*	34
14–18 y	3.3*	130	38*	ND	16*	1.6*	52
19–30 y	3.7*	130	38*	ND	17*	1.6*	56
31–50 y	3.7*	130	38*	ND	17*	1.6*	56
51–70 y	3.7*	130	30*	ND	14*	1.6*	56
>70 y	3.7*	130	30*	ND	14*	1.6*	56
Females							
9–13 y	2.1*	130	26*	ND	10*	1.0*	34
14–18 y	2.3*	130	26*	ND	11*	1.1*	46
19–30 y	2.7*	130	25*	ND	12*	1.1*	46
31–50 y	2.7*	130	25*	ND	12*	1.1*	46
51–70 y	2.7*	130	21*	ND	11*	1.1*	46
>70 y	2.7*	130	21*	ND	11*	1.1*	46
Pregnancy							
14–18 y	3.0*	175	28*	ND	13*	1.4*	71
19–30 y	3.0*	175	28*	ND	13*	1.4*	71
31–50 y	3.0*	175	28*	ND	13*	1.4*	71
Lactation							
14–18 y	3.8*	210	29*	ND	13*	1.3*	71
19–30 y	3.8*	210	29*	ND	13*	1.3*	71
31–50 y	3.8*	210	29*	ND	13*	1.3*	71

NOTE: This table presents Recommended Dietary Allowances (RDAs) in **bold** type and Adequate Intakes (AIs) in ordinary type followed by an asterisk (*). RDAs and AIs may both be used as goals for individual intake. RDAs are set to meet the needs of almost all (97 to 98 percent) individuals in a group. For healthy infants fed human milk, the AI is the mean intake. The AI for other life stage and gender groups is believed to cover the needs of all individuals in the group, but lack of data or uncertainty in the data prevent being able to specify with confidence the percentage of individuals covered by this intake.

[a] Total water includes all water contained in food, beverages, and drinking water.
[b] Based on 0.8 g/kg body weight for the reference body weight.
[c] Change from 13.5 in prepublication copy due to calculation error.

Dietary Reference Intakes (DRIs): Additional Macronutrient Recommendations
Food and Nutrition Board, Institute of Medicine, National Academies

Macronutrient	Recommendation
Dietary cholesterol	As low as possible while consuming a nutritionally adequate diet
Trans fatty acids	As low as possible while consuming a nutritionally adequate diet
Saturated fatty acids	As low as possible while consuming a nutritionally adequate diet
Added sugars	Limit to no more than 25% of total energy

SOURCE: *Dietary Reference Intakes for Energy, Carbohydrate, Fiber, Fat, Fatty Acids, Cholesterol, Protein, and Amino Acids* (2002).

Nutrition

Through the Life Cycle

FOURTH EDITION

Nutrition
Through the Life Cycle
FOURTH EDITION

Judith E. Brown
Ph.D., M.P.H., R.D.
University of Minnesota

with

Janet S. Isaacs, Ph.D., R.D.
Nutrition Consultant Raleigh, North Carolina

U. Beate Krinke, Ph.D., M.P.H., R.D.
University of Minnesota

Ellen Lechtenberg, R.D., IBCLC
Primary Children's Medical Center

Maureen A. Murtaugh, Ph.D., R.D.
University of Utah School of Medicine

Carolyn Sharbaugh, M.S., R.D.
Nutrition Consultant

Patricia L. Splett, Ph.D., R.D., M.P.H.
Nutrition Consultant

Jamie Stang, Ph.D., M.P.H., R.D.
University of Minnesota

Nancy H. Wooldridge, M.S., R.D., L.D.
University of Alabama at Birmingham

WADSWORTH
CENGAGE Learning

Australia • Brazil • Japan • Korea • Mexico • Singapore • Spain • United Kingdom • United States

WADSWORTH
CENGAGE Learning™

Nutrition Through the Life Cycle,
Fourth Edition
Judith E. Brown, Janet S. Isaacs, U. Beate Krinke,
Ellen Lechtenberg, Maureen A. Murtaugh,
Carolyn Sharbaugh, Patricia L. Splett,
Jamie Stang, Nancy H. Wooldridge

Senior Acquisitions Editor: Peggy Williams

Senior Developmental Editor: Nedah Rose

Assistant Editor: Elesha Feldman

Editorial Assistant: Alexis Glubka

Media Editor: Miriam Myers

Senior Marketing Manager: Laura McGinn

Marketing Assistant: Elizabeth Wong

Senior Marketing Communications Manager:
Linda Yip

Content Project Management: Pre-PressPMG

Creative Director: Rob Hugel

Senior Art Director: John Walker

Print Buyer: Linda Hsu

Rights Acquisitions Account Manager, Text:
Roberta Broyer

Rights Acquisitions Account Manager, Image:
Dean Dauphinais

Production Service: Pre-PressPMG

Photo Researcher: Pre-PressPMG

Cover Designer: Riezebos/Holzbaur: Bill Alexander

Cover Images: Front cover (top to bottom):
GettyImages/Image Source; GettyImages/Fabrice
LEROUGE; Corbis 1/© Randy Faris; GettyImages/
Hola Images; Corbis/Bruderer/Blend Images;
Corbis/© Juice Images; GettyImages/PhotoAlto/
Michele Constantini; Photo Library/Pierre Bourrier;
Back cover (top to bottom): Corbis/© Juice Images;
GettyImages/Image Source; Corbis/Bruderer/Blend
Images; GettyImages/Fabrice LEROUGE

Compositor: Pre-PressPMG

For product information and technology assistance, contact us at
Cengage Learning Customer & Sales Support, 1-800-354-9706

For permission to use material from this text or product,
submit all requests online at **cengage.com/permissions**
Further permissions questions can be emailed to
permissionrequest@cengage.com

Library of Congress Control Number: 2010926933

ISBN-13: 978-0-538-73341-0

ISBN-10: 0-538-73341-1

Wadsworth
20 Davis Drive
Belmont, CA 94002-3098
USA

Cengage Learning is a leading provider of customized learning solutions with office locations around the globe, including Singapore, the United Kingdom, Australia, Mexico, Brazil, and Japan. Locate your local office at: **www.cengage.com/global**

Cengage Learning products are represented in Canada by Nelson Education, Ltd.

To learn more about Wadsworth, visit **www.cengage.com/wadsworth.**

Purchase any of our products at your local college store or at our preferred online store **www.cengagebrain.com**

Printed in the United States of America
2 3 4 5 6 7 14 13 12 11

Contents in Brief

Preface   *xxi*

Chapter 1
Nutrition Basics   1

Chapter 2
Preconception Nutrition   51

Chapter 3
Preconception Nutrition   70
CONDITIONS AND INTERVENTIONS

Chapter 4
Nutrition During Pregnancy   87

Chapter 5
Nutrition During Pregnancy   134
CONDITIONS AND INTERVENTIONS

Chapter 6
Nutrition During Lactation   159

Chapter 7
Nutrition During Lactation   193
CONDITIONS AND INTERVENTIONS

Chapter 8
Infant Nutrition   222

Chapter 9
Infant Nutrition   247
CONDITIONS AND INTERVENTIONS

Chapter 10
Toddler and Preschooler Nutrition   266

Chapter 11
Toddler and Preschooler Nutrition   296
CONDITIONS AND INTERVENTIONS

Chapter 12
Child and Preadolescent Nutrition   310

Chapter 13
Child and Preadolescent Nutrition   338
CONDITIONS AND INTERVENTIONS

Chapter 14
Adolescent Nutrition   356

Chapter 15
Adolescent Nutrition   385
CONDITIONS AND INTERVENTIONS

Chapter 16
Adult Nutrition   405

Chapter 17
Adult Nutrition   428
CONDITIONS AND INTERVENTIONS

Chapter 18
Nutrition and Older Adults   454

Chapter 19
Nutrition and Older Adults   486
CONDITIONS AND INTERVENTIONS

Answers to Review Questions   AR-1

Appendix A
Summary of Research of Effects of Exercise Activities on Health of Older Adults   A-1

Appendix B
Measurement Abbreviations and Equivalents   A-3

Appendix C
Body Mass Index (BMI)   A-5

Appendix D
Carbohydrate Counting for Type 1 Diabetes   A-6

References   R-1

Glossary   G-1

Index   I-1

Contents

Preface   xxi

All chapters include Resources and References

Chapter 1
Nutrition Basics   1

Introduction   2

Principles of the Science of Nutrition   2
Essential and Nonessential Nutrients   3
Dietary Intake Standards   4
Carbohydrates   4
Protein   6
Fats (Lipids)   8
Vitamins   13
Other Substances in Food   13
Minerals   22
Water   22

Nutritional Labeling   35
Nutrition Facts Panel   35
Ingredient Label   35
Dietary Supplement Labeling   36
Herbal Remedies   36
Functional Foods   36

The Life-Course Approach to Nutrition and Health   38
Meeting Nutritional Needs Across the Life Cycle   38
Dietary Considerations Based on Ethnicity   38
Dietary Considerations Based on Religion   38

Nutritional Assessment   39
Community-Level Assessment   39
Individual-Level Nutritional Assessment   39
Dietary Assessment   39
Anthropometric Assessment   41
Biochemical Assessment   41

Monitoring the Nation's Nutritional Health   42

Public Food and Nutrition Programs   42
WIC   42

Nationwide Priorities for Improvements in Nutritional Health   43
U.S. Nutrition and Health Guidelines   44

Chapter 2
Preconception Nutrition   51

Introduction   52

Preconception Overview   52
2010 Nutrition Objectives for the Nation Related to the Preconceptional Period   52

Reproductive Physiology   53
Female Reproductive System   53
Male Reproductive System   55

Sources of Disruptions in Fertility   56

Nutrition-Related Disruptions in Fertility   56
Undernutrition and Fertility   57
Body Fat and Fertility   58

Nutrition Time Line

1621
First Thanksgiving
feast at Plymouth
colony

H. Armstrong Roberts/
ClassicStock/Alamy

1702
First coffeehouse
in America opens
in Philadelphia

Photodisc

1734
Scurvy recognized

Exercise and Infertility   59
Oxidative Stress, Antioxidant Nutrient Status, and
    Fertility   60
Multivitamin Supplement, Folate Intake,
    and Fertility   61
Caffeine and Fertility   61

Nutrition and Contraceptives   62
Oral Contraceptives and Nutritional Status   63
Contraceptive Injections   63
Contraceptive Implants   63
Contraceptive Patches   63
Emerging Forms of Contraceptives   63

Other Preconceptional Nutrition Concerns   64
Very Early Pregnancy Nutrition Exposures   64
Recommended Dietary Intakes for Preconceptional
    Women   65

Model Preconceptional Nutrition Programs   65
Preconceptional Benefits of WIC   66
Decreasing Iron Deficiency in Preconceptional
    Women in Indonesia   66
Preconception Care: Preparing for Pregnancy   66

**Nutrition Programs and Services Delivery Before
Pregnancy   66**
The Nutrition Care Process   66

Chapter 3

Preconception Nutrition   70
Conditions and Interventions

Introduction   71

Premenstrual Syndrome   71
Caffeine Intake and PMS   71
Exercise and Stress Reduction   71
Magnesium, Calcium, Vitamin D, and Vitamin B_6 Supplements
    and PMS Symptoms   72

Obesity and Fertility   72
Central Body Fat and Fertility   73
Weight Loss and Fertility   73

Hypothalamic Amenorrhea   74
Nutritional Management of Hypothalamic
    Amenorrhea   75

The Female Athlete Triad and Fertility   75
Nutritional Management of the Female
    Athlete Triad   75

Eating Disorders and Fertility   75
Nutritional Management of Women with Anorexia
    Nervosa or Bulimia Nervosa   75

Diabetes Mellitus Prior to Pregnancy   76
Nutritional Management of Type 1 Diabetes   76
Nutritional Management of Type 2 Diabetes   76
Other Components of the Nutritional Management
    of Type 2 Diabetes   77
Reducing the Risk of Type 2 Diabetes   78

Polycystic Ovary Syndrome   78
Nutritional Management of Women with PCOS   79

Disorders of Metabolism   79
Phenylketonuria (PKU)   79
Nutritional Management of PKU   81
Celiac Disease   81
Nutritional Management of Celiac Disease   81

**Herbal Remedies for Fertility-Related
Problems   82**

Chapter 4

Nutrition During   87
Pregnancy

Introduction   88

The Status of Pregnancy Outcomes   88
Infant Mortality   89
Low Birth Weight, Preterm Delivery, and Infant
    Mortality   90

Reducing Infant Mortality and Morbidity   90
Health Objectives for the Year 2010   90

Nutrition Time Line ⟶

John A. Rizzo/ Photodisc/Getty Images

1744

First record of ice cream in America at Maryland colony

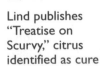
Photodisc

1747

Lind publishes "Treatise on Scurvy," citrus identified as cure

1750

Ojibway and Sioux war over control of wild rice stands

1762

Sandwich invented by the Earl of Sandwich

C Squared Studios/ Photodisc/Getty Images

Physiology of Pregnancy 91
Maternal Physiology 91
Normal Physiological Changes During Pregnancy 92
The Placenta 96

Embryonic and Fetal Growth and Development 97
Critical Periods of Growth and Development 97
Fetal Body Composition 99
Variation in Fetal Growth 99
Nutrition, Miscarriages, and Preterm Delivery 102

The Fetal-Origins Hypothesis of Later Disease Risk 102
Mechanisms Underlying the Fetal-Origins
Hypothesis 103
Limitations of the Fetal-Origins Hypothesis 104

Pregnancy Weight Gain 104
Pregnancy Weight Gain Recommendations 104
Composition of Weight Gain in Pregnancy 107
Postpartum Weight Retention 107

**Nutrition and the Course and Outcome of
Pregnancy 107**
Famine and Pregnancy Outcome 107
Contemporary Prenatal Nutrition Research Results 109

Nutrient Needs During Pregnancy 109
The Need for Energy 109
The Need for Carbohydrates 110
Alcohol and Pregnancy Outcome 110
The Need for Protein 110
Vegetarian Diets in Pregnancy 111

The Need for Fat 112
Omega-3 Fatty Acids EPA and DHA During
Pregnancy 113
The Need for Water 114

**The Need for Vitamins and Minerals During
Pregnancy 114**
Folate 114
Folate and Congenital Abnormalities 115
Choline 116
Vitamin A 116
Vitamin D 117

The Need for Minerals During Pregnancy 118
Calcium 118
Fluoride 118

Iron 119
Iodine 121
Sodium 121

Bioactive Components of Food 122
Caffeine 122

Healthy Diets for Pregnancy 122
Effect of Taste and Smell Changes on Dietary Intake
During Pregnancy 123
Pica 123
Assessment of Nutritional Status During Pregnancy 124
Dietary Assessment During Pregnancy 124
Nutrition Biomarker Assessment 125

Dietary Supplements During Pregnancy 125
Multivitamin and Mineral Prenatal Supplements 126
Herbal Remedies and Pregnancy 127

Exercise and Pregnancy Outcome 128
Exercise Recommendations for Pregnant Women 128

Food Safety Issues During Pregnancy 128
Mercury Contamination 129

Common Health Problems During Pregnancy 129
Nausea and Vomiting 129
Heartburn 130
Constipation 130

**Model Nutrition Programs for Risk Reduction in
Pregnancy 130**
The Montreal Diet Dispensary 130
The WIC Program 131

Chapter 5
Nutrition During Pregnancy   134
Conditions and Interventions

Introduction 135

Obesity and Pregnancy 135
Obesity and Infant Outcomes 136
Nutritional Recommendations and Interventions for Obesity
During Pregnancy 136

Nutrition Time Line

1771

Potato heralded
as famine food

1774

Americans drink
more coffee in
protest over
Britain's tea tax

© Stefano Bianchetti/CORBIS

1775

Lavoisier ("the father
of the science of
nutrition") discovers
the energy-producing
property of food

1816

Protein and
amino acids
identified
followed by
carbohydrates
and fats in the
mid 1800s

Hypertensive Disorders of Pregnancy 137
Hypertensive Disorders of Pregnancy, Oxidative Stress, and Nutrition 137
Chronic Hypertension 138
Gestational Hypertension 138
Preeclampsia–Eclampsia 138
Preeclampsia Case Presentation 140
Nutritional Recommendations and Interventions for Preeclampsia 140

Diabetes in Pregnancy 140
Gestational Diabetes 140
Potential Consequences of Gestational Diabetes 141
Risk Factors for Gestational Diabetes 142
Diagnosis of Gestational Diabetes 142
Treatment of Gestational Diabetes 143
Presentation of a Case Study 144
Exercise Benefits and Recommendations 144
Nutritional Management of Women with Gestational Diabetes 145
Low-Glycemic Index (GI) Foods 145
Postpartum Follow-Up 146
Prevention of Gestational Diabetes 146
Type 1 Diabetes During Pregnancy 146

Multifetal Pregnancies 147
Background Information about Multiple Fetuses 148
Risks Associated with Multifetal Pregnancy 149
Interventions and Services for Risk Reduction 149
Nutrition and the Outcome of Multifetal Pregnancy 150
Dietary Intake in Twin Pregnancy 150
Nutritional Recommendations for Women with Multifetal Pregnancy 151

HIV/AIDS During Pregnancy 151
Treatment of HIV/AIDS 152
Consequences of HIV/AIDS During Pregnancy 152
Nutritional Factors and HIV/AIDS During Pregnancy 153
Nutritional Management of Women with HIV/AIDS During Pregnancy 154

Eating Disorders in Pregnancy 154
Consequences of Eating Disorders in Pregnancy 154
Treatment of Women with Eating Disorders During Pregnancy 154

Nutritional Interventions for Women with Eating Disorders 154

Fetal Alcohol Spectrum 155
Effects of Alcohol on Pregnancy Outcome 155
The Fetal Alcohol Syndrome 155

Nutrition and Adolescent Pregnancy 156
Growth During Adolescent Pregnancy 156
Dietary and Other Recommendations for Pregnant Adolescents 156

Evidence-Based Practice 157

Chapter 6
Nutrition During Lactation   159

Introduction 160

Breastfeeding Goals for the United States 160

Lactation Physiology 161
Functional Units of the Mammary Gland 161
Mammary Gland Development 161
Lactogenesis 162
Hormonal Control of Lactation 162
Secretion of Milk 163
The Letdown Reflex 164

Human Milk Composition 164
Colostrum 164
Water 165
Energy 165
Lipids 166
Protein 166
Milk Carbohydrates 167
Fat-Soluble Vitamins 167
Water-Soluble Vitamins 167
Minerals in Human Milk 167
Taste of Human Milk 168

Benefits of Breastfeeding 168
Breastfeeding Benefits for Mothers 168
Breastfeeding Benefits for Infants 169

Nutrition Time Line

1833
Beaumont's experiments on a wounded man's stomach greatly expands knowledge about digestion

1871
Proteins, carbohydrates, and fats determined to be insufficient to support life; that there are other "essential" components

1895
First milk station providing children with uncontaminated milk opens in New York City

Breast Milk Supply and Demand 171
  Can Women Make Enough Milk? 171
  Does the Size of the Breast Limit a Woman's Ability to Nurse
    Her Infant? 171
  Is Feeding Frequency Related to the Amount of Milk a
    Woman Can Make? 171
  Pumping or Expressing Milk 172
  Can Women Breastfeed after Breast Reduction or
    Augmentation Surgery? 172
  What Is the Effect of Silicone Breast Implants on
    Breastfeeding? 172

The Breastfeeding Infant 172
  Optimal Duration of Breastfeeding 172
  Reflexes 173
  Preparing the Breast for Breastfeeding 173
  Breastfeeding Positioning 173
  Presenting the Breast to the Suckling Infant 173
  Mechanics of Breastfeeding 174
  Identifying Hunger and Satiety 175
  Feeding Frequency 175
  Vitamin Supplements for Breastfeeding Infants 175
  Identifying Breastfeeding Malnutrition 175
  Tooth Decay 176

Maternal Diet 177
  Nutrition Assessment of Breastfeeding Women 177
  Energy and Nutrient Needs 178

Maternal Energy Balance and Milk Composition 179
  Weight Loss During Breastfeeding 179
  Exercise and Breastfeeding 179
  Vitamin and Mineral Supplements 180
  Vitamin and Mineral Intakes 180
  Functional Foods 180
  Fluids 180
  Alternative Diets 180
  Infant Colic 180

**Factors Influencing Breastfeeding Initiation and
Duration** 181
  Obesity and Breastfeeding 181
  Socioeconomic 181

Breastfeeding Promotion, Facilitation, and Support 181
  Role of the Health Care System in Supporting
    Breastfeeding 182
  Prenatal Breastfeeding Education and Support 182

Lactation Support in Hospitals and Birthing Centers 183
Lactation Support After Discharge 185
The Workplace 186
The Community 187

Public Food and Nutrition Programs 187
  National Breastfeeding Policy 187
  USDA WIC Program 188

Model Breastfeeding Promotion Programs 188
  WIC National Breastfeeding Promotion Project—Loving
    Support Makes Breastfeeding Work 188
  Wellstart International 189

Chapter 7
Nutrition During Lactation   193
Conditions and Interventions

Introduction 194

Common Breastfeeding Conditions 194
  Sore Nipples 194
  Flat or Inverted Nipples 194
  Letdown Failure 195
  Hyperactive Letdown 195
  Hyperlactation 195
  Engorgement 195
  Plugged Duct 196
  Mastitis 196
  Low Milk Supply 198

Maternal Medications 198

Herbal Remedies 200
  Specific Herbs Used in the United States 201
  Milk Thistle/Blessed Thistle 203

Alcohol and Other Drugs and Exposures 203
  Alcohol 203
  Nicotine (Smoking Cigarettes) 205
  Marijuana 205
  Caffeine 206
  Other Drugs of Abuse 206
  Environmental Exposures 206

Nutrition Time Line

1896
Atwater publishes
*Proximate Composition of
Food Materials*

© Bettmann/CORBIS

1906
Pure Food and Drug Act passed
by President Theodore Roosevelt
to protect consumers against
contaminated foods

1910
Pasteurized milk
introduced

Jonelle Weaver/Photodisc/
Getty Images

1912
Funk suggested
scurvy, beriberi,
and pellagra
caused by
deficiency of
"vitamines" in
the diet

Neonatal Jaundice and Kernicterus 208
Bilirubin Metabolism 208
Physiologic Versus Pathologic Newborn Jaundice 209
Hyperbilirubinemia and Breastfeeding 209
Prevention and Treatment for Severe Jaundice 211
Information for Parents 212

Breastfeeding Multiples 212

Infant Allergies 212
Food Intolerance 213

Late-Preterm Infants 214

Human Milk and Preterm Infants 215

Medical Contraindications to Breastfeeding 216
Breastfeeding and HIV Infection 216

Human Milk Collection and Storage 218
Milk Banking 218

Model Programs 219
Breastfeeding Promotion in Physicians' Office Practices
(BPPOP) 219
The Rush Mothers' Milk Club 219

Chapter 8
Infant Nutrition   222

Introduction 223

Assessing Newborn Health 223
Birth Weight as an Outcome 223
Infant Mortality 223
Combating Infant Mortality 224
Standard Newborn Growth Assessment 224

Infant Development 224
Motor Development 225
Critical Periods 225
Cognitive Development 225
Digestive System Development 226
Parenting 227
Energy and Nutrient Needs 228

Caloric Needs 228
Protein Needs 228
Fats 228
Metabolic Rate, Calories, Fats, and Protein—How Do
They All Tie Together? 229
Other Nutrients and Non-nutrients 229

Physical Growth Assessment 230
Interpretation of Growth Data 231

Feeding in Early Infancy 231
Breast Milk and Formula 231
Cow's Milk During Infancy 233
Soy Protein-Based Formulas During Infancy 233

Development of Infant Feeding Skills 233
Introduction of Solid Foods 234
The Importance of Infant Feeding Position 236
Preparing for Drinking from a Cup 236
Food Texture and Development 236
First Foods 237
Inappropriate and Unsafe Food Choices 238
Water 238
How Much Food Is Enough for Infants? 238
How Infants Learn Food Preferences 239

Nutrition Guidance 239
Infants and Exercise 239
Supplements for Infants 239

Common Nutritional Problems and Concerns 240
Failure to Thrive 240
Nutrition Intervention for Failure to Thrive 241
Colic 241
Iron-Deficiency Anemia 241
Diarrhea and Constipation 242
Prevention of Baby-Bottle Caries and Ear Infections 242
Food Allergies and Intolerances 242
Lactose Intolerance 243

Cross-Cultural Considerations 243

Vegetarian Diets 243

Nutrition Intervention for Risk Reduction 244
Model Program: Newborn Screening 244

Nutrition Time Line ————————————————▶

C Squared Studios/
Photodisc/Getty
Images

1913

First vitamin
discovered (vitamin A)

1914

Goldberger identifies the
cause of pellagra (niacin
deficiency) in poor
children to be a missing
component of the diet
rather than a germ as
others believed

1916

First dietary guidance
material produced for
the public was released.
It was titled "Food for
Young Children."

1917

First food groups
published, The Five Food
Groups: Milk and Meat;
Vegetables and Fruits;
Cereals; Fats
and Fat Foods; Sugars
and Sugary Foods

Chapter 9
Infant Nutrition   247
Conditions and Interventions

Introduction   248

Infants at Risk   248
Families of Infants with Special Health Care Needs   249

Energy and Nutrient Needs   249
Energy Needs   249
Protein Requirements   250
Fats   250
Vitamins and Minerals   251

Growth   251
Growth in Preterm Infants   252
Does Intrauterine Growth Predict Growth
   Outside?   252
Interpretation of Growth   253

Nutrition for Infants with Special Health Care Needs   254
Nutrition Risks to Development   254

Severe Preterm Birth and Nutrition   255
How Sick Babies Are Fed   256
What to Feed Preterm Infants   256
Preterm Infants and Feeding   257

Infants with Congenital Anomalies and Chronic Illness   257
Infants with Genetic Disorders   260

Feeding Problems   261

Nutrition Interventions   261

Nutrition Services   263

Chapter 10
Toddler and Preschooler   266
Nutrition

Introduction   267
Definitions of the Life-Cycle Stage   267
Importance of Nutrition   267

Tracking Toddler and Preschooler Health   267
Healthy People 2010   267

Normal Growth and Development   267
Measuring Growth   269
The 2000 CDC Growth Charts   269
WHO Growth Standards   271
Common Problems with Measuring and Plotting Growth
   Data   271

Physiological and Cognitive Development   271
Toddlers   271
Preschool-Age Children   273
Temperament Differences   275
Food Preference Development, Appetite, and Satiety   275

Energy and Nutrient Needs   277
Energy Needs   277
Protein   278
Vitamins and Minerals   278

Common Nutrition Problems   278
Iron-Deficiency Anemia   278
Dental Caries   279
Constipation   280
Elevated Blood Lead Levels   280
Food Security   281
Food Safety   281

Prevention of Nutrition-Related Disorders   282
Overweight and Obesity in Toddlers and Preschoolers   282
Assessment of Overweight and Obesity   282
Prevention of Overweight and Obesity   283
Treatment of Overweight and Obesity Expert Committee
   Recommendations   283
Dietary Guidelines for Americans 2005   284
Nutrition and Prevention of Cardiovascular Disease in
   Toddlers and Preschoolers   284
Vitamin and Mineral Supplements   285
Herbal Supplements   286

Dietary and Physical Activity Recommendations   286
Dietary Guidelines   286
Food Guide Pyramid   287
Recommendations for Intake of Iron, Fiber, Fat, and
   Calcium   287
Fluids   289
Recommended vs. Actual Food Intake   289

Nutrition Time Line

1921
First fortified food produced: Iodized salt. It was needed to prevent widespread iodine deficiency goiter in many parts of the United States

Leonard Lessin/Photolibrary

1928
American Society for Nutritional Sciences and the *Journal of Nutrition* founded

1929
Essential fatty acids identified

Photodisc

Cross-Cultural Considerations   290
Vegetarian Diets   290
Child Care Nutrition Standards   291
Physical Activity Recommendations   291

Nutrition Intervention for Risk Reduction   291
Nutrition Assessment   291
Model Program   291

Public Food and Nutrition Programs   292
WIC   292
WIC's Farmers' Market Nutrition Program   293
Head Start and Early Head Start   293
Supplemental Nutrition Assistance Program (formerly the
    Food Stamp Program)   293

Chapter 11
Toddler and Preschooler   296
Nutrition
Conditions and Interventions

Introduction   297

**Who Are Children with Special Health Care
    Needs?   297**

**Nutrition Needs of Toddlers and Preschoolers with
    Chronic Conditions   298**

Growth Assessment   299

Feeding Problems   300
Behavioral Feeding Problems   300
Excessive Fluid Intake   301
Feeding Problems and Food Safety   302
Feeding Problems from Disabilities Involving Neuromuscular
    Control   302

Nutrition-Related Conditions   302
Failure to Thrive   302
Toddler Diarrhea and Celiac Disease   303
Autism   304
Muscle Coordination Problems and Cerebral Palsy   304
Pulmonary Problems   306
Developmental Delay and Evaluations   306

Food Allergies and Intolerance   306

Dietary Supplements and Herbal Remedies   307

Sources of Nutrition Services   307

Chapter 12
Child and Preadolescent   310
Nutrition

Introduction   311
Definitions of the Life Cycle Stage   311
Importance of Nutrition   311

Tracking Child and Preadolescent Health   311
Healthy People 2010   312

Normal Growth and Development   313
The 2000 CDC Growth Charts   313
WHO Growth References   313

**Physiological and Cognitive Development of
    School-Age Children   313**
Physiological Development   313
Cognitive Development   314
Development of Feeding Skills and Eating
    Behaviors   315

**Energy and Nutrient Needs of School-Age
    Children   317**
Energy Needs   317
Protein   317
Vitamins and Minerals   318

Common Nutrition Problems   318
Iron Deficiency   318
Dental Caries   318

**Prevention of Nutrition-Related Disorders in
    School-Age Children   319**
Overweight and Obesity in School-Age Children   319
Addressing the Problem of Pediatric Overweight and
    Obesity   321
Nutrition and Prevention of Cardiovascular Disease in
    School-Age Children   322
Dietary Supplements   324

Nutrition Time Line

1930s

Vitamin C identified
in 1932, followed
by pantothenic acid
and riboflavin in
1933, and vitamin K
in 1934

Photodisc

1937

Pellagra found
to be due to
a deficiency
of niacin

1941

First refined
grain-enrichment
standards developed

Dietary Recommendations 324
Recommendations for Intake of Iron, Fiber, Fat, Calcium,
  Vitamin D and Fluids 324
Recommended vs. Actual Food Intake 326
Cross-Cultural Considerations 327
Vegetarian Diets 327

Physical Activity Recommendations 327
Recommendations vs. Actual Activity 327
Determinants of Physical Activity 328
Organized Sports 328

**Nutrition Intervention for Risk
  Reduction 328**
Nutrition Education 328
Nutrition Integrity in Schools 329
Nutrition Assessment 330
Model Programs 330

Public Food and Nutrition Programs 331
The National School Lunch Program 332
School Breakfast Program 333
Summer Food Service Program 333
Team Nutrition 334

Chapter 13
Child and Preadolescent   338
Nutrition
Conditions and Interventions

Introduction 339

**"Children Are Children First"—What Does that
  Mean? 339**

**Nutritional Requirements of Children with Special
  Health Care Needs 340**
Energy Needs 340
Protein Needs 340
Other Nutrients 340

Growth Assessment 341
Growth Assessment and Interpretation in Children with
  Chronic Conditions 341
Body Composition and Growth 341

Nutrition Recommendations 342
Methods of Meeting Nutritional Requirements 344
Fluids 345

**Eating and Feeding Problems in Children with Special
  Health Care Needs 345**
Specific Disorders 346

Dietary Supplements and Herbal Remedies 351

Sources of Nutrition Services 351
USDA Child Nutrition Program 351
Maternal and Child Health Block Program of the U.S.
  Department of Health and Human Services (HHS) 351
Public School Regulations: 504 Accommodation and
  IDEA 353
Nutrition Intervention Model Program 353

Chapter 14
Adolescent Nutrition   356

Introduction 357

Nutritional Needs in a Time of Change 357

Normal Physical Growth and Development 358
Changes in Weight, Body Composition, and Skeletal Mass 359

Normal Psychosocial Development 360

**Health and Eating-Related Behaviors During
  Adolescence 361**
Vegetarian Diets 363
Dietary Intake and Adequacy Among Adolescents 365

**Energy and Nutrient Requirements of
  Adolescents 365**
Energy 367
Protein 368
Carbohydrates 368
Dietary Fiber 369
Fat 369
Calcium 369
Iron 370
Vitamin D 370
Folate 371
Vitamin C 372

Nutrition Time Line

1941
First Recommended
Dietary Allowances (RDAs)
announced by President
Franklin Roosevelt on radio

AP Photo

1946
National School
Lunch Act passed

David Buffington/
Photodisc/Getty Images

1947
Vitamin B_{12}
identified

Nutrition Screening, Assessment, and Intervention 372
Nutrition Education and Counseling 373

Physical Activity and Sports 378
Factors Affecting Physical Activity 378

Promoting Healthy Eating and Physical Activity Behaviors 380
Effective Nutrition Messages for Youth 380
Parent Involvement 380
School Programs 380
Community Involvement in Nutritionally Supportive Environments 383

Chapter 15
Adolescent Nutrition 385
Conditions and Interventions

Introduction 386

Overweight and Obesity 386
Health Implications of Adolescent Overweight and Obesity 387
Assessment and Treatment of Adolescent Overweight and Obesity 387

Supplement Use 389
Vitamin–Mineral Supplements 389
Ergogenic Supplements Used by Teens 390

Nutrition for Adolescent Athletes 391
Fluids and Hydration 391
Special Dietary Practices 391

Substance Use 392

Iron-Deficiency Anemia 392

Cardiovascular Disease 393
Hypertension 393
Hyperlipidemia 394

Dieting, Disordered Eating, and Eating Disorders 395
The Continuum of Eating Concerns and Disorders 395
Dieting Behaviors 395
Body Dissatisfaction 396
Disordered Eating Behaviors 397
Eating Disorders 397

Anorexia Nervosa 397
Bulimia Nervosa 398
Binge-Eating Disorder 399
Etiology of Eating Disorders 400
Treating Eating Disorders 401
Preventing Eating Disorders 402
Children and Adolescents with Chronic Health Conditions 403

Chapter 16
Adult Nutrition 405

Introduction 406
Importance of Nutrition 406

Health Objectives for the Nation 406

Physiological Changes During Adulthood 407
Body Composition Changes in Adults 408
Estimating Energy Needs in Adults 408
Energy Adjustments for Weight Change 409
Age-Related Changes in Energy Expenditure 409
Fad Diets 410
Continuum of Nutritional Health 410
States of Nutritional Health 411
Health Disparities Among Groups of Adults 412

Dietary Recommendations for Adults 413
Dietary Guidelines for Americans 413
Vegetarian Diets 414
Beverage Intake Recommendations 415
Alcoholic Beverages 415
Water Intake Recommendations 416
Effects of Caffeine Intake on Water Need 416
Dietary Supplements and Functional Foods 417

Nutrient Recommendations 417
Risk Nutrients 417

Physical Activity Recommendations 421
Physical Activity, Body Composition, and Metabolic Function 421
Physical Activity Types and Settings 422
Physical Activity and Lifestyle 422
Diet and Physical Activity 422

Nutrition Time Line

1953
Double helix structure of DNA discovered

Photodisc

1965
Food Stamp Act passed, Food Stamp program established

1966
Child Nutrition Act adds school breakfast to the National School Lunch Program

1968
First national nutrition survey in United States launched (the Ten State Nutrition Survey)

Nutrition Intervention for Risk Reduction   422
The Eating Competence Model   423
A Model Health-Promotion Program   424
Public Food and Nutrition Programs   425
Putting It All Together   425

Chapter 17
Adult Nutrition   428
Conditions and Interventions

Introduction   429

Overweight and Obesity   429
Effects of Obesity   430
Etiology of Obesity   430
Screening and Assessment   431
Recommendation for Weight-Management Therapy   432
Nutrition Assessment   432
Nutrition Interventions for Weight Management   432
Weight Loss   432
Medical Nutrition Therapy for Weight
   Management   432
Cognitive Behavioral Therapy for Weight
   Management   432
Physical Activity for Weight Management   434
The Challenge of Weight Maintenance   434
Pharmocotherapy for Weight Loss   434
Bariatric Surgery   435
Cardiovascular Disease   436
Prevalence of CVD   436
Etiology of Atherosclerosis   436
Physiological Effects of Atherosclerosis   436
Risk Factors for CVD   437
Screening and Assessment of CVD   437
Nutrition Interventions for CVD   437
Primary Prevention   437
Medical Nutrition Therapy for CVD   440
Pharmacotherapy of CVD   441

Metabolic Syndrome   442
Introduction   442
Prevalence of Metabolic Syndrome   442
Etiology of Metabolic Syndrome   442
Effects of Metabolic Syndrome   442

Screening and Assessment   443
Nutrition Interventions for Metabolic Syndrome   444

Diabetes Mellitus   444
Prevalence of Diabetes   444
Disparities in the Prevalence of Diabetes   444
Etiology of Diabetes   444
Physiological Effects of Diabetes   445
Prevention of Diabetes Complications   445
Screening and Assessment   445
Nutrition Assessment   445
Interventions for Diabetes   445
Medical Nutrition Therapy for Diabetes   446
ADA Exchange Lists   446
Carbohydrate Counting   447
Self-Monitored Blood Glucose   447
Physical Activity in Diabetes Management   447
Pharmacological Therapy of Type 2 Diabetes   448

Cancer   448
Prevalence of Cancer   448
Physiological Effects of Cancer   448
Etiology of Cancer   449
Risk Factors for Cancer   449
Screening and Assessment   449
Nutrition Interventions for Cancer   450
Alternative Medicine and Cancer Treatment   450

HIV Disease   451
Prevalence of HIV   451
Physiological Effects of HIV   451
Etiology of HIV   451
Assessment   451
Nutrition Interventions for HIV   452

Chapter 18
Nutrition and Older   454
Adults

Introduction   455
What Counts as Old Depends On Who Is Counting   455
Food Matters: Nutrition Contributes to a Long
   and Healthy Life   456

A Picture of the Aging Population: Vital Statistics   456
Global Population Trends: Life Expectancy and Life Span   456

Nutrition Time Line

1970
First Canadian national
nutrition survey launched
(Nutrition Canada
National Survey)

1972
Special Supplemental
Food and Nutrition
Program for Women,
Infants, and Children
(WIC) established

1977
Dietary Goals
for the United
States issued

1978
First Health
Objectives
for the Nation
released

1989
First national
scientific consensus
report on diet and
chronic disease
published

Nutrition: A Component of Health Objectives for
the Older Adult Population   457

Theories of Aging   457
Programmed Aging   457
Wear-and-Tear Theories of Aging   458
Calorie Restriction to Increase Longevity   459

Physiological Changes   459
Body-Composition Changes   459
Changing Sensual Awareness: Taste and Smell, Chewing and
Swallowing, Appetite and Thirst   461

Nutritional Risk Factors   462

**Dietary Recommendations: Pyramids for Older
Adults   465**

Nutrient Recommendations   467
Estimating Energy Needs   467
Nutrient Recommendations for Older Adults: Energy
Sources   469
Recommendations for Fluid   472
Age-Associated Changes: Nutrients of Concern   472
Nutrient Supplements: When, Why, Who, What, and How
Much?   474
Dietary Supplements, Functional Foods, Nutraceuticals, and
Older Adults   476
Nutrient Recommendations: Using the Food Label   478
Cross-Cultural Considerations in Making Dietary
Recommendations   478

Food Safety Recommendations   479

Physical Activity Recommendations   479
Exercise Guidelines   480

**Nutrition Policy and Intervention for Risk
Reduction   480**
Nutrition Education   480

Community Food and Nutrition Programs   482
Nutrition Programs Serving Older Adults   482
Store-to-Door: A Nongovernmental Service that Supports
Aging in Place   482
Senior Nutrition Program: Promoting Socialization and
Improved Nutrition   483
The Promise of Prevention: Health Promotion   483

Chapter 19
Nutrition and Older Adults   486
Conditions and Interventions

Introduction: The Importance of Nutrition   487

Nutrition and Health   487

**Heart Disease: Coronary Heart Disease,
Cerebrovascular Disease, Peripheral Artery
Disease   488**
Prevalence   488
Risk Factors   488
Nutritional Remedies for Cardiovascular
Diseases   488

Stroke   490
Definition   490
Prevalence   490
Etiology   490
Effects of Stroke   490
Risk Factors   490
Nutritional Remedies   491

Hypertension   491
Definition   491
Prevalence   491
Etiology   491
Effects of Hypertension   491
Risk Factors   491
Nutritional Remedies   491

Diabetes   493
Special Concerns for Older Adults   493
Effects of Diabetes   493
Nutritional Interventions   493

Obesity   494
Definition   494
Prevalence   494
Etiology, Effects, and Risk Factors of
Obesity   494
Nutritional Remedies   495

Nutrition Time Line

1997
RDAs expanded to
Dietary Reference
Intakes (DRIs)

1998
Folic acid
fortification
of refined grain
products begins

Jeff Greenberg/Alamy

2003
Sequencing of DNA in the
human genome completed.
Marks beginning of new era
of research in nutrient–gene
interactions

2009
Global epidemics
of obesity and
diabetes threaten
gains in life
expectancy

Osteoporosis 496
Definition 496
Prevalence 496
Etiology 497
Effects of Osteoporosis 497
Risk Factors 498
Nutritional Remedies 498
Other Issues Impacting Nutritional Remedies 499

Oral Health 500

Gastrointestinal Diseases 501
Gastroesophageal Reflux Disease (GERD) 501
Stomach Conditions Affect Nutrient Availability: Vitamin B_{12}
Malabsorption 501
Constipation 503

Inflammatory Diseases: Osteoarthritis 505
Definition 505
Etiology 505
Effects of Osteoarthritis 505
Risk Factors 506
Nutritional Remedies 506

Cognitive Disorders: Alzheimer's Disease 507
Definition 507
Prevalence of Dementia 507
Etiology of Cognitive Disorders 508
Effects of Cognitive Disorders 509
Nutrition Interventions for Cognitive Disorders 509

Medications and Polypharmacy 509

Low Body Weight/Underweight 509
Definition 509
Etiology 510
Nutrition Interventions 512

Dehydration 513
Definition 513
Etiology 513

Effects of Dehydration 514
Nutritional Interventions 514
Rehydrate Slowly 514
Dehydration at End of Life 514

Bereavement 514

Answers to Review Questions   AR-1

Appendix A
Summary of Research of
Effects of Exercise Activities on
Health of Older Adults   A-1

Appendix B
Measurement Abbreviations
and Equivalents   A-3

Appendix C
Body Mass Index (BMI)   A-5

Appendix D
Carbohydrate Counting
for Type 1 Diabetes   A-6

References R-1

Glossary G-1

Index I-1

Preface

It is our privilege to offer you the fourth edition of *Nutrition Through the Life Cycle*. This text was initially developed, and has been revised, to address the needs of instructors teaching, and students taking, a two- to four-credit course in life-cycle nutrition. It is written at a level that assumes students have had an introductory nutrition course. Overall, the text is intended to give instructors a tool they can productively use to enhance their teaching efforts, and to give students an engaging and rewarding educational experience they will carry with them throughout their lives and careers.

The authors of *Nutrition Through the Life Cycle* represent a group of experts who are actively engaged in clinical practice, teaching, and research related to nutrition during specific phases of the life cycle. All of us remain totally dedicated to the goals established for the text at its conception: to make the text comprehensive, logically organized, science-based, realistic, and relevant to the needs of instructors and students.

Chapter 1 summarizes key elements of introductory nutrition and gives students who need it a chance to update or renew their knowledge. Coverage of the life-cycle phases begins with preconceptional nutrition and continues with each major phase of the life cycle through adulthood and the special needs of the elderly. Each of these 19 chapters was developed based on a common organizational framework that includes key nutrition concepts, prevalence statistics, physiological principles, nutritional needs and recommendations, model programs, case studies, and recommended practices.

To meet the knowledge needs of students with the variety of career goals represented in many life-cycle nutrition courses, we include two chapters for each life-cycle phase. The first chapter for each life-cycle phase covers normal nutrition topics, and the second covers nutrition-related conditions and interventions. Every chapter focuses on scientifically based information and employs up-to-date resources and references. Each chapter ends with:

- A list of key points
- Review questions (excluding chapter 1; answers are located at the end of the book)
- A directory of internet and other resources for reliable information on topics presented in the chapter

New to the Fourth Edition

Advances in knowledge about nutrition and health through the life cycle are expanding at a remarkably high rate.

New research is taking our understanding of the roles played by nutrients, nutrient–gene interactions, body fat, physical activity, and dietary supplements to new levels. The continued escalation rates of overweight, obesity, and type 2 diabetes are having broad effects on the incidence of disease throughout the life cycle. New knowledge about nutrition and health through the life cycle requires that we understand the effects of nutrients and body fat on hormonal activity, nutrient triggers to gene expression, and the roles of nutrients in the development and correction of chronic inflammation, oxidative stress, and endothelial dysfunction. Recommendations for dietary and nutrient supplement intake and for physical activity in health and disease are changing due to these understandings.

The practice of dietetics and nutrition is changing due to the emerging emphasis on electronic medical records, evidence-based health care services, and standardization of care delivery. The American Dietetic Association is responding to these changes though the development of nutrition care process standards. These process standards are intended to provide a systematic approach to the delivery of nutrition care to patients and clients. Revised standards for nutrition assessment have also been developed by the American Dietetic Association.

You will see these emerging areas of direct relevance to nutrition and updated information about nutrition incorporated throughout the fourth edition of *Nutrition Through the Life Cycle*.

The fourth edition differs from the third in important ways. We have:

- Added 8–12 review questions at the end of Chapters 2–19.
- Expanded coverage of nutrition assessment in all chapters.
- Added nutrition biomarker values by life-cycle stage.
- Added content and case studies related to the nutrition care process (NCP).
- Expanded presentation of oxidative stress, free radicals, antioxidants, chronic inflammation, and their relationship to diet and disease states from infertility to aging.
- Expanded coverage of ethnic disparities in health status.
- Added photographs, tables, and figures to enhance instruction and student understanding of material presented.

- Increased emphasis on use of SI units; expanded list of SI unit conversion factors for nutrition biomarkers appear in Appendix C.
- Incorporated presentations of mechanisms underlying nutrition and disease relationships.

Chapter-by-Chapter Changes

In addition to the enhancements just listed for every chapter, we have extensively revised and updated the text based on the most current research. The following is a list of some of those changes.

Chapter 1: Nutrition Basics

- Expanded coverage of the life-cycle concept as it relates to nutrition and lifelong health
- New illustration that shows the similarity among the chemical structures of monosaccharides
- Updated food-sources-of-nutrients tables to reflect food manufacturers' fortification changes and new analyses of nutrient content of foods
- Updated content on cholesterol, including plant cholesterol
- Updated content on fiber
- New content on choline
- Updated summary table on the vitamins
- Revised definition of *kwashiorkor* based on new evidence of its potential cause(s)
- Added section on increasing omega-3 fatty acid intake
- Updated table on top sources of antioxidant-rich foods
- Added content on low and high folic acid intake and cancer risk
- Updated content on nutrient–gene interactions during critical stages of growth and development and later health risks
- Revised content on glycemic index and health
- Expanded coverage of nutrition and autoimmune diseases, chronic inflammation, and oxidative stress
- Updated table on examples of diseases and disorders linked to diet
- Added content on shared dietary risk factors for chronic disease
- Expanded coverage on characteristics of healthy diets
- Modified content related to issues surrounding the "good food, bad food" concept
- Expanded content on nutrition assessment, including a new table of normal levels of various nutrition biomarkers that may be used as part of a biochemical assessment of nutritional status
- New information about mypyramid.gov features, the Mediterranean diet, and the DASH diet

Chapter 2: Preconception Nutrition

- Added background information and summary table related to the Nutrition Care Process
- Added case study that uses the Nutrition Care Process procedures

- Update on success (or the lack of it) in meeting Health Objectives for the year 2010 related to preconceptional men and women
- Updated content and table on factors related to altered fertility in women and men
- Revised content on body fat, obesity, and fertility
- Substantially modified sections on factors affecting male fertility to reflect new research findings related to nutrition exposures and sperm development and function
- Updated and expanded table on caffeine content of foods and beverages
- Updated content on oral contraceptives and nutrition status
- Added content on multivitamin supplements, folate intake, and fertility

Chapter 3: Preconception Nutrition: Conditions and Interventions

- Expanded presentation of ethnic and racial disparities in nutrition-related health disorders
- Added examples of mechanisms underlying nutrition and health relationships prior to and early in pregnancy
- Substantially expanded coverage on obesity and fertility, with new sections on bariatric surgery, nutrient needs, and fertility outcomes
- Updated table on biological bases of infertility in obese men and women
- Added coverage of the effects of obesity on chronic inflammation and insulin resistance on fertility-related disorders
- Reorganized, updated section on nutritional management of type 2 diabetes before pregnancy
- Updated section on nutrition recommendations for type 2 diabetes before pregnancy
- New section on pre-diabetes
- Updated sections on PKU and celiac disease, including a section on maternal PKU

Chapter 4: Nutrition During Pregnancy

- Updated natality statistics
- Added content related to rates of preterm delivery and low birth weight by ethnic/racial background
- Added content related to effects of nutrition exposures during critical periods of growth and development on gene expression and future health; added definitions and discussion of "developmental plasticity" and "epigenetic" mechanisms
- Updated content on pre-pregnancy weight status, weight gain, and pregnancy outcomes
- Updated content on early pregnancy blood lipid levels, oxidative stress, and pregnancy outcomes
- Refined coverage of effects of diet on pregnancy outcomes

- Expanded coverage of the functions of protein, fat, and essential fatty acids during pregnancy
- Updated coverage on the effects of protein, fat, and essential fatty acids on the course and outcome of pregnancy
- New table on the EPA+DHA, vitamin A, and vitamin D
- Updated content related to folate and pregnancy; added content on choline and pregnancy course and outcome
- Updated content on vitamin D and the course and outcome of pregnancy
- New overview sections on the need for vitamins during pregnancy and the need for minerals during pregnancy
- Updated section on fluoride and enamel formation during pregnancy
- Updated content on iron and iodine and pregnancy; added section on bioactive food components
- New section on nutrition assessment during pregnancy that includes a nutrition-biomarker assessment component and a table of normal nutrition biomarker concentrations during pregnancy
- Updated section on dietary supplements (including vitamin and mineral supplements) and pregnancy; new table on the nutrient formulation of common prenatal multivitamin and mineral supplements

Chapter 5: Nutrition During Pregnancy: Conditions and Interventions

- Increased coverage of mechanisms underlying nutrition and health disorders during pregnancy
- Reorganized to make obesity and pregnancy outcome the first condition presented, with new material including a table titled "Comparative prevalence of obesity prior to pregnancy and outcomes related to pre-pregnancy weight status"
- New sections: "Nutrition Recommendations and Interventions for Obesity During Pregnancy," "Pregnancy after Bariatric Surgery," and "Nutrition Care for Pregnant Women Post Bariatric Surgery"
- New coverage of dumping syndrome
- Fully updated table on "Dietary and other environmental exposures that increase or decrease chronic inflammation and oxidative stress"
- Updated content on carbohydrates, glycemic index, and diabetes during pregnancy
- New section titled "Vitamin and Mineral Supplementation and the Risk of Preeclampsia"
- Updated content on the nutritional management of diabetes (gestational, type 1, and type 2) during pregnancy
- New illustration of chorionicity in coverage on twin pregnancy

Chapter 6: Nutrition During Lactation

- New table on nutrition assessment for breastfeeding women
- Updated information on breastfeeding rates
- New diagram of breast that reflects latest understanding of ductal structure
- New discussion of the risks of cruciferous vegetables
- New material on breastfeeding duration
- Coverage of effects of implant surgery on breastfeeding
- New discussion of infant colic and its nutritional management
- New discussion of the effect of the WIC program on breastfeeding prevalence and duration
- Added overview of the obstacles to breastfeeding
- New information on the preparation of the breast for lactation during pregnancy and common problems such as leaking prior to delivery
- Additional information on MyPyramid for Moms, specific to pregnancy and lactation

Chapter 7: Nutrition During Lactation: Conditions and Interventions

- Updated discussion of common breastfeeding conditions
- Added section on flat or inverted nipples
- Added section on hyperlactation
- Updated information in engorgement section, including instructions on use of cabbage leaves
- Updated information in plugged-duct section, including instruction on use of lecithin
- Updated section on mastitis, modifying definition and treatment options
- Additional section on low milk supply
- Updated section on maternal medications, including discussion of OTC cold remedies, antihistamines, and decongestants
- Updated herbal remedies section
- Updated alcohol section, including AAP recommendations
- Updated smoking information with potential changes to mother's milk and infant behaviors
- Updated THC/marijuana section, including discussion of new studies
- Near-term section retitled as "late preterm"
- Added/updated info in the human-milk preterm section, reflecting current studies on infection rates
- Updated HIV section with WHO policy
- Additional information on human-milk storage guidelines, including estimated volumes
- Updated statistics on milk banking in North America

Chapter 8: Infant Nutrition

- Updated to include latest information on vitamin D AAP supplement change
- Updated U.S. standing in infant mortality rates
- Updated soy formula guidelines from AAP
- Revised discussion of protein digestion in food intolerances and allergies section
- Erickson's developmental stages added to support critical periods
- New World Health Organization growth charts added
- Updated infant mortality statistics
- Updated discussion of starting spoon feeding, in light of AAP guidelines
- Updated tables with latest statistics, including mortality rates by race of mother and key nutrients in infant formulas

Chapter 9: Infant Nutrition: Conditions and Interventions

- Updated references and studies on impacts of preterm
- Additional outcome information on prematurity
- New information on vitamin A supplementation
- New information in food safety section
- New material on effectiveness of EPA
- New material on fetal programming
- Added emphasis on breast milk in feeding problems section

Chapter 10: Toddler and Preschooler Nutrition

- Updated children's health data
- Added section on WHO growth standards
- Revised terminology in dental caries section
- Revised section on elevated blood lead levels
- Updated section on overweight and obesity in toddlers and preschoolers to include the recommendations of the obesity expert committee
- Major edits to section on nutrition and prevention of cardiovascular disease in toddlers and preschoolers, with more recently published AHA guidelines
- Updated AAP recommendations for vitamin and mineral supplementation
- Added information on herbal use in WIC population
- Updated survey information on recommended vs. actual food intake
- New section on nutrition assessment and one table of biochemical indices
- Food Stamp section changed to Supplemental Nutrition Assistance Program (SNAP)

Chapter 11: Toddler and Preschooler Nutrition: Conditions and Interventions

- Autism section expanded to include prevalence and AAP clinical report

- Renamed section: "Combating Autism in HRSA"
- New section that includes AAP statement on complementary and alternative medicine in pediatrics
- New information on the National Dissemination Center for Children with Disabilities to explain need for finding services
- New information on obesity and special needs
- Clarification of the relationship between CP and preterm birth

Chapter 12: Child and Preadolescent Nutrition

- Updated statistics on child and preadolescent health
- New definitions of overweight and obesity in CDC growth chart section
- New section on WHO growth references for this age group
- Additional information about AAP recommendations and CDC recommendations for screening for iron deficiency
- Additional information about "screen time" in section on effects of TV viewing
- New coverage of expert committee's recommendations on assessment, prevention, and treatment of pediatric overweight/obesity
- Updated recommendations for CV disease prevention based on AHA references
- New AAP vitamin D recommendations in calcium section
- New section on lactose intolerance
- Updated soft-drink consumption data
- Updated consumption data using "What We Eat in America" report
- In physical activity section, added AAP recommendations, including information about the "built environment"
- Updated AAP policy statement on organized sports
- Updated information in section on nutrition integrity in schools

Chapter 13: Child and Preadolescent Nutrition: Conditions and Interventions

- Expanded description of type 1 and type 2 diabetes
- Added appendix with carbohydrate-counting example table comparing type 1 and type 2 nutrition
- Added information on ADHD treatment requiring cardiac monitoring
- Expanded PKU section to include discussion of inborn; added discussion of new PKU medication Kuvan
- Updated information on prevalence and incidence
- Updated CAM section to include special needs
- New sections on PKU and ADHD management

Chapter 14: Adolescent Nutrition

- Updated information on nutrient intake and eating behaviors (meal skipping, dieting, family meals)
- Updated dietary fiber recommendations to include current DRI and AAP recommendations
- Added information on target goals for iron deficiency for HP 2010
- Added section on vitamin D
- Updated folate section to reflect current estimates of intake and adequacy
- New weight terminology in the assessment section, using recommendations in the 2007 Guidelines for Assessment, Prevention and Treatment of Child Overweight and Obesity
- Added section on the use of technology for nutrition counseling
- Updated physical activity statistics
- Updated data on the nutrition environment, particularly the school environment, using the 2006 SHPPS data from CDC
- Updated section on school wellness policies

Chapter 15: Adolescent Nutrition: Conditions and Interventions

- Revised overweight/obesity section to reflect the new staged approach to prevention and treatment of adolescent overweight and obesity as outlined in the 2007 Guidelines for Assessment, Prevention and Treatment of Child Overweight and Obesity
- Updated information on supplement use and substance use using the most recent YRBS data (2007–2008)
- Revised eating disorders section to include standard practices for screening for disordered eating behaviors, as well as to reflect new advancements in eating-disorder treatment and prevention

Chapter 16: Adult Nutrition

- New coverage of psychosocial changes in adulthood and their nutritional implications
- New coverage of the continuum of nutritional health
- New emphasis on health disparity and a new "in-focus" on the social determinants of health
- Inclusion of the Eating Competency Model, which offers an alternative paradigm for nutrition education and dietary guidance to complement the disease and risk-factor orientation
- Inclusion of the beverage guidance system, with illustrations of the quantity and caloric contribution of beverages
- Inclusion of the Mifflin–St. Jeor energy estimation formula for healthy adults (replaces the Harris-Benedict equation)

- New tables on functional foods, cues for regulating food intake, caloric and alcohol content of beverages, and physical activity guidelines for adults
- All other tables updated

Chapter 17: Adult Nutrition: Conditions and Interventions

- New emphasis on the role that overweight and obesity plays in the development and progression of the major chronic diseases
- Metabolic syndrome expanded into a separate section
- Increased focus on primary prevention, including a table on recommended actions to change environmental factors
- Updated perspective on HIV disease as a chronic disease
- New table on areas of comprehensive nutrition assessment, with screening and assessment information pertinent to each chronic disease
- Weight-status table expanded to report prevalence by race/ethnicity
- New tables on risk factors and assessment criteria for cardiovascular disease and diabetes and on dietary risk factors for cancer and nutrition and physical activity guidelines for cancer prevention

Chapter 18: Nutrition and Older Adults

- Updated information on theories of aging
- New figures, including newest versions of older-adult food pyramids
- New table on life expectancies by country
- New table on "What We Eat in America"
- Latest statistics on physical activity and older adults

Chapter 19: Nutrition and Older Adults: Conditions and Interventions

- New information on cognition and nutrition
- Updated table on prevalence of chronic disease
- Updated information about disease and health disparities
- New table comparing BMI ranges used by practitioners and policy makers
- New table on falls and fall-related injuries
- New information on misconceptions about constipation
- New table showing conditions associated with cognitive disorders

Instructor Resources

Updated for the fourth edition is a Power Lecture DVD-ROM that contains Microsoft PowerPoint™ lecture presentations with artwork, chapter outlines, classroom activities, lecture launchers, Internet exercises, discussion questions, hyperlinks to relevant websites, and case studies. The DVD-ROM also includes a Test Bank, expanded and improved, that contains multiple-choice, true/false, matching, and discussion exercises.

Acknowledgments

It takes the combined talents and hard work of authors, editors, assistants, and the publisher to develop a new edition of a textbook and its instructional resources. The dedicated efforts of Nedah Rose, Senior Developmental Editor, in securing and managing the enhancements offered in the fourth edition are applauded by all of the authors. We are fortunate to have Peggy Williams, Senior Acquisitions Editor, heading up the team that supports the growth and development of *Nutrition Through the Life Cycle*. We appreciate the creativity and energy of Laura McGinn in marketing this text.

Two new highly qualified authors have joined the group in this edition of the text. Dr. Patricia Splett, a nutrition consultant and frequent advisor on program effectiveness for the American Dietetic Association and other nonprofit health groups, is now lead author on Chapters 16 and 17. Ellen Lechtenberg, R. D., IBCLC, joins us as first author on Chapter 7. Ms. Lechtenberg is the Lactation Program Coordinator at Primary Children's Medical Center in Salt Lake City. We would like to take this opportunity to thank Carolyn Sharbaugh for the excellent work on the first three editions as co-author of Chapter 6 and as lead author of Chapter 7. Thank you a ton, Carolyn.

Reviewers

Many thanks to the following reviewers, whose careful reading and thoughtful comments helped enormously in shaping both earlier editions and this fourth edition.

Betty Alford
Texas Woman's University

Leta Aljadir
University of Delaware

Clint Allred
Texas A&M University

Dea Hanson Baxter
Georgia State University

Janet Colson
Middle Tennessee State University

Kathleen Davis
Texas Women's University

Shelley R. Hancock
The University of Alabama

Laura Horn
Cincinnati State Community College

Dr. Mary Jacob
California State University, Long Beach

Pera Jambazian
California State University, Los Angeles

Tay Seacord Kennedy
Oklahoma State University

Younghee Kim
Bowling Green State University

Barbara Kirks
California State University, Chico

Kaye Stanek Krogstrand
University of Nebraska

Karen Kubena
Texas A&M University

Sally Ann Lederman
Columbia University

Richard Lewis
University of Georgia

Marcia Magnus
Florida International University

J. Harriett McCoy
University of Arkansas

Sharon McWhinney
Prairie View A&M University

Janis Mena
University of Florida

Robert Reynolds
University of Illinois at Chicago

Sharon Nickols-Richardson
Virginia Polytechnic Institute and State University

Lisa Roth
University of Florida

Claire Schmelzer
Eastern Kentucky University

Adria Sherman
Rutgers University

Carmen R. Roman-Shriver
Texas Tech University

Joanne Slavin
University of Minnesota

Joanne Spaide
Professor Emeritus, University of Northern Iowa

Diana-Marie Spillman
Miami University, Oxford, Ohio

Wendy Stuhldreher
Slippery Rock University of Pennsylvania

Anne VanBeber
Texas Christian University

Phyllis Moser-Veillon
University of Maryland

Janelle Walter
Baylor University

Doris Wang
University of Minnesota Crookston

Suzy Weems, Ph.D.
Stephen F. Austin State University

Kay Wilder
Point Loma Nazarene College

Nutrition
Through the Life Cycle
FOURTH EDITION

Chapter 1

Nutrition Basics

Prepared by **Judith E. Brown**

Key Nutrition Concepts

1 Nutrition is the study of foods, their nutrients and other chemical constituents, and the effects of food constituents on health.

2 Nutrition is an interdisciplinary science.

3 Nutrition recommendations for the public change as new knowledge about nutrition and health relationships is gained.

4 At the core of the science of nutrition are principles that represent basic truths and serve as the foundation of our understanding about nutrition.

5 Healthy individuals require the same nutrients across the life cycle but in differing amounts. Nutritional needs can be met by a wide variety of cultural and religious food practices.

6 Nutritional status during one stage of the life cycle influences health status during subsequent life-cycle stages.

Introduction

Need to freshen up your knowledge of nutrition? Or, do you need to get up to speed on basic nutrition for the course? This chapter presents information about nutrition that paves the way to greater understanding of specific needs and benefits related to nutrition by life-cycle stage.

Nutrition is an interdisciplinary science focused on the study of foods, *nutrients*, and other food constituents and health. The body of knowledge about nutrition is large and is growing rapidly, changing views on what constitutes the best nutrition advice. You are encouraged to refer to nutrition texts and to use the online resources listed at the end of this chapter to fill in any knowledge gaps. You are also encouraged to stay informed in the future and to keep an open mind about the best nutrition advice for many health-related issues. Scientific evidence that drives decisions about nutrition and health changes with time.

> **Nutrients** Chemical substances in foods that are used by the body for growth and health.
>
> **Food Security** Access at all times to a sufficient supply of safe, nutritious foods.
>
> **Food Insecurity** Limited or uncertain availability of safe, nutritious foods, or the ability to acquire them in socially acceptable ways.
>
> **Calorie** A unit of measure of the amount of energy supplied by food. Also known as the "kilocalorie" (kcal), or the "large Calorie."

This chapter centers on (1) the principles of the science of nutrition, (2) nutrients and other constituents of food, (3) nutritional assessment, (4) public food and nutrition programs, and (5) nationwide priorities for improvements in the *public's nutritional health*.

Principles of the Science of Nutrition

Every field of science is governed by a set of principles that provides the foundation for growth in knowledge. These principles change little with time. Knowledge of the principles of nutrition listed in Table 1.1 will serve as a springboard to greater understanding of the nutrition and health relationships explored in the chapters to come.

PRINCIPLE #1   Food is a basic need of humans.

Humans need enough food to live and the right assortment of foods for optimal health (Illustration 1.1 on the next page). People who have enough food to meet their needs at all times experience *food security*. They are able to acquire food in socially acceptable ways—without having to scavenge or steal food. *Food insecurity* exists when the availability of safe, nutritious foods, or the ability to acquire them in socially acceptable ways, is limited or uncertain. About 12% of U.S. households are food insecure.[1]

PRINCIPLE #2   Foods provide energy (calories), nutrients, and other substances needed for growth and health.

People eat foods for many different reasons. The most compelling reason is the requirement for *calories* (energy), nutrients, and other substances supplied by foods for growth and health.

Table 1.1 Principles of human nutrition

Principle #1 Food is a basic need of humans.

Principle #2 Foods provide energy (calories), nutrients, and other substances needed for growth and health.

Principle #3 Health problems related to nutrition originate within cells.

Principle #4 Poor nutrition can result from both inadequate and excessive levels of nutrient intake.

Principle #5 Humans have adaptive mechanisms for managing fluctuations in food intake.

Principle #6 Malnutrition can result from poor diets and from disease states, genetic factors, or combinations of these causes.

Principle #7 Some groups of people are at higher risk of becoming inadequately nourished than others.

Principle #8 Poor nutrition can influence the development of certain chronic diseases.

Principle #9 Adequacy, variety, and balance are key characteristics of a healthy diet.

Principle #10 There are no "good" or "bad" foods.

Illustration 1.1 The need for food is part of Maslow's hierarchy of needs.

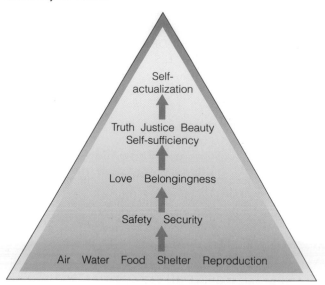

A calorie is a measure of the amount of energy transferred from food to the body. Because calories are a unit of measure and not a substance actually present in food, they are not considered to be nutrients.

Nutrients are chemical substances in food that the body uses for a variety of functions that support growth, tissue maintenance and repair, and ongoing health. Essentially, every part of our body was once a nutrient consumed in food.

There are six categories of nutrients (Table 1.2). Each category except water consists of a number of different substances.

Essential and Nonessential Nutrients

Of the many nutrients required for growth and health, some must be provided by the diet while others can be made by the body.

Essential Nutrients   Nutrients the body cannot manufacture, or generally produce in sufficient amounts, are referred to as *essential nutrients*. Here *essential* means "required in the diet." All of the following nutrients are considered essential:

- Carbohydrates
- Certain amino acids (the *essential amino acids*: histidine, isoleucine, leucine, lysine, methionine, phenylalanine, threonine, tryptophan, and valine)
- Linoleic acid and alpha-linolenic acid (essential fatty acids)
- Vitamins
- Minerals
- Water

Table 1.2 The six categories of nutrients

1. Carbohydrates   Chemical substances in foods that consist of a single sugar molecule or multiples of sugar molecules in various forms. Sugar and fruit, starchy vegetables, and whole grain products are good dietary sources.
2. Proteins   Chemical substances in foods that are made up of chains of amino acids. Animal products and dried beans are examples of protein sources.
3. Fats (Lipids)   Components of food that are soluble in fat but not in water. They are more properly referred to as "lipids." Most fats are composed of glycerol attached to three fatty acids. Oil, butter, sausage, and avocado are examples of rich sources of dietary fats.
4. Vitamins   Fourteen specific chemical substances that perform specific functions in the body. Vitamins are present in many foods and are essential components of the diet. Vegetables, fruits, and grains are good sources of vitamins.
5. Minerals   In the context of nutrition, minerals consist of 15 elements found in foods that perform particular functions in the body. Milk, dark, leafy vegetables, and meat are good sources of minerals.
6. Water   An essential component of the diet provided by food and fluid.

Nonessential Nutrients   Cholesterol, creatine, and glucose are examples of nonessential nutrients. *Nonessential nutrients* are present in food and used by the body, but they do not have to be part of our diets. Many of the beneficial chemical substances in plants are not considered essential, for example, yet they play important roles in maintaining health.

Essential Nutrients   Substances required for growth and health that cannot be produced, or produced in sufficient amounts, by the body. They must be obtained from the diet.

Essential Amino Acids   Amino acids that cannot be synthesized in adequate amounts by humans and therefore must be obtained from the diet. Also called "indispensible amino acids."

Nonessential Nutrients   Nutrients required for growth and health that can be produced by the body from other components of the diet.

Requirements for Essential Nutrients   All humans require the same set of essential nutrients, but the amount of nutrients needed varies based on:

- Age
- Body size
- Gender
- Genetic traits
- Growth
- Illness
- Lifestyle habits (e.g., smoking, alcohol intake)
- Medication use
- Pregnancy and lactation

Illustration 1.2 Theoretical framework, terms, and abbreviations used in the Dietary Reference Intakes.

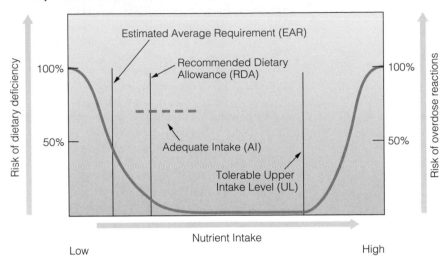

Amounts of essential nutrients required each day vary a great deal, from cups (for water) to micrograms (for example, for folate and vitamin B$_{12}$).

Dietary Intake Standards

Dietary intake standards developed for the public cannot take into account all of the factors that influence nutrient needs, but they do account for the major ones of age, gender, growth, and pregnancy and lactation. Intake standards are called Dietary Reference Intakes (DRIs).

- Dietary Reference Intakes (DRIs). This is the general term used for the nutrient intake standards for healthy people.
- Recommended Dietary Allowances (RDAs). These are levels of essential nutrient intake judged to be adequate to meet the known nutrient needs of practically all healthy persons while decreasing the risk of certain chronic diseases.
- Adequate Intakes (AIs). These are "tentative" RDAs. AIs are based on less conclusive scientific information than are the RDAs.
- Estimated Average Requirements (EARs). These are nutrient intake values that are estimated to meet the requirements of half the healthy individuals in a group. The EARs are used to assess adequacy of intakes of population groups.
- Tolerable Upper Intake Levels (ULs). These are upper limits of nutrient intake compatible with health. The ULs do not reflect desired levels of intake. Rather, they represent total, daily levels of nutrient intake from food, fortified foods, and supplements that should not be exceeded.

Daily Values (DVs) Scientifically agreed-upon standards for daily intakes of nutrients from the diet developed for use on nutrition labels.

DRIs have been developed for most of the essential nutrients and will be updated periodically. (These are listed on the inside front covers of this text.) Current DRIs were developed through a joint U.S.–Canadian effort, and the standards apply to both countries. The DRIs are levels of nutrient intake intended for use as reference values for planning and assessing diets for healthy people. They consist of the Recommended Dietary Allowances (RDAs), which specify intake levels that meet the nutrient needs of over 98% of healthy people, and the other categories of intake standards described in Illustration 1.2. It is recommended that individuals aim for nutrient intakes that approximate the RDAs or Adequate Intake (AI) levels. Estimated Average Requirements (EARs) should be used to examine the possibility of inadequate intakes in individuals and within groups. Additional tests are required to confirm inadequate nutrient intakes and status.[2]

Tolerable Upper Intake Levels (ULs) The DRIs include a table indicating levels of daily nutrient intake from foods, fortified products, and supplements that should not be exceeded. They can be used to assess the safety of high intakes of nutrients, particularly from supplements.

Standards of Nutrient Intake for Nutrition Labels The Nutrition Facts panel on packaged foods uses standard levels of nutrient intakes based on an earlier edition of recommended dietary intake levels.

The levels are known as *Daily Values (DVs)* and are used to identify the amount of a nutrient provided in a serving of food compared to the standard level. The "% DV" listed on nutrition labels represents the percentages of the standards obtained from one serving of the food product. Table 1.3 lists DV standard amounts for nutrients that are mandatory or voluntary components of nutrition labels. Additional information on nutrition labeling is presented later in this chapter.

Carbohydrates

Carbohydrates are used by the body mainly as a source of readily available energy. They consist of the simple sugars (monosaccharides and disaccharides), complex carbohydrates (the polysaccharides), most dietary sources of fiber, and alcohol sugars. Alcohol (ethanol) is closely related chemically to carbohydrates and is usually considered to be part of this nutrient category. Illustration 1.3 shows the similarity in the chemical structure of basic carbohydrate units. The most basic forms of carbohydrates are single molecules called monosaccharides.

Table 1.3 Daily Values (DVs) for nutrition labeling based on intakes of 2000 calories per day in adults and children aged 4 years and above

Mandatory Components of the Nutrition Label

Food Component	Daily Value (DV)
Total fat	65 g[a]
Saturated fat	20 g
Cholesterol	300 mg[a]
Sodium	2400 mg
Total carbohydrate	300 g
Dietary fiber	25 g
Vitamin A	5000 IU[a]
Vitamin C	60 mg
Calcium	1000 mg
Iron	18 mg

[a]g = grams; mg = milligrams; IU = International Units

Glucose (also called "blood sugar" and "dextrose"), fructose ("fruit sugar"), and galactose are the most common monosaccharides. Molecules containing two monosaccharides are called disaccharides. The most common disaccharides are:

- Sucrose (glucose + fructose, or common table sugar)
- Maltose (glucose + glucose, or malt sugar)
- Lactose (glucose + galactose, or milk sugar)

Complex carbohydrates (also called polysaccharides) are considered "complex" because they have more elaborate chemical structures than the simple sugars. They include:

- Starches (the plant form of stored carbohydrate)
- Glycogen (the animal form of stored carbohydrate)
- Most types of fiber

Illustration 1.3 Chemical structures of some simple carbohydrates.

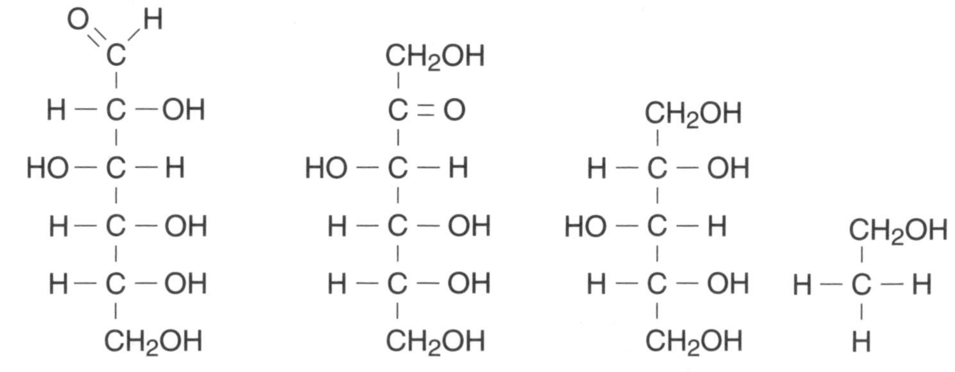

Glucose (monosaccharide)

Fructose (monosaccharide)

Xylitol (alcohol sugar)

Ethanol (alcohol)

Each type of simple and complex carbohydrate, except fiber, provides 4 calories per gram. Dietary fiber supplies two calories per gram on average, even though fiber cannot be broken down by human digestive enzymes. Bacteria in the large intestine can digest some types of dietary fiber, however. These bacteria excrete fatty acids as a waste product of fiber digestion. The fatty acids are absorbed and used as a source of energy. The total contribution of fiber to our energy intake is modest (around 50 calories), and supplying energy is not a major function of fiber.[3] The main function of fiber is to provide "bulk" for normal elimination. It has other beneficial properties, however. High-fiber diets reduce the rate of glucose absorption (a benefit for people with diabetes) and may help prevent cardiovascular disease and some types of cancer.[4]

Nonalcoholic in the beverage sense, alcohol sugars are like simple sugars, except that they include a chemical component of alcohol. Xylitol, mannitol, and sorbitol are common forms of alcohol sugars. Some are very sweet, and only small amounts are needed to sweeten commercial beverages, gums, yogurt, and other products. Unlike the simple sugars, alcohol sugars do not promote tooth decay.

Alcohol (consumed as ethanol) is considered to be part of the carbohydrate family because its chemical structure is similar to that of glucose. It is a product of the fermentation of sugar with yeast. With 7 calories per gram, alcohol has more calories per gram than do other carbohydrates.

Glycemic Index of Carbohydrates and Carbohydrates in Foods In the not-too-distant past, it was assumed that "a carbohydrate is a carbohydrate is a carbohydrate." It was thought that all types of carbohydrates had the same effect on blood glucose levels and health, so it didn't matter what type was consumed. As is the case with many untested assumptions, this one fell by the wayside. It is now known that some types of simple and complex carbohydrates in foods elevate blood glucose levels more than do others. Such differences are particularly important to people with disorders such as *insulin resistance* and *type 2 diabetes*.[5] (An "In Focus" box on page 82 expands upon the topics of insulin resistance and diabetes.)

Carbohydrates and carbohydrate-containing foods are now being classified by

Insulin Resistance A condition is which cell membranes have a reduced sensitivity to insulin so that more insulin than normal is required to transport a given amount of glucose into cells.

Type 2 Diabetes A disease characterized by high blood glucose levels due to the body's inability to use insulin normally, to produce enough insulin, or both.

Table 1.4   Glycemic Index (GI) of selected foods[8,9]

High GI (70 and Higher)		Medium GI (56–69)		Low GI (55 and Lower)	
Glucose	100	Breadfruit	69	Honey	55
French bread	95	Orange soda	68	Oatmeal	54
Scone	92	Sucrose	68	Corn	53
Potato, baked	85	Taco shells	68	Cracked wheat bread	53
Potato, instant mashed	85	Angel food cake	67	Orange juice	52
Corn Chex	83	Croissant	67	Banana	52
Pretzel	83	Cream of Wheat	66	Mango	51
Rice Krispies	82	Quaker Quick Oats	65	Potato, boiled	50
Cornflakes	81	Chapati	62	Muesli	48
Corn Pops	80	French bread with		Green peas	48
Gatorade	78	butter and jam	62	Pasta	48
Jelly beans	78	Couscous	61	Carrots, raw	47
Doughnut, cake	76	Raisin Bran	61	Cassava	46
Waffle, frozen	76	Sweet potato	61	Lactose	46
French fries	75	Bran muffin	60	Milk chocolate	43
Shredded Wheat	75	Just Right cereal	60	All Bran	42
Cheerios	74	Rice, white or brown	60	Orange	42
Popcorn	72	Blueberry muffin	59	Peach	42
Watermelon	72	Coca-Cola/cola	58	Apple juice	40
Grape-Nuts	71	Power Bar	56	Plum	39
Wheat bread	70			Apple	38
White bread	70			Pear	38
				Tomato juice	38
				Yam	37
				Dried beans	25
				Grapefruit	25
				Cherries	22
				Fructose	19
				Xylitol	8
				Hummus	6

the extent to which they increase blood glucose levels. This classification system is called the *glycemic index*. Carbohydrates that are digested and absorbed quickly have a high glycemic index and raise blood glucose levels to a higher extent than do those with lower glycemic index values (Table 1.4).

Diets providing low glycemic index carbohydrates have generally been found to improve blood glucose control in people with diabetes, as well as to reduce elevated levels of blood cholesterol and triglycerides; increase levels of beneficial HDL cholesterol; and decrease the risk of developing type 2 diabetes, some types of cancer, and heart disease.[6]

Recommended Intake Level   Recommended intake of carbohydrates is based on their contribution to total energy intake. It is recommended that 45–65% of calories come from carbohydrates. Added sugar should constitute no more than 25% of total caloric intake. It is recommended that adult females consume between 21 and 25 grams, and males 30–38 grams of total dietary fiber daily.[7]

Food Sources of Carbohydrates   Carbohydrates are widely distributed in plant foods, while milk is the only important animal source of carbohydrates (lactose). Table 1.5 on page 7 lists selected food sources by type of carbohydrate. Additional information about total carbohydrate and fiber content of foods can be found on nutrition information labels on food packages and at this Web address: www. ars.usda.gov/Services/docs.htm?docid=17477.

Protein

Protein in foods provides the body with *amino acids* used to build and maintain tissues such as muscle, bone, enzymes, and red blood cells. The body can also use protein as a source of energy—it provides 4 calories per gram. However, this is not a primary function of protein.

Glycemic Index   A measure of the extent to which blood glucose is raised by a 50-gram portion of a carbohydrate-containing food compared to 50 grams of glucose or white bread.

Amino Acids   The "building blocks" of protein. Unlike carbohydrates and fats, amino acids contain nitrogen.

Table 1.5   Food sources of carbohydrates

A. Simple Sugars (Mono- and Disaccharides)

	Portion Size	Grams of Carbohydrates		Portion Size	Grams of Carbohydrates
Breakfast Cereals			Apple	1 med	16
Raisin Bran	1 c	18	Orange	1 med	14
Corn Pops	1 c	14	Peach	1 med	8
Frosted Cheerios	1 c	13	**Vegetables**		
Bran Flakes	¾ c	5	Corn	½ c	3
Grape-Nuts	½ c	3	Broccoli	½ c	2
Special K	1 c	3	Potato	1 med	1
Wheat Chex	1 c	2	**Beverages**		
Cornflakes	1 c	2	Soft drinks	12 oz	38
Sweeteners			Fruit drinks	1 c	29
Honey	1 tsp	6	Skim milk	1 c	12
Corn syrup	1 tsp	5	Whole milk	1 c	11
Maple syrup	1 tsp	4	**Candy**		
Table sugar	1 tsp	4	Hard candy	1 oz	28
Fruits			Gumdrops	1 oz	25
Watermelon	1 pc (4" × 8")	25	Caramels	1 oz	21
Banana	1 med	21	Milk chocolate	1 oz	16

B. Complex Carbohydrates (Starches)

	Portion Size	Grams of Carbohydrates		Portion Size	Grams of Carbohydrates
Grain Products			**Dried Beans**		
Rice, white, cooked	½ c	21	White beans, cooked	½ c	13
Pasta, cooked	½ c	15	Kidney beans, cooked	½ c	12
Oatmeal, cooked	½ c	12	Lima beans, cooked	½ c	11
Cheerios	1 c	11	**Vegetables**		
Cornflakes	1 c	11	Potato	1 med	30
Bread, whole wheat	1 slice	7	Corn	½ c	10
			Broccoli	½ c	2

C. Total Fiber

	Portion Size	Grams of Total Fiber		Portion Size	Grams of Total Fiber
Grain Products			Green peas	½ c	4
Bran Buds	½ c	12	Carrots	½ c	3
All Bran	½ c	11	Potato, with skin	1 med	4
Raisin Bran	1 c	7	Collard greens	½ c	3
Bran Flakes	¾ c	5	Corn	½ c	3
Oatmeal, cooked	1 c	4	Cauliflower	½ c	2
Bread, whole wheat	1 slice	2	**Nuts**		
Fruits			Almonds	½ c	5
Raspberries	1 c	8	Peanuts	½ c	3
Avocado	½ med	7	Peanut butter	2 Tbsp	2
Mango	1 med	4	**Dried Beans**		
Pear, with skin	1 med	4	Pinto beans, cooked	½ c	10
Apple, with skin	1 med	3	Black beans, cooked	½ c	8
Orange	1 med	3	Black-eyed peas, cooked	½ c	8
Banana	1 med	3	Navy beans, cooked	½ c	6
Vegetables			Lentils, cooked	½ c	5
Lima beans	½ c	7			

Of the common types of amino acids, nine must be provided by the diet and are classified as "essential amino acids." (These are listed on page 3.) Many different amino acids obtained from food perform important functions, but since the body can manufacture these from other amino acids, they are classified as "nonessential amino acids."

Kwashiorkor A severe form of protein-energy malnutrition in young children. It is characterized by swelling, fatty liver, susceptibility to infection, profound apathy, and poor appetite. The cause of kwashiorkor is unclear.

Fatty Acids The fat-soluble components of fats in foods.

Glycerol A component of fats that is soluble in water. It is converted to glucose in the body.

Essential Fatty Acids Components of fat that are a required part of the diet (i.e., linoleic and alpha-linolenic acids). Both contain unsaturated fatty acids.

Prostaglandins A group of physiologically active substances derived from the essential fatty acids. They are present in many tissues and perform such functions as the constriction or dilation of blood vessels and stimulation of smooth muscles and the uterus.

Thromboxanes Biologically active substances produced in platelets that increase platelet aggregation (and therefore promote blood clotting), constrict blood vessels, and increase blood pressure.

Prostacyclins Biologically active substances produced by blood vessel walls that inhibit platelet aggregation (and therefore blood clotting), dilate blood vessels, and reduce blood pressure.

Food sources of protein differ in quality, based on the types and amounts of amino acids they contain. Foods of high protein quality include a balanced assortment of all of the essential amino acids. Protein from milk, cheese, meat, eggs, and other animal products is considered high quality. Plant sources of protein, with the exception of soybeans, do not provide all nine essential amino acids. Combinations of plant foods, such as grains or seeds with dried beans, however, yield high-quality protein. Amino acids found in these individual foods "complement" each other, thus providing a source of high-quality protein.

Recommended Protein Intake DRIs for protein are shown on the inside front cover of this text. In general, proteins should contribute 10–35% of total energy intake.[7] Protein deficiency, although rare in economically developed countries, leads to loss of muscle tissue, growth failure, weakness, reduced resistance to disease, kidney and heart problems, and contributes to the development of a severe form of protein-energy malnutrition known as *kwashiorkor*. Protein deficiency in adults produces a loss of body tissue protein, heart abnormalities, severe diarrhea, and other health problems.

Food Sources of Protein Animal products and dried beans are particularly good sources of protein. These and other food sources of proteins are listed in Table 1.6.

Fats (Lipids)

Fats in food share the property of being soluble in fats but not in water. They are actually a subcategory of *lipids,* but this category of macronutrient is referred to as fat in the DRIs.[7] Lipids include fats, oils, and related compounds such as cholesterol. Fats are generally solid at room temperature,

Table 1.6 Food sources of protein

	Portion Size	Grams of Protein
Meats		
Beef, lean	3 oz	26
Tuna, in water	3 oz	24
Hamburger, lean	3 oz	24
Chicken, no skin	3 oz	24
Lamb	3 oz	22
Pork chop, lean	3 oz	20
Haddock, broiled	3 oz	19
Egg	1 med	6
Dairy Products		
Cottage cheese, low fat	½ c	14
Yogurt, low fat	1 c	13
Milk, skim	1 c	9
Milk, whole	1 c	8
Swiss cheese	1 oz	8
Cheddar cheese	1 oz	7
Grain Products		
Oatmeal, cooked	½ c	4
Pasta, cooked	½ c	4
Bread	1 slice	2
Rice, white or brown	½ c	2

whereas oils are usually liquid. Fats and oils are made up of various types of triglycerides (triacylglycerols), which consist of three *fatty acids* attached to *glycerol* (Illustration 1.4). The number of carbons contained in the fatty acid component of triglycerides varies from 8 to 22.

Fats and oils are a concentrated source of energy, providing 9 calories per gram. Fats perform a number of important functions in the body. They are precursors for cholesterol and sex-hormone synthesis, components of cell membranes, vehicles for carrying certain vitamins that are soluble in fats only, and suppliers of the *essential fatty acids* required for growth and health.

Essential Fatty Acids There are two essential fatty acids: linoleic acid and alpha-linolenic acid. Because these fatty acids are essential, they must be supplied in the diet. The central nervous system is particularly rich in derivatives of these two fatty acids. They are found in phospholipids, which—along with cholesterol—are the primary lipids in the brain and other nervous system tissue. Biologically active derivatives of essential fatty acids include *prostaglandins, thromboxanes*, and *prostacyclins*.

Linoleic Acid Linoleic acid is the parent of the omega-6 (or n-6) fatty acid family. One of the major derivatives of linoleic acid is arachidonic acid. Arachidonic acid serves as a primary structural component of the central nervous system. Most vegetable oils, meats, and human milk are good sources of linoleic acid. American diets tend to

Illustration 1.4 Basic structure of a triglyceride.

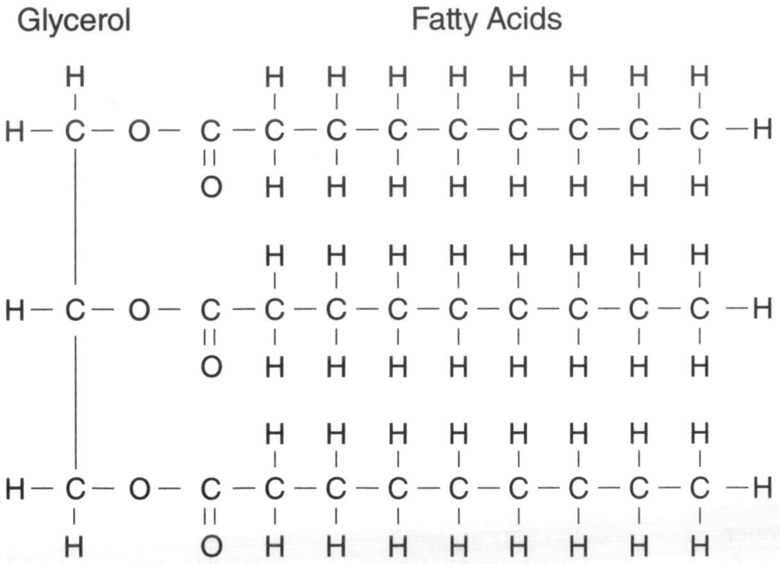

Glycerol   Fatty Acids

provide sufficient to excessive levels of linoleic acid, and considerable amounts are stored in body fat.

Alpha-Linolenic Acid Alpha-linolenic acid is the parent of the omega-3 (n-3) fatty acid family. It is present in many types of dark green vegetables, vegetable oils, and flaxseed. Derivatives of this essential fatty acid include eicosapentaenoic acid (EPA) and docosahexaenoic acid (DHA). Relatively little EPA and DHA are produced in the body from alpha-linolenic acid because the conversion process is slow.[7] EPA and DHA also enter the body through intake of fatty, cold-water fish and shellfish and human milk. The EPA and DHA content of fish provide health benefits. Regular consumption of fish (two or more meals per week of fish) not only protects against irregular heartbeat, sudden death, and stroke but also reduces high blood pressure and plaque formation in arteries.[10] DHA is found in large amounts in the central nervous system, the retina of the eye, and the testes. The body stores only small amounts of alpha-linolenic acid, EPA, and DHA.[11]

Increasing Omega-3 Fatty Acid Intake In the past it was thought that consuming high amounts of omega-6 fatty acids compared to omega-3 fatty acids could interfere with the availability of omega-3 fatty acids, particularly EPA and DHA. Although still somewhat controversial, it appears that rather high intakes of omega-6 fatty acids do not interfere with the availability of EPA and DHA.[12] What interferes most with the availability of EPA and DHA for body functions is inadequate intake. On average, adults in the United States and Canada consume 100 mg EPA plus DHA daily, far short of the recommended intake of 500 mg daily.[13]

Saturated and Unsaturated Fats Fats (lipids) come in two basic types: *saturated* and *unsaturated*. Whether

a fat is saturated or not depends on whether it has one or more double bonds between carbon atoms in one or more of the fatty acid components of the fat. If one double bond is present in one or more of the fatty acids, the fat is considered *monounsaturated*; if two or more are present, the fat is *polyunsaturated*.

Some unsaturated fatty acids are highly unsaturated. Alpha-linolenic acid, for example, contains three double bonds, arachidonic acid four, EPA five, and DHA six. These fatty acids are less stable than fatty acids with fewer double bonds, because double bonds between atoms are weaker than single bonds.

Saturated fats contain no double bonds between carbons and tend to be solid at room temperature. Animal products such as butter, cheese, and meats and two plant oils (coconut and palm) are rich sources of saturated fats. Fat we consume in our diets, whether it contains primarily saturated or unsaturated fatty acids, is generally in the triglyceride (or triacylglycerol) form.

Fats (lipids) also come in these forms:

- Monoglycerides (or monoacylglycerols), consisting of glycerol plus one fatty acid
- Diglycerides (or diacylglycerols), consisting of glycerol and two fatty acids

Although most foods contain both saturated and unsaturated fats, animal foods tend to contain more saturated and less unsaturated fat than plant foods. Saturated fatty acids tend to increase blood levels of LDL cholesterol (the lipoprotein that increases heart-disease risk when present in high levels), whereas unsaturated fatty acids tend to decrease LDL-cholesterol levels.[14]

Hydrogenation and Trans Fats Oils can be made solid by adding hydrogen to the double bonds of their unsaturated fatty acids. This process, called hydrogenation, makes some of the fatty acids in oils saturated and enhances storage life and baking qualities. Hydrogenation may alter the molecular structure of the fatty acids, however, changing the naturally occurring *cis* structure to the *trans* form. Trans fatty acids raise blood LDL-cholesterol levels to a greater extent than do saturated fatty acids. Trans fatty acids are naturally present in dairy products and meats, but the primary

Saturated Fats Fats in which adjacent carbons in the fatty acid component are linked by single bonds only (e.g., –C–C–C–C–).

Unsaturated Fats Fats in which adjacent carbons in one or more fatty acids are linked by one or more double bonds (e.g., –C–C=C–C=C–).

Monounsaturated Fats Fats in which only one pair of adjacent carbons in one or more of its fatty acids is linked by a double bond (e.g., –C–C=C–C–).

Polyunsaturated Fats Fats in which more than one pair of adjacent carbons in one or more of its fatty acids are linked by two or more double bonds (e.g., –C–C=C–C=C–).

Trans fat   A type of unsaturated fat present in hydrogenated oils, margarine, shortenings, pastries, and some cooking oils that increase the risk of heart disease. Fats containing fatty acids in the *trans* versus the more common *cis* form are generally referred to as *trans* fat.

Cholesterol   A fat-soluble, colorless liquid primarily found in animals products.

dietary sources are products made from hydrogenated fats. Due to new nutrition labeling requirements and public uproar, the **trans fat** content of bakery products, chips, fast foods, and other products made with hydrogenated fats is decreasing.

Cholesterol   Dietary *cholesterol* is a fatlike, clear liquid substance primarily found in lean and fat components of animal products. Cholesterol is a component of all animal cell membranes, the brain, and the nerves. It is the precursor of estrogen, testosterone, and vitamin D, which is manufactured in the skin upon exposure to sunlight. The body generally produces only one-third of the cholesterol our bodies use, because more than sufficient amounts of cholesterol are provided in most people's diet. The extent to which dietary cholesterol intake modifies blood cholesterol level appears to vary a good deal based on genetic tendencies.[15] Dietary cholesterol intake affects blood cholesterol level substantially less than do saturated and *trans* fat intake, however.[7] Leading sources of dietary cholesterol are egg yolks, meat, milk and milk products, and fats such as butter.

Recommended Intake of Fats   Scientific evidence and opinions related to the effects of fat on health have changed substantially in recent years—and so have recommendations for fat intake. In the past, it was recommended that Americans aim for diets providing less than 30% of total calories from fat. Evidence indicating that the *type* of fat consumed is more important to health than is total fat intake has changed this advice. The watchwords for thinking about fat have become "not all fats are created equal: some are better for you than others." Concerns that high-fat diets encourage the development of obesity have been eased by studies demonstrating that excessive caloric intakes—and not just diets high in fat—are related

to weight gain. Current recommendations regarding fat intake do not encourage increased fat consumption, but rather emphasize that healthy diets include certain types of fat and that total caloric intake and physical activity are the most important components of weight management.[7]

Fats that elevate levels of LDL-cholesterol (which increases the risk of heart disease) are regarded as "unhealthful," while those that lower LDL cholesterol and raise blood levels of HDL cholesterol (the one that helps the body get rid of cholesterol in the blood) are considered healthful. The list of unhealthy fats includes *trans* fats, saturated fats, and cholesterol. Monounsaturated fats, polyunsaturated fats, alpha-linolenic acid, DHA, and EPA are considered healthful fats.

Current recommendations call for consumption of 20–35% of total calories from fat. The AIs for the essential fatty acid linoleic acid are set at 17 grams a day for men and 12 grams for women. AIs for the other essential fatty acid, alpha-linolenic acid, are 1.6 grams per day for men and 1.1 grams for women. It is recommended that people keep their intake of *trans* fats and saturated fats as low as possible while consuming a nutritionally adequate diet. Americans are being encouraged to increase consumption of EPA and DHA by eating fish more often. They are being urged to reduce saturated fat intake in order to reduce the risk of heart disease.[7]

There is no recommended level of cholesterol intake, because there is no evidence that cholesterol is required in the diet. The body is able to produce enough cholesterol, and people do not develop a cholesterol deficiency disease if it is not consumed. Because blood cholesterol levels tend to increase somewhat as consumption of cholesterol increases, it is recommended that intake should be minimal. Cholesterol intake averages around 237 mg per day in the United States, but a more health-promoting level of intake would be less than 200 mg a day.[16]

Food Sources of Fat   The fat content of many foods can be identified by reading the nutrition information labels on food packages. The amount of total fat, saturated fat, *trans* fat, and cholesterol in a serving of food is listed

Table 1.7 Food sources of fats

			A. Total Fat			
	Portion Size	Grams of Total Fat		Portion Size	Grams of Total Fat
Fats and Oils			Hamburger, 21% fat	3 oz	15.0
Mayonnaise	1 Tbsp	11.0	Hamburger, 16% fat	3 oz	13.5
Ranch dressing	1 Tbsp	6.0	Steak, rib-eye	3 oz	9.9
Vegetable oils	1 tsp	4.7	Bacon	3 strips	9.0
Butter	1 tsp	4.0	Steak, round	3 oz	5.2
Margarine	1 tsp	4.0	Chicken, baked, no skin	3 oz	4.0
Meats, Fish			Flounder, baked	3 oz	1.0
Sausage	4 links	18.0	Shrimp, boiled	3 oz	1.0
Hot dog	2 oz	17.0			

Table 1.7 Food sources of fats (continued)

A. Total Fat (Continued)

	Portion Size	Grams of Total Fat		Portion Size	Grams of Total Fat
Fast Foods			Milk, 1%	1 c	2.9
Whopper	8.9 oz	32.0	Milk, skim	1 c	0.4
Big Mac	6.6 oz	31.4	Yogurt, frozen	1 c	0.3
Quarter Pounder with Cheese	6.8 oz	28.6	**Other Foods**		
Veggie pita	1	17.0	Avocado	½	15.0
Subway meatball sandwich	1	16.0	Almonds	1 oz	15.0
Subway turkey sandwich	1	4.0	Cashews	1 oz	13.2
Milk and Milk Products			French fries, small serving	1	10.0
Cheddar cheese	1 oz	9.5	Taco chips	1 oz (10 chips)	10.0
Milk, whole	1 c	8.5	Potato chips	1 oz (14 chips)	7.0
American cheese	1 oz	6.0	Peanut butter	1 Tbsp	6.1
Cottage cheese, regular	½ c	5.1	Egg	1	6.0
Milk, 2%	1 c	5.0			

B. Saturated Fats

	Portion Size	Grams of Saturated Fat		Portion Size	Grams of Saturated Fat
Fats and Oils			Haddock, breaded, fried	3 oz	3.0
Margarine	1 tsp	2.9	Rabbit	3 oz	3.0
Butter	1 tsp	2.4	Pork chop, lean	3 oz	2.7
Salad dressing, ranch	1 Tbsp	1.2	Steak, round, lean	3 oz	2.0
Peanut oil	1 tsp	0.9	Turkey, roasted	3 oz	2.0
Olive oil	1 tsp	0.7	Chicken, baked, no skin	3 oz	1.7
Salad dressing, thousand island	1 Tbsp	0.5	Prime rib, lean	3 oz	1.3
Canola oil	1 tsp	0.3	Venison	3 oz	1.1
Milk and Milk Products			Tuna, in water	3 oz	0.4
Cheddar cheese	1 oz	5.9	**Fast Foods**		
American cheese	1 oz	5.5	Croissant w/ egg, bacon, & cheese	1	16.0
Milk, whole	1 c	5.1	Sausage croissant	1	16.0
Cottage cheese, regular	½ c	3.0	Whopper	1	11.0
Milk, 2%	1 c	2.9	Cheeseburger	1	9.0
Milk, 1%	1 c	1.5	Bac'n Cheddar Deluxe	1	8.7
Milk, skim	1 c	0.3	Taco, regular	1	4.0
Meats, Fish			Chicken breast sandwich	1	3.0
Hamburger, 21% fat	3 oz	6.7	**Nuts and Seeds**		
Sausage, links	4	5.6	Macadamia nuts	1 oz	3.2
Hot dog	1	4.9	Peanuts, dry-roasted	1 oz	1.9
Chicken, fried, with skin	3 oz	3.8	Sunflower seeds	1 oz	1.6
Salami	3 oz	3.6			

C. Unsaturated Fats

	Portion Size	Grams of Unsaturated Fat		Portion Size	Grams of Unsaturated Fat
Fats and Oils			Haddock, breaded, fried	3 oz	6.5
Canola oil	1 tsp	4.1	Chicken, baked, no skin	3 oz	6.0
Vegetable oils	1 tsp	3.6	Pork chop, lean	3 oz	5.3
Margarine	1 tsp	2.9	Turkey, roasted	3 oz	4.5
Butter	1 tsp	1.3	Tuna, in water	3 oz	0.7
Milk and Milk Products			Egg	1	5.0
Cottage cheese, regular	½ c	3.0	**Nuts and Seeds**		
Cheddar cheese	1 oz	2.9	Sunflower seeds	1 oz	16.6
American cheese	1 oz	2.8	Almonds	1 oz	12.6
Milk, whole	1 c	2.8	Peanuts	1 oz	11.3
Meats, Fish			Cashews	1 oz	10.2
Hamburger, 21% fat	3 oz	10.9			

continued

Table 1.7 Food sources of fats (continued)

D. Trans Fats

	Portion Size	Grams Trans Fats		Portion Size	Grams Trans Fats
Fats and Oils			**Milk**		
Margarine, stick	1 tsp	1.3	Whole	1 c	0.2
Margarine, tub (soft)	1 tsp	0.1	**Other Foods**		
Shortening	1 tsp	0.3	Doughnut	1	3.2
Butter	1 tsp	0.1	Danish pastry	1	3.0
Margarine, "no trans fat"	1 tsp	0	French fries, small serving	1	2.9
Meats			Cookies	2	1.8
Beef	3 oz	0.5	Corn chips	1 oz	1.4
Chicken	3 oz	0.1	Cake	1 slice	1.0
			Crackers	4 squares	0.5

E. Cholesterol

	Portion Size	Milligrams Cholesterol		Portion Size	Milligrams Cholesterol
Fats and Oils			Ostrich, ground	3 oz	63
Butter	1 tsp	10.3	Pork chop, lean	3 oz	60
Vegetable oils, margarine	1 tsp	0	Hamburger, 10% fat	3 oz	60
Meats, Fish			Venison	3 oz	48
Brain	3 oz	1476	Wild pig	3 oz	33
Liver	3 oz	470	Goat, roasted	3 oz	32
Egg	1	212	Tuna, in water	3 oz	25
Veal	3 oz	128	**Milk and Milk Products**		
Shrimp	3 oz	107	Ice cream, regular	1 c	56
Prime rib	3 oz	80	Milk, whole	1 c	34
Chicken, baked, no skin	3 oz	75	Milk, 2%	1 c	22
Salmon, broiled	3 oz	74	Yogurt, low fat	1 c	17
Turkey, baked, no skin	3 oz	65	Milk, 1%	1 c	14
Hamburger, 20% fat	3 oz	64	Milk, skim	1 c	7

F. Omega-3 (n-3) Fatty Acids

	Portion Size	Milligrams EPA + DHA		Portion Size	Milligrams EPA + DHA
Fish and Seafood			Oysters	3.5 oz	375
Fish oil	1 tsp	2796	Snapper	3.5 oz	273
Shad	3.5 oz	2046	Shrimp	3.5 oz	268
Salmon, farmed	3.5 oz	1825	Clams	3.5 oz	241
Anchovies	3.5 oz	1747	Haddock	3.5 oz	202
Herring	3.5 oz	1712	Catfish, wild	3.5 oz	201
Salmon, wild	3.5 oz	1564	Crawfish	3.5 oz	187
Whitefish	3.5 oz	1370	Sheepshead	3.5 oz	162
Mackerel	3.5 oz	1023	Tuna, light, and in oil	3.5 oz	109
Sardines	3.5 oz	840	Lobster	3.5 oz	71
Tilefish	3.5 oz	796	**Other**		
Whiting	3.5 oz	440	Egg yolk	1	40
Flounder	3.5 oz	426	DHA-fortified egg	1	150
Trout, freshwater	3.5 oz	420	Human milk	4.0 oz	126
			DHA-fortified beverages	4 oz	32

*Mercury content <0.2 ppm as given in *Mercury levels in commercial fish and shellfish*, 2006 update, U.S. Environmental Protection Agency, www.epa.gov.

on the label. Table 1.7 lists the total fat, saturated fat, unsaturated fat, *trans* fat, cholesterol, and omega-3–fatty acid contents (EPA and DHA) of selected foods.

Vitamins

Vitamins are chemical substances in foods that perform specific functions in the body. Fourteen have been discovered so far. They are classified as either fat soluble or water soluble (Table 1.8).

The B-complex vitamins and vitamin C are soluble in water and found dissolved in water in foods. The fat-soluble vitamins consist of vitamins A, D, E, and K and are present in the fat portions of foods. (To remember the fat-soluble vitamins, think of "DEKA" for vitamins D, E, K, and A.) Only these chemical substances are truly vitamins. Substances such as coenzyme Q_{10}, inositol, provitamin B_5 complex, and pangamic acid (vitamin B_{15}) may be called vitamins, but they are not. Except for vitamin B_{12}, water-soluble vitamin stores in the body are limited and run out within a few weeks to a few months after intake becomes inadequate. Fat-soluble vitamins are stored in the body's fat tissues and the liver. These stores can be sizable and last from months to years when intake is low.

Excessive consumption of the fat-soluble vitamins from supplements, especially of vitamins A and D, produces various symptoms of toxicity. High intake of the water-soluble vitamins from supplements can also produce adverse health effects. Toxicity symptoms from water-soluble vitamins, however, tend to last a shorter time and are more quickly remedied. Vitamin overdoses are very rarely related to food intake.

Vitamins do not provide energy or, with the exception of choline, serve as structural components of the body. Some play critical roles as *coenzymes* in chemical changes that take place in the body, known as *metabolism*. Vitamin A is needed to replace the cells that line the mouth and esophagus, thiamin is needed for maintenance of normal appetite, and riboflavin and folate are needed for the synthesis of body proteins. Other vitamins (vitamins C and E, and beta-carotene—a precursor of vitamin A) act as *antioxidants* and perform other functions. By preventing or repairing damage to cells due to oxidation, these vitamins help maintain body tissues and prevent disease.

Primary functions, consequences of deficiency and overdose, primary food sources, and comments about each vitamin are listed in Table 1.9 starting on the next page.

Recommended Intake of Vitamins Recommendations for levels of intake of vitamins are presented in the tables on the inside front covers of this text. Note that Tolerable Upper Levels of Intake (ULs) for many vitamins are also given; they represent levels of intake that should not be exceeded. Table 1.10 (pages 19–21) lists food sources of each vitamin.

Other Substances in Food

"Things don't happen by accident in nature. If you observe it, it has a reason for being there."

Norman Krinsky, Tufts University

There are many substances in foods in addition to nutrients that affect health. Some foods contain naturally occurring toxins, such as poison in puffer fish and solanine in green sections near the skin of some potatoes. Consuming the poison in puffer fish can be lethal; large doses of solanine can interfere with nerve impulses. Some plant pigments, hormones, and other naturally occurring substances that protect plants from insects, oxidization, and other damaging exposures also appear to benefit human health. These substances in plants are referred to as *phytochemicals*, and knowledge about their effects on human health is advancing rapidly. Many of the phytochemicals that benefit health are pigments that act as antioxidants in the human body. Table 1.11 shows a list of the top food sources of antioxidants. Notice that most of the foods listed are colorful. That's due to the antioxidant pigments they contain.

Consumption of foods rich in specific pigments and other phytochemicals, rather than consumption of isolated phytochemicals, may help prevent certain types of cancer, cataracts, type 2 diabetes, hypertension, infections, and heart disease. High intakes of certain phytochemicals from vegetables, fruits, nuts, seeds, and whole-grain products may partially account for lower rates of heart disease and cancer observed in people with high intakes of these foods.[19,20]

Coenzymes Chemical substances that activate enzymes.

Metabolism The chemical changes that take place in the body. The conversion of glucose to energy or body fat is an example of a metabolic process.

Antioxidants Chemical substances that prevent or repair damage to cells caused by exposure to oxidizing agents such as oxygen, ozone, and smoke and to other oxidizing agents normally produced in the body. Many different antioxidants are found in foods; some are made by the body.

Phytochemicals (phyto = plants) Chemical substances in plants, some of which affect body processes in humans that may benefit health.

Table 1.8 Vitamin solubility

Water-Soluble Vitamins	Fat-Soluble Vitamins
B-complex vitamins	Vitamin A (retinol, beta-carotene)
Thiamin (B_1)	
Riboflavin (B_2)	Vitamin D (1,25 dihydroxy-cholecalciferol)
Niacin (B_3)	
Vitamin B_6	
Folate	Vitamin E (alpha-tocopherol)
Vitamin B_{12}	
Biotin	Vitamin K
Pantothenic acid	
Choline	
Vitamin C (ascorbic acid)	

Table 1.9 Summary of the vitamins

The Water-Soluble Vitamins

	Primary Functions	Consequences of Deficiency
Thiamin (vitamin B$_1$) AI[a] women: 1.1 mg men: 1.2 mg	• Coenzyme in the metabolism of carbohydrates, alcohol, and some amino acids • Required for the growth and maintenance of nerve and muscle tissues • Required for normal appetite	• Fatigue, weakness • Nerve disorders, mental confusion, apathy • Impaired growth • Swelling • Heart irregularity and failure
Riboflavin (vitamin B$_2$) AI women: 1.1 mg men: 1.3 mg	• Coenzyme involved in energy metabolism of carbohydrates, proteins, and fats • Coenzyme function in cell division • Promotes growth and tissue repair • Promotes normal vision	• Reddened lips, cracks at both corners of the mouth • Fatigue
Niacin (vitamin B$_3$) RDA women: 14 mg men: 16 mg UL: 35 mg (from supplements and fortified foods)	• Coenzyme involved in energy metabolism • Coenzyme required for the synthesis of body fats • Helps maintain normal nervous system functions	• Skin disorders • Nervous and mental disorders • Diarrhea, indigestion • Fatigue
Vitamin B$_6$ (pyridoxine) AI women: 1.3 mg men: 1.3 mg UL: 100 mg	• Coenzyme involved in amnio acid, glucose, and fatty acid metabolism and neurotransmitter synthesis • Coenzyme in the conversion of tryptophan to niacin • Required for normal red blood cell formation • Required for the synthesis of lipids in the nervous and immune systems	• Irritability, depression • Convulsions, twitching • Muscular weakness • Dermatitis near the eyes • Anemia • Kidney stones
Folate (folacin, folic acid) RDA women: 400 mcg men: 400 mcg UL: 1000 mcg (from supplements and fortified foods)	• Required for the conversion of homocysteine to methionine • Methyl (CH$_3$) group donor and coenzyme in DNA synthesis, gene expression and regulation • Required for the normal formation of red blood and other cells	• Megaloblastic cells and anemia • Diarrhea, weakness, irritability, paranoid behavior • Red, sore tongue • Increased blood homocysteine levels • Neural tube defects, low birthweight and preterm delivery (in pregnancy)
Vitamin B$_{12}$ (cyanocobalamin) AI women: 2.4 mcg men: 2.4 mcg	• Coenzyme involved in the synthesis of DNA, RNA, and myelin • Required for the conversion of homocysteine to methionine • Needed for normal red blood cell development	• Neurological disorders (nervousness, tingling sensations, brain degeneration) • Pernicious anemia • Increased blood homocysteine levels • Fatigue
Biotin AI women: 30 mcg men: 30 mcg	• Required by enzymes involved in fat, protein, and glycogen metabolism	• Seizures, vision problems • Hearing loss • Weakness

[a]AI (Adequate Intakes) and RDAs (Recommended Dietary Allowances) are for 19–30-year-olds; UL (Upper Limits) are for 19–70-year-olds, 1997–2004.

Table 1.9 Summary of the vitamins (continued)

The Water-Soluble Vitamins (Continued)

Consequences of Overdose	Primary Food Sources	Highlights and Comments
• High intakes of thiamin are rapidly excreted by the kidneys. Oral doses of 500 mg/day or less are considered safe	• Grains and grain products (cereals, rice, pasta, bread) • Ready-to-eat cereals • Pork and ham, liver • Milk, cheese, yogurt • Dried beans and nuts	• Need increases with carbohydrate intake • There is no "e" on the end of *thiamin* • Deficiency rare in the U.S.; may occur in people with alcoholism • Enriched grains and cereals prevent thiamin deficiency
• None known. High doses are rapidly excreted by the kidneys	• Milk, yogurt, cheese • Grains and grain products (cereals, rice, pasta, bread) • Liver, poultry, fish, beef • Eggs	• Destroyed by exposure to light
• Flushing, headache, cramps, rapid heartbeat, nausea, diarrhea, decreased liver function with doses above 0.5 g per day	• Meats (all types) • Grains and grain products (cereals, rice, pasta, bread) • Dried beans and nuts • Milk, cheese, yogurt • Ready-to-eat cereals • Coffee • Potatoes	• Niacin has a precursor—tryptophan. Tryptophan, an amino acid, is converted to niacin by the body. Much of our niacin intake comes from tryptophan • High doses raise HDL-cholesterol levels, decrease LDL-cholesterol, and lower triglyceride levels
• Bone pain, loss of feeling in fingers and toes, muscular weakness, numbness, loss of balance (mimicking multiple sclerosis)	• Oatmeal, bread, breakfast cereals • Bananas, avocados, prunes, tomatoes, potatoes • Chicken, liver • Dried beans • Meats (all types), milk • Green and leafy vegetables	• Vitamins go from B_3 to B_6 because B_4 and B_5 were found to be duplicates of vitamins already identified
• May cover up signs of vitamin B_{12} deficiency (pernicious anemia)	• Fortified, refined grain products (bread, flour, pasta) • Ready-to-eat cereals • Dark green, leafy vegetables (spinach, collards, romaine) • Broccoli, brussels sprouts • Oranges, bananas, grapefruit • Milk, cheese, yogurt • Dried beans	• Folate means "foliage." It was first discovered in leafy green vegetables • This vitamin is easily destroyed by heat • Synthetic form (folic acid) added to fortified grain products is better absorbed than naturally occurring folates • Very low and very high intakes of folic acid appear to increase the risk of some types of cancer
• None known. Excess vitamin B_{12} is rapidly excreted by the kidneys or is not absorbed into the bloodstream • Vitamin B_{12} injections may cause a temporary feeling of heightened energy	• Animal products: beef, lamb, liver, clams, crab, fish, poultry, eggs • Milk and milk products • Ready-to-eat cereals	• Older people and vegans are at risk for vitamin B_{12} deficiency • Some people become vitamin B_{12} deficient because they are genetically unable to absorb it • Vitamin B_{12} is found in animal products and microorganisms only
• None known. Excesses are rapidly excreted	• Grain and cereal products • Meats, dried beans, cooked eggs • Vegetables	• Deficiency is extremely rare. May be induced by the overconsumption of raw eggs

continued

Table 1.9 Summary of the vitamins (continued)

The Water-Soluble Vitamins (Continued)

	Primary Functions	Consequences of Deficiency
Pantothenic acid (pantothenate) AI women: 5 mg men: 5 mg	• Coenzyme involved in energy metabolism of carbohydrates and fats • Coenzyme in protein metabolism	• Fatigue, sleep disturbances, numbness, impaired coordination • Vomiting, nausea

Vitamin C (ascorbic acid)
RDA   women: 75 mg
          men: 90 mg
          UL: 2000 mg
	• Required for collagen synthesis
	• Acts as an antioxidant; protects LDL cholesterol, eye tissues, sperm proteins, DNA, and lipids against oxidation
	• Required for the conversion of Fe^{++} to Fe^{+++}
	• Required for neurotransmitters and steroid hormone synthesis
		• Bleeding and bruising easily due to weakened blood vessels, cartilage, and other tissues containing collagen
		• Slow recovery from infections and poor wound healing
		• Fatigue, depression

Choline AI women: 425 mg AI men: 550 mg UL: 3.5 g	• Serves as a structural and signaling component of cell membranes • Required for the normal development of memory and attention processes during early life • Required for the transport and metabolism of fat and cholesterol	• Fatty liver • Infertility • Hypertension

The Fat-Soluble Vitamins

	Primary Functions	Consequences of Deficiency

Vitamin A
RDA   women: 700 mcg
          men: 900 mcg
          UL: 3000 mcg
	• Needed for the formation and maintenance of mucous membranes, skin, bone
	• Needed for vision in dim light
		• Increased susceptibility to infection, increased incidence and severity of infection (including measles)
		• Impaired vision, xerophthalmia, blindness
		• Inability to see in dim light

Vitamin E (alpha-tocopherol)
RDA   women: 15 mg
          men: 15 mg
          UL: 1000 mg
	• Acts as an antioxidant, prevents damage to cell membranes in blood cells, lungs, and other tissues by repairing damage caused by free radicals
	• Reduces oxidation of LDL cholesterol
		• Muscle loss, nerve damage
		• Anemia
		• Weakness
		• Many adults may have nonoptimal blood levels

Table 1.9 Summary of the vitamins (continued)

The Water-Soluble Vitamins (Continued)

Consequences of Overdose	Primary Food Sources	Highlights and Comments
• None known. Excesses are rapidly excreted	• Many foods, including meats, grains, vegetables, fruits, and milk	• Deficiency is very rare
• Intakes of 1 g or more per day can cause nausea, cramps, and diarrhea and may increase the risk of kidney stones	• Fruits: oranges, lemons, limes, strawberries, cantaloupe, honeydew melon, grapefruit, kiwi fruit, mango, papaya • Vegetables: broccoli, green and red peppers, collards, cabbage, tomatoes, asparagus, potatoes • Ready-to-eat cereals	• Need increases among smokers (to 110–125 mg per day) • Is fragile; easily destroyed by heat and exposure to air • Supplements may decrease duration and symptoms of colds in some people • Deficiency may develop within 3 weeks of very low intake
• Low blood pressure • Sweating, diarrhea • Fishy body odor • Liver damage	• Beef • Eggs • Pork • Dried beans • Fish • Milk	• Most of the choline we consume from foods comes from its location in cell membranes • Lecithin, an additive commonly found in processed foods, is a rich source of choline • Choline is primarily found in animal products • It is considered a B-complex vitamin

The Fat-Soluble Vitamins (Continued)

Consequences of Overdose	Primary Food Sources	Highlights and Comments
• Vitamin A toxicity (hypervitamosis A) with acute doses of 500,000 IU, or long-term intake of 50,000 IU per day; limit retinol use in pregnancy to 5000 IU daily • Nausea, irritability, blurred vision, weakness • Increased pressure in the skull, headache • Liver damage • Hair loss, dry skin • Birth defects	• Vitamin A is found in animal products only • Liver, butter, margarine, milk, cheese, eggs • Ready-to-eat cereals	• Beta-carotene is a vitamin A precursor or "provitamin" It functions as an antioxidant • Symptoms of vitamin A toxicity may mimic those of brain tumors and liver disease. Vitamin A toxicity is sometimes misdiagnosed because of the similarities in symptoms • 1 mcg retinol equivalent = 5 IU vitamin A or 3.6 mcg beta-carotene
• Intakes of up to 800 IU per day are unrelated to toxic side effects; over 800 IU per day may increase bleeding (blood-clotting time) • Avoid supplement use if aspirin, anticoagulants, or fish oil supplements are taken regularly	• Oils and fats • Salad dressings, mayonnaise, margarine, shortening, butter • Whole grains, wheat germ • Leafy, green vegetables, tomatoes • Nuts and seeds • Eggs	• Vitamin E is destroyed by exposure to oxygen and heat • Oils naturally contain vitamin E; it's there to protect the fat from breakdown due to free radicals • Eight forms of vitamin E exist, and each has different antioxidant strengths • Natural form is better absorbed than synthetic form: 15 IU alpha-tocopherol = 22 IU d-alpha-tocopherol (natural form) and 33 IU synthetic vitamin E

continued

Table 1.9 Summary of the vitamins (continued)

The Fat-Soluble Vitamins (Continued)

	Primary Functions	Consequences of Deficiency

Vitamin D
(**Vitamin D$_2$** = ergocalciferol,
Vitamin D$_3$ = cholicalciferol)
AI   women: 5 mcg
            (200 IU)
       men: 5 mcg
            (200 IU)
       UL: 50 mcg
            (2000 IU)
	• Required for calcium and phosphorus metabolism in the intestines and bone, and for their utilization in bone and teeth formation, nerve and muscle activity
	• Inhibits inflammation
		• Weak, deformed bones (children)
		• Loss of calcium from bones (adults), osteoporosis
		• Increased risk of heart disease, type 1 diabetes, metabolic syndrome, and other inflammatory diseases

Vitamin K (**phylloquinone, menaquinone**) AI women: 90 mcg men: 120 mcg	• Regulation of synthesis of blood-clotting proteins • Aids in the incorporation of calcium into bones	• Bleeding, bruises • Decreased calcium in bones • Deficiency is rare; may be induced by the long-term use (months or more) of antibiotics

Table 1.10 Food sources of vitamins

Thiamin

Food	Amount	Thiamin (Milligrams)	Food	Amount	Thiamin (Milligrams)
Meats			Rice	½ c	0.1
Pork roast	3 oz	0.8	Bread	1 slice	0.1
Beef	3 oz	0.4	**Vegetables**		
Ham	3 oz	0.4	Peas	½ c	0.3
Liver	3 oz	0.2	Lima beans	½ c	0.2
Nuts and Seeds			Corn	½ c	0.1
Sunflower seeds	¼ c	0.7	Broccoli	½ c	0.1
Peanuts	¼ c	0.1	Potato	1 med	0.1
Almonds	¼ c	0.1	**Fruits**		
Grains			Orange juice	1 c	0.2
Bran flakes	1 c (1 oz)	0.6	Orange	1	0.1
Macaroni	½ c	0.1	Avocado	½	0.1

Riboflavin

Food	Amount	Riboflavin (Milligrams)	Food	Amount	Riboflavin (Milligrams)
Milk and Milk Products			Beef	3 oz	0.2
Milk	1 c	0.5	Tuna	3 oz	0.1
2% milk	1 c	0.5	**Vegetables**		
Yogurt, low fat	1 c	0.5	Collard greens	½ c	0.3
Skim milk	1 c	0.4	Broccoli	½ c	0.2
Yogurt	1 c	0.1	Spinach, cooked	½ c	0.1
American cheese	1 oz	0.1	**Eggs**		
Cheddar cheese	1 oz	0.1	Egg	1	0.2
Meats			**Grains**		
Liver	3 oz	3.6	Macaroni	½ c	0.1
Pork chop	3 oz	0.3	Bread	1 slice	0.1

Table 1.9 Summary of the vitamins (continued)

The Fat-Soluble Vitamins (Continued)

Consequences of Overdose	Primary Food Sources	Highlights and Comments
• Mental retardation in young children • Abnormal bone growth and formation • Nausea, diarrhea, irritability, weight loss • Deposition of calcium in organs such as the kidneys, liver, and heart • Toxicity possible with long-term dose levels over 10,000 IU per day	• Vitamin D–fortified milk and margarine • Butter • Fatty fish • Eggs • Mushrooms • Milk products such as cheese and yogurt, and breads and cereals may be fortified with vitamin D	• Vitamin D_3 is the most active form of this vitamin • Vitamin D is manufactured from cholesterol in cells beneath the surface of the skin upon exposure of the skin to sunlight • Poor vitamin D status is common in all age groups • The AI for vitamin D may be increased in 2009 • Breastfed infants with little sun exposure benefit from vitamin D supplements
• Toxicity is a problem only when synthetic forms of vitamin K are taken in excessive amounts; that may cause liver disease	• Leafy, green vegetables • Grain products	• Vitamin K is produced by bacteria in the gut; part of our vitamin K supply comes from these bacteria • Newborns are given a vitamin K injection because they have "sterile" guts and consequently no vitamin K–producing bacteria

Table 1.10 Food sources of vitamins (continued)

Niacin

Food	Amount	Niacin (Milligrams)	Food	Amount	Niacin (Milligrams)
Meats			**Vegetables**		
Liver	3 oz	14.0	Asparagus	½ c	1.5
Tuna	3 oz	10.3	**Grains**		
Turkey	3 oz	9.5	Wheat germ	1 oz	1.5
Chicken	3 oz	7.9	Brown rice	½ c	1.2
Salmon	3 oz	6.9	Noodles, enriched	½ c	1.0
Veal	3 oz	5.2	Rice, white, enriched	½ c	1.0
Beef (round steak)	3 oz	5.1	Bread, enriched	1 slice	0.7
Pork	3 oz	4.5	**Milk and Milk Products**		
Haddock	3 oz	2.7	Cottage cheese	½ c	2.6
Scallops	3 oz	1.1	Milk	1 c	1.9
Nuts and Seeds					
Peanuts	1 oz	4.9			

Vitamin B_6

Food	Amount	Vitamin B_6 (Milligrams)	Food	Amount	Vitamin B_6 (Milligrams)
Meats			**Eggs**		
Liver	3 oz	0.8	Egg	1	0.3
Salmon	3 oz	0.7	**Legumes**		
Other fish	3 oz	0.6	Split peas	½ c	0.6
Chicken	3 oz	0.4	Dried beans, cooked	½ c	0.4
Ham	3 oz	0.4	**Fruits**		
Hamburger	3 oz	0.4	Banana	1	0.6
Veal	3 oz	0.4	Avocado	½	0.4
Pork	3 oz	0.3	Watermelon	1 c	0.3
Beef	3 oz	0.2			

continued

Table 1.10 Food sources of vitamins (continued)

Vitamin B$_6$ (Continued)

Food	Amount	Vitamin B$_6$ (Milligrams)	Food	Amount	Vitamin B$_6$ (Milligrams)
Vegetables			Sweet potato	½ c	0.2
Turnip greens	½ c	0.7	Carrots	½ c	0.2
Brussels sprouts	½ c	0.4	Peas	½ c	0.1
Potato	1	0.2			

Folate

Food	Amount	Folate (Micrograms)	Food	Amount	Folate (Micrograms)
Vegetables			**Fruits**		
Garbanzo beans	½ c	141	Cantaloupe	¼ whole	100
Spinach, cooked	½ c	131	Orange juice	1 c	87
Navy beans	½ c	128	Orange	1	59
Asparagus	½ c	120	**Grains**[a]		
Brussels sprouts	½ c	116	Ready-to-eat cereals	1 c (1 oz)	100–400
Black-eyed peas	½ c	102	Oatmeal	½ c	97
Romaine lettuce	1 c	86	Rice	½ c	77
Lima beans	½ c	71	Noodles	½ c	45
Peas	½ c	70	Wheat germ	2 Tbsp	40
Collard greens, cooked	½ c	56			
Sweet potato	½ c	43			
Broccoli	½ c	43			

Vitamin B$_{12}$

Food	Amount	Vitamin B$_{12}$ (Micrograms)	Food	Amount	Vitamin B$_{12}$ (Micrograms)
Meats			**Milk and Milk Products**		
Liver	3 oz	6.8	Skim milk	1 c	1.0
Trout	3 oz	3.6	Milk	1 c	0.9
Beef	3 oz	2.2	Yogurt	1 c	0.8
Clams	3 oz	2.0	Cottage cheese	½ c	0.7
Crab	3 oz	1.8	American cheese	1 oz	0.2
Lamb	3 oz	1.8	Cheddar cheese	1 oz	0.2
Tuna	3 oz	1.8	**Eggs**		
Veal	3 oz	1.7	Egg	1	0.6
Hamburger, regular	3 oz	1.5			

Choline

Food	Amount	Choline (Milligrams)	Food	Amount	Choline (Milligrams)
Meats			Navy beans, boiled	½ c	41
Beef	3 oz	111	Collards, cooked	½ c	39
Pork chop	3 oz	94	Black-eyed peas (cowpeas)	½ c	39
Lamb	3 oz	89	Chickpeas (garbanzo beans)	½ c	35
Ham	3 oz	87	Brussels sprouts	½ c	32
Beef	3 oz	85	Broccoli	½ c	32
Turkey	3 oz	70	Collard greens	½ c	30
Salmon	3 oz	56	Refried beans	½ c	29
Eggs			**Milk and Milk products**		
Egg	1 large	126	Milk, 2%	1 c	40
Vegetables			Cottage cheese, low fat	½ c	37
Baked beans	½ c	50	Yogurt, low fat	1 c	35

[a]Fortified, refined grain products such as bread, rice, pasta, and crackers provide approximately 40 to 60 mcg of folic acid per standard serving.

Table 1.10 Food sources of vitamins (continued)

Vitamin C

Food	Amount	Vitamin C (Milligrams)	Food	Amount	Vitamin C (Milligrams)
Fruits			Watermelon	1 c	15
Orange juice, vitamin C–fortified	1 c	108	**Vegetables**		
Kiwi fruit	1 or ½ c	108	Green peppers	½ c	95
Grapefruit juice, fresh	1 c	94	Cauliflower, raw	½ c	75
Cranberry juice cocktail	1 c	90	Broccoli	½ c	70
Orange	1	85	Brussels sprouts	½ c	65
Strawberries, fresh	1 c	84	Collard greens	½ c	48
Orange juice, fresh	1 c	82	Vegetable (V-8) juice	¾ c	45
Cantaloupe	¼ whole	63	Tomato juice	¾ c	33
Grapefruit	1 med	51	Cauliflower, cooked	½ c	30
Raspberries, fresh	1 c	31	Potato	1 med	29
			Tomato	1 med	23

Vitamin A (Retinol)

Food Sources of Vitamin A (Retinol)	Amount	Vitamin A (Micrograms RE)[b]	Food Sources of Vitamin A (Retinol)	Amount	Vitamin A (Micrograms RE)[b]
Meats			2% milk	1 c	139
Liver	3 oz	9124	American cheese	1 oz	82
Salmon	3 oz	53	Whole milk	1 c	76
Tuna	3 oz	14	Swiss cheese	1 oz	65
Eggs			**Fats**		
Egg	1 med	84	Margarine, fortified	1 tsp	46
Milk and Milk Products			Butter	1 tsp	38
Skim milk, fortified	1 c	149			

Vitamin A (Beta-Carotene)

Food Sources of Beta-Carotene	Amount	Vitamin A Value (Micrograms RE)[b]	Food Sources of Beta-Carotene	Amount	Vitamin A Value (Micrograms RE)[b]
Vegetables			**Fruits**		
Pumpkin, canned	½ c	2712	Green peppers	½ c	40
Sweet potato, canned	½ c	1935	Cantaloupe	¼ whole	430
Carrots, raw	½ c	1913	Apricots, canned	½ c	210
Spinach, cooked	½ c	739	Nectarine	1 med	101
Collard greens, cooked	½ c	175	Watermelon	1 c	59
Broccoli, cooked	½ c	109	Peaches, canned	½ c	47
Winter squash	½ c	53	Papaya	½ c	20

Vitamin E

Food	Amount	Vitamin E (IU)[c]	Food	Amount	Vitamin E (IU)[c]
Oils			Collard greens	½ c	3.1
Vegetable oil	1 Tbsp	6.7	Asparagus	½ c	2.1
Mayonnaise	1 Tbsp	3.4	Spinach, raw	1 c	1.5
Margarine	1 Tbsp	2.7	**Grains**		
Salad dressing	1 Tbsp	2.2	Wheat germ	2 Tbsp	4.2
Nuts and Seeds			Bread, whole wheat	1 slice	2.5
Sunflower seeds	¼ c	27.1	Bread, white	1 slice	1.2
Almonds	¼ c	12.7	**Seafood**		
Peanuts	¼ c	4.9	Crab	3 oz	4.5
Cashews	¼ c	0.7	Shrimp	3 oz	3.7
Vegetables			Fish	3 oz	2.4
Sweet potato	½ c	6.9			

[b]RE (retinol equivalent) = 3.33 IU
[c]15 mg alpha-tocopherol = 22 IU d-alpha tocopherol (natural form) and 33 IU synthetic vitamin E

continued

Table 1.10 Food sources of vitamins (continued)

Vitamin D

Food	Amount	Vitamin D (IU)[d]	Food	Amount	Vitamin D (IU)[d]
Milk			**Organ meats**		
Milk, whole, low fat, or skim	1 c	100	Beef liver	3 oz	42
Fish and seafood			Chicken liver	3 oz	40
Salmon	3 oz	340	**Eggs**		
Tuna	3 oz	150	Egg yolk	1	27
Shrimp	3 oz	127			

[d]40 IU 5 1 mcg

Table 1.11 Top sources of antioxidant-rich foods[17,18]

Pomegranate	Red cabbage
Blackberries	Pecans
Walnuts	Cloves, ground
Blueberries, wild	Grape juice
Strawberries	Chocolate, dark
Raspberries	Cranberry juice
Artichokes	Wine, red
Cranberries	Pineapple juice
Blueberries, domestic	Guava nectar
Coffee, brewed	Mango nectar

Minerals

Humans require the 15 minerals listed in Table 1.12 (page 21). Minerals are unlike other nutrients in that they consist of single atoms and carry a charge in solution. The property of being charged (or having an unequal number of electrons and protons) is related to many of the functions of minerals. The charge carried by minerals allows them to combine with other minerals to form stable complexes in bone, teeth, cartilage, and other tissues. In body fluids, charged minerals serve as a source of electrical power that stimulates muscles to contract (e.g., the heart to beat) and nerves to react. Minerals also help the body maintain an adequate amount of water in tissues and control how acidic or basic body fluids remain.

The tendency of minerals to form complexes has implications for the absorption of minerals from food.

Table 1.12 Minerals required by humans

Calcium	Fluoride	Chromium
Phosphorus	Iodine	Molybdenum
Magnesium	Selenium	Sodium
Iron	Copper	Potassium
Zinc	Manganese	Chloride

Calcium and zinc, for example, may combine with other minerals in supplements or with dietary fiber and form complexes that cannot be absorbed. Therefore, in general, the proportion of total mineral intake that is absorbed is less than for vitamins.

Functions, consequences of deficiency and overdose, primary food sources, and comments about the 15 minerals needed by humans are summarized in Table 1.13 (pages 22–25).

Recommended Intake of Minerals Recommendations for intake of minerals are presented in the tables on the inside front covers of this text. Note that Tolerable Upper Levels of Intake for many minerals are also given in a separate table. Table 1.14 lists food sources of each mineral.

Water

Water is the last, but not the least, nutrient category. Adults are about 60–70% water by weight. Water provides the medium in which most chemical reactions take place in the body. It plays a role in energy transformation, the excretion of wastes, and temperature regulation.

People need enough water to replace daily losses from perspiration, urination, and exhalation. In normal weather conditions with normal physical activity levels, the total water requirement of adult males is 15–16 cups from foods and fluids per day. The corresponding figure for females is 11 cups. Total water intake includes drinking water, water in beverages, and water that is a part of food. Water from food provides about 19% of our total water intake.[7] The need for water is generally met by consuming sufficient fluids to satisfy thirst.[21] The need for water is greater in hot and humid climates, and when physical activity levels are high. People generally consume 75% of their water intake from water and other fluids and 25% from foods.[22] Adequate consumption of water is indicated by the excretion of urine that is pale yellow and normal in volume.[23]

Dietary Sources of Water   The best sources of water are tap and bottled water; nonalcoholic beverages such as fruit juice, milk, and vegetable juice; and brothy soups. Alcohol tends to increase water loss though urine, so beverages such as beer and wine are not as "hydrating" as water is. Caffeinated beverages are hydrating in people who are accustomed to consuming them.[24]

PRINCIPLE #3   Health problems related to nutrition originate within cells.

The functions of each cell are maintained by the nutrients it receives. Problems arise when a cell's need for nutrients differs from the amounts that are available. Cells (Illustration 1.5) are the building blocks of tissues (such as bones and muscles), organs (the heart, kidney, and liver, for example), and systems (such as the circulatory and respiratory systems). Normal cell health and functions are maintained when a nutritional and environmental utopia exists within and around cells. This state of optimal cellular nutrient conditions supports *homeostasis* in the body.

Disruptions in the availability of nutrients, or the presence of harmful substances in the cell's environment, initiate diseases and disorders that eventually affect tissues, organs, and systems. For example, folate, a B vitamin, is required for protein synthesis within cells. When too little folate is available, cells produce proteins with abnormal shapes and functions. Abnormalities in the shape of red blood cell proteins lead to functional changes that produce loss of appetite, weakness, and irritability.

PRINCIPLE #4   Poor nutrition can result from both inadequate and excessive levels of nutrient intake.

Illustration 1.6 Nutrient function and consequences by level of intake.

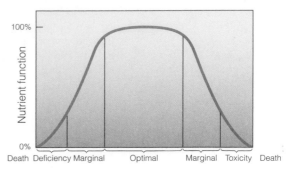

Increasing concentration or intake of nutrient ⟹

Each nutrient has a range of intake levels that corresponds to optimum functioning of that nutrient (Illustration 1.6). Intake levels below and above this range are associated with impaired functions.

> **Homeostasis**   Constancy of the internal environment. The balance of fluids, nutrients, gases, temperature, and other conditions needed to ensure ongoing, proper functioning of cells and, therefore, all parts of the body.

Inadequate intake of an essential nutrient, if prolonged, results in obvious deficiency diseases. Marginally deficient diets produce subtle changes in behavior or physical condition. If the optimal intake range is exceeded (usually by overdoses of supplements), mild to severe changes in mental and physical functions occur, depending on the amount of the excess and the nutrient involved. Overt vitamin C deficiency, for example, produces irritability, bleeding gums, pain upon being touched, and failure of bone growth. Marginal deficiency may cause delayed wound healing. The length of time a deficiency or toxicity takes to develop depends on the type and amount of the nutrient consumed and the extent of body nutrient reserves. Intakes of 32 mg/day of vitamin C, or about a third of the RDA for adults (75 mg and 90 mg per day for women and men, respectively), lower blood vitamin C levels to the deficient state within three weeks.[25] On the excessive side, too much supplemental vitamin C causes diarrhea. For nutrients, enough is as good as a feast.

Steps in the Development of Nutrient Deficiencies and Toxicities   Poor nutrition due to inadequate diet generally develops in the stages outlined in Illustration 1.7.

After a period of deficient intake of an essential nutrient, tissue reserves become depleted, and subsequently blood levels of the nutrient decline. When the blood level can no longer supply cells with optimal amounts

Illustration 1.5 Schematic representation of the structure and major components of a human cell.

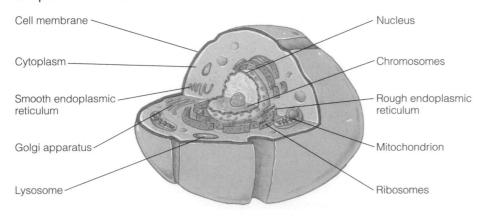

Cell membrane
Cytoplasm
Smooth endoplasmic reticulum
Golgi apparatus
Lysosome
Nucleus
Chromosomes
Rough endoplasmic reticulum
Mitochondrion
Ribosomes

Table 1.13 Summary of minerals

	Primary Functions	Consequences of Deficiency
Calcium AI* women: 1000 mg men: 1000 mg UL: 2500 mg	• Component of bones and teeth • Required for muscle and nerve activity, blood clotting	• Poorly mineralized, weak bones (osteoporosis) • Rickets in children • Osteomalacia (rickets in adults) • Stunted growth in children • Convulsions, muscle spasms
Phosphorus RDA women: 700 mg men: 700 mg UL: 4000 mg	• Component of bones and teeth • Component of certain enzymes and other substances involved in energy formation • Required for maintenance of acid-base balance of body fluids	• Loss of appetite • Nausea, vomiting • Weakness • Confusion • Loss of calcium from bones
Magnesium RDA women: 310 mg men: 400 mg UL: 350 mg (from supplements only)	• Component of bones and teeth • Needed for nerve activity • Activates enzymes involved in energy and protein formation	• Stunted growth in children • Weakness • Muscle spasms • Personality changes
Iron RDA women: 18 mg men: 8 mg UL: 45 mg	• Transports oxygen as a component of hemoglobin in red blood cells • Component of myoglobin (a muscle protein) • Required for certain reactions involving energy formation	• Iron deficiency • Iron-deficiency anemia • Weakness, fatigue • Hair loss • Pale appearance • Reduced attention span and resistance to infection • Mental retardation, developmental delay in children
Zinc RDA women: 8 mg men: 11 mg UL: 40 mg	• Required for the activation of many enzymes involved in the reproduction of proteins • Component of insulin, many enzymes	• Growth failure • Delayed sexual maturation • Slow wound healing • Loss of taste and appetite • In pregnancy, low-birth-weight infants and preterm delivery
Fluoride AI women: 3 mg men: 4 mg UL: 10 mg	• Component of bones and teeth (enamel)	• Tooth decay and other dental diseases
Iodine RDA women: 150 mcg men: 150 mcg UL: 1100 mcg	• Component of thyroid hormones that help regulate energy production and growth • Required for normal brain development	• Goiter • Cretinism (mental retardation, hearing loss, growth failure)

*AIs and RDAs are for women and men 19–30 years of age; ULs are males and females 19–70 years of age, 1997–2004.

Table 1.13 Summary of minerals (continued)

Consequences of Overdose	Primary Food Sources	Highlights and Comments
• Drowsiness • Calcium deposits in kidneys, liver, and other tissues • Suppression of bone remodeling • Decreased zinc absorption	• Milk and milk products (cheese, yogurt) • Broccoli • Dried beans • Calcium-fortified foods (some juices, breakfast cereals, bread, for example)	• The average intake of calcium among U.S. women is approximately 60% of the DRI • One in four women and one in eight men in the U.S. develop osteoporosis • Adequate calcium and vitamin D status must be maintained to prevent bone loss
• Muscle spasms	• Milk and milk products (cheese, yogurt) • Meats • Seeds, nuts • Phosphates added to foods	• Deficiency is generally related to disease processes
• Diarrhea • Dehydration • Impaired nerve activity due to disrupted utilization of calcium	• Plant foods (dried beans, tofu, peanuts, potatoes, green vegetables) • Milk • Bread • Ready-to-eat cereals • Coffee	• Magnesium is primarily found in plant foods, where it is attached to chlorophyll • Average intake among U.S. adults is below the RDA
• Hemochromatosis ("iron poisoning") • Vomiting, abdominal pain • Blue coloration of skin • Liver and heart damage, diabetes • Decreased zinc absorption • Atherosclerosis (plaque buildup) in older adults	• Liver, beef, pork • Dried beans • Iron-fortified cereals • Prunes, apricots, raisins • Spinach • Bread • Pasta	• Cooking foods in iron and stainless steel pans increases the iron content of the foods • Vitamin C, meat, and alcohol increase iron absorption • Iron deficiency is the most common nutritional deficiency in the world • Average iron intake of young children and women in the U.S. is low
• Over 25 mg/day is associated with nausea, vomiting, weakness, fatigue, susceptibility to infection, copper deficiency, and metallic taste in mouth • Increased blood lipids	• Meats (all kinds) • Grains • Nuts • Milk and milk products (cheese, yogurt) • Ready-to-eat cereals • Bread	• Like iron, zinc is better absorbed from meats than from plants • Marginal zinc deficiency may be common, especially in children • Zinc supplements may decrease duration and severity of the common cold
• Fluorosis • Brittle bones • Mottled teeth • Nerve abnormalities	• Fluoridated water and foods and beverages made with it • Tea • Shrimp, crab	• Toothpastes, mouth rinses, and other dental care products may provide fluoride • Fluoride overdose has been caused by ingestion of fluoridated toothpaste
• Over 1 mg/day may produce pimples, goiter, and decreased thyroid function	• Iodized salt • Milk and milk products • Seaweed, seafoods • Bread from commercial bakeries	• Iodine deficiency remains a major health problem in some developing countries • Amount of iodine in plants depends on iodine content of soil • Most of the iodine in our diet comes from the incidental addition of iodine to foods from cleaning compounds used by food manufacturers

continued

Table 1.13 Summary of minerals (continued)

		Primary Functions	Consequences of Deficiency
Selenium RDA women: 55 mcg men: 55 mcg UL: 400 mcg		• Acts as an antioxidant in conjunction with vitamin E (protects cells from damage due to exposure to oxygen) • Needed for thyroid hormone production	• Anemia • Muscle pain and tenderness • Keshan disease (heart failure), Kashin-Beck disease (joint disease)
Copper RDA women: 900 mcg men: 900 mcg UL: 10,000 mcg		• Component of enzymes involved in the body's utilization of iron and oxygen • Functions in growth, immunity, cholesterol and glucose utilization, brain development	• Anemia • Seizures • Nerve and bone abnormalities in children • Growth retardation
Manganese AI women: 2.3 mg men: 1.8 mg		• Required for the formation of body fat and bone	• Weight loss • Rash • Nausea and vomiting
Chromium AI women: 35 mcg men: 25 mcg		• Required for the normal utilization of glucose and fat	• Elevated blood glucose and triglyceride levels • Weight loss
Molybdenum RDA women: 45 mcg men: 45 mcg UL: 2000 mcg		• Component of enzymes involved in the transfer of oxygen from one molecule to another	• Rapid heartbeat and breathing • Nausea, vomiting • Coma
Sodium AI adults: 1500 mg UL adults: 2300 mg		• Regulation of acid-base balance in body fluids • Maintenance of water balance in body tissues • Activation of muscles and nerves	• Weakness • Apathy • Poor appetite • Muscle cramps • Headache • Swelling
Potassium AI adults: 4700 mg		• Same as for sodium	• Weakness • Irritability, mental confusion • Irregular heartbeat • Paralysis
Chloride AI adults: 2300 mg		• Component of hydrochloric acid secreted by the stomach (used in digestion) • Maintenance of acid-base balance of body fluids • Maintenance of water balance in the body	• Muscle cramps • Apathy • Poor appetite • Long-term mental retardation in infants

Table 1.13 Summary of minerals (continued)

Consequences of Overdose	Primary Food Sources	Highlights and Comments
• "Selenosis"; symptoms of selenosis are hair and fingernail loss, weakness, liver damage, irritability, and "garlic" or "metallic" breath	• Meats and seafoods • Eggs • Whole grains	• Content of foods depends on amount of selenium in soil, water, and animal feeds • May play a role in the prevention of some types of cancer
• Wilson's disease (excessive accumulation of copper in the liver and kidneys) • Vomiting, diarrhea • Tremors • Liver disease	• Bread • Potatoes • Grains • Dried beans • Nuts and seeds • Seafood • Ready-to-eat cereals	• Toxicity can result from copper pipes and cooking pans • Average intake in the U.S. is below the RDA
• Infertility in men • Disruptions in the nervous system (psychotic symptoms) • Muscle spasms	• Whole grains • Coffee, tea • Dried beans • Nuts	• Toxicity is related to overexposure to manganese dust in miners
• Kidney and skin damage	• Whole grains • Wheat germ • Liver, meat • Beer, wine • Oysters	• Toxicity usually results from exposure in chrome-making industries or overuse of supplements • Supplements do not build muscle mass or increase endurance
• Loss of copper from the body • Joint pain • Growth failure • Anemia • Gout	• Dried beans • Grains • Dark green vegetables • Liver • Milk and milk products	• Deficiency is extraordinarily rare
• High blood pressure in susceptible people • Kidney disease • Heart problems	• Foods processed with salt • Cured foods (corned beef, ham, bacon, pickles, sauerkraut) • Table and sea salt • Bread • Milk, cheese • Salad dressing	• Very few foods naturally contain much sodium; processed foods are the leading source • High-sodium diets are associated with hypertension in "salt-sensitive" people • Kidney disease, excessive water consumption are related to sodium depletion
• Irregular heartbeat, heart attack	• Plant foods (potatoes, squash, lima beans, tomatoes, plantains, bananas, oranges, avocados) • Meats • Milk and milk products • Coffee	• Content of vegetables is often reduced in processed foods • Diuretics (water pills), vomiting, diarrhea may deplete potassium • Salt substitutes often contain potassium
• Vomiting	• Same as for sodium (most of the chloride in our diets comes from salt)	• Excessive vomiting and diarrhea may cause chloride deficiency • Legislation regulating the composition of infant formulas was enacted in response to formula-related chloride deficiency and subsequent mental retardation in infants

Table 1.14 Food sources of minerals

Magnesium

Food	Amount	Magnesium (mg)	Food	Amount	Magnesium (mg)
Legumes			**Vegetables**		
Lentils, cooked	½ c	134	Bean sprouts	½ c	98
Split peas, cooked	½ c	134	Black-eyed peas	½ c	58
Tofu	½ c	130	Spinach, cooked	½ c	48
Nuts			Lima beans	½ c	32
Peanuts	¼ c	247	**Milk and Milk Products**		
Cashews	¼ c	93	Milk	1 c	30
Almonds	¼ c	80	Cheddar cheese	1 oz	8
Grains			American cheese	1 oz	6
Bran buds	1 c	240	**Meats**		
Wild rice, cooked	½ c	119	Chicken	3 oz	25
Breakfast cereal, fortified	1 c	85	Beef	3 oz	20
Wheat germ	2 tbs	45	Pork	3 oz	20

Calcium*

Food	Amount	Calcium (mg)	Food	Amount	Calcium (mg)
Milk and Milk Products			Ice milk	1 c	180
Yogurt, low fat	1 c	413	American cheese	1 oz	175
Milk shake			Custard	½ c	150
(low-fat frozen yogurt)	1¼ c	352	Cottage cheese	1½ c	70
Yogurt with fruit, low fat	1 c	315	Cottage cheese, low fat	½ c	69
Skim milk	1 c	301	**Vegetables**		
1% milk	1 c	300	Spinach, cooked	½ c	122
2% milk	1 c	298	Kale	1½ c	47
3.25% milk (whole)	1 c	288	Broccoli	½ c	36
Swiss cheese	1 oz	270	**Legumes**		
Milk shake (whole milk)	1¼ c	250	Tofu	½ c	260
Frozen yogurt, low fat	1 c	248	Dried beans, cooked	½ c	60
Frappuccino	1 c	220	**Foods Fortified with Calcium**		
Cheddar cheese	1 oz	204	Orange juice	1 c	350
Frozen yogurt	1 c	200	Waffles, frozen	2	300
Cream soup	1 c	186	Soymilk	1 c	200–400
Pudding	½ c	185	Breakfast cereals	1 c	150–1000
Ice cream	1 c	180			

Selenium

Food	Amount	Selenium (mcg)	Food	Amount	Selenium (mcg)
Seafood			Ham	3 oz	29
Lobster	3 oz	66	Beef	3 oz	22
Tuna	3 oz	60	Bacon	3 oz	21
Shrimp	3 oz	54	Chicken	3 oz	18
Oysters	3 oz	48	Lamb	3 oz	14
Fish	3 oz	40	Veal	3 oz	10
Meats			**Eggs**		
Liver	3 oz	56	Egg	1 med	37

Sodium

Food	Amount	Sodium (mg)	Food	Amount	Sodium (mg)
Miscellaneous			Sea salt	1 tsp	1716
Salt	1 tsp	2132	Ravioli, canned	1 c	1065
Dill pickle	1 (4½ oz)	1930	Spaghetti with sauce, canned	1 c	955

*Actually, the richest source of calcium is alligator meat; 3½ ounces contain about 1231 milligrams of calcium. But just try to find it on your grocer's shelf!

Table 1.14 Food sources of minerals (continued)

Sodium (Continued)

Food	Amount	Sodium (mg)	Food	Amount	Sodium (mg)
Baking soda	1 tsp	821	Fish, smoked	3 oz	444
Beef broth	1 c	810	Bologna	1 oz	370
Chicken broth	1 c	770	**Milk and Milk Products**		
Gravy	¼ c	720	Cream soup	1 c	1070
Italian dressing	2 Tbsp	720	Cottage cheese	½ c	455
Pretzels	5 (1 oz)	500	American cheese	1 oz	405
Green olives	5	465	Cheese spread	1 oz	274
Pizza with cheese	1 wedge	455	Parmesan cheese	1 oz	247
Soy sauce	1 tsp	444	Gouda cheese	1 oz	232
Cheese twists	1 c	329	Cheddar cheese	1 oz	175
Bacon	3 slices	303	Skim milk	1 c	125
French dressing	2 Tbsp	220	Whole milk	1 c	120
Potato chips	1 oz (10 pieces)	200	**Grains**		
Catsup	1 Tbsp	155	Bran flakes	1 c	363
Meats, Fish			Cornflakes	1 c	325
Corned beef	3 oz	808	Croissant	1 med	270
Ham	3 oz	800	Bagel	1	260
Fish, canned	3 oz	735	English muffin	1	203
Meat loaf	3 oz	555	White bread	1 slice	130
Sausage	3 oz	483	Whole wheat bread	1 slice	130
Hot dog	1	477	Saltine crackers	4 squares	125

Iron

Food	Amount	Iron (mg)	Food	Amount	Iron (mg)
Meat and Meat Alternates			English muffin	1	1.6
Liver	3 oz	7.5	Rye bread	1 slice	1.0
Round steak	3 oz	3.0	Whole wheat bread	1 slice	0.8
Hamburger, lean	3 oz	3.0	White bread	1 slice	0.6
Baked beans	½ c	3.0	**Fruits**		
Pork	3 oz	2.7	Prune juice	1 c	9.0
White beans	½ c	2.7	Apricots, dried	½ c	2.5
Soybeans	½ c	2.5	Prunes	5 med	2.0
Pork and beans	½ c	2.3	Raisins	¼ c	1.3
Fish	3 oz	1.0	Plums	3 med	1.1
Chicken	3 oz	1.0	**Vegetables**		
Grains			Spinach, cooked	½ c	2.3
Breakfast cereal, iron-fortified	1 c	8.0 (4–18)	Lima beans	½ c	2.2
Oatmeal, fortified, cooked	1 c	8.0	Black-eyed peas	½ c	1.7
			Peas	½ c	1.6
Bagel	1	1.7	Asparagus	½ c	1.5

Zinc

Food	Amount	Zinc (mg)	Food	Amount	Zinc (mg)
Meats, Seafood			Pork	3 oz	2.4
Liver	3 oz	4.6	Chicken	3 oz	2.0
Beef	3 oz	4.0	**Legumes**		
Crab	½ c	3.5	Dried beans, cooked	½ c	1.0
Lamb	3 oz	3.5	Split peas, cooked	½ c	0.9
Turkey ham	3 oz	2.5			

continued

Table 1.14 Food sources of minerals (continued)

Zinc (Continued)

Food	Amount	Zinc (mg)	Food	Amount	Zinc (mg)
Grains			Cashews	¼ c	1.8
Breakfast cereal, fortified	1 c	1.5–4.0	Sunflower seeds	¼ c	1.7
Wheat germ	2 Tbsp	2.4	Peanut butter	2 Tbsp	0.9
Oatmeal, cooked	1 c	1.2	**Milk and Milk Products**		
Bran flakes	1 c	1.0	Cheddar cheese	1 oz	1.1
Brown rice, cooked	½ c	0.6	Whole milk	1 c	0.9
White rice	½ c	0.4	American cheese	1 oz	0.8
Nuts and Seeds					
Pecans	¼ c	2.0			

Phosphorus

Food	Amount	Phosphorus (mg)	Food	Amount	Phosphorus (mg)
Milk and Milk Products			**Grains**		
Yogurt	1 c	327	Bran flakes	1 c	180
Skim milk	1 c	250	Shredded wheat	2 large biscuits	81
Whole milk	1 c	250	Whole wheat bread	1 slice	52
Cottage cheese	½ c	150	Noodles, cooked	½ c	47
American cheese	1 oz	130	Rice, cooked	½ c	29
Meats, Fish			White bread	1 slice	24
Pork	3 oz	275	**Vegetables**		
Hamburger	3 oz	165	Potato	1 med	101
Tuna	3 oz	162	Corn	½ c	73
Lobster	3 oz	125	Peas	½ c	70
Chicken	3 oz	120	French fries	½ c	61
Nuts and Seeds			Broccoli	½ c	54
Sunflower seeds	¼ c	319	**Other**		
Peanuts	¼ c	141	Milk chocolate	1 oz	66
Pine nuts	¼ c	106	Cola	12 oz	51
Peanut butter	1 Tbsp	61	Diet cola	12 oz	45

Potassium

Food	Amount	Potassium (mg)	Food	Amount	Potassium (mg)
Vegetables			Hamburger	3 oz	480
Potato	1 med	780	Lamb	3 oz	382
Winter squash	½ c	327	Pork	3 oz	335
Tomato	1 med	300	Chicken	3 oz	208
Celery	1 stalk	270	**Grains**		
Carrots	1 med	245	Bran buds	1 c	1080
Broccoli	½ c	205	Bran flakes	1 c	248
Fruits			Raisin Bran	1 c	242
Avocado	½ med	680	Wheat flakes	1 c	96
Orange juice	1 c	469	**Milk and Milk Products**		
Banana	1 med	440	Yogurt	1 c	531
Raisins	¼ c	370	Skim milk	1 c	400
Prunes	4 large	300	Whole milk	1 c	370
Watermelon	1 c	158	**Other**		
Meats, Fish			Salt substitutes	1 tsp	1300–2378
Fish	3 oz	500			

Illustration 1.7 Usual steps in the development of nutrient deficiencies and toxicities.

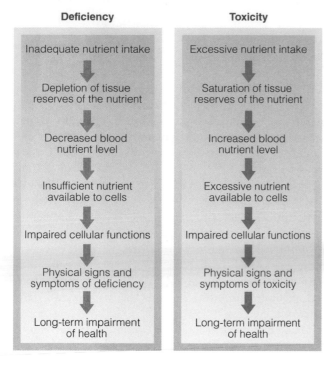

nutrients may develop from a low milk intake and an otherwise poor diet.

The "Ripple Effect"   Dietary changes affect the level of intake of many nutrients. Switching from a high-fat to a low-fat diet, for instance, generally results in a lower intake of calories and higher intake of dietary fiber and vitamins. Consequently, dietary changes introduced for the purpose of improving intake of a particular nutrient produce a "ripple effect" on the intake of other nutrients.

> **PRINCIPLE #5**   Humans have adaptive mechanisms for managing fluctuations in food intake.

Healthy humans have adaptive mechanisms that partially protect the body from poor health due to fluctuations in nutrient intake. These mechanisms act to conserve nutrients when dietary supply is low and to eliminate them when excessively high amounts are present. Dietary surpluses of nutrients such as iron, calcium, vitamin A, and vitamin B_{12} are stored within tissues for later use. In the case of iron and calcium, absorption is also regulated so that the amount absorbed changes in response to the body's need for these nutrients. The body has a low storage capacity for other nutrients, such as vitamin C and water, and excesses are eliminated through urine or stools. Fluctuations in energy intake are primarily regulated by changes in appetite. If too few calories are consumed, however, the body will obtain energy from its glycogen and fat stores. If caloric intakes remain low and a significant amount of body weight is lost, the body down-regulates its need for energy by lowering body temperature and the capacity for physical work. When energy intake exceeds need, the extra is converted to fat—and to a lesser extent, to glycogen—and stored for later use.

Although they provide an important buffer, these built-in mechanisms do not protect humans from all the consequences of poor diets. An excessive vitamin A or selenium intake over time, for example, results in toxicity disease; excessive energy intake creates health problems related to obesity; and deficient intakes of other vitamins and minerals compromise health in many ways.

of nutrients, cell processes change. These changes have a negative effect on the cell's ability to form proteins appropriately, regulate energy formation and use, protect itself from oxidation, or carry out other normal functions. If the deficiency continues, groups of cells malfunction, which leads to problems related to tissue and organ functions. Physical signs of the deficiency may then develop, such as growth failure with protein deficiency or an inability to walk as a result of beriberi (thiamin deficiency). Eventually, some problems produced by the deficiency can no longer be reversed by increased nutrient intake. Blindness that results from serious vitamin A deficiency, for example, is irreversible.

Excessively high intakes of many essential nutrients produce toxicity diseases. Excessive vitamin A, for example, produces hypervitaminosis A, and selenium overdose leads to selenosis. Signs of toxicity stem from an increased level of the nutrient in the blood and the subsequent oversupply of the nutrient to cells. The high nutrient load upsets the balance needed for optimal cell function. These changes in cell function lead to the signs and symptoms of a toxicity disease.

For both deficiency and toxicity diseases, the best way to correct the problem is at the level of intake. Identifying and fixing intake problems prevents related health problems from developing.

Nutrient Deficiencies Are Usually Multiple   Most foods contain many nutrients, so poor diets are generally inadequate in many nutrients. Calcium and vitamin D, for example, are present in milk. Deficiencies of both of these

> **PRINCIPLE #6**   Malnutrition can result from poor diets and from disease states, genetic factors, or combinations of these causes.

Malnutrition means "poor nutrition" and results from either inadequate or excessive availability of energy and nutrients. Niacin toxicity, obesity, and iron deficiency are examples of malnutrition.

Malnutrition can result from poor diets as well as from diseases that interfere with the body's ability to use the nutrients consumed.

Malnutrition   Poor nutrition resulting from an excess or lack of calories or nutrients.

Nutrient–Gene Interactions   Advances in knowledge about nutrient–gene interactions in health and disease are revolutionizing the science and practice of nutrition. This new field of nutrition science is called *nutrigenomics* and is highlighted in the "In Focus" box. Genes provide the codes for enzyme and other protein synthesis, and they consequently affect body functions in a huge number of ways. Although individuals are 99.9% genetically identical, the 0.1% difference in genetic codes makes everyone unique. Variations in gene types (or genotypes) contribute to disease resistance and development, and to the way individuals respond to various drugs.[26]

Hundreds of diseases and disorders related to single-gene defects have been identified, and many of these affect nutrient needs. Four examples of such diseases and disorders—phenylketonuria, celiac disease, lactose intolerance, and hemochromatosis—are described in Table 1.15.[27] Most diseases and disorders related to genetic makeup, however, are due to interactions among environmental factors, genotype, and gene functions.[11]

Food and nutrient intake is a prominent environmental factor that interacts with genotype and gene function. Lack of adequate nutrition during pregnancy, for example, can program gene functions for life in ways that increase or decrease the risk of chronic disease development.[28] Throughout life, components of foods consumed affect gene function by turning specific genes "on" or "off," thereby affecting what metabolic reactions occur within the body.[29] Newly identified relationships between dietary

Primary Malnutrition Malnutrition that results directly from inadequate or excessive dietary intake of energy or nutrients.

Secondary Malnutrition Malnutrition that results from a condition (e.g., disease, surgical procedure, medication use) rather than primarily from dietary intake.

Autoimmune Disease A disease related to the destruction of the body's own cells by substances produced by the immune system that mistakenly recognize certain cell components as harmful.

Primary malnutrition results when a poor nutritional state is dietary in origin. *Secondary malnutrition*, on the other hand, is precipitated by a disease state, surgical procedure, or medication. Diarrhea, alcoholism, AIDS, and gastrointestinal tract bleeding are examples of conditions that may cause secondary malnutrition.

Table 1.15 Examples of single-gene disorders that affect nutrient need

PKU (phenylketonuria)	A very rare disorder caused by the lack of the enzyme phenylalanine hydroxylase. Lack of this enzyme causes phenylalanine, an essential amino acid, to build up in the blood. High blood levels of phenylalanine during growth lead to mental retardation, poor growth, and other problems. PKU is treated by low-phenylalanine diets.
Celiac disease	An *autoimmune disease* characterized by inflammation of the small-intestine lining resulting from a genetically based intolerance to a component of gluten. The inflammation produces diarrhea, fatty stools, weight loss, and vitamin and mineral deficiencies. Also called tropical sprue and gluten-sensitive enteropathy.
Lactose intolerance	A common disorder in adults in many countries resulting from lack of the enzyme lactase. Ingestion of lactose in dairy products causes gas, cramps, and nausea due to the presence of undigested lactose in the gut. About 25% of the U.S. population and 75% of adults worldwide are reported to be lactose intolerant.[31] Rates of lactose intolerance are particularly high in African Americans, Latinos, and Asians.[32] The preferred but less commonly used term for lactose intolerance is lactose maldigestion.
Hemochromatosis	A disorder affecting 1 in 200 people that occurs due to a genetic deficiency of a protein that helps regulate iron absorption. Individuals with hemochromatosis absorb more iron than normal and have excessive levels of body iron. High levels of body iron have toxic effects on tissues such as the liver and heart. The disorder can also be produced by excessive levels of iron intake over time and frequent iron injections or blood transfusions.

components and genes are being announced regularly. Here are a few examples of effects of nutrient–gene interactions on health status:

- Consumption of high glycemic index carbohydrates appears to increase the risk of type 2 diabetes in individuals with a certain form of a gene involved in insulin production and secretion.[30]

- High alcohol intake during pregnancy in some women sharply increases the risk of fetal alcohol syndrome in her fetus, but the fetuses of other women with different genetic traits are not affected by high alcohol intake.

- Regular consumption of green tea reduces the risk of prostate cancer in certain individuals with a particular genetic trait.[6,26,27]

Genetic factors alone cannot explain the rapid rise in obesity and type 2 diabetes in the United States, but they do provide clues about needed preventive and therapeutic measures.

PRINCIPLE #7 Some groups of people are at higher risk of becoming inadequately nourished than others.

Women who are pregnant or breastfeeding, infants, children, people who are ill, and frail elderly persons have a greater need for nutrients than healthy adults and elderly people do. As a result, they are at higher risk of becoming inadequately nourished than others. Within these groups, those at highest risk of nutritional insults are the poor. In cases of widespread food shortages, such as those induced by war or natural disaster, the health of these nutritionally vulnerable groups is compromised the soonest and the most.

PRINCIPLE #8 Poor nutrition can influence the development of certain chronic diseases.

"We currently have evidence from major clinical trials that show the efficacy of nutrition practices in 'curing' diseases, slowing disease progression, and markedly decreasing the risk to a similar extent as pharmacologic therapy."

—Penny Kris-Etherton[14]

Today, the major causes of death among Americans are slow-developing, lifestyle-related *chronic diseases* (Illustration 1.8). Based on government survey data, 44% of Americans have a chronic condition such as diabetes, heart disease, cancer, *hypertension*, or high cholesterol levels, and 13% have three or more of these conditions.[36]

The leading causes of death among Americans are heart disease and cancer. Together they account for 50%

of all deaths. Western-type diets high in saturated and *trans* fats, and low in vegetables, fruits, and whole grain products, are linked to the development of heart disease.[34] Six types of cancer, including colon, pancreatic, and breast cancer, are related to obesity, habitually low intakes of vegetables and fruits, and high levels of intake of processed meats.

Diet is related to three other leading causes of death: diabetes, *stroke*, and *Alzheimer's disease*. Example of relationships between diseases and disorders and diet are overviewed in Table 1.16.

Shared Dietary Risk Factors A number of the diseases and disorders listed in Table 1.16 share the common dietary risk factors of low intakes of vegetables, fruits, and whole grains; excess calorie intake and body fat; and high animal-fat intake. These risk factors are associated with the development of *chronic inflammation* and *oxidative stress*, conditions that are strongly related to the development of heart disease, diabetes, *osteoporosis*, Alzheimer's disease, cancer, and other chronic diseases.[43]

Inadequate and excessive nutrient intakes may contribute to the development of more than one disease and produce disease by more than one mechanism. The effects of habitually poor diets on chronic disease development often take years to become apparent.

PRINCIPLE #9 Adequacy, variety, and balance are key characteristics of a healthy diet.

Healthy diets contain many different foods that together provide calories, nutrients, and other beneficial substances in amounts that promote the optimal functioning of cells and health. A variety of food is required to obtain all the nutrients

Chronic Disease Slow-developing, long-lasting diseases that are not contagious (e.g., heart disease, cancer, diabetes). They can be treated but not always cured.

Hypertension High blood pressure. It is defined as blood pressure exerted inside blood vessel walls that typically exceeds 140/90 mmHg (millimeters of mercury).

Stroke An event that occurs when a blood vessel in the brain ruptures or becomes blocked. Stroke is often associated with "hardening of the arteries" in the brain. Also called cerebral vascular accident.

Alzheimer's Disease A brain disease that represents the most common form of dementia. It is characterized by memory loss for recent events that expands to more distant memories over the course of five to ten years. It eventually produces profound intellectual decline characterized by dementia and personal helplessness.

Chronic Inflammation Low-grade inflammation that lasts weeks, months, or years. Inflammation is the first response of the body's immune system to infectious agents, toxins, or irritants. It triggers the release of biologically active substances that promote oxidation and other reactions to counteract the infection, toxin, or irritant. A side effect of chronic inflammation is that it also damages lipids, cells, and tissues.

Oxidative Stress A condition that occurs when cells are exposed to more oxidizing molecules (such as free radicals) than to antioxidant molecules that neutralize them. Over time, oxidative stress causes damage to lipids, DNA, cells, and tissues. It increases the risk of heart disease, type 2 diabetes, cancer, and other diseases.

Osteoporosis A condition in which low bone density or weak bone structure leads to an increased risk of bone fracture.

Table 1.16 Examples of diseases and disorders linked to diet[37–42]

Disease/Disorder	Dietary Connection
Heart disease	High saturated fat, *trans* fat, and cholesterol intake; low vegetable, fruit, and whole grain intake; low fish or fish oil intake; and excessive body fat
Cancer	Low vegetable and fruit intake, excessive body fat and alcohol intake, regular consumption of processed meat
Stroke	Low vegetable and fruit intake, excessive alcohol intake, high animal-fat diets
Type 2 diabetes	Excessive body fat; low vegetable and fruit intake; high saturated fat, sugary foods, and refined grain-product intake
Cirrhosis of the liver	Excessive alcohol intake, poor overall diet quality
Hypertension	Excessive sodium (salt) and low potassium intake; high intake of refined grain products and alcohol; poor vitamin D status; low vegetable and fruit intake; excessive body fat
Iron-deficiency anemia	Low iron intake
Tooth decay and gum disease	Excessive and frequent sugar consumption, inadequate fluoride intake
Obesity	Excessive calorie intake
Chronic inflammation and oxidative stress	Excessive intake of calorie-dense foods, high fat diets, low intake of antioxidant-rich vegetables and fruits and omega-3 fatty acids (DHA and EPA), poor vitamin D status

Nutrient-Dense Foods Foods that contain relatively high amounts of nutrients compared to their caloric value.

Empty-Calorie Foods Foods that provide an excess of calories relative to their nutrient content.

Illustration 1.8 Poor nutrition increases the risk of many chronic diseases.

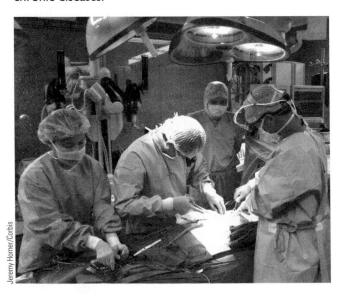

Jeremy Horner/Corbis

needed because, although no one food contains them all, many different combinations of food can make up a healthy diet.

Adequate diets are most easily obtained by consuming foods that are good sources of a number of nutrients but not packed with calories. Such foods are considered **nutrient-dense foods.** Those that provide calories and low amounts of nutrients are considered **"empty-calorie" foods.** Vegetables, fruits, lean meats, dried beans, breads, and cereals are nutrient-dense. Foods such as beer, chips, candy, pastries, sodas, and fruit drinks lead the list of empty-calorie foods.

Variety is a core characteristic of healthy diets because the essential nutrient and phytochemical content of foods differ. Consumption of an assortment of foods from each of the basic food groups increases the probability that the diet will provide enough nutrients. You could, for example, eat three servings of potatoes a day to meet the recommended "3-a day" guideline. But you would consume a much broader variety of vitamins, minerals, and plant antioxidants, for example, if you consumed spinach and tomatoes along with potatoes.

A healthy diet provides a balanced selection of food types and amounts. Regular consumption of fried foods, high-fat meats, and sweets, and infrequent intake of whole grains and colorful vegetables, for example, can knock a diet out of balance. To bring diets into balance, Americans are being urged to consume more vegetables, fruits, whole grains, dried beans, and low-fat meats and dairy products, and to consume less sugar, animal fat, and saturated fat than they do now.

PRINCIPLE #10   There are no "good" or "bad" foods.

"All things in nutriment are good or bad relatively."

Hippocrates

People tend to classify foods as being "good" or "bad," but such opinions about individual foods oversimplify the potential contribution of these foods to a diet.[44] Although opinions about which foods are good or bad vary, hot dogs, ice cream, candy, bacon, and french fries are often judged to be bad, whereas vegetables, fruits, and whole-grain products are given the "good" stamp. Unless we're talking about spoiled stew, poisonous mushrooms, or something similar, however, no food can be accurately labeled as good or bad. Ice cream can be a "good" food for physically active, normal-weight individuals with high calorie needs who have otherwise met their nutrient requirements by consuming nutrient-dense foods. Some people who only eat what they consider to be "good" foods, like vegetables, fruits, whole grains, and tofu, may still miss the healthful diet mark due to inadequate consumption of essential fatty acids and certain vitamins and minerals. All foods can fit into a healthful diet as long as nutrients needs are met at calorie-intake levels that maintain a healthy body weight.[44]

Nutritional Labeling

In 1990 the U.S. Congress passed legislation establishing requirements for nutrition information, nutrient content claims, and health claims presented on food and dietary supplement labels. This legislation, called the Nutrition Labeling and Education Act, requires that almost all multiple-ingredient foods and *dietary supplements* be labeled with a Nutrition Facts panel (Illustration 1.9). The act also requires that nutrient content and health claims appearing on package labels, such as "*trans* fat–free" and "helps prevent cancer," qualify based on criteria established by the Food and Drug Administration (FDA).

Concern about rising rates of overweight and type 2 diabetes among children and youth prompted senators in the United States to propose national menu-labeling standards for fast food and other restaurants in 2009. The legislation would require restaurant chains with 20 or more outlets to post calories on menus and menu boards and make additional information about the fat, saturated fat, carbohydrate, sodium, protein, and fiber content available in writing upon request.

Nutrition Facts Panel

For foods, the Nutrition Facts panel must list the content of fat, saturated fat, *trans* fat, cholesterol, sodium, total

Illustration 1.9 Example of a Nutrition Facts panel.

Nutrition Facts		
Serving Size 1 Entree		
Serving Per Container 1		
Amount Per Serving		
Calories 380 Calories from Fat 170		
		%Daily Value
Total Fat 19g		**29%**
Saturated Fat 10g		**50%**
Trans Fat 2g		
Cholesterol 85g		**28%**
Sodium 810mg		**34%**
Total Carbohydrate 33g		**11%**
Dietary Fiber 3g		**12%**
Sugars 5g		
Protein 20g		
Vitamin A 10%		Vitamin C 0%
Calcium 10%		Iron 15%

Percent Daily Values are based on a 2000 calorie diet. Your daily values may be higher or lower depending on your calorie needs:

		Calories	2000	2500
Total Fat	Less Than		65g	80g
Sat Fat	Less Than		20g	25g
Cholesterol	Less Than		300mg	300mg
Sodium	Less Than		2400mg	2400mg
Total Carbohydrate			300g	375g
Dietary Fiber			25g	30g

carbohydrates, fiber, sugars, protein, vitamins A and C, and calcium and iron in a standard serving. Additional nutrients may be listed on a voluntary basis. If a health claim about a particular nutrient is made for the product, the product's content of the nutrient addressed in the claim must be shown. Nutrition Facts panels contain a column that lists the % Daily Value (% DV) for each relevant nutrient. This information helps consumers decide, for example, whether the carbohydrate content of a serving of a specific food product is a lot or a little.

Dietary Supplements   Any product intended to supplement the diet, including vitamin and mineral supplements, proteins, enzymes, amino acids, fish oils, fatty acids, hormones and hormone precursors, and herbs and other plant extracts. In the United States, such products must be labeled "Dietary Supplement."

Nutrient content claims made on food package labels must meet specific criteria. Products labeled "no *trans* fat" or "*trans* fat–free," for example, must contain less than 0.5 grams of *trans* fat and of saturated fat. Products labeled "low sodium" must contain less than 140 mg sodium per serving.

Ingredient Label

Food products must list ingredients in an "ingredient label." The list must begin with the ingredient that contributes the greatest amount of weight to the product and continue with the other ingredients on a weight basis.

The FDA now requires that ingredient labels note the presence of common food allergens in products. Potential food allergens that must be listed are milk, eggs, fish, shellfish, tree nuts, wheat, peanuts, and soybeans. These eight foods account for 90% of food allergies.

Dietary Supplement Labeling

> "You can call anything a dietary supplement, even something you grow in your back yard."
>
> Donna Porter, RD, PhD, Congressional Research Service

Dietary supplements such as herbs, amino acid pills and powders, and vitamin and mineral supplements must show a "Supplement Facts" panel that lists serving size, ingredients, and % DV of essential nutrients contained. Because they do not have to be shown to be safe and effective before they are sold, labels on dietary supplements cannot claim to treat, cure, or prevent disease. They can be labeled with standardized nutrition content claims such as "high in calcium" or "a good source of fiber." They can also be labeled with health claims such as "may reduce the risk of heart disease" if the product qualifies based on nutrition labeling requirements. Dietary supplements can make other claims on product labels not approved by the FDA, such as "supports the immune system" or "helps maintain mental health," as long as the label doesn't state or imply that the product will prevent, cure, or treat disease. If a health claim is made on a dietary supplement label, the label also must present the FDA disclaimer:

> This product has not been evaluated by the FDA. This product is not intended to diagnose, treat, cure, or prevent any disease.

Enrichment and Fortification Some foods are labeled as "enriched" or "fortified." These two terms have specific definitions developed prior to the Nutrition Labeling and Education Act. *Enrichment* pertains only to refined grain products and covers some of the vitamins and one of the minerals lost when grains are refined. By law, producers of bread, cornmeal, crackers, flour tortillas, white rice, and other products made with refined grains must use flours enriched with thiamin, riboflavin, niacin, and iron.

Any food can be fortified with added vitamins and minerals, and its manufacturers most often do so on a voluntary basis to enhance product sales. However, some foods must be fortified. Refined grain flours must be fortified with folic acid, milk with vitamin D, and low-fat and skim milk with vitamin D and vitamin A. Although fortification is not required for salt, it is often fortified with iodine. *Fortification* of these foods has contributed substantially to reductions in the incidence of diseases related to inadequate intakes.[45]

Enrichment The replacement of thiamin, riboflavin, niacin, and iron that are lost when grains are refined.

Fortification The addition of one or more vitamins or minerals to a food product.

Functional Foods Generally taken to mean food, fortified foods, and enhanced food products that may have health benefits beyond the effects of essential nutrients they contain.

Herbal Remedies

The FDA considers herbal products to be dietary supplements; they are taken by many people during various stages of the life cycle. Thousands of types of herbal products are available (Illustration 1.10). Some herbal remedies act like drugs and have side effects, but they are not considered to be drugs and are loosely regulated. They do not have to be shown to be safe or effective before they are marketed. Herbs vary substantially in safety and effectiveness—they can have positive, negative, or neutral effects on health. Knowledge of the effects of herbal remedies is far from complete, making it difficult to determine appropriateness of their use in many cases.

The extent to which herbs pose a risk to health depends on the amount taken, the duration of use, and the user's age, stage, and health status.

Functional Foods

Also known as "neutraceuticals," *functional foods* include a variety of products that have theoretically been modified to enhance their contribution to a healthy diet. Foods are made "functional" by:

- Taking out potentially harmful components (e.g., cholesterol in egg yolks and lactose in milk)
- Increasing the amount of nutrients and beneficial non-nutrients (e.g., fiber-fortified liquid meals, calcium- and vitamin C–fortified orange juice)

Illustration 1.10 Herbal products are widely available on the market.

Barry Austin/Photodisc/Getty Images

- Adding new beneficial compounds to foods (e.g., "friendly" bacteria to yogurt and other milk products)

Functional foods, such as those shown in Illustration 1.11, are not regulated, and no specific standards apply to them.[46] Health claims, however, can be made for functional foods given they have been approved by the FDA. Increasingly, the list of functional foods is becoming infiltrated with sports bars, soups, beverages, and cereals spiked with vitamins, minerals, and herbs. Some of the products carry labels with unsubstantiated health claims and may be of little benefit or are potentially unsafe.[47] For these products, the label "functional food" is a marketing term. Others, such as lactose-free milk, xylitol-sweetened gum, and iodized salt, benefit the health of some people.

Prebiotics and Probiotics   The terms *prebiotics* and *probiotics* were derived from *antibiotics* due to their probable effects on increasing resistance to various diseases. Prebiotics and probiotics are in a class of functional foods by themselves. *Prebiotics* are fiberlike, indigestible carbohydrates that are broken down by bacteria in the colon. The breakdown products foster the growth of beneficial bacteria. The digestive tract generally contains over 500 species of microorganisms and 100 trillion bacteria.[48] Some species of bacteria such as *E. coli* may cause disease; others, such as strains of lactobacillus and bifidobacteria, prevent certain diseases.[49] Because they foster the growth of beneficial bacteria, prebiotics are considered "intestinal fertilizer." *Probiotic* is the term for live, beneficial ("friendly") bacteria that enter food products during fermentation and aging processes. Those that survive digestive enzymes and acids may start colonies of beneficial bacteria in the digestive tract. Table 1.17 lists foods and other sources of pre- and probiotics.

Prebiotics and probiotics have been credited with important benefits, such as the prevention and treatment of diarrhea and other infections in the gastrointestinal tract; prevention of colon cancer; decreased blood levels of triglycerides, cholesterol, and glucose; and decreased dental caries.[49] Prebiotics appear to be safe in general; however, probiotics may be harmful to individuals who may develop blood infections.[49,50] The primary side effects associated with prebiotic and probiotic use are flatulence, bloating, and constipation.[49]

Availability of foods containing prebiotics and probiotics is much more common in Japan and European countries than in Canada or the United States. However, availability of such products is increasing in these countries as research results shed light on their safety and effectiveness.[51]

> **Prebiotics**   Certain fiberlike forms of indigestible carbohydrates that support the growth of beneficial bacteria in the lower intestine. Nicknamed "intestinal fertilizer."
>
> **Probiotics**   Strains of lactobacillus and bifidobacteria that have beneficial effects on the body. Also called "friendly bacteria."

Illustration 1.11   Examples of functional foods.

Andy Crawford/Dorling Kindersley/Getty Images

Table 1.17 Food and other sources of prebiotics and probiotics

Probiotics

Fermented or aged milk and milk products
- Yogurt with live culture
- Buttermilk
- Kefir
- Cottage cheese
- Dairy spreads with added inulin

Other fermented products
- Soy sauce
- Tempeh
- Fresh sauerkraut
- Miso

Breast milk

Probiotic tablets, powders, and nutritional beverages

Prebiotics

Chicory

Jerusalem artichokes

Wheat

Barley

Rye

Onions

Garlic

Leeks

Prebiotic tablets, powders, and nutritional beverages

The Life-Course Approach to Nutrition and Health

Nutritional needs should be met at every stage of the life cycle because nutritional status at one stage influences health status in the next ones. Lack of adequate nutrition during pregnancy, for example, can program gene functions for life in ways that set the stage for life-long metabolic changes that increase the risk of chronic-disease development.[52] Iron deficiency experienced by young children can decrease intellectual capacity later in life, and adequate vitamin D status during adolescence and early adulthood decreases the risk of breast cancer in older women.[53] Disease prevention and health promotion, rather than repair of health problems, requires a focus on meeting nutritional and other health needs of individuals during every stage of the life cycle.

Meeting Nutritional Needs Across the Life Cycle

Healthy individuals require the same nutrients throughout life, but amounts of nutrients needed vary based on age, growth, and development. Nutrient needs during each stage of the life cycle can be met through a variety of foods and food practices. There is no one best diet for everyone. Traditional diets defined by diverse cultures and religions provide the foundation for meeting individuals' nutritional needs and the framework for dietary modification when needed.[33] Although it is inaccurate to say that all or most members of a particular cultural group or religion follow the same dietary practices, groups of individuals may share common beliefs about food and food-intake practices.

Dietary Considerations Based on Ethnicity

People immigrating to the United States and other countries both preserve dietary traditions of their cultural group and integrate cross-cultural adaptations into their dietary practices. The extent to which culturally based food habits change depends to some extent on income, food cost, and ethnic food availability. Immigrant families from El Salvador who live in urban areas of the United States, for example, maintain many cultural food practices from their homeland:

- Breakfast generally consists of fried beans, corn tortillas, occasionally eggs, and sweetened coffee with boiled milk.
- Lunch consists of soup, fried meat, rice or rice with vegetables, corn tortillas, and fruit juice.
- Dinner will offer fried beef or chicken, corn tortillas, rice, dried beans, fruit juice, and black coffee.

Cross-cultural adaptations made by a portion of Salvadorans immigrating to the United States include the addition of french fries, hamburgers, American cheeses, salad dressing, tacos, flour tortillas, and peanut butter to their diets.[54]

Sometimes diets of native populations change when their numbers become overwhelmed by other population groups. A primary example of this phenomenon is represented by changes in traditional dietary practices of Native Americans. In general, traditional diets of Indians in the United States consisted of foods such as buffalo, deer, wild berries and other fruits, corn, turnips, squash, wild potato, and wild rice. Loss of land and buffalo, discrimination, poverty, and food programs that offered refined flour, sugar, salt pork, and other high-fat meats drastically changed what Indians ate, how they lived, and their health status. Activities aimed at bringing back traditional foods and dietary practices are under way among many Indian groups.[55]

Food preferences of African Americans vary widely but may stem from their cultural food heritage. Historically important foods include corn bread, pork, buttermilk, rice, sweet potatoes, greens, cabbage, salt pork, and fried fish. "Soul foods" make less of the African American diet now than in the past but remain foods of choice for special occasions and are the foods most likely to be revered.[56]

Dietary Considerations Based on Religion

Many religions have special dietary laws and practices. For example:

- Hindus may not consume foods such as garlic and onions, which are believed to hinder spiritual development.
- Buddhists in certain countries tend to be vegetarian or to eat fish as their only choice of meat. In countries such as Tibet and Japan, vegetarianism is rare among Buddhists.
- Alcohol is prohibited as part of Sikhism, and meat prepared by kosher or halal methods is avoided.
- The Church of Latter Day Saints, or the Mormon Church, prohibits alcohol and discourages consumption of caffeine. Mormons may eat meat and prize wheat.
- Seventh-Day Adventists tend to follow a strict lacto-ovo vegetarian diet and exclude alcohol and caffeine. Whole grains, vegetables, and fruits are considered to be the base of diets, and dried beans, low-fat dairy products, and eggs may be consumed infrequently.
- Jewish dietary laws require that foods consumed must be kosher, or fit to eat according to Judaic law. Organizations are certified as supplying foods that are kosher. The Jewish calendar includes six fasting days that call for total abstinence from food or drink.
- The Muslim religion has dietary laws that require foods to be halal, or permitted for consumption by Muslims. Pork consumption is not allowed, nor is the consumption of animals slaughtered in the name of any god other than Allah. Slaughterhouses must be under the supervision of a halal certifier in order for meat to be considered fit to eat, although

some Muslims will eat other meats. Consuming alcohol is prohibited.[57,58]

Additional information about cultural and religious food practices and beliefs can be obtained directly by getting to know people from a variety of cultures and their dietary preferences. This information can be of great benefit in nutrition education and counseling situations.

Nutritional Assessment

Nutritional assessment of groups and individuals is a prerequisite to planning for the prevention or solution of nutrition-related health problems. It represents a broad area within the field of nutrition and is only highlighted here. Resources related to the selection of appropriate nutritional assessment techniques and their implementation are listed at the end of this chapter.

Nutritional status may be assessed for a population group or for an individual. Community-level assessment identifies a population's status using broad nutrition and health indicators, whereas individual assessment provides the baseline for anticipatory guidance and nutrition intervention.

Community-Level Assessment

A target community's "state of nutritional health" can generally be estimated using existing vital statistics data, seeking the opinions of target group members and local health experts, and making observations. Knowledge of average household incomes; the proportion of families participating in the Food Stamp Program, soup kitchens, school breakfast programs, or food banks; and the age distribution of the group can help identify key nutrition concerns and issues. In large communities, rates of infant mortality, heart disease, and cancer can reveal whether the incidence of these problems is unusually high.

Information gathered from community-level nutritional assessment can be used to develop community-wide programs addressing specific problem areas, such as childhood obesity or iron-deficiency anemia. Nutrition programs should be integrated into community-based health programs.

Individual-Level Nutritional Assessment

Nutritional assessment of individuals has four major components:

- Clinical/physical assessment
- Dietary assessment
- Anthropometric assessment
- Biochemical assessment

Data from all of these areas are needed to describe a person's nutritional status. Data on height and weight provide information on weight status, for example, and knowledge of blood iron levels tells you something about iron status. It cannot be concluded that people who are normal weight or have good iron status are "well nourished." Single measures do not describe a person's nutritional status.

Clinical/physical Assessment   A clinical/physical assessment involves visual inspection of a person by a trained *registered dietitian* or other qualified professional to note features that may be related to malnutrition. Excessive or inadequate body fat, paleness, bruises, and brittle hair are examples of features that may suggest nutrition-related problems. Physical characteristics are nonspecific indicators, but they can support other findings related to nutritional status. They cannot be used as the sole criterion upon which to base a decision about the presence or absence of a particular nutrition problem.

> **Registered Dietitian**   An individual who has acquired food and nutrition knowledge and skills necessary to pass a national registration examination and who participates in continuing professional education.

Dietary Assessment

Many methods are used for assessing dietary intake. For clinical purposes, 24-hour dietary recalls and food records analyzed by computer programs are most common. Single, 24-hour recalls and food frequency questionnaires are most useful for estimating dietary intakes for groups, whereas multiple recalls and dietary histories are generally used for assessments of individual diets.

24-Hour Dietary Recalls and Records   Becoming proficient at administering 24-hour recalls takes training and practice. Food records, on the other hand, are completed by clients themselves. These are more accurate if the client has also received some training. Generally, the purpose of assessing an individual's diet is to estimate the person's overall diet quality so that strengths and weaknesses can be identified, or to assess intake of specific nutrients that may be involved in disease states.

Information on at least three days of dietary intake (preferably two usual weekdays and one weekend day) is needed to obtain a reliable estimate of intake by food group, calories, and nutrients. A good approach is to have a trained dietary interviewer administer a 24-hour recall and then have the client record her or his own diet on two other days. The experience of thinking about what was eaten, portion sizes, ingredients, and recipes helps train people to complete their food records accurately. Completed records should be reviewed with the client during a telephone call or clinic session to make sure they are accurate.

Dietary History   Dietary histories have been used for decades and represent a quantitative method of dietary assessment. They require an interview that is about 1½ hours long and includes a 24-hour dietary recall modified to represent usual intake, careful deliberations over food types and portions, and a cross-check food frequency questionnaire that confirms 24-hour usual dietary intake information. Results must be coded, checked, and processed. Although expensive, diet histories provide more complete and accurate data than most other dietary assessment methods.[59]

Food Frequency Questionnaires   Food frequency questionnaires are often used in epidemiological studies to estimate food and nutrient intake of groups of people. These tools are considered semiquantitative because they force people into describing food intake based on a limited number of food choices and portion sizes (Illustration 1.12). Validated food frequencies are relatively inexpensive to administer and tabulate, and they provide good enough estimates of dietary intake to rank people by their food and nutrient intake levels. They tend to underestimate food intake and provide data that are more likely to fail to identify nutrient and health relationships than are quantitative assessment techniques such as the dietary history.[60]

Web Dietary Assessment Resources   Several high-quality computer programs and Internet resources are available for dietary assessment. My Pyramid Tracker, developed by the U.S. Department of Agriculture (USDA), is an example of a high-quality Internet resource. This interactive program provides an analysis of nutrient and food intake. MyPyramid Tracker can be found at www.mypyramid.gov.

USDA's Automated 5-step Multiple-Pass Method
A computerized, interactive method for collecting interviewer-administered 24-hour dietary recalls either in person or by telephone has been validated by several studies.[61,62] It's called the Automated Multiple-Pass Dietary Recall and is being used in government-sponsored nutritional studies. It utilizes a five-step multiple-pass 24-hour recall. The term *multiple pass* refers to the repeated use of

questions that hone the accuracy of information provided by interviewees about the food they ate the previous day.

The five-step interview process used by the automated multiple-pass method consists of the following:

1. **The Quick List**   Quickly collect a list of foods and beverages consumed the previous day.
2. **The Forgotten Foods List**   Probe for foods forgotten during development of the Quick List.
3. **Time and Occasion List**   Collect information on the time and eating occasion for each food.
4. **The Detail Cycle**   Collect detailed information on the description and amount of food consumed using USDA's interactive Food Model Booklet and measuring guides.
5. **Final Probe Review**   Review 24-hour recall and ask about anything else consumed.

Additional information on this advanced method for assessing dietary intake can be found at www.ars.usda.gov/Services/docs.htm?docid=7710.

The Healthy Eating Index   The HEI (Healthy Eating Index) assesses a person's reported dietary intake based on 10 dietary components that cover intake of the Pyramid's five basic food groups, and the Dietary Guidelines for Americans recommendations for fat and sodium intake, and dietary variety. The components are assigned scores based on the extent to which diets meet recommended standards of intake. The HEI is primarily used for population monitoring of dietary quality, evaluation of interventions, and research. It has been shown to be a valid tool for assessing dietary quality.[63]

Additional information about the HEI is available online from the USDA at www.cnpp.usda.gov/Healthy EatingIndex.htm.

Illustration 1.12 Example component of a food frequency questionnaire.

	Frequency of Consumption								
Food	Never or less than once per month	1–3 per month	1 per week	2–4 per week	5–6 per week	1 per day	2–3 per day	4–5 per day	6+ per day
1. a. Broth-type soups, 1 cup									
b. Tap water, 1 cup									
c. Sparkling or mineral water, 1-cup serving									
d. Decaffeinated black tea, iced or hot, 1 cup									
e. Herbal tea, (no caffine) iced or hot, 1 cup									
2. Custard or pudding, 1/2 cup									
3. Onions, 1/4 cup, alone or in combination									

SOURCE: J. Brown, University of Minnesota; Diana Project form, adapted from W. Willett's Food Frequency Questionnaire.

Anthropometric Assessment

Individual measures of body size (height, weight, percent body fat, bone density, and head and waist circumferences, for example) are useful in the assessment of nutritional status—if done correctly. Each measure requires use of standard techniques and calibrated instruments by trained personnel. Unfortunately, *anthropometric* measurements are frequently performed and recorded incorrectly in clinical practice. Training on anthropometric measures is often available through public health agencies and programs such as WIC (Special Supplemental Nutrition Program for Women, Infants, and Children), and courses and training sessions are sometimes presented at universities.

Biochemical Assessment

Nutrient and enzyme levels, DNA characteristics, and other biological markers are components of a biochemical assessment of nutritional status. Which nutrition biomarkers are measured depends on what problems are suspected, based on other evidence. For example, a young child who tires easily, has a short attention span, and does not appear to be consuming sufficient iron based on dietary assessment results may have blood taken for analyses of hemoglobin and serum ferritin (markers of iron status). Suspected inborn errors of metabolism that may underlie nutrient malabsorption may be identified through DNA or other tests. Such results provide specific information on a component of a person's nutritional status and are very helpful in diagnosing a particular condition. Table 1.18 provides examples of normal levels of various nutrition biomarkers that may be used as part of a biochemical assessment of a nutritional status, or to help diagnose a particular deficiency or nutrient-related health problem.

Anthropometry   The science of measuring the human body and its parts.

Table 1.18 Examples of normal levels of various nutrition biomarkers that may be used as part of a biochemical assessment of nutritional status[a, 64-67]

Nutrient	Women	Men
Red blood cell (erythrocyte) count, whole blood	$4.0–5.2 \times 10^6/mm^3$	$4.5–5.9 \times 10^6/mm^3$
Ferritin, serum	12–150 ng/mL	15–300 ng/mL
Folate, serum or plasma	>4.0–17.5 ng/mL (9.0–39.7 nmol/L)	3.1–17.5 ng/mL (7.0–39.7 nmol/L)
Folate, red blood cell	>187–645 ng/mL >424–1,426 nmol/L	150–450 ng/mL 340–1,020 nmol/L
Hematocrit, whole blood	36–46%	41–53%
Hemoglobin, whole blood	12–16 g/dL (7.4–9.9 mmol/L)	13.5–17.5 g/dL (8.4–10.9 mmol/L)
Iron, serum	40–150 µg/dL (7.2–27.0 µmol/L)	50–160 µg/dL (9.0–28.7 µmol/L)
Mean corpuscular hemoglobin, whole blood	26–34 pg/cell	26–34 pg/cell
Vitamin B_{12}, serum or plasma	>250 pg/mL (>185 pmol/L)	>250 pg/mL (>185 pmol/L)
Vitamin A, serum	20–100 µg/dL (0.7–3.5 µmol/L)	20–100 µg/dL (0.7–3.5 µmol/L)
Riboflavin, serum	4–24 µg/dL (106–638 nmol/L)	4–24 µg/dL (106–638 nmol/L)
Vitamin B_6, plasma	5–30 ng/mL (20–121 nmol/L)	5–30 ng/mL (20–121 nmol/L)
Vitamin C, serum	0.4–1.0 mg/dL (23–57 µmol/L)	0.4–1.0 mg/dL (23–57 µmol/L)
Vitamin D_3 (25 hydroxy-vitamin D), plasma	30–68 ng/mL (75–170 nmol/L)	30–68 ng/mL (75–170 nmol/L)
Vitamin E, serum	5–18 µg/mL (12–42 µmol/L)	5–18 µg/mL (12–42 µmol/L)
Vitamin K, serum	0.13–1.19 ng/mL (0.29–2.64 nmol/L)	0.13–1.19 ng/mL (0.29–2.64 nmol/L)

[a]SI units given in parentheses. Refer to the second page of Appendix B for factors used to convert convention units to SI units. Nutritional biomarker values and reference ranges are affected by many variables, including sample and method of analysis used and population characteristics. They change as more information becomes known about nutrient biomarker levels and health relationships.

Table 1.19 U.S. national nutrition monitoring systems

Survey	Purpose
1. National Health and Nutrition Examination Survey (NHANES)	• Assesses dietary intake, health, and nutritional status in a sample of 5000 people of all ages on a yearly basis; it oversamples low-income individuals.
2. Behavioral Risk Factor Surveillance System	• Identifies self-reported dietary behaviors, nutritional knowledge, and health status of thousands of adults by telephone interview.
3. Nationwide Food Consumption Survey (NFCS)	• Estimates food and nutrient intake and understanding of diet and health relationships among a national sample of individuals.
4. Total Diet Study	• Ongoing assessment of levels of various pesticides, contaminants, and nutrients in foods and diets.

After the fact-finding phase of nutritional assessment, the nutritionist or other professionals must "apply their brains to their clients' problems." There is no "one size fits all" approach to solving nutrition problems—each has to be figured out individually.

In the future, biochemical assessments will include a nutrigenomics profile to identify health risks due to interactions among an individual's genetic makeup, gene functions, and components of food.[68]

Monitoring the Nation's Nutritional Health

Food availability, dietary intake, weight status, and nutrition-related disease incidence are investigated regularly in the United States by the National Nutrition Monitoring System. This wide-ranging system is primarily responsible for the conduct of *nutrition surveillance* and *nutrition monitoring* studies.

The first U.S. nutrition survey began in 1936 when hunger, poor growth in children, and vitamin and mineral deficiencies were common. Today's surveys monitor rates of obesity, diabetes, and other nutrition-related disorders; the safety of the food supply (for example, the mercury content of fish and pesticide residues in vegetables and fruits); and food and nutrient intake.

The four major ongoing U.S. studies related to diet, the food supply, and nutritional health are overviewed in Table 1.19. Together with findings from studies conducted by university researchers and others, the results give direction to food and nutrition policies and programs aimed at safeguarding the food supply, improving the population's nutritional health, and maintaining a successful agricultural economy.[69]

Nutrition Surveillance Continuous assessment of nutritional status for the purpose of detecting changes in trend or distribution in order to initiate corrective measures.

Nutrition Monitoring Assessment of dietary or nutrition status at intermittent times with the aim of detecting changes in the dietary or nutritional status of a population.

Public Food and Nutrition Programs

A variety of federal, state, and local programs are available to provide food and nutrition services to families and individuals. Many communities have nutrition coalitions or partnership groups that collaborate on meeting the food and nutritional needs of community members. Programs representing church-based feeding sites, food shelves, Second Harvest Programs, the Salvation Army, missions, and others are usually a part of local coalitions. These central resources can be identified by contacting the local public health or cooperative extension agency. State-level programs are generally part of large national programs.

About 1 in 5 Americans participates in at least one of the USDA's 15 food assistance programs at some point during the year. The Food Stamp Program (now called the Supplemental Nutrition Assistance Program or "SNAP") is the nation's largest food assistance program, serving about 27 million low-income Americans in 2005. The program also serves as a source of demand for the products of American farmers and food companies.[70] Some programs, such as the School Lunch Program, benefit many children. Other programs are targeted to families and individuals in need. *Need* is generally defined as including individual and household incomes below the poverty line. Some programs, such as WIC, have eligibility standards of up to 185% of the poverty line (Table 1.20). Income guidelines change periodically and are several thousand dollars higher for people in Alaska and Hawaii.

WIC

The Special Supplemental Nutrition Program for Women, Infants, and Children (WIC) was established in 1972 and is administered by the USDA. The program provides

Table 1.20 Income eligibility standards for the WIC program (≤185% of poverty income, 2009–2010)

Household Size	Household Income per Year
1	$20,036
2	26,955
3	33,874
4	40,793
5	47,712
6	54,631
7	61,550
8	68,550
Each additional member	6,919

and a quarter of all young children in the United States participate in the WIC program.

The WIC program has been shown to have a number of positive effects on the health of participants. Infants born to women participating in WIC while pregnant weigh more, and they are less likely to be small at birth and to be born before term, than the infants of nonparticipating low-income women. Children served by WIC tend to consume more nutritious diets and experience lower rates of iron deficiency than children who are low-income but not enrolled in WIC. WIC is cost-effective: every dollar invested in WIC prenatal nutrition services saves $3.13 on Medicaid costs for infants during the first 2 years of life.[71]

Table 1.21 presents information on existing federal food and nutrition programs. You can get more information on these programs online at www.nutrition.gov.

nutrition education and counselling as well as food vouchers for low-income pregnant, postpartum, and breastfeeding women and for low-income children under the age of 5 years. Food vouchers apply to nutritious foods such as fortified breakfast cereals, iron-fortified infant cereals and formula, milk, cheese, eggs, peanut butter, dried beans whole grain products, fruits, vegetables, and 100% fruit and vegetable juices. Some WIC programs offer vouchers for the purchase of produce at farmer's markets. WIC staff also provides breastfeeding support and referrals to health-care and social-service providers.

Eligibility for WIC is based on low-income status and the presence of a nutritional risk, such as iron deficiency or being underweight. WIC services are provided through approximately 10,000 clinic sites throughout the United States and in American Samoa, Guam, Puerto Rico, and the Virgin Islands. The program serves more than 7 million women and children each year. Nearly half of all infants

Nationwide Priorities for Improvements in Nutritional Health

Public health initiatives involving population-based improvements in food safety, food availability, and nutritional status have led to major gains in the health status of the country's population. Among the important components of this success story are programs that have expanded the availability of housing; safe food and water; foods fortified with iodine, iron, vitamin D, or foliate; fluoridated public water; food assistance; and nutrition education. Since 1900, the average life span of persons in the United States has lengthened by more than 30 years, and 25 years of this gain are estimated to be due to the quiet revolutions that have taken place in public health.[71]

Table 1.21 Examples of federal food and nutrition programs

Program	Activity
Child and Adult Care Food Program (CACFP)	Reimburses child and adult care organizations in low-income areas for provision of nutritious foods.
Summer Food Service Program	Provides foods to children in low-income areas during the summer.
School Breakfast and Lunch Programs	Provide free breakfasts and reduced-cost or no-cost lunches to children from families who cannot afford to buy them.
Food Stamp Program	Subsidizes food purchases of low-income families and individuals.
WIC	Serves low-income, high-risk, and pregnant and breastfeeding women and children up to 5 years of age. Provides supplemental nutritious foods and nutrition education as an adjunct to health care.
Head Start Program	Includes nutrition education for children and parents and supplies meals for children in the program.

Although the United States spends more on health care than any other country, statistics from the World Health Organization (WHO) indicate that it ranks twenty-sixth among developed countries of the world in life expectancy. Today's priorities for improvements in the public's health and longevity center on reducing obesity, infant mortality, smoking, excessive alcohol consumption, accidents, violence, and physical inactivity. Goals for dietary changes are a central part of the nation's overall plan for health improvements. For the two out of three Americans who do not smoke or drink excessively, dietary intakes represent the major environmental influence on long-term health.[72]

Goals for improving the nutritional health of the nation are summarized in the document "Healthy People 2010: Objectives for the Nation" (Table 1.22). Because the seeds of many chronic diseases are planted during pregnancy and childhood, major emphasis is placed on dietary habits early in life.

U.S. Nutrition and Health Guidelines

In the United States, the national guidelines for diet and physical activity are called the "Dietary Guidelines for Americans" and the major educational tool for consumers "MyPyramid" (Illustration 1.13).

Dietary Guidelines for Americans   The Dietary Guidelines for Americans provide science-based recommendations to promote health and to reduce the risk for

---

Table 1.22 2010 Nutrition Objectives for the Nation

√ Increase the proportion of adults who are at a healthy weight from 42 to 60%.

√ Reduce the proportion of adults who are obese from 23 to 15%.

√ Reduce the proportion of children and adolescents who are overweight or obese from 11 to 5%.

√ Reduce growth retardation among low-income children under age 5 years from 8 to 5%.

√ Increase the proportion of persons aged 2 years and older who:

- Consume at least 2 daily servings of fruit from 28 to 75%
- Consume at least 3 daily servings of vegetables, with at least one-third being dark-green or deep-yellow vegetables, from 3 to 50%
- Consume at least 6 daily servings of grain products, with at least three being whole grains, from 7 to 50%
- Consume less than 10% of calories from saturated fat from 36 to 75%
- Consume no more than 30% of total calories from fat from 33 to 75%
- Consume 2400 mg or less of sodium daily from 21 to 65%
- Meet dietary recommendations for calcium from 46 to 75%

√ Reduce iron deficiency among young children and females of childbearing age from 4–11% to 1–7%.

√ Reduce anemia among low-income pregnant females in their third trimester from 29 to 20%.

√ Increase the proportion of worksites that offer nutrition or weight-management classes or counseling from 55 to 85%.

√ Increase the proportion of physician office visits made by patients with the diagnosis of cardiovascular disease, diabetes, osteoporosis, or hyperlipidemia that include counseling or education related to diet and nutrition from 42 to 75%.

√ Increase food security among U.S. households, and in so, doing reduce hunger rates.

Other Objectives Related to Nutrition

√ Reduce infant mortality from 7.6 to no more than 5 per 1000 live births.

√ Reduce the incidence of spina bifida and other neural tube defects from 7 to 3 per 10,000 live births.

√ Reduce the incidence of birth defects from 1.7 to 1.2 per 1000 live births.

√ Increase the proportion of women who receive preconceptional counseling.

√ Increase the proportion of pregnant women who begin prenatal care in the first trimester from 81 to 90% or more.

√ Reduce low birthweight (<2500 grams) from 7.3 to 5%.

√ Reduce preterm births (<37 weeks) from 9.1 to 7.6%.

√ Increase abstinence from alcohol use by pregnant women from 79 to 95%.

√ Reduce the incidence of fetal alcohol syndrome.

√ Increase the proportion of women who gain weight appropriately during pregnancy.

√ Increase from 60 to 75% the proportion of women who exclusively breastfeed after delivery.

Illustration 1.13 The major nutritional and physical activity guides in the United States.

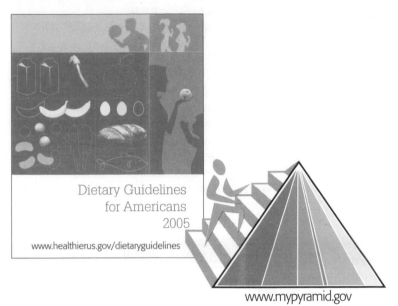

major chronic diseases through diet and physical activity. Due to its credibility and focus on health promotion and disease prevention for the public, the Dietary Guidelines form the basis of federal food and nutrition education programs and policies. Under legislative mandate, the Dietary Guidelines for Americans must be updated every five years.

The 2005 edition of the Dietary Guidelines stresses the importance of selecting nutrient-dense foods, balancing caloric intake with output, and increasing physical activity. This document includes the promise that the health of most individuals will be enhanced if the recommendations are followed. Each recommendation is part of an integrated whole—all the recommendations should be implemented for best results.

Focus Areas and Key Recommendations
The Dietary Guidelines for 2005 include nine "Focus Areas" and 23 "Key Recommendations." These are highlighted in Table 1.23. "Key Recommendations for Specific Population Groups" are also provided by the Dietary Guidelines. Special population groups addressed are infants and young children, pregnant women, older adults, and people with weakened immune systems. These and other recommendations included in the Dietary Guidelines are available online (see the resources section at the end of this chapter).

Implementation of the Dietary Guidelines The MyPyramid Food Guide is the major how-to tool intended to help the public implement the 2005 Dietary

Table 1.23 The 2005 Dietary Guidelines for Americans: focus areas and examples of key recommendations

Adequate Nutrients within Calorie Needs
- Consume the variety of foods and beverages cited in MyPyramid food guide or the DASH eating plan. (Additional information about the DASH eating plan is presented on page 46.)

Weight Management
- Achieve and maintain a healthy body weight.

Physical Activity
- Engage in at least 30 minutes of moderate-intensity physical activity, above usual activity, at work or home on most days of the week.

Food Groups to Encourage
- Choose and consume colorful vegetables, fruits, whole-grain products, and low-fat dairy products.

Fats
- Consume less than 10% of calories from saturated fatty acids, less than 300 mg/day of cholesterol, and keep consumption of *trans* fatty acids as low as possible.

Carbohydrates
- Choose fiber-rich fruits, vegetables, and whole grains often.

Sodium and Potassium
- Consume less than 2300 mg of sodium (approximately 1 tsp of salt) per day.

Alcoholic Beverages
- Alcoholic beverages, if chosen, should be consumed sensibly and in moderation.

Food Safety
- Avoid food-borne illness by cleaning hands and food-preparation surfaces and cooking foods to safe temperatures.

Guidelines. In addition, the Dietary Guidelines report identified the Dietary Approaches to Stop Hypertension (DASH) Eating Plan as being consistent with the recommendations. MyPyramid covers both food selection and physical exercise, and the DASH Eating Plan addresses only dietary intake. Both are valuable tools that provide the framework for planning nutrient-dense, calorically appropriate diets that diminish the risk of chronic disease.

MyPyramid Food Guide   Food group guides to the intake of healthful diets have been available in the United States since 1916. Known by names such as "Basic Four Food Groups" and "Food Guide Pyramid," the guides have been periodically updated. New releases of food-group guides reflect existing scientific knowledge about nutrition and health and are modified to address emerging health problems.

The U.S. Department of Agriculture (USDA) released its newest version of the food guide in 2005. Called "MyPyramid," it is very different from previous guides. It is Internet-based and resource-filled, and some of the educational resources offered are interactive. It is a popular website.

MyPyramid guidance stresses the importance of lower calorie intakes and increased physical activity levels. Because low intakes increase the risk of certain chronic diseases, the new food guide emphasizes intake of whole-grain products, colourful vegetables, low-fat dairy products, and lean meats. MyPyramid uses cups and ounces as the primary measures of how much food to consume daily and it gives recommended number of cups or ounces for food within each group.

MyPyramid.gov: The Website   A large assortment of educational and assessment tools are available at the MyPyramid website (www.mypyramid.gov). Here are a few examples of the types of resources you can find:

- The *For Professionals* link will connect you to detailed information about using MyPyramid educational materials and provides seven days of sample menus that correspond to MyPyramid recommendations for a 2000-calorie food pattern. Four of the seven days included in the menus, and the nutrient analysis for the menus, are shown in Illustration 1.14. Menus offered as examples on MyPyramid may not correspond to individual food preferences and contain relatively few ethnic foods.

- *Inside the Pyramid* explains each food group, discretionary calories, and physical activity recommendations. It provides information on which foods are within the various groups and food measure equivalents so you can convert food amounts into cups and ounces, and oils into teaspoons. Table 1.24 lists

food measure equivalents for common foods by food group shown in this section of *Inside the Pyramid*.

- *MyPyramid Tracker* is a dietary and physical activity assessment tool that provides information on your diet quality, physical activity status, and links to nutrient and physical activity information. Although the My Pyramid Food Guide is not designed for weight-loss diets, the My Pyramid Tracker tool can be used to assess and monitor caloric intake and physical activity levels.

- *MyPyramid Menu Planner* can be used to develop menus based on food preferences and calorie need. After you answer questions about an individual's age, sex, height, weight, and physical activity level, the planner calculates calorie need and food-group servings. Individuals can pick the foods that go into the menu and to see how well those choices stack up against the recommendations.

The DASH Diet   Originally published as a diet that helps control mild and moderate high blood pressure in experimental studies, the DASH Eating Plan also reduces the risk of cancer, osteoporosis, and heart disease. Improvements in blood pressure are generally seen within two weeks of starting this dietary pattern.[73]

The DASH dietary pattern emphasizes fruits, vegetables, low-fat dairy foods, whole-grain products, poultry, fish, and nuts. Only low amounts of fat, red meats, sweets, and sugar-containing beverages are included. This dietary pattern provides ample amounts of potassium, magnesium, calcium, fiber, and protein and limited amounts of saturated and *trans* fats. Recommendations for types and amounts of food included in this eating plan for a 2000-calorie diet are shown in Table 1.22.

Although one calorie level is shown in the table, DASH Eating Plans for 12 calorie levels (1600 to 3200) are available at the website listed in Table 1.25.

The Mediterranean Diet   Adherence to MyPyramid, DASH, and the Mediterranean dietary recommendations effectively lowers the risk of developing a range of diseases, from heart disease to diabetes.[14] The Mediterranean diet was originally based on foods consumed by people in Greece, Crete, southern Italy, and other Mediterranean areas where rates of chronic disease were low and life expectancy long.[16]

The Mediterranean dietary pattern, shown in Illustration 1.15, emphasizes daily consumption of bread, pasta, fruits, vegetables, olive oil, legumes, cheese and yogurt, and meats such as fish, poultry, veal, and lamb. Daily physical activity, a traditional part of life in these areas, is included in the plan. Fish, poultry, eggs, and sweets are recommended weekly, and red meat monthly. Wine with meals is part of the Mediterranean diet.

Illustration 1.14 My Pyramid sample menu (showing 4 of 7 days) and analysis for a 2000-calorie food pattern.

Sample Menus for a 2000-Calorie Food Pattern

Averaged over a week, this seven-day menu provides all of the recommended amounts of nutrients and food from each food group. (Italicized foods are part of the dish or food that precedes it, which is not italicized.)

Food Group	Daily Average Over One Week		Nutrient	Daily Average Over One Week
GRAINS	Total Grains (oz eq) 6.0 Whole Grains 3.4 Refined Grains 2.6		Calories	1994
			Protein, g	98
			Protein, % kcal	20
			Carbohydrate, g	264
			Carbohydrate, % kcal	53
VEGETABLES *	Total Veg* (cups) 2.6		Total fat, g	67
			Total fat, % kcal	30
			Saturated fat, g	16
			Saturated fat, % kcal	7.0
			Monounsaturated fat, g	23
			Polyunsaturated fat, g	23
			Linoleic acid, g	21
FRUITS	Fruits (cups) 2.1		Alpha-linolenic acid, g	1.1
			Cholesterol, mg	207
			Total dietary fiber, g	31
			Potassium, mg	4715
MILK	Milk (cups) 3.1		Sodium, mg*	1948
			Calcium, mg	1389
			Magnesium, mg	432
			Copper, mg	1.9
			Iron, mg	21
			Phosphorus, mg	1830
			Zinc, mg	14
			Thiamin, mg	1.9
MEAT & BEANS	Meat/ Beans (oz eq) 5.6		Riboflavin, mg	2.5
			Niacin equivalents, mg	24
			Vitamin B6, mg	2.9
			Vitamin B12, mcg	18.4
			Vitamin C, mg	190
			Vitamin E, mg (AT)	18.9
OILS	Oils (tsp/grams) 7.2 tsp/32.4 g		Vitamin A, mcg (RAE)	1430
			Dietary folate equivalents, mcg	558

*Vegetable subgroups	(weekly totals)
Dk-Green Veg (cups)	3.3
Orange Veg (cups)	2.3
Beans/Peas (cups)	3.0
Starchy Veg (cups)	3.4
Other Veg (cups)	6.6

* Starred items are foods that are labeled as no-salt-added, low-sodium, or low-salt versions of the foods. They can also be prepared from scratch with little or no added salt. All other foods are regular commercial products that contain variable levels of sodium. Average sodium level of the seven day menu assumes no salt added in cooking or at the table.

Sample Menus for a 2000-Calorie Food Pattern

Averaged over a week, this seven-day menu provides all of the recommended amounts of nutrients and food from each food group. (Italicized foods are part of the dish or food that precedes it.)

Day 1

BREAKFAST

Breakfast burrito
  1 flour tortilla (7˝ diameter)
  1 scrambled egg (in 1 tsp soft margarine)
  1/3 cup black beans*
  2 tbsp salsa
1 cup orange juice
1 cup fat-free milk

LUNCH

Roast beef sandwich
  1 whole-grain sandwich bun
  3 ounces lean roast beef
  2 slices tomato
  1/4 cup shredded romaine lettuce
  1/8 cup sauteed mushrooms (in 1 tsp oil)
  1 1/2 ounce part-skim mozzarella cheese
  1 tsp yellow mustard
3/4 cup baked potato wedges*
  1 tbsp ketchup
1 unsweetened beverage

DINNER

Stuffed broiled salmon
  5 ounce salmon filet
  1 ounce bread stuffing mix
  1 tbsp chopped onions
  1 tbsp diced celery
  2 tsp canola oil
1/2 cup saffron (white) rice
  1 ounce slivered almonds
1/2 cup steamed broccoli
  1 tsp soft margarine
1 cup fat-free milk

SNACKS

1 cup cantaloupe

Day 2

BREAKFAST

Hot cereal
  1/2 cup cooked oatmeal
  2 tbsp raisins
  1 tsp soft margarine
1/2 cup fat-free milk
1 cup orange juice

LUNCH

Taco salad
  2 ounces tortilla chips
  2 ounces ground turkey, sauteed in
  2 tsp sunflower oil
  1/2 cup black beans*
  1/2 cup iceberg lettuce
  2 slices tomato
  1 ounce low-fat cheddar cheese
  2 tbsp salsa
  1/2 cup avocado
  1 tsp time juice
1 unsweetened beverage

DINNER

Spinach lasagna
  1 cup lasagna noodles, cooked (2 oz dry)
  2/3 cup cooked spinach
  1/2 cup ricotta cheese
  1/2 cup tomato sauce tomato bits*
  1 ounce part-skim mozzarella cheese
1 ounce whole-wheat dinner roll
1 cup fat-free milk

SNACKS

1/2 ounce dry-roasted almonds*
1/4 cup pineapple
2 tbsp raisins

Day 3

BREAKFAST

Cold cereal
  1 cup bran flakes
  1 cup fat-free milk
  1 small banana
1 slice whole wheat toast
  1 tsp soft margarine
1 cup prune juice

LUNCH

Tuna fish sandwich
  2 slices rye bread
  3 ounces tuna (packed in water, drained)
  2 tsp mayonnaise
  1 tbsp diced celery
  1/4 cup shredded romaine lettuce
  2 slices tomato
1 medium pear
1 cup fat-free milk

DINNER

Roasted chicked breast
  3 ounces boneless skinless chicken breast*
1 large baked sweet potato
1/2 cup peas and onions
  1 tsp soft magarine
1 ounce whole-wheat dinner roll
  1 tsp soft margarine
1 cup leafy greens salad
  3 tsp sunflower oil and vinegar dressing

SNACKS

1/4 cup dried apricots
1 cup low-fat fruited yogurt

Day 4

BREAKFAST

1 whole wheat English muffin
  2 tsp soft margarine
  1 tbsp jam or preserves
1 medium grapefruit
1 hard-cooked egg
1 unsweetened beverage

LUNCH

White bean-vegetable soup
  1 1/4 cup chunky vegetable soup
  1/2 cup white beans *
2 ounce breadstick
8 baby carrots
1 cup fat-free milk

DINNER

Rigatoni with meat sauce
  1 cup rigatoni pasta (2 ounces dry)
  1/2 cup tomato sauce tomato bits *
  2 ounces extra lean cooked ground beef (sauteed in 2 tsp vegetable oil)
  3 tbsp grated Parmesan cheese
Spinach salad
  1 cup baby spinach leaves
  1/2 cup tangerine slices
  1/2 ounce chopped walnuts
  3 tsp sunflower oil and vinegar dressing
1 cup fat-free milk

SNACKS

1 cup low-fat fruited yogurt

Table 1.24 MyPyramid food measure equivalents

Grains	
Bagel	1 mini bagel = 1 oz
	1 large bagel = 4 oz
Biscuit	1–2" diameter = 1 oz
	1–3" diameter = 2 oz
Bread	1 slice = 1 oz
Cooked cereal	½ cup = 1 oz
Crackers	5 whole wheat = 1 oz
	7 square/round = 1 oz
English muffin	½ muffin = 1 oz
Muffin	1–2½" diameter = 1 oz
	1–3½" diameter = 3 oz
Pancake	1–4½" diameter = 1 oz
	2–3" diameter = 1 oz
Popcorn	3 cups = 1 oz
Breakfast cereal	1 cup flakes = 1 oz
	1¼ cups = 1 oz
Rice	½ cup = 1 oz
Pasta	½ cup = 1 oz
Tortilla	1–6" diameter = 1 oz
	1–12" diameter = 4 oz

Vegetables	
Cooked	1 cup = 1 cup
Carrots	2 medium = 1 cup
	12 baby = 1 cup
Celery	1 large stalk = 1 cup
Corn on the cob	1–6" long = ½ cup
	1–9" long = 1 cup
Green/red peppers	1 large = 1 cup
Potatoes	1 medium (3" diameter) = 1 cup
Raw, leafy greens	2 cups = 1 cup
Tomato	1 large = 1 cup

Milk	
Milk	1 cup = 1 cup
Yogurt	1 cup = 1 cup
Cheese	1½ oz hard = 1 cup
	⅓ cup shredded = 1 cup
	2 oz processed = 1 cup
	½ cup ricotta = 1 cup
	2 cups cottage cheese = 1 cup
Pudding	1 cup = 1 cup
Frozen yogurt	1 cup = 1 cup
Ice cream	1½ cup = 1 cup

Fruits	
Apple	1 small = 1 cup
	½ large = 1 cup
Banana	1 large = 1 cup
Cantaloupe	⅛ = 1 cup
Grapes	32 = 1 cup
Grapefruit	1 = 1 cup
Orange	1 = 1 cup
Peach	1 = 1 cup
Pear	1 = 1 cup
Plums	3 = 1 cup
Strawberries	8 large = 1 cup
Watermelon	1" wedge = 1 cup
Dried fruit	½ cup = 1 cup
Fruit juice	1 cup = 1 cup

Meat, Fish and Beans	
Steak	1–3½" × 2½" × ½" = 3 oz
Hamburger	1 small = 2 oz
	1 medium = 4 oz
	1 large = 6 oz
Chicken	½ breast = 3 oz
	1 thigh = 2 oz
	1 leg = 3½ oz
Pork chops	1 medium = 3 oz
Fish	1 small can tuna = 3½ oz
	1 small fish = 3 oz
	1 salmon steak = 5 oz
Seafood	5 large srimp = 1 oz
	10 medium clams = 3 oz
	½ cup crab = 2 oz
	½ cup lobster = 2½ oz
Eggs	1 small = 1 oz
	1 large = 2 oz

Table 1.25 The DASH Eating Plan for a 2000-calorie diet

	Number of servings		Number of servings
Grains*	6–8	Lean meats, poultry, fish	2 or fewer
Vegetables	4–5	Nuts, seeds, legumes	4–5/week
Fruits	4–5	Fats and oils	2–3
Fat-free or low-fat milk and milk products	2–3	Sweets	5/week

*whole-grain products primarily

SOURCE: Dietary Guidelines for Americans, Appendix A-1: The DASH Eating Plan at 1,600-, 2,000-, 2,600-, and 3,100- Caloric Levels (www.health.gov/dietaryguidelines/dga2005/document/html/appendixA.htm).

Illustration 1.15 The Mediterranean Diet Pyramid.

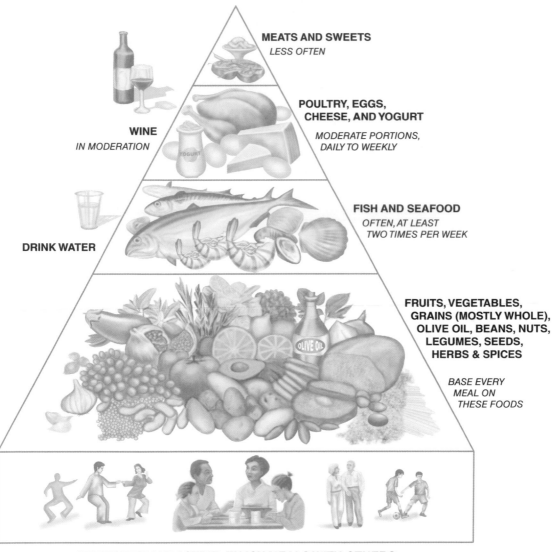

Mediterranean Diet Pyramid
A contemporary approach to delicious, healthy eating

MEATS AND SWEETS
LESS OFTEN

POULTRY, EGGS, CHEESE, AND YOGURT
MODERATE PORTIONS, DAILY TO WEEKLY

WINE
IN MODERATION

FISH AND SEAFOOD
OFTEN, AT LEAST TWO TIMES PER WEEK

DRINK WATER

FRUITS, VEGETABLES, GRAINS (MOSTLY WHOLE), OLIVE OIL, BEANS, NUTS, LEGUMES, SEEDS, HERBS & SPICES

BASE EVERY MEAL ON THESE FOODS

BE PHYSICALLY ACTIVE; ENJOY MEALS WITH OTHERS

Resources

Public Nutrition Programs

Start here if you want to identify government resources related to the Dietary Guidelines and to nutrition information, services, and programs.
Website: **www.nutrition.gov**
Learn about the WIC program at this site.
Website: **www.fns.usda.gov/wic**

Information on public food and nutrition assistance programs is presented at this site.
Website: **www.ers.usda.gov/Browse/FoodNutritionAssistance**

Dietary Assessment

View nutrient profiles for 13,000 foods commonly eaten in the United States in the What's In the Foods You Eat search tool.
Website: **www.ars.usda.gov/foodsearch**

Find the USDA's Automated Multiple-Pass Method for 24-hour dietary recalls at this address.
Website: **www.ars.usda.gov/Services/docs.htm?docid=7710**

Locate the MyPyramid one-day dietary assessment tool here.
Website: **www.mypyramidtracker.gov**

The below address leads to the home page for the Healthy Eating Index. Learn more about its methods and uses here.
Website: **www.cnpp.usda.gov/healthyeatingindex.htm**

Nutrition and Health Guides

The U.S. Department of Agriculture's MyPyramid Food Guide main page, with links to dietary and physical activity assessments, menu planning information, interactive learning tools, and much more.
Website: **www.mypyramid.gov**

This terrific site from the Food and Nutrition Center at the National Agricultural Library is stuffed with information on public food and nutrition programs, nutrition labeling, the Dietary Guidelines and MyPyramid Food Guide, and much more.
Website: fnic.nal.usda.gov/nal_display/index.php?info_center=4&tax_level=1

American Dietetic Association

Links to a wide variety of resources on nutrition and health.
Website: www.eatright.org and www.dietitians.ca

Institute of Medicine: Reports

Want more information on the Dietary Reference Intakes? You can get full reports here.
Website: http://www.iom.edu/Reports.aspx?series={508F5CFF-EE88-4FF6-92BF-8D6CAB46F52E}

Medline

Get access to scientific journal articles on a variety of nutrition topics.
Website: www.nlm.nih.gov/medlineplus

Dietary Guidelines for Americans, 2010; Canada's Food Guide

Food Guide

Check out the Meetings of the 2010 Dietary Guidelines Advisory Committee and hear the science behind the 2010 Dietary Guidelines for Americans.
Website: http://www.cnpp.usda.gov/DietaryGuidelines.htm

Get your copy of Canada's 2008 food guide for healthy living from this site.
Website: www.hc-sc.gc.ca/fn-an/food-guide-aliment/index-eng.php

Chapter 2

Preconception Nutrition

Scott Stulberg/SuperStock

Photodisc

Yuri Arcurs, 2009/Used under license from Shutterstock.Com

Prepared by **Judith E. Brown**

Key Nutrition Concepts

1   Fertility is achieved and maintained by carefully orchestrated, complex processes that can be disrupted by a number of factors related to body composition and dietary intake.

2   Oral contraceptives and contraceptive implants can adversely affect some aspects of nutritional status.

3   Optimal nutritional status prior to pregnancy enhances the likelihood of conception and helps ensure a healthy pregnancy and robust newborn.

Introduction

Human reproduction is the result of a superb orchestration of complex and interrelated genetic, biological, environmental, and behavioral processes. Given favorable states of health, these processes occur smoothly in females and males and set the stage for successful reproduction. However, less than optimal states of health, brought about by conditions such as acute undernutrition or high levels of alcohol intake, can disrupt these finely tuned processes and diminish reproductive capacity. Sometimes conception occurs in the presence of poor nutritional or health status. Such events increase the likelihood that fetal growth and development, and the health of the mother during pregnancy, will be compromised.

This chapter first highlights vital statistics related to the preconception period and presents background information on reproductive physiology. Then the focus is on (1) nutrition and the development and maintenance of the biological capacity to reproduce, (2) nutritional effects of contraceptives, (3) preconceptional nutritional status and the course and outcome of pregnancy, and (4) model programs that promote preconceptional nutritional health. The chapter ends with an overview and case study related to the "Nutrition Care Process," a newly introduced and emerging system for the delivery of nutrition services to people in every phase of the life cycle. The following chapter addresses the role of nutrition in specific conditions—such as premenstrual syndrome, diabetes, eating disorders, obesity, celiac disease, and polycystic ovary syndrome—that affect preconceptional health or very early pregnancy outcomes.

Infertility   Absence of production of children. Commonly used to mean a biological inability to bear children.

Infecundity   Biological inability to bear children after 1 year of unprotected intercourse.

Fertility   Actual production of children. The word best applies to specific vital statistic rates, but it is commonly taken to mean the ability to bear children.

Fecundity   Biological ability to bear children.

Miscarriage   Generally defined as the loss of a conceptus in the first 20 weeks of pregnancy. Also called spontaneous abortion.

Endocrine   A system of ductless glands, such as the thyroid, adrenal glands, ovaries, and testes, that produces secretions that affect body functions.

Immunological   Having to do with the immune system and its functions in protecting the body from bacterial, viral, fungal, or other infections and from foreign proteins (i.e., those proteins that differ from proteins normally found in the body).

Subfertility   Reduced level of fertility characterized by unusually long time to conception (over 12 months) or repeated early pregnancy losses.

Preconception Overview

Between 9 to 15% of all couples in developed countries are involuntarily childless.[1] They are generally considered to be *infertile,* or more correctly *infecund.*

Fertility refers to the actual production of children, whereas *fecundity* addresses the biological capacity to bear children. *Fertility* best applies to vital statistics on fertility rates, or the number of births per 1000 women of childbearing age (15–44 years in most statistical reports). For example, in 2006 the U.S. fertility rate was 68.5 births per 1000 women aged 15 to 44 years.[2] Even scientists and clinicians rarely use these terms correctly, however, and to do so in this chapter would cause undue confusion. The familiar meanings of *fertility* and *infertility* are used in this chapter.

Infertility is generally defined as the lack of conception after 1 year of unprotected intercourse. This definition leads to a "yes/no" answer about fertility that is misleading. Approximately 40% of couples diagnosed as infertile will conceive a child in 3 years without the help of technology.[3] Chances of conceiving decrease the longer infertility lasts and as women and men age beyond 35 years.[1]

Healthy couples having regular, unprotected intercourse have a 20 to 25% chance of a diagnosed pregnancy within a given menstrual cycle.[4] However, many more conceptions probably occur. Studies show that 30–50% of conceptions are lost by resorption into the uterine wall within the first 6 weeks after conception. Another 9% are lost by *miscarriage* in the first 20 weeks of pregnancy.[5]

The most common known cause of miscarriage is the presence of a severe defect in the fetus. Miscarriages can also be caused by maternal infection, structural abnormalities of the uterus, *endocrine* or *immunological* disturbances, and unknown, random events.[5]

Women who experience multiple miscarriages (variously defined as two or three), men who have sperm abnormalities (such as low sperm count or density, malformed sperm, or immobile sperm), and women who ovulate infrequently are considered *subfertile.* It is estimated that 18% of married couples in the United States are subfertile due to delayed time to conception (over 12 months) or repeated, early pregnancy losses.[3] A silver lining to subfertility is that the reproductive capacity of one individual can compensate for diminished potential in the other. In addition, subfertility can be diminished by improvements in diet and lifestyle.[6]

2010 Nutrition Objectives for the Nation Related to the Preconceptional Period

National priorities for improvement in health status prior to pregnancy include five related to nutrition (Table 2.1). Many of the health objectives set for 2010 have not yet been met and will continue to represent priority health

Table 2.1 2010 nutrition objectives for the nation related to preconception

- Increase the proportion of adults who are at a healthy weight from 42 to 60%.
- Reduce the proportion of adults who are obese from 23 to 15%.
- Reduce iron deficiency among females of childbearing age from 4–11% to 1–7%.
- Reduce the incidence of spina bifida and other neural-tube defects from 7 to 3 per 10,000 live births and birth defects from 1.7 to 1.2 per 1000 live births.
- Increase the proportion of women who receive preconceptional counseling.

concern in the 2020 update of the health objectives. Incidences of some of the priority health problems slated for reductions (the proportion of adults who are obese, for example) have increased over the past 10 years.[7]

Reproductive Physiology

The reproductive systems of females and males (Illustration 2.1 on the next page) begin developing in the first months after conception and continue to grow in size and complexity of function through *puberty*. Females are born with a complement of immature *ova* and males with sperm-producing capabilities. The capacity for reproduction is established during puberty when hormonal changes cause the maturation of the reproductive system over the course of 3 to 5 years.

Approximately 7 million immature ova, or *primordial follicles,* are formed during early fetal development, but only about one-half million per ovary remain by the onset of puberty. During a woman's fertile years, some 400–500 ova will mature and be released for possible fertilization. Very few ova remain by *menopause.*

For men, sperm numbers and viability decrease somewhat after approximately 35 years of age, but sperm are still produced from puberty onward.[8] Because females are born with their lifetime supply of ova, the number with chromosomes damaged by oxidation, radioactive particle exposure, and aging increases with time. Consequently, children born to women older than roughly 35 years of age are more likely to have disorders related to defects in chromosomes than are children born to younger women.[1]

Female Reproductive System

During puberty females develop monthly **menstrual cycles,** the purpose of which is to prepare an ovum for fertilization by sperm and the uterus for implantation of a fertilized egg. Menstrual cycles result from complex interactions among hormones secreted by the hypothalamus, the **pituitary gland,** and the ovary. Knowledge of hormonal changes during the menstrual cycle is expanding, and the process is more complex than indicated in this presentation, which focuses on nutritional effects on hormonal changes in the menstrual cycle and on fertility.

Menstrual cycles are 28 days long on average, but it is not uncommon for cycles to be several days shorter or longer. The first day of the cycle is when menses, or blood flow, begins. The first half of the cycle is called the *follicular phase;* the last 14 days is the *luteal phase.* Hormonal changes during these phases of the menstrual cycle are shown in Illustration 2.2 on page 55.

Hormonal Effects During the Menstrual Cycle   At the beginning of the follicular phase, estrogen stimulates the hypothalamus to secrete *gonadotropin-releasing hormone* (GnRH), which causes the pituitary gland to release the *follicle-stimulating hormone* (FSH) and *luteinizing hormone* (LH). (See Table 2.2 on page 56 for definitions of these hormones.) FSH prompts the growth and maturation of 6–20 follicles, or capsules in the surface of the ovary in which ova mature. The presence of FSH stimulates the production of *estrogen* by cells within the follicles. Estrogen and FSH further stimulate the growth and maturation of follicles while rising LH levels cause cells within the follicles to secrete *progesterone*. Estrogen and progesterone also prompt the uterine wall (or endometrium) to store glycogen and other nutrients and to expand the growth of blood vessels and connective tissue. These changes prepare the uterus for nourishing a conceptus after implantation. Just prior to ovulation, which usually occurs on day 14 of a 28-day menstrual cycle, blood levels of FSH and LH peak. The surge in LH level results in the release of an ovum from a follicle, and voilà! Ovulation occurs.

The luteal phase of the menstrual cycle begins after ovulation. Much of the hormonal activity that regulates biological processes during this half of the cycle is initiated by the cells in the follicle left behind when the egg was released. These cells grow in number and size and form the **corpus luteum** from the original follicle. The corpus luteum secretes large amounts of progesterone and some estrogen. These hormones now inhibit the production

Puberty   The period in life during which humans become biologically capable of reproduction.

Ova   Eggs of the female produced and stored within the ovaries (singular = *ovum*).

Menopause   Cessation of the menstrual cycle and reproductive capacity in females.

Menstrual Cycle   An approximately 4-week interval in which hormones direct a buildup of blood and nutrient stores within the wall of the uterus and ovum maturation and release. If the ovum is fertilized by a sperm, the stored blood and nutrients are used to support the growth of the fertilized ovum. If fertilization does not occur, they are released from the uterine wall over a period of 3 to 7 days. The period of blood flow is called the *menses*, or the menstrual period.

Pituitary Gland   A pea-sized gland located at the base of the brain. It is connected to the hypothalamus and produces and secretes growth hormone, prolactin, oxytocin, follicle-stimulating hormone, luteinizing hormone, and other hormones in response to signals from the hypothalamus.

Corpus Luteum   (*corpus* = body, *luteum* = yellow) A tissue about 12 mm in diameter formed from the follicle that contained the ovum prior to its release. It produces estrogen and progesterone. The "yellow body" derivation comes from the accumulation of lipid precursors of these hormones in the corpus luteum.

Illustration 2.1 Mature female and male reproductive systems.

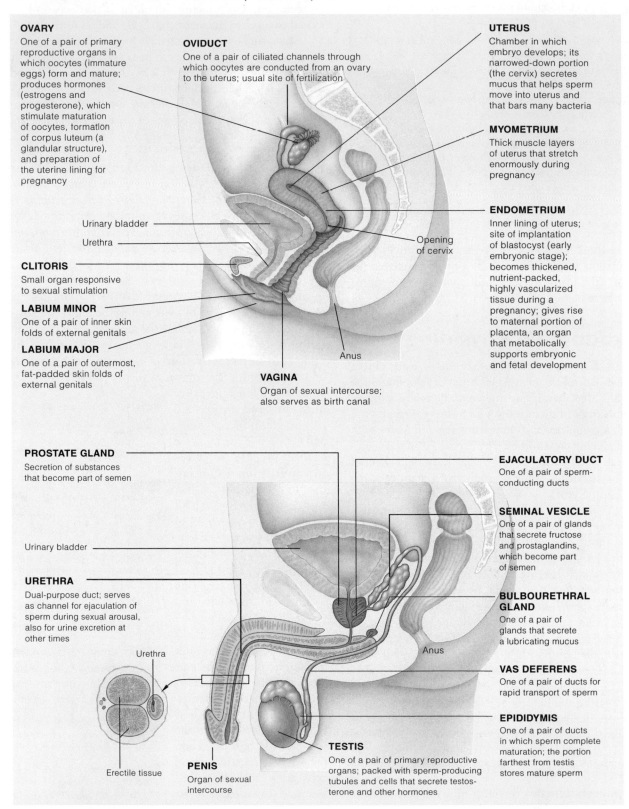

OVARY
One of a pair of primary reproductive organs in which oocytes (immature eggs) form and mature; produces hormones (estrogens and progesterone), which stimulate maturation of oocytes, formation of corpus luteum (a glandular structure), and preparation of the uterine lining for pregnancy

OVIDUCT
One of a pair of ciliated channels through which oocytes are conducted from an ovary to the uterus; usual site of fertilization

UTERUS
Chamber in which embryo develops; its narrowed-down portion (the cervix) secretes mucus that helps sperm move into uterus and that bars many bacteria

MYOMETRIUM
Thick muscle layers of uterus that stretch enormously during pregnancy

ENDOMETRIUM
Inner lining of uterus; site of implantation of blastocyst (early embryonic stage); becomes thickened, nutrient-packed, highly vascularized tissue during a pregnancy; gives rise to maternal portion of placenta, an organ that metabolically supports embryonic and fetal development

Urinary bladder

Urethra

Opening of cervix

CLITORIS
Small organ responsive to sexual stimulation

LABIUM MINOR
One of a pair of inner skin folds of external genitals

LABIUM MAJOR
One of a pair of outermost, fat-padded skin folds of external genitals

Anus

VAGINA
Organ of sexual intercourse; also serves as birth canal

PROSTATE GLAND
Secretion of substances that become part of semen

EJACULATORY DUCT
One of a pair of sperm-conducting ducts

SEMINAL VESICLE
One of a pair of glands that secrete fructose and prostaglandins, which become part of semen

Urinary bladder

URETHRA
Dual-purpose duct; serves as channel for ejaculation of sperm during sexual arousal, also for urine excretion at other times

BULBOURETHRAL GLAND
One of a pair of glands that secrete a lubricating mucus

Anus

VAS DEFERENS
One of a pair of ducts for rapid transport of sperm

Urethra

EPIDIDYMIS
One of a pair of ducts in which sperm complete maturation; the portion farthest from testis stores mature sperm

Erectile tissue

PENIS
Organ of sexual intercourse

TESTIS
One of a pair of primary reproductive organs; packed with sperm-producing tubules and cells that secrete testosterone and other hormones

of GnRH, and thus the secretion of FSH and LH. Without sufficient FSH and LH, ova within follicles do not mature and are not released. (This is also how estrogen and progesterone in some birth control pills inhibit ova maturation and release.) Estrogen and progesterone secreted by the corpus luteum further stimulate the development of the endometrium. If the ovum is not fertilized, the production of hormones by the corpus luteum declines, and blood levels

Illustration 2.2 Changes in the ovary and uterus, correlated with changing hormone levels during the follicular and luteal phases of the menstrual cycle.

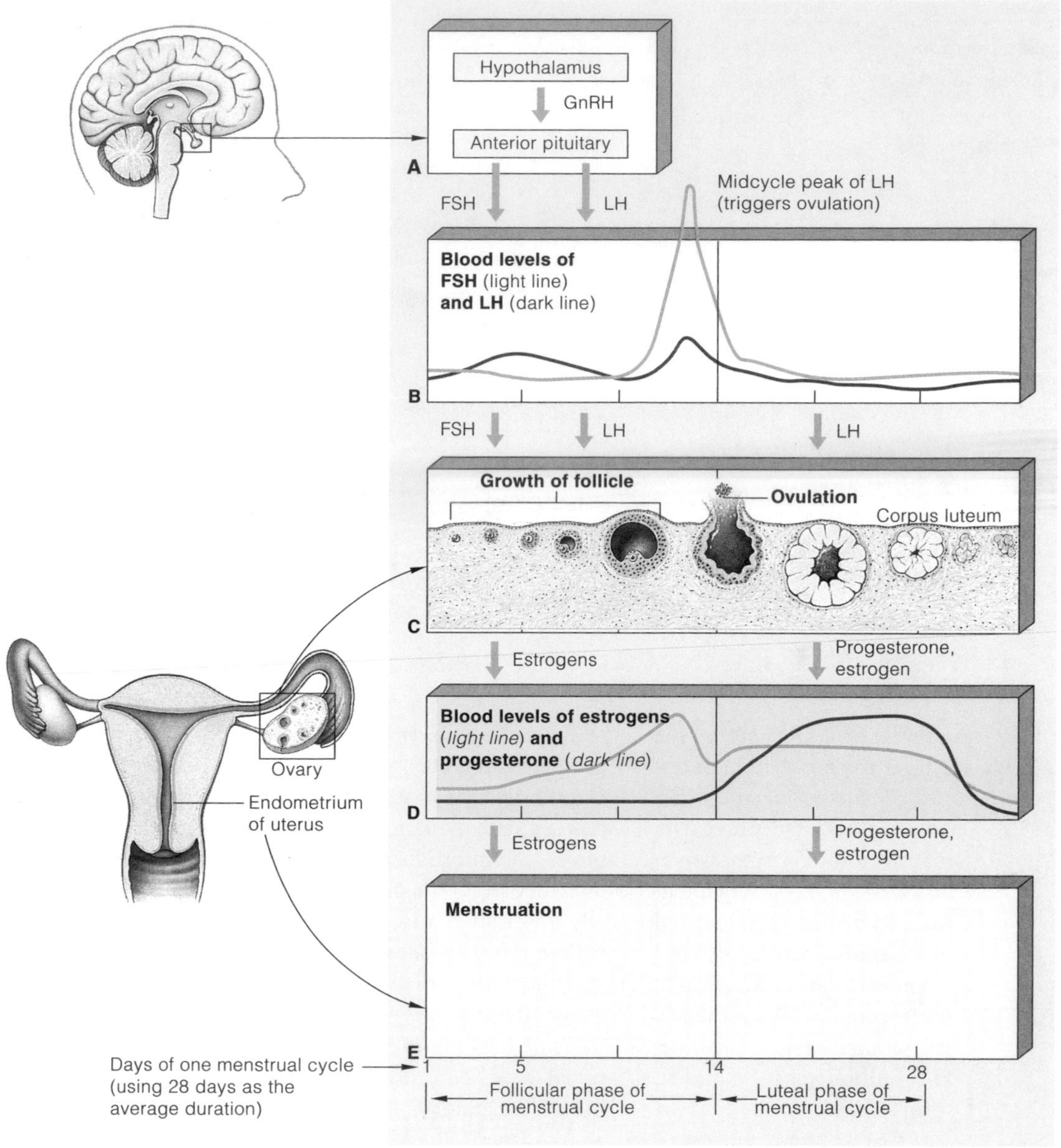

of progesterone and estrogen fall. This decline removes the inhibitory effect of these hormones on GnRH release, and GnRH is again able to stimulate release of FSH for the next cycle of follicle development, and of LH for the stimulation of progesterone and estrogen production. Decreased levels of progesterone and estrogen also cause blood vessels in the uterine wall to constrict, allowing the uterine wall to release its outer layer in the *menstrual flow*. Cramps and other side effects of menstruation can be traced to the production of *prostaglandins* by the uterus. These substances cause the uterus to contract and release the blood and nutrients stored in the uterine wall.

If the ovum is fertilized, it will generally implant in the lining of the uterus within 8 to 10 days. Hormones secreted by the dividing, fertilized egg signal the corpus luteum to increase in size and to continue to produce enough estrogen and progesterone to maintain the nutrient and blood vessel supply in the endometrium. The corpus luteum ceases to function within the first few months of pregnancy, when it is no longer needed for hormone production.

Male Reproductive System

Reproductive capacity in males is established by complex interactions among the hypothalamus, pituitary gland, and *testes*. The process in males is ongoing

Prostaglandins A group of physiologically active substances derived from the essential fatty acids. They are present in many tissues and perform such functions as the constriction or dilation of blood vessels and stimulation of smooth muscles and the uterus.

Testes Male reproductive glands located in the scrotum. Also called testicles.

Table 2.2 Hormones that affect reproduction

Hormone	Abbreviation	Source	Action
Gonadotropin-releasing hormone	GnRH	Hypothalamus	Stimulates release of FSH and LH
Follicle-stimulating hormone	FSH	Pituitary	Stimulates the maturation of ova and sperm
Luteinizing hormone	LH	Pituitary	Stimulates secretion of estrogen, progesterone, and testosterone and growth of the corpus luteum
Estrogen (most abundant form is estradiol)		Ovaries, testes, fat cells, corpus luteum, and placenta (during pregnancy)	Stimulates release of GnRH in follicular phase and inhibits in luteal phase; stimulates thickening of uterine wall during menstrual cycle
Progesterone (progestin, progestogen, and gestagon are similar)		Ovaries and placenta	"Progestational": prepares uterus for fertilized ovum and to maintain a pregnancy; stimulates uterine lining buildup during menstrual cycle; helps stimulate cell division of fertilized ova; inhibits action of testosterone
Testosterone		Mostly by testes	Stimulates maturation of male sex organs and sperm, formation of muscle tissue, and other functions

rather than cyclic. Fluctuating levels of GnRH signal the release of FSH and LH, which trigger the production of testosterone (Table 2.2) by the testes. Testosterone and other *androgens* stimulate the maturation of sperm, which takes 70–80 days. When mature, sperm are transported to the *epididymis* for storage. Upon ejaculation, sperm mix with secretions from the testes, seminal vesicle, prostate, and bulbourethral gland to form *semen.*

Just as some aspects of female reproductive processes remain unclear, scientists have yet to fully elucidate hormonal and other processes involved in male reproduction.

Sources of Disruptions in Fertility

The intricate mechanisms that regulate fertility can be disrupted by many factors, including adverse nutritional exposures, contraceptive use,

Androgens Types of steroid hormones produced in the testes, ovaries, and adrenal cortex from cholesterol. Some androgens (testosterone, dihydrotestosterone) stimulate development and functioning of male sex organs.

Epididymis Tissues on top of the testes that store sperm.

Semen The penile ejaculate containing a mixture of sperm and secretions from the testes, prostate, and other glands. It is rich in zinc, fructose, and other nutrients. Also called seminal fluid.

Pelvic Inflammatory Disease (PID) A general term applied to infections of the cervix, uterus, fallopian tubes, or ovaries. Occurs predominantly in young women and is generally caused by infection with a sexually transmitted disease, such as gonorrhea or chlamydia, or with intrauterine device (IUD) use.

Endometriosis A disease characterized by the presence of endometrial tissue in abnormal locations, such as deep within the uterine wall, in the ovary, or in other sites within the body. The condition is quite painful and is associated with abnormal menstrual cycles and infertility in 30–40% of affected women.

severe stress, infection, tubal damage and other structural problems, and chromosomal abnormalities (Table 2.3).[5,6] Conditions that modify fertility appear to affect hormones that regulate ovulation, the presence or length of the luteal phase, sperm production, or the tubular passageways that ova and sperm must travel for conception to occur. Sexually transmitted infections, for example, can result in *pelvic inflammatory disease (PID)*, which may lead to scarring and blockage of the fallopian tubes.[9] *Endometriosis* is also a common cause of reduced fertility. It develops when portions of the endometrial wall that build up during menstrual cycles become embedded within other body tissues. Endocrine abnormalities that modify hormonal regulation of fertility are the leading diagnoses related to infertility. "Unknown cause" is the second leading diagnosis, however, and is applied to about one-half of all cases of male and female infertility.[10,11]

Nutrition-Related Disruptions in Fertility

In the past few years there has been an explosion in research reports related to the effects of nutrition and other lifestyle factors on fertility. Nutrition and other lifestyle changes are now viewed as a core component of the prevention and treatment of many cases of infertility.[12,13]

Nutrient intake from food and dietary supplements, and body fat, affect fertility primarily by 1) altering the environment in which eggs and sperm develop, and 2) modifying levels of hormones involved in reproductive

Table 2.3 Factors related to altered fertility in women and men

Females and Males	Females	Males
• Weight loss >10 to 15% of normal weight	• Recent oral contraceptive use (within 2 months)	• Inadequate zinc status
• Inadequate antioxidant status (selenium, vitamins C and E)	• Anorexia nervosa, bulimia nervosa	• Heavy metal exposure (lead, mercury, cadmium, manganese)
• Inadequate body fat	• Vegan diets	• Halogen (in some pesticides) and glycol (in antifreeze, de-icers) exposure
• Excessive body fat, especially central fat	• Age >35 years	• Estrogen exposure (in DDT, PCBs)
• Extreme levels of exercise	• Metabolic syndrome	• Sperm defects (quality, motility)
• High alcohol intake	• Pelvic inflammatory disease (PID)	• Excessive heat to testes
• Endocrine disorders (e.g., hypothyroidism, Cushing's disease)	• Endometriosis	• Steroid abuse
• Structural abnormalities of the reproductive tract	• Polycystic ovary syndrome	• High intake of soy foods
• Chromosomal abnormalities in sperm and eggs	• Poor iron stores	
• Celiac disease		
• Oxidative stress		
• Severe psychological stress		
• Infection (sexually transmitted diseases)		
• Diabetes, cancer, other disorders		
• Some medications		

processes.[14] Nutrient intake and body fat before conception also influence the mother's health during pregnancy and the growth and development of the fetus. Nutritional factors generally exert only temporary influence on fertility; normal fertility returns once the problem is corrected.

Undernutrition and Fertility

Does undernutrition decrease fertility in populations? The answer depends on whether the undernutrition is long-term (chronic) or short-term (acute). Chronic undernutrition appears to reduce fertility by only a small amount.[1] Acute undernutrition due to famine or deliberate weight loss in normal-weight women clearly decreases fertility.

Chronic Undernutrition   The primary effect of chronic undernutrition on reproduction in women is the birth of small and frail infants who have a high likelihood of death in the first year of life.[15] Infant death rates in developing countries where malnutrition is common often exceed 50 per 1000 live births. By contrast, infant death rates are less than 3 per 1000 live births in places such as Hong Kong, Japan, and Singapore.[16]

The effect of chronic undernutrition on fertility is difficult to study accurately, and conclusions about relationships will change as more is learned. Investigations of relationships between chronic undernutrition and fertility are complicated by differences in the use of contraception, age of puberty and marriage, breastfeeding duration (longer periods of breastfeeding increase the time to the next pregnancy), access to induced abortions, and social and economic incentives or constraints on family size.[17] In less developed countries with poor access to contraception, births per woman average 6 to 8, whereas in developed countries (where contraception is generally available) they average at or below the replacement level of 2.1 births per woman.[18] Without careful study, it might appear that fertility is *lower*[19] in better-nourished women in developed countries than in poorly nourished women in less developed countries.

Of the environmental factors that influence fertility, education and child survival appear to be the most important. Fertility rates in poor countries decline substantially as women become educated and as child survival increases.[19,20]

Acute Undernutrition   Undernutrition among previously well-nourished women is associated with a dramatic decline in fertility that recovers when food intake does.[15] Periods of feast and famine in the nomadic Kung tribe of Botswana and among the Turkana people of Kenya, for example, are associated with major shifts in fertility.[21,22] These groups of hunter-gatherers (although relatively few survive now) experience sharp seasonal fluctuations in body weight depending on the success of their hunting and foraging for plant foods. Birthrates decline substantially during periods of famine and increase with food availability. When these hunter-gatherers become farmers and food supply is more dependable, body weight increases, activity levels decrease, and pregnancy rates go up.

Other evidence also suggests a connection between undernutrition and infertility. Food shortages in Europe in the seventeenth and eighteenth centuries were accompanied by dramatic declines in birthrates. Famine in Holland during World War II led to calorie intakes of about 1000 per day among women. One out of two women in famine-affected areas stopped menstruating, and the birthrate dropped by 53%. Fertility status improved within 4 months after the end of the famine, but for many women it took as long as a year for their menstrual cycles to return to normal.[23] Similarly, the 1974–1975 famine in Bangladesh resulted in a 40% decline in births.[24]

Famines are associated with more than disruptions in the food supply. They are usually accompanied by low availability of fuel for heating and cooking, poor living conditions, anxiety, fear, and despair. These factors also probably contribute to the declines in fertility observed with famine.

Acute reduction in food intake appears to reduce reproductive capacity by modifying hormonal signals that regulate ovulation and menstrual cycles in females.[24] It also appears to impair sperm maturation in males.

Body Fat and Fertility

Excessive and inadequate levels of body fat are related to declines in fertility in women and men. Body fat–related declines in fertility are primarily related to changes in hormone concentrations.[1,25] In obese individuals, increased levels of *oxidative stress* and exposure of eggs and sperm to oxidative damage are also related to infertility.[1]

Fat cells produce estrogen, androgens, and *leptin,* and the availability of these hormones changes with body fat content. Changes in the availability of these hormones interfere with reproductive processes such as follicular development, ovulation, and sperm maturation and production.[1]

Oxidative Stress   A condition that occurs when cells are exposed to more reactive oxygen molecules (such as free radicals) than to antioxidant molecules. Certain environmental pollutants, ozone, smoke, radiation, excess body fat, high-fat diets, and inflammation are sources of reactive oxygen molecules.

Leptin   A protein secreted by fat cells that, by binding to specific receptor sites in the hypothalamus, decreases appetite, increases energy expenditure, and stimulates gonadotropin secretion. Leptin levels are elevated by high, and reduced by low, levels of body fat.

Anovulatory Cycles   Menstrual cycles in which ovulation does not occur.

Amenorrhea   Absence of menstrual cycle.

Body Mass Index (BMI)   Weight in kg/height in m². BMIs <18.5 are considered underweight, 18.5–24.9 normal weight, 25–29.9 overweight, and BMIs of 30 and higher obesity.

Excessive Body Fat and Infertility
Most obese women and men are not infertile, but they are more likely to experience delays in the time it takes to become pregnant.[1,26] Obese women tend to have higher levels of estrogen, androgens, and leptin than non–obese women.[1,27] These hormonal changes favor the development of menstrual-cycle irregularity (it occurs in 30 to 47% of overweight and obese women), ovulatory failure and *anovulatory cycles,* and *amenorrhea.*[26] Obesity in men is associated with lower levels of testosterone and increased estrogen and leptin levels.[28]

These changes are related to reduced sperm production in 16% of obese men and higher-than-average rates of erectile dysfunction.[26,29] The existence of excessive body fat is generally indicated by *body mass index* values over 30 kg/m².[3,30] Infertility treatments, such as the use of drugs to induce ovulation, are less effective in obese than in normal-weight women.[31] Loss of body fat is related to improvements in hormone levels, reduced oxidative stress, and improved conception rates in both men and women.[1]

Approximately one in three men and women in the United States is obese (or has a body mass index of 30 kg/m² or greater), making infertility related to excess body fat a common problem and an important health and quality-of-life concern.[1] The topic of obesity and fertility is revisited in more detail in the next chapter.

Inadequate Body Fat and Fertility   It appears that a critical level of body fat (usually indicated by a body mass index over 20 kg/m²) is needed to trigger and sustain normal reproductive functions in women.[32,33] Low levels of body fat during adolescence is related to delays in the age of onset of menstruation and to reduced fertility later in life.[32] Impaired fertility in underweight women often takes the form of delayed time to conception and amenorrhea.[15,33] Lowered libido and reduced sperm production have been identified in underweight men with low levels of body fat.[34]

Weight Loss and Fertility in Normal-Weight Women and Men   In normal-weight women, weight loss that exceeds approximately 10–15% of usual weight decreases estrogen, LH, and FSH concentrations.[35] Consequences of these hormonal changes include amenorrhea, anovulatory cycles, and short or absent luteal phases. It is estimated that about 30% of cases of impaired fertility are related to simple weight loss. In the past, this form of amenorrhea was called "weight-related amenorrhea." It is now called "hypothalamic amenorrhea." Hormone levels tend to return to normal when weight is restored to within 95% of previous weight.[15] Case Study 2.1 provides an example of the effect of weight loss on fertility.

Weight gain is the recommended first-line treatment for amenorrhea related to low body weight. In many cases, however, the advice is more easily given than applied. About 10% of underweight women will not consider weight gain and may change health care providers in search of a different solution to infertility.[12]

Case Study 2.1

Cyclic Infertility with Weight Loss and Gain

Photodisc

After four years of experiencing amenorrhea, Tonya seeks medical care to help her become pregnant. She is convinced that her lack of menstrual periods is the cause of her infertility. Tonya's height is 5 feet 5 inches; her weight is 107 pounds, which she has maintained for 4 years (she previously weighed 121 pounds). Her FSH and LH levels are both abnormally low, and she is not ovulating. When the importance and methods of weight gain are explained to her, Tonya agrees to gain some weight. After she regains 7 pounds, her LH level is normal, but her FSH level is still low, and the luteal phase of her cycles is abnormally short. When her weight reaches 119 pounds, Tonya's LH and FSH levels, ovulation, and menstrual cycles are normal.

Questions

1. Was Tonya underweight or normal weight based on BMI when she weighed 107 pounds? (Use the BMI chart on the inside back cover of the text to answer this question.)
2. Can you determine Tonya's body fat content based on her BMI?
3. Why isn't Tonya ovulating?
4. What likely happened to Tonya's average estrogen level when her weight decreased from 121 to 107 pounds?
5. What are two likely reasons Tonya was advised to gain weight to improve her chances of conception rather than being given Clomid or another ovulation-inducing drug?

(Answers are located in the Instructor's Manual for the 4th edition of *Nutrition Through the Life Cycle*.)

Treatment of underweight women with Clomid (clomiphene citrate, a drug that induces ovulation) generally does not improve fertility until weight is regained. Fertility may be improved through the use of GnRH, FSH, and other hormones. However, twice as many infants born to underweight women receiving such therapy are small for gestational age compared to infants born to underweight women who gain weight and experience unassisted conception.[15]

The eating disorder anorexia nervosa is associated with similar, but more severe, changes in endocrine and hypothalamic functions than those seen with weight loss in normal-weight women. (This topic, as well as hypothalamic amenorrhea, is covered in Chapter 3.)

Weight loss decreases fertility in men just as it does in women. In the classic starvation experiments by Keys during World War II,[34] men experiencing a 50% reduction in caloric intake reported substantially reduced sexual drive early in the study. Sperm viability and motility decreased as weight reached 10 to 15% below normal, and sperm production ceased entirely when weight loss exceeded 25% of normal weight. Sperm production and libido returned to normal after weight was regained.

Exercise and Infertility

The adverse effects of intense levels of physical activity on fertility were observed over 40 years ago in female competitive athletes. Since then, a number of studies have shown that young female athletes may experience delayed age at puberty and lack menstrual cycles. Average age of menarche for competitive female athletes and ballet dancers is often delayed by 2 to 4 years. The delay in menarche increases if females begin training for events that require thinness (such as gymnastics) before menarche normally would begin. Very high levels of exercise can also interrupt previously established, normal menstrual cycles. The presence of abnormal cycles reportedly ranges from about 23% in joggers to 86% in female bodybuilders (Table 2.4).[34,35]

Delays and interruptions in normal menstrual cycles appear to result from hormonal and metabolic changes primarily related to caloric deficits rather than intense exercise.[36] Metabolic and hormonal status generally reverts to normal after high levels of training and caloric deficits end.

Some of the hormones involved in fertility impairments perform other important functions in the body,

Table 2.4 Incidence of irregular or absent menstrual cycles in female athletes and sedentary women

	Incidence of Irregular or Absent Menstrual Cycles
Joggers (5 to 30 miles per week)	23%
Runners (over 30 miles per week)	34%
Long-distance runners (over 70 miles per week)	43%
Competitive bodybuilders	86%
Noncompetitive bodybuilders	30%
Volleyball players	48%
Ballet dancers	44%
Sedentary women	13%

which may also be disrupted. Reduced levels of estrogen that accompany low levels of body fat and amenorrhea, for example, may decrease bone density and increase the risk of shortness, bone fractures, and osteoporosis.[37]

Oxidative Stress, Antioxidant Nutrient Status, and Fertility

A growing body of research suggests that intake of *antioxidants* such as vitamin E, vitamin C, beta-carotene, and selenium and antioxidant-rich pigments in vegetables and fruits plays an important role in fertility in women and men.[38] Antioxidant nutrients are needed to protect cells of the reproductive system, including eggs and sperm, from damage due to oxidative stress. Oxidative stress occurs when the production of potentially destructive reactive oxygen molecules (*free radicals*) exceeds the body's own antioxidant defenses. Reactive oxygen molecules attack polyunsaturated fatty acids in sperm membranes, and that decreases sperm motility and reduces the ability of sperm to fuse with an egg. Once the membrane surrounding sperm is damaged, reactive oxygen molecules can enter the sperm cell and damage DNA.[39] This can result in the passage of defective DNA on to the conceptus.[40] Oxidative stress is observed in approximately half of all infertile men.[41] In women, oxidative stress can harm egg and follicular development and can interfere with corpus luteum function and implantation of the egg in the uterine wall.[42]

Antioxidants Chemical substances that prevent or repair damage to molecules and cells caused by oxidizing agents. Vitamins C (see Illustration 2.3) and E, selenium, and certain components of plants function as antioxidants.

Free Radicals Chemical substances (often oxygen-based) that are missing electrons. The absence of electrons makes the chemical substance reactive and prone to oxidizing nearby molecules by stealing electrons from them. Free radicals can damage lipids, cell membranes, DNA, and tissues by altering their chemical structure and functions. They also form as a normal part of metabolism. Over time, oxidative stress causes damage to lipids, cell membranes, DNA, cells, and tissues.

Effects of Antioxidant Intake on Fertility A number of studies have shown lower intakes of antioxidant nutrients in infertile than fertile women and men. Other studies have noted that higher average intakes of antioxidants are associated with improvements in levels of oxidative stress in infertile women and improved sperm maturation, motility, concentration, and reduced DNA and chromosome damage in men.[39,43,44] Supplemental intakes of vitamin E and selenium appear to improve sperm quality in infertile men.[45,46] Regular intake of vitamin C, vitamin E, and beta-carotene supplements have been related to increased sperm number and motility.[44] Although the use of antioxidants for the treatment of infertility disorders in women and men is being advocated,[13,46] additional clinical trials are needed to confirm their beneficial effects.

Zinc Status and Fertility in Men Zinc plays important roles in the reduction of oxidative stress, in sperm maturation, and in testosterone synthesis, and it has been investigated for its potential role in infertility.[53] Lower zinc status in men has been found to be related to poorer sperm quality[54] and sperm concentrations, and to abnormal sperm shapes.[55] Zinc supplementation alone,[56] or combined with vitamin E and vitamin C supplementation,[57] has been found to improve sperm quality.

Illustration 2.3 Apple slices exposed to air will turn brown due to oxidation reactions. The oxidation reactions are prevented if the slices are coated with a vitamin C-rich solution such as lemon juice.

Christopher O Driscoll/iStockphoto

Plant Foods and Fertility Women who regularly consume plant-based, low-fat diets are more likely to have irregular menstrual cycles than omnivores. These results apply to vegans who are thin, normal weight, or overweight.[47,48] Diets providing less than 20% of calories from fat appear to lengthen menstrual cycles among women in general.[49]

Regular intake of soy foods such as tofu, soymilk, tempeh, and textured soy protein appears to be related to reduced sperm count in men,[50] and to a one–day increase in menstrual cycle length in women.[49] Effects of high plant- and soy-food diets on fertility may be related to the influence of certain phytochemicals in plant foods on levels of estradiol, progesterone, and luteinizing hormone.[51] Additionally, some of the effects may be related to the consumption of nutritionally inferior vegan diets.[52]

Multivitamin Supplement, Folate Intake, and Fertility

Multivitamin intake by preconceptional women has been associated with a lower risk of ovulatory infertility in the large Nurse's Health Study.[58] Intake of folic acid from the supplements appears to account for much of the decline in ovulatory infertility observed. Folate status may affect male fertility as well. Higher levels of dietary intakes of folate from food and supplements in healthy men have been related to the presence of fewer chromosomally abnormal sperm than identified among men with lower intake of folate.[59]

Preconception Iron Status, Fertility, and Pregnancy Outcomes Iron status prior to pregnancy is related to fertility and pregnancy outcomes. Results of a large prospective study of nurses indicate that infertility due to a lack of ovulation is related to iron intake. In this study, women who regularly used iron supplements and consumed plant sources of iron were 60% less likely to develop ovulatory infertility than women who did not. Mechanisms underlying the link between iron status and ovulation are not yet established.[110]

Iron deficiency prior to pregnancy has been shown to increase the risk that iron-deficiency anemia will occur during pregnancy and that infants will be born with low stores of iron. Iron deficiency before pregnancy is also related to increased rates of preterm delivery.[111] Iron deficiency occurs in 12% of U.S. women of childbearing age overall, and among 22% of Mexican American and 19% of African American women.[7] About one-half of women in the United State enter pregnancy with inadequate iron stores.

It is easier and more efficient to build up iron stores before pregnancy than during pregnancy.[111] Iron status can generally be improved by taking modest doses of iron supplements (18 mg a day) and by the regular consumption of vitamin C–rich fruits and vegetables along with plant sources of iron, iron-fortified cereals, and lean meats.

Caffeine and Fertility

Should women concerned about infertility consume coffee and other foods with caffeine (Table 2.5)? The previous answer to this question of "it looks like they should be concerned" is now in question.

In a study of European women, researchers found that the chance of conception within a 10-month interval of unprotected intercourse was half as likely among women who consumed over 4 cups of coffee per day (>500 mg caffeine) versus the conception rate of women who consumed little coffee.[60] Another study reported that intake of over 300 mg of caffeine daily from coffee, sodas, and tea decreased the chance of conceiving by 27% per cycle compared to negligible caffeine intake.[61] In both studies, the effect of caffeine on time to conception was stronger in women who smoked.

Other studies have failed to find effects of caffeine intake prior to pregnancy on the amount of time it takes to become pregnant,[62] on ovulatory infertility,[63] or on indicators of ovarian function.[64] Results of research on the effects of caffeine or coffee intake on fertility in women are conflicting, and overall effects are likely weak.[1] Effects of coffee or caffeine consumption on fertility may be due to one or more of the hundreds of other biologically active substances in coffee, or to characteristics of women who consume lots of coffee and other sources of caffeine.

Alcohol and Fertility Alcohol may influence fertility by decreasing estrogen and testosterone levels and by disrupting menstrual cycles and testicular functions.[64] In a study of 430 Danish couples attempting pregnancy for 6 months, consumption of from 1 to 5 alcohol-containing drinks per week by women was related to a 39% lower chance of conception. Alcoholic-beverage consumption of over 10 drinks per week was related to a 66% reduction in the probability of conception during the 6-month period.[65] Consumption of 7 or more drinks a week has been associated with a doubling of risk for infertility in women over the age of 30 only.[66]

Not all studies show an effect of alcohol intake on fertility, and some show effects only in women and men with very high intakes of alcohol.[67–69] Excess alcohol intake early in pregnancy is related to impaired fetal growth and development. Consequently, it is recommended that women restrict their alcohol intake while attempting pregnancy to avoid the possibility of alcohol-related harm to the developing fetus.[70]

Heavy-Metal Exposure Exposure to high levels of lead is related to decreased sperm production and abnormal sperm motility and shape.[71] Inhaled or ingested lead is transported to the pituitary gland, where it appears

Table 2.5 Caffeine content of foods and beverages

Foods and Beverages	Caffeine (mg)
Coffee, 1 c	
Drip	115–175
Decaffeinated	0–4
Instant	61–70
Percolated	97–140
Espresso (2 oz)	100
Tea, 1 c	
Black, brewed 5 minutes, U.S. brand	32–144
Black, brewed 5 minutes, Imported brand	40–176
Green, brewed 5 minutes	25
Instant	40–80
Soft drinks, 12 oz	
Mountain Dew	54
Coca-Cola	47
Cherry Coke	47
Diet Coca-Cola	47
Dr. Pepper	40
Pepsi-Cola	38
Diet Pepsi	37
Ginger ale	0
7-Up	0
Energy drinks	
Ripped Force, (8 oz)	120
Power Shot (1 oz)	100
Red Bull (8 oz)	80
Full Throttle (8 oz)	72
Jolt (12 oz)	72
Kick (12 oz)	56
Surge (8 oz)	35
Chocolate products	
Cocoa/chocolate milk (1 cup)	10–17
Milk chocolate (1 oz)	1–15
Chocolate syrup (2 Tbsp)	4
Nonprescription drugs, 2 tablets	
Nodoz	200
Vivarin	200
Excedrin	130
Weight-loss pills	150

mercury in Hong Kong has been associated with decreased sperm count and abnormal semen.[73] Consumption of fish from the U.S. Great Lakes does not appear to pose similar problems.[74]

Exposure to excess levels of cadmium, molybdenum, manganese, boron, cobalt, copper, nickel, silver, or tin may also affect male fertility.[75,76] These metals may build up in male reproductive systems through the inhalation of fumes or dust containing particles or through long-term use of dietary supplements, industrial pollution, or consumption of contaminated water.[75,77]

Nutrition and Contraceptives

The contraceptive revolution emerged in full force in the 1960s when use of pills with heavy doses of estradiol (the most biologically active form of estrogen) and progesterone became widespread. The adverse side effects of these oral contraceptives were plentiful and included increased risk of heart attack and stroke, elevated blood lipids, glucose intolerance, weight gain, and folate and vitamin B_6 deficiencies. New generations of oral contraceptives employed increasingly lower hormone doses, and side effects diminished substantially. Use of the current generation of oral contraceptives is associated with fewer adverse side effects than the pills used in the past, but risks remain (Table 2.6).

Fertility-control products for females also include contraceptive implants, patches, and injections. These,

Table 2.6 Nutrition-related side effects of contraceptives

Oral Contraceptives
Increased blood levels of HDL cholesterol (the "good" cholesterol)
Increased blood levels of triglycerides and LDL cholesterol
Increased risk of venous thromboembolism (blood clots), cervical cancer, and cardiovascular disease
Decreased blood levels of vitamins B_{12} and B_6
Increased blood levels of copper

Contraceptive Injections (Depo-Provera)
Weight gain
Increased blood levels of LDL cholesterol and insulin
Decreased blood levels of HDL cholesterol
Decreased bone density

Contraceptive Implants (Norplant)
Weight gain

to disrupt hormonal communications with the testes. The result is lowered testosterone levels and decreased sperm production and motility. The men most likely to be exposed to excess lead tend to be workers in smelting and battery factories.[72]

Mercury can build up in fish living in contaminated waters. Ingestion of fish from waters contaminated with too, have a number of nutritional side effects. Development of hormonal contraceptive methods for control of fertility in males lags far behind advances in female contraception.

Oral Contraceptives and Nutritional Status

Oral contraceptives increase blood levels of triglycerides by 30% and total cholesterol levels by 6% on average. HDL cholesterol—the "good" blood cholesterol fraction—is increased slightly by these contraceptives.[78]

Women taking oral contraceptives have a two fold risk of thromboembolism (blood-clot formation) and are at increased risk of cervical cancer and cardiovascular disease.[78,79] Long-term use of oral contraceptives (10 years) is associated with the benefit of a decrease in the risk of ovarian cancer.[80]

Oral contraceptives still decrease blood levels of certain nutrients. Blood levels of vitamins B_{12} and B_6 have been found to be lower in oral-contraceptive users versus nonusers in several studies.[81,82] In a study involving Canadian adolescents, serum copper levels were found to be 34 to 55% higher among oral contraceptive users versus nonusers.[81]

It is recommended that females who are obese, over the age of 35 years, and smoke; and those who have cardiovascular disease, hypertension, diabetes, or are immobilized, use nonhormonal methods of contraception due to their increased risk of venous thromboembolism. Women who take oral contraceptive pills are cautioned against consuming more than a half-ounce of licorice per day. Genuine black licorice contains glycyrrhizic acid, a substance that gives licorice its distinctive taste. Consumption of several ounces of licorice can increase blood pressure and fluid retention in women using the pill.[83] It is generally also recommended that women stop using oral contraceptive pills about 3 months prior to attempting pregnancy.[84]

Contraceptive Injections

DMPA (depot medroxyprogesterone acetate), or Depo-Provera, is the primary type of injectable contraception used in females. Injections that suppress ovulation are given every 3 months. Although highly effective, Depo-Provera has discontinuation rates that average over 50% within the first year of use. Weight gain is a leading reason for discontinuation (27%); irregular periods (24%), fatigue (23%), headache (25%), and abdominal pain (18%) are also commonly reported reasons for discontinuation.[85,86] Weight gain averages 12 pounds (5.5 kg) during one to three years of Depo-Provera use.[87,88] Long-term use of this contraceptive is related to decreased bone density and blood levels of HDL cholesterol and increased levels of LDL cholesterol and insulin.[89]

Contraceptive Implants

Norplant (levonorgestral), the leading contraceptive implant, prevents conception for up to 7 years.[90] This contraceptive method is highly effective for normal-weight and underweight women, as evidenced by a 1.9% pregnancy rate.[63] Pregnancy rates are 4.2% in obese women, however.[91,92] High rates of side effects, especially erratic bleeding (69%), weight gain (41%), and headaches (30%), lead to early removal of the implant in about half of users.[93] Average weight gain 1 year after the implant has been reported to be 9 pounds (4.1 kg).[94]

Contraceptive Patches

The contraception patch releases a type of estrogen and progesterone. Tests indicate it is highly effective and easy to use. The patch is placed on the skin for 3 weeks and then taken off for a week.

Contraceptive patches increase blood levels of cholesterol and triglycerides to a greater extent than do oral contraceptives and increase the risk of blood-clot formation.[95] On the positive side, use of the patch is related to slight increases in HDL-cholesterol levels. The most commonly reported adverse reactions to contraceptive patches are breast soreness, headaches, application-site reactions, and abdominal pain. Contraceptive patches are less effective in women weighing over 198 pounds (90 kg) than in women who weigh less. Women who should not use a contraceptive patch include those with a history of heart disease, stroke, blood clots, and reproductive cancers.[96] Pregnancy should be separated from use of the patch by at least 6 weeks. Contraceptive patches are relatively expensive. Three patches (good for a month) cost about $35, whereas generic oral contraceptive pills cost approximately $5 per month.

Emerging Forms of Contraceptives

Several types of male contraceptives are under development and testing. One of the products being tested consists of periodic injections of progestin and implanted crystallized testosterone. Initial tests show that the hormones effectively suppress sperm production and are 97% effective in preventing conception. Men receiving the progestin/testosterone contraceptive gained an average of 3% of their initial body weight during the 12-month treatment period. Additional studies on the safety of this new contraceptive are under way.[97] Contraceptive pills for males that suppress sperm production are now available but are not widely used yet.

New types of contraceptives such as the monthly vaginal ring, hormone-releasing IUDs, and contraceptive pills that limit menstrual cycles to four or fewer per year have become available.[98] Little is known about the nutrition side effects of these forms of contraceptives.

Other Preconceptional Nutrition Concerns

Approximately 8 to 10 days after an ovum is fertilized, it implants into the uterine wall. Within the first month after conception, the developing *embryo* will have grown from a single cell to millions of cells, basic structures of organs will have formed, and the blueprint for future growth and development will have been established. All this often happens before women may know they are pregnant or attend a prenatal clinic. The time to establish a state of optimal health and nutritional status is before conception.

Very Early Pregnancy Nutrition Exposures

Table 2.7 summarizes major nutritional exposures that adversely affect the growth and development of the embryo and *fetus*. It is important to be aware that any of these conditions, if present preconceptionally, may impair embryonic and fetal growth and development. This chapter touches on the importance of adequate folate intake prior to conception. Chapter 4, on nutrition during pregnancy, expands the folate discussion and returns to the other topics listed in Table 2.7.

Folate Status Prior to Conception and Neural Tube Defects   Folate status prior to conception is an important concern because inadequate folate very early in pregnancy can cause *neural tube defects* (NTDs). These defects develop within 21 days after conception—or before many women even know they are pregnant, and well before prenatal care begins.[99] Knowledge of the folate–neural tube defect relationship, and awareness that folate intake is inadequate in many women of childbearing age, prompted efforts to increase folate intake. In particular, efforts are focused on encouraging women to consume folic acid, a highly absorbable, synthetic form of this B vitamin. In 1998, that task was made easier when the Food and Drug Administration mandated that refined grain products such as white bread, grits, crackers, rice, and pasta be fortified with folic acid.

Embryo   The developing organism from conception through 8 weeks.

Fetus   The developing organism from 8 weeks after conception to the moment of birth.

Neural Tube Defects (NTDs)   Spina bifida and other malformations of the neural tube. Defects result from incomplete formation of the neural tube during the first month after conception.

Nearly 40 countries now fortify refined grain products with folic acid, and rates of NTDs have fallen in each of these countries. Rates of NTDs in Nova Scotia, Canada, for example, fell 55% in the 2 years after folic-acid fortification became mandatory. Decline in the incidence of NTDs in the United States is significant but lower than that achieved in Canada.[100] Reliable data on NTD rates for the United States are not yet available. Countries in the European Union have yet to implement folic-acid

fortification; rates of NTDs in these countries have not changed.[101]

Intake of folic acid and folate status has increased substantially in most U.S. population groups since fortification. Since 1998, the prevalence of low levels of red-blood-cell folate (a marker of long-term folate status) has declined from 45 to 7%. Rates of low red-cell folate levels continue to be high (20–21%) among non-Hispanic blacks, however.[102] Concern exists that some women are unaware of the importance of consuming folic-acid-fortified refined grain products and other sources of folate prior to pregnancy. Although women of reproductive age have increased their median intake of folic acid by at least 100 mcg a day, only 39% of white women, 26% of black women, and 28% of Mexican American women have reached the 400-mcg-a-day target for folate intake.[103] Only 17% of women of child bearing age in Canada meet the goal of 400-mcg folic acid per day.[104]

Table 2.7   Nutritional exposures before and very early in pregnancy that disrupt fetal growth and development

Weight Status
- Being underweight increases the risk of maternal complications during pregnancy and the delivery of small and early newborns.
- Obesity increases the risk of clinical complications during pregnancy and delivery of newborns with neural tube defects or excessive body fat.

Nutrient Status
- Insufficient folate intake increases the risk of embryonic development of neural tube defects.
- Excessive vitamin A intake (retinol, retinoic acid) increases the risk the fetus will develop facial and heart abnormalities.
- High maternal blood levels of lead increase the risk of mental retardation in the offspring.
- Iodine deficiency early in pregnancy increases the risk that children will experience impaired mental and physical development.
- Iron deficiency increases the risk of early delivery and development of iron deficiency in the child within the first few years of life.

Alcohol
- Regular intake of alcohol increases the risk of *fetal alcohol syndrome* and *fetal alcohol effects,* both of which include impaired mental and physical development.

Diabetes
- Poorly controlled blood glucose levels early in pregnancy increase the risk of fetal malformations, excessive infant size at birth, and the development of diabetes in the offspring later in life.

Table 2.8 MyPyramid food-group recommendations for preconceptional women with various levels of caloric need

Calorie Need	Food Group*					
	Grains (Ounces)	Vegetables (Cups)	Fruits (Cups)	Milk (Cups)	Meat & Beans (Ounces)	Oil (tsp)
1800	6	2.5	1.5	3	5	5
2000	6	2.5	2	3	5.5	5.5
2200	7	3	2	3	6	6
2400	8	3	2	3	6.5	7

*"Other" calories are allotted for desserts, sweets, and fats based on caloric need: 2000 calorie need = 195 other calories; 2100 calorie need = 250; 2200 calorie need = 290; and 2400 calorie need = 360.

Women can get enough folate by consuming a good basic diet and a fully fortified breakfast cereal (Smart Start, Total, or Product 19, for example) or a regular breakfast cereal (Cheerios, Corn Flakes, Raisin Bran, etc.) and 6 to 8 servings of refined grain products each day. Folic acid supplements (400 mcg per day) can also provide folic acid.

Potential Adverse Effect of Increased Folic Acid Although women should aim to get enough folic acid prior to pregnancy, increased levels of folic acid intake may encourage cancer development in some women (and men) and may discourage its development in others. Moderate dietary increases in folic-acid intake initiated before the development of colon cancer appear to help prevent the cancer from spreading, whereas excessive intake or increased intake after colon cancer has developed may increase tumor growth.[105]

Recommended Dietary Intakes for Preconceptional Women

Recommendations for food and nutrient intakes for women who may become pregnant differ from those for adult women in general in several ways. It is recommended that women who may become pregnant (1) consume 400 mcg of folate from fortified grain products, vegetables, fruits, or supplements; (2) take no more than 5000 IU of vitamin A (retinol or retinoic acid) from supplements daily; and (3) limit or omit alcohol-containing beverages.[99] Recommendations for nutrient intake given in the Dietary Reference Intakes (DRIs) should be applied, paying careful attention to the Tolerable Upper Intake Levels (see tables on the inside front covers of this text).

Food selection prior to pregnancy should be based on the MyPyramid recommendations (see Table 2.8). The number of servings recommended from each food group is given in cups, ounces, or teaspoons and is determined primarily by a person's caloric need. An example menu offering several options for food selection for meals and snacks is given in Table 2.9.

Model Preconceptional Nutrition Programs

This section highlights two model programs, one in the United States and one in Indonesia, related to nutrition during the preconceptional period.

Table 2.9 Example menu for preconceptional women that corresponds to MyPyramid recommendations. The menu works for men, too, if portion sizes are adjusted upward to meet caloric needs.

Breakfast

Fruit juice, 1 c
Fortified, whole-grain breakfast cereal (hot or cold), 1 c
Sliced peaches or banana, 1 c
Skim or low-fat milk, 1 c
Coffee or tea

Lunch

Pork almond ding, 1 c, or taco salad with beans, chicken, or beef, 1 c
Rice, 1 c
Skim or low-fat milk, 1 c

Dinner

Veggie burger, lean beef, poultry, or fish, 3–4 oz
Macaroni salad, pasta, or potato, ½ c
Kale, turnip greens, collards, or spinach, 1 c
Tortilla, pita bread, corn-bread muffin, or whole-grain roll, 1
Pumpkin or sweet-potato pie, 1 slice, or ice cream, 1 c
Skim or low-fat milk, 1 c

Snacks

Yogurt with fruit
Graham crackers
Dried fruit
Nuts
Seeds
Apple, pear, mango, pineapple, or other fruit

Preconceptional Benefits of WIC

Women are eligible to enter the WIC program (Supplemental Nutrition Program for Women, Infants, and Children) when pregnant. In this model program in California, women were experimentally provided WIC food supplements and nutrition education during *and* between consecutive pregnancies. Women who received WIC benefits during one pregnancy and continued through the first two months of the next pregnancy had better iron status and delivered newborns with higher birth weights and greater lengths than did women who received WIC benefits during pregnancy only.[106] The study demonstrates that low-income women at nutritional risk benefit from WIC services before, as well as during, pregnancy. Additional information about the WIC program is presented in Chapter 1.

Decreasing Iron Deficiency in Preconceptional Women in Indonesia

Approximately one in every two women in Indonesia experiences iron-deficiency anemia during pregnancy. In a unique effort to prevent this problem, the Ministry of Health initiated regulations that require a couple applying for a marriage license to receive advice on iron status from those dispensing the license. All women are now advised to take 30–60 mg of iron along with folic acid in a supplement. Of 344 women studied after the program was initiated, 98% reported that they had purchased and taken iron-folate tablets; 56% had taken at least 30 tablets. The incidence of iron deficiency in this group of women dropped by almost half.[107]

Preconception Care: Preparing for Pregnancy

"Each woman, man, and couple should make a reproductive life plan that includes whether and when they want to have children and how they will maintain their reproductive health."

Centers for Disease Control and Prevention, 2006

Increasingly, routine health care visits and educational sessions are being recommended and introduced into health care organizations. Services focus on risk assessment of behaviors such as weight status; dietary and alcohol intake; folate and iron status; and vitamin, mineral, and herbal supplement use, as well as on the presence of diseases such as diabetes, hypertension, infection, and genetic traits that may be transmitted to offspring. Psychosocial needs should also be addressed as part of preconceptional care, and referrals made to appropriate services for issues such as eating disorders, abuse, violence, or lack of food or shelter.[78] The desire of couples planning for pregnancy to have a healthy newborn makes the preconceptional period a prime time for positive behavioral changes. It presents opportunities to make lasting improvements in the health and well-being of individuals and families.

CDC's Preconception Health Initiative   Efforts to boost the availability and utilization of preconception health care services now have the backing of the Centers for Disease Control and Prevention (CDC). In 2006, the CDC released a report highlighting recommendations for improving preconception health and health care services. The report was developed in response to the slow progress the United States had made in improving rates of poor pregnancy outcomes and in achieving the 2010 Health Objectives for the Nation related to preconception health. Additionally, the recommendations address the persistent problem of higher rates of poor pregnancy outcomes in African Americans, Hispanic Americans, and other groups compared to Caucasians.

The report concludes that preconception health care should be delivered at regularly scheduled primary care visits and include education about preconception health and pregnancy outcome; screening for vaccination, weight, and iron and folate status; assessment of alcohol use; and management of disorders such as diabetes and celiac disease. The CDC recommends that preconception services include counseling to modify individual health behaviors that, if left unchanged, would negatively impact fertility and pregnancy outcomes. Initiatives such as this one are increasing public awareness of the importance of preconception health and of planning pregnancy.[108]

Starting pregnancy in the best health status possible can make an important difference to reproductive outcomes. It should be recognized, however, that even in ideal conditions, continued infertility, early pregnancy loss, fetal malformations, and maternal complications will sometimes occur.

Nutrition Programs and Services Delivery Before Pregnancy

The American Dietetic Association has recently developed nutrition care standards intended to serve as guidelines for the delivery of nutrition services.[109] The standards are called the "Nutrition Care Process" and they are part of new technology-based systems being developed to facilitate health-services delivery and cost evaluation, electronic charting, coding, and outcome measurement. The process is being evaluated and will be revised as needed on an ongoing basis.

The Nutrition Care Process

The Nutrition Care Process (NCP) focuses on the delivery of effective nutrition care through the use of:

- A common approach to, and standardized methods for, nutrition assessment.
- Specific terms to describe nutrition diagnoses identified by the nutrition assessment. Nutrition diagnoses are treated independently by a nutrition practitioner.

- Effective intervention plans and goals related to treating the specific nutrition diagnoses.
- Nutrition monitoring and evaluation techniques that identify client/patient outcomes relevant to the nutrition diagnoses and the intervention plans and goals.

The Nutrition Care Process in practice consists of four steps:

1. Nutrition assessment
2. Nutrition diagnosis
3. Nutrition intervention
4. Nutrition monitoring and evaluation

Table 2.10 summarizes components of these four steps.

The Nutrition Care Process Related to the Preconception Period Components of the Nutrition Care Process vary to some extent based on the lifecycle stage of the individuals being served. Preconception services are tailored to the nutritional needs of women before pregnancy, and to the nutrition and reproductive health needs of men. Topics such as weight status; folic acid, iron, and antioxidant intake; dietary supplement use; reproductive history; contraceptive use; and effects of existing disorders on early fetal development would be addressed as part of the nutrition care process. Case Study 2.2 addresses a preconception health problem related to nutrition and asks students to use the Nutrition Care Process to formulate responses to questions asked about the case.

Table 2.10 Summary of the components of the four steps of the Nutrition Care Process[109]

Step 1: Nutrition Assessment

- Food/Nutrition-Related History
  - Food and nutrient intake (e.g., diet history, fat intake)
  - Medicinal/dietary supplement intake (e.g., medications, supplement use)
  - Knowledge/beliefs/attitudes (e.g., level of nutrition knowledge, unscientific beliefs, attitudes)
  - Nutrition-related behaviors (e.g., binge-eating behavior, willingness to try new foods)
  - Food and supplies availability (e.g., eligibility for and utilization of government/community food and nutrition programs)
  - Physical activity (e.g., physical activity history, intensity)
- Biochemical Data, Medical Tests, and Procedures
  - Laboratory data (e.g., glucose, C-reactive protein; note: normal nutrition biomarker values for adult men and women are provided in Table 1.18 in Chapter 1)
  - Tests (e.g., resting metabolic rate, bone density)
- Anthropometric Measurements
  - Body size, growth (BMI, growth pattern)
- Client History
  - Medical history, treatments, use of alternative/complementary medicine, social history

Step 2: Nutrition Diagnosis

- Identification of nutrition-specific problem(s) based on results of the nutrition assessment. Diagnoses are classified as being related to:
  - Food and nutrient intake (e.g., food and nutrient amounts consumed relative to need, such as "excessive energy intake," "underweight," "excess fat intake")
  - Clinical (e.g., nutrition problems related to medical and physical conditions such as "swallowing difficulties due to [name the factor, e.g., lack of saliva]," "breastfeeding difficulty due to [name the factor, e.g., failure to obtain a let-down reflex]," "altered lab value, low vitamin D due to [name the factor, e.g., sun sensitivity]")
  - Behavioral-environmental (e.g., problem related to behavior, knowledge, beliefs, and access to food such as "food- and nutrition-related knowledge deficit," "limited access to food")

Step 3: Nutrition Intervention

- Identify and implement effective, individually tailored nutrition interventions that will resolve or improve the nutrition problems identified. Nutrition intervention strategies are organized into four categories:
  1. Food and/or nutrient delivery
  2. Nutrition education
  3. Nutrition counseling
  4. Coordination of nutrition care

Step 4: Nutrition Monitoring and Evaluation

  - Measure and monitor changes in the client's nutrition-related health status and evaluate the effectiveness of the intervention based on client nutrition and health outcomes.

Nutrition Care Process (NCP) Case Study 2.2

Male Infertility

Life in View/Photo Researchers, Inc.

This case study requires use of the American Dietetic Association's Nutrition Care Process.

Mr. Trigger, a 29-year old male and his healthy 32-year-old wife have been unable to conceive after two years of unprotected intercourse. Results of medical tests led Mr. Trigger to be diagnosed with male-factor infertility due to low sperm production. No other health problems were identified. His weight was assessed to be 260 lbs. (118 kg) and his height measured 5 ft. 10 in. (1.78 m). Mr. Trigger was referred to a registered dietitian for nutritional assessment, diagnoses, intervention, and follow-up.

Results of the nutritional assessment revealed that Mr. Trigger was physically inactive and consumed an average of 145 calories more per day than recommended based on his weight, height, and physical activity level. Results of the dietary and supplement intake, nutrition-focused physical findings, nutrition knowledge, behaviors, and beliefs, and food availability assessments, revealed no additional problem areas.

Questions

1. What is Mr. Trigger's BMI?
2. Name an appropriate priority nutrition diagnosis for this case.
3. Name one potential nutrition intervention that addresses the nutrition diagnosis.
4. Cite one nutrition-related indicator that could be used to monitor and evaluate the intervention.

(Answers are located in the Instructor's Manual for the 4th edition of *Nutrition Through the Life Cycle*.)

Key Points

1. Preconceptional nutritional status influences maternal health and the course and outcome of pregnancy.

2. Dietary intake, supplement use, and body-fat composition affect the development and maintenance of a person's biological capacity to reproduce.

3. The primary effect of undernutrition on reproduction in women is the birth of small and frail infants who have a high risk of mortality early in life.

4. Acute undernutrition in previously well-nourished women is associated with a dramatic decline in their biological capacity to conceive.

5. Inadequate and excessive levels of body fat, weight loss in normal-weight individuals, oxidative stress, low antioxidant intake, high alcohol intake, eating disorders, certain chronic health problems, vegan diets, high soy food diets, inadequate zinc status, and heavy-metal exposure are related to fertility.

6. Oral contraceptives increase levels of HDL-cholesterol, triglycerides, and LDL-cholesterol; they decrease blood levels of vitamin B_{12} and copper somewhat. Contraceptive injections and implants are associated with weight gain.

7. Adequate folate status prior to pregnancy substantially reduces the risk of neural-tube defects in newborns. Over 60% of preconceptional women in the United States fail to consume the recommended amount of folate.

8. Low iron stores prior to pregnancy increase the risk of iron deficiency during pregnancy, preterm delivery, and low iron stores in the infant. Iron stores can be more effectively accumulated prior to rather than during pregnancy.

9. Adequate and healthy preconceptional diets for women and men are described by the MyPyramid Food Guide.

10. Preconception health services should be a part of primary health care and would likely improve fertility and pregnancy outcomes.

11. The process by which nutrition care service should be provided to preconceptional women and men (and others) has been defined by the American Dietetic Association.

Review Questions

1. Health Objectives for the Nation for 2010 related to preconception health and nutrition have largely been achieved.
   ____ True   ____ False

2. Estrogen is produced by fat cells, the ovaries, and the testes.
   ____ True   ____ False

3. The maturation of sperm takes 70–80 days.
   ____ True   ____ False

4. Both inadequate and excessive levels of body fat affect fertility.
   ____ True   ____ False

5. Weight loss of 10–15 percent of body weight in normal-weight men, but not in normal-weight women, decreases fertility.
   ____ True   ____ False

6. Vegan diets in men and high soy food intake in women are related to subfertility.
   ____ True   ____ False

7. Studies have consistently shown that high intakes of caffeine or coffee reduce fertility.
   ____ True   ____ False

8. The current generation of oral contraceptive pills have no adverse effects on nutritional status.
   ____ True   ____ False

9. Folic acid status of preconceptional women in the United States has greatly improved since the introduction of folic-acid fortification of refined-grain products.
   ____ True   ____ False

10. List two potential effects of oxidative stress on fertility.
    1.
    2.

11. Nutrient intake and body fat affect fertility primarily by (list the two primary effects):
    1.
    2.

12. What is the definition of "anovulatory cycles"?
    _____
    _____

Resources

Vital Statistics

The National Center for Health Statistics provides information on fertility and birthrates, and other vital statistics data.
Website: cdc.gov/nchs

Medscape Women's Health Journal

Receive automatic updates on women's health, fertility, and contraception topics by subscribing to this free online journal.
Website: http://www.medscape.com/womenshealth

WIC Homepage

Learn more about the WIC program at this website.
Website: www.fns.usda.gov/wic

The March of Dimes

Visit the blog for moms-to-be, read answers to common questions about the pre-pregnancy period, and find out more about birth defects at the organization's website.
Website: www.marchofdimes.com

Journal Articles

Look up research articles related to any of the topics covered in this chapter through these sites.
Websites: www.ncbi.nlm.nih.gov/pubmed
www.nlm.nih.gov/medlineplus/femalereproductivesystem.html
www.cdc.gov/reproductivehealth/index.htm
http://familydoctor.org/x5042.xml
www.nlm.nih.gov/medlineplus/infertility.html

"Women's nutritional status before conception influences physiologic events during pregnancy, and nutrition during pregnancy sets the stage for meeting nutritional needs during lactation."

J. C. King[1]

Chapter 3

Preconception Nutrition:
Conditions and Interventions

Prepared by **Judith E. Brown**

Key Nutrition Concepts

1   Nutrition and other lifestyle changes are a core component of the treatment of a variety of common health problems of women and men prior to conception.

2   Nutritional and health status before and during the first 2 months after conception influences embryonic development and the risk of complications during pregnancy.

Introduction

This chapter addresses specific nutrition-related conditions of women before conception and during the *periconceptional period,* and the interventions that address them. Conditions presented here have important implications for health and well-being, and for reproductive outcomes. We begin with a discussion of premenstrual syndrome and progress to obesity, hypothalamic amenorrhea, the female athlete triad, eating disorders, diabetes, polycystic ovary syndrome, inborn errors of metabolism, and celiac disease. Then we address the role of herbal remedies in the treatment of disorders of menstruation and fertility.

Premenstrual Syndrome

It wasn't until 1987 that PMS, or *premenstrual syndrome,* moved from the psychogenic disorder section of medical textbooks to chapters on physiologically based problems. It is diagnosed according to criteria stipulated in the fourth version of the *Diagnostic and Statistical Manual of Mental Disorders (DSM IV).* A standard questionnaire, rather than physical examination or laboratory tests, is used to diagnose PMS. Common physical signs and psychological symptoms of PMS are listed in Table 3.1. The diagnosis of PMS is made when women exhibit one to five of the signs and symptoms in two consecutive luteal phases.[2]

PMS is characterized by life-disrupting physiological and psychological changes that begin in the luteal phase of the menstrual cycle and end with menses (menstrual bleeding). It occurs in about 40% of menstruating women.[2,3]

Premenstrual dysphoric disorder (PMDD) is a severe form of PMS. It is characterized by marked mood swings, depressed mood, disruptions of parenting and partnership relationships, decreased work productivity, irritability, anxiety, and physical symptoms (breast tenderness, headache, joint or muscle pain).[2] PMDD is diagnosed when five or more signs or symptoms of PMS occur during at least two consecutive menstrual cycles. One of the symptoms must be related to depression, anxiety, or mood swings. The treatment strategy for PMDD is basically the same as that for PMS but more aggressive.[4,5]

The cause of PMS is unknown, but it is thought to be related to abnormal serotonin activity following ovulation.[2] Almost all remedies tested for PMS show about a 30% decline in symptoms with placebo, so an effective treatment must bring relief to an even higher proportion of women receiving it in experimental studies. Serotonin reuptake inhibitors, which are the active ingredient in some types of antidepressants or oral contraceptives, effectively reduce PMS symptoms in some women. Decreased caffeine intake; exercise and stress reduction; magnesium, calcium, or vitamin B$_6$ supplements; and a number of herbal remedies may also be used to treat PMS.[5]

Caffeine Intake and PMS

It is commonly recommended that women reduce their intake of coffee and other beverages high in caffeine to decrease PMS symptoms. This recommendation is holding up with time. A study at the University of Oregon demonstrated that PMS symptoms in college-age women increased in severity as coffee intake increased from 1 cup to 8 to 10 cups a day.[6] Risk of severe symptoms was eight times higher in women consuming the highest average daily amount of coffee (8 to 10 cups) compared to non-coffee drinkers. In an additional study involving 423 nurses, the incidence of premenstrual syndrome (PMS) was found to be over twice as high among women consuming over 1 cup of coffee daily versus consumers of less coffee. This study also found that a negative attitude toward menstruation was much more common among women experiencing PMS than among women who did not.[7]

Exercise and Stress Reduction

Increasing daily physical activity and reducing daily stressors appear to diminish PMS symptoms in many women.[8] Regular physical activity tends to improve energy level,

Periconceptional Period   Around the time of conception, generally defined as the month before and the month after conception.

Premenstrual Syndrome   (*premenstrual* = the period of time preceding menstrual bleeding; *syndrome* = a constellation of symptoms) A condition occurring among women of reproductive age that includes a group of physical, psychological, and behavioral symptoms with onset in the luteal phase and subsiding with menstrual bleeding. Premenstrual dysphoric disorder (PMDD) is a severe form of PMS.

Table 3.1 Common signs and symptoms of PMS

Physical Signs	Psychological Symptoms
• Fatigue	• Craving for sweet or salty foods
• Abdominal bloating	• Depression
• Swelling of the hands or feet	• Irritability
• Headache	• Mood swings
• Tender breasts	• Anxiety
• Nausea	• Social withdrawal

mood, and feelings of well-being in women with PMS. Relieving stress—by such techniques as sitting comfortably and quietly with eyes closed while relaxing deep muscles, breathing through the nose, and exhaling while silently saying a word such as "one"—appears to decrease symptoms. When done for 15 to 20 minutes twice daily over 5 months, this exercise was associated with a 58% improvement in PMS symptoms.[8]

Magnesium, Calcium, Vitamin D, and Vitamin B₆ Supplements and PMS Symptoms

Magnesium, calcium, vitamin D, and vitamin B₆ supplements appear to decrease symptoms of PMS in some women. Mechanisms underlying improvements are incompletely understood.

Magnesium   Magnesium supplements of 200 mg per day given during two cycles have been shown to decrease swelling, breast tenderness, and abdominal bloating symptoms of PMS. The beneficial response to magnesium was seen during the second month of treatment.[9] While some studies on magnesium supplementation for PMS symptom reduction have identified benefits, others have not.[6] The 200-mg daily dose of magnesium is below the Tolerable Upper Intake Level (UL) for magnesium of 350 mg daily and is therefore considered safe.

Calcium and Vitamin D   Calcium supplements of 1200 mg per day for three cycles were found to reduce the PMS symptoms of irritability, depression, anxiety, headaches, and cramps by 48%, versus a reduction of 30% in the placebo group. The effect of calcium increased with duration of supplement use.[10] A similarly designed placebo-controlled trial using 500 mg calcium twice daily yielded similar results, with benefits noted particularly for early fatigue, appetite changes, and depressive symptoms.[11]

> **Sex Hormone Binding Globulin (SHBG)**   A protein that binds with the sex hormones testosterone and estrogen. Also called steroid hormone binding globulin, because testosterone and estrogen are produced from cholesterol and are thus considered to be steroid hormones. These hormones are inactive when bound to SHBG, but are available for use when needed. Low levels of SHBG are related to increased availability of testosterone and estrogen in the body.

A recent study of women with PMS and control women demonstrated that vitamin D in addition to calcium may affect risk.[12] Women with PMS were found to have lower blood levels of vitamin D and calcium, as well as lower intakes, than women without the syndrome. The risk of developing PMS was 40% lower in women with average vitamin D intakes of 706 IU (17.7 mcg) per day compared to women consuming 112 IU (2.8 mcg). Daily calcium intakes of 1,283 mg were related to a 30% risk reduction compared to women consuming 529 mg. The UL for vitamin D for women is 50 mcg (2000 IU), and 2500 mg for calcium.

Vitamin B6   Vitamin B6 (pyridoxine) is involved in the synthesis of neurotransmitters such as serotonin that are related to the development of PMS. Some trials have found no effect of B6 on PMS symptoms, but the others have identified benefits.[13–15] It is generally concluded that pharmacological doses of vitamin B6 in the range of 50 to 100 mg per day reduce the severity of premenstrual depressive symptoms in some women.[2]

Vitamin B₆ supplements are sometimes used in clinical practice to diminish PMS symptoms. The UL for vitamin B₆ is 100 mg/day; doses recommended should not exceed this level.

Obesity and Fertility

Rates of reproductive health problems related to excess body fat are increasing in the United States and other countries along with rising rates of obesity. Among 20- to 39-year-olds in the United States, 23.3% of men and 27.0% of women are obese. Rates of obesity in adults vary by racial/ethnic background (Table 3.2), indicating that health risks related to obesity are not equally shared by all population subgroups. The prevalence of severe obesity, assessed as BMIs of 40 kg/m² or higher—equivalent to weights that are 100 pounds higher than normal—has increased to around 5%.[16]

Men who are obese are at risk of infertility due to low levels of testosterone and *sex hormone binding globulin (SHBG),* and low sperm count (Table 3.3).[17] Obesity in women increases the risk of infertility due to highly irregular or anovulatory menstrual cycles. These conditions are likely related to high levels of estrogen and free testosterone that can accompany obesity. Irregular menstrual cycles

Table 3.2 Prevalence of obesity in subgroups of adults in the United States, 2008[16]

	Males	Females
Hispanics/Latinos	32.4%	35.4%
Whites	25.3%	24.7%
Blacks	29.0%	37.7%

Table 3.3 Biological bases of infertility in obese men and women[17, 22, 23]

Men	Women
• Low testosterone and sex hormone binding globulin levels	• High estrogen, free-testosterone, and leptin levels
• Elevated leptin, follicle stimulating hormone (FSH), and estrogen levels	• Reduced levels of sex hormone binding globulin
• Decreased sperm count, sperm motility; increased malformed sperm	• Insulin resistance
• Oxidative stress, inflammation	• Oxidative stress, inflammation

In Focus

Metabolic Syndrome

Definition: **Metabolic syndrome** is not a specific disease but rather a cluster of abnormal metabolic and other health indicators. It is diagnosed when three of the following five conditions exist:

1. Waist circumference >40" in men, >35" in women (These are an indicator of the presence of insulin resistance. Other population-based definitions of elevated waist circumference may also be used.)
2. Blood triglycerides ≥150 mg/dL
3. HDL-cholesterol <40 mg/dL in men and <50 mg/dL in women
4. Blood pressure of ≥130/85 mm Hg
5. Fasting blood glucose ≥110 mg/dL[25]

Prevalence: It is estimated that one in four U.S. adults has metabolic syndrome.[26]

Major physiological aspects and consequences: The cluster of metabolic risk factors found in people with metabolic syndrome greatly increases the risk of development of cardiovascular disease and type 2 diabetes. Metabolic syndrome is also characterized by *chronic inflammation* that promotes oxidation reactions. Over time, chronic inflammation damages cells and body functions and can impair reproductive functions in both women and men.[19, 22]

The first-line therapy for metabolic syndrome is lifestyle changes that emphasize dietary modifications, weight reduction, and exercise. Diets high in antioxidant-rich fruits, vegetables, whole-grain products, fiber, and low-fat dairy products are recommended. Such diets decrease inflammation, plasma triglyceride levels, body weight, plasma glucose levels, and blood pressure and increase HDL-cholesterol levels.[27]

Risk factors: People with metabolic syndrome are often obese, have high levels of central body fat, and are insulin resistant.

contribute to a lack of early prenatal care in some obese women. Women may not be aware that they have become pregnant because delayed onset of menstrual cycles may not be unusual.

Insulin resistance frequently occurs along with obesity and contributes to adverse hormonal changes that affect fertility. Insulin can bind to specific receptors on the ovary and stimulate testosterone production. Androgens such as testosterone suppress follicular growth, leading to ovulatory dysfunction.[18]

Obesity is also related to *chronic inflammation* that can damage developing eggs and maturing sperm.[19] Chronic inflammation appears to be sustained, in part, by poorer vitamin D status commonly identified in people with high levels of body fat.[20]

Obese women have higher rates of *metabolic syndrome, polycystic ovary syndrome*, gestational diabetes and hypertension, fetal overgrowth, cesarean delivery,

and stillbirth than women who are not obese.[21] Metabolic syndrome is a common condition in obese men as well. It is referred to again in this and other chapters and is overviewed in the "In Focus" box.

Central Body Fat and Fertility

The presence of central body obesity, indicated by a waist circumference of 35 inches or greater in women and over 40 inches in men, is a risk factor for impaired fertility.[17, 24] In a study of women attending an artificial insemination clinic due to infertility in their partners, conception occurred in just half as many women with high central body fat as in other women.[28] In general, it takes women with high central body fat stores longer to become pregnant than it does women with low levels of central fat.[29]

Weight Loss and Fertility

Weight loss should be the first therapy option for men and women who are obese and infertile.[17] (Read about one woman's experience with weight loss and fertility in Case Study 3.1 on the next page.)

Studies of both women and men have demonstrated that weight loss of 7 to 22 pounds in women with BMIs over 25 kg/m^2, and of 100 pounds in massively obese men, are related to a return of fertility in most study participants.[21,30–32] Weight loss in the studies of obese women was accomplished by diet and exercise; the studies of obese men employed weight-loss surgery.

Weight reduction through diet and exercise is considered the first therapeutic option for infertility in obese people. It is less costly than medications or surgery, has fewer complications, and has many health benefits.[17]

Diets for Weight Loss   Diets for people trying to lose weight should be healthful, balanced, and provide all required nutrients in amounts recommended. Nutrient deficits, such as inadequate vitamin D status or low calcium intake, that exist before weight loss should be addressed in the dietary plan for weight loss. Diets should be planned around foods that correspond to individual food preferences and resources.

Weight loss tends to be maintained if weight is lost

Insulin Resistance   A condition in which cells "resist" the action of insulin in facilitating the passage of glucose into cells.

Chronic Inflammation   Low-grade inflammation that lasts weeks, months, or years. Inflammation is the first response of the body's immune system to infection or irritation. Inflammation triggers the release of biologically active substances that promote oxidation and other potentially harmful reactions in the body.

Metabolic Syndrome   A constellation of metabolic abnormalities that increase the risk of type 2 diabetes, heart disease, and other disorders. It is characterized by insulin resistance, abdominal obesity, high blood pressure and triglyceride levels, low levels of HDL cholesterol, and impaired glucose tolerance. Also called Syndrome X and insulin-resistance syndrome.

Polycystic Ovary Syndrome (PCOS)   (*polycysts* = many cysts; i.e., abnormal sacs with membranous linings) A condition in females characterized by insulin resistance, high blood insulin and testosterone levels, obesity, polycystic ovaries, menstrual dysfunction, amenorrhea, infertility, hirsutism (excess body hair), and acne.

Case Study 3.1

Westend 61 GmbH/Alamy

Anna Marie's Tale

Exercise can be bad for you—or at least it is for Anna Marie. She and her husband Mark already have two delightful children, full-time jobs, and hectic schedules. Mark wants more children, but Anna Marie is dead set against it. Mark refuses to use contraception and has made Anna Marie promise not to use any, either. Anna Marie makes the promise because she thinks she can avoid pregnancy by staying at her weight of 210 pounds. At this weight, Anna Marie seldom has a menstrual period and figures the odds of conception are slim. For 2 years, Anna Marie's plan for avoiding conception has worked.

Now that the children are older, Anna Marie finds she has a bit of free time, which she uses to indulge her love of swimming. Within months of beginning her program of swimming regularly, however, Anna Marie abandons it. Her menstrual periods have become regular, and her contraception method is lost.

Anna Marie's weight at 210 pounds has remained stable during the months she has been swimming. It appears that her improved level of physical fitness and body fat has improved her fertility status.

Questions

1. What was likely the reason for Anna Marie's infertility when she was inactive?
2. Give an example of a hormonal change that may have occurred after Anna Marie began exercising regularly.
3. Name a possible health consequence related to Anna Marie's high weight and lack of menstrual cycles.

(Answers are located in the Instructor's Manual for the 4th edition of *Nutrition Through the Life Cycle*.)

---

slowly and accompanied by acceptable changes in diet, physical activity, and other lifestyle behaviors.[33] Cutting back on food intake by 100 calories per day can lead to a 10-pound loss in a year, for example. Dramatic changes in diets and physical activity tend not to last and lead to weight re-gain in most people. The challenge for individuals attempting weight loss is to identify small, acceptable changes in diet and exercise that are enjoyable and lasting. For individuals who cannot lose weight on their own, diet drugs and surgery constitute the remaining options for weight loss.[17]

Bariatric (Weight-Loss) Surgery   Weight-loss surgery represents the measure of last resort for weight loss for obese individuals.[17] The surgery is reserved for people with BMIs over 40, or over 35 kg/m[2] if a serious medical condition related to obesity exists. It generally reduces body weight to levels that qualify as overweight or somewhat obese. In most women and men, bariatric surgery is followed by a return to normal hormone levels, decreased inflammation, and improved fertility.[34] In the long term, benefits of fat loss related to the surgery include resolution of type 2 diabetes in 40% of individuals and improved glucose control in others.[35] The surgery effectively reduces the risk of developing gestational diabetes and hypertension during pregnancy.[34] Bariatric surgery is related to a number of complications, however, and some of them may compromise nutritional health of pregnant women and infant outcomes.

Bariatric surgery increases the risk that women will develop deficiencies of iron, folate, calcium, and vitamins A, B_{12}, and K.[36] The risk that such deficiencies will develop is increased in women who experience chronic vomiting or malabsorption, or who fail to take appropriate vitamin and mineral supplements. If uncorrected, nutrient deficiencies can impair fetal growth and nutrient stores of the infant.[37] Pregnancy is not recommended during the first year after bariatric surgery when weight loss is most rapid.[38]

Hypothalamic Amenorrhea

One of the most common causes of anovulation and loss of menstrual cycles is *hypothalamic amenorrhea*.[39] Women affected by hypothalamic amenorrhea, previously called "weight-related amenorrhea," have no abnormalities of

> **Hypothalamic Amenorrhea**   Cessation of menstruation related to changes in hypothalamic signals that maintain the secretion of hormones required for ovulation. Changes in hypothalamic signals appear to be triggered by an energy deficit. Also called functional amenorrhea and weight-related amenorrhea.

the hypothalamus, but its functions appear to be disrupted by an energy deficiency. Deficits in energy and possibly nutrients appear to disrupt hypothalamic signals that lead to normal secretion of gonadotropin-releasing hormone and luteinizing hormone.[40] (These hormones are described in Chapter 2.) Women with hypothalamic amenorrhea may also be leptin-deficient due to low levels of body fat. These hormonal changes prevent ovulation.

The onset of hypothalamic amenorrhea is related to being underweight, weight loss, or weight loss accompanied by intense exercise. It usually occurs in women engaged in intellectual professions or those exposed to social stress. A minority of cases of hypothalamic amenorrhea appear to be related to high levels of social stress accompanied by subtle deficits in calorie intake. It is often preceded by menstrual irregularities lasting months to years.[41]

Nutritional Management of Hypothalamic Amenorrhea

"...Let's not tread into new expensive treatments without correction of nutrient deficiencies or without first attempting to modify the dietary intake to meet all the energy and nutrient needs of the patient."

Fima Lifshitz, MD[42]

Fertility can be restored in underweight and energy-deficient women by hormonal therapy. Risks of pregnancy and newborn complications, as well as health care costs, are higher, however, if this approach is taken. The first treatment approach should be weight gain, accomplished by the consumption of a healthful diet. Weight gains of 6 to 11 pounds (3 to 5 kg) are usually sufficient to restore fertility and improve the outcome of pregnancy.[21]

The Female Athlete Triad and Fertility

Improved opportunities for female participation in sports in recent decades have been followed by an upsurge in the number of females who are competitive athletes. Although this is a healthy trend, some who are involved in sports that emphasize a lean body type are compromising their health. Very high levels of physical activity combined with negative caloric balance can place women at risk of developing the "female athlete triad." It is called a triad because it consists of three major conditions: amenorrhea, disordered eating, and osteoporosis. Amenorrhea associated with the female athlete triad appears to be triggered when energy intake is about 30% less than energy requirement.[43] This level of energy deficit leads to a loss of normal secretion of luteinizing hormone, a subsequent lack of estrogen production, and other hormonal changes also seen in hypothalamic amenorrhea. Metabolic changes triggered by hormonal shifts result in decreased bone density and an increased susceptibility to stress fractures in affected athletes.[44]

Nutritional Management of the Female Athlete Triad

Treatment of the female athlete triad focuses on correction of the negative energy balance and associated eating disorders, and on restoration of bone mass accretion. Peak bone mass is established before age 30, so it is particularly important that interruptions in bone development be short in duration. Vitamin D, calcium, and other supplements may be needed in addition to a balanced and adequate diet to facilitate bone development.[45]

Eating Disorders and Fertility

Both *anorexia nervosa* and *bulimia nervosa* are related to menstrual irregularities and infertility. These disorders affect about 3 to 5% of young women, and likely twice that many have clinically important symptoms related to these eating disorders.[46] Amenorrhea is a cardinal manifestation of anorexia nervosa, and little bleeding during menses (oligomenorrhea) or amenorrhea may occur in women with bulimia nervosa. Amenorrhea in anorexia nervosa is related to irregular release of GnRH and very low levels of estrogen. Menses generally returns upon weight gain, but some cases of infertility persist even after normal weight is attained. This effect may be related to continued low levels of body fat, low dietary fat intake, excessive exercise, or other factors.[46] Osteoporosis and short stature are other potential long-lasting effects of anorexia nervosa.[47]

> **Anorexia Nervosa** (*anorexia* = poor appetite; *nervosa* = mental disorder) A disorder characterized by extreme underweight, malnutrition, amenorrhea, low bone density, irrational fear of weight gain, restricted food intake, hyperactivity, and disturbances in body image.
>
> **Bulimia Nervosa** (*bulimia* = ox hunger) A disorder characterized by repeated bouts of uncontrolled, rapid ingestion of large quantities of food (binge eating) followed by self-induced vomiting, laxative or diuretic use, fasting, or vigorous exercise in order to prevent weight gain. Binge eating is often followed by feelings of disgust and guilt. Menstrual cycle abnormalities may accompany this disorder.

Food binges and crash diets associated with bulimia nervosa are related to low FSH and LH levels, menstrual disturbances, and infertility.[48]

Nutritional Management of Women with Anorexia Nervosa or Bulimia Nervosa

The primary therapeutic goal for anorexia nervosa is normalization of body weight, and for bulimia nervosa, normalization of eating behaviors.[49] Recommended treatment for anorexia nervosa involves long-term, multidisciplinary services. Hospitalization may be required in severe cases. Certain psychotherapeutic medications are moderately effective for treating bulimia nervosa, but cognitive-behavioral therapy is the best, established approach.

Additional coverage of eating disorders can be found in Chapters 5 and 15.

Diabetes Mellitus Prior to Pregnancy

Many women with diabetes mellitus are unaware that the disorder increases the risk of maternal and fetal complications, and they fail to get blood glucose under excellent control prior to conception.[50] High blood glucose levels during the first 2 months of pregnancy are *teratogenic;* they are associated with a two- to threefold increase in *congenital abnormalities* in newborns. Exposure to high blood glucose during the first 2 months in utero is related to malformations of the pelvis, central nervous system, and heart in newborns, as well as to higher rates of miscarriage.[51]

Management approaches to blood glucose control in diabetes depends, in part, on the type of diabetes present. Women may have *type 1 diabetes* or *type 2 diabetes.* Once thought of as a disease of older adults, type 2 diabetes is becoming increasingly common in young adults and childrens due to the obesity epidemic. Approximately 4 in 5 U.S. adults with type 2 diabetes are overweight or obese.[52] Background information on diabetes is presented in the "In Focus" box on page 77.

Nutritional Management of Type 1 Diabetes

The main goals of the management of type 1 diabetes are blood glucose control, resolution of coexisting health problems, and health maintenance. Diets are controlled in carbohydrate content because carbohydrates raise blood glucose levels and increase insulin need to a greater extent than do protein or fats. Insulin-to-carbohydrate ratios, carbohydrate counting, or the diabetes exchange system can be used to adjust mealtime insulin doses based on carbohydrate intake.[56]

People with type 1 diabetes are urged to replace simple sugars with reasonable amounts of artificial sweeteners. Foods low in glycemic index and high in fiber (especially soluble fiber such as oatmeal) are encouraged, as are brightly colored fruits and vegetables, low-fat meat and dairy products, fish, dried beans, and nuts and seeds.[57,58] Reduced-calorie diet plans should be included as part of the care for individuals with type 1 diabetes who would benefit from weight loss.

Physical activity is generally part of a diabetes care plan because it improves blood glucose levels, physical fitness, and insulin utilization. Both strength and aerobic exercises are recommended.[57] Individualized meal and physical activity plans and follow-up care for individuals with type 1 diabetes should be provided by an experienced health care team that includes a registered dietitian.[58]

Nutritional Management of Type 2 Diabetes

> "There is a substantial gap between recommended diabetes care and the care patients actually receive in the health care setting."[60]

Some people with type 2 diabetes can manage their glucose levels with diet and exercise, whereas others will need an oral medication that increases insulin production or sensitivity, or insulin to further boost glucose absorption into cells.[61]

Individualized diet and exercise recommendations and an educational and follow-up program developed and implemented by registered dietitians, certified diabetes educators (CDE), physicians, and nurses are preferred for diabetes management. Carefully planned and monitored dietary recommendations are a major component of the management of type 2 diabetes. Individual blood glucose levels vary a good deal in response to diet composition, so dietary prescriptions must be tailored for every person.

For patients with type 2 diabetes, the American Diabetes Association's guidelines recommend:

- Weight loss of 7% of body weight or more
- Percent of total calories from the energy nutrients: 15–20% protein, <30% fat, and approximately 50% carbohydrates
- Percent of total calories from saturated fat: <7%
- Percent of total calories from *trans* fat: as low as possible
- Restriction of cholesterol intake to 200 mg per day or less
- 14 g fiber per 1000 calories of food intake
- Whole grains should comprise half of all grain intake
- Low glycemic-index foods that are rich in fiber and other important nutrients should be encouraged[56]

There is no evidence to support prescribing diets such as "no concentrated sweets" or "no added sugar."

Sucrose can be substituted for other carbohydrate foods in meal plans. Restrictions on food choices for people with diabetes should only be implemented when indicated by scientific evidence.[56]

Glycemic Index   *Glycemic index (GI)* is a measure of the extent to which 50 grams (about 1¾ ounces) of carbohydrate-containing foods raise 2-hour postprandial blood glucose levels compared to 50 grams of glucose or white bread. Not all expert committees on diabetes recommend the use of low-GI

Teratogenic   Exposures that produce malformations in embryos or fetuses.

Congenital Abnormality   A structural, functional, or metabolic abnormality present at birth. Also called congenital anomalies. These may be caused by environmental or genetic factors, or by a combination of the two. Structural abnormalities are generally referred to as congenital malformations, and metabolic abnormalities as inborn errors of metabolism.

Type 1 Diabetes   A disease characterized by high blood glucose levels resulting from destruction of the insulin-producing cells of the pancreas. This type of diabetes was called juvenile-onset diabetes and insulin-dependent diabetes in the past, and its official name is type 1 diabetes mellitus.

Type 2 Diabetes   A disease characterized by high blood glucose levels due to the body's inability to use insulin normally, or to produce enough insulin. This type of diabetes was called adult-onset diabetes and non- insulin-dependent diabetes in the past, and its official name is type 2 diabetes mellitus.

Glycemic Index (GI)   A measure of the extent to which blood glucose levels are raised by a specific amount of carbohydrate-containing food compared to the same amount of glucose or white bread.

In Focus

Diabetes

Definition: There are three major types of diabetes mellitus: type 1, type 2, and gestational. All types of diabetes are characterized by abnormally high blood glucose levels, or fasting levels of 126 mg/dL (7 mmol/L) or higher. People with diabetes are considered to be "carbohydrate intolerant" because carbohydrate consumption tends to raise blood glucose levels.

Type 1 diabetes is carbohydrate intolerance resulting from destruction of insulin-producing cells of the pancreas. Individuals with type 1 diabetes require an external supply of insulin. It is considered an *autoimmune disease.* About 10% of cases of diabetes are of this type.

Type 2 diabetes is carbohydrate intolerance due to the body's inability to use insulin normally, to produce enough insulin, or both. About 90% of cases of diabetes are of this type.

Gestational diabetes is carbohydrate intolerance that begins or is first recognized during pregnancy. It is closely related to type 2 diabetes. (Additional information about gestational diabetes is presented in Chapter 5.)

Prevalence: 10% of U.S. adults have type 2 diabetes and less than 1% of youth and adults have type 1. Gestational diabetes is diagnosed in 3–7% of pregnancies.[53]

Rates of all types of diabetes are increasing worldwide. Increased rates of type 2 and gestational diabetes are related to esclating global rates of obesity.

Major physiological aspects and consequences: Major signs and symptoms of diabetes are frequent urination, increased thirst and fluid intake, increased appetite, and elevated blood glucose levels.

People with type 2 and gestational diabetes are generally obese and have insulin resistance. Insulin resistance that develops with obesity partly originates from metabolic changes initiated in fat cells.

High amounts of body fat are related to an increased release of fatty acids from fat cells into blood. Through a series of complex mechanisms, high circulating levels of fatty acids stimulate the production of proinflammaory molecules and free radicals. These substances are involved in the development of oxidative stress and chronic inflammation.[54] The resulting oxidative and chronic inflammatory states promote the development of insulin resistance in liver and muscle cells, excessive fat storage in the liver, and increased insulin output by beta cells of the pancreas. Over time, insulin resistance can lead to the exhaustion of beta cells and decreased insulin production, elevated blood glucose levels, and the onset of type 2 diabetes.[54]

Elevated blood glucose levels in people with diabetes have many adverse consequences in the short- and long-term. Such consequences can be limited or postponed by tightly managed blood glucose levels. Diabetes increases the risk of coronary heart disease, kidney disease, vision problems and blindness, nerve problems, and loss of limbs.

Type 1 and 2 diabetes are chronic diseases that require lifelong management. Patient education, nutritional support, and self-glucose monitoring are important for long-term blood glucose control. Some people with type 2 diabetes can control the disease through dietary and physical activity changes and weight loss.

Risk factors: Exposure to certain infectious diseases, drugs, other environmental agents, and vitamin D inadequacy early in life can trigger the onset of Type 1 diabetes in genetically susceptible individuals. Obesity, especially central obesity, is the main risk factor for type 2 and gestational diabetes.[55]

foods as a primary strategy in the dietary management of diabetes. However, mounting evidence indicates that low-GI diets are beneficial for the control of blood glucose and insulin levels. Diets that provide low-GI carbohydrates and approximately 30 grams of fiber daily are associated with reduced blood levels of glucose, insulin, and triglycerides versus lower-fiber (15 g/day), high-GI diets. (Table 1.4 in Chapter 1 lists the GI of a variety of foods.) Foods with high GI raise blood glucose and insulin levels more than do foods with low GI, and high-GI foods lead to more episodes of hyperglycemia (high blood glucose level) than do diets providing mainly low-GI carbohydrates.[107,108]

It is currently recommended that low-glycemic index foods rich in fiber and other important nutrients be included in the diets of individuals with diabetes.[56]

Other Components of the Nutritional Management of Type 2 Diabetes

Individuals with diabetes are at risk for heart disease due to abnormal blood lipid levels. Consequently, diets recommended include foods that improve blood lipid concentrations. Diets that help lower high-LDL cholesterol without lowering concentrations of beneficial HDL cholesterol, and that lower elevated triglyceride levels, follow these guidelines:

- Provide fat mainly in the form of monounsaturated fatty acids (vegetables, olive oil, peanuts and peanut oils, nuts, and seeds)
- Keep intake of saturated fats from animal products below 7% of total calories
- Limit cholesterol intake to less than 200 mg per day
- Minimize intake of *trans* fats from bakery products, fried foods, and snack foods

Fish intake is often encouraged for people with diabetes. Consuming 2–3 servings of fish per week (excluding commercially available fried

Autoimmune Diseases Diseases that result from a failure of an organism to recognize its own constituent parts as "self." The organism attempts to defend itself from the perceived foreign substance through actions of its immune system. These actions can damage molecules, tissues, and organs. Type 1 diabetes, lupus, and rheumatoid arthritis are examples of autoimmune diseases.

fish fillets), or taking fish oil supplements, lowers blood triglycerides in people with elevated levels.[56] Sample menus that incorporate nutritional criteria for type 2 diabetes are given in Table 5.1 of Chapter 5.

There is no clear evidence of benefit from vitamin or mineral supplementation in people with diabetes who do not have underlying deficiencies. Routine supplementation with antioxidants, such as vitamin E, vitamin C, and beta-carotene, or with chromium is not recommended because of lack of evidence of benefit.[56]

Physical Activity Recommendations for Type 2 Diabetes   Regular physical activity is an important component of the management of type 2 diabetes. In addition to facilitating weight loss, physical activity reduces insulin resistance, decreases blood pressure and body fat content, and improves blood lipid and glucose levels.[59] Moderate-intensity aerobic and strength-building activities, such as weight lifting, jogging, fast walking, aerobic dancing, and swimming, are recommended. The target generally set for the duration of physical activity is 150 minutes per week, or an average of 21 minutes per day.[59]

Weight Loss and Type 2 Diabetes   Weight loss in overweight and obese individuals with diabetes lowers blood glucose levels and blood pressure, improves blood lipid concentrations, and increases insulin sensitivity. The most effective approach to weight loss in people with diabetes tested so far combines reduced caloric intake with exercise. Both aerobic and strength-building exercises are recommended. Behavioral therapy generally includes self-monitoring of body weight, diet, and physical activities; stress management; and problem-solving skill building. Selection of acceptable foods and dietary patterns, realistic goals for weight loss, enjoyable physical activities, and helpful behavioral therapies are critical to the long-term success of weight-control efforts. Diet drugs and obesity surgery may be indicated for individuals who are unable to lose weight through behavioral changes.[62]

A concerted effort will be needed to translate new knowledge about diet and nutrition into improved treatment for and prevention of diabetes and its complications.[63]

Reducing the Risk of Type 2 Diabetes

Type 2 diabetes develops over years and is generally preceded by worsening blood glucose levels that eventually qualify as *prediabetes*.[64] Individuals with prediabetes are at high risk of developing type 2 diabetes. The onset of type 2 diabetes can be postponed and sometimes prevented

Prediabetes   A condition in which blood glucose levels are higher than normal but not high enough for the diagnosis of diabetes. It is characterized by impaired glucose tolerance, or fasting blood glucose levels between 110 to 126 mg/dL.

by interventions that reduce body weight. One large study taking place over 3 years found that losses in body weight of about 7% and 150 minutes per week of exercise reduced the risk of developing type 2 diabetes by 50%.[65]

The American Diabetes Association recommends that individuals at high risk for developing type 2 diabetes be offered structured programs that emphasize lifestyle changes that promote regular physical activity and moderate weight loss (7% of body weight). Individuals at risk should be encouraged to meet recommendations for dietary fiber intake (14 g fiber/1,000 kcal) in part by consuming whole grains (one-half of grain intake). Consumption of low-glycemic index foods that are rich in fiber and other important nutrients should be encouraged.[56]

Polycystic Ovary Syndrome

Case scenario: Lupe is a 28-year-old woman who is 5 feet 3 inches tall and weighs 208 pounds. She and her husband want to start a family, but her highly irregular periods and a failure to become pregnant as soon as desired have brought her to see her doctor. At the clinic it is determined that Lupe's waist circumference is 38 inches, and her body mass index (BMI) is 37 kg/m². Laboratory tests show that her blood levels of insulin and triglycerides are high and that she is not ovulating. Lupe is diagnosed as having polycystic ovary syndrome, or PCOS.

Between 5 and 10% of women of reproductive age have PCOS, and it is a leading cause of female infertility.[66] PCOS is not a disease but rather a syndrome that consists of a variety of clinical signs. It is considered by some experts to be a sex-specific form of metabolic syndrome.[67] Most women with PCOS are infertile due to the absence of ovulation, and they have menstrual irregularities. Characteristically, the outer layer of the ovaries of women with PCOS is thick and hard, and it may look yellowish. Due to the hard covering on the ovaries, follicles don't break open to release the egg, so ovulation does not occur.[66]

Many women with PCOS are obese, and even in women with PCOS who are not obese, levels of intraabdominal fat are usually high.[68] Some women with PCOS have excess body hair (hirsutism), acne, and high blood levels of insulin, triglycerides, and androgens; and they have low levels of HDL cholesterol (Table 3.4). PCOS is sometimes difficult to diagnose (and may therefore not be treated) because signs and symptoms of the disorder vary among individual women.[69]

The cause of PCOS is still debated, but insulin resistance plays a pivotal role in most cases regardless of body weight.[68] Less commonly, PCOS is caused by androgen-secreting tumors in the ovaries or adrenal gland, other disorders, and certain medications.[70] High blood levels of

Table 3.4 Variation in clinical signs associated with PCOS[69,72,74]

Clinical Sign	Percent of Women with PCOS Affected
Menstrual irregularities	90%
Polycystic ovaries	67–86%
Excess abdominal fat	80%
Insulin resistance	80%
Overweight, obesity	80%
Abnormal facial and body hair	70%
High testosterone levels	70%
Infertility	70%
Low HDL-cholesterol levels	64%
High triglycerides	47%

insulin stimulate the ovaries to produce androgens (such as testosterone), and excess androgens disrupt development of follicles.[71] High blood levels of androgens also lead to excess hair growth on the face and other parts of the body, while high insulin levels raise triglyceride and lower HDL-cholesterol levels.[72]

PCOS appears to have a genetic component, and its development is influenced by environment-gene interactions. It tends to run in families where females have a history of infertility, menstrual problems, type 2 diabetes, central obesity, and hirusutism. Inutero exposures that affect fetal gene programming may be a factor in its development. Although obesity does not cause PCOS, it exacerbates reproductive and metabolic problems associated with it. Rates of PCOS in teens and women increase with increasing rates of overweight and obesity in females.[62] Women with PCOS are at increased risk of spontaneous abortions, gestational and type 2 diabetes, hypertension, and cardiovascular disease.[73]

Nutritional Management of Women with PCOS

"Many women with PCOS have their symptoms ignored for years and have never received the proper diagnosis and treatment. She may have been told 'just lose weight!' and admonished when she found that very hard to do."

Martha McKittrick, RD, CDN, CDE[75]

The primary goal in the treatment of PCOS is to increase insulin sensitivity. A number of insulin-sensitizing drugs, such as metformin and rosiglitazone, can be used to lower blood insulin levels and reduce the excess production of androgens by the ovaries.[83] Other drugs may be used to stimulate ovulation. The preferred first-line treatment for women with PCOS, however, is dietory modification, weight loss and exercise.[83] Weight loss and exercise improve insulin sensitivity, benefit blood lipids and insulin levels, and lower

fasting glucose and testosterone levels in women with PCOS. Care must be taken to individualize eating and exercise plans if weight loss is to succeed. PCOS is a long-term health problem that requires a sustainable approach to weight loss and exercise. In addition, women may benefit from knowing more about PCOS, long-term health risks, and why weight loss and exercise are needed.[75]

Dietary recommendations for teens and women with PCOS emphasize marine sources of the omega-3 fatty acids eicosapentaenoic acid (EPA) and docosahexaenoic acid (DHA) or fish oils, whole grains, fruits, and vegetables high in antioxidants and fiber, regular meals, non-fat dairy products, and low-GI carbohydrates.[74,84] EPA and DHA increase insulin sensitivity. Basic foods such as whole grains and high-fiber, low-GI carbohydrates are encouraged to limit blood glucose surges and insulin production. Weight loss is recommended (if needed), as is aerobic and strength-building exercise (30 minutes or more per day). If drugs are used to treat the symptoms of PCOS, they should be used in conjunction with diet and exercise.[21]

Clinicians have observed that many women with PCOS crave sweets and, because these are highly desired, recommend that they not be totally excluded.[69] Females with PCOS may be scared to eat and may develop eating disorders out of fear of getting diabetes.[74]

Symptoms of PCOS tend to improve substantially with weight loss of 5–10% of initial body weight.[38] Symptoms generally worsen if diet, weight loss, and exercise recommendations are not followed.[69] Most women with PCOS are able to modify eating and exercise behaviors and have children.[74]

Counseling for women with PCOS should be supportive and convey a real understanding of the difficulties the woman has probably been through in getting appropriate advice and help in the past.[75]

Disorders of Metabolism

Two metabolic disorders that affect embryonic development or fertility are covered here: *phenylketonuria (PKU)* and *celiac disease.*

Phenylketonuria (PKU)

Phenylketonuria (also called hyperphenylalaninemia)isthe most frequently inherited disorder of amino-acid metabolism and one of the few preventable causes of mental retardation. It occurs in roughly 1 in 10,000 individuals.[76]

Phenylketonuria (PKU) An inherited error in phenylalanine metabolism most commonly caused by a deficiency of phenylalanine hydroxylase, which converts the essential amino acid phenylalanine to the nonessential amino acid tyrosine. Also called hyperphenylalaninemia.

Celiac Disease An autoimmune disease that causes malabsorption due to an inherited sensitivity to the gliadin portion of gluten in wheat, rye, and barley. It is often responsible for iron, folate, zinc, and other deficiencies. Also called celiac sprue and nontropical sprue.

Phenylketonuria derives its name from the characteristic presence of phenylalanine in the urine of people with this condition. PKU is an inherited problem that causes elevation in blood phenylalanine levels due to very low levels or lack of the enzyme phenylalanine hydroxylase. Lack of this enzyme diminishes the conversion of the essential amino acid phenylalanine to tyrosine, a nonessential amino acid, and causes phenylalanine to accumulate in blood.

If present during early pregnancy, high levels of phenylalanine accumulate in the embryo and fetus and impair normal central nervous system development. Elevated phenylalanine levels in the first 8 weeks of pregnancy increase the risk of heart defects. The risk increases if high blood levels of phenylalanine are combined with low-protein diets early in pregnancy.[85] Untreated women with PKU have a 92% chance of delivering a newborn with mental retardation, and a 73% chance that the infant will be born with an abnormally small head (microcephaly).[86,87]

Infants born to women with high blood levels of phenylalanine during pregnancy are at elevated risk of seizures, hyperactivity, and abnormal behavioral patterns later in life.[76]

Not all infants born to a parent with PKU inherit the disorder, and some infants inherit it from parents who do not have PKU. It is important that infants born with PKU be identified and started on low-phenylalanine formula as soon after birth as possible. Table 3.5 presents a brief case study of the journey of a woman born with PKU that was identified late and inadequately controlled. Individuals born with PKU who adhere to an adequate, low-phenylalanine diet during childhood and later in life tend to develop normally or at levels that are somewhat below normal (Illustration 3.1).[78]

People with undiagnosed PKU may self-select low-protein foods because meat and other rich sources of

---

Table 3.5 One Person's PKU Experience

Margaret was a healthy newborn but during her first year of life she started to experience seizures and difficulty in standing and walking. When she was 13 months old she was diagnosed with PKU and started on a low-phenylalanine diet. Her development improved until the age of 8 years when she was taken off the diet. It was reasoned that she no longer needed the low-phenylalanine diet because "her brain had grown." Around the age of 19 Margaret became violent, destructive, hysterical, and self-abusive. Her family put her back on the PKU diet and although she continued to have seizures and remain profoundly retarded, her behaviors and quality of life improved.[77]

Illustration 3.1 You can't pick a person with well-controlled PKU out of a crowd.

Andi Berger, 2009/Used under license from Shutterstock.com

protein make them light-headed and easily confused. They may find it difficult to comprehend all the information they receive after the inborn error is diagnosed.[81]

Maternal PKU   Twenty-five years ago, it was thought that children with PKU could safely go off the PKU diet after the brain developed. Later it was made abundantly clear that PKU diets are for life. Elevated phenylalanine levels continue to impair mental functions and health long after the growing years.[79] Women who go off the PKU diet after childhood and become pregnant are at risk for a condition called "maternal PKU."

Uncontrolled PKU in women represents risks for the fetus during pregnancy, even in fetuses that did not inherit the disorder. The extent of the harm caused to the fetus increases with increased maternal phenylalanine levels. Adverse effect on the fetus can be minimized if maternal phenylalanine levels are well controlled from the beginning of pregnancy and energy and nutrient needs of the mother are met.[76,79]

In the 1960s, the United States adopted newborn screening programs that test infants for a variety of genetic disorders including PKU.[79] States track individuals testing positive for PKU to ensure they are provided with medical and nutrition services and prescription low-phenylalanine formulas and are made aware of the potential adverse effects of uncontrolled PKU throughout pregnancy. Not all women are able to follow the diet, however, because it is expensive and unpalatable to them, while others may get confused and have trouble staying on the diet.[79]

Alternatives to diet therapy for PKU management are being developed and tested. In the future, it is likely that PKU will be managed largely by techniques such as cell transplantation or gene therapy.[80]

Nutritional Management of PKU

PKU can be successfully managed by a low-phenylalanine diet, instituted and monitored with the help of an experienced registered dietitian. PKU diets are individualized based on blood phenylalanine response to protein foods. Successful PKU diets maintain blood concentrations of phenylalanine in the range of 120–360 μmol/L (2–6 mg/dL).[79] High phenylalanine protein foods such as meat, fish, eggs, and wheat are excluded from the diet. Protein needs are met primarily though consumption of high-protein, low-phenylalanine formulas and other formulated products. The formulated products are generally fortified with tyrosine, vitamins, and minerals. Vegetables, fruits, fats, sugars and high-carbohydrate foods, and phenylalanine-free breads, flour, and pasta are included in the diet. Milk is allowed if needed to maintain a minimal blood phenylalanine level.[76]

The PKU diet should be followed throughout life, but it is critical that it be adhered to prior to conception and maintained throughout pregnancy. It usually takes about 4 to 6 months to learn and implement the PKU diet and to lower blood phenylalanine levels.[81]

Celiac Disease

Celiac disease, highlighted in the "In Focus" box on page 84 is related to somewhat higher rates of infertility and to substantially higher rates of subfertility.[88] In males, untreated celiac disease is related to alterations in the actions of androgens, delayed sexual maturation, and hypogonadism. Hypogonadism is marked by a deficiency of sex hormones and poor development and functioning of the reproductive system. In females, untreated celiac disease is associated with amenorrhea, increased rates of miscarriage, fetal growth restriction, low birthweight deliveries, and a short duration of lactation. It is hypothesized that the effects of celiac disease on reproductive functions in males and females is related to malabsorption-induced deficiencies of nutrients such as vitamins A, E, and D; folate; and iron,[89] and direct effects of inflammation on intestinal and other tissues.[90] Table 3.6 lists vitamin and mineral deficiencies and other health consequences of untreated celiac disease. Normal reproductive functions return after celiac disease has been stabilized with a nutritionally adequate gluten-free diet.[89,90]

Not all individuals with celiac disease have overt symptoms, so it may be missed as a underlying cause of infertility. It should be considered in unexplained cases of infertility, early pregnacy loss, and poor pregnancy outcomes.[91] Case Study 3.2 describes the experience of a woman ultimately diagnosed with celiac disease.

Table 3.6 Vitamin and mineral deficiencies and other health consequences that may occur in people with untreated celiac disease[83,92]

Vitamin Deficiencies	Other Potential Health Problems
Folate	Lactose maldigestion, intolerance
Vitamin B$_{12}$	Weight loss
Vitamin A	Anemia
Vitamin D	Osteoporosis
Vitamin E	Subfertility
Vitamin K	Growth failure (in children and adolescents)
	Irritable bowel disease
Mineral Deficiencies	
Calcium	
Iron	

Nutritional Management of Celiac Disease

Treatment of celiac disease centers on the goals of elimination of gluten from the diet, correction of vitamin and mineral deficiences, and the long-term maintenance of health. The cornerstone of treatment—elimination of gluten—can be challenging. Gluten is found in a variety of nongrain foods, including hot dogs, deli meats, some vitamin and mineral supplements, flavored potato chips, bouillon, and salad dressing. Table 3.7 lists foods that contain gluten and others that do not. An example of a gluten-free one-day diet for an adult with celiac disease is shown in Table 3.8.

People with celiac disease are generally avid food-label readers and eventually become skilled at selecting *gluten-free* foods. In 2008, the Food and Drug Administration started allowing gluten-free foods to be labeled "gluten-free" if they qualify according to a standard definition.[93]

Many different food-product labels announce "gluten-free," "without gluten," and "no gluten," but there are no federal inspections in place to guarantee that food so labeled actually qualifies as gluten-free. The Gluten-Free Certification Organization tests food products for gluten content and certifies them with a GF mark (Illustration 3.2) if found to contain less than 10 ppm, or 3 mg, gluten per serving. Reliable Internet resources for people with celiac disease are listed at the end of this chapter.

> **Gluten-free** A food labeling term that indicates a product does not contain any species of wheat, rye, barley, or their hybrids, or ingredients that contain these grains, or 20 or more parts per million (ppm) gluten (about 6 mg per serving). (FDA proposed definition.[83])

Case Study 3.2

joeysworld.com/Alamy

Celiac Disease

Chloe, age 30, has not had a period for over 2 years. Her gynecological exam turns up no abnormalities, but the hormones she is given to stimulate her menstrual periods do not work. Since the age of 10, Chloe has had painful stomach cramps, frequent diarrhea or constipation, and periodic iron-deficiency anemia. Multiple visits to doctors have failed to find the cause of Chloe's health problems. At around the age of 20, Chloe had begun to wonder if she was a medical anomaly or a hypochondriac.

Still bothered by her health problems and about to be married, Chloe seeks care again. This time she is seen by a nurse practitioner who has just read an article on celiac disease. The nurse sends Chloe to a registered dietitian, who advises Chloe on a gluten-free diet. After faithfully following the diet for a week, Chloe feels better. The cramps, diarrhea, and constipation are much improved, and later on, her menstrual cycles return. She returns to her doctor for a checkup and requests a test for celiac disease. By that time, however, her intestinal biopsy comes back normal because she has been on the diet for months.

Questions

1. What should have been the first clue that Chloe might have celiac disease?
2. What facts provide other clues to the possibility of celiac disease?
3. How long will Chloe have to stay on a gluten-free diet?

(Answers are located in the Instructor's Manual for the 4th edition of *Nutrition Through the Life Cycle*.)

Table 3.7 Examples of foods that do or do not contain gluten[83, 94, 95]	
Gluten-Containing Foods	**Gluten-Free Foods[a]**
Beer, ale	Fruit
Barley	Vegetables
Broth, bouillon powder/cubes	Dried beans
Brown rice syrup	Amaranth
Bulgur	Cassava, millet
Commercial soups, salad dressings	Grits, corn, cornmeal
Breads, cereals, pastas	Quinoa
Imitation seafood	Oatmeal (gluten-free)
Cakes, pies, cookies	Fats
Processed meats	Fresh meats, fish
Soy sauce	Soy flour, cereals, tofu
Wheat starch	Rice, wild rice
Pizza	Eggs
Macaroni and cheese	Nuts, seeds
Seasonings	Cheese (not processed)
Marinades, gravies	Popcorn
Rye-containing products	Milks
Vegetarian meat substitutes	Chips (100% corn, potato)
Flavored-rice packaged products	

[a]Assumes foods have not been contaminated with gluten during processing and are free of gluten-containing ingredients.

Evidence-Based Nutrition Practice Guidelines for Celiac Disease   Evidence-based nutrition practice guidelines and the Nutrition Care Process for celiac disease have been recently developed by the American Dietetic Association.[99] Major components of nutrition care services and the Nutrition Care Process are highlighted in Table 3.9. Monitoring individuals for abnormalities in nutritional status that may impact reproductive outcomes is an important part of the Nutrition Care Process for celiac disease.

Herbal Remedies for Fertility-Related Problems

Women may use herbs and supplements for chronic gynecologic conditions such as premenstrual syndrome, menstrual problems, and infertility. Therapies that carry some level of support from randomized controlled trials indicate that thiamin and vitamin E may be helpful for painful menstrual periods and that calcium, magnesium, vitamin D, and vitamin B$_6$ may help relieve premenstrual syndrome symptoms. Herbal extracts from the berries of the chaste tree (named after the plant monks were said to have chewed to inspire chasity) also show promise for safely relieving PMS symptoms. Chaste-tree berry extracts are not yet considered safe for women who may become or who are pregnant, or who are taking oral contraceptive pills.[100, 101] Evening primrose oil, which contains high

Table 3.8 Example of one day's diet and snack option for an adult with celiac disease

Breakfast	Dinner
Gluten-free bagel with nut butter Sliced bananas in yogurt Tea	Lamb stew (thickened with potato starch) with carrots and lentils Rice Gluten-free cake Low-fat milk
Lunch	**Snack Options**
Gluten-free pasta salad with chicken, broccoli, and tomatoes Oil and vinegar dressing Gluten-free roll with margarine Fresh fruit Low-fat milk	Popcorn Spring rolls with rice paper Ice cream Fruit Dark chocolate Gluten-free cookies String cheese Rice cakes

amounts of essential fatty acids, does not appear to beat a placebo in relation to PMS relief.[102]

There are too few trials involving herbs and supplement use for the treatment of infertility to warrant solid recommendations, but chaste berry, antioxidants, and Fertility Blend (a proprietary supplement containing chaste berry, green tea extracts, L-arginine, vitamins, and minerals) have some preliminary support in terms of safety and effectiveness.[100] Sperm concentrations, motility, and the percent of sperm with normal shape have been found

to improve among infertile men given 200 mcg selenium plus 600 mg N-acetyl-cysteine (a derivative of the amino acid cysteine) daily for 6 months. Both selenium and

Illustration 3.2 The Gluten-Free Certification Organization tests and certifies products as gluten-free and awards qualifying products this mark.

Table 3.9 Key features of the Nutrition Care Process for individuals with celiac disease[99]

A. Nutrition Assessment

1. Assessment of food/nutrition-related history
   - Food and nutrient intake with focus on vitamins and minerals listed in Table 3.5
   - Knowledge, skills, attitudes about celiac and dietary change
   - Access to food
2. Assessment of biochemical data and medical results
   - Severity of intestinal lining damage
   - Presence of anemia, osteoporosis, other diseases

B. Nutrition Intervention

1. Provide education and guidance on nutritionally adequate, gluten-free diet
2. Advise on the use of gluten-free multivitamin and mineral supplement as required
3. Provide resources and education on label reading, food cross-contamination, and support groups

C. Nutrition Monitoring and Evaluation

1. Monitor dietary intake, gluten intake from all sources, celiac antibody levels
2. Monitor persistent gastrointestinal symptoms not eliminated by a gluten-free diet, coordinate care
3. Monitor nutrition risks for poor pregnancy outcomes

In Focus

Celiac Disease

Definition: Celiac disease is an autoimmune disease that occurs in people with a genetic susceptibility to the protein gliadin found in the gluten component of wheat, rye, and barley. Oats do not contain gluten, but commercial oats and oat products may be cross-contaminated by grains that do. Gluten-free oat products should be selected.[96] Celiac disease is also called *celiac sprue* and *nontropical sprue*.

Diagnosis: Blood tests for celiac antibodies and genetic markers of celiac disease are used to help identify the presence of celiac disease. The gold-standard diagnostic test for celiac disease, however, is small bowel biopsy and examination of cells for signs of damage due to the disease. The test has to be undertaken while individuals are consuming their normal diet and not a gluten-free one. Removal of gluten from the diet corrects intestinal cell damage and may lead to an incorrect diagnosis.

Prevalence: Celiac disease occurs in one in every 133 people in the United States and is three times more common in women than men. The incidence of celiac disease appears to be increasing.[96] Although awareness of, and screening for, celiac disease is increasing, far more individuals have the disease than have been diagnosed with it.[97]

Major physiological aspects and consequences: The presence of gliadin in the small intestine triggers an *auto-immune response* that causes an inflammatory reaction to occur in the inside lining of the small intestine. Over time, the inflammation causes the lining of the small intestine to become flattened and to absorb nutrients poorly (Illustration 3.3). The damage produced by chronic inflammation in the small intestine may lead to a variety of vitamin and mineral deficiencies and other health consequences (Table 3.5).

Signs and symptoms of celiac disease range from very mild to severe and vary by age and sex. Diarrhea, nutrient malabsorption, bloating, weight loss, iron-deficiency anemia, infertility, fatigue, and growth failure in children often characterize the disease. Many cases of celiac disease are clinically silent—presenting no clear, related symptoms.[96] Long-standing, untreated celiac disease predisposes individuals to other autoimmune diseases. The only effective treatment is a lifelong gluten-free diet.

Risk factors: The primary risk factor for celiac disease is a genetic predisposition toward reacting to gliadin as a foreign protein. Repeated exposure to certain types of infectious agents, such as *rotavirus*, may also trigger the onset of the disease in genetically susceptible people.[98]

Autoimmune Response Chemicals released by the immune system that attack its own molecules, cells, and tissues.

Rotavirus A virus that is the most common cause of severe diarrhea among children. Diarrhea caused by rotavirus generally lasts 2 days, and recovery is full in otherwise healthy children. The rotavirus is generally spread from an infected peron's stools to food.

N-acetyl-cysteine function as antioxidants and may help prevent oxidative damage to sperm.[103] Coenzyme Q_{10} given at a dose of 200 mg per day for 6 months in males with poor sperm motility for unknown reasons increased sperm motility and pregnancy rates.[104,105] Coenzyme Q_{10} acts as an antioxidant and is a coenzyme in energy formation. It may help protect sperm membranes from oxidation and facilitate energy formation

Illustration 3.3 Sections of the small intestine showing normal villi structures of a person without celiac disease (a) and the flattened villi that develop in people with untreated celiac disease (b).

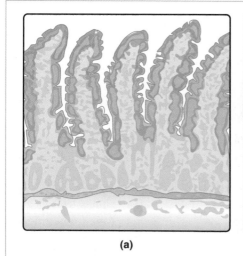
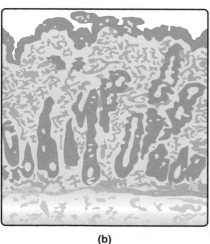

(a)   (b)

by sperm. Coenzyme Q_{10} appears to be well tolerated.[106] Bee propolis, which consists of plant resins collected by honeybees, appears to have anti-inflammatory activity and has been found to increase pregnancy rates in women with mild endometriosis. Use of the compound was not related to adverse side effects in one study.[109]

Women and men using herbs for fertility problems should inform their health care providers. They should avoid herbs and other dietary supplements that have not been demonstrated to be safe for use during early pregnancy.

Key Points

1. Dietary intake, supplement use, weight status, and exercise levels before conception affect fertility in women and men, and the course and outcome of pregnancy in women.

2. Symptoms of PMS can be modified in some females by specific changes in dietary intake and supplement use.

3. Obesity is related to a number of hormonal and metabolic changes that compromise fertility and health status in men and women.

4. Modest levels of weight loss in obese women and men, and weight gain in underweight individuals, improve fertility.

5. Rates of diseases and disorders associated with obesity such as metabolic syndrome, polycystic ovary syndrome, and type 2 and gestational diabetes are increasing in the United States and other countries. Obesity-related health problems affecting fertility and the course and outcome of pregnancy are being seen increasingly in clinical practice.

6. Chronic inflammation is an important component of disorders such as infertility, metabolic syndrome, and polycystic ovary syndrome. Antioxidant nutrients and the omega-3 fatty acids EPA and DHA may play key roles in reducing adverse effects of chronic inflammation.

7. Insulin resistance is a key feature of obesity, type 2 and gestational diabetes, metabolic syndrome, and polycystic ovary syndrome.

8. Energy deficits in individuals with hypothalamic amenorrhea, eating disorders, and the female athlete triad are related to hormonal changes that reduce fertility.

9. Energy and estrogen deficits in women with the female athlete triad lead to reduced fertility and lower bone formation in young women.

10. PKU is a genetic disorder that causes blood phenylalanine levels to rise to toxic concentrations in untreated individuals with the disorder. Untreated PKU can produce malformations, neurological disorders, and severe mental retardation in children and adults. It is treated with a low-phenylalanine diet for life.

11. Untreated celiac disease is related to multiple vitamin and mineral deficiencies and infertility, early pregnancy, impaired fetal growth, and other poor pregnancy outcomes. Standard treatment for celiac disease is a gluten-free diet.

12. Some herbal remedies and other dietary supplements appear to reduce symptoms of premenstrual syndrome and may aid fertility. Due to limited knowledge of the safety of herbs and other dietary supplements taken before conception and during early pregnancy, they are best used under medical supervision.

Review Questions

Lois was diagnosed with type 2 diabetes 2 months before she planned to become pregnant. To get her blood glucose under control before pregnancy, Lois worked with a dietitian who specializes in diabetes, and together they developed a plan to reduce Lois's blood glucose levels.

The following four questions refer to this case and appropriate components of the nutritional management of type 2 diabetes. Check *true* is the statement is correct and *false* if it is incorrect.

1. As part of the plan, Lois would have to exclude from her diet sugar and foods containing sugar.
   _____ True   _____ False

2. Assume Lois will attempt to consume 2000 calories a day. Based on that level of calorie intake, her fiber intake should total approximately 28 grams daily.
   _____ True   _____ False

3. Lois plans to perform aerobic exercise 25 minutes daily because it will help improve insulin resistance, lower blood glucose levels, and improve her blood lipid levels.
   _____ True   _____ False

4. Lois should take a chromium supplement to help lower her blood glucose levels.
   _____ True   _____ False

5. High intake of caffeine or coffee has been found to increase the risk of premenstrual syndrome.
____ True   ____ False

6. Obesity is related to infertility and chronic inflammation in men only.
____ True   ____ False

7. Obesity is associated with inadequate vitamin D status.
____ True   ____ False

8. Health problems related to anorexia nervosa resolve after the eating disorder is successfully treated.
____ True   ____ False

9. Weight loss of 7% or more of body weight and exercise in individuals with prediabetes reduce the risk of type 2 diabetes.
____ True   ____ False

10. Excess visceral fat and insulin resistance are shared characteristics among women with PCOS.
____ True   ____ False

11. Individuals with PKU can safely consume high-protein foods such as beef, chicken, or eggs once a day.
____ True   ____ False

12. Most individuals with celiac disease share the symptoms of abdominal pain, diarrhea, and bloating.
____ True   ____ False

Resources

Health Topics

High-quality information on eating disorders, PMS, and other conditions as presented in this chapter can be found here.
Website: www.healthfinder.gov

PCOS Support

Chat rooms for women with PCOS include the following:
Website: www.obgyn.net/PCOS/PCOS.asp
Website: www.PCOSupport.org

Celiac Disease Resources

Support journal articles, and other resources for people with celiac disease are available from the following:
Websites:

The Celiac Disease Foundation: www.celiac.org
Celiac Sprue Association, USA: www.csaceliacs.org
Gluten Intolerance Group: www.gluten.net
Raising Our Celiac Kids: www.celiackids.com
Resource guide for gluten-free diets: www.glutenfreediet.ca

Obtain a list of gluten-free foods at this site.
Website: www.celiac.com

Select a company's name and get a list of the gluten-free products it sells, or select a food type and receive a list of companies that sell the gluten-free version of the food.
Website: www.gfco.org/products.php

Obesity and Weight-Control Resources

The Centers for Disease Control and Prevention provide updated obesity statistics.
Website: www.cdc.gov/obesity/data/trends.html

Read about small steps changes for weight loss and weight-loss maintenance.
Website: www.healthfinder.gov/prevention/ViewTopic.aspx?topicID=25

National Institute of Diabetes and Digestive and Kidney Diseases offers information on weight control.
Website: www.nlm.nih.gov/medlineplus/weightcontrol.html

NIH's Weight Control Information network presents leading myths about weight control, weight-loss diets, food, and physical activity.
Website: www.win.niddk.nih.gov/publications/myths.htm

The FTC has developed and updates a site on bogus weight-loss advertisements and guides for industries that may produce them.
Website: http://www.ftc.gov/redflag/

Diabetes Resources

Support and other resources related to diabetes are available from the following resources:
Websites:

American Diabetes Association: www.diabetes.org
The National Institutes of Health: www.niddk.nih.gov
Canadian Diabetes Association: www.diabetes.ca

Physical Activity Resources

View a slide show on weight-training exercises produced by the Mayo Clinic staff.
Website: www.mayoclinic.com/health/weight-training/SM00041&slide=1

Find strength-building exercise plans.
Website: http://www.healthfinder.gov/getactive

NIH provides a list of physical activities for large people.
Website: win.niddk.nih.gov/publications/active.htm

Obtain a copy of the 2008 Physical Activity Guidelines for Americans, the first such guidelines to be published by the federal government. The site contains information and tools for increasing physical activity based on individual preferences, and offers physical-activity recording forms.
Website: www.health.gov/paguidelines

Healthful Diet Resources

MyPyramid.gov provides a wealth of resources related to diet and physical activity planning and evaluation.
Website: www.mypyramidtracker.gov

Chapter 4

Nutrition During Pregnancy

Prepared by **Judith E. Brown**

Key Nutrition Concepts

1   Many aspects of nutritional status, such as dietary intake, supplement use, and weight change, influence the course and outcome of pregnancy.

2   The fetus is not a parasite; it depends on the mother's nutrient intake to meet its nutritional needs.

3   Periods of rapid growth and development of fetal organs and tissues occur during specific times throughout pregnancy. Essential nutrients must be available in required amounts during these times for fetal growth and development to proceed optimally.

4   The risk of heart disease, diabetes, hypertension, and other health problems during adulthood may be influenced by maternal nutrition during pregnancy.

Introduction

The 9 months of pregnancy represent the most intense period of growth and development humans ever experience. How well these processes go depends on many factors, most of which are modifiable. Of the factors affecting fetal growth and development that are within our control to change, nutritional status stands out. At no other time in life are the benefits of optimal nutritional status more obvious than during pregnancy.

This chapter addresses the status of pregnancy outcomes in the United States and other countries. It covers physiological changes that take place to accommodate pregnancy, and the impact of these changes on maternal nutritional needs. The chapter presents the roles of nutrition in fostering fetal growth, development, and long-term health; it also covers dietary supplement use and weight-gain recommendations.

The discussion goes on to consider common problems during pregnancy that can be addressed with nutritional remedies. Chapter 5 addresses clinical conditions and nutrition interventions during pregnancy. We begin this chapter by highlighting vital statistics reports that clearly show a need for improving pregnancy outcomes in the United States.

The Status of Pregnancy Outcomes

The status of reproductive outcomes in the United States and other economically developed countries is routinely assessed through examination of a particular set of vital statistics data called *natality statistics* (*natality* means "related to birth"). Natality statistics summarize important information about the occurrence of pregnancy complications and harmful behaviors, in addition to infant mortality (death) and morbidity (illness) rates within a specific population. These data are used to identify problems in need of resolution and to identify progress in meeting national goals for improvement in the course and outcome of pregnancy.

Illustration 4.1 presents a time line of key intervals and events before, during, and after pregnancy. Specific time points and periods are labeled. Table 4.1 provides natality statistics for the periods and outcomes shown in Illustration 4.1. It also provides a comparison of the most recent rates available for outcomes such as infant mortality and low birth weight with rates reported in 1995. Terms listed in Table 4.1 are referred to frequently in this chapter. Also referred to often in this chapter are weights in grams (g) and kilograms (kg), as well as in pounds and ounces. There are 448 grams in a pound and 2.2 pounds in a kilogram. (Note: A table of measurement abbreviations and equivalents is located in Appendix B.)

Illustration 4.1 Time-related terms before, during, and after pregnancy.

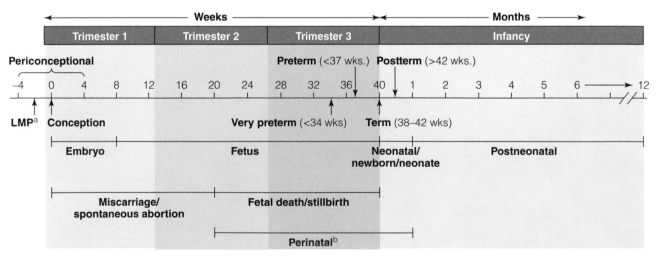

ᵃLMP = last menstrual period

ᵇPerinatal definition varies from 20 to 24 weeks gestation to 7 to 28 days after birth.

Table 4.1 Natality statistics: rates, definitions, and trends in the United States[1,2]

	Rates 1995	2006/2007	Definition
Maternal mortality	7.1	7.1	Deaths/100,000 live births
Fetal deaths (stillbirths)	7.0	6.2	Deaths/1000 pregnancies over 20 weeks gestation
Perinatal mortality	7.6	6.6	Deaths/1000 deliveries over 20 weeks gestation to 7 days after birth
Neonatal mortality	4.9	4.5	Deaths from delivery to 28 days/1000 live births
Postneonatal mortality	2.7	2.3	Deaths from 28 days after birth to 1 year/1000 live births
Infant mortality	7.6	6.8	Deaths from birth to age 1 year/1000 live births
Preterm	11.0	12.7	Births <37 weeks gestation/100 live births
Very preterm	1.9	2.0	Births <34 weeks gestation/100 live births
Low birth weight	7.3	8.2	Newborn weights <2500 g (5 lb 8 oz)/100 live births
Very low birth weight	1.4	1.5	Newborn weights <1500 g (3 lb 4 oz)/100 live births
Multifetal pregnancies			
Twins	1 in 40	1 in 32	Number of twin births/total live births
Triplets+	1 in 784	1 in 667	Number of triplets plus higher-order multiple births/total live births
Adolescent pregnancies	56.8	42.5	Births/1000 females aged 15 to 19 years

Infant Mortality

> "Infant mortality is a mirror of a population's physical health and socioeconomic status."

Infant mortality reflects the general health status of a population to a considerable degree because so many of the environmental factors that affect the health of pregnant women and newborns also affect the health of the rest of the population. The graph presented in Illustration 4.2 demonstrates this point. This historical view of infant mortality rates indicates that population-wide improvements in social circumstances, infectious disease control, and availability of safe and nutritious foods have corresponded to greater reductions in infant mortality than have technological advances in medical care.[3] Small improvements in infant mortality in the past few decades in the United States are largely due to technological advances in medical care that save ill newborns. High-level medical care has not favorably affected the need for such care, however. The United States spends more money on health care than any other

Illustration 4.2 Chronology of events related to declines in infant mortality in the United States.

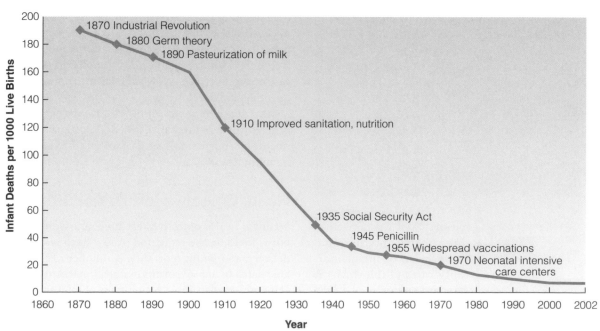

SOURCE: Judith E. Brown, 2003.

Table 4.2 Infant mortality rates (deaths per 1000 live births) for 31 countries listed from lowest to highest, 2004[5]

Country	Infant Mortality Rate
Singapore	2.0
Hong Kong	2.5
Japan	2.8
Sweden	3.1
Norway	3.2
Finland	3.3
Spain	3.5
Czech Republic	3.7
France	3.9
Portugal	4.0
Germany	4.1
Greece	4.1
Italy	4.1
Netherlands	4.1
Switzerland	4.2
Belgium	4.3
Denmark	4.4
Austria	4.5
Israel	4.5
Australia	4.7
Ireland	4.9
Scotland	4.9
England and Wales	5.0
Canada	5.3
Northern Ireland	5.5
New Zealand	5.7
Cuba	5.8
Hungary	6.6
Poland	6.9
Slovakia	6.9
United States	6.9
Puerto Rico	8.1
Chile	8.4
Costa Rica	9.0
Russian Federation	11.5
Bulgaria	11.7
Romania	16.8

Liveborn Infant   A liveborn infant is the outcome of delivery when a completely expelled or extracted fetus breathes, or shows any sign of life such as beating of the heart, pulsation of the umbilical cord, or definite movement of voluntary muscles, whether or not the cord has been cut or the placenta is still attached.

that differences in the definition or reporting of infant mortality among countries account for the relatively high ranking of the United States.[4] Two-thirds of deaths to liveborn infants occur within the first month after birth, or during the neonatal period.[6]

Low Birth Weight, Preterm Delivery, and Infant Mortality

Infants born low birth weight or preterm are at substantially higher risk of dying in the first year of life than are larger and older newborns. Low-birth-weight infants, for example, make up 8.2% of all births, yet comprise 66% of all infant deaths. The 12.7% of newborns delivered prior to 37 weeks of pregnancy similarly account for a disproportionately large number of infant deaths.[1] Low-birthweight and preterm infant outcomes are intertwined in that the shorter the pregnancy, the less newborns tend to weigh. Table 4.3 shows increases in birth weight with the duration of pregnancy and birth-weight-specific infant mortality rates.

Rates of preterm delivery and low birth weight in the United States have trended slowly upward since 1983 and remain higher in African American infants than in other infants (Table 4.4). The high rates of preterm and low birth weight in African American infants clearly represents a problem in need of resolution.

Reducing Infant Mortality and Morbidity

Deaths and illnesses associated with low-birth-weight and preterm infants can be reduced through improvements in the birth weight of newborns. Infants weighing 3500 to 4500 grams at birth (or 7 lb 12 oz to 10 lb) are least likely to die within the first year of life, as well as in the perinatal, neonatal, and postneonatal periods.[1] Newborns weighing 3500 to 4500 grams are also at an advantage as a group in relation to overall health status and subsequent mental development.[8] They are less likely to develop heart disease, diabetes, lung disease, hypertension, and other disorders later in life.[9] Reducing the proportion of infants born small or early would clearly decrease infant mortality.

Health Objectives for the Year 2010

National health objectives for pregnant women and newborns focus on the reduction of low birth weight, preterm delivery, and infant mortality. A number of the objectives are related to improvements in nutritional status (Table 4.5). Data presented in Table 4.1 on rates of infant mortality, low birth weight, and preterm delivery indicate that it is unlikely that important objectives for pregnant women and infants will be met by 2010.

nation, yet it ranks twenty-ninth in the international comparison of infant mortality rates (Table 4.2).[5] A standard definition of a *liveborn infant* is implemented internationally to identify infant live births. It is unlikely

Table 4.3 Range of birth weights by gestational age, U.S.[7]

Birth Weight		Weeks Gestation	Infant Mortality Rate
Pounds (lb) and Ounces (oz)	Grams		
<1 lb 2 oz	<500	<22	846
1 lb 2 oz–2 lb 3 oz	500–999	22–27	316
2 lb 3 oz–3 lb 5 oz	1000–1499	27–29	62
3 lb 5 oz–4 lb 6 oz	1500–1999	29–31	28
4 lb 6 oz–5 lb 8 oz	2000–2499	31–33	12
5 lb 8 oz–6 lb 10 oz	2500–2999	33–36	4.6
6 lb 10 oz–7 lb 11 oz	3000–3499	36–40	2.4
7 lb 11 oz–8 lb 13 oz	3500–3999	40+	1.7
8 lb 13 oz–9 lb 14 oz	4000–4499	40+	1.5
9 lb 14 oz–11 lb	4500–4999	40+	2.5
>11 lb	5000+	40+	—

Table 4.4 Rates of preterm delivery and low birth weight for the United States' population and by ethnic/racial background, 2007[2]

	Preterm	Low Birth Weight
All races and origins	12.7	8.2
Blacks	18.3	13.8
Whites	11.5	7.2
American Indian/ Alaskan Native	13.9	7.5
Asian/Pacific Islander	10.9	8.1
Hispanic	12.3	6.9

Table 4.5 Health objectives for the nation related to pregnant women and infants

- Reduce anemia among low-income pregnant females in their third trimester from 29 to 20%.
- Reduce infant mortality from 7.6 to no more than 5 per 1000 live births.
- Reduce the incidence of spina bifida and other neural tube defects from 7 to 3 per 10,000 live births.
- Reduce low birth weight (<2500 g) from 7.3 to 5%.
- Reduce preterm births (<37 weeks) from 9.1 to 7.6%.
- Increase abstinence from alcohol use by pregnant women from 79 to 95%.
- Reduce the incidence of fetal alcohol syndrome.
- Increase the proportion of women who gain weight appropriately during pregnancy.

Physiology of Pregnancy

Conception triggers thousands of complex and sequenced biological changes that transform two united cells into a member of the next generation of human beings. The rapidity with which structures and functions develop in mother and fetus and the time-critical nature of energy and nutrient needs make maternal nutritional status a key element of successful reproduction.

Pregnancy begins at conception; that occurs approximately 14 days before a woman's next menstrual period is scheduled to begin. Assessed from conception, pregnancy averages 38 weeks, or 266 days, in length. Most commonly, however, pregnancy duration is given as 40 weeks (280 days) because it is measured from the date of the first day of the last menstrual period (LMP). Consequently, the common way of measuring pregnancy duration includes two nonpregnant weeks at the beginning. The anticipated date of delivery is denoted by the ancient terminology of "estimated date of confinement," or EDC. Assessment of duration of pregnancy as weeks from conception is correctly termed *gestational age*, whereas time in pregnancy estimated from LMP reflects *menstrual age*. It is particularly important to get these terms straight during early fetal development, when a 2-week error in duration of pregnancy may mean miscalculating the timing of nutrient-related events in pregnancy.

Maternal Physiology

Changes in maternal physiology during pregnancy are so profound that they were previously considered abnormal and in need of correction. Doctors routinely advised pregnant women to follow low-sodium diets to reduce fluid retention, restricted their patients' weight gain and dietary intake to prevent complications at delivery, and prescribed excessive levels of iron and other supplements

Table 4.6 Sequence of tissue development and approximate gestational week of maximal rates of change in maternal systems, the placenta, and fetus during pregnancy[10]

Tissue	Sequence of Development	Gestational Week of Maximal Rate of Growth
Maternal plasma volume	1	20
Maternal nutrient stores	2	20
Placental weight	3	31
Uterine blood flow	4	37
Fetal weight	5	37

Similarly, the maximal rate of placental growth is timed to precede that of fetal weight gain. This sequence of events ensures that the placenta is fully prepared for the high level of functioning that will be needed as fetal weight increases most rapidly. Fetuses depend on the functioning of multiple systems, established well in advance of their maximal rates of growth and development. Abnormalities in the development of any of these physiological systems can modify subsequent fetal growth and development.

to bring blood nutrient levels back up to "normal." We now know that what is considered normal physiological status of nonpregnant women cannot be considered normal for women who are pregnant. Fortunately, it is now understood that attempts to bring maternal physiological changes back to nonpregnant levels may cause more harm than good to the pregnancy.

Changes in maternal body composition and functions occur in a specific sequence during pregnancy. The order of the sequence is absolute because the successful completion of each change depends on the one before it. Because maternal physiological changes set the stage for fetal growth and development, they begin in earnest within a week after conception.[10]

The sequence of physiological changes taking place during pregnancy is listed in Table 4.6. The table indicates the timing of maximal rates of change in maternal tissues, the *placenta,* and fetal weight across pregnancy. To provide the fetus with sufficient energy, nutrients, and oxygen for growth, the mother must first expand the volume of plasma that can be circulated. Maternal nutrient stores are accumulated next. These stores are established in advance of the time they will be needed to support large gains in fetal weight.

Placenta   A disk-shaped organ of nutrient and gas interchange between mother and fetus. At term, the placenta weighs about 15% of the weight of the fetus.

Normal Physiological Changes During Pregnancy

Physiological changes in pregnancy can be divided into two basic groups: those occurring in the first half of pregnancy and those in the second half. In general, physiological changes in the first half are considered "maternal anabolic" changes because they build the capacity of the mother's body to deliver relatively large quantities of blood, oxygen, and nutrients to the fetus in the second half of pregnancy. The second half is a time of "maternal catabolic" changes in which energy and nutrient stores, and the heightened capacity to deliver stored energy and nutrients to the fetus, predominate (Table 4.7). Approximately 10% of fetal growth is accomplished in the first half of pregnancy, and the remaining 90% occurs in the second half.[12]

The list of physiological changes that normally occur during pregnancy is extensive (Table 4.8), and such changes affect every maternal organ and system. Changes that are most directly related to maternal energy and nutrient needs are discussed further.

Body Water Changes   A woman's body gains a good deal of water during pregnancy, primarily due to increased volumes of plasma and extracellular fluid, as well as amniotic fluid.[14] Total body water increases in

Table 4.7 Summary of maternal anabolic and catabolic phases of pregnancy[11-13]

Maternal Anabolic Phase 0–20 Weeks	Maternal Catabolic Phase 20+ Weeks
Blood volume expansion, increased cardiac output	Mobilization of fat and nutrient stores
Buildup of fat, nutrient, and liver glycogen stores	Increased production and blood levels of glucose, triglycerides, and fatty acids; decreased liver glycogen stores
Growth of some maternal organs	Accelerated fasting metabolism
Increased appetite, food intake (positive caloric balance)	Increased appetite and food intake decline somewhat near term
Decreased exercise tolerance	Increased levels of catabolic hormones
Increased levels of anabolic hormones	

Table 4.8 Normal changes in maternal physiology during pregnancy[10,11]

Blood Volume Expansion
- Blood volume increases 20%
- Plasma volume increases 50%
- Edema (occurs in 60–75% of women)

Hemodilution
- Concentrations of most vitamins and minerals in blood decrease

Blood Lipid Levels
- Increased concentrations of cholesterol, LDL cholesterol, triglycerides, HDL cholesterol

Blood Glucose Levels
- Increased insulin resistance (increased plasma levels of glucose and insulin)

Maternal Organ and Tissue Enlargement
- Heart, thyroid, liver, kidneys, uterus, breasts, adipose tissue

Circulatory System
- Increased cardiac output through increased heart rate and stroke volume (30–50%)
- Increased heart rate (16% or 6 beats/min)
- Decreased blood pressure in the first half of pregnancy (−9%), followed by a return to nonpregnancy levels in the second half

Respiratory System
- Increased tidal volume, or the amount of air inhaled and exhaled (30–40%)
- Increased oxygen consumption (10%)

Food Intake
- Increased appetite and food intake; weight gain
- Taste and odor changes, modification in preference for some foods
- Increased thirst

Gastrointestinal Changes
- Relaxed gastrointestinal tract muscle tone
- Increased gastric and intestinal transit time
- Nausea (70%), vomiting (40%)
- Heartburn
- Constipation

Kidney Changes
- Increased glomerular filtration rate (50–60%)
- Increased sodium conservation
- Increased nutrient spillage into urine; protein is conserved
- Increased risk of urinary tract infection

Immune System
- Suppressed immunity
- Increased risk of urinary and reproductive tract infection

Basal metabolism
- Increased basal metabolic rate in second half of pregnancy
- Increased body temperature

Hormones
- Placental secretions of large amounts of hormones needed to support physiological changes of pregnancy

pregnancy range from 7 to 10 liters (approximately 7 to 10 quarts, or about 2 to 2½ gallons). About two-thirds of the expansion is intracellular (blood and body tissues) and one-third is extracellular (fluid in spaces between cells).[10] Plasma volume begins to increase within a few weeks after conception and reaches a maximum at approximately 34 weeks. Early pregnancy surges in plasma volume appear to be the primary reason that pregnant women feel tired and become exhausted easily when undertaking exercise performed routinely prior to pregnancy. Fatigue associated with plasma-volume increases in the second and third months of pregnancy declines as other compensatory physiological adjustments are made.

Gains in body water vary a good deal among women during normal pregnancy. High gains are associated with increasing degrees of *edema* and weight gain. If not accompanied by hypertension, edema generally reflects a healthy expansion of plasma volume. Birth weight is strongly related to plasma volume: generally, the greater the expansion, the greater the newborn size.[10] The increased volume of water in the blood is responsible for the "dilution effect" of pregnancy on blood concentrations of some vitamins and minerals. Blood levels of fat-soluble vitamins tend to

increase in pregnancy, whereas levels of the water-soluble vitamins tend to decrease. Vitamin supplement use can modify these relationships.[13]

Hormonal Changes   Many physiological changes in pregnancy are modulated by hormones produced by the placenta. Table 4.9 summarizes normal physiological changes that occur in pregnancy, and Illustration 4.3 presents a picture of how hormone levels change. The placenta serves many roles, but a key one is the production of *steroid hormones*, such as progesterone and estrogen. The placenta is also the main supplier of many other hormones needed to support the physiological changes of pregnancy.

Maternal Nutrient Metabolism   Adjustments in maternal nutrient metabolism are apparent within the first few weeks after conception and progress throughout pregnancy.[12] Many of the adjustments are directed toward ensuring that nutrients will be available to the fetus during periods of high nutrient need. Fetal nutrient

Edema   Swelling (usually of the legs and feet, but can also extend throughout the body) due to an accumulation of extracellular fluid.

Steroid Hormones   Hormones such as progesterone, estrogen, and testosterone produced primarily from cholesterol.

Table 4.9 Key placental hormones and examples of their roles in pregnancy[13,16]

Human chorionic gonadotropin (hCG)

Maintains early pregnancy by stimulating the corpus luteum to produce estrogen and progesterone. It stimulates growth of the endometrium. The placenta produces estrogen and progesterone after the first 2 months of pregnancy

Progesterone

Maintains the implant; stimulates growth of the endometrium and its secretion of nutrients; relaxes smooth muscles of the uterine blood vessels and gastrointestinal tract; stimulates breast development; promotes lipid deposition

Estrogen

Increases lipid formation and storage, protein synthesis, and uterine blood flow; prompts uterine and breast duct development; promotes ligament flexibility

Human chorionic somatotropin (hCS)

Increases maternal insulin resistance to maintain glucose availability for fetal use; promotes protein synthesis and the breakdown of fat for energy for maternal use

Leptin

May participate in the regulation of appetite and lipid metabolism, weight gain, and utilization of fat stores

and for fetal structures to develop. Because normal fetal tissue growth and development are genetically timed, nutrients must be available at the same time that genes controlling fetal growth and development are expressed.[15]

Carbohydrate Metabolism   Many adjustments in carbohydrate metabolism are made during pregnancy that promote the availability of glucose to the fetus. Glucose is the fetus's preferred fuel, even though fats can be utilized for energy. Continued availability of a fetal supply of glucose is accomplished primarily through metabolic changes that promote maternal insulin resistance. These changes, sometimes referred to as the *diabetogenic effect of pregnancy,* make normal pregnant women slightly carbohydrate intolerant in the third trimester of pregnancy.[17] Illustration 4.4 provides an example of the normal levels of plasma glucose and insulin during late pregnancy compared to prepregnancy levels.

Carbohydrate metabolism in the first half of pregnancy is characterized by estrogen- and progesterone-stimulated increases in insulin production and conversion of glucose to glycogen and fat. In the second half, rising levels of hCS and prolactin from the mother's pituitary gland inhibit the conversion of glucose to glycogen and fat.[13] At the same time, insulin resistance builds in the mother, increasing her reliance on fats for energy. Decreased conversion of

needs are driven by genetically timed sequences of fetal tissue growth and development. The amount and types of nutrients required depend on the type and amount of nutrients needed for specific metabolic pathways to function

Illustration 4.4 Plasma glucose and insulin levels in nonpregnant women and in women near term.

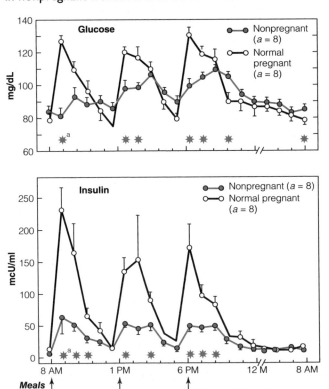

Illustration 4.3 Changes in maternal plasma concentration of hormones during pregnancy.

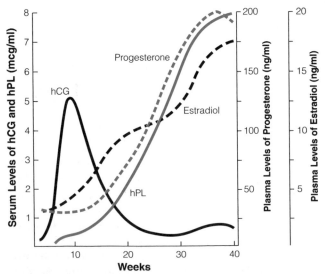

SOURCE: From Pedro Rosso, *Nutrition and Metabolism in Pregnancy: Mother and Fetus.* Copyright © 1990 by Oxford University Press, Inc. Used by permission of Oxford University Press, Inc.

SOURCE: Reprinted from *American Journal of Obstetrics and Gynecology* 140(6): 730–736. R. L. Phelps et al., © 1981, with permission from Elsevier; a * indicates statistical significance.

glucose to glycogen and fat, lowered maternal utilization of glucose, and increased liver production of glucose help to ensure that a constant supply of glucose for fetal growth and development is available in the second half of pregnancy.

Fasting maternal blood glucose levels decline in the third trimester due to increased utilization of glucose by the rapidly growing fetus. However, postmeal blood glucose concentrations are elevated and remain higher longer than before pregnancy.[17]

Accelerated Fasting Metabolism   Maternal metabolism is rapidly converted toward *glucogenic amino acid* utilization, fat oxidation, and increased production of *ketones* with fasts that last longer than 12 hours. Decreased levels of plasma glucose and insulin and increased levels of triglycerides, free fatty acids, and ketones are seen hours before they occur in nonpregnant fasting women. The rapid conversion to fasting metabolism allows pregnant women to use primarily stored fat for energy while sparing glucose and amino acids for fetal use.[17]

Although these metabolic adaptations help ensure a constant fetal supply of glucose, fasting eventually increases the dependence of the fetus on ketone bodies for energy. Prolonged fetal utilization of ketones, such as occurs in women with poorly controlled diabetes or in those who lose weight during part or all of pregnancy, is associated with reduced growth and impaired intellectual development of the offspring.[18]

Protein Metabolism   Nitrogen and protein are needed in increased amounts during pregnancy for synthesis of new maternal and fetal tissues. It is estimated that 925 grams (2 pounds) of protein is accumulated during pregnancy.[19] To some extent the increased need for protein is met through reduced levels of nitrogen excretion and the conservation of amino acids for protein tissue synthesis. There is no evidence, however, that the mother's body stores protein early in pregnancy in order to meet fetal needs for protein later in pregnancy. Maternal and fetal needs for protein are primarily fulfilled by the mother's intake of protein during pregnancy.[12]

Fat Metabolism   Multiple changes occur in the body's utilization of fats during pregnancy. Overall, changes in lipid metabolism promote the accumulation of maternal fat stores in the first half of pregnancy and enhance fat mobilization in the second half.[10] In addition to seeing increasing maternal reliance on fat stores for energy as pregnancy progresses, we see blood levels of many lipoproteins increase dramatically (Table 4.10). Plasma triglyceride levels increase first and most dramatically, reaching three times nonpregnant levels by term.[12,17] Cholesterol-containing

Table 4.10 Changes in cholesterol and triglyceride levels during pregnancy[21,22]

Trimester	Cholesterol mmol/L	(mg/dL)	Triglycerides mmol/L	(mg/dL)
1	5.78	(223)	1.19	(105)
2	6.88	(266)	1.32	(117)
3	8.14	(314)	2.58	(228)
Nonpregnant	5.11	(197)	0.80	(71)

lipoproteins, phospholipids, and fatty acids also increase, but to a lesser extent than do triglycerides. The increased cholesterol supply is used by the placenta for steroid hormone synthesis, and by the fetus for nerve and cell-membrane formation.[17] High concentrations of cholesterol and triglycerides observed during pregnancy do not promote the development of atherosclerosis (hardening of the arteries), as they may in adults.[20] Small increases in HDL cholesterol in pregnancy appear to decline within a year postpartum and remain lower than prepregnancy levels. It is speculated that declines in HDL cholesterol after pregnancy may contribute to an increased risk of heart disease in women. Other changes in serum lipids appear to revert to prepregnancy levels postpartum.[22]

By the third trimester of pregnancy, most women have a lipid profile that would be considered atherogenic, if not for pregnancy. These blood lipid changes are normal, however, which is why blood lipid screening is not recommended during pregnancy.[22] Normal changes in blood lipid levels during pregnancy appear to be unrelated to maternal dietary intake.[23]

Mineral Metabolism   Impressive changes in mineral metabolism occur during pregnancy. Calcium metabolism is characterized by an increased rate of bone turnover and reformation.[14] Elevated levels of body water and tissue synthesis during pregnancy are accompanied by increased requirements for sodium and other minerals. Sodium metabolism is delicately balanced during pregnancy to promote an accumulation of sodium by the mother, placenta, and fetus. This is accomplished by changes in the kidneys that increase aldosterone secretion and the retention of sodium. This normal change in pregnancy renders ineffective and potentially harmful any attempts to prevent and treat high blood pressure in pregnancy by reducing sodium intake. Sodium restriction may overstress mechanisms that act to conserve sodium and lead to functional and growth impairments due to sodium depletion.[24]

Glucogenic Amino Acids   Amino acids such as alanine and glutamate that can be converted to glucose.

Ketones   Metabolic by-products of the breakdown of fatty acids in energy formation. β-hydroxybutyric acid, acetoacetic acid, and acetone are the major ketones, or "ketone bodies."

Illustration 4.5 A placenta.

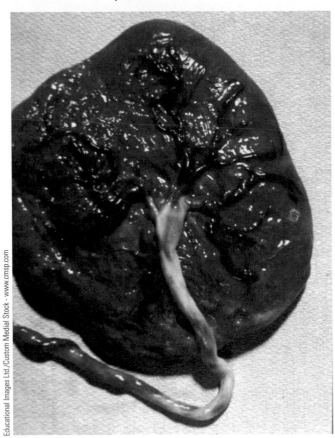

Educational Images Ltd./Custom Medial Stock - www.cmsp.com

Illustration 4.6 Structure of the placenta. Maternal arteries and veins are part of the maternal circulation, whereas umbilical arteries and veins are part of the fetal circulation. Blood enters the fetus through umbilical veins and exits through umbilical arteries.

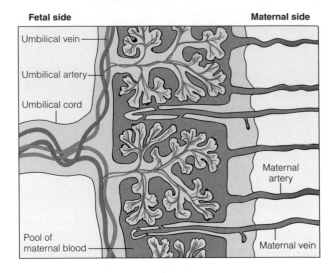

The Placenta

The word *placenta* is derived from the Latin word for *cake*. The placenta, with its round, disklike shape (Illustration 4.5), looks somewhat like a cake—the more so the more active the imagination. The placenta develops from embryonic tissue and is larger than the fetus for most of pregnancy. Development of the placenta precedes fetal development.

Functions of the placenta include:

- Hormone and enzyme production,
- Nutrient and gas exchange between the mother and fetus
- Removal of waste products from the fetus

Its structure, including a double lining of cells separating maternal and fetal blood, acts as a barrier to some harmful compounds, and it governs the rate of passage of nutrients and other substances into and out of the fetal circulation (Illustration 4.6). The barrier role of the placenta is better described as a fence than as a filter that guards the fetus against all things harmful. Many potentially harmful substances (alcohol, excessive levels of some vitamins, drugs, and certain viruses, for example) do pass through the placenta to the fetus. The placenta is a barrier to the passage of maternal red blood cells, bacteria, and many large proteins. The placenta also prevents the mixing of fetal and maternal blood until delivery, when ruptures in blood vessels may occur.

Nutrient Transfer   The placenta uses 30–40% of the glucose delivered by the maternal circulation. If nutrient supply is low, the placenta fulfills its needs before nutrients are made available to the fetus. If nutrient supplies fall short of meeting placental needs, functioning of the placenta is compromised to sustain the nutrient supply and health of the mother.[25]

Nutrient transfer across the placenta depends on a number of factors, including:

- The size and the charge of molecules available for transport
- Lipid solubility of the particles being transported
- The concentration of nutrients in maternal and fetal blood

Small molecules with little or no charge (water, for example) and lipids (cholesterol and ketones, for instance) pass through the placenta most easily, while large molecules (e.g., insulin and enzymes) aren't transferred at all. Nutrient exchange between the mother and fetus is unregulated for some nutrients, oxygen, and carbon dioxide; it is highly regulated for other nutrients. Nutrient transfer based on concentration gradients determined by the levels of the nutrient in the maternal and the fetal blood is unregulated. In these cases, nutrients cross placenta membranes by simple diffusion from blood with high concentration of the nutrient to blood with lower concentration.

Three primary mechanisms regulate nutrient transfer: facilitated diffusion, active transport, and endocytosis (or pinocytosis). Table 4.11 summarizes mechanisms of

Table 4.11 Mechanisms of nutrient transport across the placenta[13,26]

Mechanism	Examples of Nutrients
Passive diffusion (also called *simple diffusion*) Nutrients transferred from blood with higher concentration levels to blood with lower concentration levels	Water, some amino acids and glucose, free fatty acids, ketones, vitamins E and K,[a] some minerals (sodium, chloride), gases
Facilitated diffusion Receptors ("carriers") on cell membranes increase the rate of nutrient transfer	Some glucose, iron, vitamins A and D
Active transport Energy (from ATP) and cell membrane receptors	Water-soluble vitamins, some minerals (calcium, zinc, iron, potassium) and amino acids
Endocytosis (also called *pinocytosis*) Nutrients and other molecules are engulfed by placenta membrane and released into fetal blood supply	Immunoglobulins, albumin

[a]Vitamin K crosses the placenta slowly and to a limited degree.

nutrient transfer across the placenta and provides examples of nutrients transported by each specific mechanism as they are known.

The fetus receives small amounts of water and other nutrients from ingestion of *amniotic fluid*. By the second half of pregnancy, the fetus is able to swallow and absorb water, minerals, nitrogenous waste products, and other substances in amniotic fluid.[27]

The Fetus Is Not a Parasite   The fetus is not a "parasite"—it cannot take whatever nutrients it needs from the mother's body at the mother's expense. When maternal nutrient intakes fall below optimum levels or adjustment thresholds, fetal growth and development are compromised more than maternal health.[12] In general, nutrients will first be used to support maternal nutrient needs for her health and physiological changes, and next for placental development, before they become available at optimal levels to the fetus. For example:

- Underweight women gaining the same amount of weight as normal-weight women tend to deliver smaller infants and to retain more of the weight gained during pregnancy at the expense of fetal growth.[13]
- Fetal growth tends to be reduced in pregnant teenagers who gain height during pregnancy compared to fetal growth in teens who do not grow during pregnancy.[28]
- Vitamin and mineral deficiencies and toxicities in newborns have been observed in women who showed no signs of deficiency or toxicity diseases during pregnancy.[13]

If the fetus did act as a parasite, it would harm the mother for its own benefit. Rather, the fetus is generally harmed more by poor maternal nutritional status than is the mother.[26]

Embryonic and Fetal Growth and Development

The rate of human *growth* and *development* is higher during gestation than at any time thereafter. If the rate of weight gain achieved in the 9 months of gestation continued after delivery, infants would weigh about 160 pounds at their first birthdays and be 20 feet tall by age 20! Table 4.12 provides an overview of embryonic and fetal growth and development during pregnancy.

Critical Periods of Growth and Development

Fetal growth and development proceed along genetically determined pathways in which cells are programmed to multiply, *differentiate*, and establish long-term functional levels during set time intervals. Such time intervals are known as *critical periods*

Amniotic Fluid   The fluid contained in the amniotic sac that surrounds the fetus in the uterus.

Growth   Increase in an organism's size through cell multiplication (hyperplasia) and enlargement of cell size (hypertrophy).

Development   Progression of the physical and mental capabilities of an organism through growth and differentiation of organs and tissues, and integration of functions.

Differentiation   Cellular acquisition of one or more characteristics or functions different from that of the original cells.

Critical Periods   Preprogrammed time periods during embryonic and fetal development when specific cells, organs, and tissues are formed and integrated, or functional levels established. Also called *sensitive periods*.

Table 4.12 Notes on normal embryonic and fetal growth and development[13,29]

Day 1	Conception; one cell called the zygote exists.		Week 9	Embryo now considered a fetus.
Day 2–3	Eight cells have formed (called the morula) and enter the uterine cavity.		Month 3	Weighs 1 oz; primitive egg and sperm cells developed, hard palate fuses, breathes in amniotic fluid.
Day 6–8	The morula becomes fluid-filled and is renamed the blastocyst. The blastocyst is comprised of 250 cells, and cell differentiation begins.		Month 4	Weighs about 6 oz; placenta diameter is 3 inches.
Day 10	Embryo implants into the uterine wall, where glycogen is accumulating.		Month 5	Weighs about 1 lb, 11 inches long; skeleton begins to calcify, hair grows.
Day 12	Embryo is composed of thousands of cells; differentiation well under way. Utero placental circulation being formed.		Month 6	14 inches long; fat accumulation begins, permanent teeth buds form; lungs, gastrointestinal tract, and kidneys formed but are not fully functional.
Week 4 (21–28 days)	¼ inch long; rudimentary head, trunk, arms; heart "practices" beating; spinal cord and two major brain lobes present.			
Week 5 (28–35 days)	Rudimentary kidney, liver, circulatory system, eyes, ears, mouth, hands, arms, and gastrointestinal tract; heart beats 65 times per minute, circulating its own newly formed blood.		Month 7	Gains ½–1 oz per day.
			Months 8 and 9	Gains about 1 oz per day; stores fat, glycogen, iron, folate, B_6 and B_{12}, riboflavin, calcium, magnesium, vitamins A, E, D; functions of organs continue to develop. Growth rate declines near term. Placenta weighs 500 – 650 g (1–1½ lb) at term.
Week 7 (49–56 days)	½ inch long, weighs 2–3 g (less than a teaspoon of sugar); brain sends impulses, gastrointestinal tract produces enzymes, kidney eliminates some waste products, liver produces red blood cells, muscles work. (Approximately 25% of blastocysts and embryos will be lost before 7 weeks.)			

and are most intense during the first 2 months after conception, when a majority of organs and tissues form. On the whole, critical periods represent a "one-way street," because it is not possible to reverse directions and correct errors in growth or development that occurred during a previous critical period. Consequently, adverse effects of nutritional and other insults occurring during critical periods of growth and development persist throughout life.[30]

Hyperplasia Critical periods of growth and development are characterized by hyperplasia, or an increase in cell multiplication. Because every human cell has a specific amount of DNA, periods of hyperplasia can be determined by noting times during gestation when the DNA content of specific organs and tissues increases sharply. The critical period of rapid cell multiplication of the forebrain, for example, is between 10 and 20 weeks of gestation (Illustration 4.7).

The brain is the first organ that develops in humans, and along with the rest of the central nervous system, it is given priority access to energy, nutrient, and oxygen supplies. Thus, in conditions of low energy, nutrient, and oxygen availability, the needs of the central nervous system will be met before those of other fetal tissues such as the liver or muscles. The heart and adrenal glands come next after the central nervous system in the hierarchy of targets for preferential nutrient delivery.[25]

Deficits or excesses in nutrients supplied to the embryo and fetus during critical periods of cell multiplication can produce lifelong defects in organ and tissue structure and function. The organ or tissue undergoing critical periods of growth at the time of the adverse exposure will be affected most.[15] For example, the neural tube develops into the brain and spinal cord during weeks 3 and 4 after conception. If folate supplies are inadequate during this critical period of growth, permanent defects in brain or spinal cord formation occur, regardless of folate availability at other times. Other tissues—such as the pancreas, which does not undergo rapid cell multiplication until the third trimester of pregnancy—do not appear to be affected by the early shortage of folate.

Illustration 4.7 The critical period of cell multiplication of the forebrain. Growth in cell numbers is indicated by increases in DNA content of a given amount of tissue.

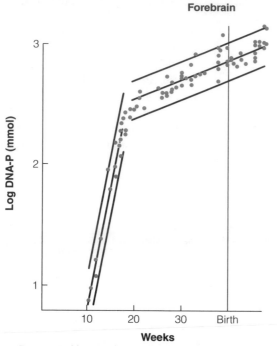

Forebrain

SOURCE: From J. Dobbing and J. Sands, "Quantitative Growth and Development of Human Brain," in *Archives of Disease of Children*, 48(10):757–767. © 1973 BMJ Publishing Group. Reprinted with permission.

Some degree of hyperplasia takes place in a number of organs and tissues in the first year or two after birth and during the adolescent growth spurt. Cells of the central nervous system, for instance, continue to multiply for about two years after birth, but at a much slower pace than early in pregnancy. Skeletal and muscle cells increase in number during the adolescent growth spurt.[31]

In utero and early life changes in DNA content of the brain have been investigated in fetuses, infants, and young children dying from non-nutritional causes and from undernutrition. Illustration 4.8 presents results from one such study that show deficits in DNA content (or cell number) in the brains of children dying of protein-energy malnutrition versus those dying from accidents. Deficits in DNA were apparent a few months after birth, indicating that severe malnutrition early in pregnancy reduced brain cell number *in utero*.[32]

Hyperplasia and Hypertrophy   Cell multiplication continues at a lower rate after critical periods of cell multiplication and is accompanied by increases in the size of cells. This phase of growth can be seen in Illustration 4.7, where it begins around 20 weeks in the forebrain when the rate of increase in DNA content slows. Cell size increases mainly due to an accumulation of protein and lipids inside of cells. Consequently, increases in cell size can be determined by measuring the protein or lipid content of cells. Specialized functions of cells, such as production of digestive enzymes by

Illustration 4.8 DNA content of the cerebellum of the human brain in young children dying from non-nutritional causes and from undernutrition.

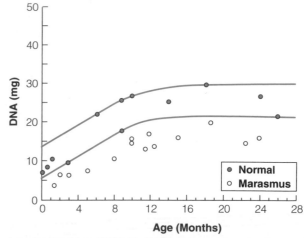

SOURCE: Reprinted from M. Winick, "Malnutrition and Brain Development," *Journal of Pediatrics* 74(6):667–679, © 1969, with permission from Elsevier.

cells within the small intestine or neurotransmitters by nerve cells, occur along with increases in cell number and size.[13]

Hypertrophy   Periods of hyperplasia-hypertrophy are followed by hypertrophy only. During this phase, cells continue to accumulate protein and lipids, and functional levels continue to grow in sophistication, but cells no longer multiply. Reductions in cell size caused by unfavorable nutrient environments or other conditions are associated with deficits in organ and tissue functions, such as reduced mental capabilities or declines in muscular coordination. Such functional changes can often be reduced or reversed later if deficits are corrected.[30]

Maturation   The last phase of growth and development is maturation—the stabilization of cell number and size. This phase occurs after tissues and organs are fully developed later in life.

Fetal Body Composition

The fetus undergoes marked changes in body composition during pregnancy (Table 4.13). The general trend is toward progressive increases in fat, protein, and mineral content. Some of the most drastic changes take place in the last 5 weeks of pregnancy, when fat and mineral content increase substantially.

Variation in Fetal Growth

Given a healthy mother and fetal access to needed amounts of energy, nutrients, and oxygen and freedom from toxins, fetal genetic growth potential is achieved.[33] However, as evidenced by the relatively high rate of low birth weight in the United States, optimal conditions required for achievement of genetic growth potential often do not exist during pregnancy. Variations in fetal growth and development

Table 4.13 Estimated changes in body composition of the fetus by time in pregnancy[13,36]

Component	10 Weeks	20 Weeks	30 Weeks	40 Weeks
Body weight, g	10	300	1667	3450
Water, g	<9	263	1364	700
Protein, g	<1	22	134	446
Fat, g	<1	26	66	525
Sodium, meq	<1	32	136	243
Potassium, meq	<1	12	75	170
Calcium, g	<1	1	10	28
Magnesium, mg	<1	5	31	76
Iron, mg	<1	17	104	278
Zinc, mg	<1	6	26	53

dSGA. If weight, length, and head circumference are less than the 10th percentile for gestational age, then the newborn is considered pSGA. Approximately two-thirds of SGA newborns in the United States are disproportionately small, and one-third are proportionately small.[33] Illustration 4.9 provides photos of newborns with differing sizes for gestational age.

dSGA Infants who are disproportionately small for gestational age look skinny, wasted, and wrinkly. They tend to have small abdominal circumferences, reflecting a lack of glycogen

are not generally due to genetic causes but rather to environmental factors such as energy, nutrient, and oxygen availability, and to conditions that interfere with genetically programmed growth and development. Insulin-like growth factor-1 (IGF-1) is the primary growth stimulator of the fetus. It promotes uptake of nutrients by the fetus and inhibits fetal tissue breakdown. Levels of IGF-1 are sensitive to maternal nutrition; its levels are decreased by undernutrition. Low levels of IGF-1 decrease muscle and skeletal mass and produce asymmetrical growth.[25] Factors such as prepregnancy underweight and shortness, low weight gain during pregnancy, poor dietary intakes, smoking, drug abuse, and certain clinical complications of pregnancy are associated with reduced fetal growth.[34]

Risk of illness and death varies substantially with size at birth and is particularly high for newborns experiencing intrauterine growth retardation (IUGR).[35] For a portion of newborns, smallness at birth is normal and may reflect familial genetic traits. Because IUGR is complicated to determine, it is usually approximated by assessment of size for gestational age using a reference standard (Table 4.14). Infants are generally considered likely to have experienced intrauterine growth retardation if their weight for gestational age or length is low. Newborns whose weight is less than the 10th percentile for gestational age are considered *small for gestational age,* or *SGA*. This determination is further categorized into *disproportionately small for gestational age (dSGA)* and *proportionately small for gestational age (pSGA)*. Newborns who weigh less than the 10th percentile of weight for gestational age but have normal length and head circumference for age are considered

Small for Gestational Age (SGA) Newborn weight is ≤10th percentile for gestational age. Also called *small for date (SFD)*.

Disproportionately Small for Gestational Age (dSGA) Newborn weight is ≤10th percentile of weight for gestational age; length and head circumference are normal. Also called *asymmetrical SGA*.

Proportionately Small for Gestational Age (pSGA) Newborn weight, length, and head circumference are ≤10th percentile for gestational age. Also called *symmetrical SGA*.

Illustration 4.9 The newborn on the top is disproportionately small for gestational age, the middle newborn is proportionately small for gestational age, and the newborn on the bottom is large for gestational age.

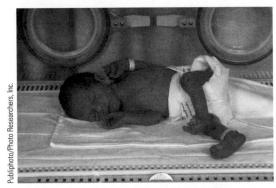

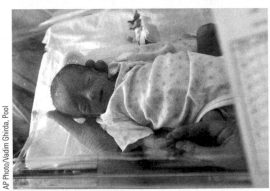

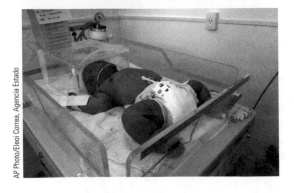

Table 4.14 Percentiles of weight in grams for newborn gestational age

Gestational Age (wk)	5th Pctl	10th Pctl	50th Pctl	90th Pctl	95th Pctl
20	249	275	412	772	912
21	280	314	433	790	957
22	330	376	496	826	1023
23	385	440	582	882	1107
24	435	498	674	977	1223
25	480	558	779	1138	1397
26	529	625	899	1362	1640
27	591	702	1035	1635	1927
28	670	798	1196	1977	2237
29	772	925	1394	2361	2553
30	910	1085	1637	2710	2847
31	1088	1278	1918	2986	3108
32	1294	1495	2203	3200	3338
33	1513	1725	2458	3370	3536
34	1735	1950	2667	3502	3697
35	1950	2159	2831	3596	3812
36	2156	2354	2974	3668	3888
37	2357	2541	3117	3755	3956
38	2543	2714	3263	3867	4027
39	2685	2852	3400	3980	4107
40	2761	2929	3495	4060	4185
41	2777	2948	3527	4094	4217
42	2764	2935	3522	4098	4213
43	2741	2907	3505	4096	4178
44	2724	2885	3491	4096	4122

NOTE: Pctl = percentile
SOURCE: From *Obstetrics and Gynecology*, Vol. 87, No. 2, 1996, pp. 163–168, table 2. Copyright © 1996. Reprinted by permission of Lippincott, Williams & Wilkins.

experienced long-term malnutrition *in utero*, due to factors such as prepregnancy underweight, consistently low rates of maternal weight gain in pregnancy and the corresponding inadequate dietary intake, or chronic exposure to alcohol.[13]

Because nutritional insults existed during critical periods of growth early in pregnancy, pSGA infants generally have a reduced number of cells in organs and tissues. These babies tend to exhibit fewer health problems at birth than do dSGA infants, but catch-up growth is poorer, even with nutritional rehabilitation. On average, pSGA infants remain shorter and lighter and have smaller head circumferences throughout life than do infants born *appropriate for gestational age (AGA)* or *large for gestational age (LGA)*.[37]

The goal of nutritional rehabilitation for pSGA infants should be catch-up in weight and length, and not just weight. This goal appears to be easier to reach if pSGA infants are breastfed. Excessive weight gain by pSGA infants appears to increase the risk of obesity and insulin-resistance-related disorders, such as hypertension and type 2 diabetes, later in life.[38]

LGA Newborns with weights greater than the 90th percentile for gestational age are considered to be large for gestational age. About 1–2% of U.S. newborns are LGA. Although it is difficult to predict LGA, it appears to be related to prepregnancy obesity, poorly controlled diabetes in pregnancy, excessive weight gain in pregnancy (over 44 pounds), and possibly other factors.

Except for infants born to women with poorly controlled diabetes during pregnancy or other health problems, LGA newborns experience far lower illness and death rates than do SGA infants, and they tend to be taller later in life.[39] Delivery and postpartum complications in mothers, however, tend to be higher with LGA newborns, and these include increased rates of operative delivery, *shoulder dystocia,* and postpartum hemorrhage.

stores in the liver, and little body fat. It appears that these infants have experienced *in utero* malnutrition in the third trimester of pregnancy and that it compromised liver glycogen and fat storage. Short-term episodes of malnutrition, such as maternal weight loss or low weight gain late in pregnancy that compromise energy, nutrient, or oxygen availability appear to be related to dSGA.[33] These infants generally have smaller organ sizes but the normal number of cells in organs and tissues.

Infants who are dSGA are at risk of developing the "hypos" after birth (hypoglycemia, hypocalcemia, hypomagnesiumenia, and hypothermia). If the period of maternal undernutrition was short, dSGA infants tend experience good catch-up growth with nutritional rehabilitation.[25] Unfortunately, disproportionately small infants tend to perform less well academically and are at greater risk than other infants for heart disease, hypertension, and type 2 diabetes in the adult years.[8]

pSGA Proportionately SGA newborns look small but well proportioned. It is believed that these infants have

Appropriate for Gestational Age (AGA) Weight, length, and head circumference are between the 10th and 90th percentiles for gestational age.

Large for Gestational Age (LGA) Weight for gestational age exceeds the 90th percentile for gestational age. Also defined as birth weight greater than 4500 g (≥10 lb) and referred to as *excessively sized for gestational age,* or *macrosomic.*

Shoulder Dystocia Blockage or difficulty of delivery due to obstruction of the birth canal by the infant's shoulders.

Nutrition, Miscarriages, and Preterm Delivery

Several other pregnancy outcomes are related in part to maternal nutrition. Highlighted here are the roles played by nutrition in miscarriage and preterm delivery.

Miscarriages   Over 30% of implanted embryos are lost by reabsorption into the uterus or expulsion before 20 weeks of conception. Roughly a third of these losses are recognized as a miscarriage. Such early losses of embryos and fetuses are thought to be primarily caused by genetic, uterine, or hormonal abnormalities, reproductive tract infections, or tissue rejection due to immune system disorders.[40]

The presence of nausea and vomiting early in pregnancy is related to a low risk of miscarriage. Nausea and vomiting may occur as a side effect of healthy changes in hormone levels.[2,34] Women who enter pregnancy underweight are at higher risk of miscarriage than are normal and overweight women.[41-43] Elevated blood cholesterol (>230 mg/dL) or triglyceride concentration (>140 mg/dL) and high levels of markers of inflammation in the first half of pregnancy have been linked to a substantial increase in the risk of miscarriage.[44] The use of multivitamins during early pregnancy has been associated with a decreased risk of miscarriage. However, it is not yet clear whether the reduction in risk is due to the vitamins or healthier habits and diets of women who use them.[45]

Preterm Delivery   Infants born preterm are at greater risk than other infants of death, neurological problems reflected later in low IQ scores, congenital malformations, and chronic health problems such as *cerebral palsy*. The risk for these outcomes increases rapidly as gestational age at birth decreases. Infants born very preterm (<34 weeks) commonly have problems related to growth, digestion, respiration, and other conditions due to immaturity.[46] Low stores of fat, essential fatty acids, glycogen, calcium, iron, zinc, and other nutrients in very preterm infants may also interfere with growth and health after delivery.[47] Additionally, breast-milk content of riboflavin and vitamins A, C, and B_{12} may be low in women who have inadequate intake of these vitamins during the third trimester of pregnancy.[48-50]

Underweight women who gain less than the recommended amount of weight during pregnancy are at particularly high risk for preterm delivery.[51] Women entering pregnancy obese are also at increased risk, but to a lesser extent than is the case for underweight women.[52]

Cerebral Palsy   A group of disorders characterized by impaired muscle activity and coordination present at birth or developed during early childhood.

Fetal-Origins Hypothesis   The theory that exposures to adverse nutritional and other conditions during critical or sensitive periods of growth and development can permanently affect body structures and functions. Such changes may predispose individuals to cardiovascular diseases, type 2 diabetes, hypertension, and other disorders later in life. Also called *metabolic programming* and *developmental origins of disease*.

Several studies, but not all, have identified a protective effect of multivitamin supplement use before pregnancy on preterm delivery.[53-55] Additionally, it appears that women who exercise during pregnancy are at lower risk of preterm delivery than are women who do not exercise.[56]

Within the past few years, a number of studies have identified increased levels of cholesterol, triglycerides, or free fatty acids and elevated levels of markers of inflammation and oxidative stress in the first half of pregnancy in women delivering preterm.[44,57,59] Higher than average cholesterol levels have been observed as early as 8 weeks of pregnancy in women delivering preterm as.[44] Elevated levels of these lipids appear to be present early in pregnancy, before major increases in blood lipid levels generally occur. This result is raising the question of whether women with high levels of lipids coming into pregnancy are at increased risk of preterm delivery. Indications of increased inflammatory markers and oxidative stress early in pregnancy suggest that chronic inflammation and oxidative stress may be involved in the development of physiological conditions that favor preterm delivery.[44] Whether diets rich in antioxidant nutrients and measures that reduce lipid levels and oxidative stress decrease the risk for preterm delivery is not currently known.

Although preterm delivery is a major health problem in the United States, its etiology remains unclear, and the search for effective prevention programs continues.[60] A portion of preterm deliveries appears to be related to genital tract infections, insufficient uterine-placental blood flow, placental abruption (bleeding into the uterus), pre-pregnancy underweight, low weight gain in pregnancy, short interpregnancy interval (<6 months), and high levels of psychological or social stress. It is also fairly common in women who have previously delivered preterm.[60] Improvements in prenatal care for women at risk of preterm delivery—such as close supervision of the pregnancy, inclusion of nutritional counseling, encouragement of adequate weight gain in underweight and normal-weight women, and home visits—appear to decrease the risk of preterm delivery somewhat.[61]

The Fetal-Origins Hypothesis of Later Disease Risk

"... Early life events play a powerful role in influencing later susceptibility to certain chronic disease."

—Peter D. Simonetti et al.[62]

In the last decade, thinking about chronic disease risk has changed substantially. In contrast to the earlier idea that disease risk begins during childhood or in the adult years, studies testing the *fetal-origins hypothesis* indicate that risks begin *in utero*. The concept that chronic disease

Table 4.15 Examples of diseases and other conditions in adults related to smallness or thinness at birth[61,63–65]

Allergies	Mood disorders
Autoimmune diseases	Obesity
Bronchitis	Osteoporosis
Cardiovascular disease	Ovarian cancer
Decreased bone mineral content	Polycystic ovary syndrome
	Schizophrenia
Gestational diabetes	Short stature
Hypertension	Stroke
Irritable bowel syndrome	Subfertility in males
Kidney disease	Suicide
Metabolic syndrome	Type 2 diabetes

risk can be established *in utero* is strongly supported by animal studies and by investigations in humans.[9] Much of the evidence that relates *in utero* exposures to later disease in humans comes from studies showing increased risk for high levels of visceral fat, obesity, heart disease, hypertension, type 2 diabetes, gestational diabetes, and chronic bronchitis in small, short, and thin newborns (Table 4.15). Maternal nutrition is hypothesized to play a key role in mechanisms that lead to later disease risk because it is a major factor affecting fetal growth and development.[64] Although smallness and thinness at birth are recognized as risk factors for later disease development, specific aspects of maternal nutrition unrelated to size at birth have also been related to disease development later in life.[61]

Relatively small reductions in weight or disproportions in newborn size have been related to increased later disease risk. Risk of cardiovascular disease (heart disease and stroke), for instance, is associated with birth weights below 7.5 pounds (3360 g)—weights that are often considered "normal." Results of the U.S. Nurses Study, which compared newborn birth weight to risk of cardiovascular disease in adults, are provided in Table 4.16 and illustrate

Table 4.16 Association of birth weight with the risk of cardiovascular disease in the U.S. Nurses Study[66]

Birth Weight	Relative Risk of:	
	Heart Disease	Stroke
<5 lb (2240 g)	1.5	2.3
5–5½ lb (2240–2500 g)	1.3	1.4
5½–7 lb (2500–3136 g)	1.1	1.3
7–8½ lb (3136–3808 g)	1.0	1.0
8½–10 lb (3808–4480 g)	1.0	1.0
>10 lb (>4480 g)	0.7	0.7

this point. Infants at risk for later disease include those born at weights below that genetically programmed, even if birth weights are considered "normal."[66]

Mechanisms Underlying the Fetal-Origins Hypothesis

The process of human growth and development *in utero* and during the first year of life is not inflexible or solely determined by genes. It is also influenced by environmental exposures. This characteristic of very early growth and development has been termed ***developmental plasticity***. Environmental exposures modify development through ***epigenetic*** mechanisms that program metabolic changes in gene activity and not DNA structure. Epigenetic mechanisms influence growth and development by silencing certain genes (or turning them off) and activating (turning on) others. For example, epigenetic mechanisms can develop in response to maternal undernutrition by reducing cell multiplication in the kidneys while sparing brain growth.[62,67] Epigenetic changes may last throughout all cell divisions for the remainder of the cell's life and for multiple generations.[68]

Epigenetic effects on gene function are initiated by adaptive responses the fetus makes to cues from the mother about her health and physical condition. The mother's health and physical condition are determined by factors such as genetic makeup, diet, level of body fat, and illnesses. Biological indicators of these conditions are transferred to the fetus by the mother's blood. The fetus adapts to the environmental cues sent by the mother by modifying the function of certain genes. These modifications can foster fetal survival *in utero* and later in life given the same set of environmental circumstances.[61]

Examples of Developmental Programming Effects Results of studies during human pregnancy, and laboratory studies involving animals, point to ways in which *in utero* energy and nutrient exposures may prompt fetal adaptations that affect gene function programming and later disease risk. Energy availability has been a major focus of many of these studies because of its importance to fetal growth and development. In human studies, energy available for fetal growth and development is often assessed by pregnancy weight change and newborn birth size.

An inadequate availability of glucose during fetal growth and development would hinder central nervous system (CNS) development and threaten fetal survival. Mechanisms are set in place, however, that triage available glucose to the central nervous system.

Developmental Plasticity The concept that development can be modified by particular environmental conditions experienced by a fetus or infant.

Epigenetics (*epi* = over, above) Biological mechanisms that change gene function without changing the structure of DNA. Epigenetic mechanisms are affected by environmental factors.

This change represents an adaptation by the body to how glucose utilization is programmed to operate.

What adaptations are made to ensure the CNS gets priority access to glucose? Animal studies indicate that the expression of genes that produce insulin receptors on muscle cell membranes may be suppressed in response to a low availability of glucose. This increases insulin resistance and decreases uptake of glucose by muscle cells, and reduces their growth. It also increases the availability of glucose for CNS development.

Adaptations that decrease muscle utilization of glucose and reduce muscle size may serve the offspring well later in life if food availability and intake are limited. If food is abundant and food intake is high, however, such adaptations may lead to elevated blood levels of glucose and insulin. These changes may increase the risk of obesity, type 2 diabetes, gestational diabetes, and other disorders associated with insulin resistance.[69]

Increased susceptibility to insulin resistance and weight gain in infants experiencing nutritional insults *in utero* has been attributed to a "thrifty phenotype," or genetic functional types programmed *in utero* that act to conserve energy.[70]

The function of genes involved in cholesterol metabolism appear to be modified in males with birth weights less than 3.2 kg (7 lb). In these individuals, production of "good" cholesterol, HDL, tends to decrease in response to high-fat, high-saturated-fat diets. HDL cholesterol production generally increases in males with higher birth weights in response to this type of diet. High blood levels of HDL-cholesterol are protective against heart disease.[71]

Some studies have shown a link between maternal nutritional exposures during pregnancy and later disease risk in infants with a wide range of birth weights. Low weight gain around mid-pregnancy, for example, has been associated with higher blood pressure in children, and low levels of maternal body fat during pregnancy with increased risk of heart disease in offspring.[72]

Limitations of the Fetal-Origins Hypothesis

The hypothesis that maternal and fetal nutritional exposures influence later disease risk is gaining support and recognition. Many questions are unanswered, however. Which specific nutritional exposures are responsible for changes in gene function and increased disease risk? When do the vulnerable periods of fetal sensitivity to poor nutrition occur? What levels of energy and nutrient availability are related to the optimal functioning of genes? How can we "rescue" or repair detrimental epigenetic changes so that they correspond to a person's actual rather than *in utero* or early-life exposures? Progress is being made toward finding out answers to each of these questions. The implications of the associations between maternal and fetal nutrition and adult disease risk are immense.

Pregnancy Weight Gain

"Any obstetrician who allows a woman to lose her attractiveness (i.e., gain too much weight) is depriving her of many things that make for her mental well-being, her husband's contentment, and her own personal satisfaction."

Loughran, *American Journal of Obstetrics and Gynecology,* 1946

Weight gain during pregnancy is an important consideration because newborn weight and health status tend to increase as weight gain increases. Birth weights of infants born to women with weight gains of 15 pounds (7 kg) for example, average 3100 grams (6 lb 14 oz). This weight is about 500 grams less than the average birth weight of 3600 grams (8 lb) in women gaining 30 pounds (13.6 kg). Rates of low birth weight are higher in women gaining too little weight during pregnancy.[26] Weight gain during pregnancy is an indicator of plasma volume expansion and positive calorie balance, and provides a rough index of dietary adequacy.[73]

Multiple studies show broad agreement on amounts of weight gain that are related to the birth of infants with weights that place them within the lowest category of risk for death or health problems.[34] Yet how much weight should be gained during pregnancy remains a hotly debated topic. Earlier in the last century, when gains were routinely restricted to 15 or 20 pounds, weight gain in pregnancy was seen as the cause of pregnancy hypertension, difficult deliveries, and obesity in women. Pregnant women would be placed on low-calorie diets and given diuretics and amphetamines and urged to use saccharin to limit weight gain.[74]

Although none of these notions have been shown to be true, weight gain during pregnancy still represents a prickly issue. Weight gain and body weight are not only a matter of health, but are also closely linked to some people's view of what is socially acceptable.

Psychological and sociological biases related to body weight and shape in women are an important reason to apply recommendations for weight gain in pregnancy based on scientific studies and consensus.

Pregnancy Weight Gain Recommendations

Current recommendations for weight gain in pregnancy are based primarily on gains associated with the birth of healthy-sized newborns (approximately 3500–4500 g or 7 lb 13 oz to 10 lb).[33] As shown in Illustration 4.10, however, prepregnancy weight status influences the relationship between weight gain and birth weight. The higher the weight before pregnancy, the lower the weight gain needed to produce healthy-sized infants. Recommended weight gains for women of all ethnicities and statures entering pregnancy underweight, normal-weight, overweight, and obese are displayed in Table 4.17.[75] This table also presents the range in weight gain provisionally recommended for twin pregnancy.[75] (Weight gain recommendations for twin pregnancy are presented in more detail in Chapter 5.)

Illustration 4.10 Pregnancy weight gain by prepregnancy weight status and birth weight.

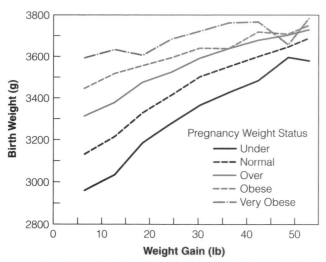

SOURCE: From *Clinical Nutrition*, Vol. 7, Fig. 1, p. 186, J. E. Brown, © 1988. Reprinted by permission from Elsevier.

Because underweight women tend to retain some of the weight gained in pregnancy for their own needs, they need to gain more weight in pregnancy than do other women. Overweight and obese women, on the other hand, are able to use a portion of their energy stores to support fetal growth, so they need to gain less.

Duration of gestation, smoking, maternal health status, *gravida,* and *parity* also influence birth weight. Consequently, gaining a certain amount of weight during pregnancy does not guarantee that newborns will be a healthy size. It does improve the chances that this will happen, however.

Approximately 40% of U.S. women gain within the recommended weight ranges during pregnancy.[76] For all except the obese, women who gain within the recommended ranges are approximately half as likely to deliver low-birth-weight or SGA newborns as are women who gain less. Rates of LGA newborns, Caesarean-section deliveries, and postpartum weight retention tend to be higher when pregnancy weight gain exceeds that recommended.[76] It is suggested that insulin resistance may be related to excessive weight gains during pregnancy and some of the adverse neonatal outcomes.[77]

Restriction of pregnancy weight gain to levels below the recommended ranges is not recommended. It does not decrease the risk of pregnancy-related hypertension and is associated with increased infant death and low birth weight, and poorer offspring growth and development.[20] In addition, low weight gain in pregnancy may increase the risk that infants will develop heart disease, type 2 diabetes, hypertension, and other types of chronic disease later in life.[35]

Gravida Number of pregnancies a woman has experienced.

Parity The number of previous deliveries experienced by a woman; *nulliparous* = no previous deliveries, *primiparous* = one previous delivery, *multiparous* = two or more previous deliveries. Women who have delivered infants are considered to be "parous."

Rate of Pregnancy Weight Gain Rates at which weight is gained during pregnancy appear to be as important to newborn outcomes as is total weight gain. Low rates of gain in the first trimester of pregnancy may down-regulate fetal growth and result in reduced birth weight and thinness.[78] For underweight and normal-weight women, rates of gain of less than 0.5 pound (0.25 kg) per week in the second half of pregnancy, and of less than 0.75 pound (0.37 kg) per week in the third trimester of pregnancy, double the risk of preterm delivery and SGA newborns. For overweight and obese women, rates of gain of less than 0.5 pound (0.25 kg) per week in the third trimester also double the risk of preterm birth.[79] Third-trimester weight gains exceeding approximately 1.5 pounds a week (0.7 kg), however, add little to birth weight in normal-weight and heavier women, and may increase postpartum weight retention.[80]

Rate of weight gain is generally highest around mid-pregnancy—which is prior to the time the fetus gains most of its weight (Illustration 4.11). In general, the pattern of gain should be within a few pounds of that represented by the weight-gain curves shown in Illustration 4.12.[34]

Illustration 4.11 Rates of maternal and fetal weight gain during pregnancy.

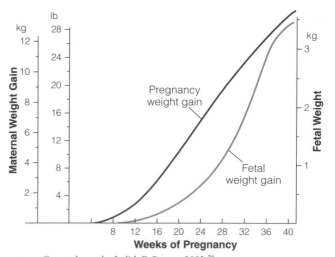

SOURCE: Curves drawn by Judith E. Brown, 2002.[78]

Table 4.17 Pregnancy weight gain recommendations[75]	
Prepregnancy Weight Status Body Mass Index	**Recommended Weight Gain**
Underweight, <18.5 kg/m²	28–40 lb (12.7–18.2 kg)
Normal weight, 18.5–24.9 kg/m²	25–35 lb (11.4–15.9 kg)
Overweight, 25–29.9 kg/m²	15–25 lb (6.8–11.4 kg)
Obese, 30 kg/m² or higher	11–20 lb (5.0–9.1 kg)
Twin pregnancy	25–54 lb (11.4–24.5 kg)

Illustration 4.12 The Institute of Medicine's prenatal weight-gain graph and weight-recording form for women.

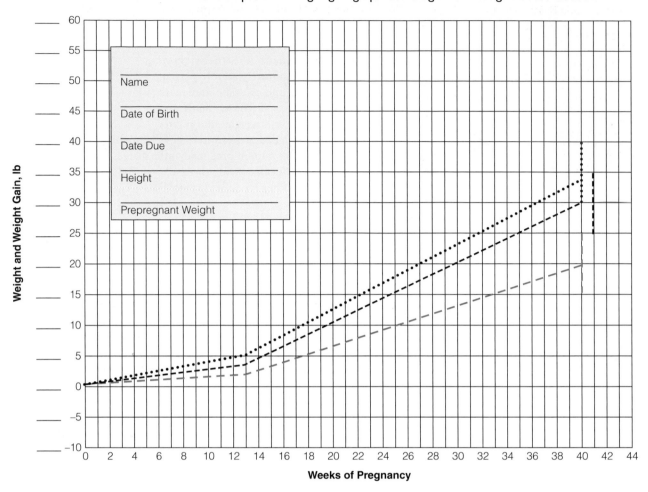

Weight Record			
Date	Weeks of Gestation	Weight	Notes

Please bring this chart with you to each prenatal visit.

SOURCE: Reprinted with permission from Prenatal Weight Gain Chart, *Nutrition During Pregnancy and Lactation: an Implementation Guide*, 1992 by the National Academy of Sciences. Courtesy of the National Academies Press, Washington, D.C.

Table 4.18 Components of weight gain during pregnancy for healthy, normal-weight women delivering a 3500-g (about 8-lb) infant at term[10,12,13,36]

Component	Weight Gain, Grams			
	10 Weeks	20 Weeks	30 Weeks	40 Weeks
Fetus	5	300	1500	3550
Placenta	20	170	430	670
Uterus	140	320	600	1120
Amniotic fluid	30	350	750	896
Breasts	45	180	360	448
Blood supply	100	600	1300	1344
Extracellular fluid	0	265	803	3200
Maternal fat stores	315	2135	3640	3500
Total weight gain at term = 14.7 kg or 32 lb				

Some weight (3 to 5 pounds) should be gained in the first trimester, followed by gradual and consistent gains thereafter. The rate of weight gain often slows a bit a few weeks prior to delivery, but as is the case for the rest of pregnancy, weight should not be lost until after delivery.[78]

Composition of Weight Gain in Pregnancy

A question often asked by pregnant women is, "Where does the weight gain go?" Where the weight gain generally goes by time in pregnancy is shown in Table 4.18. The fetus actually comprises only about a third of the total weight gained during pregnancy in women who enter pregnancy at normal weight or underweight. Most of the rest of the weight is accounted for by the increased weight of maternal tissues.

Body Fat Changes   Pregnant women store a significant amount of body fat in normal pregnancy in order to meet their own and the fetus's energy needs, and quite likely to prepare for the energy demands of lactation. Body fat stores increase the most between 10 and 20 weeks of pregnancy, or before fetal energy requirements are highest. Levels of stored fat tend to decrease before the end of pregnancy. Only 0.5 kg of the approximately 3.5 kg of fat stored during pregnancy is deposited in the fetus.[12,30]

Postpartum Weight Retention

Concern about the role of pregnancy weight gain in fostering long-term maternal obesity has increased in the United States, along with the rising incidence of obesity in adults. Increased weight after pregnancy appears to be related to a variety of factors, including excessively high weight gain in pregnancy (over 45 lb, or 20.5 kg), weight gain after delivery, and low activity levels.[81] High blood levels of insulin early in pregnancy, and levels of leptin, have been related to increased weight gain during pregnancy. Levels of both hormones are related to diet.[82,83]

Women tend to lose about 15 pounds the day of delivery, but subsequent weight loss is highly variable.[81] On average, however, women who gain within the recommended ranges of weight gain are 2.0 pounds (0.9 kg) heavier 1 year after delivery than they were before pregnancy.[84] This gain is slightly above the amount of weight women tend to gain with age.[85] Postpartum weight retention tends to be slightly less in women who breastfeed for at least 6 months after pregnancy.[86] Women who gain less than the recommended amount of weight gain in pregnancy do not retain less weight on average after pregnancy than do women who gain within the ranges.[89] Postpartum weight can be reduced by identifying high weight gainers during pregnancy and getting the women identified involved in an exercise and healthy-eating program.[87]

Nutrition and the Course and Outcome of Pregnancy

> "A mother who wishes her child to have black eyes should frequently eat mice."
>
> Prenatal diet folklore from ancient Rome[88]

The history of beliefs about the effects of maternal diet on the course and outcome of pregnancy is rife with superstition, ill-founded and hazardous conclusions, and unhelpful suggestions. Societies have shared a belief in the importance of "eating right" during pregnancy for the child's sake, but actual knowledge about maternal nutrition and the course and outcome of pregnancy has been acquired only relatively recently.

Famine and Pregnancy Outcome

Much of the scientific interest in the effects of maternal nutrition on the course and outcome of pregnancy comes from studies done in the first half of the twentieth century. Ecological studies on effects of famines in

Europe and Japan during World War II on the course and outcome of pregnancy demonstrated potential negative, as well as positive, effects of food intake on fertility and newborn outcomes.

The Dutch Hunger Winter, 1943–1944

As mentioned briefly in Chapter 2, people in many parts of Holland experienced severe food shortages for an 8-month period during World War II due to enemy occupation of major cities. Although people in Holland were generally well nourished and had a reasonable standard of living before the disaster, conditions rapidly deteriorated during the famine. In addition to intakes that averaged only about 1100 kcal and 34 grams of protein per day, fuel was in low supply and the winter harsh.

Carefully kept records by health officials showed a sharp decline in pregnancy rates of over 50% during the famine, an effect attributed to absent and irregular menstrual periods. Average birth weight declined by 372 grams (13 oz), delivery of low-birth-weight infants increased by 50%, and rates of infant deaths increased. Birth weight did not fully "catch up" in infants born to women exposed to famine early in pregnancy, even if they received enough food later in pregnancy. This result supports the notion that the fetal growth trajectory may be established early in pregnancy and that early nutritional deprivations limit fetal growth regardless of food intake later in pregnancy.[89]

Although the Dutch famine was associated with major declines in fertility and newborn health and survival, the rather good nutritional status of women prior to the famine likely protected pregnant women and their infants from more severe disruptions in health. Normal fertility status and newborn outcomes returned within a year after the famine ended.[90]

Studies undertaken in the last 30 years on adults who were born to women during the hunger winter (the Dutch famine cohort) show relationships between the timing of famine during pregnancy and adult offspring health outcomes. Examples of relationships identified are shown in Table 4.19.

The Seige of Leningrad, 1942

Unlike people in Holland, the population in Leningrad (now called St. Petersburg) had experienced moderate deprivations in nutritional status and quality of life prior to the famine. As was the case for pregnant women in Holland, the famine in Leningrad resulted in average intakes of approximately 1100 kcal per day. Infertility and low-birth-weight rates increased over 50%, infant death rates rose, and birth weights dropped by an average of 535 grams (1.2 lb) during the famine.[93] Rates of pSGA newborns also increased, suggesting that the poor nutritional

status of women coming into pregnancy and persistent undernutrition during pregnancy interfered with critical periods of fetal growth.

Food Shortages in Japan

Effects of World War II–associated food shortages on reproductive outcomes in Japan were similar to those observed in Holland. Japanese women tended to be well nourished prior to the shortages. Lack of food before and during pregnancy was reflected in decreased fertility status among women and in reductions in birth weight that averaged 200 grams.

Social and economic improvements occurring in Japan after the war led to increased availability of many foods, including animal products. This higher plane of nutrition achieved during the postwar years in Japan was accompanied by major increases in newborn size and the "growing up" of Japanese children. In a trend that continues today, subsequent generations of Japanese adults averaged 2 inches taller than the previous generation.[94] Infant mortality in Japan, which ranked among the highest for industrialized nations prior to World War II, declined incredibly after the war and remains well below rates in the United States and in a number of other developed countries.[95]

Food shortages continue to occur in various parts of the world and to adversely affect fertility and the course and outcome of pregnancy. Effects have become predictable, such that declines in fertility and newborn size and vitality are viewed as part of the consequences of such disasters. For example, the siege of Sarajevo, which decreased food availability during 1993–1994, led to reduced caloric and nutrient intakes during pregnancy, reduced maternal weight gain and newborn weights, and increased rates of perinatal mortality and congenital anomalies.[96] Birth weight did not fully catch up in infants born to women exposed to famine early in pregnancy, even if they received enough food later in pregnancy. This result supports the notion that the fetal growth trajectory may be established early in pregnancy and that early nutritional

Table 4.19 Exposure to the Dutch World War II famine by time in pregnancy and adult offspring health risks[91,92]

| | Period of Famine | |
First Trimester	First and/or Second Trimester	Second Half of Pregnancy
Schizophrenia	Antisocial personality disorder	Decreased glucose tolerance
High LDL and low HDL cholesterol		
High body weight and central body fat		
Infertility		
Neural tube defects		

deprivations limit fetal growth regardless of food intake later in pregnancy.[89]

Contemporary Prenatal Nutrition Research Results

"Good nutritional status maintained before and throughout pregnancy decreases the risk of birth defects, suboptimal fetal growth and development, and chronic health problems later in life."

Position of the American Dietetic Association[97]

Nutrient Needs During Pregnancy

Nutrient requirements during pregnancy are not static. They vary during the course of pregnancy depending on prepregnancy nutrient stores, body size and composition, physical activity levels, stage of pregnancy, and health status. For the most part, nutrient needs can be and are optimally met by consuming well balanced, adequate, and healthful diets consisting of basic foods. Healthful diets established during pregnancy can last well beyond pregnancy and benefit health for life.

Carefully conducted studies of diet and pregnancy outcome in the first half of the twentieth century began the era of scientifically based recommendations on nutrition and pregnancy. The now-classic studies conducted by Bertha Burke at Harvard in the 1940s were particularly influential.[98] These studies showed that diet quality during pregnancy, assessed using diet histories, was strongly related to newborn health status. Newborns assessed as having optimal physical condition by pediatricians were found to be much more common among women consuming high-quality diets, whereas those with the poorest physical condition were born to women with the poorest-quality diets. Average birth weight of newborns assessed as being in optimal physical condition was 7 pounds, 15 ounces in females, and 8 pounds, 8 ounces in males.[99] Although Burke's studies did not show that high-quality pregnancy diets by themselves were responsible for robust newborn health, they provided some of the first evidence that prenatal diet quality may strongly influence pregnancy outcome.

Thousands of other studies on the effects of nutrition on the course and outcome of pregnancy are now available. The following sections highlight research results and recommendations related to calories, key nutrients, and other substances in food that influence the course and outcome of pregnancy.

The Need for Energy

Energy requirements increase during pregnancy, mainly due to increased maternal body mass and fetal growth. The additional requirements can be allocated to different

Illustration 4.13 Components of increased oxygen consumption in normal pregnancy.

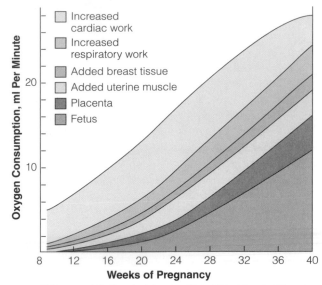

SOURCE: F. Hytten and G. Chamberlain, eds., *Clinical Physiology in Obstetrics.* Reprinted by permission of Blackwell Science Ltd.

maternal and fetal tissues by estimating the amount of oxygen used (or "consumed") by the various tissues. Illustration 4.13 shows the results of work on oxygen consumption during pregnancy undertaken by Hytten and Chamberlain. Approximately one-third of the increased calorie need in pregnancy is related to increased work of the heart, and another third to increased energy needs for respiration and accretion of breast tissue, uterine muscles, and the placenta. The fetus accounts for about a third of the increased energy needs of pregnancy.[10]

The increased need for energy in pregnancy averages 300 kcal a day, or a total of 80,000 kcal.[10] The DRIs for energy intake for pregnancy are +340 kcal per day for the second trimester and +452 kcal per day for the third trimester of pregnancy. Caloric intake recommendations represent a rough estimate that by no means applies to every woman.[100]

Additional energy requirements of women have been found by different studies to range from 210–570 kcal a day.[12] The need for additional calories during pregnancy may be a good deal lower in women who perform little exercise, and higher in women who are very active. Low levels of energy expenditure from physical activity are common in the first trimester of pregnancy, and the energy savings may produce a positive caloric balance even though a woman's caloric intake hasn't changed much. Contrary to a previous belief, energy needs of pregnant women do not appear to be affected by "metabolic efficiencies" of pregnancy that decrease caloric need.[101]

Illustration 4.14 shows the difference between caloric (kcal) intake and estimated caloric balance throughout pregnancy in a group of women served by a health

Illustration 4.14 Estimated caloric balance in pregnancy through 6–8 weeks postpartum.

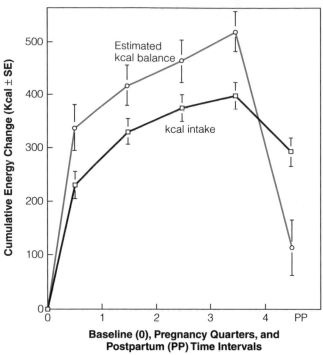

Estimated kcal balance

kcal intake

Cumulative Energy Change (Kcal ± SE)

Baseline (0), Pregnancy Quarters, and Postpartum (PP) Time Intervals

SOURCE: From *Clinical Perinatology*, 24(2):433–449, by J. E. Brown and E. S. B. Khan, © 1997. Reprinted by permission of W. B. Saunders Co.

maintenance organization.[101] The graph indicates that estimated caloric balance is higher than caloric intake throughout pregnancy and becomes negative postpartum. The positive caloric balance observed during pregnancy is due to the fact that women consumed more calories than they expended in physical activity and basal metabolism.

Adequacy of calorie intake is most easily assessed in practice by pregnancy weight gain. Rates of gain in women who do not have noticeable edema are a good indicator of caloric balance.

The Need for Carbohydrates

Approximately 50–60% of total caloric intake during pregnancy should come from carbohydrates. Women should consume a minimum of 175 grams carbohydrates to meet the fetal brain's need for glucose. On average, women in the United States consume 53% of calories (269 g) from carbohydrates during pregnancy.[102] Basic foods such as vegetables, fruits, and whole-grain products containing fiber and a variety of other nutrients are good choices for high-carbohydrate foods. These foods provide beneficial phytochemicals, such as plant antioxidants, and protection against constipation.[103] In addition, sources of carbohydrates that do not contain added sugars and fat tend to be less energy-dense than foods that do and may help women manage pregnancy weight gain.[104]

Artificial Sweeteners There is no evidence that consumption of aspartame (Nutrasweet) or acesulfame K (Sunette) is harmful in pregnancy.[105] Diet soft drinks and other artificially sweetened beverages and foods are often poor sources of nutrients, however, and may displace other, more nutrient-dense foods in the diet.

Alcohol and Pregnancy Outcome

Far fewer pregnant women consume alcohol than do non-pregnant women. Approximately 12.2% of pregnant women in the United States consume an alcohol-containing beverage once in a month during pregnancy, whereas about 54% of women who are not pregnant drink that amount.[60] Most women appear to be aware of the message that not drinking during pregnancy is good for the developing fetus.

Alcohol ingested by a pregnant woman readily passes through the placenta to the fetus where it can interrupt normal growth and development. Adverse effects of high amounts of alcohol intake (such as several drinks per day or more) are strongly related to abnormal mental development and growth in the offspring, and the deficits are lifelong. Adverse effects of alcohol intake during pregnancy are mild or undetectable when intakes are low or when alcohol intake exists but is infrequent.

There is no clearly defined safe level of alcohol intake during pregnancy; therefore it is strongly advised that women who are pregnant do not drink. It is further recommended that women who *may* become pregnant not drink alcohol. *In utero* alcohol exposure during the first, critical months of pregnancy may impair organ development.[106]

Frequent consumption of high amounts of alcohol from early pregnancy onward is related to the development of fetal alcohol syndrome. This topic is addressed in the next chapter.

The Need for Protein

The recommended protein intake for pregnancy is +25 grams per day, or 71 grams daily, and as 1.1 gram/kg body weight, for females aged 14 and older. On average, pregnant women in the United States consume 78 grams of protein daily.[102] Physiological adaptations in protein metabolism during pregnancy shift in the direction of meeting maternal and fetal needs for protein. Consequently, less protein is used for energy and more is used for protein synthesis.[107]

Protein requirements increase during pregnancy primarily due to protein tissue accretion. Of the approximately 925 grams of protein (2 pounds) accumulated in protein tissues during pregnancy, 440 grams are taken up by the fetus, 216 grams are used for increases in maternal blood and extracellular fluid volume, 166 grams are consumed by the uterus, and 100 grams are accumulated by the placenta. Additional protein is also required to maintain the protein tissue developed.[102] Protein supplements

Table 4.20 Tool for estimating protein intake

Food	Protein, Grams	How much protein is there in this usual day's diet?	
Milk, 1 c	8	2 slices toast	6
Cheese, 1 oz	7	1 c milk	8
Egg, 1	7	3 oz tuna	21
Meat, 1 oz	7	2 sl bread	6
Dried beans, 1 c	13	2 oz chicken	14
Bread, 1 slice or oz	3	1 oz cheese	7
		2 tortillas	6
		½ c refried beans	7
		Total g protein	= 75

Table 4.21 Vegetarian food guide adapted for pregnant women[110–112]

Food Group	Servings per Day
A. Grains	
Whole-grain bread, 1 slice	6–11
Cooked grains, ½ c	
Fortified cold cereal, 1 oz	
Fortified cooked cereal, ½ c	
Corn, ½ c	
Pasta, ½ c	
Tortilla, 1 small	
Crackers, 4 small	
B. Legumes, Nuts, Seeds, Dairy	5–7
Dried beans, cooked, ½ c	
Peas, ½ c	
Soy products, ½ c or 2–3 oz	
Soynuts, ¼ c	
Nut and seed butter, 2 Tbsp	
Nuts and seeds, ¼ c	
Eggs, 1	
Cow's milk, 1 c	
Cheese, 1 oz	
Yogurt, ½ c	
Fortified soymilk, 1 c	
C. Vegetables	4
Cooked vegetables, ½ c	
Raw vegetables, 1 c	
Vegetable juice, ½ c	
D. Fruits	2
Medium-sized fruit, 1	
Cut-up raw or cooked, ½ c	
Fruit juice, ½ c	
Dried fruit, ¼ c	
E. Fats, Oils, and Sweets	2+ depending on caloric need
Mayonaise, oil, margarine, 1 Tbsp	
Honey, syrup, jams, jellies, sugar, 1 Tbsp	

do not benefit the course or outcome of pregnancy in well-nourished women.[108]

Protein content of nonvegetarian diets can be simply estimated by evaluating women's usual daily intake of major sources of protein. A tool for estimating protein intake is shown in Table 4.20.

Vegetarian Diets in Pregnancy

"The topic of vegetarian dietary practices often brings with it a variety of images and attitudes regarding those who follow such practices. Those attitudes may have limited, if any, basis in actual fact."

Patricia Johnston, 1988[109]

Nutrient needs in pregnancy may be met by many different types of diets, including those that omit animal products.[97] It is the type and amount of food consumed, not the label placed on it, that determines the appropriateness for dietary intake during pregnancy.

A food guide for pregnant women who exclude animal products from their diet can be found in Table 4.21. Diets of pregnant vegetarians are sometimes low in vitamins B_{12} and D, calcium, iron, zinc, and omega-3 fatty acids eicosapentaenoic and docosahexaenoic acids due to the lack of consumption of rich food sources of these nutrients. Vitamin B_{12} deficiency during pregnancy may not become apparent until after delivery. Two cases of neurological impairment and growth failure due to maternal B_{12} deficiency were identified in 4- to 8-month-old infants in Georgia in 2001. Both infants were born to women who followed a vegetarian diet during pregnancy.[113]

Protein intake is generally adequate in the vegetarian diet, but it may be low in vegans. Protein needs are met by vegetarians who regularly consume a variety of plant sources of protein and meet energy needs. In pregnant women who consume no animal products, the variety of plant protein sources needs to include complementary sources of protein daily. Protein sources that complement each other, or provide a complete source of protein,

include legumes (such as lentils, chickpeas, black-eyed peas, black beans, and lima beans) and grains (corn, rice, bulgur, and barley, for example). Protein requirements in vegetarians whose main source of protein is cereals and legumes may be 30% higher than for non-vegetarians due to the low digestibility of protein in these foods.[110]

Availability of vegetarian food products in large grocery and organic-food stores has expanded substantially in the past few years. Vegetarians can now select veggie burgers, meat analog entrees, meals-in-a-cup, and frozen desserts from food-store shelves. Fortified juice, soymilks, breakfast cereal, and meat substitutes are available and can contribute substantially to vegetarians' intake of vitamins B_{12} and D and calcium. DHA derived from algae can be used to provide a source of this omega-3 fatty acid

Case Study 4.1

chris stock photography/Alamy

Vegan Diet During Pregnancy

Ms. Lederman, a healthy 32-year-old woman entering her thirteenth week of pregnancy, asks her doctor for a referral to a dietitian to discuss her vegan diet. She receives the referral, and while making an appointment with the nutrition consulting service, she is asked to record her food intake for 3 days prior to the appointment. Ms. L follows the instructions she was given and carefully completes a 3-day food record. Prior to the appointment, she sends her food record to the dietitian she will be seeing.

During the appointment, the dietitian learns that Ms. L started pregnancy at normal weight, has gained 3 pounds so far in pregnancy, has no history of iron or another nutrient deficiency, and is experiencing a normal course of pregnancy. Ms. L has been a vegan since the age of 16, and although she believes it is good for her health, she worries that her baby may not be getting the nutrients she or he needs. Ms. L wears sunscreen whenever she goes outside, so she makes little or no vitamin D in her skin. She makes sure to combine plant sources of protein (usually dried beans and grains), so she'll consume complete sources of protein every day.

Results of the dietary analysis performed by the dietitian showed the following average calorie and nutrient intake levels:

Kcal: 2237
Protein, g: 71
Linoleic acid (n-6 fatty acids), g: 15.2
Alpha-linolenic acid (n-3 fatty acids), g: 0.54
Vitamin B_{12}, mcg: 2.1
Vitamin D, mcg: 3 (120 IU)
Zinc, mg: 15

Questions

1. Is Ms. L consuming enough protein?
2. Based on the information presented, which nutrients are consumed in amounts that are below the DRI standard for pregnancy?
3. Suggest three types of food Ms. L could consume to bring up her intake of the nutrients identified in question 2.

(Answers are located in the Instructor's Manual for the 4th edition of *Nutrition Through the Life Cycle*.)

in diets of vegetarian pregnant women who do not consume fish or seafood.[110] (Additional information on the omega-3 fatty acids follows.)

Computerized nutrient analysis of several days of usual food intake may be especially helpful in vegetarian diets due to the variability of dietary practices. If indicated by the results of a nutrition assessment, vegetarian pregnant women should be counseled about dietary modifications to meet individual needs and provided with information about vegetarian food sources of specific nutrients. Evaluation of rate of weight gain in pregnancy is generally a good way to assess the adequacy of energy intake. Case Study 4.1 is related to the dietary assessment results of a pregnant vegan woman.

The Need for Fat

It is estimated that pregnant women consume, on average, 33% of total calories from fat.[114] Fat consumed in foods is used as an energy source for fetal growth and development and serves as a source of fat-soluble vitamins. Fat also provides essential fatty acids that are specifically required for components of fetal growth and development. It is recommended that pregnant women consume 13 grams of the essential fatty acid linoleic acid daily, and 1.4 grams of the other essential fatty acid, alpha-linolenic acid. Diets in the United States tend to provide sufficient amounts of linoleic acid but too little alpha-linolenic acid and other fatty acids related to it.[114–115]

Illustration 4.15 The structure of the omega-3 fatty acid DHA, showing the "alpha" end on the left and the "omega" end on the right.

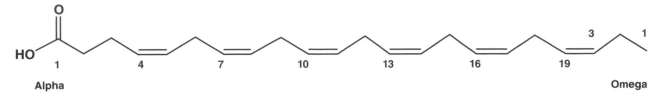

Alpha   **Omega**

Rich food sources of linoleic acid include safflower, corn, sunflower, and soy oil. Alpha-linolenic acid is found in good quantities in flaxseed, walnut, soybean, and canola oils, and leafy green vegetables.

Linoleic acid is the primary fatty acid of the n-6, or the omega-6, fatty acid family, and alpha-linolenic acid is the major n-3, or omega-3, fatty acid. The term "omega-6" or "omega-3" is assigned to these fatty acids based on the location of the first double bond from the end of the carbon chain of the fatty acid (Illustration 4.15). Linoleic acid and alpha-linolenic acid are considered to be long-chained, polyunsaturated fatty acids (LCPUFA) and are sometimes referred to as such.

Linoleic and alpha-linolenic acids serve as structural components of cell membranes. The brain, retina, and other neural tissues of the fetus are particularly rich in these fatty acids.[116] Derivatives of linoleic acid and alpha-linolenic acid serve as precursors for *eicosanoids* that regulate numerous cell and organ functions.[117] Two members of the alpha-linolenic acid family of fatty acids, eicosapentaenoic acid (EPA) and docosahexaenoic acid (DHA), play particularly important roles in pregnancy.

EPA and DHA are highly unsaturated molecules: EPA contains 5 double bonds between carbons, and DHA has 6. The double bonds tend to break down upon exposure to light, heat, or oxygen in the air. They are the components of fish that become oxidized with time and release a "fishy" odor.[118]

Omega-3 Fatty Acids EPA and DHA During Pregnancy

EPA and DHA can be derived from food sources of alpha-linolenic acid, but only in limited quantities. In pregnant women, 9% of alpha-linolenic acid is converted to EPA and DHA.[119] Even relatively high intakes of alpha-linolenic acid during pregnancy fail to increase maternal blood content of EPA and DHA.[120] Consequently, adequate intake of these two omega-3 fatty acids depends on the consumption of food sources of EPA and DHA, or the use of supplements. Fish and seafood are by far the richest food sources of EPA and DHA. (Food sources of EPA+DHA are listed in Table 1.7 in Chapter 1.) Intakes of 500 mg to 3 grams per day of EPA+DHA do not appear to be related to excessive bleeding.[121,122] Intakes of 500 mg per day are considered safe for consumption by healthy women in pregnancy.[121]

EPA and DHA are selectively transported through the placenta to the fetus and, given adequate maternal intake of these fatty acids, concentrations of EPA and DHA become higher in fetal blood than in maternal blood in the third trimester of pregnancy. Maternal stores of EPA and DHA may become depleted during pregnancy due to their increased use by the developing fetus. Preterm infants may be born with low stores of EPA and DHA.[122]

Eicosanoid derivatives of EPA reduce inflammation, dilate blood vessels, and reduce blood clotting. DHA is a major structural component of phospholipids in cell membranes in the central nervous system, including retinal photoreceptors. High amounts of DHA are also found in sperm. Optimal functioning of the central nervous system appears to depend on the availability of sufficient amounts of DHA during critical phases of growth and development when central nervous system tissues are being formed.[123]

Women who consume adequate amounts of EPA and DHA during pregnancy and lactation tend to deliver infants with somewhat higher levels of intelligence, better vision, and otherwise more mature central nervous system functioning than do women who consume low amounts of these fatty acids. Sufficient intake of EPA and DHA during pregnancy has been found to prolong gestation by an average of 4 days, and to decrease the risk of preterm delivery.[124–126]

> **Eicosanoids** Molecules synthesized from essential fatty acids. They exert complex control over many bodily systems, mainly in inflammation and immunity, and act as messengers in the central nervous system.

Dietary Intake Recommendations for EPA and DHA An adequate intake of EPA and DHA during pregnancy is estimated to be 250 mg or more per day.[121] Most women in the United States consume less than a third of this amount,[114] and vegan women are at particularly high risk of poor EPA and DHA status.[127] The Food and Drug Administration recommends that intake of EPA and DHA does not exceed 3 grams per day.

EPA and DHA are found together in fish, fish oils, and seafood (it turns out that fish really is "brain food"). Fish liver oils may contain relatively high amounts of vitamins A and D. Fish oils generally contain low levels of these vitamins (Table 4.22). DHA is available from egg yolk and DHA-fortified eggs, orange juice, bars, and other products, and in certain types of algae that produce it. Human milk from women with adequate intakes is an excellent source of DHA. Prenatal supplements are becoming

Table 4.22 EPA + DHA, vitamin A, and vitamin D content of 4.5 g (1 tsp) cod liver oil and of fish oil[128]

	EPA + DHA	Vitamin A	Vitamin D
Cod liver oil, 1 tsp	810 mg	4500 IU	450 IU
Fish oil, salmon, 1 tsp	1410 mg	0	15 IU
Fish oil, sardines, 1 tsp	940 mg	0	0
Fish oil, herring, 1 tsp	470 mg	0	0

a source of these fatty acids. Many types of prenatal supplements provide DHA or EPA+DHA in addition to vitamins and minerals.

Due to the presence of mercury and other contaminants in some types of fish, it is recommended that women who are pregnant or breastfeeding consume no more than 12 ounces of fish per week. Fish consumed should be good sources of EPA and DHA and contain low amounts of mercury and other contaminants. Fish known to generally contain high levels of mercury (swordfish, king mackerel, tilefish, and shark) should not be consumed. No more than 6 ounces per week of albacore tuna (labeled as "white tuna" on cans) should be consumed each week.[129] The Environmental Protection Agency provides information on the safety of locally caught fish at the website www.epa.gov/waterscience/fish/states.htm.

Many pregnant women avoid eating fish during pregnancy due to a concern that its content of mercury and other pollutants may harm the baby. Avoidance of modest fish consumption due to confusion regarding risks and benefits could result in suboptimal neurodevelopment in children.[130,131] For women who do not like fish, fish oil supplementation appears safe and beneficial.[132]

> **Preeclampsia**   A pregnancy-specific condition that usually occurs after 20 weeks of pregnancy (but may occur earlier). It is characterized by increased blood pressure and protein in the urine and is associated with decreased blood flow to maternal organs and through the placenta.

The Need for Water

The large increase in water need during pregnancy is generally met by increased levels of thirst. On average, women consume about 9 cups of fluid daily during pregnancy.[133] Women who engage in physical activity in hot and humid climates should drink enough to keep urine light-colored and normal in volume. Water, diluted fruit juice, iced tea, and other unsweetened beverages are good choices for staying hydrated.

The Need for Vitamins and Minerals During Pregnancy

Requirements for most vitamins and minerals increase during pregnancy due to metabolic demands associated with placental and fetal growth, expansion of maternal tissues and plasma volume, and increased nutrient needs for tissue maintenance. Maternal physiological adaptations involve changes in vitamin and mineral absorption and utilization that respond to the changing needs for these nutrients by time in pregnancy.[12]

Folate

Inadequate folate during pregnancy has long been associated with anemia in pregnancy and reduced fetal growth.[134] Only during the last two decades, however, has the broad spectrum of effects of folate been recognized. Discoveries of the multiple effects of inadequate folate intake on the development of congenital abnormalities and clinical complications of pregnancy represent some of the most important advances in our knowledge about nutrition and pregnancy.

Folate Background   The term *folate* encompasses all compounds that have the properties of folic acid and includes monoglutamate and polyglutamate forms of the vitamin. The monoglutamate form of folate is represented primarily by folic acid, a synthetic form of folate used in fortified foods and supplements. A similar monoglutamate form of folate naturally occurs in a few foods. Food sources of folate contain primarily the polyglutamate form of folate. The two major types of folates are often distinguished by referring to the monoglutamates as folic acid and the polyglutamates as dietary folate.

Bioavailability of folic acid and dietary folate differs substantially. Folic acid is nearly 100% bioavailable if taken in a supplement on an empty stomach, and 85% bioavailable if consumed with food or in fortified foods. Naturally occurring folates are 50% bioavailable on average.[135]

Folate requirements increase dramatically during pregnancy due to the extensive organ and tissue growth that takes place.

Functions of Folate   Folate is a methyl group (CH_3) donor and enzyme cofactor in metabolic reactions involved in the synthesis of DNA, gene expression, and gene regulation. Deficiency of folate impairs these processes, leading to abnormal cell division and tissue formation.[136] Folate serves as a methyl donor in the conversion of homocysteine to the amino acid methionine. The conversion of homocysteine to methionine depends primarily on three enzymes and folate, vitamin B_{12}, and vitamin B_6 cofactors. Lack of folate in particular, and less commonly a lack of vitamin B_{12}, as well as genetic abnormalities in the enzymes can lead to an accumulation of homocysteine. This may result in methionine shortage at a crucial stage of fetal development. High cellular and plasma levels of homocysteine may increase the risk of rupture of the placenta, stillbirth, preterm delivery, *preeclampsia,* structural abnormalities (congenital defects) in the newborn, and reduced birth weight. Folic acid supplements (500–600 mcg per day) in the second and third trimesters of pregnancy decrease homocysteine levels and improve pregnancy outcomes.

A common genetic defect has been identified in the enzyme 5,10-methylene tetrahydrofolate reductase (MTHFR). This variant of the normal enzyme reduces the level of activity of MTHFR by about half. Variant forms of MTHFR and defects in methionine synthase are thought to be present in approximately 30% of the population.[137]

Folate and Congenital Abnormalities

> "There is a widespread belief that congenital malformations are always the result of defective genes. Perhaps a nutritional deficiency resulting in a defective gene leads to the same congenital abnormality."
>
> R. D. Mussey, 1949[88]

Researchers have known since the 1950s that low and high intakes of certain vitamins and minerals cause congenital abnormalities in laboratory animals. They have also known that neural-tube defects, brain and heart defects, and cleft palate can be caused by feeding pregnant rats folate-deficient diets.[138] Firmly held beliefs that only severe malnutrition affects fetal growth and that genetic errors are the sole cause of congenital abnormalities delayed recognition of the importance of folate to human pregnancy.[139]

Neural-tube defects (NTDs) are malformations of the spinal cord and brain. There are three major types of NTDs:

- Spina bifida is marked by the spinal cord failing to close, leaving a gap where spinal fluid collects during pregnancy (see Illustration 4.16). Paralysis below the gap in the spinal cord occurs in severe cases.
- Anencephaly is the absence of the brain or spinal cord.
- Encephalocele is characterized by the protrusion of the brain through the skull.

Illustration 4.16 A newborn child with spina bifida.

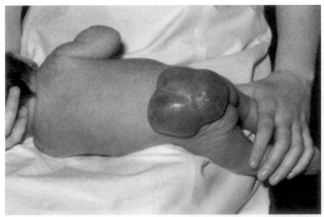

It is now well accepted that inadequate availability of folate between 21 and 27 days after conception (when the embryo is only 2–3 mm in length) can interrupt normal cell differentiation and cause NTDs.[140] Neural-tube defects are among the most common types of congenital abnormalities identified in infants, with approximately 4000 pregnancies affected each year in the United States.[141] NTDs are among the most preventable types of congenital abnormalities that exist.[142] Approximately 70% of cases of NTDs can be prevented by consumption of adequate folate before and during very early pregnancy.[139]

Folate Status of Women in the United States Folate status is assessed by serum and red cell folate levels. Of the two measures, red cell folate levels are the preferred indicator because they represent long-term folate intake, whereas serum folate levels reflect only recent intake. Levels of red cell folate of over 300 ng/mL (or 680 nmol/L) are associated with very low risk of NTDs.[143] These levels of red cell folate can generally be achieved by folic acid intakes that average 400 mcg daily.[144] As shown in Illustration 4.17, red cell folate levels are higher among women who consume folic-acid-fortified cereals or supplements compared to women consuming folate from food only.

Folate status in women of childbearing age in the United States has improved since the advent of folic-acid fortification of refined grain products in 1998. Average levels of red cell folate in U.S. women have increased from 181 to 235 ng/mL since fortification began.[142] Low levels of intake of folic-acid-fortified grain products and breakfast cereals still leave some women with too little folate, however.[145] Low red cell folate levels are disproportionately found in black

Illustration 4.17 Mean red cell folate level in preconceptional women by level of intake of various sources of folate.

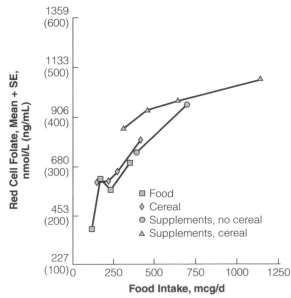

SOURCE: Brown JE et al., *JAMA*, Vol. 277, No. 6, p. 551 (Feb 19, 1997).

Table 4.23 Food sources of folate

	Amount	Folic Acid (mcg)
A. Foods		
Orange	1	40
Orange juice	6 oz	82
Pineapple juice	6 oz	44
Papaya juice	6 oz	40
Dried beans	½ c	50
B. Fortified Foods		
Highly fortified breakfast cereal[a]	1 c or 1 oz	400
Breakfast cereal	1 c or 1 oz	100
Bread, roll	1 slice or 1 oz	40
Pasta	½ c	30
Rice	½ c	30

[a]Includes Product 19, Smart Start, Special K, and Total.

women in the United States compared to white or Mexican American women.[146]

Dietary Sources of Folate   Many vegetables and fruits are good sources of folate (see Table 1.9 in Chapter 1), but only a few foods contain the highly bioavailable form of folate. Table 4.23 lists some foods that naturally contain the highly bioavailable, monoglutamate form of folate and foods that provide folic acid through fortification.

Adequacy of folic acid intake before and during pregnancy can be estimated by adding up the amount of folic acid in foods typically consumed in the daily diet using the data in the table. Whole-grain products including breads and pastas, brown rice, oatmeal, shredded wheat, and organic grain products may or may not be fortified with folic acid. You have to check food labels to find that out.

Recommended Intake of Folate   Due to variation in folate bioavailability, the DRI for folate takes into consideration a measure called *dietary folate equivalents,* or DFE. One DFE equals any of the following:

- 1 mcg food folate
- 0.6 mcg folic acid consumed in fortified foods or a supplement taken with food
- 0.5 mcg of folic acid taken as a supplement on an empty stomach

Folic acid taken in a supplement without food provides twice the dietary folate equivalents as does an equivalent amount of folate from food.

It is recommended that women consume 600 mcg DFE of folate per day during pregnancy and include 400 mcg folic acid from fortified foods or supplements.[135] The remaining 200 mcg DFE should be obtained from vegetables and fruits. These nutrient-dense foods provide an average of 40 mcg of folate per serving.[141] Because

NTDs develop before women may realize they are pregnant, adequate folate should be consumed several months prior to, as well as throughout, pregnancy.

Women who have previously delivered an infant affected by an NTD are being urged to take 4000 mcg (4.0 mg) of folic acid in a supplement to reduce the risk of recurrence.[147] This dose, however, may be much higher than needed based on results of clinical trials.[148,149] The upper limit for intake of folic acid from fortified foods and supplements is set at 1000 mcg per day. There is no upper limit for folate consumed in its naturally occurring form in foods. The 1000 mcg level represents an amount of folic acid that may mask the neurological signs of vitamin B_{12} deficiency. If left untreated, vitamin B_{12} deficiency leads to irreversible neurological damage.[141]

Choline

Choline is a B-complex vitamin that humans can produce, but not in high enough amounts to meet needs when dietary intake of choline is very limited. The need for choline increases during pregnancy due to its roles as a component of phosopholipids in cell membranes and a precursor of intracellular messengers.[150] Choline can be converted to betaine, which, like folate, serves as a source of methyl groups used to regulate gene function, neural-tube and brain development, and the conversion of homocysteine to methionine.[151] Large amounts of choline are transported via the mother's blood to the embryo and fetus during pregnancy.[150]

The RDA for choline in pregnancy is 450 mg daily. Average intake of choline by women between the ages of 20 and 39 years in the United States, however, is 274 mg per day.[114] Some experts are concerned that pregnant women may not be getting enough choline for optimal fetal brain growth and intellectual development.[152] Although this is a much discussed and exciting area of research, it is not yet clear whether low dietary choline availability in women during pregnancy represents a risk factor for fetal brain and intellectual development.[153]

Choline status tends to be adequate in women who regularly consume eggs and meat, the two major sources of this vitamin.[150]

Vitamin A

Vitamin A is a key nutrient in pregnancy because it plays important roles in reactions involved in cell differentiation. Deficiency of this vitamin is rare in pregnant women in industrialized countries, but it is a major problem in many developing nations. Vitamin A deficiency that occurs early in pregnancy can produce malformations of the fetal lungs, urinary tract, and heart.[154]

Of more concern than vitamin A deficiency in the United States are problems associated with excessive intakes of vitamin A in the form of retinol or retinoic acid (but not beta-carotene). Intakes of these forms of vitamin A

Illustration 4.18 An 8-month-old infant exposed to high levels of retinoic acid *in utero*. Note the high forehead, flat nasal bridge, and malformed ear.

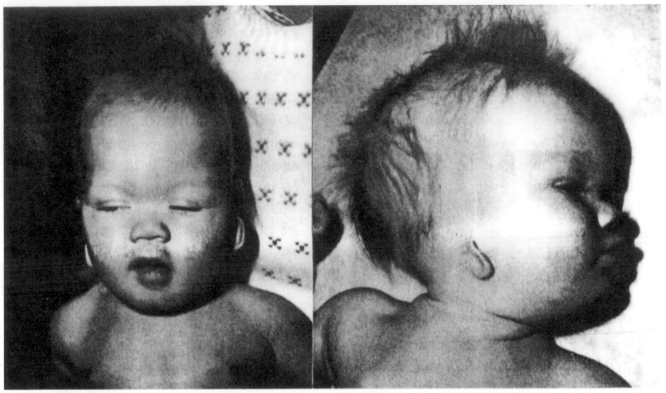

SOURCE: Used by permission of Harcourt Health Sciences, Inc. Lott IT et al., "Fetal hydrocephalus and ear abnormalities associated with maternal use of isotretinoin," *J Pediatr* 1984;105:597–600.

of over 10,000 IU per day, and the use of medications such as Accutane and Retin-A for acne and wrinkle treatment, increase the risk of fetal abnormalities. Effects are particularly striking in infants born to women using Accutane or Retin-A early in pregnancy (see Illustration 4.18). Fetal exposure to the high doses of retinoic acid in these drugs tends to result in "retinoic acid syndrome." Features of this syndrome include small ears or no ears, abnormal or missing ear canals, brain malformation, and heart defects.[155]

Due to the potential toxicity of retinol, it is recommended that women take no more than 5,000 IU of vitamin A as retinol from supplements during pregnancy.[24] Most supplements made today contain beta-carotene rather than retinol. High intakes of beta-carotene have not been related to birth defects.[55] Although women were issued strong warnings not to take Retin-A or Accutane if pregnancy was possible, ill-timed use continue to occur, as did the retinoic acid syndrome. Accutane is no longer on the market, but other retinoic acid-based medications for acne remain available.[166]

Vitamin D

Vitamin D supports fetal growth, the addition of calcium to bone, and tooth and enamel formation. Lack of a sufficient supply of vitamin D during pregnancy compromises fetal as well as childhood bone development.[157] Infants born

to women with poor vitamin D status tend to be smaller than average, more likely to have low blood calcium levels (hypocalcaemia) at birth, and more likely to have poorly calcified bones and abnormal enamel. They are also more likely to develop dental caries in childhood.[157–160]

Vitamin D also supports normal functioning of the immune system and can inhibit inflammation. There are indications that vitamin D deficiency during pregnancy may be related to adverse outcomes based on these roles. Miscarriage, preeclampsia, preterm birth, maternal infection, and the development of type 1 diabetes and asthma in children are under investigation for their relationship to maternal vitamin D insufficiency.[161]

Prevalence of Vitamin D Inadequacy Vitamin D inadequacy during pregnancy is common. A study of 206 pregnant women in Toronto found vitamin D deficiency in 35%.[158] Of 400 women tested in Pittsburgh, vitamin D deficiency or insufficiency was identified in 83% of black and 47% of white pregnant women.[161] These high rates of inadequate vitamin D status during pregnancy were mirrored in the vitamin D status of infants. During the first month of life, 93% of black and 66% of white infants were vitamin D insufficient.[161] Intakes of 10 mcg (400 IU) vitamin D daily are related to low serum levels of 25 hydroxyvitamin D (vitamin D_3, the nutrition biomarker of vitamin D status).[157]

Table 4.24 Risk factors for vitamin D inadequacy during pregnancy

Vegan diet

Consumption of small amounts of vitamin D–fortified milk or of raw milk

Limited exposure of the skin to the direct rays of the sun

Consistent use of sun block

Dark skin

Obesity

Obese women appear to be at increased risk for inadequate vitamin D status due to low levels of release of stored vitamin D from fat cells. As many as 61% of obese women have been identified as having low serum levels of vitamin D compared to approximately 36% in women who are not obese.[161] Vegan women are at risk for poor vitamin D status because vitamin D is naturally present only in animal products. Risk factors for inadequate vitamin D status during pregnancy are summarized in Table 4.24.

Recommendations for Vitamin D Intake During Pregnancy   An intake of 5 mcg (200 IU) vitamin D daily is officially recommended for pregnancy. This amount of vitamin D can be obtained by consuming 3 cups of vitamin D–fortified milk a day, or by exposing the skin to sunshine. Two 15-minute sunbathing sessions per week lead to the production of about 1250 mcg (50,000 IU) vitamin D and a low risk of sunburn in most people. Individuals with dark skin need two to five times this length of sun exposure to produce that much vitamin D. Winter sunlight in northern climates is too weak to produce vitamin D formation in the skin. There is no evidence that vitamin D overdose occurs due to sun exposure.[162]

Some experts assert that more vitamin D than 5 mcg daily during pregnancy is needed. Vitamin D intake recommendations are being revised and it would not be surprising to see recommended intake levels for vitamin D increase. The current upper limit for vitamin D intake from foods and supplements is 50 mcg (2000 IU) daily.

The Need for Minerals During Pregnancy

Requirements for minerals during pregnancy are generally higher than for women who are not pregnant due to increased need for them as components of maternal and fetal tissues, and to facilitate metabolic processes that occur during pregnancy. This section addresses minerals of most concern during pregnancy.

Calcium

Calcium is primarily needed in pregnancy for fetal skeletal mineralization and maintenance of maternal bone health. Approximately 30 grams of calcium (a little over an ounce) is transferred from the mother to the fetus during pregnancy. Fetal demand for calcium peaks in the third trimester when fetal bones are mineralizing at a high rate.[163]

Calcium metabolism changes meaningfully during pregnancy. Absorption of calcium from food increases, excretion of calcium in urine decreases, and bone mineral turnover takes place at a higher rate.[164] The additional requirement for calcium in the last quarter of pregnancy is approximately 300 mg per day and may be obtained by increased absorption and by release of calcium from bone.[165] (Calcium is not taken from the teeth, however.[166]) Calcium lost from bones appears to be replaced after pregnancy in women with adequate intakes of calcium and vitamin D.[165] Inadequate calcium intake has been related to increased blood pressure during pregnancy, decreased subsequent bone remineralization, increased blood pressure of infants, and decreased breast-milk concentration of calcium.[163]

Calcium and the Release of Lead from Bones   Lead in maternal blood can cross the placenta and be taken up by the fetus.[168] Elevated blood lead levels in pregnancy are a cause for concern because they are related to miscarriage, preterm birth, low-birth-weight infants, impaired central nervous system development, and subsequent developmental delays in children.[167] Poor, urban, and immigrant populations are at greater risk for exposure to lead-based paint and environmental contamination than are other groups in the United States.[169]

Pregnant women who do not consume enough calcium show greater increases in blood lead levels than women who consume 1000 mg (the DRI for calcium) or more per day. Bone tissues contain about 95% of the body's lead content, and the lead is released into the bloodstream when bones demineralize. Bone tissues demineralize to a greater extent in pregnant women who fail to consume adequate calcium.[167]

Calcium needs during pregnancy can be met by drinking 3 cups of milk or calcium-fortified soymilk, or 2 cups of calcium-fortified orange juice plus a cup of milk, or by choosing a sufficient number of other good sources of calcium daily. (Table 1.14 in Chapter 1 lists food sources of calcium.)

Fluoride

Teeth begin to develop *in utero*, so why isn't it recommended that pregnant women consume sufficient fluoride so that the fetus builds cavity-resistant teeth? A limited amount of fluoride is transferred from the mother's blood to the developing enamel of the fetus. Major gains in the fluoride composition of enamel, however, occur in the years after birth when enamel in primary and permanent teeth fully develops and hardens.[170] Children of pregnant women given fluoride supplements during pregnancy have the same rates of dental caries as do children of women who did not receive supplements.[171]

Iron

Iron status is a leading topic of discussion in prenatal nutrition because the need for iron increases substantially; women require about 1000 mg (1 g) of additional iron for pregnancy.

- 300 mg is used by the fetus and placenta.
- 250 mg is lost at delivery.
- 450 mg is used to increase red blood cell mass.

Maternal iron stores get a boost after delivery when iron liberated during the breakdown of surplus red blood cells is recycled.[172]

Approximately 12% of women enter pregnancy with *iron deficiency* and little stored iron, and consequently are at risk of developing *iron-deficiency anemia* in pregnancy.[173]

Iron-Deficiency Anemia in Pregnancy   Over recent decades, rates of iron-deficiency anemia in pregnancy have remained high in women in developing as well as developed countries (Table 4.25). Iron-deficiency anemia at the beginning of pregnancy increases the risk of preterm delivery and low-birth-weight infants by two to three times.[174] Iron deficiency during pregnancy is related to lower scores on intelligence, language, gross motor, and attention tests in affected children at the age of 5 years.[175] The mechanisms underlying these effects are unknown, but they may be related to decreased oxygen delivery to the placenta and fetus, increased rates of infection, or adverse effects of iron deficiency on brain development.[174] Iron deficiency often occurs toward the end of pregnancy even among women who enter pregnancy with some iron stores. It is far more common than iron-deficiency anemia.

Iron deficiency and iron-deficiency anemia are related to reduced iron stores in newborns. A fetus from a well-nourished mother is able to store a 6 to 8 month supply of

iron during the last two months *in utero*. Preterm infants are at risk for iron deficiency in infancy because they have less time to accumulate iron in late pregnancy.[176]

Assessment of Iron Status   Red cell mass increases substantially (30%) in pregnancy. However, plasma volume expands more (by about 50%). The higher increase in plasma volume compared to red cell mass makes it appear that amounts of hemoglobin, ferritin, and packed red blood cells have decreased. They have not decreased but rather have become diluted by the large increase in plasma volume. Hemoglobin concentration normally decreases until the middle of the second trimester and then rises somewhat in the third. It is not necessary to prevent normal declines in hemoglobin level during pregnancy.[177]

Due to the dilution effects of increased plasma volume, changes in hemoglobin levels tend to be more indicative of plasma volume expansion than of iron status.[174] Low levels of hemoglobin or serum ferritin may be associated with high plasma volume expansion (hypervolemia), and high hemoglobin levels are related to low plasma volume expansion (hypovolemia). Low levels of plasma volume expansion are associated with reduced fetal growth, whereas newborns tend to be larger in women with higher levels of plasma volume expansion.[177]

The Centers for Disease Control have developed standard hemoglobin levels to be used in the identification of iron-deficiency anemia in pregnant women. These standards (shown in Table 4.26) represent levels below the 5th percentile of hemoglobin values in pregnancy.[178]

> **Iron Deficiency**   A condition marked by depleted iron stories. It is characterized by weakness, fatigue, short attention span, poor appetite, increased susceptibility to infection, and irritability.
>
> **Iron-Deficiency Anemia**   A condition often marked by low hemoglobin level it is characterized by the signs of iron deficiency plus paleness, exhaustion, and a rapid heart rate.

Table 4.25 Estimates of the incidence of iron-deficiency anemia in women in developing and developed countries[172,173]

	% with Iron-Deficiency Anemia	
	Developing Countries	Developed Counties
Nonpregnant	43	12
Pregnant	56	18

Table 4.26 CDC's gestational age-specific cutoffs for anemia in pregnancy[178]

Gestational Weeks	Hemoglobin (g/dL) Indicating Anemia[a]
12	<11.0
16	<10.6
20	<10.5
24	<10.5
28	<10.7
32	<11.0
36	<11.4
40	<11.9

[a]For women living in high altitudes, hemoglobin values should be increased by 0.2 g/dL for every 1000 feet above 3000 and by 0.3 g/dL for every 1000 feet above 7000. For cigarette smokers, hemoglobin values should be adjusted upward by 0.3 g/dL.

By trimester, hemoglobin levels indicative of iron-deficiency anemia are:

- <11.0 g/dL in the first and third trimesters
- <10.5 g/dL in the second trimester

Serum ferritin cut-points indicative of iron-deficiency anemia in pregnancy have also been developed:[178]

	Serum Ferritin, ng/mL
Normal	>35
Depleted Stores	<20
Iron Deficiency	≤15

Hemoglobin and serum ferritin are the most commonly employed measures of iron status in pregnant women.[178]

The diagnosis of iron-deficiency anemia is more complicated than often thought. No single test of iron status is totally accurate because (1) many factors, including infection and inflammation, affect iron status; and (2) each test measures a different aspect of iron status. It is best to base the diagnosis of iron-deficiency anemia on results of several tests.[178]

Women entering pregnancy with adequate iron stores tend to absorb about 10% of total iron ingested; those with low stores absorb more—about 20% of the iron consumed. The largest percentage of iron absorption, 40%, occurs in women who enter pregnancy with iron-deficiency anemia.

Iron absorption from foods and supplements is enhanced in women with low iron stores during pregnancy, and absorption increases as pregnancy progresses.[34] Absorption is highest after the thirtieth week of pregnancy, when the greatest amount of iron transfer to the fetus occurs. Maternal iron depletion in pregnancy decreases fetal iron stores, increases the risk that infants will develop iron-deficiency anemia, and is associated with development of maternal postpartum depression.[55]

Pros and Cons of Iron Supplementation   Absorption of iron from multimineral supplements is substantially lower than is iron absorption from supplements containing iron only. For example, women given a multimineral supplement containing iron, calcium, and magnesium absorb less than 5% of the iron, whereas women given a similar dose of iron in a supplement containing iron only absorb over twice that much.[239]

The amount of iron absorbed from supplements depends primarily on women's need for iron and the amount of iron in the supplement. As can be seen in Illustration 4.19, the amount of iron absorbed from supplements decreases substantially as the dose of iron increases. Acceptance of high levels of iron supplementation by women is often poor. Nausea, cramps, gas, and constipation are associated with the presence of free iron in the intestines, and these side effects increase as doses of supplemental iron increase (Table 4.27). Side effects experienced in using iron supplements are a major reason that women fail to take them.[173] Unused iron supplements, when stored and later found by young children, pose a risk of iron poisoning. Difficulties related to building iron stores during pregnancy provide a strong rationale for screening women for iron status prior to pregnancy and establishing good levels of stored iron before pregnancy if needed.[173]

A relatively new concern about high-dose iron supplements is emerging. Iron supplements providing 60 mg or more iron per day regularly expose the intestinal mucosa to free iron radicals. The oxidizing effects of iron radicals cause inflammation and mitochondrial damage in cells.[242] In addition, iron doses over 30 mg per day may decrease zinc absorption status.[34]

Illustration 4.19 Effect of does of supplemental iron on iron absorption in women during pregnancy.

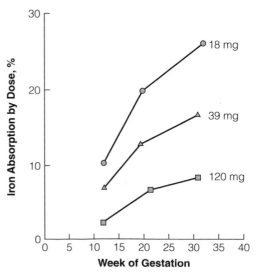
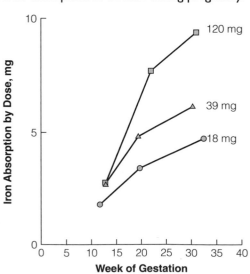

SOURCE: From Pedro Rosso, *Nutrition and Metabolism in Pregnancy: Mother and Fetus.* Copyright © 1990 by Oxford University Press, Inc. Used by permission of Oxford University Press, Inc.

Table 4.27 Increased occurrence of side effects in women by supplemental iron dose[240, 241]

Dose of Iron, mg/day	Side Effects
60	32%
120	40%
240	72%

Table 4.28 Percent of elemental iron by weight in various types of iron supplements

Supplement Type	Iron Content
Ferrous sulfate	20%
Ferrous gluconate	12%
Ferrous fumarate	32%

Amounts of elemental iron in supplements vary depending on the form of the iron compound in the supplement (Table 4.28). The proportion of iron absorbed from a constant amount of iron from each type of supplement listed in Table 4.28 is approximately equal.[243]

Recommendations Related to Iron Supplementation in Pregnancy   It is generally recommended that pregnant women in the United States take a 30-mg iron supplement daily after the twelveth week of pregnancy.[244] Women with iron-deficiency anemia are often given 60–180 mg of iron per day.[245]

It has been suggested that women's iron status be assessed at the first prenatal visit to determine if there is a need for iron supplements. A 30-mg iron supplement would be indicated when hemoglobin levels are <11 g/dL, or if serum ferritin levels are <30 ng/mL. Women with higher values would be monitored for iron status but not given a supplement.[244]

Recommended Intake of Iron During Pregnancy   The increased need for iron can be met by intakes that lead to an additional 3.7 mg absorbed iron per day on average throughout pregnancy. This is a large increase, especially considering that nonpregnant women consuming the DRI for iron (18 mg) absorb only around 1.8 mg of iron daily. Given an ongoing need for 1.8 mg of absorbed iron a day, and the additional need of 3.7 mg of iron daily for pregnancy, the total need for absorbed iron during pregnancy is 5.5 mg daily. Assuming 20% of iron consumed is absorbed, average iron consumption of 27 mg per day (the RDA for iron for pregnancy) will meet the iron needs of pregnancy. The Upper Limit for iron intake during pregnancy is set at 45 mg per day.

Iodine

Iodine is required in pregnancy by the mother and fetus for thyroid function and energy production, and for fetal brain development.[179] Deficiency of iodine early in pregnancy can lead to *hypothyroidism* in the offspring. Hypothyroidism in infants is endemic in parts of southern and eastern Europe, Asia, Africa, and Latin America.[180] The incidence of infant hypothyroidism has been found to decrease by over 70% when at-risk women in developing countries are given iodine supplements before or in the first half of pregnancy.

Rates of infant deaths are also substantially improved, as is the psychomotor development of the offspring.[181] Iodine supplementation in the second half of pregnancy does not improve infant outcomes.[180]

About half of pregnant women in the United States consume less than the recommended 220 mcg of iodine daily, and 7% have low urinary iodine levels.[182] The most reliable source of iodine is iodized salt. One teaspoon contains 400 mcg iodine. Fish, shellfish, seaweed, and some types of tea provide iodine. Women who consume iodized salt are not likely to need supplemental iodine.[183] Kelp and seaweed should not be used as a source of iodine because they vary too much in iodine content.[179] Iodine can also be provided through supplements containing iodine. The American Thyroid Association recommends that prenatal supplements contain 150 mcg of iodine.[184] Usual iodine intake should not exceed 1100 mcg daily during pregnancy.

Sodium

Sodium plays a critical role in maintaining the body's water balance. Requirements for it increase markedly during pregnancy due to plasma volume expansion. But the need for increased amounts of sodium in pregnancy hasn't always been appreciated. Thirty years ago in the United States, it was accepted practice to put all pregnant women on low-sodium diets. It was then thought that sodium increased water retention and blood pressure, and that sodium restriction would prevent edema and high blood pressure. We now know this isn't accurate and that inadequate sodium intake can complicate the course and outcome of pregnancy.[184] Sodium restriction during pregnancy may exhaust sodium conservation mechanisms and lead to excessive sodium loss.[190]

Sodium restriction is not indicated in normal pregnancy or for the control of edema or high blood pressure that develops in pregnancy. Women should be advised to consume salt "to taste" unless contradicted by a medical condition related to salt intake.[191]

Hypothyroidism   A condition characterized by growth impairment and mental retardation and deafness when caused by inadequate maternal intake of iodine during pregnancy. Used to be called *cretinism*.

Bioactive Components of Food

Foods contain thousands of biologically active substances that are not considered essential nutrients but nonetheless influence health. These food substances are generally referred to as *bioactive food components* and include hundreds of phytochemicals (plant chemicals) that may influence maternal health. Antioxidant pigments in plant foods and caffeine are two primary examples of bioactive food components that have been investigated for effects on maternal and newborn health. Many of the beneficial effects of antioxidant consumption in foods are not found when the antioxidants are taken as supplements.[185]

Normal pregnancy is a pro-oxidative state and is accompanied by an increased requirement for antioxidants. Many plant pigments act as antioxidants and help protect fetal DNA from damage due to exposure to oxygen and other oxidizing chemicals produced in the body. Antioxidants also reduce maternal tissue damage associated with inflammation and oxidation. Vitamins C and E (also found in plant foods) likewise perform important antioxidant roles during pregnancy.[186] High intakes during pregnancy of foods rich in vitamin E, for example, appear to reduce the risk of asthma in children. Children born to women consuming diets providing 26 mg of vitamin E daily on average are less likely to develop wheezing and asthma during early childhood than are children born to women who consume 14 mg of vitamin E daily.[187] It is suggested that vitamin E intake from food during pregnancy may reduce asthma by decreasing lung inflammation in the offspring.[188]

Foods rich in antioxidants advertise that fact by their color. Red, orange, dark green, deep yellow, and blue-purple fruits and vegetables generally provide good amounts of antioxidants. Many of these same foods are rich in vitamin C. Women who are pregnant should consume at least 5 cups of vegetables and fruits daily.

Caffeine

Caffeine has long been suspected of causing adverse effects in pregnant women because it increases heart rate, acts as a diuretic, and stimulates the central nervous system. It easily passes from maternal to fetal blood and lingers in the fetus longer than in maternal blood because the fetus excretes it more slowly.[192]

Due to its high caffeine level, coffee has generally been the target of investigations on the effects of caffeine intake on pregnancy outcome. Coffee, however, contains hundreds of substances, and some of these have effects similar to those of caffeine. So,

Bioactive Food Components Constituents in foods or dietary supplements other than those needed to meet basic human nutritional needs that are responsible for changes in health status.

although conclusions about caffeine's effects on pregnancy are largely based on coffee intake, it is possible that other components of coffee are responsible for effects observed. Coffee is by far the largest contributor to caffeine intake in most people,[193] and pregnant women consume on average 144 mg caffeine from coffee per day in pregnancy.[194] (Table 2.5 in Chapter 2 provides a list of the caffeine content of beverages and foods.)

Despite the possibilities, evidence from well-controlled trials of caffeine or coffee intake and fetal growth restriction, malformations, preterm birth, and delivery complications does not support a relationship.[195,196] No long-term consequences of coffee intake during pregnancy have been observed in children 7 years later. Children of women who drank coffee during pregnancy have been found to have similar levels of intellectual and neuromotor development when compared to children of non-coffee drinkers.[196] Avoidance of caffeine and coffee during pregnancy does not appear to improve pregnancy outcomes or infant birth weight.[195] It is generally concluded that intake of up to 4 cups of coffee per day during pregnancy is safe.[196]

Healthy Diets for Pregnancy

Healthy diets for women during pregnancy are described in terms of calories and nutrient intake, and by food choices. Such diets have a number of characteristics in common (Table 4.29).

Adequacy of caloric intake during pregnancy is generally based on rate of weight gain, but for nutrients it is based

Table 4.29 Basics of a good diet for normal pregnancy

Good pregnancy diets:

1. Provide sufficient calories to support appropriate rates of weight gain.
2. Follow the MyPyramid food group recommendations.
3. Provide all essential nutrients at recommended levels of intake from the diet (with the possible exception of iron).
4. Include 600 mcg folate, of which 400 mcg is folic acid, daily.
5. Include 250 mg EPA and DHA daily.
6. Provide sufficient dietary fiber (28 g/day).
7. Include 9 c fluid daily.
8. Include salt "to taste."
9. Exclude alcohol and limit coffee intake to ≤4 cups per day.
10. Are satisfying and enjoyable.

Table 4.30 Recommended Dietary Allowances and Upper Limits for pregnant and nonpregnant women aged 19–30 years*

	Pregnant	Nonpregnant	Upper Limit (UL)
Energy, kcal			
2nd trimester	+350	2403	—
3rd trimester	+452		
Protein, gm	71	46	—
Linoleic acid, g	13	12	—
Alpha-linolenic acid, g	1.4	1.1	—
Vitamin A, mcg	770	700	3000
Vitamin C, mg	85	75	2000
Vitamin D, mcg[a]	5	5	50
Vitamin E, mg	15	15	1000[c]
Vitamin K, mcg	90	90	—
Thiamin, mg	1.4	1.1	—
Riboflavin, mg	1.4	1.1	—
Niacin, mg	18	14	35[c]
Vitamin B_6	1.9	1.3	100
Folate, mcg[b]	600	400	1000[c,d]
Vitamin B_{12}, mcg	2.6	2.4	—
Pantothenic acid, mcg	6	5	—
Biotin, mcg	30	30	—
Choline, g	450	425	3.5
Calcium, mg	1000	1000	2500
Chromium, mcg	30	25	—
Copper, mcg	1000	900	10,000
Fluoride, mg	3	3	10
Iodine, mcg	220	150	1100
Iron, mg	27	18	45
Magnesium, mg	350	310	350[c]
Manganese, mg	2	1.8	11
Molybdenum, mcg	50	45	2000
Phosphorus, mg	700	700	3500
Selenium, mcg	60	55	400
Zinc, mg	11	8	40

* DRIs for females <19 and >30 years are listed inside the front covers of this book.

[a] 1 mcg = 40 IU vitamin D; DRI applies in the absence of adequate sunlight.

[b] As Dietary Folate Equivalent (DFE). 1 DFE = 1 mcg food folate = 0.6 mcg folic acid from fortified food or supplement consumed with food = 0.5 mcg of a supplement taken on an empty stomach.

[c] UL applies to intake from supplements or synthetic form only.

[d] Applies to intake of folic acid.

on the DRIs (Table 4.30). Nutrient intakes during pregnancy should approximate those given in the DRI table, and food intake should correspond to the recommended types and quantity of food recommended in MyPyramid.

Effect of Taste and Smell Changes on Dietary Intake During Pregnancy

No inner voice directs women to consume foods that provide needed nutrients during pregnancy. Pregnant women may, however, develop food preferences and aversions due to changes in the sense of taste and smell, and they may experience *pica*.

Changes in the way certain foods taste, and the odor of foods and other substances, affect two out of three women during pregnancy. If asked to recall, many previously pregnant women could tell you which foods tasted really good to them, and which odors made them feel queasy to even think about. Increased preference for foods such as sweets, fruits, salty foods, and dairy products are common.[197] The odors of meat being cooked, coffee, perfume, cigarette smoke, and gasoline are common nasal offenders and may stimulate episodes of nausea.[198] The biological bases for such changes are not known, but they are suspected of being related to hormonal changes of pregnancy.

Pica

"Pica permits the mind no rest until it is satisfied."

F. W. Craig, 1935

Classified as an eating disorder, pica affects over half of pregnant women in some locations of the southern part of the United States. It is more common in African Americans than in other ethnic groups, and it is common enough to be considered a normal behavior in some countries. Historically, one type of pica—*geophagia*—was thought to provide women with additional minerals and to ease gastrointestinal upsets. The cause of pica remains a mystery.[199]

Nonfood items most commonly craved and consumed by pregnant women with pica include ice or freezer frost (*pagophagia*), laundry starch or cornstarch (*amylophagia*), baking soda and powder, and clay or dirt (geophagia). Women experiencing pica are more likely to be iron deficient than those who don't, and iron-deficiency anemia is especially common among pregnant women who compulsively consume ice or freezer frost.[200] It is

Pica An eating disorder characterized by the compulsion to eat substances that are not food.

Geophagia Compulsive consumption of clay or dirt.

Pagophagia Compulsive consumption of ice or freezer frost.

Amylophagia Compulsive consumption of laundry starch or cornstarch.

not clear, however, whether iron deficiency leads to pica or if pica leads to iron deficiency.

Pica does not appear to be related to newborn weight or preterm delivery. It can, however, complicate control of gestational diabetes if starch is eaten, and it has caused lead poisoning, intestinal obstruction, and parasitic infestation of the gastrointestinal tract.[200] Women with amylophagia sometimes accept powdered milk as an alternative to laundry starch or cornstarch, and treating anemia often stops the craving for ice or freezer frost.

Cultural Considerations   People tend to be attached to existing food preferences, many of which may have deep cultural roots. Dietary recommendations will differ for Native Alaskans accustomed to a diet based on wild game; for Cambodians, Vietnamese, and Somalis who may think no meal is complete without rice; and for lactose-intolerant individuals.

The belief that consumption of certain foods "marks" the baby is common in many cultures. People may think, for example, that a woman who loves mangos and eats lots of them during pregnancy may have a baby born with a "mango-shaped" birthmark. Some cultures would hold that the baby will also have learned to love mangos because its mother ate them often while pregnant.

Dietary recommendations that are not consistent with a person's usual dietary practices and beliefs, or that are not viewed as acceptable or even preferred by the woman, are least likely to be effective. For best results, dietary adjustments recommended for each individual pregnant woman should take into account her usual practices and preferences.

Assessment of Nutritional Status During Pregnancy

A comprehensive approach to nutrition assessment in pregnancy includes an evaluation of dietary intake, weight status, biomarkers of nutrient status, food preferences and resources, previous pregnancy and health history, and dietary supplement use.[201] In this chapter we highlight two of these components: assessment of dietary intake and nutrition biomarkers.

Dietary Assessment During Pregnancy

Routine assessment of dietary practices is recommended for all pregnant women to determine the need for an improved diet or vitamin and mineral supplements.[34] Dietary assessment in pregnancy should cover usual dietary intake, dietary supplement use, and weight-gain progress. For best results, several days of accurately recorded, usual intake should be used for each assessment.

Several levels of dietary assessment can be undertaken. (Internet resources for dietary assessment are listed at the end of the chapter.) Which assessment level is best primarily depends on the skill level of the health professional responsible for interpreting the results. Results of food-based assessments are rather straightforward to interpret, whereas computerized assessments of levels of nutrient intake are more complex.

The MyPyramid food guide provides a good way to assess a typical diet during pregnancy. Table 4.31 presents this guide based on a caloric need of 2400, which is not unusual for pregnant women in the United States. Daily amounts

Table 4.31 MyPyramid Food Guide for pregnant women[a]

Food Group	Ounces/Cups Recommended per Day	Examples of Equivalent Measures
Grains	8 oz (includes 4 oz whole grain products)	• 1 slice bread = 1 oz • 1 c cold cereal = 1 oz • 1 c cooked rice, pasta, or cereal = 2 oz
Vegetables	3 c (includes dark green and orange vegetables)	• 2 c tossed salad = 1 c
Fruits	2 c	• 1 c fruit juice = 1 c
Milk	3 c	• ⅓ c shredded cheese = 1 c • 2 slices American cheese = 1 c • 1½ oz hard cheese = 1 c • 1½ c ice cream = 1 c
Meat and beans	6½ oz	• 1 small egg = 1 oz • 1 Tbsp peanut butter = 1 oz • ¼ c dried beans = 1 oz • ½ oz nuts = 1 oz
Oils	7 tsp	• 1 Tbsp mayonnaise = 2½ tsp oil • 1 Tbsp salad dressing = 1 tsp oil

[a] The MyPyramid Plan for Moms shown here is for a 32-year-old female, 5 feet 6 inches tall, 130 pounds before pregnancy, who is physically active for 30 to 60 minutes a day. She is in her second trimester of pregnancy.

Table 4.32 An example of one day's typical diet for a pregnant women based on MyPyramid food intake recommendations for a 2400-calorie diet, and results of an energy and nutrient assessment[a] of the day's diet

One-Day Typical Diet		Energy (kcal) in the Diet	2376	
			Nutrient Analysis	Pregnancy RDA/AI
Grains, 8 oz	Cheerios, 1 c			
	Craked wheat bread, 2 slices	Protein, g	108	71
	Brown rice, 1 c	Fiber, g	26	28
	Roll, 2 oz	Total fat, g	24	
	Corn tortilla, 1 oz	Vitamin A, mcg	2120	770
Vegetables, 3 c	Carrots, cooked, 1 c	Vitamin C, mg	125	85
	Tomato slices, 1 c	Vitamin E, mg	8.8	15
	Potatoes, boiled, 1 c	Thiamin (B_1), mg	1.8	1.4
Fruit, 2 c	Orange, 1	Riboflavin (B_2), mg	2.4	1.4
	Banana, 1	Niacin (B_3), mg	43	18
Milk, 3 c	1% milk, 2 c	Folate, mcg	488	600
	American cheese, 1½ oz	Vitamin B_6, mg	3.4	1.9
Meat and beans	Light tuna in water, 2½ oz	Vitamin B_{12}, mcg	3.4	2.6
	Chicken, baked, no skin, 4 oz	Calcium, mg	1350	1000
Fats and oils, 7 tsp	Soybean oil, 4 tsp	Magnesium, mg	421	350
	Italian dressing, 2 Tbsp	Iron, mg	18	27
Discretionary	Soft drink, 12 oz	Zinc, mg	13	11
calories (360 kcal)	Sugar, 2 tsp	Selenium, mcg	177	60
	Margarine, 4 tsp	Potassium, mg	3720	4700
Energy and nutrient analysis of the day's diet[a]				

[a]Analyzed using the mypyramid.gov program.

recommended for each food group can be compared to that recorded by a pregnant woman and the results used to identify the general quality of the diet. Table 4.32 shows how closely the example of a day's diet based on MyPyramid recommendations matches recommended levels of nutrient intake for pregnancy. With the exception of vitamin E and iron (which are supplied in rather low amounts by the example diet), nutrient levels correspond to recommended intakes for pregnancy.

Computerized analysis, given accurate records and entry of dietary intake and a high-quality nutrient database, provides results useful for estimating the quantity of calories and nutrients consumed. Detailed knowledge of dietary intake is particularly useful for women at risk of nutrient inadequacies or excesses, and for women with conditions such as gestational diabetes, food intolerances, and multifetal pregnancy.

Nutrition Biomarker Assessment

Nutrition assessment of pregnant women usually includes laboratory tests of iron status, and will include tests to determine the status of other nutrients as indicated. Due to the normal physiological changes occurring during pregnancy, such as hemodilution, that affect blood nutrient concentrations, assessment of nutrition biomarkers should employ standards developed for pregnancy.[202] Blood nutrient concentrations change with time during pregnancy, so no one value per nutrient for all of pregnancy accurately reflects status.

Studies reporting reference values for nutrition biomarkers during pregnancy are beginning to appear in the scientific literature. These values are shown in Table 4.33. The concentrations listed by week of gestation consists of values from the 2.5 percentile to the 97.5 percentile of the distribution of values within a sample of well-nourished women with healthy, uncomplicated pregnancies. These values are assumed to reflect normal ranges of nutrition biomarker concentrations during pregnancy. They are intended to assist clinicians in distinguishing between physiological changes and pathological states during pregnancy.[202]

Dietary Supplements During Pregnancy

Dietary supplements used by pregnant women primarily include vitamins, minerals, and herbs. Multivitamin and mineral supplements are routinely recommended to pregnant women by most clinicians in the United States, and many women make the decision to use nutrient and

Table 4.33 Reference values for nutrition biomarkers during normal pregnancy in healthy women[a, 158,178,179,202]

Nutrient	Weeks Gestation	Reference Values
Calcium, mmol/L	7–17	2.18–2.53
	24–28	2.04–2.40
	34–38	2.04–2.41
Chloride, mmol/L	7–17	100–107
	24–28	99–108
	34–38	97–109
Ferritin, μg/L	7–17	7.1–106.4
	24–28	3.8–49.8
	34–38	4.8–43.5
Hemoglobin, g/dL	0–14	>11.0
	14–26	>10.5
	26–40	>110.0
Hematocrit, %	0–14	>33.0
	14–26	>32.0
	26–40	>33.0
Iodin, urinary, μg/L	0–40	150–249
Iron, μmol/L	7–17	8.7–37.0
	24–28	8.0–50.0
	34–38	7.6–34.5
Magnesium, mmol/L	7–17	0.70–0.96
	24–28	0.63–0.91
	34–38	0.57–0.87
Potassium, mmol/L	7–17	3.24–4.86
	24–28	3.27–4.62
	34–38	3.32–5.09
Sodium, mmol/L	7–17	133.2–140.5
	24–28	129.2–139.3
	34–38	127.0–140.2
Transferrin, g/L	7–17	1.92–3.85
	24–28	2.72–4.36
	34–38	2.88–5.12
Triglycerides, mmol/L	7–17	0.55–3.08
	24–28	1.09–3.63
	34–38	1.62–5.12
Vitamin D, nmol/L (25-hydroxyvitamin D)	0–40	≥80 (optimum) <35 (deficient)

[a] See Appendix B for a table of factors used to convert SI Units to conventional units. Nutrition biomarkers considered to be in the normal range vary based on percentile cut-points used. The 5th to 95th percentiles are sometimes used and not the 2.5 to 97.5 percentiles reported in this table. Reference values and blood nutrient concentrations considered "normal" or "abnormal" during pregnancy change based on advances in knowledge. The symbol "μg" means "micrograms," sometimes abbreviated as mcg.

herbal supplements on their own.[203] Clinicians in general support the use of certain dietary supplements. Of 900 doctors and 277 nurses included in a recent survey, 73 to 89% personally used supplements, and 79 to 82% recommended that their patients use them.[204] There is little evidence supporting the safety and effectiveness of many of the dietary supplements available on the market and used by pregnant women, however.

Multivitamin and Mineral Prenatal Supplements

With the exception of iron, nutrient needs during pregnancy should be met by the consumption of a well-balanced and adequate diet.[97] This approach to meeting nutrient needs should be considered first because foods provide antioxidants, fiber, and other beneficial bioactive substances. Healthy diets also provide adequate amounts of protein, health-promoting sources of dietary fat, and nutrient-rich sources of carbohydrates.[97]

Multivitamin and mineral prenatal supplements may benefit women who:

- Do not ordinarily consume an adequate diet
- Have multifetal pregnancy
- Smoke, drink, or use drugs
- Are vegans
- Have iron deficiency anemia
- Have a diagnosed nutrient deficiency[34, 206, 205]

Standard prenatal multivitamin and mineral supplements taken before and during pregnancy appear to benefit women in need of them. Prenatal vitamin and mineral supplement use by low-income pregnant women has been found to decrease the risk of preterm, low birth weight, and certain congenital malformations.[208,209]

About 95% of all pregnant women and 75% of women in WIC take a vitamin and mineral supplement regularly during pregnancy.[45,210] Hundreds of types of prenatal supplements are available by prescription (those that contain over 1 mg folic acid), over the counter, or on the Internet. They contain an amazing array of nutrients, from vitamins and minerals to seaweed, borage, and don guai. Some of the prenatal supplements sold over the Internet contain ingredients that are not considered safe for use in pregnancy, and others provide unreasonably high levels of vitamins or minerals. Table 4.34 summarizes the range in nutrient amounts found in 12 examples of prenatal supplements and compares the amounts to mean nutrient

Table 4.34 Range of daily dose levels of nutrients in 12 prenatal supplements and comparison with recommended intake levels during pregnancy and mean intakes of nutrients for women age 20–29 years[a]

Nutrient	Range in Amounts	RDA	Mean Intake, 20–29-Year-Old Women	UL
Vitamin A	3000–8000 IU	2564 IU	1572 IU	9990 IU
Vitamin E	4–60 IU	22 IU	9.7 IU	1490 IU
Vitamin B_6	2.6–25 mg	1.9 mg	1.6 mg	100 mg
Folate	800–1000 mcg	600 mcg	474 mcg	1000 mcg
Vitamin B_{12}	4–100 mcg	2.6 mcg	4.0 mcg	—
Vitamin C	60–120 mg	85 mg	82 mg	2000 mg
Vitamin D	400–610 IU	200 IU	—	2000 IU
Calcium	68–1300 mg	1000 mg	806 mg	2500 mg
Magnesium	20–200 mg	350 mg	242 mg	350 mg
Iron	21–51 mg	27 mg	13.8 mg	45 mg
Iodine	0–290 mcg	220 mcg	200 mcg	1100 mcg
Zinc	15–30 mg	11 mg	10.8 mg	40 mg
DHA/DHA+EPA	0–440 mg	—	80 mg	—

[a]Prenatal nutrient supplement content determined from company website information, 5/09. United States population-wide information on average nutrient intake of pregnant women is unavailable, so data for women age 20 to 29 are used.[114] RDAs listed are for 19–30-year-old pregnant women, and Tolerable Upper Intake Levels (ULs) for pregnant women age 19–50 years. Table developed by Judith E. Brown, 7/09.

intakes of women in the Untied States, recommended intake levels during pregnancy, and the Tolerable Upper Intake Levels (ULs) of nutrients for pregnancy.

Supplements provided to pregnant women should contain the essential nutrients most likely to be lacking in their diets. These nutrients include vitamin B_6, folic acid, vitamin D, iron, iodine, and EPA+DHA.[206] Nutrient amounts should approximate recommended intake levels and not exceed Tolerable Upper Intake Levels for pregnant women. Supplement use should be accompanied by nutritional counseling that helps women select and consume foods that add up to a healthful diet.

Herbal Remedies and Pregnancy

Herbs are generally regarded by the public and by some health professionals as helpful, safe, and gentle. It is estimated that in the eastern United States, 45% of pregnant women use herbal products during pregnancy. Women may not report use of herbs to their health care provider based on concerns about the provider's knowledge about herbs or a bias against them.[203]

The active ingredients of herbal products are often similar to those in medications that may not be approved for use in pregnancy.[58] About one-third of commonly used herbal supplements have been deemed unsafe for use by pregnant women.[207] Table 4.35 lists some of these herbs. Women who use herbs should be provided respectful counseling about effectiveness and safety, and directed toward reliable sources of information about them.[203]

Advice to use herbal remedies during pregnancy appears to be based primarily on their traditional use in different societies. This strategy for assessing the safety of herbs doesn't always work. Some herbs considered safe based on traditional use have been found to produce malformations in animal studies.[211] Others, such as blue cohosh, which was previously thought to safely induce uterine contractions, may increase the risk of heart failure in the baby.[212] Ginseng, the most commonly used herb in the world, has been found to cause malformations in rat embryos,[211] and ginkgo may promote excessive bleeding.[213] Peppermint tea and ginger root, taken for nausea, appear to be safe.[211]

Table 4.35 Herbs to avoid in pregnancy[211,212]

Aloe vera	Ergot
Anise	Feverfew
Black cohosh	Ginkgo
Black haw	Ginseng
Blue cohosh	Juniper
Borage	Kava
Buckthorn	Licorice
Comfrey	Pennyroyal
Cotton root	Raspberry leaf
Dandelion leaf	Saw palmetto
Ephedra, ma huang	Senna

Ginger, given in oral doses of 1 gram daily for 4 days, has been found to decrease the severity of nausea and vomiting during pregnancy in a majority of women. Ginger use in this study involving 70 women was not related to complications of pregnancy or poor pregnancy outcomes.[214]

Manufacturers of herbal remedies do not have to prove they are safe for use by pregnant women. However, the FDA does advise that claims related to pregnancy not be made for herbal supplements.[207]

Exercise and Pregnancy Outcome

> "Exercise is no longer simply being allowed during pregnancy, it is actively being encouraged."
>
> K. Johnson[215]

There is no evidence that moderate or vigorous exercise undertaken by healthy women consuming high-quality diets and gaining appropriate amounts of weight is harmful to mother or fetus. The bulk of evidence indicates that exercise during pregnancy benefits both the mother and her fetus. Women who exercise regularly during pregnancy feel healthier and have an enhanced sense of well-being, and their labors appear to be somewhat shorter than is the case for women who do not exercise.[216]

Researchers and practitioners are beginning to focus more on the advantages of exercise than on possible disadvantages. Women who exercise regularly during pregnancy reduce their risk of developing gestational diabetes, pregnancy-induced hypertension, low back pain, excessive weight gain, and blood clots.[217]

Exercise during pregnancy can reduce fetal growth in women who are poorly nourished and gain little weight in pregnancy. It is also important for women to avoid dehydration by drinking plenty of fluids while exercising and not to become overheated during physical activity.[218]

Is it safe to begin an exercise program during pregnancy? Not only is it generally safe, it is being encouraged.[219] Beginning an exercise program during pregnancy may improve fetal growth. This effect was shown in a study involving nonexercising pregnant women who began to exercise at 8 weeks of pregnancy. Women participated in three to five weight-bearing exercise sessions a week until delivery. Placenta function was better, and newborn weight and length greater, in exercising women compared to women who did not exercise.[220]

L. Monocytogenes, or Listeria   A foodborne bacterial infection that can lead to preterm delivery and stillbirth in pregnant women. Listeria infection is commonly associated with the ingestion of soft cheeses, unpasteurized milk, ready-to-eat deli meats, and hot dogs.

T. Gondii, or Toxoplasmosis   A parasitic infection that can impair fetal brain development. The source of the infection is often hands contaminated with soil or the contents of a cat litter box; or raw or partially cooked pork, lamb, or venison.

Table 4.36 Target heart rates for healthy pregnant women[220]

Age, Years	Heart Rate
<20	140–155
20–29	135–150
30–39	130–145
40+	125–140

Exercise Recommendations for Pregnant Women

Exercise recommendations for pregnant women are similar to those for other healthy women. Pregnant women should exercise three to five times a week for 20–30 minutes at a heart rate that achieves 60–70% VO_2 max (Table 4.36). Exercise should begin with about 5 minutes of warm-up movements and end with the same length of cooldown activities. Recommended types of exercise include walking, cycling, swimming, jogging, and dancing. Better left until after pregnancy are activities such as water and snow skiing, surfing, mountain climbing, scuba diving, and horseback riding. Switching to non-weight-bearing exercises is advised toward the end of pregnancy.[216]

Food Safety Issues During Pregnancy

Certain foodborne illnesses can be devastating during pregnancy. Increased progesterone levels that normally occur decrease pregnant women's ability to resist infectious diseases, so they are more susceptible to the effects of foodborne infections.[221] One particularly important foodborne illness is caused by *Listeria monocytogenes*. The placenta does not protect the fetus from listeria infection in the mother. Listerosis during pregnancy is associated with spontaneous abortion and stillbirth in one-third of fetuses and mild infection in mothers.[222–223] To prevent this foodborne infection, pregnant women should not eat raw or smoked fish, oysters, unpasteurized cheese, raw or undercooked meat, or unpasteurized milk. Luncheon meats, hot dogs, and other processed meats should be stored correctly, and foods such as hot dogs heated thoroughly.[227]

The protozoan *Toxoplasma gondii* also causes serious effects in pregnant women and their fetuses. This protozoan can be transferred from mother to fetus and cause mental retardation, blindness, seizures, and death.[224] Sources of *T. gondii* include raw and undercooked meats, the surface of fruits and vegetables, and cat litter. Cats that eat wild animals and undercooked meats can become infected and transfer the infection through the air and via stools left in their litter boxes.[225]

Mercury Contamination

Fish have come under fire as a potential source of mercury overload due to contamination of waters and fish by fungicides, fossil fuel exhaust, and products used in smelting plants, pulp and paper mills, leather-tanning facilities, and chemical manufacturing plants. Mercury, which passes from the mother's blood to the fetus, is a fetal neurotoxin that can produce mild to severe effects on fetal brain development. Fetuses exposed to high amounts of mercury can develop mental retardation, hearing loss, numbness, and seizures. Pregnant women are generally only slightly affected by the mercury overload. However, it accumulates in the mother's tissues and may increase fetal exposure to mercury during pregnancy and lactation.[226]

High levels of mercury are most likely to be present in the muscles of large, long-lived predatory fish such as sharks, swordfish, tilefish, albacore tuna, walleye, pickerel, and bass. Mercury content of bottom feeders, such as carp, channel catfish, and white sucker, is generally less than half the amount found in predatory fish. Other fish that tend to have low mercury content include "light" (not white) tuna, haddock, tilapia, salmon, cod, pollack, and sole. Shrimp, lobster, and crab generally have low mercury content, too. Recommendations for fish intake during pregnancy are given on pages 113–114.

Common Health Problems During Pregnancy

Some of the physiological changes that occur in pregnancy are accompanied by side effects that can dull the bliss of expecting a child by making women feel physically miserable. Common ailments of pregnancy, such as nausea and vomiting, heartburn, and constipation, are generally more amenable to prevention than to treatment, but often can be relieved through dietary measures.

Nausea and Vomiting

Nausea occurs in about eight in ten pregnancies, and vomiting in five of ten.[227] Symptoms of nausea generally begin around week 5 of gestation and generally disappear by week 12. Up to 15% of pregnant women will experience some nausea and vomiting throughout pregnancy. The conditions are so common that they are considered a normal part of pregnancy. Unless severe or prolonged, nausea and vomiting during pregnancy are associated with a reduction in risk of miscarriage of greater than 60% and with healthy newborn outcomes. The cause of nausea and vomiting is not yet clear, but they are thought to be related to increased levels of human chorionic gonadotropin, progesterone, estrogen, or other hormones early in pregnancy.[228]

In the past, the nausea and vomiting of pregnancy was called "morning sickness," because it was thought to occur mostly after waking up. It actually occurs at all times of day—a mere 17% of women experience nausea and vomiting only in the morning.[228] Iron supplements may aggravate nausea and vomiting when taken in the first trimester of pregnancy.[173]

Hyperemesis Gravidarum   Between 1 and 2% of pregnant women with nausea and vomiting develop *hyperemesis gravidarum* (more commonly called hyperemesis).[228] Hyperemesis is characterized by severe nausea and vomiting that last throughout much of pregnancy. It can be debilitating. In addition to the mother feeling very sick, frequent vomiting can lead to weight loss, electrolyte imbalances, and dehydration. Women with hyperemesis who gain weight normally during pregnancy (about 30 pounds total) are not at increased risk of delivering small infants, but women who gain less (21–22 pounds) are.[229]

Management of Nausea and Vomiting   Many approaches to the treatment of nausea and vomiting are used in clinical practice, but only a few are considered safe and effective. Dietary interventions represent a safe method, primarily because the short- and long-term safety of many drugs and herbal remedies early in pregnancy is unclear.[230] Here are some general recommendations for women experiencing nausea and vomiting:

- Continue to gain weight.
- Separate liquid and solid food intake.
- Avoid odors and foods that trigger nausea.
- Select foods that are well tolerated.

Many women find that hard-boiled eggs, potato chips, popcorn, yogurt, crackers, and other high-carbohydrate foods are well tolerated. Personal support and understanding are important components of counseling women with nausea and vomiting. Care should be taken to individualize dietary advice based on each woman's food preferences and tolerances. Women with hyperemesis may require rehydration therapy to restore fluids and electrolyte balance.[227]

Periodically, articles will appear in the popular press claiming that nausea and vomiting are caused by certain foods, and that women should avoid them to protect their fetus from harmful substances in the foods. Not too long ago it was claimed that bitter-tasting vegetables, for example, should be avoided. When put to the test, this notion was found to be groundless.[231] Theoretical claims that certain foods elicit nausea and vomiting in order to protect the fetus from harmful effects of the food should be considered unreliable until proven in scientific studies.

Dietary Supplements for the Treatment of Nausea and Vomiting   Three types of dietary supplements have been found to decrease the symptoms of nausea and vomiting in pregnancy:

- Vitamin B$_6$ (pyridoxine) supplements given in a 10–25 mg dose every 8 hours reduce the severity of nausea in many women.[232] The upper limit for vitamin B$_6$ intake during pregnancy is 100 mg per day.
- Multivitamin supplements taken prior to and early in pregnancy may decrease the occurrence of nausea and vomiting.[233]
- Ginger in doses of 1 gram a day for 4 days may decrease nausea and vomiting.[214]

Use of moderate doses of vitamins in a multivitamin supplement, and vitamin B$_6$ in doses under the Tolerable Upper Intake Level, appear safe. Further research is needed before a definitive statement can be made regarding the safety of ginger use during pregnancy.

Heartburn

Pregnancy is accompanied by relaxation of gastrointestinal tract muscles. This effect is attributed primarily to progesterone. Relaxation of the muscular valve known as the cardiac or lower esophageal sphincter at the top of the stomach is thought to be the principal reason for the 40–80% incidence of heartburn in women during pregnancy.[234] The loose upper valve may allow stomach contents to be pushed back into the esophagus.[235]

Management of Heartburn   Dietary advice for the prevention and management of heartburn includes:

- Ingest small meals frequently.
- Do not go to bed with a full stomach.
- Avoid foods that seem to make heartburn worse.

Elevating the upper body during sleep, and not bending down so your head is below your waist, also reduce gastric reflux. Antacid tablets, which act locally in the stomach, are often recommended, but sodium bicarbonate (baking soda) and heartburn pills usually are not.[234]

Constipation

Relaxed gastrointestinal muscle tone is thought to be primarily responsible for the increased incidence of constipation and hemorrhoids in pregnancy. One way to prevent these maladies is to consume approximately 30 grams of dietary fiber daily.[234] (Food sources of fiber are listed in Table 1.5c in Chapter 1.) Laxative pills are not recommended for use by pregnant women, but soluble fiber in products such as Metamucil, Citrucel, and Perdiem are considered safe and effective for the prevention and treatment of constipation.[234] Women should drink a cup or more of water along with the fiber supplement.

Model Nutrition Programs for Risk Reduction in Pregnancy

"Pregnancy may be the most sensitive period of the life cycle in which intervention may reap the greatest benefits."

A. Prentice[164]

Two programs that have been shown to substantially improve pregnancy outcomes are highlighted in this section. First is the intervention program offered by the Montreal Diet Dispensary (MDD); second is the Supplemental Nutrition Program for Women, Infants, and Children (WIC).

The Montreal Diet Dispensary

The Montreal Diet Dispensary (MDD) has served low-income, high-risk pregnant females with nutritional assessment and intervention services since the early 1900s. Part of the rationale for the WIC program in the United States was based on the successes of the MDD program. The program is located in a large, comfortable house (see Illustration 4.20) in urban Montreal. Clients are warmly welcomed into a nonthreatening, relaxed setting.

Developed as an adjunct to routine prenatal care, the MDD intervention strategy has four major components:

1. Assess the usual dietary intake and risk profile of each pregnant woman, including calories, protein, and selected vitamin and mineral adequacy; also assess stress level.
2. Determine individual nutritional rehabilitation needs based on results of the assessment.

Illustration 4.20 The Montreal Diet Dispensary.

Judith Brown

3. Teach clients the importance of optimal nutrition and about changes that should be made through practical examples.

4. Provide regular follow-up and supervision.

The MDD dietitians are carefully trained and hold the interests of their clients first in their hearts. They treat clients with respect, openness, and affection; they also address client needs, such as transportation or emergency food or housing. Staff interactions with clients are nonjudgmental in nature and include positive feedback and praise for dietary changes and other successes of clients.

The initial client visit to the MDD takes about 75 minutes, and follow-up visits are scheduled at 2-week intervals for 40 minutes each. Women are identified as undernourished if their protein intake falls below that recommended for pregnancy, and an additional protein allowance is added to the diet. Women who are underweight are given an additional daily allowance of 20 grams protein and 200 calories for each additional pound of weight gain needed to achieve a maximum of 2 pounds per week. Women identified as being under excessive stress (such as having a partner in jail, being homeless, or being abused) receive an additional allowance of protein and calories and lots of positive attention. Food supplements, including milk and eggs, and vitamin supplements are provided to women who need them.

Impact of MDD Services   Multiple studies have shown that women receiving MDD services have higher-birth-weight infants (+107 grams), fewer low-birth-weight infants (−50%), and infants with lower rates of perinatal mortality than is the case for similar women not receiving MDD services.[236,237]

The program is cost-effective in relation to savings on newborn critical care, and programs based on MDD services have spread across Canada. Expenditures per client average $450. The program is primarily supported by Centraide of Greater Montreal, provincial and federal programs, and other contributions.[238]

The WIC Program

The WIC program represents an outstanding example of a successful public program intended to serve the nutritional needs of low-income women and families. It is cited as a model program in several other chapters and is described in Chapter 1.

In operation since 1974, WIC provides nutritional assessment, education and counseling, food supplements, and access to health services to over 6 million participants. WIC serves low-income pregnant, postpartum, and breastfeeding women, and children up to 5 years of age who are at nutritional risk. Supplemental food provided to women includes milk, ready-to-eat cereals, dried beans, fruit juice, and cheese; some programs offer vouchers for farmer's markets.

Participation in WIC is related to reduced rates of iron-deficiency anemia in pregnancy, higher-birth-weight infants, decreased low-birth-weight infants, and lower rates of iron-deficiency anemia in women after delivery. For each dollar invested in WIC, approximately $3 in health care costs are saved. Internet addresses leading to additional information about WIC are listed in the resources section at the end of this chapter.

Key Points

1. Nutritional status before and during pregnancy can modify the health of women during pregnancy, as well as the current and future health of infants.

2. The United States spends more money on health care than any other country, yet its birth outcomes are far from the best internationally. Improved maternal nutrition could help improve the health status of U.S. newborns.

3. A woman's body prepares in advance for upcoming physiological events related to placental growth and fetal growth and development (such as proliferation of cells in organs and tissues of the placenta and fetus, and rapid increases in fetal weight). Consequently, nutritional needs must be met *prior* to the physiological changes.

4. Functions of the placenta include hormone and enzyme production, nutrient and gas exchange between mother and fetus, and removal of waste products from the fetus.

5. The placenta does *not* block all harmful substances from entering the fetus.

6. The fetus is *not* a parasite. It cannot take whatever nutrients it needs from the mother's body.

7. Variations in fetal growth and development are generally *not* due to genetic causes but rather to environmental factors such as energy, nutrient, and oxygen availability, and to conditions that interfere with genetically programmed growth and development.

8. Energy and nutrient availability is considered the major intrauterine environmental factor that alters expression of fetal genes. This phenomenon represents the major mechanism that underlies the relationship between maternal nutrition and later disease risk.

9. Pregnancy weight gain affects birth weight and long-term health outcomes. Weight gain recommendations are based on prepregnancy weight status.

10. Excessive weight gain is related to postpartum weight retention.

11. Caloric adequacy during pregnancy can be estimated by weight gain.

12. High-quality vegetarian diets promote a healthy course and outcome of pregnancy.

13. Consumption of the omega-3 fatty acids, EPA and DHA, promote visual and intellectual development in offspring, and increase gestational duration somewhat. Most U.S. women consume too little EPA and DHA during pregnancy.

14. Key nutrients of particular importance during pregnancy are folate, vitamin D, calcium, iron, iodine, and EPA and DHA. Antioxidants from plant food also play key roles in maintaining maternal and fetal health.

15. Not all pregnant women need a multivitamin and mineral supplement during pregnancy. Women at risk of deficiencies do.

16. In general, exercise is beneficial to the course and outcome of pregnancy.

17. Certain foodborne illnesses in pregnant women can threaten fetal survival.

18. Some of the common discomforts of pregnancy, such as nausea and vomiting and constipation, can be ameliorated by nutritional measures.

Review Questions

A. The next four questions pertain to the following case:

Tony was born at 35 weeks of pregnancy and weighed 2075 grams. His waist circumference was low relative to his weight and length. Tony was hospitalized for an infection at 4 months of age.

1. Tony was born preterm.
   ___ True   ___ False

2. How much did Tony weigh in pounds and ounces?
   ___ lbs   ___ oz

3. Tony was proportionately small for gestational age.
   ___ True   ___ False

4. Tony was hospitalized during the perinatal period.
   ___ True   ___ False

5. High intakes of regular coffee or caffeine during pregnancy are strongly related to preterm delivery.
   ___ True   ___ False

6. Vegan diets are NOT recommended for pregnancy because they generally fail to provide sufficient protein to support normal growth and development of the fetus.
   ___ True   ___ False

7. The amount of iron absorbed from supplements decreases as the amount of iron in the supplement increases.
   ___ True   ___ False

8. The Institute of Medicine recommends that all pregnant women take a multivitamin and mineral supplement during pregnancy.
   ___ True   ___ False

B. Use the conversion factors listed in the "Conventional Units to SI Units" table in Appendix C to convert the following conventional unit measures to SI units.

9. 52 ng/mL vitamin D (25 hydoxyvitamin D) = ___ nmol/L

10. 86 ng/mL ferritin = ___ pmol/L

11. 402 ng/mL red cell folate = ___ nmol/L

Resources

Pregnancy Resources and Information

Visit the Women's Health Resource Center for access to journal articles and summaries; information on pregnancy, growth, and development; and health care and diversity information.
Website: **www.medscape.com**

The Bureau of Maternal and Child Health website provides information on programs for pregnant women, hot topics, and announcements of new publications.
Website: **http://mchlibrary.info**

The National Library of Medicine website offers extensive coverage of scientific journal articles, summaries, and educational resources from a variety of reputable organizations on pregnancy, nutrition, diet, and disorders of pregnancy.
Website: **www.nlm.nih.gov/medlineplus**

Fish Advisories

This site links to local freshwater fish advisories.
Website: **www.epa.gov/waterscience/fish/states.htm**

Continuing Education Presentations

American Society for Nutrition members can sign in at www.nutrition.org and go to this page to access a presentation at the 2009 Experimental Biology meetings on "Nutritional Experiences in Early Life as Determinants of the Adult Metabolic Phenotype."

Website: **www.nutrition.org/education-and-professional-development/videotaped-symposia-from-asn-annual-meeting-at-experimental-biology-2009**

Food Safety

Food safety information for mothers-to-be is available from the FDA.
Website: **www.cfsan.fda.gov/~pregnant/pregnant.html**

Pregnancy Information

The American Pregnancy Association offers a search tool for information about pregnancy by topic (e.g., pregnancy calendar, ovulation calendar, paternity testing information, finding a health care professional).
Website: **http://www.americanpregnancy.org**

Pregnancy Food Guides

Obtain a "MyPyramid for Moms" menu plan and dietary analyses at this site.
Website: **www.mypyramid.gov/mypyramidmoms/ pyramidmoms_plan.aspx**

"Eating Well with Canada's Food Guide" provides women with the information they need to eat well during pregnancy.
Website: **www.hc-sc.gc.ca/fn-an/nutrition/prenatal/index-eng.php**

Natality Statistics

The Centers for Disease Control and Prevention offers "Wonder," a single-point access to public health reports and health statistics.
Website: **wonder.cdc.gov**

U.S. Government

The U.S. government's site provides information on food and nutrition programs and eligibility; links to scientific references; and information about the nutrition needs of infants, children, adults, and seniors.
Website: **www.nutrition.gov**

WIC

The USDA provides access to information about the WIC program, the WIC Works Food Safety Resource List, and other resources.
Website: **www.fns.usda.gov/wic/aboutwic**

Dietary Analysis

Select "MyPyramid Tracker" and run dietary intake records one day at a time. Analyzes diet by food groups and selected nutrients.
Website: **www.mypyramid.gov**

This USDA site is the best one for food composition data.
Website: **www.ars.usda.gov/nutrientdata**

"Women at greater risk of adverse birth outcomes benefit the most from educational health care messages."

M. D. Kogan et al.

Chapter 5

Nutrition During Pregnancy:
Conditions and Interventions

Photodisc

Ayleen and Jeremy Perez-Marty

Brand X Pictures

Prepared by **Judith E. Brown**

Key Nutrition Concepts

1 Some complications of pregnancy are related to women's nutritional status.

2 Nutritional interventions for a number of complications of pregnancy can benefit maternal and infant health outcomes.

3 Nutritional interventions during pregnancy should be based on scientific evidence that supports their safety, effectiveness, and affordability.

Introduction

> "Practice is science touched with emotion."
>
> Stephen Paget, 1909, *Confessio Medici*

Almost all healthy women expect that their pregnancies will proceed normally and that they will be rewarded at delivery with a healthy newborn. For the vast majority of pregnancies, this expectation is fulfilled. For other women, however, the path to a healthy newborn is strewn with obstacles in the form of health problems that women bring into or develop during pregnancy. This chapter addresses a number of these health conditions and the role of nutrition in their etiology and management. The specific health conditions presented include hypertensive disorders of pregnancy, preexisting and gestational diabetes, obesity, multifetal pregnancy, HIV/AIDS, eating disorders, fetal alcohol spectrum, and adolescent pregnancy. Excess body fat alters many metabolic processes and influences the development of two of the clinical conditions presented in this chapter. The subject of obesity during pregnancy is presented first.

Obesity and Pregnancy

Obesity prior to pregnancy is associated with higher rates of gestational diabetes and hypertensive disorders of pregnancy (Table 5.1).[1]

The increased risk of these disorders is associated with unfavorable metabolic changes initiated by excessive body fat, such as:

- Increased blood glucose levels
- High C-reactive protein levels (a key marker of inflammation)
- Increased blood concentration of insulin
- Insulin resistance
- Increased blood pressure
- High blood levels of total cholesterol, LDL-cholesterol, and triglycerides
- Low levels of HDL-cholesterol[2]

Approximately 70% of obese persons and 23% of normal-weight individuals have two or more of these metabolic abnormalities that increase disease risk.[3]

Metabolic effects associated with obesity are closely related to the presence of high amounts of visceral fat. Visceral fat lies beneath skin and muscles of the abdomen (Illustration 5.1). It is much more metabolically active than subcutaneous fat (fat that lies beneath the skin) and is more strongly related to disease risk.[4,5] Metabolic processes initiated by visceral fat produce chronic inflammation, free-radical generation, and oxidative stress. These disruptions promote the development of insulin resistance, elevated blood glucose, insulin, and triglyceride concentrations, and increased blood pressure.[6] These changes, in turn, increase the risk of gestational diabetes, hypertensive disorders, and other clinical conditions during pregnancy.

Table 5.1 Comparative prevalence of obesity prior to pregnancy and outcomes related to prepregnancy weight status

Weight status	Underweight	Normal	Overweight	Obese	Very obese	Extremely obese
BMI, kg/m^2	<18.5	18.5–24.9	25–29.9	30–34.9	35–39.9	≥40
Preterm	11.6%	8.1%	8.4%	10.6%	8.9%	12.4%
Gestational diabetes	3.5%	3.8%	4.7%	7.0%	9.6%	11.0%
Preexisting diabetes	0%	0.8%	1.7%	2.4%	6.9%	9.7%
Hypertensive disorder	5.8%	9.1%	13.3%	20.7%	23.3%	31.7%

SOURCE: Data are from Chu et al.

Illustration 5.1 Subcutaneous fat is located under the skin (left photo). The illustration on the right shows the location of visceral fat.

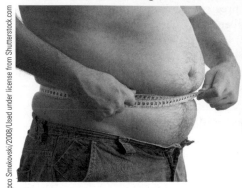

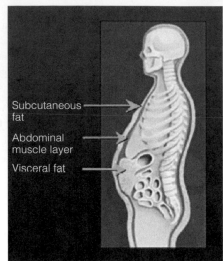

Subcutaneous fat

Abdominal muscle layer

Visceral fat

Ljupco Smokovski/2008/Used under license from Shutterstock.com

Dumping Syndrome A condition characterized by weakness, dizziness, flushing, nausea, and palpitation immediately or shortly after eating and produced by abnormally rapid emptying of the stomach, especially in individuals who have had part of the stomach removed.

Normal-weight and overweight individuals with excessive visceral fat deposits are also at increased risk of metabolic abnormalities and diseases associated with them.[7]

The prevalence of disorders such as gestational diabetes and hypertension during pregnancy is increasing due to higher rates of obesity. Approximately 28.4% of women 20 to 34 years old in the United States are obese now compared to 10% in 1980.[8]

Obesity and Infant Outcomes

Rates of stillbirth, large-for-gestational newborns, and Cesarean-section delivery tend to be higher in obese than in non-obese women.[9] Some infants born to women who enter pregnancy obese are at high risk of becoming overweight during childhood and of developing type 2 diabetes later in life.[2] The increased risks appear to be partly related to fetal exposure to high levels of insulin and a propensity toward developing insulin resistance.[10]

Nutritional Recommendations and Interventions for Obesity During Pregnancy

Nutrition and other health care services should be provided to obese women in the same nonjudgmental and nondiscriminatory manner as for other groups of women.[11] Many nutrition recommendations for pregnancy in obese women, including meeting nutrient needs through consumption of a variety of basic foods, participating in physical activity, and maintaining appropriate rates of weight gain, are the same as those for women of other sizes.

Nutrition assessment, counseling and development of interventions to address identified nutrition problems, and monitoring and evaluation of the results of interventions are core components of nutrition service delivery.[12] Changes in calorie intake and physical activity levels may be needed to adjust rates of pregnancy weight gain.

Pregnancy After Bariatric Surgery The use of bariatric surgery for weight loss in severe obesity has increased dramatically over the past 10 years. About half of individuals who undergo this surgery are women of childbearing age.[17]

Women lose weight rapidly after bariatric surgery due to limited food intake, fat malabsorption, and also due to the presence of *dumping syndrome,* a common side effect of the surgery.[14]

Nausea, vomiting, and other symptoms of dumping syndrome may persist into pregnancy. The rapid weight loss and limited food and nutrient intake lead to depletion of tissue stores of many nutrients. Thiamin deficiency can develop within 20 days after the surgery,[18] and deficiencies of vitamins D and B_{12}, iron, calcium, and folate are common.[15] Vitamin and mineral supplementation and monitoring of nutrient biomarkers are routine components of care for individuals undergoing weight-loss surgery. It is recommended that pregnancy be postponed for a year or two after bariatric surgery, when body weight is stable and nutrient stores have been established.[9]

Maternal and neonatal health outcomes in women who have had bariatric surgery tend to be acceptable as long as adequate maternal nutrition is maintained.[16] Maternal complications such as gestational diabetes and preeclampsia are lower in obese women who had weight-loss surgery compared to obese women who did not.[17]

Nutrition Care for Pregnant Women Post-Bariatric Surgery Nutrition care services are recommended for pregnant women with a history of bariatric surgery. These services should include assessment of dietary intake, supplement use, nutrient biomarker status, weight gain, physical activity, and gastrointestinal symptoms.[12] Nutrition interventions are then implemented as needed to solve nutrition-related problems identified.[9]

Nutrient deficiencies and supplementation requirements will vary during pregnancy based on the type of bariatric surgery performed.[18] Roux-en-Y bypass and biliopancreatic diversion surgery severely restrict food intake and are associated with greater weight loss and nutrient deficits than is "lap band" (vertical banded gastroplasty) surgery.[15] Lap band surgery involves placement of a band around the top of the stomach. The band limits the size of the stomach and restricts the amount of food that can be consumed. The lap band is sometimes adjusted during pregnancy to help regulate food intake and weight gain.[14]

Women with a history of bariatric surgery generally qualify as "at risk" for gestational diabetes, and blood glucose assessment will be ordered. Many women will experience dumping syndrome if given the standard 100-gram glucose load for blood glucose testing. Alternatives to this test such as fasting glucose levels or home blood glucose values should be used instead of the oral glucose load.[19]

Hypertensive Disorders of Pregnancy

Hypertensive disorders of pregnancy are the second leading cause of maternal mortality in the United States. They affect 6 to 10% of pregnancies and contribute significantly to stillbirths, fetal and newborn deaths, and other adverse outcomes of pregnancy. The causes of most cases of hypertension during pregnancy remain unknown, and cures for these disorders remain elusive.[13]

Several types of hypertensive disorders in pregnancy have been identified (Table 5.2). In the past, the major types of hypertensive disorders in pregnancy were grouped under the heading "pregnancy-induced hypertension," or PIH. This terminology is being phased out in favor of the classification scheme for hypertensive disorders of pregnancy presented in Table 5.2.

Hypertensive Disorders of Pregnancy, Oxidative Stress, and Nutrition

All forms of hypertension in pregnancy (as well as other disorders such as diabetes) are related to chronic inflammation, *oxidative stress,* and damage to the *endothelium* of blood vessels throughout the body.[21,22] Over time, oxidative stress within the endothelium leads to endothelial dysfunction. Consequences of endothelial dysfunction include impaired blood flow, an increased tendency of blood to clot, and plaque formation.[22]

A number of nutritional and other environmental factors are related to chronic inflammation and oxidative stress (Table 5.3) Except for weight loss, pregnant women with hypertensive disorders may benefit from lifestyle and dietary changes that lower inflammation and oxidative stress.

> **Oxidative Stress** A condition that occurs when cells are exposed to more oxidizing molecules (such as free radicals) than to antioxidant molecules that neutralize them and help repair cell damage. Over time, oxidative stress causes damage to lipids, DNA, cells, and tissues.
>
> **Endothelium** The layer of cells lining the inside of blood vessels.

Table 5.2 Definitions and features of hypertensive disorders of pregnancy[*2]

Chronic Hypertension

Hypertension that is present before pregnancy or diagnosed before 20 weeks of pregnancy. Hypertension is defined as blood pressure ≥140 mm Hg systolic or ≥90 mm Hg diastolic blood pressure.

Hypertension first diagnosed during pregnancy that does not resolve after pregnancy is also classified as chronic hypertension.

Gestational Hypertension

This condition exists when elevated blood pressure levels are detected for the first time after mid-pregnancy. It is not accompanied by proteinuria. If blood pressure returns to normal by 12 weeks postpartum, the condition is considered to be transient hypertension of pregnancy. If it remains elevated, then the woman is considered to have chronic hypertension.

Women with gestational hypertension are at lower risk for poor pregnancy outcomes than are women with preeclampsia.

Preeclampsia–Eclampsia

A pregnancy-specific syndrome that usually occurs after 20 weeks gestation (but that may occur earlier) in previously normotensive women. It is determined by increased blood pressure during pregnancy to ≥140 mm Hg systolic or ≥90 mm Hg diastolic and is accompanied by proteinuria. In the absence of proteinuria, the disease is highly suspected when increased blood pressure is accompanied by headache, blurred vision, abdominal pain, low platelet count, and abnormal liver enzyme values.

- Proteinuria is defined as the urinary excretion of ≥0.3 grams of protein in a 24-hour urine specimen. This usually correlates well with readings of ≥30 mg/dL protein, or ≥2 on dipstick readings taken in samples from women free of urinary tract infection. In the absence of urinary tract infection, proteinuria is a manifestation of kidney damage.
- Eclampsia is defined as the occurrence of seizures that cannot be attributed to other causes in women with preeclampsia.

Preeclampsia Superimposed on Chronic Hypertension

This disorder is characterized by the development of proteinuria during pregnancy in women with chronic hypertension. In women with hypertension and proteinuria before 20 weeks of pregnancy, it is indicated by a sudden increase in proteinuria, blood pressure, or abnormal platelet or liver enzyme levels.

---

*Blood pressure values used to determine status should be based on two or more measurements of blood pressure in relaxed settings.

Table 5.3 Dietary and other environmental exposures that increase or decrease chronic inflammation and oxidative stress[13,20]

1. Decrease
   - Regular intake of colorful fruits and vegetables, dried beans, and whole-grain products
   - Adequate intake of the omega-3 fatty acids EPA and DHA
   - Vitamin D sufficiency
   - Physical activity
2. Increase
   - Frequent intake of processed and high-fat meats
   - Regular intake of baked products and snack foods with *trans* fats
   - Frequent consumption of soft drinks, other high-sugar beverages
   - Physical inactivity
   - High levels of body fat, especially visceral fat
   - Smoking

Chronic Hypertension

The incidence of chronic hypertension—or that diagnosed prior to pregnancy or before 20 weeks after conception—ranges from 1 to 5% depending on the population studied. The condition is more likely to occur in African Americans, obese women, women over 35 years of age, and women who experienced high blood pressure in a previous pregnancy.[23]

Prostacyclin A potent inhibitor of platelet aggregation and a powerful vasodilator and blood pressure reducer derived from n-3 fatty acids.

Thromboxane The parent of a group of thromboxanes derived from the n-6 fatty acid arachidonic acid. Thromboxane increases platelet aggregation and constricts blood vessels, causing blood pressure to increase.

Women with mild hypertension may be taken off antihypertension medications preconceptionally or early in pregnancy, because the drugs do not appear to improve the course or outcome of pregnancy.[24] Mild hypertension in healthy women that does not become worse during pregnancy appears to pose few risks to maternal and newborn health. Pregnancies among women with blood pressures ≥160/110 mm Hg—either or both values—are associated with an increased risk of fetal death, preterm delivery, and fetal growth retardation. Selection of the proper antihypertension medicines for women during pregnancy reduces these risks somewhat.

Nutritional Interventions for Women with Chronic Hypertension in Pregnancy Preconceptional and pregnancy diets of women with hypertension should be carefully monitored with the aim of achieving adequate and balanced diets for pregnancy. Weight-gain recommendations are the same as for other pregnant women.

Women with salt-sensitive hypertension, or hypertension that responds to dietary sodium intake, must be managed along a fine line between consuming too much sodium for good blood pressure control and consuming too little at the potential cost of impaired fetal growth.[24] For women with hypertension that was managed successfully in part by a low-sodium diet prior to pregnancy, continuing that dietary approach is generally recommended.[25]

Gestational Hypertension

Gestational hypertension is usually diagnosed after 20 weeks of pregnancy. Unlike women with preeclampsia, women with gestational hypertension do not have proteinuria. Women with this disorder are at greater risk for hypertension and stroke later in life. Women with gestational hypertension tend to be overweight or obese and have excess central body fat.[26]

Preeclampsia–Eclampsia

Preeclampsia–eclampsia represents a syndrome characterized by:

- Oxidative stress, inadequate antioxidant defenses, inflammation, and endothelial dysfunction
- Platelet aggregation and blood coagulation due to deficits in *prostacyclin* relative to *thromboxane*
- Blood vessel spasms and constriction, restricted blood flow
- Increased blood pressure
- Insulin resistance
- Adverse maternal immune system responses to the placenta [21,27,28]
- Elevated blood levels of triglycerides, free fatty acids, and cholesterol

Many of the metabolic abnormalities observed in preeclampsia are present before it is diagnosed and are the same as those for cardiovascular disease.[29] Occurrence of preeclampsia during pregnancy is predictive of later development of cardiovascular disease.[30]

Virtually all maternal organs can be affected in preeclampsia. Organs most affected by small blood clots, vasoconstriction, and reduced blood flow are the placenta and the mother's kidney, liver, and brain.[31]

Eclampsia can be a life-threatening condition and one that is difficult to predict. Eclamptic seizures appear to be related to hypertension, the tendency of blood to clot, and spasms of and damage to blood vessels in the brain. It complicates about 1 in 2000 pregnancies.[32]

Signs and symptoms of preeclampsia range from mild to severe (Table 5.4), as do the health consequences (Table 5.5). The cause of preeclampsia is unknown but appears to originate from abnormal implantation and vascularization of

Table 5.4 Signs and symptoms of preeclampsia[33]

- Hypertension
- Increased urinary protein (albumin)
- Decreased plasma volume expansion (hemoglobin levels .13 g/dL)
- Low urine output
- Persistent and severe headaches
- Sensitivity of the eyes to bright light
- Blurred vision
- Abdominal pain
- Nausea

Table 5.5 Outcomes related to the existence of preeclampsia during pregnancy[23,35]

Mother
- Early delivery by cesarean section
- Acute renal (kidney) dysfunction
- Increased risk of gestational diabetes, hypertension, and type 2 diabetes later in life
- Abruptio placenta (rupture of the placenta)

Newborn
- Growth restriction
- Respiratory distress syndrome

Table 5.6 Risk factors for preeclampsia[23,38-40]

- First pregnancy (nulliparous)
- Obesity, especially high levels of central body fat
- Underweight
- Mother's smallness at birth
- African Americans, American Indians
- History of preeclampsia
- Preexisting diabetes mellitus
- Age over 35 years
- Multifetal pregnancy
- Insulin resistance
- Abnormally high blood triglyceride levels
- Chronic hypertension
- Renal disease
- Poor vitamin D status
- Poor calcium status
- Consumption of a pro-inflammatory, pro-oxidative stress diet

the placenta, and poor blood flow through the placenta.[31] Abnormal blood flow through the placenta is an important characteristic of preeclampsia because it decreases the delivery of nutrients and gases to the fetus. It appears to be related to oxidative stress, reduced antioxidant defenses, and endothelial dysfunction.[13] Insulin resistance is also a common characteristic of preeclampsia and contributes to some of the negative consequences observed.[27] The only cure is delivery.[25] Signs and symptoms of preeclampsia generally disappear rapidly after delivery, but eclampsia may occur within 12 days following delivery.[34]

Women with preeclampsia are at increased risk for developing gestational diabetes during pregnancy, and type 2 diabetes, hypertension, heart disease, and stroke later in life.[36] About 15% of women with gestational diabetes and 30% of those with type 2 diabetes prior to pregnancy will develop preeclampsia.[32] A history of preeclampsia increases the risk that it will occur in subsequent pregnancies.[38]

Risk Factors for the Development of Preeclampsia The roots of preeclampsia lie very early in pregnancy, but as yet there is no reliable means of identifying women who will develop it before the condition is established.[38] However, women with insulin resistance, obesity, abnormally high triglyceride levels, or other characteristics listed in Table 5.6 are at increased risk for developing the disease.

Increased rates of preterm delivery and low-birth-weight in infants born to women with preeclampsia are partly related to clinical decisions to deliver fetuses early in order to treat the disease. Most infants born to women with this disorder are normal weight, however, and some newborns are large for gestational age. Variations in birth weight associated with preeclampsia appear to be related to the severity of the disease in individual women.[41]

The risk of developing preeclampsia is higher in women who were born small for gestational age (SGA). It appears that growth restriction *in utero* may impair mechanisms involved in the regulation of blood pressure and increase the probability that high blood pressure will develop with the physiological stresses of pregnancy.[42]

Vitamin and Mineral Supplementation and the Risk of Preeclampsia Oxidative stress and a lack of antioxidant defenses appear to play key roles in the development of preeclampsia. Based on this knowledge, it was theorized that therapeutic doses of vitamins C and E (which function as antioxidants) would decrease oxidative stress and the risk of preeclampsia. Results from early studies suggested that this did happen, but later, better-designed clinical trials failed to identify a true relationship between supplemental vitamin C and E intake and preeclampsia. It is now concluded that vitamin C and E supplements should not be used to prevent preeclampsia.[43] Supplemental vitamin D, however, has been related to a reduced risk of preeclampsia in women with poor vitamin D status.[40,44]

Use of multivitamin and mineral supplement in the months before and early in pregnancy have been related to reduced risk of preeclampsia in normal-weight women. [45,46]

Calcium Supplements and Preeclampsia   Calcium performs a number of functions related to endothelial function and maintenance of normal blood pressure, and intake has been found to be low in many groups of women. Consequently, potential roles of calcium supplementation during pregnancy in the prevention of preeclamspia have been investigated. Overall, studies have found that daily doses of one gram or more of supplemental calcium during pregnancy reduce the risk for preeclampsia by half. The extent of reduction in risk for preeclampsia was greatest in women with low calcium intakes prior to supplementation. Children born to supplemented women have been found to have lower blood pressure than children whose mothers were not supplemented. [47]

Dietary Intake and the Risk of Preeclampsia   Certain patterns of dietary intake during the first 22 weeks of pregnancy have been related to the risk of preeclampsia. Diets characterized by high intake of plant foods that tend to decrease chronic inflammation and oxidative stress have been linked to a decreased risk of preeclampsia compared to diets that regularly include processed meat, sweet drinks, and salty snacks. [13]

Prepregnancy and early-pregnancy diets containing relatively high amounts of fiber (over 21 grams a day) have been related to a significant reduction in the risk of preeclampsia. High-fiber diets may modify the risk of preeclampsia by reducing abnormally high blood concentrations of triglycerides and cholesterol that may contribute to the development of oxidative stress. [21,48]

Sodium (Salt) Intake and the Risk of Preeclampsia   In the past it was thought that high sodium intakes were related to the development of preeclampsia and that low salt intakes would help prevent it. These clinical assumptions have not been found to be accurate in clinical studies. Salt restriction during pregnancy does not prevent preeclampsia, hypertension, or other complications of pregnancy. Routine salt restriction is not recommended. Rather, it is recommended that salt consumption during pregnancy remain a matter of personal preference. [49]

Preeclampsia Case Presentation

Signs, symptoms, severity, and causes of preeclampsia vary from woman to woman. Therefore, appropriate interventions for women presenting differing aspects of the syndrome are best designed on a case-by-case basis. [41] By way of example, Case Study 5.1 describes the course of preeclampsia in one woman experiencing the condition.

Gestational Diabetes Carbohydrate intolerance with onset of, or first recognition in, pregnancy.

Nutritional Recommendations and Interventions for Preeclampsia

In the best of circumstances, dietary interventions for preeclampsia would begin prior to pregnancy. This approach might give women the opportunities to decrease body weight and stores of central body fat, become physically fit, and consume a diet that reduces inflammation and oxidative stress. Short of those circumstances, dietary recommendations and interventions should begin in at-risk women as early in pregnancy as possible.

Nutritional and physical activity recommendations that may benefit women at risk of preeclampsia include:

- 1000–2000 mg per day of dietary or supplemental calcium
- Adequate vitamin D status
- Use of a multivitamin-mineral supplement
- Five or more servings of colorful vegetables and fruits daily
- Consumption of the assortment of other basic foods recommended in MyPyramid
- Moderate exercise (for example, walking, swimming, noncompetitive tennis, or dancing for 30 minutes) daily unless medically contraindicated
- Weight gain that follows recommendations based on prepregnancy weight status

Iron supplements, especially if taken in high doses, may aggravate inflammation by increasing the body's free-radical load. [50] Women with preeclampsia should not be given high-dose iron supplements.

Diabetes in Pregnancy

Diabetes is a leading complication in pregnancy. It has several forms:

- Gestational diabetes
- Type 2 diabetes
- Type 1 diabetes
- Other specific types [51]

Gestational diabetes and type 2 diabetes are part of the same disease process. This chapter focuses on gestational diabetes, which develops during pregnancy. Information related to type 1 diabetes, or insulin-dependent diabetes, is presented as well.

Gestational Diabetes

> "The difference between gestational and type 2 diabetes may be the moment of detection."
>
> Branchtein [52]

Approximately 7.5% of pregnant women develop *gestational diabetes* during pregnancy, and the incidence is increasing along with obesity. Gestational diabetes accounts for 88% of all cases of diabetes in pregnancy. [53]

A Case of Preeclampsia

Susan is a 19-year-old "meat-and-potato" type eater who rarely consumes vegetables, fruits, or dairy products. She likes monosyllabic vegetables (beans, corn, and peas), bananas and oranges, and chocolate milk. She generally consumes one of these vegetables and fruits each day, and always has a glass of chocolate milk. Susan consumes sweetened iced tea throughout the day, and twice a week she eats rice with meat rather than potatoes. She finds this type of diet satisfying and rarely consumes foods other than those mentioned.

Her first 17 weeks of pregnancy were uneventful. At week 18 she was found to have proteinuria. By week 22, her blood pressure had increased to 150/100 mm Hg, and she was diagnosed with preeclampsia. Laboratory studies indicated that her blood glucose level was on the high side of normal and that she was insulin resistant. She was lost to follow-up after her week 22 visit.

A bit overweight prior to pregnancy, Susan did not gain weight and restricted her salt intake after midpregnancy. She believed these actions would help lower her weight and blood pressure. Although she was given a supply of prenatal vitamin and mineral supplements early in pregnancy, she rarely remembered to take them. Her baby, weighing 5 pounds 5 ounces (2380 grams), was delivered by cesarean section at week 36.

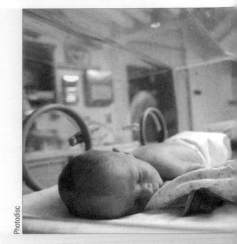
Photodisc

Questions

1. List three ways in which Susan's dietary intake likely contributes to oxidative stress.
2. Identify two other characteristics of her diet that are contraindicated for women with preeclampsia. Answers should be different than those for question 1.
3. List three health problems Susan is at increased risk of developing due to her history of preeclampsia.

(Answers are located in the Instructor's Manual for the 4th edition of *Nutrition Through the Life Cycle*.)

Gestational diabetes in underweight and normal-weight women appears to be related to insulin resistance combined with a deficit in insulin production, whereas insulin resistance—not inadequate insulin production—may underlie it in obese women.[54,55] Women who develop gestational diabetes appear to enter pregnancy with a predisposition to insulin resistance and type 2 diabetes that is expressed due to physiological changes that occur during pregnancy.

The insulin resistance brought into pregnancy, or the tendency to develop it, may be clinically silent in that glucose levels may not be elevated and blood pressure may be normal. However, high blood levels of glucose and other signs related to increased insulin resistance develop as pregnancy progresses. Women with gestational diabetes develop elevated levels not only of blood glucose but also of triglycerides, fatty acids, and sometimes blood pressure. Gestational diabetes appears to be related to exaggerated metabolic changes favoring oxidative stress and elevated blood glucose levels.[54,56]

High maternal blood glucose levels reach the fetus (Illustration 5.2) and cause the fetus to increase insulin production to lower those levels. The higher the level of blood glucose received, the larger the fetal output of insulin.

Potential Consequences of Gestational Diabetes

Potential consequences associated with gestational diabetes are summarized in Table 5.7. Elevated *hemoglobin A1c* levels, a long-term marker of blood glucose concentrations, indicate poor glucose control and higher risk of adverse outcomes. Specifically, hemoglobin A1c levels over 8% are associated with higher rates of spontaneous abortion, stillbirth, neonatal death, and *congenital anomalies* than are values

Hemoglobin A1c A form of hemoglobin used to identify blood glucose levels over the lifetime of a red blood cell (120 days). Glucose molecules in blood will attach to hemoglobin (and stay attached). The amount of glucose that attaches to hemoglobin is proportional to levels of glucose in the blood. The normal range of hemoglobin A1c is 4 to 5.9%. Also called glycosylated hemoglobin and glycated hemoglobin.

Congenital Anomalies Structural, functional, or metabolic abnormalities present at birth. Also called congenital abnormalities.

Illustration 5.2 Concentrations of fetal blood glucose following an intravenous dose of glucose to the mother.

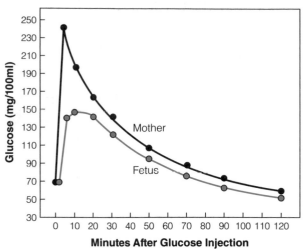

SOURCE: F. Hytten and G. Chamberlain, eds., *Clinical Physiology in Obstetrics,* Blackwell, 1980. Reprinted by permission. Also based on data from Coultart, et al. Reprinted by permission.

Table 5.7 Adverse outcomes associated with gestational diabetes[62]

Mother
- Cesarean delivery to prevent shoulder dystocia
- Increased risk for preeclampsia during pregnancy
- Increased risk of type 2 diabetes, hypertension, and obesity later in life
- Increased risk for gestational diabetes in a subsequent pregnancy

Offspring
- Stillbirth
- Spontaneous abortion
- Congenital anomalies
- Macrosomia (>10 lb or >4500g)
- Neonatal hypoglycemia, death
- Increased risk of insulin resistance, type 2 diabetes, high blood pressure, and obesity later in life

below 7%.[57] Exposure to high insulin levels *in utero* leads to increased glucose uptake into cells and the conversion of glucose to triglycerides. These changes increase fetal formation of fat and muscle tissue and may program metabolic adaptations, increasing the likelihood that insulin resistance, type 2 diabetes, high blood pressure, and obesity will develop later in life. The chances that these disorders will occur increase with higher maternal levels of glucose and triglycerides during pregnancy.[58,59]

Effects of high maternal levels of glucose and triglycerides are particularly striking in the Pima Indians of Arizona. Fetal exposure to poorly controlled maternal diabetes incurs a tenfold increase in the risk that children will develop type 2 diabetes. Offspring of diabetic Pima mothers are heavier at birth, have higher body mass index (BMI) throughout childhood, and have 7–20 times greater incidence of type 2 diabetes in early adulthood. Although risks of these conditions increase in offspring of women with poorly controlled diabetes in general, the pronounced effect in Pima Indians is likely due to a strong genetic tendency toward insulin resistance and obesity.[60]

The end of pregnancy initially restores insulin sensitivity in most women with gestational diabetes. However, a degree of insulin resistance often remains.[37] Close to half of women with gestational diabetes in a previous pregnancy will develop it in a subsequent pregnancy.[61] Women with weight gain after pregnancy and repeated pregnancies continue to experience insulin insufficiency and resistance; this group is at even higher risk of developing type 2 diabetes later in life. Among women who have experienced gestational diabetes, those requiring insulin therapy have higher blood pressure than women whose gestational diabetes was controlled with diet and exercise.[37]

Risk Factors for Gestational Diabetes

Both type 2 and gestational diabetes are linked to multiple inherited predispositions and their environmental triggers, such as excess body fat and low physical activity levels.[64] Results of a large prospective study indicate that the risk for gestational diabetes decreases 26% for each 10 grams of fiber consumed daily from plant sources. Diets both low in cereal fiber and high in glycemic load are associated with a 2.15-fold increased risk compared to diets high in cereal fiber and low in glycemic load.[63] About half of women who develop gestational diabetes have no identified risk for the disease.[64] Risk factors for gestational diabetes are outlined in Table 5.8.

Diagnosis of Gestational Diabetes

The diagnosis of gestational diabetes is based on abnormal blood glucose levels. Glucose screening is recommended for women at high risk of gestational diabetes at the initial visit or as soon as possible thereafter. High risk is identified in women who have one or more of the following:

- Marked obesity
- Diabetes in a mother, father, sister, or brother
- History of glucose intolerance
- Previous macrosomic infant
- Current glucosuria

A 50-gram oral glucose challenge test is generally used for blood glucose screening. This test can be done without fasting. Blood is collected 1 hour after the glucose load is consumed and tested for glucose content. This test should be followed by an oral glucose tolerance test if

Table 5.8 Risk factors for gestational diabetes[61,63,64]

- Obesity, especially high levels of central body fat
- Weight gain between pregnancies
- Underweight
- Age over 35 years
- Native American, Hispanic, African American, South or East Asian, Pacific Islander, indigenous Australian ancestry
- Family history of gestational diabetes
- History of delivery of a macrosomic newborn (>4500 g or >10 lb)
- Chronic hypertension
- Mother was SGA at birth
- History of gestational diabetes in a previous pregnancy
- Diabetes in pregnant women's mothers during the pregnancy with them and LGA at birth
- Low fiber intake, high-glycemic-load diets

Table 5.9 Comparison of outcomes of unrecognized and diet-treated gestational diabetes

Outcome	Gestational Diabetes Unrecognized	Diet-Treated	Controls
LGA (>90th percentile)	44%	9%	5%
Macrosomia (>4500 g)	44%	15%	8%
Shoulder dystocia	25%	3%	3%
Birth trauma	25%	0%	0%

SOURCE: Data from Adams, 1998.[67]

- Normal prepregnancy weight and weight gain during pregnancy
- No history of glucose intolerance
- No prior poor obstetrical outcomes[62]

Women with gestational diabetes may notice an increased level of thirst (especially in the morning), an increased volume of urine, and other signs related to high blood glucose levels.[66] Urinary glucose cannot be used to diagnose nor monitor gestational diabetes, because the results do not correspond to blood glucose levels.[67]

Treatment of Gestational Diabetes

A team approach to caring for women with diabetes in pregnancy is advised. Such teams often consist of an obstetrician, a registered dietitian who is also a certified diabetes educator, a nurse educator, and an endocrinologist. The main stay of treatment is medical nutrition therapy that begins with attempts to normalize blood glucose levels with diet and exercise.[65]

Management of blood glucose concentrations with diet and exercise is considered successful when fasting blood glucose values remain at 95 mg/dL or less, or when 1-hour postprandial values are 140 mg/dL or less and 2-hour postprandial levels are 120 mg/dL or less. Insulin is recommended when fasting glucose levels or when 1- and 2-hour postprandial glucose values exceed these cut-points. [65]

Medical nutrition therapy has been shown to effectively normalize blood glucose levels and to decrease the risk of adverse perinatal outcomes.[68] Results shown in Table 5.9 demonstrate the effect and the usefulness of identifying and intervening upon women with gestational diabetes. It can also be noted from the results that a higher proportion of large newborns occurs even with medical nutrition therapy, but that the incidence is substantially less than in women with untreated gestational diabetes.

Blood glucose levels can be brought down by low caloric intakes. However, accelerated rates of starvation metabolism during pregnancy, as well as potentially

glucose level is high, or ≥130 mg/dL (7.2 mmol/L).[65] (You can convert mg/dL to millimoles per liter, or mmol/L, by multiplying mg/dL by 0.05551.)

The oral glucose tolerance test (OGTT) is the basis for the diagnosis of most cases of gestational diabetes. It can be bypassed among women with very high glucose screening results and treatment started. A 100-gram glucose 3-hour test is used for the OGTT. A diagnosis of gestational diabetes is made when two or more values for venous serum or plasma glucose concentrations exceed these levels:

Overnight fast	95 mg/dL
One hour after glucose load	180 mg/dL
Two hours after glucose load	155 mg/dL
Three hours after glucose load	140 mg/dL[65]

Because of their increased risk for preeclampsia, women with gestational diabetes should be closely monitored for preeclampsia.[62]

A plasma glucose screening between 24 and 28 weeks of pregnancy is recommended for women at "average risk" and for high-risk women not determined by glucose screen to have elevated glucose levels earlier. Average risk is defined as women who fit neither the low- nor the high-risk profile.

Glucose screens are not recommended for women at low risk, defined as:

- Age <25 years
- Member of a low-risk ethnic group (those other than Hispanic, African American, South or East Asian, Pacific Islander, Native American, or indigenous Australian)
- No diabetes in first-degree relatives

Case Study 5.2

Photodisc

Elizabeth's Story: Gestational Diabetes

Elizabeth is a 36-year-old who entered pregnancy with a BMI of 23.5 kg/m². She began receiving prenatal care at 32 weeks gestation and was screened for gestational diabetes the next day. Results of her oral glucose tolerance test revealed the following blood glucose levels:

Fasting:	90 mg/dL
1-hour:	195 mg/dL
2-hour:	163 mg/dL
3-hour:	135 mg/dL

Elizabeth's health care provider advised her to consume a no-sugar, low-carbohydrate diet and to keep her weight gain low throughout the rest of pregnancy. She delivered a large infant (4750 grams) at 39 weeks gestation.

Questions

1. Did Elizabeth have gestational diabetes?
2. Was she insulin resistant?
3. What's the most likely reason Elizabeth delivered an abnormally large newborn?
4. What was wrong with the dietary advice Elizabeth was given?
5. List three components of appropriate dietary advice for women with gestational diabetes.

(Answers are located in the Instructor's Manual for the 4th edition of *Nutrition Through the Life Cycle*.)

deleterious effects of resulting ketonemia on fetal development, exclude this approach to blood glucose control.[65] Correspondingly, restriction of pregnancy weight gain to below recommended amounts is not advised.[71] Aggressive treatment of gestational diabetes that excessively limits caloric intake and weight gain increases the risk of SGA newborns.[75] On the other hand, excessively high caloric balances and weight gains are of concern because they increase the risk of macrosomia.[69]

Type 2 diabetes in nonpregnant individuals is often treated with sulfonylurea oral medications. These drugs cannot be used in pregnancy because they cross the placenta and stimulate fetal insulin production. Other types of oral medications such as glyburide and metformin are being tested for use among women with gestational diabetes.[70] Oral agents that reduce blood glucose levels are not yet recommended for use in pregnancy due to the lack of randomized, controlled clinical trials that demonstrate their safety.[65]

Presentation of a Case Study

No two women with gestational diabetes share the same history, risks, needs, and response to treatment. Case Study 5.2 represents an individual's experience with the disorder.

Exercise Benefits and Recommendations

Insulin resistance is decreased and blood glucose control enhanced by regular aerobic exercise such as walking, jogging, biking, golfing, hiking, swimming, and moderate weight lifting. This appears to be the case as well in women with gestational diabetes. Weight lifting with the arms 3 days a week for 20 minutes per session for 6 weeks, and exercising on a recumbent bicycle at 50% VO_2 max for 45 minutes three times a week, have been found to normalize blood glucose levels in some women.[71]

Levels of exercise that approximate 50–60% of VO_2 max, or maximal oxygen uptake, are most often recommended for women with gestational diabetes. These levels are estimated in practice using a formula for heart rates associated with various levels of VO_2 max. The formula is $220 - age \times 0.50$ (for 50% of VO_2 max) = heartbeats per minute. In the case of a 29-year-old, the estimated heart rate at 50% of VO_2 max would be $220 - 29 \times 0.50$, or 96 beats per minute. This would be the maximum heart rate she should experience while exercising. Levels of exercise should make women become slightly sweaty but not overheated, dehydrated, or exhausted.[71]

Nutritional Management of Women with Gestational Diabetes

A primary outcome goal for women with gestational diabetes is well-controlled blood glucose levels. Other goals include the normalization of carbohydrate metabolism and a reduction in the mother's and offspring's subsequent risk of diabetes, hypertension, heart disease, and obesity.[65] For most women, diet and exercise changes will be the primary way to achieve these goals. In other women, supplemental insulin will be needed to help achieve glucose goals.

The following are components of the nutritional management of women with gestational diabetes:

- Assessing dietary habits and exercise habits
- Developing an individualized diet and exercise plan for blood glucose control
- Monitoring weight gain, dietary intake
- Interpreting blood glucose and urinary ketone results
- Ensuring follow-up during pregnancy and postpartum[71]

Women with type 2 diabetes coming into pregnancy are managed in much the same fashion as are women with gestational diabetes, only nutritional care begins earlier. Ideally, normal blood glucose levels should be established prior to conception and then maintained in good control through pregnancy. Diet and exercise plans for women with type 2 diabetes can often be based on what has worked in the past, thus simplifying planning for needs associated with pregnancy.[72]

The Diet Plan   In general, diets developed for women with gestational diabetes emphasize:

- Whole-grain breads and cereals, vegetables, fruits, and high-fiber foods
- Limited intake of simple sugars and foods and beverages that contain them
- Low-GI foods, or high fiber carbohydrate foods that do not greatly raise glucose levels
- Unsaturated fats
- Three regular meals and snacks daily[72]

Dietary planning is based around a calculated level of caloric need. These initial estimates of caloric need are intended to meet both maternal and fetal demand for energy while limiting increases in blood glucose levels. They are based on the pregnant woman's weight status and weight gain goals for pregnancy. Estimated levels of caloric need according to women's current weight status are shown in Table 5.10.

Women's allotment of calories are generally spread across three meals and several snacks, including a bedtime snack to help prevent nighttime hypoglycemia. Proportions of daily calorie intake generally assigned to meals and snacks are:

Table 5.10 Estimating levels of caloric need in women with gestational diabetes[65]

Current Weight Status	BMI, kg/m^2	Calories per kg Body Weight, kcal/kg
Underweight	<18.5	up to 40
Normal weight	18.5–24.9	30
Overweight, obese	25–34	25
Morbidly obese	≥34	20 or less

- 10-20% for breakfast
- 20-30% for lunch
- 30-40% for dinner
- 30% for snacks[65]

Caloric levels and meal and snack plans are considered to be starting points and often require modifications after results of blood glucose home monitoring tests are known.

Dietary management of gestational diabetes calls for control of carbohydrate intake because carbohydrates raise blood glucose values more than protein or fats do. The following percent distributions of total calories from carbohydrate, protein, and fat have been established for gestational diabetes:

- Carbohydrates: 40-50%

  Carbohydrate calories should be obtained from complex carbohydrate foods that are high in fiber.

- Protein: 20%
- Fat: 30-40%

  Fat calories should be obtained primarily from food sources of unsaturated fats.[65]

The relatively low-carbohydrate, high-fat diet decreases the need for insulin by lowering the amount of glucose absorbed from food, and blunts postprandial increases in blood glucose and insulin levels. The addition of high-fiber foods to diet plans further enhances blood glucose control. These changes in turn reduce fetal overgrowth and other adverse effects of insulin resistance and high blood levels of glucose during pregnancy.[73]

Low-Glycemic Index (GI) Foods

Whether low-GI foods benefit women with diabetes in pregnancy has been much debated and is somewhat controversial. Low-GI foods help women with gestational diabetes sustain modest improvements in blood glucose levels and decrease insulin requirements.[74] Illustration 5.3 demonstrates this point by showing blood glucose levels after a meal containing white bread (GI = 70) or spaghetti (GI = 48) is consumed.

Illustration 5.3 Blood glucose response in people with type 2 diabetes to meals containing white bread or spaghetti.

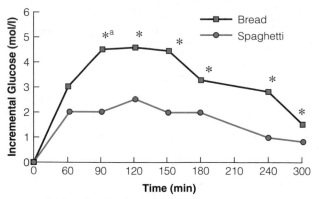

SOURCE: G. Riccardi and A. A. Rivellese, "Diabetes: Nutrition in Prevention and Management." *Nutr Metab Cardiovasc Dis* 1999; 9:33–6. Reproduced with permission of Medikal Press S.r.l.

* denote statistical significance.

Example Meal Plans   Individualized diet plans for women with gestational diabetes include a variety of foods that correspond to the preferences and needs of women. Two examples of such diet plans are shown in Table 5.11. One menu provides approximately 2200 calories, the other 2400. Both menus include low-GI food sources of carbohydrate and meet nutrient needs of women during pregnancy.

Urinary Ketone Testing   Women with gestational diabetes may be instructed to monitor urinary ketone levels using dipsticks. The presence of ketones indicates a negative calorie balance that is likely related to inadequate calorie intake or skipped meals. A positive ketone urine test can be used to help women monitor and adjust calorie intake. When interpreting results of urinary ketone tests, keep in mind that 10–20% of pregnant women spill ketones after an overnight fast.[71] This means the severity and consistency of positive findings for urinary ketones should be considered.

Postpartum Follow-Up

About 15% of women with gestational diabetes will remain glucose intolerant postpartum, and 10–15% will develop type 2 diabetes within 2–5 years. Most women who managed their gestational diabetes with diet and exercise will not require monitoring of blood glucose levels after pregnancy. Women requiring insulin for glucose management should be tested for fasting and 2-hour postprandial blood glucose values before hospital discharge. A 75-gram oral glucose tolerance test is recommended between 6 to 12 weeks postpartum in women who were diagnosed with gestational diabetes during pregnancy but tested negative for glucose intolerance postpartum. Negative results should be followed by repeated glucose testing every 3 years.[65]

Prevention of Gestational Diabetes

Reducing overweight and obesity, increasing physical activity, and decreasing insulin resistance prior to pregnancy are important components of reducing the risk of gestational diabetes.[77] The risk of type 2 diabetes after pregnancy can be reduced substantially by healthful eating, aerobic and resistance exercise, and maintenance of normal weight.[65]

Type 1 Diabetes During Pregnancy

Women with type 1 diabetes have deficient insulin output and must rely on insulin injections or an insulin pump to meet their need for insulin. Type 1 diabetes represents a potentially more hazardous condition to mother and fetus than do most cases of gestational diabetes.

Type 1 diabetes places women at risk of kidney disease, hypertension, and other complications of pregnancy.[78] Newborns of women with this type of diabetes are at increased risk of mortality, of being SGA or LGA, and of experiencing hypoglycemia and other problems within 12 hours after birth. Hypoglycemia occurs in about half of macrosomic infants.[58] Coming into pregnancy with this type of diabetes increases (by 2–3% to 6–9%) the risk of congenital malformations of the pelvis, central nervous system, and heart in offspring. Good control of blood glucose levels reduces the risk of malformations. Maintenance of normal glucose levels from the start of pregnancy decreases the risk of fetal malformations and macrosomia substantially.[79]

Blood glucose control from the beginning of pregnancy is also important because the fetal growth trajectory may be largely determined in the first half of pregnancy. Exposure to high amounts of glucose and insulin when the fetal growth trajectory is being established may set the "metabolic stage" for fetal accumulation of fat and lean tissue later in pregnancy.[80] Even relatively low elevations in blood glucose levels can meaningfully increase birth weight.[81] Unfortunately, only 10% of women with type 1 diabetes receive preconceptional care.[80]

Availability of a variety of new insulins, the insulin pump, and self-monitoring technology has revolutionized the care of type 1 diabetes during pregnancy.

Nutritional Management of Type 1 Diabetes in Pregnancy   Primary goals for the nutritional management of type 1 diabetes in pregnancy are continual control of blood glucose levels, nutritional adequacy of dietary intake, achievement of recommended amounts of weight gain, and a healthy mother and newborn. Careful home monitoring of glucose levels and adjustments in dietary intake, exercise, and insulin dose based on the results are key events that increase the likelihood of reaching these goals. Monitoring urinary ketones is

Table 5.11 Examples of three-meal, three-snack one-day menus at two caloric and carbohydrate levels for women with gestational diabetes

2200-Calorie Diet	Carbohydrates, g	Calories	2400-Calorie Diet	Carbohydrates, g	Calories
Breakfast			**Breakfast**		
All Bran, ½ c	22	80	Complete Wheat		
2% milk, ½ c	6	61	Bran Flakes, ¾ c	23	90
Mozzarella cheese			2% milk, ½ c	6	61
stick, 1 oz	1	78	Egg, 1	1	74
Black coffee, tea			Black coffee, tea		
Morning Snack			**Morning Snack**		
Oat-bran bagel, ½	19	98	Peanuts, 2 oz	10	326
Sugar-free, low-fat			Carrot, 1	7	31
yogurt, 1 c	17	155	Graham crackers,		
Lunch			4 small or 1 sheet	11	59
Tuna salad, ½ c	19	192	**Lunch**		
Whole-grain			Beef or chicken		
bread, 2 slices	24	130	burrito, 1	33	255
Carrot and celery			Salsa, ½ c	7	33
sticks, 1 c	7	31	Black beans, 1 c	40	228
Potato salad, ½ c	14	179	Apple, 1	21	81
Orange, 1	15	62	Black coffee, tea,		
Black coffee, tea,			water, or diet soda		
water, or diet soda			**Midday Snack**		
Midday Snack			Banana, ½	28	55
Peaches canned in	29	109	2% milk, 1 c	12	121
juice, ½ c			**Dinner**		
2% milk, 1 c	12	121	Lean pork chop, 4 oz	0	263
Dinner			Pinto beans, 1 c	22	116
Lean roast beef, 4 oz	0	235	Corn bread, 1 oz	12	92
Broccoli, 1 c with:	10	50	Margarine, 1 tsp	1	33
Cheese (melted), 1 oz	1	105	Garden salad, 2 cup	0	10
Whole-grain			Feta cheese, 1 oz	1	74
roll, 2 oz	30	167	Salad dressing,		
Margarine, 2 tsp	0	67	2 Tbsp	3	104
Grapes, 15	14	53	Black coffee, tea,		
Black coffee, tea, water,			water, or diet soda		
or diet soda			**Bedtime Snack**		
Bedtime Snack			Peanut butter,		
Hard-boiled egg, 1	1	74	2 Tbsp	12	190
Crackers, 4 sq	8	50	Rice cake, 1	8	35
2% milk, 1 c	12	121	2% milk, 1 c	12	121
Total:	**261 g**	**2218**	**Total:**	**270 g**	**2442**
	or 48% of total calories			or 44% of total calories	

SOURCE: Developed by Judith Brown.

particularly important in women with type 1 diabetes because they are more prone to developing ketosis than are women with gestational diabetes.[51] Inclusion of ample amounts of dietary fiber (25–35 g per day) reduces insulin requirements in many women with type 1 diabetes in pregnancy.[73]

Multifetal Pregnancies

Rates of multifetal pregnancy in the United States have increased markedly since 1980. Twin births, which accounted for 1 in 56 births in 1980, constituted 1 in 32 births in 2006. Rates of triplet and higher-order multiple births

Assisted Reproductive Technology (ART)   An umbrella term for fertility treatments such as *in vitro* fertilization (IVF, a technique in which egg cells are fertilized by sperm outside the woman's body), artificial insemination, and hormone treatments.

(referred to as *triplet* + births) increased from 1 in 2941 to 1 in 653 births.[83] The leading reason for the increased prevalence of multifetal pregnancies in the United States and other developed countries is the use of ***assisted reproductive technology***. Rates of twin and higher-order births are highest by far in women 45–54 years old (1 in 5 births), the age group most likely to receive assisted reproductive technological interventions to achieve pregnancy.[83]

The progressively older ages at which U.S. women are bearing children also contribute to rising rates of multifetal pregnancies. The chances of a spontaneous multifetal pregnancy increase with age after about 35 years. Rates of spontaneous multifetal pregnancy also increase with increasing weight status. For example, the rate of twin pregnancy is about two times higher in obese than in underweight women.[84] Rates of triplet + pregnancies are headed downward due to improved assisted reproductive technologies that reduce higher-order multifetal pregnancies.[82]

Upward trends in low birth weight and preterm delivery in the United States over recent years have been strongly influenced by the upsurge in multiple births. Only 3% of newborns are from multifetal pregnancies, yet they account for 21% of all low-birth-weight newborns, 14% of preterm births, and 13% of infant deaths.[85]

Background Information about Multiple Fetuses

The most common type of multifetal pregnancies, those with twin fetuses, come in several types and levels of risk. Twins are dizygotic (DZ) if two eggs were fertilized, and

monozygotic (MZ) if one egg was. Monozygotic twins result when the fertilized and rapidly dividing egg splits in two within days after conception. The term *identical* is often used to describe MZ twins, and *fraternal* denotes DZ twins. These terms are misleading, so the preferred terms are *monozygotic* and *dizygotic*.[86] About 70% of twins are DZ, and 30% are MZ.

Monozygotic twins are always the same sex, whereas DZ twins are the same sex half the time. Monozygotic twins are genetically identical in almost all ways, but they are seldom absolutely identical. Genetic differences in pairs of MZ twins can result from chromosome abnormalities in one twin, unequal genetic expression of maternally and paternally derived genes, and environmental effects on gene expression. Rates of MZ twins are remarkably stable across population groups and do not appear to be influenced by heredity.[87]

Dizygotic twins represent individuals with differing genetic "fingerprints." The incidence of DZ twin pregnancies is influenced both by inherited and environmental factors. Rates of DZ twins vary among racial groups and by country. Rates tend to decrease in populations during famine and to increase when food availability and nutritional status improve.[88] Periconceptional vitamin and mineral supplement use has also been related to an increased incidence of DZ twin pregnancy.[89]

Twins also vary in the number of placentas; some are born having used the same placenta, but more commonly each fetus has its own. Twins may share a common amniotic sac and one of the membranes around the sac (the chorion), or have separate amniotic sacs and membranes (Illustration 5.4). Twins at highest risk of death, malformations, growth retardation, short gestation, and other serious problems are those that share the same

Illustration 5.4 Variations in amniotic sacs, chorions, and placentas in twin pregnancy. Drawing (a) shows twins with two amniotic sacs, two chorions, and two placentas. Drawing (b) represents twins sharing one amniotic sac, chorion, and placenta, and drawing (c) shows twins with two amniotic sacs, one chorion, and two placentas that have grown into one.

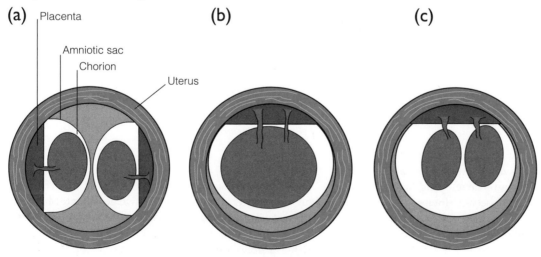

SOURCE: Schematic representations drawn by the author with the help of Scott Strachan, 2009.

Illustration 5.5 Rates of fetal weight gain in singleton, twin, and triplet fetuses.

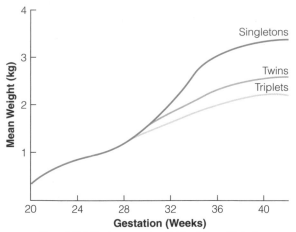

SOURCE: From "Multifetal pregnancies: Epidemiology, Clinical Characteristics, and Management," by M. Smith-Levitin et al., in Maternal-Fetal Medicine: Principles and Practice, 3rd Ed., R.K. Creasy and R. Resnick (Eds.), p. 589–601. Copyright © 1994. Reprinted by permission of W. B. Saunders Co.

amniotic sac and chorion, and to a lesser degree, MZ twins in general.[87]

Determining twin type is not always an easy task during or after pregnancy. Definitive diagnoses of tough cases can be made through DNA fingerprinting.[90]

In Utero **Growth of Twins and Triplets**   Fetal growth patterns of twins and triplets compared to singleton fetuses are shown in Illustration 5.5. Rates of weight gain for each group of fetuses are the same until about 28 weeks of gestation. Rates of weight gain begin to decline in twin and triplet fetuses after that point, however, and remain lower until delivery. Variations in birth weight of twin and triplet newborns appear to be related to factors that affect fetal growth after 28 weeks of pregnancy.[90]

The Vanishing Twin Phenomenon   The disappearance of embryos within 13 weeks of conception is not unusual. It has been estimated that 6 to 12% of pregnancies begin as twins, but that only about 3% result in the birth of twins. Most fetal losses silently occur by absorption into the uterus within the first 8 weeks after conception. The prognosis for continued viability of a pregnancy associated with a vanishing twin tends to be good.[91]

Risks Associated with Multifetal Pregnancy

Singleton pregnancy is the biological norm for humans, so it may be expected that multifetal pregnancy would be accompanied by increasing health risks (Table 5.12).[82] Multifetal pregnancies present substantial risks to both mother and fetuses, and the risks increase as the number of fetuses increases (Table 5.13). Newborns from twin pregnancies at lowest risk of death in the perinatal period weigh between 3000 and 3500 grams (6.7 to 7.8 lb) at

Table 5.12 Risks to mother and fetuses associated with multifetal pregnancy[92]

Pregnant Women
- Preeclampsia
- Iron-deficiency anemia
- Gestational diabetes
- Hyperemesis gravidarum
- Placenta previa
- Kidney disease
- Fetal loss
- Preterm delivery
- Cesarean delivery

Newborns
- Neonatal death
- Congenital abnormalities
- Respiratory distress syndrome
- Intraventricular hemorrhage
- Cerebral palsy

Table 5.13 Average birth weight and gestational age at delivery, and low-birth-weight rates, of singleton, twin, and triplet newborns[92,93]

	Mean Birth Weight	Mean Gestational Age	Low-Birth-Weight Rate
Singletons	3440 g (7.7 lb)	39–40 weeks	6%
Twins	2400 g (5.4 lb)	37 weeks	54%
Triplets	1800 g (4.0 lb)	33–34 weeks	90%

birth and are born at 37–39 weeks gestation. Triplets tend to do best when they weigh over 2000 grams (4.5 lb) and are born at 34–35 weeks gestation.[94]

Unfortunately, these outcomes do not represent the usual. Data presented in Table 5.14 on the next page show that median weights of twins born at 37, 38, and 39 weeks gestation fall below the 3000- to 3500-gram range. However, the 3000- to 3500-gram birth weight range for twins, and the >2000-gram mark for triplet newborns, can serve as goals for the provision of nutrition services.

Interventions and Services for Risk Reduction

Special multidisciplinary programs that offer women with multifetal pregnancy a consistent, main provider of care; preterm prevention education; increased attention to nutritional needs; and intensive follow-up achieve better pregnancy outcomes than does routine prenatal care.[93] Rates of very low birth weight (≤1500 g or ≤3.3 lb) have

Table 5.14 Median birth weight for gestational age at delivery of twins	
Gestational Age, Weeks	Birth Weight
28	995 g (2.2 lb)
29	1145 g (2.6 lb)
30	1300 g (2.9 lb)
31	1445 g (3.2 lb)
32	1580 g (3.5 lb)
33	1750 g (3.9 lb)
34	1905 g (4.3 lb)
35	2165 g (4.8 lb)
36	2275 g (5.1 lb)
37	2430 g (5.4 lb)
38	2565 g (5.7 lb)
39	2680 g (6.0 lb)
40	2810 g (6.3 lb)
41	2685 g (6.0 lb)

SOURCE: Data from Cohen SB, 1997.[96]

Table 5.15 Prepregnancy weight status and weight-gain relationships in twin pregnancy[84]	
Prepregnancy Weight Status	Weight Gain Related to Birth Weights of >2500 g (5.5 lb)
Underweight	44.2 lb (20.1 kg)
Normal weight	40.9 lb (18.6 kg)
Overweight	37.8 lb (17.2 kg)
Obese	37.2 lb (16.9 kg)
Very obese	29.2 lb (13.3 kg)

Rate of Weight Gain in Twin Pregnancy A positive rate of weight gain in the first half of twin pregnancy is strongly associated with increased birth weight.[92] On the other hand, weight loss after 28 weeks of pregnancy increases the risk of preterm delivery by threefold.[100]

Reasonable rates of weight gain for women with twin pregnancy are:

- 0.5 pounds (0.2 kg) per week in the first trimester
- 1.5 pounds (0.7 kg) per week in the second and third trimesters[92]

Weight Gain in Triplet Pregnancy Several studies have examined the relationship between weight gain and birth weight in women with triplets. The general result is that weight gains of about 50 pounds (22.7 kg) correspond to healthy-sized triplets. Rates of gain related to a total weight gain of 50 pounds in women who will average 33 to 34 weeks of gestation are 1.5 pounds (0.7 kg) per week or more, starting as early in pregnancy as possible.[92]

Dietary Intake in Twin Pregnancy

Ensuring "adequate nutrition" is widely acknowledged to be a key component of prenatal care for women with multifetal pregnancy. However, it is not clear what constitutes adequate nutrition. Energy and nutrient needs clearly increase during multifetal pregnancy due to increased levels of maternal blood volume, extracellular fluid, and uterine, placental, and fetal growth. The normally high expansion in extracellular volume and its side effect of leg and ankle edema can be seen in the healthy woman with a twin pregnancy shown in Illustration 5.6. Increases in energy and nutrient needs place demands on the mother in terms of the nutritional costs of building and maintaining these tissues. Although their newborns are smaller, women with twins still produce around 5000 g (11.2 lb) of fetal weight, and women with triplets 5400 g (13.4 lb) or more.

Evidence of higher caloric need for tissue maintenance and growth in multifetal than singleton pregnancy comes from studies that show increased weight gain and a quicker onset of starvation metabolism in women expecting more than one newborn. Reduced rates of twin

been reported to be substantially lower (6% versus 26%), neonatal intensive care admissions three times lower (13% versus 38%), and perinatal mortality strikingly lower (1% versus 8%) among women who receive such services.[95] Interventions offered by the Montreal Diet Dispensary, which focuses on improving the nutritional status and well-being of the pregnant women served, have been shown to substantially reduce poor outcomes compared to those for similar women not receiving the services. Improvements include a 27% reduction in the rate of low birth weight, 47% decline in very low birth weight, 32% lower rate of preterm delivery, and a 79% drop in mortality during the first 7 days after birth.[97]

Nutrition and the Outcome of Multifetal Pregnancy

Nutritional factors are suspected of playing a major role in the course and outcome of multifetal pregnancy, but much remains to be learned. Of the nutritional factors that influence multifetal pregnancy, weight gain during twin pregnancy has been studied most.

Weight Gain in Multifetal Pregnancy As with singleton pregnancy, weight gain in multifetal pregnancy is linearly related to birth weight, and weight gains associated with newborn weight vary based on prepregnancy weight status (Table 5.15).[98] The Institute of Medicine recommends that women with twins gain 25 to 54 pounds (11.4 to 24.5 kg). It is provisionally advised that normal weight women gain toward the upper end of this range, and overweight and obese women closer to the lower end.[99]

Illustration 5.6 The woman pregnant with twins shown in this photo labeled it "My poor feet!" The leg and ankle swelling are an expected and normal part of a healthy twin pregnancy.

SOURCE: Photo courtesy of Alyeen and Jeremy Perez-Marty.

deliveries, as well as the higher incidence of twins in overweight and obese women, imply that energy status is an important factor in multifetal pregnancy.[92] Whereas it is obvious that energy and nutrient needs are higher in multifetal than singleton pregnancy, levels of energy balance and nutrient intake associated with optimal outcomes of multifetal pregnancy have not been quantitated.

Results of a large prospective study indicate that women with twins enter pregnancy with higher average caloric intakes (2030 versus 1789 cal per day) and consume an average of 265 cal more per day during pregnancy than women with singleton pregnancy. Nutrient intakes during pregnancy are also higher in women bearing twins than with singleton pregnancy.[84]

Several studies have concluded that the need for specific nutrients is increased during multifetal pregnancy. The need for essential fatty acids (linoleic and alpha-linolenic acid) appears to be increased in multifetal pregnancy. Poor essential fatty acid status is related to neurologic abnormalities and vision impairments in twin offspring.[101] Requirements for iron and calcium have also been found to be increased based on the magnitude of physiological changes that take place in multifetal pregnancy. Levels of essential fatty acids, iron, or calcium required by women to meet these increased needs are unknown, however.[92] Case Study 5.3 addresses a twin pregnancy. This case study requires use of the American Dietetic Association's Nutrition Care Process.

Vitamin and Mineral Supplements and Multifetal Pregnancy   Benefits and hazards of multivitamin and mineral supplement use in multifetal pregnancy have not been reported. Consequently, the extent to which they may be required is unknown. Levels of nutrient intake exceeding the DRI Tolerable Upper Intake Levels should be avoided.

Nutritional Recommendations for Women with Multifetal Pregnancy

Due to the lack of study results, nutritional recommendations for women with multifetal pregnancy are largely based on logical assumptions and theories (Table 5.16). It is reasoned, for example, that caloric needs for twin pregnancy can be extrapolated from weight gain. Theoretically, to achieve a 40-pound (18.2 kg) weight gain, or 10 pounds (4.5 kg) more than in singleton pregnancy, women with twins would need to consume approximately 35,000 cal more during pregnancy than do women with singleton pregnancies. This increase would amount to about 150 cal per day above the level for singleton pregnancy, or an average of 450 cal more per day than prepregnancy. To achieve higher rates of gain, underweight women may need a higher level of intake, and overweight and obese women lower levels. Energy needs will also vary by energy expenditure levels. As for singleton pregnancy, adequacy of caloric intake can be estimated by weight-gain progress.[92]

Food-intake recommendations for women with multifetal pregnancy are primarily estimated based on assumptions related to caloric and nutrient needs. Women with multifetal pregnancy likely benefit from diets selected from the MyPyramid groups and nutrient intakes that somewhat exceed the RDAs/AIs.

Although twin pregnancies are higher-risk than singleton pregnancies, outcomes of twin pregnancy can be excellent. Illustration 5.7 shows healthy, term newborn twins. Their mother remained in good health during pregnancy while consuming the type of diet and supplements and gaining weight as recommended by her health care providers.

Recommendations from the Popular Press   Websites, books, and pamphlets are available that provide ample amounts of scientifically unsupported "guesses" about food and nutrient requirements of women with multifetal pregnancy. Even if presented with steely resolution, any advice that strays from current scientifically based wisdom about nutritional needs of women during pregnancy should be sidestepped.

HIV/AIDS During Pregnancy

The world first became aware of acquired immunodeficiency syndrome (AIDS) in the summer of 1981. It was caused by a newly recognized microbe, the human immunodeficiency virus (HIV). Since then, over 100,000 women of childbearing

Table 5.16 "Best practice" recommendations for nutrition during multifetal pregnancy[92]

Weight Gain

Twin pregnancy: Overall gain of 35–45 lb (15.9–20.5 kg). Underweight women should gain at the upper end of this range, and overweight and obese women at the lower end.

- First trimester: 4–6 lb (1.8–2.7 kg)
- Second and third trimesters: 1.5 lb (0.7 kg) per week

Triplet pregnancy: Overall gain of approximately 50 lb (22.7 kg)

- Gain of 1.5 lb (0.7 kg) per week through pregnancy

Daily Food Intake

Twin pregnancy (2400–2800 calories a day)

- Grains: 8–10 oz
- Vegetables: 3–3.5 c
- Fruits: 2–2.5 c
- Meat and beans: 6.5–7 oz
- Milk: 3 c
- Oil: 7–8 tsp
- Discretionary calorie allowance: 362–426

Triplet pregnancy

- Food intake from the MyPyramid groups should be consumed at a level that promotes targeted weight gain.

Caloric Intake

Twin pregnancy

- 450 calories above prepregnancy intake, or the amount consistent with targeted weight-gain progress

Triplet pregnancy

- Caloric intake levels should promote targeted weight-gain progress.

Nutrient Intake

Twin and triplet pregnancy

- RDA or AI levels or somewhat more than these levels
- Intakes should be lower than ULs.

Vitamin and Mineral Supplements

Twin pregnancy

- Use a prenatal vitamin and mineral supplement.

Triplet pregnancy

- Provide prenatal vitamin and mineral supplements; avoid excessively high amounts of nutrients.

Illustration 5.7 The outcome of a healthy twin pregnancy: Isa weighed 6 pounds, 8 ounces (2912 grams) and Manu, 6 pounds, 7 ounces (2884 grams). The twins were delivered by a scheduled cesarean section at 38 weeks, 4 days, and were above average weight for their gestational age.

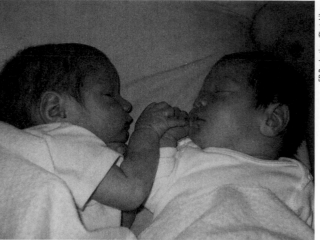

SOURCE: Photo courtesy of Alyeen and Jeremy Perez-Marty.

age in the United States have been diagnosed with AIDS,[92] and 33 million worldwide.[103] Transmission of the virus during pregnancy and delivery is a major route to the spread of the infection. Approximately 20% of children with HIV/AIDS are infected during pregnancy or delivery, and 14–21% during breastfeeding.[104]

Treatment of HIV/AIDS

The primary focus of care for pregnant women with HIV/AIDS is the prevention of transmission of the virus to the fetus and infant. The transmission of the AIDS virus from mother to child is negligible if treatment is provided before, during, and after pregnancy.[104] In developing countries where there is not enough money to purchase the drugs, transmission rates can be substantially reduced by giving the mother a short course of a specific anti-HIV drug before delivery.[105]

Consequences of HIV/AIDS During Pregnancy

Disease processes such as compromised immune system functions related to HIV/AIDS progress during pregnancy, but it does not appear that the infection itself is related to adverse pregnancy outcomes. Although adverse pregnancy outcomes such as preterm delivery, fetal growth retardation, and low birth weight tend to be higher in women with HIV/AIDS, differences are most closely related to poverty, poor food availability, compromised health status, and the coexistence of other infections.[104]

Case Study 5.3

Nutrition Care Process Case Study 5.3

This case study involves use of the American Dietetic Association's Nutrition Care Process.[76]

Twin Pregnancy

At 37 years of age, Señora Mendez was in her 23rd week of pregnancy and expecting her second child. Or so she thought, until an ultrasound scan detected twins. Her prepregnancy weight was 142 pounds (64.5 kg) and her BMI 23 kg/m². Señora Mendez's weight-gain progress has been poor due to nausea and vomiting experienced in the first half of pregnancy. Otherwise, Sra. Mendez was experiencing a normal pregnancy for women expecting twins. Concerns about her weight-gain progress and the nutritional needs of women with twin gestation prompted her certified-nurse midwife to prescribe a prenatal vitamin and mineral supplement and to refer her to a registered dietitian/certified diabetes educator.

A nutritional assessment completed during week 25 of pregnancy identified that Sra. Mendez had gained 14 pounds since conception, and that her typical dietary intake excluded food sources of EPA and DHA. Her plasma 25-hydroxyvitamin D level was below 24 nmol/L, indicating low vitamin D status. No other nutrition-related problems were identified.

(Answers to the case study are located in the Instructor's manual.)

Questions

1. Assume Sra. Mendez will deliver toward the end of week 37 and is now in the beginning of week 25. How many pounds (or kilograms) should be set as a goal for weekly weight gain if a total pregnancy gain at the midpoint of the provisonally recommended range for weight gain in twin pregnancy is to be achieved?
2. State three appropriate nutrition diagnoses for this case.
3. Name one potential nutrition intervention that would address each of the nutrition diagnoses stated.
4. Cite a nutrition-related indicator that could be used to monitor and evaluate each of the nutrition interventions stated.

(Answers are located in the Instructor's Manual for the 4th edition of *Nutrition Through the Life Cycle*.)

Nutritional Factors and HIV/AIDS During Pregnancy

HIV/AIDS is related to poor nutritional status that further compromises the body's ability to fight infections. The disease can lead to nutrient losses and fat malabsorption due to diarrhea, and inflammatory responses to the infection cause the loss of lean muscle mass. Loss of calcium from bones and decreased bone density is a common finding in individuals with HIV/AIDS.[106] Risks of inadequate nutrient status of a wide variety of vitamins and minerals increase as the disease progresses. Nutritional needs increase the most during the later stages of HIV/AIDS as diarrhea, wasting, and reductions in CD4 counts (a measure of white blood cells that help the body fight infection) increase. New drugs used to treat HIV/AIDS are associated with increased insulin resistance and the accumulation of central body fat.[107]

The compromised immune status of women with HIV/AIDS, and further decreases in immune response during pregnancy, mean that women with the disease are at high risk of developing foodborne infections during pregnancy. Risk of infection originating from foods can be decreased if raw or uncooked meats and seafood and unpasteurized milk products and honey are not consumed. Safe food-handling practices at home can also reduce the risk of foodborne infection.[106]

Nutritional Management of Women with HIV/AIDS During Pregnancy

Goals for the nutritional management of women with HIV/AIDS include:

- Maintenance of a positive nitrogen balance and preservation of lean muscle and bone mass
- Adequate intake of energy and nutrients to support maternal physiological changes and fetal growth and development
- Correction of elements of poor nutritional status identified by nutritional assessment
- Adoption of safe food-handling practices
- Delivery of a healthy newborn[106]

Nutrient requirements of HIV-infected pregnant women are not known, but it is suspected that energy and nutrient needs will be somewhat higher due to the effects of the virus on the body. Good nutritional status prior to pregnancy and use of multivitamin mineral supplements during pregnancy are associated with better outcomes in women with HIV/AIDS[111,112]

Insufficient information exists to provide specific standards for nutritional care for women with HIV/AIDS during pregnancy. Consequently, nutritional recommendations for women with HIV/AIDS are consistent with recommendations for pregnant women in general. As is the case for other pregnant women, foods should be the primary source of nutrients in women with HIV/AIDS.[106]

Eating Disorders in Pregnancy

Eating disorders represent relatively rare conditions in pregnancy because many women with such disorders are subfertile or infertile. Such disorders can have far-reaching effects on both mother and fetus, however, when they do occur. The eating disorder most commonly observed among pregnant women in the United States is bulimia nervosa, or a condition marked by both severe food restriction and bingeing and purging.[109] It is estimated that 1–3% of adolescents and young women in the United States have this condition.[108] Women with bulimia nervosa exhibit poorly controlled eating patterns marked by recurrent episodes of binge eating. To prevent weight gain, women will induce vomiting, use laxatives, exercise intensely, or fast after binges. Self-worth in women with bulimia nervosa is usually closely tied to their weight and shape. A history of sexual abuse is common among women with this eating disorder, as well as in women with anorexia nervosa.[108]

Pregnancy is rarely suspected in women with anorexia nervosa because amenorrhea is a diagnostic criteria for the disorder. Nonetheless, women with anorexia nervosa may occasionally ovulate and become at risk for conception. To women with anorexia nervosa, body weight is of utmost importance, and they are generally fully dedicated to achieving extreme thinness. Adolescents and women with this condition will refuse to eat, even when ravenously hungry; limit their food choices to low-calorie foods only; and exercise excessively.[108]

Eating disorder symptoms often subside during the second and third trimesters of pregnancy, but they rarely vanish altogether. Symptoms tend to return after delivery, sometimes to a more severe extent than was the case prior to pregnancy.[110] Information on eating disorders in nonpregnant individuals is included in Chapter 15.

Consequences of Eating Disorders in Pregnancy

Women with eating disorders during pregnancy are at higher risk for spontaneous abortion, hypertension, and difficult deliveries than are women without an eating disorder. Pregnancy weight gain is generally below the recommended amounts, and newborns tend to be smaller and to experience higher rates of neonatal complications.[109]

Treatment of Women with Eating Disorders During Pregnancy

It is recommended that pregnant women with eating disorders be referred to an eating disorders clinic or specialist. Most large communities have special clinics and programs for women with eating disorders, and they commonly use a team approach to problem-solving around the eating disorder. Nutritionists or dietitians often participate in these services because they are knowledgeable about the woman's individual nutritional needs and those of pregnancy.

Health professionals serving women with eating disorders in pregnancy can facilitate open communication and behavioral change by gently encouraging women to talk about their eating disorder, fears, and concerns.[110]

Nutritional Interventions for Women with Eating Disorders

Behavioral changes required for improvements in nutritional status and weight gain in women with eating disorders are most likely to work when the changes are considered acceptable to the women with the disorder. Frequently, the health professional presents the types of changes that need to be made and explains why, and then works with the woman to develop specific plans accomplishing these changes.

Fetal Alcohol Spectrum

New terminology related to the harmful effects of alcohol on fetal growth and development has recently been introduced by the Centers for Disease Control and Prevention (CDC).[113] The term *fetal alcohol spectrum* is now being used to describe the range of effects of fetal alcohol exposure on mental development and physical growth. This range includes behavioral problems, mental retardation, aggressiveness, nervousness, short attention span, and growth-stunting and birth defects.[114] A diagnosis of *fetal alcohol syndrome* is made when infants or young children exhibit a specific set of characteristics included in the fetal alcohol spectrum.

Fetal exposure to alcohol during pregnancy is a leading, preventable cause of birth defects, mental retardation, and developmental disorders in children and adults. Approximately 12% of women in the United States consume alcohol once a month during pregnancy, and 2% consume 5 or more drinks on at least one occasion.[115] It is estimated that 1 in every 1000 newborns in the United States is affected by the fetal alcohol syndrome.[113]

Effects of Alcohol on Pregnancy Outcome

Alcohol consumed by a woman easily crosses the placenta to the fetus. Because the fetus has yet to fully develop enzymes that break it down, alcohol lingers in the fetal circulation. This situation, combined with the fact that the fetus is smaller and has far less blood than the mother does, increases the harmful effects of alcohol on the fetus as compared to the mother. Alcohol exposure during critical periods of growth and development can permanently impair organ and tissue formation, growth, health, and mental development.

Poor dietary intakes of some women who consume alcohol regularly in pregnancy, as well as the negative effect of alcohol on the availability of certain nutrients, may also contribute to the harmful effects of alcohol exposure during pregnancy.[116]

Consumption of 4 or more drinks a day, or occasional episodes of consumption of 5 or more drinks in a row, is considered to represent heavy alcohol intake during pregnancy. Heavy drinking during pregnancy increases the risk of miscarriage, stillbirth, and infant death within the first month after delivery.[117] Approximately 40% of the fetuses born to women who drink heavily early in pregnancy will develop fetal alcohol syndrome (FAS). The likelihood that the fetus will be affected by FAS increases as the number of drinks consumed early in pregnancy increases (Table 5.17). Because a "safe" dose of alcohol consumption during pregnancy has not been

identified, it is recommended that women do not drink alcohol while pregnant.[119]

The Fetal Alcohol Syndrome

FAS was first described in 1973 and consists of pSGA, mental retardation, and a set of common malformations (Illustration 5.8). Diagnosis of FAS is difficult, however, because many of the facial features are not unique to the syndrome. Short noses, flat nasal bridges, and thin upper lips, for example, are also normal facial features. New criteria developed by the CDC are aimed at limiting this

Table 5.17 Approximate incidence of structural abnormalities of the brain in 5- to 14-week-old fetuses exposed to alcohol

Maternal Alcohol Intake, Drinks	Abnormal Brain Structure
13 to 31/day	100%
6.3 to 13/day	83%
2 to 6.3 occasionally	29%
≤2/day	0%

SOURCE: Table is based on information presented by Konovalov et al., 1997.[118]

Illustration 5.8 Features of FAS in children.

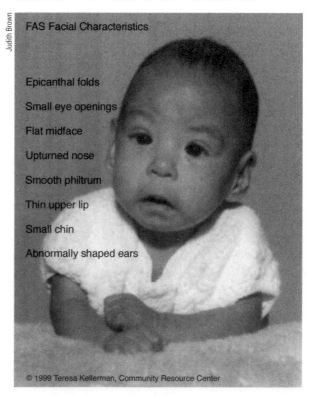

Judith Brown

FAS Facial Characteristics

Epicanthal folds

Small eye openings

Flat midface

Upturned nose

Smooth philtrum

Thin upper lip

Small chin

Abnormally shaped ears

© 1999 Teresa Kellerman, Community Resource Center

Philtrum The vertical groove between the bottom of the nose and the upper lip. The philtrum is smooth, or flat, when there is no groove.

Vermillion Border The exposed pink or reddish margin of a lip. A thin vermillion border in FAS denotes a thin upper lip.

Palpebral Fissure The space between the top and bottom eyelid when the eye is open. This opening is small in children with FAS.

confusion. To be diagnosed with FAS, children must have the characteristic smooth *philtrum,* thin *vermillion border,* small *palpebral fissures,* and a documented neurological disorder (for example, a small head circumference, problems with physical movements, seizures, or cognitive delay), and reduced growth.[113]

Children with FAS tend to have poor coordination, are hyperactive or have a short attention span, have behavioral problems, and remain small for their age. Adults with FAS often find it difficult to hold a job or live independently.[117,120]

Approximately 10 times the number of infants born with FAS have lesser degrees of alcohol-related damage. These effects are characterized by mental and behavioral abnormalities but not malformations.[114]

Nutrition and Adolescent Pregnancy

Between 1991 and 2004, rates of pregnancy among teens 15–19 years old fell 28% but they are now trending upward. The United States continues to have one of the highest rates of adolescent pregnancy of all developed countries.[121] Adolescents are at higher risk for a number of clinical complications and other unfavorable outcomes compared to adult women (Table 5.18). The downward shift in birth weight observed in newborns of adolescent mothers is related to increased rates of perinatal mortality, and suggests the existence of unfavorable nutrition and health conditions during pregnancy.[122]

The extent to which increased rates of poor outcomes in pregnant teens are associated with biological immaturity or with lifestyle factors such as drug use, smoking, and poor dietary intakes (that influence health status)

Table 5.18 Risks associated with adolescent pregnancy[122-124]

- Low birth weight
- Perinatal death
- Cesarean delivery
- Cephalopelvic disproportion (head too large for birth canal)
- Preeclampsia
- Iron-deficiency anemia
- Delayed, reduced educational achievement
- Low income

---

is unclear. Age-related differences in outcome diminish substantially when potentially harmful lifestyle factors are taken into account, diminishing the theory that biological immaturity accounts for the differences. Iron deficiency and vitamin D insufficiency have been identified in 30 to 60% of pregnant adolescents.[127,128]

Very young adolescents becoming pregnant within a few years after the onset of menstruation may be at risk due to biological immaturity. They tend to have shorter gestations and a higher likelihood of cephalopelvic disproportion.[122] Poorly nourished, growing adolescent mothers may compete with the fetus for calories and nutrients—and win.[123]

Growth During Adolescent Pregnancy

Young teens who are growing when pregnancy occurs continue to gain height and weight during pregnancy—but at the expense of fetal growth. Teens who continue to grow during pregnancy give birth, on average, to infants that weigh 155 grams less than infants of adult women, even if they gain more weight than adults do.[123] Rates of spontaneous abortion, preterm birth, and low birth weight are also higher in growing than nongrowing adolescents.[125] Young adolescents gain more maternal fat tissue during the last trimester of pregnancy and retain more weight postpartum than do nongrowing teens. Growing teens experience a surge in blood leptin levels during the last trimester, which may decrease maternal use of fat stores and increase utilization of glucose by the mother. Increased use of glucose by the mother appears to decrease energy "availability" to the fetus.[126]

Obesity, Excess Weight Gain, and Adolescent Pregnancy Increasing rates of overweight status and obesity among adolescents appear to be placing additional teens at risk of poor pregnancy outcomes. Adolescents entering pregnancy overweight or obese are at increased risk for cesarean delivery, hypertensive disorders of pregnancy, gestational diabetes, and the delivery of excessively large newborns.[124] In obese adolescents, weight gains during pregnancy that exceed those recommended are related to decreased placental growth and the birth of infants disproportionately small for gestational age.[125]

Dietary and Other Recommendations for Pregnant Adolescents

Recommendations for pregnant adolescents are basically the same as for older pregnant women, with a few exceptions. Recommendations for weight gain and protein intake are the same, but young adolescents may need more calories to support their own growth as well as that of the fetus. Caloric need should be met by a nutrient-dense diet and lead to rates of weight gain that follow those recommended. Pregnant adolescents have a higher requirement for calcium. The AI for pregnant teens for calcium is 1300 mg per day, or 300 mg higher than for adult pregnant women. This increased need can be met by the consumption of 4 daily servings of milk and milk products, combined with a varied, basic diet.

The importance of lifestyle and other environmental factors to pregnancy outcome in teens emphasizes the need for special, comprehensive teen pregnancy health care programs. Nutrition counseling is an important component of the multidisciplinary services that should be offered to pregnant adolescents. Nutrition services that include individualized nutrition assessment, intervention, education, guidance on weight gain, and follow-up enhance birthweight outcomes. Additional specialized services that focus on the psychosocial needs of pregnant adolescents, support/discussion groups, and home visits also contribute to improved maternal and infant outcomes for adolescents.[129]

Because most pregnant adolescents have low income, referral to appropriate food and nutrition programs and other assistance related to health care, housing, and education should be core components of services.

Evidence-Based Practice

"Enormous amounts of new knowledge are barreling down the information highway, but they are not arriving at the doorsteps of our patients."

Claude Lenfant, National Institutes of Health[117]

The clinical nutritional management of the conditions covered in this chapter, as well as other complications during pregnancy, is not entirely evidence-based. Such practices are a problem when they burden women and families with costs, call for dietary changes not known to work, or potentially cause harm. Practices not based on evidence that likely pose little risk or burden and may potentially be of help should nonetheless be carefully evaluated. Outdated practices often linger far too long, at the expense of missed opportunities for real improvements.

Use of practices not supported by scientific evidence should always be questioned and confirmed to represent "best practice" insofar as that can be determined. To know what best practice is requires vigilant attention to scientific developments related to the nutritional management of clinical conditions during pregnancy.

Key Points

1. Maternal diet, weight status, and physical activity levels influence the development of healthy pregnancy outcomes and a number of conditions that adversely effect the course and outcome of pregnancy.

2. Oxidative stress and endothelial dysfunction appear to be related to the development and progression of hypertensive disorders of pregnancy.

3. Oxidative stress occurs when the body's exposure to oxidizing agents is greater than its supply of antioxidants. Oxidative stress damages the endothelium (the inside lining of blood vessels), and that contributes to endothelial dysfunction. Endothelial dysfunction impairs blood flow, causes an increased tendency for blood to clot, increases plaque formation, and causes other problems.

4. Excess body fat, low levels of physical activity, *trans* fats, lack of antioxidants, and elevated blood glucose increase oxidative stress. It is lowered by weight loss in overweight individuals, increased levels of exercise, exclusion of *trans* fats from the diet, normal blood glucose levels, and sufficient intake of antioxidant defenses.

5. Nutritional management of gestational diabetes focuses on individually based dietary and exercise plans that help maintain blood glucose levels within the normal range and that foster maternal health and appropriate weight gain.

6. Multifetal pregnancies are classified as "high risk" because of above-average rates of complications and less-than-optimal pregnancy outcomes. Energy, nutrient, and weight-gain needs are somewhat higher in multifetal than singleton pregnancies.

7. Pregnant women with HIV/AIDS may benefit from nutrition interventions that conserve lean muscle and bone mass, correct nutrient deficiencies, and lead to healthful dietary intake and weight gain in pregnancy. Foodborne illnesses can severely affect people with HIV/AIDS, making food safety a priority concern.

8. Women with eating disorders should be closely monitored during pregnancy to facilitate appropriate dietary intake and weight gain.

9. High levels of alcohol intake during pregnancy are associated with a broad range of mental and physical disorders. The range of disorders is called the *fetal alcohol spectrum*. It is recommended that women do not drink alcohol-containing beverages during pregnancy.

10. Growing adolescents are at risk of poor pregnancy outcomes. Pregnancy outcomes of adolescents who are not growing primary depend on the health status of the teen. Dietary quality and weight status are important components of the health status of adolescents and influence the course and outcome of their pregnancies.

Review Questions

1. Obesity prior to pregnancy is associated with metabolic disorders that promote the development of hypertensive disorders of pregnancy and gestational diabetes.
   ____ True   ____ False

2. Bariatric surgery prior to pregnancy decreases the risk of gestational diabetes and preeclampsia during pregnancy.
   ____ True   ____ False

3. Chronic inflammation and oxidative stress occur in preeclampsia but not in other hypertensive disorders of pregnancy.
   ____ True   ____ False

4. Multivitamin and mineral supplementation before and early in pregnancy has been associated with a decreased risk of preeclampsia during pregnancy.
   ____ True   ____ False

5. Salt restriction during pregnancy decreases the incidence of all forms of hypertension during pregnancy and is recommended for all pregnant women.
   ____ True   ____ False

6. Gestational diabetes is characterized by abnormally high blood triglyceride and glucose concentrations.
   ____ True   ____ False

7. It is recommended that women with diabetes gain 5 lb (2.3 kg) less during pregnancy than the weight gains recommended for women who do not have diabetes.
   ____ True   ____ False

The next three questions refer to the one-day, 2200-calorie diet listed on the left-hand side of Table 5.12.

8. What percent of total calories are provided by the snack foods listed in this one-day diet?

9. Does the percent of total calories provided at lunch correspond to the recommendation for calorie distribution at lunch for women with gestational diabetes?

10. Assume the 2200-calorie-per-day menu was developed for a woman with gestational diabetes who weighs 132 pounds. Is she underweight, normal-weight, overweight, or obese?

11. It is estimated that women with twin pregnancy need 150 calories more each day than do women with singleton pregnancy.
    ____ True   ____ False

12. Women with HIV/AIDS who are well-nourished prior to pregnancy tend to have better pregnancy outcomes than do poorly nourished women.
    ____ True   ____ False

13. Women with bulimia nervosa and anorexia nervosa tend to improve their eating habits dramatically during pregnancy.
    ____ True   ____ False

14. Iron and vitamin D status may be compromised in many pregnant adolescents.
    ____ True   ____ False

15. Clinical decisions regarding nutrition and health relationships during pregnancy should be evidence-based and not rely on clinical assumptions or previous experiences.
    ____ True   ____ False

Resources

Clinical Conditions of Pregnancy

Visit the Women's Health Resource Center for access to journal articles and summaries and health care information.
Website: **www.medscape.com**

The National Library of Medicine's website offers extensive coverage of scientific journal articles, summaries, and educational resources from a variety of reputable organizations on disorders of pregnancy.
Website: **www.nlm.nih.gov/medlineplus**

The Canadian Diabetes Association's website section "About Diabetes" clearly presents facts on gestational diabetes.
Website: **www.diabetes.ca**

The United States Department of Agriculture (USDA) provides a nutrient composition of foods resource, and a source of information on nutrient content of individual foods.
Website: **www.ars.usda.gov/nutrientdata**
Website: **www.ars.usda.gov/Main/docs.htm?docid=15869**

As a mother, one of the best things that only you can do for your baby is to breastfeed. Breastfeeding is more than a lifestyle choice — it is an important health choice. Any amount of time that you can do it will help both you and your baby. While breastfeeding isn't the only option for feeding your baby, every mother has the potential to succeed and make it a wonderful experience.

U.S. Department of Health & Human Services, 2009

Chapter 6

Nutrition During Lactation

Prepared by **Maureen A. Murtaugh, Ellen Lechtenberg,** and **Carolyn Sharbaugh** with **Denise Sofka**

Key Nutrition Concepts

1 Human milk is the best food for newborn infants for the first year of life or longer.

2 Maternal diet does not significantly alter the protein, carbohydrate, fat, and major mineral composition of breast milk, but it does alter the fatty acid profile and the amounts of some vitamins and trace minerals.

3 When maternal diet is inadequate, the quality of milk is preserved over the quantity for the majority of nutrients.

4 Health care policies and procedures and the knowledge and attitudes of health care providers affect community breastfeeding rates.

Introduction

The benefits of breastfeeding to mothers and infants are well established. Federal breastfeeding promotion efforts and greater understanding of the advantages of breastfeeding have contributed to the resurgence of breastfeeding in the United States since the 1970s. Nevertheless, racial and ethnic disparities in breastfeeding initiation rates remain, and despite the knowledge that the benefits increase with longer duration, there has been little increase in the duration of lactation among all women.

The health care system, the workplace, and the community can either hinder or facilitate the initiation and continuation of breastfeeding. Health programs can play a significant role in increasing breastfeeding rates to optimize maternal and infant nutrition. Health care professionals who wish to manage and promote breastfeeding should understand the physiology of lactation, the composition of human milk, and the benefits to mothers and infants. Helping women achieve appropriate nutritional status to optimize breastfeeding requires consideration of energy and nutrient needs, weight goals, effects of exercise during breastfeeding, and vitamin and mineral supplement needs.

Multilevel (health care system, community, workplace, and family) support is critical for women who suffer from common breastfeeding challenges and medical conditions. Human milk is the preferred food for premature and sick newborns. It is rarely necessary to discontinue breastfeeding to manage medical problems or medication use. However, adequately experienced and informed health care professionals are needed to provide support for successful breastfeeding. Common breastfeeding conditions and interventions are discussed in Chapter 7.

Breastfeeding Goals for the United States

"During the twentieth century, infant feeding practices have undergone dramatic changes that reflect shifts in values and attitudes in the U.S. society as a whole. They have tended to occur first among those women at the forefront of changes in dominant social values and among those with resources (whether it is time, energy or money) to permit adoption of new feeding practices."

Institute of Medicine, Subcommittee on
Nutrition During Lactation, 1991[1]

In the early 1900s, almost all infants in the United States were breastfed. As safe human milk substitutes (HMSs) became widely available, breastfeeding rates steadily declined, reaching levels below 30% in the 1950s and 1960s, and then rose dramatically in the 1970s.[2] In the early 1980s levels peaked above 60%. In recognition of the health and economic benefits of breastfeeding, national goals for breastfeeding rates have been established and revised since 1980.[3] Nonetheless, breastfeeding rates declined until the early 1990s.[2]

Longer duration of breastfeeding is associated with higher education and socio-economic status, living in the western region of the U.S., and being older and married. Healthy People 2010 contains broad-reaching national health goals for the new decade, focusing on two major themes: (1) increasing the quality and years of healthy life and (2) eliminating racial and ethnic disparities in health status.[3] In addition to adding specific breastfeeding goals for black or African Americans and Hispanic or Latina Americans, Healthy People 2010 places increased emphasis on the duration of breastfeeding (Table 6.1).

Recent changes to the Healthy People 2010 objectives include an objective to increase the proportion of mothers who exclusively breastfeed their infants through 3 months to 60% and through 6 months to 25%.[4] Despite evidence suggesting progress toward meeting the Healthy People 2010 goals for breastfeeding initiation, breastfeeding duration, and exclusive breastfeeding[3], rates still fall below desired levels. Seventy-seven percent of infants born in 2005 and 2006 are reported to have been breastfed in the early postpartum period.[5] Despite the American Academy of Pediatrics' recommendation that an infant be exclusively breastfed until 6 months of age[6], rates of exclusive breastfeeding continue to be lower than 50% (Tables 6.1 and 6.2). In the early postpartum period (7 days), exclusive breastfeeding is 59.4%. Rates fall to 44.3% at 2 months, to 29% by 4 months, and 13.9% at 6 months.

Gaps in breastfeeding remain—geographically, and among racial groups, educational groups, income groups, and by marital status.[5,7,8] Similar disparities are evident in breastfeeding rates among first-time mothers.[9] Breastfeeding rates are almost 25% lower among mothers participating in the Special Supplemental Nutrition Program for Women, Infants, and Children (WIC) compared with women who do not participate in WIC.[10] Breastfeeding rates are highest in the western states, with California, Washington, Oregon, and Hawaii meeting the 75% breastfeeding initiation rate. Some of the lowest rates persist among southern states and among African American and American Indian women. Rates of breastfeeding among Hispanic or Latina women now equal or exceed rates among white women. Rates remain lower among younger and unmarried women and among women with lower income.

If the United States is to reach these breastfeeding goals, health professionals must take an active role in promoting and supporting lactation. Becoming a breastfeeding advocate requires a thorough understanding of lactation physiology and thorough knowledge of clinical and community resources for support.

Lactation Physiology

Functional Units of the Mammary Gland

The functional units of the **mammary gland** are the **alveoli** (Illustration 6.1). Each alveolus is composed of a cluster of cells (**secretory cells**) with a duct in the center, whose job it is to secrete milk. The ducts are arranged like branches of a tree, each smaller duct leading to six to ten larger collecting ducts. These branchlike collecting ducts lead to the nipple. **Myoepithelial cells** surround the secretory cells. Myoepithelial cells can contract under the influence of **oxytocin** and cause milk to be ejected into the ducts.

Mammary Gland The source of milk for offspring, also commonly called the breast. The presence of mammary glands is a characteristic of mammals.

Alveoli A rounded or oblong-shaped cavity present in the breast.

Secretory Cells Cells in the acinus (milk gland) that are responsible for secreting milk components into the ducts.

Myoepithelial Cells Specialized cells that line the alveoli and that can contract to cause milk to be secreted into the duct.

Oxytocin A hormone produced during letdown that causes milk to be ejected into the ducts.

Mammary Gland Development

During puberty, the ovaries mature and the release of estrogen and progesterone increases (Table 6.3). The cyclic release of these two hormones governs pubertal breast development (Illustration 6.2). The mammary gland develops its lobular structure (**lobes**)

Lobes Rounded structures of the mammary gland.

Table 6.1 Healthy People 2010 breastfeeding objectives for the nation

Objective: Increase the proportion of mothers who breastfeed their babies.

	1998 Baseline Percent (%)	2010 Goal Percent (%)
In early postpartum period		
All women	64	75
Black or African American	45	75
Hispanic or Latino	66	75
White	68	75
At 6 months		
All women	29	60
Black or African American	19	50
Hispanic or Latino	28	50
White	31	50
At 1 year		
All women	16	25
Black or African American	9	25
Hispanic or Latino	17	25
White	19	25

SOURCE: U.S. Department of Health and Human Services. Healthy People 2010: Conference Edition—volumes I and II. Washington, D.C.: U.S. Department of Health and Human Services, Public Health Service, Office of the Assistant Secretary for Health, January 2000.[3,4]

Table 6.2 Breastfeeding rates in birth cohorts

	Infants Born 2000–2003	Infants Born 2004–2006	HP 2010 Goal
Early postpartum	70.9	73.8	75
Exclusive breastfeeding through 3 months	30.5	60	
Any breastfeeding at 6 months	34.2	41.5	50
Exclusive breastfeeding through 6 months	1.3	25	
Any breastfeeding at 12 months	15.7	20.9	25

SOURCE: Breastfeeding Trends and Updated National Health Objectives for Exclusive Breastfeeding. *Morb Mortal Wkly Rep* 2007; 56(3), 760–3.[4]

Lactogenesis Another term for human milk production.

Prolactin A hormone necessary for milk production.

under the cyclic production of progesterone and is usually complete within 12 to 18 months after menarche. As the ductal system matures, cells that can secrete milk develop, the nipple grows, and its pigmentation changes. Fibrous and fatty tissues increase around the ducts.

In pregnancy, the luteal and placental hormones (placental lactogen and chorionic gonadotropin) allow further preparation for breastfeeding (Illustration 6.2). Estrogen stimulates development of the glands that will make milk. Progesterone allows the tubules to elongate and the cells that line the tubules (epithelial cells) to duplicate.

Illustration 6.1 Breast of a lactating female.
This cut away view shows the mammary glands and ducts.

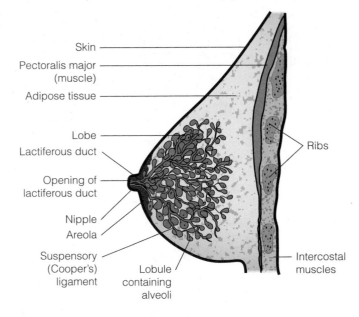

Skin
Pectoralis major (muscle)
Adipose tissue
Lobe
Lactiferous duct
Opening of lactiferous duct
Nipple
Areola
Suspensory (Cooper's) ligament
Lobule containing alveoli
Ribs
Intercostal muscles

Lactogenesis

Breast milk production, or *lactogenesis*, is classically described as occurring in three stages.[11] The first stage, or lactogenesis I, begins during the last trimester of pregnancy; the second and third stages (lactogenesis II and III) occur after birth. Lactogenesis I may be impacted by premature delivery, method of delivery, and other factors. These may explain why mothers who deliver prematurely are often unable to develop a full milk supply (25–35 ounces per day).

- *Lactogenesis I.* During the first stage of milk production, milk begins to form, and the lactose and protein content of milk increase. This stage extends through the first few days postpartum.
- *Lactogenesis II.* This stage begins 2–5 days postpartum and is marked by increased blood flow to the mammary gland. Clinically, it is considered the onset of copious milk secretion, or "when milk comes in." Significant changes in both the milk composition and the quantity of milk that can be produced occur over the first 10 days of the baby's life.
- *Lactogenesis III.* This stage of breast milk production begins about 10 days after birth and is the stage in which the milk composition becomes stable.[12]

Hormonal Control of Lactation

Prolactin and oxytocin are necessary for establishing and maintaining a milk supply. *Prolactin* is a hormone that stimulates milk production. Suckling is a major stimulator of prolactin secretion: prolactin levels double with suckling.[13] Stress, sleep, and sexual intercourse also stimulate prolactin levels. To prevent milk production in the last 3 months of pregnancy, prolactin activity is suppressed by a prolactin-inhibiting factor that is released by the hypothalamus. This inhibition of prolactin allows the

Table 6.3 Hormones contributing to breast development and lactation

Hormone	Role in Lactation	Stage of Lactation
Estrogen	Ductal growth	Mammary gland differentiation with menstruation
Progesterone	Alveolar development	After onset of menses and during pregnancy
Human growth hormone	Development of terminal end buds	Mammary gland development
Human placental lactogen	Alveolar development	Pregnancy
Prolactin	Alveolar development and milk secretion	Pregnancy and breastfeeding (from the third trimester of pregnancy to weaning)
Oxytocin	Letdown: ejection of milk from myopithelial cells	From the onset of milk secretion to weaning

Illustration 6.2 Breast development from puberty to lactation.

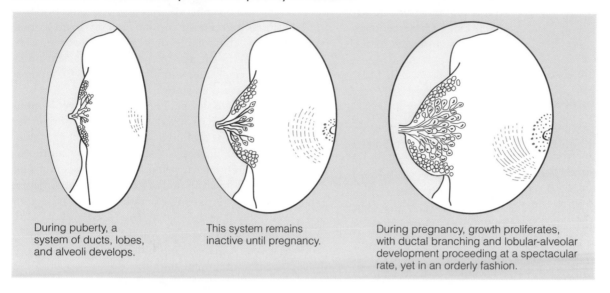

During puberty, a system of ducts, lobes, and alveoli develops.

This system remains inactive until pregnancy.

During pregnancy, growth proliferates, with ductal branching and lobular-alveolar development proceeding at a spectacular rate, yet in an orderly fashion.

Illustration 6.3 The pathways for secretion of milk components.

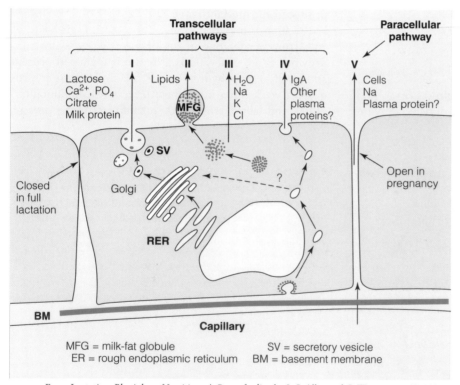

MFG = milk-fat globule
ER = rough endoplasmic reticulum
SV = secretory vesicle
BM = basement membrane

SOURCE: From *Lactation: Physiology, Nutrition, & Breastfeeding* by J. C. Allen and C. Watters, pp. 49–102. Copyright © 1983 Klumer Academics. Reprinted with permission.

of milk from the milk gland (acinus) into the milk ducts. Women may experience tingling or sometimes sharp shooting pain that lasts about a minute and corresponds with contractions in the milk ducts. Oxytocin also acts on the uterus, causing it to contract, seal blood vessels, and shrink its size.

Secretion of Milk

Although the process of milk production is complex, understanding the basic mechanisms of milk secretion is important to understanding how factors such as nutritional status, supplementation, medications, and disease may affect breastfeeding or milk composition. As described by Neville et al., the secretory cell in the breast uses five pathways for milk secretion (Illustration 6.3).[15] Briefly, some components like lactose are made in the secretory cells and secreted into ducts. Water, sodium, potassium, and chloride are able to pass through alveolar cell membranes in either direction (passive diffusion). Milk fat comes from triglycerides from the mother's blood and from new fatty acids produced in the breast. Fats are made soluble in milk by addition of a protein carrier to form milk-fat globules.[16] The milk-fat globules are then secreted into the ducts. Immunoglobulin A and other plasma proteins are

mother's body to prepare for milk production during pregnancy. The actual level of prolactin in the blood is not related to the amount of milk made, but prolactin is necessary for milk synthesis to occur.[14]

Oxytocin release is also stimulated by suckling or nipple stimulation. Its main role is in letdown, or the ejection

Colostrum The milk produced in the first 2–3 days after the baby is born. Colostrum is higher in protein and lower in lactose than milk produced after a milk supply is established.

captured from the mother's blood and taken into the alveolar cells. These proteins are then secreted into the milk ducts.

The Letdown Reflex

The letdown reflex stimulates milk release from the breast. The stimuli from the infant suckling are passed through nerves to the hypothalamus, which responds by promoting oxytocin release from the posterior pituitary gland (Illustration 6.4). The oxytocin causes contraction of the myoepithelial cells surrounding the secretory cells. As a result, milk is released through the ducts, making it available to the infant. Other stimuli, such as hearing a baby cry, sexual arousal, and thinking about nursing, can also cause letdown, and milk will leak from the breasts.

Human Milk Composition

> "Thus, the complexity of milk as a system designed to deliver nutrients and nonnutritive messages to the neonate has increased."
>
> R. G. Jensen, *Handbook of Milk Composition*[17]

Human milk is an elegantly designed natural resource. It is the only food needed by the majority of healthy infants for approximately 6 months. The composition of milk is designed not only to nurture, but also to protect infants from infectious and certain chronic diseases. Human milk composition is changeable over a single feeding, over a day, according to the age of the infant or gestation at delivery, presence of infection in the breast, with menses, and maternal nutritional status.

As our ability to measure and identify novel components increases, we recognize that the composition of human milk is complex. Hundreds of components of human milk have been identified, and their nutritive and non-nutritive roles are under investigation. The basic nutrient composition of colostrum and mature milk is provided in Table 6.4, and a comparison of mature human milk with cow's milk follows in Table 6.5. *The Handbook of Milk Composition*[17] and *Breastfeeding: A Guide for the Medical Professional*[13] provide more detailed descriptions of the composition of human and other milks.

Colostrum

The first milk, *colostrum*, is a thick, often yellow fluid produced during lactogenesis II (days 1–3 after infant birth). Infants may drink only 2 to 10 mL (1.5–2 tsp) of colostrum per feeding in the first 2–3 days. Colostrum provides about 580–700 kcal/L and is higher in protein and lower in carbohydrate and fat than mature milk (produced 2 weeks after infant birth). Secretory immunoglobulin A and lactoferrin are the primary proteins present

Illustration 6.4 The letdown reflex.

An infant suckling at the breast stimulates the pituitary to release the hormones prolactin and oxytocin.

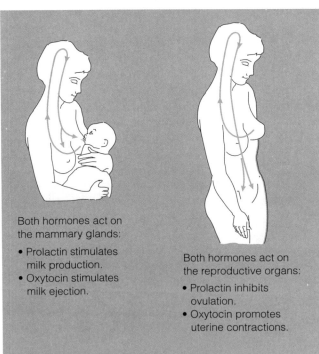

Both hormones act on the mammary glands:
- Prolactin stimulates milk production.
- Oxytocin stimulates milk ejection.

Both hormones act on the reproductive organs:
- Prolactin inhibits ovulation.
- Oxytocin promotes uterine contractions.

Table 6.4	Compositions of 100 mL colostrum (days 1–5 postpartum) and mature milk (day 15 postpartum)	
Contents	**Colostrum**	**Mature Milk**
Calories (kcal)	55	67
Fat (g)	2.9	4.2
Lactose (g)	5.3	7.0
Total protein (g)	2.0	1.1
Secretory IgA	0.5[a]	0.1
Lactoferrin	0.5	0.2
Casein	0.5	0.4
Calcium (mg)	28	30
Sodium (mg)	48	15
Vitamin A (μg retinol equivalents)	151	75
Vitamin B$_1$ (μg)	2	14
Vitamin B$_2$ (μg)	30	40
Vitamin C (μg)	6	5

[a]Concentration is considerably higher at 1–3 days postpartum than at days 4 and 5.

SOURCE: Adapted from Prentice A. Constituents of human milk. *Food and Nutrition Bulletin* 17(4), The United Nations University Press. December 1996.[16]

Table 6.5 Human and cow's milk composition

Nutrient	Units	Human Milk (1 fl oz)	Whole Cow's Milk (1 fl oz)	Nutrient	Units	Human Milk (1 fl oz)	Whole Cow's Milk (1 fl oz)
Water	g	26.95	26.94	**Lipids**			
Energy	kcal	22	19	Saturated Fatty acids, total	g	0.619	0.569
Energy	kj	90	78				
Protein	g	0.32	0.96	4:0	g	0.000	0.023
Total lipids (fat)	g	1.35	0.99	6:0	g	0.000	0.023
Carbohydrate	g	2.12	1. 46	8:0	g	0.000	0.023
Fiber, total dietary	g	0.0	0.0	10:0	g	0.019	0.023
Sugars, total	g	2.12	1.60	12:0	g	0.079	0.023
				14:0	g	0.099	0.091
Minerals				16:0	g	0.283	0.000
Calcium, Ca	mg	10	34	18:0	g	0.090	0.253
Iron, Fe	mg	0.01	0.01	Monounsaturated Fatty acids, total	g	0.511	0.000
Magnesium, Mg	mg	1	3				
Phosphorus, P	mg	4	26	Polyunsaturated Fatty acids, total	g	0.153	0.000
Potassium, K	mg	16	40				
Sodium, Na	mg	5	13	Cholesterol	mg	4	0.000
Zinc, Zn	mg	0.05	0.11				
Copper, Cu	mg	0.016	0.008	**Amino acids**			0.000
Manganese, Mn	mg	0.008	0.001	Tryptophan	g	0.005	0.022
Selenium, Se	mcg	0.6	1.1	Threonine	g	0.014	0.043
				Isoleucine	g	0.017	0.049
Vitamins				Leucine	g	0.029	0.079
Vitamin C, total ascorbic acid	mg	1.5	0.0	Lysine	g	0.021	0.042
				Methionine	g	0.006	0.022
Thiamin	mg	0.004	0.014	Cystine	g	0.006	0.005
Riboflavin	mg	0.011	0.052	Phenylalanine	g	0.014	0.044
Niacin	mg	0.055	0.027	Tyrosine	g	0.016	0.045
Pantothenic acid	mg	0.069	0.114	Valine	g	0.019	0.057
Vitamin B_6	mg	0.003	0.011	Arginine	g	0.013	0.022
Folate, DFE	mcg_DFE	2	2	Histidine	g	0.007	0.022
Vitamin B_{12}	mcg	0.02	0.14	Alanine	g	0.011	0.031
Vitamin A, RAE	mcg_RAE	19	14	Aspartic acid	g	0.025	0.071
Vitamin E (alpha tocopherol)	mg	0.02	0.02	Glutamic acid	g	0.052	0.193
				Glycine	g	0.008	0.022
Vitamin D	IU	1	16	Proline	g	0.025	0.102
Vitamin K (phylloquinone)	mcg	0.1	0.1	Serine	g	0.013	0.032

SOURCE: Adapted from USDA National Nutrient Database for Standard Reference, Release 22, 2009.[17]

in colostrum, but other proteins present in mature milk are not present. The concentration of mononuclear cells (a specific type of white blood cell from the mother that provides immune protection) is highest in colostrum. Colostrum has higher concentrations of sodium, potassium, and chloride than more mature milk.

Water

Breast milk is isotonic with plasma. This biological design of milk means that babies do not need water or other fluids to maintain hydration, even in hot climates.[20] As a major component of human milk, water allows suspension of the milk sugars, proteins, immunoglobulin A, sodium, potassium, citrate, magnesium, calcium, chloride, and water-soluble vitamins.

Energy

Human milk provides approximately 0.65 kcal/mL, although the energy content varies with its fat (and, to a lesser degree, protein and carbohydrate) composition.

Breastfed infants consume fewer calories than infants fed HMS.[21,22] It is not known whether this difference in energy intake of breastfed infants has to do with the composition of human milk, the inability to see the volume of feedings when providing human milk, the differences in the suckling at the breast compared to an artificial nipple, or other factors. Infants who are breastfed are thinner for their weight at 8–11 months than infants fed HMS, but these differences disappear by 12–23 months of age and few differences are notable by 5 years of age.[23]

Lipids

Lipids are the second largest component of breast milk by concentration (3–5% in mature milk). Lipids provide half of the energy of human milk.[17] Human milk fat is low at the beginning of a feeding in foremilk, and higher at the end in the hindmilk that follows.

Effect of Maternal Diet on Fat Composition   The fatty acid profile of human milk varies with the diet of the mother.[24] When diets rich in polyunsaturated fats are consumed, more polyunsaturated fatty acids are present in the milk.[17] When a mother is losing weight, the fatty acid profile of her fat stores is reflected in the milk.[25] When very low-fat diets with adequate calories from carbohydrate and protein are fed, more medium-chain fatty acids are synthesized in the breast.

Clinicians use the value of 20 kcal/oz of human milk to calculate energy provided to an infant. However, fat content may vary considerably from mother to mother and may vary diurnally as well as within a feeding. Fat values in foremilk may range from 39.7–46.7% energy (17.9–23.6 kcal/oz) and hindmilk from 60.7–80.1% energy (23.5–33.2 kcal/oz).[26] Overall, the energy ranged from 20.9 to 26.2 calories per ounce.

DHA   Milk DHA levels are increased by maternal supplementation.[27] Recent interest in lipids in human milk stems from studies showing developmental advantages provided by docosahexaenioic acid (DHA).[28] DHA is essential for retinal development and accumulates during the last months of pregnancy. The advantages of human milk seem particularly important to premature infants born before 37 weeks, perhaps because the concentrations of DHA are higher in the milk of mothers delivering preterm infants as compared to full-term infants.[29] Advantages for term infants have been demonstrated as well. For example, a Norwegian study suggests that cod liver oil supplementation during pregnancy was associated with higher IQ scores at 4 years of age in breastfed versus HMS-fed infants.[30] Cod liver oil contains high levels of DHA as well as high levels of vitamin A and vitamin D, so it should be used with caution.

Trans Fatty Acids   *Trans* fatty acids stemming from the mother's diet are present in human milk.[31] *Trans* fat concentrations are similar in American and Canadian women, but lower in the milk of women from European and African countries. Removal of *trans* fatty acids from many food products in Canada led to lower levels of *trans* fat in human milk.[32] Similar trends are expected in the United States.

Cholesterol   Cholesterol, an essential component of all cell membranes, is needed for growth and replication of cells. Cholesterol concentration ranges from 10–20 mg/d and varies depending on the time of day.[17] Breastfed infants have higher intakes of cholesterol and higher levels of serum cholesterol than infants fed HMS.[33] Early consumption of cholesterol through breast milk appears to be related to lower blood cholesterol levels later in life.[34]

Protein

The protein content of mature human milk is relatively low (0.8–1.0%) compared to other mammalian milks, such as cow's milk (Table 6.5). The concentration of proteins synthesized in the breast are more affected by the age of the infant (time since delivery) than maternal intake and maternal serum proteins. Proteins synthesized by the breast are more variable because hormones that regulate gene expression and guide protein synthesis change with time.[35] Despite the relatively low concentration, human milk proteins have important nutritive and non-nutritive value. Proteins and their digestive products, such as peptides, exhibit a variety of antiviral and antimicrobial effects.[36] Enzymes present in human milk might also provide protection by facilitating actions that prevent inflammation.

Casein   Casein is the major class of protein in mature milk from women who deliver either at term or preterm.[37] Casein, calcium phosphate, and other ions such as magnesium and citrate appear as an aggregate and are the source of milk's white appearance.[38] Casein's digestive products, casein phosphopeptides, keep calcium in soluble form and facilitate its absorption.

Whey Proteins   Whey proteins are the proteins that remain soluble in water after casein is precipitated from milk by acid or enzymes. Whey proteins include milk and serum proteins, enzymes, and immunoglobulins, among others. Several mineral-, hormone-, or vitamin-binding proteins are also identified as components of whey proteins. These include lactoferrin, which carries iron in a form that is easy to absorb and has bacteriostatic activity. The enzymes present in whey proteins aid in digestion and protection against bacteria.

Nonprotein Nitrogen   Nonprotein nitrogen provides 20–25% of the nitrogen in milk.[38] Urea accounts for 30–50% of nonprotein nitrogen, and nucleotides for 20%, depending on the stage of lactation and the diet of the mother. Some of this nitrogen is available for the infant to use for producing nonessential amino acids. Some of the nonprotein nitrogen is used to produce other proteins with biological roles such as hormones, growth factors, free amino acids, nucleic acids, nucleotides, and carnitine. The role of individual nucleotides in human milk is under investigation; however, nucleotides appear to play important roles in growth and disease resistance.

Milk Carbohydrates

Lactose is the dominant carbohydrate in human milk. Other carbohydrates—including monosaccharides (such as glucose), polysaccharides, oligosaccharides, and protein-bound carbohydrates—are also present.[39] Lactose enhances calcium absorption. As the second largest carbohydrate component, oligosaccharides contribute calories at low *osmolality*, stimulate the growth of *bifidus* bacteria in the gut, and inhibit the growth of *E. coli* and other potentially harmful bacteria.

Oligosaccharides   Oligosaccharides are medium-length carbohydrates containing lactose on one end. Oligosaccharides can be free, or bound to proteins as glycoproteins, or bound to lipids as glycolipids, or they can bind to other structures. The conjugated and unconjugated oligosaccharides are classified as glycans. Over 130 different oligosaccharides are present as functional ingredients of human milk.[40,41] Oligosaccharides in human milk prevent the binding of pathogenic microorganisms to the gut, thereby preventing infection and diarrhea.

Fat-Soluble Vitamins

Vitamin A   Colostrum has approximately twice the concentration of vitamin A as mature milk does. Some of the vitamin A in human milk is in the form of beta-carotene. Its presence is responsible for the characteristic yellow color of colostrum. In mature milk, vitamin A is present at 75 µg/dl or 280 IU/dl.[42] These levels are adequate to meet infant needs.

Vitamin D   Vitamin D is present in both lipid and aqueous (water) compartments of human milk. Most vitamin D is in the form of $25\text{-}OH_2$ vitamin D and vitamin D_3. Vitamin D levels of human milk vary with maternal diet and exposure to sunshine.[43] Maternal exposure to sunlight has been reported to increase the vitamin D_3 level in milk tenfold.[44] It is yet unknown how much maternal vitamin D supplement is needed to ensure adequate maternal and infant vitamin D status when sunlight exposure is insufficient, though researchers are actively pursuing the answer.[45]

Vitamin E   The level of total tocopherols in human milk is related to the milk's fat content. Human milk contains 40 µg of vitamin E per gram of lipid in the milk.[46] Levels of alpha-tocopherol decrease from colostrum to transitional milk and to mature milk, whereas beta and gamma tocopherols remain stable throughout each stage of lactation. The level of vitamin E present in human milk is adequate to meet the needs of full-term infants for muscle integrity and resistance of red blood cells to hemolysis (breaking of red blood cells). The levels of vitamin E in preterm milk have been reported to be the same[47] and higher[48] than in term milk. However, in both reports, the levels present were not considered adequate to meet the needs of preterm infants.

> **Osmolality**   A measure of the concentration of particles in solution.
>
> **Monovalent Ion**   An atom with an electrical charge of +1 or −1.

Vitamin K   Vitamin K is present in human milk at levels of 2.3 µg/dL.[46] Approximately 5% of breastfed infants are at risk for vitamin K deficiency based on vitamin K-dependent clotting factors. There are cases of vitamin K deficiency among exclusively breastfed infants who did not receive vitamin K at birth.

Water-Soluble Vitamins

Water-soluble vitamins in human milk are generally responsive to the content of the maternal diet or supplements (vitamin C, riboflavin, niacin, B_6, and biotin). Clinical problems relating to water-soluble vitamins are rare in infants nursed by mothers with inadequate diets.[13] Vitamin B_6 is considered most likely to be deficient in human milk; levels of B_6 in human milk directly reflect maternal intake.[49]

Vitamin B_{12} and Folic Acid   Vitamin B_{12} and folic acid are bound to whey proteins in human milk; therefore, their content in milk is less influenced by maternal intake of these vitamins than are the other water-soluble vitamins. Factors that influence protein secretion (hormones and the age of the infant, or time since delivery) are more likely to alter the human milk levels of B_{12} and folate than is dietary intake.[1,50] Infant illness associated with low folate levels in milk has not been reported. Folate levels increase with the duration of lactation despite a decrease in maternal serum and red blood cell folate levels.[1] B_{12} deficiency, or low levels of B_{12}, in milk has been reported for women who have had gastric bypass surgery, have hypothyroidism, consume vegan diets, have latent pernicious anemia, or are generally malnourished.[51]

Minerals in Human Milk

The minerals in human milk contribute substantially to the osmolality of human milk. *Monovalent ion* secretion is managed closely by the alveolar cells, in balance with lactose, to maintain the isosmotic composition of human milk.

Mineral content in milk is related to the growth rate of the offspring. The mineral content of human milk is much lower than the concentration in cow's milk and the milk of other animals whose offspring grow faster. With the exception of magnesium, the concentration of minerals decreases over the first 4 months postpartum. This decline in the mineral content of milk during the period of rapid growth is not what one would expect, but infant growth is well supported.[52] The lower mineral concentration of human milk is easier for the kidneys to handle. This reduced load on the kidneys is considered a significant benefit of human milk.

Bioavailability An important feature of several of the minerals (magnesium, calcium, iron, zinc) in human milk is the packaging that makes them highly available (bioavailable) to the infant.[53] Packaging minerals so that the infant can use them efficiently also reduces the burden to the mother because less of the mineral is needed in the milk. For example, zinc is 49% available from human milk, but only 10% available from cow's milk and cow's milk–based HMS.[54] Exclusively breastfed infants have little risk of anemia,[55] despite the seemingly low concentration of iron in human milk. One study suggests that infants who are exclusively breastfed for 6.5 months are less likely to be anemic than those nursed exclusively for 5.5 months.[56]

Zinc The importance of zinc to human growth is well established. Human milk zinc is bound to protein and is highly available, in comparison to cow's milk and cow's milk–based HMS. Both the zinc intake (per kg) and the zinc requirements of infants decline after the first few months.[57] Normally, zinc homeostasis and human milk zinc levels are maintained even in the face of low maternal zinc intake.[58] Rare cases of zinc deficiency, which appears as intractable diaper rash, have been noted in exclusively breastfed infants, however.[54] A defect in the mammary gland uptake of zinc has been described as the cause of low milk concentration when maternal serum zinc concentrations are normal. In these cases, infants seem to respond to zinc supplementation.

Trace Minerals Trace minerals (copper, selenium, chromium, manganese, molybdenum, nickel, and fluorine) are present in the human body in small concentrations and are essential for growth and development. Less is known about trace minerals and infant health than about other nutrients. In general, however, the levels of trace minerals in human milk are not altered by the mother's diet or supplement use, excepting fluoride. The DRI for fluoride is 0.1 mg daily for infants less than 6 months.[59] Fluoride provided in community water is safe for breastfeeding women and their infants. Most infants who live in areas with fluoridated water do not need an additional supplement.[5,60] If bottled water is used, water with fluoride added should be purchased.

Taste of Human Milk

> "...too full o' th' milk of human kindness to catch the nearest way."
>
> Shakespeare's *Macbeth*, Act I, Scene V

This line from Shakespeare reflects the centuries-old belief that a breastfeeding woman's diet influences the composition of her milk and has a long-lasting influence on the child. The flavor of human milk is an important taste experience for newborn infants, but flavor of human milk is often ignored when the benefits of human milk or its composition is considered. Human milk is slightly sweet[61] and it carries the flavors of compounds ingested, such as mint, garlic, vanilla, and alcohol.[62]

The transfer of flavor compounds appears to occur selectively and in relatively low amounts.[63] Infant responses to flavors in milk seem to depend on the length of time since the mother consumed the food, and the amount and frequency of the flavor that the mother consumed (new versus repeated exposure). Infants seem more interested in their mother's milk when flavors are new to them. Researchers found that infants nursed at the breast longer if a flavor (garlic) was new to them than if the mother had taken garlic tablets for several days.[64] Infants who were exposed to carrot juice flavor in their mother's milk ate less of a carrot-flavored cereal and spent less time feeding at the breast than infants who had not been exposed to the carrot flavor. Thus, exposing infants to a variety of flavors in human milk may contribute to their interest in and consumption of human milk as well as their acceptance of new flavors in solid foods.[65]

Benefits of Breastfeeding

Breastfeeding Benefits for Mothers

Breastfeeding women experience hormonal, physical, and psychosocial benefits.[66] Breastfeeding immediately increases levels of oxytocin, a hormone that stimulates uterine contractions, minimizes maternal postpartum blood loss, and helps the uterus to return to nonpregnant size.[67]

After the birth, the return of fertility (through monthly ovulation) is delayed in most women during breastfeeding, particularly with exclusive breastfeeding.[67] This delay in ovulation results in longer intervals between pregnancies. Breastfeeding alone, however, is not as effective as other available birth control methods. Consequently, many health care professionals in the United States do not offer breastfeeding as an option for birth control.

Many women experience psychological benefits, including increased self-confidence and facilitated bonding with their infants.[68] Many still consider faster return to prepregnancy weight a benefit of breastfeeding; however, women may lose or gain weight while nursing. The impact of breastfeeding on maternal weight is discussed in more detail later in this chapter. In addition to these short-term benefits, women who nurse at a younger age and for longer duration have lower risk of breast and ovarian cancer[69-70] and rheumatoid arthritis.[71]

Breastfeeding Benefits for Infants

"Breastfeeding—the main source of active and passive immunity in the vulnerable early months and years of life—is considered to be the most effective preventive means of reducing the death rate of children under five."[72]

Lubbock, M. H., Clark, D., and Goldman, A. S. Breastfeeding: maintaining an irreplaceable immunological resource. *Nature Reviews Immunology.* 2004; 4(Jul):565–572.

Nutritional Benefits   The value of the composition of human milk is widely recognized. Companies that make HMS often use human milk as the standard, recognizing the many unique properties of human milk:

- With its dynamic composition and the appropriate balance of nutrients, human milk provides optimal nutrition to the infant.[13,17]
- The balance of nutrients in human milk matches human infant requirements for growth and development closely; no other animal milk or HMS meets infant needs as well.
- Human milk is isosmotic (of similar ion concentration; in this case human milk and plasma are of similar ion concentration) and therefore meets the requirements for infants without other forms of food or water.
- The relatively low protein content of breast milk compared to cow's milk meets the infant's needs without overloading the immature kidneys with nitrogen.
- Whey protein in human milk forms a soft, easily digestible curd.
- Human milk provides generous amounts of lipids in the form of essential fatty acids, saturated fatty acids, medium-chain triglycerides, and cholesterol.
- Long-chain polyunsaturated fatty acids, especially docosahexaenoic acid (DHA), which promotes optimal development of the central nervous system, are present in human milk and are present in only some of the HMS marketed in the United States.
- Minerals in breast milk are largely protein-bound and balanced to enhance their availability and meet infant needs with minimal demand on maternal reserves.

Immunological Benefits   One of the most important realizations about breastfeeding in the last few decades is the ability of human milk to protect against infections. Cells (T- and B-lymphocytes), *secretory immunoglobins* (sIgA, sIgG, sIgM, sIgE, sIgD), histocompatibility antigens, T-cell products, many nonspecific factors (e.g., complement, bifidus factor), carrier proteins (lactoferrin, transferring, vitamin B_{12}-binding protein, and corticoid-binding protein), and enzymes (lysozyme, lipoprotein lipase, leukocyte enzymes) are components of milk that confer immunological benefits.

Cellular components in human milk (*macrophages, neutrophils, T- and B-lymphocytes,* and *epithelial cells*) are especially high in colostrum but are also present for months in mature human milk in lower concentrations. The function of macrophages in human milk includes phagocytosis of fungi and bacteria, killing of bacteria, and production of the complement proteins, lysozyme, and lactoferrin and immunoglobin A and G.[13]

Leukocyte function appears to offer more protection to the breast than to immunocompetence of the infant. Neutrophils, however, appear to be activated and contribute to phagocytosis at the mucosa of the infant's gastrointestinal tract.[73] Both T- and B-lymphocytes provide the infant with protection against organisms in the gastrointestinal tract. This protection may extend beyond acute infection to allergy, necrotizing enterocolitis, tuberculosis, and neonatal meningitis.[13]

Immunoglobins are thought to be transported from maternal plasma across secretory epithelium to create secretory immunoglobins.[74] The predominant (90%) immunoglobulin in human milk, *secretory immunoglobin A(sIgA),* also appears to be most important in terms of the protection conferred to the infant. sIgA and sIgM protect the infant by blocking colonization with pathogens and limiting the number of antigens that cross the mucosal barrier. sIgA protects against enteroviruses, cytomegalovirus, herpes simplex virus, respiratory syncytial virus, rubella, retrovirus, and rotavirus,[13] and sIgM protects against cytomegalovirus, respiratory syncytial virus, and rubella.

Bifidus factor is a growth factor (probably a carbohydrate) that supports growth of *Lactobacillus bifidus. Lactobacillus* is a probiotic bacterium that stimulates antibody production and enhances phagocytosis of antigens.[13]

Lysozyme protects against enterobacteria and other gram-positive bacteria. Lysozyme is secreted by neutrophils and macrophages.

Immunoglobin   A specific protein that is produced by blood cells to fight infection.

Macrophages   A white blood cell that acts mainly through phagocytosis.

Neutrophils   Class of white blood cells that are involved in the protection against infection.

T-lymphocyte   A white blood cell that is active in fighting infection. (May also be called T-cell; the t in T-cell stands for thymus.) These cells coordinate the immune system by secreting hormones that act on other cells.

B-lymphocytes   White blood cells that are responsible for producing immunoglobulins.

Epithelial Cells   Cells that line the surface of the body.

Secretory Immunoglobin A   A protein found in secretions that protect the body's mucosal surfaces from infections. The mode of action may be by reducing the binding of a microorganism with cells lining the digestive tract. It is present in human colostrum but not transferred across the placenta.

Binding proteins in human milk bind iron and vitamin B_{12}, making the nutrients unavailable for pathogens to grow in the infant's gastrointestinal tract. Such factors are also responsible for some of the differences in intestinal flora (natural bacteria of the gastrointestinal tract) of breastfed infants versus HMS-fed infants.

Individual fatty acids and other milk components (oligosaccharides, gangliosides, and glycoconjugates) resulting from digestion of human milk are antimicrobial.[75] The digestive products of triacylglycerides and the lipid globule appear to protect against *Escherichia coli* 0157:H7, *Campylobacter jejuni*, *Listeria monocytogenes*, and *Clostridium perfringens*.[76] Monoacylglycerides are able to lyse enveloped viruses, bacteria, and protozoa. Glycoconjugates (glycoproteins, glycolipids, glycoaminoglycans, and oligosaccharides) may bind pathogens directly, thus preventing infection. Nucleotides are reported to increase resistance to *Staphylococcus aureus* and *Candida albicans* and may increase response to vaccine antigens.[77]

Growth factors and hormones in human milk, such as insulin, enhance the maturation of the infant's gastrointestinal tract. These substances also help to protect the infant, especially neonates, against viral and bacterial pathogens.

Lower Infant Mortality in Developing Countries   In the developing world, 10 million children die each year, and 60% are believed to be preventable deaths.[78] Improving breastfeeding practices could save approximately 1.3 million lives annually, and continuing breastfeeding with complementary foods could save an additional 600,000. This protection of lives is at the center of the World Health Organization (WHO) and UNICEF's joint efforts called the Global Strategy for Infant and Young Child Feeding to remind the international community of the impact of feeding practices (including breastfeeding) on children's health outcomes.[78]

Breastfeeding may also play a role in reducing the risk of sudden infant death syndrome (SIDS), but this is still under investigation. Researchers disagree about whether breastfeeding has a primary effect in reducing risk of SIDS. An analysis of available studies found that bottle-feeding increases the risk of SIDS, but other factors related to feeding choice may be responsible for this finding.[79]

Fewer Acute Illnesses   Reduced infant illness is evident in countries with high infant illness (*morbidity*) and death (*mortality*) rates, poor sanitation, and questionable water supplies. Even in the United States and other developed nations, where modern health care systems, safe water, and proper sanitation are commonplace, there is a clear relationship between breastfeeding and reduced rates of illness in infants. In U.S. samples, the *incidence* of diarrhea is estimated to be

Morbidity   The rate of illnesses in a population.

Mortality   Rate of death.

50% lower in exclusively breastfed infants.[80] Internationally, gastrointestinal infection was lower among infants exclusively breastfed for 6 months when compared to those exclusively breastfed for only 3 months.[80,81] Ear infections are 19% lower, and the number of prolonged episodes of ear infection was 80% lower among breastfed infants than among infants fed HMS. In a U.S. population study, breastfed infants experienced 17% less coughing and wheezing and 29% less vomiting than did infants fed HMS.[82]

Reductions in Chronic Illness   In addition to the lower rate of acute illnesses in breastfed children, breastfeeding also seems to protect against chronic childhood diseases. Breastfeeding may reduce the risk of celiac disease,[83] inflammatory bowel disease,[84] and neuroblastoma.[85] HMS feeding results in an increase in the risk of allergy (30%) and asthmatic disease (25%).[86] These reductions in acute and chronic infant illness increase with greater use of human milk.[87] For example, infants who receive some human milk and some HMS are at 60% greater risk of ear infection than those fed exclusively human milk. The risks, particularly for allergy and asthmatic disease, are reduced for the duration of breastfeeding and for months to years after weaning.[86]

Breastfeeding and Childhood Overweight   Considerable attention has been paid to the role of breastfeeding in preventing obesity, but this relationship is still a topic of controversy. Breastfed infants typically are leaner than HMS-fed infants at 1 year of age without any difference in activity level or development.[88] A large body of literature suggests that there is a small reduction in risk of overweight in children older than 3 years of age who were breastfed.[89] The effect of breastfeeding on the incidence of overweight was greater with longer duration of breastfeeding in some studies,[90] but not all.[91] Several potential mechanisms have been identified for the modest reduction in obesity in children who were breastfed, including metabolic programming, possibly related to leptin, ghrelin, and other neurometabolic messengers delivered in human milk.[92, 93] Other contributors could include learned self-regulation of energy intake and other characteristics of the families or parents, such as healthy lifestyle. Given the epidemic of obesity in the United States and beyond, it is likely that the research in this area will continue.

Cognitive Benefits   Several reports have linked breastfeeding, and especially duration of breastfeeding, with cognitive benefits, assessed by IQ.[30,94] The increases in cognitive ability associated with breastfeeding are significant even after adjusting for family environment.[95] Cognitive development gains increase with the duration of breastfeeding.[94] In addition, higher intelligence quotients (IQ) of infants breastfed for 6 months appear to be greater among infants born small for gestational age (11 points) than among infants born appropriate weight for age (3 points).[96] The differences in *cognitive function* are also greater in premature infants fed human milk than in those fed HMS.[97] Recognition that the fatty acid composition of milk plays an important role in neuropsychological development bolsters the credibility of psychological or cognitive benefits from breastfeeding.

Analgesic Effects   Breastfeeding seems to work as an analgesic in infants. Breastfeeding during venipuncture seems to reduce infant pain as well as a 30% glucose solution followed by pacifier use.[98] However, breastfeeding before a heel prick[99] did not seem to reduce infant pain response. Breastfeeding may be used to reduce infant discomfort during minor invasive procedures.[100]

Socioeconomic Benefits   A decrease in medical care for breastfed infants is the primary socioeconomic benefit of breastfeeding. Medicaid costs for breastfed WIC infants in Colorado were $175 lower than for infants who were fed HMS.[101] Never-breastfed infants have an excess of care for lower respiratory tract illness, otitis media, and gastrointestinal disease compared to infants breastfed for at least 3 months.[102] Each 1000 never-breastfed infants had 2033 more sick care visits, 212 days of hospitalization, and 609 more prescriptions. In addition, in one study of two companies with established lactation programs, the one-day maternal absenteeism from work due to infant illness was approximately two-thirds lower in breastfeeding women than nonbreastfeeding women.[103] Companies benefit through lower medical costs and greater employee productivity.

Breast Milk Supply and Demand

Can Women Make Enough Milk?

Typical milk production averages approximately 600 mL (240 mL = 1 cup, or 8 ounces) in the month postpartum and continues to increase to approximately 750–800 mL per day by 4–5 months postpartum.[1] Milk production can range from 450–1200 mL per day in women who are nursing one infant.[22] Infant weight, the caloric density of milk, and the infant's age contribute significantly to the infant demand for milk. Milk increases to meet the demand of twins, triplets, or infants and toddlers suckling simultaneously; it can also be increased by pumping the milk.

Traditionally, factors such as how vigorously an infant nurses, how much time the infant is at the breast, and how many times he or she nurses in a day were thought to control milk production. We now know that milk synthesis (rate of accumulation of milk in the breast) is related to infant demand.[104] That is, the removal of milk from the breast seems to be the signal to make more milk, and most women are able to increase their milk production to meet infant demand.[105]

An average of 24% of milk is left in the breast after feeding.[106] Thus, the short-term milk storage of the breast does not seem to be a limiting factor to infant milk intake. The average rate of synthesis in a day is only 64% of the highest rate of milk synthesis, suggesting that milk synthesis could be increased considerably. Comparisons of milk production between mothers of singletons and twins shows that the breasts have the capacity to synthesize much more than a singleton infant usually drinks.[107]

Cognitive Function   The process of thinking.

Does the Size of the Breast Limit a Woman's Ability to Nurse Her Infant?

The size of a woman's breast does not determine the amount of milk production tissue (clusters of alveoli containing secretory cells that produce the milk).[108] Much of the variation in breast size is due to the amount of fat in the breast. The size of the breast does limit storage because of limitations in the expansion of the ducts. Daily milk production is not related to the total milk storage capacity within the breast, however.[106] This means that women with small breasts can produce the same amount of milk as women with large breasts, although the latter woman may be able to feed her infant less frequently to deliver the same volume of milk compared with a woman with smaller breasts.

Is Feeding Frequency Related to the Amount of Milk a Woman Can Make?

Feeding frequency is not consistently related to milk production. The rate of milk synthesis is highly variable between breasts and between feedings.[109] However, the amount of milk produced in 24 hours and the total milk withdrawn in that 24-hour period are highly related.[106] Milk synthesis is able to quickly respond to infant demand.

The breast responds to the degree of emptying during a feeding, and this response is a link between maternal milk supply and infant demand. Daly proposed that the breast responds to the infant's need by measuring how completely the infant empties the breast.[102,106,109] For example, if a lot of milk is left in the breast, then milk synthesis will be low to prevent engorgement; if the breast is fully emptied, synthesis will be high to replenish the milk supply.

Exact mechanisms of milk supply and demand are not well understood, but they seem to be related to a protein called feedback inhibitor of lactation (FIL).[109] FIL is an active whey protein that inhibits milk secretion. This protein inhibits all milk components equally according to their concentration in milk. Therefore, this protein seems to affect milk quantity only, not milk composition.

Pumping or Expressing Milk

Pumping or expressing milk may be needed for many reasons, including maternal or infant illness or separation. Women can express milk using several different methods: manually, hand pumps, commercial electric pumps, or hospital-grade electric pumps. A pump that allows mothers to pump both breasts at the same time (10 minutes per session) can save time over single pumping (20 minutes per session).[110] Electric pumps are efficient and may increase prolactin more than hand expression or hand pumping.[110,111] Insufficient milk production is a common problem among women who express milk. Researchers working with women who pumped their breasts report that 8 to 12 or more milk expressions per day were necessary to stimulate an adequate production of milk.[110] The optimal number of expressions in a 24-hour period is likely to differ for women according to how well they empty the breast and the storage capacity of the breast. Women who are able to establish an adequate volume of milk (>500 mL per day) in the first 2 weeks postpartum are more likely to still have enough milk for their infant at 4–5 weeks postpartum.[110] This recommendation is consistent with the advice to nurse the infant (or pump) early and often to build a good milk supply. See Chapter 7 for information on storage of milk.

Can Women Breastfeed after Breast Reduction or Augmentation Surgery?

Information regarding breastfeeding rates after breast surgery is scarce. Accumulating evidence does suggest that women who undergo breast reduction surgery may be at risk for unsuccessful lactation, as evidenced by lower breastfeeding rates and duration and greater perception of insufficient milk supply than among women without prior breast surgery. Almost all (91.8%) women with prior breast reduction surgery reported problems with breastfeeding.[112] The type of surgery, including the location and amount of breast tissue removed and the damage to remaining tissue, appears to be an important determinant of ability to breastfeed. Women may choose to have incision around the lower part of the breast to avoid damage to the ductal system caused by incision in the midst of the breast. Women with peri-areolar (around the nipple) incisions experience greater difficulties with breastfeeding because of damage to the ductal system.[113] After augmentation surgery, compression of the ducts in the breast may lead to poor milk production. Lactation consultants recommend that the surgery date, type, and incision used, as well as prior breastfeeding experience, be ascertained. Infants should be closely followed to prompt intervention when needed.

What Is the Effect of Silicone Breast Implants on Breastfeeding?

Nearly half of the adverse maternal-child reports to the U.S. Food and Drug Administration were related to problems with or concerns about the safety of the implants with respect to breastfeeding.[114] The American Academy of Pediatrics does not consider silicone implants a reason not to breastfeed.[115] Nearly a million women in the United States have breast implants containing silicone. Early reports introduced concern of esophageal dysfunction in children of women who had silicone implants, but more recent research found no evidence to support such claims, and silicone concentrations in milk from women with implants are not elevated[116] and, in fact, are lower than those in formula and cow's milk.[117] It is possible that immunological responses could cause unfavorable effects; however, there is no evidence suggesting direct toxicity to the infant. Therefore, the most likely influence on breastfeeding is similar to the effects of saline implants: compression of ducts leading to poor milk production.

The Breastfeeding Infant

Optimal Duration of Breastfeeding

The gap between "best breastfeeding practice" and the norms for breastfeeding in the United States makes study of the optimal duration particularly important for health care professionals. The health benefits to the mother–child pair should be the primary criteria to determine the optimal duration of breastfeeding—and not simply whether the cultural environment makes such duration practical. The American Academy of Pediatrics (AAP) has taken a clear stand on this issue, saying that breastfeeding should continue for a year or longer.[6] The U.S. Surgeon General recommends human milk feeding exclusively for 6 months, noting further that it is better to breastfeed for 6 months and best to breastfeed for 12 months, with solid foods being introduced at 4–6 months.[66]

Infants who are breastfed for 6 months experience fewer illnesses from gastrointestinal infection than do infants who are given HMS and breast milk at 3 or 4 months of age. Deficits in growth have not been demonstrated among infants in developing or developed countries who are exclusively breastfed for 6 months or longer.[116] Breastfeeding can prevent intestinal blood loss in infants—a factor that should be considered when determining its optimal duration. Infants fed cow's milk before the age of 6 months suffer nutritionally significant losses of iron via intestinal blood[117]—an observation that supports the AAP's recommendation.[6] Through 1 year, breastfed infants suffer fewer acute infections than formula-fed infants do, a finding that supports breastfeeding beyond the introduction of solids.

Reflexes

Healthy term infants are born with several reflexes that enable newborns to nourish themselves. Observations show that 18-week-old fetuses start sucking. By 34 weeks gestation, the suck has adequate pace and rhythm to be nutritive. The gag reflex is the reflex that prevents taking food and fluids into the lungs. This reflex is developed by 28 weeks gestation. These reflexes allow term infants to suck and swallow in a coordinated pattern that protects the airways.

Two other reflexes describe the infant's ability to position herself to breastfeed. The *oral search reflex* is described as the infant opening his or her mouth wide in proximity to the breast while thrusting the tongue forward. The *rooting reflex* results in the infant turning to the side when stimulated on the side of the upper or lower lip. Infants come forward, open their mouth, and extend the tongue when the center of either upper or lower lip is stimulated.

The presence of these reflexes is important to the success of breastfeeding. However, successful nursing also requires appropriate positioning of the infant at the breast and adequate maternal letdown and milk production. Appropriate positioning and maternal assessment of infant nursing behaviors must be learned. Support from lactation consultants and/or other health care professionals who are trained in lactation may be necessary.

Preparing the Breast for Breastfeeding

Breasts and nipples begin to be sore in the first trimester, but the tenderness usually subsides by the end of the first trimester. Enlargement of the breast and nipple are evident by the end of the first trimester and continue through pregnancy. By the third trimester, Montgomery glands, sebaceous glands that produce oils to lubricate the nipple and areola, become pronounced and the nipple darken. Rubbing the breast with a towel to toughen the breast for breastfeeding is no longer recommended, as it may remove the natural lubrication provided by the Montgomery glands. Gentle massage is recommended by the La Leche League to get women accustomed to handing their breasts and prepare them for expressing milk. Women with flat or inverted nipples may be instructed on the Hoffman technique to break up adhesions. The thumbs are placed opposite each other at the base of the nipple and thumbs are simultaneously pressed in and away from each other. This procedure is repeated with the position of the thumbs varied around the nipple. Women should understand that breasts may leak milk prior to delivery. Anecdotal reports include milk leaking as early as 20 weeks of pregnancy.

Breastfeeding Positioning

Proper positioning of the infant at the breast is important to breastfeeding success.[118,119] Mothers need to learn from health professionals experienced in optimal positioning, because improper positioning causes pain and possible damage to nipple and breast tissue. The mother may need to use cushions, pillows, or a footstool to be comfortable and positioned well to nurse the infant (see Illustration 6.5).

Presenting the Breast to the Suckling Infant

Women use their hand to shape and position the breast so that the infant can easily latch. The grasp should allow the infant to place a sufficient amount of the areola into his or her mouth. For palmar grasp, the mother places her thumb above the areola, and remaining fingers are placed

Illustration 6.5 Positions for breastfeeding.

a. Cradle hold

b. Football or clutch hold

c. Cross-cuddle hold

Illustration 6.6 Attachment.

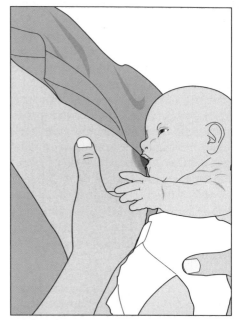

a. Touching the baby's bottom lip with the nipple stimulates the oral search reflex.

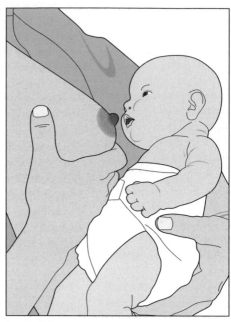

b. The infant opens his mouth wide.

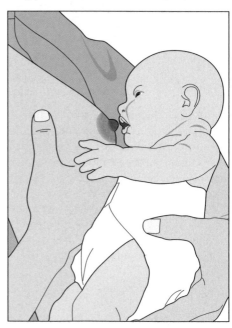

c. The infant is brought to the breast with the nipple centered in his mouth.

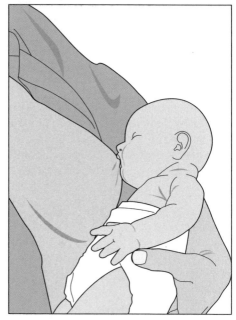

d. The infant is properly latched at the breast and has all of the areola in his mouth.

nipple and stimulate the oral search reflex by touching the baby's bottom lip with her nipple. Then the infant will open his or her mouth wide and should be brought to the breast with the nipple centered in the mouth (Illustration 6.6). This process is called *latching on*. Infants who are properly attached at the breast have all or most of the areola (the dark pigmented skin around the nipple) in their mouth. If the mother pulls down the infant's lower lip, she should see the tongue lying around the lower gum line. The baby's nose should be close to the breast, with breathing unrestricted. The mother should hear swallowing, but not smacking, clicking, or slurping. Women who consistently have pain when the infant is suckling should consult a professional trained in lactation to correct the attachment position.[118,119] Infants who are positioned correctly at the breast start to suckle almost immediately, change from quick short sucks to slow deep sucks, and remain relaxed.

Mechanics of Breastfeeding

During feeding, suction created within the baby's mouth causes the mother's nipple and areola to elongate and form a teat. The baby's jaw moves her tongue toward the areola, compressing it and causing milk to travel from the milk ducts to the infant's mouth. The baby then raises the anterior portion of the tongue to complete the process. Afterward, the baby depresses and retracts the posterior portion of her tongue in peristaltic motions, forming a groove in the tongue that channels milk to the back of the oral cavity. This backward movement of the tongue creates a negative pressure, allowing milk to be delivered into the baby's mouth. Receptors in the back of the oral cavity are stimulated and initiate the swallowing reflex.

under the breast to form a C or V. A scissor grasp, placement of her thumb and index finger above the areola with the remaining 3 fingers below, may also be used. It is important to note that the nipple is not tipped upward when the breast is presented to the infant to avoid improper latch-on and nipple abrasion.

Once the mother is comfortable, she should hold the baby so the mouth is directly in front of the

A correctly placed nipple does not move in the baby's mouth during suckling.

Identifying Hunger and Satiety

When infants are hungry, they begin to bring their hands to their mouth, suck on them, and start moving their head from side to side with their mouth open (rooting reflex). Infants should be fed when these signs of hunger are displayed rather than waiting for crying, a late sign of hunger. Recognizing early hunger behaviors and initiating feeding before the infant becomes very upset helps mothers and infants who have difficulty nursing.

Nutritive and non-nutritive sucks are different. Feedings begin with non-nutritive sucking. The infant sucks quickly and not particularly rhythmically. Nutritive sucking is slower and more rhythmic as the infant begins to suck and swallow. A mother can hear the infant suck in a quiet room.

Infants should be allowed to nurse as long as they want at one breast. Infants who are fed for shorter periods from both breasts can get larger amounts of foremilk. The high lactose content of foremilk can cause diarrhea.[120] Allowing the infant to nurse at one breast until satisfied creates a pattern that assures that the infant gets both foremilk and hindmilk. Infants who fall asleep before they empty the breast can be kept awake by gently tickling the feet, rubbing the head, and talking to the baby.

The higher fat content of hindmilk may help in signaling satiety. Infants will stop nursing when full. If they are still hungry, after burping, they can be offered the other breast.

Feeding Frequency

Stomach emptying occurs in about 1 1/2 hours for breastfed infants. Ten to 12 feedings per day are normal for newborn infants. Different feeding patterns can meet infant needs. In one study, infants who did not feed from midnight through early morning consumed more in the other feedings, particularly in the morning. Milk intake and weight gain of these infants in the first 4 months of life were similar to those of infants whose feedings were distributed over 24 hours.[121]

Vitamin Supplements for Breastfeeding Infants

All infants, whether they are fed human milk or HMS, are vulnerable to vitamin K–deficiency bleeding (VKDB). All infants in the United States receive a vitamin K supplement (1.0 mg by injection) at birth because it is known to decrease the risk of VKDB.[6] Questions regarding potential health risks, including cancer, related to vitamin K injections in the newborn persist. Therefore, in 2003, the American Academy of Pediatrics (AAP) called for further research regarding the "efficacy, safety, and bioavailability of oral formulations and optimal dosing regimens of vitamin K to prevent late VKDB."

Exclusively breastfed infants should be given a supplement of 400 IU of vitamin D per a day beginning in the first 2 months of life.[122] A minimum intake of 400 IU of vitamin D supplementation should continue through adolescence. This recommendation by the AAP was made in response to an increasing incidence of vitamin D–deficiency rickets among infants who are exclusively breastfed.

However, the AAP no longer recommends routine fluoride supplements during the first 6 months of life. After 6 months, the decision to supplement with fluoride should be made based on individual situations.[6] If the water supply contains 0.7 to 1.0 ppm of fluoride, no supplement is needed. If the water, food, and toothpaste contains less than 0.3 ppm fluoride, then 0.25 mg of fluoride is recommended. When supplementation is indicated for breastfed infants, maternal supplementation may be the best route.

Breastfed infants do not need iron-fortified HMS or supplements[13] because they rarely experience iron deficiency. The excess iron in HMS might bind with lactoferrin in human milk, resulting in a loss of the protective activity of lactoferrin.

Identifying Breastfeeding Malnutrition

A normal newborn weight loss of up to 7% can occur in the first week postpartum. A loss of 10% should trigger an evaluation of milk transfer to the infant by a lactation consultant or other trained professional who can provide support needed to maintain breastfeeding. Malnourished infants become sleepy and nonresponsive, and have a weak cry and few wet diapers. The clinician can use the diagnostic flowchart in Illustration 6.7[13] to help diagnose failure to thrive in breastfed infants. By the fifth to seventh day postpartum, infants who are getting adequate nourishment have wet diapers approximately 6 times a day and have 3–4 soft, yellowish stools per day. Their urine is pale yellow and dilute while stools are loose and seedy (some small particles are present in the stool). Infants who are slow gainers and not malnourished are alert, bright, responsive, and develop normally. In contrast, infants who are failing to thrive are apathetic, hard to arouse, and have a weak cry. They have few wet diapers, and their urine is concentrated. Their stools are infrequent.

Case Study 6.1

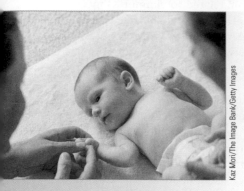

Breastfeeding and Adequate Nourishment

Molly G. is a 24-year-old office manager and part-time aerobics instructor who has delivered vaginally, without complications, a healthy, full-term son, Daniel. With a birth weight of 3200 grams (7 lb), Daniel is the first child for Molly and her husband. Molly is 162 cm (5 ft 4 in) tall with a prepregancy weight of 56.8 kg (125 lb). She gained 25 kg (55 lb) during her uncomplicated pregnancy and has been a lacto-ovo vegetarian for 5 years. After a 12-hour stay in a birthing center, Molly and her husband bring Daniel home.

At 4 days postpartum, Molly, her husband, and her mother-in-law bring the baby to the health care center for his first follow-up visit. Molly and her husband are very concerned about whether their son is getting adequate nourishment, so the dietitian is called to see the family. During nutrition assessment, the following information is documented. Daniel weighs 3000 grams (6 lb 6 oz). The parents report that Daniel nurses vigorously about every 1 1/2 to 2 hours and never sleeps for more than a couple of hours. Molly says that her milk "came in" on the second postpartum day and that she feels like all she does is nurse. Her nipples are tender, but not uncomfortably sore. She reports that Daniel has at least six to eight wet diapers and two to three very loose stools each day. She wonders if she has enough milk and worries about how she will ever return to work in 2 months. She also wants to lose the excess weight she gained during the pregnancy and is eager to return to her aerobics classes. Her husband and mother-in-law are supportive, but they worry about the baby.

The dietitians's nutrition assessment of the infant concludes that Daniel is a healthy infant with no nutritional problems. The family is referred to their pediatrician for further follow-up.

Questions

1. What factors put Molly at high risk for early termination of breastfeeding?
2. What factors indicate that Daniel is getting adequate nourishment?
3. What concerns do you have about Molly's diet? What advice would you give her about her weight-loss plans and eagerness to return to exercise? Do Molly or Daniel need any vitamin–mineral supplements?
4. If Molly lived in your community, what resources would be available for help and support for breastfeeding mothers?
5. What steps can Molly take to continue successful breastfeeding when she returns to work in 2 months?

(Answers are located in the Instructor's Manual for the 4th edition of *Nutrition Through the Life Cycle*.)

Mothers of slow-gaining infants in particular should be advised to let the infant nurse at one breast until it is empty or the infant stops nursing, rather than switching breasts after a specific amount of time. This regimen assures that the infant gets hindmilk with a higher fat and calorie content. Lawrence[13] recommends evaluation of slow-gaining infants when the slow-gaining pattern is first recognized (Illustration 6.7).

Tooth Decay

Human milk has infection-fighting components that inhibit the formation of dental caries. Nevertheless, caries can occur in children who are breastfed.[123] Frequent nursing at night after 1 year of age is a risk factor for dental caries. Nevertheless, the prevention of dental caries is not justification for advising early weaning. Rather,

Illustration 6.7 Diagnostic flowchart for failure to thrive.

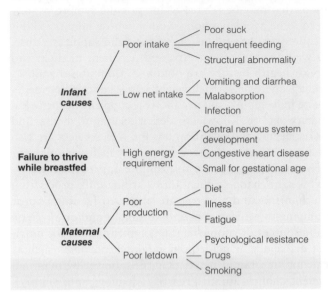

SOURCE: Lawrence, R. A. and Lawrence, R. M. *Breastfeeding: A Guide for the Medical Profession*, 5th ed. St. Louis: Mosby, 2006.[11]

the mother should be instructed on the prevention and treatment of early childhood dental caries. All children should be seen by a qualified dentist 6 months after the first tooth erupts or by 12 months of age. Mothers are the primary source of bacteria that cause early childhood caries. Therefore, mothers or primary care givers should

be given information on oral hygiene, diet, fluoride, caries removal, and prevention of caries.

Breastfed babies have straighter teeth due to the development of a well-rounded dental arch.[124] A well-rounded arch may also help prevent sleep apnea later in life.

Maternal Diet

The U.S. Department of Agriculture's MyPyramid Food Guide (presented in Chapter 1) has been adapted for pregnant and breastfeeding women. The Dietary Guidelines indicate that moderate weight reduction can be achieved by the breastfeeding mother without compromising the weight gain of the nursing infant.[125]

Diets formed around a MyPyramid Moms food plan provide a healthy assortment of nutrients at specified caloric levels appropriate for the stage of breastfeeding (www.mypyramid.gov/mypyramidmoms/) (Illustration 6.8).

Nutrition Assessment of Breastfeeding Women

Nutrition assessment for adults is outlined in Table 2.10. Assessment for breastfeeding women includes similar parameters, with a focus on commonly encountered problems, delineated in Table 6.6. Attention should be paid to appropriate maternal energy intake to achieve or maintain

Illustration 6.8 MyPyramid for Moms menu planning resource.

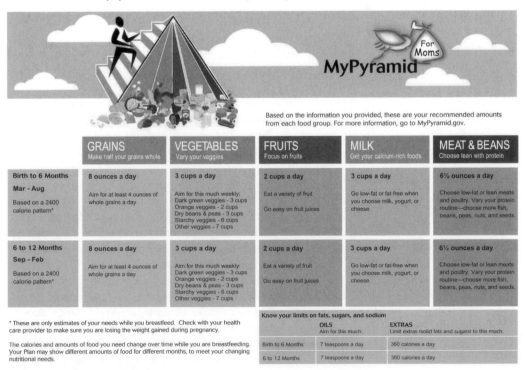

Table 6.6 Summary of the components of the four steps of the Nutrition Care Process for Breastfeeding Women (ADA)

Step 1. Nutrition Assessment for Breastfeeding Women

- Food and nutrient intake
  - Diet history
- Knowledge/attitudes/beliefs
  - Foods that may increase risk of infant colic
  - Adequate energy intake for return to pre-pregnant weight
- Medications, herbs, and supplements
  - Check for safety and compatibility with breastfeeding (see more on herbs and medications in Chapter 7)
  - Need for supplements based on inadequate intake, missing food groups, or vegan dietary pattern
- Nutrition-related behaviors
  - Excessive dieting or calorie restriction
- Physical activity
  - Return to physical activity after adequate recovery from delivery
- Biochemical data, medical tests, and procedures
  - Gradual return to non-pregnant, non-lactating values
- Anthropometric measures
  - BMI to return to pre-pregnant or ideal levels
- Client history
  - Medical history, treatments (beast surgeries), use of alternative/complementary medicine, social history

Step 2. Common Nutrition Diagnoses for Breastfeeding Women

- Altered maternal BMI
  - Obesity, underweight, or related to rate of weight loss
- Nutrient inadequacy or excess
  - Vegan
  - Lactose intolerance
- Perceived or real inadequate milk production
- Behavioral-environmental
  - Knowledge deficit
  - Need/qualify for WIC services

Step 3. Common Nutrition Interventions for Breastfeeding Women

- Alter energy intake to achieve ideal weight or weight goal
- Reinforce principles of milk production, early signs of infant hunger, and signs of adequate infant milk intake
- Recommend vitamin B_{12} for vegans or other supplements if intake is inadequate

Step 4. Common Nutrition Monitoring and Evaluation Plan Components

- Weight change/BMI
- Infant growth
- B_{12} status

ideal or healthy target maternal body weight. Specific foods that may be related to infant colic symptoms may need to be avoided among women whose infants exhibit symptoms. A careful history of food and supplement use, as well as physical activity, and pertinent medical and breast health history are obtained. Common diagnoses include altered maternal weight status, concern for inadequate milk supply, and need for B_{12} supplement for vegan dietary pattern. Knowledge deficit relating to dieting and breastfeeding and qualifications for WIC services are also common. Nutrition intervention may include a plan for adjusting energy intake to promote appropriate maternal weight change, education regarding milk production mechanisms and appropriate maternal feeding and/or pumping to achieve appropriate supply, and/or alternate food sources or supplement recommendations for nutrient deficiency or inadequate intake of nutrients such as calcium or vitamin B_{12}. A plan for monitoring maternal weight change, infant growth (adequate milk supply), or nutrient status rounds out the nutrition assessment process.

Energy and Nutrient Needs

The DRIs for normal-weight lactating women assume that the energy spent for milk production is 500 cal per day in the first 6 months and 400 cal afterward.[126] The 2002 DRI is 330 additional cal to support 0.8 kg per month weight loss (170 cal per day) for the first 6 months and 400 cal per day afterward. However, a recent review of a state-of-the-art study of energetics in exclusively breastfeeding women calls for a review of recommendations, citing a total energy cost of approximately 623 cal per day assuming 750 grams of milk produced at 0.67 kcal/g and 80% efficiency.[127] With mobilization of approximately 170 kcal per day, net energy needs were estimated at approximately 450 kcal per day.

We now understand that women use several mechanisms to meet the energy needs of lactation. Adjustments in energy intake and energy expenditure must be balanced to meet those needs. Goldberg et al.[128] found that women increased food intake (56% of the need for milk production) and decreased physical activity (44% of the energy need for milk production) to meet the increase in energy needs for lactation. Doubly labeled water studies suggest that the components of energy expenditure vary greatly, and measurements of dietary intake can be unreliable.[129] Therefore, a single recommendation for energy for lactating women could never address all of the individual ways that women meet their energy needs. Assessment of adequacy of energy intake of breastfeeding women should always be made within the context of the mother's overall nutritional status and weight changes and the adequacy of the infant's growth.

Maternal Energy Balance and Milk Composition

The composition of breast milk depends on maternal nutritional status. Protein–calorie malnutrition results in an energy deficit that reduces the volume of milk produced but does not usually compromise the composition of the milk. Several studies show that milk production is maintained when there is a modest level of negative calorie balance. Animal models first identified a potential threshold effect of energy restriction. Baboons fed 60% of their voluntary intake significantly reduced milk production.[130] Yet, baboons fed 80% of their usual intake maintained milk production. Randomized studies such as the one performed on baboons cannot ethically be done with humans. However, a series of human studies on weight loss during breastfeeding (discussed below) support a threshold effect of energy limitations on lactation.

Weight Loss During Breastfeeding

Current DRIs are written assuming a weight loss of 0.8 kg/month.[126] In addition, mechanisms that favor use of maternal fat stores and delivery of nutrients to the breast seem to occur during lactation. Despite these mechanisms that should favor weight loss in breastfeeding women, loss by 12 months postpartum is on average less than the amount needed to return to prepregnancy weight.[131] Even more surprising, postpartum weight changes are smaller in developing countries (–0.1 kg/mo) than in industrialized nations (–0.8 kg/mo). The failure to return to prepregnancy weight may be due to changes in energy intake, energy expenditure, and fat mobilization that easily meets energy needs.

A number of small studies in the United States have addressed the issue of whether weight loss influences milk production. Strode et al.[132] first observed women who voluntarily reduced their energy intake to 68% of their estimated needs for 7 days. No differences in infant intake or milk composition were observed. In the week following the diet, women who consumed fewer than 1500 cal tended to experience a decrease in milk volume.

Women who consumed over 1500 cal per day experienced no decrease. In addition, 22 healthy postpartum women who participated in a 10-week weight-loss program, which reduced energy intake by 23%, maintained milk production.[133] The women lost an average of about a pound a week during the 10 weeks.

Studies of weight loss during lactation also suggest that modest energy restriction (500 cal per day) can be accomplished without large decreases in the quality of the maternal diet, but that the macronutrient content of the diet may influence maternal weight loss and fat content of the milk differentially. Maternal reduction of energy intake through lowered intake of sugary and high-fat foods resulted in similar micronutrient intakes to women who did not decrease intake, with the exceptions of calcium and vitamin D.[134] None of the women consumed adequate vitamins C or E. Therefore, careful attention to consumption of calcium and vitamin D–rich foods such as low-fat and fat-free dairy products as well as fruit and vegetable and whole-grain consumption is needed. A small recent study challenges conventional wisdom that suggests that diet does not substantively alter human milk composition. When the women consumed a high-fat diet (55% fat, 30% carbohydrate, 15% protein), they had similar milk volume and similar protein and carbohydrate content of milk, but a greater energy deficit through higher milk-fat content milk to when they consumed a high carbohydrate diet (60% carbohydrate, 15% protein, 25% fat).[135]

Exercise and Breastfeeding

Studies[136,137] examining the effect of increasing energy expenditure on weight and lactation suggest that it is safe. In a *cross-sectional study*, vigorous exercise increased energy expenditure, but these women also increased their energy intake so that the calorie deficit was similar in the two groups.[136] There were no significant differences in milk volume between groups, although the group of exercising women tended to have higher milk volumes. A later 12-week randomized exercise intervention trial studied women who were 6 weeks postpartum and exclusively breastfeeding, to examine the effects of exercise on body composition and energy expenditure during breastfeeding.[137] One group followed an aerobic exercise regimen for 45 minutes, 5 days a week. The other group did no exercise. The exercising women increased energy expenditure by 400 kcal per day and increased energy intake to compensate for this use. Both groups experienced similar weight changes, milk volume, composition, and infant weight gain and serum prolactin levels. An additional study suggests that immunological factors in milk are not reduced by exercise in breastfeeding women.[138] These studies suggest that lactating women efficiently balance their energy intake to support energy expenditure.

The available evidence suggests that modest energy restriction combined with increases in activity may be effective at helping women to lose weight, while improving

> **Cross-Sectional Study**   A study that measures the current disease and exposure status of all individuals in a sample at a single time point (i.e., a study that measures the prevalence of breastfeeding at 6 months of age).

their metabolic profile and increasing fat losses. Despite the still small numbers of women who have been studied, a consistent lack of effect on milk production (infant intake), milk energy output, and infant growth is encouraging.

Vitamin and Mineral Supplements

A 1991 Institute of Medicine report, *Nutrition during Lactation,* stated that well-nourished breastfeeding women do not need routine vitamin or mineral supplementation.[1] Instead, supplementation should target specific nutritional needs of individual women. Supplementation strategies should take into account how nutrients are secreted into human milk and the potential for nutrient–nutrient interactions in mothers and their infants. For example, women who avoid dairy products completely should use calcium (1200 mg) and vitamin D (10 µg) supplements.

Vitamin and Mineral Intakes

Vitamin and mineral intakes that do not meet recommended levels (of folate, thiamin, vitamin A, calcium, iron, and zinc) have been reported for lactating women.[42,59,139] Ten percent of lactating women have thiamin intakes below the recommended levels,[139] whereas fewer than 5% of nonlactating women have intakes below the 1998 DRI. However, these reports of inadequate intake have not been followed by reports of deficits in nutritional status of the mother–infant pair. Nor is there a recommendation for vitamin and mineral supplements for all breastfeeding women. A careful balance is needed between concern over inadequate maternal dietary intake and causing women not to breastfeed because they do not have an optimal diet.

Functional Foods

Concern has been expressed about possible ill effects of high intakes of fortified foods in addition to supplements. Although this is an important issue, studies to date have not identified adverse reactions related to fortified-food consumption and RDA levels of nutrients in supplements.

Fluids

There is no evidence that increasing fluid intake will increase milk production or that a short-term fluid deficit results in a decrease in milk production. Fluid demands rise during breastfeeding, however, so women should drink fluids to thirst. Once a mother and her infant have the nursing routine down, she may find it convenient to have something to drink while she nurses. Although many women want to know how many glasses to drink per day, the amount needed varies depending on climate, milk production, body size, and other factors. Therefore, women are advised to drink enough fluids to keep their urine pale yellow. The current RDI for water for lactating women sums the recommendation for nonpregnant and nonlactating women (2.7 L for women 19–30) and the water content of the average milk output during the first 6 months (0.78 L milk ¥ 87% water = 0.68 L) for a total need of 3.4 L for women 19–30.[140]

Alternative Diets

Breastfeeding women can follow alternative eating patterns and be well nourished. The goal is to adequately nourish the mother and child, not to force women to use supplements and/or products that are not part of their normal eating patterns. Incorporation of soy products, vegetarian diets of various sorts, and other alternative diet choices can be followed as long as they meet maternal nutritional needs. Vegans who do not consume dairy products and eggs, however, may need to plan carefully to consume adequate amounts of calories, protein, calcium, vitamin D, vitamin B_{12}, iron, and zinc. Vegetarians' intakes of protein are generally adequate as long as energy intake is adequate. Breastfeeding women who consume no animal products should use plant foods with bioavailable B_{12} from sources such as yeasts, seaweed, and fortified soy products. Women who are unable to get adequate B_{12} from foods should take a vitamin B_{12} supplement. The 1991 Institute of Medicine[1] report recommends a multiple vitamin–mineral supplement for vegetarians because human milk may be low in vitamin B_{12} even when mothers do not exhibit deficiency symptoms. See Chapter 7 for information on the use of herbals during breastfeeding.

Infant Colic

Infant colic is defined as crying for more than three hours a day when the cause is not a medical problem. It is widely believed that components of maternal diet are related to infant colic. Information is growing to support this idea.[141,142] An observational trial suggested that a mother's consumption of cow's milk, onions, cabbage, broccoli, and chocolate were associated with greater likelihood of colic in the infant.[141] A randomized trial assessing maternal avoidance of cow's milk, eggs, peanuts, tree nuts, wheat, soy, and fish resulted in a reduction in colic symptoms of their infants in the first 6 weeks of life.[142] Women should be encouraged to exclude only those foods that seem to cause problems and to be careful to replace nutrients that might be lost by avoiding classes of foods. For example, excluding dairy foods may limit calcium and vitamin D intake.

Factors Influencing Breastfeeding Initiation and Duration

Obesity and Breastfeeding

Overweight and obesity prior to pregnancy and excess weight gain during pregnancy are associated with shorter duration of breastfeeding.[143] This association is independent of socioeconomic status and other factors also known to affect breastfeeding duration. This influence of obesity may be related to lower prolactin responses early postpartum and resulting difficulty in establishing adequate milk supply.[144,145] Therefore, maintenance of ideal body weight is important for lactation success.

Socioeconomic

All new mothers, both low-income and more affluent, need support for breastfeeding. However, low-income women often lack the education, support, and confidence to interpret the abundant and pervasive mixed messages on infant feeding practices. Consider the strikingly different context for pregnancy, birth, and parenting for low-income women and their more affluent counterparts:

Profile of a Low-income Pregnant Woman   She says she wants to do what is best for her baby and, in fact, knows the breast is best. However, she is afraid breastfeeding will cause her baby to be too "clingy." She feels uncomfortable about nursing around family, much less in public. She is certainly not up for the pain she has heard breastfeeding causes. To make things more difficult, in the hospital, she is separated from her baby soon after delivery and is given little assistance for getting started. She is sent home from the hospital with free samples of formula.

Profile of an Affluent Pregnant Woman   The affluent expectant mother has friends who have breastfed and have helped build her confidence that she can breastfeed successfully. She may have been able to choose her birth setting and select a hospital with knowledgeable staff who allow mother and baby to stay together around the clock. Because she knows there may be bumps in the road getting started, she seeks out support from friends or the doctor after discharge. At home, she has a supportive husband who is proud of her for offering the best for their baby. If she returns to work, she knows she can still breastfeed to keep that special closeness with her baby even after returning to work.

Common barriers to breastfeeding initiation expressed by expectant mothers:

- Embarrassment
- Time and social constraints, and concerns about loss of freedom (particularly issues of working mothers)
- Lack of support from family and friends
- Lack of confidence
- Concerns about diet and health practices
- In adolescents, fear of pain[146-148]

Additional obstacles to the initiation and continuation of breastfeeding include:

- Insufficient prenatal breastfeeding education
- Health care provider apathy and misinformation
- Inadequate health care provider lactation management training
- Disruptive hospital policies
- Early hospital discharge
- Lack of routine follow-up care and postpartum home health visits
- Maternal employment, especially in the absence of workplace facilities and support for breastfeeding
- Lack of broad societal support
- Media portrayal of bottle-feeding as the norm
- Unfounded concern that breastfeeding causes breast ptosis (sagging)[149]
- Commercial promotion of infant formula through distribution of hospital discharge packs, coupons for free or discounted formula, and television and general magazine advertising[6,150]

Breastfeeding Promotion, Facilitation, and Support

"Significant steps must be taken to increase breastfeeding rates in the United States and to close the wide racial and ethnic gaps in breastfeeding. This goal can only be achieved by supporting breastfeeding in the family, community, workplace, health care sector, and society."

Health and Human Services Blueprint for Action on Breastfeeding, 2000[66]

The support a woman receives from those around her directly impacts on her capability to breastfeed optimally.[66] A number of factors in the health care system, workplace, and community facilitate the initiation and continuation of breastfeeding.

Role of the Health Care System in Supporting Breastfeeding

Health care providers and facilities exert tremendous influence over the mother–infant dyad, with the power to promote and model optimal breastfeeding practices during prenatal care, at delivery, and after discharge. There is evidence for the effectiveness of lay[150] and professional support[151] on the duration of any breastfeeding.

The AAP is working in partnership with the Maternal and Child Health Bureau (MCHB), maternal and pediatric health professionals, residency programs, public health representatives, and other breastfeeding personnel to strengthen the AAP/MCHB Breastfeeding Promotion in Physicians Office Practices (BPOPIII). This program aims to (1) provide training in breastfeeding promotion to pediatric practices and individuals though webcasts or teleconferencing technology; (2) develop model residency program curricula for obstetrics, pediatrics, and family medicine; (3) provide technical assistance and resources to physicians, residents, public health representatives, and families; (4) strengthen and expand national collaborative networks and action groups at local, state, and regional levels to implement effective breastfeeding strategies and initiatives in underserved populations; and (5) assess changes in breastfeeding rates in physicians' practices after the practice implements breastfeeding education, counseling interventions, and ongoing support for mothers.

Doula An individual who surrounds, interacts with, and aids the mother at any time within the period that includes pregnancy, birth, and lactation; may be a relative, friend, or neighbor and is usually but not necessarily female. One who gives psychological encouragement and physical assistance to a new mother.[11]

Prenatal Breastfeeding Education and Support

Culturally competent prenatal breastfeeding education that is given frequently in person can have a significant positive influence on breastfeeding rates.[152,153] Best Start Social Marketing developed an effective three-step counseling strategy (Table 6.7) that quickly identifies a woman's particular barriers to choosing breastfeeding and provides targeted education while affirming the woman's ability to breastfeed.[154]

This Best Start approach replaces forced-choice questions such as "Are you going to breast- or bottle-feed?" with open-ended questions such as "What have you heard about breastfeeding?" or "What questions do you have about breastfeeding?" This provides the women with an opportunity to begin a dialogue with her provider about the infant feeding decision. The use of follow-up probes can help the counselor understand the woman's specific concern. Counselors should avoid the common pitfall of overwhelming women with too much

Table 6.7 The Best Start three-step breastfeeding counseling strategy[147]

1. Ask open-ended questions to identify the woman's concerns.
   - Dietitian: "What have you heard about breastfeeding?"
   - Client: "I hear it's best for my baby, but all my friends say it really hurts!"
2. Affirm her feelings by reassuring her that these feelings are normal.
   - Dietitian: "You know, most women worry about whether it will hurt."
3. Educate by clarifying how other women like her have dealt with her concerns. Avoid overeducating or giving the impression that breastfeeding is hard to master.
   - Dietitian: "Did you know that it is not supposed to be painful, and if you are having discomfort, there are people who can help make it better?"

information, which can give the impression that breastfeeding is difficult.[148]

Another effective strategy utilizes peer counselors and peer group discussions with at least one or two women who have successfully breastfed.[155,156] Exposure to mothers nursing their babies increases a woman's level of comfort with breastfeeding and provides a forum for informal discussion with family and friends. Discussion of an individual's personal experience can be an effective way to help women see and believe that others like her share her concerns.[146]

Toward the end of pregnancy, women need information on what to expect in the hospital or birthing center and practical tips for initiating breastfeeding. Because fathers,[157–159] grandmothers,[160] *doulas*, friends, and social networks[161] all have a powerful role on infant feeding decisions, it is important to include these influential people as often as possible in breastfeeding promotion efforts. An Italian study demonstrates the power of educating fathers.[157] Higher breastfeeding rates (25%) were achieved at 6 months in the group where fathers received education on their role in supporting breastfeeding than in the control group (15%). Differences were even greater when women reported difficulties with breastfeeding (24.5% versus 4.5%). Several key points are shown in Table 6.8.

The environment for the delivery of prenatal care is an important barrier or facilitator of breastfeeding. It should provide positive messages about breastfeeding, such as posters that promote breastfeeding and magazines and literature in the waiting room; there should be no advertisements or promotions of formula. Patient education

Table 6.8 Key teaching points prior to birth[147]

In the hospital or birthing center, mothers should:

Request early first feeding

Practice frequent, exclusive breastfeeding

Ask to be taught swallowing indicators

Learn indicators of sufficient intake

Ask for help if it hurts

Know sources for help

Understand postpartum rest and recovery needs

Avoid supplements unless medically indicated

SOURCE: Adapted from S. Page-Goertz and S. McCamman, Breast-feeding Success: You Can Make the Difference. *The Perinatal Nutrition Report* 4 (Winter 1998).

Table 6.9 World Health Organization's International/ UNICEF Code on the marketing of breast milk substitutes

- No advertising of breast milk substitutes
- No free samples or supplies
- No promotion of products through health care facilities
- No company sales representative to advise mothers
- No gifts or personal samples to health workers
- No gifts or pictures idealizing formula feeding, including pictures of infants, on the labels of the infant milk containers
- Information to health workers should be scientific and factual
- All information on artificial feeding, including labels, should explain the benefits of breastfeeding and the costs and hazards associated with formula feeding
- Unsuitable products should not be promoted for babies
- Manufacturers and distributors should comply with the Code's provisions even if countries have not adopted laws or other measures

SOURCE: Adapted from "World Health Organization: Contemporary patterns of breast-feeding," Report on the WHO Collaborative Study on Breastfeeding, Geneva; World Health Organization, 1981.[165]

materials that include formula advertising, samples, and business reply cards for free formula[162-163] are in direct violation of the World Health Organization's International Code on the Marketing of Breast Milk Substitutes (Table 6.9). Women who have been exposed to materials and products from formula companies prenatally are more likely to stop breastfeeding in the first 2 weeks.[163] Use of these materials provides a subtle message that infant formula is equivalent to breast milk.

Recent evidence stemming from focus groups intended to inform a national campaign to promote breastfeeding in the United States suggests that breastfeeding is a learned skill.[165] The process associated with success is called "confident commitment." The components are 1) confidence in the process of breastfeeding, 2) confidence in the ability to breastfeed, and 3) commitment to making breastfeeding work despite obstacles. Mothers who achieved confident commitment before birth were successful when faced with common challenges of breastfeeding. In contrast, among mothers without "confident commitment," the decision to breastfeed weakened with challenges.

Although not all women will choose to breastfeed, the goal of prenatal breastfeeding education is to empower every woman with sufficient knowledge to make an informed decision about how to feed her baby. Some professionals view breastfeeding as a personal choice rather than a public health issue and voice concern that breastfeeding promotional efforts may cause women who choose HMS feeding to feel guilty. On the other hand, some women who formula-feed report feeling angry about not getting enough breastfeeding information during their pregnancy.[147] In recognition of the benefits of breastfeeding and the important role of health professionals in promoting and supporting breastfeeding, the leading health and professional organizations in the United States that provide perinatal care have established policies in support of breastfeeding as the preferred infant feeding method.[6,166]

Successful maternal education program examples are highlighted in *The CDC Guide to Breastfeeding Interventions*.[152] These include health insurance plans that provide breastfeeding education for their members, Baby Friendly Hospital Initiatives offering patient infant-feeding classes, and health departments offering training programs for persons who provide breastfeeding education.

Lactation Support in Hospitals and Birthing Centers

Hospital policies and routines significantly impact on critical early experience with breastfeeding, with effects extending far beyond the short stay.[168] Illustration 6.9 provides examples of practices that influence this pivotal initiation experience. As in prenatal care settings, the distribution of free samples of infant formula coupons and hospital discharge packs is discouraged because of the detrimental effects of this practice on breastfeeding success, particularly among vulnerable groups such as new mothers and low-income women.[168] Model policies for hospitals and physicians are available from the Breastfeeding Coalition (www.inlandempirebreastfeedingcoalition.org/index.htm) and the Academy of Breastfeeding Medicine (http://bfmed.org).

In an effort to promote, protect, and support breastfeeding in hospitals and birthing centers worldwide, the World Health Organization (WHO) and UNICEF

Illustration 6.9 Hospital practices that influence breastfeeding initiation. [61,147]

	Hospital Practices That Influence Breastfeeding Initiation			
	← **Strongly Encouraging** →	← **Encouraging** →	← **Discouraging** →	← **Strongly Discouraging** →
Physical Contact	• Baby put to breast immediately in delivery room • Baby not taken from mother after delivery • Woman helped by staff to suckle baby in recovery room • Rooming-in; staff help with baby care in room, not only in nursery	• Staff sensitivity to cultural norms and expectations of woman	• Scheduled feedings regardless of mother's breastfeeding wishes	• Mother-infant separation at birth • Mother-infant housed on separate floors in postpartum period • Mother separated from baby due to bilirubin problem • No rooming-in policy
Verbal Communication	• Staff initiates discussion re: woman's intention to breastfeed pre- and intrapartum • Staff encourages and reinforces breastfeeding immediately on labor and delivery • Staff discusses use of breast pump and realities of separation from baby, re: breastfeeding	• Appropriate language skills of staff, teaching how to handle breast engorgement and nipple problems • Staff's own skills and comfort re: art of breastfeeding and time to teach woman on one-to-one basis	• Staff instructs woman "to get good night's rest and miss the feed" • Strict times allotted for breastfeeding regardless of mother/baby's feeding "cycle"	• Woman told to "take it easy," "get your rest" . . . impression that breastfeeding is effortful/tiring • Woman told she doesn't "do it right," staff interrupts her efforts, corrects her re: positions, etc.
Nonverbal Communication	• Pictures of woman breastfeeding • Staff (doctors as well as nurses) give reinforcement for breastfeeding (respect, smiles, affirmation) • Nurse (or any attendant) making mother comfortable and helping to arrange baby at breast for nursing • Woman sees others breastfeeding in hospital	• Literature on breastfeeding in understandable terms • Closed-circuit TV show in hospital on breastfeeding	• Pictures of woman bottle-feeding • Staff interrupts her breastfeeding session for lab tests, etc. • Woman doesn't see others breastfeeding	• Woman given infant formula kit and infant food literature • Woman sees official-looking nurses authoritatively caring for babies by bottle-feeding (leads to woman's insecurities re: own capability of care)
Experiential	• If breastfeeding not immediately successful, staff continues to be supportive • Previous success with breastfeeding experience in hospital			• Previous failure with breastfeeding experience in hospital

Table 6.10 The Baby-Friendly Hospital Initiative 10 steps to successful breastfeeding

1. Have a written breastfeeding policy that is regularly communicated to all health care staff.
2. Train all staff in skills necessary to implement this policy.
3. Inform all pregnant women about the benefits and management of breastfeeding.
4. Help mothers initiate breastfeeding within half an hour of birth.
5. Show mothers how to breastfeed and how to sustain lactation, even if they should be separated from their infants.
6. Feed newborn infants nothing but breast milk, unless medically indicated, and under no circumstances provide breast milk substitutes, feeding bottles, or pacifiers free of charge or at low cost.
7. Practice rooming-in, which allows mothers and infants to remain together 24 hours a day.
8. Encourage breastfeeding on demand.
9. Give no artificial pacifiers to breastfeeding infants.
10. Help start breastfeeding support groups and refer mothers to them.

SOURCE: World Health Organization, "Protecting, Promoting and Supporting Breast-feeding: The Special Role of Maternity Services," A Joint WHO/UNICIF Statement. Geneva, Switzerland, 1989.[169]

Table 6.11 Important elements of worksite lactation support programs

- Prenatal lactation education tailored for working women
- Corporate policies providing information for all employees on the benefits of breastfeeding and on why their breastfeeding co-workers need support
- Education for personnel about the services available to support breastfeeding women
- Adequate breaks, flexible work hours, job sharing, and part-time work
- Private "Mother's Rooms" for expressing milk in a secure and relaxing environment
- Access to hospital-grade, autocycling breast pumps at the workplace
- Small refrigerators for the safe storage of breast milk
- Subsidization or purchase of individually owned portable breast pumps for employees
- Access to lactation professional on-site or by phone to give breastfeeding education, counseling, and support during pregnancy, after delivery, and when the mother returns to work
- Coordination with on-site or near-site child care programs so the infant can be breastfed during the day
- Support groups for working mothers with children

SOURCE: U.S. Department of Health and Human Services. HHS Blueprint for Action on Breastfeeding. Washington, D.C.: U.S. Department of Health and Human Services, Office on Women's Health, 2000.[66]

established the Baby Friendly Hospital Initiative in 1992.[169] This initiative focuses on 10 evidence-based components of hospital care that impact on breastfeeding success (Table 6.10). The Baby Friendly USA program designates facilities within the United States who meet the guidelines. As of September 2006, there were 55 hospitals and birthing centers in the United States that had met all of the criteria in Table 6.11 and were designated as Baby Friendly (www.babyfriendlyusa.org/eng/03.html), whereas in 2003 there were only 38. Evidence suggests that the Baby Friendly Hospital Initiative is responsible for an increase in breastfeeding rates in Switzerland.[170] Further monitoring of the Baby Friendly Hospital Initiative is needed to document its successes.

In 2002 WHO and UNICEF came together to try to revitalize the international community in breastfeeding promotion with the Global Strategy for Infant and Young Child Feeding.[78] This report builds on the *Innocenti Declaration*[171] and the Baby Friendly Hospital Initiative by recognizing the importance of feeding in all children, including those in difficult circumstances such as emergency situations, low-birth-weight infants, and infants of mothers with HIV. The initiative includes (1) a call on governments to develop and implement policy on infant and child feeding within the context of the national policy for nutrition, child, and reproductive health as well as poverty reduction; (2) access to skilled support for initiation and maintenance of breastfeeding exclusively for 6 months and with safe complementary foods for up to 2 years or beyond; (3) empowerment of health care professionals to provide breastfeeding support and extend their services into the community; (4) review of progress in implementation of the International Code of Marketing of Breast Milk Substitutes and consideration of new measures to protect families from commercial interests and influence; and (5) enactment of legislation to protect the breastfeeding rights of working women in accordance with international labor standards. This report provides an important framework for acceleration of support of appropriate feeding for infants and children worldwide by linking resources and intervention areas available in many sectors.

Innocenti Declaration On the Protection, Promotion, and Support of Breastfeeding: Policy statement adopted by participants at the World Health Organization UNICEF policymakers' meeting on breastfeeding,, a global initiative held in Italy in 1990. The policy established exclusive breastfeeding from birth to 4–6 months of age as a global goal for optimal maternal and child health.

Lactation Support After Discharge

Breastfeeding support is essential in the first few weeks after delivery, as lactation is being established.[66] Younger women and women with lower socioeconomic status are more likely to stop breastfeeding by 4 weeks postpartum and cite sore nipples, inadequate milk supply, feeling that the infant is not satisfied, and infant problems as reasons for stopping.[172] A study of inner-city Baltimore WIC program participants[146] provides strong evidence that 7 to 10 days postpartum is the critical window for providing breastfeeding support; 35% of mothers who initiated breastfeeding in the hospital had stopped by 7 to 10 days. Lactation consultant services provided on a one-to-one basis in a randomized intervention for WIC participants in New Jersey demonstrated that there was greater initiation of breastfeeding, but no significant increase in the rate of exclusive breastfeeding at 3 months[173].

A pediatrician, nurse, or other knowledgeable health care practitioner (home visit or in the office) should see all breastfeeding mothers and their newborns when the newborn is 2 to 4 days of age. Breastfeeding should be observed and evaluated for evidence of successful breastfeeding behavior. This is also an important time to revisit the major concerns the mother identified during pregnancy, as well as discuss any new concerns. Mothers should be armed with information on sources of trained, skilled, and available help in the community, such as *lactation consultants,* peer counselors, the WIC program, or *La Leche League,* should questions or complications arise.[13,146] Follow-up telephone calls, as necessary, provide additional support to mothers who are not fully confident in their ability to breastfeed successfully.

Lactation Consultant   Health care professional whose scope of practice is focused on providing education and management to prevent and solve breastfeeding problems and to encourage a social environment that effectively supports the breastfeeding mother–infant dyad. Those who successfully complete the International Board of Lactation Consultant Examiners (IBLCE) certification process are entitled to use the IBCLC (International Board Certified Lactation Consultant) after their names (www.iblee.org).

La Leche League   International, nonprofit, nonsectarian organization dedicated to providing education, information, support, and encouragement to women who want to breastfeed. Founded in 1956 by seven women who had learned about successful breastfeeding while nursing their own babies, it currently has approximately 7100 accredited lay leaders to facilitate more than 3000 monthly mother-to-mother breastfeeding support group meetings around the world (www.lalecheleague.org).

The Workplace

The increase in the proportion of women working that began after World War II has been one of the most significant social and economic trends in modern U.S. history. In 1940, one in four U.S. workers was a woman; by 1998 almost one in two workers was a woman. In 2005, nearly 60% of mothers with children younger than 3 years of age worked.

Barriers to breastfeeding and employment have been recognized by the Surgeon General for over 20 years.[174] These barriers include lack of on-site day care, insufficiently paid maternity leave, rigid work schedules, and employers who lack knowledge about breastfeeding.

Current law ensures a woman's right to breastfeed her infant anywhere on federal property that she and her child are authorized to be. Still in process is further legislation to require that women cannot be fired or discriminated against if they breastfeed or express milk during their own lunchtime or break time, to provide a tax credit for employers who provide lactation services, and to develop minimum standards for breast pump safety.

Planning to return to work full-time does not appear to impact on breastfeeding initiation rates substantially.[2] Breastfeeding duration, however, is adversely affected by employment.[175,176] At 6 months, 35% of nonemployed women are still breastfeeding, compared with 33.4% of mothers working part-time, and only 23% of those employed full-time. Compared to women who do not return to work, short maternity leaves of 6 weeks or less or 6–12 weeks were associated with a fourfold or twofold increase in early cessation of breastfeeding, respectively.[177] Part-time work is more conducive to breastfeeding; the number of hours mothers work per day is inversely associated with the likelihood that the mother will continue to breastfeed.[178] Women in professional occupations breastfeed longer than do women in sales, clerical, or technical occupations.[179] A Los Angeles-area WIC program provided electric pumps upon request to participants and found a 5.5-time greater chance that women did not request formula at 6 months if they received an electric pump when requested than if they did not, suggesting that breastfeeding was more successful after return to work when a breast pump was available.[180] A lactation program (including an employee's choice of a class on the benefits of breastfeeding), services of a lactation consultant, and a private room in the workplace with equipment for pumping resulted in an average weaning age of 9 months.[177] Importantly, more than 50% continued to breastfeed to 6 months, and the majority were working full-time (84.2%).

Studies indicate that women who continue to breastfeed once returning to work miss less time from work because of baby-related illnesses, and have shorter absences when they do miss work, compared with women who do not breastfeed.[101] Worksite programs that support breastfeeding facilitate the continuation of breastfeeding after mothers return to their jobs and offer additional advantages to employers: employee morale and loyalty, image as family-friendly, recruiting for personnel, and retention of employees after childbirth all improved.[181,182] Companies that have adopted breastfeeding support programs have noted cost savings of $3 per $1 invested in breastfeeding support in addition to lower health care costs over the first year.[183] Key elements of worksite lactation support programs are presented in Table 6.11.

Women planning to return to work have several choices. Breast milk can be expressed during the day into sterile containers, refrigerated or frozen, and then used for subsequent bottle feedings when the mother is at work. (See Chapter 7 for storage guidelines.) With on-site child care, it is possible to breastfeed during breaks and lunch hours. Another possibility is to train the body to produce milk only when the mother is home during the evenings and during the night. To accomplish this, a woman should omit one feeding at a time during the periods of the day when she will not be feeding or expressing milk. This will help her to reduce her milk supply without experiencing engorgement. She gradually weans to the feedings at the appropriate time of the day. This method works because removal of milk is the stimuli for milk production. Generally, at least two feedings per day are needed for women to continue making milk. No evidence indicates that it is necessary to introduce a bottle sooner than 10 days before returning to work.[184] Unless a mother is returning to work immediately after delivery, a bottle should not be introduced before lactation is well established, which is usually at least 4 weeks. Information about hospital-grade breast pumps is readily available on the Internet.

The Community

To increase breastfeeding rates in a community, it is important to identify community attitudes and obstacles to breastfeeding, and to solicit the support for breastfeeding from community leadership. Establishment of a multidisciplinary breastfeeding task force with representatives from physicians, hospitals and birthing centers, public health, home visitors, La Leche League, government, industry, school boards, and journalists can be an effective vehicle for assessing community breastfeeding support needs and sponsoring collaborative efforts to overcoming community obstacles to breastfeeding.[146] Barriers to breastfeeding may include lack of access to reliable and culturally appropriate sources of information and social support, cultural perception of bottle feeding as the norm, aggressive marketing of breast milk substitutes, and laws that prohibit breastfeeding in public. Public outrage in response to a 2006 image of an infant feeding at a woman's breast on the front of a magazine about babies demonstrates that breastfeeding is not perceived as a cultural norm.

In the past decade, legislative efforts have been made to protect a woman's right to breastfeed. States vary widely in not only whether there is legislation relating to breastfeeding, but also the depth and breadth of the legislation (www.lalecheleague.org/LawBills.html). Legislation is used to protect a woman's right to breastfeed, to consider breastfeeding in family law situations, to regulate breast pumps, and to provide incentives to employers who provide breastfeeding support. Legislation addresses issues such as a woman's right to breastfeed in public and on federal property, express milk at work, and to be exempt from jury duty.

To facilitate breastfeeding support advocacy among health professionals in health care facilities and in the community, the National Alliance of Breastfeeding Advocacy (NABA) maintains a searchable database on the Web.

The database includes a state-by-state listing of all known breastfeeding coalitions and task forces, AAP chapter presidents, International Lactation Consultant Association affiliates, La Leche League International contacts, WIC breastfeeding directors and Lamaze state contacts, and Title V Maternal and Child Health Directors.

Public Food and Nutrition Programs

National Breastfeeding Policy

The U.S. Department of Health and Human Services (DHHS) is the lead federal agency for policy development to advance the promotion, protection, and support of breastfeeding for families in the United States. Over the past several decades, the Office of the Surgeon General and the Maternal and Child Health Bureau have highlighted the public health importance of breastfeeding though numerous workshops and publications (Table 6.12).

Many HHS agencies have breastfeeding initiatives. The Title V Maternal and Child Health programs of the Health Resources and Services Administration provide substantial support services, training, and research for breastfeeding (www.mchb.hrsa.gov). The Centers for Disease Control and Prevention play a major role in supporting breastfeeding

Table 6.12 Landmark U.S. breastfeeding policy statements and conferences

- Report of the Surgeon General's Workshop on Breastfeeding and Human Lactation[1,175]
- Follow-up Report: Surgeon General's Workshop on Breastfeeding and Human Lactation[185]
- Healthy People 2010 Breastfeeding Goals for the Nation[3]
- DHHS Maternal and Child Health Bureau National Workshop: Call to Action: Better Nutrition for Mothers, Children, and Families. Washington, D.C.: National Center for Education in Maternal and Child Health[186]
- Second Follow-up Report: Surgeon General's Workshop on Breastfeeding and Human Lactation[187]
- National Breastfeeding Policy Conference. Presented by the UCLA Center for Healthier Children, Families and Communities, Breastfeeding Resource Program, in cooperation with the U.S. Department of Health and Human Services, Health Resources and Services Administration, Maternal and Child Health Bureau, and the Centers for Disease Control and Prevention.[188]
- Healthy People 2010 Breastfeeding Goals for the Nation[3]
- DHHS Blueprint for Action on Breastfeeding[66]
- U.S. Breastfeeding Committee Strategic Plan[189]

nationally through applied research, program evaluation, and surveillance (www.cdc.gov).

USDA WIC Program

WIC, the Special Supplemental Nutrition Program for Women, Infants, and Children, is a federal program operated by the U.S. Department of Agriculture Food and Nutrition Service in partnership with state and local health departments. Created in 1972, WIC is designed to provide nutrition education, supplementary foods, and referrals for health and social services to economically disadvantaged women who are pregnant, postpartum, or are caring for infants and children under 5. WIC operates through a network of state health departments, Indian tribal organizations, U.S. territories, and local agencies providing services to more than 8 million program participants per year.

In 1989, Congress mandated (Public Law 101-147) a specific portion of each state's WIC budget allocation to be used exclusively for the promotion and support of breastfeeding among its participants and authorized the use of WIC administrative funds to purchase breastfeeding aids such as breast pumps. Reauthorization legislation, the Healthy Meals for Healthy Americans Act of 1994 (P.L. 103-448), increased the budget allocation to $21 for each pregnant and breastfeeding woman in support of breastfeeding promotion. Through this legislation, each state has a breastfeeding coordinator and a plan to coordinate operations with local agency programs for breastfeeding promotion. In 2005, WIC expenditures were more than 3 billion dollars (www.fns.usda.gov/pd/wisummary).

The USDA Food and Nutrition Information Center supports a WIC Works website to serve health and nutrition professionals working in the WIC Program. The WIC Works site, www.nal.usda.gov/wicworks, includes an e-mail discussion group, links to training materials on breastfeeding promotion, and information on how to share resources and recommendations.

Social Marketing Combines the principles of commercial marketing with health education to promote a socially beneficial idea, practice, or product.[147]

Focus Group Interviews Small group discussions guided by a trained moderator that provide insights into the participant's perceptions, attitudes, and opinions on a designated topic.[145]

Model Breastfeeding Promotion Programs

WIC National Breastfeeding Promotion Project—Loving Support Makes Breastfeeding Work

In 1995, the USDA Food and Nutrition Service entered into a cooperative agreement with Best Start *Social Marketing* (a not-for-profit organization assisting public health, education, and social service organizations with social marketing services) to develop a national WIC breastfeeding promotion project to implement at the state level. The campaign has five goals: (1) to increase breastfeeding initiation rates among WIC participants, (2) to increase breastfeeding duration among WIC participants, (3) to increase referrals to WIC clinics for breastfeeding support, (4) to increase general public acceptance and support of breastfeeding, and (5) to provide support and technical assistance to WIC professionals in promoting breastfeeding.

Qualitative research data were collected in 10 pilot states through a series of observations, personal and telephone interviews, and *focus group interviews* with WIC participants and individuals who might influence their infant feeding decisions such as mothers, boyfriends and husbands, health care providers, and WIC staff.[190] Motivations and perceived barriers related to breastfeeding as well as social network influences on feeding choice were identified.

Results from the formative research were used to develop a marketing plan and program material. In contrast to the traditional public health approach of addressing breastfeeding as a medical health decision, the marketing plan repositioned breastfeeding as a way for a family to establish a special relationship with their child from the very onset of his or her life.[148,189] The campaign slogan, "Loving Support Makes Breastfeeding Work," capitalizes on the concept that everyone is important to women's breastfeeding success—family, friends, doctors, and the community. Campaign materials and a counseling program were developed to help mothers work through individual barriers and constraints to breastfeeding. The key messages are (1) helping women feel comfortable with breastfeeding, (2) tips on how breastfeeding can work around a busy schedule, and (3) the involvement of family and friends to make breastfeeding a success. Since 1997, the Loving Support Makes Breastfeeding Work campaign has expanded to 72 state agencies, Indian tribal organizations, and territories participating at various levels.[191, 192]

Prior to initiation of the Loving Support campaign in 1997, Mississippi ranked fiftieth in the nation in breastfeeding initiation and duration rates. In 1999, the state moved into forty-eighth place nationally in breastfeeding initiation rates at hospital discharge. Duration rates at 6 months, which climbed from 7.0% to 15.4% from 1998 to 1999, moved Mississippi WIC from fiftieth to thirty-third in the country. Building on the idea that fathers are important, a father-to-father breastfeeding promotion program in Texas resulted in increased breastfeeding rates at WIC clinics using peer dads.[193] Based on data from the Ross Breastfeeding Survey,[2] breastfeeding initiation rates in Iowa increased from 57.8% to 65.1% after a year of the campaign. The rates of women still nursing at 6 months after birth also increased from 20.4% to 32.2%. Increased breastfeeding support from relatives and friends was also documented from data collected in a mail survey. Black women in New York say that WIC program staff influence their decision to breastfeed.[195] Women reported that free formula influenced formula feeding, but that personalized breastfeeding promotion and trusting relationships with WIC staff encouraged breastfeeding. In New Jersey, a breastfeeding initiation program increased breastfeeding initiation and exclusive breastfeeding at 3 months.[173]

Wellstart International

Wellstart International is an independent, nonprofit organization headquartered in San Diego, California, that is dedicated to supporting the health and nutrition of mothers and infants worldwide through the promotion of breastfeeding. Wellstart focuses on the development of local leadership and teamwork by providing education and technical assistance to perinatal health care providers and educators around the world, enabling them to promote maternal and child health in their own settings through the support of breastfeeding. Wellstart faculty and staff offer in-depth clinical and programmatic expertise and both domestic and international experience to hospitals, clinics, and university schools of medicine, nursing, and nutrition, as well as to a wide variety of governmental and nongovernmental health and population agencies. Wellstart is a designated World Health Organization Collaborating Center on Breastfeeding Promotion, with international and domestic efforts actively taking place.

The combined approach of community outreach activities, coupled with the use of Information, Education, and Communication (IEC) materials targeting key behaviors, and coordination and referral to trained service providers, produced significant behavior changes at the local, regional, and national levels:[192]

- *Initiation of Breastfeeding.* Behaviors related to initiation of the first breastfeed improved (p < 0.001), with the optimal behavior of initiating breastfeeding within the first hour increasing from 46% to 72% within the demonstration areas.
- *Exclusive Breastfeeding.* Exclusive breastfeeding for all age groups increased (p < 0.001), with levels increasing from 24% to 74% for the 0–3 month age group, from 8% to 55% for the 4–6 month age group, and from 15.5% to 65% for the combined 0–6 month age group.
- *Infants receiving water and herbal teas.* The use of water or herbal teas within the 0–3 and 4–6 month age groups declined (p < 0.001), as

would be expected if exclusive breastfeeding was increasing.

- *Bottle and Pacifier Use.* For infants 0–12 months of age, bottle use decreased from 49% to 12%; pacifier use decreased from 42% to 13.6% (p < 0.001).
- *Complementary Feeding.* Data related to timely introduction of complementary foods, as well as types of complementary foods received by infants, also showed substantial improvements (p < 0.001). The percentage of infants 7–9 months of age receiving no complementary food decreased from 16% to 3%.

In 1996, the Centers for Disease Control launched lactation support programs in federal workplaces. Components of the program include a two-hour prenatal breastfeeding class for expectant employees, breastfeeding counselling via e-mail and telephone by a board-certified lactation consultant from the baby's birth until breastfeeding ceases, individual consultation with a lactation consultant two weeks prior to return to work to make breastfeeding and work easier, a double electric pump, refrigerator, telephone, nursing stool and cleaning supplies in a dedicated lactation room, and discussion groups three times per year. The program boasts a large percentage of mothers who successfully breastfeed for longer than 6 months. The Office of Women's Health as created a program to educate employers about the value of supporting breastfeeding in the workplace called the Business Case for Breastfeeding (http://www.womenshealth.gov/breastfeeding/programs/business-case/index.cfm). The kit provides a sample policy for supporting breastfeeding employees, a breastfeeding program assessment form, and templates for flyers and feedback as well as a resource guide.

On a global level, we must do everything we can "to increase women's confidence in their ability to breastfeed. Such empowerment involves the removal of constraints and influence that manipulate perceptions and behavior towards breastfeeding, often by subtle and indirect means. Furthermore, obstacles to breastfeeding within the health system, the workplace, and the community must be eliminated." These words are from the WHO/UNICEF Innocenti Declaration on the Protection Promotion and Support of Breastfeeding (1990).[171] As we have seen, breastfeeding is best for the vast majority of infants and is physiologically possible for the vast majority of women. The challenge is to overcome barriers and provide support systems at the local, national, and international level so that the initiation and duration of breastfeeding will continue to increase.

Key Points

1. With rare exceptions, human milk is the optimal food for infants, exclusively for 6 months and with supplemental foods for a year or longer. Benefits to infants include protection from iron deficiency, better gains in cognitive ability, fewer acute respiratory and gastrointestinal illnesses, and lower risk of sudden infant death syndrome, celiac disease, inflammatory bowel disease, neuroblastoma, allergies, and asthma.

2. A thorough understanding of the anatomy and physiology of lactation is key to enabling health care providers in providing effective lactation support.

3. Maternal benefits of breastfeeding include minimization of postpartum blood loss, delayed fertility, greater self-confidence and bonding with the baby, and reduced risk of ovarian and breast cancers.

4. Newborns who are getting adequate human milk have about six wet diapers and three to four soft, yellowish stools per day by 5–7 days postpartum. Even slow gainers are alert, bright, and responsive, whereas infants who are failing to thrive are apathetic, hard to arouse, and have a weak cry.

5. Removal of milk from the breast is the stimulus for making more milk. Most women can make enough milk for their infant, although storage capacity (size of the breast) differences may determine how often the infant feeds in 24 hours.

6. Breastfed infants should be supplemented with vitamin K at birth and with 400 IU of vitamin D per day beginning at 2 months.[122] In areas where water is not fluoridated, infant supplementation after the age of 6 months may be the best choice.

7. Breastfeeding women can lose modest amounts of weight while breastfeeding by choosing a diet following the MyPyramid food guide. Breastfeeding women need an additional 330 calories per day in the first 6 months, and 400 cal thereafter. Energy intake should be adjusted for activity level and achievement and maintenance of healthy weight.

8. Maternal diet does not significantly alter the protein, carbohydrate, and major mineral composition of breast milk, but it does affect the fatty acid profile, amounts of some vitamins, and some, but not all, trace minerals. For a majority of nutrients, the quality of the milk is preserved over the quantity of milk when maternal diet is inadequate.

9. Maternal diet may be associated with infant colic. Avoidance of cow's milk and cow's-milk products, eggs, peanuts, tree nuts, wheat, soy, and fish have been associated with a reduction in infant colic symptoms in the first 6 weeks of life.

10. Successful breastfeeding is possible for women who follow vegetarian diets. Careful attention to vitamin B_{12} supplementation is important for women who are vegan.

11. Support for breastfeeding women from husbands, mothers, sisters, health care providers, communities, employers, and policy makers is critical to breastfeeding success and impacts breastfeeding rates in the community.

Review Questions

1. The composition of human milk changes
   a. over the time of day, but stays constant through a single feeding, and over time from birth
   b. over the time from birth, but stays constant through a single feeding and over the day
   c. over a single feeding, and across the day, but stays constant over time from birth
   d. over the time of day, within a single feeding, and over the time since birth

2. Colostrum differs from mature human milk in the following ways:
   a. It is higher in protein mononuclear cells and is thick and yellow in color
   b. It is higher in protein, but lower in sodium, potassium, and chloride, and is yellow in color
   c. It is lower in protein, but higher in fat, sodium, potassium, and chloride, and is bluish in color
   d. It is lower in protein, higher in carbohydrates, and similar in sodium, potassium, and chloride

3. The benefits of breastfeeding include
   a. hormonal, physical, and psychological benefits for the mother, but no differences for the infant.
   b. no hormonal physical, or psychological benefits for the mother, but the infant gets heightened protection from infections and several chronic diseases.
   c. hormonal, physical, and psychological benefits for the mother, and the infant gets heightened protection from infections and several chronic diseases.
   d. no hormonal benefits, but decreased breast cancer in the mother, and the infant gets heightened protection from infections and several chronic diseases.

4. The best way to increase human milk production is:
   a. To reduce fat in the mothers' diet
   b. To remove milk from the breast and feed frequently

c. To feed the infant at both breasts in a single feeding

d. To use a pump to express milk at the beginning of a feeding

5. Nipple pain is common in breastfeeding. Women should be counseled to

a. get used to it; it will eventually go away.

b. use antibiotic creams to prevent mastitis.

c. see a lactation consultant for proper positioning of the infant at the breast.

d. use a pump instead of having the infant at the breast until the pain goes away.

6. When counseling a healthy woman about what to eat during breastfeeding, the best advice is:

a. Eat and drink whatever you want; it doesn't change the milk composition or hurt the baby as long as you take a multivitamin.

b. Eat a variety of foods, following the MyPyramid for Moms.

c. Avoid chocolate, mint, tomatoes, onions, and cabbage because they cause colic in the baby.

d. Avoid wheat, milk, eggs, and nuts to avoid allergies in the infant.

7. Which of the following is true about maternal weight loss while breastfeeding?

a. Women should not try to lose weight while breastfeeding because it changes the milk composition.

b. Women should not try to lose weight while breastfeeding because milk production will suffer.

c. Women can try to lose weight while breastfeeding as long as it is modest (500 kcal/day deficit).

d. Women can try to lose as much weight as they want while breastfeeding because ketones in breast milk are not harmful to the baby.

8. A vegan breastfeeding woman should be advised to

a. continue her vegan diet, if she chooses, while making sure that her intake of vitamin B_{12} and other nutrients is adequate from food and/or supplements.

b. continue her vegan diet with no concerns about adequate nutrients in her milk.

c. become a lacto-ovo vegetarian to ensure adequate calcium and vitamin D intake.

d. begin adding meat and other iron-rich foods to her diet to ensure her milk has adequate iron.

9. Fear of inadequate milk supply is a contributor to early breastfeeding cessation. What would be key components of breastfeeding advice to give to a new mother in the early postpartum period to help her develop confidence in her milk supply?

a. Buy an infant scale and weigh the infant before and after each feeding.

b. Feed the infant every four hours even if the infant doesn't seem hungry, and if she has about 6 wet diapers a day, she is eating adequately.

c. Feed the infant when she displays early signs of hunger, and if the infant has about 6 wet diapers and about 3 stools daily, she is eating adequately.

d. Nurse the infant at both breasts at each feeding and offer a formula supplement if the infant still seems hungry.

10. Describe the stages of lactogenesis, including the timeframe of breastfeeding, where it occurs, major milk composition changes (if any), and milk production changes.

11. Describe four of the nutritional benefits of human milk to the infant.

Resources

The National Women's Health Information Center
Website: www.womenshealth.gov/breastfeeding

MyPyramid for Pregnancy & Breastfeeding
Website: http://www.mypyramid.gov/mypyramidmoms/index.html

DHHS HRSA Maternal and Child Health Bureau
Telephone: 301-443-0205
Website: http://mchb.hrsa.gov

Food and Nutrition Information Center, Lifecycle, Nutrition
Website: http://fnic.nal.usda.gov/nal_display/index.php?info_center=4&tax_level=2&tax_subject=257&topic_id=1357&placement_default=0

USDA Women, Infants, and Children (WIC) Program
Telephone: 703-305-2736
Website: www.fns.usda.gov/wic/

United States Breastfeeding Committee
Website: www.usbreastfeeding.org

La Leche League International
Telephone: 847-519-7730
Website: www.llli.org/
International Lactation Consultants Association
Telephone: 919-861-5577
Website: www.ilca.org

Wellstart International
Telephone: 619-295-5192
Website: www.wellstart.org

Academy of Breastfeeding Medicine
Telephone: 800-990-4ABM (toll-free)
Website: www.bfmed.org

Baby-Friendly USA Hospital Initiative
Telephone: 508-888-8044
Website: www.babyfriendlyusa.org

Best Start Social Marketing
Telephone: 813-971-2119
Website: www.beststartinc.org

Chapter 7

Nutrition During Lactation:
Conditions and Interventions

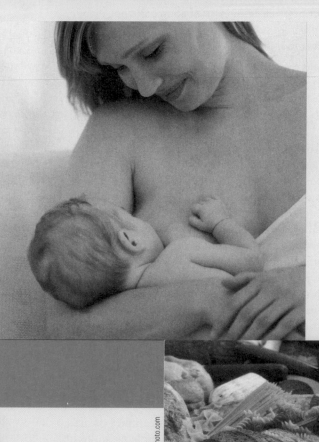

Prepared by Ellen Lechtenberg and Maureen A. Murtaugh with Denise Sofka and Carolyn Sharbaugh

Key Nutrition Concepts

1 Human milk is the preferred feeding for all premature and sick newborns with rare exceptions.

2 Breastfeeding women need consistent, informed, and individualized care in the hospital and at home after discharge.

3 It is usually not necessary to discontinue breastfeeding to manage medical problems of the mother or infant; any medical decision to limit a mother's breastfeeding must be justified by the fact that the risk to her baby clearly outweighs the benefits of breastfeeding.

4 Feeding infants early in the post delivery period whenever possible is important to successful breastfeeding. Early intervention to address questions or problems is equally important for maintaining breastfeeding.

5 Most medications (including over-the-counter as well as prescription drugs), drugs of abuse, alcohol, nicotine, and herbal remedies taken by nursing mothers are excreted in breast milk.

6 Twins and other multiples can be successfully breast-fed without formula supplementation.

Introduction

The key to successful breastfeeding management is for the mother–infant breastfeeding dyad to receive support and informed, consistent, and individualized care from health care professionals both in the hospital and after discharge. The vast majority of women do not experience significant problems with breastfeeding, and many of the more common problems that do arise can be prevented through prenatal breastfeeding education and a positive, supportive breastfeeding initiation period.

This chapter discusses prevention and treatment of common breastfeeding conditions. Issues related to maternal use of medications, herbal remedies, drugs of abuse, and environmental contaminants are addressed. The chapter presents important considerations for breastfeeding multiples, preterm infants, and infants with medical problems. It provides information on the safe collection and storage of human milk and milk banks. The chapter concludes with case studies providing examples of management of challenging breastfeeding problems and with examples of model programs promoting support for breastfeeding in the health care system.

Common Breastfeeding Conditions

Sore Nipples

Early, mild nipple discomfort is common among women initiating breastfeeding. In most women, the discomfort is transient and usually subsides by the end of the first week.[2]

Severe nipple pain, the presence of nipple cracks or fissures, pain that persists throughout a feeding, or pain that is not improved by the end of the first week should not be considered normal and requires evaluation.[3] Painful nipples and nipple trauma can lead women to become discouraged. As many as one third of women who experience nipple pain and trauma switch to bottle feeding within the first 6 weeks postpartum.[3,7]

The best prevention of nipple pain and soreness is proper positioning of the baby on the breast. In order to feed effectively, the baby needs to draw the breast deeply into the mouth, so that the mother's nipple approximates to the junction of the hard and soft palate.[4] This enables the baby to use its tongue smoothly and rhythmically against the undersurface of the breast and remove milk from the ducts. With a good latch, the mother's nipple is so far back in the baby's mouth that it is beyond the reach of the compression wave of the tongue, so no pain is caused, and no damage is done. If a woman is experiencing pain, a lactation consultant or a health care professional well trained in lactation should observe the mother nursing her baby. The lactation consultant can determine whether the pain is simply related to early breastfeeding, or if a problem exists.

The main causes of persistent nipple pain are trauma from poor positioning of the infant at the breast, poor latch, improper release of suction after a feed, infection (thrush or *staphyloccus aureus*), pumping with too much suction or incorrect breast flange size, a disorganized or dysfunctional suck, and dermatologic abnormalities.[3] Nipple damage can occur even prior to discharge from the hospital.[5] Breast care and cleaning rituals can also contribute to nipple soreness. Proper cleaning of the lactating breast involves only daily washing with warm water. Soaps and other cleansing products can irritate the nipple, and some creams and lotions can cause an allergic reaction and skin irritation. Plastic-backed breast pads used to prevent milk leakage can trap moisture and inhibit air flow to the nipple.[3]

Women can take simple steps to manage nipple pain. Recommended strategies include letting breasts air-dry after nursing, rubbing expressed milk or an all-purpose ointment (not petroleum-based) on nipples, and using warm compresses on sore nipples.[2,3] The common belief that limiting the frequency or duration of feedings will prevent or heal sore nipples is not substantiated by the literature.[5] Use of a pump to express milk can help to maintain supply if the nipples are so sore that the mother cannot nurse.[4] However, the suction on the breast pump should be adjusted carefully. High suction can make nipples sore and red and even cause blisters.

Flat or Inverted Nipples

Some women have flat or inverted nipples which do not extend very far in the baby's mouth. This should not impact breastfeeding if the latch is correct. If the baby is struggling with latch, instruct the mother to roll her nipple between her fingers or use a breast pump prior to feeding to help draw out the nipple.

Letdown Failure

"After she latches on, take deep, long breaths—think yoga not Lamaze when you nurse. As you exhale, visualize the milk letting down through your breasts into the baby's mouth."

C. Martin and N. F. Krebs[6]

Letdown failure is not common[6], but because letdown is necessary to successful breastfeeding, it is important to address the matter. Stress may inhibit oxytocin as well as alcohol and distractions. Oxytocin nasal spray can be prescribed by a physician for letdown failure. The synthetic oxytocin is sprayed into the nose and stimulates letdown, but it can only be used for a few days to help women get through a tough time. Other methods should be used at the same time to stimulate letdown. Prolonged letdown failure will cause lactation suppression.[7] Martin and Krebs[6] recommend a number of techniques to help women relax and enhance letdown:

- Play soothing music that the mother can focus on while nursing.
- Have the partner rub his knuckles down her spine.
- Try different nursing positions.
- Get out of the house. Most babies enjoy a walk.
- Arrange for some time alone (a few hours).
- Decrease number of caffeinated beverages and increase water consumption for a few weeks.

Hyperactive Letdown

Hyperactive letdown can also be a problem, especially among first-time mothers. When letdown is overactive, milk streams from the breast as feeding begins. Milk may also leak from the breast that the infant is not being nursed from. The milk streams quickly, and the infant may be overwhelmed by the volume. The infant may choke, cough, or gulp to keep up with the flow. When the infant gulps, the infant may take in air, develop gas pain, and then become fussy.

Management includes removing the infant from the breast when letdown occurs and waiting for the milk flow to slow down before putting the infant back to the breast.[7] The mother can also express the milk until the flow slows. (Expressed milk can be frozen for later use.) Expressing milk also allows infants to get hind milk and prevents gas and colic that may result from a large volume of relatively low-fat milk.

Hyperlactation

Hyperlactation occurs when the milk volume being produced by the mother far exceeds the intake of the baby. A hyperactive letdown may be a sign of overproduction. Other signs include breasts that are not drained completely during a feeding, chronic plugged ducts, leaking in between feedings, and pain with letdown or deep in the breast. Symptoms in the baby include those mentioned above with hyperactive letdown as well as milk leaking from the breast, spitting up, poor weight gain due to high volume of low-fat milk, or good initial weight gain followed by poor weight gain. During the feeding, the baby may have difficulty maintaining latch and may also arch back off of the breast.[8] Other symptoms in the baby may include excessive gas and green frothy explosive stools from rapid transit time. These symptoms often lead the mother to incorrectly limit her intake of milk or other foods, believing the symptoms are caused by an allergy instead of a fore-milk/hind-milk imbalance. If intake of high volume of low-fat fore-milk continues over time, the baby may develop colitis.

Management in the mother is to reduce production. This can be done by having the baby nurse on only one side during feedings and having the mother express milk on the other side only for comfort. Cabbage leaves or cold compresses may also be used to decrease production.

Engorgement

Engorgement occurs when breasts are overfilled with milk. This is common in first-time mothers. Engorgement occurs when the supply-and-demand process is not yet established, and the milk is abundant. It also occurs with infrequent or ineffective removal of milk from the breast because of mother–infant separation, a sleepy baby, sore nipples, or improper breastfeeding technique.[3] The best way to prevent engorgement is to nurse the infant frequently. (Newborn infants will often nurse every hour and a half.) If the infant is not available to nurse, expressing milk every few hours will prevent engorgement while helping to build and maintain a milk supply.

The peak time for engorgement varies among women and can occur any time from day 2 through day 14 but is most common on day 2–3. Engorgement can result in discomfort, difficulty in establishing milk flow, and difficulty in latch-on.[7] Severe engorgement inhibits milk flow because the swollen tissue is compressing the milk ducts—not because the mother is failing to experience the letdown or milk ejection reflex. Once engorgement occurs, there are several simple treatments to help ease the discomfort. It is important for the mother to express milk until her breasts are no longer hard before putting the infant to breast. This will make it more comfortable for her and easier for her baby to latch on. When an infant is unable to extract milk effectively, the use of hand expression or an electric breast pump can help establish milk flow and soften the breast to make it easier for the infant to attach properly and further extract milk. Women can use analgesics to reduce pain from engorgement. A warm shower, warm compresses with massage before feedings, and expressing milk will help to relieve pressure and help trigger milk flow. Application of cold compresses or gel packs between feedings helps to reduce pain and swelling.[3,8]

Cabbage Leaves   Many believe that cabbage leaves (either cool or at room temperature) reduce discomfort and swelling associated with engorgement, although it is not known how the effects are mediated. In recent randomized trials, cabbage leaves and gel packs were equally effective in the treatment of engorgement, as were cabbage extract and a placebo cream.[9] Raw leaves are applied directly to the breast until they wilt, which is approximately 20 minutes. Cabbage leaves should be used only 2–3 times per 24 hours to prevent reduction in milk supply.[10]

Plugged Duct

An obstructed, or "plugged," duct is a localized blockage of milk resulting from milk stasis (milk remaining in the ducts).[3] The mother may feel a painful knot in one breast and usually does not have a fever or other signs of illness. Treatment for plugged ducts is gentle massage, warm compresses, and complete emptying of the breast.[3,6] Women should consider changing nursing positions to facilitate emptying of the breast. For example, if the woman is nursing while lying down, she may try a sitting position, or she may switch the position of holding the infant (see Illustration 6.5 in Chapter 6). If the woman is pumping, the flange size should be assessed. In most cases, using a larger size flange will improve efficiency of pumping. When plugging occurs repeatedly, a gentle manual massage before nursing often results in the plug being expelled. Consider use of lecithin for chronic plugged ducts. Lecithin is a phospholipid used as an emulsifier to keep fat dispersed and suspended in water instead of building up in the ducts. One tablespoon per day of oral granular lecithin has been reported an effective therapy. If not resolved, plugged ducts lead to mastitis.[11]

Mastitis

Mastitis is an inflammation of the breast most commonly found in breastfeeding women. It can be infective or non-infective. It occurs in about 3–20% of breastfeeding women[8,11] with the highest incidence 2 to 6 weeks after birth, although it may occur at anytime.[3] Some women get mastitis after having cracked or sore nipples, and some get it without any noticeable problem on the surface of the breast. Missing a feeding or the infant sleeping through the night may precipitate engorgement, plugged ducts, and then mastitis. Restriction from a tight bra or clothing may also increase risk of developing mastitis. Symptoms of mastitis are similar to those of a plugged duct (Table 7.1). In both conditions, there is a tender, hot, enlarged, hard, wedge-shaped area in the breast, and often an area of redness on the surface of the breast. Cases of mastitis are usually accompanied by a fever and flu-like symptoms.

It is important for the mother to continue nursing through mastitis. Effective milk removal is the most important management step in mastitis. If nursing is too painful, the mother should express milk from the breast by hand or pump. The techniques used to minimize pain from engorgement may also be used for mastitis. Ibuprofen is commonly recommended to help with the pain and inflammation.[12] Adequate rest, fluids, and nutrition are also important. In mild cases of mastitis or when symptoms have been present for less than 24 hours, efficient milk emptying via frequent nursing or pumping may be sufficient to resolve the mastitis. If symptoms do not improve within 24 hours, antibiotics should be started.[12] In a randomized trial, half of 55 women treated only with antibiotics had breast abscess, recurrent mastitis, or symptoms lasting longer than 2 weeks, compared to only 2 of 55 who also emptied their breasts (breasts can be emptied by feeding the baby or pumping).[14] Significant delays in seeking treatment for mastitis are associated with the development of abscess and recurrent mastitis.[3] Abrupt weaning is not recommended and increases risk of abscess development (see also Case Study 7.1). Vertical transmission of human immunodeficiency virus (HIV) from mother to child increases during mastitis infection. The World Health Organization recommends avoidance of breastfeeding on the affected side until the infection resolves.[8]

Table 7.1 Comparison of symptoms of engorgement, plugged duct, and mastitis

Characteristics	Engorgement	Plugged Duct	Mastitis
Onset	Gradual, immediately postpartum	Gradual, after feedings	Sudden, after 10 days
Site	Both breasts	One breast	Usually one breast
Swelling and Heat	Generalized	May shift, little or no heat	Localized, red, hot, swollen area on breast
Pain	Generalized	Mild, but localized	Intense, but localized
Body Temperature	No fever	No fever	Fever (>101°F)
Other Symptoms	Feels well	Feels well	Flu-like symptoms

SOURCE: Reprinted from Breastfeeding: A Guide for the Medical Professions, 5th Ed. by R.A. Lawrence and R.M. Lawrence, Table 8.5, © 1999, with permission from Elsevier.

Case Study 7.1

Chronic Mastitis

This was the first and an unremarkable pregnancy for 29-year-old Barbara Ann. Barbara Ann has reported experiencing "a little" breast enlargement during her pregnancy.

Photodisc

Her infant is first put to the breast at 2 hours postpartum, and the infant latches well according to mom, and suckles vigorously. The infant nurses every 2 hours over the first 3 to 4 days postpartum. Barbara Ann's breasts became noticeably fuller during the third postpartum day, and by the fourth postpartum day they are painfully engorged. In addition, Barbara Ann reports painful, burning, cracked nipples. The engorgement makes it difficult for her baby to latch at the breast. The baby becomes irritable, and Barbara Ann experiences a significant amount of pain. A lactation consultant gives Barbara Ann guidelines for engorgement management.

On day 5, the engorgement is still causing discomfort. Barbara Ann's nipples have become more cracked and painful. The lactation consultant notes that the infant's latch has become shallow and tight, probably in an attempt to control the flow of milk. However, the infant shows all the signs of adequate intake, including 10 very wet and 5 soiled diapers during the 24 hours prior to the consultation.

By day 7 postpartum, Barbara Ann has mastitis. She is treated with a 7-day course of dicloxacillin. A lactation consultant assists her in achieving a proper infant latch.

By day 14, Barbara Ann is feeling much better. The mastitis has resolved, and her nipples are healing. She still has tenderness during infant feedings and a healing crack on the right side. Her breasts are still uncomfortably full and are occasionally swollen and tender.

At 3 weeks postpartum, Barbara Ann develops an inflamed area on the right breast that remains red and tender despite applying warmth and massage to the area. The lactation consultant helps Barbara Ann to position the infant in a way that allows drainage of the inflamed area and recommends she pump the affected side to relieve the discomfort. The crack on the right nipple has improved, but is still not completely healed. Barbara Ann continues to show signs of oversupply, such as breasts feeling uncomfortably full, even after feeding, and excessive milk leakage between feedings. The lactation consultant provides Barbara Ann with techniques to decrease her overproduction.

After 10 days of persistent burning pain in the nipple area, Barbara Ann is treated with fluconazole for a yeast infection. Seven days after starting the fluconazole, a topical nystatin ointment is prescribed for her nipples and an oral suspension for her infant.

At 7 weeks postpartum, Barbara Ann calls the lactation consultant to report another flare-up of mastitis. Her health care provider prescribes a 10-day course of dicloxacillan. Barbara Ann is still treating her nipples with nystatin ointment. At 8 weeks postpartum her mastitis resolved; her nipple pain is still present, but improving. Barbara Ann is nursing the infant on one side only per feeding and reports that the infant latches better when she is in a more reclined position.

SOURCE: Adapted from: Anonymous. Case management of a breastfeeding mother with persistent oversupply and recurrent breast infections. *J Hum Lact*, 2000, 16:221–5.

Questions

1. Name the causes of engorgement.
2. List at least 2 possible nutrition diagnosis for this case.
3. Identify at least one nutrition intervention for each of the diagnosis listed.
4. Name potential indicators for each intervention listed.

(Answers are located in the Instructor's Manual for the 4th edition of *Nutrition Through the Life Cycle*).

Low Milk Supply

Real or perceived insufficient milk supply is the most common reason for cessation of breastfeeding. Low milk supply is usually caused by the mother not breast-feeding or pumping often enough or inefficient empty-ing of the breast caused by a poor latch or incorrect flange size while pumping. Stress may also contribute to a low milk supply.[13] The mother may wish to use a galactogogue, which is a drug or herb used to increase milk supply. Galactogogues should be used only after evaluating maternal causes for low supply and when increasing breastfeeding and/or pumping has not been successful.[15] Encourage the mother to nurse or pump using a hospital-grade electric breast pump every 2 to 3 hours during the day and once at night. If the baby is not nursing effectively, the mother may need to pump after breastfeeding to improve her supply. Make sure her diet is adequate and her fluid intake is appropriate. En-courage resting and relaxation techniques. Review cur-rent medication and hormonal contraceptives. Estrogen is known to inhibit lactation. Progesterone-only birth control pills have been used without problems on milk supply.[16] Common medications used as galactogogues are Metoclopramide (Reglan) and Domperidone (Mo-tilium). Metoclopramide is the most commonly used medication for improving milk supply in the United States.[15] It is prescribed by a physician, who should also follow the mother for potential side effects. The usual dose is 10 mg taken 3–4 times per day for 14 days with a taper over 4–5 days. Metoclopramide may cause fa-tigue and drowsiness as well as diarrhea. Increased de-pression, anxiety, confusion, dizziness, or headache may also occur. Side effects should be reported to prescrib-ing physician. It may be necessary for the mother with these side effects to discontinue use of the medication. Domperidone is commonly used outside the United States. It has been approved for use in most countries in the devel-oped world. The U.S. Food and Drug Administration (FDA) issued a warning let-ter in 2004 against use of Domperidone, citing safety issues with the IV form. Risks of importation are also discussed in the warn-ing letter.[17] Side effects are not common. Domperidone has been shown to be safe and effective in improving milk production in a randomized control study.[15] Both medications increase prolactin levels through dopamine receptors.

Fenugreek (*Trigonella foenum-graceum*), goat's-rue (*Galega officinalis*), and milk thistle (*Silybum marianum*)/blessed thistle (*Cnicus benedictus*) are common herbal ga-lactogogues.[15] It is important to inform the mother that most herbal or natural galactogogues have not been evalu-ated scientifically and are not regulated by the U.S. Food and Drug Administration. Galactogogues are used in the short-term (1–3 weeks). Studies are lacking on potential side effects of longer-term use. These products are discussed further in the "Herbal Remedies" section of this chapter.

Maternal Medications

"It is equally inappropriate to discontinue breastfeeding when it is not medically necessary to do so as it is to continue breastfeeding while taking contraindicated drugs."

R. A. Lawrence[18]

The single most common medical issue health care pro-viders face in managing breastfeeding patients is maternal medication use.[18] Ninety to ninety-nine percent of breast-feeding women receive some type of medication during their first week postpartum.[19] Most medications taken by nursing mothers are excreted in breast milk. These include both over-the-counter medications and prescription drugs. Unfortunately, data on drug safety may not be readily available to women and health care providers.[20] There is data which suggests mothers who discontinue breastfeed-ing prematurely, reported concern about the use of medi-cations as a major reason.[19] Recommending that a mother discontinue breastfeeding to take a medication is almost never required and should only be done as a last resort. For most maternal conditions, required drug therapy choices are available that will not cause harm to the infant.[19,21]

Two key questions to address in the analysis of the risk of an infant's exposure to a drug excreted in breast milk are: How much of the drug is excreted in milk, and at the level of excretion, what is the risk of adverse effects?[22,23] Among the numerous variables to examine to answer these questions are:

- The pharmacokinetic properties of the drug
- Time-averaged breast *milk/plasma ratio* of the drug
- The drug *exposure index*
- The infant's ability to absorb, detoxify, and excrete the agent
- The dose, strength, and duration of dosing
- The infant's age, feeding pattern, total diet, and health[18,22]

Additional considerations are the well-established intereth-nic and racial differences in drug responsiveness, exposure

Milk to Plasma Drug Concentration Ratio (M/P Ratio) The ratio of the concentration of drug in milk to the concentration of drug in maternal plasma.[19] Since the ratio varies over time, a time-averaged ratio provides more meaningful information than data obtained at a single time point. It is helpful in understanding the mechanisms of drug transfer and should not be viewed as a predictor of risk to the infant, as it is the concentration of the drug in milk, and not the M/P ratio, that is critical to the calculation of infant dose and assessment of risk.[19,23,24]

Exposure Index The average infant milk intake per kilogram body weight per day × (the milk-to-plasma ratio divided by the rate of drug clearance) × 100. It is indicative of the amount of the drug in the breast milk that the infant ingests and is expressed as a percentage of the therapeutic (or equivalent) dose for the infant.[14]

of the infant to the drug during pregnancy, whether the drug can be safely given to the infant directly,[18] and the relative infant dose.[19] The ultimate test of drug safety is the measurement of the infant's plasma drug concentration and any side effects from the drug on the infant.[22] Carefully controlled studies on large enough samples to validate the results are rare but have increased during the last decade. Use of the risk-benefit algorithm for assessment of drug use in breastfeeding may also be helpful.

Fortunately, numerous resources (Table 7.2) based on a thorough evaluation of available evidence can assist the health care provider and mother in identifying which drugs are safe and which are not. The American Academy of Pediatrics (AAP) Committee on Drugs publishes guidelines for practitioners.[24-26] The guidelines provide a list of drugs divided into the following seven categories according to risk factors in relationship to breastfeeding:

1. Cytotoxic drugs that may interfere with the cellular metabolism of the nursing infant

2. Drugs of abuse for which adverse effects on the infant during breastfeeding have been reported

3. Radioactive compounds that require temporary cessation of breastfeeding

4. Drugs for which the effect on nursing infants is unknown, but may be of concern

5. Drugs that have been associated with significant effect on some nursing infants and should be given to nursing mothers with caution

6. Maternal medications usually compatible with breastfeeding

7. Food and environmental agents having no effect on breastfeeding

The list, which is updated periodically, includes only those drugs about which there is published information. Other useful monographs and review articles provide additional information on a wide array of medications.[19-22,26]

The Breastfeeding and Human Lactation Study Center at the University of Rochester (see Table 7.2) continually updates its database of more than 3000 references on drugs, medications, and contaminants in human milk and is a resource for complex questions on the risks to the breastfed infant. The TOXNET LactMed database is available online (http://toxnet.nlm.nih.gov/). There are also textbooks with information on drugs and breastfeeding. "Medications and Mother's Milk" by Thomas Hale (see Table 7.2) is updated regularly. It includes information on prescription and over-the-counter medications as well as common herbal remedies.[19] The **Physician's Desk Reference (PDR)** is not a good source for information about drugs and breastfeeding because the information is derived directly from pharmaceutical companies whose first concern is avoiding liability. When there are no studies that prove beyond a doubt that a drug is safe for nursing mothers, the drug companies must advise against use while breastfeeding—even if what is known about the drug suggests that there is little cause for concern.

Table 7.2 Resources on drugs, medications, and contaminants in human milk

- American Academy of Pediatrics, Committee on Drugs. The Transfer of Drugs and Other Chemicals Into Human Milk (RE9403). *Pediatrics* 2001; 108:776.[24] *Website*: www.aap.org.http://aappolicy.aappublications.org/cgi/content/full/pediatrics;108/3/776
- Briggs, G. G., Freeman, R. K., and Yaffe, S. J. *Drugs in Pregnancy and Lactation*, 6th ed. Baltimore, MD: Williams and Wilkins, 2001.[21]
- Hale, T. W. *Medications and Mothers' Milk*, 13th ed. Amarillo, TX: Pharmasoft Medical Publishing, 2008.[19] Dr. Hale will answer questions from health professionals posed on the Pharmasoft website: www.iBreastfeeding.com.
- *Breastfeeding and Maternal Medication: Recommendations for Drugs in the Eleventh WHO Model List of Essential Drugs*.[30] UNICEF World Health Organization, 2003. *Website*: www.who.int/child-adolescent-health/New_Publications/NUTRITION/BF_Maternal_Medication.pdf.
- Blumenthal, M., Busse, W., Goldberg, A., et al., eds. *The Complete German Commission E Monographs: Therapeutic Guide to Herbal Medicines*. Boston: Integrative Medicine Communications, 1998.[31]
- The Breastfeeding and Human Lactation Center, University of Rochester. This service is available for complex medication questions (9:30 a.m. to 4 p.m. EST, Monday to Friday, at 585-275-0088).
- HerbMed (www.herbmed.org): an interactive, electronic herbal database with links to scientific publications.
- Humphrey, S. *The Nursing Mother's Herbal*. MN: Fairview Press, 2003.
- TOXNET: National Library of Medicine Drug and Lactation database, containing summaries of published literature on the effects of over 400 drugs on lactation.
- REPROTOX: An online proprietary reproductive toxicology database, http://reprotox.org.

Medications contraindicated during breastfeeding include antineoplastic agents, radioactive isotopes, drugs of abuse and drugs that suppress lactation.[18,19,24] Fortunately, safer alternative medications can be recommended as a substitute for most other drugs with known adverse effects on infants. Specific knowledge about a medication's safety during breastfeeding will allow proper treatment and avoid unnecessary maternal anxiety and undue risk. For example, many cold remedies, antihistamines, and decongestants are listed as usually compatible with breastfeeding, but they may suppress lactation. In a study of women who took a single 60 mg dose of Sudafed (psuedoephedrine), 24-hour milk production was decreased by 24%.[19] This medication should be avoided in mothers with marginal or low production. Milk production should be closely monitored by the mother taking such medications.

Many women have questions on the safety of oral contraceptive use during lactation. There is currently no evidence of harm, but few women have been studied. There is evidence that combined oral contraceptives may reduce the volume of breast milk.[19,27] The American College of Obstetricians and Gynecologists (ACOG) and the World Health Organization recommend against using combined oral contraceptives in the first 6 weeks postpartum. If lactation is well established at 6 weeks, ACOG recommends monitoring the infant's nutritional status if combined oral contraceptives are initiated.[27] WHO does not recommend using combined oral contraceptives from 6 weeks to 6 months unless other forms of contraceptives are not available. The La Leche League International recommends avoiding combined oral contraceptives in breastfeeding because other forms on contraception are available.[27] Progestin-only oral contraceptives and implants are safe and effective during lactation.[19,28] Implants that deliver orally active steroids should only be used after 6 weeks postpartum to avoid transferring of steroids to the newborn.[28] The Depo-provera shot is also recommended at 6 weeks postpartum.[19]

If a drug or surgery is elective, a mother may be able to delay it until the baby is weaned. If a breastfeeding mother needs a specific medication, and the hazards to the infant are minimal, she should be instructed to take the medication after breastfeeding, at the lowest effective dose, and for the shortest duration.[18,29] Other important steps can be taken to further minimize the effects (Table 7.3). It is also sometimes possible to choose alternative routes for administration of a medication to reduce exposure. For example, prescribing an inhalant instead of a drug taken by mouth, or a topical application rather than oral dosing, reduces infant exposure. If a drug is to be taken for diagnostic testing (such as a radioactive agent), a mother may need to withhold breastfeeding for a short period of time, pumping and discarding her milk. The mother can plan ahead in these cases and pump prior to the procedure or surgery. She can freeze her expressed milk for use while breastfeeding is withheld. Discontinuing breastfeeding due to maternal

Table 7.3 Minimizing the effect of maternal medication[18]
1. Avoid long-acting forms: Accumulation in the infant is a genuine concern because the infant may have more difficulty excreting a long-acting form of a drug, which usually requires detoxification in the liver.
2. Schedule doses carefully: Check usual absorption rates and peak blood levels of the drug, and schedule the doses so that the least amount possible gets into the milk. In order to minimize milk levels of most drugs, the safest time for a mother to take the drug is usually immediately after her infant nurses.
3. Evaluate the infant: Watch for any unusual signs or symptoms, such as changes in feeding pattern or sleeping habits, fussiness, or rash.
4. Choose the drug that produces the least amount in the milk.

medications is a last resort but may be necessary for the health and well-being of the mother (for example, if she needs chemotherapy or radioactive treatment). Any decision to limit a mother's breastfeeding must be justified by the fact that the risk to her baby clearly outweighs the benefits of breastfeeding.

Herbal Remedies

Numerous herbs have been used in folk and traditional systems of healing to affect the flow of milk (Table 7.4), or to treat mastitis, infant colic, and thrush.[31–34] However, scientific information about herb use during lactation, particularly from recent studies, is sparse. The limited pertinent safety data are based on traditional use, animal studies, and knowledge of the pharmacologic activities of the products' constituents. Medicinal herbs should be viewed as drugs, with evaluation of both their pharmacological and toxicological potential.[1,9,35]

A mother may perceive herbs as natural and therefore safe and even preferable to conventional over-the-counter medicines, or prescription drugs. However, the risks of using some herbal remedies may outweigh the potential benefits. The same risk factors that apply to drugs also apply to herbs. "Medications and Mother's Milk" (2008) contains a systematic lactation risk analysis of numerous herbal remedies. Many herbs are far from benign, and many are contraindicated during lactation (Table 7.5). Because little is known about the amount secreted in human milk or the effects on preterm or term infants, herbs that are central nervous system stimulants, herbs that destroy cells (cytotoxic), or herbs that are laxatives, hepatotoxic, carcinogenic, mutogenic, or contain potentially toxic essential oils, are not recommended during lactation.[32,36]

Table 7.4 Herbs traditionally used to affect milk production*[32,33]

Herbs to Promote Milk Flow

Anise
Astralagus
Milk thistle
Caraway
Celery root and seed
Chaste tree berry (chasteberry) or monk's pepper
Fennel
Fenugreek
Goat's-rue
Hollyhoc
Hibiscus flower
Lemongrass
Marshmallow
Stinging nettle
Raspberry
Rauwolfia
Verbena

Herbs that Reduce Milk Flow

Castor bean
Jasmine flower
Fresh parsley
Sage

SOURCE: JOURNAL OF THE AMERICAN PHARMACEUTICAL ASSOCIATION: APHA by M. L. Hardy. Copyright 2000 by AMERICAN PHARMACISTS ASSOCIATION (J). Reproduced with permission of AMERICAN PHARMACISTS ASSOCIATION (J) in the format Textbook and in the format extranet posting via Copyright Clearance Center.

*Many traditional galactologues are not currently considered appropriate during lactation (see Table 7.5).

The toxic effects of herbs are often not due to the herb itself, but are caused by products containing misidentified plants or contaminants such as heavy metals, synthetic drugs, microbial toxins, and toxic botanicals.[37] *Medicinal herbs* in the United States are regulated as dietary supplements and are not tested for safety or efficacy.

Herbal teas that are safe for the infant and mother during lactation are presented in Table 7.6. Lawrence[18] recommends using only herbal teas "that are prepared carefully, using only herbs for essence (e.g., Celestial Seasonings brand tea) and avoiding heavy doses of herbs with active principles." Careful attention should also be given to preparation, avoiding long steeping times.

Some culinary herbs may lead to problems when used extensively. In lactation and herbal texts, sage has a folk reputation for lowering milk supply,[32] as do parsley and peppermint, especially if the oil is taken by mouth in large doses.[38] Consumed on occasion, however, in small amounts as part of a reasonably varied diet, peppermint, parsley, sage, and other culinary herbs currently have no documented negative effect on lactation.

Although various herbal gels, ointments, or creams are often suggested for use on the nipples, any substance applied to the breast or nipples could easily be ingested by the nursing child. The use of herbal oils is not recommended.[38] In one infant, severe breathing difficulties were documented after the mother used menthol, a significant component of peppermint oil, on her nipples.[39]

Although health care practitioners may wish otherwise, some mothers may refuse prescription drugs and insist on using herbal alternatives. If a mother is consuming a large amount of any herbal product, its contents should be checked. Important information to obtain on the product includes its name, a list of all ingredients, the names of the plants or other components (include the plants' Latin names if possible), details of the preparation, and the amount consumed.[38] Reliable sources of herbal information[30,31,40,41] (Table 7.2) or the regional poison-control center may be able to identify potentially harmful pharmacological and toxicological ingredients.

In balancing the risks and benefits in a given situation, consideration should be given to the benefits of continued breastfeeding to the baby and the mother. It is also important to consider the varied nature of lactation: newborns face different risks than do older babies or toddlers because of immaturity; infants consume varying amounts of human milk; mothers may be looking forward to many months of lactation yet need or desire the benefits of medicinal herbs. A few of the widely used herbs in the United States are discussed below.

Medicinal Herbs Plants used to prevent or remedy illness.[34]

Specific Herbs Used in the United States

Echinacea Echinacea is used for the common cold and to enhance the immune system. Insufficient reliable data is available on its entry into breast milk and effects on the infant. It is available in many forms; the tincture form contains 15–90% alcohol.[36] Gastrointestinal distress in some women has been reported. Consumption of echinacea during lactation is not recommended.[42]

Ginseng Root Ginseng root, widely believed to increase capacity for mental work and physical activity and to reduce stress, contains dozens of steroid-like glycosides, sterols, coumarins, flavonoids, and polysaccharides.[18] It is reported to have estrogen-like effects in some women, with breast pain common with extended use and mammary swelling also reported. While there has been considerable animal experimentation with ginseng root, human data is not extensive. There is no information on transfer to human milk.[36] The lack of standardized preparations, information on dosage, and accurate recording of side effects is a problem. Because of the reported breast effects and occasional reports of vaginal bleeding, the use of ginseng during lactation may not be advisable.[5,28]

St. John's Wort St. John's wort is widely used in the United States and Europe as a mood stabilizer and antidepressant. It has been shown to have fewer side effects

202   Nutrition Through the Life Cycle

Table 7.5 Medicinal herbs considered not appropriate for use during pregnancy or lactation[18,36,42,]

Agnus castus	Ephedra (ma huang)	Motherwort
Alkanet	Eucalyptus	Myrrh
Aloes	Eupatorium	Nettle
Angelica	Euphorbia	Osha
Apricot kernel	Fennel	Passionflower
Aristolchia	Feverfew	Pennyroyal
Asafoetida	Foxglove	Petasites
Avens	Frangula	Plantain
Basil	Fucus	Pleurisy root
Bladderwrack	Gentian	Pokeroot
Blue flag	German chamomile	Poplar
Bogbean	Germander	Prickly ash
Boldo	Ginkgo biloba	Pulsatilla
Bonese	Ginseng, eleuthero	Queen's delight
Borage	Ginseng, panax	Ragwort
Broom	Golden seal	Red clover
Buchu	Ground ivy	Roman chaparral
Buckthorn	Groundsel	Sassafras
Bugleweed	Guarana	Senna
Burdock	Hawthorne	Shepherd's purse
Calamus	Heliotropium	Skullcap
Calendula	Hops	Skunk cabbage
Cascara	Horehound, black	Squill
Cayenne pepper	Horehound, white	St. John's wort
Chamomile	Horsetail	Stephania
Chasteberry	Hydrocotlye	Stillingia
Chinese rhubarb	Jamaica dogwood	Tansy
Cohosh, black	Joe-pye weed	Tonka bean
Cohosh, blue	Juniper	Uva-ursi
Coltsfoot	Kava kava	Valerian root
Comfrey	Licorice	Vervain
Cornsilk	Liferoot	Wild carrot
Cottonroot	Lobelia	Willow
Crotalaria	Male fern	Wormwood
Darniana	Mandrake	Yarrow
Devil's claw	Mate	Yellow dock
Dogbane	Meadowsweet	Yohimbine
Dong quai	Melilot	
Echinacea	Mistletoe	

*Exclusion from this list should not be a recommendation for safety.

Table 7.6 Herbal teas considered safe during lactation[18]

Tea	Origin/Use
Chicory	Root/caffeine-free coffee substitute
Orange spice	Mixture/flavoring
Peppermint	Leaves/flavoring (limit duration use)
Raspberry	Fruit/flavoring
Red bush tea	Leaves, fine twigs/beverage
Rose hips	Fruits/vitamin C

SOURCE: Reprinted from Breastfeeding: A Guide for the Medical Professions, 5th Ed. by R.A. Lawrence and R.M. Lawrence, Table 7.6, © 1999, with permission from Elsevier.

in trials compared to prescription antidepressants.[19] Although the plant contains at least 10 classes of biologically active compounds, hypericin and hyperforin, and their metabolites pseudohypericin and adhyperforin, seem to be the most important for their neuropharmocologic properties.[43] There is growing evidence that hyperforin may be the key constituent responsible for the antidepressant property of this herb. Composition of St. John's wort preparations may vary based on climate and other variables related to growth, harvesting, and processing. Many products available in the United States and Canada may be of poor quality and contain little or no St. John's wort. The mother taking St. John's wort should only purchase from reputable sources. Large doses decrease prolactin levels, which could potentially reduce breast milk supply.[18] There are also concerns about the numerous potential herb–drug interactions because of the ability of St. John's wort to induce the metabolic activity of cytochrome P450 (CYP), particularly for drugs used to treat human immunodeficiency virus (HIV).[43]

There is limited information on the excretion of active components in breast milk and effect on the infant. In a study of 5 mothers who were taking 300 mg of St. John's wort 3 times daily, hyperforin was excreted into breast milk at very low levels and the relative infant doses were 0.9–2.5%. This level of infant exposure to hyperforin through milk is comparable to levels reported in most studies assessing antidepressants or neuroleptics. No side effects were seen in the mothers or infants.[43] A recent clinical trial compared 33 breastfeeding women taking St. John's wort with 101 disease-matched controls and 33 age- and parity-matched controls. No differences between groups were found in maternal adverse side effects, maternal report of decreased milk production, or in infant weight in the first year. In the group taking St. John's wort, 5 of the infants reported either colic, drowsiness, or lethargy, compared with only 1 infant in each of the control groups.[44] The symptoms were not severe, and specific medical treatment of the affected infants was not required. St. John's wort may be a reasonable choice for treatment of postpartum depression using high-quality products.[19] Mothers who are breastfeeding while taking St. John's wort should be alert to changes in infant behavior and should be closely supervised by a pediatrician. If the mother is taking other medications or herbal supplements, she should be informed that drug–herb interactions are possible.

Fenugreek   This spice is used as an artificial flavor for maple syrup, in teas, poultices, and ointments, and as an ingredient in East Indian cooking. It is also the most commonly used herbal galactogogue (mik production stimulate). While there is limited scientific evidence to back this claim, there are anecdotal reports of its successful use to increase milk supply. In one account of 1200 women taking fenugreek, almost all reported an increase in milk production within 24 to 72 hours.[45] Rare maternal adverse effects include diarrhea; a maple-like aroma in urine, breast milk, or sweat; and exacerbation of asthmatic symptoms.[46] This herb is derived from a plant in the same family as peanuts and chickpeas and has potential for allergy in sensitive infants. There are reports of colic, abdominal upset, and diarrhea among babies whose mothers took fenugreek.[36] Transfer into milk is assumed because the infant's urine smells of maple syrup. The usual dosage (2 to 3 capsules three times daily or one cup of tea three times a day) is considered compatible with breastfeeding.[46]

Goat's-Rue   This is a widely used galactogogue in Europe and South America and is becoming increasingly popular in the United States. In the 1900s, cows fed goat's-rue were observed to have increased milk supply. No controlled human studies have been done. Only one side effect has been reported, according to the Academy of Breastfeeding Medicine. "Maternal ingestion of a lactation tea containing extracts of licorice, fennel, anise, and goat's-rue was linked to drowsiness, hypotonia, lethargy, emesis and poor suckling in two breastfed neonates. An infection work-up was negative, and symptoms and signs resolved on discontinuation of the tea and a 2-day break from breastfeeding."[15] No other side effects have been reported. The usual dose is one teaspoon dried leaves steeped in eight ounces of water for ten minutes and used as a tea with one cup taken three times per day.

Milk Thistle/Blessed Thistle

Blessed thistle has been used in Europe and has become increasingly popular in the United States. It is often combined with fenugreek. There have been no randomized controlled studies done on this galactogogue. It is taken as a tea 2–3 times per day. The tea is usually prepared by simmering one teaspoon crushed seeds in eight ounces of water. It is also available in capsules.

Many mothers may be taking herbal products for low milk supply. There are many other commonly used herbal galactogogues (see Table 7.4). Keep in mind that there are few human studies done on most of these products. Safety, proper dosing, and intended and unintended effects may be unknown.[11]

Alcohol and Other Drugs and Exposures

> "Avoid prescribing or proscribing it [alcohol] and . . . assist the mother in appropriately adjusting her alcohol consumption in both timing and volume."
>
> R. A. Lawrence[18]

Alcohol

The harmful effects of alcohol consumption during pregnancy are well documented, and drinking during pregnancy is clearly not recommended. In the past, recommendations on alcohol consumption during lactation were ambiguous and somewhat controversial. The AAP policy statement on "Breastfeeding and the Use of Human Milk" states that "breastfeeding mothers should avoid the use of alcoholic beverages because alcohol is concentrated in breastmilk and its use can inhibit milk production."[47] Alcohol consumed by the mother passes quickly into her breast milk, and the effects on the breastfeeding baby are directly related to the amount the mother consumes.[48,49]

The level of alcohol in breast milk matches the maternal plasma levels at the time of the infant feeding. Peak maternal plasma and breast milk levels are reached 30 to 60 minutes after alcohol consumption and approximately 60 to 90 minutes when taken with food.[18] As the alcohol clears from a mother's blood, it clears from her milk. It takes a 120-pound woman about 2 to 3 hours to eliminate from her body the alcohol in one serving of beer or wine (Table 7.7).[50] The common practice of pumping the breasts and then discarding the milk immediately after drinking alcohol does not hasten the disappearance of alcohol from the milk, as the newly produced milk will still contain alcohol as long as the mother has measurable blood alcohol levels.[48]

In many cultures, folklore passed down for generations encourages the use of alcohol as a galactologue that facilitates milk letdown and rectifies milk insufficiency as well as sedating and calming the "fussy" infant.[51] In contrast to this folklore, there is now strong evidence of a negative dose-related impact of alcohol on milk supply and the milk letdown reflex. In 2005, a study[51] found that during the immediate hours after alcohol consumption by lactating women, the hormonal milieu underlying lactation performance is disrupted. Oxytocin levels significantly decrease, whereas prolactin levels increase significantly. The diminished oxytocin response was significantly related to decreases in milk yield and milk ejection (letdown). In contrast, changes in prolactin were related to self-reported feelings of drunkenness. Recommending alcohol to women as an aid to lactation may be counterproductive; mothers may feel more relaxed, but the hormonal disruption diminishes the infant's milk supply.

Maternal alcohol consumption affects the odor of breast milk and the volume consumed by the infant within a half-hour to an hour after consumption.[8] Breast-fed infants consumed, on average, 20% less breast milk during the 3 to 4 hour period following their mothers' consumption of an alcoholic beverage (0.3 g ethanol/kg).[48,52] Compensatory increases in intake were then observed during the 8 to 16 hours after exposure when mothers refrained from drinking.[52]

Recent studies on the impact of maternal alcohol ingestion during lactation on infant sleep patterns and psychomotor development have raised concerns about regular consumption of alcohol while lactating. In one study, 11 of 13 breastfed infants had a reduction of more than 40% in active sleep after consuming their mothers' expressed breast milk flavored with alcohol (32 mg) on one testing day and expressed breast milk alone on the other.[53] All infants spent significantly less total time sleeping after consumption of the breast milk with alcohol (56.8 minutes with alcohol compared to 78.2 minutes without). A follow-up study replicated this finding and showed that infants can compensate for this deficit by increasing the amount of time spent in active (rapid eye movement) sleep during the 20.5 hours after the sleep-deficit period.[54] The investigators concluded that short-term exposure to small amounts of alcohol in breast milk produces distinctive changes in the infant's sleep–wake patterning. Since both the observed reduced milk consumption and the infant sleep deficits occur only when breastfeeding follows shortly after the mother's alcohol consumption, a nursing woman who drinks occasionally can limit her infant's exposure to alcohol by timing breastfeeding in relation to her drinking.[55]

Epidemiologic data on the effects of moderate drinking throughout the lactation period on the human infant are limited. In a study of 400 infants born to members of a health insurance plan, no differences in the infant's cognitive development scores were found at 1 year of age between infants whose mothers consumed alcohol while nursing and those that did not. However, Bayley Psychomotor Development Index scores at 1 year of age were slightly lower among infants who were exposed to alcohol through breast milk than among those who were not exposed.[56] Since current research does not show that occasional use (one to two drinks) of alcohol is harmful to the baby, La Leche League continues to support the opinion that the occasional use of alcohol in limited amounts is compatible with breastfeeding.[57] The American Academy of Pediatrics places alcohol in the category "Maternal Medication Usually Compatible with Breastfeeding."[24,25] It lists possible side effects if consumed in large amounts, including drowsiness, deep sleep, weakness, and abnormal weight gain in the infant. The Institute of Medicine Subcommittee on Nutrition During Lactation recommends that lactating women should be advised that if alcohol is consumed, intake should be limited to "no more than 0.5 grams of alcohol per kilogram of maternal body weight per day."[58] For a 60-kilogram (132-pound) woman, 0.5 grams of alcohol per kilogram of body weight corresponds to approximately 2 to 2.5 ounces of liquor, 8 ounces of table wine, or 2 cans of beer.[59] Many feel that nursing mothers are already placed under too many restrictions and may be discouraged from initiating or

Table 7.7 Alcohol and breastfeeding: Time (h:min) until zero level in milk is reached for women at different body weights[50]

| Maternal Body Weight | | Time Alcohol Remains | | |
lb	(kg)	1 Drink	2 Drink	3 Drink
100	(45.4)	2:42	5:25	8:08
120	(54.4)	2:30	5:00	7:30
140	(63.5)	2:19	4:38	6:58
160	(72.6)	2:10	4:20	6:30
180	(81.6)	2:01	4:03	6:05

NOTE: Time is calculated from beginning of drinking. Assumptions made: alcohol metabolism is constant at 15 mg/dl; height of the women is 162.56 cm (5 feet, 4 inches). 1 drink = 12 oz of 5% beer or 5 oz of 11% wine or 1.5 oz of 40% liquor. Example: For a 100-lb woman who consumed 2 drinks in 1 h, it would take 5 h 25 min for there to be no alcohol in her breast milk, but for a 180-lb woman drinking the same amount, it would take 4 h 3 min.
SOURCE: Adapted from E. Ho and A. Collantes et al., Alcohol and breastfeeding: calculation of time to zero level in milk. Biol Neonate, 2001; 80:219–22.

continuing to breastfeed if alcohol is prohibited because they feel they will face too many limitations.

If a mother does choose to have a drink or two, she can wait for the alcohol to clear her system before nursing according to the times given in Table 7.7. She can plan ahead and have alcohol-free expressed milk stored for the occasion. If she becomes engorged, she can pump her breasts as a means of comfort, and discard her alcohol-containing milk. Drinking water, resting, or pumping and discarding breast milk will not hasten the removal of alcohol from the milk, as the alcohol content of milk matches the maternal plasma alcohol levels. Mothers who are intoxicated should not breastfeed until they are completely sober.

Nicotine (Smoking Cigarettes)

Regardless of feeding choice (breast or bottle), the health risks for infants posed by having a mother who smokes are many, including otitis media, exacerbations of asthma, respiratory infections, and gastrointestinal dysregulation such as colic and acid reflux.[60] It is not ideal to smoke and breastfeed, but it is worse to smoke and not breastfeed. Well-documented data provide clear evidence that children of smoking mothers do better if breastfed in regard to general health, respiratory illness, and risk of sudden infant death syndrome (SIDS)[18] than if bottle-fed. Unfortunately, women who smoke cigarettes are less likely to breastfeed than nonsmokers, are less likely to seek help with breastfeeding difficulties than nonsmokers, and are at increased risk for stopping breastfeeding by 3 months.[61] While lower milk output has been reported among smoking mothers,[61,62] it is unknown which components of cigarette smoke are responsible for the reduced milk production, and several studies provide evidence that smoking does not necessarily hinder breastfeeding.[61] Lower fat content in breast milk has been reported. Breastfeeding infants whose mothers smoke show poorer growth. The reasons for this are not clear.[15] Changes in the odor and flavor of the breast milk might affect breast-milk intake in the breastfed infant.[15,62] Mennella et al. found that sleep patterns were impacted by smoking. They documented a dose-response effect with the greatest impact on sleep disruptions in the infants receiving the largest dose of nicotine via human milk.[63] Substantial epidemiological evidence suggests that social and behavioral factors and not physiological factors are largely responsible for the lower rates of breastfeeding found among smokers.[61] Some women believe that smoking is a barrier to breastfeeding; they do not believe they could, or should, adhere to the kinds of healthy practices they think are required of mothers to breastfeed.

Nicotine levels in breast milk of women who smoke are between 1.5 and 3.0 times higher than the level in the mother's blood,[24,61,63] and the mean 24-hour nicotine concentrations in breast milk rise as cigarette consumption increases. There is no evidence to document whether this amount of nicotine presents a health risk to the nursing infant, and because breastfeeding and smoking may be less detrimental to the child than bottle-feeding and

smoking, the AAP Committee on Drugs removed nicotine (and thus smoking) from its 2001 list of drugs of abuse with adverse effects on the infant during breastfeeding.[24]

Dahlstrom et al.[64] estimated that the dose of nicotine in breastfeeding infants was 1 µg per kilogram per feeding, based on data on nicotine concentrations in breast milk within 30 minutes after smoking. Women who smoke 10 to 20 cigarettes per day have 0.4 to 0.5 mg of nicotine/L in their milk. The total daily systematic infant dose from breast milk is estimated at less than 0 µg/kg/day, or 50 times less than the exposure of a 70-kg adult smoking 20 cigarettes per day or using a 21-mg nicotine patch.[65] With gradual intake over a day's time, the infant can metabolize nicotine in the liver and excrete the chemical in the kidney. Numerous studies of nicotine and cotinine concentrations in the nursing mother and her infant confirm that although bottle-fed infants born to smoking mothers and raised in a smoking environment have significant levels of nicotine and metabolites in their urine, breastfed infants have higher levels.[62]

Tobacco smoking also increases the exposure of infants to organochloride pesticides, PCBs, and hexachlorobenzene through breast milk and secondhand smoke.[66] Women should be counseled not to smoke while nursing or in the infant's presence. Mothers who are not willing to stop smoking should cut down, consider low-nicotine cigarettes, and delay feedings as long as possible after smoking. The half-life of nicotine is 95 minutes.

When used as directed, smoking cessation aids that replace nicotine do not appear to pose any more problems for the breastfeeding infant than maternal smoking does.[61,67] Since transdermal nicotine (nicotine patch) provides a steady level of nicotine in plasma and thus in breast milk, the mother cannot control the level of nicotine in breast milk except by changing the strength of the patch. As the mother progresses through to lower patch strengths during smoking cessation therapy, the transfer of nicotine to the infant via milk decreases as much as 70 percent.[61] Mothers who use nicotine replacement therapy intermittently (gum, nasal spray, or inhalation) might minimize the nicotine in their milk by prolonging the duration between nicotine administration and breastfeeding.[65]

Marijuana

Delta-9-tetrahydrocannabinol (THC), an active ingredient in marijuana, transfers and concentrates in breast milk and is absorbed and metabolized by the nursing infant. One study showed that one hour after ingestion, there was an eightfold accumulation of THC in the breast milk compared to the maternal plasma level.[15,19] There is evidence from animal studies of structural changes in the brain cells of newborn animals nursed by mothers whose milk contained THC. Impairment of deoxyribonucleic acid (DNA) and ribonucleic acid (RNA) formation and neurotransmitter systems essential for proper growth and development has been described.[68] Significant absorption and metabolism has been shown in studies.[19] In one study following breastfeeding mothers and their infants for 12 months, marijuana exposure via the mother's milk during the first month postpartum appeared to be associated with a decrease in infant motor development at 1 year of age.[69] Concerns about marijuana use during lactation include the amount of the drug the infant ingests while nursing and the amount inhaled from the environment. Exposure in infants to marijuana via breast milk will provide positive urine tests for up to 2–3 weeks.[19] The possible effect on DNA and RNA metabolism should discourage any maternal use, especially since brain cell development is still taking place in the first months of life. The American Academy of Pediatrics classifies THC as a drug of abuse that is contraindicated during lactation.[24]

Persistant Organic Pollutants (POPs) A family of chemicals manufactured either for a specific purpose (e.g., pesticides or flame retardants in electrical equipment or furniture) or produced as byproducts of incinerated waste. The POP family includes dioxins, polychlorinated biphenyuls (PCBs), polybrominated didpheyl ether (PBDE), and organochlorine pesticides.

Caffeine

Although caffeine ingestion is a frequent concern of breastfeeding mothers, moderate intake causes no problems for most breastfeeding mothers and babies. A dose of caffeine equivalent to a cup of coffee results in breast milk levels of 1% of the level in maternal plasma and, consequently, low levels in the infant.[18] However, because the infant's ability to metabolize caffeine does not fully develop until 3 to 4 months of age, caffeine does accumulate in the infant. Cases of caffeine excess in breastfed infants have been documented.[70] Symptoms, which include infants being wakeful, hyperactive, and fussy, did not require hospitalization and disappeared over a week's time after caffeine was removed from the maternal diet. No long-term effects of caffeine exposure during lactation have been documented.[71]

While most breastfed infants can tolerate a maternal caffeine intake equivalent to five or fewer 5-ounce cups of coffee per day, or less than 750 mL per day, some babies may be more sensitive than others. If a mother suspects her baby is reacting to caffeine, she may try avoiding caffeine from all sources (coffee, tea, soft drinks, over-the-counter medications, chocolate) for 2 to 3 weeks.[57]

Other Drugs of Abuse

Amphetamines, cocaine, heroin, and phencyclidine hydrochloride (angel dust, PCP) are classified by the AAP[24] as drugs of abuse that are contraindicated during

lactation. The AAP guidelines strongly state that these compounds and all other drugs of abuse are hazardous not only to the nursing infant, but also to the mother's physical and emotional health. In addition to their adverse pharmacological effects on the mother and infant, street drugs lack standardization and may be contaminated with other active ingredients, bacteria, heavy metals, or pesticides.[59]

Environmental Exposures

"The advantages of breastfeeding far outweigh the potential risks from environmental pollutants. Taking into account breastfeeding's short- and long-term health benefits for infants and mothers, the World Health Organization (WHO) recommends breastfeeding in all but extreme circumstances."

World Health[72]

There is now unambiguous data that breast milk accumulates and harbors potentially toxic environmental pollutants.[66, 72–75] Persistant organohalogens, including *persistent organic pollutants (POPs),* heavy metals, and volatile solvents, are among the toxic chemicals most often found in breast milk. A woman comes in contact with environmental chemicals as a matter of course in daily life,[74] through air pollution, drinking water, and diet. A woman also comes in contact with a wide range of environmental chemicals in the home that have the potential to appear in her breast milk, such as household cleaning and personal care products, paints, furniture strippers, and pesticides. Exposure to environmental chemicals can also be occupational.

While the presence of low levels of environmental chemicals, such as those in Table 7.8, in most human milk samples is well documented, the significance of the

Table 7.8 Environmental pollutants that may be found in human milk

Pollutant	Potential Health Effect
DDT, DDE	Estrogenic, antiandrogenic activity
PCB/PCDF	Ectodermal defects, developmental delay
TCDD (Dioxin)	Choloracne
Chlordane	Neurotoxicity
Heptachlor	Neurotoxicity
Hexachlorobenzine	Hypotonia, seizures, rash
Volatile organic compounds	
Tetracholorethylene	Hepatotoxicity
Trichlorethylene	Hepatotoxicity
Halothane	Hepatotocicity
Carbon disulfide	Neurotoxicity
Lead	Renal, central nervous system impairment
Mercury	Central nervous system impairment
Brominated flame retardants	Thyroid disorders, brain development

SOURCE: Adapted from U.S. Department of Health and Human Services, Blueprint for action on breastfeeding.[78]

presence of these contaminants on the well-being of the mother and the infant is unknown. Several recent studies[66,72-75] address concerns about potential impact on duration of lactation, neurodevelopmental and immunologic outcomes, and carcinogenic effects. While the body of research is growing, huge gaps in the current knowledge of any ill effects remain. The body of evidence supports the benefits of breastfeeding over any potential risks from environmental chemicals. In fact, other factors in breast milk may have a protective effect on normal neurologic development and immunologic outcomes.[63,66] At this time, there are no established "normal" or "abnormal" levels in breast milk for clinical interpretation, and breast milk is not routinely tested for environmental exposures.[66,75,76]

Unless the mother has a high level of occupational exposures, extreme dietary exposures (e.g., from fish in contaminated waters), or unusual residential exposures to hazardous or toxic chemicals, breastfeeding remains overwhelmingly the preferred choice compared with breast-milk substitutes.[78-80] The World Health Organization,[79] the American Academy of Pediatrics,[80] the U.S. Department of Health and Human Services,[78] and other major health organizations overwhelmingly support the importance of breastfeeding even in a contaminated world. The benefits of breastfeeding, which include high levels of antioxidants, may prove to be essential to compensate for and outweigh the risk of toxic effects from the environment.

Currently the focus of scientific concerns is being directed toward removing potentially toxic substances from the environment and establishing a U.S. Breast Milk Monitoring Program to track trends in exposure levels over time.[75,81] Encouragingly, data from several countries show a decline in the level of DDT and dioxin metabolites in human milk.[81] Unfortunately, the presence of other persistent chemicals, such as flame-retardant chemicals, is increasing.[74]

Women should be advised about how to reduce exposures that may affect breast milk quality rather than abandoning breastfeeding for artificial methods (Table 7.9). Women should avoid fish such as swordfish and shark or freshwater fish from waters reported as contaminated by local health agencies and limit exposure to chemicals such as pesticides and solvents found in paints, non-water-based glues, furniture strippers, nail polish, and gas fumes.

Table 7.9 Steps a breastfeeding mother can take to reduce exposure to environmental chemicals

1. Avoid smoking cigarettes and drinking alcohol. Some studies have found levels of contaminants are higher in those who smoke and drink alcoholic beverages.

2. Be aware in purchasing homes that some houses or apartments, especially those built before 1978, might have lead-based paints.

3. In general, eat a variety of foods low in animal fats; remove skin and excess fat from meats and poultry. Avoiding high-fat dairy products may reduce the potential burden of fat-soluble contaminants. Avoid processed foods made from ground meat and animal parts such as sausage and hot dogs.

4. Increase consumption of grains, fruits, and vegetables. Thoroughly wash and peel fruits and vegetables to help eliminate the hazard of pesticide residues on the skin. When available, eat food grown without fertilizer or pesticide application. Eat organically grown food, if available.

5. Avoid fish that may have high mercury levels, such as swordfish, shark, tuna, king mackerel, tilefish, and locally caught fish from areas with fish advisories.

6. Limit exposure to chemicals such as solvents found in paints, non-water-based glues, furniture strippers, nail polish, and gasoline fumes.

7. If you work with solvents (in the workplace or at home), postpone breastfeeding for several hours after exposure.

8. Run tap water through a home filter before drinking. Filters can reduce levels of common tap water pollutants.

9. Remove the plastic cover of dry-cleaned clothing. Air out the garments in a room with open windows, or hang dry-cleaning outside for 12–24 hours.

10. Attempt to avoid occupational exposure to chemical contaminants in the workplace, and seek improved chemical safety standards for all employees, especially pregnant and lactating women. Workers should be diligent in following their workplaces' safety recommendations.

11. Alert other family members to be sensitive to contaminant residue they may inadvertently bring into the home. It is possible, for example, to carry PCBs home on clothes, body, or tools. If this is the case, the individual should shower and change clothing before leaving work and keep and launder work clothes separate from other clothing.

12. Review additional suggestions for avoiding chemical exposures in and around the home from the report from the Enviornmental Working Group (see www.ewg.org/reports/mothersmilk/part5.php).

Neonatal Jaundice and Kernicterus

"The AAP discourages the interruption of breastfeeding in healthy term newborns and encourages continued and frequent breastfeeding (at least eight to twelve times every 24 hours)."

American Academy of Pediatrics[82]

Neonatal jaundice is a yellow discoloring of the skin caused by too much bilirubin in the blood (**hyperbilirubinemia**). It is a common and usually benign condition that resolves on its own or with minimal intervention. At least 60% of full-term and 80% of preterm infants will become visibly jaundiced[83] with their serum bilirubin levels exceeding 5 to 7 mg/dl (85 to 199 µmol/L).[84] If hyperbilirubinemia does not resolve and becomes sufficiently severe, the elevated bilirubin levels can cause permanent neurological damage.[83-86]

> **Hyperbilirubinemia** Elevated blood levels of bilirubin, a yellow pigment that is a byproduct from the breakdown of fetal hemoglobin.

In recent years, the overall incidence of infant jaundice has risen,[85] and hyperbilirubinemia is the most frequent cause for hospital readmission during the first two weeks of life in the United States.[87] More infants are becoming jaundiced, and their jaundice is more severe. The Joint Commission on Accreditation of Healthcare Organizations (JCAHO),[88] the Centers for Disease Control, and the American Academy of Pediatrics[82] have all noted the increasing rates and the need for prevention, early detection, and prompt treatment. Risk factors for the development of severe hyperbiliruminia have been identified (see Table 7.10). Higher rates of breastfeeding in conjunction with shorter postpartum hospital stays is the leading explanation for the higher prevalence of neonatal jaundice.[85,89]

It is important for all health professionals to understand the causes of, risk factors for, and early signs of hyperbilirubinemia. Preventing toxicity from excessive jaundice and protecting and ensuring successful breastfeeding require an understanding of the normal and abnormal patterns and mechanisms of jaundice in the newborn period, particularly the mechanisms related to human milk intake.[90]

Bilirubin Metabolism

Bilirubin is a byproduct of the normal physiologic degradation of hemoglobin. Most hemoglobin in the neonate is derived from fetal erythrocytes. Since higher levels of hemoglobin are necessary *in utero* to carry the oxygen delivered to the fetus by the placenta, the normal full-term infant has a hematocrit of 50–60%. As soon as the infant is born and begins to breathe room air, the need for high levels of hemoglobin is gone, and excess erythrocytes

Table 7.10 Risk factors for severe hyperbilirubinemia[82]

Major Risk Factors

Predischarge total serum bilirubin (TSB) or transcutaneous bilirubin (TcB) in the high-risk zone

Jaundice observed in the first 24 hours of life

Blood group incompatibility with positive direct antiglobulin test, other known hemolytic disease, elevated ETCO

Gestational age 35–36 weeks

Previous siblings received phototherapy

Cephalohematoma or significant bruising

Exclusive breastfeeding, particularly if nursing is not going well and weight loss is excessive

East Asian race

Minor Risk Factors

Predischarge total serum bilirubin (TSB) or TcB in the high intermediate risk zone

Gestational age 37–38 weeks

Jaundice observed before discharge

Previous sibling with jaundice

Macrosomic infant of a diabetic mother

Maternal age ≥25 years

Male gender

Decreased Risk Factors

TSB or TcB level in the low-risk zone

Gestational age ≥41 weeks

Exclusive bottle feeding

Black race

Discharge from hospital after 72 hours

Oxytocin used in labor

ABO incompatibility

SOURCE: Reproduced with permission from Pediatrics, 114:297–316, Copyright © 2004 by the AAP.

are destroyed. The released hemoglobin is broken down by the reticuloendothelial system; bilirubin, an insoluble byproduct of the breakdown of hemoglobin, is released into the circulation bound to albumin or another transport protein. The insoluble form of bilirubin is removed from circulation by the liver, which conjugates bilirubin to a water-soluble form and excretes it via the bile to the stool. The balance between liver cell uptake of bilirubin, the rate of bilirubin production, and the rate of bilirubin resorption through the intestines determines the total serum bilirubin (TSB) level.

Before birth, the maternal liver is responsible for metabolism and clearance of fetal bilirubin. After birth, unique developmental factors that control the production, conjugation, and excretion of bilirubin predispose the neonate to hyperbilirubinemia:[11,85,91,92]

- Bilirubin production in the neonate is double that of an adult because of breakdown of fetal erythrocytes.
- Uptake of insoluble bilirubin by the liver is limited because of a reduction in the concentration of ligandin, a bilirubin-binding protein in the liver cell.
- Conjugation to a water-soluble form is limited in the liver because of deficient activity of uridine diphosphoglucuronosyl transferase (UDPGT), a liver enzyme responsible for bilurubin conjugation.
- Excretion of bilirubin is delayed because of an enzyme present in the intestine of the newborn, beta glucuronidase, which converts conjugated bilirubin back into its unconjugated state, which is reabsorbed.

Physiologic Versus Pathologic Newborn Jaundice

After the first 24 hours, rising bilirubin levels in healthy term infants are reflective of the physiological breakdown of fetal hemoglobin, the increased resorption of the bilirubin from the intestines, and the limited ability of the newborn's immature liver to process large amounts of bilirubin as effectively as a mature liver. Neonates tend to produce more bilirubin than they can eliminate. Prematurity magnifies this imbalance. Retention of unconjugated bilirubin by the newborn is known as normal newborn jaundice or physiologic jaundice of the newborn.[90] Excessive bilirubin is deposited in various tissues, including the skin, muscles, and mucous membranes of the body, causing the skin to take on a yellowish color. In healthy newborns, this condition is temporary and usually resolves within a few days without treatment. In the typical newborn population, bilirubin levels rise steadily in the first three to four days of life and peak around the fifth day of life, and then decline. Bilirubin levels of healthy preterm infants peak later (day 6 to day 7) and take longer to resolve. Bilirubin levels in physiologic jaundice are usually less than 12 mg of bilirubin per dl of blood of infants of white or black mothers, and average 10 to 14 mg in infants of Asian ancestry, including Chinese, Japanese, and Korean, and Native Americans.[93] Levels in white and black mothers also peak earlier than in Asian or Native American mothers.[94]

In contrast to physiologic jaundice of the newborn, pathologic jaundice begins earlier (sometime it is observed before 24 hours of age), rises faster, and lasts longer. A TSB greater than 8 mg/dl in the first 24 hours should be investigated for pathologic origin. Causes of pathologic jaundice include the following:[84]

- Hemolytic disease (immune disorders, Rh isoimmunization, ABO or minor blood type incompatibility)
- Erythrocyte disorders (glucose-6-phosphate-dehudrogenase deficiency, hereditary spherocytosis)

- Extravasation of blood (cephalohematoma, subgaleal hemorrhage, bruising)
- Inborn errors of metabolism/conjugation defects (galactosemia, Crigler-Najjar syndrome types I and II, Gilbert's syndrome, Lucey-Driscoll syndrome)
- Hypothyroidism
- Polycythemia
- Macrosomic infant of diabetic mother
- Intestinal obstruction; delayed passage of meconium
- Sepsis

In most cases of pathological jaundice, frequent breastfeeding (8 to 12 times every 24 hours) can continue during diagnosis and treatment of pathological jaundice.[82,89] An advantage of colostrum and mature human milk is stimulation of bowel movements, speeding elimination of bilirubin. However, in jaundice caused by galactosemia, breastfeeding is contraindicated.[11,18]

Since bilirubin is a cell toxin, concern arises when TSB elevates to levels with the potential to cause permanent damage. The brain and brain cells, if destroyed by bilirubin deposits, do not regenerate.[85] ***Bilirubin encephalophathy,*** or ***kernicterus,*** has a mortality rate of 50%, and survivors usually are burdened with severe problems including cerebral palsy, hearing loss, paralysis of upward gaze, and intellectual and other handicaps.[74] While full-scale kernicterus is rare, there has been an increase in reported cases in the last two decades.[85,86] There is also concern that mild effects of bilirubin on the brain may be manifested clinically in later life with symptoms such as incoordination, excessive contractions of muscles (hypertonicity), and mental retardation, or perhaps learning disabilities.[18,86]

> **Kernicterus or Bilirubin Encephalopathy** The chronic and permanent clinical sequelae that are the end result of very high untreated bilirubin levels. Excessive bilirubin in the system is deposited in the brain, causing toxicity to the basal ganglia and various brainstem nuclei.[82]

Hyperbilirubinemia and Breastfeeding

Jaundice in the breastfed infant has been divided into types, early or late, based on the age of onset (Table 7.11). It is important to differentiate between the two to establish effective prevention and treatment. Early onset of elevated unconjugated bilirubin unexplained by other pathologic factors is associated with inadequate feeding and is called "breastfeeding jaundice" or, more precisely, "breast-nonfeeding jaundice."[90] Late onset (after day 5) prolonged elevation of unconjugated bilirubin associated with the ingestion of breast milk is called breast milk jaundice.[90,95]

Breast-Nonfeeding Jaundice The optimally breastfed infant initiates breastfeeding in the first hours, followed by at least 8 to 12 breastfeeds per day for the first 1 to

Table 7.11 Comparison of early and late jaundice associated with hyperbilirubinemia while breastfeeding

	Early Jaundice	Late Jaundice
Onset	Occurs 2–5 days of age	Occurs 5–10 days of age
Duration	Transient: 10 days	Persists >1 month
Incidence	More common with first child; approximately 60% of U.S. newborns become clinically jaundiced	All children of a given mother
Feeding	Infrequent feeds Receiving water or dextrose water	Milk volume not a problem
		May have abundant milk supply
		No supplements
Stools	Stools delayed and infrequent	Normal stooling
Total Serum Bilirubin	Peaks <15 mg/dL None or phototherapy	May be >20 mg/dL Treatment: phototherapy Discontinue breastfeeding temporarily Rarely, exchange transfusion.
Associations	Low Apgar scores, water or dextrose water supplement, prematurity	None identified

SOURCE: Adapted from **Breastfeeding: A Guide for the Medical Profession**, 6th ed., by R. A. Lawrence and R. M. Lawrence, 2005, with permission from Elsevier.

Table 7.12 Management outline for early jaundice while breastfeeding (breast-nonfeeding jaundice)

1. Monitor all infants for initial stooling. Stimulate stool if no stool in 24 hours.
2. Initiate breastfeeding early and frequently. Frequent, short feeding is more effective than infrequent, prolonged feeding, although total time may be the same.
3. Discourage water, dextrose water, or formula supplements.
4. Monitor weight, voidings, and stooling in association with breastfeeding pattern.
5. When bilirubin level approaches 15 mg/dl, stimulate stooling, augment feeds, stimulate breast milk production with pumping, and use phototherapy if this aggressive approach fails and bilirubin exceeds 20 mg/dl.
6. Be aware that no evidence suggests early jaundice is associated with "an abnormality" of the breast milk, so withdrawing breast milk as a trial is only indicated if jaundice persists longer than 6 days or rises above 20 mg/dl or the mother has a history of a previously affected infant.

SOURCE: Reprinted from Breastfeeding: A Guide for the Medical Professions, 5th Ed. by R.A. Lawrence and R.M. Lawrence, Table 14-4 © 1999, with permission from Elsevier.

2 weeks without any water or other food supplementation, and with good positioning that assures effective milk transfer to the infant, and weight loss of less than 8% of birth weight.[88,89] Differences in bilirubin levels between adequately breastfed infants and formula-fed infants have not been found to be significant.[90] In contrast, infants who nurse infrequently or inefficiently ingest fewer calories and lose more weight than formula-fed infants are at risk for elevated bilirubin levels. Suboptimal breastfeeding can delay passage of meconium and reduce fecal weight, increasing the hepatic circulation of bilirubin. In addition, reduced milk intake produces a state of partial starvation in the infant, which further increases the intestinal absorption of bilirubin.[96] Delay in initiation of breastfeeding beyond the first hour of life, and administration of water to infants either before initiation of breastfeeding or in addition to breastfeeding, significantly reduce the frequency of breastfeeding and increase bilirubin concentrations. Excessive hyperbilirubinemia causes lethargy and poor feeding in some infants, further reducing breastfeeding frequency and duration and milk production, feeding a vicious cycle of increasing bilirubin levels.

It is now well established that early-onset breastfeeding jaundice results from reduced volume of milk transfer to the infant, limiting caloric intake and producing a state of partial starvation and weight loss equivalent of the adult disorder known as starvation jaundice. Lawrence outlines treatment guidelines (Table 7.12) aimed at treating the actual cause—that is, failed breastfeeding or inadequate stooling or underfeeding.[18] The goal is to evaluate the breastfeeding for frequency, length of suckling, and apparent supply of milk, and then adjust the breastfeeding to solve the problem.[96] If stooling is an issue, the infant should be stimulated to stool. If starvation is the problem, the infant should receive temporary supplemental feeding by cup or bottle while the milk supply is being increased by better breastfeeding techniques.

Early discharge of infants from the hospital at less than 72 hours of life has raised concerns about the ability to evaluate breastfeeding and the opportunity to evaluate infants for jaundice.[82,85,89,92] Formal observation of breastfeeding with evaluation of effectiveness of breastfeeding and milk transfer at regular intervals throughout the first days of life can identify breastfeeding problems sufficiently early to ensure correction of problems. The American Academy of Pediatrics strongly recommends that all breastfed infants be evaluated by a trained observer within two to three days after discharge from the hospital.[82] Particular attention must be paid to infants who are <38 weeks gestation and infants at moderate or high risk for severe hyperbilirubinemia (Table 7.10).[82,92]

Breast Milk Jaundice Syndrome   In contrast to physiological newborn jaundice, which peaks in the third day and then begins to drop, breast milk jaundice syndrome becomes apparent after the third day, and bilirubin levels may peak any time from the seventh to the tenth day, with untreated cases being reported to peak as late as the fifteenth day.[18] In breast milk jaundice syndrome, no correlation exists with weight loss or gain, and stools are normal. Initially breast milk jaundice syndrome was thought to be an unusual and distinct type of newborn jaundice affecting 1% of all breastfed neonates. More recent research reports demonstrated that at least one-third of all breastfed infants are clinically jaundiced in the third week of life and that two-thirds have significant unconjugated hyperbilirubinemia in the third week, in contrast to the absence of hyperbilirubinemia in the third week in full-term artificially fed infants.[11,95] What was once believed to be a clinical disorder is now recognized as a normally occurring extension of physiolgic jaundice of the newborn.[94,95] However, at this time there is insufficient evidence to support the popular theory that breast milk jaundice may provide protective effects for newborns by the antioxidant effects of bilirubin, compensating for the relative deficiency of endogenous antioxidants in newborns.[90]

The undisputed cause of breast milk jaundice syndrome is unresolved. It is believed to be caused by a combination of factors: a substance in most mothers' milk that increases intestinal absorption of bilirubin and individual variations in the infant's ability to process bilirubin.[11,18,90] To treat severe elevations in unconjugated bilirubin in breast milk jaundice syndrome, the AAP Clinical Practice Guidelines for Hyperbilirubinemia[82] are applied with the goal of promptly lowering TSB bilirubin levels substantially to limits set for the infant's age and level of risk. To establish the diagnosis of breast milk jaundice firmly when the bilirubin level is above 16 mg/dl for more than 24 hours, a short, temporary interruption of breastfeeding (12 to 24 hours) while monitoring bilirubin levels is recommended.[18,90]

The belief that severe hyperbilirubinemia in breastfed infants cannot result in kernicterus is erroneous and dangerous; 98% of the 105 cases of kernicterus in the U.S. Kernicterus Registry were breastfed infants.[96] Kernicterus, while rare, can develop in otherwise healthy full-term breastfed newborns or breastfed infants with sepsis.

Breastfed infants need to be followed closely, supported effectively, and evaluated appropriately in order to avoid rare cases of severe hyperbilirubinemia and kernicterus.[82]

Interrelationships between Breast-Nonfeeding Jaundice and Breast Milk Jaundice Syndrome   While breast-nonfeeding jaundice and breast milk jaundice are two separate entities, they can have an interactive effect on each other. Infants with breast milk jaundice who manifest higher levels of bilirubin in the second and third weeks of life, often over 15 mg/dl, have been noted to have had relatively high serum bilirubin concentrations during the first 3 to 5 days of life due to breast-nonfeeding jaundice, hemolysis, or unknown etiology.[95] Gartner postulates that these early, elevated bilirubin levels may produce an enlarged bilirubin pool. Then the ingestion of mature milk and a consequent enhancement of the enterohepatic circulation may enlarge the pool even further.[95]

Prevention and Treatment for Severe Jaundice

The American Academy of Pediatrics guidelines for the management of hyperbilirubinemia in healthy term newborns include a detailed algorithm for the management of jaundice in the newborn nursery and guidelines for initiating phototherapy.[82] Phototherapy involves placing the newborn under special fluorescent lights that, like sunlight, assist in removing jaundice from the skin. The light is absorbed by the bilirubin, changing it to a water-soluble product, which can then be eliminated without having to be conjugated by the liver.

Historically, treatment for jaundice in American hospitals involved phototherapy and discontinuing breastfeeding either permanently or until the bilirubin levels were acceptable. In addition, many health professionals believed that newborn infants would become dehydrated if they were not supplemented with water or formula during the first days of breastfeeding. Recent research shows these practices are counterproductive.[82,90] The benefits of early and frequent breastfeeding in the first days of life for prevention of hyperbilirubinemia through maintaining hydration and stimulating the passage of stool are now well documented. The passage of stool in the newborn is important because there are 450 mg of bilirubin in the intestinal tract *meconium* of the average newborn.[18] To avoid reabsorption of bilirubin from the gut into the serum, passing meconium in the stool is critical. The current AAP hyperbilirubinemia management guidelines discourage the interruption of breastfeeding in healthy term newborns and encourage continued and frequent breastfeeding (at least 8 to 12 feedings every 24 hours). The AAP recommends against routine supplementation of nondehydrated breastfed infants with water

Meconium   Dark green muclaginous material in the intestine of the full-term fetus.

or dextrose water, as this practice will not prevent hyperbilirubinemia or decrease total serum bilirubin levels.[82]

Information for Parents

Health professionals should convey a balanced approach when communicating with parents about jaundice. Parents need to know that most breastfed infants will become jaundiced and that the overwhelming majority of cases will be benign. Only a small fraction of these infants are at risk for developing extreme hyperbilirubinemia and kernicterus. However, parents also need to fully understand the serious consequences of extremely elevated bilirubin levels and should have their infant evaluated by a health professional if jaundice develops. Health professionals need to understand that feelings of guilt are common among mothers of jaundiced infants as many mothers feel that they caused the jaundice by breastfeeding.[97] By providing accurate information and encouragement to breastfeed, health professionals have great impact on whether a mother continues breastfeeding after her experience with neonatal jaundice.

Breastfeeding Multiples

Since 1980, the birthrate of twins has increased by 59%, and the birthrate of higher-order multiples (triplets and more) has increased by over 400%. In the United States multiples currently represent over 3% of live births. The benefits of breastfeeding to mother and infant are multiplied with twins and higher-order multiples, who often are born at risk.[99] History and numerous case reports[100-102] provide ample evidence that an individual mother can provide adequate nourishment for more than one infant. In seventeenth-century France, wet nurses in foundling homes fed 3 to 6 infants, who were often of differing ages with different daily requirements.[18,103] Breastfeeding initiation rates of nearly 70% have been reported by surveys of members of Mothers of Super Twins (MOST), Parents of Multiple Births Association (POMBA) of Canada, and Double Talk, a newsletter for parents of multiples. Some mothers of triplets and quadruplets have fully breastfed their babies.[11,101]

Frequency and effectiveness of breastfeeding are the keys to building a plentiful milk supply. The more often a baby nurses, the more milk there will be.[18,99] Mothers who exclusively breastfeed twins or triplets can produce 2 to 3 liters/day, although this involves nursing an average of 15 or more times per day.[104] The main obstacle to nursing multiples is not usually the milk supply, but time and fatigue of the mother. Parents of twins and higher-order multiples need support in four major areas: organization, feeding, individualization, and stress management.[99]

Mothers of twins or higher-order multiples often face special challenges in the establishment of lactation after birth. Approximately 60% of twins and 90% of higher-order multiples in the United States are born at less than 37 weeks gestation,

Allergic Diseases Conditions resulting from hypersensitivity to a physical or chemical agent.

and premature multiples often experience medical complications that potentially interfere with breastfeeding.[106] Breastfeeding initiation may take place in the neonatal intensive care unit, usually because of prematurity and low birth weight. Mothers of multiples may be coping with the effects of a more physically demanding pregnancy and birth or complications of pregnancy. Mothers may experience exaggerated postpartum sleep deprivation related to round-the-clock care of two or more newborns, concern for sick newborns, or staggered infant discharge, which results in time divided between infants at home and in the hospital. In addition, every aspect of breastfeeding management is affected by the dynamics that multiple newborns create.[18,99] In a recent study, breastfeeding initiation rates as high as 73% were found among full-term multiples compared with initiation rates of only 57% in preterm multiples. Breastfeeding duration was also lower in preterm multiples than in term multiples (12 weeks versus 24 weeks).[11,106]

Health care professionals can help the mothers of multiples face the many challenges of breastfeeding by offering consistent, informed, individualized care and support in the hospital and after discharge.[104] Knowledgeable care providers can help parents distinguish between multiples-specific issues, normal variations in an individual infant's breastfeeding abilities and patterns, and actual breastfeeding problems. Mothers need information on when and how to initiate simultaneous feedings, practical tips for managing night-time nursing and fatigue, and how to assure herself that her babies are getting ample nourishment.[99] Parents need to be informed of resources for parenting multiples and for receiving support for breastfeeding in their community (see Resources at the end of the chapter).

A well-defined plan for the health care of the lactating woman that includes screening for nutritional problems and providing dietary guidance is also important.[58] Mothers should be encouraged to drink to satisfy their thirst, to eat nutritious foods, and to sleep when the babies sleep. As in singleton nursing, women nursing multiples should be encouraged to obtain their nutrients from a well-balanced, varied diet rather than from vitamin–mineral supplements.

Infant Allergies

Protection from *allergic diseases* is one of the most important benefits of breastfeeding. There are numerous studies in this area being done at the present time. Many studies have shown exclusive breastfeeding for at least 4 months can protect against ectopic dermatitis and wheezing illnesses in children for the first 10 years of life.[16,107–109] Other studies have shown no effect on allergy.[16] Elimination of major food allergens from the diet of breastfeeding mothers of infants at high risk for atopic disease was previously recommended to delay or prevent some food allergies and atopic dermatitis. However, recent expert reviews concluded that there is no strong evidence to support this recommendation.[107,109,110] A recent well-controlled study confirmed that the presence of food proteins in human milk is common, but this can be highly variable between women consuming the same challenge (dose) food.[110] A number of other studies measuring proteins in human milk following challenge doses of eggs, cow's milk, gluten, and wheat protein also found variable response between women.[108,109,110]

The development of infant *food allergy* is influenced by genetic risk for allergy, duration of breastfeeding, time for introduction of other foods, maternal cigarette smoking during pregnancy and parental smoking, air pollution, exposure to infectious disease, and by maternal diet and immune systems.[108] Several mechanisms (Table 7.13) are thought to contribute to the protective effect of breastfeeding. There is preliminary evidence that increasing the breastfeeding mother's dietary omega-3 (n-3) polyunsaturated fatty acids may offer protection from some childhood allergies.[110]

Common pediatric food allergens include:

- Cow's milk
- Wheat
- Eggs
- Peanuts
- Soybeans
- Tree nuts (e.g., almonds, Brazil nuts, walnuts, hazelnuts)

Infants with a positive family history of allergies should be exclusively breastfed for at least 4 to 6 months with continuance of breastfeeding for as long as possible.[18,25,108] Advice to breastfeeding mothers with a family history of allergies regarding elimination of common allergens in their own diets should be individualized. Although several studies have documented the presence of food allergens—particularly milk and peanut protein—in breast milk sufficient to induce food reactions in infants, the amounts are variable. The American Academy of Pediatrics specifically suggests restricting peanuts and tree nuts in the maternal diet of high-risk infants.[25] If there is no history of allergy to a specific food in the mother's or father's family, avoiding a food because it is a potential allergen is an unnecessary precaution. Only if there is a family history of

Table 7.13 Possible reasons for allergy preventive effects of breastfeeding[101,102]

- Low content of allergens
- Transfer of maternal immunity
- Long-chain fatty acids and IGA in breast milk protect against inflammation and infections
- Regulation of infant immunity
- Influence on gut microbial flora

allergies, or if a baby shows allergic symptoms, should a mother consider avoiding certain foods. If a mother avoids certain foods, care must be taken to ensure that her diet remains nutritionally adequate.[111]

Food Allergy (Hypersensitivity) Abnormal or exaggerated immunologic response, usually immunoglobulin E (IgE) mediated, to a specific food protein.

Food Intolerance An adverse reaction involving digestion or metabolism but not the immune system.

Food Intolerance

Although infants may have sensitivities to certain foods, there is no scientific basis for the concern about gassy foods, such as cabbage or legumes, causing gas in the breastfed baby. In mothers, the normal intestinal flora produces gas from the action on fiber in the intestinal tract. Neither the fiber nor the gas is absorbed from the intestinal tract, and neither enters the mother's milk. Likewise, the acid content of the maternal diet does not affect the breast milk because it does not change the pH of the maternal plasma.

It is a common practice for the mother who is breastfeeding her child with colic to believe her child's symptoms are caused by a *food intolerance*. The role of diet in infantile colic, which affects up to 28% of infants in the first months of life,[112] is controversial. While breastfeeding is not protective against infantile colic, several studies report a reduction in persistent crying after elimination of cow's milk and other food proteins from the maternal diet. In a recent randomized, controlled trial, a low-allergen maternal diet was associated with a reduction in distressed behavior among breastfed infants with colic.[112] The mothers excluded cow's milk, eggs, peanuts, tree nuts, wheat, soy, and fish from their diet.

Characteristic essential oils in foods such as garlic and spices may pass into the milk, and an occasional infant objects to their presence. Studies by Mennella and Beauchamp confirm that the diet of lactating women alters the sensory qualities of her milk.[113,114] Extensive clinical experience also suggests that some infants are sensitive to certain foods in the mother's diet. According to Lawrence, garlic, onions, cabbage, turnips, broccoli, beans, rhubarb, apricots, or prunes may be bothersome to some infants, making them colicky for 24 hours.[18] In the summer, a heavy diet of melon, peaches, and other fresh fruits may cause colic and diarrhea in the infant. Red pepper has been reported to cause dermatitis in the breastfed infant within an hour of milk ingestion.[115] Contrary to popular belief, chocolate rarely causes problems and can be consumed in moderation without causing colic, diarrhea, or constipation in most infants.[18]

If a mother suspects that her baby reacts to a specific food, it may be helpful for her to keep a record of foods eaten, along with notes on the baby's symptoms or behavior. If highly allergic or sensitive, infants may react to foods their mothers have eaten within minutes, although symptoms generally show up between 4 and 24 hours after exposure. While symptoms will improve in most infants after the offending food has been removed from

Case Study 7.2

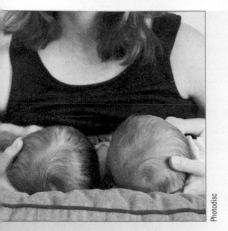

Photodisc

Breastfeeding Premature Infants

Thirty-five year-old Stacey delivers twin boys at 30 weeks gestation: baby Andrew is 2 pounds, 9 ounces and 13.5 inches long; baby Mark is 1 pound, 13 ounces and 14 inches long. Stacey had a difficult pregnancy that included severe nausea and vomiting, heartburn, preeclampsia, and preterm labor.

Stacey is very committed to breastfeeding and was able to use a hospital-grade electronic breast pump approximately 6 hours after delivery. She pumps every 2–3 hours or 8–9 times daily in order to establish her milk supply. At the end of the first week, Stacey is pumping about 14–16 oz/day per baby, and by 3 1/2 weeks she is pumping 4–5 oz per breast at each pumping. Stacey wakes at night to pump when she is full. She has placed the pumping equipment by the bed and become adept at pumping, getting out of bed only to put the milk in the refrigerator. At 2 weeks postpartum, Stacey experiences a plugged duct and has difficulty emptying the right breast for 2 days.

The twin boys have suffered the usual preterm difficulties with breathing, apnea, and bradycardia. Initially the twins are tube-fed. As their condition improves, baby Andrew is first put to breast 3 weeks after his birth, and baby Mark several weeks later. Baby Mark has more difficulty learning to latch on and suck and is growing more slowly than is his brother. Multiple interventions are used to achieve breastfeeding success. On advice from a lactation consultant, a nipple shield helps baby Mark latch on. In response to slow weight gain in baby Mark, the lactation consultant recommends that the baby receive hindmilk, which is often higher in fat and calories.

Mark and Andrew are released from the hospital a day after their due date. Stacey continues using the nipple shield for several weeks with Mark, trying without the nipple shield every few days. After 3 weeks at home, baby Mark is able to latch on without the nipple shield. The twins are breastfed and also receive up to 3 bottles of fortified expressed breast milk or premature infant formula per day for the first 2 months at home. The babies take feedings equally well from bottle or breast.

Questions

1. How often should the mother be pumping to establish and maintain a full milk supply?
2. List at least two possible nutrition diagnoses for this case.
3. Identify at least one nutrition intervention for each diagnosis listed.
4. Name potential indicators for each intervention listed.

(Answers are located in the Instructor's Manual for the 4th edition of *Nutrition Through the Life Cycle*.)

---

the mother's diet for 5 to 7 days, it may take 2 weeks to totally eliminate all traces of the offending substance from both the mother and baby[116] (see Case Study 7.2).

Late-Preterm Infants

Infants born between 34 and 37 weeks are considered late-preterm. They account for as high as 75% of all preterm singleton births.[117] These infants too often are treated as full-term; they often have subtle immaturity that makes establishing breastfeeding difficult and places the infant

at risk for insufficient milk intake, hypoglycemia, jaundice, and poor weight gain.[16,118] Late-preterm infants have higher rates of readmission during the neonatal period compared to term infants.[119] Infants may have cardio-respiratory instability, especially in upright position; poor temperature control; lower glycogen and fat stores to prevent hypoglycemia; an immature immune system; and the suck–swallow coordination may be poorly coordinated and result in poor latch-on and milk transfer. The main emphasis in postpartum care should be on building and maintaining the milk supply and feeding the infant. If an infant

Illustration 7.1 Near-term infant breastfeeding cascade.

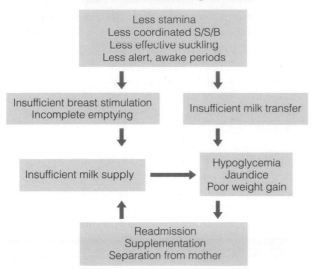

Near-Term Breastfeeding Cascade

SOURCE: Reprinted with permission from N. E. Wight, "Breastfeeding the borderline (near-term) preterm infant," in *Pediatrics Annals* 32:329–336. Copyright © 2003 Slack Inc.

is not sucking vigorously, mothers of late-preterm infants should pump after each feeding attempt, or at least every 3 hours, to build the milk supply. A specific feeding plan, including the plan to wake a sleepy infant every 2–4 hours, is recommended to avoid the late-preterm breastfeeding cascade (Illustration 7.1). Mothers and their late-preterm infants should have a feeding assessment by a trained lactation professional who can provide intervention for positioning, potential suck problems, and other breastfeeding issues. Weighing an infant before and after feedings may be useful to determine milk transfer. Discharge planning should include follow-up care, home visits, and lactation counseling as needed. Close follow-up should be continued until the infant is gaining weight and the mother is comfortable.[118]

Human Milk and Preterm Infants

"Hospitals and physicians should recommend human milk for premature and other high-risk infants either by direct breastfeeding and/or using the mother's own expressed milk. Maternal support and education on breastfeeding and milk expression should be provided from the earliest possible time. Mother-infant skin-to-skin contact and direct breastfeeding should be encouraged as early as feasible. Fortification of expressed human milk is indicated for many very low birth weight infants."

American Academy of Pediatrics[47]

The benefits of breastfeeding may be most visible among preterm infants who are born immature and without adequate stores of nutrients. The nutritional benefits include

ease of protein digestion, fat absorption, and improved lactose digestion.[120–123] The known health and developmental benefits include better visual acuity, greater motor and mental development at 1.5 years of age, greater verbal intelligence quotient at 7–8 years of age, and a lower incidence of serious infectious disease, including necrotizing enterocolitis and sepsis, even among infants who also receive some human milk substitutes.[122–124] Nosocomial infection rates may also be decreased by as much as 50% in premature infants who receive at least 50% of their daily feedings from mother's milk.[124] Lower weight and length gains and poorer bone mineralization has been reported for preterm infants fed human milk.[125]

The composition of milk from women who deliver preterm infants is higher in protein, slightly lower in lactose, and higher in energy content (58–70 Kcal/100 mL) compared to the milk of women who deliver full-term infants (approximately 62 Kcal/100 mL). Once growth is established, the nutritional needs of the preterm infant exceed the content of human milk for protein, calcium, phosphorus, magnesium, sodium, copper, zinc, riboflavin, pyridoxine, folic acid, and vitamins C, D, E, and K.[25] Human milk fortifiers are available that provide additional protein, minerals, and vitamins.[25] Infants who receive fortified human milk do not need additional supplementation unless a specific nutritional problem is identified. The health benefits, particularly reduction in sepsis and necrotizing enterocolitis, outweigh the slightly lower rate of weight gain and length gain observed among preterm infants fed fortified human milk compared to those who receive a human milk substitute for premature infants.[125]

Early feeding seems to be important for preterm infants (when medically appropriate). Early feeding may be important to the ability to digest[121] and to the development of the infant's digestive system.[126] Often, milk must be expressed and stored for preterm infants. Establishing a milk supply is important to the mother's ability to maintain a milk supply that will meet her infant's demands after several weeks (see Chapter 6, "Can Women Make Enough Milk?"). A woman who is pumping less than 750 mL of milk by 2 weeks may need additional support to establish a milk supply that will meet her infant's needs beyond the first month.

Despite the health benefits of feeding human milk to preterm infants, the incidence and duration of breastfeeding of preterm infants in the United States is about 30% lower than rates for full-term infants.[123] The challenges of feeding low-birth-weight infants include provision of adequate calorie and nutrient intake, establishing and maintaining an adequate milk supply, and transitioning from gavage feeding to feeding from the breast.[123] Strategies to improve breastfeeding rates in this vulnerable population include providing the parents with information necessary to make an informed decision to breastfeed; assisting the mother with the establishment and maintenance of a milk supply; ensuring correct breast milk management (storage and handling) techniques; providing skin-to-skin (kangaroo) care and opportunities for non-nutritive sucking at the breast; managing the transition to the breast; measuring milk transfer; preparing the infant and family for discharge; and providing appropriate follow-up care.[127] Counseling mothers of very low-birth-weight infants on the benefits of breast milk and providing breastfeeding support increases breastfeeding initiation and breast milk feeding without increasing maternal anxiety and stress.[128]

In addition to improving breast-milk supply, skin-to-skin contact provides protection to the premature infant in the NICU. The lactating mother produces specific antibodies in her milk against pathogens in the infant's environment through the enteromammary system.[16]

Medical Contraindications to Breastfeeding

Few medical problems in the mother or baby are absolute contraindications to breastfeeding (Table 7.14). There are very few infectious pathogens that pose a risk to the newborn that outweighs the potential benefits of breastfeeding.[129] Even infants with metabolic disorders such as pheynylketonuria can continue to breastfeed in combination with a specialized formula to meet calorie and protein needs. When mothers or infants have medical or other problems that cause a poor suck or other feeding problems, early identification and appropriate support from a lactation consultant is necessary for successful breastfeeding. In some cases, pumping milk may be necessary to maintain a supply of milk while problems with the infant suck are addressed and corrected. Whenever a medical situation presents a potential risk for breastfeeding infants, the theoretical risk must be carefully measured against the projected benefits of breastfeeding.[130]

Breastfeeding and HIV Infection

Every year approximately 750,000 children worldwide become infected with **HIV**, mostly through mother-to-child transmission during pregnancy, delivery, or breastfeeding.[131] The transmission of HIV type 1 from mother to child through breastfeeding is well documented. Reports of transmission rates range between 5% and 20%, with prolonged breastfeeding doubling the rate to 35–40%.[131] Factors contributing to these variable rates include strain of HIV, maternal illness, immune status and viral load, duration of breastfeeding (timing of transmission), primary infection of the mother during the breastfeeding period, exclusive breastfeeding versus mixed feeding, mastitis,

HIV Human immunodeficiency virus.

AIDS Acquired immunodeficiency syndrome.

maternal vitamin deficiencies (A, C, E, or B vitamins), and the availability of antiretroviral therapy.[129–133]

In developed countries, where safe and affordable breast-milk substitutes are available, HIV-infected women should be counseled strongly not to breastfeed.[134] The U.S. Department of Health and Human Services' Blueprint for Action on Breastfeeding states that "HIV-infected women in the United States should not breastfeed or provide their breast milk for the nutrition of their own or other infants because of the risk of HIV transmission to the child" (p. 12).[78]

The choice for women with HIV in developing countries is not so clear. In most cases, breast milk substitutes are not affordable to families or to government-sponsored public health programs, and they pose a serious health risk to infants both with and without HIV.[131,133] Women and infants in developing countries who are most at risk for HIV also face poor water quality and sanitation and are at high risk for diarrheal diseases and other infections. Boiling of water for decontamination presents another obstacle, as fuel to boil water is either difficult to obtain or expensive. A WHO-sponsored meta-analysis documented a sixfold increase in mortality due to diarrheal disease in the first 6 months of life for infants who were not breastfed when compared with breastfed infants and twice the risk of pneumonia deaths.[135] In certain populations, the benefits of breastfeeding may outweigh the risks of HIV transmission. Breastfeeding is thought to be responsible for about 300,000 HIV infections per year, while UNICEF estimates that not breastfeeding is responsible for 1.5 million child deaths per year.[136] A mathematical modeling exercise that looked at the risks of infant morbidity and mortality by feeding choices suggests that exclusive breastfeeding for 6 months would be the best option in locations where infant mortality is over 40 per 1000.[136] Locations in most of sub-Saharan Africa fall into this category. Recent studies have provided strong evidence that breastfeeding does not pose any mortality or other health risks to the HIV-infected mother.[136]

Originally, the World Health Organization (WHO) recommended that in developing countries or areas where the risk of infant mortality from infection is great, breastfeeding is recommend even if the mother has **AIDS**.[137] This policy was clarified at a meeting of the WHO, the Children's Charity UNICEF, and UNAIDS in Geneva in October 2000. The policy now states: "Exclusive breastfeeding is recommended for HIV-infected women for the first six months of life unless replacement feeding is acceptable, feasible, affordable, sustainable and safe for them and their infants before that time."[138] Exclusive breastfeeding by the HIV-infected mother has been shown to reduce risk of transmission in the early months when compared to partial breastfeeding or mixed feeding.[139] A recent review of the advantages and disadvantages of replacement feeding options concluded that

Table 7.14 Summary of medical contraindications to breastfeeding in the United States

Problem	OK to Breastfeed in U.S.?	Condition
Infectious Diseases		
Acute infectious disease	Yes	Respiratory, reproductive, gastrointestinal infection
HIV	No	HIV-positive in developed countries
Active tuberculosis	Yes	After mother has received 2 or more weeks of treatment
Hepatitis		
A	Yes	As soon as mother receives gamma globulin
B	Yes	After infant receives HGIB; first dose of hepatitis B vaccine should be given before hospital discharge
C	Yes	If no co-infections (e.g., HIV)
Veneral warts	Yes	
Herpes viruses		
Cytomegalovirus	Yes	
Herpes simplex	Yes	Except if lesion on breast
Varicella-zoster (chickenpox)	Yes	As soon as mother becomes noninfectious
Epstein-Barr	Yes	
Toxoplasmosis	Yes	
Mastitis	Yes	
Lyme disease	Yes	As soon as mother initiates treatment
HTLV-I	No	
Over-the-Counter/Prescription Drugs and Street Drugs		
Antimetabolites	No	
Radiopharmaceuticals		
Diagnostic dose	Yes	After radioactive compound has cleared mother's plasma
Therapeutic dose	No	
Drugs of abuse	No	Exceptions: cigarettes, alcohol
Other medications	Yes	Drug-by-drug assessment
Environmental Contaminants		
Herbicides	Usually	Exposure unlikely (except workers heavily exposed to dioxins)
Pesticides		
DDT, DDE	Usually	Exposure unlikely
PCBx, PBBs	Usually	Levels in milk very low
Cyclodiene pesticides	Usually	Exposure unlikely
Heavy metals		
Lead	Yes	Unless maternal level >40 mg/dl
Mercury	Yes	Unless mother symptomatic and levels measurable in breast milk
Cadmium	Usually	Exposure unlikely
Radionuclides	Yes	Risk greater to bottle-fed infants
Metabolic Disorders		
Galactosemia	No	
Pheynylketonuria	Yes	Human milk supplemented with phyenylalanine-free formula

SOURCE: Adapted from **Breastfeeding: A Guide for the Medical Profession**, 6th ed., by R. A. Lawrence and R. M. Lawrence, copyright 2005, with permission from Elsevier.

microbicide treatment of breast milk shows the greatest promise. This option has been shown to be broad-spectrum, quick, inexpensive, capable of preserving the milk's nutritional and immune status, and it does not require a heat source.[141]

Detailed instructions on counseling HIV-infected women are available from WHO. All women should be encouraged to know their HIV status and seek early prenatal care. Women need to be aware of the risks of HIV transmission during pregnancy and lactation.

Human Milk Collection and Storage

"Human milk is the most appropriate food for infants, and is also used as medical therapy for older children and adults with certain medical conditions. Human milk has a long history and proven track record both as nutrition and therapy."

Human Milk Banking Association of North America

The appropriate collection and storage of human milk is important whether the milk is for the mother's own infant or to be donated. All of the collection containers used should be cleaned by dishwasher or sterilized by boiling. Hand pumps, electric handheld pumps, hospital-grade electric breast pumps, and manual expression can be used to extract the milk. The American Academy of Breastfeeding Medicine has published evidence-based guidelines for the collection and storage of human milk for home and human milk banking.[142] The Human Milk Banking Association of North America has published "Best Practice for Expressing, Storing and Handling Human Milk in Hospitals, Homes and Child Care Settings."[143] Table 7.15 presents current recommendations for milk storage for home use. How much volume should the mother be pumping? Volumes vary according to initiation and frequency of pumping. Prematurity and stress also impact production, as discussed earlier. The pumping frequency in the early days after delivery correlates to production goals. The mother who pumps 8–10 times per day on days 2 and 3 post-delivery is more likely to achieve production goals than the mother who doesn't initiate pumping until day 3 and only pumps 3–4 times per day. The mother should begin pumping 6 hours after delivery, if possible. Typical volumes to produce the first few days are 1–10cc/ pumping. By day 5–8, those volumes should increase to 1 ½ to 2 ½ ounces, and by day 10–14, the mother should be producing 2 ½ to 4 ounces/day. A full milk supply is considered 25–35 ounces per day.[144–145]

Milk Banking

"Banked human milk may be a suitable feeding alternative for infants whose mothers are unable or unwilling to provide their own milk. Human milk Banks banks in North America adhere to national guidelines for quality control of screening and testing of donors and pasteurize all milk before distribution. Fresh human milk from unscreened donors is not recommended because of the risk of transmission of infectious agents."

American Academy of Pediatrics[47]

Human milk banks provide human milk to infants who cannot be breastfed by their mother. Premature and sick infants are most likely to receive banked milk. A woman can donate milk once or on a continuing basis if her supply exceeds the demands of her infant. There is a long history of providing human milk to infants by persons other than the biological mother.[146] Wet nurses were the main source of human milk until the early 1900s for infants not fed by their biological mothers. Milk banks began in Europe and followed in the United States. Some neonatal intensive care units had informal milk banks until the 1980s. As a result of the human immunodeficiency virus, the resurgence of tuberculosis, and risks related to donors who might abuse drugs, human milk banks are now scarce in North America, but because of recognition of the importance of human milk, demand is increasing. Approximately 166,336 ounces of milk were distributed in 2007. This is an increase of 185% in distribution since 2000.

Table 7.15 Guidelines for storage of human milk for home use

Breast Milk	Room Temperature	Refrigerator	Freezer
Freshly expressed into a closed container	6–8 hr @ 78° F or lower	3–5 days @ 39° F or lower	2 weeks in freezer compartment inside refrigerator 3–6 months in freezer section of a refrigerator with separate door 6–12 months in deep freezer at 0° F
Previously frozen—thawed in refrigerator but not warmed or used	4 hr or less	Store in refrigerator 24 hr	Do not refreeze
Thawed outside refrigerator in warm water	For completion of feeding	Hold for 4 hr or until next feeding	Do not refreeze
Infant has begun feeding	Use only for completion of feeding and then discard	Discard	Discard

SOURCE: Adapted from Breastfeeding: A Guide for the Medical Professions 6th Ed., by R.A. Lawrence and R.M. Lawrence, © 2005, with permission from Elsevier.

Presently, there are 11 milk banks in North America.[147] A network of milk banks meets and shares information through the Human Milk Banking Association of North America (HMBANA). A copy of the association's guidelines for milk storage is available for a fee.

Human milk donors are chosen by their health profile. Women are carefully screened before they can donate extra milk to milk banks. Milk banks that belong to the Milk Banking Association of North America require telephone screening, a written health and lifestyle history, and verification of the health of the mother and baby by the health care provider of each. Blood samples are tested for hepatitis B, hepatitis C, HIV, HTLV, and syphilis by the milk bank. Women are not accepted if they are acutely ill, have had a blood transfusion or an organ transplant within a year, drink more than 2 ounces of liquor daily, regularly use medications or megavitamins, smoke, or use street drugs. Additionally, women who eat no animal products must take vitamin supplements with B_{12} to be eligible to donate.

Human milk is carefully pasteurized to kill any potential pathogens while preserving the nutrients and active immune properties of the milk. The North American Human Milk Banking Association communicates closely with the Food and Drug Administration to follow guidelines for use of human tissues and fluids. Human milk for milk banks is stored frozen to preserve the immunologic and nutritional components. Rigid plastic (polypropylene) containers are recommended for keeping the milk composition stable. White blood cells stick to glass, but not to plastic containers.[148]

A prescription from a physician or a hospital is needed to order milk for an infant from one of the North American Milk Banking Association milk banks. Costs are approximately $3.50 per ounce before shipping charges, significantly more than the cost of human milk substitutes.[146] Some insurance companies and Medicaid programs cover the fees when it is demonstrated that donor milk is the most appropriate therapy for a specific patient.

Model Programs

Breastfeeding Promotion in Physicians' Office Practices (BPPOP)

The American Academy of Pediatrics (AAP) receives funding support from the Maternal and Child Health Bureau (MCHB), USDHHS, and the AAP Friends of the Children Fund for this innovative program designed to boost breastfeeding promotion and support in underserved populations. Initiated in 1997, BPPOP's original mission was to improve the ability of AAP members to support new mothers and their breastfeeding infants and to encourage pediatricians to collaborate with others to develop breastfeeding promotion programs. Pediatricians enrolling in

Nick White & Carrie Beecroft/Photodisc/Getty Images

the program received a resource kit of educational materials and other strategies to more effectively promote, support, and manage breastfeeding with all families in their practice. In addition, pediatricians were provided technical assistance by telephone and e-mail regarding breastfeeding concerns from AAP staff and were encouraged to participate in community and regional collaborative action groups. After over 700 pediatricians nationwide joined the program, BPPOP was expanded in 2002 (BPPOP II) to include obstetricians, family physicians, and other health care providers and to specifically target office practices working with racially and ethnically diverse populations. A speaker's kit and materials targeting underserved populations were added to the resource kit, along with the newest strategies and opportunities for multidisciplinary networking for community breastfeeding promotion. Physicians joining the program complete a self-assessment questionnaire at the beginning and end of the program and measure the impact of breastfeeding promotion efforts by tracking the breastfeeding initiation and duration rates within their practices.

BPPOP II concluded in 2004. Gaps in the program were identified—including the need for focused training related to breastfeeding support and management. It also became evident that this training should occur well before physicians are in practice, preferably in residency and medical school. To address this shortcoming, the BPPOP program entered its third phase (BPPOP III), which aims to educate pediatric, obstetric/gynecologic, and family medicine residents. This will be done through a pilot-tested breastfeeding residency curriculum that will be distributed to residency programs in need, especially those that are composed of racially and ethnically diverse populations of residents and patients.

The Rush Mothers' Milk Club

The Rush Mothers' Milk Club at Rush Presbyterian–St. Luke's Medical Center is an evidence-based program of breastfeeding interventions for the neonatal intensive-care unit (NICU).[133] The program uses a team approach to feeding very low-birth-weight infants their own mothers' milk (OMM). The infants' mothers work in partnership with neonatologists, neonatal nurse practitioners, bedside nurses, and other health care professionals to ensure that the latest research is applied to an infant's OMM feeding plan. Interventions to sustain lactation for program participants have evolved from the evidence about barriers to providing OMM to NICU infants. They include preventing and treating low milk volume, achieving adequate infant growth on OMM feedings, and making the transition to at-breast-feedings in the NICU and post-discharge periods. Additionally, research about the effectiveness of peer support to sustain lactation has been incorporated into the program.[149]

Major program components include: (1) providing information for mothers to make an informed decision, (2) providing access to a hospital-grade electric breast pump, (3) providing skin-to-skin care and suckling at the empty breast as practice for the newborn, (4) babies feeding at the breast as soon as they are able to suck and swallow effectively, and (5) nursery staff helping the mother prepare for breastfeeding after discharge (www.rush.edu/patients/children/publications/notes/preemies.html). The club serves as a place for mothers to discuss their goals and concerns about breastfeeding. Mothers learn the value of their milk to their high-risk baby

from the Special Care Nursery Staff and from the Rush Mothers' Milk Club. The mothers learn to measure the amount of fat and calories in their own milk and learn to capture the highest-calorie portion of milk, which is usually produced during the last 10 minutes of pumping. To create a bond between the mother and baby, mothers are encouraged to use the breast pumps at the baby's bedside. Family members and friends are also encouraged to participate in the weekly Mothers' Milk Club meetings to learn the importance of breastfeeding to the high-risk infant.

The success of the Rush Mothers' Milk Club is measured by its breastfeeding initiation rates. Between 95 and 97% of all mothers who deliver high-risk infants at Rush Presbyterian–St. Luke's Medical Center begin to nurse, compared with national rates for high-risk infants of only 30 to 40%. A group of low-income African American mothers who delivered babies below 1500 grams had OMM initiation rates of 63.4%, the highest reported rates for this population in the nation.[149] The program has evidence that these high rates of initiation are due to two primary interventions: the clarity of the message that the mothers received about the importance of OMM from health care providers and their immediate access to electric breast pump rental.

Key Points

1. The majority of mothers and infants do not experience significant problems with breastfeeding. Many of the more common problems can be prevented through prenatal breastfeeding education and from informed, consistent, and individualized care and support from health professionals both in the hospital and after discharge.

2. Most medications (prescription or over-the-counter) and herbal supplements taken by the mother are excreted in her breast milk and should not be ingested until the risks to the infant are established. For most maternal conditions, required drug therapy choices are available that will not cause harm to the breastfeeding infant; recommending that a mother discontinue breastfeeding to take a medication is rarely required.

3. The level of alcohol in breast milk matches the maternal plasma alcohol levels at the time of the infant feeding; a nursing woman who drinks occasionally can limit her infant's exposure to alcohol by timing breastfeeding in relation to her drinking.

4. Regardless of feeding choice, maternal smoking presents significant health risks for infants.

5. While low levels of environmental pollutants are present in most human milk, their impact on the well-being of the mother and infant is unknown.

The World Health Organization and other scientific groups state that the advantages of breastfeeding far outweigh the potential risks from environmental pollutants and recommend breastfeeding in all but extreme circumstances.

6. A thorough understanding of the normal and abnormal patterns and mechanisms of jaundice (hyperbilirubinemia) in the newborn period is important for all health professionals to prevent toxicity from excessive jaundice and for protecting and ensuring successful breastfeeding. Early and frequent breastfeeding (at least 8 to 12 times in 24 hours) in the first days of life helps prevent hyperbilirubinemia through maintaining infant hydration and stimulating the passage of stool. The AAP recommends against routine supplementation of nondehydrated breastfed infants with water or dextrose water, as this practice will not prevent jaundice.

7. Twins and other multiples can be successfully breastfed without supplementation.

8. Exclusive breastfeeding for at least 4 months can protect against ectopic dermatitis and wheezing illnesses in children up to age 6. Advice to breastfeeding mothers with a family history of allergies regarding elimination of common allergens in their own diet should be individualized. If there is no

history of allergy to a specific food in the mother's or father's family, avoiding a food because it is a potential allergen is an unnecessary precaution.

9. Human milk is the preferred food for all premature and sick newborns, with rare exceptions.

10. In developed countries, where safe and affordable breast milk substitutes are available, HIV-infected women should be counseled strongly not to breastfeed to prevent mother-to-child transmission of HIV through breast milk.

11. Health professionals should provide breastfeeding mothers with current evidence-based

guidelines for the collection (through hand pumps, electric handheld pumps, hospital-grade electric breast pumps, or manual expression) and storage of human milk for home use or human milk banking.

12. In most situations, the medical problems of the mother or infant can be managed without discontinuing breastfeeding. Any medical decision to limit breastfeeding must be justified by the fact that the risk to the infant clearly outweighs the benefits of breastfeeding.

Review Questions

1. List three common causes of persistent nipple pain.
2. Signs of a hyperactive let-down are:
   a. Milk streams quickly
   b. Infant may be overwhelmed by the volume
   c. Infant may choke, cough, or gulp
   d. All of the above
3. List three reasons a breastfeeding mother may develop mastitis.
4. True or false: Before a mother uses galactogogues to improve supply, she should be pumping with a hospital-grade electric pump every 2–3 hours during the day and once at night or she should be pumping after each feeding.
5. True or false: The alcohol level in breast milk matches maternal plasma levels.
6. True or false: Smoking can lower milk production and decrease fat content.

7. List at least three major risk factors for developing severe hyperbilirubinenia.
8. True or False: The main obstacle to nursing multiples is milk supply.
9. Preterm infants have higher readmission rates because of:
   a. Hypoglycaemia
   b. Suck/swallow coordination
   c. Hypothermia
   d. All of the above
10. Marijuana is excreted into breast milk at what ratio compared to maternal plasma levels?
    a. Twofold
    b. Fourfold
    c. Sixfold
    d. Eightfold

Resources

The Academy of Breastfeeding Medicine
Contains evidence-based protocols.
Website: www.bfmed.org

Best Start Social Marketing
Website: www.Bestartinc.org

The Human Milk Banking Association of North America, Inc.
Represents all of the North American human milk banks that collect, pasteurize, and distribute donated mother's milk.
Website: www.hmbana.org

International Lactation Consultants Association
Website: www.ilca.org

La Leche League
Website: www.lalecheleague.org

C. Martin and N. F. Krebs
The Nursing Mother's Problem Solver. New York: Fireside Publishing, 2000.

Resources for the Parents of Multiples
La Leche League International
Website: http://www.llli.org/NB/NBmultiples.html
Granada, K. K. Mothering Multiples: Breastfeeding and Caring for Twins and More!!! (Revised Edition). Schaumburg, IL: La Leche League International, 1999.
Website: www.tripletconnection.org

Chapter 8

Infant Nutrition

Prepared by **Janet Sugarman Isaacs**

Key Nutrition Concepts

1   The dynamic growth experienced in infancy is the most rapid of any age. Inadequate nutrition in infancy, however, leads to consequences that may be lifelong, harming both future growth and future development.

2   Progression in feeding skills expresses important developmental steps in infancy that signal growth and nutrition status.

3   Nutrient requirements of term newborns have to be modified for preterm infants. Knowing the needs of newborn infants who are ill or smaller than normal results in greater understanding of the complex nutritional needs of all newborns and infants.

4   Changing feeding practices, such as the care of infants outside the home and the early introduction of foods, markedly affects nutritional status of infants.

Introduction

This chapter is about healthy *full-term infants* born at 37 weeks of gestation or later, and healthy *preterm infants* born at 34 weeks or later. These newborns are expected to have typical growth and development. The term *normal* is not used much in this chapter, because its opposite, *abnormal,* is an emotionally laden term, particularly when describing babies to their parents. *Typical* is used in place of *normal* when possible.

This chapter discusses how nutrition is an important contributor to the complex development of infants. Both biological and environmental factors interact during infant growth and development. Models about the interaction of biological and environmental factors are often incomplete. They are not always adequate for describing complex interactions, such as how mealtime stimulates language development and how food preferences develop during infancy. The complexity of infant development contributes to our individuality later on.

The Healthy People 2010 objectives involving infants are concerned with reducing infant mortality, preterm birth rates, incidence of spina bifida and neural-tube defects, fetal alcohol syndrome, and other birth defects (see also Chapter 9 for discussion of these topics). Table 8.1 shows progress towards the 2010 objective. The 2020 objectives and progress towards them are tracked by the Centers for Disease Control and Health Promotion, as more infants are surviving with conditions previously considered fatal who require prolonged use of high-technology health care.

Assessing Newborn Health

Birth Weight as an Outcome

The weight of a newborn is a key measure of health status during pregnancy. The average gestation for a full-term infant is 40 weeks, with a range from 37 to 42 weeks. Full-term infants usually weigh 2500–3800 grams (5.5 to 8.5 lb) and are 47–54 centimeters (18.5–21.5 inches) in length.[1] There were over 4 million births in the United States in 2006, and 88% of these newborns were full term.[2] Infants with normal birth weights are less likely to require intensive care and are usually healthy in the long run. Conversely, pre-term births, regardless of birth weight, are those born at 37 weeks or less. Preterm means incomplete development has taken place.[3]

> **Full-Term Infants**   Infants born between 37 and 42 weeks of gestation.
>
> **Preterm Infants**   Infants born before 37 weeks of gestation.
>
> **Infant Mortality**   Death that occurs within the first year of life.

Infant Mortality

Worldwide *infant mortality* rates rank the United States at twenty-eighth place, far worse than many other countries.[3] The reasons for the high rate of mortality are many, but the prevalence of low birth weight is a major factor. In 2006, 8.3% of live births in the United States were low

Table 8.1 United States infant mortality rates by race of the mother and progress in 2010 Healthy People Objective 16 to reduce infant mortality

	Deaths per 1000 Live Births[3]	Trend Over Last Decade
2010 goal	4.5	
All newborns	6.8	Slow decline
White	5.8	Plateau
Black	13.6	Declining
Asian/Pacific Islander	4.9	Variable
American Indian or Alaska Native	8.0	Declining
Hispanic/Mexican	5.5	Variable

birth weight, or less than 2500 grams (5.5 lb).[2] Preterm birth (births at less than 37 completed weeks of gestation) is a key risk factor for infant death since it is continuing to increase.[3] The three leading causes of infant mortality in 2006 were congenital malformations, complications related to preterm births, and sudden infant death syndrome.[2,3] The higher incidence of infant mortality, low birth weight, and preterm births in African American infants is of particular concern, as seen in Table 8.1.[3] The basis for racial disparity and preterm birth is a major focus in federal initiatives to combat infant deaths in the United States. Despite efforts to lower neonatal and infant deaths, rates are still too high.

Combating Infant Mortality

Efforts to improve newborn health are under way on many levels. In the United States, improved access to specialized care for mothers and infants has been credited in part for a decline in the infant mortality rate.[2] This is a multifaceted problem, however, affected by:

- Social and economic status of the families and women
- Access to health care
- Medical interventions
- Teenage pregnancy rates
- Availability of abortion services
- Failure to prevent preterm and low-birth-weight births

Resources have been concentrated on the proportion of newborns identified at risk. Some of these resources are major payers of health services, such as Medicaid and the Child Health Initiatives Program (CHIP).[4] The following concepts underscore the commitment of resources for infants:

- Recognition that birth weight is important for long-term health outcomes
- Understanding that prevention and treatment of complications for at-risk infants are investments for the future

The emphasis on prevention is seen in various programs. The Early Periodic Screening, Detection, and Treatment Program (*EPSDT*) is a major source of preventive and routine care for infants in low-income families. Immunizations during infancy are another example of a prevention approach.

Nutrition is included in some national prevention programs. The Special Supplemental Food Program for Women, Infants, and Children (WIC) and the Centers for Disease Control (CDC) collaborate to track infant growth as a part of the Nutrition Surveillance Program.[5] The Bright Futures program promotes and improves the health, education, and well-being of infants and children. Nutrition is one component of the program's guidelines about common issues and concerns. Bright Futures is an example of a comprehensive approach to health supervision by a collaboration of government and professional groups.[6]

Standard Newborn Growth Assessment

Newborn health status is assessed by various indicators of growth and development taken right after birth. Indicators include birth weight, length, and head circumference for gestational age. The designation "small for gestational age" (SGA)—also called "small for dates," *"intrauterine growth retardation (IUGR),"* or "intrauterine growth restriction"—means the newborn's birth weight falls below the 10th percentile of weight for gestational age.[7] Infants above the 90th percentile are considered large for gestational age (LGA). Those in between are appropriate for gestational age (AGA).[8]

Infant Development

Monitoring infants' nutritional status requires an understanding of their overall development. Full-term newborns have a wider range of abilities than previously recognized; they hear and move in response to familiar sounds, such as the mother's voice.[9] Newborns demonstrate four states of arousal, ranging from sleeping to fully alert, and responsiveness differs in part on their state of arousal.[9] Recognizing the state of arousal is a part of nursing successfully.

Organs and systems developed during gestation continue to increase in size and complexity during infancy. The newborn's central nervous system is immature; that is, the neurons in the brain are less organized compared to those of the older infant. As a result, the newborn gives inconsistent or subtle cues of hunger and other needs, compared to the cues given later. The fact that newborns can *root, suckle,* and coordinate swallowing and breathing within hours of birth shows that feeding is directed by reflexes and the central nervous system.[9,10] Newborn *reflexes* are protective for them. These reflexes fade as they are replaced by purposeful movements during the first few months of life.[9] Table 8.2 on the next page lists major reflexes in newborns.[9-11]

EPSDT   The Early Periodic Screening, Detection, and Treatment Program is a part of Medicaid and provides routine checkups for low-income families.

Intrauterine Growth Retardation (IUGR)   Fetal under-growth from any cause, resulting in a disproportionality in weight, length, or weight-for-length percentiles for gestational age. Sometimes called *intrauterine growth restriction.*

Reflex   An automatic (unlearned) response that is triggered by a specific stimulus.

Rooting Reflex   Action that occurs if one cheek is touched, resulting in the infant's head turning toward that cheek and the infant opening his mouth.

Suckle   A reflexive movement of the tongue moving forward and backward; earliest feeding skill.

Table 8.2 Major reflexes found in newborns

Name	Response	Significance
Babinski	A baby's toes fan out when the sole of the foot is stroked	Perhaps a remnant of evolution from heel to toe.
Blink	A baby's eyes close in response to bright light or loud noise	Protects the eyes
Moro	A baby throws its arms out and then inward (as if embracing)	May help a baby cling to its mother in response to loud noise or when its head falls
Palmar	A baby grasps an object placed in the palm of its hand	Precursor to voluntary grasping
Rooting	When a baby's cheek is stroked, it turns its head toward the cheek that was stroked and opens its mouth	Helps a baby find the nipple
Stepping	A baby who is held upright by an adult and is then moved	Precursor to voluntary walking forward begins to step rhythmically
Sucking	A baby sucks when an object is placed in its mouth.	Permits feeding
Withdrawal	A baby withdraws its foot when the sole is pricked with a pin	Protects a baby from unpleasant stimulation

SOURCE: From KAIL/CAVANAUGH. Human Development, 2E. © 2000 Wadsworth, a part of Cengage Learning, Inc. Reproduced by permission. www.cengage.com/permissions.

Motor Development

Motor development reflects an infant's ability to control voluntary muscle movement. There are several models for describing infant development, but none provide a complete description and explanation of the rapid advances in motor skills achieved during infancy.[9,11] Illustration 8.1 on page 244 depicts motor development during the first 15 months.[7] It is a great source of pride for parents when their baby first rolls over or sits up. The development of muscle control is top-down, meaning head control is the start, and last comes lower legs.[10] Muscle development is also from central to peripheral, meaning the infant learns to control the shoulder and arm muscles before muscles in the hands.[10,12] Motor development influences both the ability of the infant to feed and the amount of calories expended in the activity.[13] An example of how motor development affects feeding is the ability to sit in a high chair. Only when an infant has achieved the motor development of head control and sitting balance and certain reflexes have disappeared can oral feeding with a spoon be achieved.[12] The development of motor skills slowly increases the caloric needs of infants over time because increased activity requires more energy.[13] Infants who crawl expend more calories in physical activity than do younger infants who cannot roll over.

Critical Periods

The concept of critical period is based on a fixed time period during which certain behaviors emerge. Piaget's stages of cognitive development and Erickson's psychological stages of development are examples of theories of development in which time periods, or window of development, occur when certain skills must be learned in order for subsequent learning to occur.[11] A critical period for the development of oral feeding may explain some later feeding problems in infancy.[12] In the typical healthy newborn, the mouth is a source of pleasure and exploring, an important form of early learning. When a newborn has a prolonged period of respiratory support, for example, the baby may not associate mouth sensations with pleasure, but rather with discomfort. Under such circumstances the critical period for associating mouth sensations with pleasure and exploration may have been missed. After discharge, such an infant may be a reluctant feeder and have difficulty learning to enjoy food from a spoon.

Cognitive Development

The concept of biological and environmental systems interacting is seen in Illustration 8.2 on page 245, showing *sensorimotor* development.[9] These skills influence feeding in important ways. For example, the stage during which infants are very sensitive to food texture is also when their speech skills are emerging.[9] The interaction between the environment and stimulating the senses in the developing brain is now seen as structuring the nervous system in the long term.[14] The latest research suggests that access to adequate calories and protein may not be sufficient for maximizing brain maturation if the social and emotional growth of the infant are not stimulated simultaneously.[14] Cognitive

Sensorimotor   An early learning system in which the infant's senses and motor skills provide input to the central nervous system.

Ilustration 8.1 Gross motor skills.

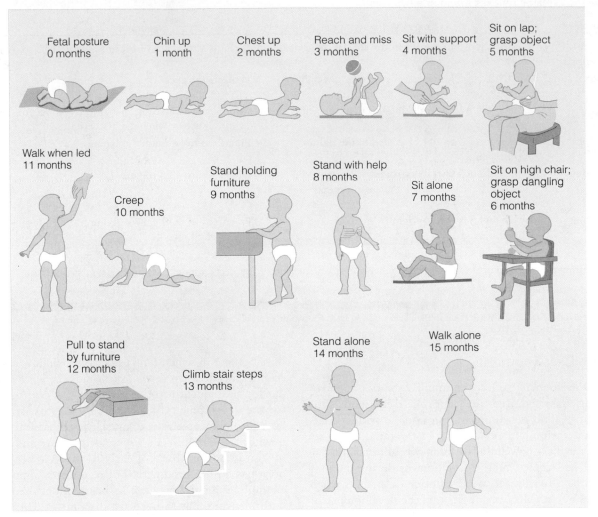

Based on Shirley, 1931, and Bayley, 1969.

development is also subjected to genetic controls, which turn genes on and off in different time frames and sites within the body.[15] Specific vitamins being needed at specific time frames of development is an example of the interaction of genetics and environment.

Digestive System Development

"Now good digestion wait on appetite, and health on both."

William Shakespeare, *Macbeth*

A healthy digestive system is necessary for successful feeding. Parents may worry about gastrointestinal problems, in part because of misinformation about nutrition in infancy.[16] For example, an infant with soft, loose stools may be considered by the parents to have diarrhea if they do not know that this is typical for breastfed infants.[17] Another common example is parents being concerned that

their infant's gastrointestinal discomfort may interfere with weight gain, even though growth usually progresses well. It takes up to 6 months for the infant gastrointestinal tract to mature, and the time required varies considerably from one individual to another.[3,6]

During the third trimester, the fetus swallows amniotic fluid, and this stimulates the lining of the intestine to grow and mature.[7] At birth, the healthy newborn's digestive system is sufficiently mature to digest fats, protein, and simple sugars and to absorb fats and amino acids. Although healthy newborns do not have the same levels of digestive enzymes or rate of stomach emptying as older infants, the gut is functional at birth.[7]

After birth and through the early infancy period, the coordination of peristalsis within the gastrointestinal tract improves. Maturation of peristalsis and rate of passage are associated with some forms of gastrointestinal

Illustration 8.2 Sensorimotor stage of development.

SUBSTAGES DURING THE SENSORIMOTOR STAGE OF DEVELOPMENT

Substage	Age (months)	Accomplishment	Example
1	0–1	Reflexes become coordinated.	Sucking a nipple 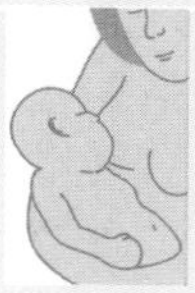
2	1–4	Primary circular reactions appear —an infant's first learned reactions to the world.	Thumb sucking 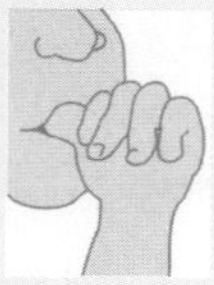
3	4–8	Secondary circular reactions emerge, allowing infants to explore the world of objects.	Shaking a toy to hear a rattle 
4	8–12	Means–end sequencing of schemes is seen, marking the onset of intentional behavior.	Moving an obstacle to reach a toy 
5	12–18	Tertiary circular reactions develop, allowing children to experiment.	Shaking different toys to hear the sounds they make 
6	18–24	Symbolic processing is revealed in language, gestures, and pretend play.	Eating pretend food with a pretend fork 

SOURCE: From KAIL/CAVANAUGH. Human Development, 2E. © 2000 Wadsworth, a part of Cengage Learning, Inc. Reproduced by permission. www.cengage.com/permissions.

discomfort in infants.[3] Infants often have conditions that reflect the immaturity of the gut, such as colic, *gastroesophageal reflux (GER)*, unexplained diarrhea, and constipation.[6,8] Such conditions do not interfere with the ability of the intestinal villa to absorb nutrients, and typically do not hinder growth. Other factors influence the rate of food passage through the colon and the gastrointestinal discomfort seen in infants. These include:

- *Osmolarity* of foods or liquids (which affects how much water is in the intestine)
- Colon bacterial flora
- Water and fluid balance in the body[7]

Parenting

"A babe in the house is a well-spring of pleasure."

Martin Farquhar Tupper, *On Education*

Even though the newborn can breast- or bottle-feed after birth, skills of new parents develop slowly. The parents' ability to recognize and respond to infant cues of hunger and satiety improve over time. New parents have to learn the temperament of their infants. Temperament has a biological basis and includes the infant's style or patterns of behavior.[9] Temperament includes the infant's emotional reactions to new situations, activity level, and sociability. The fit between the infant's temperament and that of the parents can increase or decrease feeding problems. For example, new parents may take a while to recognize that the newborn is more comfortable nursing at one noise level in the home than at another. Infants who are 6 months of age or older are better able to let parents know their needs and temperament. Conflicts in temperament may escalate as time progresses; they may become a factor in failure to thrive or other growth and feeding problems in later years.[12]

Gastroesophageal Reflux (GER) Movement of the stomach contents backward into the esophagus due to stomach muscle contractions. The condition may require treatment depending on its duration and degree. Also known as *gastroesophageal reflux disease (GERD)*.

Osmolarity Measure of the number of particles in a solution, which predicts the tendency of the particles to move from high to low concentration. Osmolarity is a factor in many systems, such as in fluid and electrolyte balance.

Energy and Nutrient Needs

The 2005 Dietary Guidelines for Americans address the needs of children aged 2 and older—not infants. Recommendations for infants are from the Dietary Reference Intakes (DRI), which are based on research over decades on infant growth and health and from professional groups such as the National Academy of Medicine, the American Academy of Pediatrics, and the American Dietetic Association.[13,19–21]

Caloric Needs

The caloric needs of typical infants are higher per pound of body weight than at any other time of life. The range in caloric requirements for individual infants is broad, ranging from 80 to 120 calories per kg (2.2 lb) body weight.[13] The average caloric need of infants in the first 6 months of life is 108 cal per kg body weight, based on growth in breastfed infants.[13] From 6 to 12 months of age, the average caloric need is 98 cal/kg.[13] Factors that account for the range of caloric needs of infants include the following:

- Weight
- Growth rate
- Sleep/wake cycle
- Temperature and climate
- Physical activity
- Metabolic response to food
- Health status and recovery from illness

Current recommendations for infants of 108 and 98 cal are considered about 15% too high, based on new study results.[13] However, too few energy-expenditure studies have been done on infants to reach a consensus about changing energy requirements.

Protein Needs

Recommended protein intake from birth up to 6 months averages 2.2 g of protein per kg of body weight, and from 6 to 12 months the need is for 1.6 g of protein per kg.[13] Protein needs of individual infants vary with the same factors listed for calorie needs. Protein needs are influenced more directly by body composition than calorie needs are, because metabolically active muscles require more protein for maintenance.

Short-Chain Fats Carbon molecules that provide fatty acids less than 6 carbons long as products of energy generation from fat breakdown inside cells. Short-chain fatty acids are not usually found in foods.

Medium-Chain Fats Carbon molecules that provide fatty acids with 6–10 carbons, again not typically found in foods.

Long-Chain Fats Carbon molecules that provide fatty acids with 12 or more carbons, which are commonly found in foods.

Most young infants who breastfeed or consume the recommended amounts of infant formula meet protein needs without added foods. Infants may exceed their protein needs based on the DRI when they consume more formula than recommended for age and when protein sources such as baby cereal are added to infant formula.[22] Inadequate or excessive protein intake can result for infants who are offered formula that is not made correctly, such as when less or more water is used than appropriate in preparation. Essential amino acids required by healthy infants are constant across the first year of life.

Fats

There is no specific recommended intake level of fats for infants. Fat restriction is not recommended. Breast milk provides 55% of its calories from fat, and this percentage reflects an adequate intake of fat by infants.[17] The main source of fat in most infant diets is breast milk or formula. Cholesterol intake should not be limited in infancy because infants have a high need for it and its related metabolites in gonad and brain development.[23] The percentage of fat in the diet drops after the infant accepts baby foods, since most baby foods are low in fat. Infants need fat, which is a concentrated source of calories, to support their high need for calories. Fat requirements in infancy are complicated by the differences in digestion and transport of fats based on fatty acid chain length.[23] *Short-* and *medium-chain fats* such as those in breast milk are more readily utilized than *long-chain fats,* such as in some infant formulas. Long-chain fatty acids are the most common type in food, but they are more difficult for young infants to utilize. Examples of long-chain fatty acids are C16–C18 and include palmitic (C16:0), stearic (C18:1), and linoleic (C18:2) acids.[23,24]

Infants use fats to supply energy to the liver, brain, and muscles, including the heart. The fact that infants have high caloric needs compared to those for older children means that infants use fats more regularly for generating energy. Young infants cannot tolerate fasting for long because it quickly uses up both carbohydrate and fat energy sources. This effect of fasting explains in part why infants cannot sleep through the night. In rare cases, some infants cannot metabolize fats due to a genetic condition that blocks specific enzymes needed to generate energy. Such infants may get sick suddenly; a few have been identified with this rare condition of fat metabolism only after dying of what appeared to be sudden infant death syndrome (SIDS).[25]

Fats in food provide the two essential fatty acids, linoleic and alpha-linolenic acid. Essential fatty acids are substrates for hormones, steroids, endocrine, and neuroactive compounds in the developing brain.[23] The long-chain polyunsaturated fatty acids docosahexaenoic acid (DHA) and eicosapenaenoic acid (EPA) are derived to some extent from an essential fatty acid.[25] Full-term breastfed babies do not need supplemental fat components or essential fatty acids.[26]

Metabolic Rate, Calories, Fats, and Protein— How Do They All Tie Together?

The metabolic rate of infants is the highest of any period after birth.[7] The higher rate is primarily related to infants' rapid growth rate and the high proportion of infant weight that is made up of muscle. The usual body fuel for metabolism is glucose. When sufficient glucose is available, growth is likely to proceed. When glucose from carbohydrates is limited, amino acids will be converted into glucose and used for energy and therefore are made unavailable for growth. The conversion of amino acids into glucose is a more dynamic process in infants as compared to adults. The breakdown of amino acids for use as energy occurs during illness in adults, but it can occur daily in fast-growing infants. Circulating amino acids in the blood from ingested foods will be used for glucose production, and if these are not sufficient, the body will release amino acids from muscles. This process of breaking down body protein to generate energy is known as catabolism. If catabolism goes on too long, it will slow or stop growth in infants. The precise site of all this metabolic activity is inside organs such as the liver, and in *mitochondria* within cells. If carbohydrates are not provided in sufficient amounts, growth will plateau because ingested protein and fats will be used for meeting energy needs.

Other Nutrients and Non-nutrients

Fluoride   The DRI for fluoride is 0.1 mg daily for infants less than 6 months of age, and 0.5 mg daily for 7- to 12-month-olds.[27] Fluoride is incorporated into the enamel of forming teeth, including those not yet erupted. Dental caries in early childhood are more frequent if an infant does not meet the DRI for fluoride. If an infant has more fluoride than recommended, tooth discoloration may result later. Community water fluoridation is safe for breastfeeding women and for infants.[6] Most infants who live in areas with fluoridated water do not require another source of fluoride. Fluoride is low in breast milk.[27] In areas in which fluoridated water is not available, prescribed fluoride is recommended for breastfed infants. If families routinely purchase bottled water, they should select water that has fluoride added.[6]

Vitamin D   Vitamin D, or preformed forms of vitamin D such as cholecalciferol, is required for bone mineralization with calcium.[27] Vitamin D is recommended for all infants starting shortly after birth at the level of 400 IU per day.

This recommendation for exclusively or partially breastfed infants is as supplemental vitamin D daily. For non-breastfed infants who are consuming one quart of vitamin D-fortified formula, no additional vitamin D supplement is needed. For those infants who consume less than one quart of vitamin D-fortified formula, supplemental vitamin D may be needed to meet the 400 mg IU daily recommendation. The identification of babies with rickets has resulted in this recommendation for breastfed infants from the American Academy of Pediatrics (AAP).[20] Vitamin D is not supplied by human milk in sufficient amounts, but it is added to infant formulas. Vitamin D is discussed in Chapter 6, on page 167. When outside, parents are encouraged to apply ultraviolet-light skin-protective lotion to their older infant's exposed skin, which blocks vitamin D formation from sunlight.

Sodium   Sodium is a major component of extracellular fluid and an important regulator of fluid balance. Estimated minimum requirements for sodium are 120 mg for 0- to 5-month-olds and 200 mg for 6- to 12-month-olds.[19] Breast milk's content of sodium was used as the basis for setting the sodium requirements for infants, and infant formula is supplemented with sodium to match the amount in breast milk. Typical infants do not have difficulty maintaining body fluids and electrolytes, even though they may not show thirst as a separate signal from hunger. Young infants do not sweat as much as older children, so losses from sweating are not major losses for infants. Illnesses such as diarrhea or vomiting cause the loss of sodium and water and increase the risk of dehydration. Infants do not need salt added to foods to maintain adequate sodium intake.

Fiber   Although there are dietary fiber recommendations for toddlers and children, there are no dietary fiber recommendations for infants.[28] Commercial and homemade baby foods are generally not significant sources of dietary fiber because preparation methods reduce dietary fiber.[28] However, fiber-containing foods such as fruits, vegetables, and grains are appropriate foods for older infants.[29]

> **Mitochondria**   Intracellular unit in which fatty acid breakdown takes place and many enzyme systems for energy production inside cells are regulated.

Lead   Although lead is not a nutrient, it can be associated with iron and calcium status during infancy. Elevated blood lead levels can be toxic to the developing brain, interfere with calcium and iron absorption, and bring about slowed growth and shorter stature.[15] Infants may inadvertently be exposed to environmental sources of lead. Lead may be a contaminant in water from lead pipes, particularly if the house was built before 1950. Older homes may contain lead-based paints that taste sweet to infants. Screening for lead poisoning is recommended starting

Table 8.3 Typical gains in weight and height for age in infancy[31]

Age	Weight Gain Grams	Weight Gain Grams (Pounds)	Length Gain mm	Length Gain mm (Inches)
	Per Day	Per Month	Per Day	Per Month
0–3 months	20–30	600–900 (1.3–2)	1	30 (1.2)
3–6 months	15–21	450–630 (1–1.4)	0.68	20 (0.8)
6–12 months	10–13	300–390 (0.7–0.9)	0.47	14 (0.6)

at 9 to 12 months of age.[6] If siblings have been found to have lead poisoning, screening for lead may be started at 6 months. Infancy is not the peak age for lead poisoning, but infants can be exposed if their parents work with lead-containing products. For example, if the father is a truck driver who uses leaded gasoline, his work clothes may have lead dust on them. If these clothes are mixed with the rest of the household laundry, or children play in the laundry room where lead dust has settled, an infant in the home may become exposed.

Physical Growth Assessment

Tracking growth in length and in weight helps identify health problems early, preventing or minimizing slowing of the growth rate. Parents understand that a sign of health is growth of their babies. By the time children reach school age, most families have a wall in their homes that proudly displays many height measurements over time. Healthy newborns double their birth weight by age 4–6 months and triple it by 1 year.[1] Growth reflects nutritional adequacy, health status, and economic and other environmental influences on the family. There is a wide range of growth attainment considered normal, however, and healthy babies may follow different patterns of growth. Often, healthy infants have short periods when their weight gain is slower or faster than at other times. Slight variations in growth rate can result from illness, teething, inappropriate feeding position, or family disruption. The overall growth pattern is important, and each assessment is compared to the whole picture. Table 8.3 shows typical growth rates during the first year of life.

Accurate assessment of growth and interpretation of growth rates are important components of health care for infants (Illustration 8.3). Accuracy requires calibrated scales, a recumbent-length measurement board with an attached right-angled headpiece, and a nonstretch tape for measurement of head circumference. Table 8.4 shows how to avoid common errors that interfere with accurate measurements. Makers of measuring equipment recommend checking their accuracy periodically, such as once a month. Calibration of measuring equipment is carried out by using standard weights (or lengths) to confirm accuracy and precision over the range that the equipment measures.

Standard techniques should be used to measure growth; these require practice and consistency. Equipment needed to measure infant growth is different from equipment for measuring children and adults. The scale bed must be long enough to allow the infant to lie down. Length is measured

Illustration 8.3 Infant being measured on length board and scale.

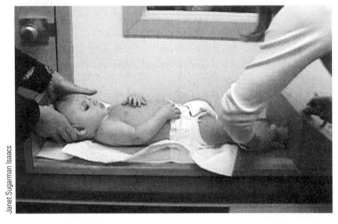

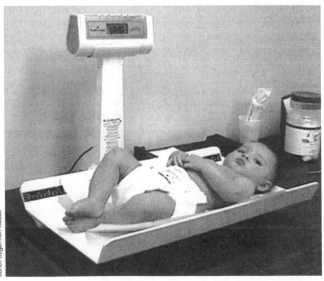

Janet Sugarman Isaacs

Table 8.4 Measuring growth accurately in infants

To Avoid Measurement Errors

- Use measuring equipment that was calibrated recently
- Confirm that the scale is on zero before starting
- Make sure the infant is not holding or wearing anything that adds weight or length
- Confirm the position of the infant for length measurements:

  Head position—the infant's eyes are looking straight up and the head is in midline, touching the head board

  Neither hips nor knees are bent

  Heel is measured with foot flat against the foot board

- Head circumference measure is at the widest part of the head

To Avoid Growth Plotting Errors

- Calculate the age accurately in months after confirming the date of birth
- Confirm plotting on the metric scale if kilograms were measured, not the pound scale
- Confirm that the plotted weight and length are marked well enough to read easily without being so large as to change percentiles

with the infant lying down with a head and foot board at right angles to the firm surface. Positioning the baby quickly and carefully is a skill needed for accurate measurement of recumbent length. In weighing, clothing, hair ornaments, and how much the baby jiggles the scale are examples of factors that could add error to measurements.

Interpretation of Growth Data

The CDC's National Center for Chronic Disease Prevention and Health Promotion 2000 infant growth charts are based on five national surveys and represent a larger sample of infants than that used for previous growth charts.[1] The infant growth charts are based on infants weighed nude on calibrated scales. The charts take into account differences in the growth patterns for formula-fed compared to breastfed infants, regardless of race or ethnic background.[1] Growth charts for 0- to 36-month-olds consist of a prepared graph for each gender, showing:

- Weight for age
- Length for age
- Weight for length
- Head circumference for age

After the CDC 2000 growth charts' publication, the World Health Organization (WHO) published growth charts covering infancy in 2006. The WHO growth charts include charts for the age range of 0–6 months and 6 months to 2 years for each gender. The WHO growth charts are based on data from 6 countries and only include infants who have breastfed for 6 months and live with a non-smoking mother. The infant WHO charts offer body mass index for age for each gender from birth to age five years, and cover a wider range than the percentiles of the CDC growth charts for weight for age and length for age.[30]

The more times a baby has been measured and the growth plotted, the more likely the growth trend will be clear, in spite of minor errors. Measures over time can identify a change in rate of weight or length gain and the need for intervention. Growth is so fast during infancy that it may be easier to determine growth problems during infancy than later. Every month in infancy there is an increase in both weight and length, which is not expected in older children. Warning signs of growth difficulties are lack of weight or height gain; plateau in weight, length, or head circumference for more than one month; or a drop in weight without regain within a few weeks.[7] Head circumference increases as a result of brain growth. If head circumference is not increasing typically, neither weight nor height increases are likely to track on standard growth percentiles. In the rare circumstance that an infant has a rapid increase in head circumference, this is not a sign of good nutrition or normal growth, but it may signal a condition that requires immediate attention to protect brain development. The rate of weight gain during infancy is not necessarily predictive of future growth patterns after infancy, nor a risk for long-term overweight, compared to the weight-gain pattern later in childhood.

Feeding in Early Infancy

"Food is the first enjoyment of life."

Lin Yutan, *The Importance of Living*

Breast Milk and Formula

The American Academy of Pediatrics and the American Dietetic Association recommend exclusive breastfeeding for the first 6 months of life and continuation of breastfeeding for the second 6 months as optimum nutrition in infancy.[32,33] Infants who are born preterm benefit from breastfeeding, too. Encouragement of breastfeeding right after birth, before the mother's milk supply is available, is an example of a birthing practice that is endorsed.[34] Other recommended practices include teaching safe handling and storage of expressed human milk.[34] The nutrient composition of breast milk is presented in Chapter 6. For young infants less than 6 months old, no other liquids or foods are recommended in addition to breast milk and

Table 8.5 Typical daily volumes for young infants not being breastfed

Age of Infant	Typical Intake of Formula per Day (24 hours)
Birth to 1 month	16–20 fl oz per day, 8–12 feedings/day, 1–2.5 fl oz per feeding
1 to 2 months	18–26 fl oz per day, 8–10 feedings/day, 2–4 fl oz per feeding
2 to 3 months	22–30 fl oz per day, 6–8 feedings/day, 3–5 fl oz per feeding
3 to 4 months	24–32 fl oz per day, 4–6 feedings/day, 4–8 fl oz per feeding

formula.[6] Recommendations for formula intake of young infants are shown in Table 8.5.

The growth rate and health status of an infant are better indicators of the adequacy of the baby's intake than is the volume of breast milk or formula. Infant formulas for full-term newborns are typically 20 cal/fl oz when prepared as directed. Formulas for infants born prematurely provide 22 or 24 cal/fl oz. Some health providers recommend further increasing the caloric density of formula for some preterm infants, but such recommendations are not appropriate for most infants. Most infants can be quite flexible in accepting formula, lukewarm or cold, or changes in formula brands.

Table 8.6 gives an overview of the composition of commercially available infant formulas compared to

Table 8.6 How infant formulas are modified compared to breast milk

Macronutrients	Breast Milk	Cow's Milk–Based Formula	Soybean-Based Formula
Protein	7% of calories	9–12%	11–13%
Carbohydrates	38% of calories	41–43%	39–45%
Fats	55% of calories	48–50%	45–49%

| Other ways infant formulas are modified compared to breast milk ||||

What Is Modified	How It Is Modified	Examples from Two Major Manufacturers
Calorie level	Increase in calories from 20 calories/fl oz to 22 or 24 calories/fl oz (for preterm infants).	EnfaCare Lipil is 22 calories/fl oz. Similac with Iron 24 is 24 calories/fl oz.
Form of protein	Protein is broken down to short amino acid fragments (hydrolyzed protein) or into single amino acids. Source of protein changed.	Similac Neosure Advance has amino acids. Enfamil Nutramigen has hydrolyzed milk protein. Prosobee has hydrolyzed soy protein in place of milk-based protein.
Type of sugar	Lactose is replaced by other sugars, such as sucrose or glucose polymers from various carbohydrate sources.	Enfamil LactoFree has lactose replaced by corn syrup solids (which provides glucose). Prosobee has carbohydrates from corn syrup solids. Neither has sucrose or lactose.
Type of fat	Long-chain fatty acids partially replaced with medium-chain fatty acids (MCT) and source of fat changed.	Pregestimil has about half of the long-chain fats, replaced by a mixture of vegetable oils. Enfamil Nutramigen has no MCT oil, but has vegetable oils in place of animal-based fats.
Allergy/intolerance	Replacement of milk-based protein with protein from soybeans or replacement of whole proteins with amino acid fragments or single amino acids.	Similac Isomil and Enfamil Prosobee have milk protein replaced by soy protein.
Micronutrients	Increased calcium and phosphorus concentration for preterm infants. Decreased minerals related to renal function. Added essential fatty acids (see above). Lower supplemental iron.	Enfamil Premature Lipil Similac PM 60/40 is modified in calcium, phosphorus, and is low in iron. Similac Special Care Advance 24 is a low-iron formula sold only to hospitals for preterm infants. Enfamil Low Iron and Similac Low Iron have lower levels of iron than the standard formula.
Thickness	Added rice or fiber for gastrointestinal problems.	Similac Isomil D.F. (D.F. = diarrhea free) for short-term use; it has added fiber from soy. Enfamil A.R. has added rice.
Age of infant	Target age 0–12 months	Similac Isomil Advance
	Target age 9–24 months	Similac Isomil 2

breast milk, and Table 8.7 compares various formulas to one another.[35-38] Some formulas have been developed for common conditions of healthy infants, such as gastroesophageal reflux (GER) or frequent diarrhea. The specialty formula market appears to be growing, such as follow-up formulas, hypoallergenic formulas, and "organic" formulas. Selenium and nucleotides for preterm infants are examples of recent formula additives.[25] Table 8.7 shows how little different types of commonly used infant formulas vary in key nutrients, as they are all based on the same nutrient guidelines.[22]

Cow's Milk During Infancy

The American Academy of Pediatrics recommends that whole cow's milk, skim milk, and reduced-fat milks not be used in infancy.[39, 40] Iron-deficiency anemia has been linked to early introduction of whole cow's milk. Low iron availability may come about as a result of gastrointestinal blood loss, low absorption of other minerals (calcium and phosphorus), or the lack of other iron-rich foods in the diet.[39] Studies on infants who were 7.5 months of age confirmed earlier findings that blood loss with whole cow's milk is more likely if the infant had been breastfed earlier rather than fed infant formula.[39] The high cost of infant formulas may result in families selecting cow's milk for older infants who are not breastfed.

Soy Protein-Based Formulas During Infancy

The American Academy of Pediatrics recommends limited use of formulas in which soy protein has been substituted for milk protein, when maternal breast milk is not available.[41] The basis for the recommendation is that there is little scientific evidence for the increasing use of soy formulas in infancy, which accounted for almost 25% of the formula market in 2008. The differences in cow's milk-based infant formula and soy protein-based infant formulas extend beyond the protein component to the fat and carbohydrate components of the formulas.[41] Soy protein formulas contain plant origin hormone-like components and dietary fiber components that can alter mineral absorption in healthy infants. The use of soy formula is not recommended for routine use for managing infantile colic or as offering an advantage over cow's-milk formula for preventing allergy in healthy at-risk infants.

Development of Infant Feeding Skills

Infants are born with reflexes that prepare them to feed successfully. As noted earlier, these reflexes include rooting, mouthing, head turning, gagging, swallowing, and coordinating breathing and swallowing.[9] Infants are also born with food-intake regulation mechanisms that adjust over time with development of the infant.[14] In early infancy, self-regulation of feeding is mediated by the pleasure of the sensation of fullness. Inherent preferences are in place for a sweet taste, which is also a pleasurable sensation.[26] After the first 4–6 weeks, reflexes fade and infants learn to purposely signal wants and needs. However, it is not until much later—about age 3—that children can verbalize that they are hungry. In between the reflexes fading and the child speaking, appetite and food intake are regulated by biological and environmental factors interacting with one another.

Table 8.8 shows infant developmental milestones and readiness for feeding skills.[9,10,12] The interaction of biology and environment prevails here, too. For example, depression in a caregiver may be an underestimated variable in the development of infant feeding. Maternal depression may bring about a lower level of interaction between the parent and infant during feeding, reducing the number or volume of feedings and increasing the risk of slower weight gain.[40] Media influences and changes in social practices also affect how babies are fed. Examples are cultural and ethnic perceptions of breastfeeding and the availability of quality child care for infants.

Several models help assess readiness for a breastfed infant to begin eating from a spoon at around 6 months. The developmental model is based on looking for signs of readiness, such as being able to move the tongue from side to side without moving the head.[10,12] The infant must be able to keep her head upright and sit with little support before initiating spoon-feeding. Models based on chronological age as well as those based on cues from the infant are considered outdated. Most infants adapt to a variety of feeding regimens, and various feeding practices can be healthy for them. The parents' ability to read the infant's cues of hunger, satiation, tiredness, and discomfort influence feeding-skill progression. The cues infants give may include:

- Watching the food being opened in anticipation of eating
- Tight fists or reaching for the spoon as a sign of hunger
- Showing irritation if the feeding pace is too slow or if the feeder temporarily stops
- Starting to play with the food or spoon as the infant begins to get full
- Slowing the pace of eating, or turning away from food when they want to end the meal
- Stopping eating or spitting out food when they have had enough to eat

Infants relate positive and pleasurable attributes to satisfaction of their hunger as part of a successful feeding experience. If there are long episodes of pain from gastroesophageal reflux or constipation, these can become the basis for later feeding problems as the association of eating and pleasure is replaced by an association of eating and discomfort.[3] An infant who makes the association

Table 8.7 Key nutrients in popular infant formulas per 100 calories as served (about 5 fl oz)

Major Nutrients	Units	Ross Similac Advance	Mead-Johnson Enfamil Lipil	Nestle Good Start DHA+ARA
Protein	g	2.07	2.1	2.2
Fat	g	5.4	5.3	5.1
DHA*	mg	8	17	17
Linoleic acid	mg	1000	860	900
Carbohydrates	g	10.8	10.9	11.2
Thiamine B$_1$	micrograms	100	80	100
Vitamin B$_{12}$	micrograms	0.25	0.3	0.33
Calcium	mg	78	78	64
Iron	mg	1.8	1.8	1.5
Zinc	mg	0.75	1	0.8

*docosahexaenoic acid

Illustration 8.4 Infant reaching with her tongue for a spoon.

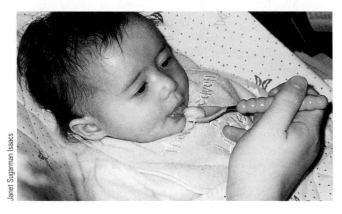

Janet Sugarman Isaacs

between eating and discomfort is likely to be seen as an irritable baby. This may set up a cycle of the infant being difficult to calm and the parent being frustrated. If this cycle is not replaced by the more positive association of eating and pleasure, the feeding difficulty in infancy may later be characterized by pickiness, food refusals, and difficult mealtime behavior in an older child.[10] A negative association of pain and eating may persist and appear as a behavioral problem at mealtimes.

Introduction of Solid Foods

Infants begin with food offered on a spoon in a small portion size of 1–2 tablespoons for a meal, with one or two meals per day. The purpose of offering food on a spoon to infants at 6 months is to stimulate mouth muscle development, and less for nutritional needs, which are met from breast milk. Watching a baby learn how to eat from a spoon is fun for new parents. If the baby has achieved the developmental milestones in Table 8.8, it may take him only a few days of practice to start spoon-feeding and to

learn to consume 1 tablespoon of semisoft food as a meal. Spoon-feeding is really two new experiences for a baby: a spoon is not soft and warm in the mouth like the breast, and whatever food is selected does not feel the same as breast milk on the tongue. At first the baby tries to suck food on a spoon like it is a liquid, so some food will come out of the mouth.

Babies respond strongly to new tastes or smells, regardless of the first food. Introducing a baby to food on a spoon includes these recommendations (in addition to those in the following section on infant feeding position):

- Time the first spoon-feeding experiences for when the baby is not overly tired or hungry, but active and playful.
- Offer a small spoon with a shallow bowl. The temperature of the spoon may have to be considered if it can conduct hot or cold readily.
- Give the baby time to open his or her mouth and extend the tongue toward the food. If the baby cannot extend the tongue farther out than the lower lip, the baby is not ready for spoon-feeding.
- Place the bowl of the spoon on the tongue with slight downward pressure toward the front of the mouth. Touching the back of the tongue may elicit a gag response.
- The spoon should be almost level. It is not a good practice to scrape the food off the spoon with the baby's gums by tilting the spoon handle up too high. The baby's chin should be slightly down to protect the airway.
- The pace of eating should be based on watching for the baby to swallow. Rushing will increase the risk of choking and of the infant having an unpleasant experience.

Table 8.7 (continued)

BPM Bright** Beginnings Ultra	BPM Bright** Beginnings Organic	BPM Bright** Beginnings Soy	Ross Isomil Advance (Soy)	Mead-Johnson Prosobee (Soy)	Nestle Good Start DHA+ARA Soy
2.2	2.2	2.7	2.45	2.5	2.5
5.3	5.3	5.3	5.465	5.3	5.1
19	19	19	8	17	17
500	750	750	1000	860	920
10.6	10.6	10.4	10.3	10.6	11.1
100	100	101	60	80	60
0.20	0.2	0.3	0.45	0.3	0.3
63	63	90	105	105	105
1.8	1.8	1.8	1.8	1.8	1.8
0.8	0.75	0.75	0.75	1.2	0.9

**often renamed as various grocery store brands

Table 8.8 Development of infant feeding skills

Chronological Age	Developmental Milestone	Feeding Skills
Birth to 1 month	Vision is blurry; hears clearly. Head is oversized for muscle strength of the neck and upper body.	Suckling and sucking reflexes. Frequent feedings of 8–12 per 24 hr. Only thin liquids tolerated.
1–3 months	Cannot separate movement of tongue from head movements. Head control emerges. Smiles and laughs. Puts hands together.	Volume increases up to 6–8 fl oz per feeding, so number of feedings per day drops to 4–8 per 24 hr. Sucking pattern allows thin liquids to be easily swallowed. Learns to recognize bottle (if bottle-fed).
4–6 months	Able to move tongue from side to side. Working on sitting balance with stable sitting emerging. Drooling is uncontrolled. Disappearance of newborn reflexes allows more voluntary movements. Teething and eruption of upper and lower central incisors.	Interest in munching, biting, and new tastes. Before 6 months, cannot easily swallow lumpy foods, but pureed foods swallowed. 6–8 fl oz per feeding and 4–5 feedings per day after 6 months (may be variable if breastfeeding). Holds bottle (if bottle-fed).
7–9 months	Hand use emerges, with pincer grasp and ability to release. Stable independent sitting. Crawling on hands and knees. Starting to use sounds, may say "mama" and "dada".	Self-feeding with hands emerges. Munching and biting emerges. Indicates hunger and fullness clearly. Prefers bottle, but little loss from a held, open cup.
10–12 months	Can pull to stand; standing alone emerges. Enjoys making sounds as if words. Can pick up small objects, such as a raisin. Can bang toys together with two hands. Has consistent routines about bedtime, diaper changing. Usually does not drool anymore.	Likes self-feeding with hands. Spoon self-feeding emerges. Drinks from an open cup as well as from a bottle. Uses upper and lower lip to clear food off a spoon. Enjoys chopped or easily chewed food or foods with lumps. Sitting position for eating Enjoys table foods even if some baby foods still used

- First meals may be small in volume—only 5 or 6 baby spoons—and last about 10 minutes, based on the baby's interest.

After mastering the new skill of eating from a spoon, babies quickly teach their parents how to feed them by indicating the rate of eating they like. Common mistakes happen if the person feeding the baby does not read the signs that the baby is giving.

The Importance of Infant Feeding Position

Positioning infants for feeding with a bottle and for eating from a spoon are important because improper positioning is associated with choking, discomfort while eating, and ear infections.[12] The semi-upright feeding position as exemplified in car seats or infant carriers is recommended for the first few months.[10] Unsafe feeding positions, such as propping a bottle or placing the baby on a pillow, increase the risk for choking and overfeeding. The recommended sleeping position for young infants is lying on the back without elevating the head on a pillow. This position is not recommended for feeding, which reinforces why feeding an infant in bed with a bottle is not generally recommended.[6,10]

Spoon-feeding also has a recommended infant feeding position. The infant can better control his mouth and head in a seated position with good support for the back and feet. The person offering the spoon should sit directly in front of the infant and make eye contact without requiring the baby to turn his head.[10,12] A high chair is an appropriate feeding chair when the infant can sit without assistance. The infant should be kept in a sitting position by use of a seat belt so that the hips and legs are at 90 degrees. This position assists the infant's balance and digestion. If the infant is sliding out from under the tray of the high chair with the hips forward, the stomach is under more pressure, and spitting up is more likely.

Some apparently healthy infants show resistance to learning feeding skills or react to the introduction of foods in an unusual manner. These problems in early feeding experiences are sometimes warning signs of more general health or developmental difficulties.[18] They may signal emerging problems that cannot be diagnosed until later. Families who call attention to early feeding problems may assist their infants in the long run by having the problem recognized earlier. For example, some infants who start and stop feeding frequently, but then do not feed for several minutes in a row, may later be discovered to have heart problems. The coordination of eating and breathing may have been the basis for the starting and stopping. Some infants are very reluctant feeders and are later diagnosed with a milk protein intolerance. Case Study 8.1 describes a baby who refuses to eat when her mother thinks she must be hungry.

Weaning Discontinuation of breastfeeding or bottle-feeding and substitution of food for breast milk or infant formula.

Preparing for Drinking from a Cup

The process of *weaning* starts in infancy, and usually is completed in toddlerhood. The recommended age for weaning the infant from the breast or from a bottle to drinking from a cup is from 12 to 24 months.[12] Breast-fed infants may make the transition to drinking from a cup without ever having any liquids in a bottle. If breastfeeding is continued as recommended for the first year of life, introducing a cup for water and for juices after 6 months is recommended, near the time that foods are offered on a spoon. By the time weaning from breast-feeding is planned, the one-year-old infant will be skilled enough at drinking from a cup to meet fluid needs without a bottle.

Infants who are not exclusively breastfed, or are breastfed for less than 12 months, need to have additional fluids offered in a bottle because their ability to meet their fluid needs by drinking from a cup are not sufficiently developed. Developmental readiness for a cup begins at 6–8 months.[12] Eight-month-old infants enjoy trying to mimic drinking from open cups that they see in the home. However, the ability to elevate the tongue and control the liquid emerges later, at closer to a year. The 10- to 12-month-old infant enjoys drinking from a held cup and trying to hold his own cup, even though the main feeding method is the breast or bottle. Infants are likely to decrease total intake of calories from breast milk or infant formula if served in a cup, because they are less efficient in the mouth skills needed. At first the typical portion size of fluid from a cup is 1 to 2 ounces. The infant who is weaned too soon may plateau in weight because of decreased total calorie intake. The drop in total fluids consumed may result in constipation. Changing from a bottle to a covered "sippy" cup with a small spout is not the same developmental step as weaning to an open cup.[12] The mouth skills needed in controlling liquids with the tongue are more advanced with an open cup. The skills learned in drinking from an open cup also encourage speech development.

Food Texture and Development

> "They say fingers were made before forks and hands before knives."
>
> Jonathan Swift

Weaning is not complete until the caloric intake from breast milk is provided from foods and liquids. Infants advance from swallowing only fluids to pureed soupy foods at 6 months.[6,12] Before that they can move liquids only from the front to the back of the mouth. The mouth is exquisitely sensitive to texture. If food with soft lumps is presented too soon, it causes an unpleasant sensation of choking. When infants are less than 6 months old, they can move their tongues from side to side. At 6–8 months, they are ready for foods with

Case Study 8.1

Baby Samantha Will Not Eat

Photodisc

Nutrition Assessment of the nutrition care process, based on data sources of the parent interview of Kathy, and physician records for Samantha.

Samantha is a healthy 8-month-old girl who lives with her mother, Kathy, her father, and her older sister, who is almost 3 years old. Both parents now work full-time, and both children attend day care full-time. Kathy nursed Samantha exclusively before she returned to work and built up a supply of frozen breast milk. She nurses her twice per day now, early in the morning and before Samantha goes to sleep. Samantha gets breast milk offered in bottles at day care. Samantha is reported by the day care staff to be a good baby. However, when Kathy picks her up after work, Samantha wants to be held and will not sit in her high chair or eat dinner. She cries if she is not held. Samantha's sister wants to eat as soon at they get home. Kathy has so much to do at home after work, she finds it difficult to hold Samantha at such a busy time. Kathy thinks that Samantha must be hungry and that she would be less irritable if she ate her dinner.

These questions pertain to whether there is an appropriate nutrition diagnosis or intervention from this assessment.

1. What signs is baby Samantha giving to show that she needs to be comforted rather than fed?
2. How might Kathy change her routine to give baby Samantha more attention and meet the needs of her older daughter?
3. At 8 months, is Samantha too young to overeat due to unmet emotional needs?
4. Should Kathy stop or continue breastfeeding to improve Samantha's eating?

(Answers are located in the Instructor's Manual for the 4th edition of *Nutrition Through the Life Cycle*).

---

a lumpy but soft texture to elicit munching and jaw movements;[12] these movements simulate chewing. By 8–10 months, infants are able to chew and swallow soft mashed foods without choking. It is important to offer infants foods that do not require much chewing, because infants do not develop mature chewing skills until they are toddlers.

First Foods

The first food generally recommended for infants at 6 months is baby cereal, such as iron-fortified cereal mixed with water or breast milk. Rice cereal is a common first food because it is easily digested and *hypoallergenic.* When to add baby cereal or other food to an infant's diet is determined not only by developmental milestones as recommended, but by other reasons, such as these:

- Some parents add baby foods because they think this will make the baby sleep longer. This practice is neither recommended nor effective for most infants. This common belief may result in introduction of baby cereal before the infant has developed the skills to eat from a spoon.

- Some families are instructed by pediatricians to add dry rice cereal as part of the treatment for gastrointestinal problems, because it tends to thicken infant formula.

Fruits and vegetables, such as pears, applesauce, or carrots, are also sometimes first foods for infants. What are considered healthy first foods for infants vary in different cultures and ethnic groups. Regardless of what foods are offered first, the timing and spacing of new foods can be used to identify any negative reactions. Common recommendations for parents of 6-month-olds are to add only one new food at a time and to offer it over 2 or 3 days. There are specific recommendations regarding the timing and spacing of foods known to trigger food allergies in families with histories of this problem (discussed later in the "Food Allergies and Intolerances" section).

Hypoallergenic Foods or products that have a low risk of promoting food or other allergies.

Commercial baby foods are not a necessity for infants. Parents can prepare baby foods at home using a blender or food processor, or by mashing with a fork. Care must be taken, however, to provide a soupy texture and to avoid contamination of home-prepared baby food by bacteria on food or from unsanitary storage methods. The nutrient content of home-prepared baby foods can vary widely depending on how they are prepared and stored. Adding salt and sugar to home-prepared baby foods are examples of variables that can decrease nutritional quality. The advantage of home-prepared baby food is that a wider variety of foods may be introduced that are likely to be a part of the diet later. Additionally, money not spent on commercial baby foods may be significant savings for some families.

Commercial baby foods are commonly selected because of their sanitation and convenience. Families who pack food for day care or who travel with infants find commercial baby foods convenient. Parents have a lot of choices to make in selecting baby foods. Selection should be based on the nutritional needs of the infant, not on what is available in local stores and the eating habits of the adult shoppers. Examples of baby foods that may reflect shopper's selections rather than baby needs are fruits with added tapioca or baby food desserts and snack foods. They are not recommended for most infants. Jar and plastic tub serving sizes of baby foods are based on industry standards, not necessarily recommended portion sizes. Portion sizes for infants should be based on appetite. Finishing an opened container of baby food may encourage overeating if parents do not pay attention to signs from the infant.

Many foods eaten by other family members are appropriate foods for infants who are 9–12 months of age. Examples are regular applesauce, yogurt, soft-cooked green beans, mashed potatoes, cooked hot cereal, and Cheerios.

Inappropriate and Unsafe Food Choices

New parents may inadvertently select foods for infants based on their own likes and dislikes, rather than on the infant's needs. Such choices are problematic when they increase the risk of choking. Here are some examples of unsafe foods for infants:[6,10,12]

- Popcorn
- Peanuts
- Raisins, whole grapes
- Uncut stringy meats
- Gum and gummy-textured candies
- Hard candy, jelly beans
- Hot dog pieces
- Hard raw fruits or vegetables, such as apples, green beans

Some foods present a choking risk for infants because of their lower chewing skills. Under-chewed pieces of food can obstruct the infant's airway because voluntary coughing and clearing the throat are skills not yet learned.[10]

Moreover, the infant may not be able to clear food from the roof of the mouth. A sticky food such as peanut butter against the hard palate may fall to the back of the mouth and present a risk of choking. Foods that do not easily stay together, such as potato chips, also can cause choking. A chip breaks apart in the mouth, but little pieces may stay crunchy. Small pieces may present to the back of the mouth before the infant can use the tongue to move the pieces to the sides and initiate munching.

Water

Breast milk or formula generally provides adequate water for healthy infants for 6 months.[6,17] Infant drooling generally does not increase the need for water. In hot, humid climates, infants have increased needs for water, but water should not replace breast milk or formula. Added water can be used to meet fluid needs, but not caloric and nutrient needs. All forms of fluids contribute to meeting the infant's water needs. Often parents are reluctant to say they offered their infants sips from their own glasses containing soft drinks or drinks containing caffeine or alcohol. This may be important information to include in a food intake record, especially if the contents are not recommended for babies. The replacement of an infant formula with a less nutritionally rich alternative such as juice, "sports drinks," cola, or tea has been found to be a contributor to lower-quality diets for infants.[42]

Water needs of infants are a concern because dehydration is such a common response to illness in infancy. The infant has limited ability to signal thirst, especially when sick. Vomiting and diarrhea result in dehydration more rapidly in infants than in older children, with symptoms that are more difficult to interpret.[10,43] Replacement of electrolytes has been the basis for a variety of over-the-counter fluid replacement products, such as Pedialyte, "sports drinks," and Gatorade, that are marketed to parents. These products contain some glucose (dextrose) along with sodium, potassium, and water. The amount of glucose provides significantly lower calories than do breast milk or formula—usually 3 cal per fl oz compared to 20 cal per fl oz. Such liquids can be overused, and they may result in weight loss even in healthy infants. Juice is not needed to meet the fluid needs of infants. The American Academy of Pediatrics recommends that juices need not be introduced into the diet before 6 months of age, and never at bedtime.[42] There is a recommendation for juice after age 1, but not before. If juice is offered to an infant over 6 months, it is recommended to be offered in a cup and not in a bottle. Infancy may be the time in which a habit of excessive juice intake starts, so limiting juice volume is a way to avoid problems later.

How Much Food Is Enough for Infants?

Parents improve in understanding infant feeding behavior as the infant–parent interaction matures from early to late infancy. During early infancy while the sleep/wake cycle of the infant is irregular, it is common for new parents to interpret all discomfort as signs of hunger. The infant's ability to self-calm is a developmental step that plays out differently with different temperaments and parenting styles.[44] Infants who are quite sensitive to what is happening around them are likely to be viewed as irritable and hungry if they cry frequently.[45] In contrast, infants who sleep through usual household noises and are less reactive to their immediate environment are likely to be offered food fewer times per day. As a result of different responses to temperament, a pattern of excessive or inadequate food and formula intakes may result.

The following is an example of excessive intake for a 3-month-old not being breastfed. Total formula intake: 33 fl oz (seven bottles per day, ranging from 3 to 5 fl oz per bottle), offered at 7:30 a.m., 11:00 a.m., 12:45 p.m., 2:30 p.m., 5:30 p.m., 8:45 p.m., and 11:30 p.m., and one serving of baby-food applesauce, fed on a spoon at 9:00 a.m. This is overfeeding because excessive formula is being offered, along with premature offering of spooned food. The frequency of the bottles being offered suggest that the parents are interpreting the baby's signs of discomfort as hunger when she may have other needs, such as for being held, changed, or calmed by movement or touch. Overfeeding is less likely with breastfeeding.

In the first few months, the oral need to suck is easily confused with hunger by new parents. The typical forward and backward tongue movements of the infant's first attempts to eat from a spoon may seem to be a rejection of food by new parents.[12] The infant appears to spit out the food, but this is a sign of learning to swallow and not necessarily a taste preference. The same food that appears to be rejected will be accepted as the infant learns to move the food from the front to the back of the mouth.[46] It may appear to a parent that the infant does not like a food if he or she appears to choke. This choking response is more likely based on the position of the spoon on the tongue.[10,12] The mouth is very sensitive, particularly toward the back. If the bowl of the spoon is too far back, it will cause a gagging reaction, regardless of the taste of the food.[47]

How Infants Learn Food Preferences

Infants learn food preferences largely based on their experiences with food. Breastfed infants may be exposed to a wider variety of tastes within breast milk than are infants offered formula.[48] The different foods that the breastfeeding mother eats may result in some flavor compounds being transmitted to the nursing infant.[49] Studies on infants in the age range of 4–7 months showed that acceptance of new foods was more rapid than acceptance of new foods after the first year of life.[46] In the 1920s and 1930s, pediatrician Clara Davis conducted studies on the self-selection of food by weaning infants.[50] Dr. Davis was able to demonstrate that older infants are able to select and consume amounts of food needed to sustain normal growth.[50] Her classic studies were interpreted by later generations as supporting the concept that infants and children will instinctively select a well-balanced diet. However, these studies were subject to misperception because careful attention was not paid to the original methods, in which only nutritious and unsweetened foods were available.[51]

Food preferences of infants are largely learned, but genetic predisposition toward sweet tastes and against bitter foods may modify food preferences. Food preferences developed in infancy set the stage for lifelong food habits. The development of trust and security for an infant are crucial, but this need not be linked to overfeeding the infant.[14] Recognizing an infant's specific needs and responding to them appropriately is important. If offered only a limited variety of foods with little interaction during the meal, infants may learn to refuse to eat as a method of gaining attention. For example, 10-month-old infants enjoy throwing food on the floor just to have someone bring them more. They may enjoy the sound of banging a spoon on a high chair more than eating. Infants who learn to get attention by not eating are likely to manipulate the behavior of adults even more successfully as toddlers.

Nutrition Guidance

Nutrition guidance materials have been developed for parents from many sources, such as the WIC program, makers of infant foods, and professionals such as those in the Bright Future in Practice initiative.[6,28,35] The need for nutrition education was demonstrated in a study of mothers and pregnant women, which showed that misunderstanding about infant nutrition was common.[16] Infant feeding recommendations from nutrition education materials are sampled in Table 8.9.

Infants and Exercise

The exercise and fitness benefits for adults do not apply to infants. Providing a stimulating environment is recommended, so infants can explore and move as a part of their developmental milestones. The American Academy on Pediatrics Committee on Sports Medicine policy statement recommends that structured infant exercise programs should not be promoted as being therapeutically beneficial for healthy infants.[52] Infants do not have the strength or reflexes to protect themselves, and their bones are more easily broken than those of older children and adults.

Supplements for Infants

Specific supplements are recommended for breastfed infants in the United States and Canada, under certain circumstances:

- Fluoride supplements are recommended if the family lives in a place that does not provide fluoridated water.
- If breast milk is the only form of nutrition after 6 months, fluoride is recommended.[32]

Table 8.9 Infant feeding recommendations

Topics	Nutrition Education Sample Content
Appropriate use of infant formula (if not breastfeeding)	Mixing instructions for diluting concentrated formula, keeping formula sanitary by refrigeration, and monitoring how long offered formula is left out. Feeding positions for the infant and bottle, and burping the infant during feedings.
Baby food and sanitation	Serving sizes for infants of different ages, refrigeration and sanitation for opened jars of baby food, problems from mixing different baby foods.
Preparing baby foods at home	Avoiding spices, salt, and pepper in baby foods. Using safe food-handling techniques in preparing and storing servings.
Prevention of dental caries	Recommendations for bedtime and nap time to avoid sugary liquids pooling in the mouth. Identifying liquids that may promote dental caries.
Feeding position	Feeding positions for starting food on a spoon. How to tell if the infant's high chair is safe for feeding.
Signs of hunger and fullness	Identifying early signs compared to later signs of hunger. How infants of various ages communicate at mealtime. Reinforcing and rewarding infant signs of hunger and fullness.
Preventing accidents and injury	Checking temperature of baby foods and liquids. Use of appropriate car seats and safety belts.
Spitting up—when to be concerned	Typical feeding behaviors in young infants. Signs of overfeeding. Spitting up and signs of illness. Discussing signs and symptoms with health providers.

Failure to Thrive (FTT) Condition of inadequate weight or height gain thought to result from a caloric deficit, whether or not the cause can be identified as a health problem.

Organic Failure to Thrive Inadequate weight or height gain resulting from a health problem, such as iron-deficiency anemia or a cardiac or genetic disease.

Nonorganic Failure to Thrive Inadequate weight or height gain without an identifiable biological cause, so that an environmental cause is suspected.

Developmental Disabilities General term used to group specific diagnoses together that limit daily living and functioning and occur before age 21.

- Elemental iron (at 3 mg per kg body weight of the infant) may be prescribed if the mother was anemic during pregnancy.[6]
- Vitamin B$_{12}$ may be prescribed if the mother is a vegan.[48,53]
- Vitamin D supplements may be needed if the infant is partially or exclusively breastfeeding.

Supplements may also be prescribed for infants who were born early at low birth weight. They may need vitamins A and E and iron due to low stores of these nutrients usually accumulated late in pregnancy. A liquid multivitamin and mineral with fluoride is a common prescription for the healthy premature baby, regardless of breastfeeding status.[6]

Common Nutritional Problems and Concerns

Common nutritional problems during infancy are failure to thrive, colic, iron-deficiency anemia, constipation, dental caries, and food allergies. Parents often overestimate the association between eating and these common health problems. Parents should discuss their concerns with the infant's health care providers.

Failure to Thrive

Failure to thrive (FTT) is a diagnosis that can be made during infancy or later. Various terms are used to refine FTT, such as *organic* (meaning a diagnosed medical illness is the basis), *nonorganic* (meaning not based on a medical diagnosis), and mixed type. Although growth failure may be brought about by a variety of medical and social conditions, FTT is primarily used to describe conditions in which a calorie deficit is suspected.[8,54] FTT is an emotionally loaded diagnosis for parents, because the term implies that someone failed. Examples of nonorganic or environmental factors are maternal depression, mental illness, alcohol or drug abuse in the home, feeding delegated to siblings or others unable to respond to the infant, and overdilution of formula. The relationship of FTT to poverty has been well documented.[55] Examples of organic reasons for FTT commonly found are untreated GER, chronic illnesses such as ear infection or respiratory illness, and *developmental disabilities*.[56] (The connection between FTT and developmental disabilities is further discussed in Chapter 9.) If there is a medical basis for expecting that the infant will not fit standard growth projections, FTT is not an appropriate term to use. For example, growth for an infant born with IUGR should be based on this medical history and related testing near birth. If this infant at 11 months of age is taken to a new health care setting, FTT may be suspected because of the infant's small size—unless the true cause, IUGR, is revealed.

Table 8.10 provides an example of an assessment that can be used to determine if FTT is present. The assessment of FTT depends on tracking growth. Once

Table 8.10 Complete nutritional assessment of an infant to rule out failure to thrive

- Review records of weight, length, head circumference, fetal or maternal risk factors such as rate of weight gain during pregnancy, newborn screening results, Apgar scores, and physical exams after birth.
- Interpret all available growth records from providers, WIC, and emergency-room visits.
- Interpret current growth measurements and indicators of body composition, such as fat measurements.
- Review family structure, education, and social supports with attention to access to food and formula (if not breastfed).
- Analyze and interpret current food and fluid intakes as reported by the primary caregiver(s).
- Rule out a biological basis for FTT from available records and laboratory results.
- Observe and interpret parent–child interactions, feeding duration, and the feeding skills of the infant.

FTT is suspected, review of medical records often indicates that growth measurements have been taken in a variety of health care settings with different equipment and personnel, at times when the infant was both well and sick. These records may produce an irregular growth pattern that is difficult to interpret.

Nutrition Intervention for Failure to Thrive

Failure to thrive may be a basis for referral to a registered dietitian. Correction of FTT usually is not as simple as just feeding the baby, but increasing caloric and protein intake is the first step.[49,50] The registered dietitian's role is to assess the growth and nutritional adequacy, establish a care plan, and provide follow-up as part of a team approach. She may work with other specialists concerning medical or psychological aspects. Nutrition interventions may establish caloric and protein intake goals and a feeding schedule to assure adequate nutrition is being provided.[35] Other interventions may include:

- Gaining agreement from the caregivers about how and when intake and weight monitoring will be done
- Enrolling the infant in an early intervention program in the local area
- Arranging for transportation or solving other barriers to follow-up care
- Assessing social supports to assure a constant supply of food and formula (if used)

- Assisting the family in advocating for the infant within the health care delivery system, such as locating a local pediatrician and getting prescriptions filled

FTT is one reason that social service agencies become involved with families. Most new parents handle stress without hurting their infants, but a few react in ways that result in infants presenting with FTT or worse. After investigation, FTT may be determined to be a form of child abuse, as a result of neglect. Some infants diagnosed with FTT become at risk for child abuse and need foster care if the home situation does not improve.[45]

Colic

Colic is the sudden onset of irritability, fussiness, or crying in a young infant.[57] Parents usually think that the infant has abdominal pain. Episodes may have a pattern of onset at the same time of day, for about the same duration every day, with all symptoms disappearing by the third or fourth month. The association of colic symptoms, gastrointestinal upset, and infant feeding practices has been studied, but no definitive cause has been shown.[57]

The response to colic is often to change baby formulas if the infant is not breastfeeding, although the change in formulas usually does not change the pattern of colic. Recommendations to relieve colic may include rocking, swaddling, bathing, or other ways of calming the infant, positioning the baby well for eating, or burping to relieve gas.[6,35] One theory about why infants have colic points to the mother's diet while breastfeeding, particularly her consumption of milk or specific foods such as onions. Identifying the origin of colic requires more research.

Iron-Deficiency Anemia

Iron deficiency in infants is less common than iron deficiency in toddlers. Iron reserves in full-term infants reflect the prenatal iron stores of the mother.[58] Women with iron-deficiency anemia during pregnancy pass on less iron to the fetus, a condition that may increase risk of anemia during infancy. Infants who have iron deficiency may be exposed to other risk factors to their overall development, such as low birth weight, elevated lead levels, or generalized undernutrition.[58] Family income at or below the poverty level is also a risk for iron deficiency.[58,59] Research in infants who have long-term and severe iron-deficiency anemia suggests inadequate iron contributes to long-term learning delays from its role in central nervous system development.[15,29] Treatment of iron-deficiency anemia in infancy is generally by prescribed oral elemental iron administered as a liquid.[6,59]

Colic A condition marked by a sudden onset of irritability, fussiness, or crying in a young infant between 2 weeks and 3 months of age who is otherwise growing and healthy.

Breastfed infants may be prescribed elemental oral iron and also receive iron through iron-fortified baby cereal at 6 months of age. For infants who are not breastfed, a usual source of iron for formula-fed babies is iron-fortified infant formula. Iron from this source improves iron status measured during the first year and is well accepted.[53] In a randomized study of healthy infants, those who received iron-fortified infant formula had, by the end of the first year, significantly improved biochemical measures of iron status compared to those who received infant formula without added iron. However, there were no differences in the developmental scores of the two groups by 15 months of age.[54]

The level of iron in iron-fortified formula has been 15 mg per liter, or 11.5 mg per quart, based on the RDA of 6 mg of iron for infants up to 6 months and 10 mg for infants from 6 to 12 months.[18] New infant formulas with a lower level of supplementation are also marketed in part as a result of gastrointestinal side effects that have been attributed to iron added to formula.[43] The "low-iron" formula has 8 mg iron per liter, or 4.5 mg of iron per quart.[35] Some manufacturers are not recommending the low-iron formula beyond 4 months of age, because it would not meet the RDA for iron.

Diarrhea and Constipation

Diarrhea and constipation may be attributed to dietary components such as breast milk or use of an iron supplement. Parents think that diarrhea and constipation are related to the infant's diet and want to change the diet or feeding plan to lower gastrointestinal upset. In fact, diarrhea can result from viral and bacterial infections, food intolerance, or changes in fluid intake.[43] Typically, young infants have more stools per day than do older infants, and have them soon after oral intake.[55] The number of stools varies widely from two per day to six per day, decreasing as the infant matures.[55] Parents of breastfed infants generally do not have concerns about constipation, because the infant's stool is generally soft. Infants fed soybean-based infant formulas may have more constipation than those fed a cow's milk–based formula. Recommendations for avoiding constipation are to assure that the infant is getting sufficient fluids and to avoid medications unless prescribed for the infant. Some parents use prune or other juices that have a laxative effect for an older infant, but there is a risk of creating a fluid imbalance and subsequent diarrhea.[43] Foods with high dietary fiber are generally not recommended for infants with constipation because many sources, such as whole wheat crackers or apples with peels, present a choking hazard and are not recommended for infants.

The cause of diarrhea during infancy may or may not be identifiable. Diarrhea in an infant may become a serious problem if the infant becomes dehydrated or less responsive.[43] Most infants have 1 or 2 days of loose stools without weight loss or signs of illness, such as after getting immunizations. General recommendations are to continue to feed the usual diet during diarrhea.[6,35] Breast milk does not cause diarrhea. During a bout of diarrhea, continuing adequate intake of fluids such as breast milk or infant formula is generally sufficient to prevent dehydration.

Prevention of Baby-Bottle Caries and Ear Infections

Baby-bottle caries are found in children older than 1 year, but are initiated by feeding practices during infancy. Infants have high oral needs, which means they love to suck and to explore by putting things in their mouths. They derive comfort from sucking and may relax or fall asleep while sucking. The use of a bottle containing formula, juice, or other high-carbohydrate foods to calm a baby enough to sleep, however, may set her up for dental caries.[56] During sleep the infant swallows less, allowing the contents of the bottle to pool in the mouth. These pools of formula or juice create a rich environment for the bacteria that cause tooth decay to proliferate, increasing the risk for tooth decay.

Risk for ear infections is also correlated with excessive use of a baby bottle as a bedtime practice, as a result of the feeding position.[56] The shorter and more vertical tubes in the ears of infants are under different pressure during the process of sucking from a bottle.[35] If the infant is feeding by lying down while drinking, the liquid does not fully drain from the ear tubes. The buildup of liquid in the tubes increases the risk of ear infections. In a study of over 200 infants, pacifiers and bottle-feeding were correlated with greater prevalence of ear infections.[60] Infants who were breastfed did not have as high a rate of ear infections.[60]

Here are some good feeding practices to limit baby-bottle caries and ear infections related to baby bottles:

- Limit the use of a bottle as part of a bedtime ritual.
- Offer juices in a cup, not a bottle.
- Put only water in a bottle if offered for sleep.
- Examine and clean emerging baby teeth to prevent caries from developing.

Food Allergies and Intolerances

The prevalence of true food allergies is higher in younger than in older children. About 6–8% of children under 4 years of age have allergies that started in infancy.[61,62]

An infant may develop a food allergy to the protein in a cow's milk–based formula over time. Often such a problem follows a gastrointestinal illness. When the infant is well, protein digestion and absorption is of groups of two or three amino acids linked together. After an illness, small patches of irritated or inflamed intestinal lining may allow protein fragments of larger lengths of amino acids to be absorbed. Such fragments are hypothesized to trigger a reaction as if a foreign protein had invaded, setting up a local immune or inflammatory response.[63] This absorption of intact protein fragments is the basis for allergic reactions. When this happens with cow's-milk protein, it is likely that soy-based formulas will also cause the same allergic reaction.

The most common allergic reactions are respiratory and skin symptoms, such as wheezing or skin rashes.[62] Food allergies are confirmed by specific laboratory tests after infancy.[57] True allergies can present as an array of reactions building up over time, so that it may take several years for the initiating cause to be identified.

Food intolerances are frequently suspected in infants. Families may consider skin rashes, upper airway congestion, diarrhea, and other forms of gastrointestinal upset to be food allergies, but often they are not.[57] Infants are not usually given all the tests for food intolerance and allergies, in part because their immune systems are not mature. The infant with suspected protein intolerance may be changed to a specialized formula composed of *hydrolyzed protein.*[58] Such formulas are expensive and have a taste that many infant reject, so treating only by symptoms usually fail. With true food allergies, the protein of a hydrolyzed formula is already broken down, so it does not trigger the same response as intact protein fragments do. A family with a known allergy or intolerance may lower the risk of the allergy occurring in their infant by breastfeeding, and by postponing into the second or third year introduction of allergy-causing foods such as wheat, eggs, and peanut butter. It is important for families not to overly restrict such foods thought to cause allergies from an infant's diet unless required. If many foods are being avoided, there may be consequences such as decreasing the nutritional adequacy of the diet and reinforcing behaviors of rejecting foods and limiting variety. Allergy and intolerance symptoms are more common in response to nonfood items such as grasses and dust, so many different sources of symptoms must be considered.

Lactose Intolerance

Lactose intolerance is a food intolerance in infancy characterized by cramps, nausea, and pain, and by alternating diarrhea and constipation. Infants who are breastfeeding may develop lactose intolerance, because breast milk has lactose in it.[17] Usually diagnostic gastrointestinal testing is needed to confirm true lactose intolerance. In infants, true lactose intolerance is uncommon and tends to be overestimated, since it is easy to confuse with colic, and it can appear after a variety of minor illnesses. Gastrointestinal infections may temporarily cause lactose intolerance because the irritated area of the intestine interferes with lactose breakdown.[26] The ability to digest lactose generally returns shortly after the illness subsides. *Lactose* is in all dairy products, so it is in cow's milk–based infant formulas.[17,35–38] Lactose-free infant formulas are those in which lactose has been replace with other carbohydrate sources, such as modified corn starch or sucrose, or formulas in which the disaccharide lactose is broken down into monosaccharides, such as in lactose-free cow's milk. Lactose intolerance is less common during infancy than at older ages in groups that are susceptible to it. An infant who was fed a lactose-free formula is likely to be able to eat dairy products later. Because dairy products are such an important source of calcium, introducing foods with low lactose is recommended for older infants who appeared to be lactose intolerant when younger.

Cross-Cultural Considerations

Commercial baby foods reflect the bias of the dominant American culture. There is little ethnic diversity in baby foods—no collards or Mexican beans. Many successful avenues to nourish a healthy infant are available, and room ought to be made for different cultural patterns in the development of feeding practices. Some cultural practices are clearly unsafe and must be discouraged, such as a mother pre-chewing meats for a baby. Cultural practices that support the development of competence in parents can be encouraged, even if not part of the dominant culture. Examples of practices that may reflect cultural choices are swaddling an infant, or having an infant sleep in the parent's bed or in a room with a certain temperature. Practices based on family traditions may be forms of social support for new parents. Only if new parents have not considered the safety of the infant or have little knowledge of other, safer alternatives should cultural practices be discouraged. For example, it may be a cultural practice to offer meats to adults but not to infants. Parents should be informed that older infants may safely eat meats that are cut up or soft-cooked to avoid causing choking. Some cultures consider meat-based soups as infant foods.

Cultural considerations may affect the family's willingness to participate in assistance programs such as WIC or early-intervention programs. The dignity of the family unit, including extended relatives, has to be considered when counseling families about infant feeding practices. Food-based cultural patterns may be part of a religious tradition, so sensitivity to the family unit would include recognizing and understanding such beliefs.

> **Hydrolyzed Protein Formula**   Formula that contains enzymatically digested protein, or single amino acids, rather than protein as it naturally occurs in foods.
>
> **Lactose**   A form of sugar or carbohydrate composed of galactose and glucose.

Vegetarian Diets

Studies have found that infants who receive well-planned vegetarian diets grow normally.[48] The most restrictive diets, vegan and macrobiotic diets, have been associated with slower growth rates in infancy, particularly if infants do not receive enough breast milk.[48,59] Breastfed vegan infants should receive supplements containing vitamin D, vitamin B$_{12}$, and possibly iron and zinc.[59] The composition of the breast milk from vegan mothers may differ in small ways from standard breast milk.[48] An example is in the ratio of types of fat, although the total fat in breast milk from vegan mothers is the same. Impacts of these differences on the health of infants are generally not known.

Vegetarian diets range from adequate to inadequate, depending on the degree to which the diet is restricted, just as diets for omnivores range from adequate to inadequate.[59] Either food sources, such as fortified infant cereals and soymilk, or supplements can be used to assure adequate vitamin and mineral intake. Vegetarian families who avoid all products of animal origin, including milk and eggs, require carefully selected fortified foods or a higher degree of supplementation for their infants.[59] Periodic assessments of dietary intake, growth, and health status can be used to monitor the infant fed a vegetarian diet. Vegetarian infants have similar risk for developing food allergies from soy products, wheat, and nuts as do other infants.

Galactosemia   A rare genetic condition of carbohydrate metabolism in which a blocked or inactive enzyme does not allow breakdown of galactose, causing serious illness in infancy.

Hypothyroidism   Condition in which thyroid hormone is not produced in sufficient quantities, interfering with growth and mental development if untreated in infants.

Nutrition Intervention for Risk Reduction

The Early Head Start program is an example of a federal program that is focused on preventing and reducing risks to infant development.[6] The Early Head Start program was developed to work with infants and their families, especially new families at risk due to drug abuse, infants with disabilities, or teenage mothers. Nutrition services are among a wide range of services typically offered in an Early Head Start program. Other services may include home-based early childhood education, case management, and mental health support services, as well as health and socialization services. The Early Head Start program assists families in coordinating WIC participation with food stamps, routine well-baby visits, and day care, as needed.

Nutrition intervention programs may include a nutrition assessment (see Table 8.10).

Model Program: Newborn Screening

In the United States and many other countries, all newborns are screened for rare conditions that may cause disability or death. Such screening, from a small dried blood spot, was initiated in the 1960s after early treatment of phenylketonuria (PKU) was shown to prevent later mental retardation in young children.[31] Most states screened for three to six different conditions, such as PKU, *galactosemia, hypothyroidism,* and sickle-cell disease.[31] New technology has resulted in expanded newborn screening, so states are now testing for as many as 57 different conditions from the same dried-blood test.[31] Many of the disorders that can be detected by expanded newborn screening are treated by diet. Dietary treatment avoids the substance that has a metabolic block and replaces other dietary components that are usually provided in the foods that are avoided.[31] Expanded infant screening for genetic disorders is likely to continue to expand nutrition knowledge overall.

Key Points

1. Infants born full-term and preterm infants born between 34 weeks and 38 weeks of gestation are the same in their milestones of growth, development, and feeding in the first year of life.

2. In addition to access to adequate nutrition, environmental and societal factors have been credited with decreasing infant mortality.

3. The ability of infants to feed and eat is based on developmental skills that show readiness for the next step; parents learn to read the signals of readiness from their infants over time.

4. Energy and nutrient needs of infants are modulated by individual differences in sleep/wake cycle, exposure to temperatures, and state of health, among other factors.

5. The priority is energy needs first; protein and carbohydrates will be converted to meet energy needs if sufficient calories are not consumed, slowing growth over time.

6. Limiting micronutrients are vitamin D and fluoride in some environments, so supplementation is recommended.

7. Growth as weight, length, and head circumference accretion is monitored and interpreted over the first year.

8. Introduction of solid foods is also a developmental stage for parents in learning to read signs of hunger, fullness, and food preferences in their infants and to know safe food choices.

9. Common nutrition problems in the first year such as failure to thrive, colic, iron deficiency anemia, and baby bottle caries are usually solved by combining parent educational, nutritional, and medical approaches.

10. Infants can thrive with many different cultural and parenting styles; nutrition guidance for new parents is available from reliable sources in every community.

Review Questions

1. Which is a true statement about limiting nutrients in infancy?
   a. Lead comes from infant formula overfeeding.
   b. Vitamin D supplementation is needed in infants fed exclusively breast milk but not in those fed exclusively formula.
   c. Dietary fiber is low in infant diets in the same manner as in diets of toddlers and children.
   d. Fluoride is not limiting until teeth start emerging in later infancy.

2. What is an accurate statement about newborn infants?
   a. The large for gestational age infant is born more than 42 weeks after conception.
   b. The newborn rooting reflex makes the infant start suckling.
   c. Newborn reflexes promote the coordination of feeding and breathing.
   d. The ability to sit is achieved with the help of newborn reflexes.

3. Which food intake carries the lowest risk for choking for an 8 month old infant?
   a. Bottle of infant formula, 1.5 TB parent's omelette of Canadian bacon, green peppers and scrambled egg, pinched off buttered toast pieces.
   b. Bottle of infant formula, 0.5 jar infant stage 2 chicken and vegetable dinner, one soda cracker with peanut butter and 3 slices fresh apple.
   c. Bottle of infant formula, 1.5 tablespoon of chicken salad containing grapes, celery and nuts and 1.5 TB applesauce.
   d. Bottle of infant formula, 1.5 TB cooked carrot slices, 3 pieces cut-up fresh peach and shredded deli ham

4. The digestive system during infancy can be accurately described by which statement?
   a. Gastrointestinal problems often interfere with health and growth in young infants.
   b. The digestive tract is not completely developed for digestion for months after birth.
   c. Infant stools change based on the ability of the intestinal villa to absorb nutrients.

   d. Regarding hunger and satiety cues, parenting skills to read them develop more slowly than infants' skills to give them.

5. Which is a true statement concerning fats?
   a. Infants need a higher percent of total calories as dietary fats than recommended for adults.
   b. Breast milk is lower in total dietary fat than infant formula.
   c. Saturated long-chain fatty acids are not needed for healthy infant feeding.
   d. Fats of short and medium chain lengths are in breast milk and standard infant formula.

6. What is a disadvantage of feeding an infant a larger volume of formula or breast milk than recommended? (short answer)

7. Which of the following statements is accurate?
   a. Newborns who weigh 6 lbs and no ounces are in the low birth weight range.
   b. Infants with normal birth weights reflect a healthy pregnancy.
   c. Infants with low gestational age of 34 weeks are the leading cause of infant deaths
   d. Hispanic/Mexican infants have a higher rate of infant deaths than all newborns

8. This question is about a 4 month old infant girl who is growing well but not sleeping through the night. The mother thinks the baby may be sick since she cries so much? What is _not_ a possible explanation for an infant not sleeping through the night?
   a. The infant is fasting while sleeping.
   b. This can be a sign of gastrointestinal upset that is not a sign of hunger, but colic.
   c. This shows that the infant needs an additional nutrition source such as baby cereal.
   d. This means the frequency of breast feeding each day needs to be increased.

9. Identify at least 2 characteristics of baby foods appropriate to be first foods? (short answer)

10. Identify at least 2 reasons failure to thrive may not be a result of insufficient energy intake in an infant? (short answer)

Resources

American Academy of Pediatrics

Reliable and credible sources of pediatric medical expertise in position papers, with sections for consumers and health care providers.
Website: www.pediatrics.org

American Dietetic Association

Consumer and provider information that includes child nutrition and health information and access to credible resources.
Website: www.eatright.org

Food Allergy Network

Reliable and scientifically based information for families with diagnosed allergies. This site routinely includes recipes.
Website: www.foodallergy.org

Health/Infants

Get the latest news on children's health from CNN.
Website: http://www.cnn.com/HEALTH/livingwell/children

National Center for Growth Statistics (source of growth charts)

Site for obtaining growth charts and guidelines for their use.
Website: www.cdc.gov/nchs

Canadian Perinatal Surveillance System

The Public Health Agency of Canada produced this good source of growth charts for various gestational ages.
Website: www.phac-aspc.gc.ca/rhs-ssg/bwga-pnag

Chapter 9

Infant Nutrition:
Conditions and Interventions

Prepared by **Janet Sugarman Isaacs**

Key Nutrition Concepts

1 Infants who are born preterm or who are sick early in life require nutritional assessment and interventions to meet their nutritional needs for growth and development.

2 Early nutrition services and other interventions can improve long-term health and growth among infants born with a variety of conditions.

3 The number of infants requiring specialized nutrition and health care is increasing due to the improved survival rates of small and sick newborns.

Introduction

Most infants are born healthy and then grow and develop in the usual manner. This chapter addresses the nutritional needs of infants who have health problems before or shortly after birth and are at risk for health or developmental difficulties. Within the first year of life, most infants have minor illnesses that do not interfere with growth and development. However, infants who were sick or small as neonates are likely to have conditions that may change the course of growth or development. **Children with special health care needs** is a broad term that includes the infants discussed in this chapter. As in Chapter 8, this chapter models sensitive communication with families by avoiding the word *normal* and, by implication, *abnormal* when referring to infants with special health care needs. Similarly, the designation *normal growth* is replaced by the phrase *standard growth* when referring to the CDC growth charts. Language such as "she is below normal on the growth charts" is replaced with "family-friendly" language, such as "she is the weight of a typical 6-month-old." Most families use the word *premature* (or *prematurity*) comfortably, but *preterm birth* is the conventional usage in maternity care. Both terms refer to infants born before 37 weeks of gestation, and both are used in this chapter.

Children with Special Health Care Needs A federal category of services for infants, children, and adolescents with, or at risk for, physical or developmental disability, or with a chronic medical condition caused by or associated with genetic/metabolic disorders, birth defects, prematurity, trauma, infection, or perinatal exposure to drugs.

Low-Birth-Weight Infant (LBW) An infant weighing <2500 g or <5 lb 8 oz at birth.

Very Low-Birth-Weight Infant (VLBW) An infant weighing <1500 g or <3 lb 5 oz at birth.

Extremely Low-Birth-Weight Infant (ELBW) An infant weighing <1000 g or 2 lb 3 oz at birth.

Neonatal Death Death that occurs in the period from the day of birth through the first 28 days of life.

Perinatal Death Death occurring at or after 20 weeks of gestation and through the first 28 days of life.

Infants at Risk

The overall U.S. infant mortality rate decreased 45% between 1980 and 2006, but it has plateaued in this new century.[1] The health care system has been more successful at saving ill infants than in preventing preterm birth, low birth weight, or chronic conditions.[2,3] The costs of preterm birth are not only the longer stay for care after birth but also costs over the first year of life, which are more than doubled from higher rates of physician and hospital visits.[3] The number of infants requiring nutritional services is increasing in large measure because of advances in neonatal intensive care. Small preterm infants who did not survive in the past are now being "saved." These are *low-birth-weight infants*, *very low-birth-weight infants*, and *extremely low-birth-weight infants*, who require the most intensive resources to support life and account for many *neonatal* and *perinatal deaths*. The smallest living newborns, who weigh 501–600 grams (1 lb 2 oz to 1 lb 5 oz), have a 31% chance of survival at birth.[2] This birth-weight range corresponds to about 23 weeks of gestation in the second trimester of pregnancy. Infants with birth weights of 901–1000 grams (2 lb to 2 lb 3 oz) are in the 29-week range of gestation and have an 88% change of survival.[2] The outcomes of those infants who survive preterm birth include 60% rate of disability for those with extremely low birth weights and 31% rate of disability for those with very low birth weights. Nutrition needs from preterm infants continue to differ from those of full-term infants well past infancy, and they are found to correlate with health into adulthood[2]. Infants with genetic disorders, malformations, or birth complications have also benefited from advances in treatment and are less likely to die in infancy. However, they are much more likely to have chronic conditions, with increased need for medical, nutritional, and educational services later.

Regardless of what condition is involved, these nutrition questions are likely:

- How is the baby growing?
- Is the diet providing all required nutrients?
- How is the infant being fed?

In-depth nutrition assessments make sure nutrition is not limiting an infant's growth and development. Such assessments are needed by three main groups of infants:

- Infants born before 34 weeks of gestation. Preterm infants are born at less than 37 weeks of gestation, but generally only those born before 34 weeks of gestation have higher nutritional risks.
- Infants born with consequences of abnormal development during pregnancy, such as infants born with heart malformations as a result of the heart not forming correctly or exposure to toxins

during gestation. Exposure to alcohol during gestation may interfere with brain formation and result in permanent changes in brain function. This second category includes infants with genetic syndromes, such as *Down syndrome*.

- Infants at risk for chronic health problems. Risks may come from the treatment needed to save their lives, or from the home environment that the baby enters. Examples of conditions that increase risks are *seizures* or cocaine withdrawal symptoms. Long-term consequences, such as later learning problems, may not be known for years.

Families of Infants with Special Health Care Needs

Every parent's wish is that his or her baby will be healthy. When parents find out their newborn has medical problems, they grieve for the loss of the perfect child of their dreams. The emotional impact of having a sick newborn

Illustration 9.1 Infant girl with Down syndrome after her heart surgery.

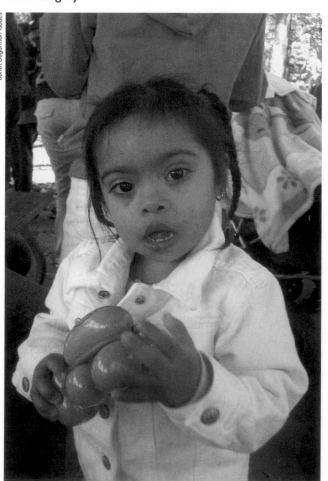

Janet Sugarman Isaacs

overwhelms many parents, and providers of services for these families must be sensitive to parents' emotional needs. Coping styles of various family members vary, even if they are well prepared. It may take over a year for some family members to understand how the baby is doing and to adjust to the special needs of the child. For conditions with long-term consequences, the first year may not be long enough for parents to see that their infant is developing differently from other babies.

Energy and Nutrient Needs

Nutrient requirements for infants with health conditions are based on the recommendations for healthy infants.[3] Specific nutrients may be adjusted higher or lower based on the health condition involved. Adjusting the diet to changing conditions and close monitoring of growth and development may result in changing recommendations quite frequently in the first year. Scientific frontiers of medicine, genetics, nutrition, and technology interact in caring for sick infants. Nutrition requirements are not known for every condition, and individuals respond at their own pace of growth and development, so many nutrition recommendations are based on the best judgment under the circumstances.

Energy Needs

For infants with special health care requirements, RDA may not be appropriate, since they are based on the needs of generally healthy infants.[3] DRI are used instead, so that the estimated energy requirements (EER) and adequate intakes (AI) can be considered as appropriate to the circumstances. For infants with special health care needs, caloric needs may be the same, less, or more than the DRI for infants (570 kcal/day).[3] Some conditions in newborns that have caloric needs based on the DRI are cleft lip and palate or phenylketonuria (PKU). The more common situation is that caloric needs are increased. Caloric needs of sick infants can be estimated with measurements such as *indirect calorimetry*.[4] Machines that measure indirect calorimetry are often available in intensive care nurseries. Estimated energy needs can change, however, with medications, activity, health conditions, and growth. Such estimates show that sick and small infants vary in their caloric needs more widely than expected, and that caloric deficits in preterm infants may be more common than previously

> **Down Syndrome** Condition in which three copies of chromosome 21 occur, resulting in lower muscle strength, lower intelligence, and greater risk for overweight.
>
> **Seizures** Condition in which electrical nerve transmission in the brain is disrupted, resulting in periods of loss of function that vary in severity.
>
> **Indirect Calorimetry** Measurement of energy requirements based on oxygen consumption and carbon dioxide release.

known.[4] Extra calories are needed in circumstances such as the following:

- Infections
- Fever
- Difficulty breathing
- Temperature regulation
- Recovery from surgery and complications

Infants who are born preterm at less than 34 weeks of gestation particularly need higher energy intakes. The American Academy of Pediatrics suggests that premature infants need 120 cal/kg.[4] Intakes for recovering premature babies may be even higher, and the range of caloric need can be wide. The European Society for Gastroenterology and Nutrition gives a caloric intake range of 95–165 cal/kg.[5] The amount of calories needed to gain 15 grams per day is recommended.[5] Infants born with VLBW or ELBW may still be weak and have difficulty feeding when they come home. Their higher calorie needs may be more difficult to meet given the small volume that they can consume. Over time, the recovering infant may increase intake so that even 180 cal/kg per day may be consumed.

Some infants need less energy than the RDA. These are infants born with smaller muscles or lower activity as a result of the inability to move certain muscles. An infant who has Down syndrome or one with repaired spina bifida needs fewer calories than the DRI of 108 cal/kg body weight.[4] Too many calories would interfere with her efforts to crawl; her weak muscles would have more body weight to move.

Protein Requirements

As noted for energy needs, protein requirements of infants with special health care needs may be higher, lower, or the same as other infants. Protein requirements based on the DRI of 1.52 grams of protein per kg body weight are recommended if the condition does not affect growth or digestion.[4] Protein recommendations are sufficient if total calories are high enough to meet energy needs.[5,6] The concept of protein sparing is important in fast-growing infants. If enough energy is available from glucose generated from foods containing fats and carbohydrates, then the amino acids generated from protein-containing foods are spared; that is, the amino acids are available for growth. If glucose from foods is not sufficient for meeting energy needs, however, amino acids from digestion of protein foods will be used to meet energy needs, and therefore less will be available for growth. When total caloric intake is low, protein-rich foods become an energy source. In this circumstance, providing the DRI for protein may be inadequate and result in slow growth. With preterm infants, a sign of inadequate protein may be slow head growth, which is an indicator of brain growth.[6]

Conditions that could slow growth may require higher protein levels than the DRI for protein for infants. Higher protein recommendations are common in early infancy for conditions such as recovery from surgery or LBW. Protein intakes of 3.0–3.5 g/kg are appropriate for premature and recovering infants.[4,6] For recovering from some complications of ELBW, high protein intakes—as much as 4 g/kg—appear safe with adequate fluids, and without kidney problems.[6] The importance of protein to growing neonates is hard to overemphasize. Protein deficits in preterm infants have been shown even when high protein is provided, depending on how small and sick they are and how soon after birth the infant is being fed.[5,6]

Protein recommendations lower than the DRI are unusual in infancy. Infants with lower muscle activity as a result of smaller-sized muscles generally need lower protein. One example is Down syndrome. Conditions that lower physical activity and movement are often not identified until the infant is old enough to be moving around, closer to the end of the first year. During infancy, muscle tone is known to change over time, so muscle coordination and movement problems are usually confirmed later.

Form of Protein   Many illnesses interfere with the functioning of the gastrointestinal tract and digestion, even though newborns are born with intact enzymes for protein digestion. Protein and fat digestion depend on liver and pancreatic enzymes for intestinal absorption. However, many conditions associated with preterm birth and illness stress the liver and reduce its ability to function, causing changes in protein and fat digestion. Sick infants may require forms of protein in which amino acids are in short chains, such as in hydrolyzed protein, or single amino acids.[5,6] Other examples of infants needing protein that has been broken down are those with metabolic disorders.[7] Total protein may be limited and partially replaced by mixtures of specific amino acids. For infants with PKU, meat and dairy-product intake have to be limited, because these foods contain too much of the amino acid phenylalanine.

Fats

Infants need a high-fat diet compared to older people, because fats provide energy. Up to 55% of calories from fats may be recommended.[6] The need for calories provided by fats is especially important in sick or recovering infants. Low-fat diets are generally not recommended for infants. Conditions that require limiting fats for infants are uncommon. One example is very sick infants who require heart-lung bypass machines as a part of major surgery. Such infants are given low-fat diets after age 2, without a specific level of fat restriction.[6] Fats are more difficult to absorb for infants with VLBW or ELBW because they require pancreatic and liver enzymes.[6] These enzyme systems may also be impaired in sick infants. Naturally occurring long-chain fats in breast milk may be supplemented with medium-chain fatty acids for sick infants. Medium-chain triglycerides do not require bile for absorption, so they are preferred.[6] Making sufficient bile for digesting long-chain fats requires healthy livers, like those in full-term infants, that are not likely in preterm births. **MCT oil** can be added to ensure calories from fats are available. Additionally, the essential fatty acids—alpha-linolenic and linoleic acid, as well as docosahexaenoic acid (DHA) and arachidonic acid (AA)—are provided in breast milk, human-milk fortifier, or special formulas.[6,8] The impact of adding these supplements to infant formulas has been reported for outcomes such as growth and measures of cognitive development, but results depend on the age and health of infants being studied, measures used, and methods of feeding. So far, improved cognitive development for LBW infants provided DHA and ARA in human milk was found when the infants were tested at 6 months of age. [8]

Vitamins and Minerals

DRIs for vitamins and minerals are appropriate for many infants with health conditions because recommendations are set with a safety margin.[9] However, DRIs are based on growth of typical infants, not those in which *catch-up growth* is required.[5,6] Vitamin and mineral requirements are affected by various health conditions, particularly those involving digestion. Prescribed medications may increase the turnover of specific vitamins.[9] Some infants with special health care needs have restrictions in volume consumed or activity that increase or decrease needs for specific vitamins or minerals. For example, limited volume of liquids may rule out vitamin-rich juices in the diet of infants with breathing problems.

High-potency vitamin and mineral supplements are usually prescribed for sick or recovering infants. Calcium is a potentially limiting nutrient in sick infants because calcium imbalance and *hypocalcemia* occur with a variety of conditions.[6] Iron, vitamin B$_{12}$, vitamin D, and fluoride are limiting in some specific situations.[6,9] Even after infants who were LBW or VLBW are eating well, vitamin and mineral requirements may be higher than the DRI, depending on specific health conditions. After preterm birth and discharge to the home, deficiencies in copper, zinc, and vitamin D are rare; they may be checked by blood tests.[4] The early signs of rickets as seen by X-ray are considered a sign of needing more vitamin D, above the 400 IU daily recommendation for term and preterm infants.[4]

Human milk fortifiers are used to boost calories as well as to provide additional vitamins and minerals in some infants in neonatal intensive care units.[6] In order to meet the higher requirements of specific vitamins and minerals for VLBW infants, such products can be added to breast milk. Such products are used under specific conditions, such as when an infant can only tolerate a low volume per feeding. They are intended to bridge the gap between breast milk and the extra needs of a VLBW infant.[8] Major ingredients are vitamins A, D, and C, and minerals such as calcium, phosphorus, sodium, and chloride.[9] Iron is not provided in human milk fortifiers, and it is prescribed as needed.[9]

Some infant formula products also provide vitamins and minerals in concentrated amounts in formulas for premature infants. Such formulas are supplemented with extra calcium, phosphorus, copper, and zinc compared to standard formulas.[4,6] (The high levels of vitamins and minerals in premature infant formulas are shown in Table 9.2 later in this chapter.) For some conditions, vitamins are used not only as dietary components, but as pharmaceuticals. An example is a condition in which vitamin B$_{12}$ is injected as a part of therapy for a rare genetic disorder of protein metabolism.[7] Vitamin A supplementation has been proven to be effective for low-birth-weight infants to lower lung and breathing problems during recovery.[10]

Growth

Growth in infancy is usually a reassuring sign that sufficient calories and nutrients are provided. The Centers for Disease Control's 2000 growth charts are a good starting point for monitoring of growth in infants with health risks.[11] The first goal of nutritional care is to maintain growth for age and gender. Later, this approach may be modified if there is a growth pattern typical for a specific condition that is identified after the first year. A steady accretion of weight or height is a sign of adequate growth, even if gains are not at the typical rate. Plateaus in weight or height, or weight gain followed by weight loss, are signs of inadequate growth. Growth may be assessed reliably using each infant as his or her own control regardless of health conditions. As noted in Chapter 8, the methods of assessing growth require consistency and accuracy in order to make sure growth is interpreted correctly. Errors such as confusing pounds and kilograms in plotting interfere with interpretation no matter what growth chart is being used.

Usually, providing sufficient calories and nutrients results in good growth, but not in all cases. Sometimes slow growth is a symptom of an underlying condition, rather than a sign of inadequate nutrition. For example, infants who are born with genetic forms of kidney disease are short even when they consume adequate diets during the first year. Refinements in the usual methods and

MCT Oil A liquid form of dietary fat used to boost calories; composed of medium-chain triglycerides.

Catch-Up Growth Period of time shortly after a slow growth period when the rate of weight and height gains is likely to be faster than expected for age and gender.

Hypocalcemia Condition in which body pools of calcium are unbalanced, and low levels are measured in blood as a part of a generalized reaction to illnesses.

interpretation of growth are needed in conditions known to influence growth and development. These include:

- Using growth charts for specific diagnoses, such as Down syndrome growth charts[12] (a list of specialty growth charts is included in Chapter 11)
- Looking for biochemical indicators of tissue stores of nutrients such as iron or protein, and of electrolytes such as potassium and sodium
- Noting indicators of body composition, such as taking body-fat measurement (these can be used to show caloric intake is not limiting growth because fat stores are adequate)
- Special attention to indicators of brain growth, such as measuring head circumference, which may help to explain short stature or other unusual growth patterns
- Using treatment guidelines or published protocols, including disorder-specific weight-gain graphs and recommendations, instead of standard growth charts
- Considering medications that change weight gain, appetite, or body composition (side effects of medications can explain rapid changes in weight)

Growth in Preterm Infants

Growth charts developed by the CDC can be used to assess growth progress of preterm infants with birth weights over 2500 grams.[11] The body composition of infants born preterm is not the same as that of term infants, in part because these infants have missed part of the third trimester, when fat is added rapidly.[6] In fact, body fat buildup is a late sign of recovery from preterm delivery. Body composition at various gestational ages is used to adjust growth expectations based on age. Treatment of the infant's medical condition may also affect growth expectations; for instance, fluid accumulation may artificially increase weight. As a result of such considerations, VLBW and ELBW infants are not represented by the standard growth charts.[11] Growth for newborns with birth weights between 501 and 1501 grams can be tracked by the Neonatal Research Network Growth Observational Study Research Network, sponsored by the National Institutes of Health, National Institute of Child Health and Human Development (NICHD).[13,14] Its website (https://neonatal.rti.org/birth_curves/dsp_BirthCurves.cfm) has a VLBW Postnatal Growth Chart component that projects growth from whatever birth weight, birth length, or birth head circumference is entered.[14] The management of preterm birth is changing so rapidly that recently published growth charts may not reflect growth patterns achieved with current practices.

Growth charts for preterm infants are needed in hospitals for determining the age of gestation at birth, and for tracking growth until discharge. Standard growth charts often replace them when the child goes home. Growth charts for preterm births are being updated routinely and cover more of fetal life, starting at 22 weeks of gestation and ending usually at 10 weeks after the expected 40 weeks of gestation. All preterm growth charts show that head circumference is a main indicator of growth approaching that in term infants; both weight and length lag at discharge even when 40 weeks gestational age is reached[13,14]

Correction for Gestational Age   Gestation-adjusted age is calculated by subtracting gestational age at birth from 40 weeks (the length of a full-term pregnancy). The resulting number of weeks is divided by four to obtain months. The result in months is then subtracted from the infant's current age. For example, if an infant was born at 30 weeks gestation, she is 10 weeks early. This equals 2.5 months preterm. When she is 3 months old, her gestation-adjusted age is 2 weeks, or 0.5 months. This age of 0.5 months would be used in plotting her growth on the growth chart as part of assessing her growth and development.

Does Intrauterine Growth Predict Growth Outside?

"Apples don't fall from a pear tree."

French saying

The answer is yes, no, and maybe. Fetal monitoring during pregnancy and in-depth knowledge about the development of various organ systems provides a clear pattern of growth at various gestational ages. However, many factors during and after pregnancy are known to affect growth rate. In summary, these are:

- Intrauterine environment, particularly the adequacy of the placenta in delivering nutrients; the presence of toxins such as viruses, alcohol, or maternal medications; or the depletion of a needed substance, such as folic acid
- Fetal-origin errors in cell migration or formation of organs, whether or not a cause is known; various nutrients, such as vitamin A, have been implicated in such errors
- Unknown factors that cause preterm birth, such as environmental toxins in air pollution

Research is emerging that shows some infants were born prematurely due to conditions originating during the intrauterine period.[6] As discussed in Chapter 8, SGA and intrauterine growth retardation are terms used to describe infants who are smaller than expected at birth. SGA is the more general term because it is based on the population of infants of the same gestational age.[6] Both predict higher medical risks and need for close growth monitoring.

If the intrauterine insult was early in gestation, body weight, length, and head size (brain size) and parent–child interaction are affected.[6] There has been a change in the number and size of fetal cells. The abnormal fetal growth pattern may persist despite adequate medical and nutrition support after birth. Examples of conditions causing early insults are infants born after cocaine or alcohol exposure, both of which are associated with IUGR and preterm birth.[15] Later exposure in the second or early part of the third trimester may result in preservation of head size and body length, but low weight. Some genetic conditions characterized by small size are not diagnosed until childhood, but have IUGR noted in the medical history.[7]

Intrauterine growth may not predict growth for some infants whose birth removes them from adverse exposure within the intrauterine environment. Examples of this situation are maternal uncontrolled diabetes, smoking, phenylketonuria, or maternal seizures treated by medications. In such cases the rate of growth after birth may be improved and normalized during the first year of life.[9] In general, the earlier the exposure to the toxin, the worse the effects on later growth.[16] Sometimes marijuana, alcohol, tobacco, and crack/cocaine have been used at various times during pregnancy, so growth impacts may be based on amount, timing, and interactions of toxins.

Much evidence supports the idea that what happens in early life during critical or sensitive periods can have lifetime consequences; a hypothesis is called fetal programming or fetal origins.[17] Many studies on the outcome of preterm birth, including those on feeding, growth, and types of nutrition, support the fetal programming hypothesis.[17] Preterm birth at every level, LBW, VLBW, or ELBW categories, has been correlated with lifelong impacts from measures such as school performance, drop-out rate, adult size, adult chronic diseases, and mental health status.[18,19,20] Nutrition outcomes regarding growth and feeding fit the same pattern as seen in motor skills, cognition and developmental outcomes.[19] For example, studies suggest that the most important risks for later growth may be neonates born with smaller head circumferences, regardless of level of nutrition support.[16,19]

The fetal origins hypothesis fits how early medical treatment can change the rate of intrauterine growth. If the intrauterine growth was fine, preterm birth or its complications may slow or plateau growth early in infancy. This may mask whether or not typical growth goals are appropriate all during infancy. Various studies which have measured recovery from early growth problems have found that medical factors can overwhelm nutrition benefits.[19–21] For example, a study of infants who were short for age and were provided nutritional supplements for 2 years found that the infants were still short for age at 11 and 12 years.[16] Whether or not adequate nutrition is provided early in life, of course, depends on how *adequate nutrition* is defined. Methods of feeding and managing the administration of nutrients for preterm infants are

constantly improving, but persistent growth failure after preterm birth continues, regardless of the best nutritional practices.[21]

Conditions that affect growth may be time-limited. Catch-up growth may be seen during recovery in many infants, resulting in changing growth interpretation. Surgery, when required for heart conditions, may delay growth for a short time. Growth may slow and then catch up with respiratory illnesses that resolve with medications. Usually, increased access to adequate nutrition improves growth. Only close monitoring over time may show signs of catch-up growth early, such as an increase in fat stores or length.

The amount of time needed for catch-up growth for premature infants differs based on gestational age at birth and subsequent complications.[22] Clinical conventions are to provide 1 year for catch-up growth for infants born 32 weeks or later, and 3 years for catch-up growth for VLBW or ELBW infants. Catch-up growth after low birth weight is usually encouraged, but rapid growth in infancy may be raising risks for later chronic conditions, such as cardiovascular disease and diabetes if the rate of weight gain is excessive.[23] It is likely such infants will have growth percentiles in the lower end of the normal range during childhood.

Interpretation of Growth

Hospital discharge after preterm birth may be based on a pattern of weight gain, such as 20–30 grams per day.[9] Strong emphasis is placed on growth as a sign of improving health in monitoring small and sick infants after discharge, but complications make this difficult to achieve. An example of a common diagnosis in preterm babies and difficulty in interpretation of growth is the lung condition bronchopulmonary dysplasia (BPD). Rates of growth among infants with BPD are different from those in full-term infants and preterm infants who do not have BPD. While infants with BPD are recovering, the growth pattern of these infants is affected for the entire first year of life.[15] The reasons for the slower growth are higher nutritional requirements, changes in endocrine and pulmonary systems, and perhaps interaction among these systems.

Growth-rate changes are closely associated with the frequency of illness, hospitalizations, and medical history.[13] Conditions acquired as a result of preterm birth that make growth difficult to interpret include:

- Symptoms related to intestinal absorption that can temporarily or permanently change nutrient requirements
- *Microcephaly* (small head size) or *macrocephaly* (large head size) compared to other growth indicators, which may be a sign that growth may

Microcephaly Small head size for age and gender as measured by centimeters (or inches) of head circumference.

Macrocephaly Large head size for age and gender as measured by centimeters (or inches) of head circumference.

be affected as a result of neurological consequences; both large and small head size in infancy can affect muscle mass, body composition, and subsequent growth

- Variable rates of recovery and growth, which are seen for many infants; infants are as different from each other as the rest of us, but these differences are hard to see soon after birth

Nutrition for Infants with Special Health Care Needs

Infants who are small or sick near birth may have major growing and feeding problems.[23] Infancy is such a vulnerable time of life that most health conditions occurring this early interfere with growth and development. Over time, most of these problems resolve, although some become chronic, and a few result in death. Nutrition plays an important part in preventing illness, maintaining health, and treating conditions in infancy. Nutrition tends to become more important over time to maintain

Developmental Delay Conditions represented by at least a 25% delay by standard evaluation in one or more areas of development, such as gross or fine motor, cognitive, communication, social, or emotional development.

growth if conditions are chronic and impact feeding (Illustration 9.2).[22] Table 9.1 shows nutrition problems in infants with special health care needs. Nutrition assessment documents these concerns, and nutrition services are provided based on the assessment.

Nutrition Risks to Development

Many health conditions change the infant's rate of development. *Developmental delay* describes the interaction of

Illustration 9.2 Baby girl wearing a heart monitor at home.

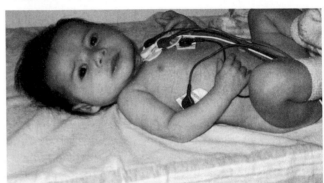

Janet Sugarman Isaacs

Table 9.1 Nutrition concerns in infants with special health care needs

Growth	Slow rate of weight gain
	Fast rate of weight gain
	Slow rate of gain in length
	Disproportionate rate of weight to height gain
	Unusual growth pattern with plateau in weight or length gain
	Altered body composition that decreases or increases muscle size or activity
	Altered brain size that decreases or increases muscle size or activity
	Altered size of organs or skeleton, such as an enlarged liver or shortened leg length
Nutritional Adequacy	Calorie needs are higher or lower
	Nutrient requirements higher or lower overall
	Specific nutrients, such as protein or sodium, are required in higher or lower amounts
	Vitamins, minerals, or cofactors (such as carnitine) are required in higher or lower amounts
Feeding	Disruption of the delivery of nutrients as a result of:
	• Structure or functioning of the mouth or oral cavity
	• Structure or functioning of the gastrointestinal tract, including diarrhea, vomiting, and constipation
	• Appetite suppression by constipation or medications
	• Disrupted interaction of the infant with the parent, such as infant cues being so subtle that parent responses are delayed
	• Posture or position that promotes or interferes during meal times
	• Timing of nursing, meals, and snacks throughout the day
	• Inappropriate food choices or methods of preparation
	• Interruptions in adequate shelter for feeding and sleeping
	• Instructions unclear or too complicated for the parent to follow

a chronic condition with development. The terms *children with special health care needs* and *developmental delay* are general terms used to allow nutritional, medical, and developmental services to be provided for infants.

Developmental delay is used to describe a wide range of symptoms that reflect slow development. Symptoms that relate to nutrition are common. These symptoms include infants who are growing slower than expected for age, or have difficulty in feeding, such as refusing food from a spoon by 8 months of age. An example is a 2-month-old girl who does not breastfeed for more than a few minutes per side. At first this may appear to be a problem of breastfeeding position or frequency. By 4 months, weight gain is slower than expected, so now growth concerns and feeding concerns are interacting; it is not clear if they are separate or related problems. These concerns are sufficient to request an evaluation for eligibility for intervention services. Several months later, after various services have been put in place, it may still be unclear if these nutrition problems are from development, a health condition such as a heart murmur, or the interaction of both. In any case, the girl fits the category of a child with a special health care need. This allows the family to benefit from nutritional, medical, and developmental interventions, without requiring a specific diagnosis. Infants generally are not old enough to have a specific diagnosis related to development, such as mental retardation or *autism*.[23]

Down syndrome is an example of a condition in which developmental delay is noted in infancy. Down syndrome prevalence is about 13 cases per 10,000 live births.[1] Nutrition concerns with infants who have Down syndrome are feeding difficulties related to weak muscles in the face, and overall; high risk of overweight; and constipation. Heart and intestinal conditions are more common in infants with Down syndrome, so their nutrient needs may be increased if surgery is required. This is also an example of a chronic condition in which nutrition problems such as overweight increase over time if prevention and maintaining health are not addressed. Growth requires close monitoring to identify and prevent overweight starting in infancy. Infants with Down syndrome love to suck and have things in their mouths so much that it is easy to overfeed them. Development of movement occurs at a slower rate, with lower physical activity, which also can contribute to overweight. Giving parents their own copy of Down syndrome growth charts for infants is recommended after the diagnosis is confirmed.[16] These charts may be helpful in recognizing typical growth and preventing overweight early. These special growth charts are available from places that serve children with Down syndrome, such as developmental or genetics clinics in major medical centers.[16]

Not all children with developmental delay in infancy have developmental disabilities later. For example, an infant with breathing problems may be slower to grow and to crawl as a result of his higher caloric needs during the first year of life. Such an infant may show developmental delay in motor skills, but by age 3 he will have improved in overall health and have caught up to others in motor skills. He would not have a developmental disability. Other examples are infants from high-risk pregnancies, such as those born large for gestational age as a result of poorly controlled gestational diabetes. Many infants require short stays in intensive care units for glucose regulation; some may have long-term risks for their development. Some infants with developmental delays continue to have slower development over time. After infancy, when standard testing and evaluation can be performed, the term *developmental delay* may be replaced with a more specific type of medical or developmental diagnosis.[24]

Severe Preterm Birth and Nutrition

The yearly incidence of VLBW in the United States is approximately 60,000 infants—about the same as the population of a small city, such as Iowa City, Iowa. Infants with birth weights near 1500 grams (3 lb 4 oz) have gestational ages from 28 to 32 weeks and a survival rate of almost 90%.[25] Each infant requires immediate intensive care hospitalization, and if they survive, these infants continue to have high nutritional needs throughout infancy. ELBW infants weigh less than 1500 grams and have gestational ages ranging from 23 to 28 weeks. Despite advances in the care of such infants, disability such as delayed development is a common outcome of ELBW.[26] Attention and learning problems in school-age children are at higher rates than in children who were born at term, although the majority of children who were ELBW do not have disabilities.[21] Some outcome studies are demonstrating lifelong consequences of low birth weight, such as impact on later employment as adults.[27]

Nutrition problems resulting from VLBW and ELBW preterm births are addressed as they present. The initial problem after birth is that the newborn cannot nurse like a full-term infant, and most require respiratory support to breathe. Getting adequate calories and nutrients into the preterm infant requires *nutrition support*, usually first *parenteral feeding* and then *enteral feeding* methods.[9,27] Feeding problems of preterm infants are discussed later in this chapter. All newborns have high metabolic rates, and they will use fat stores and protein in tissues and muscles to meet glucose needs if consumed calories and

Autism   Condition of deficits in communication and social interaction with onset generally before age 3, in which mealtime behavior and eating problems occur along with other behavioral and sensory problems.

Nutrition Support   Provision of nutrients by methods other than eating regular foods or drinking regular beverages, such as directly accessing the stomach by tube or placing nutrients into the bloodstream.

Parenteral Feeding   Delivery of nutrients directly to the bloodstream.

Enteral Feeding   Method of delivering nutrients directly to the digestive system, in contrast to methods that bypass the digestive system.

nutrients are not sufficient. This happens sooner in infants than in adults.[4,5] Providing sufficient calories and nutrients to meet requirements and preserve ingested protein and calories for growth is the goal, but it may be difficult and take more time than expected in sick and recovering infants.

How Sick Babies Are Fed

Gastrointestinal upset is a response to many conditions in newborns, whether the intestines are the initial problem or not. VLBW, ELBW, and sick infants are especially vulnerable to problems related to the gastrointestinal tract. Such problems directly affect how calories and nutrients are provided, as well as the composition of the diet. For example, if a newborn gets an infection, an early sign may be inflammation of the intestine. As a response, the method of feeding the infant has to be adjusted. Inflamed or damaged areas may slow or interrupt typical intestinal muscle movements, resulting in signs of increasing illness.[9,13] Blood loss from the intestines is a sign of *necrotizing enterocolitis (NEC)*, a serious condition in the neonate. When this occurs, oral feeding is stopped and replaced by parenteral nutrition.

> **Necrotizing Enterocolitis (NEC)**   Condition with inflammation or damage to a section of the intestine, with a grading from mild to severe.
>
> **Oral-Gastric (OG) Feeding**   A form of enteral nutrition support for delivering nutrition by tube placement from the mouth to the stomach.
>
> **Transpyloric Feeding (TP)**   Form of enteral nutrition support for delivering nutrition by tube placement from the nose or mouth into the upper part of the small intestine.
>
> **Gastrostomy Feeding**   Form of enteral nutrition support for delivering nutrition by tube placement directly into the stomach, bypassing the mouth through a surgical procedure that creates an opening through the abdominal wall and stomach.
>
> **Jejunostomy Feeding**   Form of enteral nutrition support for delivering nutrition by tube placement directly into the upper part of the small intestine.

Many gastrointestinal conditions interfere with infant feeding, such as gastroesophogeal reflux, constipation, spitting up, and vomiting. In small and sick newborns, these gastrointestinal conditions may represent slow or uncoordinated movements of the intestinal muscles.[9] These conditions do not rule out enteral feeding, which stimulates the intestines and keeps them healthy. Feeding methods are selected based on the length of time before it is expected the baby can nurse or feed without help. Gavage feedings may be used. These are slow feedings sent from the mouth or nose into the stomach though a tube. Infants who are too weak to breastfeed may be offered the comfort of the breast or pacifier along with gavage feeding. *Oral-gastric (OG) feeding* is also used. Other enteral methods are *transpyloric feeding (TP), gastrostomy feeding*, and *jejunostomy feeding*.[9] These methods are used when nutrition support is expected to be needed for several months.[23]

Food Safety   Preterm babies with immature immunological systems are prone to infection, so every effort is made to ensure their feedings do not become contaminated. The rate of feeding preterm infants is often much slower than that for full-term infants, and formula or breast milk is at room temperature for a longer time. Contamination of feeding equipment increases with time; consequently, hospitals have policies requiring them to change the feedings often, such as every 4 hours.[28–30] Hospitals avoid using powdered formulas as much as possible since liquid formulas can be heat-treated to higher temperatures.[28] Frozen and then thawed breast milk is tracked so that it is given to the correct baby (usually too weak to suck) inside set time frames. The purity and safety of infant formula that is available to hospitals or sold to the public is regulated by the Food and Drug Administration Center for Food Safety Administration.

What to Feed Preterm Infants

Breast milk is the recommended source of nutrition for preterm infants. Colostrum and breast milk are produced even when the mother delivers very early.[31] Preterm human milk has increased protein content compared to term milk.[31] Hospital protocols and policies for having mothers pump and freeze breast milk for later use by their preterm infants are highly recommended.[32] Staff training to encourage new mothers at home to rest enough and pump enough to stimulate breast milk production is also recommended.[33] Barriers to breastfeeding small and sick newborns are partially based on their abilities, on how sick they are, and on the care system. Promoting breast milk for preterm infants is recommended as hospital policy by the American Academy of Pediatrics, but hospitals differ in practices.[31] Medical conditions in the infant may undermine breastfeeding. Infants are generally able to nurse successfully at about 37 weeks of gestation. Prior to that age, they may benefit from being put to the breast to stimulate non-nutritive sucking, or sucking that does not deliver milk to be swallowed. There are a few conditions in which human milk is unsafe for preterm or sick infants. Breastfeeding is contraindicated when breast milk contains harmful medications, street drugs, viruses, or other infective agents, or when the infant has a specific type of gastrointestinal tract malformation or inborn errors of metabolism.[31]

Depending on the infant's birth weight and health status, breast milk may be insufficient in nutrients unless supplemented by human milk fortifier and/or other sources of calories, such as MCT oil. If not fed modified breast milk or nursing, the infant's source of nutrition may be cow's milk- or soybean-based formulas.[32] Whey as the predominant form of protein from cow's milk is recommended because its amino acid profile is closer to that of human milk.[8]

Infant formulas for preterm infants are available for home use after hospital discharge, if breast milk is not available. They provide the higher calories and nutrient levels that small infants need compared to term infants.[34] Standard formula that is 20 calories per fluid ounce can also be used for preterm infants, modified in a manner similar to breast milk to boost calories and nutrients. High-calorie formulas, such as 28 cal/fl oz, may be appropriate for some infants. Such high-calorie formulas are not routinely used because they have high osmolarity, which may affect fluid and electrolyte balance. Table 9.2 shows a comparison of premature and standard formulas.[34] If the infant easily fatigues or is too weak to suck enough volume, 22 or 24 cal/fl oz formulas may be recommended.[9] The sources to add extra calories and nutrients are selected based on the infant's gastrointestinal tolerance and volume requirements. They may include MCT oil; polycose; rice baby cereal; and, rarely, human milk fortifier. Routine nutritional assessment of the infant's growth tracks the diet's effectiveness in providing adequate nutrients and calories.

Preterm Infants and Feeding

VLBW or ELBW infants usually progress at their own rate of development regarding feeding skills. The goal is the same as for all infants—to achieve good nutritional status, as indicated by growth and feeding skills progression. Most families enjoy feeding their infants and experience few long-term feeding problems. Infants who were born

Table 9.2 Selected nutrient composition of term and preterm formulas (at normal dilution; per 100 calories or 4.5 fl oz)

Nutrients	20 Cal/ Fl Oz	22 Cal/ Fl Oz	24 Cal/ Fl Oz
Protein	2.1 g	2.8 g	3 g
Linoleic acid	860 mg	950 mg	1060 mg
Vitamin A	300 IU	450 IU	1250 IU
Vitamin D	60 IU	80 IU	270 IU
Vitamin E	2 IU	4 IU	6.3 IU
Thiamin (B_1)	80 mcg	200 mcg	200 mcg
Riboflavin (B_2)	140 mcg	200 mcg	300 mcg
Vitamin B_6	60 mcg	100 mcg	150 mcg
Vitamin B_{12}	0.3 mcg	0.3 mcg	0.25 mcg
Niacin	1000 mcg	2000 mcg	4000 mcg
Folic acid	16 mcg	26 mcg	35 mcg
Pantothenic acid	500 mcg	850 mcg	1200 mcg
Biotin	3 mcg	6 mcg	4 mcg
Vitamin C	12 mg	16 mg	20 mg
Inositol	6 mg	30 mg	17 mg
Calcium	78 mg	120 mg	165 mg
Copper	75 mcg	120 mcg	125 mcg

Table 9.3 Preterm and term infant feeding differences

Preterm Infant	Term Infant
Central nervous system does not signal hunger	Signals hunger; has supportive newborn feeding reflexes
Unstable feeding position, such as a forward head position	Stable and facilitating feeding position from newborn reflexes
Oral hypersensitivity	Readily accepts food by mouth

preterm, however, may be hard to feed. There are several reasons for this:[32]

- Fatigue: Low levels of arousal of weak or sick infants may lessen feeding duration.

- Low tolerance of volume: Abdominal distention due to feeding may result in changes in breathing and heart rate, so that the infant stops feeding.

- "Disorganized feeding" may result from the infant having experienced defensive and unpleasant reactions to feedings or procedures, so anything coming to the mouth causes a stress reaction rather than a pleasurable reaction.[9]

Regardless of the associated conditions, certain feeding characteristics of preterm infants are distinct from those of term infants (Table 9.3). Most recovering infants improve in their feeding abilities with time. Anxiety decreases as parents become more comfortable with caring for their infant at home. The underlying reflexes that associate pleasure with feeding reemerge, and the interaction of the infant with the feeder improves. There is a lot of room for hope in the feeding process after discharge.

Major advances in understanding the nutritional needs of preterm babies have come about from working with smaller and smaller infants. Table 9.4 presents an example of a typical diet for a premature baby who was not breastfed and had a 3-month hospital stay after birth. The infant has gastroesophageal reflux and prescribed medications that are included in his feeding instructions. This example shows that providing an adequate diet is an important part of the infant's growth and development and his recovery from preterm birth complications (see Case Study 9.1 on the next page).

Infants with Congenital Anomalies and Chronic Illness

Infants who are not preterm but require neonatal intensive care may be at risk for chronic illness. About half of babies in neonatal intensive care units have normal birth

Case Study 9.1

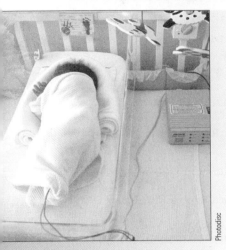

Photodisc

Premature Birth in an At-Risk Family

Nutrition Assessment uses data sources from hospital records, and foster-mother interview. Previous nutrition diagnoses include altered nutrition-related laboratory values, swallowing difficulty, imbalance of nutrients, and inadequate protein-energy intake related to birth at 30 weeks of gestation. The case study is about appropriate nutrition interventions.

A baby named Eric is born, appropriate for gestational age, at 1.4 kilograms (3 lb). Like his mother, he tests positive for cocaine. Eric receives routine intensive care services, which include ruling out sepsis, and is given head ultrasound studies. By 33 weeks he is being fed OG and has only transient respiratory difficulties. Prior to discharge at 37 weeks, he appears to be developing normally. He drinks 22 fl oz of formula per day at the rate of about 1.5–2 fl oz per feeding, with 10 feedings per day. He is placed in foster care with an experienced foster mother who has two older children.

Eric's custody is reconsidered when his biological mother expresses interest when he is about 9 months of age. Shortly after that, his mother's parental rights are terminated based on criminal charges. His foster mother reports that he has been a colicky baby, and he has had at least three ear infections during his first year of life. He is enrolled in an *early intervention program* based on his prematurity and intrauterine drug exposure. Eric's initial developmental testing was within normal limits at 6 months.

Eric's foster mother expresses concern about his intervals of periodic crying, during which he accepts neither soothing nor a bottle. Eric is diagnosed with gastroesophageal reflux (GER) and slow gastric emptying, and he is treated medically starting at 8 months. He is slow to accept foods on a spoon, with gagging and spitting up. His growth is near the 25th percentile for weight and height. His head circumference is at the 5th percentile. Eric does not sit up unassisted until 8.5 months, but he turns over from stomach to back and vice versa easily. He is sent to a genetic specialist because he appears to have some facial features consistent with fetal alcohol syndrome, such as low-set ears, wide nasal bridge, and thin upper lip. The diagnosis is not confirmed because he is too young, but it is reported as a possibility to be reevaluated after he is older and can be given developmental tests. If this diagnosis were confirmed, standard growth charts would not fit Eric, because short stature and low weight are part of the diagnosis, even when adequate nutrition is provided. In infancy Eric's growth was within normal limits after correction for prematurity, with the same trend of a lower head circumference.

Eric's foster mother expresses an interest in adoption when he is almost 1 year old. She is pleased that he needs no medication and is growing well. His early intervention services are continued based on his at-risk status, because no specific 25% delay has been documented at 1 year. (Later, at 34 months of age, he is diagnosed with mixed developmental delay based on cognitive and speech delays.) Eric is adopted by the foster family.

Early Intervention Program Educational intervention for the development of children from birth up to 3 years of age.

Questions

1. Did Eric's early birth account for his slow growth later?
2. Did nutrition affect when Eric's developmental delay probably started? Note that it was diagnosed at about age 3.
3. What are the signs that Eric can outgrow his problems?

(Answers are located in the Instructor's Manual for the 4th edition of *Nutrition Through the Life Cycle*.)

Table 9.4 Example diet for a VLBW infant at 8 months of age, with a corrected age of 4.5 months (weight 4.5 kg, 5–25% on VLBW Premature Boys growth chart)

Food and Formula	Feeding Instructions
Five feedings per day of formula, each 5 fl oz, High-calorie formula (24 cal/fl oz) with 2 Tbsp rice cereal added	Provide support for semi-reclining feeding position during and up to 30 minutes after feeding
Medications for stomach added to 2 fl oz of 50% diluted apple juice two times per day	Encourage use of pacifier for comfort between feedings
Two meals of pureed baby foods fed with spoon, total intake one 2 oz jar	Offer bottles every 3 hours except overnight if no signs of hunger before then
Liquid vitamin/mineral supplement	Keep scheduled appointments for WIC; weigh in at MD office, bringing diet log book

weights and experience lower mortality than LBW infants. They tend to have a higher rate of *congenital anomalies* (22%) and often require rehospitalization.[35] These infants need more nutrition services than typical infants because their growth and feeding development require close monitoring and intervention.

Congenital anomalies are recorded in the Birth Defects Monitoring Program (BDMP).[36] Prevalence data is published by the Centers for Disease Control based on states and hospitals that participate voluntarily in surveillance programs. The United States keeps prevalence data not just for infants but also for children of various ages. *Infant mortality attributable to birth defects (IMBD)* accounts for approximately 2 deaths per 1000 live births.[36] Whereas infant mortality is decreasing in the United States, the proportion of infants who die from IMBD is increasing.[36] The major types of birth defects associated with death are, first, heart malformations and, next, central nervous system defects.[36] Examples of central nervous system congenital anomalies are spina bifida and *anencephaly*. Since folic acid has been added as a supplement in grains and flours, rates of spina bifida and related conditions have declined 26%.[37]

Infants with congenital anomalies, genetic syndromes, and malformations fit in the category of children with special health care needs, so they are eligible for a wide range of medical, nutritional, and educational services to maximize their growth and development.[25] All babies born with such conditions have risks to maintaining good nutrition status as they receive treatment. The nutrition concerns are growth, adequacy of the diet in providing required nutrients, and feeding development, as shown in Table 9.1. Nutrition services range from temporary to long term, and they are as diverse as the many different types of conditions involved. Several examples of disorders with minor and major nutritional consequences follow.

Major nutritional impacts are exemplified by disorders that involve the gastrointestinal tract. Infants with *diaphragmatic hernia* or *tracheoesophageal atresia* cannot safely eat by mouth and require nutrition support

and several surgeries during infancy.[32,36] Diaphragmatic hernia occurs in 1 in 4000 live births due to failure of the diaphragm to form completely. Tracheoesophageal atresia occurs in 1 in 4500 live births due to an error in development of the trachea. These examples of conditions treated by neonatal surgery and intensive care have been credited with lower infant mortality for such congenital anomalies.[36] Both conditions change the motility of the gastrointestinal tract, so sufficient calories, nutrients to maintain growth, and oral feeding are important parts of the treatment plan.[32] Such infants miss the windows of development when oral feeding is pleasurable and may have residual feeding problems, such as disliking eating by mouth, well into early childhood. Financing such intensive health care and maintaining the child's normal social and emotional development are major issues for the family. The families of such infants have many specialty health care providers, and their complex financial and emotional repercussions can also influence the infant's development. Both of these conditions eventually result in children being able to eat like everyone else (Illustration 9.3).

Common examples of congenital anomalies are those for infants with *cleft lip and palate*. Major feeding difficulties occur before and after corrective surgeries, and they sometimes interfere with growth in infancy and early childhood.[38,39] Assistance in feeding by registered

Congenital Anomaly Condition evident in a newborn that is diagnosed at or near birth, usually as a genetic or chronic condition, such as spina bifida or cleft lip and palate.

Infant Mortality Attributable to Birth Defects (IMBD) Category used in tracking infant deaths in which specific diagnoses have a high mortality.

Anencephaly Condition initiated early in gestation of the central nervous system in which the brain is not formed correctly, resulting in neonatal death.

Diaphragmatic Hernia Displacement of the intestines up into the lung area due to incomplete formation of the diaphragm *in utero*.

Tracheoesophageal Atresia Incomplete connection between the esophagus and the stomach *in utero*, resulting in a shortened esophagus.

Cleft Lip and Palate Condition in which the upper lip and roof of the mouth are not formed completely and are surgically corrected, resulting in feeding, speaking, and hearing difficulties in childhood.

Illustration 9.3 Infant in hospital being encouraged to eat with a spoon.

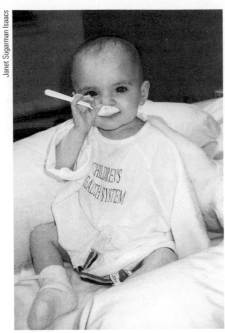

Janet Sugarman Isaacs

dieticians as part of a team approach is needed, because hearing, speech, and language problems are associated with cleft lip and palate. Positions for eating are adapted, and use of special feeding devices for the infant's cleft are needed. Cleft lip and palate may occur alone or as a part of various rare genetic conditions, so growth and feeding problems should also be assessed after corrective surgery.[38]

Infants with Genetic Disorders

Infants diagnosed with genetic disorders near birth are a small subset of infants with congenital anomalies or chronic conditions. They also fit in the category of infants with special health care needs. The expanded use of prenatal genetic tests has the consequence that some families know ahead of time that the baby has a specific condition at birth. When the condition is treated by a special diet, nutritional therapy may begin right after birth. The mother can plan ahead to breastfeed by pumping and freezing her milk. The newborn is fed the appropriate diet while confirmatory testing proceeds. Once the newborn's testing confirms the prenatal test results, therapy can set the amounts of breast milk to be mixed with a special formula at home.

Maple Syrup Urine Disease Rare genetic condition of protein metabolism in which breakdown by-products build up in blood and urine, causing coma and death if untreated.

The number of genetic disorders that can be identified in newborns is increasing rapidly, particularly through expanded infant screening programs. Population-based prevalence estimates keep decreasing, such as 1 in 800 live births for those with screened metabolic conditions compared to 1 in 2500–5000 live births.[36] The nutritional implication of expanded genetic screening and diagnosis is that more newborns require special diets immediately. Infants with rare genetic conditions such as galactosemia or *maple syrup urine disease* need the diet started within days of birth—waiting even one week can result in irreversible brain damage or death.[10] The infants who are identified in newborn screening with metabolic or genetic conditions usually require special formulas. During the diagnostic process, relatives of the affected newborn may be identified with later-onset, milder forms of the same disorder. Genetic centers for inborn errors of metabolism clinics are notified when a newborn screening result needs follow-up. Immediate action is taken to locate the family and confirm the diagnosis. In such circumstances, newborn screening results in early diagnosis and avoids a costly stay in the hospital intensive care unit. For example, an infant with galactosemia, if not picked up by the initial abnormal screening result, is likely to have been hospitalized with possible sepsis or liver problems by the time a second screen is to be collected. If the baby with galactosemia receives supportive measures, such as sugar solutions, and then soy-based infant formulas without galactose, recovery is usually rapid. Some of the disorders picked up by newborn screening do not make the baby sick in early infancy, but later. It is difficult for parents of a healthy-appearing newborn to be told that a special diet is needed to prevent illness later. An example of a condition that can be picked up by newborn screening before illness presents is cystic fibrosis. Although the prevalence of spina bifida and anencephaly has decreased since supplementation of the food supply with folic acid, genetic forms of these disorders have also been more clearly identified and account for 50–70% of these defects.[37]

Other genetic conditions managed by diet or requiring restriction or supplemental levels of nutrients identified in infancy are usually the more severe form of the same condition that can be found in children and adults.[39] Conditions include the following:

Urea cycle disorders requiring protein restriction (example: citrullinemia)

Fat-related disorders requiring restrictions on specific fatty acids (example: "LCHAD," long-chain hydroxyacyl-CoA dehydrogenase deficiency)

Carbohydrate-related disorders requiring restrictions of type or timing of carbohydrates (example: glycogen-storage disease)

Disorders sensitive to high-dose vitamins (example: B_{12} responsive methylmalonic acidemia; also may require a protein-modified diet)

Overgrowth disorders presenting as early obesity (example: Bardet-Biedl syndrome)

Renal genetic disorders managed with a protein-restricted diet to delay end-stage renal disease (example: polycystic kidney disease)

Bone genetic disorders responsive to calcium and vitamin D supplementation (example: osteogenesis imperfecta)

Increased use of genetic tests is exemplified by the condition called *VCFS*. This test may be ordered for any infant with a heart defect. VCFS is a condition in which a small piece of chromosome 22 is deleted.[40] Recent incidence estimates place it at 1 in every 4000 births.[40] This makes VCFS relatively common for a genetic condition—more common than PKU or cystic fibrosis, and second only to Down syndrome as a cause of mental retardation. Infants with VCFS may have a wide range of conditions affecting the heart, immune system, and calcium balance, and later speech and learning problems. Only when the genetic probe became available was it understood that three separate disorders involved the same deletion. As a result, the incidence of VCFS was underreported before the probe was available. Nutrition services may be required based on short stature, heart malformations, heart surgery, and resulting feeding problems (see Case Study 9.2).

Feeding Problems

Infants who were born preterm or have chronic health problems tend to be more irritable and less able to signal their wants and needs compared to healthy infants. Feeding difficulties are reported in 40–45% of families with VLBW infants.[32] Children with developmental disabilities have more frequent feeding problems, as high as 70%, that may or may not be identified in infancy.[6] Table 9.5, on page 281, gives signs of feeding difficulties in high-risk infants that can be caught early.

By the time feeding problems require interventions to prevent further growth and developmental problems, families and infants may be frustrated from their feeding experiences. Infants who are difficult to feed are at risk for failure to thrive (FTT), child abuse, and neglect.[23] Infant feeding guidelines for term infants are appropriate for many preterm infants if they were healthy and had gestational ages such as 35 weeks. Preterm infants who were VLBW or ELBW need infant feeding guidelines based on their adjusted gestational age. As an example, the recommendation for adding food on a spoon at 6 months would be adjusted to 8 months for an infant who was born at 32 weeks of gestation. Even with this adjustment, feeding problems are common because preterm infants may be extra sensitive. The emphasis on weight gain and catch-up growth inadvertently results in overfeeding and signs of gastrointestinal discomfort, such as spitting up. Some preterm infants by late infancy have learned to get attention by devices such as refusing to eat, dropping food off the high chair, or throwing a cup.

> **VCFS (also known as DiGeorge Syndrome and 22q11 microdeletion)** Condition in which chromosome 22 has a small deletion, resulting in a wide range of heart, speech, and learning difficulties.

Nutrition Interventions

When feeding problems are identified in infancy, interventions are required to assure growth and development. Interventions may include any or all of these:

- Assess growth more frequently or in more depth, such as by measuring body-fat stores to identify a change in rate of weight or length gain. This would include head growth measurement.
- Monitor the infant's intake of all liquids and foods by a diet analysis to document that enough calories and nutrients are being consumed. The infant's intake may be variable due to illness, congestion, or medications that lower the appetite.
- Change the frequency or volume of feedings as needed to meet calorie and nutrient needs.
- Adjust the timing of nursing, snacks, or meals as needed to fit medication or sleeping schedules.
- Assess the infant's feeding position and support as needed. This may be important if the infant cannot sit without support.
- Change the diet composition to improve the nutrient density, so that the infant has to expend less effort to meet energy or nutrient needs.
- Provide parent education or support services as needed, so that the feeding environment is positive and low in stress.
- Observe the interaction of the infant and mother (or whoever is routinely feeding the infant) at home or in a developmental program to make sure that signs of hunger and comfort result in a positive feeding experience for the pair.
- Adjust routine nutrition guidelines to the developmental abilities of the infant even if different from the chronological age or gestation-corrected age.

Often, attempts to improve the feeding experience are successful in meeting the infant's calorie and nutrient needs. However, when calorie and nutrient needs are higher than usual, additional steps are needed to make sure that the diet is enriched. Table 9.6 shows some of the special

Case Study 9.2

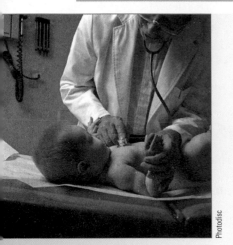

Photodisc

Noah's Cardiac and Genetic Condition

Nutrition Assessment uses data sources of hospital records, physician records, and parent interviews. Several nutrition diagnoses are involved, including feeding difficulty, inadequate protein-energy intake related to open heart surgery and VCFS. This case study is about what other nutrition diagnoses are appropriate and which nutrition interventions are needed for baby Noah.

His mother is successful at expressing her milk and maintaining her breast milk supply by using the breast pump provided by the medical center during hospitalization. Pumping and freezing the breast milk for the baby is important to the family. Noah is a reluctant eater by breast or bottle, with intake usually only 1 or 2 ounces per feeding, when allowed to feed. He is too weak to feed throughout most of the hospitalization.

At discharge the family is referred to a local early intervention program, WIC, Supplemental Social Insurance (SSI), and the state program for children with special needs. Specialty clinic and local follow-up appointments are made. Feeding difficulties concern the family, and both a lactation consultant and a registered dietician are involved at home.

Noah nurses frequently, but briefly, due to fatigue. Growth is slower than expected, but the cardiologist does not think Noah's slow weight gain is a result of the cardiac problem. Noah tolerates only small volumes, even after his family implements the recommendations from the lactation consultant. The consultant recommends offering breast milk in a bottle so that its caloric density can be increased by added rice cereal and MCT oil. The family perceives this recommendation as undermining the mother's effort to breastfeed, and they are unwilling to try it. Noah likes to nurse, but for such a short time that the richer hindmilk may not be available. Over time, the family offers food on a spoon and continues nursing. Weight and height gain do not fit the expected rate, but appear to be fairly consistent.

The family increasingly enjoys parenting, and thinks Noah is a beautiful infant. They have not contacted WIC or the early intervention program. They regard the baby as getting better after surgery. They consider his small size a result of heart surgery. Many people they meet assume he is a premature baby, but his small size is not as much a matter of concern to his parents as it is to health providers.

This case example demonstrates inadequate growth rate, feeding problems, and questionable adequacy of the diet. The genetic syndrome, heart condition, nutritional inadequacy of the baby's diet, and the impact of stress and coping on the mother's milk supply all could explain his slow growth. Growth expectations are unclear because there is no growth chart for this genetic syndrome, or for infants with cardiac anomalies. The standard growth chart is the only one available, but it may not be appropriate to predict future growth.

Important parts of Noah's growth assessment include that his fat stores are good, showing he has access to enough calories, and that his head circumference percentile is low, which suggests that his brain is not growing at the typical rate. This could be due to neurological damage during surgery or afterward, or to the underlying genetic syndrome. It is probably not due to inadequate nutrition in early infancy, because the body tries to preserve brain growth, but there is no way to rule that out. Noah's feeding problems are subtle signs of his developmental delay, although this is clear only in hindsight.

Questions

1. How does breastfeeding benefit Noah?
2. How do you know what is good growth in such a case, when the standard growth chart may not fit?
3. Why doesn't the family want to have WIC or early intervention services, although they are eligible?

(Answers are located in the Instructor's Manual for the 4th edition of *Nutrition Through the Life Cycle*.)

Table 9.5 Signs of feeding problems in high-risk infants

In Early Infancy (Under 6 Months of Age)

- Baby has a weak suck and cannot make a seal on the nipple; breast milk or formula runs out of the mouth on whatever side is lower, with obvious fatigue after a few minutes of sucking.
- Baby appears to be hungry all the time due to low volume consumed per feeding, and/or time between feedings does not appear to increase from one month to the next.
- Extended feeding times are seen, with the baby napping during the feeding despite efforts to keep the baby interested in the feeding.
- The mother is not sure that the baby is swallowing, although she is appearing to suck.

In Later Infancy (Over 6 Months of Age)

- The baby cannot maintain good head control while being fed from a spoon.
- The baby resists spoon-feeding by not opening her mouth when food is offered.
- The baby drinks from a bottle but does not accept baby foods after repeated attempts.
- The baby resists anything in the mouth except a bottle, breast, nipple, or pacifier.
- The baby does not explore the mouth with fingers or try to mouth toys.
- The baby resists lumpy and textured foods; she may turn her face away or push food away.
- The baby does not give signs to the parents that clearly indicate hunger or fullness.

Table 9.6 Examples of infant formula for special needs

Condition	Example of Special Infant Formula
Pulmonary problems such as bronchopulmonary dysplasia or cardiac defect	Breast milk or standard infant formula with polycose and MCT oil to provide 28 cal/fl oz (high calories in a low volume)
Phenylketonuria (genetic disorder of protein metabolism)	Mixture of amino acids, carbohydrates, fats, vitamins, and minerals without the amino acid phenylalanine
Maple syrup urine disease (genetic disorder of protein metabolism)	Mixture of amino acids, carbohydrates, fats, vitamins, and minerals without the amino acids leucine, isoleucine, and valine
VLBW infant who required surgery after Necrotizing enterocolitis	Mixture of amino acids, carbohydrates, fats, vitamins, and minerals
Gastroesophageal reflux and swallowing problem	Standard infant formula with baby rice cereal (increased thickness is to lower risk of choking and vomiting)
Chronic renal failure (hereditary kidney disease)	Concentrated natural protein, fats, and carbohydrates providing 40 cal/fl oz

formulas that may be used by infants who have feeding problems or chronic conditions that increase their nutrient requirements.

Nutrition Services

Infants who were born preterm or with special health care needs have access to more nutrition services than do other infants. The following programs are sources of nutrition services or finances to pay for nutrition services:[41]

- Federal disability programs
- Individuals with Disabilities Education Act (IDEA), Part C
- Early Head Start
- WIC
- State funding from the Maternal and Child Health (MCH) Block Grant

Infants with disabling conditions are eligible for Supplemental Social Insurance (SSI), a federal program within the Social Security Administration.[41] SSI provides the family with a disability check and access to health insurance if its income meets federal guidelines.

Nutrition services are part of educational programs in IDEA, including services for children 0–2 years of age.[47] Early Head Start programs enroll infants with special health care needs. Infants with a nutrition-related diagnosis such as PKU or a cardiac problem would be eligible. Early Head Start staff are trained to feed infants special diets.

Each state has to designate a portion of the federal MCH Block Grant for children with special health care needs.[41] Services differ from state to state, but all states provide care for infants with chronic conditions. The following are some examples of how nutrition services are provided:

- Specialty clinic services, such as having a nutrition consultant attend a cystic fibrosis clinic
- Contractual services for providing special formulas or therapy for groups of patients who need more nutrition care than usually provided

- Visiting at schools or programs to conduct nutrition assessments or coordinate follow-up recommendations, such as making sure that a specific diet is being offered or that mealtime behavior is being monitored
- Transporting teams of specialists to rural or isolated areas for direct care
- Development and distribution of nutrition education materials for staff training

Every state also has a program funded by MCH to identify and advocate for children with special needs, such as the Developmental Disabilities Council.[41]

Key Points

1. The number of infants at risk for having special health care needs is increasing even though survival rates of preterm infants are improving over decades.

2. Nutrition guidance for infants with special health care needs has to be adjusted to fit their individual energy and nutrient needs.

3. Infants born severely preterm temporarily require modified forms of protein, fats, carbohydrates, vitamins, and minerals that are not in breast milk or typical infant formulas.

4. Growth in infants with special health care needs reflects nutritional intake and many other factors, such as intrauterine environment, developmental delay, and underlying medical conditions.

5. Infants with severe preterm birth or congenital anomalies can be fed directly into the stomach or blood stream when oral feeding is not safe.

6. Feeding and eating difficulties are common in infants who require intensive medical care, so nutrition services track growth patterns closely.

7. Educational and developmental services for infants at risk for special health care needs are encouraged; such services include attention from nutrition experts and assistance in feeding and eating problems.

8. Among the many genetic conditions identified in infancy, some require special formulas and close nutrition management to promote normal development.

Review Questions

1. Why are infant feeding problems more common in 6 month old infants who were born at 32 weeks of gestation than in 6 month old infants born full-term?

   a. Early birth makes mothers anxious which can lower her milk supply after hospital discharge.

   b. Infants do not have skills for nursing until they reach a corrected age of 40 weeks of gestation.

   c. Neonatal intensive care units have to use nutrition support which slows oral feeding development.

   d. High energy needs means a high volume of breast milk is needed, even if the infant has weak muscles.

2. What factors make nutritional requirements higher for an infant now 8 months of age who was born at 29 weeks of gestation?

   a. Being born so early resulted in less body fat and nutrient reserves

   b. The persistence of newborn reflexes

   c. More frequent illnesses due to the immaturity of the immune system.

   d. The immaturity of the gastrointestinal track lessens nutrient absorption.

Questions 3, 4, 5 and 6 concern an infant who was born 3 months ago at 34 weeks of gestation weighing 2 pounds and 3 ounces. Answer if each statement is true or false.

3. This infant is 3 months chronological age and one month of age corrected for prematurity.

4. This infant is large for gestational age.

5. This infant's nutritional needs are met exclusively with breast milk.

6. This infant would be eligible within IDEA for nutrition services if needed.

7. Infants with genetic disorders may or may not be identified by newborn screening. Identify a

genetic condition that is not found by maternal or newborn screening.

8. Describe at least 2 factors other than nutrition that could impact growth of an infant with special health care needs.

9. Describe 2 forms of feeding an infant other than oral feeding such as nursing.

10. Describe 2 factors that can explain how intrauterine growth rate does not match growth in early infancy.

Resources

Emory University Pediatrics Department

This website includes information on preterm births and risks for providers and parents, with sections on nutrition in various health conditions.
Website: www.emory.edu/PEDS

National Association of Councils on Developmental Disabilities

This website includes public policy and advocacy resources for providers and parents all over the United States concerning various disabilities, including services for infants.
Website: www.nacdd.org/

National Center for Growth Statistics (source of growth charts)

Site for obtaining growth charts and guidelines for their use.
Website: www.cdc.gov/nchs

National Early Childhood Technical Assistance Center

This website provides publications and other resources about programs for infants and children. Staff training materials are also available for those who are working with infants and young children.
Website: http://www.nectac.org/

Neonatology on the Web

This website provides resources for health care professionals, including practice guidelines and consensus statements.
Website: www.neonatology.org

Preemieparents.com

This website lists sites for recommended reading and parent resources.
Website: www.preemieparents.com

Gaining and Growing: Assuring Nutritional Care of Preterm Infants

This site provides solid information and case studies about nutrition and feeding problem interventions for outpatient follow-up of infants who had been born early. The case-study approach gives practical suggestions, "red flags," and progress over time.
Website: http://depts.washington.edu/growing/

United Cerebral Palsy Associations

This website provides links to service sites in the United States.
Website: www.ucp.org

"Enough is as good
as a feast."
Proverbs

Chapter 10

Toddler and Preschooler Nutrition

Prepared by Nancy H. Wooldridge

Key Nutrition Concepts

1 Children continue to grow and develop physically, cognitively, and emotionally during the toddler and preschool-age years, adding many new skills rapidly with time.

2 Learning to enjoy new foods and developing feeding skills are important components of this period of increasing independence and exploration.

3 Children have an innate ability to self-regulate food intake. Parents and caretakers need to provide children nutritious foods and let children decide how much to eat.

4 Parents and caretakers have tremendous influence on children's development of appropriate eating, physical activity, and other health behaviors and habits formed during the toddler and preschool years. These lessons are mainly transferred by example.

Introduction

This chapter describes the growth and development of toddlers and preschool-age children and their relationships to nutrition and the establishment of eating patterns. Growth during the toddler and preschool-age years is slower than in infancy but steady. This slowing of *growth velocity* is reflected in a decreased appetite; however, young children need adequate calories and nutrients to meet their nutritional needs. The eating and health habits established at this early stage of life may impact food habits and subsequent health later in life. The development of new skills and increasing independence mark the toddler and preschool stages. Learning about and accepting new foods, developing feeding skills, and establishing healthy food preferences and eating habits are important aspects of this stage of development.

Definitions of the Life-Cycle Stage

Toddlers are generally defined as children between the ages of 1 and 3 years. This stage of development is characterized by a rapid increase in *gross* and *fine motor skills* with subsequent increases in independence, exploration of the environment, and language skills. *Preschool-age children* are between 3 and 5 years of age. Characteristics of this stage of development include increasing autonomy; experiencing broader social circumstances, such as attending preschool or staying with friends and relatives; increasing language skills; and expanding ability to control behavior.

Importance of Nutrition

Adequate intake of energy and nutrients is necessary for toddlers and preschool-age children to achieve their full growth and developmental potential. Undernutrition during these years impairs children's cognitive development as well as their ability to explore their environments.[1] Long-term effects of undernutrition, such as failure to thrive and cognitive impairment, may be prevented or reduced with adequate nutrition and environmental support.

Tracking Toddler and Preschooler Health

The percent of children living in poverty is a widely used indicator of child well-being, and this rate has increased dramatically in recent years.[2, 3] The poverty rate for children in 2006 was 18%.[2] Thirty-three percent of children had no parent with full-time, year-round employment in 2006, and 14% of these children lacked health insurance.[2] In 2006, 33% of children (or about 22 million) lived in single-parent families, which makes them more likely to live in poverty.[2, 3]

When evaluating young children's nutritional status and offering nutrition education to parents, it is important to consider children's home environments. Establishing healthy eating habits may not be high on a family's priority list when the home environment is one of poverty and food insecurity. Disparities in child health indicators, including those of nutrition status in this age group, exist among races and ethnicities.

> **Growth Velocity** The rate of growth over time.
>
> **Toddlers** Children between the ages of 1 and 3 years.
>
> **Gross Motor Skills** Development and use of large muscle groups as exhibited by walking alone, running, walking up stairs, riding a tricycle, hopping, and skipping.
>
> **Fine Motor Skills** Development and use of smaller muscle groups demonstrated by stacking objects, scribbling, and copying a circle or square.
>
> **Preschool-Age Children** Children between the ages of 3 and 5 years, who are not yet attending kindergarten.

Healthy People 2010

Healthy People 2010—objectives for the nation for improvements in health status by the year 2010—includes a number of objectives that directly relate to toddlers and preschoolers.[4] These are listed in Table 10.1 with available progress to date noted. Healthy People 2020 is under development (www.hhs.gov) and, according to the proposed framework, will retain most of these objectives with some modifications.

Normal Growth and Development

An infant's birth weight triples in the first 12 months of life, but growth velocity slows thereafter until the adolescent growth spurt. On average, toddlers gain 8 ounces (0.23 kg) per month and 0.4 inches (1 cm) of height per month, while preschoolers gain 4.4 pounds (2 kg) and 2.75 inches (7 cm) per year.[5] This decrease in rate of

Table 10.1 Healthy People 2010 objectives related to toddlers and preschool age children[4, CDC, 2009]

Healthy People 2010 Objective	Baseline	Progress	Target
19.4: Reduce growth retardation among low-income children under age 5	6%	6% (2006)	4%
American Indian or Alaska native	5%	5% (2006)	4%
Asian or Pacific Islander	7%	6% (2006)	4%
Hispanic or Latino	5%	6% (2006)	4%
Black or African American, not Hispanic or Latino	7%	7% (2006)	4%
White, not Hispanic or Latino	6%	7% (2006)	4%
19.5: Increase the proportion of persons aged 2 years and older who consume at least two daily servings of fruit	39%	40% (2004)	75%
Mexican American	40%	46% (2004)	75%
Black or African American, not Hispanic or Latino	35%	42% (2004)	75%
White, not Hispanic or Latino	39%	38% (2004)	75%
19.6: Increase the proportion of persons aged 2 years and older who consume at least three daily servings of vegetables, with at least one-third being dark green or deep yellow vegetables	4%	4% (2004)	50%
Black or African American, not Hispanic or Latino	6%	3% (2004)	50%
White, not Hispanic or Latino	3%	4% (2004)	50%
19.7: Increase the proportion of persons aged 2 years and older who consume at least six daily servings of grain products, with at least three being whole grains	4%	3% (2004)	50%
White, not Hispanic or Latino	4%	3% (2004)	50%
19.8: Increase the proportion of persons aged 2 years and older who consume less than 10% of calories from saturated fat	36%	34% (2004)	75%
Mexican American	37%	43% (2004)	75%
Black or African American, not Hispanic or Latino	31%	37% (2004)	75%
White, not Hispanic or Latino	35%	30% (2004)	75%
19.9: Increase the proportion of persons aged 2 years and older who consume no more than 30% of calories from fat	33%	29% (2004)	75%
Mexican American	33%	37% (2004)	75%
Black or African American, not Hispanic or Latino	26%	27% (2004)	75%
White, not Hispanic or Latino	33%	26% (2004)	75%
19.10: Increase the proportion of persons aged 2 years and older who consume 2400 mg or less of sodium daily	15%	13% (2004)	65%
Black or African American, not Hispanic or Latino	17%	17% (2004)	65%
White, not Hispanic or Latino	15%	12% (2004)	65%
19.11: Increase the proportion of persons aged 2 years and older who meet dietary recommendations for calcium	31%	42% (2004)	74%
Mexican American	28%	37% (2004)	74%
Black or African American, not Hispanic or Latino	16%	19% (2004)	74%
White, not Hispanic or Latino	35%	48% (2004)	74%
19.12a: Reduce iron deficiency among young children (aged 1 to 2 years)	9%	9% (2002)	5%
Mexican American	17%	14% (2002)	5%
Black or African American, not Hispanic or Latino	10%	7% (2004)	5%
White, not Hispanic or Latino	6%	7% (2004)	5%
19.12b: Reduce iron deficiency among young children	4%	6% (2002)	1%
Mexican American	6%	7% (2002)	1%

growth is accompanied by a reduced appetite and food intake in toddlers and preschoolers. A common complaint of parents of children this age is that their children have a much lower appetite and a lower interest in food or eating compared to their appetite and food intake during infancy. Parents need to be reassured that a decrease in appetite is part of normal growth and development for children in this age group.

Illustration 10.1 Measuring the recumbent length of a toddler is a two-person job!

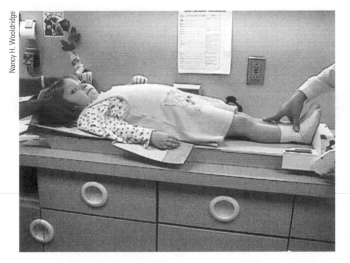

Measuring Growth

In monitoring a child's physical growth, it is important for children to be accurately weighed and measured at periodic intervals. Toddlers less than 2 years of age should be weighed without clothing or a diaper. The *recumbent length* of toddlers should be measured on a length board with a fixed head board and moveable foot board. Proper measurement of recumbent length requires two adults—one at the child's head making sure the crown of the head is placed firmly against the head board, and the other making sure that the child's legs are fully extended and placing the foot board at the child's heels. Proper positioning for measurement of a child's recumbent length is shown in Illustration 10.1. Preschool-age children should be weighed and measured without shoes and in light-weight clothing. Calibrated scales should be used and a height board should be used for measuring *stature*. Illustrations 10.2 and 10.3 further demonstrate the proper techniques for weighing and measuring young children. It is important that both weight and height be plotted on the appropriate growth charts, such as the 2000 CDC Growth Charts discussed next.

Recumbent Length Measurement of length while the child is lying down. Recumbent length is used to measure toddlers <24 months of age and those between 24 and 36 months who are unable to stand unassisted.

Stature Standing height.

Illustration 10.2 Young child being weighed.

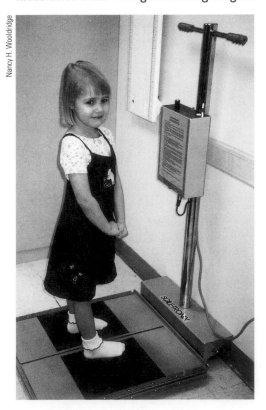

Illustration 10.3 Measuring the stature of a preschool-age child.

The 2000 CDC Growth Charts

A full set of the "CDC Growth Charts: United States" can be found on the CDC website (See the "Resources" section at the end of this chapter or www.cdc.gov/nchs).[6] The growth charts are based on data from cycles 2 and 3 of the National Health and Examination Survey (NHES) and the National Health and Nutrition Examination Surveys (NHANES) I, II, and III, and they provide a reference for how children in the United States are growing.[7] Illustration 10.4 shows an example of one of the charts, which depicts the growth of a healthy child. Charts are gender-specific and are available for birth to 36 months and for 2 to 20 years. With these charts, the health care professional can plot and monitor weight-for-age, length- or stature-for-age, head circumference-for-age,

Illustration 10.4 Birth to 36 months: Girls' length-for-age and weight-for-age percentiles.[6]

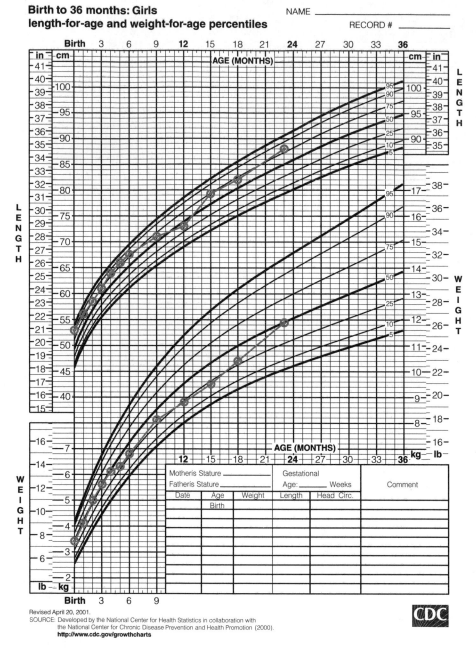

Birth to 36 months: Girls
length-for-age and weight-for-age percentiles

NAME _____

RECORD # _____

Revised April 20, 2001.
SOURCE: Developed by the National Center for Health Statistics in collaboration with
the National Center for Chronic Disease Prevention and Health Promotion (2000).
http://www.cdc.gov/growthcharts

CDC

fairly steady percentile range. It is important to monitor a child's growth over time and to identify any deviations in growth. It is the pattern of growth that is important to assess rather than any one single measurement. A weight measurement without a length or stature measurement doesn't indicate how appropriate the weight is for the child's length or stature. There are Internet-based programs and personal digital assistant devices available that will compute exact growth percentiles, but tracking and assessing growth curves over time are still essential.

Body mass index, or BMI, provides a guideline for assessing underweight and overweight in children and adults. Body mass index is predictive of body fat for children over 2 years of age, as BMI normative values are not available for children less than 2 years of age.[8] For children 2 years of age or older, a BMI in the 85th percentile or greater but less than the 95th percentile indicates overweight, and a BMI in the 95th percentile or greater indicates obesity.[8] For children less than 2 years of age, a weight-for-length greater than the 95th percentile is considered overweight.[8] A weight-for-length or a BMI for age percentile less than the 5th percentile indicates underweight. BMI fluctuates throughout childhood. BMI increases in infancy; it decreases during preschool years, hitting its lowest point at approximately 4–6 years of age; and then it increases into adulthood. Because of this normal fluctuation of BMI, the only way to know if a child's BMI is within a normal range is to plot BMI-for-age on the appropriate growth curve. In pediatrics, the goal is to strive for a BMI-for-age in the normal range and not a specific BMI range, as is the goal in adults.

Growth charts visually aid parents by demonstrating the expected slowing of the growth velocity during the toddler and preschool stage of development. Although the curves for weight-for-age and length- or stature-for-age continue to increase during the toddler and preschool-age years, the slope of the curve is not as steep as it is during the first year of life.

weight-for-length, weight-for-stature, and BMI-for-age. There is overlap between the two sets of growth charts for children between 24 and 36 months of age. If the child's recumbent length is measured, then the birth-to-36-months growth chart is the appropriate one to use. If the child over 2 years of age is measured standing, the 2-to-20-years growth chart is the correct choice. Children's growth usually "tracks" within a

Body Mass Index An index that correlates with total body fat content or percent body fat and is an acceptable measure of adiposity or body fatness in children and adults.[8] It is calculated by dividing weight in kilograms by the square of height in meters (kg/m[2]).

WHO Growth Standards

In 2006, the World Health Organization (WHO) published growth standards for children from birth to 5 years of age, based on growth data collected over time on healthy breastfed infants and young children in six different countries. The mothers of these children were non-smokers and the children had adequate diets and were free of infections. These international growth standards indicate how children should grow under optimal environmental conditions regardless of their ethnicity or socioeconomic status. WHO growth standards are available for boys and girls from birth to 5 years of age for length/height-for-age, weight-for-age, weight-for-length, weight-for-height, and BMI-for-age.[9] (See the "Resources" section at the end of this chapter or www.who.int/childgrowth.)

Common Problems with Measuring and Plotting Growth Data

Incorrectly measured or plotted growth parameters in young children can lead to errors in health status assessment. Standard procedures should be followed, calibrated and appropriate equipment should be used, and plotting should be double-checked—including checking the age of the child—to avoid such errors. Choosing the appropriate growth chart based on how the child was measured (recumbent length versus stature) and on the child's gender is as important as using the current growth charts.

Physiological and Cognitive Development

Toddlers

An explosion in the development of new skills occurs during the toddler years. Most children begin to walk independently at about their first birthday. At first the walking is more like a "toddle" with a wide-based gait.[5] After practicing for several months, the toddler achieves greater steadiness and soon will be able to stop, turn, and stoop without falling over. Gross motor skills, such as sitting on a small chair and climbing on furniture, develop rapidly at this age, and with practice, great improvements in balance and agility take place. At about 15 months, children can crawl up stairs; by about 18 months, they can run stiffly. Most toddlers can walk up and down stairs one step at a time by 24 months, and jump in place. At about 30 months, children have advanced to going up stairs by alternating their feet. By 36 months of age, children are ready for tricycles.

Children become increasingly mobile and independent with improvements in gross motor skills. Toddlers are fascinated with these newfound skills, showing a readiness to put the skills into practice and to develop new skills.

However, toddlers have no sense of dangerous situations. At this age, children are especially vulnerable to accidental injuries and ingestion of harmful substances. In fact, the leading cause of death among young children is unintentional injuries.[4] Parents and caregivers have to constantly watch over toddlers, preferably in environments made "child-safe."

Cognitive Development of Toddlers   With toddlers' newly acquired physical skills, exploring the environment accelerates, and exerting their newfound independence becomes very important to them. Toddlers now have the power to control the distance between themselves and their parents. *Nelson's Textbook of Pediatrics*[5] describes how toddlers often "orbit" around their parents like planets, moving away, looking back, moving farther, and then returning.

From a socialization standpoint, the child moves from being primarily self-centered to being more interactive. The toddler now possesses the ability to explore the environment and to develop new relationships. Fears of certain situations, such as separation, darkness, loud sounds, wind, rain, and lightning, commonly emerge during this period as the child learns to deal with changes in the environment. Children develop rituals in their daily activities in an attempt to deal with these fears.

Social development also involves imitating others, such as parents, caretakers, siblings, and peers, during this time. The child in this stage begins to learn about the family's cultural customs, including those related to meals and food.

Dramatic development of language skills occurs from age 18 to 24 months. Once a child realizes that words can stand for things, his vocabulary erupts from 10 to 15 words at 18 months to 100 or more words at 2 years of age. The toddler soon begins combining words to make simple sentences. By 36 months, the child uses three-word sentences.[5]

An important social change for toddlers is increased determination to express their own will. This expression often comes in the form of negativism and the beginning of temper tantrums, which give this stage of development its label of "the terrible twos." With an increase in motor development coupled with an increasing quest for independence, toddlers try to do more and more things, pushing their capabilities to the limit. Thus the toddler can become easily frustrated and negative. The child seeks more independence and at the same time needs the parents and caretakers for security and reassurance. Toddler behavior uncannily parallels the same type of behavior commonly seen in adolescents!

Development of Feeding Skills in Toddlers   Many babies begin to wean from the breast or the bottle at about 9 to 10 months of age, when their solid food intake increases, and they learn to drink from a cup.[10] Parents need to pay attention to cues of readiness for weaning, such as disinterest in breastfeeding or bottle feeding. The time it takes to wean is variable and depends on both the child and the mother. Weaning will be easier for those babies who adapt well to change. Weaning is a sign of the toddler's growing independence and is usually complete by 12 to 14 months of age, although the age varies from child to child.

Gross and fine motor development during the toddler years enhances children's ability to chew foods of different textures and to self-feed. Between 12 and 18 months, toddlers are able to move the tongue from side to side (or laterally) and learn to chew food with rotary, rather than just up and down, movements. Toddlers can now handle chopped or soft table food.

At about 12 months, children have a refined pincer grasp that enables them to pick up small objects, such as cooked peas and carrots, and put them into their mouths. Children will be able to use a spoon around this age, but not very well. At 18 to 24 months, toddlers are able to use the tongue to clean the lips and have well-developed rotary chewing movements. Now the toddler can handle meats, raw fruits and vegetables, and multiple food textures.

A strong need for independence in self-feeding emerges during the toddler age. "I do it!" and "No, no, no!" are commonly heard phrases in households where toddlers reside. As toddlers practice their newly found skills, they become easily distracted. Parents need to realize that their toddler's sometimes-fierce independence is part of normal growth and development and represents an ongoing process of separation from dependency on the parents and caretakers.

Increasing fine motor and visual motor coordination skills allow toddlers to use cups and spoons more effectively. Although toddlers' skill with a spoon increases during the second year, they prefer to eat with their hands. Initial attempts at self-feeding are inevitably messy, as Illustration 10.5 depicts, but represent an important stage of development. It is important that parents and caretakers keep distractions, such as television, to a minimum during mealtimes, and allow their toddlers to practice self-feeding skills and to experience new foods and textures. The child derives pleasure in self-feeding and exploring new tastes. Learning to self-feed allows the child to develop mastery of an important part of everyday life.

Adult supervision of eating is imperative due to the high risk of choking on foods at this age. Toddlers should always be seated during meals and snacks, preferably in a high chair or booster seat with the family, and not allowed to "eat on the run." Foods that may cause choking, such as hard candy, popcorn, nuts, whole grapes, and hot dogs, should not be served to children less than 2 years of age.[10]

Feeding Behaviors of Toddlers   The toddler's need for rituals, a hallmark of this stage of development, may

Illustration 10.5 Toddler enjoying mealtime!

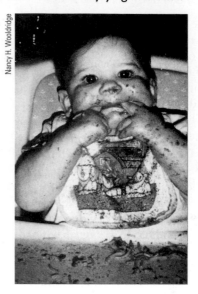

be linked to the development of food jags. Many toddlers demonstrate strong food preferences and dislikes. They can go through prolonged periods of refusing a particular food or foods they previously liked. The intensity of the refusal or the negative attitude toward a particular food will be influenced by the child's temperament (see the "Temperament Differences" section on page 275). To circumvent food jags, parents can serve new foods along with familiar foods. New foods are better accepted if they are served when the child is hungry and if she sees other members of the family eating these foods. Eventually, toddlers' natural curiosity will get the best of them. Toddlers are great imitators, which includes imitating the eating behavior of others.

Mealtime is an opportunity for toddlers to practice newly acquired language and social skills and to develop a positive self-image. It is not the time for battles over food or "force-feedings." Establishing the habit of eating breakfast is an important part of healthy eating behaviors. Family mealtime provides an opportunity for parents and caretakers to model healthy eating behaviors for the young child.

Appetite and Food Intake of Toddlers   Parents need to be reminded that toddlers naturally have a decreased interest in food because of slowing growth, and a corresponding decrease in appetite. Besides, with all of their newfound gross and fine motor skills, they have places to go and new environments to explore! It is a part of normal growth and development for toddlers to have a decreased interest in food and to be easily distracted at mealtime.

Toddlers need toddler-sized portions. One rule of thumb for serving size is 1 tablespoon of food per year of age. Applying this rule, a serving for a 2-year-old child

Case Study 10.1

Making Mealtime Pleasant

Lindsey, a 24-month-old girl, lives with her parents. She stays at a child care center during the week while both of her parents are at work. On the weekends, her parents enjoy their time with Lindsey, although a lot of their time is spent running errands and catching up on household chores. Partly to appease Lindsey, her parents allow her to have as much of her favorite beverage, apple juice, from a "sippy" cup as she wants between meals. Lindsey also has free access to snacks such as crackers, slices of cheese, and cookies. When the family sits down to have a meal together, Lindsey plays with her food and usually doesn't eat much. She tells her parents that she doesn't like the food being served and wants "something else." She soon becomes fussy and wants to get down from her booster seat. To try to keep her at the table with them, her parents turn on the television or play Lindsey's favorite cartoons. If that does not quiet Lindsey, her mother offers to prepare another food item of Lindsey's choice. Mealtime has become an unpleasant experience for the family.

Questions

1. Identify some of the inappropriate eating habits that Lindsey's parents have allowed her to develop.
2. Considering Lindsey's stage of development, what advice would you give her parents in their attempts to increase the number of foods she will eat?
3. In what types of food-preparation activities would it be appropriate for Lindsey's parents to have her participate? Why is this important?
4. What suggestions do you have for snack food items for Lindsey?
5. Would you advise Lindsey's parents to give her a daily multivitamin supplement?
6. What advice would you give the family regarding physical activity?

(Answers are located in the Instructor's Manual for the 4th edition of *Nutrition Through the Life Cycle*.)

---

would be about 2 tablespoons. It is better to give the child a small portion and allow him to ask for more than to serve large portions. Parents often overestimate portion sizes needed by their young child, which may contribute to labeling the child as a "picky" eater. Because toddlers can't eat a large amount of food at one time, snacks are vital in meeting the child's nutritional needs. It is important, however, that toddlers not be allowed to "graze" throughout the day on sweetened beverages and foods such as cookies and chips. These foods can "kill" their limited appetite for basic foods at meal and snack times. In considering the toddler's need for rituals and limit setting, parents and caretakers need to establish regular but flexible meal and snack times, allowing enough time between meals and snacks for the toddler to get hungry. It is important that toddlers be allowed to control the amount of food eaten by hunger rather than by parental pressure to eat more. Case Study 10.1 illustrates some of these points.

Preschool-Age Children

Preschool-age children continue to expand their gross and fine motor capabilities. At age 4, the child can hop, jump on one foot, and climb well. The child can ride a tricycle, or a bicycle with training wheels, and can throw a ball overhand.[5]

Cognitive Development of Preschool-Age Children

Magical thinking and egocentrism characterize the preschool period.[5] Egocentrism does not mean that the child is selfish, but rather that the child is not able to accept another's point of view. The child is beginning to interact with a widening circle of adults and peers. During the preschool years, children gradually move from primarily relying on external behavioral limits, such as those demanded by parents and caregivers, to learning to limit behavior internally. This transition is a prerequisite to functioning in a school classroom.[5] Also during this time, children's play starts to become more cooperative, such as building a tower of blocks together. Toward the end of the preschool years, children move to more organized group play, such as playing tag or "house."

Control is a central issue for preschool children. They will test their parents' limits and still resort to temper tantrums to get their way. Temper tantrums generally peak between the ages of 2 and 4 years.[5] The child's challenge is to separate, and the parent's challenge is to appropriately set limits and at the same time to let go, another parallel with adolescence. Parents need to strike an appropriate balance for setting limits. Too-tightly controlled limits can undermine the child's sense of initiative and cause him or her to act out, whereas loose limits can cause the child to feel anxious and that no one is in control.

Language develops rapidly during the preschool years and is an important indicator of both cognitive and emotional development. Between ages 2 and 5, children's vocabularies increase from 50 to 100 words to more than 2000 words, and their language progresses from two- to three-word sentences to complete sentences.[5]

Development of Feeding Skills in Preschool-Age Children

The preschool-age child can use a fork and a spoon and uses a cup well. Cutting and spreading with a knife may need some refinement. Children should be seated comfortably at the table for all meals and snacks. Eating is not as messy a process during the preschool years as it was during toddlerhood. Spills still do occur, but they are not intentional. Foods that cause choking in young children should be modified to make them safer, such as cutting grapes in half lengthwise and cutting hot dogs in quarters lengthwise and then cutting into small bites. Adult supervision during mealtime is still imperative.

Feeding Behaviors of Preschool-Age Children

As during the toddler years, parents of preschool-age children need to be reminded that the child's rate of growth continues to be relatively slow, with a relatively small appetite and food intake. Growth occurs in "spurts" during childhood. Appetite and food intake increase in advance of a growth spurt, causing children to add some weight that will be used for the upcoming spurt in height. Therefore, the appetite of a preschool-age child can be quite variable.

Preschool-age children want to be helpful and to please their parents and caretakers. This characteristic makes the preschool years a good time to teach children about foods, food selection, and preparation by involving them in simple food-related activities. For instance, outings to a farmers' market can introduce children to a variety of fresh vegetables and fruit. Allowing children to

> **Table 10.2 Meal preparation activities for young children[11]**
>
> Let children select and help prepare a whole-grain side dish.
>
> Let children help shop for, clean, peel, or cut up vegetables and fruits, depending on their age.
>
> Let children decide on the dinner vegetable or what goes into salads.

be involved in meal-related activities, such as those listed in Table 10.2 from MyPyramid for Kids, can be quite instructive.[11] Families of preschool-age children need to continue to be encouraged to eat together, like the family in Illustration 10.6.

Innate Ability to Control Energy Intake   An important principle of nutrition for young children, and one with direct application to child feeding, is children's ability to self-regulate food intake. If allowed to decide when to eat and when to stop eating without outside interference, children eat as much as they need.[12,13] Children have an innate ability to adjust their caloric intake to meet energy needs. The preschool-age child's intake may fluctuate widely from meal to meal and day to day. But over a week's time, the young child's intake remains relatively stable.[14] Parents who try to interfere with the child's ability to self-regulate intake by forcing the child to "clean her plate" or using food as a reward are asking the child to overeat or undereat.

Although children can self-regulate caloric intake, no inborn mechanisms direct them to select and consume a well-balanced diet.[15] Children learn healthful eating habits.[16] Parents give up some control over what their preschool child eats if the child spends more time away from home in a child care center or with extended family members. Preschool-age children continue to learn about food and food habits by observing their parents, caretakers, peers, and siblings, and they begin to be influenced by what they see on television and through other forms of media. Their own food habits and food preferences are established at this time.

Appetite and Food Intake of Preschoolers   Parents of preschool-age children often describe their children's appetite by calling the child "a picky eater." One reason a child may want the same foods all of the time is because familiar foods may be comforting to the child. Another reason is that the child may be trying to exert control over this aspect of her life. The child's eating and food selection can easily become a battleground between parent and child; this scenario should be avoided. Some practical suggestions for parents and caretakers of children

Illustration 10.6 Sharing family meals is an important aspect of development in young children.

this age include serving child-sized portions and serving the food in an attractive way. Young children often do not like their foods to touch or to be mixed together, such as in casseroles or salads. They typically do not like strongly flavored vegetables and other foods, or spicy foods, at this young age. Just as with toddlers, parents of preschool-age children should not allow their children to eat and drink indiscriminately between meals and snacks. This behavior often "kills" the appetite at mealtime. Children should not be forced to stay at the table until they have eaten a certain amount of food as determined by the parent.

Temperament Differences

> "Better is a dinner of herbs where love is, than a fatted ox and hatred with it."
>
> Proverbs 15:17

Temperament is defined as the behavioral style of the child, or the "how" of behavior. Three temperamental clusters have been defined: the "easy" child (about 40% of children), the "difficult" child (10%), and the "slow-to-warm-up" child (15%). [17] The remaining children, classified as "intermediate-low" or "intermediate-high," demonstrate a mixture of behaviors but gravitate toward one end of the spectrum. [17]

Children's temperaments affect feeding and mealtime behavior. The "easy" child is regular in function, adapts easily to regular schedules, and tries and accepts new foods readily. The "difficult" child, on the other hand, is characterized by irregularity in function and slow adaptability. This child is more reluctant to accept new foods and can be negative about them. The "slow-to-warm-up" child exhibits slow adaptability and negative responses to many new foods with mild intensity. With repeated exposure to new foods, this child can learn to accept them over time with limited complaining. [17]

The "goodness of fit" between the temperaments of the child and the parent or caretaker can influence feeding and eating experiences. [17] A mismatch can result in conflict over eating and food. Parents and caretakers need to be aware of the child's temperament when attempting to meet nutritional needs. The difficult or slow-to-warm-up child may pose special challenges that need to be addressed by gradually exposing the child to new foods and not hurrying him or her to accept them. [17]

Food Preference Development, Appetite, and Satiety

Food preference development and regulation of food intake have been studied extensively by Leann Birch and associates. [18,19] It is clear that children's food preferences do determine what foods they consume. Children naturally prefer sweet and slightly salty tastes and generally reject sour and bitter tastes. These preferences appear to be unlearned and present in the newborn period. Children eat foods that are familiar to them, a fact that emphasizes the importance environment plays in the development of food preferences. Children tend to reject new foods but may learn to accept a new food with repeated exposure to it. It may, however, take eight to ten exposures to a new food before it is accepted. Children who are raised in an environment where all members of the family eat a variety of foods are more likely to eat a variety of foods. One study showed that 5-year-old girls' fruit and vegetable intakes were related to their parents' fruit and vegetable intakes.[20]

Children also appear to have preferences for foods that are energy dense due to high levels of sugar and fat.[18,19] This preference may develop because children associate eating energy-dense foods with pleasant feelings of satiety, or because these types of foods may be associated with special social occasions such as birthday parties. The context in which foods are offered to a child influences the child's food preferences. Foods served on a limited basis but used as a reward become highly desirable. Restricting a young child's access to a palatable food may actually promote the desirability and intake of that food.[21] Coercing or forcing children to eat foods can have a long-term negative impact on their preference for these foods.[18-20]

Media Influence Young children are also influenced by media. One study of advertisements during programming aimed specifically at toddlers and preschool-age children on three different networks found that more than half of all food advertisements were aimed specifically at children, and the majority of these advertisements were for fast-food chains or sweetened cereals. The ads associated the advertised product with fun and/or excitement and energy. Fast-food ads seemed to focus on building brand recognition through the use of licensed characters, logos, and slogans and were less likely to show food during the ads.[22]

Appetite, Satiety Children's energy intake regulation has been studied by giving children *preloads* of food or beverage of varying energy content followed by self-selected meals. In one such study, children ages 3 to 5 years were given either a low-energy preload beverage made with aspartame (Nutrasweet), a low-calorie sugar substitute, or a high-energy preload beverage made with sucrose. Fat and protein content of the preloads did not differ. Children were then allowed to self-select their lunches. Children who had

Preloads Beverages or food such as yogurt in which the energy/macronutrient content has been varied by the use of various carbohydrate and fat sources. The preload is given before a meal or snack and subsequent intake is monitored. This study design has been employed by Birch et al. in their studies of appetite, satiety, and food preferences in young children.[19]

the low-calorie beverage before lunch consumed more calories at lunch, while those who had the higher-calorie beverage consumed fewer calories. These results indicate that young children are able to adjust caloric intake based on caloric need.[19, 23] Similar studies were conducted in 2-to-5-year-olds using foods with dietary fat or Olestra, a nonenergy fat substitute. Results indicate that children compensated for the lower level of calories in food when olestra was substituted for dietary fat.

The preloading protocol just described was also used to study children's responsiveness to caloric content of foods in the presence or absence of common feeding advice from adults. In one group, teachers were trained to minimize their control over how much the children ate. In the other group, teachers were trained to focus the children on external factors to control their intake, such as rewarding the children for finishing the portions served to them or encouraging them to eat because it was "time to eat." Results of this investigation show that when the adults focused the children on external cues for eating, children lost their ability to regulate food intake based on calories. It appears that children's innate ability to regulate caloric intake can be altered by child-feeding practices that focus on external cues rather than the child's own hunger and satiety signals.[23]

The effects of portion size on children's intakes were compared between classes of 3-year-old and 5-year-old children. The children were served either a small, medium, or large portion of macaroni and cheese along with standard amounts of other foods in their usual lunchtime setting. Analysis of amount of food eaten showed that portion size did not affect the younger children's intakes; their intakes remained constant despite the amount of food served to them. In contrast, the 5-year-old children's intakes increased significantly with the larger portion sizes. The researchers conclude that by 5 years of age, children are influenced by the size of portions served to them, another external factor that influences intake.[24] A similar study in preschoolers with an average age of 4 years found that doubling an age-appropriate portion size of an entrée increased entrée and total energy intake by 25% and 15% respectively.[25] These investigators raise the question of what effect large portion sizes have on overeating and, consequently, on the development of childhood overweight, and their results point to the possible benefits of allowing children to self-select their portion sizes.

Another study of 5-year-old girls and their parents looked at the effects of parents' restriction of palatable foods on their children's consumption of these foods. After a self-selected standard lunch, these 5-year-old girls were given free access to snack foods, such as ice cream, potato chips, fruit-chew candy, and chocolate bars. The daughters of parents who reported restricting access to snack foods indicated to the investigators that they ate "too much" of the snack foods and also reported negative emotions about eating the snack foods. Parents' restriction of foods actually promoted the consumption of these foods by their young daughters and, of even more concern, the daughters reported feeling badly about eating these "forbidden" foods.[26] A related study found a lower self-concept in 5-year-old girls with high body weight.[27] Daughters of parents who restricted access to food and expressed concern about their daughter's weight status tend to have negative self-evaluations.[26] Mothers in particular seem to have more influence over their young daughters' beliefs about food and dieting.[28] Satter describes the optimal "feeding relationship" as one in which parents and caretakers are responsible for what children are offered to eat and the environment in which the food is served, while children are responsible for how much they eat or even whether they eat at a particular meal or snack.[29] According to Satter, if this feeding relationship is respected, then feeding and potential weight problems can be prevented.[29] Parenting includes influencing what is served to children and the environment in which it is served, at home and in child care settings, as well.

What implications does all this research have on child feeding practices? Based on the results of these studies, it appears that by late preschool age, children are more responsive to external cues than to their innate ability to self-regulate intake. Table 10.3 sums up the practical applications of Birch's work.[19] The importance of appropriate parenting skills in helping children learn to self-regulate food intake and possibly avoid problems with obesity is echoed by a panel of obesity experts.[8, 30] Birch's research also reinforces the important role that parents and caretakers play in modeling healthy eating behaviors for young children.

Energy and Nutrient Needs

Dietary Reference Intakes (DRIs) were developed from 1997 through 2004. The series of reports presents a comprehensive set of reference values for nutrient intakes for healthy individuals and populations in the U.S. and Canada. Reference intake values are available for females and males aged 0.0 – 0.5 year, 0.5 to 1 year, 1-3, 4-8, and 9-13 years of age. Information about the various DRI publications can be found on the National Academy Press website (see the "Resources" section at the end of this chapter). Published DRI tables are provided on the inside front cover of this book.

Energy Needs

DRIs have been established for the energy needs of young children.[31] The formula for Estimated Energy Requirements (EER) for children ages 13–36 months is (89 × weight of child [kg] – 100) + 20 (kcal for energy deposition).

Table 10.3 Practical applications of child-feeding research[19]

- Parents should respond appropriately to children's hunger and satiety signals.
- Parents should focus on the long-term goal of developing healthy self-controls of eating in children and should look beyond their concerns regarding composition and quantity of foods children consume or fears that children may eat too much and become overweight.
- Parents should not attempt to control children's food intakes by attaching contingencies ("No dessert until you finish your rutabagas") and coercive practices ("Clean your plate, children in Bangladesh are starving").
- Parents should be cautioned not to severely restrict "junk foods," foods high in fat and sugar, as that may make these foods even more desirable to the child.
- Parental influence should be positively focused on the child developing food preferences and selection patterns of a variety of foods consistent with a healthy diet. Parental modeling of eating a varied diet at family mealtime will have a strong influence on children.
- Children have an unlearned preference for sweet and slightly salty tastes; they tend to dislike bitter, sour, and spicy foods.
- Children tend to be wary of new foods and tastes, and it may take repeated exposures to new foods before these are accepted.
- Children need to be served appropriate child-sized servings of food.
- Child feeding experiences should take place in secure, happy, and positive environments with adult supervision.
- Children should never be forced to eat anything.

For example, a healthy 24-month-old girl who weighs 12 kg would have an EER of (89 × 12 kg – 100) + 20 = 988 kilocalories. Beginning at age 3, the DRI equations for estimating energy requirements are based on a child's gender, age, height, weight, and physical activity level (PAL). Categories of activity are defined in terms of walking equivalence. Table 10.4 depicts estimated energy requirements for reference boys and girls at selected ages. Energy needs of toddlers and preschool-age children reflect the slowing growth velocity of children in this age group.

Dietary Reference Intakes (DRIs) Quantitative estimates of nutrient intakes, used as reference values for assessing the diets of healthy people. DRIs include Recommended Dietary Allowances (RDAs), Adequate Intakes (AI), Tolerable Upper Intake Levels (UL), and Estimated Average Requirements (EAR).

Table 10.4 Estimated energy requirements (in kcals) for reference boys and girls at selected ages and varying physical activity levels (PAL)[31]

Age/Gender	Reference Weight (kg [lbs])	Reference Height (m [in])	Sedentary PAL (Kcal/d)	Low Active PAL (Kcal/d)	Active PAL (Kcal/d)	Very Active PAL (Kcal/d)
3-year-old boy	14.3 (31.5)	0.95 (37.4)	1162	1324	1485	1683
4-year-old boy	16.2 (35.7)	1.02 (40.2)	1215	1390	1566	1783
5-year-old boy	18.4 (40.5)	1.09 (42.9)	1275	1466	1658	1894
3-year-old girl	13.9 (30.6)	0.94 (37.0)	1080	1243	1395	1649
4-year-old girl	15.8 (34.8)	1.01 (39.8)	1133	1310	1475	1750
5-year-old girl	17.9 (39.4)	1.08 (42.5)	1189	1379	1557	1854

Protein

The DRIs for protein for the toddler/preschool age groups can be found in Table 10.5.[31] *Recommended Dietary Allowances (RDAs)* have been established for protein. These recommendations are easily met with typical American diets as well as with vegetarian diets. Adequate energy intake to meet an individual child's needs has a protein-sparing effect; that is, with adequate energy intake, protein is used for growth and tissue repair rather than for energy. Ingestion of high-quality protein, such as milk and other animal products, lowers the amount of total protein needed in the diet to provide the essential amino acids.

Recommended Dietary Allowances (RDAs) The average daily dietary intake levels sufficient to meet the nutrient requirements of nearly all (97% to 98%) healthy individuals in a population group. RDAs serve as goals for individuals.

Anemia A reduction below normal in the number of red blood cells per cubic mm in the quantity of hemoglobin, or in the volume of packed red cells per 100 mL of blood. This reduction occurs when the balance between blood loss and blood production is disturbed.

Vitamins and Minerals

DRIs for vitamins and minerals have been established for the toddler and preschool-age child. Most children from birth to 5 years are meeting the targeted levels of consumption of most nutrients, except for iron, calcium, and zinc (see the "Recommended vs. Actual Intake" section on pages 289–290). The DRIs for these key nutrients are listed in Table 10.6.[32,33]

Table 10.5 Dietary Reference Intakes for protein[31]

Age	RDA* g/kg/d
1–3 years	1.1 g/kg/d or 13 g/day*
4–8 years	0.95 g/kg/d or 19 g/day*

SOURCE: From the Institute of Medicine, Food & Nutrition Board.
*RDA based on average weight for age (reference individual)

Table 10.6 Dietary Reference Intakes for key nutrients for toddlers and preschoolers[32,33]

Age	Recommended Dietary Allowances		Adequate Intake
	Iron (mg/d)	Zinc (mg/d)	Calcium (mg/d)
1–3 years	7	3	500
4–8 years	10	5	800

Common Nutrition Problems

Iron-Deficiency Anemia

Iron deficiency and iron-deficiency *anemia* are prevalent nutrition problems among young children in the United States, although the prevalence is decreasing. A rapid growth rate coupled with frequently inadequate intake of dietary iron places toddlers, especially 9- to 18-month-olds, at the highest risk for iron deficiency.[34] According to the 1999–2000 National Health and Nutrition Examination Survey (NHANES 1999–2000), 7% of toddlers age 1 to 2 years are iron deficient (down from 9% in NHANES III); of these, 2% have iron-deficiency anemia (down from 3% in NHANES III).[35] In numbers, these percentages translate to approximately 700,000 toddlers with iron deficiency, and of these, 240,000 have iron-deficiency anemia.[35] The full impact of this nutrition problem is profound. Iron-deficiency anemia in young children appears to cause long-term delays in cognitive development and behavioral disturbances.[1,34]

Table 10.7 depicts the progressing signs of iron deficiency. Iron deficiency can be defined as absent bone marrow iron stores, an increase in hemoglobin concentration of <1.0 g/dl after treatment with iron, or other abnormal lab values, such as serum ferritin concentration, the

Table 10.7 Progression of iron deficiency

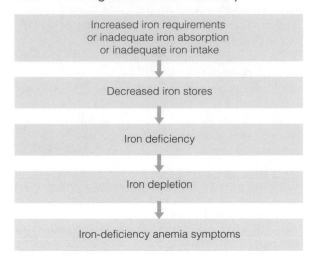

storage form of iron.[34] The definition of iron-deficiency anemia is less than the 5th percentile of the distribution of *hemoglobin* concentration or *hematocrit* in a healthy reference population. Age- and sex-specific cutoff values for anemia are derived from NHANES III data. For children 1 to 2 years of age, the diagnosis of anemia would be made if the hemoglobin concentration is <11.0 g/dl and hematocrit <32.9%. For children ages 2 to 5 years, a hemoglobin value <11.1 g/dl or hematocrit <33.0% is diagnostic of iron-deficiency anemia.

Not all anemias are due to iron deficiency. Other causes of anemia include other nutritional deficiencies (such as folate or vitamin B_{12}), chronic inflammation, or recent or current infection.[34]

One Healthy People 2010 objective is to reduce iron deficiency in children ages 1 to 2 years from 9% to 5% and in children ages 3 to 4 years from 4% to 1%.[4] Part of reaching this goal will mean reducing or eliminating disparities in iron deficiency by race and family income level. The prevalence of iron deficiency is higher in African Americans than in white children (10% versus 6% for children ages 1 to 2 years) and is highest in Mexican American children (17% of children ages 1 to 2 years).[4] Children of families with incomes <130% of the poverty threshold have a higher incidence of iron deficiency than those of families with a higher income (12% versus 7%).

Preventing Iron Deficiency   The Centers for Disease Control published recommendations for preventing iron deficiency in the United States.[34] It is recommended that children 1 to 5 years of age drink no more than 24 ounces of cow's milk, goat's milk, or soy milk each day because of the low iron content of these milks. Larger intakes may displace high-iron foods. For detecting iron deficiency, it is recommended that children at high risk for iron deficiency, such as low-income children and migrant and recently arrived refugee children, be tested for iron

deficiency between the ages of 9 and 12 months, 6 months later, and then annually from ages 2 to 5 years. For children who are not at high risk for iron deficiency, selective screening of children at risk only is recommended. Children at risk include those who have a low-iron diet, consume more than 24 ounces of milk per day, have limited access to food because of poverty or neglect, and who have special health care needs, such as an inborn error of metabolism or chronic illness.

Nutrition Intervention for Iron-Deficiency Anemia   Treatment of iron-deficiency anemia includes supplementation with iron drops at a dose of 3 mg/kg per day, counseling of parents or caretakers about diets that prevent iron deficiency, and repeat screening in 4 weeks. An increase of >1 g/dl in hemoglobin concentration, or >3% in hematocrit, within 4 weeks of initiation of treatment confirms the diagnosis of iron deficiency. If the anemia is responsive to treatment, dietary counseling should be reinforced, and the iron treatment should be continued for 2 months. At that time, the hemoglobin and hematocrit should be rechecked, and the child should be reassessed in 6 months. If the hemoglobin and hematocrit do not increase after 4 weeks of iron treatment, further diagnostic tests are needed. Iron status will not improve with iron supplements if the cause of the anemia is not directly related to a need for iron.[34]

Dental Caries

Approximately one in five children aged 2 to 4 years has decay in the primary or permanent teeth.[4] One of the Healthy People 2010 objectives is to reduce the proportion of children and adolescents who have dental caries experience in their primary or permanent teeth.[4] A primary cause of dental decay is habitual use of a bottle or a no-spill training cup with milk or fruit juice at bedtime or throughout the day. Prolonged exposure of the teeth to these fluids can produce *early childhood caries*, formerly called "nursing bottle caries" or "baby bottle tooth decay."[36] Upper front teeth are most severely affected by decay, which is where fluids pool when toddlers fall asleep while drinking from a bottle. Toddlers with baby-bottle tooth decay are at increased risk for caries in the permanent teeth.[37] The incidence of baby-bottle tooth decay is highest among Hispanic, American Indian, and Alaska Native children, and among children whose parents have less than a high school education.[4]

Hemoglobin   A protein that is the oxygen-carrying component of red blood cells. A decrease in hemoglobin concentration in red blood cells is a late indicator of iron deficiency.

Hematocrit   An indicator of the proportion of whole blood occupied by red blood cells. A decrease in hematocrit is a late indicator of iron deficiency.

Early Childhood Caries (ECC)   The presence of one or more decayed (noncavitated or cavitated lesions), missing (due to caries), or filled tooth surfaces in any primary tooth in a child 71 months of age or younger.[36]

Food sources of carbohydrates such as milk and fruit juice can have direct effects on dental caries development because *Streptococcus mutans,* the main type of bacteria that cause tooth decay, use carbohydrates for food. Bacteria present in the mouth excrete acid that causes the tooth decay.[37] Consequently, the more often and longer teeth are exposed to carbohydrates, the more the environment in the mouth is conducive to the development of tooth decay. Foods containing carbohydrates that stick to the surface of the teeth, such as sticky candy like caramel, are strong caries promoters. Rinsing the mouth with water or brushing teeth to get rid of the carbohydrate stuck to teeth reduces caries formation. Young children allowed to "graze" or indiscriminately eat or drink throughout the day likely expose their teeth to carbohydrates for a longer period of time, which encourages bacteria proliferation and tooth decay. Crunchy foods such as carrot sticks and apple slices, when age-appropriate, are good choices for snacks because they are less likely to promote tooth decay than are sticky candies.

Fluoride   Children need a source of fluoride in the diet, preferably from fluoridated water and the use of fluoridated toothpaste. If the water supply is not adequately fluoridated, a fluoride supplement is recommended. The American Dental Association, the American Academy of Pediatrics, and the American Academy of Pediatric Dentistry have devised a fluoride supplementation schedule, which is based on the child's age and the fluoride content of the local water supply.[38] Children ages 6 months to 3 years need 0.25 mg of fluoride per day if their local water supply has <0.3 ppm of fluoride. Children 3 to 6 years of age need 0.5 mg fluoride per day if their water supply has <0.3 ppm, but only 0.25 mg fluoride per day if the local water has 0.3 to 0.6 ppm of fluoride.[38] Excessive fluoride supplementation, consumption of toothpaste with fluoride, and natural water supplies high in fluoride can cause *fluorosis.*

> **Fluorosis**   Permanent white or brownish staining of the enamel of teeth caused by excessive ingestion of fluoride before teeth have erupted.[37]

Although otherwise harmless, fluorosis produces permanent staining of the enamel of teeth, particularly permanent teeth. Because of the risk of fluorosis, fluoride supplements are only available by prescription. Few foods contain much fluoride, but fluoridated water used in beverages and food preparation does provide fluoride.

Constipation

Constipation, or hard and dry stools associated with painful bowel movements, is a common problem of young children. Sometimes "stool holding" develops when the child does not completely empty the rectum, which can lead to chronic overdistension so that eventually the child is retaining a large fecal mass.[39] Then having a bowel movement can become painful to the child, which leads to more "stool holding," and a vicious cycle ensues. A pediatrician should manage the treatment of "stool holding."

Diets providing adequate total or dietary fiber for age (see fiber recommendations on page 288) guard against constipation. Some of the best food sources of dietary fiber for toddlers and preschoolers are whole-grain breads and cereals, legumes, and fruits and vegetables appropriate for age. Too much fiber should be avoided, however. Young children easily develop diarrhea from high amounts of fiber, and high-fiber foods may displace other energy-dense foods and may decrease the bioavailability of some minerals, such as iron and calcium.

Elevated Blood Lead Levels

According to the latest surveillance data, approximately 2.2% of children 1 to 5 years of age have high blood lead levels, exceeding 10 mcg/dl.[40] The number of children with high blood lead levels continues to decline throughout the United States, and a Healthy People 2010 goal is to eliminate blood lead levels ≥10 mcg/dl in young children.[4,40] The major sources of lead exposure for young children are airborne lead, which has decreased in recent decades with the elimination of lead from gasoline and with the enforcement of industrial emissions standards, and leaded chips and dust, mainly from deteriorating lead paint.[41] Young children are particularly at risk for developing high levels of lead because, in exploring their environment, they enjoy putting things into their mouths. Depending on their surroundings, some of these objects may be high in lead. Damage caused by lead exposure may begin during pregnancy as lead is transported across the placenta to the fetus. Blood lead levels peak at about 2 years of age.[41] There are racial, ethnic, and socioeconomic disparities in children with high lead levels, with higher rates found in children living in poverty, children of minority groups, and recent immigrants.[41,42]

High blood lead levels affect the functioning of many tissues in the body, including the brain, blood, and kidneys. Low-level exposure to lead is associated with decreases in IQ and impaired motor, behavioral, and physical abilities.[41,43] Elevated blood lead levels may decrease growth in young children.[44] Historically, a blood lead level of 10 mcg/dl prompted action. However, some more recent research indicates that the physical and mental development of children may be affected by blood lead levels <10 mcg/dl, and no safe level of blood lead in children has been established.[41,45] About 25% of children still live in housing with deteriorated lead-based paint.[41] Children living in housing built before 1978 are at increased risk of high lead levels because lead-based paint may have been used on these houses.[45] Lead-based paint chips taste sweet, tempting children to consume them. As the age of housing decreases nationally, so does the incidence of high lead levels in children.[4] Lead can enter the food supply through lead-soldered water pipes, contaminated water supplies, and from certain canned goods from other countries that contain lead-solder seals. Nonfood items containing lead include contaminated dirt and lead weights. Some parental occupations can be a source of lead. In these cases, parents should remove work clothing at work and wash work clothes separately. Other sources of lead include ceramic glazes and pewter used in some folk remedies and hobbies.[41] The CDC published guidelines for screening children for lead poisoning in 1997.[46] The American Academy of Pediatrics endorsed these guidelines in a policy statement published in 1998 and one updated in 2005.[41,47] The AAP advocates for a shift in focus from case identification and management to primary prevention.[41] With the decrease in prevalence of elevated blood lead levels, a shift toward targeted screening has begun and continues to be developed.[41,46] Federal policy requires lead screening of children who are enrolled in Medicaid. Screening is also recommended for children enrolled in other public assistance programs, such as WIC.[10,40,46,47] Most local and state health departments have established recommendations for targeted screening based on the risk factors in the community. When indicated, lead screening should be obtained at 9–12 months of age and again around 24 months of age, when blood lead levels peak. Besides the age of the housing, other risk factors for high blood lead levels include living in poverty and having a sibling or playmate who has had high blood lead levels.

Nutritional Considerations Some of the risk factors for elevated blood lead levels are also the same risk factors for iron-deficiency anemia, such as young age, poor nutrition, and low socioeconomic status.[43] Iron-deficiency anemia is associated with pica, the ingestion of non-food items, such as paint chips, which is a risk factor for lead ingestion. Some studies suggest that adequate iron intake may decrease lead absorption, which reinforces the benefits of treating iron-deficiency anemia in young children (see the "Iron-Deficiency Anemia" section on pages 278–279).[43] Some studies suggest that vitamin C may increase lead excretion. Although the evidence is not strong enough to recommend for or against vitamin C supplementation for children with elevated blood lead levels, it is important for young children to have sources of vitamin C in their diets for the prevention of iron deficiency.[34,43] There is good evidence that dietary calcium competitively decreases lead absorption, but there is no clinical evidence that supplementing calcium beyond the adequate intake for age has a clinical effect on elevated blood lead level.[43] Although animal and epidemiological studies point to an association between dietary fat intake and elevated blood lead levels, low-fat diets are not recommended for the treatment of elevated blood lead levels in young children. No clinical data exist showing the benefits, and fat is an important source of calories for young children.[41,43] To summarize, eliminating sources of lead in the child's environment is the most important step toward eliminating elevated blood lead levels in children. In addition, preventing iron deficiency and promoting a well-balanced diet that includes good sources of calcium and vitamin C help to prevent this problem in young children.

Food Security

One of the Healthy People 2010 objectives is to increase *food security* among U.S. households to 94% from a baseline of 88%.[4] However, in 2007, only 84.2% of U.S. households with children reported being food secure indicating a decrease in food security.[48] Food insecurity was reported in the remaining 15.8% of households with children with about 8.3% of these households reporting one or more children being food insecure at some time during the year.[48] Food insecurity is more likely to exist among American Indian or Alaska Native, African American, and Hispanic or Latino people than among whites; in households with children, particularly those headed by single women; and in lower-income-level households (<130% of poverty threshold).[4]

Food security is particularly important for young children because of their high nutrient needs for growth and development. Young children are a vulnerable group because they must depend on their parents and caretakers to supply them with adequate access to food. It appears that children who are hungry and have multiple experiences with food insufficiency are more likely to exhibit behavioral, emotional, and academic problems as compared to other children who do not experience hunger repeatedly.[49]

Food Safety

Young children are especially vulnerable to foodborne illnesses because they can become ill from smaller doses of organisms. Key foodborne pathogens include *Campylobacter* species and *Salmonella* species, which are the most frequently reported foodborne illnesses in the United States, and the emerging pathogen *E. coli 0157:H7*.[4] The highest rate of *Campylobacter* species infections is seen in children under age 1.[4] *Campylobacter* is transmitted by handling raw poultry, eating undercooked poultry, drinking raw milk or nonchlorinated water, or handling infected animal or human feces.[50] The most common cause of *Salmonella* food poisoning is consumption of foods containing undercooked or raw eggs, such as raw cookie dough containing eggs. Children younger than 10 years of age account for a disproportionate percent of cases of *E. coli*-related illness. It is a serious disease and can cause bloody diarrhea and *hemolytic uremic syndrome (HUS)*. Outbreaks of *E. coli* have been associated with ingestion of contaminated,

Food Security Access at all times to a sufficient supply of safe, nutritious foods.

Hemolytic Uremic Syndrome (HUS) A serious, sometimes fatal complication associated with illness caused by *E. coli 0157:H7*, which occurs primarily in children under the age of 10 years. HUS is characterized by renal failure, **hemolytic anemia**, and a severe decrease in **platelet** count.[4]

undercooked hamburger meat, unpasteurized apple cider and juice, and unpasteurized milk. Employing proper food storage and preparation techniques at home, in child care centers, and in retail food establishments is essential for decreasing the incidence of foodborne illnesses in young children. Contamination of food products can occur at any point along the way from production to consumption. Therefore, risk reduction and controls can be targeted at various steps in food processing. One food safety education program, called FightBAC™, was developed by the Partnership for Food Safety Education, a partnership of industry, state and consumer organizations, and government agencies, including the CDC and EPA. FightBAC has four food-safety practice messages:[4,51]

- Clean: Wash hands and surfaces often.
- Separate: Don't cross-contaminate.
- Cook: Cook to proper temperatures.
- Chill: Refrigerate promptly.

Under the Dietary Guideline on Food Safety, there is a key recommendation for young children. The guideline recommends that young children do not eat or drink raw (unpasteurized) milk or any products made from unpasteurized milk, raw or undercooked meat and poultry, raw or undercooked fish or shellfish, unpasteurized juices, and raw sprouts.[52] The Dietary Guidelines offer additional strategies for keeping food safe. Child care workers as well as family members and other caretakers of children need to be well educated in food-safety issues.

The U.S. Environmental Protection Agency is in the process of evaluating all existing standards for pesticides.[4] A major objective of this evaluation is to ensure that the current levels of pesticides in the food supply and drinking water are safe for young children.

Prevention of Nutrition-Related Disorders

The prevalence of *overweight* and *obesity* among children, adolescents, and adults in the United States has increased and represents a major public health problem. High-energy, high-fat diets coupled with sedentary lifestyles are thought to be major contributors to the increase in weight. Cardiovascular disease, a major cause of death and morbidity in the United States today, is also thought to be influenced by diets and sedentary lifestyles. Food habits, preferences, and behaviors established during the toddler and preschool ages logically influence dietary habits later in life and subsequent health status. Behaviors associated with risk factors for cardiovascular disease, including dietary habits, physical activity behaviors, and the use of tobacco, can be acquired in childhood.[53] The American Heart Association strongly advocates that primary prevention of atherosclerotic disease begin in childhood.[53] Families are encouraged to adopt dietary and exercise patterns that promote a healthy lifestyle.

Overweight and Obesity in Toddlers and Preschoolers

According to the NHANES 2007–2008 data, 10.4% of children ages 2 to 5 years had BMI-for-age percentiles greater than or equal to the 95th percentile. No significant difference in prevalence of overweight and obesity was found between male and female children, but differences do exist by race/ethnicity. Hispanic male children had significantly greater odds of having high BMI than non-Hispanic white male children. Non-Hispanic black female children were significantly more likely to have high BMI than non-Hispanic white female children.[54]

Obesity is a multifaceted problem that is difficult to treat, making prevention the preferred approach. The American Medical Association (AMA) in collaboration with the Department of Health and Human Services (DHHS) convened a multidisciplinary committee of experts to develop evidence-based recommendations on the assessment, prevention, and treatment of child and youth overweight and obesity. The expert committee's work culminated in the publication of four articles: a summary report and three separate articles on the topics of assessment, prevention, and treatment, which were endorsed by twelve professional organizations including the American Academy of Pediatrics (AAP), the American Dietetic Association (ADA), and the American Heart Association (AHA).[8,30,55,56]

Assessment of Overweight and Obesity

Body mass index-for-age percentile is recommended as the screening tool for assessment of pediatric overweight and obesity. A BMI-for-age percentile of 85th to 94th is defined as *overweight*, and a BMI-for-age ≥95th is defined as *obesity*.[8,55] BMI normative values are not available for children <2 years of age. For these young children, a weight-for-length >95th percentile is considered to be overweight.[8]

During the preschool years, a decrease in body mass index (BMI), or weight-for-height squared $[wt(kg)/ht(m)^2]$, is a normal part of growth and development. BMI usually reaches its lowest point at approximately 4 to 6 years of age and then increases gradually in the period called *adiposity rebound* or *BMI rebound*.[57] Early adiposity rebound in children increases the risk of adult obesity.[58]

Other essential components of assessment include 1) evaluation of the child's medical risk, including parental obesity, family medical history, and evaluation of

Overweight Body mass index-for-age between the 85th and 94th percentiles.[8,55]

Obesity BMI-for-age greater than or equal to the 95th percentile.[8,55]

Adiposity or BMI Rebound A normal increase in body mass index that occurs after BMI declines and reaches its lowest point at 4 to 6 years of age.[57]

weight-related problems such as sleep and respiratory problems, and 2) behavior risk assessment, including dietary and physical activity behaviors.[8,55] Another aspect of the behavioral assessment is evaluation of the child's and/or family's attitudes toward and capacity to change some behaviors.[8,55]

Prevention of Overweight and Obesity

Prevention is the best approach for overweight and obesity. All children should be targeted for prevention of overweight and obesity from birth by instituting lifestyle behaviors that prevent obesity.[8,30]

The expert panel has identified the following target behaviors in the prevention of pediatric overweight and obesity based on available evidence and on analysis of available data and expertise: [8, 30]

- Limiting sugar-sweetened beverages
- Encouraging consumption of recommended amounts of fruits and vegetables
- Limiting television and other screen time by allowing a maximum of 2 hours of screen time per day and removing televisions and other screens from children's bedrooms
- Eating breakfast every day
- Limiting eating out at restaurants, especially fast food restaurants
- Limiting portion sizes
- Eating a diet rich in calcium
- Eating a diet high in fiber
- Eating a diet that follows the Dietary Reference Intakes for macronutrients (carbohydrates, protein, and fat)
- Promoting moderate to vigorous physical activity for at least 60 minutes each day
- Limiting energy-dense foods

Parenting techniques, such as finding reasons to praise the child's behavior but never using food as a reward, foster the development of healthy eating behaviors in children and help children to self-regulate food intake (see the "Food Preference Development, Appetite, and Satiety" section on page 275).

Treatment of Overweight and Obesity Expert Committee Recommendations

The goal of overweight and obesity treatment is the improvement of long-term physical health through permanent healthy lifestyle habits and behavior modification.[8,56] Improvement is measured by a decrease in BMI-for-age percentile, but this is difficult to see in the short-term. Weight measurements on a regular basis can be used to measure progress in the short-term. Maintaining weight while gaining height can be the best treatment for obese children between the ages of 2 and 5. This approach allows the obese child to "grow into his or her weight" and to lower BMI. If weight loss does occur, it should not exceed 1 pound per month in children this age, whether they fall into the overweight or obese category.[8,56]

The expert committee recommends a staged approach to pediatric overweight and obesity treatment. These stages are depicted in Illustration 10.7, and a brief description of each stage follows.

Stage 1: Prevention Plus – This stage focuses on the behaviors identified in the prevention section. The health care provider can identify dietary and physical activity behaviors in the individual child and family that would be appropriate to target and, using motivational interviewing techniques, can assist the family in making appropriate changes. Targeted behaviors can be achieved in steps, with the ultimate goal being an improvement in BMI-for-age percentile. This stage involves more frequent follow-up based on the individual child and family's needs.[8,56]

Stage 2: Structured Weight Management (SWM) – This stage of treatment is more structured and requires more frequent follow-up. Some of the elements of the Structured Weight Management Plan include a planned diet or daily eating plan, further reduction of screen time to <1 hour per day, and planned supervised physical activity or active play for 60 minutes per day. This stage includes monitoring behavior through the use of logs and planned reinforcement for achieving targeted behaviors. A registered dietitian or a clinician who has received additional training is required to work with the child and her family on an eating plan. Staff members also need to be trained in motivational interviewing. Depending on the family, a counselor may need to be involved to address parenting skills or to help resolve family issues that may be barriers to healthy lifestyle behaviors. Monthly visits are recommended, some of which may be group sessions.[8,56]

Stage 3: Comprehensive Multidisciplinary Intervention – In this stage, the intensity of behavior change is increased and a multidisciplinary team, including a registered dietitian, an exercise specialist, a behavioral counselor, and the primary care provider, is needed to maximize support for behavior change. Weekly visits are recommended for a minimum of 8–12 weeks, some of which may be group visits. This stage is a structured program in behavior modification that includes, at a minimum, food monitoring, short-term diet resulting in a negative energy balance, and physical activity goal setting. Parents are involved in behavior modification for their child and in parental training for improving the home environment. There is a systematic evaluation of body measurements, diet, and physical activity at specified time intervals. If weight loss occurs, it should not be more than 1 pound per month in the young child.[8,56]

Stage 4: Tertiary Care Intervention – This stage is offered to severely obese adolescents who have failed

Illustration 10.7 Suggested staged treatment for overweight and obesity for 2- to 5-year old children.

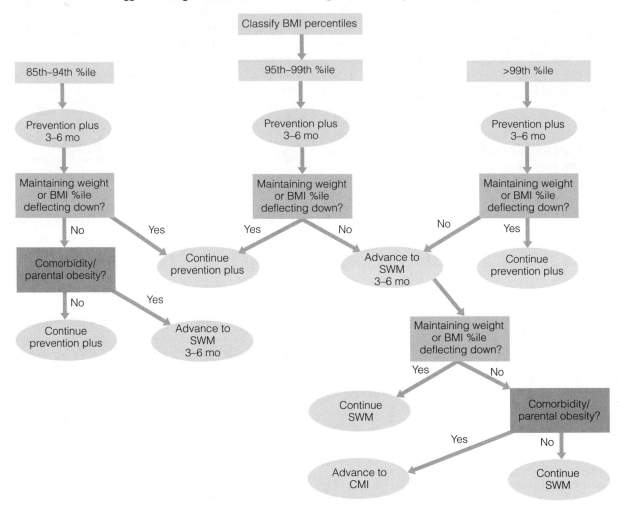

SOURCE: Reproduced with permission from Pediatrics, Vol. 120, December, Supplement 4, Figure 1, p. S273, © 2007 American Academy of Pediatrics.

other interventions. This stage is not appropriate for the obese toddler or preschool-age child.[8, 56]

Dietary Guidelines for Americans 2005

The Dietary Guidelines for Americans 2005 also include some recommendations related to pediatric overweight and obesity. The weight management guideline includes a key recommendation that the rate of body-weight gain be reduced for overweight children while care is taken to allow enough calories for normal growth and development. It is further recommended that a health care provider be consulted before placing a child on a weight-reduction diet.[52]

Another key recommendation of the Dietary Guideline on Weight Management is targeted at overweight children with chronic diseases and/or on medication, indicating that gradual weight loss may be warranted. The recommendation is that a health care provider be consulted about weight-loss strategies prior to starting a weight-reduction program to ensure appropriate management of other health conditions.[52] Reducing weight at this young age is tricky because sufficient nutrients must be provided for children to reach their full height potential and to remain healthy.

Nutrition and Prevention of Cardiovascular Disease in Toddlers and Preschoolers

Heart disease is the number-one cause of death in the United States today.[4] A leading risk factor for cardiovascular disease, which includes diseases of the heart and blood vessels, is elevated levels of *LDL cholesterol*.

Heart Disease The leading cause of death and a common cause of illness and disability in the United States. Coronary heart disease is the principal form of heart disease and is caused by buildup of cholesterol deposits in the coronary arteries, which feed the heart.

LDL Cholesterol Low-density lipoprotein cholesterol, the lipid most associated with atherosclerotic disease. Diets high in saturated fat, *trans* fatty acids, and dietary cholesterol have been shown to increase LDL-cholesterol levels.

Children with *familial hyperlipidemias* and obese children can have high levels of LDL cholesterol. High intakes of saturated fat, *trans fatty acids*, and, to a lesser extent, dietary cholesterol elevate LDL-cholesterol levels in children and adults alike. Other nutrition-related risk factors for cardiovascular disease include high triglyceride levels and high body mass index.[53] Fatty streaks, which can be precursors to the buildup of fat deposits in blood vessels, have been found in the arteries of young children. Some experts believe that these streaks can represent the beginning of *atherosclerosis* and cardiovascular disease.[53]

The American Heart Association (AHA) has published guidelines for the primary prevention of atherosclerotic cardiovascular disease beginning in childhood.[53] In these guidelines, the AHA recommends that all children be screened for risk factors of developing future cardiovascular disease. Furthermore, the AHA recommends that all children have an overall healthy eating pattern while maintaining an appropriate body weight, desirable lipid profile, and desirable blood pressure. Avoidance of smoking is encouraged as well as daily physical activity and the reduction of sedentary time.[53] Like the obesity expert panel, the AHA has also published strategies for implementing the recommended guidelines that employ behavior change theory and motivational interviewing.[59] The American Heart Association has published dietary recommendations for children, which have been endorsed by the American Academy of Pediatrics and which are consistent with the U.S. Dietary Guidelines (see the "Dietary Guidelines" section on pages 286–287).[52,60,61] The recommended diet includes fruits and vegetables, whole-grain breads and cereals, the use of non-fat or low-fat dairy products, and two servings of fish weekly. The use of vegetable oils and soft margarines low in saturated fat and *trans* fatty acids instead of butter or most other animal fats is recommended. The recommendations also include reducing intake of sugar-sweetened beverages and foods and reducing salt intake.[60-62] For children ages 2 to 3 years, 30–35% of total energy from fat is recommended. For children 4 years of age or older, the recommendation is 25–35% of total energy from fat.[60] A *trans* fatty acid intake <1% of total calories is also recommended.[60,61] Studies have shown that such dietary restrictions are safe and effective for reducing risk factors in childhood without negatively impacting growth.[63]

According to the new DRIs,[31] the acceptable macronutrient distribution ranges (AMDRs) for fat are 30–40% for children 1 to 3 years and 25–35% for children 4 to 18 years. No specific recommendations for total fat per day in the diets of young children have been made. Adequate intake levels for the essential fatty acids, linoleic acid and alpha-linolenic acid, have been determined.[31]

The American Academy of Pediatrics endorses the population approach to a healthful diet for all children older than 2 years of age. Children with a positive family history of dyslipidemia or premature cardiovascular disease should be screened for high lipid levels. The individual approach to screening with a fasting lipid profile is also recommended for children whose family history is not known and for those children with other cardiovascular risk factors such as overweight/obesity, hypertension, or diabetes mellitus. Screening for these children should take place after 2 years of age, but not later than 10 years of age.[64]

Dietary recommendations are different for children who are at increased risk of developing premature cardiovascular disease or are found to have high lipid levels. These children need periodic screening for blood cholesterol levels and close follow-up. If LDL-cholesterol levels are high, restriction of total calories from saturated fat to less than 7% and of dietary cholesterol to no more than 200 mg per day is recommended. These children need to be closely monitored by a physician and registered dietitian.[53]

Vitamin and Mineral Supplements

Children who consume a variety of basic foods can meet all of their nutrient needs without vitamin or mineral supplements. Eating a diet of a variety of foods is the preferred way to get needed nutrients because foods contain many other substances, such as phytochemicals, in addition to nutrients that benefit health.

The American Academy of Pediatrics recommends vitamin and mineral supplementation for children who are at high risk of developing or have one or more nutrient deficiencies.[38] Children at risk of nutrient deficiency according to the AAP include those:

1. With anorexia or an inadequate appetite or who follow fad diets
2. With chronic disease
3. From deprived families or who suffer from parental neglect or abuse
4. Who participate in a dietary program for managing obesity
5. Who consume a vegetarian diet without adequate intake of dairy products
6. With failure to thrive

Despite these recommendations, data from the NHANES III indicate that children 1 to 5 years of age

Familial Hyperlipidemia A condition that runs in families and results in high levels of serum cholesterol and other lipids.

Trans Fatty Acids Fatty acids that have unusual shapes resulting from the hydrogenation of polyunsaturated fatty acids. *Trans* fatty acids also occur naturally in small amounts in foods such as dairy products and beef.

Atherosclerosis A type of hardening of the arteries in which cholesterol is deposited in the arteries. These deposits narrow the coronary arteries and may reduce the flow of blood to the heart.

are major users of supplements.[65] Approximately one in two 3-year-olds in the United States is given a vitamin and mineral supplement by parents.[66] Considering characteristics of mothers who give their children supplements, such as having health insurance and a greater household income, children most likely to receive a supplement are those at low risk of developing nutrient deficiencies—children who would most likely benefit from supplements are less likely to receive them.

If given to children, vitamin and mineral supplement doses should not exceed the DRI for age. Parents and caretakers should be warned against giving high amounts of vitamins and minerals to children, particularly vitamins A (retinol) and D. The *Tolerable Upper Intake Levels* shown in the DRI tables should serve as a guide to excessive levels of nutrient intake from fortified foods and supplements.

Tolerable Upper Intake Levels   Highest level of daily nutrient intake that is likely to pose no risk of adverse health effects to almost all individuals in the general population; gives levels of intake that may result in adverse effects if exceeded on a regular basis.

Herbal Supplements

The use of herbal remedies for various disorders is increasing in the United States today, as is the use of complementary and alternative medicine practices in general. Parents and caretakers who take herbs are likely to give these products to their children. In one survey with a 59.4% response rate, participants of the Special Supplemental Nutrition Program for Women, Infants, and Children (WIC) in the states of Kansas and Wisconsin indicated herbal use by either caregivers, children, or both in nearly half of the returned surveys. Herb use was statistically greater among Latino children than non-Latino children, 48.4% vs. 31.4% respectively. Of the 1363 children reported to use herbs, 820 of them were younger than 5 years of age.[67]

Few definitive studies exist on the effectiveness of these substances in preventing disease and promoting health in adults, much less in children. In spite of the lack of scientific evidence, anecdotal reports of benefits abound. However, some reports have linked herbal preparations to adverse effects.[68] Information on herb use should be obtained during the nutrition assessment of a child to rule out herbs as a source of health problems. Currently, herbal supplements are not regulated and using the products can lead to uncertain results. Children given various herbs are the "test subjects" in these uncontrolled studies. Parents should be advised of the potential risks of herbal therapies and the need for close monitoring of their child if they choose to give herbs to their child. The National Institutes of Health's (NIH) National Center for Complementary and Alternative Medicine (NCCAM) website is listed in the "Resources" section at the end of this chapter and provides reports on the known safety and effectiveness of various herbal remedies and alternative medical practices.

Dietary and Physical Activity Recommendations

"... Children ages 2 to 11 years should achieve optimal physical and cognitive development, attain a healthy weight, enjoy food, and reduce the risk of chronic disease through appropriate eating habits and participation in regular physical activity."[69]

The American Dietetic Association

Taking into consideration the energy and nutrient needs of young children and the common nutritional problems and concerns of this age group, it is easy to understand the importance of dietary recommendations for toddlers and preschoolers. A primary recommendation is that young children eat a variety of foods. This recommendation is more easily achieved if healthful food preferences and eating habits are acquired during the early years. Food preferences in conjunction with food availability form the foundation of the child's diet. Limited food selection, therefore, will influence the adequacy of the child's diet by decreasing variety. Parents and caretakers cannot expect a child to "do as I say, but not as I do." Nutrition education aimed at the adults in the child's life becomes as important as nutrition education directed at the child, if not more so.

Dietary recommendations have been developed and disseminated by the federal government and professional organizations. Two sets of guidelines for young children's diets are available: the Dietary Guidelines for Americans and the Food Guide Pyramid.[11,52] Recommendations for caloric and nutrient intake are represented in the DRIs.

Dietary Guidelines

The Dietary Guidelines for Americans 2005 (discussed in Chapter 1) include some key recommendations for specific population groups, including children. The guidelines emphasize that children should be offered a variety of foods, including grain products (at least half of which should be whole grains), vegetables and whole fruits, and low-fat dairy products.[52] The guidelines also recommend reducing the amount of added sugar in children's diets. Children 2 to 8 years should drink 2 cups per day of fat-free or low-fat milk or equivalent milk products. With regard to fats, the guidelines recommend keeping total fat intake between 30 and 35% of calories for children 2 to 3 years of age and between 25 and 35% of calories for children and adolescents 4 to 18 years of age. Fat is an important source of calories, fat-soluble vitamins, and essential fatty acids for young children. Most fats in children's diets should come from sources of polyunsaturated and monunstaurated fatty acids, such as fish, nuts, and vegetable oils. Some fish may contain higher levels of mercury that may harm a young child's developing nervous system. The Food and Drug Administration (FDA) and the Environmental Protection Agency (EPA) advise that young children should eat fish and shellfish that are lower in mercury. Information about the mercury content of fish in a specific area can be obtained from the FDA (see "Resources").

The guidelines also recommend that beans, lean meats, and poultry be added as appropriate for the child. Foods high in fat and sugar, such as candy, cookies, and cakes, should be limited in children's diets. The Dietary Guidelines also emphasize the importance of parents modeling this type of diet for their children, or encouraging them "to do as I say and as I do." The guidelines emphasize the importance of physical activity. Parents are advised to encourage their children to engage in at least 60 minutes of physical activity on most (preferably all) days of the week and to limit the time spent in sedentary activities, such as watching TV and playing computer and video games, that replace physical activity.[52] It is important that parents model for their children a lifestyle that includes a varied diet and regular physical activity. There are also some key recommendations regarding weight management for overweight children (see the "Dietary Guidelines for Americans 2005" section on page 284) and for food safety (see "Food Safety," pages 281–282).

As mentioned previously, the American Heart Association dietary recommendations are consistent with the U.S. Dietary Guidelines. The AHA places special emphasis on adequate intake of omega-3 fatty acids and recommends introducing and regularly serving fish as an entrée to children. The recommendations also emphasize physical activity and balancing intake with physical activity.[53, 60, 61]

Food Guide Pyramid

The USDA has developed the MyPyramid for Kids (Illustration 10.8) targeted at young children.[11] The color-coded pyramid encourages children to consume a variety of foods, to make wise food choices from each of the food groups, and to limit foods high in fat and sugars. To emphasize the importance of physical activity, illustrations of children being active are depicted around the pyramid. MyPyramid for Kids Tips for Families can be found in Table 10.8. Illustration 10.9 depicts a MyPyramid Steps to a Healthier You food guide chart for a 4-year-old boy who is physically active for at least 60 minutes every day.

Illustration 10.8 MyPyramid for Kids.[11]

Table 10.8 MyPyramid for Kids Tips for Families[11]

Eat Right	Exercise
1. Make half your grains whole.	1. Set a good example.
2. Vary your veggies.	2. Take the President's Challenge as a family.
3. Focus on fruits.	3. Establish a routine.
4. Get your calcium-rich foods.	4. Have an activity party.
5. Go lean with protein.	5. Set up a home gym.
6. Change your oil.	6. Move it!
7. Don't sugarcoat it.	7. Give activity gifts.

Recommendations for Intake of Iron, Fiber, Fat, and Calcium

Adequate iron intake is necessary to prevent iron deficiency and iron-deficiency anemia in toddlers and preschoolers. Appropriate fiber intake is needed to prevent constipation and may provide long-term disease prevention. Fat is an important source of calories, essential fatty acids, and fat-soluble vitamins in young children's diets. Adequate calcium intake is important for children to achieve peak bone mass, and yet about 20% of children do not meet the DRIs for calcium.[4]

Iron Adequate iron intake is important in this age group to prevent iron deficiency. Good sources of dietary

Illustration 10.9 MyPyramid Steps to a Healthier You—Individual plan for a 4-year-old boy who is physically active for at least 60 minutes every day.[11]

GRAINS 5 ounces	VEGETABLES 2 cups	FRUITS 1 ½ cups	MILK 3 cups	MEAT & BEANS 5 ounces
Make half your grains whole	**Vary your veggies** Aim for these amounts each week:	**Focus on fruits**	**Get your calcium-rich foods**	**Go lean with protein**
Aim for at least **3 ounces** of whole grains a day	**Dark green veggies** = 2 cups **Orange veggies** = 1 ½ cups **Dry beans & peas** = 2 ½ cups **Starchy veggies** = 2 ½ cups **Other veggies** = 5 ½ cups	Eat a variety of fruit Go easy on fruit juices	Go low-fat or fat-free when you choose milk, yogurt or cheese	Choose low-fat or lean meats and poultry Vary your protein, routine— choose more fish, beans, peas, nuts, and seeds

Find your balance between food and physical activity
Be physically active for at least **60 minutes** every day, or most days.

Know your limits on fats, sugars, and sodium
Your allowance for oils is **5 teaspoons a day.**
Limit extras-solid fats and sugars - to **130 calories a day**

Your results are based on a 1600 calorie pattern.   **Name:** _____

This calorie level is only an estimate of your needs. Monitor your body weight to see if you need to adjust your calorie intake.

iron can be found in Chapter 1 on page 29. Meats, which are good sources of iron, can be ground or chopped to make them easier for toddlers to chew. Fortified breakfast cereals and dried beans and peas are also good sources of iron.

"Toddler" milks, or iron-fortified commercial formulas for toddlers, are available. Healthy children who consume a variety of foods, and whose milk intake is less than 24 ounces daily, obtain adequate iron without these special products. Other commercial beverages being marketed to parents include formulas that were originally designed for children with illnesses or who had to be fed complete nutrition through a feeding tube. Such special products are expensive and are unnecessary for healthy children. It would be better for parents of healthy children to spend their food dollars on a variety of healthy foods rather than on these special products.

Fiber   Ample dietary fiber intake has been associated with the prevention of heart disease, certain cancers, diabetes, and hypertension in adults. Whether fiber helps prevent these problems as young children become adults is not known, but it is clear that fiber in a child's diet helps prevent constipation and is part of a healthy diet. Too much fiber in a child's diet can be detrimental because

high-fiber diets have the potential of reducing the energy density of the diet, which could impact growth.[70] High-fiber diets could also impact the bioavailabilty of some minerals, such as iron and calcium.

The recommendations for total fiber intake based on the DRIs can be found in Table 10.9.[31] Total fiber is the sum of dietary fiber and functional fiber. Earlier recommendations were based on dietary fiber alone. Including fruits, vegetables, and whole-grain breads and cereal products in the diet can increase the dietary fiber intake of children. Those who meet the recommendation consume more breads and cereals, fruits, vegetables, legumes, nuts, and seeds than those who do not. Children with adequate fiber intake tend to have lower intakes of fat and cholesterol, and higher intakes of dietary fiber, vitamins A and E, folate, magnesium, and iron, than do those children who have low dietary fiber intakes.[71]

Table 10.9 Adequate intake of total fiber for children[31]

1–3 years of age	19 g/day of total fiber
4–8 years of age	25 g/day of total fiber

Fat An appropriate amount of fat in a young child's diet can be achieved by employing the principles of the Dietary Guidelines and the Food Guide Pyramid that promote a diet of whole-grain breads and cereals, beans and peas, fruits and vegetables, low-fat dairy products after 2 years of age, and lean meats.[11,52] Foods high in fat are used sparingly, especially foods high in saturated fat and *trans* fatty acids. However, an appropriate amount of dietary fat is necessary to meet children's needs for calories, essential fatty acids, and fat-soluble vitamins. As discussed in Chapter 1, good sources of the essential fatty acid linoleic acid are peanut, canola, corn, safflower, and other vegetable oils. Flaxseed, soy, and canola oils are good sources of the essential fatty acid alpha-linolenic acid.

It is important to include sources of fat-soluble vitamins in the diets of young children. Good sources of vitamin A include whole eggs and dairy products. Sources of vitamin D include exposure to sunlight and vitamin D-fortified milk. The American Academy of Pediatrics recommends a daily intake of 400 IU of vitamin D for all healthy children.[72] Corn, soybean, and safflower oils are excellent sources of vitamin E. Vitamin K is widely distributed in both animal and plant foods.

Calcium Adequate calcium intake in childhood affects peak bone mass. A high peak bone mass is thought to be protective against osteoporosis and fractures later in life.[73] However, many children do not consume adequate calcium. An estimated 21% of children 2 to 8 years of age consume less than their DRI for calcium.[4] The recommendation for daily calcium intake in the DRI table is 500 mg/day for children ages 1 to 3 years and 800 mg/day for children ages 4 to 8 years. An important aspect of adequate calcium intake in toddlers and preschoolers is the development of eating patterns that will lead to adequate calcium intake later in childhood.[73]

Dietary sources of calcium are listed in Chapter 1 on page 28. Dairy products are good sources of calcium, as are canned fish with soft bones such as sardines, dark green leafy vegetables such as kale and bok choy, tofu made with calcium, and calcium-fortified foods and beverages such as calcium-fortified orange juice. Nonfat and low-fat dairy products are low in saturated fat while still serving as a good source of calcium.

Fluids

Healthy toddlers and preschoolers will consume enough fluid through beverages, foods, and sips and glasses of water to meet their needs. Fluid requirements increase with fever, vomiting, diarrhea, and when children are in hot, dry, or humid environments.

Consumption of milk has decreased among young children since the late 1970s, but consumption of carbonated soft drinks has increased by about the same amount. According to food consumption surveys, young children consume large amounts of sweetened beverages, including fruit juice, soft drinks, and sweetened iced tea, to the detriment of the overall nutritional balance of their diet and oral health. Consumption of sugar-sweetened beverages and 100% fruit juice begins at a young age and has increased in recent years.[74] According to the Feeding Infants and Toddlers Study 2002 (FITS), 28% of 12- to 14-month-olds, 37% of 15- to 18-month-olds, and 44% of 19- to 24-month-olds consume sweetened beverages, primarily as fruit-flavored drinks, at least once a day.[75] Approximately 50% of children ages 2 to 5 years consume soft drinks.[76] Today sugar-sweetened beverages contribute 10 to 15% of total calorie intake for children.[74] Children with high consumption of regular soft drinks (more than 9 ounces per day) consume more calories and less milk and fruit juice than those with lower consumptions of regular soft drinks. Water is a good and underused "thirst quencher" for toddlers and preschoolers, as long as milk (2 cups) is part of their regular diet and fruit-juice consumption is <4–6 ounces, as recommended by the American Academy of Pediatrics.[77] Parents and caretakers can offer children water to drink between meals and snacks.

Recommended vs. Actual Food Intake

Ongoing national surveys examine food and nutrient intakes of Americans, including young children. The report "What We Eat in America" (WWEIA), a joint project of the U.S. Department of Agriculture (USDA) and the U.S. Department of Health and Human Services (DHHS), is the dietary intake component of the National Health and Nutrition Examination Survey (NHANES). In these national surveys, dietary data for children younger than 6 years of age is provided by an adult caretaker.[78]

According to the "What We Eat in America" report, which utilizes data from NHANES 2005–2006,[78] young children meet their energy needs. Mean percentages of total energy from carbohydrate, protein, total fat, saturated fat, and cholesterol intake in the diets of toddlers and preschoolers are shown in Table 10.10. Toddlers who participated in the Feeding Infants and Toddlers Study 2002 (FITS) exceeded estimated energy requirements by 31%.[79] The total fat intake of about 31% of total calories is within the target range for children this age. Two- to five-year-olds have sodium intakes of about 2400 mg per day for boys and about 2150 mg per day for girls, with the sodium intake recommendation at 2300 mg per day.

In general, young children consume more than enough protein and fat. In a longitudinal study of the nutrient and food intakes of preschool children ages 24 to 60 months, mean intakes of zinc, folic acid, and vitamins D and E were consistently below the recommended levels.[80] Low intakes of zinc, vitamin E, and iron were found in toddlers ages 12 to 18 months, a time of dietary transition.[81] Vitamin E intakes less than the estimated average requirement were found in 58% of toddlers who participated in

Table 10.10 Mean percentages of total calories from carbohydrate, protein, total fat, saturated fatty acids, and cholesterol intake[74]

2–5 years of Age	Carbohydrate (%)	Protein (%)	Total fat (%)	Saturated Fatty Acids (%)	Cholesterol (mg/d)
Males	55.9	13.9	31.4	11.5	174
Females	56.0	14.3	31.1	11.3	164

the Feeding Infants and Toddlers Study 2002 (FITS).[79] The means for nutrient intakes often hide problems at the extremes, however. They fail to indicate the percentage of children with low nutrient intakes of less than 66% of the recommended levels, and children with high nutrient intakes that exceed the Tolerable Upper Intake Levels.

The diets of children who ate fast food were found to be higher in total energy, total fat, total carbohydrate, added sugars, and sugar-sweetened beverages, and to have less fiber, less milk, and fewer fruits and non-starchy vegetables than the diets of children who did not eat fast food.[82] Using this same data set, it was found that 11% of children ages 2 to 3 years and 12% of children ages 4 to 5 years consumed greater than 25% of total energy from added sugar. Increased added sugar consumption was associated with decreased nutrient and food-group intakes and increased percentage of children not meeting the DRIs.[83] According to the FITS Study, by 19 to 24 months, 62% of toddlers consumed some type of baked dessert, 20% consumed candy, and 44% consumed sweetened beverages in a day.[75]

Children's portions sizes have remained constant over the years except for meat portions, which have decreased. This stability in portion sizes of young children over time reinforces the hypothesis that young children are capable of self-regulating energy intake. Portion sizes were positively related to both body-weight percentiles and energy intake. It seems that young children self-regulate energy intake by adjusting portion size.[84]

Cross-Cultural Considerations

When working with families from different cultures, it is important to learn as much as possible about the culture's food-related beliefs and practices. Ask the parents and caretakers about their experiences with food, including foods used for special occasions. It is also helpful to know whether foods are used for home remedies or to promote certain aspects of health. Cultural beliefs influence many child feeding practices, such as what foods are best for young children, which cause digestive upsets, or which help relieve illnesses. It is important for the health care provider to build on the cultural practices and to reinforce the positive practices while attempting to affect change that could be more beneficial to the young child.

For example, peanut or polyunsaturated oils can be recommended to Chinese Americans for stir-frying instead of the more traditional lard or chicken fat.

Vegetarian Diets

Young children can grow and develop normally on vegetarian or vegan diets, as long as their dietary patterns are intelligently planned. Vegetarian diets are rich in fruits, vegetables, and whole grains, the consumption of which is encouraged for the general population. However, young children in particular need some energy-dense foods to reduce the total amount of food required. The amount of vegetarian foods needed to meet nutrient needs may be more food than young children can eat. Young children need to eat several times a day to meet their energy needs because their stomachs cannot hold a lot of food at one time.

Children who are fed **vegan** and **macrobiotic diets** tend to have lower rates of growth, although still within normal ranges, during the first 5 years of life compared to children given a mixed diet.[85] Strict vegan diets, which exclude all foods of animal origin, may be deficient in vitamins B_{12} and D, zinc, and omega-3 fatty acids, and may also be low in calcium, unless fortified foods are consumed. Protein needs are usually met if the diet is adequate in energy and a variety of foods are included.[86] Children on vegan diets should receive vitamin B_{12} supplements or consume fortified breakfast cereals, textured soy protein, or soy milk fortified with vitamin B_{12}. The vitamin B_{12} status of children following vegetarian and vegan diets should be monitored on a regular basis, as vitamin B_{12}–deficiency may cause vitamin B_{12}–deficiency anemia. Iron-deficiency anemia is an infrequent problem among children consuming a vegetarian diet.

Vitamin D adequacy can be achieved by diet or by sun exposure. Good sources of vitamin D for children include fortified soy milk, fortified breakfast cereals, and fortified margarines. Zinc is found in foods of animal origin. Plant sources of zinc include legumes, nuts, and whole grains. Vegetable products are also lacking in omega-3 fatty

Vegan Diet The most restrictive of vegetarian diets, allowing only plant foods.

Macrobiotic Diet This diet falls between semivegetarian and vegan diets and includes foods such as brown rice, other grains, vegetables, fish, dried beans, spices, and fruits.

acids. Therefore, including a source of these fatty acids, such as canola or soybean oils, is advisable.[86] Foods containing phytates, such as unrefined cereals, may interfere with calcium absorption; if the child's diet contains a lot of unrefined cereals, higher calcium intakes may be needed.[86] Good sources of calcium for children on strict vegetarian diets include fortified soy milk, calcium-fortified orange juice, tofu processed with calcium, and certain vegetables, such as broccoli and kale.[10] Supplements may be necessary for some children with inadequate intakes that are not remedied by dietary means.

Guidelines recommended for vegetarian diets for young children have been developed and are given here:[10]

- Allow the child to eat several times a day (i.e., three meals and two to three snacks).
- Avoid serving the child bran and an excessive amount of bulky foods, such as bran muffins and raw fruits and vegetables.
- Include in the diet some sources of energy-dense foods such as cheese and avocado.
- Include enough fat (at least 30% of total calories) and a source of omega-3 fatty acids, such as canola or soybean oils.
- Include sources of vitamin B_{12}, vitamin D, and calcium in the diet, or supplement if required.

Child Care Nutrition Standards

"All child care programs should achieve recommended standards for meeting children's nutrition and nutrition education needs in a safe, sanitary, supportive environment that promotes healthy growth and development."[87]

The American Dietetic Association

An estimated 23 million children in the United States require child care while their parents work, making foods children eat away from home a major contribution to their overall intake. Nutrition standards for child care services exist and specify minimum requirements for amounts and types of foods to include in meals and snacks, as well as food-service safety procedures.[87,88] These standards also address nutrition learning experiences and education for children, staff, and parents as well as the physical and emotional environment in which meals and snacks are served. It is recommended that children in part-day programs (4 to 7 hours per day) receive food that provides at least one-third of their daily calorie and nutrient needs in at least one meal and two snacks or two meals and one snack. A child in a full-day program (8 hours or more) should receive foods that meet one-half to two-thirds of the child's daily needs based on the DRIs in at least two meals and two snacks or three snacks and one meal. Food should be offered at intervals of not less than 2 hours and not more than 3 hours and should be consistent with the Dietary Guidelines for Americans.[52]

Physical Activity Recommendations

Physical activity is an important component of a healthy lifestyle. Physical activity helps to maintain energy balance while strengthening muscles. Inactivity is thought to be a major contributor to the increasing prevalence of obesity. The Dietary Guidelines for Americans 2005 recommend that young children engage in play activity for at least 60 minutes on most, preferably all, days of the week.[52] Some suggested activities from MyPyramid for Kids[11] include:

- Taking a nature walk
- Riding a tricycle or bicycle
- Walking, skipping, or running
- Most important—having fun while being active!

This advice is reinforced by the American Academy of Pediatrics (AAP).[89] The AAP suggests that toddlers, under the supervision of an adult caregiver, engage in activities such as walking in the neighborhood, park, or zoo and free play outdoors. For the preschool-age child, the AAP lists appropriate activities as running, swimming, tumbling, throwing, and catching, under adult supervision. No television viewing is recommended for children less than 2 years of age, while screen time should be limited to less than 2 hours per day for all other age groups.[89] Removing screens from children's bedrooms is also recommended.[30] Parents are encouraged to set a good example for their children by being physically active themselves and limiting the amount of time that the family spends watching TV and playing computer and video games.

Nutrition Intervention for Risk Reduction

Nutrition Assessment

Components of a nutrition assessment include a food/nutrition-related history, pertinent biochemical measurements, anthropometric measurements such as weight, height, body mass index percentile, and a medical history. Table 10.11 lists biochemical indices that are pertinent to the young child. Based on this information, the nutrition professional can identify any nutrition diagnoses, design a nutrition intervention plan with family input, and make a plan for monitoring and evaluation (see Table 2.10 for the four steps of the Nutrition Care Process).

Model Program

"Bright Futures in Practice: Nutrition" is an example of a model program for nutrition intervention for risk reduction.[10] This guide is a component of the larger project "Bright Futures Guidelines for Health Supervision of Infants, Children, and Adolescents."[90]

Table 10.11 Normal values of biochemical nutritional parameters[38,64,90]

Test Iron Status	Normal Values
Hematocrit, %	39
Hemoglobin, g/dL	14
Serum ferritin, ng/mL	>15
Serum iron, mcg/dl	>60
Serum total iron binding capacity, mcg/dl	350–400
Serum transferring [revert to original] saturation, %	>16
Serum transferrin, mg/dl	170–250
Erythrocyte prtoporphyrin, mcg/dl red blood cells	>70
Lead Screening	
Blood lead levels	<10 mcg/dl
Dyslipidemia Screen	
Total cholesterol	<170 mg/dl
LDL-cholesterol	<110 mg/dl

Refer to page A-4 of the Appendix for a table for converting Conventional Units shown in the table to SI units.

The purpose of the Bright Futures program is to foster trusting relationships between the child, health professionals, the family, and the community to promote optimal health for the child.[90] Bright Futures guidelines are developmentally based and address the physical, mental, cognitive, and social development of infants, children, adolescents, and their families.

Many different "user-friendly" materials and tools are available from this program to assist in the implementation of the guidelines. In addition to the "Bright Futures Guidelines for Health Supervision," implementation guides have been published for oral health, general nutrition, physical activity, mental health, and families. "Bright Futures in Practice: Nutrition" is based on three critical principles:[10]

1. Nutrition must be integrated into the lives of infants, children, adolescents, and families.

2. Good nutrition requires balance.

3. An element of joy enhances nutrition, health, and well-being.

The program is based on the premise that optimal nutrition for children should be approached from the perspective of development of the child and put in the context of the environment in which the child lives.[10] It emphasizes the development of healthy eating and physical activity behaviors. Nutrition supervision guidelines are given for each age group, and within each broad age group, interview questions, screening and assessment, and nutrition counseling topics are provided. The program information lists desired outcomes for the child and discusses the role of the family, as well as answering frequently asked questions. For example, by utilizing the guidelines, health care providers will be able to provide anticipatory guidance to parents of toddlers for the proper advancements of toddler diets based on growth and development. The implementation guide also addresses special topics related to pediatric nutrition, including oral health, vegetarian eating practices, iron-deficiency anemia, and obesity. The "Bright Futures in Practice: Nutrition" guidelines are a valuable resource to anyone who is interested in promoting healthy eating and physical activity behaviors in children. Ordering information for Bright Futures materials is available on the program's website, which is listed in the "Resources" section.

Public Food and Nutrition Programs

Young children and their families can benefit from a number of federally sponsored food and nutrition programs. Four example programs are presented here.

WIC

The Special Supplemental Nutrition Program for Women, Infants, and Children,[91] previously described in Chapter 8, is administered by the Food and Nutrition Service of the U.S. Department of Agriculture (USDA). It is one of the most successful federally funded nutrition programs in the United States. In FY 2001 through FY 2003, an average of 5.7 million children and 1.8 million mothers received WIC services each month. Participation in WIC services improves the growth, iron status, and the quality of dietary intake of nutritionally at-risk infants and children up to age 5 years.[92]

As in infancy, children must live in a low-income household, 185% or less of the federal poverty level, and be at "nutrition risk" to be eligible for WIC services. "Nutrition risk" means a child has a medical- or dietary-based condition that places the child at increased risk. Such conditions include iron-deficiency anemia, underweight, overweight, a chronic illness such as cystic fibrosis, or the child consumes an inadequate diet.[91] Children receive nutrition assistance, education, and follow-up services by specially trained registered dietitians and nutritionists. Vouchers for food items such as milk, juice, eggs, cheese, peanut butter, and fortified cereals are given to eligible families. These vouchers are exchanged for the food items at authorized retailers.

WIC's Farmers' Market Nutrition Program

The Farmers' Market Nutrition Program is a special seasonal program for WIC participants. This program provides vouchers for the purchase of locally grown produce at farmers' markets. The program is designed to help low-income families increase their consumption of fresh fruits and vegetables.

Head Start and Early Head Start

Administered by the U.S. Department of Health and Human Services, Head Start and Early Head Start are comprehensive child development programs, serving children from birth to 5 years of age, pregnant women, and their families. Nearly one million U.S. children participate in this program. The overall goal is to increase the readiness for school of children from economically disadvantaged families. A range of individualized, culturally appropriate services are provided through Head Start and related agencies, including educational, health, nutritional, social, and other services.[93] More specific information about Early Head Start can be found in Chapter 9.

Supplemental Nutrition Assistance Program (formerly the Food Stamp Program)

The Supplemental Nutrition Assistance Program (SNAP), administered by the USDA, is designed to help adults in low-income households buy food. In 2008, about 28.4 million people living in 12.7 million households received assistance in the United States. The monetary amount of food vouchers provided to an eligible household depends on the number of people in it and the income of the household. Income eligibility criteria for this and a number of other federal programs can be found at the USDA's Food and Nutrition Service website (www.fns.usda.gov/fsp). The average monthly amount of benefits received through SNAP in 2008 was $226.59 per household, enough to help families and individuals pay for a portion of the food they need. Each state must develop a food stamp nutrition education plan based on federal guidance.[94] Participation in the Food Stamp Program is associated with increased intakes of a number of nutrients.[92]

Key Points

1. Periodic and accurate measurements of a young child's growth are important indicators of the child's nutritional status. Adequate and appropriate growth is the ultimate outcome indicator of adequate nutrition and health.

2. The types of foods offered to children and methods of feeding are based on an individual child's growth and development.

3. Food habits are learned behaviors.

4. Common nutritional problems such as iron-deficiency anemia, dental caries, and constipation in a healthy child can be addressed through adjustments in the child's diet.

5. The prevalence of and risk for overweight and obesity are increasing even among young children in the United States.

6. The U.S. Dietary Guidelines recommend that children eat a variety of foods and increase physical activity.

7. The MyPyramid for Kids reinforces eating a variety of foods and increasing physical activity.

8. Surveys of food intake indicate that most children meet their nutritional needs, but that their intakes exceed recommendations for energy, sodium, total fat, and saturated fat.

9. A well-planned vegetarian diet can meet the needs of a growing child.

10. Public food and nutrition programs are important resources for many young children in the United States.

Review Questions

1. Which of the following growth parameters is used as a screening tool for assessing underweight and overweight or obesity in young children?

   a. Weight-for-age percentile

   b. Length- or stature-for-age percentile

   c. Body mass index-for-age percentile

   d. Actual body mass index (weight in kg/height in meters²)

   e. Actual weight in kg

2. Which of the following strategies would be appropriate for parents to use in teaching their young child healthy eating behaviors?

   a. Give a dessert as a reward for eating fruits and vegetables.

   b. Expect the child to eat foods that parents will not eat themselves.

   c. Coax the child to eat, especially new foods.

d. Serve child-sized portions or allow the child to self-serve portions.

e. Make the child stay at the table until she has cleaned her plate.

3. Obesity is defined as:

a. Weight-for-age percentile ≥85th to 94th

b. Weight-for-age percentile ≥95th

c. Weight-for-length percentile ≥95th

d. BMI-for-age percentile ≥85th to 94th

e. BMI-for-age percentile ≥95th

4. Which of the following statements regarding fat in children's diets is correct?

a. Dietary fat is an important source of calories in a young child's diet.

b. The amount of saturated fat is not restricted in a young child's diet.

c. The amount of *trans* fatty acids is not restricted in a young child's diet.

d. Restricting fat in a young child's diet has been shown to decrease growth.

e. Dietary fat recommendations are the same for children regardless of cardiovascular disease risk.

5. True or false: According to food consumption surveys, young children exceed estimated needs for calories.

6. Which of the following would be the best choice for a between-meal snack for a preschooler? Choose the one best answer.

a. 100% fruit juice

b. Raisins

c. Fortified gummy bears

d. Peanut butter crackers

e. Oatmeal cookies

7. Several health and professional organizations and agencies, such as the American Academy of Pediatrics, the American Heart Association, and the USDA, have published similar recommendations for a healthy diet for young children. Briefly describe the food components of such a diet.

8. Describe the role of physical activity in a healthy lifestyle for young children. Give examples of appropriate activities for children.

Resources

American Academy of Pediatrics

The American Academy of Pediatrics website contains consumer information on current news topics affecting the pediatric population. Policy statements can be found on this page, and consumer publications can be ordered through the book store.

Website: www.aap.org

American Dietetic Association

Besides information for members of the American Dietetic Association, this website contains information for consumers on various topics of interest. There is a "Daily Tip" and a feature article, plus consumer information on topics such as food safety and healthy lifestyles.

Website: www.eatright.org

Bright Futures

The Bright Futures publications are developmentally based guidelines for health supervision and address the physical, mental, cognitive, and social development of infants, children, adolescents, and their families. Four implementation guides have also been published to date, addressing oral health, mental health, general nutrition, and physical activity. Bright Futures is a collaborative project of the American Academy of Pediatrics and the Maternal and Child Health Bureau, Health Resources and Services Administration, Department of Health and Human Services.

Website: www.brightfutures.aap.org

Celebrating Diversity

The purpose of this publication is to assist health professionals in learning to communicate effectively with a diverse clientele. Topics covered in the book include using food to create common ground, changing food patterns, understanding how food choices are made, communicating with clients and families, and working within the community.

Graves, D. E. and Suitor, C. W. *Celebrating Diversity: Approaching Families Through Their Food*, 2nd ed. Arlington, VA: National Center for Education in Maternal and Child Health, 1998.

Centers for Disease Control and Prevention, National Center for Health Statistics

The CDC website provides background information on its growth charts (CDC Growth Charts: United States. May 30, 2000). Also, individual growth charts can be downloaded and printed.

Website: www.cdc.gov/nchs/about/major/nhanes/ growthcharts/charts.htm

Diabetes Care and Education, Dietetic Practice Group of the American Dietetic Association

This series of booklets addresses food practices, customs, and holiday foods of various ethnic groups. Examples for incorporating traditional foods of various groups of people into dietary recommendations are given. Booklets describing the following ethnic groups are available: Alaska Native, Chinese

American, Filipino American, Hmong American, Jewish, and Navajo.

Ethnic and Regional Food Practices, A Series. Chicago, IL: The American Dietetic Association, 1994–1999.

FitSource: A Web Directory for Providers

This resource, produced by the National Child Care Information and Technical Assistance Center and the Child Care Bureau, contains a wide variety of tools for incorporating age-appropriate nutrition and physical activity into child care and after-school programs. It is designed for use by program administrators, directors, technical assistance providers, and others interested in promoting proper nutrition and physical activity.

Website: **http://nccic.acf.hhs.gov/fitsource/**

Food and Drug Administration

The FDA provides information about the safety of the food supply, including information about the mercury content of fish and shellfish in certain areas. Information on dietary supplements is also available.

Call FDA's food information line toll-free at 1-888-SAFE-FOOD, or visit www.fda.gov.

MyPyramid.gov

As described in Chapter 1, MyPyramid.gov is Internet-based and resource-filled. As the name suggests, the MyPyramid for Kids is designed for children and includes such resources as a coloring page, a kids' worksheet, and classroom materials.

Website: **www.MyPyramid.gov**

National Academy Press, Dietary Reference Intakes

The website of the National Academy Press, the publisher for the National Academies, contains descriptions of available books. Over 2000 books are available online free of charge. The current Dietary Reference Intakes are also available on this website and can be read online for free.

Website: **www.nap.edu**

National Center for Complementary and Alternative Medicine, National Institutes of Health

The National Center for Complementary and Alternative Medicine is dedicated to science-based information on complementary and alternative healing practices. The "Herbs at a Glance" section provides information and available research findings about specific herbs and botanicals. This website provides information for consumers and practitioners as well as information about related news and events.

Website: **www.nccam.nih.gov**

National Network for Child Care

This website, hosted by Iowa State University Extension, provides articles, resources, and links on a variety of topics of interest to professionals and families who care for children and youth. Topics include child development, nutrition, and health and safety.

Website: **www.nncc.org**

Partnership for Food Safety Education

This partnership was formed in 1997 for the purpose of educating the public about safe food handling to reduce foodborne illnesses. This website promotes the partnership's four food-safety practices to the educator, consumer, and the media.

Website: **www.fightbac.org**

USDA/Agricultural Research Service Child Nutrition Research Center

The Child Nutrition Research Center is housed at Baylor College of Medicine in the Texas Medical Center in Houston, TX. In addition to information about the ongoing research at the center, this website includes useful tools such as the "Kids' BMI Calculator" and the "Kids' Energy Needs Calculator," which is for healthy children 2 to 20 years of age. By entering the child's age, height, weight, and activity level, users can get an estimate of calorie needs and how much of the different food groups are needed to meet these needs through a healthy diet, based on the Dietary Reference Intakes.

Website: **www.bcm.edu/cnrc**

USDA, Center for Nutrition Policy and Promotion

This website provides information on USDA materials as well as resources, including the Dietary Guidelines for Americans and the MyPyramid for Kids.

Website: **www.cnpp.usda.gov/Resources.htm**

WHO Growth Standards

Charts and tables of z-scores and percentiles of these international growth standards for boys and girls from birth to 5 years of age can be downloaded from this website.

Website: **www.who.int/childgrowth**

Chapter 11

Toddler and Preschooler Nutrition:
Conditions and Interventions

Prepared by **Janet Sugarman Isaacs**

Key Nutrition Concepts

1 Nutrition problems in young children with special health care needs include underweight, overweight, feeding difficulties, and higher nutrient needs as a result of chronic health conditions.

2 Feeding difficulties in preschoolers and toddlers appear as food refusals, picky appetites, and concerns about growth.

3 Nutrition services for toddlers and preschoolers with chronic health problems are provided in various settings, including schools and other educational programs and specialty clinics.

4 Toddlers and preschoolers at risk for chronic conditions have the same nutritional problems, concerns, and needs as other children.

Introduction

Most toddlers and preschoolers are healthy and develop as expected. This chapter discusses children who do not fit the typical pattern, *children with special health care needs* associated with a *chronic condition* or disability, or children who are at risk. Sometimes no diagnosis has been made, and yet parents, health care providers, or preschool teachers have a nagging feeling that something is not right about how the child is growing and developing. This chapter covers nutrition needs and services for young children with food allergies, breathing or *pulmonary* problems, feeding and growth problems, or developmental delays, and those at risk for needing nutrition support. The Healthy People 2010 objectives (in Chapter 10, Table 10.1) are not appropriate for children with special health care needs if their underlying health care condition affects any of the nutrients involved. Similarly, the 2005 U.S. Dietary Guidelines are not customized to the underlying conditions of children with special health care needs.

Who Are Children with Special Health Care Needs?

The child who does not see, hear, or walk is easily recognized as having a chronic condition. It can be difficult and expensive, however, to identify some other children with special health care needs. Criteria for labeling chronic conditions in children vary from state to state. More than 40 different federal definitions describe the term *disability*.[1] Criteria used for identifying disabilities in adults do not fit children, because the criteria are related to a person's ability to work or perform household chores. *Chronic condition* and *disability* mean the same thing in referring to toddlers and preschoolers. Prevalence estimates for disabilities range from 5% to 31% of children.[1,2] Whatever the number, nutrition problems are common in children with disabilities. Up to 90% of children with disabilities have some type of nutritional problem.[3]

The difficulty in finding nutrition-related services for children with special health care needs has resulted in a variety of Internet resources targeted at families. Services for toddlers and preschoolers with chronic illnesses or those at risk are grouped with services for infants and those for older children, depending on location and funding source. Educational services categorize toddlers and preschoolers differently than do medical services. One of the best resources for providers and families to find services is the National Dissemination Center for Children with Disabilities, funded by the U.S. Department of Education. It identifies disabilities for toddlers and preschoolers and has outcome studies on the effectiveness of educational interventions.

What is important is that toddlers and children are entitled to the same services as older people with chronic illnesses are, with additional help. They are covered by the Americans with Disabilities Act, the Social Security Disability Program, the Supplemental Social Security Insurance (SSI) Program, and services for families without health insurance coverage.[4] Additional help comes from educational regulations ensuring that all children with disabilities have a free, appropriate public education. Nutrition services are funded within education regulations in the Individuals with Disabilities Education Act (IDEA).[4] Most children start school at age 5 or 6 years, but children at risk or who have special needs may attend well before that, as soon as the need is identified. The sooner special educational, nutritional, and health care interventions are started, the better for the overall development of the child. Parents of a typical child choose and pay for day care or a preschool program. For a child with special health care needs, day care or educational programs are selected based on nutrition and other types of therapy provided by state and federal resources. Nutrition services can be provided to young children within special educational programs and services both as preschoolers (3 to 5 years old) and from birth up to 3 years old.[4,5] Services have to be culturally appropriate for various ethnic groups, reflecting food preferences, religious beliefs, and sensitivity to dress and language; otherwise, they are likely to be rejected.

Eligibility for services does not require a specific diagnosis. *Early intervention services* are based on the following:[4,5]

● Developmental delays in one or more of the following areas: cognitive, physical, language and speech, psychosocial, or self-help skills

Children with Special Health Care Needs *A general term for infants and children with, or at risk for, physical or developmental disabilities or chronic medical conditions from genetic or metabolic disorders, birth defects, premature births, trauma, infection, or prenatal exposure to drugs.*

Chronic Condition *Disorder of health or development that is the usual state for an individual and unlikely to change, although secondary conditions may result over time.*

Pulmonary *Related to the lungs and their movement of air for exchange of carbon dioxide and oxygen.*

Early Intervention Services *Federally mandated evaluation and therapy services for children in the age range from birth to 3 years under the Individuals with Disabilities Education Act.*

- A physical or mental condition with a high probability of delay, such as Down syndrome
- At risk medically or environmentally for substantial developmental delay if services are not provided

A number of chronic conditions are suspected, but not obvious, in the first year of life. The diagnosis often becomes clear in the toddler and preschool years, however. Standardized developmental screening, evaluation, and testing for these ages show more reliability than they do for infants. Parents who were told about possible disabilities during infancy move beyond coping by denial or disbelief and are willing to seek out services. The tendency is to resist labeling a young child with a diagnosis, so some suspected conditions are not confirmed until school age if the delay in diagnosis will not harm the child.

Nutrition Needs of Toddlers and Preschoolers with Chronic Conditions

Toddlers or preschoolers with chronic health conditions are at risk for the same nutrition-related problems and concerns as are other children.[6] Consequently, every attempt should be made to meet their overall nutritional needs and to ensure normal growth and development. The DRIs for toddlers and preschoolers provide a good starting point for setting protein, vitamin, and mineral needs for children with chronic conditions.[7] (DRI tables are located on the inside front cover of this text.) The recommendations for typical children concerning dietary fiber, prevention of lead poisoning, and iron-deficiency anemia apply to children at risk or already diagnosed with special health care needs.[5]

Locating appropriate nutrition-related services can be difficult because educational categories of disabilities are not well aligned with nutrition diagnoses. For example, there is no category of special education for obesity or feeding problems. A toddler who is identified by genetic testing with Prader-Willi syndrome has a high need for nutrition services, but educational categories that fit Prader-Willi syndrome include developmental delay, speech or language impairment, or rare disorders. Early intervention programs and medical providers use different terminology for the same child with Prader-Willi syndrome due to health insurance coverage and special educational local resources. This is why a young boy with Prader-Willi syndrome in special education may not be identified for nutrition services until obesity is evident, and part of his condition is that he eats so well.

For children with various specific conditions, standard nutrition guidance does not apply. The DRI for dietary fiber for children may be too low for some conditions and too high for others.[8] Children with sickle-cell disease have more specific blood iron and lead testing than indicated by the usual guidelines. Iron-rich foods to increase their iron stores may not be appropriate when iron also comes with blood transfusions. Consequently, nutritional needs must be customized to the child.[6,8]

Chronic conditions may result in poor appetite, although there are increased caloric needs.[8] Table 11.1 gives examples of conditions in which caloric needs may be high or low.[6,8] Each child must be assessed to confirm caloric needs. A child may have an interval of needing additional calories based on the course of the chronic condition. Changes in caloric needs may explain why both underweight or overweight are more common in children with chronic conditions than in other children.[6] Overweight and obesity are common in Down syndrome and spina bifida in part because of lower caloric needs due to low muscle mass, lower mobility, and short stature.[6] Overall health status is worsened by excessive body weight, so matching caloric intake to needs is important no matter how difficult. Underweight results in part from the chronic illness and its treatment. Children with chronic illnesses may be more likely to experience weight loss with any illness. Underweight children with a chronic condition may or may not benefit from food choices for weight gain. In underweight children, it is inappropriate to make some of the usual recommendations, such as reducing fat intake.[5]

Recommendations regarding food intake, vitamin and mineral supplementation, and mealtime behaviors also should be customized to the individual child. Children who are frequently sick or have low energy levels and appetites may dislike eating foods that are hard to chew or take a long time to eat. Some food-intake problems related to chronic illness may result from the children's

Cystic Fibrosis Condition in which a genetically changed chromosome 7 interferes with all the exocrine functions in the body, but particularly pulmonary complications, causing chronic illness.

Diplegia Condition in which the part of the brain controlling movement of the legs is damaged, interfering with muscle control and ambulation.

Pediatric AIDS Acquired immunodeficiency syndrome in which infection-fighting abilities of the body are destroyed by a virus.

Prader-Willi Syndrome Condition in which partial deletion of chromosome 15 interferes with control of appetite, muscle development, and cognition.

Bronchopulmonary Dysplasia (BPD) Condition in which the underdeveloped lungs in a preterm infant are damaged so that breathing requires extra effort.

Table 11.1 Chronic conditions generally associated with high and low caloric needs	
Higher Caloric Need Conditions	**Lower Caloric Need Conditions**
Cystic fibrosis	Down syndrome
Renal disease	Spina bifida
Ambulatory children with diplegia	Nonambulatory children with *diplegia*
Pediatric AIDS	*Prader-Willi syndrome*
Bronchopulmonary dysplasia (BPD)	Nonambulatory children with short stature

behavior. It is age-appropriate for children to express their food likes and dislikes, insist on their independence, and go on food jags. It can be difficult but important to distinguish between food-intake problems related to the chronic condition and those related to "growing up" in the toddler and preschool years.

Growth Assessment

Most toddlers and preschoolers with chronic conditions are provided an assessment of nutritional status as a first step in determining whether more intensive levels of nutrition services are needed. The need for nutrition services is identified by answers to these sorts of questions:

- Is the child's growth on track?
- Is his or her diet adequate?
- Are the child's feeding or eating skills appropriate for the child's age?
- Does the diagnosis affect nutritional needs?

A variety of nutrition screening tools exist for assessing the nutritional status of children with chronic conditions.[6,8] Such tools are useful for children at risk as well as those already diagnosed with conditions such as asthma, HIV infection, allergies, and cerebral palsy. After assessment, nutrition intervention services provide methods to improve nutritional status. Several conditions that require nutrition intervention services include failure to thrive, celiac disease, breathing problems, and muscle coordination problems.

Children with special health care needs often have conditions that affect growth even when adequate nutrients are provided. In such cases, the 2000 CDC growth charts require interpretation based on the child's previous growth pattern.[9] If a thin and small-appearing child has adequate fat stores, adding calories may be harmful; it is important to recognize the growth pattern as healthy for that child. Trying to add calories in such a case may promote overweight in the form of excess fat stores. Growth patterns in children with special health care needs are also affected by earlier feeding practices and nutrient adequacy, as well as prescribed medications, particularly those that change body composition, such as steroids.[10,11]

Specific growth charts developed for chronic conditions, when available, are recommended to be used. Often both the CDC chart and the specialty chart give the best perspective of a growth pattern. For children up to age 38 months who are born low birth weight (LBW) or very low birth weight (VLBW), the Infant Health and Development Program (IHDP) growth percentile charts are appropriate.[12] Correction for prematurity, as discussed in Chapter 9, makes the charts useful for preterm babies as well as toddlers. For a child born 3 months early, for example, the IHDP growth charts could be used at a chronological age of 41 months. Plotting on both the 2000 CDC growth charts and the IHDP chart documents catch-up growth. When catch-up growth happens, the child's growth pattern crosses channels on the growth chart, for weight as well as for length.

Special health care providers commonly use a special head growth chart.[13] It provides head circumference percentiles from birth to age 18 years and is used to determine whether head growth falls within normal limits or indicates a neurological condition, such as *Rett syndrome*.[13] Rett syndrome is a rare disorder, characterized by a reduced rate of head growth beginning in the toddler years (see Illustration 11.1).[14] Later, over time, the rate of

Rett Syndrome Condition in which a genetic change on the X chromosome results in severe neurological delays, causing children to be short, thin-appearing, and unable to talk.

Illustration 11.1 Nellhaus head circumference growth chart plotted for girl with Rett syndrome.

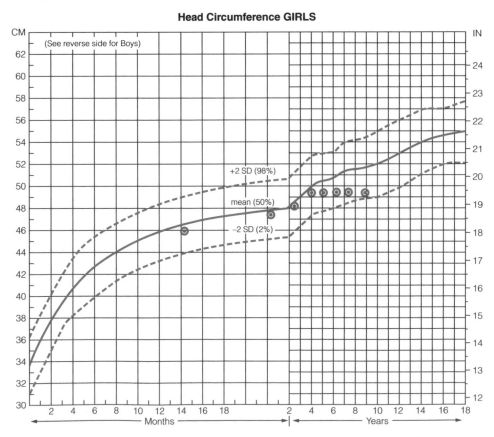

Head Circumference GIRLS

Table 11.2 Food choices of a 2.5-year-old child with suspected developmental delay

Likes	Dislikes
3 packets instant Cream of Wheat with added sugar and margarine (refused offered apple slices)	Hamburger meats, or any other kind of meat
Macaroni and cheese (refused offered sandwich with lettuce and bologna)	Green beans or any kind of vegetables
Banana, with peel removed (refused other cut-up fruits offered)	Vegetable soup
Pudding, only chocolate	Salads of all kinds
Cheese puffs (refused corn chips)	Casseroles or any mixtures of foods
Juices of all kinds, in a sippy cup	Milk, and milk with any flavoring added

Meningitis Viral or bacterial infection in the central nervous system that is likely to cause a range of long-term consequences in infancy, such as mental retardation, blindness, and hearing loss.

Attention Deficit Hyperactivity Disorder (ADHD) Condition characterized by low impulse control and short attention span, with and without a high level of overall activity.

weight and height accretion slows in girls with Rett syndrome.[14] Decreased rate of head growth in toddlers and preschoolers may be indicative of problems from infancy, such as prematurity, or consequences of infection such as *meningitis*. Some clinics plot head circumference on the back of the CDC 2000 growth chart (up to age 3 years) as well as on the special head circumference growth chart.

Feeding Problems

Children with special health care needs have many of the same developmental feeding issues as other children, such as using food to control their parents' behavior at mealtime and going on food jags. Feeding problems that are part of underlying health conditions may emerge in the toddler and preschool years on top of the usual feeding difficulties and require extra attention. Some feeding problems during the toddler years are typical in children who are later diagnosed with a chronic condition. Examples of such conditions include gastroesophageal reflux, asthma (pulmonary problems in general), developmental delay, cerebral palsy, attention deficit hyperactivity disorder, and autism.[3,8] As toddlers, these children tend to display signs of feeding problems, such as low interest in eating, long mealtimes (over 30 minutes), preferring liquids over solids, and food refusals. Children at risk for developmental delay often prove more difficult to feed as toddlers and preschoolers.[3] The child may drink liquids excessively, or eat foods usually preferred by younger children (see Case Study 11.1 on the next page). Recognizing that the child needs to be treated as younger than current chronological age may be a necessary step. Offering the child food textures that he or she can eat successfully within a monotonous diet, or continuing to offer a bottle, may be appropriate choices in these circumstances.

Table 11.2 shows an example of the likes and dislikes of a 2.5-year-old child. The child likes only a few foods, which are not especially nutritious. Usual recommendations are to add variety to the child's diet and to ensure intake of meats, milk, and vegetables. This recommendation is appropriate for a typical child, but this child's eating pattern suggests a feeding problem. The soft textures and mild tastes of preferred foods characterize a child closer to 1 year of age. The foods the child dislikes require higher oral skills. An evaluation of the child's overall level of functioning will likely indicate a developmental delay of the child's feeding skills.

Behavioral Feeding Problems

"Every mouth prefers its own soup."

Sephardic saying

Mealtime behavioral problems and food refusals are common in children with behavioral and attention disorders. These concerns often bring parents to nutrition experts for solutions. Behavioral disorders that affect nutritional status are autism and attention-focusing problems, such as *attention deficit hyperactivity disorder (ADHD)*. ADHD may be suspected during the preschool years, but it is primarily treated during school years. (ADHD is discussed further in Chapter 13.) Table 11.3 shows the intake of a 2-year-old

Table 11.3 Dietary intake of 2-year-old child with suspected autism

Dry Fruit Loops cereal
10 fl oz calcium-supplemented orange juice drink
Chicken fingers from a specific fast-food restaurant
French fries
10 fl oz calcium-supplemented orange juice drink
Waverly crackers
Pringles potato chips
10 fl oz calcium-supplemented orange juice drink
Oatmeal cake
10 fl oz calcium-supplemented orange juice drink

Case Study 11.1

A Picky Eater

Photodisc

Nutrition assessment sources are the parent interview and physician records for Greg. Greg does not yet have a nutrition or medical diagnosis. The case study is about whether or not nutrition diagnoses for feeding difficulty, biting/chewing (masticatory) difficulty, or those related to weight are appropriate:

Greg is a well-groomed boy almost 3 years old. He has been growing as expected, but he does not talk. He can walk and move about well, but he prefers to play alone. Favorite foods are juices in his sippy cup, which he likes to carry around; macaroni and cheese; white bread without crusts; mashed potato; Honeycomb cereal; and crackers. Greg cries and throws food that he does not like, such as hamburgers, fruits, most vegetables, and any food combinations. He will periodically eat cheese pizza, scrambled eggs, and applesauce.

His mother tries talking to the pediatrician about his picky appetite, but the pediatrician reassures her that Greg will eat when he is hungry and she should not worry. Greg's mother is frustrated that he is so difficult to take out to eat because of the tantrums he throws in restaurants and friends' homes. He sometimes eats a large portion of a food he likes. Most of the time, he is satisfied just drinking juices all day from his sippy cup, and he is rarely interested in eating when others eat. He is able to eat with a spoon, but he does not like to touch foods with his hands. Greg has been referred for speech therapy, but his therapy does not address his eating. His medical history shows that he was born full-term and has had three ear infections but no major illnesses.

Nutrition assessment shows that Greg is consuming adequate calories at 1350 calories/day, or 85 cal/kg. His diet is excessive in vitamin C and B vitamins, with adequate protein at the RDA for his age. His sources of protein are mainly his starchy foods of bread, crackers, and dry cereal.

Questions

1. What are the signs that Greg's feeding problem may be related to his speech?
2. Because Greg is growing well and meeting his calorie needs, why not just wait for him to mature to accept other foods?
3. Was his pediatrician wrong to say that Greg will eat when he is hungry?

(Answers are located in the Instructor's Manual for the 4th edition of *Nutrition Through the Life Cycle*.)

with a feeding problem resulting in a self-restricted diet typical of autism. The child refuses to eat many foods and is rigid in what he will eat. He does not respond to feeling hungry, as other 2-year-old children do. When he is not given foods he likes, he refuses to eat at all and has temper tantrums during which he can injure himself. His self-restricted diet is a part of the condition, which affects how he senses everything in his environment. He prefers to drink rather than eat foods, so a high proportion of his total calories come from one type of drink. Interventions to improve the diet for this child may include a complete vitamin and mineral supplement, and adding one new food by offering it many times (15–20 times) over 1 or 2 months. Nutrition interventions should be incorporated into this child's

overall treatment plan, provided within a special education program. (See Case Study 11.1.)

Excessive Fluid Intake

The food intake noted in Table 11.3 highlights a common issue related to excessive fluids. Many young children prefer to drink rather than eat solid foods, especially when they are not feeling well. Families of chronically ill children tend to offer juices and lower-nutrient beverages in an effort to achieve growth when eating is difficult. The American Academy of Pediatrics recommendation to limit juice intake to 4 to 6 fluid ounces per day for ages 1–6 years applies to all children.[16]

Neuromuscular Disorders   Conditions of the nervous system characterized by difficulty with voluntary or involuntary control of muscle movement.

Hypotonia   Condition characterized by low muscle tone, floppiness, or muscle weakness.

Hypertonia   Condition characterized by high muscle tone, stiffness, or spasticity.

Gastrostomy   Form of enteral nutrition support for delivering nutrition by tube directly into the stomach, bypassing the mouth through a surgical procedure that creates an opening through the abdominal wall and stomach.

Medical Neglect   Failure of parent or caretaker to seek, obtain, and follow through with a complete diagnostic study or medical, dental, or mental health treatment for a health problem, symptom, or condition that, if untreated, could become severe enough to present a danger to the child.

Calcium-fortified juices may be appropriate if other sources of calcium are limited, but these juices can also be over-consumed. In a child who already has gastrointestinal problems, it may not be clear if problems are caused by excessive juice intake.[16] Excess juice resulting in a pattern of low milk intake has been documented to result in smaller stature and lower bone density.[17] In young children who may be less active due to chronic conditions, the negative impact may be larger. Intake of high-sucrose beverages for young children also raises concerns related to dietary adequacy, as other more nutritious beverages are replaced.[18]

Feeding Problems and Food Safety

Toddlers and preschoolers with chronic conditions are at greater risk for food-contamination problems. Some feeding problems result in prolonged needs for soft, easy-to-eat food textures well past the age when baby foods are eaten. Fork-mashing or blending foods may invite bacterial contamination or spoilage over time.[6] Similarly, complete nutritional supplements and formulas are subject to contamination, particularly in the tubing used to give them. How often tubing and devices to deliver formulas are changed can be a food-safety issue. Some families, aware of the high cost of such devices, tend to use them longer than recommended.

Feeding Problems from Disabilities Involving Neuromuscular Control

Children who have feeding problems related to muscle control of swallowing or control of the mouth or upper body may choke or cough while eating or refuse foods that require chewing.[3] These types of feeding problems result from conditions such as cerebral palsy or other *neuromuscular disorders* and genetic disorders such as Down syndrome. These signs of feeding and swallowing problems in toddlers or preschoolers generally appear more severe than the reactions of infants who are learning how to munch and chew foods.[8] The decrease in appetite expected in toddlers and preschoolers may be pronounced in children who find eating difficult and unpleasant. These feeding problems may require further study to make sure

eating is safe for a child, and not related to frequent illness such as bronchitis or pneumonia.

A child with *hypotonia* or *hypertonia* in the upper body may experience difficulty sitting for a meal and self-feeding with a spoon.[8] If these feeding problems are not resolved by providing therapy in early-intervention programs or schools, children are likely to resist eating over time. They may then need a form of nutrition support, such as placement of a *gastrostomy*.

Nutrition-Related Conditions

Failure to Thrive

Failure to thrive (FTT) is a condition in which a caloric deficit is suspected.[19] FTT has a slightly different basis in toddlers and preschoolers, who may have grown adequately during the first year. Their decrease in growth rate occurs at the age when appetite typically decreases and control issues at mealtime are expected, making identifying the cause of FTT more difficult. Generally FTT is suspected when a child's growth declines more than two growth percentiles, placing him near or below the lowest percentile in weight-for-age, weight-for-length, and/or BMI. FTT may result from a complex interplay of medical and environmental factors, such as the following:[4,6]

- Digestive problems such as gastrointestinal reflux or celiac disease
- Asthma or breathing problems
- Neurological conditions such as seizures
- Pediatric AIDS

Children who have chronic illnesses or were born preterm have a higher risk of FTT as a result of abuse or *medical neglect*.[19] They have greater needs than other children do, and they may be more irritable and demanding, which places them at risk. Often a specific nutrient or group of nutrients is suspected of being inadequate in the diet of children with FTT, when the more appropriate emphasis should be placed on energy and protein. Copper and zinc in the blood of toddlers with FTT was reported to be the same as age-matched controls, although protein intake was lower.[20]

Recovery from FTT can include catch-up growth, which is an acceleration in growth rate for age.[3] If calories are provided at a higher level than for a typical child of the same age, catch-up growth is likely (see Illustration 11.2). The length of time needed for catch-up growth varies, but some weight gain should be documented within a few weeks. For example, recovery from FTT for one 3-year-old was a gain of 6 pounds—more weight than is typical to gain in 1 year—within the first 3 months of living in a new home.

The opposite of FTT, obesity, is usually not a concern in the toddler and preschool age range. However, there is a growing group of overgrowth or obesity syndromes identified in the toddler and preschool age range that are quite different than typical childhood obesity. An example is Wiedemann-Beckwith syndrome, in which an usually high rate of weight gain is typical in toddlers. Such conditions are not necessarily a result of excessive energy intake but of endocrine or metabolic abnormalities, body composition changes, or drug side effects. Many conditions have behavioral or developmental components that are found at older ages than the early obesity. Such conditions are good reminders of how nutrition services and recommendations have to be adjusted for children with special needs.

Toddler Diarrhea and Celiac Disease

Toddlers are likely to develop diarrhea. The condition is called toddler diarrhea, in which otherwise healthy growing children have diarrhea so often that their parents bring them for a checkup.[3] Testing shows no intestinal damage and normal blood levels, without FTT or weight loss. The dietary culprit is likely to be excessive intake of juices that contain sucrose or sorbitol. The diarrhea results from excess water being pulled into the intestine, so limiting juices intake may be recommended.[16]

Illustration 11.2 Growth chart for a girl with failure to thrive before and after intervention.

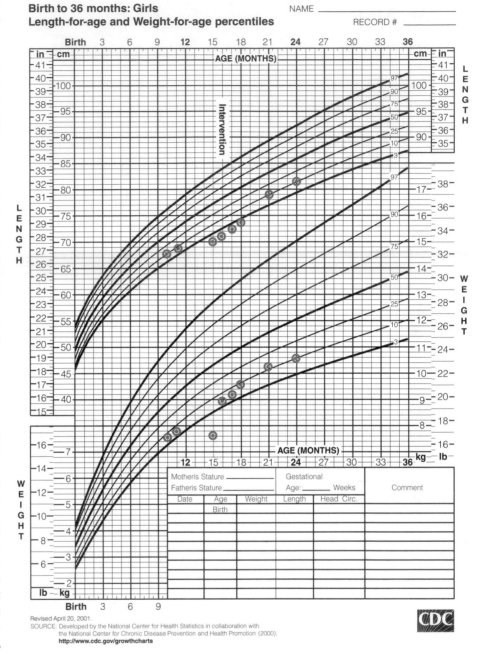

Celiac disease occurs in people who are sensitive to gluten, a component of wheat, rye, and barley. It has a prevalence of 1 in 3000 people within certain ethnic groups, such as those of Middle Eastern or Irish ancestry.[21] Celiac may be related to other chronic conditions, such as Down syndrome, and management of cancer by certain chemotherapy regimens. Symptoms of diarrhea and other digestive problems usually develop by 2 years of age. Confirmation of the condition is based on testing blood for the antibodies to gluten. Dietary management requires complete restriction of any foods with gluten.

This list includes everything made with flour, such as bread and pasta, as well as foods with wheat, barley, or rye as an additive.[21] The allowed foods include rice, soy, corn, and potato flours. Oats are gluten-free but may be contaminated with gluten from wheat mixed in. Meats, fruits, and vegetables are not restricted, but many processed foods use wheat flour for thickening. After instituting dietary restrictions, the intestinal damage heals, and the digestive symptoms disappear. The parents of preschoolers with celiac disease learn to be expert readers of food labels because intestinal damage recurs if gluten is eaten by mistake.

Autism

The toddler and preschool years are when behavioral signs of autism are noted by families. Early screening and diagnostic evaluations are recommended by the American Academy of Pediatrics for those with suspected speech delays, repetitive behaviors and social skill deficits.[22] The average age of diagnosis of autism has been dropping as a result of efforts for early diagnosis and is now 30 months of age. Autism is the third most common developmental disability in the U.S., with an incidence of one in every 166 children. Preschoolers with autism are sensitive to sensory information of all kinds. This sensitivity usually results in a rigid, self-restricted range of food choices that makes parents and caregivers worry about the adequacy of the child's diet (Table 11.3). Diet-related interventions for autism such as a gluten- and casein-free diet have not been endorsed by professional societies, but they are well known to families from autism support groups. Milk substitutes to avoid casein may or may not meet the child's need for calcium, vitamin D, protein, or other nutrients. Families of children with autism report a variety of gastrointestinal problems, but dietary recommendations for autism are the same as for any other child of the same age who has feeding problems. Funding in the 2006 federal program called Combating Autism encourages nutrition experts to receive training in managing those with autism because feeding problems in toddlers and preschoolers are so prevalent.

> **Spastic Quadriplegia**   A form of cerebral palsy in which brain damage interferes with voluntary muscle control in both arms and legs.

Muscle Coordination Problems and Cerebral Palsy

The prevalence of cerebral palsy, 2.0 to 2.5 per 1000 children, is skewed because 40–50% of those diagnosed were LBW, VLBW, and ELBW infants.[2,15] Toddlers and preschoolers who were born preterm account for much of the increase in muscle coordination problems seen before cerebral palsy can be diagnosed. Those at risk for or confirmed with cerebral palsy need nutrition assessments that include body composition indexes, such as fat stores.[23] Nutrition interventions are then based on these findings, and they may include encouraging weight gain if body fat stores are low. Some toddlers have muscle coordination problems combined with other developmental delays, rather than the diagnoses of cerebral palsy or spastic quadriplegia. The general term "developmental delay" makes some families see nutrition and growth problems as something the child will outgrow. The lack of a clear diagnosis means that appropriate nutrition services are delayed and growth and feeding problems worsen without interventions. The child with muscle coordination problems may appear thin as a result of small muscle size and not low fat stores; weight gain is not needed. Their growth tracking may not fit the CDC growth charts, yet may also not fit on the spastic quadriplegia growth charts for toddlers or preschoolers.[24] In such circumstances, the Infant Health and Development Program growth charts may be appropriate. Part of the growth assessment for a preschooler with cerebral palsy may include an estimate of caloric needs for activity, which may be higher or lower than expected. A girl may expend higher energy in her efforts to coordinate walking while receiving physical therapy three days per week at school. Her activity may be lower if she is in a wheelchair most of the time.

Feeding assessment for a child with severe cerebral palsy (also called *spastic quadriplegia*) may be necessary as part of the overall nutritional assessment.[8] The assessment may include an observation of eating to determine any restrictions in the type of foods that the child can eat, and whether coordinating muscles for chewing, swallowing, and/or using a spoon or fork are working well. Table 11.4 provides a food-intake record for a 4-year-old girl with spastic quadriplegia who does not walk and is receiving nutritional

Table 11.4   Meal pattern and recommended foods for an underweight girl with feeding problems as a result of weakness

Meal Pattern: Small, Frequent Meals and Snacks to Prevent Tiredness at Meals	Recommended Foods That Are Easy to Chew, with Small Portions
Breakfast at home	**Breakfast:** 1/2 c oatmeal with added soft fruit, margarine, and brown sugar
Mid-morning snack	**Snacks:** 1 slice deli meat with 6 fl oz whole milk with Carnation Instant Breakfast added
Lunch (at preschool)	
Afternoon snack (at preschool)	1/2 c soft-cooked sliced apples with added margarine
After-school snack	Cake-type cookie (frosting allowed)
Dinner	**Dinner:** 1/2 c mashed potato with added margarine
Bedtime snack	3 Tbsp meat loaf
	3 Tbsp soft-cooked carrots with added margarine
	Bedtime snack: Chocolate cake with frosting and 4 fl oz whole milk

Early Intervention Services for a Boy at Risk for Nutrition Support

The Nutrition Assessment is based on hospital and physician medical records, and parent interviews. Several nutrition diagnoses and interventions have already been documented prior to Robert's enrollment in an early intervention program, including inadequate energy intake and feeding difficulty related to his preterm birth at 30 weeks gestation. This case study is about additional nutrition concerns that need to be assessed and diagnosed as Robert gets older.

Robert is 2.3 years old. His premature birth was related to exposure to an intrauterine infection. All in the family agree that he is small, but their main concern is that he is difficult to feed. He cries and refuses to eat when offered meals, even those with his favorite foods.

The registered dietician who consults at the early intervention program meets with the family, assesses Robert, and reviews his medical records. Nutrition services are first planned to boost calories to stimulate weight gain. Observing Robert being fed by his mother is part of the nutrition services. Other therapists at the early intervention center are involved in making sure that Robert is positioned well to eat, so that he is sitting up without extra effort. The nutritionist and occupational therapist are concerned that Robert is choking so easily, and they talk to the family about contacting his pediatrician. They send a fax to the pediatrician's office recommending tests to study Robert's swallowing.

Robert does not attend the early intervention program for the next 3 weeks. The tests demonstrate that he is aspirating some of his liquids into his lungs, so oral feeding is unsafe. He requires a gastrostomy for feeding and is hospitalized for surgery. His parents learn how to feed him through the gastrostomy.

When Robert returns to the early intervention program, his pediatrician asks the early intervention staff to monitor his weight and to reinforce the discharge feeding instructions with the family. Nutrition services provided in the early intervention program are changed from working on Robert's oral feeding to monitoring and documenting his growth as adjustments are made in his gastrostomy feeding schedule. Over the next 6 months, Robert gains weight. He starts being more interactive with the staff at the early intervention center and makes some developmental progress in his walking and speaking. He is still a small child, but his improved nutritional status is confirmed by his adequate body fat measurements. His ability to return to eating by mouth will be reassessed later in the year.

Questions

1. What are the signs that Robert needs gastrostomy feeding?
2. Could the gastrostomy placement have been prevented if Robert had gained weight?
3. Can Robert enjoy life if he cannot eat?

(Answers are located in the Instructor's Manual for the 4th edition of *Nutrition Through the Life Cycle*.)

services for weight gain. Her meal pattern was adjusted because she tires easily while eating. She does not like to eat too much at a time, and she refuses to be fed by another person (which is appropriate for her age). She can chew foods such as fresh apple, but then is too tired to eat something else. She eats a larger portion if the food is soft and does not require her to work so hard. She has gained weight at a slow rate, and her fat stores are low. The first plan is to use regular foods that are easy for her to eat to meet her nutritional needs, including cooked rather than fresh vegetables and fruits, and to avoid hard-to-chew foods, such as roast beef or corn on the cob. If she does not gain weight by eating foods such as those suggested in Table 11.4, she may need nutritional supplementation to assure her nutritional needs are met within her feeding limitations (see Case Study 11.2).

Pulmonary Problems

Breathing conditions are examples of common problems in children with special health care needs with major nutritional consequences. Breathing problems increase nutritional needs, lower interest in eating, and can slow growth rate (see Case Study 11.2).[25] Infants who were born preterm are especially likely as toddlers to have breathing problems. Up to 80% of 1000-gram infants can develop chronic lung disease.[25] Examples of pulmonary diseases or chronic lung disease are bronchopulmonary dysplasia (BPD) and *asthma*. Asthma is self-reported in 58 of every 1000 children under 5 years of age.[26] Asthma results in more emergency-room visits for children under 5 years—at 121 visits per 10,000 people—than it does in older children with asthma.[26] Asthma does not necessarily require nutrition services, but some children have asthma as a result of food allergies.[27]

> **Asthma**   Condition in which the lungs are unable to exchange air due to lack of expansion of air sacs. It can result in a chronic illness and sometimes unconsciousness and death if not treated.
>
> **Work of Breathing (WOB)**   A common term used to express extra respiratory effort in a variety of pulmonary conditions.
>
> **Mental Retardation**   Substantially below-average intelligence and problems in adapting to the environment, which emerge before age 18 years.
>
> **Anaphylaxis**   Sudden onset of a reaction with mild to severe symptoms, including a decrease in ability to breathe, which may be severe enough to cause a coma.

Toddlers and preschoolers with BPD have a positive long-term prognosis because new lung tissue can grow until about 8 years of age.[25] Toddlers and preschoolers with serious breathing problems generally need extra caloric intake due to the extra energy expended in breathing. Increased *work of breathing (WOB)* occurs with different pulmonary conditions and generally leads to low interest in feeding, partially as a result of tiredness.[26] Feeding difficulties have several causes in a toddler treated for BPD:[25]

- The normal progression of feeding skills is interrupted.
- Medications and their side effects contribute to high nutrition needs.
- Interrupted sleep and fatigue make hunger and fullness cues harder to interpret.

By the preschool years, the impact of BPD on slowing the rate of weight gain is usually clear. Exposure to common respiratory illnesses, which are minor in typical children, can require a trip back to the hospital for some children with BPD. Increased frequency of infections adds another limitation to catch-up growth. Neither the CDC growth chart nor the IHDP preterm growth chart may be helpful in predicting the child's growth pattern, but periods of good health are usually accompanied by an increase in weight and appetite.

Dietary recommendations for toddlers with BPD are similar to those for children with weakness (see Table 11.4). Small, frequent meals with foods that are concentrated sources of calories are needed. Easy-to-eat foods may still be recommended so that fatigue from meals is low. If the toddler with breathing problems does not gain weight as a result of dietary recommendations such as those in Table 11.4, the next step will be to add complete nutritional supplements to meet the higher caloric needs. The supplements, such as Pediasure, are also a source of vitamins and minerals.

Developmental Delay and Evaluations

Developmental delay may be suspected when specific nutrients are consumed in inadequate or excessive amounts. Iron deficiency and lead toxicity are risk factors for developmental problems.[5,6] Developmental evaluations are recommended for young children who have been sick for a long time and isolated from other children. Standardized testing aids in finding a definitive diagnosis and appropriate educational programs. Developmental delay is a specific diagnosis that may be replaced by *mental retardation* when the child is 6 or 7 years old.[4] Changes in growth rate are typical in children with developmental delay.[23] Short stature is common and part of the unusual growth pattern that often prompts referrals for genetic testing.[28] The evaluation of growth from a genetic expert may include more in-depth analyses, such as measurements of hand and foot size and bone age.[28] Genetic syndromes also can be associated with unusually fast growth. Soto's syndrome is a rare disorder in which the child is tall and large, but has delayed development.[28]

Food Allergies and Intolerance

True food allergies are estimated to be present in 2% to 8% of children.[27] Food allergies are usually identified in toddlers and preschoolers because allergy testing in infancy is not useful due to the incomplete development of the immune system. True food allergies can result in life-threatening episodes of *anaphylaxis*.[27] Examples of food allergies that may result in anaphylaxis for some children include the following:[27]

- Milk
- Eggs
- Wheat
- Peanuts
- Walnuts
- Soy
- Fish

Cow's milk protein allergy rarely persists into the toddler and preschool years. However, when cow's milk protein allergy does persist, symptoms in the toddler and preschool years may appear as more general allergy symptoms, such as asthma or skin rashes.[29] Other food allergies are often present in a child with confirmed cow's milk protein allergy, with, for example, 35% reacting also to oranges or 47% reacting also to soy milk.

Strict and complete avoidance of the food that causes the allergy is required. This abstinence includes all settings, such as eating nothing prepared at bake sales when the food ingredients are unknown. If the preschool child is on an extensively restricted diet, the quality of the diet may not meet all her nutritional needs. Such restrictions are also likely to result in mealtime behavioral problems. The parents may become overprotective, or the child may quickly learn to use restricted foods to get a parent's concern and attention. Diagnosed food allergies can greatly affect the family. For children at risk for anaphylaxis, parents and caregivers should be given instruction in emergency lifesaving procedures and use of an injectable form of epinephrine.[27]

Dietary Supplements and Herbal Remedies

"It is better to take food into the mouth than to take worries into the heart."

Yiddish saying

Families who are concerned that something may be wrong with their young children may be attracted to health and nutritional claims targeted and packaged for adults. The family that is having difficulty finding effective treatment for a child is most at risk for inappropriate or ineffective alternative products. Use of dietary supplements in children with chronic conditions is more common than in healthy children, and most take dietary supplements not prescribed nor discussed with health care providers.[30] Examples of dietary supplements are vitamins, minerals, botanicals, amino acids, and fatty acids.[30] Parent coalitions and advocacy groups are excellent sources of networking for families, but they can also be sources of nutritional claims for products and dietary regimens that have no scientific testing behind them. Down syndrome, for example, is a disorder for which nutritional supplementation has been marketed to parents. No specific nutrients, combinations of nutrients, or herbal remedies have been shown to improve the intellectual functioning of individuals with Down syndrome.[31,32] The National Down Syndrome Society cautions parents about the ineffectiveness of nutrient and herbal supplements to discourage their use, but interest continues.[31,32] What is really being marketed is hope, which families always want and need.

Constipation remedies are examples of over-the-counter products used often for children with special health care needs. Constipation is a common condition in children with various neuromuscular conditions in which muscles are weak.[6] Parents tend to try over-the-counter remedies, dietary methods, and home remedies for constipation management. The effectiveness of dietary fiber may be low when muscle weakness is an underlying problem.[3] Both overtreatment and undertreatment can get the child in trouble by worsening the constipation problem. A young child died as a result of poisoning by a laxative product administered at a higher dose than recommended.[33] Effective prescription medications for constipation management are available, but the family has to bring the problem to the attention of the health care provider. Encouraging the family to discuss the problem with a physician before trying over-the-counter products is important for many children.

Sources of Nutrition Services

Infants and toddlers who have chronic conditions are served by a variety of resources. Registered dieticians who have training in pediatrics are qualified to provide services to toddlers and preschool children with chronic conditions.

Programs in which nutrition care may be accessed include the following:[34]

- State programs for children with special health care needs
- Early intervention programs (age 0 up to 36 months)
- Early childhood education programs (IDEA, ages 3–5 years)
- Head Start; regular program or special-needs category (ages 3–5 years)
- Early Head Start; regular program or special-needs category (0 up to 36 months)
- WIC
- Low-birth-weight follow-up programs
- Child care feeding programs (USDA)

These programs are described in Chapters 8 and 10. Efforts to increase program accessibility come from state and federal governmental offices, toll-free outreach services, and websites. Specific outreach programs to locate toddlers and preschoolers at risk are funded in each state, under names such as "Child Find."[4] Because every child at risk is eligible for a screening, contacting a neighborhood public school is a good starting place to locate services, even if the child is not old enough to attend the school.

Key Points

1. Toddlers and preschoolers with special health care needs may require medication for their underlying condition that may interfere with their growth, appetite, and meal pattern. Early intervention programs and early childhood education programs include adjusting the timing of meals and snacks.

2. Nutritional, educational, and developmental providers for toddlers and preschoolers demonstrate how parents are to advocate for their children's special health care needs later on in schools and job sites.

3. Families of children with special health care needs are targets of invalid nutritional claims by those hoping to sell nutritional supplements. Over-the-counter nutrition-related products, such as vitamin or constipation remedies for adults, may be dangerous for such young children.

4. Toddlers and preschoolers with special health care needs are quite varied in their nutrition needs, but the basic concepts of supporting growth, typical feeding skill development, and meeting nutrition needs for age and activity still apply.

5. Failure to thrive is often the reason children with special health care needs enter medical, educational, and developmental services. Such cases of failure to thrive cannot be corrected with additional energy, as it is in children without special health care needs. Unusual growth patterns can be signs of conditions that are not directly related to nutrition.

6. Examples of conditions likely to appear in toddlers and preschoolers with special health needs are autism, Rett syndrome, spastic quadriplegia, asthma, developmental delay, and true food allergies.

Review Questions

1. What is an accurate statement about services for a boy with Down syndrome who has difficulty chewing when he turns 3 years old?

   a. He is now eligible to start receiving supplementary social security if his family meets the income requirements.

   b. He needs speech therapy to address his difficulty chewing.

   c. He is no longer eligible for early childhood programs under IDEA.

   d. His chewing difficulty cannot be included in his nutrition services, as it is a typical component of his underlying syndrome.

2. What is an accurate statement about a preschool girl who has a diagnosed milk-protein allergy?

   a. If she attends a public day care that participates in the USDA Child Nutrition Program, her family has to provide her lunch for her safety.

   b. Only if this allergy is a feature of an underlying condition is she considered a child with a special health care need.

   c. Her allergy would still let her have ice cream as a special treat on her birthday.

   d. It is likely, since she is a preschool-age child, that she will be diagnosed with other food-related reactions, such as skin rashes.

3. A 4-year-old with asthma does not have food allergies. Her asthma routine requires a home breathing machine and two prescribed medications. Which statement is accurate about her nutrition needs?

   a. Her diagnosis is one of the more common types of special needs and is best managed by a reduced-calorie diet to compensate for restricted activity.

   b. Compliance with her nutrition recommendations can prevent emergency-room visits for flare-ups.

   c. Her prescribed medications could impact her appetite and growth and so require nutrition services.

   d. Asthma can interrupt the normal progression of eating skills in a young child.

4. Feeding and eating problems do <u>not</u> require nutrition services under which of these conditions?

   a. When a 3-year-old boy has a food jag and refuses to eat foods he usually likes.

   b. When the parents or caregivers are concerned that a toddler wants to drink milk or juice but refuses solid foods such as fruit.

   c. When a 3-year-old child has been diagnosed with autism and is entering a behavioral management program.

   d. When a 3-year-old child is diagnosed with gastroesophageal reflux and prescribed medications again.

Questions 5, 6, and 7 concern growth assessment. Determine whether each is true or false.

5. Any child with a chronic condition should have a growth assessment as part of determining nutrition status.

6. A thin-appearing 4-year-old child may be harmed if weight gain is a goal after adequate fat stores are measured.

7. A 3-year-old boy has a growth pattern that plots lower on the standard CDC growth chart than on the IHDP growth charts; his growth is plotted incorrectly.

8. Briefly describe why measuring head size is a part of a nutritional assessment for a toddler or preschooler with a suspected special health care need.

9. Identify two examples of when standard nutrition guidance for healthy toddlers and preschoolers must be modified for a preschool-age child with special health care needs.

10. Identify two examples of when a vitamin or mineral supplement has to be a part of a nutrition services plan.

Resources

Celiac Disease Foundation
This organization identifies food and lifestyle restrictions for gluten-free diets.
Website: http://www.celiac.org/

Federal Interagency Coordinating Council Site for Families with Children with Disabilities
This website identifies, by state and city, resources for finding local intervention programs.
Website: www.fed-icc.org

Federation for Children with Special Needs
The online material for this organization includes support for families of children with special needs.
Website: www.fcsn.org

Food Allergy
A credible source of recommendations for preventing food allergy reactions, this newsletter provides recipes to avoid foods that cause reactions.
Website: www.foodallergy.org

National Center for Growth Statistics (source of growth charts)
Site for obtaining growth charts and guidelines for their use.
Website: www.cdc.gov/nchs

National Information Center for Children and Youth with Disabilities
This is a useful site for parents and providers who are looking for intervention services for children with special needs. It is targeted mainly toward educational programs.
Website: www.NICHCY.org

National Organization for Rare Diseases (NORD)
This is a credible source for parents and providers of information and resources about rare "orphan" diseases.
Website: www.rarediseases.org

Quackwatch
This website includes information on dietary supplements and products that are claimed to benefit health and nutrition.
Website: www.quackwatch.com

Studies to Advance Autism Research and Treatment (STAART) Network
The U.S. National Institutes of Health (NIH) has clinical studies on autism to define the characteristics of different subtypes and possible treatments. Nutrition in toddlers who show signs of developmental delay and dietary changes as a treatment for autism are examples from the eight STAART sites. One study, "Diet and Behavior in Young Children with Autism," is testing if a gluten- and casein-free diet has specific benefits for children with autism in the preschool-age group. This research is needed, since many families with a child with autism use Internet advocacy sites as their source of nutrition information because this therapy is not accepted by most health care providers.

Pacific West Maternal and Child Health Distance Learning Network
Personnel who can work in early intervention programs and schools on nutrition problems are limited in part because qualified nutrition providers are not familiar with children with special health care needs. Web-based distance learning resources for training about nutrition needs for children with special health care needs are vital tools for recruiting and advancing the skills of nutrition providers. This training network originated due to the distance from Hawaii and Alaska to the mainland, but it is now used internationally.
Website: http://washington.edu/pwdlearn

Chapter 12
Child and Preadolescent Nutrition

Prepared by **Nancy H. Wooldridge**

Key Nutrition Concepts

1 Children continue to grow and develop physically, cognitively, and emotionally during the middle childhood and preadolescent years in preparation for the physical and emotional changes of adolescence.

2 Children continue to develop eating and physical activity behaviors that affect their current and future states of health.

3 Although children's families continue to exert the most influence over their eating and physical activity habits, external influences, including teachers, coaches, peers, and the media, begin to have more impact on children's health habits.

4 With increasing independence, children begin to eat more meals and snacks away from home and need to be equipped to make good food choices.

Introduction

This chapter focuses on the growth and development of school-age and preadolescent children and their relationships to nutritional status. Children continue to grow physically at a steady rate during this period, but development from a cognitive, emotional, and social standpoint is tremendous. This period in a child's life is preparation for the physical and emotional demands of the adolescent growth spurt. Having family members, teachers, and others in their lives who model healthy eating and physical activity behaviors will better equip children for making good choices during adolescence and later in life.

Definitions of the Life Cycle Stage

Middle childhood is a term that generally describes children between the ages of 5 and 10 years. This stage of growth and development is also referred to as school-age, and the two terms are used interchangeably in this chapter. *Preadolescence* is generally defined as ages 9 to 11 years for girls and ages 10 to 12 years for boys. "School-age" is also used to describe preadolescence.

Importance of Nutrition

Adequate nutrition continues to play an important role during the school-age years in ensuring that children reach their full potential for growth, development, and health. Nutrition problems can still occur during this age, such as iron-deficiency anemia, undernutrition, and dental caries. Regarding weight, both ends of the spectrum are seen during this age. The prevalence of obesity is increasing, but the beginnings of eating disorders can also be detected in some school-age and preadolescent children. Therefore, adequate nutrition and the establishment of healthy eating behaviors can help to prevent immediate health problems as well as promote a healthy lifestyle, which in turn may reduce the risk of the child developing a chronic condition, such as obesity, type 2 diabetes, and/or cardiovascular disease later in life.[1] Adequate nutrition, especially eating breakfast, has been associated with improved academic performance in school and reduced tardiness and absences.[2] Meeting energy and nutrient needs, addressing common nutrition problems, and preventing nutrition-related disorders while establishing healthy eating and physical activity habits will be discussed later in this chapter.

Tracking Child and Preadolescent Health

In 2006, there were an estimated 73.7 million children in the United States under the age of 18 years, representing 25% of the total population.[3,4] Of these, 46% were between the ages of 10 and 17.[3] The statistics regarding the environments in which many of these children are growing up are alarming. From 2000 to 2006, the child poverty rate, a widely used predictor of child well-being, increased 6%, representing 1 million additional children living in poverty during that time span.[3] In 2006, approximately 8% of U.S. children lived in extreme poverty, where the household income was below 50% of the poverty level, and 40% of children lived in low-income families, where the household income was below 200% of the poverty level.[3] The rate of children living in poverty is higher for black/African American, American Indian/Alaskan Native, and Hispanic/Latino children.[3,4]

The environment in which a child lives affects the child's health status, including nutrition, and education. In 2006, 11.7% of children younger than 18 years of age, translating to 8.7 million children, did not have health insurance.[4] In 2007, only 32% of fourth-grade students scored at or above proficient reading level, while only 39% scored at or above proficient math level.[3] Lack of transportation is a significant limitation for many families. In the discussions that follow regarding nutrition during childhood, the recommendations must always be considered in the context of the individual child's environment.

Disparities in nutrition status indicators exist among races and ethnicities. For example:

- The prevalence of overweight and obesity as measured by body mass index (BMI) is significantly greater in Hispanic male children, including Mexican Americans, than in non-Hispanic white male children.[5]

Middle Childhood Children between the ages of 5 and 10 years; also referred to as "school-age."

Preadolescence The stage of development immediately preceding adolescence; 9 to 11 years of age for girls and 10 to 12 years of age for boys.

- Non-Hispanic black female children have significantly greater BMI than non-Hispanic white female children.[5]
- African-Americans have higher percentages of total calories from dietary fat.[6]

Healthy People 2010

A number of objectives in the Healthy People 2010 document are specific to children's health and well-being. Table 12.1 lists the specific Healthy People 2010 objectives that are pertinent to a discussion of middle childhood

Table 12.1 Healthy People 2010 Objectives related to school-age children[6, CDC 2009]

Healthy People 2010 Objective		Baseline	Progress	Target
19.3a:	Reduce overweight or obesity in children (aged 6 to 11 years)	11%	17% (2006)	5%
	Mexican American	16%	24% (2006)	7%
	Black or African American, not Hispanic or Latino	15%	21% (2006)	5%
	White, not Hispanic or Latino	10%	15% (2006)	5%
19.5:	Increase the proportion of persons aged 2 years and older who consume at least two daily servings of fruit	39%	40% (2004)	75%
	Mexican American	40%	46% (2004)	75%
	Black or African American, not Hispanic or Latino	35%	42% (2004)	75%
	White, not Hispanic or Latino	39%	38% (2004)	75%
19.6:	Increase the proportion of persons aged 2 years and older who consume at least three daily servings of vegetables, with at least one-third being dark green or deep yellow vegetables	4%	4% (2004)	50%
	Black or African American, not Hispanic or Latino	6%	3% (2004)	50%
	White, not Hispanic or Latino	3%	4% (2004)	50%
19.7:	Increase the proportion of persons aged 2 years and older who consume at least six daily servings of grain products, with at least three being whole grains	4%	3% (2004)	50%
	White, not Hispanic or Latino	4%	3% (2004)	50%
19.8:	Increase the proportion of persons aged 2 years and older who consume less than 10% of calories from saturated fat	36%	34% (2004)	75%
	Mexican American	37%	43% (2004)	75%
	Black or African American, not Hispanic or Latino	31%	37% (2004)	75%
	White, not Hispanic or Latino	35%	30% (2004)	75%
19.9:	Increase the proportion of persons aged 2 years and older who consume no more than 30% of calories from fat	33%	29% (2004)	75%
	Mexican American	33%	37% (2004)	75%
	Black or African American, not Hispanic or Latino	26%	27% (2004)	75%
	White, not Hispanic or Latino	33%	26% (2004)	75%
19.10:	Increase the proportion of persons aged 2 years and older who consume 2400 mg or less of sodium daily	15%	13% (2004)	65%
	Black or African American, not Hispanic or Latino	17%	17% (2004)	65%
	White, not Hispanic or Latino	15%	12% (2004)	65%
19.11:	Increase the proportion of persons aged 2 years and older who meet dietary recommendations for calcium	31%	42% (2004)	74%
	Mexican American	28%	37% (2004)	74%
	Black or African American, not Hispanic or Latino	16%	19% (2004)	74%
	White, not Hispanic or Latino	35%	48% (2004)	74%
22.8:	Increase the proportion of the nation's public and private schools that require daily physical education for all students (middle and junior high schools)	6.4%	Not available	9.4%
22.14:	Increase the proportion of trips made by walking	31%	36% (2003)	50%
22.15:	Increase the proportion of trips made by bicycling	2.4	1.5% (2001)	5.0

and preadolescence with available progress to date by race and ethnicity. According to the proposed framework for Healthy People 2020 (www.healthypeople.gov/hp2020), many of the objectives are being retained with some modification because the 2010 objectives were not met.

Normal Growth and Development

During the school-age years, the child's growth is steady, but the growth velocity is not as great as it was during infancy or as great as it will be during adolescence. The average annual growth during the school years is 7 pounds (3–3.5 kg) in weight and 2.5 inches (6 cm) in height.[7] Children of this age continue to have spurts of growth that usually coincide with periods of increased appetite and intake. During periods of slower growth, the child's appetite and intake will decrease. Parents should not be overly concerned with this variability in appetite and intake in their school-age children.

Periodic monitoring of growth continues to be important in order to identify any deviations in the child's growth pattern. Children should continue to be weighed on calibrated scales without shoes and in lightweight clothing. The child's stature or standing height should be measured without shoes and utilizing a height board (see Illustration 10.3 in Chapter 10). A height board consists of a non-stretchable tape on a flat surface like a wall with a moveable right-angle head board. The child's heels should be up against the wall or flat surface, and the child should be instructed to stand tall, looking straight ahead with arms by the sides, during the measurement. Both weight and height should be plotted on the appropriate 2000 CDC growth charts, discussed next.

The 2000 CDC Growth Charts

The "CDC Growth Charts: United States" are excellent tools for monitoring the growth of a child.[8] The growth charts, which are pertinent to the school-age child, cover weight-for-age, stature-for-age, and body mass index (BMI)-for-age for boys and girls and can be downloaded from the CDC website www.cdc.gov/nchs (see the "Resources" section). As described in Chapter 10, the growth charts are based on data from the second and third cycles of the National Health and Examination Survey (NHES) and the National Health and Nutrition Examination Surveys (NHANES) I, II, and III. However, weight data for children greater than 6 years of age who participated in NHANES III were not included because there was a known higher prevalence of overweight for these ages. Incorporating this information into the growth charts would reflect an unhealthy standard.[9] Gender-specific BMI-for-age greater than or equal to the 85th percentile but less than or equal to the 94th percentile defines overweight, while BMI-for-age values greater than or equal to the 95th percentile define obesity.[10, 11]

Illustrations 12.1 and 12.2 depict the growth of a healthy child. A chart for weight-for-stature up to a height of 48 in (122 cm) is also available for the younger school-age child. As with the toddler and preschooler, it is the child's pattern of growth over time that is important rather than any single measurement. The tracking of BMI-for-age percentile is an important screening tool for overweight and obesity as well as undernutrition. Making sure to use the correct age of the child when plotting on the growth charts and using the most current growth curves will help to avoid errors.

WHO Growth References

In 2007, the World Health Organization (WHO) released growth references for older school-age children and adolescents, including height-for-age, weight-for-age, and BMI-for age.[12] Unlike the WHO Child Growth Standards for 0 to 5 years, which were based on prospective growth data, the growth reference for school-age children and adolescents was constructed using existing historical data, including the 1977 NCHS/WHO growth reference from 5 to 19 years, which was a non-obese sample.[12] Applying state-of-the-art statistical methods, the 2007 curves were matched to the WHO under-five curves and BMI values at age 19 years compare to the BMI cutoff used for adults for identifying overweight (BMI >25.0 kg/m^2) and obesity (BMI >30.0 kg/m^2).[12] The WHO growth reference can be downloaded from the WHO website www.who.int/childgrowth (see the "Resources" section).

Physiological and Cognitive Development of School-Age Children

Physiological Development

During middle childhood, muscular strength, motor coordination, and stamina increase progressively.[7] Children are able to perform more complex pattern movements, therefore affording them opportunities to participate in activities such as dance, sports, gymnastics, and other physical activities.

During the early childhood years, percent body fat reaches a minimum of 16% in females and 13% in males. Percent body fat then increases in preparation for the adolescent growth spurt. This increase in percent body fat, which usually occurs on average at 6.0– 6.3 years of age, is called *adiposity rebound* or *BMI rebound* and is reflected in the BMI-for-age growth charts.[13] The increase in percent body fat with puberty is earlier and greater in females

Illustration 12.1 2 to 20 years: Girls' stature-for-age and weight-for-age percentiles.[8]

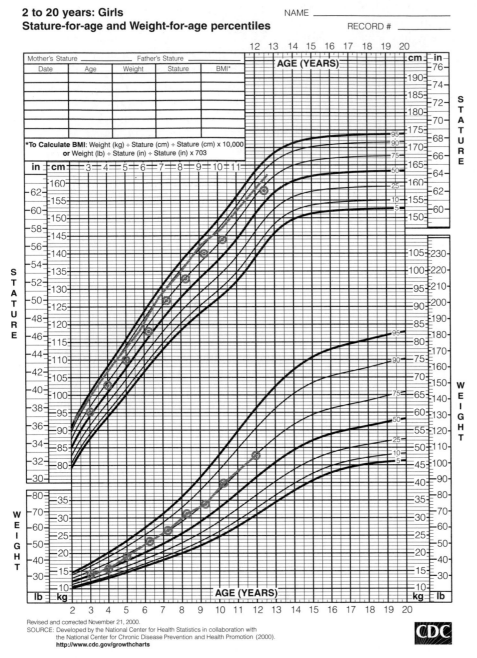

2 to 20 years: Girls
Stature-for-age and Weight-for-age percentiles

NAME _____

RECORD # _____

*To Calculate BMI: Weight (kg) ÷ Stature (cm) ÷ Stature (cm) x 10,000
or Weight (lb) ÷ Stature (in) ÷ Stature (in) x 703

Revised and corrected November 21, 2000.
SOURCE: Developed by the National Center for Health Statistics in collaboration with
the National Center for Chronic Disease Prevention and Health Promotion (2000).
http://www.cdc.gov/growthcharts

CDC

values, as it is in adults, but rather to have a BMI-for-age percentile within the normal range.

With the increase in body fat, preadolescents, especially girls, may be concerned that they are becoming overweight. Parents need to be aware that an increase in body fat during this stage is part of normal growth and development. Parents need to be able to reassure their child that these changes are most likely not permanent; parents also need to be careful not to reinforce a preoccupation with weight and size. Boys may become concerned about developing muscle mass and need to understand that they will not be able to increase their muscle mass until middle adolescence (see Chapter 14).[1]

Cognitive Development

The major developmental achievement during middle childhood is self-efficacy, the knowledge of what to do and the ability to do it. During the school-age years, children move from a preoperational period of development to one of "concrete operations."[7] This stage is characterized by being able to focus on several aspects of a situation at the same time; being able to have more rational cause/effect reasoning; being able to classify, reclassify, and generalize; and a decrease in egocentrism, which allows the child to see another's point of view. Schoolwork becomes increasingly complex as the child gets older. School-age children also enjoy playing strategy games, displaying growing cognitive and language development.

During this stage, the child is developing a sense of self. Children become increasingly independent and are learning their roles in the family, at school, and in the community.[1] Peer relationships become increasingly important, and children begin to separate from their own families by spending the night at a friend's or relative's house. More and more time is spent watching television and playing video games. Older children may be able to walk or ride a bicycle to a neighborhood store and purchase snack items. Thus, influences outside the home

than in males (19% for females versus 14% for males). During middle childhood, boys have more lean body mass per centimeter of height than girls do. These differences in body composition become more pronounced during adolescence.[1] It is important to understand that BMI is not constant throughout childhood. This is because height is a component of the formula for BMI (weight in kg divided by height in meters squared), and a child's height is constantly increasing as he grows. Plotting BMI-for-age percentile on the growth charts is the only way to know if a child's BMI is outside the normal range for his age. The goal for children is not to strive for a certain range of BMI

Illustration 12.2 2 to 20 years: Girls' body mass index-for-age percentiles.[8]

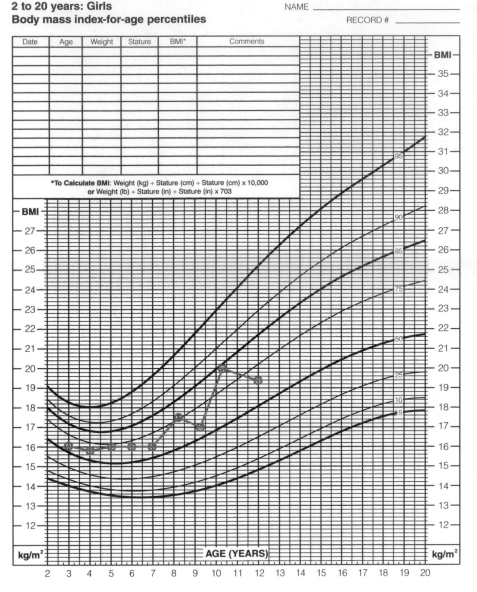

2 to 20 years: Girls
Body mass index-for-age percentiles

NAME _____

RECORD # _____

Date	Age	Weight	Stature	BMI*	Comments

*To Calculate BMI: Weight (kg) ÷ Stature (cm) ÷ Stature (cm) x 10,000
or Weight (lb) ÷ Stature (in) ÷ Stature (in) x 703

AGE (YEARS)

SOURCE: Developed by the National Center for Health Statistics in collaboration with
the National Center for Chronic Disease Prevention and Health Promotion (2000).
http://www.cdc.gov/growthcharts

CDC

environment play an increasing role in all aspects of the child's life.

Development of Feeding Skills and Eating Behaviors

With increased motor coordination, school-age children develop increased feeding skills. During childhood, the child masters the use of eating utensils, can be involved in simple food preparation, and can be assigned chores related to mealtime such as setting the table. By performing these tasks, the child learns to contribute to the family, which boosts developing self-esteem. The complexity of the tasks can be increased as the child grows older.

At the same time, the child is learning about different foods, simple food preparation, and some basic nutrition facts.

Eating Behaviors Parents and older siblings continue to have the most influence on a child's attitudes toward food and food choices during middle childhood and preadolescence. The eating behaviors and cultural food practices and preferences of parents will influence the child's food likes and dislikes. The feeding relationship between parent and child, as described in Chapter 10, still applies to the school-age child. Parents are responsible for the food environment in the home, what foods are available, and when they are served. The child is responsible for how much she eats.[14] Parents need to continue to be positive role models for their children in terms of healthy eating behaviors. They also need to provide the necessary guidance so that the child will be able to make healthy food choices when away from home.

Family Mealtime Families should try to eat meals together. When children are involved in school-related activities, eating together is often difficult for families to achieve because of the family members' hectic schedules. But eating together as a family should be encouraged as a goal, allowing time for conversation (see Illustration 12.3). Excessive reprimanding and arguments should be avoided during mealtime.

One study of 9- to 14-year-old children of participants of the Nurses' Health Study II found a positive relationship between families eating dinner together and the overall quality of the children's diets.[15] Children who ate dinner with their families had higher energy intakes as well as higher intakes of nutrients such as fiber, calcium, folate, iron, and vitamins B_6, B_{12}, C, and E. These children also reported eating more fruits and vegetables, eating less fried food when away from home, and drinking fewer soft drinks. The percentage of children who reported eating family dinner decreased with the age of the child, indicating that family dinner becomes more of a challenge as children get older.[15]

Illustration 12.3 A family enjoying mealtime together.

Janet Sugarman Isaacs

Outside Influences School-age children spend more and more time away from home, which is an important part of normal growth and development. Peer influence becomes greater as the child's world expands beyond the family. The increased peer influence extends to attitudes toward foods and food choices. Children may suddenly request a new food or refuse a previous favorite food, based on recommendations from a peer.

Teachers and coaches have an increasing influence on the child's attitudes toward food and eating behaviors. Nutrition should be part of the health curriculum, and what is learned in the classroom should be reinforced by foods available in the school cafeteria. Vending machines present in school as a source of extra funding can also reinforce good nutrition with appropriate choices, or these can be a source of high-fat, high-sugar foods and beverages (see the "Nutrition Integrity in Schools" section).

Media Influence In their expanding world, children come under the influence of the media. Children want to try foods they see advertised on television. One study analyzed the commercials aired during Saturday morning television programming and found that 49% of all advertisements were for food.[16] Of these food advertisements, 91% were for foods or beverages high in fat, sodium, or added sugars or were low in nutrients, which is not in line with the recommendations of the U.S. Department of Agriculture's MyPyramid (see Chapter 1).[16,17] Fast-food establishments, with their playgrounds and giveaways, are also attractive to children. With more and more children having access to the Internet, food companies are finding new ways to market their products to children. The new forms of marketing include "advergames," online games that feature the company's product or brand character; "viral marketing," in which children are encouraged to send e-mails to their friends about a product; television advertising online, which blurs the lines between advertising and entertainment; and "advercation," a combination of advertising and education. The impact of online marketing on children's food choices needs further study.[18]

Snacking Snacks continue to contribute significantly to a child's daily intake. During middle childhood, children cannot consume large amounts of food at one time and therefore need snacks to meet their nutrient needs. Many children prepare their own breakfasts or after-school snacks. These children need to have a variety of foods available to them, be equipped with nutrition education for making their own food choices, and have some age-appropriate knowledge and skill in food preparation—assuming, of course, that the family has adequate access to food.

Food Preference Development, Appetite, and Satiety Food preference development, appetite, and satiety in young children were thoroughly discussed in Chapter 10. Researchers have described the innate ability of young children to internally control their energy intake and their responsiveness to energy density. The internal controls can be altered by external factors, such as child-feeding practices. Studies in 9- to 10-year-old children found that these older children were not as responsive to energy density as young preschool-age children were.[19]

External factors such as the time of day, the presence of other people, and the availability of good food begin to override the internal controls of hunger and satiety as children get older.

Birch and associates, who have performed extensive research in the area of the development of food preferences and appetite control in children, have also examined the relationships among children's adiposity, child-feeding practices, and children's responsiveness to energy density.[20] These researchers found that children of parents who imposed authoritarian controls on their children's eating were less likely to be responsive to energy density. In other words, these children were not able to listen to internal cues in energy regulation. In girls, regulation of energy intake was inversely related to their adiposity. Heavier girls were less likely to be able to regulate their intake based on internal cues.

Body Image/Excessive Dieting Parents who had difficulty controlling their own intakes seemed to impose more restrictions on their children. A study of mothers and their 5-year-old daughters found that this transfer of "restrictive" eating practices may begin as early as the preschool age.[21] The more the mother is concerned with her own weight and with the risk of her daughter becoming overweight, the more likely she is to employ restrictive child-feeding practices. These researchers hypothesize that chronic dieting and dietary restraints, which are commonly seen in adolescent girls and young women, may have their beginnings in the early regulation of energy intake and may be related to the amount of parental control exerted over the child's eating.[20]

Young girls seem to have a preoccupation with weight and size at an early age. With the normal increase in body mass index or body fatness in preadolescence, many girls and their mothers may interpret this phenomenon of normal growth and development as a sign that the child is developing a weight problem. By imposing controls and restrictions over their daughters' intakes, mothers may actually be promoting the intake of the forbidden or restricted foods.[22] Similar results were found in 5-year-old girls whose parents restricted palatable snack foods.[23] Not only did parental restriction promote the consumption of these forbidden foods by the young girls, but these children reported feeling badly about eating these foods. Early "dieting" may actually be a risk factor for the development of obesity.[24] Dieting, which imposes restrictions, is similar to controlling child-feeding practices, which restrict children's intake. Both methods ignore internal cues of hunger and satiety. These types of child-feeding practices not only contribute to the onset of obesity and possibly a nutritionally inferior diet but may also be contributing to the beginnings of eating disorders. Eating disorders are discussed in more detail in Chapter 14.

Many studies of ethnic differences in body image and body-size preferences have been conducted hypothesizing that there are ethnic and gender differences in these parameters. The research conducted to date is not conclusive. One study of men and women of four ethnicities/races—black, Hispanic, Asian, and white—found that ethnicity alone did not influence the preference for body shapes or tolerance for obesity.[25] In working with individual families, however, it is important to try to assess their health beliefs and their preference for body size, which may impact their readiness for nutrition counseling.

Energy and Nutrient Needs of School-Age Children

Dietary Reference Intakes (DRIs), based on available scientific data on the amount of nutrients needed to maintain health and prevent chronic disease for healthy individuals and populations in the U.S. and Canada, have been developed (1997–2005). DRI tables are provided on the inside front cover of this book. Children need a variety of foods that provide enough energy, protein, carbohydrate, fat, vitamins, and minerals for optimal growth and development.[1]

Energy Needs

Energy needs of school-age children reflect the slow but steady growth rate during this stage of development. Energy needs of an individual child are dependent on the child's activity level and body size. Equations for estimating energy requirements have been developed as part of the Dietary Reference Intakes, based on a child's gender, age, height, weight, and physical activity level (PAL).[26] Estimated energy expenditure (EER) has been defined as total energy expenditure plus kilocalories for energy deposition. Categories of activity are defined in terms of walking equivalence. For example, an 8-year-old girl who weighs 56.4 pounds (25.6 kg) and is 50.4 inches (128 cm) tall will require 1360 kilocalories per day if sedentary, 1593 kcal/day if she is low-active, 1810 kcal/day if active, and 2173 kcal/day if very active. Energy allowances based on body weight are lower for school-age children than for toddlers and preschoolers. The decrease in the energy requirement per kilogram of body weight is a reflection of slowing growth rate. In addition to the DRI equations for estimating calorie requirements for children, online tools are available on websites such as MyPyramid.gov and the website of the USDA/Agriculture Research Service's Children's Nutrition Research Center (see the "Resources" section).

Protein

Based on the DRIs, the recommended protein intake for school-age children is 0.95 gram of protein per kg body weight per day for 4- to 13-year-old girls and boys.[26] School-age children can meet this recommendation by consuming diets that follow the MyPyramid recommendations

Table 12.2 Dietary Reference Intakes for key nutrients for school-age children[28,29]

Age	Recommended Daily Allowances		Adequate Intake
	Iron (mg/d)	Zinc (mg/d)	Calcium (mg/d)
4–8 years	10	5	800
9–13 years	8	8	1300

and the Dietary Guidelines for Americans.[17,27] Vegetarian diets are also appropriate for school-age children if they provide sufficient energy, complementary protein foods, a variety of foods, and adequate levels of intake of vitamins and minerals.[1] By meeting an individual child's energy needs, protein is spared for tissue repair and growth.

Vitamins and Minerals

Dietary Reference Intakes (DRIs) for vitamins and minerals have been established for the school-age and pre-adolescent child. According to food-consumption surveys, children's mean intakes of most nutrients meet or exceed the recommendations. Still, certain subsets of children do not meet their needs for key nutrients such as iron and zinc, which are important for growth, and calcium, needed to achieve peak bone mass (see the "Recommended vs. Actual Intake" section on pages 326–327). According to NHANES III data, calcium intakes are declining in 6- to 11-year-old children.[6] Dietary Reference Intakes for these key nutrients are listed in Table 12.2.[28,29]

Common Nutrition Problems

During the last century, common nutrition problems have shifted from problems of nutrient deficiencies to problems of excess nutrition, such as energy, fat, and salt. During middle childhood, some children still experience problems such as iron-deficiency anemia and dental caries, especially with easy accessibility to high-sugar foods. These nutrition problems are addressed here, followed by a thorough discussion of prevention of nutrition-related disorders.

Iron Deficiency

Iron deficiency is not as common a problem in middle childhood as it is in the toddler age group. According to the NHANES 1999–2000 survey data, 4% of 6- to 11-year-old children were found to be iron deficient, compared to 7% of toddlers.[30] Although the prevalence of iron-deficiency anemia is decreasing, these rates are still above the 2010 national health objectives. The American

Academy of Pediatrics recommends selective screening for iron-deficiency anemia for children who consume a strict vegetarian diet without taking an iron supplement.[31] The CDC recommends screening children who have known risk factors for iron-deficiency anemia, such as low iron intake and a previous diagnosis of iron-deficiency anemia.[32]

Age- and gender-specific cutoff values for anemia are based on the 5th percentile of hemoglobin and hematocrit for age from NHANES III. For children 5 to less than 8 years of age, the diagnosis of anemia is made if the hemoglobin concentration is <11.5 g/dl and hematocrit <34.5%. For children from 8 to less than 12 years of age, a hemoglobin value <11.9 g/dl or hematocrit <35.4% is diagnostic of iron-deficiency anemia.[30,32] Just as in early childhood, the treatment of iron-deficiency anemia in the school-age child consists of an oral iron trial for 4 weeks.[31] General dietary recommendations for preventing iron-deficiency anemia are the inclusion of iron-rich foods, such as meat, fish, poultry, and fortified breakfast cereals, and vitamin C-rich foods, such as citrus juice or fruit, which enhance iron absorption.

Dental Caries

Approximately one in two children age 6 to 8 years has decay in primary or permanent teeth.[6] Children with poor oral health may also develop periodontal disease. The amount of time that children's teeth are exposed to carbohydrates influences the risk of dental caries or tooth decay (see the explanation of the carogenic process in Chapter 10). Complex carbohydrates such as fruits, vegetables, and grains are better choices than simple sugars, such as soft drinks or sports drinks and candy, in relation to oral health and nutrition. Sticky carbohydrate-containing foods, such as raisins and gummy candy, are strong caries promoters. Fats and proteins may have a protective effect on enamel. Choosing snacks that are combinations of carbohydrates, proteins, and fats may decrease the risk of developing dental caries. Having regular meal and snack times versus continually snacking throughout the day is also beneficial. Rinsing the mouth after eating—or, better yet, brushing teeth regularly—also decreases the development of caries.[33] It is important that the school-age child continue to have a source of fluoride, either from the water supply or through supplementation. Details about fluoride supplementation were reported in Chapter 10.

During middle childhood, children lose their primary or baby teeth and begin to get their permanent teeth. If several teeth are missing, children may experience difficulty in chewing some foods, such as meat. Also, the orthodontic appliances commonly worn by school-age children may interfere with their ability to eat certain foods. Modifying food, such as by chopping meat or slicing fresh fruit, can help.[1]

Prevention of Nutrition-Related Disorders in School-Age Children

The prevalence of overweight and obesity among children is increasing at an alarming rate. Increases in prevalence of overweight and obesity are present in the adult population and in populations of other countries, indicating that social and environmental factors may be having an impact. Despite the increase in the prevalence of overweight, analysis of dietary data from NHANES I, II, and III indicates no corresponding increase in energy intake among children over the years. This finding suggests that physical inactivity may be a significant contributing factor to the increased prevalence of overweight.[34] The problem of increasing overweight in the United States needs to be addressed from a public health perspective.[9] Furthermore, children who are overweight are at increased risk for developing risk factors for chronic conditions, such as cardiovascular disease and type 2 diabetes mellitus.[35]

Overweight and Obesity in School-Age Children

Prevalence   According to the NHANES 2007–2008 data, approximately 19.6% of children ages 6 through 11 years are obese, with BMIs-for-age greater than or equal to the 95th percentile. [5,10,11] These percentages have increased significantly since the NHANES III study (1988–1994), when 11% of children ages 6 through 11 years were in the obese category. According to the 2007–2008 data, the proportion of children ages 6 through 11 years who are obese with BMIs greater than or equal to the 95th% ranges from 17.4% in non-Hispanic white girls to 28.3% for Hispanic males, including Mexican Americans. The prevalence of overweight and obesity in Hispanic male children was significantly greater than in non-Hispanic white male children. Prevalence of overweight among non-Hispanic white male children did not differ significantly from that in non-Hispanic black male children. Non-Hispanic black female children were significantly more likely to be overweight compared with non-Hispanic white female children.[5]

The prevalence of overweight among children has increased over time. In fact, an intrasurvey increase of about 2 to 6 percentage points occurred for most of the gender, age, and racial-ethnic groups during the six years of the NHANES III survey. The most recent survey data indicate that the prevalence of high BMI has appeared to have reached a plateau between 1999 and 2006.[5,9] As mentioned in the description of the growth charts, BMI data for children older than 6 years of age were not included in the revised growth charts because of the known increased prevalence of overweight for these ages in NHANES III. Inclusion of this information in the revised growth charts

would reflect a heavier population and would not be a healthy standard. Further analysis shows that the heaviest children are getting heavier. Extreme obesity is increasing in prevalence and has been associated with high risk for cardiovascular disease risk factors.[36] An expert committee on pediatric obesity has reported 99th-percentile cutoff points to identify obese children who are at increased risk for obesity-related complications.[10,11]

> **Bone Age**   Bone maturation; correlates well with stage of pubertal development.

Characteristics of Overweight Children   Overweight children are usually taller, have advanced *bone ages,* and experience sexual maturity at an earlier age than their non-overweight peers. From a psychosocial standpoint, overweight children look older than they are, and often adults expect them to behave as if they were older. Health consequences of obesity, such as hyperlipidemia, higher concentrations of liver enzymes, hypertension, and abnormal glucose tolerance, occur with increased frequency in obese children than in children of normal weight.[37] Analysis of data from the Bogalusa Heart Study, a community-based study of adverse risk factors in early life in a biracial population, confirms an increase in chronic disease risk factors with increasing BMI-for-age.[35] Increasing insulin levels show the strongest association with increasing BMI-for-age. Additionally, overweight children are more likely to have more than one chronic disease risk factor, especially those with BMI-for-age percentiles >99th.[36]

Type 2 diabetes mellitus, typically considered to be a disease of adults, is increasing in children and adolescents in the United States today, with up to 85% of affected children being either overweight or obese at diagnosis.[38] According to the recommendations of a panel of experts in diabetes in children, any child who is overweight, which is defined by this group as having a BMI above the 85th percentile, and who has other risk factors, should be monitored for type 2 diabetes beginning at age 10 or at puberty. Other risk factors include a family history of type 2 diabetes, belonging to certain race and ethnic groups, including African American, Hispanic American, Asian and South Pacific Islander, and Native American, and having signs of insulin resistance.[38]

It is still unclear what effect an early onset of obesity in childhood has on the risk of adult morbidity and mortality.[37] But consequences of obesity and the precursors of adult disease do occur in obese children. More studies have been performed on the relationship between obesity during adolescence and the risks of obesity in adulthood than have been performed on the relationship between childhood obesity and obesity and type 2 diabetes in adulthood (see Chapter 14).[39]

Predictors of Childhood Obesity   Dietz[13] describes critical periods in childhood for the development of obesity: gestation and early infancy, the period of BMI rebound, and adolescence. BMI rebound is the normal increase in body mass index that occurs after BMI declines and reaches its lowest point, at about 4 to 6 years of age, and is reflected in the BMI-for-age growth chart. Studies suggest that the age at which BMI rebound occurs may have a significant effect on the amount of body fat that the child will have during adolescence and into adulthood. Early BMI rebound is defined as beginning before 5.5 years of age, while the average age of BMI rebound is 6.0–6.3 years. BMI rebound after age 7 is considered late. Studies have shown that adolescents and adults who had an early BMI rebound as children have higher BMI than those subjects who had an average or late BMI rebound. Several possible mechanisms may explain the relationship between BMI rebound and subsequent obesity.[40] The period of BMI rebound may be when children are beginning to express learned behaviors related to food intake and activity. Early BMI rebound may be related to infants who were exposed to gestational diabetes during fetal development and consequently have high birth weights. Although more study is needed, the conclusion is that preventive efforts need to focus on these developmental stages.[13]

Another predictor of childhood obesity is the child's home environment. Children from birth to 8 years of age were followed over a 6-year period as part of the National Longitudinal Survey of Youth.[41] Associations between the home environment and socioeconomic factors and the development of childhood obesity were examined. Maternal obesity was found to be the most significant predictor of childhood obesity, followed by low family income and lower cognitive stimulation.

Parental obesity is associated with an increased risk of obesity in children.[42] In one study, parental obesity doubled the risk of adult obesity for both obese and nonobese children less than 10 years of age. An analysis of data from NHANES III indicated a higher percentage of overweight youth who had one obese parent as compared to those children who had no obese parents. The percentage of overweight youth increased further if both parents were obese.[43] The connection between parental obesity and obesity in children is likely due to genetic as well as environmental factors.[42]

Effects of Television Viewing and Screen Time on the Incidence of Overweight   One of the Healthy People 2010 objectives is to increase the proportion of children and adolescents who view television 2 or fewer hours per day from 60% to 75%. In 2005, this proportion was reported to be 63% for students in grades 9 to 12.[44] Related data analyzed by race and ethnicity, gender, and family income level are depicted in Table 12.3.[6,44] Since this Healthy People 2010 goal was written, young children more commonly have ready access to computers, video games, and DVDs in addition to television. "Screen

Table 12.3 Percentage of children and adolescents viewing television 2 or fewer hours per day by race/ethnicity, gender, and family income level[6,44]

Children and Adolescents Aged 8 to 16 Years	Television 2 or Fewer Hours Per Day	
	1988–1994	2005
Race and Ethnicity		
Mexican American	53%	54%
Black or African American	42	36
White	65	71
Gender		
Female	64	64
Male	54	62
Family Income Level		
Poor	53	Not available
Near poor	54	Not available
Middle/high income	64	Not available

time" is the term now used to describe the time spent in these sedentary activities, which directly competes with time spent in physical activity. To date, there is limited research on "screen time" as a contributor to pediatric overweight and obesity.[45] Most of the studies reported in the literature are specific to television viewing. However, any of these sedentary activities will likely have similar effects. The American Academy of Pediatrics recommends that children have no more than 2 hours each day of "screen time" and that televisions and other screens be removed from the child's primary sleeping area.[46]

Analysis of data collected during cycles II and III of the National Health Examination Survey (NHES) revealed significant associations between the time spent watching television and the prevalence of obesity in children and adolescents.[47] A dose–response relationship was detected. For each additional hour of television viewed in the 12- to 17-year-old group, the prevalence of obesity increased by 2%.

A strong dose–response relationship between TV viewing time and the prevalence of overweight was also found in the National Longitudinal Survey of Labor Market Experience, Youth Cohort (NLSY), which consisted of a nationally representative sample of youths age 10 to 15 years.[48] The odds of having a BMI above the 85th percentile for age and gender are significantly greater for those youths who view more than 5 hours of television per day as compared to those who watch 2 or fewer hours of television daily. Approximately 33% of the youth report watching more than 5 hours of television per day, while only 11% watch 2 or fewer hours of daily television, which is a Healthy People 2010 objective.[6,48]

According to NHANES III data, children age 11 through 13 years have the highest rates of daily television viewing.[49] Children, both males and females, who watch 4 or more hours of television daily have greater body fat and BMI than those who watch less television.[49,50] A school-based intervention program aimed at reducing third- and fourth-grade children's television, video, and video game use was shown to be effective in reducing television viewing and meals eaten in front of the television. In addition, decreases in BMI and waist circumference were seen.[51]

The proposed mechanisms by which television viewing contributes to obesity include reduced energy expenditure by displacing physical activity and increased dietary intake by eating during viewing or as a result of food advertising.[51] Analysis of NHANES III data showed a positive correlation between intake and number of hours of television watched.[50] One study found that energy expenditure during television viewing was actually significantly lower than *resting energy expenditure* in 15 obese children and 16 normal-weight children who ranged in age from 8 to 12 years.[52] Based on these findings, it is hypothesized that television viewing does contribute to the prevalence of obesity, and that treatment for childhood obesity should include a reduction in the number of hours spent watching television and videos and playing video and computer games.

Addressing the Problem of Pediatric Overweight and Obesity

"An ounce of prevention is worth a pound of cure."

Recognizing the increase in the prevalence of childhood overweight and its associated chronic health problems, the American Academy of Pediatrics (AAP) issued a policy statement in 2003 on the prevention of pediatric overweight and obesity.[53] The policy statements advocates for (1) early recognition, using BMI-for-age as a screening tool, and providing anticipatory and appropriate guidance regarding healthy eating and physical activity; and (2) advocacy for opportunities for physical activity, improvements in foods available to children, research, and third-party reimbursement for treatment of overweight.[53] To further emphasize the role of increased physical activity in the prevention of childhood obesity, the Council on Sports Medicine and Fitness and Council on School Health of the AAP released another policy statement in 2006 recommending that increased physical activity for children be encouraged, monitored, and advocated for by pediatric health care providers and public health officials.[54]

These same principles for addressing pediatric overweight and obesity were espoused in the expert committee recommendations on pediatric obesity published in 2007 and endorsed by the AAP.[10] The expert committee's evidence-based recommendations address the topics of assessment, prevention, and treatment of pediatric obesity.[10,11,45,55]

Assessment of Overweight and Obesity Body mass index-for-age percentile is recommended as the screening tool for assessment of pediatric overweight and obesity. A BMI-for-age percentile of greater than or equal to the 85th but less than or equal to the 94th is defined as overweight, and a BMI-for-age percentile greater than or equal to the 95th is defined as obesity.[10,11] As has already been discussed, 99th-percentile cutoff points for BMI have been reported to identify those children who are at increased risk of obesity-related health consequences.[10,11] Other components of assessment include evaluation of the child's medical risk, including parental obesity, and behavior risk, including dietary and physical activity behaviors. It is also essential to evaluate the child's and/ or family's attitude toward and willingness to make behavior changes.[10,11]

> **Resting Energy Expenditure** The amount of energy needed by the body in a state of rest.

Prevention of Overweight and Obesity All children should be targeted for prevention of overweight and obesity throughout their lives. The expert committee's recommendations for targeted behaviors in overweight and obesity prevention are described in detail in Chapter 10. The targeted behaviors address healthy eating and increased physical activity.[10,55]

Treatment of Overweight and Obesity The expert committee's recommendations for a four-stage approach to treatment of overweight and obesity in pediatrics are described in detail in Chapter 10.[10,45] Illustration 12.4 depicts the staged treatment for 6- to 11-year-old children.[10,45] The four stages include:

Stage 1: Prevention Plus
Stage 2: Structured Weight Management (SWM)
Stage 3: Comprehensive Multidisciplinary
          Intervention (CMI)
Stage 4: Tertiary Care Intervention (reserved for
          severely obese adolescents)

The overall goal of treatment is for the child and family to develop healthy eating and physical activity behaviors for a lifetime. For children who fall into the overweight category based on a BMI-for-age percentile of 85th to 94th, the goal of treatment should be weight maintenance or a slowing of the rate of weight gain until a BMI-for-age percentile <85th is achieved. For children with

Illustration 12.4 Staged Obesity Treatment for 6- to 11-year-old youth.

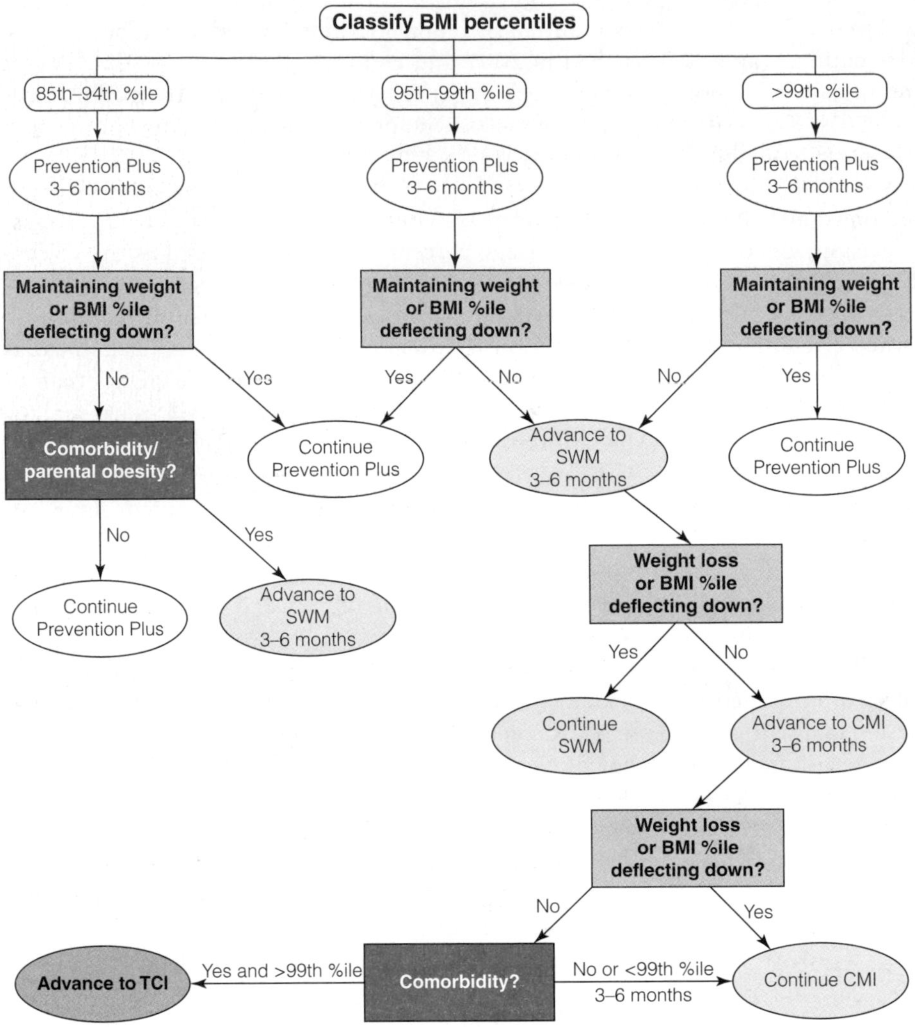

SOURCE: Reproduced with permission from *Pediatrics*, Volume 120, Supplement 4, Page S274. Copyright 2007, American Academy of Pediatrics.

BMI-for-age percentiles of 95th to 98th, weight maintenance or gradual weight loss of no more than 1 pound per week is the goal until the BMI-for-age percentile drops to <85th. Weight loss not to exceed to 2 pounds per week is the goal of treatment for those children with BMI-for-age percentiles ≥99th until a BMI-for-age percentile of <85th is achieved.[10,45] See Case Study 12.1 on the next page.

An evidence-based analysis of intervention literature showed positive effects of multicomponent, family-based programs for children between the ages of 5 and 12 years. Recommended components include parent training, dietary counseling/nutrition education, physical activity and addressing sedentary behaviors, and behavioral counseling.[56]

Some potential consequences of a weight-loss program in childhood are a slowing of linear growth and the beginnings of eating disorders. To reduce the risks associated with weight loss in childhood, the program must ensure nutritional adequacy of the diet, a nonjudgmental approach, and attention to the child's emotional state.[10,45]

Nutrition and Prevention of Cardiovascular Disease in School-Age Children

In the new DRIs, no recommendations for total grams of fat per day in the diet of children have been made.[26] Studies have shown that as long as enough energy is provided for growth, no effect of fat intake on growth has been found. In addition, the evidence is still insufficient to be able to define the optimal fat intake for promoting growth while also preventing obesity and other chronic diseases. According to the DRIs, the Acceptable Macronutrient Distribution Range (AMDR) for fat is 25 to 35% of energy for children 4 to 18 years of age.[26]

The new DRIs do stress the importance of including sources of linoleic acid (omega-6 fatty acid) and alpha-linolenic acid (omega-3 fatty acid). Adequate intake levels for these essential fatty acids have been determined and can be found in Table 12.4. Sources of linoleic acid include vegetable oils, seeds, nuts, and whole-grain breads and cereals. Fish, as well as flaxseed, soy, and canola oils, are good sources of alpha-linolenic acid. (See Chapter 1 for a complete list of food sources of linoleic acid and alpha-linolenic acid.) A diet that emphasizes fruits, vegetables, low-fat dairy products, whole-grain breads and cereals, nuts, seeds, fish, and lean meats is recommended for promoting nutrition and preventing cardiovascular disease in school-age children.[26]

The American Heart Association and the American Academy of Pediatrics have jointly issued guidelines for cardiovascular health promotion in all children and adolescents and dietary recommendations.[57–60] The diet and lifestyle recommendations were further revised in 2006.[61] The recommended diet is consistent with the Dietary Guidelines for Americans, the My Pyramid, and the DRIs, including a recommended total fat intake of 25–35% of total calories.[17,26,27] Emphasis is placed on adequate intakes of omega-3 fatty acids, with a recommendation of

Pediatric Overweight

Seven-year-old Timothy's mother takes him to his pediatrician for his annual checkup. His weight is 68 pounds (31 kg), plotted at the 95th percentile, and his height is 50 inches (127 cm), between the 75th and 90th percentiles for his age. His body mass index of 19.25 kg/m² plots at the 95th percentile for his age. His growth percentiles have been increasing over the last several years.

Timothy's mother expresses concern to the pediatrician about her son's weight. His older and younger brothers are both thinner than Timothy. Timothy's mother is obese, but his father is a normal weight for height. Timothy is in the second grade. He rides the school bus to and from school. He participates in the School Lunch Program at his school, but his parents gave him extra money in case he wants to buy some additional à la carte food items from the cafeteria or items from the vending machines. After school, Timothy and his brothers stay in their home with a babysitter until one of their parents returns home from work. Timothy usually watches TV or plays video games after school. His parents leave snack foods—chips, cookies, and sodas—in the house for their sons to have after school. His mother usually prepares their evening meal, which consists of a meat, starch, vegetables, and a dessert item. After dinner, Timothy does his homework and then usually watches more TV with his parents. He usually has a dish of ice cream before going to bed.

Questions

1. What is your assessment of Timothy's body size based on his weight-for-age, height-for-age, and BMI-for-age percentiles?

2. What suggestions do you have for Timothy's parents about improving his eating habits?

3. What suggestions do you have for Timothy's parents for increasing his physical activity level?

4. Is it significant that Timothy's mother also has a weight problem?

(Answers are located in the Instructor's Manual for the 4th edition of *Nutrition Through the Life Cycle*.)

Table 12.4 Adequate intake of linoleic acid and alpha-linolenic acid[26]*		
Gender and Age	Linoleic Acid g/day	Alpha-Linolenic Acid g/day
Children 4–8 years	10	0.9
Boys 9–13 years	12	1.2
Girls 9–13 years	10	1.0

*See Chapter 1 for a list of food sources.

at least 2 servings of fish each week. Limiting the intake of fruit juice, sugar-sweetened beverages and foods, and salt is also recommended. For children over 2 years of age, these guidelines recommend limiting foods high in saturated fats (<7 percent of total calories per day), cholesterol (<300 mg per day), and *trans* fatty acids (<1 percent of total calories per day).[57,58,61,62]

A discussion of a recommended approach to screening children for high lipid levels can be found in Chapter 10. Children with hyperlipidemias require further dietary restrictions to help control LDL cholesterol. Further restriction of dietary cholesterol to 200 mg per day and restricting *trans* fatty acids to a level as low as possible are recommended.[57,58,63] Increasing soluble fiber intake, emphasizing weight management and physical activity, and follow-up by a registered dietitian are also treatment recommendations.[57,58,63] Employing behavior change theory and motivational interviewing may be useful strategies when counseling children and families on these dietary recommendations.[62]

Dietary Supplements

Children who are healthy and consume a diet of a variety of foods do not require a vitamin and mineral supplement to meet their nutrient needs. The American Academy of Pediatrics recommends vitamin and mineral supplementation for children who are at high risk of developing nutrient deficiencies or have one or more.[31] (See Chapter 10 for a list of children at risk for nutrient deficiency.)

If vitamin and mineral supplements are given to school-age children, the supplement should not exceed the Dietary Reference Intakes for age. Parents should be warned against giving amounts of vitamins and minerals that exceed the Tolerable Upper Intake levels designated in the DRI tables.[28,29] It is not clear to what extent herbal supplements are given to school-age children. Herbal supplements are used in some cultures as home remedies. One study of children and their families who received care at a pediatric emergency department found that 12% of families reported use of at least one form of complementary and alternative medicine (CAM) to treat any of their children. Children who were treated with CAM were likely to have a caretaker who also used CAM.[64] It is important to obtain this information from parents and caretakers as part of the child's health history. The use of herbal supplements, botanicals, and vitamin/mineral supplements may be a more prevalent practice by parents of children with special health care needs (see Chapters 11 and 13).

Total Fiber   Sum of dietary fiber and functional fiber.

Dietary Fiber   Complex carbohydrates and **lignins** naturally occurring and found mainly in the plant cell wall. Dietary fiber cannot be broken down by human digestive enzymes.

Functional Fiber   Nondigestible carbohydrates including plant, animal, or commercially produced sources that have beneficial effects in humans.

Dietary Recommendations

The basic dietary recommendation for school-age and preadolescent children is to eat a diet of a variety of foods, which is why it remains so important throughout these school years for children to have a variety of foods available to them. The available food environment will affect children's food choices. Parents and other adult role models need to continue to model appropriate eating behaviors for children.

Dietary recommendations, as outlined by the USDA in the Dietary Guidelines for Americans and MyPyramid, apply to school-age children as well as to other segments of the population.[17, 27] Professional organizations, such as the American Heart Association, the American Academy of Pediatrics, and the American Dietetic Association, have also published positions on dietary guidance for healthy children, supporting the federal guidelines.[59-61,65]

Recommendations for Intake of Iron, Fiber, Fat, Calcium, Vitamin D and Fluids

Adequate iron nutrition is still important during middle childhood and preadolescence to prevent iron-deficiency anemia and its consequences. According to food-consumption surveys, children are not eating the recommended amounts of fiber in their diets, but they are exceeding the recommendations of total calories from fat and saturated fat. Calcium requirements increase during the preadolescent years, but calcium intake decreases with age.

Iron   Although iron deficiency is not as prevalent during the school-age years as it was during the toddler and preschool-age years, adequate intake of iron is still important. The inclusion of iron-rich foods—such as meats, fortified breakfast cereals, and dry beans and peas—in children's diets is important. A good vitamin C source, such as orange juice, will enhance the absorption of iron. (See Chapter 1 for a more complete list of high-iron foods.)

Fiber   As reported in Chapter 10, many health effects of fiber intake have been identified—including prevention of chronic disease in adulthood, such as heart disease, certain cancers, diabetes, and hypertension. The new recommendations for *total fiber* intake based on the DRIs can be found in Table 12.5. Total fiber is the sum of *dietary fiber* and *functional fiber*. Earlier recommendations were based on dietary fiber.[26]

To increase the dietary fiber in children's diets, parents and caretakers can begin by increasing the amount of fresh fruits and vegetables and whole-grain breads and cereals being offered. High-fiber fruits, such as apples with peels, have about 3 grams per serving, while fruit juices are low in fiber. High-fiber vegetables, such as broccoli, have about 2.5 grams per serving. Whole-grain breads, cereals, and brown rice have about 2.5 grams per serving. High-fiber cereals, such as bran flakes and raisin bran, have about 8 to 10 grams per serving. Served alone, these high-fiber cereals may not be well accepted by young children, but they can be mixed with other cereals or used in recipes for food items such as muffins. Dried beans and peas are also excellent sources of fiber, providing 4 to 7 grams of fiber per ½-cup serving.[66]

Table 12.5 Adequate intake of total fiber[26]

Gender and Age	Total fiber, g/day
Children 4–8 years	25
Boys 9–13 years	31
Girls 9–13 years	26

SOURCE: From the Institute of Medicine, Food & Nutrition Board

Fat Food intakes that follow the recommendations of the Dietary Guidelines for Americans and MyPyramid provide an appropriate amount of fat for school-age and preadolescent children.[17, 27] As discussed in the "Nutrition and Prevention of Cardiovascular Disease in School-Age Children" section, healthy diets include whole-grain breads and cereals, beans and peas, fruits and vegetables, low-fat dairy products, and lean meats, fish, and poultry. Foods high in fat, especially those high in saturated fat and *trans* fatty acids, should be kept to a minimum. However, an appropriate amount of dietary fat is necessary to meet children's needs for calories, essential fatty acids, and fat-soluble vitamins.

Calcium and Vitamin D The recommendations for adequate daily intakes of calcium are 800 milligrams for children aged 4 to 8 years and 1300 milligrams for children 9 through 18 years.[28] The higher recommendation for older children reflects the fact that most bone formation occurs during puberty. Adequate calcium intake during this time is necessary to achieve peak bone mass, which may prevent osteoporosis later in life.[67]

Good sources of calcium are listed in Chapter 1. It is difficult to meet the higher recommendations of calcium without the inclusion of dairy products, preferably low-fat dairy products. One cup of skim or low-fat milk contains about 300 mg calcium. Calcium-fortified foods such as fruit juice and soy milk are also available for children such as those on a vegan diet.

Adequate vitamin D is needed for calcium absorption. The American Academy of Pediatrics recently doubled the recommended amount of vitamin D for all healthy infants, children, and adolescents from 200 IU per day to 400 IU per day[68]. Serum 25-OH-D concentrations in children should be 20 ng/mL. Main sources of vitamin D include exposure to sunlight, vitamin D–fortified foods such as fortified cereals, and vitamin D–fortified milk (100 IU per 8 ounces).[68] Children who are at risk for vitamin D deficiency include those with increased skin pigmentation, including African-Americans and Hispanics, and those with limited sunlight exposure. For children whose calcium and vitamin D intakes are inadequate or who are at increased risk for vitamin D deficiency, supplements need to be given under the guidance of a physician or registered dietitian.

Lactose Intolerance Lactose intolerance, more commonly seen in older children than in younger children, is a common cause of abdominal pain. Lactose intolerance is a clinical syndrome of one or more gastrointestinal symptoms, such as abdominal pain, diarrhea, nausea, flatulence or bloating, after consumption of lactose or lactose-containing foods or beverages. Lactose malabsorption, the disorder that manifests itself as lactose intolerance, is caused by reduced digestion of lactose because of the low availability of the enzyme lactase. Lactase breaks down lactose, the disaccharide in milk and milk products.[69] Lactose intolerance can be caused by a primary lactase deficiency, which is the relative or absolute absence of the enzyme. Primary lactase deficiency is especially common in certain racial and ethnic groups, including Hispanics (50 to 80%), black and Ashkenazi Jewish people (60 to 80%), and Asians and Native Americans (nearly 100%).[69] Affected individuals have varying degrees of lactose intolerance, and dairy products should be included in their diets as individually tolerated. Having lactose-containing foods in small amounts, spaced throughout the day, and eaten with other foods may be tolerated by many people with lactose intolerance. Additionally, yogurts, cheeses, and lactose-reduced dairy products have lower lactose content and may be well tolerated. Secondary lactase deficiency can be caused by injury to the small bowel, such as from an acute infection. The underlying condition should be treated, and often the elimination of lactose from the diet is not necessary at all, but milk products should definitely be resumed once the underlying condition has resolved. The concern with both primary and secondary lactase deficiency is to avoid the total elimination of dairy products from the diet when it may not be necessary, as these foods are important sources of calcium, vitamin D, and other nutrients.[69]

Fluids It is of particular importance for school-age children to drink enough fluids to prevent dehydration during periods of exercise and during participation in sports, because children are at risk for dehydration and heat-related stress. Preadolescent children need to be more careful about staying hydrated than do adults and adolescents, for several reasons.[70] Children sweat less, and they get hotter during exercise. Some sports, such as football and hockey, require special protective gear that may prevent the body from being able to cool off. Children should never deprive themselves of food or water in order to meet a certain weight category, such as in wrestling.

Adults who are supervising children's physical activities need to make sure that children drink fluids before, during, and after exercise. The thirst mechanism may not work as well during exercise, and children may not realize that they need fluids. Cold water is the best fluid for children. However, children may be more likely to drink more fluids if they are flavored. Sports drinks, which contain 4 to 8% carbohydrate and diluted fruit juice, are appropriate for children. Children should not be given soft drinks or undiluted juice, because the carbohydrate load is too high to be hydrating and could cause stomach cramps, nausea, and diarrhea.[70]

Soft Drinks School-age children consume more soft drinks than preschool-age children do, but not as much as adolescents—indicating an increase in consumption

Table 12.6 Mean percentages of food energy from carbohydrate, protein, total fat, saturated fatty acids, and cholesterol intake of 6- to 11-year-old children[78]

Gender and Age	Carbohydrate (%)	Protein (%)	Total Fat (%)	Saturated Fatty Acids (%)	Cholesterol (mg/d)
Males:					
6–11 years	53.7	13.8	33.8	12.1	223
Females:					
6–11 years	53.8	13.5	33.9	12.0	237

with age. Children's consumption of all sugar-sweetened beverages has increased over time, with a 20% increase among children age 6 to 11 years from 1988 to 2004. On average, sugar-sweetened beverages add 229 calories per day to the school-age child's overall energy intake, with most of the sugar-sweetened beverage calories being consumed at home.[71] Children with high consumption of regular soft drinks (more than 9 ounces per day) consume less milk and fruit juice than do those with lower consumptions of regular soft drinks. According to analysis of NHANES III data, overweight children have a higher proportion of their energy intake from soft drinks than non-overweight children do.[34] A study of 548 ethnically diverse school-age children, with an average age 11.7 years, showed that BMI and the frequency of high BMI greater than 95th percentile increased along with increased consumption of sugar-sweetened beverages.[72] Soft drinks can contribute significantly to children's overall calorie intake while contributing little to the overall nutritional value of their diets and displacing more nutritious foods. A study of 30 children 6 to 13 years of age found that sweetened-drink consumption displaced milk from children's diets and resulted in intakes that were lower in protein, calcium, magnesium, phosphorus, and vitamin A but higher in calories.[73] According to USDA food-consumption data, children's soft-drink consumption increases with age while milk consumption decreases at a time when calcium requirements are increasing.[28,74] Children who ate fast food consumed more carbonated non-diet beverages and less milk than children who did not eat fast food.[75] Soft drinks can also contribute to a child's caffeine intake, providing 35–50 mg caffeine per 12-ounce serving.[76] Diet soft drinks do not provide sugar, and the aspartame content of diet sodas does not appear to pose a risk to healthy children.[77] Soft drinks in excess are not recommended for school-age children because they provide empty calories, promote tooth decay, and are not a good fluid choice for hydration.

Recommended vs. Actual Food Intake

The composition of children's diets, based on data from the report "What We Eat in America" (WWEIA), a

joint project of the U.S. Department of Agriculture (USDA) and the U.S. Department of Health and Human Services (DHHS), can be found in Table 12.6.[78] This table depicts the mean percentages of total energy from carbohydrate, protein, total fat, saturated fatty acid, and cholesterol intake for 6- to 11-year-old males and females. According to the data, both boys' and girls' percent of total calories from fat is within the recommended range of 25–35%. However, both boys and girls are exceeding the recommendation for total calories from saturated fat of less than 7%.[61,62] Cholesterol intake is well below the recommendation of 300 mg per day. Analysis of NHANES III data shows that the percentage of energy from fat is higher for black and Mexican American girls and black boys than for white girls and boys.[79] These differences are seen by 6 to 9 years of age in black and Mexican American girls and by 10 to 13 years of age in black boys.

Mean vitamin and mineral intakes for school-age children exceeded the recommendations except for that of vitamin E, folic acid, and calcium.[28,29,78] Table 12.7 depicts a further analysis of children's diets in relation to dietary fiber, sodium, and caffeine intake. School-age children's caffeine intake has risen dramatically for both males and females from the preschool years, indicating an increased consumption of caffeine with age. This coincides with an increase in soft-drink consumption. Thirty-three percent of U.S. children report consuming fast food on a typical day. Children who eat fast food consume more total energy, more total fat, more total carbohydrates, more added sugars, more sugar-sweetened beverages, less fiber, less milk, and fewer fruits and non-starchy vegetables than children who do not eat fast food.[75] Analysis of food consumption data indicates that snacking among children has increased over the years, and the contribution of snacks to energy intake has increased from 20% in 1977 to 25% in 1996.[80]

Table 12.7 Mean dietary fiber, sodium, and caffeine intake of 6- to 11-year-old

Gender and Age	Dietary Fiber (g)	Sodium (mg)	Caffeine (mg)
Males:			
6–11 years	14.1	3202	19.7
Females:			
6–11 years	12.0	2966	17.0

According to the NHANES III data, children age 6 to 11 years obtain about 20% of their total energy intake from beverages, with milk, soft drinks, and juice drinks being the largest contributors.[34] Drinking whole milk contributes significantly to children's saturated fat intakes.

A measure of diet quality is provided by the Healthy Eating Index (HEI; available at www.usda.gov/cnpp), which utilizes data from the ongoing National Health and Nutrition Examination Surveys.[81] The HEI measures the degree to which an individual's diet conforms to the Dietary Guidelines for Americans and uses the food-group standards found in MyPyramid.[17,27] The index also measures saturated fat intake, sodium intake, and extra calories from solid fats and added sugars. According to the most recent data, the average HEI score for children age 6 to 11 years was 54.7 out of 100, indicating that their diets needed improvement.[81] Data indicate that children need to increase consumption of whole fruit, whole grains, dark-green and deep-yellow vegetables, and legumes and need to decrease consumption of saturated fat, sodium, and extra calories from solid fats and added sugars to improve their diets.[81]

Cross-Cultural Considerations

Healthy People 2010 has as one of its major goals the elimination of health disparities among different segments of the population.[6] The reasons for the health disparities are complex but may include genetic variations, environmental factors, and health behaviors, including diet. Access to community-based, culturally competent, linguistically appropriate preventive health care is needed to eliminate these disparities.[6]

A unique characteristic of every ethnic group in America is its culturally based foods and food habits.[82] As discussed earlier, children learn food habits within the context of their family's culture. It is important for a health care professional to try to learn as much as possible about the foods and diets of the ethnic groups served, including where food is purchased and how it is prepared. The next step is to evaluate the diet within the context of the culture. Which foods or food habits have positive health benefits and should be encouraged? Which food behaviors have harmful effects on health and should be limited or modified? For example, in working with the Latino population, it is important to first of all establish the country of origin. Latino immigrants may be from Mexico, Central America, South America, or the Caribbean. Food habits are unique for each of these ethnic groups. For example, Central Americans eat a lot of legumes, rice, and corn. Fruits and vegetables are also included in the diet. These dietary practices form the basis of a healthy diet. However, lard is the most commonly used fat. Encouraging Central Americans to use a vegetable oil instead of lard is an example of a modification of a food practice to make it healthier.

Vegetarian Diets

Young children who are consuming vegetarian diets are usually following their parents' eating practices. Preadolescents, on the other hand, may choose to follow a vegetarian diet independently of the family, motivated by concerns about animal welfare, ecology, and the environment.[1] A vegetarian diet is a socially acceptable way to reduce caloric intake and may be adopted by adolescents with eating disorders (see Chapter 14). A Vegetarian Food Guide Pyramid with suggested numbers of servings from different food groups has been developed.[83] Providing adequate calories, protein, calcium, zinc, iron, omega-3 fatty acids, vitamin B_{12}, riboflavin, and vitamin D is essential in planning vegetarian diets for children.[84]

Physical Activity Recommendations

Physical activity has many proven health benefits, including prevention of chronic diseases such as coronary heart disease, building muscle strength, and controlling energy balance. Physical activity is one of the health behaviors that is important to establish in childhood as part of a healthy lifestyle that will continue into adolescence and adulthood. With the increased prevalence of childhood obesity, increasing physical activity and decreasing sedentary behaviors become important factors in controlling childhood overweight.[50]

Recommendations vs. Actual Activity

It is recommended that children engage in at least 60 minutes of physical activity every day.[10,27] Strategies for parents include:

- Set a good example by being physically active themselves and joining their children in physical activity.
- Encourage children to be physically active at home, at school, and with friends.
- Limit television and video/DVD watching, computer and video game playing, time at the computer, and other inactive forms of play by alternating with periods of physical activity.

As has already been mentioned, it is recommended that "screen time" for children be limited to not more than 2 hours per day, and televisions and other screens, such as computers, should be removed from children's bedrooms.[46] Additionally, physical activity and daily physical education should be encouraged at schools and during after-school care programs. The American Academy of Pediatrics recommends that schools establish policies that promote physical activity in safe settings, implement physical education and health education curricula, provide extracurricular physical activity, and include parents and caretakers in these activities.[85] As of 2006, only 7.9% of middle and junior high schools required daily physical education for all students.[44] Healthy People 2010 objectives include increasing the proportion of trips that school-age children make by walking and by bicycling. As of 2001, only about 36% of children and adolescents ages 5 to 15 years take walking trips to school less than 1 mile, and only about 1.5% of children and adolescents ages 5 to 15 years take bicycle trips to school of less than 2 miles.[44]

In order to meet these goals, communities need to ensure that there are safe places for children to walk and ride their bicycles. The "built environment" is the term used to describe the overall structure of the physical environment of a child's community.[86] The physical environment can be conducive to a healthy lifestyle by providing safe places such as parks for children to play in, sidewalks for walking to school, or bike paths. Or the physical environment can be a barrier to a healthy lifestyle. In addition to the distances children must walk to school, parents are also concerned about traffic danger, followed by concerns about crime and weather.[86] Programs such as the "walking school bus" and the national Safe Routes to School Program help concerned parents, health professionals such as pediatricians, and other community leaders establish safe "built environments" in their communities (see the "Resources" section). Safety measures, such as wearing a helmet when biking, need to be employed. Communities can also offer youth sports and recreation programs that are developmentally appropriate and fun for all young people. To achieve such a community environment, partnerships need to be established among federal, state, and local governments, nongovernment organizations, and private entities. The CDC has proposed strategies for promoting physical activity for children in family, school, and community settings.[87]

Determinants of Physical Activity

It is important to understand children's physical activity patterns and determinants of physical activity, so that vulnerable groups can be identified and appropriate intervention programs designed. Potential determinants of physical activity behaviors among children include physiological, environmental, psychological, social, and demographic factors.[88] Childhood physical activity has been difficult to assess and to track into adulthood. Many of the studies have identified correlates of physical activity behavior rather than predictors. The determinants of childhood physical activity are probably multidimensional and interrelated. More work needs to be done in this area, but some generalities resulting from existing studies are listed here:

- Girls are less active than boys.
- Physical activity decreases with age.
- Seasonal and climate differences are seen in children's activity levels.
- Physical education in schools has decreased.

School and neighborhood safety is an important issue in promoting physical activity. In addition, parents have direct and indirect effects on children's physical activity levels.

Organized Sports

Many school-age and preadolescent children participate in organized sport activities, through schools or other community organizations. An analysis of NHANES III data indicates that children who participate in team sports and exercise programs are less likely to be overweight as compared to nonparticipants.[43] The American Academy of Pediatrics (AAP) recognizes the fact that organized sports for children have become widely accepted in our society and do provide opportunities for physical activity for children. The AAP recommends that children be involved in sports that are appropriate for their physical and cognitive development.[89] Emphasis should be placed on having fun and on family participation rather than on being competitive. Organized sports should not take the place of regular physical activity, such as physical education in school and free play.[89] The proper use of safety equipment, such as helmets, pads, mouth guards, and goggles, should be encouraged. The AAP warns against intensive, specialized training for children. Coaches should be educated about developmental and safety issues.[89]

Recommendations for sports participation by children stress the importance of proper hydration. As noted earlier, children are at greater risk for dehydration and heat-related stresses than are adults (see the "Fluids" section).

Nutrition Intervention for Risk Reduction

"It is the position of the American Dietetic Association, the Society for Nutrition Education, and the American School Food Service Association that comprehensive nutrition services must be provided to all of the nation's preschool through grade-twelve students.[90]

The American Dietetic Association

Nutrition Education

Eating a healthy diet and participating in physical activity are important components of a healthy lifestyle that may prevent chronic disease in childhood and into adolescence and adulthood. School age is a prime time for learning about healthy lifestyles and incorporating them into daily behaviors. Schools can provide an appropriate environment for nutrition education and learning healthy lifestyle behaviors. Nutrition education studies have been conducted in school settings as well as outside of schools. Some of these programs have been knowledge-based nutrition education programs, with the focus on improving the knowledge, skills, and attitudes of children in regard to food and nutrition issues.[91] Other nutrition education programs have been more behaviorally focused, emphasizing disease risk reduction as well as enhancing health. The CDC has published "Guidelines for School Health Programs to Promote Lifelong Healthy Eating" (see Table 12.8).[92]

Nutrition Integrity in Schools

"The schools and the community have a shared responsibility to provide all students with access to high-quality foods and school-based nutrition services as an integral part of the total education program.[93]

The American Dietetic Association

Nutrition integrity in schools is defined as ensuring that all foods available to children in schools are consistent with the U.S. Dietary Guidelines for Americans and the Dietary Reference Intakes.[6,27] School nutrition programs are vital

---

Table 12.8 Recommendations for school health programs promoting healthy eating[92]

1. **Policy:** Adopt a coordinated school nutrition policy that promotes healthy eating through classroom lessons and a supportive school environment.

2. **Curriculum for nutrition education:** Implement nutrition education from preschool through secondary school as part of a sequential, comprehensive school health education curriculum designed to help students adopt healthy eating behaviors.

3. **Instruction for students:** Provide nutrition education through developmentally appropriate, culturally relevant, fun, participatory activities that involve social learning strategies.

4. **Integration of school food service and nutrition education:** Coordinate school food service with nutrition education and with other components of the comprehensive school health program to reinforce messages on healthy eating.

5. **Training for school staff:** Provide staff involved in nutrition education with adequate preservice and ongoing in-service training that focuses on teaching strategies for behavioral change.

6. **Family and community involvement:** Involve family members and the community in supporting and reinforcing nutrition education.

7. **Program evaluation:** Regularly evaluate the effectiveness of the school health program in promoting healthy eating, and change the program as appropriate to increase its effectiveness.

---

to reinforcing healthy eating habits in school-age children. Sound nutrition policies need the support of the community and school environments, and these must involve students in order to be successful. Preparing community leaders for involvement in policy development is one of the nutrition integrity core concepts.[93] Training food-service personnel, teachers, administrators, and parents is an integral part of this process. The school environment must be one that supports healthy eating and exercise patterns. Foods sold from vending machines and snack bars often do not support healthy eating and may undermine sound nutrition programs. However, in some underfunded schools, vending machine proceeds are important sources of revenue. Some schools have *pouring rights* contracts with soft-drink companies and receive a percentage of the profits.[90] Many schools sell à la carte items in addition to standard school lunches to increase revenue. One study showed that seventh-grade students in schools with à la carte items ate more fat and fewer fruits and vegetables than did students at schools without an à la carte program.[94] In the same study, for each snack vending machine, students' mean intake of fruit servings declined by 11%. Data from the third School Nutrition and Dietary Assessment study in 2005 of 395 U.S. public schools in 38 states indicated that while there were vending machines in only 17% of elementary schools, there were vending machines in 82% of middle schools; à la carte items were sold in 71% of elementary schools and 92% of middle schools. The majority of the time, vending machines and à la carte items were sources of low-nutrient, energy-dense foods and beverages such as sugar-sweetened beverages and candy.[95] It is against USDA regulations to sell *competitive foods* of minimal nutritional value. However, it is not against USDA regulations to sell these foods to students at times other than mealtimes or in other areas of the school, outside of food-service areas.

Adequate time allotted for meals is another important component of a sound nutrition program. Students can be involved in a nutrition advisory council, providing feedback about menu preferences and meal environment and serving as a communication link with other students.

The American Academy of Pediatrics (AAP) Committee on School Health has issued a policy statement about nutrition concerns regarding soft-drink consumption in schools. The AAP advocates for the elimination of sweetened beverages in schools and their replacement with beverages such as real fruit and vegetable juices, water, and low-fat white or flavored milk. Vended food or drink contracts are discouraged, and the health and nutritional interests of students should form the foundation of nutritional policies in schools.[96] The AAP has also issued a policy

Pouring Rights Contracts between schools and soft-drink companies whereby the schools receive a percentage of the profits of soft drink sales in exchange for the school offering only that soft-drink company's products on the school campus.

Competitive Foods Foods sold to children in food service areas during meal times that compete with the federal meal programs.

on physical fitness and activity in schools (see the "Physical Activity Recommendations" section).[85]

The School Health Index (SHI) for Physical Activity, Healthy Eating and a Tobacco-free Lifestyle is a self-assessment and planning tool for schools that is offered by the National Center for Chronic Disease Prevention and Health Promotion, Centers for Disease Control and Prevention.[97] The SHI helps schools:

- Identify strengths and weaknesses in health promotion policies and strategies.
- Develop an action plan for improving student health.
- Involve all stakeholders, including teachers, parents, students, and the community, in improving school policies and programs.[97]

The SHI has eight different modules for elementary schools and middle and high schools in the self-assessment, which correspond to the eight components of a coordinated school health program as depicted in Illustration 12.5. Each module consists of a score card, a questionnaire with guidance for arriving at a score, planning questions, and recommendations for implementation. Illustration 12.6 on the following page is an example of items on the Nutrition Services score card. Health topics covered by SHI include safety, physical activity, nutrition, tobacco use, and asthma.[97]

Nutrition Assessment

A nutrition assessment of a school-age child can identify nutrition-related concerns that pertain to the child's overall health status. Components of a nutrition assessment

Illustration 12.5 SHI figure.[97]

Monkey Business Images/2008/ Used under license from Shutterstock.com

include a food/nutrition-related history, pertinent biochemical measurements, anthropometric measurements such as weight, height, body mass index percentile, and medical history. According to *Bright Futures: Guidelines for Health Supervision of Infants, Children, and Adolescents,* school-age children should be routinely screened for anemia and dyslipidemia by measuring pertinent biochemical indices, as listed in Table 12.9.[98] Based on this information, the nutrition professional can identify any nutrition diagnoses, design a nutrition intervention plan with family input, and make a plan for monitoring and evaluation (see Table 2.10 for the four steps of the Nutrition Care Process).

Model Programs

The National Fruit and Vegetable Program, formerly the "5 A Day" program, is a public-private partnership of the Centers for Disease Control and Prevention, Produce for Better Health, and other health organizations. Its public health initiative is called Fruits & Veggies—More Matters and reflects the U.S. Dietary Guidelines in terms of recommendations for fruit and vegetable consumption. Available educational resources include "Explore the World with Fruits and Vegetables," which provides fun ideas for encouraging school-age children and their parents to increase their fruit and vegetable intake.[99]

Model Program: High 5 Alabama   The purpose of this study was to evaluate the effectiveness of a school-based dietary intervention program in increasing fruit and vegetable consumption among fourth-graders.[100] Twenty-eight elementary schools in the Birmingham, Alabama, metropolitan area were paired within three school districts based on ethnic composition and the proportion of students receiving free or reduced-price meals through the National School Lunch Program. One school in each pair was randomly assigned to an intervention group or a usual care control group. Assessments were completed at baseline (at the end of third grade), after year 1 (at the end of fourth grade), and after year 2 (at the end of fifth grade).

The intervention consisted of three components: classroom, parent, and food service. The classroom component of the intervention included 14 lessons, taught biweekly by trained curriculum coordinators with assistance from the regular classroom teachers. The parent component consisted of an overview during a kickoff meeting and completion of seven homework assignments by the parent and the child. Parents were also asked to encourage and support behavior change in their children. The food-service component consisted of food-service managers and workers receiving half-day training

Illustration 12.6 School Health Index Nutrition Services Score Card for Elementary Schools.[97]

School Heath Index – Elementary School

Module 4: Nutrition Services

Score Card (photocopy before using)

Instructions

1. Carefully read and discuss the Module 4 Questionnaire, which contains questions and scoring descriptions for each item listed on this Score Card.
2. Circle the most appropriate score for each item.
3. After all questions have been scored, calculate the overall Module Score and complete the Module 4 Planning Questions located at the end of this module.

		Fully in place	Partially in place	Under Development	Not in place
4.1	Breakfast and lunch programs	3	2	1	0
4.2	Variety of foods in school meals	3	2	1	0
4.3	Low-fat and skim milk available	3	2	1	0
4.4	Meals include appealing, low-fat items	3	2	1	0
4.5	A la carte offerings include appealing, low-fat items	3	2	1	0
4.6	Sites outside the cafeteria include appealing, low-fat items	3	2	1	0
4.7	Food purchasing and preparation practices to reduce fat content	3	2	1	0
4.8	Promote healthy cafeteria selections	3	2	1	0
4.9	Clean, safe, pleasant cafeteria	3	2	1	0
4.10	Preparedness for food emergencies	3	2	1	0
4.11	Collaboration between food service staff and teachers	3	2	1	0
4.12	Degree and certification for food service manager	3	2	1	0
4.13	Professional development for food service manager	3	2	1	0

Column Totals: For each column, add up the numbers that are circled and enter the sum in this row

Total points: Add the four sums above and enter the total to the right.

Module Score = (Total points/39) x 100   %

by High 5 nutritionists in purchasing, preparing, and promoting fruit and vegetables within the High 5 guidelines. Data analyzed included 24-hour recalls from the students, cafeteria observations, psychosocial measures, and parent measures.[100]

Results indicate that mean daily consumption of fruits and vegetables was higher at year 1 follow-up (3.96 vs. 2.28) and year 2 follow-up (3.2 vs. 2.21) for the intervention group as compared to the controls. At year 1 follow-up, the mean daily consumption of fruits and vegetables was higher for the intervention parents as compared to control parents, but no difference was found at year 2 follow-up. The intervention was found to be effective in subsamples, suggesting that the program can be used with boys and girls; African Americans and European Americans; low-, middle-, and higher-income families; and parents of low, medium, and high educational levels. The intervention was found to be effective at changing the fruit and vegetable consumption of fourth-grade students. Future studies are recommended to enhance the effectiveness of the intervention in changing parents' consumption patterns and to test the effectiveness of the intervention when delivered by the regular classroom teachers.[100]

Public Food and Nutrition Programs

"It is the position of the American Dietetic Association that all children and adolescents, regardless of age, sex, socioeconomic status, racial diversity, ethnic diversity, linguistic diversity, or health status, should have access to food and nutrition programs that ensure the availability of a safe and adequate food supply that promotes optimal physical, cognitive, social, and emotional growth and development. [101]

The American Dietetic Association

National child nutrition programs, which have had a federal legislative basis since 1946, contribute significantly to the food intake of school-age children. The purpose of the child nutrition programs is to provide nutritious meals to all children. These programs can also reinforce nutrition education, which takes place in the classroom. Increasing the proportion of children and adolescents ages 6 to 19 years whose overall dietary quality is enhanced by meals and snacks at schools is addressed in one of the Healthy People 2010 objectives.[6] Since the

Table 12.9 Normal Values of Biochemical Nutritional Parameters[31,98]

Test	Normal Values
Iron Status	
Hematocrit, %	39
Hemoglobin, g/dl	14
Serum ferritin, ng/mL	>15
Serum iron, mcg/dl	>60
Serum total iron binding capacity, mcg/dl	350–400
Serum transferrin saturation, %	>16
Serum transferring, mg/dl	170–250
Erythrocyte prtoporphyrin, mcg/dl red blood cells	>70
Dyslipidemia Screen	
Total cholesterol	<170 mg/dl
LDL-cholesterol	<110 mg/dl

SOURCE: From the Institute of Medicine, Food & Nutrition Board

school year 2006–2007, schools are required to develop a wellness plan that includes specific nutrition guidelines for all foods in the school, including competitive foods.[102]

Child nutrition programs include the National School Lunch Program, School Breakfast Program, Summer Feeding Program, Special Milk Program, After-School Snack Program, Team Nutrition, and the National Food Service Management Institute at the University of Mississippi.[102] Descriptions of several of these programs follow.

Commodity Program A USDA program in which food products are sent to schools for use in the child nutrition programs. Commodities are usually acquired for farm price support and surplus-removal reasons.[93]

The National School Lunch Program

The federal government provides financial assistance to schools participating in the National School Lunch Program (NSLP) through cash reimbursements for all lunches served, with additional cash for lunches served to needy children, and through commodities.[102,103] Schools must meet five major requirements in order to participate in the NSLP:

1. Lunches must be based on nutritional standards.
2. Children who are unable to pay for lunches must receive lunches for free or at a reduced price, with no discrimination between paying and nonpaying children.
3. The programs operate on a nonprofit basis.
4. The programs must be accountable.
5. Schools must participate in the *commodity program*.

School lunches must provide one-third of the Dietary Reference Intakes for the age/grade group of children

being served for energy, protein, calcium, iron, vitamin A, and vitamin C and must be consistent with the most recent version of the U.S. Dietary Guidelines for Americans, especially in regard to total fat and saturated fat, sodium, and cholesterol, when analyzed over a week's time.[27,102,103] Special emphasis is placed on serving a variety of foods and having menus that contain a variety of fruits and vegetables, low-fat dairy products, and lean meats. In addition, school food-service personnel must make food safety a priority. These programs must also meet the needs of children with disabilities and special health care needs (see Chapter 13). Although not federally mandated, schools are encouraged to allow adequate time for children to eat their lunches. Schools receive payments from the federal government based on the number of meals served by category (paid, free, or reduced-price).

Schools participating in the National School Lunch Program can choose from among four menu-planning approaches to plan school lunches:[104]

1. *Traditional Food-Based Menu-Planning Approach* Schools using this approach must plan menus with specific component and quantity requirements by offering five food items from four food components (meat/meat alternate, vegetables and/or fruits, grains/breads, and milk). Minimum portion sizes are established by ages and grade groups. Table 12.10 shows the meal pattern for lunches.
2. *Enhanced Food-Based Menu-Planning Approach* A variation of the traditional food-based menu-planning approach, this approach increases calories from low-fat food sources and increases the weekly servings of vegetables and fruits and grains/breads, in order to meet the Dietary Guidelines. The five food components are retained.
3. *Nutrient-Standard Menu-Planning Approach* This is a computer-based menu-planning system that analyzes the specific nutrient content of menus, using USDA-approved computer software. This system allows more flexibility in planning, at the same time assuring that the nutrient standards are being met.
4. *Assisted Nutrient-Standard Menu-Planning Approach* This variation of the nutrient-standard menu-planning approach is for schools that do not have the necessary technical resources to conduct nutrient analyses. This approach allows schools to have an outside source plan and analyze menus based on the school's needs and preferences.

An additional provision of the National School Lunch Program allows states and school districts to use any reasonable approach to menu planning as long as the method is reviewed and approved by the state agency or the USDA.

Table 12.10 Traditional food-based menu planning approach: meal pattern for lunches[104]

Food Components and Food Items	Minimum Quantities Ages 9 and Older Grades 4–12	Recommended Quantities Ages 12 and Older Grades 7–12	Sample Menu Minimum Quantities for Ages 9 and Older
Milk (as a beverage)	8 fl oz	8 fl oz	8 fl oz Low-fat milk
Meat or Meat Alternative (quantity of the edible portion as served)			
Lean meat, poultry, or fish	2 oz	3 oz	2 oz Hamburger patty
Alternate Protein Products	2 oz	3 oz	
Cheese	2 oz	3 oz	
Large egg	1	½ c	
Cooked dry beans or peas	½ c	¾ c	
Peanut butter or other nut or seed butters	4 Tbsp	6 Tbsp	
Yogurt, plain or flavored, unsweetened or sweetened	8 oz or 1 c	12 oz or 1½ c	
The following may be used to meet no more than 50% of the requirement and must be used in combination with any of the above: peanuts, soy nuts, tree nuts, or seeds, as listed in program guidance, or an equivalent quantity of any combination of the above meat/meat alternative (1 oz of nuts/seeds = 1 oz of cooked lean meat, poultry, or fish)	1 oz = 50%	1½ oz = 50%	
Vegetable or Fruit			
2 or more servings of vegetables, fruits, or both	¾ c	¾ c	Lettuce, tomato, Carrot/raisin salad Fresh apple
Grains/Breads (servings per week)			
Must be enriched or whole grain. One serving is 1 slice of bread or an equivalent serving of biscuits, rolls, etc., or 1/2 c of cooked rice, macaroni, noodles, other pasta products, or cereal grains.	8 servings per week, (For the purposes of this table, a week = 5 days.) minimum of 1 serving per day	10 servings per week, minimum of 1 serving per day	Enriched hamburger bun Condiments

School Breakfast Program

First authorized as a pilot program in 1966, the School Breakfast Program is a voluntary federal program. Many state legislatures have mandated breakfast programs for their districts, especially in schools serving needy populations.[103] In general, the NSLP rules also apply to the School Breakfast Program. School breakfasts must provide one-fourth of the Dietary Reference Intakes for the children being served, based on age or grade groups, and comply with the U.S. Dietary Guidelines for Americans when analyzed over a week's time. Table 12.11 shows the traditional, food-based meal pattern for breakfast. It is a special challenge for schools to allow enough time for school breakfasts before school when most of the participating children arrive at about the same time. Currently, universal breakfast programs in elementary schools are being tested as pilot programs.

Summer Food Service Program

The Summer Food Service Program provides meals to children from needy areas when school is not in session. Schools, local government agencies, or other public and private nonprofit agencies operate these programs. The federal government gives financial assistance to these programs for providing meals in areas where 50% or more of the participating children are from families whose incomes are lower than 185% of the poverty level.[103] The Summer Food

Table 12.11 Traditional food-based menu planning approach—meal pattern for breakfasts[104]

Food Components and Food Items	Grades K–12	Sample Menu
Milk (fluid)		
As a beverage, on cereal, or both	8 fl oz	8 fl oz low-fat milk
Juice/Fruit/Vegetable		
Fruit and/or vegetable; or full-strength fruit juice	½ c	½ c Orange juice or vegetable juice
Select one serving from each of the following components, two from one component, or an equivalent combination.		
Grains/Breads		
Whole-grain or enriched bread	1 slice	1 slice Enriched toast, butter, jelly
Whole-grain or enriched biscuit, roll, muffin, etc.	1 serving	
Whole-grain, enriched or fortified cereal	¾ c or 1 oz	¾ c Raisin Bran cereal
Meat or Meat Alternatives	1 oz	
Meat/poultry or fish	1 oz	
Alternate protein products	1 oz	
Cheese	1 oz	
Large egg	½	
Peanut butter or other nut or seed butters	2 Tbsp	
Cooked dry beans and peas	4 Tbsp	
Nuts and/or seeds (as listed in program guidance)	1 oz	
Yogurt, plain or flavored, unsweetened or sweetened	4 oz or ½ c	

Service Program is an important source of food for many children from food-insecure families.

Team Nutrition

Team Nutrition is a program of the USDA's Food and Nutrition Service. The program is aimed at improving children's lifelong eating and physical activity habits through application of information in the Dietary Guidelines for Americans and MyPyramid.[17,27,105] Team Nutrition is a partnership of public and private organizations, including private sector companies, nonprofit organizations, and advocacy groups that are interested in improving the health of the nation's children. Team Nutrition is an excellent example of a program that addresses the establishment of healthy eating and physical activity patterns for children on multiple fronts.

Team Nutrition operates through three behavior-oriented strategies:

- Provide training and technical assistance for child-nutrition food-service professionals to help them serve meals that meet nutrition standards while tasting and looking good.
- Provide integrated nutrition education for children and their parents with the goal of establishing healthy food and physical activity choices as part of a healthy lifestyle.
- Provide support for healthy eating and physical activity by involving community partners, including school administrators and other school and community partners.

Six communication channels are utilized: (1) food-service initiatives, (2) classroom activities, (3) school-wide events, (4) home activities, (5) community programs and events, and (6) media events and coverage.[105] Schools are recruited to become Team Nutrition schools with the benefit of receiving a resource kit of materials. Additional information is available on the Team Nutrition website at www.fns.usda.gov/tn. The site includes activities for educators, parents, and students.[105]

Key Points

1. School-age and preadolescent children continue to grow at a slow, steady rate until the adolescent growth spurt.
2. Monitoring BMI-for-age percentiles is important for screening for overweight or underweight.
3. Family mealtimes should be encouraged, as there is a positive relationship between families eating together and the overall quality of the child's diet.

4. A child's food choices are being influenced by peers, teachers, coaches, the media, and the Internet.

5. The prevalence of overweight and obesity continue to increase in the school-age and preadolescent groups.

6. Complications of overweight and obesity in children and adolescents, such as type 2 diabetes mellitus, are increasing.

7. Sedentary lifestyles and limited physical activity are contributing factors to the increase in childhood overweight.

8. School-age and preadolescent children are encouraged to eat a variety of foods and increase physical activity as outlined by the U.S. Dietary Guidelines for Americans and MyPyramid.

9. Consumption of sweetened soft drinks, which increases as children get older, is associated with increased calorie consumption and poorer diet quality.

10. Schools play an important community role in promoting healthy nutrition and physical activity patterns for children and adolescents.

Review Questions

Timothy in Case Study 12.1 returns to his pediatrician 6 months later at 7½ years of age for a weight check. He has gained 10 kg in 6 months and now weighs 90 pounds (41 kg). His height has increased to 51 inches (130 cm).

1. Calculate Timothy's BMI.

2. What is Timothy's BMI-for-age percentile? Use CDC growth curves to determine percentile.

3. Which of the following classifies Timothy's BMI-for-age percentile?
   a. Underweight
   b. Normal weight
   c. Overweight
   d. Obese
   e. >99th percentile cutoff point

4. Based on the staged treatment plan for pediatric obesity, describe the recommended approach to treatment for Timothy.

5. Regarding fluid intake in school-age children, which of the following statements is most correct?
   a. Soft drinks are good choices for hydration during exercise because the high carbohydrate content provides quick energy.
   b. Children's caffeine consumption decreases as they get older.
   c. Children who drink more soft drinks have higher intakes of calcium than do children who drink more milk.
   d. Diluted fruit juice is a good choice for hydration during exercise, as children are more likely to drink it than water.
   e. Adults who are supervising children during exercise can rely on the children to let them know when they are thirsty.

6. Name two strategies for increasing a school-age child's physical activity.

7. Name two strategies for decreasing a school-age child's sedentary behaviors.

8. Regarding child nutrition programs, which of the following statements is most correct?
   a. The nutritional analysis of the School Lunch Program includes à la carte items.
   b. "Pouring rights" means that children who participate in the School Lunch Program have the right to serve themselves beverages.
   c. Meals served as part of the School Lunch Program must provide 1/3 of the Dietary Reference Intakes of key nutrients for the age of the children being served.
   d. Meals served as part of the School Breakfast Program must provide 1/3 of the Dietary Reference Intakes of key nutrients for the age of the children being served.
   e. Team Nutrition refers to the component of the national child nutrition program that provides meals for student athletes.

Resources

The American Dietetic Association, Diabetes Care and Education Dietetic Practice Group

This series of booklets addresses the food practices, customs, and holiday foods of various ethnic groups.

Ethnic and Regional Food Practices: A Series. Chicago, IL: The American Dietetic Association, 1994–1999.

Bright Futures

The Bright Futures publications are developmentally based guidelines for health supervision that address the physical, mental, cognitive, and social development of infants, children, and adolescents and their families and provide practical guidelines for health care professionals who care for this population.

Four implemenation guides have also been published to date, addressing oral health, mental health, general nutrition, and physical activity.
Website: **www.brightfutures.aap.org**

Celebrating Diversity

The purpose of this publication is to assist health professionals in learning to communicate effectively with a diverse clientele.

Graves, D. E. and Suitor, C. W. *Celebrating Diversity: Approaching Families Through Their Food,* 2nd ed. Arlington, VA: National Center for Education in Maternal and Child Health, 1998.

The Center for Health and Health Care in Schools

This center, housed at the George Washington University School of Public Health and Health Services, is committed to achieving better health outcomes for children and adolescents through school health programs and services. Resources and fact sheets on a variety of nutrition topics, including childhood obesity, are available for educators and families.
Website: **www.healthinschools.org**

Centers for Disease Control and Prevention, National Center for Health Statistics

The CDC website provides background information on the growth charts. Also, individual growth charts can be downloaded and printed from this web page.
Website: **www.cdc.gov/nchs/about/major/nhanes/growth-charts/charts.htm**

Child Nutrition Programs

The national child nutrition programs website provides information on all programs, including the National School Lunch Program, SchoolBreakfast Program, Special Milk Program, Summer Food Service Program, and Child and Adult Care Food Program.
Website: **www.fns.usda.gov/cnd**

Dietary Reference Intakes: National Academy Press

Over 2000 books are available online at this site, free of charge, including material on the current Dietary Reference Intakes.
Website: **www.nap.edu**

Food and Culture

This book covers culturally related food and nutrition topics. The traditional food habits of key ethnic, religious, and regional groups are comprehensively reviewed. Information about traditional health beliefs and practices is also given.

Kittler, P. G. and Sucher, K. P. *Food and Culture,* 3rd ed. Belmont, CA: Wadsworth/Thomson Learning, 2001.

Institute of Medicine Committee on Prevention of Obesity in Children and Youth

This website provides several reports on progress in preventing childhood obesity, including the 2005 report "Preventing Childhood Obesity: Health in the Balance." These reports provide comprehensive national strategies that recommend specific actions for families, schools, industry, communities, and government. They can be found under "Child Health" and "Reports" on the IOM website.
Website: **www.iom.edu**

Kidnetic

This colorful and interactive website is geared toward teaching older school-age children about healthy eating and exercise habits. Information for parents and health professionals and educators is also available. The website is supported by the International Food Information Council (IFIC) Foundation and was developed in partnership with professional organizations including the American Dietetic Association.
Website: **www.kidnetic.com and www.ific.org/kidnetic (leader and parents' guide)**

Kids Count

This website provides data on critical issues affecting at-risk children and their families. The publication *Kids Count* by the Annie E. Casey Foundation is also available online at this address.
Website: **www.kidscount.org**

KidsHealth

This website, supported by the Nemours Foundation, has sections for parents, kids (older school-age), and teens. It addresses all types of health concerns, including healthy eating and exercise.
Website: **www.KidsHealth.org**

National Center for Complementary and Alternative Medicine, National Institutes of Health

This website provides science-based information for consumers and practitioners as well as information about related news and events. Information is available on specific herbs, including available scientific research.
Website: **www.nccam.nih.gov**

National Safe Routes to School Program

This program assists communities in helping children safely walk or bike to school. The report entitled *Many Steps...One Tomorrow*, available at this website, gives examples of Safe Routes to School programs throughout the United States.
Website: **www.saferoutesinfo.org**

Pediatric Pulmonary Centers, Cross-Cultural Health Care Case Studies

The national group of Pediatric Pulmonary Centers, Maternal and Child Health Bureau-funded graduate-level training programs, developed this online independent study that explores issues related to cultural competence in health care through case studies. The case studies can be accessed free of charge at the following website.
Website: **www://ppc.mchtraining.net**

Team Nutrition

This website provides information on the USDA's Team Nutrition program, which is designed to help implement the Dietary Guidelines in child nutrition programs.
Website: **www.fns.usda.gov/tn**

U.S. Department of Agriculture/Agriculture Research Service Children's Nutrition Research Center

The Child Nutrition Research Center is housed at Baylor College of Medicine in the Texas Medical Center in Houston, TX. In addition to information about the ongoing research at the center, this website also includes some useful tools such as the "Kids' BMI Calculator" and the "Kids' Energy Needs Cal-

culator," which is for healthy children 2-20 years of age. When details about a child's age, height, weight, and activity level are entered, the calculator will estimate calorie needs and number of servings of different food groups needed for a healthy diet based on the Dietary Reference Intakes
Website: www.bcm.edu/cnrc

U.S. Department of Health and Human Services, Maternal and Child Health Bureau

The chartbook "Overweight and Physical Activity Among Children: A Portrait of States and the Nation 2005" presents national and state-level data on the prevalence of overweight in children and adolescents (ages 10–17) within the context of family structure, poverty level, parental health and habits, and community surroundings.
Website: www.mchb.hrsa.gov/overweight/

Walking School Bus

A "walking school bus" is a group of children walking to school with one or more adults. Basics of starting a walking school bus, including picking a safe route, can be found at this website. A good time to kick off the program is during International Walk to School Month in October.
Website: www.walkingschoolbus.org

WHO Growth References

Gender-specific charts of height-for-age and BMI-for-age z-scores from 5 to 19 years of age and weight-for-age z-scores from 5 to 10 years of age can be downloaded from this website.
*Website: **www.who.int/childgrowth***

Chapter 13
Child and Preadolescent Nutrition:
Conditions and Interventions

Prepared by **Janet Sugarman Isaacs**

Key Nutrition Concepts

1 Children are children first, even if they have conditions that affect their growth and nutritional requirements.

2 Common nutrition problems in children with chronic conditions are underweight and overweight, and difficulties in eating enough to meet nutrient requirements.

3 Children who have special health conditions receive more intensive nutrition services in schools and health care settings than other children do.

4 Family meal patterns and routines affect nutrition for children, so providing adequate support for families improves the nutritional status of children with chronic health conditions.

Introduction

Nutrition services need to be part of the goal to help a child reach his or her full potential. This chapter discusses the nutrition needs of children with chronic conditions, such as cystic fibrosis, diabetes mellitus, cerebral palsy, phenylketonuria (PKU), and behavioral disorders. Nutrition recommendations are based on those for children generally, but they may be modified by the condition and its consequences on growth, nutrient requirements, and/or eating abilities. Other factors such as activity that increase or decrease caloric needs are also discussed. Expanding school and community resources include nutrition services for children with special health care needs or those with developmental disabilities. Advocates for those with disabilities prefer "people-first language," which this chapter models. This convention names the person first and then the condition. An example is "a girl with Down syndrome" rather than "the Down's girl." Advocates for those with disabilities also prefer the word *disabilities* to the word *handicapped*. (Use of the term *handicap* comes from the practice of using a cap to beg.)

"Children Are Children First"—What Does that Mean?

Children with special health care needs are children first, even if their conditions change their nutrition, medical, and social needs. Children are expected to become more independent in making food choices, assisting with meal preparation, and participating at mealtime with other family members. These same expectations are appropriate for children with special health care needs. For example, the child with spina bifida should be encouraged to make a salad or set the table. Modifications may be needed to

help the child be successful, such as storing plates and utensils on low shelves, or lowering counter heights to accommodate a wheelchair at a kitchen sink. The cognitive developmental gains of childhood and participation in meal preparation and mealtime are the same for children with chronic conditions.

This concept has been acknowledged in schools through federal legislation in the Individuals with Disabilities Education Act (IDEA). This law requires the least restrictive environment and is resulting in inclusive settings for more children with disabilities.[1] This concept of inclusion has major ramifications for how children receive all types of services, such as schools providing alternative foods as required in the main cafeteria with all the other children. As a result of inclusion, children in wheelchairs or with Down syndrome spend time in regular classrooms with others of the same age. Nutritional problems related to food refusals, mealtime behavior, or special diets are being addressed in the neighborhood school, in the regular classroom, as often as possible.

This same concept of treating the child as a child first is recommended at home, too. A special diet is part of an overall treatment plan that incorporates normal developmental steps of childhood. Children with diabetes or PKU do not benefit from being treated in a special manner at mealtime. As soon as possible, children are taught to take responsibility for making food choices consistent with their diet plans. Consistency and structure in the home support a child's normal development. This structure includes regular meal and snack times, and the child accepting increasing levels of responsibility in preparing foods. These approaches lower the chance of the child being overprotected or manipulating adults because of her illness or its treatment.

The number of children with chronic conditions is increasing over time, with 15–18% having a condition expected to last 12 months or more at diagnosis.[2] Numbers of children with conditions such as ADHD, obesity, and autism are increasing more rapidly than those for conditions such as lead poisoning and simple iron-deficiency anemia, which are decreasing in incidence. General nutrition guidelines for children, school nutrition educational materials, and nutrition prevention strategies may or may not be applicable to children with specific conditions. Many nutrition education curricula appropriately target the goals of preventing overweight, lowering fat, and increasing fruit and vegetable intake. These curricula provide appropriate education, for the most part, to children with conditions such as diabetes and Down syndrome. Conditions in which slow weight gain and underweight are common, such as cerebral palsy, may not fit such curricula.[3] Nutrition education may not address such children's need for high-fat foods as part of high caloric needs. A child with PKU, for example, cannot ever have protein-rich meats or dairy products.[4] When the usual nutrition guidelines are discussed and encouraged, the child with

PKU may feel isolated and confused about whether his diet is healthy. When possible, children with special health care needs are encouraged to participate in school nutrition education programs, with modifications as needed.

"Train a child in the way he should go, and when he is old, he will not depart from it."

Proverbs

Nutritional Requirements of Children with Special Health Care Needs

Caloric and protein needs are lower on a body-weight basis in childhood than during the preadolescence, toddler years, and preschool years.[5] Children with special health care needs have a wide range of nutritional requirements and more variability than other children do, based on these factors:[6,7]

- Low caloric intake may be appropriate with small muscle size.
- High protein is needed with high protein losses, such as skin breakdown.
- High fluid volume is needed with frequent losses from vomiting or diarrhea.
- High fiber may be needed for chronic-constipation management.
- Long-term use of prescribed medications may increase or decrease vitamin or mineral requirements, or change the balance of vitamins and minerals needed as a result of medication side effects.
- Routine illness is more likely to result in hospitalization or resurgence of symptoms of the underlying disorder.

These factors may replace the U.S. Dietary Guidelines and the Healthy People 2010 objectives for most children with special health care needs.

Energy Needs

Children with special health care needs may need more, less, or the same caloric intake as other children of the same age. Energy needs are amazingly complex in children, let alone those with special needs. Under ideal conditions, the caloric needs are measured by indirect calorimetry, but they are usually estimated using standard calculations that cannot take into account the specific conditions involved.[8] Machines that measure indirect calorimetry are becoming available, but they still can give only estimates of energy needs at rest, without considerations of energy needed for activity and growth. Conditions that slow growth or decrease muscle size generally result in lower caloric needs.[6,9,10] Caloric needs in a child with Prader-Willi

syndrome may be only 66% of the caloric needs of a child of the same age and gender without the syndrome.[6]

Other factors that change energy requirements are related to activity level and frequency of illnesses.[5,6] Children with a chronic condition are encouraged to participate in age-appropriate sports activities. Conditions in which activity may be especially beneficial include diabetes and mild cerebral palsy. The level of activity of children with chronic conditions may be higher or lower than activity levels in other children. Children who are very active may appear thin as a result of low caloric intake. Children with autism and attention deficit hyperactivity disorder are generally more active than other children are, and/or they may sleep less.[11,12] Such a range in level of activity is addressed as a part of a thorough nutrition assessment. Questions such as "Is the child receiving physical therapy one or three times per week?" and "How much time does the child use a walker compared to a wheelchair?" are examples of how activity can be assessed in determining caloric needs.

Protein Needs

Protein needs also can be higher, lower, or the same as those for other children, based on the condition. Healing burns and cystic fibrosis are examples of disorders with high protein needs—at 150% of the DRI.[13] Conditions such as PKU and other protein-based inborn errors of metabolism require greatly reduced amounts of natural protein in the diet.[4] Children with diabetes mellitus do not have modified protein needs.[14] The importance of protein for wound healing and for maintaining a healthy immune system makes protein requirements key for various conditions with frequent illnesses or surgeries. For example, a child with cerebral palsy who is scheduled to have hip surgery would have protein needs evaluated in a complete nutritional assessment. Higher protein may be recommended for wound healing and the prevention of skin breakdown while in a cast after surgery.

Healthy children eat foods providing intact protein, but some conditions require hydrolyzed amino acids or specific amino acid mixtures rather than intact protein. Children with severe egg, soy, or milk allergies, chronic gastrointestinal conditions, or with inborn errors may be prescribed complete nutritional supplements because they fit modified protein diets.

Other Nutrients

DRIs are good starting places to assess the need for vitamins and minerals in chronic conditions.[15] (DRIs are listed inside the back cover of this textbook.) As in all children, if the diet provides sufficient foods to meet the needs for protein, fats, and carbohydrates, it is likely the vitamin and mineral needs are also met. However, children with chronic conditions may have more difficulty meeting the DRI for vitamins and minerals as a result of these considerations:[15,16]

- Eating or feeding problems may restrict intake of foods requiring chewing, such as meats, so that certain minerals may be low in the diet.
- Prescribed medications and their side effects can increase turnover for specific nutrients, raising the recommended amount needed.
- Food refusals are common with recurrent illness, so total intake may be more variable day to day than in other children of the same age.
- Treatment of the condition necessitates specific dietary restrictions, so that vitamins and minerals usually provided in restricted foods have to be supplemented.

Nutrients such as calcium that are low in the general population of children are also problem nutrients for children with chronic conditions.[17] The American Academy of Pediatrics statement on calcium applies to children with chronic conditions. This statement recommends good-quality food sources of calcium for all children. Food sources of calcium may avoid lead contamination that has been reported in a variety of over-the-counter calcium supplements.[18] Taking high levels of supplemental calcium does not clearly benefit children with chronic conditions.[17]

Growth Assessment

The Centers for Disease Control (CDC) 2000 growth charts are a good starting place for assessing the growth of any child. Identifying children at risk for overweight and preventing long-term cardiovascular risks are important purposes underlying growth assessments of children. Such concerns may or may not apply to children with chronic conditions, but a nutrition assessment can tailor nutrition goals to specific conditions. Families dealing with conditions that shorten life, such as cystic fibrosis, would not benefit from including overweight and its long-term risks as part of the child's nutrition assessment.

Families of children with severe disabilities who have wheelchairs may be apprehensive about growth as a goal for the child. They may have long-term concerns about caring for the child at home. Families may not want the child to grow at the usual rate when activities such as lifting him or her from the bathtub or out of a wheelchair may become more difficult. Also, children with rare degenerative conditions such as *spinal muscular atrophy* have such major decreases in muscle size that growth may not occur.[10]

Most children with chronic conditions do grow, and assessing growth is an important component of nutrition services. If the child's condition is known to change the rate of weight or height gain—either slowing or accelerating it—the following signs need attention regardless of what growth chart is used:

- A plateau in weight
- A pattern of gain and then weight loss
- Not regaining weight lost during an illness
- A pattern of unexplained and unintentional weight gain

Growth Assessment and Interpretation in Children with Chronic Conditions

Factors that affect growth assessment and interpretation in childhood may not have been detectable earlier in younger children. These factors are the age of onset of the condition, *secondary conditions*, and activity. The child's age when the condition started may influence whether CDC growth charts are applicable. Early onset is more likely than later onset to affect growth in conditions such as *seizures*.[10] If the seizures started in middle childhood, the standard growth chart may be appropriate because the child's growth pattern is already established. Onset of seizures in the neonatal period may reflect more severe brain damage, which markedly slows growth rate. Then the child's own growth record over time would be the best indicator of future growth.

Toddlers and preschoolers with cerebral palsy usually do not develop secondary conditions until childhood or later. *Scoliosis* is a secondary condition that interferes with accurate measurement of stature.[6] It may develop as a result of muscle incoordination and weakness in some forms of cerebral palsy in preadolescence. If a child with cerebral palsy has stature measurements that plateau or decline, it may be a result of cerebral palsy, scoliosis, lack of adequate nutritional intake, or a combination of these three factors.[10] Nutritional interventions cannot prevent scoliosis, although nutritional consequences of its treatment may arise. Children may be provided custom-fitted back braces, so weight gain means the brace needs to be replaced. Children with scoliosis braces also may become less active because the brace restricts some types of movement. If scoliosis surgery is performed, the child may become slightly taller immediately, again showing that stature measurements have to be interpreted with care.

Body Composition and Growth

Children with special health care needs may or may not be typically proportioned in

Spinal Muscular Atrophy Condition in which muscle control declines over time as a result of nerve loss, causing death in childhood.

Secondary Condition Common consequence of a condition, which may or may not be preventable over time.

Seizures Condition in which electrical nerve transmission in the brain is disrupted, resulting in periods of loss of function that vary in severity.

Scoliosis Condition in which the vertebral bones in the back show a side-to-side curve, resulting in a shorter stature than expected if the back were straight.

muscle size, bone structure, and fat stores. Some children with good nutritional status may plot at or below the lowest percentile on a standard growth chart for height.[19] In fact, low-percentile heights are usual for a child with Down syndrome if growth is plotted on the CDC chart rather than the special growth chart for Down syndrome.[20] Short stature, low muscle tone, and low weight compared to age-matched peers are not attributed to caloric intake. They characterize the natural consequences of the *neuromuscular* changes within Down syndrome. Similarly, a child with low muscle size could have a low weight and short stature. It would be unfair to assume that the child's diet is inadequate because of the low weight. A thorough assessment that includes body composition is necessary. For example, a thin-appearing child needs to have body fat stores measured before diet recommendations are made. If body fat stores are fine, adding calories is more likely to contribute to overweight.

Children with small muscle size will have lower weights than those with regular-sized muscles.[8,10] Conditions with altered muscle size may be described using terms such as *hypotonia* or *hypertonia*. Examples include cerebral palsy, Down syndrome, and spina bifida.[10,20] Not all muscles are affected. For example, some children with spina bifida have larger muscle size in the upper body and smaller muscle size in the lower body. Variation in size of muscles may make growth interpretation more difficult. Any assessment must address risks for overweight, such as body mass index (BMI) and adiposity rebound. Standard interpretation may suggest a risk of overweight, but it may not accurately reflect that short stature is part of the child's condition. By standard interpretation, every child with Down syndrome or spina bifida could be overweight (see Case Study 13.1, page 343). For now, no established BMI tables cover specific conditions or the appropriate time for adiposity rebound.

Measuring body fat is another indicator of body composition. Skinfold fat measurements and their interpretation have to be based on consistent and repeatable standard methods.[21,22] Measuring fat stores in children is not like measuring fat stores in adults, because of the changes in body composition that come with age and growth. Calculated formulas and methods for evaluating body composition in children are not the same as those for adults.[8,23] Estimates of body composition for children with chronic conditions may be based on smaller sample sizes than for other children, but still such information is helpful. Identified low fat stores trigger recommendations to boost calories.

In-depth growth assessment may include head circumference measurement for all ages, with plotting and interpretation based on the Nellhaus head circumference growth chart, as discussed in Chapter 9.[24] Head circumference is important because

children with unusually small heads have smaller brains, a characteristic associated with short stature. Even with adequate diet and no documented eating problems, children with various genetic disorders tend to be shorter than age-matched peers.[20,25]

Special Growth Charts   Special growth charts have been published for a variety of genetic conditions.[20] Table 13.1 includes examples of these special growth charts. The number of children reported in such growth charts is not as large, nor as representative, as in the CDC 2000 growth charts. Special growth charts are revised often, based on new information emerging about the natural course of rare conditions. Some special growth charts are based on only the most severe forms of the condition, such as for children living in residential care. Many chronic conditions do not have special growth charts because they present with a wide range of severity. Conditions without a specialty growth chart, for which individuals may or may not match the standard growth charts, include the following:

- *Juvenile rheumatoid arthritis*
- Cystic fibrosis
- Rett syndrome
- Spina bifida
- Seizures
- Diabetes

Nutrition Recommendations

Children with chronic conditions require nutrition assessments to determine whether they are meeting their nutrient and caloric needs, whether eating problems such as food refusals or mealtime behavior are interfering with

Neuromuscular   Term pertaining to the central nervous system's control of muscle coordination and movement.

Juvenile Rheumatoid Arthritis   Condition in which joints become enlarged and painful as a result of the immune system; generally occurs in children or teens.

Table 13.1 Examples of specialty growth charts[2,20]

Conditions with Special Growth Charts	Comment
Achrondroplasia	Form of dwarfism
Down syndrome Trisomy13 Trisomy 18	Short stature, variable weight
Fragile X syndrome	Short stature, primary in males
Prader-Willi syndrome	Short stature, overweight
Rubinstein-Tabyi syndrome	Short stature
Sickle-cell disease	Short stature
Turner syndrome	Short stature
Spastic quadriplegia	Short stature, low weight
Marfan syndrome	Tall stature

Case Study 13.1

Adjusting Caloric Intake for a Child with Spina Bifida

This case study is about appropriate nutrition interventions, monitoring, and evaluation within the Nutrition Care Process. The nutrition diagnosis is excess energy intake related to spina bifida as evidenced by weight gain over the last year.

Sam is a third-grader in regular classes at his public school. He uses a wheelchair all the time and can transfer from his wheelchair to a chair by himself. He is on a toileting schedule at school with the assistance of a nurse. He participates in modified physical education as part of his physical therapy treatment. He likes to eat with his friends at school. His mother tries to make him cut back at the evening meal and has stopped buying some of his favorite snacks. He is mad at his mother because he likes his snacks after school when he is bored.

Nutrition assessment from Sam's last visit at the spina bifida clinic at the local hospital showed that he was overweight by measuring his fat stores. Because he cannot stand, his stature was estimated by measuring his length lying down and comparing it with his last length measurement. Standard methods could not be used to measure him, which limits the interpretation of his growth using the CDC growth chart. The chart showed Sam at the 75th percentile in weight for his age, which is not overweight for his age. His rate of weight gain of 8 pounds per year, typical for a boy his age, is too fast for his low level of activity. His estimated calorie needs are 1100 per day due to low activity and short stature, or about two-thirds of the caloric needs of others his age. Sam says he does not care about his size or being overweight. His mother is quite concerned that she would not be able to assist him if he fell or needed to be lifted.

Intervention Recommendations: The nutritionist at the clinic completes a school lunch prescription to reduce Sam's caloric intake from 650 calories to 350 calories per lunch. His meal pattern is adjusted to two meals (breakfast and lunch) and two snacks per day at home, which better fits his low caloric needs. Sam is allowed to choose his favorite snack foods to replace his evening meal. Giving him choices about his snacks increases Sam's sense of being in control and lowers the instances of expressing anger at his mother about snack foods. The clinic nutritionist calls the school to review Sam's level of activity and confirm that the lunch changes are being implemented. The physical therapist at school has found after-school swimming lessons and recommends them to Sam's mother as a way to increase his activity and socialization.

Nutrition Monitoring Recommendations: To motivate Sam to pay attention to his eating and weight gain, his teacher and his mother set up a monthly non-food reward for him if he does not gain any weight. The effectiveness of the plan to cut Sam's caloric intake and increase his activity will be assessed at his next clinic visit, when he will be weighed and have his fat stores measured.

Questions

1. Since Sam does not care about his size or being overweight, why is a diet plan necessary?

2. What are the risks from Sam's weight, since he is only at the 75th percentile for his age on the standard growth chart?

3. Will Sam grow taller when he goes though puberty and be able to eat more calories each day?

(Answers are located in the Instructor's Manual for the 4th edition of *Nutrition Through the Life Cycle*.)

meeting nutritional needs, and whether growth is on target for their age and gender. Then nutrition interventions are provided based on the assessment. The goal is for the child to maintain good nutritional status, and to prevent nutrition-related problems from being superimposed on the primary condition. (See Case Study 13.1.)

Children with special health care needs benefit from the same nutritional recommendations other children do, particularly in general areas such as dietary fiber or appropriate use of soft drinks. However, children with special health care needs may require particular formulas and nutrition support not needed for most children. Most children develop feeding skills during the toddler and preschool years; by childhood, abilities and/or disabilities that limit self-feeding and using utensils may require more aggressive support. Nutritional supports common for children are enteral supplements, when oral feeding of regular foods is insufficient in quality or amount to maintain health and to assure growth is not being limited. Table 13.2 provides a list of commonly used complete nutritional supplements and examples of their use. Children under 10 years of age are generally provided with a pediatric formula, but adult formulas may be used for children.

Methods of Meeting Nutritional Requirements

Children with special health care needs who cannot meet their nutrient requirements from regular foods may receive complete nutritional supplements in addition to meals or for partial or complete replacement for meals. The first choice is that required supplements are drunk or eaten in the usual way. If this method does not work out, complete nutritional supplements can be administered by placement of a feeding *gastrostomy*.[6] Gastrostomy feeding may be required in children with kidney diseases, some forms of cancer, and severe forms of cerebral palsy and cystic fibrosis.[26–28] Many families experience difficulty accepting a gastrostomy for meeting nutritional requirements because feeding is such an important aspect of parenting.[29] Aside from emotional aspects, insurance coverage and financial questions of paying for formulas fed by gastrostomies are major concerns for some families.

Children fed by gastrostomy can have many different schedules, such as eating orally during school and being fed by gastrostomy overnight. Table 13.3 gives an example of a feeding plan that includes gastrostomy feeding and oral feeding. If medications are required, they can be given through the gastrostomy also. For example, for children with pediatric AIDS who require many medications during the day, compliance with taking the drugs

Table 13.2 Examples of nutritional supplements and formula for children[26]

Formula	Comments
Pediatric versions of complete nutritional supplements, such as Pediasure	Generally recommended for children under 10 years of age; can be used for gastrostomy or oral nutrition support
Adult complete nutritional supplements, such as Ensure	Generally 1 calorie per milliliter is recommended for children
Enrichment of beverages, such as Carnation Instant Breakfast added to milk	Requires that milk is tolerated
Predigested formula with amino acids and medium-chain fatty acids, such as Peptamen Junior	For conditions in which intestinal absorption may be impaired
Special formulas for inborn errors of metabolism (PKU), such as Phenex-2	Usually a powder that is mixed as a beverage, but other forms such as bars and capsules are available
High-calorie booster for cystic fibrosis, such as Scandishake	Generally 2.5 calories per milliliter to concentrate calories in small volume

Table 13.3 Example of a feeding and eating schedule for an 8-year-old who eats by mouth and by gastrostomy

Daily Schedule	Comments
6:30 a.m. Night feeding pump turned off	Overnight feeding by gastrostomy runs from 9:30 p.m. until 6:30 a.m., providing about 3 fl oz per hour, so no hunger in the morning is common
7:15 a.m. Breakfast: refused	
8:00 a.m. Bus to school	
11:30 a.m. School lunch offered and about half is eaten: ½ chicken sandwich, all of french fries, with ⅓ pint of whole milk	Child has slow eating pace and is easily distracted by school lunchroom sounds
3:30 p.m. After-school snack at home of 4 oz pudding cup, two plain cookies, and 4 fl oz orange drink	Mealtime behavior at home includes many attempts to leave the table, with prompting to eat from parents
6:30 p.m. Evening meal at home: ½ cup mashed potato, 6 fl oz whole milk, refused vegetable and meat	Parents hook up night feeding pump while the child is sleeping
8:30 p.m. Bedtime	

improved after gastrostomies were placed. The parents spend less time trying to administer the medications to their children, and some children's health improved as a result of taking all of the required medications. Another example is a child who cannot safely drink liquids as a result of cerebral palsy. The child could have fluids given by gastrostomy but eat solid foods by mouth. Children with gastrostomies can swim, bathe, and do any activities they could do before the gastrostomy was placed.

Most formulas fed by gastrostomy can be consumed as beverages. Even regular foods can be blended in a recipe for gastrostomy feeding for some children. Such "home brews" for gastrostomy feeding have to be carefully monitored because they are a rich medium for bacterial contamination. Part of the decision-making process about use of special formulas and gastrostomy feeding often hinges on prior rejection of other feeding methods. When possible, gastrostomy feeding is planned as a temporary measure, with a return to oral eating later. For example, a child who has a gastrostomy because of a kidney condition may have a kidney transplant that allows removal of the gastrostomy after recovery.

Other nutrition supplements fed by gastrostomy have specific components that are unusual in beverages because they have such a strange taste. For example, formulas that contain individual amino acids generally are accepted only by those who have had them from infancy, as in the formulas for PKU. If a child required a new formula with amino acids, gastrostomy feeding would be more successful than oral feeding in most cases.

Vitamin and Mineral Supplements for Chronic Conditions   Children's complete vitamin and mineral supplements are recommended for a variety of chronic conditions to assure that the DRIs for essential nutrients are provided. However, most over-the-counter supplements are in the form of chewable tablets, so children who cannot chew well may require a liquid form of vitamins and minerals. The composition of vitamin and mineral supplements may be important because some have added ingredients not recommended for certain chronic conditions. Examples are vitamin and mineral brands with added carbohydrates, which are not allowed on a *ketogenic diet*, or those made with an artificial sweetener containing phenylalanine not recommended for children with PKU. (The ketogenic diet and PKU are discussed later in this chapter.)

The underlying diagnosis can make specific nutrients so important in the diet that they may be prescribed as pharmaceuticals. Cystic fibrosis treatment (discussed later in this chapter) requires fat-soluble vitamin supplements due to poor intestinal absorption of these nutrients. Vitamins A, D, E, and K are needed in cystic fibrosis. Vitamin B_{12} injections are needed for some protein-based inborn errors of metabolism.[4,13] Vitamin C may be prescribed above the DRI for some children with spina bifida who have frequent bladder infections.[6] The high dose of vitamin C functions as a medication rather than as a nutrient in this instance.

Use of excessive levels of vitamins and minerals can be risky, especially for underweight children. For example, a child may be counseled to add Carnation Instant Breakfast to a diet, while another provider adds a complete nutritional supplement and is unaware that the child takes a chewable children's vitamin/mineral tablet, too. Determination of the total intake of supplemented nutrients is part of a nutrition assessment. All medications, prescribed and over-the-counter, are identified within the nutrition and medical care plans.

Children with chronic conditions that limit activity or require medications that affect bone growth need special attention regarding calcium and vitamin D requirements.[17] *Galactosemia*, in which dairy products are eliminated from the diet, and cerebral palsy are two examples of conditions affecting calcium. Some children with these conditions are like older women with *osteoporosis* in that their calcium may move out of bones faster than it goes in. Providing additional calcium, phosphorus, and vitamin D may be recommended. Selecting calcium supplements that can be taken for years or decades by children raises concerns about the lead content found in some calcium supplements.[18]

Ketogenic Diet   High-fat, low-carbohydrate meal plan in which ketones are made from metabolic pathways used in converting fat as a source of energy.

Galactosemia   A rare genetic condition of carbohydrate metabolism in which a blocked or inactive enzyme does not allow breakdown of galactose. It can cause serious illness if not identified and treated soon after birth.

Osteoporosis   Condition in which low bone density or weak bone structure leads to an increased risk of bone fracture.

Fluids

Guidelines for fluids for all children are appropriate. Particular considerations for children with special health care needs are high fluid losses—for example, from drooling as a result of cerebral palsy—or behaviors that result in low fluid intakes.[6] Because constipation is common in children with neuromuscular disorders, adequate fluids are often stressed as part of a bowel management program. Children with limited ability to talk may have more difficulty indicating thirst. Many chronic health conditions carry higher risks for dehydration due to side effects of prescribed medications. A chronic condition generally does not change the fluid requirement when the child is well.

Eating and Feeding Problems in Children with Special Health Care Needs

Eating and feeding problems are diagnosed when children have difficulty accepting foods, chewing them safely, or ingesting sufficient foods and beverages to meet their nutritional requirements. About 70% of children with developmental delays have feeding difficulties, independent of whether neuromuscular problems have been identified.[6] Examples of these feeding problems include the following situations:

- Self-feeding skills are lower than the child's chronological age, requiring assistance and supervision to ensure adequate intake.
- Meals take so long or so much food is lost in the process of eating that the actual food intake is too low.
- The condition requires adjustment in the timing of meals and snacks at home and at school.

In children who do not have developmental delay, the impact of chronic conditions on eating may include behavioral problems at mealtimes, conflicts about control over food choices, and variability in appetite. Families of children with chronic conditions may focus on mealtimes and foods as methods of coping with their concerns about the child's future. For example, families may be overprotective and restrict a child from eating at friends' homes, when such activities may be appropriate for social development.

Specific Disorders

Cystic Fibrosis   Cystic fibrosis (CF) is one of the most common lethal genetic disorders, with an incidence of 1 in 1500–2000 live births.[13] It is highest among Caucasians, with an incidence of 1 in 17,000 live births for African Americans. The CF gene is located on the long arm of chromosome 7, and it has many different genetic versions. The most common genetic mutation characterizes 67% of the cases. CF affects all the exocrine functions in the body, with lung complications often causing death. Its major nutrition-related consequence is malabsorption of various nutrients due to the lack of pancreatic enzymes. This can result in a slower rate of weight and height gain, and higher energy needs due to chronic lung infections. Children with cystic fibrosis are likely to develop malnutrition as the condition progresses. Intensive nutrition interventions may be required to meet higher caloric needs.[26,27]

Nutrition interventions for CF include monitoring growth, assessing dietary intake, and increasing calories and protein by two to four times the usual recommendations to compensate for malabsorption. Every time a child with CF eats a meal or snack, he must take pills containing enzymes. Frequent eating and large, calorie-dense meals are encouraged. Gastrostomy feeding at night to boost calories is sometimes required. Vitamin and mineral supplementation, particularly fat-soluble vitamins, is a part of daily management. Children with CF are at risk for developing diabetes because the pancreas is a target organ of CF damage.[13] In recent years children diagnosed with CF have achieved longer life expectancies, and many have survived into the young adult years.

Nutrition experts working with children who have CF struggle to balance the children's high nutrient needs and frequent illnesses. Many children with CF have slow growth and are lower in weight and shorter than expected. Even with nutrition support, decline in pulmonary function over time continues. Some children with CF have lung transplants if they meet strict eligibility requirements. CF is on the leading edge of gene therapy research, giving hope to families with young children.

Diabetes Mellitus   Diabetes mellitus is a disorder of glucose metabolism and *insulin* regulation in which dietary management is crucial.[15,30] Early signs of glucose regulation dysfunction, such as insulin resistance, have been found in the childhood years in those born preterm and in those with unusual patterns of catch-up growth.[31] Type 1 diabetes is related to immune function and results in virtually no insulin production. Children with type 1 diabetes have both high and low blood sugars during diabetes management, not just high blood sugars, as in type 2 diabetes.[15] Increasingly, children with type 1 diabetes mellitus are managed with insulin pumps set to inject small amounts of insulin continuously and in response to meals rather than several insulin injections each day.[15] Use of continuous insulin pumps changes dietary management guidelines from those for individuals with type 1 diabetes managed with conventional treatment and also from those for type 2 diabetes. For type 2 diabetes, which is more common in older children and teens, some insulin may be produced in the body.[15,30] Children of Pima Indians in Arizona have a high incidence of type 2 diabetes, at 22 per 1000 children as compared to the 7.2 cases per 100,000 children seen in Ohio.[14]

Treatment for diabetes is regulation of the timing and composition of meals and exercise, along with insulin injections or medications.[14] Type 1 diabetes requires families and children to master a carbohydrate counting system.[14] This is not the same as the diabetes exchange system, but a refinement focused only on the carbohydrate content of foods, as that is what is used to adjust insulin doses accompanying meals. Appendix D illustrates the carbohydrate counting system for a child with diabetes type 1 managed with an insulin pump. A third-grader with type 1 diabetes requires a school diabetes management plan describing oversight and modification of school breakfast, school lunch, and snack time based on physical activity in school and after-school activities. If the child is invited to a birthday party, the timing of meals and snacks can be adjusted to allow the child to attend the party and eat most of the foods there. Common colds, or foods a child refuses to eat, can cause wide variation in blood

Insulin   Hormone usually produced in the pancreas to regulate movement of glucose from the bloodstream into cells within organs and muscles.

sugar, contributing to irritability, sleepiness, or difficulty with schoolwork.

In the summertime, many localities organize summer camps for children with diabetes; diet education and controlled access to a diabetic diet are provided along with the usual camp activities. Such disease-specific camps are good for breaking the social isolation that children experience when they feel they are the only ones required to follow special diets.

Seizures   Seizures are uncontrolled electrical disturbances in the brain. Epilepsy and seizures are the same disorder. Seizures in children are a relatively common condition, with an incidence of 3.5 per 1000 children.[32] Seizure activity has a range of outward signs, from uncontrollable jerking of the whole body to mild blinking. Currently, no known nutrients bring on seizures. Children who have seizures are usually treated by medications that prevent them. After some types of seizures, the child may have a period of semiconsciousness called a *postictal state* and appear to be sleeping, but he or she is difficult to wake.[32] Feeding or eating during the postictal state is not recommended, because the child may choke. Some children have long enough postictal states to miss meals. In this case, adding other eating times is needed to make up for the lost calories and nutrients.

When seizures are controlled by medications, growth usually continues at the rate typical for that child. Dietary consequences of controlled seizures are primarily related to drug–nutrient side effects, such as change in hunger or sleepiness. Some drugs should be taken without food, and others may be offered with snacks or meals. Most drugs have to be taken on a strict schedule and are not stopped without medical supervision.

Some children have uncontrolled seizures that may cause further brain damage over time. For reasons that remain unknown, seizures decrease when brain metabolism is switched from the usual fuel, glucose, to *ketones* from fat metabolism.[32] Some specialty clinics administer the *ketogenic diet* for uncontrolled seizures. The diet severely limits carbohydrates and increases calories from fat, but is adequate in calories and protein. Vitamins and minerals have to be added as supplements because the allowed food sources of carbohydrates are not sufficient to meet vitamin and mineral requirements. The ketogenic diet may allow seizure medications to be reduced or eliminated over time. However, many difficulties, such as measuring growth, blood glucose, and ketones in urine, lie in monitoring the body's reaction to such severe carbohydrate restriction. Growth during the time on a ketogenic diet may be different from that seen in the child's previous pattern. Some children improve in both weight and height when seizure activity declines. The ketogenic diet is so high in fat that some children gain weight faster than expected. The diet is generally recommended for 2 years, if it shows demonstrated effectiveness.

Cerebral Palsy   Cerebral palsy (CP) is one of the most common conditions in children with severe disabilities (Illustration 13.1). Overall incidence of cerebral palsy is about 1.4–2.4 per 1000 children.[34] *Cerebral palsy* is a general term well understood by the public; it covers a broad range of conditions resulting from brain damage. Causes of CP all involve damage to the brain early in life, either before or after birth. The initial site of brain damage does not progress, but progression of secondary effects occurs over time. Secondary effects may include contractures, scoliosis, gastroesophageal reflux, and constipation.[6,10] Many children with CP have constipation because coordinated muscle movements are part of bowel emptying, including the muscles in and over the intestines. Muscle coordination problems most easily seen in movements of the arms and legs may occur in muscles all over the body, including the abdominal muscles that assist in bowel evacuation.[34]

The form of cerebral palsy that presents the most nutrition problems is spastic quadriplegia, involving all limbs.[2] Most children with spastic quadriplegia appear thin, but this appearance may be a result of brain damage or muscle size. Children with cerebral palsy often have other forms of brain damage as well: 39–44% have mental retardation, 26–36% have seizures, and 14–18% have severe visual impairment. Causes of CP are unknown for more than one-third of affected children, and the condition may or may not be related to preterm birth. About half the time after a preterm birth, a basis for CP can be identified

Postictal State   Time of altered consciousness after a seizure; appears to be like a deep sleep.

Ketones   Small two-carbon chemicals generated by breakdown of fatty acids for energy.

Illustration 13.1   Boy with CP in a walker.

Janet Sugarman Isaacs

Illustration 13.2 Growth chart for gastrostomy feeding for a boy with spastic quadriplegia and scoliosis.

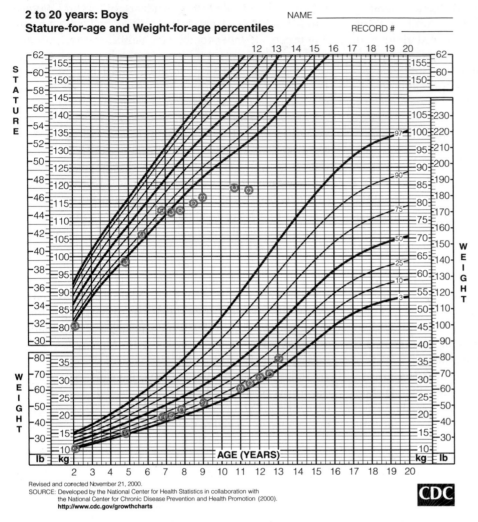

2 to 20 years: Boys
Stature-for-age and Weight-for-age percentiles

NAME _____

RECORD # _____

Revised and corrected November 21, 2000.
SOURCE: Developed by the National Center for Health Statistics in collaboration with
the National Center for Chronic Disease Prevention and Health Promotion (2000).
http://www.cdc.gov/growthcharts

CDC

even with an in-depth growth assessment. Children with small or weak muscles have lower caloric needs because they are less active as a result of little voluntary muscle control. In contrast, types of CP characterized by increased uncontrolled movement require extra calories as a result of a higher activity level. *Athetosis* is an example of this less common form of CP, in which increased energy needs have been documented.[35] Altered body composition affects many aspects of the child's nutrition and eating abilities.[7] Eating or feeding problems may appear in the forms of spilling food, long mealtimes, fatigue at mealtime, and/or requiring assistance to eat. Difficulty in controlling muscles such as those in the neck and back, those used in head position and sitting, and those in the jaw, tongue, and lips and used in swallowing may contribute to feeding and eating problems.[34]

Nutrition experts who provide services for children with CP monitor their growth and then make recommendations for food choices that fit the children's abilities for eating, for nutritional supplements if food and beverages are not providing sufficient nutrients, and for nutrition support if needed. Nutrition interventions may include the following:

- Stimulating oral feeding
- Promoting healthy eating at school
- Adjusting menus and timing of meals and snacks at home or school for meeting nutrient needs from foods that minimize fatigue during meals
- Assessing and adjusting the child's diet over time
- Using adapted self-feeding utensils or other types of feeding equipment

from the perinatal period.[34] The prevalence of CP in children born with very low birth weight has been increasing, but this trend may be a consequence of the overall increase in survivors of severe prematurity.[6] Children with CP can enjoy many activities, attend school, and later contribute to society. Persons with CP display a wide range of abilities. As indicated in the growth chart in Illustration 13.2, children with spastic quadriplegia grow, but their growth is slower than that in others, with or without gastrostomy feeding.[2]

Nutritional consequences of spastic quadriplegia are slow weight gain and other growth concerns, difficulty with feeding and eating, and changes in body composition. No specific vitamins or minerals are known to correct CP. Problem nutrients are likely to be those related to bone density, calcium, and vitamin D, or nutrients needed in higher amounts as a result of medication side effects. Recommendations for caloric needs are difficult to determine,

Athetosis Uncontrolled movements of the large muscle groups as a result of damage to the central nervous system.

Phenylketonuria (PKU) and Inborn Errors of Metabolism

PKU is a well-known example of a group of genetic disorders called inborn errors of metabolism.[4] More children are found to have inborn errors of metabolism as a result of new genetic disorders being identified and better survival after early diagnosis from newborn screening. (Newborn screening and genetic screening is discussed in Chapter 9.) These disorders require interventions to manage breakdown products from dietary protein, fats, and carbohydrates being metabolized incompletely or inadequately. Inborn errors of metabolism involve molecular- or cellular-level blocks and have little or nothing to do with the gastrointestinal level of digestion and absorption of nutrients. Examples of inborn errors of metabolism diagnosed during childhood are glycogen-storage diseases (inborn errors of carbohydrate metabolism) and medium-chain fatty acid disorders (inborn errors of fat metabolism) diagnosed in older siblings or family members after the birth an infant identified with the same disorder by newborn screening.

PKU best demonstrates the importance of diet in management of an inborn error of metabolism. PKU is neither the most common nor most typical in developmental outcome of the inborn errors group. PKU has a prevalence of 1 in 12,000 live births.[4] The main treatment is lifelong dietary management, in which more than 80% of protein intake from foods and beverages is replaced by a mixture of amino acids from which phenylalanine has been removed. The enzyme that uses phenylalanine as a substrate is either not working at all or only partially active in the liver of a person with PKU. For children with specific types of blood phenylalanine elevations, a prescribed medication can adjust the degree of dietary management required.[33] Both medication and dietary treatment reduce intake of this amino acid to the minimum amount needed as an essential amino acid. This strategy limits toxic breakdown products of accumulated phenylalanine, which the body has difficulty clearing. How excess phenylalanine causes mental retardation is not known. The PKU diet is required throughout life (Illustration 13.3).

If foods with protein are consumed in too-high amounts, PKU slowly becomes a degenerative disease affecting the brain at whatever age the treatment is stopped. A woman with PKU has to continue strict dietary adherence because high levels of phenylalanine affect every pregnancy, even if the infant did not inherit PKU.

When their diets are managed correctly, children with PKU appear to be eating meals providing less food than the meals of other children. The diet is adequate in all vitamins, minerals, protein, fats, and calories, but more nutrients are in liquid rather than solid forms. Foods to be avoided completely are protein-rich foods such as meats, eggs, regular dairy products, peanuts, and soybeans in all forms. Allowed natural sources of protein are limited amounts of regular crackers, potato chips, rice, and potatoes. Many fruits and vegetables are encouraged, if offered without added sources of protein. Some foods that are high in fats and/or sugars and generally low in natural protein, such as fried vegetables or candy canes, are safe for children with PKU. Illustration 13.4 shows a MyPyramid with the base of the pyramid as the special medical food or formula that replaces the protein in foods.

The phenylalanine-deficient protein is generally served as a liquid, called a medical food or formula. The vitamins and minerals required to meet the RDA are in the phenylalanine-deficient protein powder. If the child does not drink enough of the PKU formula, foods that the child eats to meet her vitamin, mineral, and calorie needs will elevate the blood phenylalanine. Table 13.4 is a dietary recall from a child in good control of PKU. The phenylalanine-deficient protein is also available as bars and pills. It can be expensive to buy substitute low-protein

Illustration 13.3 This girl does not appear to have a chronic illness, but she has PKU.

Illustration 13.4 Modified Food Pyramid for diet education about PKU.

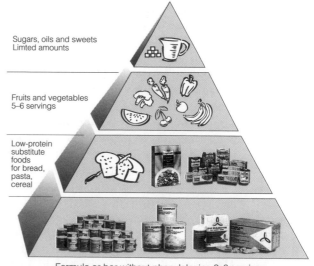

Sugars, oils and sweets Limted amounts

Fruits and vegetables 5–6 servings

Low-protein substitute foods for bread, pasta, cereal

Formula or bar without phenylalanine 2–3 servings

SOURCE: © 2000 SHS North America. Reprinted with permission.

Table 13.4 Dietary recall for a 5-year-old child with well-controlled PKU

Breakfast
2 slices low-protein bread with jelly and margarine
6 cut-up orange pieces
8 fl oz PKU formula

Lunch
½ c fruit cocktail in heavy syrup
1 c tossed salad (lettuce, tomato, celery, cucumber only) with 2 Tbsp ranch dressing
17 french fries with ketchup
6 fl oz apple juice

After-school snack
½ c microwave popcorn
8 fl oz PKU formula

Dinner
Pickle spears (dill, 3 wedges)
1 c low-protein imitation rice containing 1.5 Tbsp margarine
½ c grilled onions, green peppers, and mushrooms (on rice)
1 c canned peaches in heavy syrup
8 fl oz PKU formula

Snack
Skittles candy (small snack-size)
4 fl oz apple juice

alternative foods, such as low-protein pizza crusts, low-protein cheese, and low-protein baking mixes. Successful compliance requires use of low-protein foods to allow variety in the diet, such as low-protein pasta.

Attention Deficit Hyperactivity Disorder Attention deficit disorder and attention deficit hyperactivity (ADHD) disorder are the most common *neurobehavioral* conditions in children. The incidence of ADHD is estimated at 5% to 8% of school-age children and adolescents.[36,37] Children diagnosed with ADHD require cardiovascular monitoring before and, depending on results, during ADHD drug treatment. Children suspected of having ADHD may have a chaotic meal and snack pattern and the inability to stay seated for a meal. They may be given fewer opportunities to use kitchen appliances and get their own snacks due to impulsiveness. Theories about specific foods or nutrients causing ADHD have not been proven scientifically, but high interest in nutrition as a cause and treatment continues. The sale of herbal medicines and nutritional supplements to families with children with ADHD is common. One large survey found that 64% of children with ADHD had tried at least one type of alternative therapy and that 13% of children took some sort of multivitamin supplement.[38] Megavitamins were found ineffective in treating ADHD in a double-blind crossover trial, but such information has not stopped claims made for them or marketing and sales.[38] Often health care providers, including nutritionists, are not told about the use of these supplements by families.

Recommended procedures to confirm ADHD include at least two sites completing observation checklists about behavior. Treatment with the following two approaches has been most effective:

1. A structured behavioral approach that may also include mental health counseling and support, such as parenting classes
2. A prescribed *psychostimulant* medication; examples are Ritalin or Adderall

Nutritional concerns in ADHD include medication side effects that decrease appetite, maintenance of growth while being medicated, and mealtime behavior. Low appetite as a result of treatment of ADHD is quite variable; it depends on the timing of the medications compared to meals, dosage, and medication schedule and on how long the child has had to adjust to the medication. Less interference with appetite and growth is likely if the child does not take the medication during school holidays.

Regardless of the child's dosage schedule, ADHD medication peak activity is aimed for school hours, which includes school lunch. Nutrition interventions for children on psychostimulant medications call for timing meals and snacks around the medication's action peaks. For example, adding a large bedtime snack when the medication's effects are low is a typical recommendation. Monitoring weight and height carefully over time helps identify growth plateaus. Education for the school's lunchroom supervisors and teachers may be helpful to deal with food refusals and mealtime behavior for the child with ADHD.[37]

Pediatric HIV Most children with HIV were infected at around the time of birth. HIV in children under age 13 is classified differently than it is in adults, by age-based categories as well as level of immunosuppression from the virus. Only in the last few years have affected infants lived long enough to benefit from the combination highly active antiretroviral therapy that became available in the mid-1990s.[39] As experience has accumulated with these potent medications, growth of children with low viral loads has improved, and opportunistic infections have occurred at a lower rate.

Nutrition is an important component of HIV management. Failure to thrive is common in infected infants regardless of drug treatment, due to the dampening effect

Neurobehavioral Pertains to control of behavior by the nervous system.

Psychostimulant Classification of medication that acts on the brain to improve mental or emotional behavior.

of antiretroviral therapy on appetite and food intake. For children too young to be in charge of their own medication and eating schedules, educating the family and arranging support may be part of the nutrition intervention. Other nutrition concerns include food-related infection-control measures, assuring access to complete nutritional supplements, and referrals to food banks. If weight-gain and medication compliance problems are unresolved, gastrostomy placement for medications and supplemental feedings may be needed.[39]

Working with children with HIV is complicated and demanding, and dietary approaches have to be customized to the behavioral and developmental realities of each child. For example, an 11-year-old girl is being treated for HIV and its related illnesses. Her diet prescription includes a high-protein/high-calorie diet, with one complete vitamin/mineral supplement daily and three meals and three snacks. Her family members are to check her weight at home weekly and call in if they observe weight loss and low appetite. The girl takes four kinds of HIV-related medications, totaling 17 capsules per day. Two medications have no food-related restrictions, one is best taken with food at two different times per day, and one is best taken on an empty stomach (30 minutes before a meal or 2 hours after a meal). She also gets an injection every other week to strengthen her blood counts.

Dietary Supplements and Herbal Remedies

Children with special health care needs are found to use complementary and alternative medicine at a higher rate, 30–70%, than healthy children.[40] This includes various types of nutritional supplements and vitamins and minerals.[41] Families with children in a lengthy process of diagnosis—where the diagnosis does not lead to a definite treatment and when expense, insurance-coverage, and administrative problems tax their ability to cope—are more likely to seek alternative therapies. Some of these alternatives have questionable effectiveness and are perhaps even harmful. No herbal remedies or nutritional supplements have been found effective to prevent or treat the conditions covered in this chapter; however, nutritional claims abound for various chronic conditions. Families hear from one another about micronutrients—such as magnesium, zinc, and B_6—sold with various combinations of amino acids for Down syndrome and autism.[42] Restrictive diets, such as avoiding dairy products or gluten, have been researched for one condition and then extrapolated for another. Sports drinks and high-protein products marketed for athletes may attract families with children who have difficulty gaining weight.

Strategies to counter unscientific nutritional claims for various products include the following:

- Recognize the benefits of support for families, such as advocacy groups.
- Improve communication with health care providers, so that families ask more questions about nutrition claims of alternative treatments.
- To give them some control over decision making for their children, give families reliable information, such as scientific literature or fact sheets, without endorsing any claim.

Sources of Nutrition Services

Children with chronic conditions that interfere with their ability to function may be eligible for Supplemental Social Insurance (SSI). Low-income families are eligible for SSI depending on the child's condition. Examples of conditions usually qualifying for SSI are chromosomal disorders; mental retardation; and severe forms of seizures, cerebral palsy, and CF. A child with treated PKU is generally not eligible for SSI because treatments prevent decline in learning abilities. Also, the Americans with Disabilities Act applies to all ages. It requires, for example, that school cafeteria lines accommodate wheelchairs.

USDA Child Nutrition Program

The U.S. Department of Agriculture Child Nutrition Program, as described in previous chapters, requires that school breakfast and lunch menus be modified for children with diagnosis-specific diets or changes in the texture of foods. Parents who want their children to participate in the Child Nutrition Program cannot be charged an additional fee for providing a special diet for the child. A registered dietician or another health provider completes a prescription ordering special breakfasts or lunches. Examples of diet prescription orders are a reduced-calorie school lunch and breakfast, a pureed diet, or a nutrient-modified diet, such as a PKU diet (see Case Study 13.2). If families do not want to participate in the Child Nutrition Program and prefer to pack lunches, they can change their decision at any time during the school year. Formulas administered by gastrostomy are not required to be supplied by the Child Nutrition Program.

Maternal and Child Health Block Program of the U.S. Department of Health and Human Services (HHS)

Every state has a designated portion of federal funding for children with special health care needs.[43,44] A wide range of services can be provided based on state planning as

Case Study 13.2

Photodisc

Dealing with Food Allergies in School Settings

Nutrition diagnoses are an age-appropriate self-monitoring deficit, inability to manage self-care, and a need for a severe restriction related to a peanut allergy as evidenced by her medical history. This case study is about the appropriate nutrition interventions, monitoring, and evaluation within the Nutrition Care Process.

Judy is to start regular kindergarten. When she was 2 years old, she was diagnosed with a peanut food allergy after many episodes of asthma and hives. Her health has improved as a preschooler with avoidance of peanuts in all forms. The family has carefully watched what she eats. However, at age 4, she had an episode of breathing difficulty that required an emergency-room visit. This incident makes the family quite concerned about Judy's eating at school. She is generally not allowed to go to friends' homes to play; friends come to her house so the family can watch out for her. She has been instructed not to take any food from anyone. She has not been in day care or preschool, so starting school is a big step for the family.

Nutrition Intervention Recommendations: A meeting was held at the school after receiving the physician's statement and diagnosis. Judy's mother met with the school staff to discuss plans for Judy at school to avoid exposure to peanuts in any form. The family did not want to participate in the school lunch program. Her mother proposed to pack Judy a lunch from home, although most children eat food provided at school. With the school staff, Judy's mother discussed snack time for Judy and her eating at the cafeteria, which periodically serves food cooked in peanut oil, or food containing peanuts. The snack at kindergarten was discussed, as it is provided by parents based on a rotation schedule. It is usually milk or juice with cookies or fruit.

Monitoring Recommendations: A plan was put in place for the teacher to check the snack foods and offer a replacement snack provided by Judy's mother if she is unsure whether that day's snack contains peanuts. The school working group had written out a 504 accommodation plan for Judy's peanut allergy. It included making sure that tables where children eat peanut-containing foods are washed well, with signs in the cafeteria with Judy's picture to make sure she does not inadvertently get peanut-containing food from another child or in a food activity.

After Judy has been in school for 1 month, the family meets with the school group. In that month, two episodes resulted in what may have been hives, and her family is worried that Judy is not adjusting well. At snack time Judy did not recognize some of the foods; she refused to eat a snack most days. She appeared hungry at home after school. Her mother says she would like to send to school a snack that she knows her daughter will eat, and she wants to attend school during snack time to make sure Judy is not being teased.

Questions

1. Why is it the school's job to check for peanuts when other parents are sending snacks?
2. The parents seem overprotective. Can the teacher transfer Judy to another classroom?
3. What are the chances that Judy will outgrow the peanut allergy by next year?

(Answers are located in the Instructor's Manual for the 4th edition of *Nutrition Through the Life Cycle.*)

reported back to HHS. Nutrition services may be in specialty clinics or county health departments, or they can be contracted for providing care, ensuring access to resources such as formulas, foods, and nutrition education. Nutrition experts work with children in various settings, including schools, early-intervention programs, homes, clinics, and facilities. Also, a program in every state identifies and advocates for children with special needs. An example is the Developmental Disabilities Council.

Public School Regulations: 504 Accommodation and IDEA

Two sets of regulations guide how schools provide nutrition services in addition to the Child Nutrition Program. Nutrition services in schools are generally more available to younger children than they are to older children. Children in regular education have different access to services than do children eligible for special educational services (see Case Study 13.2).

504 Accommodation   The USDA Child Nutrition Program requires that elementary and junior and senior high school breakfast and lunch fit individual special diets when the child's physician orders them. This is called the "504 accommodation" when the child has a regular curriculum. If the child requires special education services, this requirement becomes included in his or her individualized education plan. For a child with MSUD, diabetes, or another chronic condition, the Child Nutrition Program school district director works with the child's hospital-based dietician, who prescribes the diet, if a parent requests that the child receive school meals. Child Nutrition Program regulations do not allow the parents of the child to be charged any additional cost for the modified school lunch. The child's dietary requirements also have to be met for classroom birthday parties or when food is used in classroom projects. These rules do not apply in private school, although the Americans with Disabilities Act may protect children from their special dietary needs being the basis for discrimination against them.

Individuals with Disabilities Education Act   Children eligible for special education are covered by regulations within the Individuals with Disabilities Education Act (IDEA).[1] It requires each child to have an individualized

education plan (IEP) that may include nutrition-related goals and objectives as needed. The school staff must involve the parent in developing the IEP. For some diagnoses, it would be appropriate for a nutritionist to attend the IEP meeting to make sure the teacher, teacher's aide, and other staff understand what the child needs. Nutrition, eating, and feeding problems may be a part of that plan, and it may apply to food offered in the classroom as well as that served in the regular school cafeteria. An example of IEP goals and objectives can be found in Table 13.5. For this particular plan, the child's education includes learning to eat by mouth with prompting and assistance. Nutritional supplements may be purchased as part of an education intervention called for in the child's IEP.

Nutrition Intervention Model Program

The Maternal and Child Health Bureau (MCH) is a part of the Department of Health and Human Services and funds nutrition services for chronically ill children.[44] MCH develops and promotes model programs by funding competitive grants that emphasize training health care providers, including nutrition experts. Training programs vary in length from short, intensive courses to year-long traineeships. Topics vary from nutrition for infants receiving intensive care services to nutrition problems of adolescence, such as warning signs of anorexia nervosa. Examples of such federal grant programs are the Pediatric Pulmonary Centers, Bright Futures Guidelines.[43,44]

Table 13.5   Example of nutrition objectives in an individualized education plan for an 8-year-old boy with limited oral feeding skills

1. In three of five trials, J. R. will hold food on the spoon as he moves it to his mouth without hand-over-hand assistance from his aides during three meals per week.

2. J. R. will point to what he wants to eat with his left hand in three trials after two prompts per meal three days each week.

3. J. R. will cooperate in having his gastrostomy site checked at feedings by pulling up his shirt three days in a row each week.

Key Points

1. Children with chronic health conditions still want to fit in with everyone else and be treated like others their age. Paying too much attention to the special health care needs may not help the individual become independent over time.

2. Nutrition requirements have to be customized to the individual: guidelines for healthy children may not be appropriate.

3. Energy needs (calories) are based on the child's activity as affected by the underlying condition;

needs may be higher or lower than those for others of the same age.

4. Feeding and eating problems are likely to interfere with appetite and meal patterns for conditions that require medications that have side effects.

5. The goal for meeting nutritional needs is to eat by mouth, if this is enjoyable and safe for the individual. An alternative may be to add complete nutritional supplements taken orally or by overnight feedings.

6. Vitamins and minerals are needed for some conditions at levels that are higher or lower than usually recommended.

7. Dietary management of cystic fibrosis and diabetes mellitus are examples of lifelong treatments that are combined with other medical approaches for children with special health care needs.

8. Cerebral palsy is one of the more severe conditions in which growth and eating are impacted; alternatives to using the standard growth chart and eating guidelines are necessary based on the abilities of the individual.

9. Children with special health care needs attend school like everyone else, but they may need to have a modified school lunch or eat different foods than others based on their underlying condition.

10. Nutrition and medical providers are encouraged to help families find in their community the educational and developmental services suitable for their children.

Review Questions

1. Which child requires an interpretation of growth modified from the usual CDC growth chart interpretation?
   a. A child with ADHD
   b. A child with type 1 diabetes
   c. A child with mild cerebral palsy with scoliosis
   d. A child with PKU

2. Which child requires in-depth nutrition assessment due to changes in body composition?
   a. A child with drug-nutrient interactions that increase sleepiness
   b. A child who appears thin and uses a wheelchair most of the time
   c. A child with seizures with a BMI at the 50th percentile
   d. A child with autism who looks thin and refuses a variety of foods

3. When a child has a chronic condition that limits activity, such as spina bifida, what is the greatest long-term nutrition concern?
   a. Protein limiting growth
   b. Modified food textures to avoid eating difficulties
   c. Excessive calorie intake further limiting mobility
   d. Intake of calcium and vitamin D to prevent bone fractures

4. An 8-year-old child with diabetes is eating birthday cake at school. What is a likely explanation?
   a. The child has type 2 diabetes and is obese.
   b. The child has type 1 diabetes, and he has adjusted his insulin pump to cover the cake.
   c. The child's chronic condition is not managed well, as cake is not allowed.
   d. The child had eaten a high-protein food with little carbohydrate at lunch.

   Sue has a chronic condition with a pattern of weight gain and then loss, without growth over the last year. Identify as true or false statements 5, 6, and 7.

5. Sue's condition can modify her energy needs compared to a healthy child.

6. The family may be worried about handling Sue if she grows as expected for her age.

7. Sue's nutritional intake must not be sufficient, since her growth is not as expected.

8. What are two examples of family support that would assist a child with a chronic condition?

9. When a child with ADHD refuses to eat, what are two possible explanations?

10. Explain why a child could need an IEP that includes nutrition services.

Resources

Ability OnLine Support Network
Connects young people with disabilities and chronic illnesses to peers and mentors with and without disabilities.
Website: **www.ablelink.org**

American Diabetes Association
Allows searches about diabetes in children around the world and provides professional and consumer publications.
Website: **www.diabetes.org**

Cystic Fibrosis Foundation
This large organization has information about research, services, and policy related to cystic fibrosis, including nutritional products and recommendations.
Website: **www.cff.org**

Exceptional Parent
This magazine is an excellent resource for a wide variety of conditions and includes tips for parents on how to work with care providers and educators.
Website: **www.eparent.com**

Ketogenic Diet
This website is maintained by the Packard Children's Hospital at Stanford, which provides credible information for parents and providers about the ketogenic diet and places that use it.
Website: **www.stanford.edu/group/ketodiet**

National Center for Growth Statistics (source of growth charts)
Site for obtaining growth charts and guidelines for their use.
Website: **www.cdc.gov/nchs**

National Down Syndrome Society
Includes information about its policy regarding nutritional products and directs parents to local resources for working with schools.
Website: **www.ndss.org**

Chapter 14

Adolescent Nutrition

Prepared by **Jamie Stang**

Key Nutrition Concepts

1 Nutrition needs should be determined by the degree of sexual maturation and biological maturity (biological age) instead of by chronological age.

2 Unhealthy eating behaviors common among adolescents include frequent dieting, meal skipping, use of unhealthy dieting practices, and frequent consumption of foods high in fat and sugar, such as fast foods, soft drinks, and savory snacks.

3 Concrete thinking and abstract reasoning abilities do not develop fully until late adolescence or early adulthood; therefore, education efforts need to be highly specific and based on concrete principles.

4 Adolescent eating behaviors are influenced by a variety of factors, including peer influences, parental modeling, food availability, food preferences, cost, convenience, personal and cultural beliefs, mass media, and body image.

5 Family meals decline during adolescence, but they are an important factor in improving the nutritional quality of adolescents' diets.

6 A small number of adolescents are meeting nutritional requirements for fruits, vegetables, whole grains, and calcium intake. At the same time, many adolescents exceed daily energy requirements.

7 Nutrition messages for adolescents need to focus on what is important to their lives. Focusing on the present and how good nutrition can positively impact appearance, sports performance, or academic performance is likely to have greater impact than focusing on long-term disease prevention.

8 Calcium intake is important during adolescence for development of peak bone mass, but most adolescents consume less than half of the recommended intake. Participation in physical activity during adolescence also plays a role in the development of bone mass.

Introduction

Adolescence is defined as the period of life between 11 and 21 years of age. It is a time of profound biological, emotional, social, and cognitive changes during which a child develops into an adult. Physical, emotional, and cognitive maturity is accomplished during adolescence (Illustration 14.1). Many adults view adolescence as a tumultuous, irrational phase that children must go through. However, this view does disservice to its developmental importance. The tasks of adolescence, not unlike those experienced during the toddler years, include the development of a personal identity and a unique value system separate from parents and other family members, a struggle for personal independence accompanied by the need for economic and emotional

Illustration 14.1 Average ages of pubertal, cognitive, and psychosocial maturation.

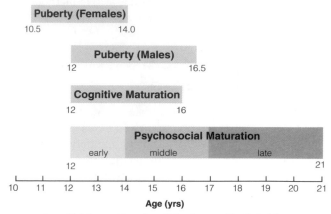

SOURCE: From "Adolescent Growth and Development" by R. L. Johnson, in Adolescent Medicine 2/e by A. Hofmann and D. Greynaus, Fig 2.1, p. 9. Copyright © 1988 McGraw-Hill Companies. Reprinted with permission.

family support, and adjustment to a new body that has changed in shape, size, and physiological capacity. When the seemingly irrational behaviors of adolescents are reframed as essential endeavors and viewed in light of these developmental tasks, adolescence can and should be viewed as a unique, positive, and integral part of human development.

Nutritional Needs in a Time of Change

The biological, psychosocial, and cognitive changes associated with adolescence have direct effects on nutritional status. The dramatic physical growth and development experienced by adolescents significantly increases their needs for energy, protein, vitamins, and minerals. However, the struggle for independence that characterizes adolescent psychosocial development often leads to the development of health-compromising eating behaviors, such as excessive dieting, meal skipping, use of unconventional nutritional and nonnutritional supplements, and the adoption of fad diets. These disparate situations create a great challenge for health care professionals. The challenging behaviors of adolescents can become opportunities for change at a time during which adult health behaviors are being formed. The search for personal identity and independence among adolescents can lead to positive, health-enhancing behaviors such as adoption of healthful eating practices, participation in competitive and non-competitive physical activities, and an overall interest in developing a healthy lifestyle. These interests and behaviors provide a good foundation on which nutrition education can build.

This chapter provides an overview of normal biological and psychosocial growth and development among adolescents and how these experiences affect the nutrient needs and eating behaviors of teens. Common concerns related to adolescent nutrition and effective methods for educating and counseling teens are also discussed.

Normal Physical Growth and Development

Early adolescence encompasses the occurrence of *puberty,* the physical transformation of a child into a young adult. The biological changes that occur during puberty include sexual maturation, increases in height and weight, accumulation of skeletal mass, and changes in body composition. Even though the sequence of these events during puberty is consistent among adolescents, the age of onset, duration, and tempo of these events vary a great deal between and within individuals. Thus, the physical appearance of adolescents of the same chronological age covers a wide range. These variations directly affect the nutrition requirements of adolescents. A 14-year-old

Puberty   The time frame during which the body matures from that of a child to that of a young adult.

Secondary Sexual Characteristics   Physiological changes that signal puberty, including enlargement of the testes, penis, and breasts and the development of pubic and facial hair.

Menses   The process of menstruation.

male who has already experienced rapid linear growth and muscular development will have noticeably different energy and nutrient needs than a 14-year-old male peer who has not yet entered puberty. For this reason, sexual maturation (or biological age) should be used to assess biological growth and development and the individual nutritional needs of adolescents rather than chronological age.

Sexual Maturation Rating (SMR), also known as "Tanner Stages," is a scale of *secondary sexual characteristics* that allows health professionals to assess the degree of pubertal maturation among adolescents, regardless of chronological age (Table 14.1). SMR is based on breast development and the appearance of pubic hair among females, and on testicular and penile development and the appearance of pubic hair among males.[1] SMR stage 1 corresponds with prepubertal growth and development, while stages 2 through 5 denote the occurrence of puberty. At SMR stage 5, sexual maturation has concluded. Sexual maturation correlates highly with linear growth, changes in weight and body composition, and hormonal changes.[1,2]

The onset of *menses* and changes in height relative to the development of secondary sexual characteristics that occur in females during puberty are shown in Illustration 14.2. Among females, the first signs of puberty are the development of breast buds and sparse, fine pubic

Table 14.1 Sexual maturity rating for girls and boys

Girls

Stage	Breast Development	Pubic Hair Growth
1	Prepubertal; nipple elevation only	Prepubertal; no pubic hair
2	Small, raised breast bud	Sparse growth of hair along labia
3	General enlargement of raising of breast and areola	Pigmentation, coarsening, and curling, with an increase in amount
4	Further enlargement with projection of areola and nipple as secondary mound	Hair resembles adult type, but not spread to medial thighs
5	Mature, adult contour, with areola in same contour as breast, and only nipple projecting	Adult type and quantity, spread to medial thighs

Boys

Stage	Genital Development	Pubic Hair Growth
1	Prepubertal; no change in size or proportion of testes, scrotum, and penis from early childhood	Prepubertal; no pubic hair
2	Enlargement of scrotum and testes; reddening and change in texture in skin of scrotum; little or no penis enlargement	Sparse growth of hair at base of penis
3	Increase first in length, then width of penis; growth of testes and scrotum	Darkening, coarsening, and curling; increase in amount
4	Enlargement of penis with growth in breadth and development of glands; further growth of testes and scrotum, darkening of scrotal skin	Hair resembles adult type, but not spread to medial thighs
5	Adult size and shape genitalia	Adult type and quantity, spread to medial thighs

SOURCE: From J. M. Tanner, Growth at Adolescence. Copyright © 1962 Blackwell Publishers. Reprinted with permission.

Illustration 14.2 Sequence of physiological changes during puberty in females.

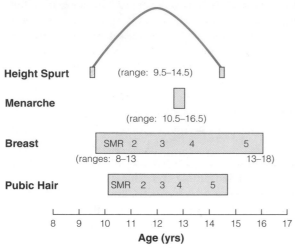

SOURCE: From J. M. Tanner, Growth at Adolescence. Copyright © 1962 Blackwell Publishers. Reprinted with permission.

Illustration 14.3 Sequence of physiological changes during puberty in males.

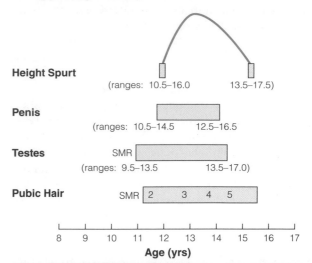

SOURCE: From J. M. Tanner, Growth at Adolescence. Copyright © 1962 Blackwell Publishers. Reprinted with permission.

hair occurring on average between 8 to 13 years of age (SMR stage 2). *Menarche* occurs 2 to 4 years after the initial development of breast buds and pubic hair, most commonly during SMR stage 4. The average age of menarche is 12.4 years, but menarche can occur as early as 9 or 10 years or as late as 17 years of age. Menarche may be delayed in highly competitive athletes or in girls who severely restrict their caloric intake to limit body fat.

Onset of the linear growth spurt occurs most commonly during SMR stage 2 in females, beginning between the ages of 9.5 and 14.5 years in most females (Illustration 14.2). Peak velocity in linear growth occurs during the end of SMR stage 2 and during SMR stage 3, approximately 6 to 12 months prior to menarche. As much as 15% to 25% of final adult height will be gained during puberty, with an average increase in height of 9.8 inches (25 cm).[3] During the peak of the adolescent growth spurt, females gain approximately 3.5 inches (8–9 cm) a year. The linear growth spurt lasts 24 to 26 months, ceasing by age 16 in most females. Some adolescent females experience small increments of growth past age 19 years, however. Linear growth may be delayed or slowed among females who severely restrict their caloric intake.

Enlargement of the *testes* and change in *scrotal* coloring are most often the first signs of puberty among males (Illustration 14.3), occurring between the ages of 10.5 and 14.5 years, with 11.6 years being the average age. The development of pubic hair is also common during SMR stage 2. Testicular enlargement begins between the ages of 9.5 and 13.5 years in males (SMR 2 to 3) and concludes between the ages of 12.7 and 17 years. The average age of "spermarche" is approximately 14 years among males. Clearly, males show a great deal of variation in the chronological age at which sexual maturation takes place.

On average, peak velocity of linear growth among males occurs during SMR stage 4, coinciding with or just following testicular development and the appearance of faint facial hair. The peak velocity of linear growth occurs at 14.4 years of age, on average. At the peak of the growth spurt, adolescent males will increase their height by 2.8 to 4.8 inches (7 to 12 cm) a year. Linear growth will continue throughout adolescence, at a progressively slower rate, ceasing at about 21 years of age.

> **Menarche** The occurrence of the first menstrual cycle.
>
> **Testes** One of the two male reproductive glands located in the scrotum.

Changes in Weight, Body Composition, and Skeletal Mass

As much as 50% of ideal adult body weight is gained during adolescence. Among females, peak weight gain follows the linear growth spurt by 3 to 6 months. During the peak velocity of weight change, which occurs at an average age of 12.5 years, girls will gain approximately 18.3 lb (8.3 kg) per year.[3] Weight gain slows around the time of menarche, but will continue into late adolescence. Adolescent females may gain as much as 14 lb (6.3 kg) during the latter half of adolescence. Peak accumulation of muscle mass occurs around or just after the onset of menses.

Body composition changes dramatically among females during puberty, with average lean body mass falling from 80% to 74% of body weight while average body fat increases from 16% to 27% at full maturity. Females experience a 44% increase in lean body mass and a 120% increase in body fat during puberty.[4] Adolescent females gain approximately 2.5 lb (1.14 kg) of body fat mass each year during puberty. Adolescent body fat levels peak among females between the ages of 15 and 16 years. Research by Rose Frisch suggests that a level of 17% body fat is required for menarche to occur and that 25% body fat is required for the development and maintenance of regular ovulatory cycles.[5] Normal changes in body fat mass can be mediated by excessive physical activity and/ or severe caloric restriction.

Even though the accumulation of body fat by females is obviously a normal and physiologically necessary process, adolescent females often view it negatively. Weight dissatisfaction is common among adolescent females during and immediately following puberty, leading to potentially health-compromising behaviors such as excessive caloric restriction, chronic dieting, use of diet pills or laxatives, and, in some cases, the development of body image distortions and eating disorders.

Among males, peak weight gain coincides with the timing of peak linear growth and peak muscle mass accumulation.[3] During peak weight gain, adolescent males gain an average of 20 lb (9 kg) per year. Body fat decreases in males during adolescence, resulting in an average of approximately 12% by the end of puberty.

Almost half of adult peak bone mass is accrued during adolescence. By age 18, more than 90% of adult skeletal mass has been formed.[6] A variety of factors contribute to the accretion of bone mass, including genetics, hormonal changes, weight-bearing exercise, cigarette smoking, consumption of alcohol, and dietary intake of calcium, vitamin D, protein, phosphorus, boron, and iron. Because bone is comprised largely of calcium, phosphorus, and protein and because a great deal of bone mass is accrued during adolescence, adequate intakes of these nutrients are critical to support optimal bone growth and development.

Normal Psychosocial Development

During adolescence, an individual develops a sense of personal identity, a moral and ethical value system, feelings of self-esteem or self-worth, and a vision of occupational aspirations. Psychosocial development is most readily understood when it is divided into three periods: early adolescence (11 to 14 years), middle adolescence (15 to 17 years), and late adolescence (18 to 21 years). Each period of psychosocial development is marked by the mastery of new emotional, cognitive, and social skills (Table 14.2).

During early adolescence, individuals begin to experience dramatic biological changes related to puberty. The development of body image and an increased awareness of sexuality are central psychosocial tasks during this

Table 14.2 Psychosocial processes and the substages of adolescent development

Substage	Emotional and Social Development	Cognitive Development
Early adolescence	• Alterations in body image secondary to dramatic changes in body shape and size • Increased awareness of sexuality • Strong need for social acceptance by peers • Strong sense of impulsivity	• Concrete thinking processes are dominant, often with limited abstract thought capacity
Middle adolescence	• Development of greater autonomy from parents and family • Continued need for peer acceptance, which may lead to risk-taking behaviors • Increased opportunities for employment outside of home, resulting in more decision making and the beginning of economic independence • Increased awareness of moral and social issues	• Development of abstract reasoning continues • May revert to concrete thinking when under stress
Late adolescence	• Further development of personal set of morals and values • Increased impulse control • Greater social, emotional, and economic independence from family • Reduced need for peer acceptance • Development of personal and vocational goals	• Abstract thought capacity fully develops

SOURCE: From G. M. Ingersoll, "Psychological and Social Development," in Textbook of Adolescent Medicine. Copyright © 1992 Saunders. Reprinted with permission.

period of adolescence. The dramatic changes in body shape and size can cause a great deal of ambivalence among adolescents, leading to the development of poor body image and eating disturbances if not addressed by family or health care professionals.

Peer influence is very strong during early adolescence. Young teens, conscious of their physical appearance and social behaviors, strive to "fit in" with their peer group. The need to fit in can affect nutritional intake among adolescents. Focus groups conducted with adolescent females revealed that situational factors, such as who students ate with and where they ate, were important factors in the food choices they made.[7] Consequently, teens express their ability and willingness to fit in with a group of peers by adopting food preferences and making food choices based on peer influences and by refuting family preferences and choices. In some cases, choices based on peer pressure can lead to improved dietary intake, such as reduced intake of animal protein due to animal welfare concerns or choosing foods with a lower carbon footprint. In other cases, the choices based on peer pressure may lead to poor dietary intake, such as consumption of fast foods, convenience foods, sweetened beverages, and other highly processed foods that are high in added fats and sugars.

The wide chronological age range during which pubertal growth and development begins and proceeds can become a major source of personal dissatisfaction for many adolescents. Males considered to be "late bloomers" often feel inferior to their peers who mature earlier and may resort to the use of anabolic steroids and other supplements in an effort to increase linear growth and muscle development. Females who mature early have been found to have more eating problems and poorer body image than their later-developing peers.[8] They are also more likely to initiate "grown-up" behaviors such as smoking, drinking alcohol, and engaging in sexual intercourse at an earlier age.[8,9] Education of young adolescents on normal variations in tempo and timing of growth and development can help to facilitate the development of a positive self-image and body image and may reduce the likelihood of early initiation of health-compromising behaviors.

Cognitively, early adolescence is a time dominated by concrete thinking, egocentrism, and impulsive behavior. Abstract reasoning abilities are not yet developed to a great extent in most adolescents, limiting their ability to understand complex health and nutrition issues. Young adolescents also lack the ability to see how their current behavior can affect their future health status or health-related behaviors.

Middle adolescence marks the development of emotional and social independence from family, especially parents. Conflicts over personal issues, including eating and physical activity behaviors, are heightened during midadolescence. Peer groups become more influential, and

their influence on food choices peaks. Physical growth and development are mostly completed during this stage. Body image issues are still of concern, especially among males who are late to mature and females. Peer acceptance remains important, and the initiation of and participation in health-compromising behaviors often occurs during this stage of development. Adolescents may believe they are invincible during this stage of development.

The emergence of abstract reasoning skills occurs rapidly during middle adolescence; however, these skills may not be applied to all areas of life. Adolescents will revert to concrete thinking skills if they feel overwhelmed or experience psychosocial stress. Teens begin to understand the relationship between current health-related behaviors and future health, even though their need to "fit in" may supplant this understanding.

Late adolescence is characterized by the development of a personal identity and individual moral beliefs. Physical growth and development is largely concluded, and body image issues are less prevalent. Older teens become more confident in their ability to handle increasingly sophisticated social situations, which is accompanied by reductions in impulsive behaviors and peer pressure. Adolescents become increasingly less economically and emotionally dependent on parents. Relationships with one individual become more influential than the need to fit in with a group of peers. Personal choice emerges.

Abstract thinking capabilities are realized during late adolescence, which assists teens in developing a sense of future goals and interests. Adolescents are now able to understand the perspectives of others and can fully perceive future consequences associated with current behaviors. This capability is especially important among adolescent females who plan to have children or who become pregnant.

Health and Eating-Related Behaviors During Adolescence

Eating patterns and behaviors of adolescents are influenced by many factors, including peer influences, parental modeling, food availability, food preferences, cost, convenience, personal and cultural beliefs, mass media, and body image. Illustration 14.4 presents a conceptual model of the many factors that influence eating behaviors of adolescents. The model depicts three interacting levels of influence that impact adolescent eating behaviors: personal or individual, environmental, and macrosystem. Personal factors that influence eating behavior include attitudes, beliefs, food preferences, *self-efficacy*, and biological changes. Environmental factors include

Self-Efficacy The ability to make effective decisions and to take responsible action based on one's own needs and desires.

Illustration 14.4 Conceptual model for factors influencing eating behavior of adolescents.

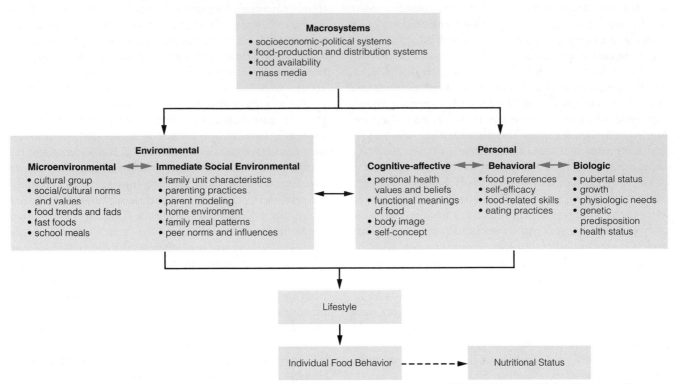

SOURCE: From M. Story and I. Alton, "Becoming a Woman: Nutrition in Adolescence," in D. A. Krummel and P. M. Kris-Etherton (Eds.), in Nutrition in Women's Health. Copyright © 1996 Aspen Publishers. Reprinted with permission.

the immediate social environment such as family, friends, and peer networks, and other factors such as school, fast-food outlets, and social and cultural norms. Macrosystem factors, which include food availability, food production and distribution systems, and mass media and advertising, play a more distal and indirect role in determining food patterns, yet can exert a powerful influence on specific food choices. To improve the eating patterns of youth, nutrition interventions should be aimed at each of the three levels of influence. In addition to these influencing factors, socioeconomic status and racial/ethnic background also play a role in shaping adolescent eating patterns. Adolescents of a lower socioeconomic status and from minority backgrounds have been found to be more likely to engage in less healthy eating behaviors, and this often leads to higher prevalence of overweight.[10]

Eating habits of adolescents are not static; they fluctuate throughout adolescence in relation to psychosocial and cognitive development. Longitudinal data of adolescent females suggest that even though body-weight percentiles track throughout adolescence, little consistency guides the intake of energy, nutrients, vitamins, and minerals from early to late adolescence.[11] Health professionals must therefore refrain from jumping to conclusions about the dietary habits of adolescents (even if they have been evaluated for nutritional status at an earlier age) and

should take the time to assess the current dietary intake of the individual.

Adolescents lead busy lives. Many are involved in extracurricular sports or academic activities, others are employed, and many must care for younger children in a family for part of the day. These activities, combined with the increased need for social and peer contact and approval, and increasing academic demands as they proceed through school, leave little time for adolescents to sit down to eat a meal. Snacking and meal skipping are commonplace among adolescents. Snacks account for up to 39% of daily food intake among adolescents, with 35% of discretionary calories and 43% of added sugars provided by snacks alone.[12] National data suggest that the proportion of calories and nutrients from foods consumed as snacks has risen during the past decade.[13,14] The average size of snacks has remained approximately the same, but the number of snacks consumed has increased, which accounts for the increased caloric intake.[13] The increased calorie intake from snacks has paralleled the increase in consumption of foods outside of the home. Adolescents are consuming a larger proportion of calories from snacking away from home, often at fast-food restaurants.[14]

Unfortunately, the food choices made by adolescents while snacking tend to favor foods high in sugar, sodium, and fat while relatively low in vitamins and minerals. Soft drinks are the most commonly chosen snacks for adolescent females, and these account for about 6% of total caloric intake.[13-15] This trend causes significant concern because high consumption of soft drinks may reduce the consumption of healthier beverage choices that are lower in energy and/or higher in calcium. Over time, these changes in nutrient intake could lead to an increased risk for chronic health problems such as osteoporosis and obesity. Health practitioners working with adolescents need to understand that snacking is a commonplace behavior among adolescents and should work with adolescents to improve food choices rather than discouraging snacking. Due to the frequency of snacking, an improvement in food choices during snacking toward more nutrient-dense foods has the potential to significantly impact the dietary intake and adequacy of adolescents.

The occurrence of meal skipping increases as adolescents mature. Breakfast is the most commonly skipped meal; as few as 21% of adolescents have been found to consume breakfast daily.[15-17] Skipping breakfast can dramatically decrease intakes of energy, protein, fiber, calcium, and folate due to the absence of breakfast cereal or other nutrient-dense foods commonly consumed at breakfast. Lunch is skipped by almost 25% and dinner is skipped by up to 8% of teens.[16-18] As with breakfast, skipping lunch and/or dinner reduces intakes of energy, protein, and other nutrients. Adolescents who skip meals should be counselled on convenient, portable, and healthy food choices that can be taken with them and eaten as meals or snacks.

As adolescents mature, they spend less time with family and more time with their peer group. Eating away from home becomes prevalent: female adolescents, for example, eat almost one-third of their meals away from home, and the average teen eats at a fast-food restaurant twice a week. Data from NHANES 2003–2004 found that 59% of 12–19-year-olds surveyed had eaten fast food on at least one of the two days during which dietary intake was assessed. Fast food accounted for 16% of the energy intake by females and 17% by males in the total sample; among the highest tertile of intake fast foods provided almost 50% of daily energy consumed.[18] Fast-food restaurants and food courts are favorite eating places of teens because they offer an informal setting with inexpensive food choices. These venues also have a high percentage of teen employees, thus increasing their social value.

Eating at fast-food restaurants has direct bearing on the nutritional status of adolescents. Many fast foods are high in total and saturated fat and low in fiber, vitamins, and minerals. However, specific choices can be made to increase the nutrient content of fast-food meals and decrease the fat content. Adolescents can be counseled to ask for juice, water, or milk instead of soft drinks, order small sandwiches instead of larger choices, choose a salad or baked potato as a side dish instead of french fries, order grilled items as opposed to fried sandwiches, and avoid "super-sizing" meals even if it seems to offer a better economic deal. Fruit cups, pancakes, fruit-and-yogurt parfaits, and pre-sliced, packaged fruit are better choices for fast-food breakfasts than breakfast sandwiches.

As adolescents spend more time with peers, participation in family meals tends to steadily decline.[19-21] This is an unfortunate occurrence, as more frequent family meals are associated with improved dietary intake among adolescents, including higher intakes of grains, fruits, and vegetables, and decreased intake of soft drinks. Vitamin and mineral intakes, including those of calcium, folate, fiber, iron, and vitamins A, C, E, B_6, and B_{12}, are higher among adolescents who frequently consume family meals.[19-22] Similar improvements in dietary intake are found among older adolescents living away from their parents who eat meals with others compared to those who eat alone.[19] A greater number of families that are eating together do so in front of the television, with a national sample of adolescents indicating that 64% of 11- to 18-year-olds had the TV on during meals.[23] Adolescents eating meals in front of the TV, even with family, may be more susceptible to food advertisements and product placements, which tend to be for higher-calorie and low-nutrient-dense foods.

Vegetarian Diets

The term *vegetarian* is used quite broadly and can consist of many different eating patterns. Table 14.3 lists the most common vegetarian diet patterns along with

Table 14.3 Types of vegetarian diets and foods excluded

Type of Vegetarian Diet	Foods Excluded
Semi- or partial-vegetarian	Red meat
Lacto-ovo-vegetarian	Meat, poultry, fish, seafood
Lacto-vegetarian	Meat, poultry, fish, seafood, eggs
Vegan (total vegetarian)	Meat, poultry, fish, seafood, eggs, dairy products (may exclude honey)
Macrobiotic	Meat, poultry, eggs, dairy, seafood, fish (fish may be included in the diets of some macrobiotic vegetarians)

SOURCE: Reprinted with permission. Hadded E. Johnston P. Vegetarian Diets and Pregnant Teens. In: Story M, Stang J, eds. Nutrition and the Pregnant Adolescent: A Practical Reference Guide. Minneapolis. MN: Center for Leadership, Education, and Training in Maternal and Child Nutrition, University of Minnesota; 2000.

the foods most commonly excluded. Among low-literacy populations or those who do not speak English as a first language, *vegetarian* may be thought to refer to a person who eats vegetables. Therefore, health professionals should ask adolescents to define what type of vegetarian diet they consume and elicit a complete list of foods that are avoided.

The prevalence of vegetarianism among adolescents is small—approximately 4% of adolescents report currently consuming a vegetarian diet, but almost 11% report having identified themselves as vegetarian at some point in time.[24] Adolescents adopt vegetarian eating plans for a variety of reasons, including cultural or religious beliefs, moral or environmental concerns, health beliefs, as a means to restrict calories and/or fat intake, and as a means of exerting independence by adopting eating behaviors that differ from those of the teen's family. Regardless of the reason for consuming a vegetarian diet, the adolescent's diet should be thoroughly assessed for nutritional adequacy. As a rule, the more foods that are restricted in the diet, the more likely it is that nutritional deficiencies will result.

Vegetarian adolescents have been found to be shorter and leaner than omnivores during childhood and to enter puberty at a later age. On average, menarche occurs 6 months later in vegetarians than among omnivores.[25] After puberty, vegetarian adolescents are as tall as or taller than omnivores and are generally leaner, although final adult height may be reached at a later age.[25,26]

Well-planned vegetarian diets can offer many health advantages to adolescents, such as a high intake of fiber and relatively high intake of the vitamins and minerals found in plant-based foods. Data suggest that vegetarian adolescents consume more fruits and vegetables, fewer sweets, fewer salty snack foods, and less fat compared to omnivorous teens.[24] When well-planned, vegetarian diets can provide adequate protein to promote growth and development among pubescent adolescents, particularly if small amounts of animal-derived foods, such as milk or cheese, are consumed at least two times per week. If vegetarian diets restrict intake of all animal-derived food products such as in vegan diets, however, careful attention must be paid to ensure adequate intakes of protein, calcium, zinc, iron, and vitamins D, B_6, and B_{12}.[26] Supplements of vitamins B_{12} and D and calcium are often required among vegans unless fortified foods are routinely consumed. A suggested dietary food guide for adolescent vegetarians is listed in Table 14.4.

Adolescents who consume vegan diets must also be assessed for adequacy of total fat and essential fatty acid intakes. Docosahexaenoic acid (DHA) is derived from alpha-linolenic acid. Although it is found in soy products, flaxseed, nuts, eggs, and canola oil, intake is very low in the diets of vegans. Diets that are low in fat may not supply an adequate ratio of linoleic acid to alpha-linolenic acid (5:1 to 10:1) in order to facilitate the metabolism of alpha-linolenic acid to DHA.[26] Therefore,

Table 14.4 Suggested daily food guide for lacto-ovo and vegan vegetarians 11 years and older requiring 2200–2800 Kcals per day

Food Groups	Lacto-Ovo Vegetarians	Vegan vegetarians
Breads, grains, cereal	9–11	10–12
Legumes	2–3	3 or more
Vegetables	4–5	5 or more
Fruits	4	4 or more
Nuts, seeds	1	4–6
Milk, yogurt, cheese	4	—
Soy, almond or rice milk (fortified with calcium and vitamin D)	—	4
Eggs	½–1	—

SOURCE: Data used with permission from E. H. Haddad, "Development of a Vegetarian Food Guide," in American Journal of Clinical Nutrition, 1994:59:307–16; and M. Story, et al. (Eds.), Bright Futures in Practice: Nutrition. © 2000 National Center for Education in Maternal and Child Health.

Table 14.5 Plant sources of alpha-linolenic acid

Food Source	Alpha-Linolenic Acid, g
Flaxseed, 2 Tbsp	4.3
Walnuts, 1 oz	1.9
Walnut oil, 1 Tbsp	1.5
Canola oil, 1 Tbsp	1.6
Soybean oil, 1 Tbsp	0.9
Soybeans, ½ c cooked	0.5
Tofu, ½ c	0.4

SOURCE: Reprinted with permission from E. Haddad and P. Johnston, "Vegetarian Diets and Pregnant Teens," in M. Story and J. Strang (Eds.), Nutrition and the Pregnant Adolescent: A Practical Reference Guide. Minneapolis, MN: Center for Leadership, Education, and Training in Maternal and Child Nutrition, University of Minnesota, 2000.

particular attention should be paid to sources of fat in the diets of vegans and other vegetarians with low fat intakes. Some plant sources of alpha-linolenic acid are shown in Table 14.5.

Adolescents who consume a vegetarian diet, particularly if they report doing so for health- or weight-related reasons, should be carefully assessed for the presence of eating disorders, chronic dieting, and body-image disturbances. Surveys of adolescents have shown that those who consume vegetarian diets are somewhat more likely to report binge eating, almost twice as likely to report frequent or chronic dieting, four times more likely to report purging, and up to eight times more likely to report laxative use than non-vegetarian peers.[24,27] These results seem to reflect the fact that many individuals who are chronic dieters or who have disorder eating patterns adopt a vegetarian diet as a means of restricting fat intake or practicing self-denial rather than the vegetarian diet causing the restrictive behaviors. Adolescent vegetarians, especially adolescent males, have been found to be at high risk for engaging in unhealthy and often extreme weight-loss behaviors.[27] Therefore, it is imperative that practitioners working with adolescents who consume a vegetarian diet explore reasons for adopting this eating style and encourage adolescents follow a nutritionally balanced and healthy diet.

Dietary Intake and Adequacy Among Adolescents

Data on food intakes of U.S. adolescents suggest that many adolescents consume diets that do not match the Dietary Guidelines for Americans or the MyPyramid recommendations.[28,29] Inadequate consumption of dairy products, grain products, fruits, and vegetables is commonplace among adolescents (Table 14.6).

Few adolescents meet recommendations for fruit or vegetable consumption.[30] Data from NHANES suggest that 35% of adolescent males and 34% of adolescent females met recommendations for vegetable intake, while 23% of males and 27% of females met recommendations for fruit intake.[30] Approximately 30% of adolescents consumed less than one serving of vegetables each day, with white potatoes making up half of the vegetables consumed by teens. Almost two-thirds (62%) of male and 57% of female teens consumed less than one serving of fruit per day. Fruit and vegetable intake by teens is not adequate to promote optimal health and reduce risk of chronic diseases.

Adequate intake of servings of grains were reported by 64% of teenage males and 48% of teen females; however, intake of whole grains was below recommended levels.[30] Intake of meat/meat alternatives was

low among adolescents surveyed in NHANES.[30] Ten percent of males and 18% of females reported less than one serving per day, with only 50% of males and 17% of females meeting recommended intakes. Intake of dairy products was especially low among adolescents. One-third of teen males and 17% of teen females met recommendations for dairy intake, according to NHANES data.[30] More than one-third (39%) of female and 29% of male teens reported intake of less than one serving of dairy per day.

It was reported that fat made up 32% of all energy consumed by adolescents.[30] Added sugars accounted for 21% of energy intake by teens. Male teens consumed 35 tsp of added sugar each day, while female teens consumed 26 tsp per day. Clearly, adolescents do not consume diets that comply with the national nutrition recommendations or that provide the recommended level of intakes for all food groups.

Energy and Nutrient Requirements of Adolescents

Increases in lean body mass, skeletal mass, and body fat that occur during puberty result in energy and nutrient needs that exceed those at any other point in life. Energy and nutrient requirements of adolescence correspond with the degree of physical maturation that has taken place. Unfortunately, there is little available data that define optimal nutrient and energy intakes during adolescence. Most existing data are extrapolated from adult or child nutritional requirements. Recommended intakes of energy, protein, and some other nutrients are based on adequate growth as opposed to optimal physiological functioning. The Dietary Reference Intakes (DRIs) provide the best estimate of nutrient requirements for adolescents (Table 14.7). It should be noted, however, that these nutrient recommendations are classified according to chronological age, as opposed to individual levels of biological development. Thus, health care professionals must use prudent professional judgment based on SMR status, and not solely on chronological age, when determining the nutrient needs of an adolescent.

Nutrient intakes of U.S. adolescents suggest that many consume inadequate amounts of vitamins and minerals; this trend is more pronounced in females than in males. It is not surprising, given the fact that most adolescents do not consume diets that comply with MyPyramid or the Dietary Guidelines for Americans. On average, adolescents consume diets inadequate in several vitamins and minerals, including folate; vitamins A, B$_6$, C, and E; and iron, zinc, magnesium, phosphorus, and calcium

Table 14.6 Percentage of adolescents meeting the recommended number of MyPyramid servings for select food groups

	Male (%)	Female (%)
Dairy products	33	17
Fruits	23	27
Vegetables	35	34
Grains	64	48
Meat	50	17

SOURCE: Pyramid Serving Intakes in the U.S. 1999–2002, 1 Day. Community Nutrition Research Group. Beltsville Human Nutrition Research Center. Agricultural Research Service, USDA. Available online at www.ba.urs.usda.gov/enrg (accessed on June 20, 2009).

Table 14.7 Dietary reference intakes of selected nutrients for preadolescents and adolescents

Life-Stage Group	Calcium (mg/d)	Phosphorus (mg/d)	Magnesium (mg/d)	Vitamin D (mg/d)[a,b]	Fluoride (mg/d)	Thiamin (mg/d)	Riboflavin (mg/d)	Niacin (mg/d)[c]
Males								
9–13 years	1300*	1250	240	5*	2*	0.9	0.9	12
14–18 years	1300*	1250	410	5*	3*	1.2	1.3	16
19–30 years	1000*	700	400	5*	4*	1.2	1.3	16
Females								
9–13 years	1300*	1250	240	5*	2*	0.9	0.9	12
14–18 years	1300*	1250	360	5*	3*	1.0	1.0	14
19–30 years	1000*	700	310	5*	3*	1.1	1.1	14
Pregnancy								
<18 years	1300*	1250	400	5*	3*	1.4	1.4	18
19–30 years	1000*	700	350	5*	3*	1.4	1.4	18
Lactation								
<18 years	1300*	1250	360	5*	3*	1.4	1.6	17
19–30 years	1000*	700	310	5*	3*	1.4	1.6	17

Life-Stage Group	Vitamin B_6 (mg/d)	Folate, (mg/d)	Vitamin B_{12} (mg/d)	Pantothenic Acid (mg/d)	Biotin (mg/d)	Choline (mg/d)[d]	Vitamin C (mg/d)	Vitamin E (mg/d)	Selenium (mg/d)
Males									
9–13 years	1.0	300	1.8	4*	20*	375*	45	11	40
14–18 years	1.3	400	2.4	5*	25*	550*	75	15	55
19–30 years	1.3	400	2.4	5*	30*	550*	90	15	55
Females									
9–13 years	1.0	300	1.8	4*	20*	375*	45	11	40
14–18 years	1.2	400[e]	2.4	5*	25*	400*	65	15	55
19–30 years	1.3	400[e]	2.4	5*	30*	425*	75	15	55
Pregnancy									
<18 years	1.9	600	2.6	6*	30*	450*	80	15	60
19–30 years	1.9	600	2.6	6*	30*	450*	85	15	60
Lactation									
<18 years	2.0	500	2.8	7*	35*	550*	115	19	70
19–30 years	2.0	500	2.8	7*	35*	550*	120	19	70

NOTE: This table presents Recommended Dietary Allowances (RDAs) in **bold type** and Adequate Intakes (AIs) in ordinary type followed by an asterisk (*). RDAs and AIs may both be used as goals for individual intake. RDAs are set to meet the needs of almost all (97–98%) individuals in a group. For healthy and breastfed infants, the AI is the mean intake. The AI for other life stage and gender groups is believed to cover needs of all individuals in the group, but lack of data or uncertainty in the data prevent being able to specify with confidence the percentage of individuals covered by this intake.

[a] As cholecalciferol. 1 mg cholecalciferol = 40 IU vitamin D.

[b] In the absence of adequate exposure to sunlight.

[c] As niacin equivalents (NE). 1 mg of niacin = 60 mg tryptophan; 0–6 months = preformed niacin (not NE).

[d] Although AIs have been set for choline, there are few data to assess whether a dietary supplement of choline is needed at all stages of the life cycle, and it may be that the choline requirement can be met by endogenous synthesis at some of these stages.

[e] In view of evidence linking folate intake with neural tube defects in the fetus, it is recommended that all women capable of becoming pregnant consume 400 mg from supplements or fortified food until their pregnancy is confirmed and they enter prenatal care, which ordinarily occurs after the end of the periconceptional period—the critical time for formation of the neural tube.

SOURCE: Reprinted with permission from Dietary Reference Intakes: Recommended Intakes for Individuals, © by the National Academy of Sciences. Courtesy of the National Academy Press, Washington, D.C.

(Table 14.8).[31] Dietary fiber intake among adolescents is also low. Diets consumed by many teens exceed current recommendations for total and saturated fats, cholesterol, sodium, and added sugar. Data on nutrient intakes of adolescents taken from *What We Eat in America 2001–2002*

suggest that more than half of teens consume less than the Estimated Average Requirement (EAR) for vitamins A and E and magnesium. More than a quarter of adolescents consume less than the EAR for vitamin C, with more than a quarter of females also consuming less than the EAR for

Table 14.8 Mean intakes of selected nutrients compared to DRIs, adolescent males and females

	9- to 13-Year-Old Males				14- to 18-Year-Old Males			
	Mean Intake	RDA/AI	EAR	% < EAR	Mean Intake	RDA/AI	EAR	% < EAR
Vitamin A	670	600	445	13	638	700	630	55
Thiamin	1.78	0.9	0.7	<3	1.96	1.2	1.0	<3
Riboflavin	2.51	0.9	0.8	<3	2.57	1.3	1.1	<3
Niacin	22.5	12	9	<3	27.0	16	12	<3
Vitamin B_6	1.87	1.0	0.8	<3	2.17	1.3	1.1	<3
Vitamin B_{12}	6.0	1.8	1.5	<3	6.69	2.4	2.0	<3
Folate	644	300	250	<3	683	400	330	4
Vitamin C	80.2	45	39	8	100.0	75	63	26
Vitamin E	6.0	11	9.0	97	7.3	15	12	>97
Calcium	1139	1300	NA	NA	1142	1300	NA	NA
Phosphorous	1431	1250	1055	9	1575	1250	1055	9
Magnesium	250	240	200	14	284	410	340	78
Iron	17.0	8	5.9	<3	19.1	11	7.7	<3
Zinc	13.0	8	7.0	<3	15.1	11	8.5	4
Sodium	3549	1500	NA	NA	2806	1500	NA	NA
Fiber	14.2	31	NA	NA	15.3	38	NA	NA

	9- to 13-Year-Old Females				14- to 18-Year-Old Females			
	Mean Intake	RDA/AI	EAR	% < EAR	Mean Intake	RDA/AI	EAR	% < EAR
Vitamin A	536	600	420	34	513	700	485	54
Thiamin	1.44	0.9	0.7	<3	1.4	1.0	0.9	12
Riboflavin	1.94	0.9	0.8	<3	1.80	1.0	0.9	6
Niacin	18.5	12	9	<3	18.6	14	11	6
Vitamin B_6	1.52	1.0	0.8	<3	1.48	1.2	1.0	16
Vitamin B_{12}	4.4	1.8	1.5	<3	4.16	2.4	2.0	8
Folate	512	300	250	<3	500	400	330	19
Vitamin C	81.0	45	39	9	75.6	65	56	42
Vitamin E	5.6	11	9	95	5.6	15	12	>97
Calcium	865	1300	NA	NA	804	1300	NA	NA
Phosphorous	1141	1250	1055	42	1099	1250	1055	49
Magnesium	215	240	200	44	206	360	300	91
Iron	13.7	8	5.7	<3	13.3	15	7.9	16
Zinc	9.8	8	7.0	10	9.5	9	7.3	26
Sodium	2806	1500	NA	NA	2799	1500	NA	NA
Fiber	12.3	26	NA	NA	11.7	26	NA	NA

Data from: Moshfegh A, Goldman J, Cleveland L. 2005. NHANES 2001–2002: Usual Nutrient Intakes from Food Compared to Dietary Reference Intakes. Accessed online at www.ars.usda.gov/foodsurvey, June 20, 2009.

phosphorus and zinc.[32] More than one-third of females consume inadequate amounts of all of these nutrients.[31]

Based on growth and development of adolescents, as well as national findings on dietary intakes of foods and nutrients, the dietary intakes of adolescents should be assessed for adequacy of intake of vitamins, minerals, energy, protein, carbohydrates, and fiber during routine health care visits. Nutrients of particular concern for teens are discussed in greater detail in the following sections.

Energy

Energy needs of adolescents are influenced by activity level, basal metabolic rate, and increased requirements to support pubertal growth and development. Basal metabolic rate (BMR) is closely associated with the amount of lean body mass of individuals. Because adolescent males experience greater increases in height, weight, and lean body mass, they have significantly higher caloric

Table 14.9 Recommended intakes of macronutrients based on IOM daily recommended intakes

	Estimated Energy Requirements (Kcals)	Carbohydrate (g)	% of Daily Energy from Carbohydrate	Fiber (g)	% of Daily Energy from Fat	Linoleic acid (g)	Alpha-linolenic acid (g)	Protein (g)	% of Daily Energy from Protein
Males									
9–13	2279	130	45–65	31	25–35	21	1.2	34	10–30
14–18	3152	130	45–65	38	25–35	16	1.6	52	10–30
Females									
9–13	2071	130	45–65	26	25–35	10	1.0	34	10–30
14–18	2368	130	45–65	26	25–35	11	1.1	46	10–30

requirements than do females. The estimated energy requirements for adolescents are listed in Table 14.9. Due to the great variability in the timing of growth and maturation among adolescents, the determination of energy needs based on velocity of growth will provide a better estimate than one based on chronological age.

The DRI for energy is based upon the assumption of a light to moderate activity level. Therefore, adolescents who participate in sports, those who are in training to increase muscle mass, and those who are more active than average may require additional energy to meet their individual needs. Approximately 35% of U.S. teenagers report participating in moderate to strenuous activity for at least 60 minutes per day at least 5 times per week.[33] Conversely, adolescents who are not physically active or those who have chronic or handicapping conditions that limit their mobility will require less energy to meet their needs. Physical activity has been found to decline throughout adolescence, with approximately 25% of adolescents reporting no moderate to strenuous physical activity.[33] Therefore, caloric needs of older adolescents who have completed puberty and are less active may be significantly lower than those of younger, active, still-growing adolescents.

Physical growth and development during puberty is sensitive to energy and nutrient intakes. When energy intakes fail to meet requirements, linear growth may be retarded and sexual maturation may be delayed. The standard way to gauge adequacy of energy intake is to assess height, weight, and body composition. If, over time, height as well as weight-for-height continuously fall within the same percentiles when plotted on gender-appropriate National Center for Health Statistics growth charts, it can be assumed that energy needs are being met. If percentile of weight-for-height measurements begin to fall or rise, a thorough assessment of energy intake should be done, and adjustments in energy intake should be made accordingly. The use of body-fat measurements, such as triceps and subscapular skinfold measurements, can provide useful information when weight-for-height does not remain consistent. Remember, however, that transient increases and decreases in body fat are commonly noted among adolescents

during puberty due to the variation in timing of increases in height, weight, and accumulation of body fat and lean body mass. Repeated measurements of weight, height, and body composition over a several-month period are needed to accurately assess adequacy of growth and development.

Protein

Protein needs of adolescents are influenced by the amount of protein required for maintenance of existing lean body mass, plus allowances for the amount required to accrue additional lean body mass during the adolescent growth spurt. The estimated protein need for adolescents is 0.85 g/kg body weight/day, slightly higher than that of adults.[34] Because protein needs vary with the degree of growth and development, requirements based on developmental age will be more accurate than absolute recommendations based on chronological age.

Recommended protein intakes are shown in Table 14.9. Protein requirements are highest for females at 11 to 14 years and for males at 15 to 18 years, when growth is at its peak. Similar to energy needs, estimation of protein needs based on timing of growth rather than chronological age is most accurate. As with energy, growth is affected by protein intakes. When protein intakes are consistently inadequate, reductions in linear growth, delays in sexual maturation, and reduced accumulation of lean body mass may be seen. Subgroups of adolescents may be at risk for marginal or low protein intakes, including those from food-insecure households, those who severely restrict calories, and those who consume vegetarian diets, most notably vegans.

Carbohydrates

Carbohydrates provide the body's primary source of dietary energy. Carbohydrate-rich foods such as fruit, vegetables, whole grains, and legumes are also the main source of dietary fiber. The recommended intake of carbohydrate among teens is 130 g/day or 45–65% of daily energy needs (Table 14.9). Sweeteners and added sugars provide approximately 21% of energy intake by teens. Males consume 35 tsp and female teens consume 26 tsp of added sugar per day.[30] Soft drinks, candy, baked goods, and sweetened beverages are major sources of added sweeteners in the diets of adolescents.

Dietary Fiber

Dietary fiber is important for normal bowel function and may play a role in the prevention of chronic diseases such as certain cancers, coronary artery disease, and type 2 diabetes mellitus. Adequate fiber intake is also thought to reduce serum cholesterol levels, moderate blood sugar levels, and reduce the risk of obesity. The DRIs set the recommended intake of dietary fiber for adolescent females at 26 g/day, for males <14 years of age at 31 g/day, and for older adolescent males at 38 g/day.[34]

National data indicate that adolescent males consume 14.2 grams of fiber per day, while adolescent females consume 12.3 g/day, well short of AAP and DRI recommendations.[32] During adolescence, fiber intake among males increases slightly with age, while it decreases with age among females. The low intake of fruits and vegetables, combined with an average intake of less than one serving of whole grains per day among adolescents, are contributing factors affecting fiber intake among adolescents.[30]

Fat

The human body requires dietary fat and essential fatty acids for normal growth and development. Current recommendations suggest that children over the age of 2 years consume no more than 25–35% of calories from fat, with no more than 10% of calories derived from saturated fat.[34, 35] Data on energy and macronutrient intakes among adolescents suggest that approximately 32% of total calories consumed are derived from fat.[30] Approximately two-thirds of teens meet the recommendations for fat intake. National dietary guidelines also suggest that adolescents consume no more than 300 mg of dietary cholesterol per day.[34,35] The DRIs recommend specific intake of linoleic and alpha-linolenic acid to support optimal growth and development (Table 14.9).

Calcium

Achieving an adequate intake of calcium during adolescence is crucial to physical growth and development. Calcium is the main constituent of bone mass. Because about half of peak bone mass is accrued during adolescence, calcium intake may be of great importance for the development of dense bone mass and the reduction of the lifetime risk of fractures and osteoporosis. Additionally, calcium needs and absorption rates are higher during adolescence than any other time except infancy.[6, 37] Female adolescents appear to have the greatest capability to absorb calcium at about the time of menarche, with calcium absorption rates decreasing from then on.[6,37] Calcium absorption rates in males also peak during early adolescence, a few years later than in females. Young adolescents have been found to retain up to four times as much calcium as young adults. Males accrue more bone mass than females at all ages during puberty, possibly due to lower intake of calcium, less weight-bearing stress on bone tissue, or hormonal influences. Clearly, an adequate intake of calcium is of paramount importance during adolescence, particularly among females.

The DRI for calcium for 9- to 18-year-olds is 1300 mg per day (Table 14.7). National data suggest that many adolescents, most notably females, do not consume the DRI for calcium. Adolescent females consume 849 mg calcium per day, while adolescent males have been found to consume about 1186 mg calcium each day.[38] These levels of dietary intake are not adequate to support the development of optimal bone mass, particularly for females. Supplements may be warranted for adolescents who do not consume adequate calcium from dietary sources.

Research suggests that adolescents are not able to meet daily calcium needs in diets that do not include dairy products without the use of calcium-fortified foods.[39] Adolescents increasingly consume their calcium in the form of fortified foods. A longitudinal study of individuals followed from childhood into young adulthood found that breads/grains, vegetables, non-alcoholic beverages, and cheese are common sources of calcium, with milk supplying less than 25% of daily intake.[40] Soy beverages are another source of calcium consumed by youth that are being considered for inclusion in school meal programs.[41] The availability of calcium from soy beverages appears low, however, and the equivalency of soy versus dairy as a calcium source is highly debated.[42] Other food sources of calcium must be carefully chosen when dairy intake is not adequate to meet daily needs.

The consumption of soft drinks by adolescents may displace the consumption of more nutrient-dense beverages, including milk and fortified juices. Studies have shown an inverse relationship between the intake of sweetened and carbonated beverages and the intake of milk and juice.[43,44] Because milk and fortified juices are significant sources of calcium in the diets of adolescents, interventions aimed at reducing consumption of soft drinks may be warranted.

Calcium consumption drops as age increases among both male and female adolescents; however, males consume greater amounts of calcium at all ages than do females.[38] Calcium intakes among adolescents are highly correlated with energy intakes. When dietary calcium intake is adjusted for energy intake, no differences in calcium density of diets are found between males and females. This fact suggests that females who restrict calories in an effort to control their weight are at particularly high risk for inadequate calcium intakes. Some variation in calcium intake follows race categories among females: Cuban, Asian, and black females consume less calcium on average than do Mexican American, Puerto Rican, and white, non-Hispanic females.

Serum Iron, Plasma Ferritin, and Transferrin Saturation Measures of iron status obtained from blood plasma or serum samples.

Heme Iron Iron contained within a protein portion of hemoglobin that is in the ferrous state.

Nonheme Iron Iron contained within a protein of hemoglobin that is in the ferric state.

In addition to calcium, physical activity also plays a role in bone development during adolescence. Physical activity patterns and participation during adolescence have been shown to be a strong predictor of adult bone density. Participation in weight-bearing activities may lead to increased bone mineral density as compared to that accrued by more sedentary adolescents.[45,46]

Iron

The rapid rate of linear growth, the increase in blood volume, and the onset of menarche during adolescence increase a teen's need for iron. The DRIs for iron for male and female adolescents are shown in Table 14.7. These recommendations are based on the amount of dietary iron intake needed to maintain a suitable level of iron storage, with additional amounts of iron added to cover the rapid linear growth and onset of menstruation that occur in male and female adolescents, respectively. Note that even though DRIs are based on chronological age, the actual iron requirements of adolescents are based on sexual maturation level. Iron needs of an adolescent will be highest during the adolescent growth spurt in males, and after menarche in females.

The age-specific hemoglobin and hematocrit values used to determine iron-deficiency anemia are listed in Table 14.10. Hemoglobin and hematocrit levels, although commonly used to screen for the presence of iron-deficiency anemia, are actually the last serum indicators of depleted iron stores to drop. More sensitive indicators of iron stores include *serum iron*, *plasma ferritin*, and *transferrin saturation*. These measures are expensive and not commonly used in the traditional medical setting, however. Estimates of iron deficiency among adolescents are 9% of 12- to 15-year-old females, 5% of 12- to 16-year-old males, 11% of 15- to 19-year-old females, and 2% of 15- to 19-year-old males.[47] While iron deficiency occurs more frequently in all adolescents, iron-deficiency anemia occurs almost exclusively in females, with a prevalence of <1% of males and 2% among females.[47] Therefore, it is assumed that although the prevalence of iron-deficiency anemia may be relatively low among adolescents, a larger proportion may have inadequate iron stores. Rates of iron deficiency and anemia are twice as high among black and Mexican American females compared to white females and are three times higher than the target prevalence goals set by Healthy People 2010.[48] This finding is particularly relevant among adolescents from low SES homes, because rates of iron deficiency tend to be higher in adolescents from low-income families.

The availability of dietary iron for absorption and utilization by the body varies by its form. The two types of dietary iron are *heme iron*, which is found in animal products, and *nonheme iron*, which is found in both animal and plant-based foods. Heme iron is highly bioavailable, while nonheme iron is much less so. More than 80% of the iron consumed is in the form of nonheme iron. Bioavailability of nonheme iron can be enhanced by consuming it with heme sources of iron or vitamin C. This point is particularly salient for adolescents who avoid animal foods as a means of restricting calories and those who consume few animal-based foods (semivegetarian) or vegetarian diets for moral or cultural reasons.

Dietary intakes of iron are estimated at 19.6 mg/day among 12–19-year-old males and 13.3 mg/day among 12–19-year-old females.[38] Data suggest that while <3% of adolescent males and young adolescent females consume less than the DRI for iron, the prevalence of very low iron intake is 16% among older adolescent females.[32] In light of the frequency of iron deficiency among adolescents, particularly females, nutrition education and counseling for teens to promote higher iron consumption is warranted.

Vitamin D

Vitamin D is a fat-soluble vitamin that plays an essential role in facilitating intestinal absorption of calcium and phosphorus that is required to maintain adequate serum levels of these minerals.[34] Vitamin D can be synthesized by the body

Table 14.10 Common biochemical indices

	Normal	Borderline	High
Hyperlipidemia			
Total cholesterol (mg/dl)	<170	170–199	≥200
LDL cholesterol (mg/dl)	<110	110–129	≥130
HDL cholesterol (mg/dl)	≥40		
Triglycerides (mg/dl)	≥200		
Iron Status			
Hemoglobin (g/dl)	Males ≥12.5 (12–15 yr) ≥13.3 (16–18 yr) ≥13.5 (18+ yr)		
	Females ≥11.8 (12–15 yr) ≥12.0 (16+ yr)		
Hematocrit (%)	Males ≥37.3 (12–15 yr) ≥39.7 (16–18 yr) ≥39.9 (18+ yr)		
	Females ≥35.7 (12–15, 18+ yr) ≥35.9 (16–18 yr)		

through exposure of skin to ultraviolet B rays of sunlight. However individuals who live in northern latitudes may not receive adequate exposure to sunlight in the winter months to facilitate synthesis of adequate amounts of vitamin D.[49] Individuals with dark skin pigmentation may also experience limited vitamin D production by the body.[34,50]

Vitamin D is essential for optimal bone formation. There appears to be an inverse relationship between serum parathyroid hormone (PTH) and vitamin D levels.[34,50] Even in the earliest states of vitamin D deficiency when no overt symptoms of deficiency are seen, PTH is elevated in order to maintain serum calcium levels through demineralization of bone. As deficiency progresses, absorption of calcium from the gastrointestinal tract is reduced, resulting in even higher levels of PTH being released into circulation and further bone demineralization.

Longitudinal data from NHANES surveys has demonstrated that serum 25 (OH) vitamin D levels decreased among adolescents during the past few decades.[49] Data from NHANES III (1988–1994) found mean serum 25 (OH) vitamin D levels of 32 ng/mL among 12–19-year-olds, while mean levels measured among the same age group during NHANES 2001–2004 were 24 ng/mL. Decreases in serum vitamin D status were seen among all age and race/ethnicity groups, but were particularly pronounced among black Americans. Females showed greater decreases in serum vitamin D status than did males. This is particularly concerning for teenage females, given their low intake of calcium during adolescence and higher risk for osteoporosis later in life. Using criteria of <10 ng/mL for deficiency and ≥30 ng/mL for sufficiency, it is believed that less than 1% of white adolescents are vitamin D deficient, but 39% of females and 29% of males do not have sufficient vitamin D status.[49] Approximately 2% of Mexican American female and <1% of male teens in the U.S. are vitamin D deficient; however, 59% of male and 76% of female Mexican American adolescents were vitamin D insufficient, according to the study. Black teens had the highest rates of vitamin D deficiency and insufficiency, with 4% of male and 10% of female teens found to be deficient and 75% of black male and 92% of black female adolescents found to be vitamin D insufficient. Clearly, vitamin D insufficiency among adolescents of color should be considered a major public health-nutrition issue.

Vitamin D is naturally found in very few foods—namely, fatty fish, fish oils, and egg yolks of hens provided with vitamin D-fortified feed.[34] The majority of dietary vitamin D intake in North America comes from fortified foods such as milk, breakfast cereals, margarines, and some juices.[34] The AI for vitamin D among adolescents is quite low, at only 200 IU (5 μg) per day; the suggested upper limit is 2000 IU (50 μg) per day. The American Academy of Pediatrics (AAP) has recommended that all adolescents who do not consume at least 400 IU of vitamin D per day through dietary sources

receive a supplement of 400 IU per day. The AAP has recommended that the criterion for deficiency be 20 ng/mL, which is lower than that used in the NHANES surveys.[50] Vitamin D intake and adequacy should be assessed for all adolescents of high-risk groups, particularly those who live in northern climates, who have limited sun exposure, who have lactose intolerance or milk allergy, who have developmental disabilities that may limit outdoor activities, or who have darkly pigmented skin.

Folate

Folate is an integral part of DNA, RNA, and protein synthesis. Thus, adolescents have increased requirements for folate during puberty. The DRI for folate is listed in Table 14.7. Folate in the form of folic acid is twice as bioavailable as other forms of folate. For this reason, dietary folate equivalents (DFEs) are used in the DRIs. One microgram of folic acid is equivalent to approximately 2 DFEs, while 1 microgram of other forms of folate is approximately equivalent to 1 DFE. Folic acid is the form of folate added to fortified cereals, breads, and other refined grain products.

Severe folate deficiency results in the development of megaloblastic anemia, which is rare among adolescents. Evidence, however, indicates that a significant proportion of adolescents have inadequate folate status. Red blood cell and serum folate levels drop during adolescence as sexual maturation proceeds, suggesting that increased folate needs during growth and development are not being met. Serum levels were found to drop significantly between childhood and adolescence in data gathered through NHANES 2005–2006, with levels of 16.1 ng/mL measured in children 4–11 years of age and levels of 11.6 ng/mL measured in adolescents.[51] According to NHANES 2005–2006 data, adolescents had the lowest red blood cell folate levels of any age group in the U.S., with levels of 233 ng/mL among males and 237 ng/mL among females. For this reason, sexual maturation level should be used to identify folate needs as opposed to chronological age.

Poor folate status among adolescent females also presents an issue related to reproduction. Studies show that adequate intakes of folate prior to pregnancy can reduce the incidence of spina bifida and selected other congenital anomalies and may reduce the risk of Down syndrome among offspring.[51] The protective effects of folate occur early in pregnancy, often before a woman knows she is pregnant. Thus, it is imperative that all women of reproductive age (15–44 years old) consume adequate folic acid, preferably through dietary sources, or if needed, through supplements.

Despite the low serum and red blood cell levels of folate, national data suggest that many adolescents consume adequate amounts of folate. Mean intakes of folate among adolescent males average 658 μg/day, while females consume 482 μg/day.[38] The DRI for folate among adolescents is 400 µg/day. However, up to 4% of male and 26% of female adolescents consume less than the EAR for folate, suggesting that there is a wide variation in folate intake among teens and that some groups may be particularly vulnerable to low folate status.[32,38] Teens who skip breakfast or do not commonly consume orange juice and ready-to-eat cereals are at an increased risk for having a low consumption of folate.

Vitamin C

Vitamin C is involved in the synthesis of collagen and other connective tissues. For this reason, vitamin C plays an important role during adolescent growth and development. The DRIs for 9- to 13-year-old and 14- to 18-year-old adolescents are shown in Table 14.7. Vitamin C intakes are generally adequate within the adolescent population. Mean intakes are estimated at 97 mg/day among teenage males and 75 mg/day among teenage females.[38] The prevalence of intakes below the EAR ranged from 8% in young adolescent males and 9% in young adolescent females to 26% of older adolescent males and 42% of older adolescent females.[32]

Vitamin C acts as an antioxidant. Smoking increases the need for this antioxidant within the body because it consumes vitamin C in oxidative reactions. Consequently, smoking results in reduced serum levels of vitamin C. Recommended levels of vitamin C intake are higher among smokers. Adolescents who use tobacco and other substances may have lower concern for health, leading to poorer-quality diets and consumption of fewer fruit and vegetables, which are primary sources of vitamin C.

Nutrition Screening, Assessment, and Intervention

The AAP has recommended ongoing review of dietary intake and nutritional status of children and adolescents.[52] The American Medical Association's Guidelines for Adolescent Preventive Services recommends that all adolescents receive annual health guidance related to healthy dietary habits and methods to achieve a healthy weight.[53] This health guidance begins by annually screening all adolescents for indicators of nutritional risk. Common concerns that should be investigated during nutrition screening include overweight, underweight, eating disorders, hyperlipidemia, hypertension, iron deficiency and/or anemia, food insecurity, and excessive intake of high-fat or high-sugar foods and beverages. Pregnant adolescents should also be assessed for adequacy of weight gain and compliance with prenatal vitamin–mineral supplement recommendations.

Nutrition screening should include an accurate measurement of height and weight, and calculation of BMI (body mass index). These data, plotted on age- and gender-appropriate National Center for Health Statistics 2000 growth charts, indicate the presence of any weight or other growth problems. Indicators of height and weight status are listed in Table 14.11. Teens below the 5th percentile of weight-for-height or BMI-for-age are considered to be underweight and should be referred for evaluation of metabolic disorders, chronic health conditions, or eating disorders. Adolescents with a BMI above the 85th percentile but below the 95th percentile are considered to be overweight. They should be evaluated to determine the presence or absence of obesity-related co-morbid complications and referred for treatment as necessary. Teenagers with a BMI greater than 95th percentile are considered to be obese and should be referred for a full medical evaluation. Referral to a weight-management program specially designed to meet the needs of adolescents may also be warranted for overweight adolescents who have completed physical growth (see Chapter 15 for additional information).

Nutrition screening should include at least a brief dietary assessment. Food frequency questionnaires, 24-hour recalls, and food diaries or food records are all appropriate for use with adolescents. Table 14.12 lists the advantages and disadvantages of each dietary assessment method. Less formal dietary assessment questionnaires that target specific behaviors, such as consumption of savory snacks and high-sugar beverages, can also be used for initial nutrition screening. These rapid assessment questionnaires and screening tools can be completed quickly and may be used to identify those adolescents in need of additional dietary assessment and nutrition counseling.

Nutrition risk indicators that may warrant further nutrition assessment and counseling are listed in Table 14.13.

Table 14.11 Indicators of height and weight status for adolescents

Indicator	Body Size Measure	Cutoff Values
Stunting (low height-for-age)	Height-for-age	<3rd percentile
Thinness (low BMI-for-age)	BMI-for-age	<5th percentile
At risk for overweight	BMI-for-age	>85th percentile, but <95th percentile
Overweight	BMI-for-age	>95th percentile

SOURCE: From Haddad, E. © American Journal of Clinical Nutrition, 1943; 59: 1248S–1254S; and Story M. Holt, K., Sofka D. (eds.) 2000. BRIGHT FUTURES IN PRACTICE: Nutrition. Reprinted by permission of the American Journal of Clinical Nutrition.

Table 14.12 Strengths and limitations of various dietary assessment methods used in clinical settings

	Strengths	Limitations	Applications
24-Hours Recall	• Does not require literacy • Relatively low respondent burden • Data may be directly entered into a dietary analysis program • May be conducted in person or over the telephone	• Dependent on respondent's memory • Relies on self-reported information • Requires skilled staff • Time consuming • Single recall does not represent usual intake	• Appropriate for most people as it does not require literacy • Useful for the assessment of intake of a variety of nutrients and assessment of meal patterning and food group intake • Useful counseling tool
Food Frequency	• Quick, easy, and affordable • May assess current as well as past diet • In a clinical setting, may be useful as a screening tool	• Does not provide valid estimates of absolute intake of individuals • Can't assess meal patterning • May not be appropriate for some population groups	• Does not provide valid estimates of absolute intake for individuals, thus of limited usefulness in clinical settings • May be useful as a screening tool; however, further development research is needed
Food Record	• Does not rely on memory • Food portions may be measured at the time of consumption • Multiple days of records provide valid measure of intake for most nutrients	• Recording foods eaten may influence what is eaten • Requires literacy • Relies on self-reported information • Requires skilled staff • Time consuming	• Appropriate for literate and motivated population groups • Useful for the assessment of intake of a variety of nutrients and assessment of meal patterning and food group intake • Useful counseling tool
Diet History	• Able to assess usual intake in a single interview • Appropriate for most people	• Relies on memory • Time consuming (60 to 90 minutes) • Requires skilled interviewer	• Appropriate for most people as it does not require literacy • Useful for assessing intake of nutrients, meal patterning, and food group intake • Useful counseling tool

SOURCE: Used with permission. Story M, Stang I, eds. Nutrition and the Pregnant Adolescent: A Practical Reference Guide. Minneapolis, MN: Center for Leadership, Education, and Training in Maternal and Child Nutrition, University of Minnesota; 2000.

Adolescents who have a poor-quality diet characterized by an excessive intake of high-fat or high-sugar foods and beverages or meal skipping should be provided with nutrition counseling that provides concrete examples of ways to improve dietary intake. Adolescents who have been found to have a nutrition-related health risk, such as hyperlipidemia, hypertension, iron-deficiency anemia, overweight, or eating disorders, should be referred for in-depth medical assessment and nutrition counseling. Pregnant adolescents may also benefit from in-depth nutrition assessment and counseling.

In-depth nutrition assessment should include a review of the full medical history, a review of psychosocial development, and evaluation of all laboratory data available. A complete and thorough dietary assessment should be performed, preferably using two dietary assessment methods. Most commonly, a food frequency questionnaire or a 3- to 7-day food record is combined with a 24-hour recall to provide accurate dietary intake data. Specific areas of nutrition concern can be identified during a complete nutrition assessment, and recommendations for nutrition education and counseling can be made accordingly.

Nutrition Education and Counseling

Providing nutrition education and counseling to teenagers requires a great deal of skill and a good understanding of normal adolescent physical and psychosocial development. Health professionals need to remember that developmentally, adolescents may look like young adults, but they are not. Their psychosocial development can vary tremendously and they may not respond favorably to traditional adult counseling methods. It is also important to not provide nutrition education messages or materials that are too childish in nature. Adolescence falls between young adulthood and childhood and requires a unique approach to education and counseling. When working with teens, it is important to treat them as individuals with unique needs and concerns. This builds upon their

Table 14.13 Key indicators of nutrition risk for adolescents

Indicators of Nutrition Risk	Relevance	Criteria for Further Screening and Assessment
FOOD CHOICES		
Consumes fewer than 2 servings fruit or fruit juice per day	Fruits and vegetables provide dietary fiber and several vitamins (such as A and C) and minerals. Low intake of fruits and vegetables is associated with an increased risk of many types of cancer. In females of childbearing age, low intake of folic acid is associated with an increased risk of giving birth to an infant with neural tube defects.	Assess the adolescent who is consuming less than 1 serving of fruit or fruit juice per day.
Consumes fewer than 3 servings of vegetables per day		Assess the adolescent who is consuming fewer than 2 servings of vegetables per day.
Consumes fewer than 6 servings of bread, cereal, pasta, rice, or other grains per day	Grain products provide complex carbohydrates, dietary fiber, vitamins, and minerals. Low intake of dietary fiber is associated with constipation and an increased risk of colon cancer.	Assess the adolescent who is consuming fewer than 3 servings of bread, cereal, pasta, rice, or other grains per day.
Consumes fewer than 3 servings of dairy products per day	Dairy products are a good source of protein, vitamins, and calcium and other minerals. Low intake of dairy products may reduce peak bone mass and contribute to later risk of osteoporosis.	Assess the adolescent who is consuming fewer than 2 servings of dairy products per day. Assess the adolescent who has a milk allergy or is lactose intolerant. Assess the adolescent who is consuming more than 20 ounces of soft drinks per day.
Consumes fewer than 2 servings of meat or meat alternatives (e.g., beans, eggs, nuts, seeds) per day	Protein-rich foods (e.g., meats, beans, dairy products) are good sources of B vitamins, iron, and zinc. Low intake of protein-rich foods may impair growth and increase the risk of iron-deficiency anemia and of delayed growth and sexual maturation. Low intake of meat or meat alternatives may indicate inadequate availability of these foods at home. Special attention should be paid to children and adolescents who follow a vegetarian diet.	Assess the adolescent who is consuming fewer than 2 servings of meat or meat alternatives per day or who consumes a vegan diet.
Has excessive intake of dietary fat	Excessive intake of total fat contributes to the risk of cardio-vascular diseases and obesity and is associated with some cancers.	Assess the adolescent who has a family history of premature cardiovascular disease. Assess the adolescent who has a body mass index (BMI) greater than or equal to the 85th percentile.
EATING BEHAVIORS		
Exhibits poor appetite	A poor appetite may indicate depression, emotional stress, chronic disease, or an eating disorder.	Assess the adolescent if BMI is less than the 15th percentile or if weight loss has occurred. Assess if irregular menses or amenorrhea have occurred for 3 months or more. Assess for organic and psychiatric disease.
Consumes food from fast food restaurants three or more times per week	Excessive consumption of convenience foods and foods from fast food restaurants is associated with high fat, calorie, and sodium intakes, as well as low intake of certain vitamins and minerals.	Assess the adolescent who is at risk for overweight/obesity, or who has diabetes mellitus, hyperlipidemia, or other conditions requiring reduction in dietary fat.

Table 14.13 Key indicators of nutrition risk for adolescents (continued)

Indicators of Nutrition Risk	Relevance	Criteria for Further Screening and Assessment
Skips breakfast, lunch, or dinner/supper three or more times per week	Meal skipping is associated with a low intake of energy and essential nutrients and, if it is a regular practice, could compromise growth and sexual development. Repeatedly skipping meals decreases the nutritional adequacy of the diet.	Assess the adolescent to ensure that meal skipping is not due to inadequate food resources or unhealthy weight-loss practices.
Consumes a vegetarian diet	Vegetarian diets can provide adequate nutrients and energy to support growth and development if well planned. Vegan diets may lack calcium, iron, and vitamins D and B12. Low-fat vegetarian diets may be adopted by adolescents who have eating disorders.	Assess the adolescent who consumes fewer than 2 servings of meat alternatives per day. Assess the adolescent who consumes fewer than 3 servings of dairy products per day. Assess the adolescent who follows a low-fat vegetarian diet and experiences weight loss for eating disorder and adequacy of energy intake.
FOOD RESOURCES		
Has inadequate financial resources to buy food, insufficient access to food, or lack of access to cooking facilities	Poverty can result in hunger and compromised food quality and nutrition status. Inadequate dietary intake interferes with learning.	Assess the adolescent who is from a family with low income, is homeless, or is a runaway.
WEIGHT AND BODY IMAGE		
Practices unhealthy eating behaviors (e.g., chronic dieting, vomiting, and using laxatives, diuretics, or diet pills to lose weight)	Chronic dieting is associated with many health concerns (fatigue, impaired growth and sexual maturation, irritability, poor concentration, impulse to binge) and can lead to eating disorders. Frequent dieting in combination with purging is often associated with other health-compromising behaviors (substance use, suicidal behaviors). Purging is associated with serious medical complications.	Assess the adolescent for eating disorders. Assess for organic and psychiatric disease. Screen for distortion in body image and dysfunctional eating behavior, especially if adolescent desires weight loss, but BMI <85th percentile.
Is excessively concerned about body size or shape	Eating disorders are associated with significant health and psychological morbidity. Eighty-five percent of all cases of eating disorders begin during adolescence. The earlier adolescents are treated, the better their long-term prognosis.	Assess the adolescent for distorted body image and dysfunctional eating behaviors, especially if adolescent wants to lose weight, but BMI is less than the 85th percentile.
Has exhibited significant weight change in past 6 months	Significant weight change during the past 6 months may indicate stress, depression, organic disease, or an eating disorder.	Assess the adolescent to determine the cause of weight loss or weight gain (limited to too much access to food, poor appetite, meal skipping, eating disorder).
GROWTH		
Has BMI less than the 5th percentile	Thinness may indicate an eating disorder or poor nutrition.	Assess the adolescent for eating disorders. Assess for organic and psychiatric disease. Assess for inadequate food resources.
Has BMI greater than the 95th percentile	Obesity is associated with elevated cholesterol levels and elevated blood pressure. Obesity is an independent risk factor for cardiovascular disease and type 2 diabetes mellitus in adults. Overweight adolescents are more likely to be overweight adults and are at increased risk for health problems as adults.	Assess the adolescent who is overweight or at risk for becoming overweight (on the basis of present weight, weight-gain patterns, family weight history).

continued

Table 14.13 Key indicators of nutrition risk for adolescents (continued)

Indicators of Nutrition Risk	Relevance	Criteria for Further Screening and Assessment
PHYSICAL ACTIVITY		
Is physically inactive: engages in physical activity fewer than 5 days per week	Lack of regular physical activity is associated with overweight, fatigue, and poor muscle tone in the short term and a greater risk of heart disease in the long term. Regular physical activity reduces the risk of cardiovascular disease, hypertension, colon cancer, and type 2 diabetes mellitus. Weight-bearing physical activity is essential for normal skeletal development during adolescence. Regular physical activity is necessary for maintaining normal muscle strength, joint structure, and joint function; contributes to psychological health and well-being; and facilitates weight reduction and weight maintenance throughout life.	Assess how much time the adolescent spends watching television/videotapes and playing computer games. Assess the adolescent's definition of physical activity.
Engages in excessive physical activity	Excessive physical activity (nearly every day or more than once a day) can be unhealthy and associated with menstrual irregularity, excessive weight loss, and malnutrition.	Assess the adolescent for eating disorders.
MEDICAL CONDITIONS		
Has chronic diseases or conditions	Medical conditions (diabetes mellitus, spina bifida, renal disease, hypertension, pregnancy, HIV infection/AIDS) have significant nutritional implications.	Assess adolescent's compliance with therapeutic dietary recommendations. Refer to dietitian if appropriate.
Has hyperlipidemia	Hyperlipidemia is a major cause of atherosclerosis and cardiovascular disease in adults.	Refer adolescent to a dietitian for cardiovascular nutrition assessment.
Has iron-deficiency anemia	Iron deficiency causes developmental delays and behavioral disturbances. Another consequence is increased lead absorption.	Screen adolescents if they have low iron intake, a history of iron-deficiency anemia, limited access to food because of poverty or neglect, special health care needs, or extensive menstrual or other blood losses. Screen annually.
Has dental caries	Eating habits have a direct impact on oral health. Calcium and vitamin D are vital for strong bones and teeth, and vitamin C is necessary for healthy gums. Frequent consumption of carbohydrate-rich foods (e.g., lollipops, soda) that stay in the mouth a long time may cause dental caries. Fluoride in water used for drinking and cooking as well as in toothpaste reduces the prevalence of dental caries.	Assess the adolescent's consumption of snacks and beverages that contain sugar, and assess snacking patterns. Assess the adolescent's access to fluoride (e.g, fluoridated water, fluoride tablets).
Is pregnant	Pregnancy increases the need for most nutrients.	Refer the adolescent to a dietitian for further assessment, education, and counseling as appropriate.
Is taking prescribed medication	Many medications interact with nutrients and can compromise nutrition status.	Assess potential interactions of prescription drugs (e.g., asthma medications, antibiotics) with nutrients.

Table 14.13 Key indicators of nutrition risk for adolescents (continued)

Indicators of Nutrition Risk	Relevance	Criteria for Further Screening and Assessment
LIFESTYLE		
Engages in heavy alcohol, tobacco, and other drug use	Alcohol, tobacco, and other drug use can adversely affect nutrient intake and nutrition status.	Assess the adolescent further for inadequate dietary intake of energy and nutrients.
Uses dietary supplements	Dietary supplements (e.g., vitamin and mineral preparations) can be healthy additions to a diet, especially for pregnant and lactating women and for people with a history of iron-deficiency anemia; however, frequent use or high doses can have serious side effects. Adolescents who use supplements to "bulk up" may be tempted to experiment with anabolic steroids. Herbal supplements for weight loss can cause tachycardia and other side effects. They may also interact with over-the-counter prescription medications.	Assess the adolescent for the type of supplements used and dosages.
Assess the adolescent for use of anabolic steroids and megadoses of other supplements.

SOURCE: Adapted from M. Story, et al., Bright Futures in Practice: Nutrition. Copyright © 2000 National Center for Education in Maternal and Child Health.

developmental process of becoming autonomous and helps to foster a sense of respect and self-reliance.

The initial component of the counseling session should involve getting to know the adolescent, including personal health or nutrition-related concerns. After establishing a rapport with the teen, the counselor should provide an overview of the events of the counseling session, including which specific nutrition topics will be discussed. Once again, the adolescent should be encouraged to add his or her own nutrition concerns to the list of topics to be discussed during the education session. After agreeing on a list of topic areas to be covered during the nutrition education session, a complete nutrition assessment should be performed. Upon completion of the assessment, the counselor and teen should work together to establish goals for improving dietary intake and reducing nutrition risk.

It is important to involve the adolescent in decision-making processes during nutrition counseling. Allowing teens to provide input as to what aspects of their eating habits they think need to be changed and what changes they are willing to make accomplishes several important tasks during the counseling session. First, the importance of the adolescent in the decision-making process is stressed, and she or he is encouraged to become involved in personal decisions about health. Second, a good rapport established between the health professional and the adolescent may lead to greater interaction between both parties. Finally, behavior change is more likely when the adolescent has suggested ways to change, thus becoming engaged in the education process and owning a willingness to change.

One or two goals during a counseling session is a reasonable number to work toward. Setting too many goals reduces the probability that the adolescent can meet all of the goals and may seem overwhelming. For each goal set, several behavior-change strategies should be mutually agreed upon for meeting that goal. These strategies should be concrete in nature and instigated by the teen. The adolescent and the counselor should also work together to decide how to determine when a goal is met. The MyPyramid online system can be utilized by teens as a means of assessing how changes in food choices affect changes in nutrient intake across time. Frequent follow-up sessions also help to provide feedback and monitor progress toward individual goals.

The use of technology to facilitate nutrition education and counseling for adolescents should be implemented whenever possible to engage adolescents and to provide nutrition information between client visits. Electronic communications methods such as text messaging, podcasts, YouTube, and online social networking sites (e.g., Facebook, Twitter) are commonly accessed by adolescents and can serve as a means to convey nutrition information in a highly engaging way. In a study of health screening prior to a well visit delivered through the use of personal digital assistants, a significantly greater number of discussions related to fruit and vegetable intake occurred compared to well visits for teens without the use of technology.[54]

Physical Activity and Sports

Regular physical activity leads to many health benefits. Physical activity improves aerobic endurance and muscular strength, may reduce the risk of developing obesity, and builds bone mass density.[37,52] Physical activity among adolescents is consistently related to higher levels of self-esteem and self-concept and lower levels of anxiety and stress. Thus, physical activity is associated with both physiological and psychological benefits, especially during adolescence, which offers opportunities to positively influence the adoption of lifelong activity patterns. Increasing physical activity among adolescents is an important goal because regular physical activity declines during adolescence and many American teens are inactive.

Physical activity is defined as any bodily movement produced by skeletal muscles which results in energy expenditure. This definition is distinguished from exercise, which is a subset of physical activity that is planned, structured, and repetitive and is done to improve or maintain physical fitness. Physical fitness is a set of attributes that are either health- or skill-related. The Physical Activity Guidelines for Americans recommends that all adolescents be physically active daily, or nearly every day, as part of play, games, sports, work, transportation, recreation, physical education, or health promotion.[55] Further, it is recommended that adolescents engage in 60 minutes or more of physical activity at least 3 days of the week, with muscle- and bone-strengthening activities included at least 3 days per week.[28,55]

Despite common knowledge about the importance and benefits of physical activity, only 35% of U.S. adolescents meet physical activity guidelines, while 25% report no moderate to vigorous physical activity on any day in a week.[33] Moreover, physical activity has been shown to decline steadily throughout adolescence, especially among females. More males than females meet daily physical activity guidelines during adolescence. Racial and ethnic differences are also noted in physical activity; white teens are more likely to meet physical activity targets, while black students are more likely to report no physical activity.[33]

Factors Affecting Physical Activity

Individual, social, and environmental factors are associated with physical activity among adolescents. Females are less active than males, and among adolescent females, blacks are less active than whites.[33] Individual factors positively associated with physical activity among young people include confidence in one's ability to engage in exercise (i.e., self-efficacy), perceptions of physical or sports competence, having positive attitudes toward physical activity, enjoying physical activity, and perceiving positive benefits associated with physical activity (i.e., excitement, fun, adventure, staying in shape, improved appearance, weight control, improving skills). Social factors associated with engaging in physical activity are peer and family support. Environmental factors associated with physical activity are having safe and convenient places to play, sports equipment, and transportation to sports or fitness programs.

Schools offer an ideal setting for promoting physical activity through physical education classes. About half of U.S. adolescents attend a physical education class at least once a week, but less than one third of U.S. students attend physical education classes daily. Physical education classes may not offer significant health benefits to some students, as many students are not physically active for 20 minutes or more during physical education classes.[33] Community programs that offer non-competitive activities where all teens get an opportunity to participate for at least 30 minutes or more are essential to meeting physical activity goals for youth, because most physical activity among adolescents occurs outside of schools, particularly among females.

The Centers for Disease Control and Prevention published *Guidelines for School and Community Programs to Promote Lifelong Physical Activity Among Young People,* which provides a developmental framework for comprehensive school and community physical activity programs to be used by school districts, educators, health professionals, and policymakers.[56] The guidelines include recommendations (Table 14.14) to promote lifelong physical activity, including school policies, physical and social environments that encourage and enable physical activity, developmentally appropriate physical education curricula and instruction, personnel training, family and community involvement, and program evaluation.

High levels of physical activity, combined with growth and development, increase adolescents' needs for energy, protein, and select vitamins and minerals. Participation in competitive sports often means an adolescent will participate in intense training and competition during an athletic season. If the athlete competes in several sports, energy and nutrient needs will remain relatively stable throughout the year. If an athlete participates in only one sport and does not maintain a training routine off-season, energy and nutrient needs may fluctuate based on the timing of the sports season. Therefore, adolescents must be assessed for seasonal and yearly physical activity when energy and nutrient needs are determined.

The energy and nutrient needs of adolescent athletes vary widely. Many of the recommendations available are based on needs of young adult athletes or are extrapolated from usual nutrient needs of adolescents. The best method of assessing the nutrient needs of athletes is to begin with general dietary needs based on Sexual Maturation Rating (SMR), adding additional allowances based

Table 14.14 Recommendations for school health programs promoting healthy eating and physical activity

Healthy Eating	Physical Activity
• Adopt a coordinated school nutrition policy that promotes healthy eating through classroom lessons and a supportive school environment.	• Establish policies that promote enjoyable lifelong physical activity among young people.
• Implement nutrition education from preschool through secondary school as part of a sequential, comprehensive school health education curriculum designed to help students adopt healthy eating behaviors.	• Provide physical and social environments that encourage and enable safe and enjoyable physical activity.
• Provide nutrition education through developmentally appropriate, culturally relevant, fun participatory activities that involve social learning strategies.	• Implement physical education curricula and instruction that emphasize enjoyable participation in physical activity and that help students develop the knowledge, attitudes, motor skills, behavioral skills, and confidence needed to adopt and maintain physically active lifestyles.
• Coordinate school food service with other components of the comprehensive school health program to reinforce messages on healthy eating.	• Provide extracurricular physical activity programs.
• Involve family members and the community in supporting and reinforcing nutrition education.	• Include parents and guardians in physical activity instruction and in extracurricular and community physical activity programs, and encourage them to support their children's participation in enjoyable physical activities.
• Evaluate regularly the effectiveness of the school health program in promoting healthy eating, and change the program as appropriate to increase the effectiveness.	• Provide training for education, coaching, recreation, health care, and other school and community personnel that imparts the knowledge and skills needed to effectively promote enjoyable, lifelong physical activity among young people.
	• Assess physical activity patterns among young people, counsel them about physical activity, refer them to appropriate programs, and advocate for physical activity instruction and programs for young people.
	• Provide a range of developmentally appropriate community sports and recreation programs that are attractive to all young people.
	• Regularly evaluate school and community physical activity instruction, programs, and facilities.

SOURCE: Centers for Disease Control and Prevention. "Guidelines for School and Community Programs to Promote Lifelong Physical Activity among Young People, MMWR 45:1996; and Guidelines for School Health Programs to Promote Lifelong Healthy Eating," MMWR 45:1996.

upon the unique needs of the individual and the intensity of physical activity he or she engages in. In order to assess individual nutrient needs, health care professionals must gather information such as:

- What sport(s) does the adolescent engage in, and what is the duration of the competition season?
- What is the level of competition of the adolescent? Is participation recreational, competitive, or highly competitive?
- What kind of training does the adolescent engage in? The method(s), intensity, and duration of training activities should be noted.
- Does the athlete typically sweat profusely or lose body weight during competition?
- Does the athlete follow a special diet or take supplements to improve athletic performance? The type, amount, and frequency of supplement use should be noted and counseling provided as necessary.

General energy and protein needs are shown in Table 14.8. These guidelines should provide the foundation for calculating protein and energy needs for athletes. Competitive athletes may require 500–1500 additional calories per day to meet their energy needs. Athletes and their parents should be encouraged to monitor weight stability throughout the sports season. During the season, particularly during intense training phases or at the beginning of a season, athletes should weigh themselves before and after practice and sporting events. Any change in body weight during the activity signals a loss of body water, which could lead to dehydration. Any weight loss that is not transient (transient losses are often due to dehydration) signifies that the caloric intake is inadequate to support growth and development. A thorough assessment of energy and protein intakes, accompanied by measurements of body composition, should be taken when unexpected weight loss occurs. Protein should supply no more than 30% of calories in the diet. Groups at risk for inadequate intake would include athletes who follow vegan diets or restrict caloric intake to maintain a particular weight. When the main sources of protein are plant-based, additional protein intake may be needed because plant-based sources of protein may be less bioavailable.

Dietary intakes of athletes should follow the MyPyramid recommendations, with the realization that the increased energy needs of athletes may require them to consume the upper limit of food-group recommendations. Athletes should be encouraged to eat a pre-event meal at least 2 to 3 hours prior to exercise; eating too close to exercise may lead to indigestion and physical discomfort. Foods that are high in fat, high in protein, and high in dietary fiber should be avoided for at least 4 hours prior to exercise, because they take longer to digest and may cause physical discomfort during exercise. Protein and fat also displace complex carbohydrates, which are the most readily available source of energy during athletic events. Post-event meals should contain approximately 400–600 calories and should be comprised of high-carbohydrate foods and adequate amounts of non-caffeinated fluids.

Calcium intakes have been shown to be below the DRIs in a significant proportion of adolescents, especially females. Athletes' increased risk for bone fractures makes adequate calcium intake extremely important. Although the mechanism responsible for this tendency has not been identified, female adolescent athletes with low calcium consumption appear to be the highest-risk group of all adolescents for bone fractures, and they therefore should make every effort to consume adequate calcium in their diets. Teen athletes who cannot or will not consume calcium from dietary sources should be counseled to take a daily calcium supplement that meets their daily needs.

Promoting Healthy Eating and Physical Activity Behaviors

Meeting the challenge of improving the nutritional health of teenagers requires the integrated efforts of teenagers, parents, educators, health care providers, schools, communities, the food industry, and policymakers all working together to create more opportunities for healthful eating.

Effective Nutrition Messages for Youth

Health professionals need to rethink how they frame messages to youth. Years ago, Leverton pointed out that too often, teenagers have been given the message that good nutrition means "eating what you don't like because it's good for you."[57] Rather, they should be told to "eat well because it will help you in what you want to do and become." Teenagers are present-oriented and tend not to be concerned about how their eating will affect them in later years. However, they are concerned about immediate, socially relevant issues such as their physical appearance, achieving and maintaining a healthy weight, and having

lots of energy. Many are also interested in optimizing sports performance. Others are concerned with environmental or moral aspects of food. Even though adolescents need to be aware of the long-term risks of an unhealthy diet and benefits of a more healthful one, focusing on the short-term or tangible benefits will have more appeal to them and is more likely to result in behavior change.

Parent Involvement

Parents should be targets for nutrition education as well as teenagers, because they fill the role of gatekeepers of foods and serve as role models for eating behavior. Even though parents may have little control over what their teenagers are eating outside the home, they have more control in the home environment. Teenagers tend to eat what is available and convenient. Parents can capitalize on this by stocking the kitchen with a variety of nutritious ready-to-eat foods and limiting the availability of high-sugar, high-fat foods within the home. Focus groups of parents of teenagers suggest that parents have concerns over whether or not they should involve teens in choosing foods served at meals, prepare alternative foods for teens when they don't like what is served, or restrict intake of specific foods.[58] In general, adolescents should be involved in food purchasing and preparation as often as possible to provide them with food-preparation and decision-making skills. Parents should refrain from offering specially prepared foods for adolescents who do not like what is served, and should instead be encouraged to provide a balanced meal. Research has found that serving adolescents fruits, vegetables, and dairy at dinner time is related to higher intakes of these foods both during adolescence and five years later, during young adulthood.[59] While parental modeling of fruit, vegetable, and dairy intake has not been found to predict adolescent intake of these foods, it has been found to have a longer-term impact, improving intakes of these foods during young adulthood. The use of different creative settings and outlets to deliver innovative nutrition education programs to parents to encourage recommended behaviors—including such settings as work sites, places of worship, community centers, libraries, supermarkets, hair and nail salons, shopping centers, housing complexes, and restaurants—should be explored.

School Programs

School-based programs can play important roles in promoting lifelong healthy eating and physical activity. Efforts to promote physical activity and healthful eating should be part of a comprehensive, coordinated school health program and should include school health instruction (curriculum), school physical education, school food service, health services (screening and preventive counseling), school-site health promotion programs for faculty and staff, and integrated community efforts.[56,60]

The Centers for Disease Control and Prevention published two complementary reports, *Guidelines for School Health Programs to Promote Lifelong Healthy Eating*[60] and *Guidelines for School and Community Programs to Promote Lifelong Physical Activity Among Young People,*[56] which provide a developmental framework for comprehensive school nutrition and physical activity programs to be used by school districts, educators, health professionals, and policymakers. The guidelines (Table 14.14) include recommendations to promote healthy eating and lifelong physical activity, including school policies, physical and social environments that encourage and enable physical activity and healthy eating, developmentally appropriate nutrition and physical education curricula and instruction, personnel training, family and community involvement, and program evaluation. The CDC has also published the School Health Index, which can be used to assess the strengths and weaknesses of schools.[61]

Nutrition Education   Nutrition education is mandated in 67% of middle schools and 72% of high schools.[62] Approximately 75% of middle school and high school students received instruction in utilizing national food guidance such as MyPyramid and preparing healthy meals and snacks, according to the 2006 School Health Policies and Practices Survey (SHPPS).[62] Data from the 2004 School Health Profile survey show that most nutrition instruction is offered as part of a required health education course.[63] An average of 5 hours of nutrition and dietary behavior instruction is provided to high school students. The nutrition topics taught in health education classes can vary widely. The percentage of health education teachers provided with opportunities for nutrition education training increased from 43% in 2000 to 65% in 2006.[62] Teacher training in basic nutrition and instructional, motivational, and behavioral change strategies increases the success of a nutrition education curriculum. Training may be most effective if teachers have the opportunity to examine their own nutrition and health beliefs, identify their personal body image, and assess their eating behaviors. Teacher training typically increases the time spent on teaching nutrition in the classroom.

By senior high, students are in the process of cognitive and social development changes that permit more advanced nutrition education concepts and activities. The ability for more abstract thinking coupled with the changing psychosocial terrain of young adolescents provides both a challenge and a unique opportunity for educators to offer new learning and teaching strategies to encourage them to make healthful food choices. Young adolescence is an ideal time to teach students how to assess their own behavior and set goals for change. As adolescents begin the social process of individuation, they become ready and eager to make their own decisions and show their individuality. Nutrition education often fails to take advantage of the social and cognitive transitions of adolescence to promote the adoption of more healthful behaviors. In addition, obstacles to implementing nutrition education programs persist, ranging from insufficient funding to teacher ambivalence to competition with other high-priority health concerns, such as HIV and substance-use prevention.[60]

Because knowledge alone is inadequate when students must decide which foods to eat and how to deal with peer and social influences, as well as with a widely available supply of high-fat, high-sugar foods, the focus of nutrition education and teaching methods should be on behavior-change strategies and skill acquisitions to make healthful food decisions. Characteristics of teaching methods found to be most effective in school health education curricula include use of discovery learning; use of student learning stations, small work groups, and cooperative learning techniques; cross-age and peer teaching; positive approaches that emphasize the intrinsic value of good health; use of personal commitment to change and goal setting; and provision of opportunities to increase self-efficacy in modifying health behaviors. Most important, adolescents need to be given repeated opportunities to develop, demonstrate, practice, and master the skills needed to make informed decisions and cope with social influences. To be effective, programs also must take into account cultural factors as well as the developmental processes of adolescents. The integration of technology as a means of assessing and monitoring eating behaviors needs to be encouraged within schools and school districts.

Nutrition Environment of the School   The National School Lunch Program (NSLP) and School Breakfast Program (SBP) are federally sponsored nutrition programs administered by the Department of Agriculture (USDA) in conjunction with state and local education agencies. Youth from households with incomes between 130% and 185% of the poverty level receive meals at reduced rates; youth from households with incomes 130% of poverty and below receive meals free. The school lunch and breakfast programs can complement and reinforce what is learned in the classroom and serve as a learning laboratory for nutrition education. The synergistic interaction between the school program and classroom learning should enhance the likelihood that adolescents will adopt healthful eating practices.

The school environment provides multiple food and nutrition activities and influences not only classroom nutrition education and school meals, but also food sold in vending machines, school stores, and snack bars; fundraising events; food rewards by teachers; corporate-sponsored nutrition education materials; and in-school advertising of food products.[64] The result can be inconsistent nutrition messages. The growing stream of commercial messages, food advertisements, and easy access to high-fat, high-sugar, high-sodium food products in schools are at cross purposes and in direct conflict with the goals of nutrition education and may negate the efforts in the classroom and lunchroom to foster healthful eating practices.

Most schools offer students the opportunity to purchase foods that are not part of the NSLP or SBP, through the option of à la carte foods, canteens or school stores, or vending machines. Data collected as part of the 2006 SHPPS demonstrated that 3% of elementary schools, 71% of middle/junior high schools, and 89% of high schools sold foods other than the NSLP offerings from vending machines, school stores, or canteens or snack bars during lunch.[62] This same study found that 12% of elementary, 19% of middle, and 24% of high schools offered brand-name fast foods for sale, such as Pizza Hut®, Dominos®, Taco Bell®, and Subway®.

Some improvements in school food service have been noted in the last decade. Policy changes at the school district level have resulted in healthier food-preparation practices and more control over the types of foods served as à la carte options. The proportion of schools offering deep-fried potatoes fell by more than half, from 40% in 2000 to 19% in 2006. Significant increases in the number of schools that used low-fat dairy products in cooking, removed skin from poultry, offered low-fat salad dressing, and steamed or baked foods as opposed to frying them were also seen during this time period.[62] More than 70% of middle and high schools offered two or more fruit, non-fried vegetable, main entrée, and 100% fruit juice choices, and 46% of schools sold bottled water in vending machines in 2006.

State education policies are often ahead of school district policies with regard to food sales and offerings. As reported in SHPPS 2006, 39% of school districts and 42% of states had policies prohibiting the sale of foods of low nutrient value ("junk foods") during mealtimes. The sale of junk foods in vending machines was prohibited by 32% of state and 30% of district policies. The majority of states (86%) provided model nutrition policies to districts and individual schools to assist them in establishing a healthy school environment and to promote healthy eating among students. The Institute of Medicine Committee on Nutrition Standards for Foods in Schools has developed a set of recommendations to guide state education departments, school districts, and schools in setting policies regarding the availability, sale, and nutrient composition of foods and beverages served at schools.[65] State-and-district level administrators should be encouraged to adopt these guidelines as part of a comprehensive school health program.

School Wellness Policies In the Child Nutrition and WIC Reauthorization Act of 2004, the United States Congress included a provision requiring that all school districts with a federally funded school meals program develop and implement wellness policies that address nutrition and physical activity by the beginning of the

2006–2007 school year (P.L. 108–265). At a minimum, each local policy was to include:

- Goals for *nutrition education, physical activity, and other school-based activities* that are designed to promote student wellness in a manner that the local educational agency determines is appropriate
- *Nutrition guidelines* selected by the local educational agency for all foods available on each school campus under the local educational agency during the school day, with the objectives of promoting student health and reducing childhood obesity
- *Guidelines for reimbursable school meals*
- A *plan for measuring implementation* of the local wellness policy, including designation of one or more persons within the local educational agency or at each school, as appropriate, charged with operational responsibility for ensuring that each school fulfils the district's local wellness policy
- *Community involvement*—including parents, students, and representatives of the school food authority, the school board, school administrators, and the public—in the development of the school wellness policy[66]

An analysis of local wellness policies from the 100 largest school districts in the United States showed the following:

- 99% address school meal nutrition standards.
- 93% address nutrition standards for à la carte foods and beverages.
- 92% address nutrition standards for foods and beverages available in vending machines.
- 65% address nutrition standards/guidelines for fundraisers held during school hours.
- 63% address nutrition standards/guidelines for classroom celebrations or parties.
- 65% address nutrition standards/guidelines for teachers using foods as rewards in the classroom.
- 50% of school districts address a recess requirement for at least elementary grade.
- 96% require physical activity for at least some grade levels.
- 97% require nutrition education for at least some grade levels.
- 95% outlined a plan for implementation and evaluation, utilizing the superintendent, school nutrition director, or wellness policy task force as the entity responsible for monitoring the policy.[67]

Guidelines to assist schools and school districts in development of wellness policies as well as specific programs that can achieve policy objectives are available from a variety of governmental and nonprofit groups (see the "Resources" section at the end of the chapter).

Community Involvement in Nutritionally Supportive Environments

Promoting lifelong healthy eating and physical activity behaviors among adolescents requires attention to the multiple behavioral and environmental influences in a community. Adolescents are most likely to adopt healthy behaviors when they receive consistent messages through multiple channels (e.g., community, home, school, and the media) and from multiple sources (e.g., parents, peers, teachers, health professionals, and the media). Because most physical activity occurs outside the school setting, community parks and recreation programs are essential for promoting physical activity among young people. Healthy eating can be integrated into these efforts by providing nutritious snacks, food-preparation activities, and other skill-based learning activities. Community coalitions or task forces can be established to assess community needs and to develop, implement, and evaluate physical activity and nutrition programs for young people.

Model Nutrition Program One example of a model community-based nutrition intervention program is the California Adolescent Nutrition and Fitness (CANFIT) program.[68] For more than 15 years, CANFIT has been working with teens and community leaders from low-income communities and communities of color to develop, implement, and evaluate culturally competent policies and programs. The program has successfully enhanced capacity and leadership of low-income communities and communities of color by involving youth, community leaders, and health professionals in collaborative processes to expand healthy eating and physical activity opportunities for youth. Through competitive grants, the CANFIT program supports and empowers adolescent-serving, community-based organizations to develop and implement nutrition education and physical activity programs for ethnic adolescents from low-income communities. Using a capacity-building model, the CANFIT program attempts to change the community context by improving access to healthier food choices and safe, affordable physical activity opportunities and by enabling adolescents to have the decision-making skills and social support necessary for making healthy nutrition and fitness choices. Examples of CANFIT grantees' projects include (1) development of after-school programs for African American girls that focuses on self-esteem, body image, healthy eating, cooking, and physical activity (e.g., hip-hop dance); (2) a nutrition and physical activity program for adolescents and their parents attending Saturday Korean language schools in Los Angeles; (3) Latino adolescents in a soccer league working with a local health department to train team coaches and parents in sports nutrition; and (4) a statewide media contest for youth (the MO project) where teens can provide ideas for how to improve their community through video and other media outlets. Innovative programs using capacity-building models, such as the CANFIT program, can provide numerous benefits to other communities.

Review Questions

1. Which of the following is considered a risk factor for low vitamin D status?
   a. Dark skin pigmentation
   b. Light skin pigmentation
   c. Low calcium intake
   d. High calcium intake

2. Sexual Maturation Rating is a method of determining when adolescents have completed menarche or spermarche.
   True   False

3. Which of the following factors contributes to increased iron needs among adolescents?
   a. Increased growth velocity during puberty
   b. Low dietary intake of iron
   c. Onset of menarche
   d. All of the above

4. School wellness policies are only required for schools that offer à la carte foods for sale.
   True   False

5. The nutrients most likely to be missing from the diets of vegan adolescents are:
   a. Vitamins A, D, E, and zinc
   b. Vitamins B_6, D, B_{12}, and calcium
   c. Vitamins B_6, B_{12}, E, and choline
   d. Vitamins A, D, E, and iron

6. Which of the following is NOT a requirement of a school wellness policy?
   a. Goals for nutrition education
   b. Goals for physical activity
   c. Development of a school nutrition advisory council
   d. Community involvement in policy development

7. What is biological age and why should it be used instead of chronological age to determine adolescent nutrient needs?

8. Why don't adolescent BMI charts use a single cutoff point for determining obesity in the same way adult BMI charts do?

9. Which of the following is NOT part of national guidelines for physical activity for adolescents:
   a. Engage in at least 60 minutes of strenuous aerobic physical activity at least 5 days a week
   b. Engage in bone-strengthening activities at least 3 days a week
   c. Engage in muscle-strengthening activities at least 3 days a week
   d. Engage in at least 60 minutes of physical activity almost every day

Resources

California Adolescent Nutrition and Fitness (CANFIT) Program
Information about CANFIT activities and links to adolescent health resources are available online.
Website: www.canfit.org

Bright Futures
View and download Bright Futures publications, including *Bright Futures in Practice: Nutrition* and *Bright Futures in Practice: Physical Activity*.
Website: www.brightfutures.org

Centers for Disease Control and Prevention, Division of Nutrition, Physical Activity and Obesity
This CDC website offers facts on physical activity and nutrition among U.S. adults and youth, model programs, program guidelines, and links to pediatric growth charts.
Website: http://www.cdc.gov/nccdphp/dnpao/index.html

National Center for Education in Maternal and Child Health
Search databases on adolescent health anad view and download publications and reports online.
Website: www.ncemch.org

The Vegetarian Resource Group
This website provides information on choosing a healthy vegetarian diet and links to other resources and websites.
Website: www.vrg.org/nutrition/teennutrition.htm

United States Department of Agriculture, Food and Nutrition Information Center
Information on Dietary Guidelines for Americans, MyPyramid, dietary supplements, food safety, the Healthy School Meals Resource System, and other sources of nutrition information can be found online.
Website: www.nal.usda.gov/fnic

School Nutrition Association
This website includes information and updates on school nutrition programs, nutrition education in schools, and school wellness policies.
Website: www.schoolnutrition.org

National Alliance for Nutrition and Activity
Website: www.schoolwellnesspolicies.org/

United States Department of Agriculture, Food and Nutrition Service, Team Nutrition
Website: www.fns.usda.gov/tn/Default.htm

Chapter 15

Adolescent Nutrition:
Conditions and Interventions

Prepared by **Jamie Stang**

Key Nutrition Concepts

1   Overweight adolescents are at increased risk for medical and psychosocial complications such as hypertension, hyperlipidemia, insulin resistance, type 2 diabetes mellitus, hypoventilation and orthopedic disorders, depression, and low self-esteem.

2   Competitive adolescent athletes require an additional 500–1500 kcals per day to meet their energy needs. Additional protein may be required among adolescent athletes who are still growing.

3   Adolescents are concerned about their weight; a national study suggests that on any given day, approximately half of adolescent females and 15% of adolescent males were attempting weight loss.

Introduction

Multiple factors influence the nutritional needs and behaviors of adolescents. This chapter presents specific nutrition concerns that affect significant numbers of adolescents, including overweight, participation in competitive sports, substance abuse, vegetarian diets, eating disorders, hypertension, and hyperlipidemia. Because overweight, sports participation, and eating disorders affect a larger group of adolescents than other conditions, they are presented in greater detail.

Overweight and Obesity

The increase in the prevalence of overweight and obesity among adolescents has nearly doubled during the past two decades. Exact reasons for this increase have not been identified. Although genetics is known to contribute to the occurrence of obesity, and having one or more overweight parent(s) increases a teen's risk of developing obesity, it alone clearly cannot account for the dramatic increase in overweight during the past two decades.[1] Environmental factors, or interactions between genetic and environmental factors, are the most likely causes of the dramatic rise in overweight and obesity. Risk factors for the development of overweight and obesity among children and adolescents include having at least one overweight parent; coming from a low-income family; being of African American, Hispanic, or American Indian/Native Alaskan race/ethnicity; and being diagnosed with a chronic or disabling condition that limits mobility.[1] Inadequate levels

of physical activity and consuming diets high in total calories and added sugars and fats are additional risk factors among a significant proportion of adolescents.[2] These environmental factors increase the risk of developing overweight if an adolescent is genetically predisposed to obesity.

Weight status among adolescents should be assessed by calculating body mass index (BMI). BMI is calculated by dividing a person's weight (kg) by their height^2 (m^2). BMI values are compared to age- and gender-appropriate percentiles to determine the appropriateness of the individual's weight for height. Youth with BMI values greater than the 85th but lower than the 95th percentile are considered overweight; those with BMI values above the 95th percentile are considered obese.[2] Growth curves based on BMIs for children and adolescents are available from the National Center for Health Statistics. An example of a BMI growth curve is shown in Illustration 15.1.

Illustration 15.1 CDC Growth Charts: United States.

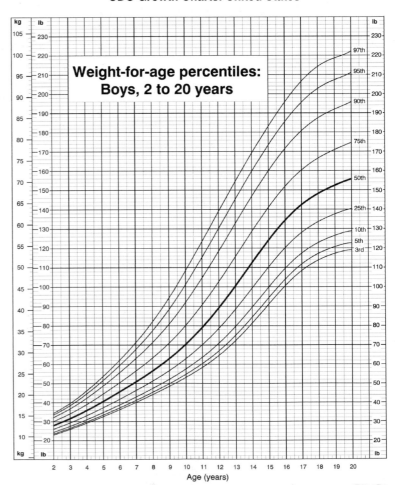

Developed by the National Center for Health Statistics in collaboration with the National Center for Chronic Disease Prevention and Health Promotion; 2000.

Table 15.1 Prevalence of at-risk-for overweight and overweight by race and gender among 12–19-year-olds in NHANES 2003–06

	Overweight	Obese
Males		
White	34.5	17.3
Black	32.1	18.5
Mexican American	30.5	19.9
Male Total	34.9	18.2
Females		
White	31.7	14.5
Black	44.5	27.7
Mexican American	37.1	19.9
Female Total	33.3	16.8

Data from the 2003–2006 NHANES suggest that 34% of U.S. adolescents are overweight and 18% are obese.[3] Table 15.1 provides prevalence estimates of overweight among adolescents in the United States, by gender and race/ethnicity. In general, the prevalence of being overweight or obese among females is highest among black teens, while among males the prevalence is highest among Mexican American teens. While American Indian students are not reported separately in the Youth Risk Behavior Survey (YRBS) or NHANES, regional data suggest that the prevalence of overweight among American Indian youth is higher than among other racial/ethnic groups.[4,5]

The persistence of overweight from childhood throughout adulthood has not been well quantified. Research suggests that the persistence of obesity from infancy to adulthood increases with age. As many as 90% of overweight adolescents can be expected to remain overweight into adulthood.[6,7] Identification of overweight at an early age is important, as research data suggest that children with BMI above the 85th percentile are more than twice as likely as children with BMI below the 50th percentile to continue to gain weight and reach overweight status by adolescence.[1,7] The risk of persistence of obesity from childhood into adulthood increases if at least one parent is overweight.[7] Risk of persistence of overweight is also higher among the most overweight individuals, especially those whose weight is more than 180% of ideal weight.

Health Implications of Adolescent Overweight and Obesity

A range of medical and psychosocial complications accompanies overweight among adolescents, including hypertension, dyslipidemia, insulin resistance, type 2 diabetes mellitus, sleep apnea and other hypoventilation disorders, orthopedic problems, hepatic diseases, body image disturbances, and lowered self-esteem.[8,9] Cardiorespiratory fitness is lower in male and female adolescents who are overweight than in those who are normal weight. Cardiorespiratory fitness is also lower in adolescents who have low levels of physical activity and high levels of sedentary behaviors.[10]

Assessment and Treatment of Adolescent Overweight and Obesity

All adolescents should be screened for appropriateness of weight-for-height on a yearly basis. Teens determined to be at risk for overweight require an in-depth medical assessment to diagnose any obesity-related complications. Illustration 15.2 provides recommended screening and referral procedures for adolescents with a BMI ≥ 85th percentile for age and gender.

National guidelines for the treatment of child and adolescent overweight and obesity recommend a staged care process based on BMI, comorbid conditions, age, and motivation.[11] The four stages include (1) Prevention Plus, (2) structured weight management, (3) comprehensive multidisciplinary intervention, and (4) tertiary care intervention. Adolescents advance through the stages based on age, biological development, level of motivation, presence of comorbid conditions, and success with previous stages of treatment (see Table 15.2). A brief overview of the stages is included below.

Stage 1: Prevention Plus
Adolescents with BMI of ≥ 85th but < 95th percentile may start out in Stage 1 if they do not exhibit significant comorbid conditions and/or have not completed their adolescent growth spurt. This level of treatment builds upon

Illustration 15.2 Primary care assessments based on adolescent BMI.

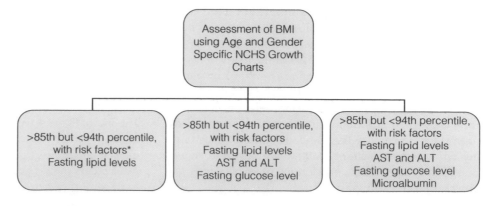

Table 15.2 Overview of staged treatment based on BMI status

BMI Percentile	Staged Treatment Protocol	Weight Goal
85th–94th	Begin with the Prevention Plus stage. If improvements in weight and/or comorbid conditions are not seen after 3–6 months, advance stage 2.	Maintenance of body weight until BMI is <85th percentile or until a downward deflection in the BMI curve is noted
95th–98th	Begin at Stage 1 or Stage 2 depending on age, developmental stage, degree of obesity, comorbidities, and motivation. Advance to Stage 3 if not responsive to treatment after 3–6 months or if an increase in comorbidities is noted.	Maintenance of body weight until BMI is <85th percentile, with a weekly weight loss of no more than 2 lb*
≥99th	Being with Stage 2 or Stage 3 depending on age, degree of obesity, comorbidities, and motivation. Evaluate for advance to Stage 4 after 3–6 months if no signs of improvement are seen.	Weight loss of no more than 2 lb/week until BMI is <85th percentile*

SOURCE: Spear, B. A., Barlow, S. E., Ervin, C., et al. Recommendations for treatment of child and adolescent overweight and obesity. Reproduced with permission from Pediatrics 120:254–258. Copyright © 2007 by the AAP.

*If weight loss is >2 lb/week in this age group, teens should be evaluated for excessive energy restrictions by the parent or self as well as for unhealthy forms of weight loss or disordered eating.

*RISK factors include family history of hypertension, early cardiovascular death, stroke, or hyperlipidemia; presence of hypertension in adolescent; adolescent tobacco use.

basic nutrition and physical activity guidance recommended to promote health and prevent disease. Specific topics that should be included as components of Stage 1 obesity treatment include consumption of at least 5 servings of fruits and vegetables per day, limiting sweetened-beverage consumption, achieving at least 60 minutes of physical activity per day, and limiting screen time (including DVDs, Internet, television, and computer or video games) to no more than 2 hours per day. Additional nutrition issues that may be addressed in Stage 1 include the importance of daily breakfast consumption, limiting take-out and restaurant meals (including fast food), participating in family meals at least 5 times per week, and encouraging youth to self-regulate their intake. Stage 1 treatment can be provided by a single health care provider, including physicians, nurses, dieticians, and other allied health care providers who have training in pediatric weight management.

Stage 2: Structured Weight Management
The second stage of pediatric weight management addresses the same behaviors as Stage 1, but does so in a more structured manner. Monitoring of food and nutrition behaviors by the adolescent and/or their parent(s) is a key component of this stage. All of the goals of Stage 1 should be reinforced, but several are modified in Stage 2. Screen time is limited to <1 hour per day in this stage, and a meal plan is introduced to emphasize nutrient-dense food choices while minimizing energy-dense foods. Journals or log books may be provided for monitoring target behaviors. Achievement of goals should be rewarded with non-food items such as new clothing or jewelry or tickets to a concert or event.

Stage 2 can be offered by a health care provider with training in behavioral pediatric weight management. Skills required to successfully implement this stage of treatment include motivational counseling, monitoring and reinforcement, and family conflict resolution. Referrals for physical therapy, mental health counseling, and medical nutrition therapy may be necessary for some adolescents with significant co-morbidities. Monthly follow-up and assessment of progress is suggested.

Stage 3: Comprehensive Multidisciplinary Intervention
Stage 3 targets the same behavioral goals as Stage 2, but does do in a more structured, multidisciplinary format with more frequent client contact. This treatment stage is provided by a team of health care professionals who specialize in pediatric obesity management. A more structured eating and physical activity plan that is designed to lead to negative caloric balance is implemented in this phase. A structured behavior-modification program is recommended, with weekly visits for at least 8–12 weeks followed by bimonthly or more frequent contact with the adolescent and his or her family. The recommended membership of the multidisciplinary team includes a physical or pediatric nurse practitioner, a behavioral or mental health counselor, a registered dietician, an exercise physiology or physical therapist, and a nurse.

Stage 4: Tertiary Care Intervention
The use of Stage 4 treatment is appropriate with severely obese youth or those who have significant, chronic comorbid conditions that necessitate intensive intervention. Adolescents should be evaluated for their level of maturity to be sure they are able to understand the high level of

commitment required as well as the potential risks associated with Stage 4 treatment. This level of treatment is provided through a tertiary weight-management center that specializes in adolescent obesity. In addition to diet and activity counseling and behavior modification, more intensive treatments such as meal replacement, a very-low-energy diet, medication, and surgery may be implemented.

Few data on the effectiveness of very-low-energy diets or meal replacements are available; however, these measures appear to be safely used for short periods of time.[11] The use of very-low-calorie diets or protein-sparing modified fasts should only be done under continuous medical supervision, as these diets have been associated with many health risks, including orthostatic hypotension, diarrhea, hyperuricemia, cholelithiasis, electrolyte imbalance, and reduced serum protein levels.[12] The use of these diets should not exceed 12 weeks in duration. There are currently only two medications that have been FDA-approved for use by adolescents: sibutramine, which is a serotonin reuptake inhibitor that increases weight loss by decreasing appetite, and orlistat, an enteric lipase inhibitor that causes fat malabsorption. Sibutramine is approved for use in adolescents >16 years of age, while orlistat is approved for adolescents >12 years of age. Side effects of weight-loss medications include insomnia, headache, hypertension, cardiac arrhythmia, depression, dizziness, edema, nausea, anxiety, steatorrhea, flatulence, fecal incontinence, blurred vision, and fat-soluble vitamin deficiencies.[12]

Gastric bypass has been used for several decades to treat severely obese adolescents who were not successful with behavior modification and lifestyle changes. Guidelines for the use of bariatric surgery among adolescents have been developed (Table 15.3).[13] In order to be considered as a candidate for bariatric surgery, adolescents must have a BMI of >40 with medical complications or a BMI of >50 without medical complications. In addition, teens should have completed the majority of their adolescent growth spurt before undergoing bariatric surgery in order to minimize potential side effects such as stunting of growth. Once the adolescent growth spurt is completed, nutrient needs are reduced, so it is less likely that nutrient deficiencies will occur among adolescents who have limited food intakes following bariatric surgery. Recently, adolescent obesity treatment experts have suggested that the BMI criteria for bariatric surgery among adolescents be reduced to a BMI of ≥ 35 in the presence of comorbid conditions; however, this is not yet standard practice.[14]

The long-term success rate of bariatric surgery among teens is not well established. Research suggests that adolescents reduce their BMI values by 19 units and lose an average of 36–47% of their excess body fat following bariatric surgery.[15,16] In a recent study of 31 teens, 39% of patients undergoing bariatric surgery experienced complications, with 25% experiencing moderate complications, 13% experiencing severe complications, and 1 teen dying within a year of the surgery.[16] Of the 30 adolescents included in follow-up, 2 began regaining weight within the first year following surgery, with 1 adolescent regaining more than 50% of lost body weight.[65]

Supplementation with multivitamin–mineral preparations following bariatric surgery is imperative for adolescents. While all nutrients may require supplementation when food intake is severely limited following surgery, calcium, vitamin B_{12}, folate, and thiamin are especially important.[13–16] Iron status of menstruating females must be monitored closely as well, with supplementation provided as needed.

Supplement Use
Vitamin–Mineral Supplements

Supplements may be used by adolescents for a variety of reasons, including improving health, treating iron deficiency, increasing energy, building muscle, and losing weight. National data suggest that more than one-third of U.S. children and adolescents consume vitamin–mineral supplements, while studies in Canada show a prevalence of vitamin/mineral use of 43%.[17,18] More than half of adolescents who report using vitamin–mineral supplements take them occasionally, with slightly less than half using them daily.[17–19] Approximately half of vitamin–mineral supplements consumed by adolescents are multivitamins without minerals, 34% are individual vitamins or minerals, 18% are multivitamins with minerals, and 17% are iron with vitamin C tablets.[18,19] Among the individual nutrient supplements used, vitamin C is the most common, followed by calcium, iron, vitamin E, and B-vitamin complex.[19]

Data on demographic differences in adolescent supplement use are apparent; supplement use is positively correlated with household income, high food-security

Table 15.3 Guidelines for consideration of bariatric surgery in adolescents

- Failure to obtain adequate weight loss after minimum of 6 months of intensive weight loss program participation
- SMR/Tanner at stage III or higher
- BMI ≥40 with medical complications or ≥50 without medical complications
- Participation in psychological and medical counseling before surgery with agreement to continue counseling after surgery
- Must have adequate support of family and a home environment conducive to long-term dietary change
- Capability to follow medical nutrition therapy protocol after surgery
- Agreement to prevent pregnancy for at least 1 year after surgery

SOURCE: Based on American Pediatric Surgical Association Clinical Task Force on Bariatric Surgery 2004.

status, having some form of health insurance, and parental education.[17–19] Adolescents who take vitamin–mineral supplements tend to consume a more nutritionally adequate diet than those who don't.[17,19] Supplement use is also positively correlated with health behaviors such as meeting physical activity goals, consuming more fruits and vegetables, and spending less than 2 hours per day watching television, playing video games, or using a computer. BMI status and intakes of total and saturated fat and cholesterol are negatively correlated with supplement use.

Few data are available to quantify the use of non-nutritional supplements such as herbs (including herbal weight-loss products) among adolescents. Data from Monroe County, New York suggest that as many as 29% of U.S. teens may use herbal products.[20] A small study of 78 Australian teens found that 18% has used herbal supplements, 5% had taken creatine and guarana, and 1% had used coenzyme Q.[21] A study of 353 teens from Canada found that 4.1% of adolescents used herbal weight-control products, 6% used energizers (e.g., bee pollen), 1.6% used L-carnitine, and 5.3% used creatine.[18] Adolescents who use herbal supplements have been found to be more likely to engage in health-compromising behaviors such as the use of cigarettes, marijuana, alcohol, and other street drugs.[20]

The use of herbs and supplements by youth is highly controversial. Adolescents may take herbal supplements for several reasons including weight loss, treatment of attention deficit disorder, and to increase energy and stamina. Youth with special health care needs, such as autism spectrum disorders, attention deficit disorder, and cystic fibrosis, may use supplements more frequently than other adolescents. Teens may also use herbal supplements to help them lose weight, build muscle, or improve athletic performance. Studies are needed to determine exactly what types of herbal products are used by adolescents, because many herbs are known to have potentially dangerous side effects, and few recommendations are available to guide the use of herbs by children or adolescents.

Ergogenic Supplements Used by Teens

YRBS data suggest that 3% of adolescents report having used illegal steroids.[4] Steroid use is reported more frequently among male (5%) than female (3%) adolescents.[4] Hispanic and white males report the highest prevalence of steroid use (5%). Steroid use appears to decrease with age, peaking during ninth grade. Supplements used by adolescent athletes include creatine; individual amino acids or protein powders; carnitine; anabolic-androgenic steroids; anabolic steroid precursors, including dehydroepiandrosterone (DHEA) and androstenedione; beta-hydroxy-beta-methylbutyrate; growth hormone; Xenadrine; and ephedra.[22,23] Steroids and other ergogenic supplements are taken orally, injected, or absorbed through transdermal patches.[22,23] They are most often used outside of the sport season to avoid detection of use. Steroids and ergogenic supplements are often taken in 1–3-month periods and are "stacked" so that the peak dose of one substance may overlap the introduction of another substance.[22] While the use of steroids and ergogenic aids is forbidden by national and NCAA regulations, few high school athletic programs test athletes for their use.[24]

Anabolic-androgenic steroids are controlled substances used to increase lean body mass and improve strength.[22,23] The use of anabolic steroids appears to be highest among competitive athletes, particularly those who participate in football, as well as gymnastics, weight training, basketball, and baseball.[23] Among non-athletes, reasons for steroid use have been to attempt to improve appearance or increase self-esteem. The use of these steroids has been linked to infertility, hypertension, physeal closure, depression, aggression, and increased risk of atherosclerosis.[22]

DHEA and androstenedione are precursors of testosterone and estrogen. Androstenedione is also a controlled, substance while DHEA is widely available as a supplement.[22] Naturally produced in the human body by the adrenal glands, DHEA levels fall in humans as age increases. Its reputed effects include reducing body fat, decreasing insulin resistance, increasing immune system function, increasing lean body mass, and decreasing risk of osteoporosis; however, no scientific evidence backs such claims.[22] As steroid precursors, andostenedione and DHEA may induce many of the same side effects as steroids, such as irreversible gynecomastia (breast enlargement) and prostate enlargement among males and hirsuitism (facial hair) among females.[22,23] As many as 4% of adolescents take androstenedione.[23]

Growth hormone (GH) has been shown to decrease subcutaneous body fat and may strengthen ligaments and tendons, resulting in fewer injuires.[22] Side effects of its use include physeal closure, hyperlipidemia, glucose intolerance, and myopathy.[22] Few data are available on the use of GH by adolescents, as it is a substance available only by prescription for the treatment of growth disorders in youth. The extent of illegal use of GH among adolescents is unknown. Given the possibility of significant side effects in pubescent adolescents undergoing hormonal changes, steroids or their precursors and GH should not be used by adolescents.

Creatine is sold as a nutritional supplement to increase lean body mass. Creatine, formed in the liver and kidney of the human body, can be obtained in more than adequate amounts from the consumption of meat. Eight percent or more of male adolescents and up to 2% of female adolescents report the use of creatine; however, the prevalence among male athletes has been found to be as high as 51%.[23] Creatine use has been found to be most prevalent in athletes who are involved with football, gymnastics, hockey, wrestling, and baseball. Studies of creatine use in adults show mixed results; data on adolescent performance are sparse.[24] It appears to be of minimal benefit to endurance athletes, and marginal benefit to those involved in short-duration, anaerobic, strength-related sports.[22] Side effects of creatine use, which seem to be dose-related, include abdominal pain and cramping, nausea, diarrhea, headache, dehydration, reduced renal function, increased tendency toward muscle strains, and muscle soreness.[22,24] No available data document the long-term health effects related to creatine use; however, chronic use may be associated with renal damage.[22,24]

Ephedra was sold as an over-the-counter supplement until 2004, when its sale was banned by the FDA. While it has been proven to increase metabolic rate, no known benefits on athletic performance have been documented.[22] Ephedra was removed from the market due to side effects including cardiac arrhythmia, hypertension, increased risk of myocardial infarction and cerebral vascular accidents, and, in extreme cases, death. Litigation to return ephedra to the marketplace in lower doses than previously sold is ongoing.[22] Ephedrine use has been reported to be at high as 26% among female and 12% among male adolescents.[23]

Nutrition for Adolescent Athletes

More than half (56%) of U.S. adolescents report playing on one or more organized sports teams.[4] Participation is higher among male adolescents than female adolescents (62% versus 50%) and reduces with age. White and black adolescents report participation in organized sports more often than do Hispanic students.

Fluids and Hydration

Fluid intake is an important issue in sports nutrition for adolescents. Young adolescents and those who are prepubertal present a particular vulnerability to heat illnesses because their bodies do not regulate body temperature as well as those of older adolescents. Adolescents can become so mentally and physically involved in physical activities that they do not pay attention to physiological signals of fluid loss, such as excessive sweating and thirst. Some athletes commonly assume they do not need additional fluids if they are not actively moving all of the time during exercise. Other factors, such as ambient temperature and humidity levels and weight of equipment (helmets, padding, etc.) worn or utilized during exercise also play a role. For instance, hockey goalies do not skate for great distances during a match, yet they may lose 5 or more pounds of body weight due to the weight of the padding and equipment they wear. Therefore, all athletes should be counseled to regularly consume fluids, even if they do not feel thirsty.

Athletes should consume 6–8 oz of fluid prior to exercise, 4–6 oz every 15–20 minutes during physical activity, and at least 8 oz of fluid following exercise. Recommendations encourage athletes to weigh themselves periodically before and after exercise or competition to determine whether they have lost body weight. Each pound of body weight lost during an activity requires ingestion of 16 oz of fluid following the activity to maintain proper hydration. Athletes should drink no more than 16 oz of fluid each 30 minutes, however, to avoid potential side effects, such as nausea.

The type of fluid an athlete drinks is affected more by peer pressure and mass media than by actual physiological need. Sports drinks and energy drinks are very popular among teens, even those who do not participate in sports. Data on children suggest that even though water is an economical, easily available fluid, it may not provide optimal benefits for athletes who participate in physically intense events or those of great duration.[25] In such events, juice diluted at a ratio of 1:2 with water, or sports drinks that contain no more than 6% to 8% carbohydrate, may allow for better hydration and physical performance. Undiluted juices, fruit drinks, carbonated beverages, energy drinks, and sports drinks that contain more than 6% to 8% carbohydrate are not recommended during exercise because they may cause gastric discomfort. Their high carbohydrate content may also delay gastric emptying. Some carbonated soft drinks and many energy drinks contain significant amounts of caffeine, which promotes diuresis.

Special Dietary Practices

Adolescent athletes may follow special diets or consume nutritional and non-nutritional supplements in an effort to improve physical performance and increase lean body mass. Special diets that are noted among adolescent athletes include carbohydrate-loading regimens and high-protein diets. Distance runners and other endurance athletes traditionally used carbohydrate loading to improve the glycogen content of muscle. It involves the manipulation of training intensity and duration along with the carbohydrate content of meals to improve glycogen formation in muscle tissue. Carbohydrate loading is traditionally a week-long process that begins with intense training 1 week prior to competition. For the first 3 days of a carbohydrate-loading week, athletes choose low-carbohydrate foods, but continue to exercise in an attempt to deplete muscle glycogen stores. During the 3 days prior to competition, athletes rest, or exercise minimally, while consuming a high-carbohydrate diet to promote glycogen formation and storage. Many athletes follow a modified version of this regimen rather than the full traditional method.

High-protein diets may take many forms for teen athletes. In general, athletes who follow high-protein diets may consume three to four times the recommended protein intake, accompanied by a relatively low intake of carbohydrate. High-protein diets should be discouraged as pre-performance dietary regimens among athletes for several reasons. First, many dietary protein sources are also sources of total and saturated fats, which may increase lifetime risk of coronary artery disease. Second, high protein and fat intakes result in reduced carbohydrate intake and may delay digestion and absorption, limiting the amount of energy available for use during physical activity. Finally, more water is required for the breakdown of protein than for either fat or carbohydrate due to the increased water loss that accompanies the excretion of nitrogen. This factor places an athlete at increased risk for dehydration, often accompanied by a decrease in physical performance. High protein intake appears to be more effective in recovering from intense physical activity than in preparing for an event.

Substance Use

The use of substances such as tobacco, alcohol, and recreational drugs can affect the nutritional status of adolescents. YRBS data suggest that 20% of adolescents self-report cigarette smoking, with 12% smoking at least 1 cigarette each day.[4] Almost 11% of teens reported smoking >10 cigarettes per day. Traditionally, male adolescents were more likely to smoke than females; however, that tendency is no longer true. Current data suggest that smoking rates in male and female adolescents are similar.[4] White students are the most likely to smoke cigarettes (23%), followed by Hispanic students (17%) and black students (12%). Twenty-two percent of white females report smoking, compared to 24% of white males. The lowest prevalence of smoking was reported among black females (8%). Males are more likely to use smokeless tobacco products, with 13% of male and 2% of female adolescents using these products.[4] Smokeless tobacco use is highest among white males (18%), followed by Hispanic males (7%) and Hispanic females (3%). Adolescents who use tobacco have been shown to have higher vitamin C requirements compared to peers who do not use tobacco.

Alcohol intake and substance use among adolescents increases with age. Data from the YRBS suggested that by the twelfth grade, 83% of adolescents had tried alcohol.[4] Almost half (45%) of teens reported current alcohol use; 26% reported binge drinking (drinking five or more alcoholic drinks during one occasion) at least 1 day during the past month. Alcohol use is significantly higher among white (30%) and Hispanic (27%) youth compared to black youth (13%). The consumption of alcohol may replace nutritious foods and beverages in the diet, compromising nutritional status. Thiamin and other B-vitamin requirements may be higher among adolescents who frequently consume large quantities of alcohol.

According to YRBS data, illicit drug use is reported by a significant number of adolescents.[3] Twenty percent reported current marijuana use, 3% reported cocaine use, 13% had used inhalants, 8% had used hallucinogens, 2.3% had used heroin, and 2% had used other injectible drugs. Six percent of teens reported using ecstasy, and 4% used methamphetamines.[4] Illicit drug use was higher

Table 15.4 Potential effects of substance use on nutrition status

Appetite suppression
Reduced nutrient intake
Decreased nutrient bioavailability
Increased nutrient losses/malabsorption
Altered nutrient synthesis, activation, and utilization
Impaired nutrient metabolism and absorption
Increased nutrient destruction
Higher metabolic requirements of nutrients
Inadequate weight gain/weight loss
Iron deficiency anemia
Decreased financial resources for food

SOURCE: Reprinted with permission. Alton I. Substance Abuse During Pregnancy. In: Story M, Stang J, eds. Nutrition and the Pregnant Adolescent: A Practical Reference Guide. Minneapolis, MN: Center for Leadership, Education, and Training in Maternal and Child Nutrition, University of Minnesota; 2000.

among males than females and was more prevalent among Hispanic youth than other races and ethnicities, with the exception of marijuana, which was used slightly more often by black youth.

Recent data on the effects of substance use on eating behaviors have focused exclusively on the risk for disordered eating behaviors, particularly on bulimia nervosa and binge-eating disorder. Disordered eating is seen more frequently among females who report smoking cigarettes, drinking alcohol, and using inhalants.[26] Among males, the use of marijuana, steroids, and inhalants were related to higher risk for disordered eating. It is believed that substance use may result in depleted stores of vitamins and minerals, including thiamin, vitamin C, and iron. Chronic ingestion of alcohol and drug use can result in a reduced appetite, leading to low dietary intakes of protein, energy, vitamins A and C, thiamin, calcium, iron, and fiber (see Table 15.4).

Iron-Deficiency Anemia

Iron-deficiency anemia is the most common nutritional deficiency noted among children and adolescents. Several risk factors are associated with its development among adolescents, including rapid growth, inadequate dietary intake of iron-rich foods or foods high in vitamin C, vegan diets, calorie-restricted diets, meal skipping, participation in strenuous or endurance sports, and heavy menstrual bleeding.[27,28] The effects of iron-deficiency anemia include delayed or impaired growth and development, fatigue, increased susceptibility to infection secondary to depressed immune system function, reductions in physical performance and endurance, and increased susceptibility to lead poisoning. Pregnant teens that are iron deficient in the early stages of gestation are at increased risk of preterm delivery and delivery of a low-birth-weight infant.

Assessment of iron-deficiency anemia compares individual hemoglobin and hematocrit levels to standard reference values. Table 15.5 lists the Centers for Disease Control and Prevention criteria for determining anemia, based on age and gender. Adjustments to these values must be made for individuals who live at altitudes greater than 3000 feet and for smokers. An adjustment of +0.3 g/dl is required for adolescents who smoke.[28] Because adolescent males are not at high risk for iron-deficiency anemia, they do not need to be screened unless they exhibit one or more of the risk criteria listed. All adolescent females should be screened every 5 years for anemia; those with one or more risk factors for anemia should be screened annually.

Treatment that follows a diagnosis of iron-deficiency anemia needs to include increased dietary intake of foods rich in iron and vitamin C as well as iron supplementation. Adolescents under the age of 12 should be supplemented with 60 mg of elemental iron per day, and teenagers over the age of 12 should receive 60 to 120 mg of elemental iron per day.[28] These recommendations spark some controversy, however, given the high doses of elemental iron. Adolescents often report gastrointestinal side effects from iron supplementation, such as constipation, nausea, and cramping. These side effects can be lessened by giving smaller doses of iron more frequently throughout the day and counseling the adolescent to take the iron supplement at mealtimes or with food sources of vitamin C. Calcium supplements, dairy products, coffee, tea, and high-fiber foods may decrease absorption of iron supplements; these foods should be avoided within 1 hour of taking an iron supplement.

Cardiovascular Disease
Hypertension

Criteria for the detection and diagnosis of hypertension are shown in Table 15.6. Adolescents are considered hypertensive if the average of three systolic and/or diastolic blood pressure readings exceeds the 95th percentile, based on age, sex, and height.[29] Blood pressure levels for the 95th percentiles for males and females are shown in Table 15.7. Classifications of blood pressure based on the average of three readings are:

- Normal blood pressure: <90th percentile
- Prehypertensive: >90th and <95th percentiles
- Stage 1 hypertension: >95th and <99th percentile + 5 mm Hg
- Stage 2 hypertension: >99th percentile + 5 mm Hg

Table 15.5 Maximum hemoglobin concentration and hematocrit values for iron-deficiency anemia[a]

Sex/Age[a]	Hemoglobin (<g/dL) Less Than:	Hematocrit (<%) Less Than:
8–12 years	11.9	35.4
Males		
12–15 years	12.5	37.3
15–18 years	13.3	39.7
18+ years	13.5	39.9
Females[b]		
12–15 years	11.8	35.7
15–18 years	12	35.9
18+ years	12	35.7

SOURCE: Abridged from Centers for Disease Control and Prevention. Recommendations to Prevent and Control Iron Deficiency Anemia in the United States. Morb Mortal Wkly Rep. 2002;51(40):897–899.

[a]Age and sex-specific cutoff values for anemia are based on the 5th percentile from the third National Health and Nutrition Examination Survey.
[b]Nonpregnant and lactating adolescents.

Table 15.6 Consensus statement guidelines for detection and diagnosis of hypertension and hyperlipidemia

	Guidelines
Hypertension	
Prehypertension	Systolic or diastolic blood pressure >90th percentile for age and gender or 120/80 mm Hg, whichever is less
Stage 1 Hypertension	Systolic or diastolic blood pressure >95th percentile for age and gender on 3 consecutive visits or 140/90 mm Hg, whichever is less
Stage 2 Hypertension	Systolic or diastolic blood pressure >99th percentile +5 mm Hg for age and gender or 160/110 mm Hg, whichever is less
Hyperlipidemia	
Total cholesterol, mg/dL	
borderline	≥ 170
abnormal	≥200
LDL cholesterol, mg/dL	
borderline	≥100
abnormal	≥130
HDL cholesterol, mg/dL	
abnormal	<40
Triglycerides, mg/dL	
abnormal	≥200

SOURCE: Adapted from American Heart Association. Dietary Recommendations for Children and Adolescents. A Guide for Practitioners. Consensus statement from the American Heart Association. Circulation. 2005;112:2061–2075.

Table 15.7 Blood pressure levels for the 90th and 95th percentiles of blood pressure for boys and girls, aged 10 to 17 years

| | | Systolic BP (mm Hg), by Height Percentile from Standard Growth Curves | | | | | | | | | | | | | | Diastolic BP (mm Hg), by Height Percentile from Standard Growth Curves | | | | | | | | | | | | | |
| | | Boys | | | | | | | Girls | | | | | | | Boys | | | | | | | Girls | | | | | | |
Age	BP Percentile*	5%	10%	25%	50%	75%	90%	95%	5%	10%	25%	50%	75%	90%	95%	5%	10%	25%	50%	75%	90%	95%	5%	10%	25%	50%	75%	90%	95%
10	90th	110	112	113	115	117	118	119	112	112	114	115	116	117	118	73	74	74	75	76	77	78	73	73	73	74	75	76	76
	95th	114	115	117	119	121	122	123	116	116	117	119	120	121	122	77	78	79	80	80	81	82	74	74	75	75	76	77	78
11	90th	112	113	115	117	119	120	121	114	114	116	117	118	119	120	74	74	75	76	77	78	78	74	74	75	75	76	77	77
	95th	116	117	119	121	123	124	125	118	118	119	121	122	123	124	78	79	79	80	81	82	83	78	78	79	79	80	81	81
12	90th	115	116	117	119	121	123	123	116	116	118	119	120	121	122	75	75	76	77	78	78	79	75	75	76	76	77	78	78
	95th	119	120	121	123	125	126	127	120	120	121	123	124	125	126	79	79	80	81	82	83	83	79	79	80	80	81	82	82
13	90th	117	118	120	122	124	125	126	118	118	119	121	122	123	124	75	76	76	77	78	79	80	76	76	77	78	78	79	80
	95th	121	122	124	126	128	129	130	121	122	123	125	126	127	128	79	80	81	82	83	83	84	80	80	81	82	82	83	84
14	90th	120	121	123	125	126	128	128	119	120	121	122	124	125	126	76	76	77	78	79	80	80	77	77	78	79	79	80	81
	95th	124	125	127	128	130	132	132	123	124	125	126	128	129	130	80	81	81	82	83	84	85	81	81	82	83	83	84	85
15	90th	123	124	125	127	129	131	131	121	121	122	124	125	126	127	77	77	78	79	80	81	81	78	78	79	79	80	81	82
	95th	127	128	129	131	133	134	135	124	125	126	128	129	130	131	81	82	83	83	84	85	86	82	82	83	83	84	85	86
16	90th	125	126	128	130	132	133	134	122	122	123	125	126	127	128	79	79	80	81	82	82	83	79	79	79	80	81	82	82
	95th	129	130	132	134	136	137	138	125	126	127	128	130	131	132	83	83	84	85	86	87	87	83	83	83	84	85	86	86
17	90th	128	129	131	133	134	136	136	122	123	124	125	126	128	128	81	81	82	83	84	85	85	79	79	79	80	81	82	82
	95th	132	133	135	136	138	140	140	126	126	127	129	130	131	132	85	85	86	87	88	89	89	83	83	83	84	85	86	86

SOURCE: Adapted from the National Heart, Lung, and Blood Institute, National High Blood Pressure Education Working Group on Hypertension Control in Children and Adolescents. Fourth Report on the Diagnosis, Evaluation, and Treatment of High Blood Pressure in Children and Adolescents. Bethesda, MN: National Institutes of Health; 2005.

*Blood pressure percentile determined by a single movement.

Risk factors for hypertension among adolescents include a family history of hypertension, high dietary intake of sodium, overweight, hyperlipidemia, inactive lifestyle, and tobacco use.[27,29] Adolescents who display one or more of these risk factors should be routinely screened for hypertension. Nutrition counseling to decrease sodium intake, to limit fat intake to 30% or less of calories, and to consume adequate amounts of fruits, vegetables, whole grains, and low-fat dairy products should be provided when hypertension is diagnosed.[30] Table 15.8 outlines the dietary recommendations suggested for adolescents to promote health and reduce cardiovascular risk factors. Weight loss is recommended for adolescents who are hypertensive in the presence of overweight. If medications are prescribed, teens still must adhere to general dietary recommendations and should still be encouraged to reach and maintain a healthy weight for their height.

Hyperlipidemia

Approximately one in four adolescents in the United States has an elevated cholesterol level.[27,30] Table 15.6 provides the classification criteria for elevated cholesterol levels in children and adolescents. Risk factors for hypercholesterolemia include a family history of cardiovascular disease or high blood cholesterol levels, cigarette smoking, overweight, hypertension, diabetes mellitus, and low level of physical activity. Adolescents who exhibit these risk factors should be screened to determine causes of hyperlipidemia and should be referred for treatment as required.[27,30] Early intervention among adolescents who have high cholesterol levels may reduce their risk of coronary artery diseases later in life.

Dietary recommendations indicate that youth over the age of 5 years should obtain less than 35% of their calories from fat, with no more than 10% of calories derived from saturated fat.[30] Dietary cholesterol intakes of 300 mg per day or less have also been recommended. Counseling adolescents with hyperlipidemia to follow these guidelines can be challenging, given their frequent consumption of fast foods and their preferred food choices. Table 15.8 outlines American Heart Association recommendations for dietary intake to promote health and reduce cardiovascular disease risk. Suggested dietary changes should take into account the eating habits of adolescents and emphasize healthier food choices as opposed to restriction of favorite foods of teens. Health professionals can work with adolescents to make healthier choices at fast food restaurants, to limit their portion sizes of high-fat food items such as high-fat snacks, and to consume adequate amounts of fruits, vegetables, grains, and low-fat dairy products.

Table 15.8 Dietary recommendations to promote health and prevent cardiovascular disease

	9–13 years	14–18 years
Calories†		
Female	1600 Kcal	1800 Kcal
Male	1800 Kcal	2200 Kcal
Fat	25–35% Kcal	25–35% Kcal
Milk/Dairy Products‡	3 c	3 c
Meat/Meat Alternatives	5 oz	
Female		5 oz
Male		6 oz
Fruits	1.5 c	
Female		1.5 c
Male		2 c
Vegetables		
Female	2 c	2.5 c
Male	2.5 c	3 c
Grains‖		
Female	5 oz	6 oz
Male	6 oz	7 oz

SOURCE: Abridged from: American Heart Association. Dietary Recommendations for Children and Adolescents. A Guide for Practitioners. Consensus statement from the American Heart Association. Circulation. 2005;112:2061–2075.

†Calorie estimates are based on a sedentary lifestyle. Increased physical activity will require additional calories: by 0–200 Kcal/d if moderately physically active; and by 200–400 Kcal/d if very physically active.
‡Milk listed is fat-free (except for children under the age of 2 years). If 1%, 2%, or whole-fat milk is substituted, this will utilize, for each cup, 19, 39, or 63 kcal of discretionary calories and add 2.6, 5.1, or 9.0 g of total fat, of which 1.3, 2.6, or 4.6 g are saturated fat.
‖Half of all grains should be whole grains.

Dieting, Disordered Eating, and Eating Disorders

The Continuum of Eating Concerns and Disorders

Eating concerns and disorders lie on a continuum ranging from mild dissatisfaction with one's body shape to serious eating disorders such as *anorexia nervosa*, *bulimia nervosa*, and *binge-eating disorder*. Along the continuum, between these endpoints, lie normative dieting behaviors and more severe disordered eating behaviors such as self-induced vomiting and binge eating (Illustration 15.3). Although engagement in anorexic behaviors and unhealthy dieting may not be frequent or intense enough to meet the formal criteria for being defined as an eating disorder, these behaviors may negatively impact health and may lead to the development of more severe eating disorders. All eating disorders present a serious public health concern in light of their prevalence and their potentially adverse effects on growth, psychosocial development, and physical health outcome.

Dieting Behaviors

Dieting behaviors among adolescents, and in particular among adolescent girls, tend to be alarmingly high. National data suggest that 60% of female and 30% of male adolescents have dieted in the past month to lose weight.[4] Dieting was once considered to be a phenomenon of white, middle-class females. Current data suggest that dieting is as prevalent among Hispanic females as white females (62%), with black females dieting at a significantly lower, although still high, rate (50%).[4] More than one-third of Hispanic males (39%) report dieting, compared to 29% of white and 25% of black males. The prevalence of dieting drops slightly with age among males but increases with age among females. Dieting remains a significant issue for adolescents of all ages, races, and ethnicities that persists into adulthood.

Dieting and the use of unhealthy weight-control behaviors may also place adolescents at increased likelihood of being overweight in the future. Neumark-Sztainer and colleagues found that over a 5-year period, adolescents who initially reported dieting or using unhealthy weight-control behaviors were more likely to be overweight 5 years later than were peers who did not report using weight-control behaviors.[31] Effective nutrition messages aimed at adolescents should focus on making healthy lifestyle changes, rather than focusing on short-term dieting behaviors that are often difficult to sustain. Shifting focus toward long-term behavior changes is needed for prevention of both eating disorders and overweight.

Dieting behaviors among youth are of concern in that they are often used by youth who are not overweight. Furthermore, unhealthful dieting behaviors in which meals are

Anorexia Nervosa An eating disorder characterized by extreme weight loss, poor body image, and irrational fears of weight gain and obesity.

Bulimia Nervosa A disorder characterized by repeated bouts of uncontrolled, rapid ingestion of large quantities of food (binge eating) followed by self-induced vomiting, laxatives or diuretic use, fasting, or vigorous exercise in order to prevent weight gain.

Binge-Eating Disorder An eating disorder characterized by periodic binge eating, which normally is not followed by vomiting or the use of laxatives. People must experience eating binges twice a week on average for over six months to qualify for this diagnosis.

Illustration 15.3 The continuum of weight-related concerns and disorders.

Body dissatisfaction → Dieting behaviors → Disordered eating → Clinically significant eating disorders

skipped, energy intake is severely restricted, or food groups are lacking are common. Dieting behaviors have been found to be associated with inadequate intakes of essential nutrients. Restricting behaviors, leading adolescents to experience hunger or cravings for specific foods, may place them at risk for binge-eating episodes. Finally, dieting behaviors may be indicative of increased risk for the later development of eating disorders; research has found that during a 3 year period, restrained eating was a significant predictor of eating-disorder risk among female adolescents.[32] Therefore, dieting should not be viewed as a normative and acceptable behavior, in particular among adolescents.

Body Dissatisfaction

During adolescence, body image and self-esteem tend to be closely intertwined; therefore, body-image concerns should not be viewed as acceptable and normative components of adolescence. Furthermore, body dissatisfaction is a main contributing factor to dieting behaviors, disordered eating behaviors, and clinical eating disorders.[34] Body dissatisfaction appears to increase dramatically following the body-weight increase that normally occurs in females around the time of menarche and remains a significant concern for females for the next 1–2 years. Binge eating and/or purging has been found to occur within 6–12 months of menarchial weight changes.[34] Adolescents with low levels of body satisfaction are also at greater risk for using other unhealthy weight-control behaviors, as well as more likely to participate in less physical activity.[33] Although actual weight status is directly associated with perceived weight status, a considerable number of teens who are not overweight perceived of themselves as overweight, particularly around the time of puberty.

In working with overweight youth who express body dissatisfaction, health professionals are challenged to help them improve their body image while simultaneously working toward weight control. All adolescents, including overweight youth, should be encouraged to appreciate the positive aspects of their bodies. Overweight adolescents may need help in accepting the fact that they may never achieve the thin ideal portrayed in the media, but they may strive toward a leaner and healthier body that is realistic for them. Tips for fostering a positive body image among adolescents, regardless of their weight, are shown in Table 15.9.

Table 15.9 Tips for fostering a positive body image among children and adolescents

Child or Adolescent	Parents	Health Professional
• Look in the mirror and focus on your positive features, not the negative ones. • Say something nice to your friends about how they look. • Think about your positive traits that are not related to appearance. • Read magazines with a critical eye, and find out what photographers and computer graphic designers do to make models look the way they do. • If you are overweight and want to lose weight, be realistic in your expectations and aim for gradual change. • Realize that everyone has a unique size and shape. • If you have questions about your size or weight, ask a health professional.	• Demonstrate healthy eating behaviors, and avoid extreme eating behaviors. • Focus on non-appearance-related traits when discussing yourself and others. • Praise your child or adolescent for academic and other successes. • Analyze media messages with your child or adolescent. • Demonstrate that you love your child or adolescent regardless of what he weighs. • If your child or adolescent is overweight, don't criticize her or his appearance—offer support instead. • Share with a health professional any concerns you have about your child's or adolescent's eating behaviors or body image.	• Discuss changes that occur during adolescence. • Assess weight concerns and body image. • If a child or adolescent has a distorted body image, explore causes and discuss potential consequences. • Discuss how the media negatively affects a child's or adolescent's body image. • Discuss the normal variation in body sizes and shapes among children and adolescents. • Educate parents, physical education instructors, and coaches about realistic and healthy body weight. • Emphasize the positive characteristics (appearance- and non-appearance related) of children and adolescents you see. • Take extra time with an overweight child or adolescent to discuss psychosocial concerns and weight-control options. • Refer children, adolescents, and parents with weight-control issues to a dietitian.

SOURCE: Story M, Holt K, Sofka D, eds. Bright Futures in Practice: Nutrition. Arlington, VA: National Center for Education in Maternal and Child Health; 2000.

Disordered Eating Behaviors

Some adolescents engage in restricting or binge/purge eating behaviors, but with less frequency or intensity than required for a formal diagnosis of an eating disorder. Behaviors typically considered in this category include self-induced vomiting, fasting or extreme dieting, binge eating, compensatory physical activity, and the use of laxatives, diuretics, or diet pills. The heterogeneity of these behaviors makes it more difficult to estimate their prevalence. Data from the YRBS show that 12% of adolescents had gone for 24 hours or longer without eating (fasting) as a means of losing weight.[4] Fasting was more commonly reported among females (16%) than among males (7%). Seventeen percent of Hispanic and white females reported fasting to lose weight, compared to 13% of black females, 7% of black males, 6% of white males, and 11% of Hispanic males.

Six percent of students surveyed in YRBS reported using diet pills or other diet formulae to lose weight.[4] The use of diet pills was reported by 8% of white and Hispanic females and 4% of black females, compared to 5% of Hispanic and 4% of black and white males. The use of vomiting and/or laxative use to lose weight was reported by 7% of Hispanic and white females, 4% of black females, 4% of Hispanic males, 3% of black males, and 1% of white males.[4] Data on compensatory exercise, or exercise for the purpose of purging calories eaten rather than for health benefits, is difficult to find. Adolescents who report more than 60 mins of physical activity per day should be carefully assessed for the purpose of excessive exercise. Teens who work out until they have "burned off" all of the calories eaten that day should be considered at high risk for disordered eating.

The types of questions used to assess disordered eating behaviors may influence prevalence estimates and may account for the disagreement on how common this issue is among teens. Based on research findings, it can be reasonably estimated that between 10% and 20% of adolescents have engaged in disordered eating behaviors. These behaviors are often overlooked in overweight adolescents, but overweight adolescents have reported the use of unhealthy and extreme weight-control behaviors, which can include purging, laxative use, self-restriction, and excessive exercise.[35] Disordered eating behaviors such as self-induced vomiting and binge eating have serious implications for health and may be precursors to a diagnosed eating disorder. Therefore, interventions aimed at their prevention are essential.

Eating Disorders

An awareness of the prevalence of eating disorders is critical in effective planning for interventions aimed at their treatment and prevention. Conditions prevalent among youth, such as disordered eating behaviors and obesity, warrant interventions such as community-based and school-based programs that have the potential to reach large numbers of youth. The small percentage of the adolescent population affected by eating disorders requires more intensive individual or small-group interventions. Estimates as to the prevalence of each of the eating disorders on the continuum are presented in Table 15.10.

Anorexia Nervosa

Anorexia nervosa and its impact on morbidity and mortality make it the most severe condition on the continuum of eating disorders. Among adolescent girls and young women, prevalence estimates of anorexia nervosa range from 0.2% to 1.0%.[36,37] Anorexia nervosa presents more frequently among females than among males; about 9 out of 10 individuals with anorexia nervosa are female. Only in recent years has attention been directed toward males with this condition; they may not be suspected of having anorexia nervosa and therefore may be diagnosed at later stages of the disease, when treatment is more difficult.

Characteristics of anorexia nervosa include preoccupation with food, self-starvation, and strong fears of being fat.[36] An adolescent may begin with dieting behaviors due to social pressures to be thin, comments by others about

Table 15.10 Estimated prevalence and brief description of weight-related concerns/disorders among adolescents

Disorder	Estimated Prevalence
Anorexia nervosa	Approximately 0.2% to 1.0% of adolescent females and young women
Bulimia nervosa	Approximately 1% to 3% of adolescent females and young women
Binge eating	Estimated 30% of population currently dieting; 2% of general population
Disordered eating behaviors	Estimated 10% to 20% of adolescents although estimates vary
Dieting behaviors	Estimates vary and range from 44% of adolescent females, 15% adolescent males, to 50% to 60% of all adolescent females are attempting to lose weight
Body dissatisfaction	Estimates vary in accordance with type of measurement used and age, gender, and ethnicity of population: approximately 60% of girls and 35% of boys are not satisfied with their weight

weight, or as a result of their discomfort with the normal pubescent weight gain. Weight loss may result in adolescents feeling more in control of their body or other aspects of their life, which further reinforces the restricting behavior. If the weight loss and accompanying body-image and self-esteem issues are not addressed early on, anorexia nervosa develops. Diagnostic criteria for anorexia nervosa are shown in Table 15.11. Key features of anorexia nervosa are refusal to maintain body weight over a minimal normal weight for age and height; intense fear of gaining weight or becoming fat, even though underweight; a distorted body image; and amenorrhea (in females).

The two subtypes of anorexia nervosa are restricting and nonrestricting. In the restricting subtype, the individual does not regularly engage in binge-eating or purging behaviors. The nonrestricting subtype exhibits regular episodes of binge-eating and purging behaviors. However, both subtypes present with a refusal to maintain a minimally normal body weight, which differentiates them from other types of eating disorders.

An estimated 10–15% of patients with anorexia nervosa die from their disease, although difficulties arise in assessing mortality rates from anorexia nervosa.[36,38]

Table 15.11 Diagnostic criteria for anorexia nervosa

- Refusal to maintain body weight at or above a minimally normal weight for age and height (e.g., weight loss leading to maintenance of body weight less than 85% of that expected; or failure to make expected weight gain during period of growth, leading to body weight less than 85% of that expected)
- Intense fear of gaining weight or becoming fat, even though underweight
- Disturbance in the way in which one's body weight or shape is experienced, undue influence of body weight or shape on self-evaluation, or denial of the seriousness of the current low body weight
- Amenorrhea in postmenarchal women; that is, the absence of at least three consecutive menstrual cycles (a woman is considered to have amenorrhea if her menstrual periods occur only following hormone-estrogen-administration)

Restricting Type
During the episode of anorexia nervosa, the person has not regularly engaged in binge eating or purging behavior (i.e., self-induced vomiting or the misuse of laxatives, diuretics, or enemas).

Binge-Eating/Purging Type
During the episode of anorexia nervosa, the person has regularly engaged in binge-eating or purging behavior (i.e., self-induced vomiting or the misuse of laxatives, diuretics, or enemas).

SOURCE: Reprinted with permission from the Diagnostic and Statistical Manual of Mental Disorders, Fourth Edition, Text Revision, Copyright 2000. American Psychiatric Association.

Reasons for fatality from anorexia include a weakened immune system due to undernutrition, gastric ruptures, cardiac arrhythmias, heart failure, and suicide. The adolescent or the family commonly denies the condition, which delays the diagnosis and treatment, resulting in a poorer prognosis for recovery. Early recognition of possible signs of anorexia nervosa and seeking out professional help significantly affect the time and intensity of treatment and improve chances for a successful recovery. Full recovery rates are estimated at <50% of individuals with anorexia nervosa; 33% show improvements; and 20% are chronically affected by this mental illness.[36,38]

Bulimia Nervosa

Bulimia nervosa is an eating disorder characterized by the consumption of large amounts of food with subsequent purging by self-induced vomiting, laxative or diuretic abuse, enemas, and/or obsessive exercising.[36,38] Whereas anorexia nervosa is characterized by severe weight loss, bulimia nervosa may show weight maintenance or extreme weight fluctuations due to alternating binges and fasts. In some individuals, anorexia and bulimia nervosa overlap. Reliable estimates of bulimia nervosa range from 1.0% to 3.0%.[36,38] As with anorexia nervosa, the vast majority of individuals with bulimia nervosa are female.

Diagnostic criteria for bulimia nervosa are shown in Table 15.12. Key features of bulimia nervosa include recurrent episodes of binge eating (rapid consumption of a large amount of food in a discrete period of time), a feeling of lack of control over eating during the binge, some form of purging food and calories from the body, and a persistent overconcern with body shape and weight.[36,38]

There are two categories of bulimia nervosa: purging and non-purging. Individuals with the purging subtype of bulimia nervosa regularly engage in self-induced vomiting and/or the use of laxatives, diuretics, or enemas to purge calories from the body. Individuals with the non-purging subtype may fast in between binge episodes and utilize compensatory exercise as a means of compensating for caloric intake. People with bulimia nervosa can be overweight, underweight, or of average weight for their height and body frame. Bulimia nervosa may be preceded by a history of dieting or restrictive eating, which are thought to contribute to the binge–purge cycle.

Mortality for bulimia nervosa appears to be lower than for anorexia nervosa. Based on a review of the existing literature in this area, it has been estimated that approximately 2% to 3% of patients die of their disease.[36,38] Recovery rates for bulimia nervosa are estimated at 48% for full recovery, 26% for improvement, and 26% for chronicity.[38] Early diagnosis and treatment and less severe behaviors are associated with better outcomes for bulimia nervosa, while having a comorbid psychological condition such as borderline personality disorder, anxiety, or alcohol abuse is associated with poorer outcomes.

Table 15.12 Diagnostic criteria for bulimia nervosa

A. *Recurrent episodes of binge eating.* An episode of binge eating is characterized by both of the following:
- eating, in a discrete period of time (e.g., within any 2-hour period), an amount of food that is definitely larger than most people would eat during a similar period of time and under similar circumstances.
- a sense of lack of control over eating during the episode (e.g., a feeling that one cannot stop eating or control what or how much one is eating).

B. Recurrent inappropriate compensatory behavior in order to prevent weight gain, such as self-induced vomiting; misuse of laxatives, diuretics, enemas, or other medications; fasting; or excessive exercise.

C. The binge eating and inappropriate compensatory behaviors both occur, on average, at least twice a week for 3 months.

D. Self-evaluation is unduly influenced by body shape and weight.

E. The disturbance does not occur exclusively during episodes of anorexia nervosa.

Purging Type: During the current episode of bulimia nervosa, the person regularly engages in self-induced vomiting or the misuse of laxatives, diuretics, or enemas.

Nonpurging Type: During the current episode of bulimia nervosa, the person has used other inappropriate compensatory behaviors, such as fasting or excessive exercise, but has not regularly engaged in self-induced vomiting or the misuse of laxatives, diuretics, or enemas.

SOURCE: Reprinted with permission from the Diagnostic and Statistical Manual of Mental Disorders, Fourth Edition, Text Revision, Copyright 2000. American Psychiatric Association.

Table 15.13 Diagnostic criteria for binge-eating disorder

A. *Recurrent episodes of binge eating.* An episode of binge eating is characterized by both of the following:
- eating, in a discrete period of time (e.g., within any 2-hour period), an amount of food that is definitely larger than most people would eat in a similar period of time and under similar circumstances.
- a sense of lack of control over eating during the episode (e.g., a feeling that one cannot stop eating or control what or how much one is eating).

B. The binge-eating episodes are associated with three (or more) of the following:
- eating much more rapidly than normal
- eating until feeling uncomfortably full
- eating large amounts of food when not feeling physically hungry
- eating alone because of being embarrassed by how much one is eating
- feeling disgusted with oneself, depressed, or guilty after overeating
- experiencing marked distress regarding binge eating
- occurring, on average, at least 2 days a week for 6 months

C. The method of determining frequency differs from that used for bulimia nervosa; future research should address whether the preferred method of setting a frequency threshold is counting the number of days on which binges occur or counting the number of episodes of binge eating.

D. The binge eating is not associated with the regular use of inappropriate compensatory behaviors (e.g., purging, fasting, excessive exercise) and does not occur exclusively during the course of anorexia nervosa or bulimia nervosa.

SOURCE: Reprinted with permission from the Diagnostic and Statistical Manual of Mental Disorders, Fourth Edition, Text Revision, Copyright 2000. American Psychiatric Association.

Binge-Eating Disorder

Binge-eating disorder (BED) is a condition in which an individual engages in eating large amounts of food and feels that these eating episodes are not within one's control.[36] BED is defined by recurrent episodes of binge eating at least 2 days a week for at least 6 months (Table 15.13). In addition, the person feels a subjective sense of a loss of control over binge eating, which is indicated by the presence of three of the following five criteria: eating rapidly, eating when not physically hungry, eating when alone, eating until uncomfortably full, and feeling self-disgust about bingeing. BED differs from bulimia nervosa in that binge eating is not followed by compensatory behaviors such as self-induced vomiting, as occurs in bulimia nervosa.

Dieting may be a risk factor for BED; however, 35% to 55% of women may experience binging before dieting.[39] Females who report dieting before bingeing are more likely to have experienced sexual abuse, which may lead to feelings of loss of control and the desire to participate in restricting behaviors to regain a sense of control. Females who experienced stressful situations, such as the death of someone close to them, were more likely to report bingeing prior to dieting, consistent with an emotional eating response. Age of onset of BED is somewhat lower for women who report bingeing first (20 years) versus those who report dieting first (25 years).

BED appears to be more prevalent among overweight clinical populations (30%) than among community samples (5% of females and 3% of males).[36] Studies on adolescents that assess the prevalence of binge eating include few that document prevalence rates of BED. In a college-student sample, the rate of BED was 2.6%. In contrast to other weight-related conditions, significant differences were not found between male and female students. Further study of the prevalence and etiology of BED among adolescents seems critical in light of the high rates of obesity among youth.

Etiology of Eating Disorders

The etiology of eating disorders is multifactorial—that is, many factors contribute to their onset. Some of the major contributory factors include social norms emphasizing thinness, being teased about one's weight, familial relations (e.g., chaotic lifestyles, lack of boundaries between family members, poor patterns of communication), physical and sexual abuse experiences, personal body shape and size, body image, and self-esteem (Table 15.14). A genetic component to eating disorders has been proposed in response to studies that have shown a higher prevalence of these disorders within families and/or twins. These factors do not operate in isolation, but rather an interaction between genetics and environmental risk factors may be necessary to increase an adolescent's risk for engaging in potentially harmful eating and dieting behaviors.[39]

In considering etiological issues, it is essential to realize that different etiological pathways may lead to weight-related disorders in different adolescents. For some adolescents, family issues may be major factors, while for others social norms may be the key factors leading to the onset of a condition. Furthermore, different conditions tend to be influenced by different factors. Potential contributory factors for eating disorders can be categorized into environmental, familial, interpersonal, and personal domains.

Environmental factors include:

- Media influences
  - Media messages and images
  - Body images portrayed in media
- Societal and cultural norms
  - Frequency of eating and snacking
  - Food preferences
  - Attitudes toward body weight
  - Roles of women and children within families and communities
- Food availability and accessibility
  - Food security or insecurity
  - Types of food easily accessible
  - Amounts of food served and consumed

Table 15.14 Screening elements and warning signs for individuals with eating disorders

Screening	Warning Signs
Body image and weight history	• Distorted body image • Extreme dissatisfaction with body shape or size • Profound fear of gaining weight or becoming fat • Unexplained weight change or fluctuations greater than 10 lbs
Eating and related behaviors	• Very low caloric intake; avoidance of fatty foods • Poor appetite; frequent bloating • Difficulty eating in front of others • Chronic dieting despite not being overweight • Binge-eating episodes • Self-induced vomiting; laxative or diuretic use
Meal patterns	• Fasting or frequent meal skipping to lose weight • Erratic meal pattern with wide variations in caloric intake
Physical activity	• Participation in physical activity with weight or size requirement (e.g., gymnastics, wrestling, ballet) • Overtraining or "compulsive" attitude about physical activity
Psychosocial assessment	• Depression • Constant thoughts about food or weight • Pressure from others to be a certain shape or size • History of physical or sexual abuse or other traumatizing life event
Health history	• Secondary amenorrhea or irregular menses • Fainting episodes or frequent light-headedness • Constipation or diarrhea unexplained by other causes
Physical examination	• BMI <5th percentile • Varying heart rate, decreased blood pressure after arising suddenly • Hypothermia; cold intolerance • Loss of muscle mass • Tooth enamel demineralization

SOURCE: The Society for Nutrition Education. L. B. Adams and M. B. Shafer, "Early Manifestations of Eating Disorders in Adolescents: Defining Those at Risk," J Nutr Educ 20 (1988); and American Medical Association. K. Perkins, N. Ferrari, and A. Rosas et al., "You Won't Know Unless You Ask: The Biopsychosocial Interview for Adolescents," Clin Pediatr 36(2), 1997; and Guidelines for Adolescent Preventive Services (GAPS): Recommendation Monogragh, 2nd ed. Chicago: American Medical Association, 1995.

Familial factors include:

- Family dynamics
  - Communication styles and patterns between family members
  - Appropriateness of parental expectations
  - Appropriateness of personal boundaries between teens and other family members
- Weight-related behaviors modeled by parents and siblings
- Feeding behaviors reinforced during childhood and adolescence

Interpersonal factors include:

- Peer norms and behaviors
  - Dieting beliefs and behaviors
  - Food preferences and behaviors
  - Weight preoccupation
- Abuse experiences
  - Verbal abuse
  - Physical or sexual abuse

Personal factors include:

- Biological factors
  - Genetic predisposition
  - Level of physical development and sexual maturation
  - Chronological age
- Psychological factors
  - Self-esteem and self-efficacy levels
  - Personal body image
  - Depression and/or anxiety
  - Coping mechanisms
- Knowledge, attitudes, and behaviors
  - Nutrition beliefs and practices
  - Health beliefs and practices
  - Physical activity beliefs and practices
  - Consequences of dieting, fasting, purging, obesity

An understanding of the etiology of eating disorders is essential to the development of effective interventions aimed at their treatment and prevention. An individual clinical setting needs to allow time to assess the factors leading to the onset of the condition for that particular adolescent. In developing prevention programs to reach larger groups of adolescents, it is more feasible to identify and address factors that may be contributing to the onset of weight-related behaviors and conditions for a broad sector of the targeted population. Although not all factors may be addressed within one intervention, it is important to be aware of the broad range of factors coming into play and the interactions between them.

Treating Eating Disorders

The complex etiology of eating disorders and their potentially life-threatening psychosocial, physical, and behavioral consequences highlight the need for a multidisciplinary treatment approach. The health care team caring for an adolescent with an eating disorder will often include a physician, dietician, nurse, psychologist, and/or psychiatrist. The role of the dietician is paramount to the treatment of eating disorders at the stages of assessment, treatment, and maintenance. Initially an adolescent may be more willing to discuss his or her concerns with a dietician than with a psychologist.

The treatment of eating disorders may take many forms. Individuals with eating disorders who are medically and psychologically stable are generally treated through outpatient programs. The frequency of contact with the health care team is usually weekly, but it may be more frequent if required. Day treatment programs, often referred to as partial-inpatient programs, may be recommended for individuals who require daily contact with the health care team and whose body weight is sufficient to remain treated as an outpatient. Day treatment programs may vary in the number of weekly visits from 3–7 days per week, depending upon the facility and the client. Inpatient programs are required for individuals with life-threatening comorbidities, unstable medical or psychological status, or severely low body weight. The criteria for inpatient care of adolescents are listed in Table 15.15.

The goal of eating-disorder treatment programs is to restore body weight, to improve social and emotional well-being, and to normalize eating behaviors. While programs

Table 15.15 Suggested criteria for inpatient treatment of eating disorders

Medical Criteria	Psychosocial Criteria
Failure to thrive (BMI <3rd percentile)	Social isolation
Rapid and dramatic weight loss	Depression
Very low caloric intake	Obsessive-compulsive disorder
Refusal to eat or drink	Suicidal thoughts or tendencies
Hypokalemia	Lack of parental support
Alkalosis	Poor family communication and dynamics
Bradycardia	Poor response to outpatient or day treatment
Pancreatic dysfunction	
Liver dysfunction	

SOURCE: Based on Herpetz-Dahlmann, B. and Slaback-Andrae, H. Overview of treatment modalities in adolescent anorexia nervosa. Child Adolesc Psychiatr Clin N Am 2008; 18:131–145.

vary, the core components of any eating-disorder treatment program include:[40]

- Treatment of medical comorbidities
- Restoration of body weight to a normal level
- Nutrition education and counseling to normalize food-related thoughts and beliefs
- Individualized psychotherapy to improve social well-being and emotional health
- Family therapy to improve communication and family function
- Group therapy

During treatment, a major role of the dietician is to help the adolescent normalize eating patterns and to feel comfortable with these changes. Some of the key goals of the nutritional care include the following:

- Thoroughly assess dietary intake and adequacy
- Recommend nutrition-related therapeutic interventions based on nutrition-assessment data
- Provide counseling to client to establish a regular pattern of nutritionally balanced meals and snacks
- Monitor dietary intake and physical activity levels to determine adequate but not excessive levels of energy intake or physical activity, with the goal of reaching and maintaining a healthy body weight
- Counsel clients to consume adequate dietary fat and fiber intake to promote satiety
- Provide counseling in conjunction with the psychologist or other mental health care provider on strategies to help clients avoid dieting behaviors and excessive exercise
- Assist clients with strategies to gradually include formerly forbidden foods into the diet, which may include role-modeling appropriate food intake at mealtimes, arranging for groups of clients to practice eating out at restaurants, and assisting clients in preparing meals
- Periodically evaluate effectiveness of interventions
- Monitor client nutritional status

For some adolescents, denial of the condition or a lack of motivation for change make nutrition counseling quite challenging. It is important for the nutritionist to work in close conjunction with other members of the health team to ensure that roles of different members of the team are clearly defined.

Preventing Eating Disorders

The high prevalence of eating disorders and their potentially harmful consequences point to a need for interventions aimed at their prevention. One of the most pressing current public-health issues that needs to be addressed concerns the prevention of eating disorders described in previous sections and the prevention of obesity. Even the prevention of a small percentage of these conditions, at a population level, returns huge benefits in terms of reducing physical, emotional, and financial burdens.

In the development of interventions aimed at the prevention of eating disorders, it is essential to address factors that contribute to the onset of these conditions for a large proportion of the targeted population, factors that are potentially modifiable, and factors suitable for addressing within the designated setting. For example, media awareness and advocacy has been suggested as a suitable approach toward preventing eating concerns and disorders. Participants may learn about how the media influence one's body image and about techniques used within the media to improve the appearance of models, and then take action toward making changes in the media. This approach is suitable in that media influences, and the internalization of media messages, may contribute to weight concerns among a large sector of the adolescent population. These adolescent perceptions are potentially modifiable and suitable for addressing within clinical, community, and school-based settings where interventions may be implemented.

Any efforts toward prevention first must consider the target audience. An important question is whether to direct interventions to all adolescents or to adolescents at increased risk for eating disorders. Reasons for providing interventions for all adolescents include the high prevalence of eating concerns among adolescents, difficulties inherent in identifying and targeting high-risk individuals, and the advantages of developing positive social norms regarding eating issues within the peer group. Taking a more targeted approach offers the advantages of better use of limited resources, more intensive interventions, and interventions developed for specific high-risk groups (e.g., ballet dancers, youth with diabetes, or overweight girls). In order to be most effective at preventing eating disorders, both types of interventions seem necessary; more general approaches address the issues of the general adolescent population, while more refined approaches can better meet the needs of specific high-risk groups.

Prevention interventions may be implemented within clinical, community, and school-based settings that serve adolescents. A recent meta-analysis of eating-disorder prevention programs suggests that more than half of published eating-disorder prevention programs reduced at least one risk factor for disordered eating, and 29% reduced the severity of disordered eating among youth.[41] Programs that focused on changing weight-related attitudes of youth and promoted healthy weight-control strategies were found to be the most effective, with effects lasting up to 2 years. Other characteristics of successful eating-disorder prevention programs included:

- Selective targeting of high-risk groups rather than all youth
- Programs targeting adolescents >15 years of age

- Programs with information provided by trained interventionists rather than counselors, teachers, or health care providers
- Programs that included multiple sessions rather than a single encounter
- Integrated interactive learning (role-playing, computer technology, etc.) rather than providing only didactic learning experiences

Most eating-disorder prevention programs have focused on females. There is a need to develop prevention programs for males and mixed-gender audiences, as well.

Children and Adolescents with Chronic Health Conditions

Approximately 18% of children and adolescents have a chronic condition or disability. These children and adolescents are at increased risk for nutrition-related health problems because of (1) physical disorders or disabilities that may affect their ability to consume, digest, or absorb nutrients; (2) biochemical imbalances caused by long-term medications or internal metabolic disturbances; (3) psychological stress from a chronic condition or physical disorder that may affect a child's appetite and food intake; and/or (4) environmental factors, often controlled by parents who may influence the child's access to and acceptance of food.

Nutrition reports of children and adolescents with special health care needs estimate that as many as 40% have nutrition risk factors that warrant a referral to a dietitcan. Common nutrition problems in children and adolescents with special health care needs include the following:

- Altered energy and nutrient needs (e.g., inborn errors of metabolism, spasticity of movement, enzyme deficiencies)
- Delayed growth

- Oral-motor dysfunction (e.g., neurological disorders, swallowing disorders)
- Elimination problems
- Drug/nutrient interactions
- Appetite disturbances
- Unusual food habits (e.g., rumination)
- Dental caries, gum disease

Malnutrition has been implicated as a major factor contributing to poor growth and short stature in adolescents with a variety of diseases (e.g., chronic inflammatory bowel disease, cystic fibrosis). Factors such as inadequate nutrient and energy intakes, excessive nutrient losses, malabsorption, and increased nutrient requirements all lead to the chronic malnourished state. Studies have shown that the energy requirements for adolescents with cystic fibrosis or inflammatory bowel disease may be 30–50% higher than the RDA for adequate growth. In addition to the increased energy needs caused by malabsorption (or in the case of adolescents with cystic fibrosis, the increased work of breathing), fever, infection, and inflammation also increase energy requirements. Whereas undernourishment is frequently seen in adolescents with chronic illnesses, obesity is common among youth with gross motor limitations or immobility. Because of limited activity, caloric requirements are lower, and the balance between intake and expenditure is often difficult, resulting in obesity.

Consideration of nutrition needs of children with chronic disabling conditions or illnesses is complex and requires specialized, individualized care by an interdisciplinary team. Assessment of nutrition status followed by nutrition intervention, when necessary, and monitoring will help ensure the health and well-being of adolescents with chronic and disabling conditions. Also, during adolescence, issues of personal responsibility and independent-living skills related to food purchasing and preparation may need to be addressed.

Review Questions

1. Stage 2 treatment for adolescent obesity can be offered by a single provider.

   True   False

2. At which state of obesity treatment is a structured eating plan designed to create a caloric deficit?
   a. Stage 1
   b. Stage 2
   c. Stage 3
   d. Stage 4

3. Which of the following nutrients may be required in greater amounts by adolescents who use alcohol and tobacco?
   a. Thiamin
   b. Iron

   c. Vitamin C
   d. All of the above

4. Which is not a criterion for inpatient treatment of eating disorders?
   a. Social isolation or depression
   b. Excessive parental expectations
   c. Weight loss of more than 10 lb in 3 months
   d. Unwillingness to eat or drink

5. Which of the following is a normal lipid profile for an adolescent?
   a. Total cholesterol = 185, HDL = 90
   b. Total cholesterol = 158, HDL = 37
   c. Total cholesterol = 179, HDL = 39
   d. Total cholesterol = 167, HDL = 57

6. All of the following criteria are required for adolescent bariatric surgery, except:
   a. Failure to lose weight during intensive lifestyle program
   b. BMI > 40 with medical complications
   c. Family history of obesity-related deaths
   d. Supportive home environment

7. How does binge-eating disorder differ from bulimia nervosa?
8. How does the use of substances such as alcohol, tobacco, and recreational drugs affect the nutritional status of adolescents?

Resources

National Institute for Children's Health Care Quality
Information on childhood obesity, including a childhood obesity intervention guide, and details on children with special health care needs.
Website: http://www.nichq.org/

Guidelines for Adolescent Nutrition Services, University of Minnesota
Information on adolescent growth and development, nutrition needs, and nutrition interventions.
Website: http://www.epi.umn.edu/let/pubs/adol_book.shtm

Adolescent Nutrition: Eating Disorders and Interventions, University of Washington
Information and tools related to identifying and treating disordered eating.
Website: http://faculty.washington.edu/jrees/adolescentnutrition.html

Life Cycle Nutrition: Adolescence, USDA Food and Nutrition Service
Information and data on adolescent nutrition with links to publications and other resources.
Website: http://fnic.nal.usda.gov/nal_display/index.php?info_center=4&tax_level=2&tax_subject=257&topic_id=1354

Child and Adolescent Nutrition Knowledge Path, The Maternal and Child Health Library
Knowledge path with links to resources about obesity in children and adolescents.
Website: http://www.mchlibrary.info/KnowledgePaths/kp_childnutr.html

The World Bank Adolescent Nutrition Resources
Information about global adolescent health issues.
Website: http://web.worldbank.org/WBSITE/EXTERNAL/TOPICS/EXTHEALTHNUTRITIONANDPOPULATION/EXTNUTRITION/0, contentMDK:20206757~menuPK:483704~pagePK:148956~piPK:216618~theSitePK:282575,00.html

Child and Adolescent Health and Development, The World Health Organization
Data, reports, and resource links about global issues related to adolescent nutrition.
Website: http://www.who.int/child_adolescent_health/en/

Nutrition for Children with Special Health Care Needs, University of Washington
Information and links to resources on nutrition needs and services for youth with special health care needs.
Website: http://depts.washington.edu/cshcnnut/resources/cshcn.html

Healthy Youth! National Center for Chronic Disease Prevention and Health Promotion, Centers for Disease Control and Prevention
Statistics, reports, and other resources related to improving the nutrition and physical-activity behaviors of youth.
Website: http://www.cdc.gov/HealthyYouth/nutrition/

Society for Adolescent Medicine, Teen and Family Resources
Links to resources on a full array of adolescent health topics, including nutrition, obesity, and eating disorders.
Website: http://www.adolescenthealth.org/categories.htm

Chapter 16

Adult Nutrition

Prepared by Patricia L. Splett
and U. Beate Krinke

Key Nutrition Concepts

1   Alterations in nutrition and health status progress over a long period and are reversible up to a point, but with continued poor nutrition, permanent damage occurs.

2   Many nutrition risk factors faced by adults are modifiable.

3   The aim of dietary recommendations is to reduce risks of specific diseases and ensure that the population consumes adequate levels of required nutrients.

4   Current food and nutrient intakes of adults often fail to match recommendations.

5   Eating competence includes enjoyment of food.

Introduction

> "Healthy living is the best revenge."
>
> Earl S. Ford[1]

Adulthood marks a long period between the active growth and development phases of infancy, childhood, and adolescence and the older adult phases, where a concern is sustaining physical and mental capacity. Adulthood is subdivided into the following segments.

Early Adulthood: The twenties generally involve becoming independent, leaving the parental home, finishing formal schooling, entering regular employment and starting a career, developing relationships, and choosing a partner. Planning, buying, and preparing food are newly developing skills for many. The thirties could be characterized by increasing responsibilities to and for others, including having children, providing for and caring for family, building a career, and involvement in community and civic affairs. There may be renewed interest in nutrition at this time "for the kids' sake."

Midlife: The forties are a period of active family responsibilities (that may include nurturing children and teenagers and, for some, building new relationships and blending families), as well as expanding work and professional roles. Managing schedules and meals becomes a challenge. Sociologists say that this is a time of reviewing life's accomplishments and beginning to recognize one's mortality.

Sandwich Generation Refers to middle-aged adults, usually women, who are multigenerational caregivers dealing with the complex roles of wife, mother, daughter, caregiver, and employee.

Chronic Diseases Slow-developing, long-lasting diseases that are not contagious (e.g., heart disease, cancer, diabetes). They can be treated but not always cured.

Healthy Weight A weight range compatible with normal function and long, healthy life.

The phase around the fifties is referred to as the *"sandwich" generation*. Many, especially women, are multigenerational caregivers who juggle the roles of caring for children and aging parents while maintaining a career. Work and career continue to be priorities for most adults. In the fifties, health concerns frequently are added to the picture. Dealing with a *chronic disease* or managing identified risk factors to prevent diseases is an added responsibility.

Old Age: By their early sixties, many adults are making the transition to retirement, have more leisure time, and are able to give greater attention to physical activity and nutrition. While many are "empty-nesters," significant numbers have children living at home and/or have responsibilities as guardians and caretakers of grandchildren, parents, or others. Food choices and lifestyle factors may take on added significance for those who are dealing with a chronic disease.

This chapter explores the nutritional needs of adults and nutrition guidance aimed at helping meet those needs. The importance of preventing injury at the cellular level, recognizing modifiable risk factors, and maintaining a *healthy weight* are emphasized.

Importance of Nutrition

The span of years between ages 20 and 64 is a time when diet, physical activity, smoking, and body weight strongly influence the future course of health and wellness. During these forty-four years, lifestyle choices interact with genetic endowment, social forces, and environmental factors to determine years of life and quality of life.[2]

The onset and severity of five of the ten leading causes of death in adults (cancer, heart disease, stroke, diabetes, and hypertension) have risk factors that can be modified through changes in nutrition and physical activity. In addition, Table 16.1 characterizes diets that lead to chronic diseases and summarizes modifiable nutritional risk factors related to short-term and chronic diseases.

Health Objectives for the Nation

The Health Objectives for the Nation address multiple adult health improvement goals. In the last decade progress was made. Rates of heart disease, cancer, and stroke deaths declined; and fat intake (as percentage of total calories) declined. However, rates of obesity and diabetes have increased, sugar intake has risen, and health care disparities still exist.[3] Data on dietary goals for disease prevention and health promotion for adults are shown on Table 16.2. These areas are ongoing objectives for 2020. A priority of national public health goals is obesity.

Table 16.1 Nutritional risk factors for chronic diseases*

Cancer — *Carcinogenic diet*
- Low fruit and vegetable intake
- Low level of antioxidants (especially vitamins A, C)
- Low intake of whole grains and fiber
- High dietary fat intake
- Nitrosamines, burnt and charred food
- High intakes of pickled and fermented food
- Alcohol consumption
- High animal-, low plant-food intake

Heart Disease — *Atherogenic diet*
- High saturated fat (>10% calories)
- *Trans*-fatty acid intake
- Dietary cholesterol intake >300 mg
- Low antioxidants, low fruit and vegetable intake
- Low intake of whole grains
- No or excess alcohol**
- High waist circumference (men >40, women >35 inches)
- Elevated plasma Apo B levels
- High levels of LDL cholesterol, especially in men
- Low levels of HDL cholesterol, especially in women
- High blood triglycerides
- Glucose intolerance
- Hypertension, Stroke
- High sodium
- Low potassium
- Low milk and dairy foods
- Excess alcohol
- Low levels of antioxidants

Obesity — *Obesogenic diet*
- Caloric intake exceed needs
- Unstructured eating
- Frequent fast-food consumption
- High fat intake
- Sugar-sweetened beverage consumption
- Energy-dense, low-nutrient food choices

Diabetes — Atherogenic diet / Obesogenic diet

*Obesity (BMI >30) and physical inactivity are independent risk factors for all of the chronic conditions.
**In middle-aged adults, moderate alcohol intake reduces risk of heart disease (for men, after age 45; and for women, after age 55).

Physiological Changes During Adulthood

For the most part, individuals have stopped growing by the time they reach their twenties. Men and women continue to develop bone density until roughly age 30. Muscular strength peaks around 25 to 30 years of age, although regular use of muscles and weight training affects strength as well as muscle size and retention. The type and amount of physical activity has a significant impact on body composition, including **lean body mass** (musculature), fat accumulation and relocation, and bone density.[4] By middle adulthood, physical changes become more apparent with the decline in size and mass of muscles and an increase in body fat. Dexterity and flexibility decline, as well as sensory and perceptual abilities. Hearing loss begins as early as age 25 (or earlier with exposure to loud music), and vision changes often become noticeable by

Carcinogenic Diet A pattern of eating and food choices that increases the risk of some cancers.

Atherogenic Diet A pattern of eating and food choices that promotes deposits of plaque in arterial walls and leads to the development of cardiovascular disease.

Obesogenic Diet A pattern of eating and food choices that leads to excessive energy intake and accumulation of body fat.

Lean Body Mass Sum of fat-free body tissue: muscle, mineral (as in bone), and water.

Table 16.2 Progress toward Healthy People 2010: Focus Area Nutrition and Overweight

Objective[a] *and 2008 Status*[b]		Target	Population Percentage Baseline, Age 20+	
			Females	Males
Increase the number who are at a healthy weight	*Worse*	60	45	38
Decrease the number who are obese (BMI 30+)	*Worse*	15	25	20
The following baseline information is for age 20–39. (Baselines for ages 40–59 are in parentheses.) Increase the proportion of persons who:				
• Eat at least two daily servings of fruit	*No progress*	75	20 (26)	23 (28)
• Eat at least three daily servings of vegetables (at least one-third dark green or deep yellow)	*No progress*	50	4 (4)	3 (4)
• Eat at least six daily servings of grain products (at least three of those as whole grains)	*No progress*	50	4 (4)	10 (10)
• Eat less than 10% of calories from saturated fat[c]	*No progress*	75	41 (42)	32 (33)
• Eat no more than 30% of calories from fat[c]	*Worse*	75	38 (33)	29 (28)
The following baseline information is for ages 20–49. Meet dietary recommendations (1000 mg) for calcium	*Improved*	75	40	64
The following baseline is for all adults, age 20+. Eat 2400 mg or less of sodium daily	*No progress*	65	30	5
Increase the proportion of physician office visits made by patients with a diagnosis of cardiovascular disease, diabetes, or hyperlipidemia that include counseling or education related to diet and nutrition	*Worse*	75	39	44

SOURCE: Food and Drug Administration and National Institutes of Health. *Healthy People 2010: National Health Promotion and Disease Prevention Objectives.* Washington, D.C.: U.S. Department of Health and Human Services, 2000.

a = source of objectives.

b = 2003–2004 NHANES data compared to 1988–1994 (baseline) data. Available at http://www.cdc.gov/nchs/hphome.htm, accessed September 2009.

c = Updated to recommend 20–35% of calories from fat, minimizing saturated and *trans*-fatty acid intake. [2002; IOM]

Climacteric Change Point in life where crucial changes occur; refers to the loss of reproductive activity, marked by menopause in women and reduction in testosterone production in men.

Menopause The end of menstruation and a marking point for increased risk of cardiovascular disease and other chronic conditions for women.

Energy Balance An equilibrium state when the number of calories consumed equals the number of calories expended.

Basal Metabolic Rate (BMR) Total energy expenditure by the body at rest. It is measured in an individual who has been awake less than 30 minutes and is still at absolute rest, has fasted for 10 hours or more, and is in a quiet room with normal, comfortable temperature.

Thermic Effect of Food (TEF) Energy required for the digestion, absorption, and metabolism of food; approximately 10% of energy needs.

age 40. Body composition slowly shifts in tandem with hormonal shifts.

Hormonal and Climacteric Changes The decline of estrogen production in women leads to *menopause,* the end of menstruation. Menopause is associated with an increase in abdominal fat and significant increase in risk of cardiovascular disease and accelerated loss of bone mass.[5] Men experience a gradual decline in testosterone level and muscle mass. Physical activity and weight training to increase muscle mass result in small and transient increases in testosterone level. Obesity is associated with higher estrogen levels in both men and women.

Body Composition Changes in Adults

The years between ages 20 and 64 are typically associated with a positive *energy balance* with an increase in weight and adiposity and a decrease in muscle mass. A redistribution of fat occurs with gains in the central and intra-abdominal space and away from subcutaneous fat. This redistribution of body fat is associated with increased risk for hypertension, insulin resistance, diabetes, stroke, gallbladder disease, and coronary artery disease.[6] These risks increase with the accumulation of additional body fat.[7]

Estimating Energy Needs in Adults

Energy needs are based on an individual's *basic metabolic rate (BMR)*, the *thermic effect of food (TEF),* and activity thermogenesis (which includes energy expended through exercise and nonexercise activity such as fidgeting). The largest component of daily energy expenditure, 60–75% for most adults, is the involuntary process of internal chemical activities that maintain the body.[8] While the brain, liver, gastrointestinal tract, heart, and kidney make up less than

5% of body weight, the active metabolic processes and functions of these organs account for about 60% of BMR. Additional energy is required for the digestion, absorption, and metabolism of food–referred to as the thermic effect of food (TEF). This amounts to approximately 10% of energy needs but varies with diet composition and across individuals. TEF is lower in some obese individuals, suggesting that more efficient digestion and absorption of food may be a factor in obesity.[9] The most variable component of energy expenditure is activity thermogenesis which accounts for 20-40% of total energy expenditure.

Calorimetry is the measurement of the amount of heat given off or absorbed by a reaction or group of reactions (as by an organism). In humans, we are interested in the energy utilized for the body's metabolic processes, which is the major component of the total daily energy expenditure. Indirect calorimetry is used to determine *resting energy expenditure (REE)*, a measure closely related to basal metabolic rate. Indirect calorimetry is done by measuring the exchange of gases during respiration, for a specific period of time, using a metabolic cart in hospitals or newer portable technology and hand-held devices in clinics and gyms. The respiratory quotient (CO_2/O_2) is used to estimate 24-hour energy expenditure.

REE can also be calculated using a validated estimation formula. The Mifflin-St. Jeor formula was developed for healthy, normal-weight and moderately overweight men and women. It uses the routinely available measures of height (cm) and weight (kg).[10]

Mifflin–St. Jeor Energy Estimation Formula
Males: REE = (10 × wt) + (6.25 × ht)
         − (5 × age) + 5
Females: REE = (10 × wt) + (6.25 × ht)
           − (5 × age) − 161

After the REE has been determined, the value is multiplied by an activity factor (1.2 sedentary, 1.55 moderately active, or 1.725 very active) to arrive at the estimated daily calorie expenditure.[11] Numerous energy calculators are available online. Many are based on the older Harris-Benedict equation, which is less accurate.[12] A simple calculation can also be used to "ballpark" caloric levels for weight maintenance, weight loss, or weight gain. Approximately 15 calories per pound per day are needed to maintain weight. Cutting to 13 calories/lb per day can result in weight loss, and increasing to 17 calories/lb per day can produce weight gain.

Energy Adjustments for Weight Change

A pound of body weight is the equivalent of approximately 3500 calories. To lose 1 lb a week, an adult would need to create a negative calorie balance of 500 calories daily. These 500 calories can be generated from a combination of decreased calorie intake and increased physical activity. For instance, brisk walking two extra miles uses roughly 200 calories, and eliminating 12 ounces of regular cola and a single-serving bag of potato chips (1 oz) allows subtraction of 300 calories, summed to total 500 calories. Seven days of burning 300 extra calories and eating 200 fewer calories leads to a weight loss of approximately 1 lb. On the other hand, a positive balance of just 100 extra calories per day will result in a gain of 10 lbs in a year!

> **Resting Energy Expenditure (REE)** Measured energy expenditure in an individual who has fasted, had no vigorous physical activity prior to the test, has been given time to relax (e.g., rest) for 30 minutes before starting measurement, and is in a quiet room with comfortable temperature.

Age-Related Changes in Energy Expenditure

Metabolic rate and energy expenditure begin to decline in early adulthood at a rate of about 2% per decade.[13] These reductions generally correspond to declines in physical activity and lean muscle mass. Between ages 25 and 65, physical working capacity (measured by VO_2 max) declines 5% to 10% per decade. The presence of musculosketetal disease, obesity, and other conditions can accelerate declines in energy expenditure and physical capacity.

In young, healthy adults, there is compensatory adjustment between physical activity and calorie intake. Early studies indicate this is also true for older adults.[14] In the mid-1960s, researchers at the Baltimore Longitudinal Study on Aging (BLSA) reported that caloric intake in men decreased 22%, from 2700 to 2100 calories, between age 30 and age 80.[15] They suggested that the decrease was due to lowered metabolic rates as well as decreased activity levels. This study also found that fat-free mass levels in women depend on energy expenditure.

National data indicate that adult caloric intake declines with age. Table 16.3 shows reported caloric intake

Table 16.3 Comparing caloric intakes of adults, by gender, from CSFII (Continuing Survey of Food Intake by Individuals, 1994–1995) and NHANES III (National Health and Nutrition Examination Survey 1988–1994) and NHANES 1999–2000

Age	Daily Caloric Intake, CSFII		Daily Caloric Intake, NHANES III		Daily Caloric NHANES 99–00	
	Males	Females	Males	Females	Males	Females
20–29	2844	1828	3025	1957	2828	2028
30–39	2702	1676	2872	1883	2828	2028
40–49	2411	1680	2545	1764	2590	1828
50–59	2259	1583	2341	1629	2590	1828
60–69	2100	1496	2110	1578	2123	1596

collected by the U.S. Department of Agriculture's Continuing Survey of Food Intake by Individuals (CSFII) and by the Department of Health and Human Services in the National Health and Nutrition Examination Survey (NHANES). Nutrition scientists suggest that underreporting by overweight people and overreporting by underweight individuals skews these numbers. However, a decline is evident for age, and an increase in caloric intake from 1994–1995 to 1999–2000 is also evident.

Evidence suggests that increased calorie intake, rather than increased sedentary activity, is primarily related to the obesity epidemic. Per capita food availability has increased 16% since 1970, and the types of foods consumed have changed dramatically (Table 16.4). Per capita figures apply to individuals of all ages and correspond to the increase in weight of the American public.[16,17]

Fad Diets

Dieting is so pervasive, and weight loss is so difficult to maintain, that weight-loss efforts support a multimillion-dollar industry. It has been estimated that 71% of females and 42% of males are dieting at any time.[18] *Weight cycling* with weight regains larger than weight losses, also called *yo-yo dieting*, is considered more harmful than persistent overweight. Weight cycling is associated with higher cardiovascular and all-cause mortality.[19]

Diet books frequently make the top-ten bestseller lists. Checking the *New York Times* and Amazon.com will reveal the current diet fad. Diets are also popular-

Weight Cycling/Yo-yo Dieting   A pattern of weight loss and regain, with weight gain exceeding the amount lost. Weight cycling is associated with higher mortality.

Fad Diets   Popular weight-loss approaches; many, but not all, are unhealthy.

ized and promoted by TV personalities, via Web advertisements and print and social media. Popular approaches to weight loss are commonly referred to as *fad diets*. Most should be viewed with caution, but some popular diets, such as the Mediterranean diet, are actually very healthful. Fad diets generally promise quick results—something easy and "guaranteed to get you into those tight new jeans." They focus on fast weight loss promoted with testimonials, alluring before-and-after pictures, and deceptive marketing tactics. Many make claims of scientific rationale based on simplistic interpretation of animal or other studies and anecdotal information. The following features flag unhealthy and potentially harmful fad-diet strategies:

- Inadequate nutrient supply (restrictions deplete tissue reserves of important nutrients)
- Severe energy restriction (diets with 800 or fewer calories a day require medical supervision)
- Unusual food restriction (for example, no carbohydrates)
- Food combinations (for example, grapefruit with all meals)
- Strict limitations (avoiding certain food groups, such as no dairy or never eating potatoes)
- Gimmicks (don't eat after 7 p.m. in the evening, eat only popcorn for lunch, drink caffeine or guarana, or eat hot peppers to speed up your metabolism)

The popular but competing approaches of fat reduction (e.g., the Mediterranean Diet and National Cholesterol Education Program recommendations), carbohydrate restriction (e.g., Atkins and The Zone) and calorie-restricted diets have been compared in several scientific studies. After six months, individuals in the low-carbohydrate program lost more weight, but at 12 months, weight losses were not significantly different.[20] Fad diets offer strategies to initiate weight loss, but most fail to offer realistic or healthful strategies to maintain weight loss, which is the real test of any weight-loss diet. The cornerstone of successful weight management is long-term lifestyle change. In weight loss, the aim is to lose fat—adipose tissue, not water, muscle, or tissue from essential organs—while consuming a variety of low-calorie, nutrient-dense foods to meet nutritional needs and maintain body stores of nutrients. Healthful weight-loss regimens also include physical activity to enhance the energy deficit and to shape and firm the body.

Continuum of Nutritional Health

With good genes, good habits, good environment, and good luck, nutritional and physical health can be maintained throughout adulthood. More likely, the interaction of those factors over the years results in "nutritional injury," which leads to alteration or loss of function at the cellular level. Nutritional injury may be minor, of short

Table 16.4 Per capita increases in food available for consumption, adjusted for plate waste and spoilage

Commodity Group	1970, lb	2003, lb	Change in Daily Calories (kcal)
Fats and oils	53	86	+216
Grains	136	194	+188
Sugars, sweeteners	119	142	+76
Meat, eggs, nuts	226	242	+24
Vegetables	337	418	+16
Fruits	242	275	+14
Dairy	564	594	−11
TOTAL	1675	1950	+523

Adapted from Farah, H. and Buzby, J. U.S. food consumption up 16% since 1970. *Amber Waves*, 11/05 (32). Available at www.ers.usda.gov/AmberWaves/November05/Findings/USFoodConsumption/htm, accessed 10/3/2006.

duration, and reversible, or, if it continues, permanent changes in cells and tissues can develop.[21] The *continuum of nutritional health* is shown in Illustration 16.1 and described below. But first, several principles of human nutrition, presented in Chapter 1, bear repeating here to emphasize their relevance to nutrition during the adult years.

- Health problems related to nutrition originate within cells.
- Poor nutrition can result from both inadequate and excessive levels of nutrient intake.
- Humans have adaptive mechanisms for managing fluctuation in food intake.
- Malnutrition can result from poor diets and from disease states, genetic factors, or combinations of these causes.
- Poor nutrition can influence the development of certain chronic diseases.
- Adequacy and balance are key characteristics of healthy diets.
- There are no "good" or "bad" foods.

Nutritional health can be viewed as a continuum ranging from "healthy" and resilient to the terminal state in which the body systems shut down and life ceases. According to this continuum, adapted from nutritional-injury models put forth by Arroyave and Leyse-Wallace,[22,23]

changes occurring at the cellular level are initially insidious and unnoticed. Alterations progress over a long period, and are reversible up to a point. But in the face of continued poor nutrition, permanent damage occurs. Altered nutrient intakes produce early changes in metabolic process that are preclinical stages of illness. This "injury" may not manifest itself until permanent damage has occurred. In the absence of signs and symptoms and awareness of a "problem," adults might not be especially concerned about food choices or motivated to adjust lifestyle behaviors.

> **Continuum of Nutritional Health** Stages of nutritional status that range from optimal to unable to sustain life. The stages are resilient and healthy, altered substrate availability, nonspecific signs and symptoms, clinical conditions, chronic conditions, and terminal illness and death.

States of Nutritional Health

The continuum of nutritional health can be represented in six states or stages.

Resilient and "Healthy"   In this state, metabolic systems are in homeostasis, and organs are functioning at optimum level. The body's defenses and immune system can counter assaults from toxins, pathogens, and stress. In this stage, nutritional guidance and education are used to encourage adequate intake—not too much, not too little—of a variety of healthful foods. The mantra is "moderation, variety, and balance." Anticipatory guidance is used to

Illustration 16.1 Continuum of nutritional status.

Nutritional state	Resilient and healthy	Altered substrate availability	Nonspecific signs and symptoms	Clinical condition	Chronic condition	Terminal illness and death
Metabolic, physiological, functional status	Metabolic homeostasis Able to defend against injury	Reduction of nutrient stores or accumulation of excess Subclinical changes	Metabolic and physiologic alterations are observable	Evident illness and medical diagnosis	Altered metabolism and structural changes in tissues become permanent	Complications advance, body systems shut down
Focus of nutrition guidance, education, or therapy	Dietary guidance to support adequate intake and anticipate risks	Dietary guidance and education to inform about risks and encourage healthy eating and lifestyle choices	Nutrition education and counseling to reduce or reverse specific risk factors	Intensive medical nutrition therapy or therapeutic lifestyle change programs to delay progression	Medical nutrition therapy and patient education to enable self-management of the condition and prevent complications	Comfort care

← *Continuum of nutritional health and intervention* →

enable healthy individuals to anticipate and plan for possible risks so they are able to make informed choices that sustain resilience and prevent nutritional injury.

Altered Substrate Availability   This early, subclinical state of nutritional harm occurs when intake does not meet needs. There is a loss of reserves and/or accumulation of excesses. Nutrients are drawn out of other body compartments, such as protein out of muscle or lung tissue and calcium from bones. There may be a buildup of by-products resulting from inefficient or altered metabolism. When substrates are not available in appropriate amounts, adaptive mechanisms kick in, but they reach limits. If measured, blood markers could show subclinical changes, but without physical signs or risk indicators, such laboratory testing normally is not done. Nutrition education and *dietary guidance* directed at the public attempts to inform people about common risks and encourages healthful diets and lifestyle choices to minimize or reverse subclinical changes.

Nonspecific Signs and Symptoms   Eventually, insufficient or excessive intake of nutrients or energy leads to observable changes. Examples include the accumulation of subcutaneous fat and central adiposity, elevated blood cholesterol, and insulin resistance. These changes are well-recognized risk factors for the development of chronic diseases. By this stage, immune function is affected and there is reduced resistance to pathogens, chemical exposures, radiation, and stress, including the continued stress of nutrient imbalance. Screening should identify these changes and signal the need for intervention. Dietary guidance, nutrition counseling, and medical nutrition therapy, delivered individually or in groups, are potential interventions to assist individuals in making changes at this stage. Goals of intervention target specific risk factors and observable signs and symptoms. These can be measured and monitored over time to assess progress in halting or reversing nutritional injury and risk factors for disease.

Clinical Condition   If changes aren't made and the nutritional injury persists, frank signs and symptoms of illness are now present and a medical diagnosis is made. Examples are atherosclerosis, osteoporosis, cancer, type 2 diabetes, and depression. Genetic predisposition, interacting with dietary components and other environmental factors, influences if and when the clinical condition develops. A clear medical diagnosis is the turning point for serious lifestyle change for some adults. But change is difficult, and intensive intervention such as medical

Dietary Guidance   Providing concise recommendations to guide daily food choices and consumer information to translate guidelines into daily food choices.

Health Disparity   Significant differences in the incidence, prevalence, mortality, and burden of disease and other adverse conditions that exist among specific population groups.

nutrition therapy or therapeutic behavior-change programs (presented in detail in Chapter 17) may be necessary to manage the disease and prevent or delay its progression and the development of side effects and complications.

Chronic Condition   At this stage, altered metabolism and structural changes in tissues become permanent and irreversible. Examples are structural damage to coronary arteries, invasive and metastatic cancer, loss of kidney function, or blindness. Major adjustments of life are necessary to self-manage the chronic disease and accommodate conditions that have significant impact on quality of life. Intervention at this stage is aimed at managing the condition, preventing further complications, reducing the degree of disability, and optimizing quality of life.

Terminal Illness and Death   At the final stage in the continuum, complications advance, body systems shut down, and life ceases.

Health Disparities Among Groups of Adults

Some population groups are more likely to experience "nutritional injury" and its consequences than are others. This is illustrated by comparisons of disease prevalence in adults of different racial/ethnic backgrounds:[24,25,26]

- High blood pressure is approximately 50% higher in African Americans compared to whites, Hispanics, and Asians (31.7% compared to 22.2%, 20.6%, and 19.5%, respectively).

- Physician-diagnosed diabetes in Mexican-American females in 2006 was 14.2% compared to 6.1% in white females.

- Blacks ≥18 years of age were more likely (35.1%) to be obese than American Indians or Alaska Natives (32.4%), whites (25.4%), and Asians (8.9%).

- The proportion of obese adults who were told they were overweight by their health care provider was significantly lower for blacks (61.1%) and Mexican Americans (56.5%) than for whites (68.8%). This example illustrates disparity in health-care delivery.

Some groups have a genetic predisposition for certain diseases. American Indians have a predisposition for diabetes; Asians develop cardiovascular disease at lower BMI and smaller waist circumference; and African Americans have greater salt sensitivity and hypertension.[26] However, genetics and environment interact to determine the actual development of the disease. Groups experiencing *health disparity* not only have higher prevalence of certain conditions and experience worse health but also tend to have less access to environmental conditions that support health, such as healthy food, good housing, quality education, and safe neighborhoods.

Table 16.5 Social determinants of health

The social gradient	People lower on the social ladder have significantly higher risk of serious illness and premature death than those near the top.
Stress	Continuing anxiety, insecurity, low self-esteem, social isolation, and lack of control over work and home life have powerful effects on health.
Early life	Slow growth and poor emotional support raise the lifetime risk of poor physical heath and reduced physical, cognitive, and emotional functioning in adulthood.
Social exclusion	The unemployed, many ethnic minority groups, guest workers, disabled persons, refugees, and homeless people are at particular risk.
Work	Stress at work contributes to differences in health, sickness absence, and premature death.
Unemployment	After accounting for other factors, unemployed people and their families suffer substantially higher risk of premature death.
Social support	Belonging to a social network makes people feel cared for, loved, esteemed, and valued, which has a powerful protective effect on heath.
Addiction	Drug use is both a response to social disparity and an important factor in worsening the resulting inequalities in health. Smoking is a drain on income and a cause of ill heath and premature death.
Food	A shortage of food and lack of variety causes malnutrition and deficiency diseases. Excess intake contributes to cardiovascular diseases, cancer, diabetes, degenerative eye diseases, obesity, and dental caries.
Transportation	Cycling, walking, and the use of public transportation promote health in four ways: by providing exercise, reducing fatal accidents, increasing social contact, and reducing air pollution.

Adapted from Wilkinson, R. and Marmot, M. *Social Determinants of Health: The Solid Facts*. Copenhagen: WHO Regional Office for Europe, 2003.

The impact of racism and other forms of discrimination is also recognized as a significant contributor to health disparity.[26] These kinds of factors are referred to as *social determinants of health* and are elaborated on Table 16.5.[27,28] Strategies directed to the social determinants of health are necessary to get at the fundamental causes behind health disparity.[29] Reducing health disparities is a priority within Healthy People 2020. Goals include: eliminate health disparities and improve health for all groups; and create social and physical environments that promote good health for all.[30] Social and environmental risk factors can be changed, but to do so requires policy and environmental change strategies which are being used in many federal initiatives.[26, 30]

Dietary Recommendations for Adults

"We see a path for a healthy diet and a healthy relationship with food and perhaps a real chance for preventing escalating obesity with less of the emotional trauma."

Susan Welsh and Etta Saltos[31]

The aim of dietary recommendations for adults is to reduce risks of specific diseases and ensure that the population consumes adequate levels of required nutrients by helping the public to understand what, and how much, to eat. *Dietary guidance systems* are sets of dietary and lifestyle recommendations, based on the latest scientific information, that are developed to promote health and prevent disease. They focus on nutrients of current concern, translate nutritional risk factors into dietary recommendations, and offer guidance on types and amounts of food to eat. They respond to current nutrient intakes and eating habits of the population and take into account developments in the food industry. Key components of the U.S. dietary guidance system are the *Dietary Guidelines* for Americans and MyPyramid. Additional background is provided in Chapter 1.

Dietary Guidelines for Americans

Recommendations in the Dietary Guidelines for Americans translate and integrate national goals for health, such as Health Objectives for the Nation from the U.S. Department of Health and Human Services, Dietary Reference Intakes (DRIs) from the National Academy of Sciences, Food and

Social Determinants of Health   Socioeconomic and environmental factors that are powerful determinants of health and are largely outside of the control of individuals and groups.

Dietary Guidance System   A comprehensive set of dietary and lifestyle recommendations, based on the latest scientific information, that are developed to promote heath and prevent disease or its complications, ensure adequate intake of nutrients of concern, and offer guidance on what and how much to eat.

Dietary Guidelines   A report, including scientific information and rationale, on dietary information and guidelines for the general public or a defined subpopulation. The guidelines provide a cohesive set of recommendations that are adopted by the government or organization. They represent policy and are integrated into food, nutrition, and health programs.

Nutrition Board, and food-consumption data from the U.S. Department of Agriculture.[32] Voluntary (i.e., nonprofit) health organizations also make science-based dietary recommendations for healthy adults related to the organization's mission. For example, the American Cancer Society has Nutrition and Physical Activity Guidelines,[33] and the American Heart Association has Diet and Lifestyle Recommendations.[34] The three mentioned guidelines are generally consistent in their advice to adults. They encourage:

- consuming greater amounts of fruits and vegetables, fiber, and low-fat dairy products;
- limiting saturated fat intake, avoiding *trans* fats, and selecting lean meats and meat alternatives;
- getting regular physical activity; and
- balancing energy intake with energy expenditure to maintain or achieve a healthy weight.

The MyPyramid *food guide* helps adults use the Dietary Guidelines for Americans to make smart choices, find balance between food and physical activity, and get the most nutrition out of their calories and stay within daily calorie needs. Table 16.6 shows food groups and recommended number of servings or amounts per day by calorie levels appropriate for adults. Calculations are based on selection of the most nutrient-dense foods of that group to leave some additional "discretionary" calories that may be used for fats or sugars. Additional guidance, also shown on Table 16.6, is available to encourage consumption of a variety of fruits and vegetables over the period of a week.

Food Guides Consumer-oriented information that concisely communicates what to eat in a day, usually using graphics. Food guides classify foods into relevant groups, describe what constitutes a serving, and identify the number of servings needed per day.

Vegetarian Diets

Among individuals aged 18 and older, 2.3% of adults report eating no meat, fish, or poultry.[35] A little over 1% eat no animal products at all. There are four common types of vegetarian diets (Table 16.7).

Individuals can achieve a high-quality diet whether or not they eat meat. A well-chosen vegetarian diet is associated with decreased mortality and morbidity.[36] Potential health benefits of vegetarian diets include:

- Disease-specific benefits
    Decreased risk of mortality
    Lower incidence of heart disease symptoms
    Lower incidence of hypertension
    Lower incidence of type 2 diabetes
    Improved risk profile for kidney disease (lower glomerular filtration rate)
    Decreased risk of prostate and colorectal cancers
- General diet quality
    Higher intake of vegetables and fruits improves diet quality
    Lower saturated fat and cholesterol intake
    Diet may also be lower in calories
    Higher intake of dietary fiber, magnesium and potassium, vitamins C and E, folate, carotenoids, flavonoids, and other phytochemicals

Table 16.6 Food intake amounts for adults by calorie level, from MyPyramid

Calorie Level	1600	2000	2400	2800	3200
Daily Food Group Amounts					
Fruits	1.5 cups	2 cups	2 cups	2.5 cups	2.5 cups
Vegetables	2 cups	2.5 cups	3 cups	3.5 cups	4 cups
Grains[1]	5 oz	6 oz	8 oz	10 oz	10 oz
Meat and beans[2]	5 oz	5.5 oz	6.5 oz	7 oz	7 oz
Milk	3 cups	3 cups	3 cups	3 cups	3 cups
Oils	5 tsp	6 tsp	7 tsp	8 tsp	11 tsp
Discretionary Calorie Allowance	132	267	362	426	648
Weekly Vegetable Subgroup Amounts					
Dark green vegetables	2 c/wk	3 c/wk	3 c/wk	3 c/wk	3 c/wk
Orange vegetables	1.5 c/wk	2 c/wk	2 c/wk	2.5 c/wk	2.5 c/wk
Legumes	2.5 c/wk	3 c/wk	3 c/wk	3.5 c/wk	3.5 c/wk
Starchy vegetables	2.5 c/wk	3 c/wk	6 c/wk	7 c/wk	9 c/wk
Other vegetables	5.5 c/wk	6.5 c/wk	7 c/wk	8.5 c/wk	10 c/wk

[1] 1 slice of bread, 1 cup of ready-to-eat cereal, or 1/2 cup of cooked rice, pasta, or cooked cereal is considered equivalent to 1 ounce. At least half of all grains consumed should be whole grains.

[2] 1 egg, 1 Tbsp peanut butter, 1/4 cup cooked dry beans, or 1/2 ounce of nuts or seeds is considered equivalent to 1 ounce. Meat, poultry, and fish should be lean.

Table 16.7 Classification of vegetarians by food groups consumed

Type of Vegetarian & Foods Not Eaten	Foods Eaten
Vegan No animal foods of any kind, no honey	Grains, nuts, seeds, nut butters, fruits, vegetables, legumes, sugar, molasses, oils, margarine, soda, alcohol, soy milk and soy analogs (e.g., textured vegetable protein "meats")
Lacto-vegetarian No meat, poultry (no eggs), fish	Above, plus milk and other dairy products, cheese, yogurt, butter
Lacto-ovo vegetarian No meat, poultry, fish	Above, plus eggs
Vegetarian No "red meat" (beef, pork, lamb, venison, buffalo, or other red meat)	Depends on individual interpretation: may include fish, both fin fish and shellfish, and poultry and game birds

- Environmental benefits

  Vegetable foods are lower on the food chain than meat, fish, and poultry and use fewer resources for production

Several nutrients are of concern because of possible shortfall in vegetarian diets. These include vitamins B_{12} and D, calcium, zinc, and omega-3 fatty acids.

Following a systematic review of the evidence, the American Dietetic Association offered the following guidelines for helping vegetarians plan healthful diets.[36]

- Choose a variety of foods, including whole grains, vegetables, fruits, legumes, nuts, seeds, and, if desired, dairy products and eggs.
- Minimize foods that are highly sweetened, high in sodium or high in fat—especially saturated fat and *trans* fat.
- Choose a variety of fruits and vegetables.
- If animal foods such as dairy products and eggs are used, choose lower-fat dairy products and use dairy products and eggs in moderation.
- Use a regular source of vitamin B_{12} and, if sunlight is limited, of vitamin D.

Beverage Intake Recommendations

The Dietary Guidelines and MyPyrimid do not give specific advice about beverage consumption. Yet 21% of calories come from beverages, and beverage choices have shifted considerably in the past decades.[37] Calories consumed in liquid form have less satiety value and are not compensated for through adjustments in food intake. Except for milk and fruit juices, beverages contribute little to essential nutrient needs. Beverage intakes of U.S. adults in ounces of fluid and calories consumed are shown in Illustration 16.2. A group of nutrition experts has proposed a beverage guidance system directed at obtaining daily fluid needs from water and making other beverage choices to minimize caloric intake. The system groups beverages into six categories and considers each category's energy and nutrient density, effect on energy intake and body weight, effect on daily intake of essential nutrients, evidence of beneficial health effects, and evidence of adverse health effects.[37]

The beverage guidance system recommends a range of 10% to 14% of energy intake from beverages. Suggested and acceptable consumption amounts are given in levels. Categories and consumption ranges follow.

Level I	water	20–50 fl oz
Level II	tea or coffee, unsweetened	0–40 fl oz
Level III	low-fat milk and soy beverage	0–16 fl oz
Level IV	noncalorically sweetened beverages	0–32 fl oz
Level V	fruit juices	4–8 fl oz
	alcoholic beverages	0–8 fl oz
Level VI	calorically sweetened beverages	0–8 fl oz

Total caffeine is limited at 400 mg/day. The system has been criticized because the effort was funded by industry and because alcohol and juice are placed in the same level.

Alcoholic Beverages

Alcohol is a popular beverage with significant social and cultural significance, but it is also a psychoactive drug with potential for abuse. Moderate alcohol intake is a recognized contributor to heart health, but alcohol also increases the risk of oral, esophageal, liver, and colorectal cancers, and breast cancer in women.

Approximately 61% of adults in the United States drink alcohol, with the highest rates among persons aged 25 to 44 years old (76% of males, 63% of females).[38] Rates of alcohol consumption decline with age. Alcohol-consumption guidelines vary by country. France and Italy have the highest levels (3-5 drinks per day for men and 2-3 drinks per day for women), while Canada (7 per week for men and women) and the United Kingdom

Illustration 16.2 Beverages consumed, by quantity and calories.

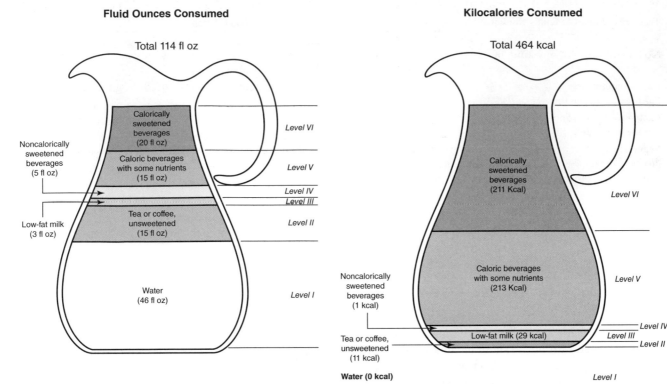

SOURCE: A new proposed guidance system for beverage consumption in the United States. *Am J Clin Nutr*, 2006:83:529–42. © 2006 American Society for Nutrition.

(1-2 per day for men and for women) have the most stringent guidelines.[39] United States guidelines are "If you drink, do so in moderation," defined as no more than 2 drinks per day for males, and no more than 1 drink per day for females.[32]

What is a drink? See Table 16.8 for drink serving size, alcohol content, and calories of selected beverages. A drink contains roughly 13–15 grams of alcohol or 0.5 oz of ethanol. (These are translated into ounces of beverage on blood-alcohol charts describing legal limits for drivers).

Water Intake Recommendations

The Food and Nutrition Board defines the adult adequate intake (AI) level for water based on median total water intake (from fluids and food) from NHANES III data for young adults age 19 to 30 years.[40] The panel did not set an upper level for water; however, water toxicity can occur.

Total Water AI for Adults

Men	3.7 liters	(125 oz)
Women	2.7 liters	(91 oz)

According to the NHANES survey, water and other beverages supplied 81% of total water needs of adults, and the moisture in food provided the remaining 19%.[41] So, on average, the quantity of fluids that men and women need to drink each day is approximately 12 cups and 9 cups, respectively, of water and other beverages (including coffee, milk, juice, soda, etc.).

Effects of Caffeine Intake on Water Need

What about fluids containing caffeine? Occasionally someone suggests that coffee, tea, or other caffeine-containing beverages not be counted as fluid intake because they have a diuretic effect on the body. Caffeine is a stimulant that relaxes the esophageal sphincter (leads to acid reflux),

Table 16.8 Alcoholic beverages: alcohol and calorie content

Amount	Beverage	Alcohol grams	Calories
12 oz	Regular beer	13	153
12 oz	Light beer	12	100
1.5 oz	Distilled spirits: gin, whiskey, rum, vodka	15	90
5 oz	Red wine	15	125
5 oz	Dry white wine	14	107
2 oz	Sherry	9	80

has a laxative effect, and temporarily increases urine production at high doses. But no evidence connects drinking coffee, tea, or other caffeine-containing fluids with dehydration.[42] The Food and Nutrition Board concluded that caffeine-containing beverages contribute to daily total water intake.[40]

Dietary Supplements and Functional Foods

There are circumstances in which dietary supplements are indicated to assure nutritional adequacy, such as during pregnancy, with certain illness, or when low calorie or nutrient restricted diets are consumed. Survey data indicate that about a third of adult males and about half of adult females take a vitamin or mineral supplement daily, nearly every day, or "every so often."[43] Supplemental vitamin and mineral use gradually increases with age. At perimenopause, many women begin to take calcium or calcium with vitamin D supplements to prevent bone loss.

Herbal and other botanical products are grouped with dietary supplements. Some of the more popular herbals and their potential uses and side effects are listed in Table 16.9.

Functional Foods   Numerous products are available that straddle the line between conventional food, supplement, and therapeutic agent. These cover many categories, as defined in Table 16.10.[44] *Functional foods* is the term used for food products that have a physiological benefit or reduce the risk of chronic disease beyond basic nutritional functions. The growing body of evidence linking diet and biologically active components of food to chronic-disease risk reduction is fueling expansion of functional foods. The aging population and consumer interest in controlling health are fueling demand. In the highly competitive food industry, functional foods are an area of tremendous growth.[45] The Food and Drug Administration regulates supplements as food and medical foods as drugs. Regulation of products in other categories depends on how the manufacturer intends to market them and if health claims are made on the package labels. Consumer interest in these products is high, and the effect of functional foods on nutritional adequacy and health is an area for investigation.

Nutrient Recommendations

"Poison is in everything, and no thing is without poison. The dosage makes it either a poison or a remedy."

Paracelsus, Swiss alchemist, 1493–1541

The Institute of Medicine's recommendations for macronutrient intake are expressed in ranges of total calorie intake and account for the fact that various eating patterns can be healthful.[46]

Acceptable Macronutrient Distribution Ranges for total calorie intake from fat, carbohydrate, and protein for adults are:

- Fat   20–35% of calories
- Carbohydrate   45–65% of calories
- Protein   10–35% of calories

Table 16.11 lists recommended nutrient intakes for adults using the DRIs or the Daily Values (DV) used in food labeling[57] and shows average intakes based on the 2005-2006 NHANES survey.[48] Note that protein, total fat, saturated fat and sodium exceed recommended levels, and that men consume too much cholesterol and sugar. Intakes are below the recommendation for fiber and several vitamins and minerals, and women have low intakes of iron and calcium. Vitamin K intake appears to be age-related, with middle-age but not young adults reaching the RDA.

Risk Nutrients

As shown in Table 16.11, intakes of certain nutrients by adults exceed or fall short of recommendations.

Fiber   Average daily dietary fiber recommendations are 38 grams for men and 25 grams for women, or 14 grams of dietary fiber per 1000 calories.[46] Daily median intakes for adult males and females barely reach half of the recommended level (Table 16.11). Fiber, especially soluble or viscous fiber, reduces absorption of cholesterol from the gut and has positive effects on blood lipids, glucose metabolism, and weight management. If dietary recommendations to increase fruit, vegetable, and whole-grain intakes were followed, the fiber recommendation could be met.

Functional Food   A food product that has a physiological benefit or reduces the risk of chronic disease beyond basic nutritional functions.

Vitamin A   Vitamin A is important for epithelial cell growth and development and for activity of immune cells. It keeps skin and mucous membrane cells healthy and resistant to infection. Top sources of vitamin A include fortified milk and cheddar cheese. Sources of beta-carotene, a precursor of vitamin A, include sweet potatoes, carrots, pumpkin, cantaloupe, broccoli, apricots, spinach, and collard greens.

Vitamin D   Exposure of the skin to sunlight is an important but inconsistent source of vitamin D. Needs may be met with 5–30 minutes of sun exposure between 10 a.m. and 3 p.m. at least twice a week to the face, arms, legs, or back without sunscreen. However, geographic latitude, season, time of day, cloud cover, and air pollution all make a difference. If above 43° latitude (a line from northern California to Boston), UV rays are not sufficient

Table 16.9 Supplements and herbal remedies, proposed claims, side effects, and potential medication interactions

Supplement, Herbal or Other Remedy	Proposed Claims	Potential Side Effects	Potential Interaction with Prescription Medications
Cholestin	Maintains desirable blood cholesterol levels	Safety of some ingredients unknown	Unknown
Creatine	Sport supplement (increased performance in short, high-intensity events); inconclusive claims regarding role in congestive heart failure	Kidney disease; side effects possible and include vomiting and diarrhea	Unknown
DHEA	Improves physical well-being throughout aging	Increases risk of breast and endometrial cancer in women and prostate cancer in men	Unknown
Echinacea	Prevents and treats colds and sore throat; anti-inflammatory	Allergies to plant components in daisy family; affects oral cavity integrity; depresses immune system if taken longer than 6 weeks	Corticosteroids
Ephedra (Ma Huang)	Promotes short-term weight loss	Insomnia, headaches, nervousness, heart palpitations, seizures, death	Corticosteroids, digoxin, monoamine oxidase inhibitors (MAOI); oral hypoglycemics
Garlic	Lowers blood cholesterol; relieves colds and other infections, asthma	Heartburn, gas, blood thinner	Oral hypoglycemics; blood thinners[a]
Ginger	Calms stomach upset, fights nausea	Central nervous system depression, heart rhythm disturbances if using very large doses	Oral hypoglycaemic medications; blood thinners
Ginkgo biloba	Increases mental skills, delays progression of Alzheimer's disease, increases blood flow, decreases depression	Nervousness, headache, stomachache, interacts with blood thinners	Thiazides; blood thinners (e.g., warfarin); MAOI
Ginseng	Increases energy, normalizes blood glucose, stimulates immune function, relieves impotence in males	Insomnia, hypertension, low blood glucose, menstrual dysfunction	Estrogens, insulin, oral hypoglycemics, blood thinners;[a] MAOI
Glucosamine and chondroitin	Slows progression of joint-space narrowing with long-term use; may relieve joint pain and improve mobility	Gastrointestinal upset (i.e., gas, soft stools, diarrhea, nausea, indigestion, heartburn) and allergic reactions in individuals with shellfish allergy	Blood thinners (chondroitin only)
Peppermint	Treats indigestion and flatulence, spasmolytic (relaxes muscles), antibacterial agent	Heartburn, allergic reactions	Unknown
St. John's wort	Relieves depression	Dry mouth, dizziness, interacts with many drugs	Cyclosporin, digoxin, iron supplements, oral contraceptives, selective serotonin re-uptake inhibitors; MAOI
SAMe	Relieves mild depression, pain relief for arthritis	May trigger manic excitement, nausea	Unknown
Saw palmetto	Improves urine flow, reduces urgency of urination in men with prostate enlargement	Nausea, abdominal pain	Estrogens

SOURCE: Table compiled from Shekelle, P. G., Hardy, M. L., Morton, S. C., et al. Efficacy and safety of ephedra and ephedrine for weight loss and athletic performance: A meta-analysis. *J Am Med Assoc* 2003; 289(12):1537–45; AACE Guidelines: Medical guidelines for the clinical use of dietary supplements and nutraceuticals. *Endocrine Prac* 2003; 9(5):418–470, Table 7; Brown, J. E. *Nutrition Now*, 3rd ed. Belmont, CA: West/Wadsworth Publishing Company, 1999: Table 24.4; Stupay, S. and Sivertsen, L. Herbal and nutritional supplement use in the elderly. *Nurs Pract* 2000; 25(9):56–67.
[a]Blood thinners include aspirin, warfarin, coumarin.

Table 16.10 Definition and examples of supplements and functional food categories

Category	Definition	Examples
Supplement	Vitamin, mineral, amino acid, herbal or other botanical, or combination ingested in pill, capsule, or liquid form	Multiple vitamin Calcium Glucosamine and chondroitin Amino acids
Functional Food Category		
Conventional foods (whole foods)	Unmodified food rich in bioactive components that have health benefits	Nuts reduce risk of cardiac death Cruciferous vegetables reduce risk of cancer Flaxseed oil for omega-3 fatty acids
Modified foods • Fortified • Enriched • Enhanced	• Bioactive component added • Replacement or addition of component naturally in the food • Food product formulated with bioactive components	• Calcium-fortified orange juice, iodized salt • Enriched flour with added vitamins and iron • Margarine with plant stanols or sterol esters for cholesterol lowering • Energy bars, yogurt, or bottled water formulated with amino acids, lutein, fish oil, probiotics, ginkgo biloba
Medical foods (available only by prescription)	A food formulated to be used under the supervision of a physician, intended for the specific dietary management of a disease or condition	Phenylketonuria (PKU) formula free of phenylalanine
Foods for special dietary use (available at the retail level)	Food designed to meet a special dietary need due to physical, physiological, pathological, or other condition; supplement the diet; or replace a meal or daily food intake	Gluten-free foods Sports drinks with electrolytes Nutrient-dense beverage for protein and calorie supplementation Meal replacements for weight-reducing diets

SOURCE: Adapted from Hasler, C. M. et al. Position of the American Dietetic Association: functional foods. *J Am Diet Assoc* 2009; 109:735–746.

from November to February to synthesize vitamin D in the skin. Food sources of vitamin D include fish and fish oils. Vitamin D_3 (cholecalciferol) has been added to milk since the 1930s to prevent rickets. Vitamin D is also one of many vitamins and minerals commonly added to breakfast cereals.

The AI for vitamin D is 5 micrograms/day up to age 50, and then it doubles to 10. The current tolerable upper intake level (UL) is 2000 IU. In recent years, vitamin D has been identified as protective in a number of conditions, including cardiovascular disease, hypertension, immune function, and cancer.[49] Vitamin D deficiency is a highly prevalent condition, present in approximately 30% to 50% of the general population.[50]

Vitamin E   Vitamin E is found naturally in some foods, added to others, and available as a dietary supplement. "Vitamin E" is the collective name for a group of fat-soluble compounds with distinctive antioxidant activities. The adult RDA for vitamin E is 15 mg TE (alpha-tocopherol equivalents), which converts to 22.4 IU.

Research suggests that vitamin E may be protective for prostate and esophageal cancer, but is mixed with regard to heart disease and eye-disorder risk.[51] A 2005 meta-analysis found that high-dosage (400 IU or more)

supplements were associated with increased all-cause mortality.[52] Food sources of vitamin E include vegetables oils (e.g., corn, safflower, and olive oils), sunflower seeds, walnuts, wheat germ, mayonnaise and salad dressings made with oil, avocado, and leafy greens.

Folic Acid   The evidence for folic acid in the prevention of neural-tube defects is so strong that the Food and Drug Administration mandated folic-acid fortification of grains and grain products beginning in 1998. Dietary surveys show that fortification is working; mean intake has increased to 508 micrograms for 20- to 39-year-old women and 641 mcg for 20- to 39-year-old men, compared to the 400 mcg DRI.[48]

Choline   Choline was first defined as an essential nutrient and assigned an Adequate Intake (AI) level in 1998 for its role in preventing fatty liver. It has many functions, including lipid transport, and along with folate and vitamin B_{12} is involved in converting homocysteine to methionine. Liver, eggs, beef, brussels sprouts and broccoli, shrimp and salmon, milk, and peanuts are food sources of choline.

Calcium   Calcium plus vitamin D and magnesium help develop and maintain bone density, which delays

Table 16.11 Selected nutrient intakes of adults, NHANES 2005–2006, compared to recommendations

	INTAKE				
	Actual				Recommended[a]
	20–29 years		40–49 years		31–50 years
Nutrient	Males	Females	Males	Females	Males/Females
Energy, kcal	2821	1959	2753	1873	—
Protein, g	106	72	107	76	56/46
Total fat, g	101	74	105	72	65 (DV)[b]
Saturated fat, g	34	26	35	24	20 (DV)[b]
Alpha-linolenic acid	1.7	1.3	1.8	1.4	1.1/1.1
Cholesterol, mg	340	238	388	255	≤ 300 (DV)[b]
Total carbohydrate, g	344	246	313	221	300 (DV)[b]
Fiber, g	17	13	18	14	38/25
Total sugar, g	160	118	141	103	125 (DV)[b]
Vitamin A, mcg RE	560	520	728	623	900/700
Vitamin C, mg	104	81	91	70	90/75
Vitamin E, mg TE	8.5	6.1	8.8	7.0	15
Folate, mcg	641	508	614	448	400
Vitamin K	96	72	107	106	120/90
Choline	401	257	448	290	550/425
Calcium, mg	1103	933	1099	923	1000
Iron, mg	19.3	14.4	19.0	13.9	8/18
Magnesium, mg	327	251	373	285	420/320[d]
Potassium, mg	2951	2205	3311	2443	3500
Sodium, mg	4476	3107	4350	3059	Up to 2300
Zinc, mg	15.4	8.3	17.5	11.1	11/8
Caffeine, mg	133	82	264	220	Up to 250[c]
Alcohol, g	19.7	6.4	23.0	9.3	20/8[d]

SOURCE: *What We Eat in America*, NHANES, 2005–2006. U.S. Department of Agriculture, Agriculture Research Service, 2008.
[a]Recommended intake according to DRI 2002 or to DV nutrient reference amount used on food labels, relevant to a 2000 kcal diet.[47]
[b]Based on a 2000 kcal diet; for other caloric levels, recommendation is no more than 30% of calories from fat and 10% from saturated fat, and 25% of calories from sugar.
[c]250 mg is an average moderate intake; 800 mg is considered excessive.
[d]Based on Dietary Guidelines
Available from www.ars.usda.gov/ba/bhmrc/fsrg, accessed 9/10/2009.

osteoporosis and reduces risk of bone fractures. Adequate calcium is a potential contributor to lowered risk of colon cancer. Milk, as part of the DASH diet (see Chapter 19), has been successfully used as part of blood-pressure reduction intervention.[53] Calcium's crucial role in metabolism is supported by physiological mechanisms to keep calcium in tight balance. Absorption of calcium from the gut increases or decreases in relation to the amount ingested and the body's need. Like many nutrients, numerous interacting mechanisms affect the absorption and utilization of calcium.

Adults could meet calcium needs by adding an 8 oz glass of milk, 8 oz container of yogurt, or 8 oz glass of calcium-fortified orange juice daily. Other good sources are calcium-fortified soy beverage, tofu, and salmon. Supplemental calcium is often recommended for adults, especially women, with low intake of dairy foods and risk of osteoporosis.

Magnesium Poor diet, diabetes, or prolonged illness can increase the risk of magnesium deficiency, with adverse effects on bone strength, heart health, and metabolic syndrome. Magnesium deficiency tends to be more problematic in older adults. Good magnesium sources are sunflower seeds, almonds, beans, milk, whole grains, spinach, and bananas.

Potassium Potassium has many functions in the body, including electrolyte balance and muscle contraction. Potassium-rich diets tend to reduce blood pressure. Good plant-based sources of potassium include dried apricots and raisins, wheat bran and wheat germ, potatoes, bananas, and orange juice.

Case Study 16.1

Run, Kristen, Run

Kristen, who was active in competitive sports throughout high school, has decided to run a marathon with some of her college friends. She is 25 years old, 5 ft 8 in tall, and weighs 135 pounds. She eats all sorts of foods, likes fruits and vegetables, but tries to avoid greasy foods. She says coffee is her downfall—she drinks 4 to 6 cups a day. She doesn't like sweets, although she keeps ice cream in her freezer. A family history notes that her mother needed angioplasty to treat occluded arteries shortly after menopause and that her father is not at risk for any chronic conditions. Although she would eventually like to have children, Kristen is not pregnant now.

An analysis of a 24-hour dietary recall shows the following:

2090 calories	34 mg iron
377 g carbohydrate (41 g total fiber)	1170 mcg RE vitamin A
98 g protein	158 mg vitamin C
33 g fat (7 g saturated fat, 1 g trans fat,	213 IU vitamin D
1.5 g omega-3 fatty acid, 99 mg cholesterol)	35 IU vitamin E
3343 mg sodium	1548 mcg folic acid
958 g calcium	6 mg pantothenic acid

Questions

1. How many calories does Kristen need to maintain her weight?
2. Is she eating enough to support daily workouts?
3. Describe three health-promoting aspects of Kristen's diet.
4. Make three suggestions that could improve Kristen's diet.

(Answers are located in the Instructor's Manual for the 4th edition of *Nutrition Through the Life Cycle*.)

Iron Iron deficiency can be one of the nutritional causes of anemia, although there are non-nutritional causes as well. The benefits of sufficient iron are healthy blood cells that carry oxygen for metabolism and overall energy; a drawback of excessive iron is greater need for antioxidants. High-iron foods include red meats, fortified cereals, and dried beans (as in baked beans, bean soup, bean dip, and chili). Enriched breads and most fruits and vegetables also contribute iron. For women, iron needs drop from 18 to 8 mg per day at menopause. Changing from a high-iron to a low-iron cereal during the menopausal transition can be an easy way to adjust for the downward shift in iron needs.

Physical Activity Recommendations

Healthy eating and increased physical activity are the featured duo for combating obesity at the individual and population level and are primary and secondary prevention strategies for several chronic diseases, including cardiovascular disease, diabetes, hypertension, osteoporosis, and colon and breast cancer.[54]

Any physical activity is better than none; even bouts of 10 minutes can have positive health benefits. The degree of benefit increases as the duration and intensity increase.[55] Government agencies and health organizations around the world recommend that adults get a minimum of 30 minutes of moderate-intensity physical activity (that increases breathing or heart rate) every day.[56] This is the amount of exercise that has been well accepted and documented in numerous studies to produce benefits. Table 16.12 lists current U.S. recommendations.

Physical Activity, Body Composition, and Metabolic Function

Favorable body composition changes (reduced fat mass and increased lean mass) occur with the adoption of regular physical activity.[55] Even in the absence of caloric restriction, aerobic physical activity equivalent to walking at 4 miles per hour for 150 minutes a week or

Table 16.12 Physical Activity Guidelines for Americans (2008): adults (aged 18–64)

- **Basic recommendation:** 2 hours and 30 minutes a week (150 minutes) of moderate-intensity, or 1 hour and 15 minutes a week (75 minutes) of vigorous-intensity aerobic physical activity, or an equivalent combination. Aerobic activity should be performed in episodes of at least 10 minutes, preferably spread throughout the week.
- **For additional health benefits:** increase to 5 hours (300 minutes) a week of moderate-intensity aerobic physical activity or 2 hours and 30 minutes (150 minutes) of vigorous-intensity physical activity.
- Adults should also do muscle-strengthening activities that involve all major muscle groups, performed on 2 or more days per week.

Moderate or vigorous intensity?

Take the talk test:

If you can talk, but not sing, you are doing moderate-intensity activity. If you can't say more than a few words without pausing for a breath, you are doing vigorous-intensity activity.

Moderate-intensity activity examples:

Baseball, brisk walking, cycling, golf (carrying bag), raking leaves, vacuuming, water aerobics

Vigorous-intensity activity examples:

Aerobic exercise classes, basketball, fast dancing, hiking with backpack, jumping rope, running, swimming laps

Ergogenic Aids Nutritional products that are purported to enhance performance. Examples range from caffeine and protein powders to sports drinks and energy gels and bars.

jogging at 6 miles per hour for 75 minutes a week results in decreases in total and abdominal adiposity that are consistent with improved metabolic function. When more physical activity is done, intra-abdominal fat losses are greater. Moreover, research indicates that abdominal fat loss is greater during exercise-induced weight loss and among those with the greatest level of adiposity.[55]

Physical Activity Types and Settings

Healthy People 2010 Objectives address three types of physical activity: aerobic activity to condition the cardiovascular and respiratory systems and increase endurance, weight-bearing and resistance exercises to strengthen muscles and bone, and stretching and flexibility activities.[3] To encourage physical activity, Healthy People objectives also include objectives for access to physical activity and fitness opportunities at worksites and community walking and

biking trails in communities. Changing the environment to foster a variety of opportunities for physical activity and accommodate a wide range of preferences and abilities has expanded into the "active living" movement and national recommendations to modify the environment to enable greater physical activity in all aspects of life.[57]

Physical Activity and Lifestyle

Regular physical activity enables adults to meet the physical demands of work and leisure comfortably; and it supports physical and mental health, helps manage weight, and enhances quality of life. Case Study 16.1 deals with physical activity and nutrition needs. The percent of adults meeting physical-activity recommendations is increasing; however, many segments of the adult population get little or no physical activity (see Table 16.13). Socioeconomic constraints, cultural preferences, and baseline levels of sedentariness or obesity are barriers to physical activity. Low-intensity social-environmental interventions, such as buddy systems, walking groups, and accessible walking trails and building exercise facilities, have been feasible, sustainable, and effective in many racial/ethnic minority groups.[57]

Diet and Physical Activity

A general healthful diet supports physical activity. However, adults engaged in competitive sports may have increased nutrient needs to meet demands of training, competition, and recovery. Nutritional *ergogenic aids*—nutritional products that are purported to enhance performance—range from caffeine and protein powders to sport drinks and energy gels and bars. Few improve performance, and some may be harmful. In spite of the popularity of energy and sports drinks, water is sufficient for hydration for routine physical activity.[58]

Nutrition Intervention for Risk Reduction

Nutrition intervention for adults takes place at many levels and in many settings. Education and counseling can be directed to the individual to increase knowledge and encourage behavior change. Discussions about BMI and eating habits are becoming a more common part of primary health care. Environmental changes, such as point-of-purchase signage about calorie or fat content of foods, or healthier food options in vending and convenience outlets, are implemented to enable healthier food choices.[59] Other interventions affect availability and access to healthful foods, such as menu guidelines for worksite cafeterias, community gardens, and limits on the number of fast-food outlets in low-income neighborhoods.[60] Health-promotion programs that include healthy eating and physical activity are sponsored by

Table 16.13 Amount of physical activity reported by adults in the Behavioral Risk Factor Surveillance Survey

	Subgroup	2001 Met recommended PA in usual week* %	2001 No leisure-time PA in past month %	2007 Met recommended PA in usual week* %	2007 No leisure-time PA in past month %
Gender	Female	43.0	28.7	47.0	26.2
	Male	47.9	23.7	50.7	21.7
Race	White	48.2	22.5	51.7	20.4
	Black	35.3	34.7	40.4	30.9
	Hispanic	38.9	39.8	42.1	35.7
	Other	42.2	28.1	45.3	23.6
Age groups	18–24	55.7	20.7	59.0	18.4
	25–34	49.6	23.2	53.2	20.8
	35–44	47.0	24.3	49.6	22.3
	45–64	41.9	27.7	46.6	24.8
	65+	36.8	34.2	39.3	32.7

*Recommended physical activity is defined as reported moderate-intensity activities in a usual week (i.e., brisk walking, bicycling, vacuuming, gardening, or anything else that causes small increases in breathing or heart rate) for at least 30 minutes per day, at least 5 days per week; or vigorous-intensity activities in a usual week (i.e., running, aerobics, heavy yard work, or anything else that causes large increases in breathing or heart rate) for at least 20 minutes per day, at least 3 days per week, or both.

churches, community organizations, and employers.[56] Community campaigns promote healthy eating messages to the general public or specific messages tailored to identified needs and risk factors within subgroups. Enrichment of commonly eaten foods with nutrients lacking in the diet is a policy intervention that changes nutrient intake without requiring nutrition education or behavior change. Reducing risk and improving the nutritional and health status of adults requires multiple strategies that combine individual, social, organizational, and policy-level changes such as those mentioned.

The Eating Competence Model

The *Eating Competence Model* offers a paradigm for nutrition education and dietary guidance that is different from but complementary to dietary guidelines that are focused on risk reduction. This model recognizes the special value of the social and sensory aspects of food and eating. The goal is to encourage "competent eaters who are positive, comfortable, and flexible with eating and are matter-of-fact and reliable about getting enough to eat of enjoyable nourishing food."[61]

The model has four components:

- **Eating attitudes:** Includes positive interest in foods and eating; self-trust about managing food and eating; and finding harmony among food desires, food choices, and amounts eaten.

- **Food acceptance:** Recognizes that enjoyment and pleasure are primary motivators for food selection, and that nutritional excellence is supported by enjoyment of a variety of food, including nutritious

food. Food acceptance means being comfortable eating a preferred food, even if it isn't the most nutritious, but it also means being able to settle for less-preferred food when necessary to satisfy caloric or other nutritional needs.

- **Regulation of food intake:** Emphasizes internally regulated eating and attention to sensations of hunger and fullness. See Table 16.14 for hunger-related cues.[62] One who has learned self-regulation has the ability to tolerate hunger when there is confidence that adequate, rewarding food will be available; the ability to stop when satisfied; and the ability to accept the body weight that evolves from internally regulated eating.

- **Eating context:** Puts priority on structure and meal planning. Meals are a predictable time to eat adequate amounts of preferred food. Eating is intentional and deliberate, and it requires discipline. The model teaches that going to some trouble to procure rewarding food, scheduling eating times, and setting aside time to eat are important. Intentional, deliberate eaters are able to postpone snacking and grazing when they are confident that there will be satisfying food at the next meal.

> **Eating Competence Model** A new paradigm for nutrion education and nutrition guidance that considers four components: eating attitudes, food acceptance, regulation of food intake, and eating context. A comptent eater is postive, comfortable, and flexible with eating and is matter-of-fact and reliable about getting enough to eat of enjoyable, nourishing food.

Within the eating competence model, nutrition education and dietary guidance help people be more attuned to

Table 16.14 Cues for regulating food intake

Appetitive cues	Physical sensation	Emotional feeling or cognition
Famished	Extreme hunger, pronounced discomfort: shakiness, crankiness, headache	Urgency and desperation to eat Often results from food insecurity (no assurance of being able to get enough to eat) May result from extreme self-restraint
Hunger, increased appetite	Physical experience of emptiness may include mild discomfort	Tolerable anticipation of eating Awareness that adequate amounts of rewarding food will soon be available
Hunger goes away	Physical feeling of emptiness subsides along with discomfort from energy deficit	Sense of relief increases; however, most people are reluctant to stop eating at this point because eating is still rewarding
Appetite goes away	Satiety: positive experience of readiness to stop eating This is a more sustaining and rewarding endpoint to eating for most people than when hunger goes away	Food stops tasting good; a subjective experience of losing interest in eating
Feeling full	For most, this is a pleasant, if occasional, endpoint to eating; it is a positive state of feeling filled up	Eating past satiety is rewarding if it follows a deliberate decision to eat more than usual, perhaps on a ceremonial occasion, because food tastes exceptionally good or because energy needs have suddenly increased
Feeling Stuffed	Negative physical state including extreme fullness, lethargy, physical discomfort, perhaps nausea; virtually universally experienced as being a negative endpoint to eating	Accompanied by a sense of chagrin at over-eating and self-indulgence; often arrived at as an unthinking or impulsive suspension of self-restraint

SOURCE: Based on Satter Eating Competency Model.

their needs and feelings; encourage them to try, and learn to like, a wide variety of nutritious foods; and remind them of the importance of taking time and having the self-discipline to plan and prepare satisfying meals.

A Model Health-Promotion Program

"Sisters Together: Move More, Eat Better" is a health awareness program that encourages African-American women 18 years and older to maintain a healthy weight by becoming more physically active and by eating nutritious foods. It is a national initiative of the Weight-control Information Network (WIN). The program was originally developed and piloted in Boston with funding from the National Institute of Diabetes and Digestive and Kidney Diseases (NIDDK) in response to the high rates of obesity among African American women. Sisters Together is a culturally relevant, community-based program that is flexible and can be tailored to African American women of all ages, communities and demographics. The program design combines social marketing with community-building strategies.

Dedicated individuals and organizations can start a program at the local level using the Sisters Together Program Guide and materials. The manual outlines the steps to creating a successful program[63]:

- identify community resources and partners,
- set realistic goals and objectives,
- work with media to raise awareness,
- decide on core activities and events, and
- measure success.

Sisters Together initiatives bring together respected leaders and organizations in the community that have goals similar to Sisters Together, address women's issues and concerns, and have credibility with African American women in community. Programs are based in settings that ranging from churches to beauty salons. They partner with community centers and YWCAs for space for group meetings or exercises classes, and link up with public programs such as WIC and SNAP.

The media, including radio, newspapers and newsletters and social media, are used to increase the visibility of the Sisters Together program, cover special events and raise awareness about relevant issues.

Core activities are determined at the local level and are designed to fill knowledge gaps, reduce barriers and increase opportunity, provide social support, and to complement other community resources. Activities are selected to match expressed needs of women in the community such as desires to have more energy, relieve stress, look good, and feel better about themselves. Program content emphasizes how physical activity and nutrition help achieve these needs. Successful activities include classes where food is prepared and tasted, featuring easy-to-prepare recipes that are low in calorie, fat and salt; and organized groups and buddy systems for physical activity such as walking clubs and classes.

Approaches encourage physical activity versus exercise and feature traditional ethnic foods and flavors but with more health–conscious ingredients and preparation methods.

Positive changes in fruit and vegetable intake and physical activity have been documented through pre- and post-evaluations. Ongoing and periodic "booster" programs are strategies to maintain changes over time.

The model of raising awareness and improving health through the establishment of community-based groups and campaigns is applicable in a wide range of adult nutrition and health issues.

Public Food and Nutrition Programs

One of the aims of national health goals is to minimize health disparities; another is to improve the food security of the population (see Table 16.15). Living in poverty is linked to poor diets and to adverse health outcomes, such as increased rates of obesity.[3] According to Census Bureau estimates for 2008, the poverty rate for the U.S. is 13.2%; however, there is great disparity. The rate is 9.9% for

whites, compared to 24.7% for blacks, 23.2 for Hispanics, and 10.2% for Asians. For adults aged 18-64 years, the rate is 11.7%.[64] The poverty threshold is derived from calculating the cost of foods needed for basic dietary requirements (according to the Thrifty Food Plan) and multiplying that cost by 3. This poverty index was developed in the early 1960s by the Social Security Administration; at that time, food costs made up about one-third of the average household budget. Currently, households spend 9–34% of after-tax income on food, depending on household income. Poor households spend disproportionately higher levels of their income on food.

For 2009, a person was considered to live in poverty if his or her annual income was $10,830 or less.[65] Poverty guidelines are calculated according to household size and adjusted for higher living costs in Alaska and Hawaii. For a four-person household, the poverty guideline is $22,050 in the contiguous states and Washington, D.C. The federal poverty guideline is published annually in the Federal Register. It is used to determine eligibility for food and nutrition assistance programs that are administered by the U.S.D.A.

By far the largest of all nutrition assistance programs is Supplemental Nutrition Assistance Program (SNAP), the new name for the federal Food Stamp Program. The updated program reflects an expanded focus on nutrition, rather than just food, and it serves 1 out of every 11 Americans each month. Dollars allocated to SNAP fund supplemental food as well as nutrition education.

Many other programs work to help hungry individuals and families gain food security. For example:

- Government extension programs teach budgeting, shopping, meal planning, and food-safety skills.
- The Second Harvest food bank network keeps food items out of the waste stream, coordinates charitable giving programs, and supplies food shelves and community kitchens.
- Soup kitchens and shelters provide hot meals and snacks for hungry and homeless people.
- Meals-on-Wheels programs serve homebound adults. Some are funded through the Agency on Aging, but many operate as voluntary community efforts.

Together, governmental and private organizations help individuals and families gain consistent access to safe, wholesome foods that are culturally acceptable. Such access is the basis of food security.

Putting It All Together

Adults need access to a variety of healthful foods, knowledge to guide food choices, and positive attitudes about food and eating, balanced with discipline. Food choices and nutrient adequacy matter throughout the adult years.

Table 16.15 Progress toward Healthy People 2010 food-security goals for adults, 2008, compared to baseline

The following applies to all adults, age 20+:	Target	Baseline	2008
Increase food security, and in so doing, decrease hunger			
All households at or <130% of poverty	94	69	66
All households >185% poverty	94	94	94.5

SOURCES: Food and Drug Administration and National Institutes of Health. *Healthy People 2010: National health promotion and disease prevention objectives.* Washington, DC: U.S. Department of Health and Human Services, 2000.
Household Food Security in the United States, 2007 / ERR-66, Economic Research Service, USDA.

Good choices, a variety of nutritious foods, and not too much food intake in early adulthood years affect health and nutritional status in future years. Food and nutrient intakes, along with other lifestyle factors, genetics, and environment, determine one's ability to maintain or restore health and minimize the development and advancement of chronic disease. The overarching message is to follow the principles of variety, moderation, and balance in choosing a diet that will be satisfying and that will achieve a healthy body weight and maintain health.

Key Points

1. Dietary habits during adulthood can raise or lower risk of chronic diseases.

2. Adulthood signals a change in nutritional focus from growth and development to maintenance of physiological health.

3. Excessive calorie intake is contributing to the increased prevalence of obesity. Balancing energy intake and physical activity helps to maintain a healthy body weight and increase the chance of a long and healthy life.

4. Subclinical nutritional injury begins long before observable signs and symptoms emerge. Early alterations in nutritional state can be reversed with good nutrition.

5. Dietary guidance tools are designed to reduce risks of specific diseases, ensure that the population consumes adequate levels of required nutrients, and help the public understand what, and how much, to eat.

6. Beverages, including alcohol, contribute a large percentage of calories to the diet and provide few nutrients. Recommendations are emerging regarding the type and amount of beverages that are part of a healthful diet.

7. Adult nutrient intake is excessive in energy, protein, fat and sodium, and is below recommended levels for fiber and several vitamins and minerals.

8. Individuals can achieve a high-quality diet with or without meat.

9. Dietary guidance and nutrition education should include appreciation for the social and sensory aspects of eating.

Review Questions

1. What percent of adults are overweight or obese?
   a. 66%
   b. 15%
   c. 24%
   d. 42%

Select the answer to questions 2, 3, 4, and 5 from the following sets of nutrients.

   a. Fat, cholesterol, and sodium
   b. Calcium, vitamin D, and magnesium
   c. Vitamins C, E, and K
   d. Vitamin A, potassium, and magnesium
   e. Vitamins B_{12} and D, and zinc

2. Which nutrients are a concern for men due to low intake?

3. Which nutrients are important for bone health?

4. Which nutrients are concern because of excessive intake?

5. Which nutrients are of concern in the diets of vegetarians?

6. Define "risk factor," and list at least four dietary risk factors and how they could be modified in adults.

7. Describe reasons behind health disparities.

8. When this chapter was written, the most popular "fad" diet was the Acai Berry Diet alone or in combination with colon cleansing. What diet fads are currently being promoted by books, websites, and print media? Critique them using information in the chapter. Which one, if any, would you recommend as healthful? Which ones would you not recommend, and why?

9. What would you consider in developing a dietary guidance system tailored to young adults?

Resources

American Dietetic Association
Website: www.eatright.org

American Heart Association
Website: www.americanheart.org

Calorie Control Council
Website: www.caloriecontrol.org

Centers for Disease Control and Prevention
Website: www.cdc.gov

Consumer Lab
Website: www.consumerlab.com

Food and Drug Administration
Website: www.fda.gov

Dietary Guidelines for Americans
Website: www.healthierus.gov/dietaryguidelines

MyPyramid
Website: www.MyPyramid.gov

National Center for Complementary and Alternative Medicine
Website: www.nccam.nih.gov

U.S. Census Bureau
Website: www.census.gov

U.S. Department of Agriculture
Website: www.fns.usda.gov

U.S.D.A. Agricultural Research Service
Website: www.ars.usda.gov/Services/docs.htm?docid=15044

U.S. Government Gateway for Nutrition Info
Website: www.nutrition.gov

Weight-control Information Network
Website: www.win.niddk.nih.gov

Vegetarian Resource Group
Website: www.vrg.org

Chapter 17

Adult Nutrition:
Conditions and Interventions

Prepared by **Patricia L. Splett
and U. Beate Krinke**

Key Nutrition Concepts

1 Dietary intake, body weight and body composition, and physical activity influence changes in the health status of adults.

2 Nutrition-related chronic diseases are interrelated and share common risk factors.

3 Nutrition interventions to modify early risk factors for nutrition-related diseases can change the course of disease progression.

4 Nutritional intervention practiced in the context of lifestyle change can lead to sustainable improvements that affect health status and quality of life.

5 The cognitive behavioral change approach incorporates strategies to build skills and to modify thinking and behavior.

Introduction

"And this I know, moreover, that to the human body it makes a great difference whether the bread be fine or coarse; of wheat with or without the hull, whether mixed with much or little water, strongly wrought or scarcely at all, baked or raw—and a multitude of similar differences....Whoever pays no attention to these things, or paying attention, does not comprehend them, how can he understand the diseases which befall a man?"

Hippocrates, 400 B.C.

Chronic diseases are now recognized as a leading heath concern of the nation, but they are also the most preventable.[1] Diseases that develop during the adult years fully or partially result from the cumulative effects of excessive energy intake, diets high in saturated fats and low in vegetables, fruits, and fiber, tobacco use, and alcohol. The lifestyle factors of physical inactivity and poor diet insidiously influence health on a day-to-day basis, and obesity has emerged as a major contributor to lifestyle-related diseases. For example, in 2007, cancer, heart disease, cerebrovascular diseases, diabetes, chronic liver disease, and cirrhosis accounted for 55% of all deaths for adults aged 20–64 years in the United States.[2] Table 17.1 shows the leading causes of death for adults aged 20–65 years.

Living with a chronic condition takes its toll. In addition, heart disease and diabetes are among the chronic conditions that significantly limit the quality of life of working-age adults.[2] On the positive side, the most prevalent diseases of adulthood can be prevented, in part, by healthful changes in eating and physical activity. Interventions targeted at prevention and early identification and treatment are taking center stage. This chapter starts with obesity and then addresses the three nutrition-related

Table 17.1 Ten leading causes of death by age group[2]

	20–24 Years	25–44 Years	45–65 Years
1	Accidents	Accidents	Cancer
2	Homicide	Cancer	Heart disease
3	Suicide	Heart disease	Accidents
4	Cancer	Suicide	Chronic respiratory disease
5	Heart disease	Homicide	Diabetes
6	Congenital anomalies	HIV/AIDS	Stroke
7	HIV/AIDS	Liver disease	Liver disease
8	Stroke	Stroke	Suicide
9	Diabetes	Diabetes	Septicemia
10	Chronic respiratory disease	Septicemia	Kidney disease

SOURCE: Centers for Disease Control and Prevention. Preliminary 2007 data from: National Vital Statistics Reports, Vol 58, No 1, August 19, 2009.

diseases that are highest contributors to premature death among adults—cancer, cardiovascular diseases, and diabetes—and concludes with HIV/AIDS, a significant cause of death for young and middle-age adults.

Overweight and Obesity

"The second day of a diet is always easier than the first. By the second day you're off it."

Jackie Gleason

We're a growing nation! Nearly one-third of U.S. adults are obese. Obesity is defined as having an excess accumulation of adipose tissue. It results from a long-term energy-in/energy-out imbalance involving excess calorie consumption and/or low energy output through physical activity. Obesity has become an "epidemic" in the U.S. and throughout most of the developed and developing world.[3] In 1990, the obesity rate across states ranged from <10% to 15%. By 2008, the obesity rate exceeded 30% in six states, while only one state had an obesity prevalence rate less than 20%. The dramatic increase in obesity is shown in Illustration 17.1.[4]

Obesity and overweight vary across age, gender, race, and income categories, as shown in Table 17.2.[5] Compared to whites, the obesity rate for blacks and Hispanics is 51% and 21% higher, respectively.[6] No adult population group met the Healthy People 2010 target of 15% obesity prevalence. Objectives for 2020 are to reduce the prevalence of adults who are obese and to increase the prevalence of adults who are at a healthy weight.

Illustration 17.1   Obesity trends in the United States.

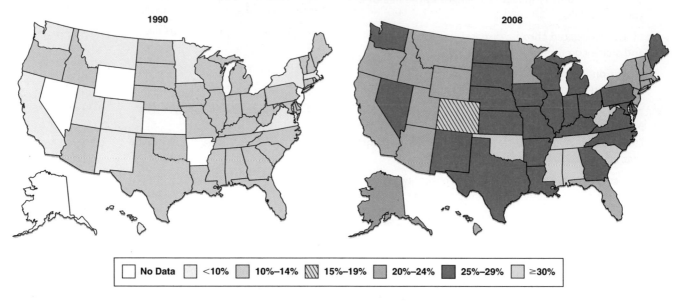

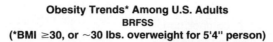

| **Insulin Resistance**   A condition in which cells "resist" the action of insulin in facilitating the passage of glucose into cells. |

Effects of Obesity

Adipocytes (fat cells) are not passive deposits of excess fat. They comprise an active endocrine organ that secretes hormone-like factors associated with chronic low-grade inflammation and *insulin resistance*.[7] These mechanisms play a role in the development or progression the major chronic diseases. Being overweight or obese increases the risk of hypertension, dyslipidemia, coronary heart disease, type 2 diabetes, stroke, gallbladder disease, osteoarthritis, sleep apnea and respiratory problems, back problems, and endometrial, breast, prostate, and colon cancers. These risks rise as the degree of excess weight rises. Persons who are obese face psychosocial complications including low-self esteem and depression, social and job discrimination, and social stigma.[8] Obesity is also associated with a shorter life expectancy.[9]

Etiology of Obesity

Overweight and obesity are not simply a matter of intake exceeding output. They are complex and chronic conditions, stemming from numerous interacting physiological, individual, and environmental factors that affect the type, frequency, and quantity of food and beverages consumed and the body's metabolic processes. Physiological processes that protect the body from starvation favor fat storage and adjust to a lower metabolic rate during periods of prolonged energy deficit (i.e., dieting). Internal mechanisms that govern energy metabolism and appetite

Table 17.2   Weight status of adults by race/ethnicity and gender

	Overweight %	Obese %	Healthy Weight %
Males	40.3	32.7	27.0
Females	25.5	35.0	39.5
American Indian or Alaska Native			
Adults ≥18 yr	34.7	32.4	32.9
Asian or Pacific Islander			
Adults ≥18 yr	29.2	8.9	61.9
Black or African American ≥20 yr			
Men	36.9	36.8	26.3
Women	24.8	52.9	22.3
Mexican American ≥20 yr			
Men	48.0	26.8	25.2
Women	31.1	41.9	27.0
Hispanic or Latino			
Adults ≥18 yr	40.3	27.5	32.2
White			
Men	40.1	32.3	27.6
Women	24.8	32.7	42.5

regulation are altered by obesity. Loss of appetite control may be due to disturbances in hormone signals that rise and fall in response to eating patterns and adjust hunger and fullness sensations, such as leptin from adipose tissue and ghrelin from the gut. Another hormonal regulatory system, the hedonic system, responds to the cognitive, motivational, and emotional aspects of food intake (perceived pleasantness, liking, and wanting).[10] The role of genetics in these mechanisms is an area of current research.

At the individual level, psychological, socioeconomic, lifestyle, and cultural factors all play a role in attitudes, behaviors, and lifestyle patterns related to the development of obesity. Life-changing events such a transition to college life[11] and pregnancy are associated with weight gain.

Environmental factors such as food advertising, the easy availability of highly palatable snack foods and beverages, larger retail package sizes, larger portions served in fast-food and sit-down restaurants, and the relatively low cost of fast food and energy-dense foods work against healthy eating and contribute to the obesity epidemic. Technological advances in every dimension of life—work, leisure time, transportation—have replaced physical activity with sedentary activity. Numerous expert groups agree that environmental and policy interventions at the national and community level, in addition to individual interventions, are necessary strategies to curb the obesity epidemic.[12]

Screening and Assessment

Classification of Obesity Body mass index (BMI) is significantly correlated with total body fat. BMI, calculated from height and weight, is used internationally for classifying overweight and obesity.[3] The formula is BMI = kg body weight divided by height in meters, squared. BMI calculators using metric or English measurements can be found online at www.cdc.gov. To calculate BMI using English measurements, following these steps:

1. Multiply body weight in pounds by 703.
2. Divide that number by height in inches.
3. Divide that number by height once more.

Overweight in adults is defined by BMI of 25.0–29.9 and obesity by BMI of 30 or greater.[7] A BMI of 30 is roughly equivalent to being 30 or more pounds overweight for a 5'4" person. Although BMI approximates body fat for most healthy individuals, there are exceptions:

- Athletes or others with greater-than-average percentages of muscle mass
- Individuals with little muscle mass
- Individuals with dense, large bones
- Dehydrated and over-hydrated individuals

Asian and Pacific Islanders experience obesity–related health problems at lower BMIs, and experts suggest that BMI cutoffs should be lowered for this population.[7,13,14]

Clinically, obesity is further classified as I, II, and III using BMI (see Table 17.3). Extreme obesity (also called morbid obesity because of its high correlation with premature death), defined as a BMI of 40 or more, is a dangerous condition that places the person at extremely high risk for cardiovascular and other diseases as well as physical disability and severely impaired quality of life.

Central Adiposity Body fat content and its distribution is a more important indicator of health than BMI.[15] Overweight, obesity, and abdominal fat are correlated with disease risk in a dose-response manner. Increased waist circumference is associated with higher risk even in persons of normal weight.[7] Waist circumference is measured and compared to sex-specific cutoffs of >40 in (102 cm) for men and >35 in (>88 cm) for women. (Cutoffs for Asians are lower: >35 in (88 cm) for Asian men and >31 in (>79 cm) for Asian women.[7]) To measure waist circumference, place a tape measure around the abdomen just above the hip bone, level with the navel and parallel to the floor. The tape should be snug but not compressing the skin. Measure after exhaling.[7,16] Refer back to Chapter 3 for more information on central adiposity.

Table 17.3 Classification of overweight and obesity by BMI, waist circumference, and associated disease risk[7]

	BMI (kg/m²)	Obesity Class	Men ≤102 cm (<40 in) Women ≤88 cm (<35 in)	Men >102 cm (>40 in) Women >88 cm (>35 in)
Underweight	<18.5		—	—
Normal	18.5–24.9		—	—
Overweight	25.0–29.9		Increased	High
Obesity	30.0–34.9	I	High	Very high
	35.0–39.9	II	Very high	Very high
Extreme Obesity	≥40	III	Extremely high	Extremely high

*Disease risk for type 2 diabetes, hypertension, and cardiovascular diseases.

Recommendation for Weight-Management Therapy

Table 17.3 shows the disease risk associated with increasing BMI and high waist circumference. The Dietary Guidelines for Americans recommend weight reduction for those who are even mildly overweight.[17] The national guidelines for treatment of overweight and obesity use BMI, waist circumference, and presence of disease conditions or cardiovascular risk factors (discussed later in this chapter) and patient motivation to prioritize individuals for treatment.[7]

Nutrition Assessment

After the need for weight management is identified, a comprehensive assessment is used to understand the individual's experience with overweight, current eating and physical-activity patterns, psychosocial and medical factors, and his/her motivation and readiness to change and goals. Table 17.4 outlines possible factors to consider during an assessment for weight management. A patient-centered interview approach allows the client's priorities and perceptions to be expressed and provides information for jointly planning appropriate goals and treatment strategies.[18]

Motivation   Several factors contribute to understanding the client's motivation to engage in a weight-loss program: reasons and motivation for weight reduction, previous weight-loss attempts, patient's understanding of causes of obesity and how obesity contributes to several diseases, attitude toward physical activity, capacity to engage in physical activity, time available for weight-loss intervention, and financial considerations. Patient's nutrition knowledge, food access, food selection, and functional capacity to prepare food and engage in physical activity are all important for individualized treatment planning.

> **Meal Replacement**   A nutritionally balanced beverage, meal bar, or packaged meal used to replace a meal in weight management.
>
> **Cognitive Behavioral Therapy**   Programs designed to build knowledge, modify beliefs and attitudes, and integrate new behaviors through a combination of skills training and analysis of behavior and thought processes over a period of several weeks. Key features are cognitive restructuring and stimulus control.

Nutrition Interventions for Weight Management

Treatment of obesity requires clinician- and patient-devised goals and treatment strategies for weight loss and reduction of obesity-related factors.[7] A successful weight-loss plan includes an eating plan that reduces caloric intake relative to calories burned, provides for nutritional needs at a safe level, incorporates physical activity, and is compatible with the individual's lifestyle. A variety of effective treatment options exist, including dietary therapy with behavioral modification techniques, altering physical-activity patterns, pharmacotherapy, surgery, and combinations of these.

Weight Loss

The good news is that relatively small amounts of weight loss (5% to 10% of body weight) can reduce or prevent the health risks associated with obesity.[19,20]

Goals of weight management are to (1) prevent further weight gain, (2) reduce body weight, and (3) maintain a lower body weight for the long term. Accomplishing this requires sustainable lifestyle changes.

A moderate rate of weight loss of ½ to 1 lb per week is recommended, but accelerated weight loss of up to 2 lb per week can also be used and may provide positive reinforcement to continue.[7,20] A deficit of 3500 kcal is necessary to lose one pound of weight; thus a weight loss of ½, 1, or 2 lb per week requires a calorie deficit of 300, 500, or 1000 kcal/day, respectively.

Medical Nutrition Therapy for Weight Management

The nutrition prescription for weight loss is an eating plan that is deficient in calories but otherwise meets guidelines for healthy eating. A balanced vitamin/mineral supplement may be recommended. Strategies for reducing caloric intake include following a menu plan and tracking calories, fat grams, or carbohydrate grams; portion control; eating 4 to 5 meals/snacks, including breakfast; and using *meal replacements* (liquid meals, meal bars, or packaged meals) for one or two meals a day. Very-low-energy diets (800 kcal/day) can produce rapid weight loss but should only be used under close medical supervision.[21]

Research comparing low-fat and low-carbohydrate diets has shown that both approaches can be successful in initial weight loss, and carbohydrate restriction had a slight advantage in weight maintenance.[22-23] However, carbohydrate restriction produces a rise in blood lipids (LDL cholesterol and triglycerides), and there is concern that the high fat intake may be associated with progression of atherosclerotic plaque.

Cognitive Behavioral Therapy for Weight Management

Successful programs developed for weight management, diabetes education, and other lifestyle changes utilize *cognitive behavioral therapy*. Programs are

Table 17.4 Nutrition assessment of adults with overweight or obesity and obesity-related diseases and conditions

Anthropometrics
- Height, weight, BMI, waist circumference, waist-hip ratio

Food and Nutrition History
- Weight history: age of onset, highest/lowest adult weights, patterns of weight gain and loss, environmental triggers to weight gain, triggers to excessive or disordered eating
- Dieting experience: number and types of diets, weight-loss medications, success of previous efforts
- Current eating patterns: meal and snack patterns (skipped meals, largest meal, snacks/grazing)
- Eating location and environment: meals eaten out (cafeteria, fast food, restaurant, carry lunch), family meals, television on at mealtime
- Types and amounts of food typically eaten: 24-hour recall or food frequency, food preferences, ethnic foods, cultural practices
- Nutritional intake: assessment of reported intake for energy and adequacy of key nutrients
  - Total caloric intake
  - Type and amount of fat (saturated, mono saturated, *trans* fats, omega-3 fatty acid)
  - Sources of key nutrients: fruits and vegetables (vitamins A, C, antioxidants and phytochemicals, potassium, fiber), bread and grains (fiber, B-vitamins, iron, folic acid), milk and dairy (calcium, vitamin D), fish, meat, beans, nuts (protein, iron, omega-3 fatty acid)
  - Energy-dense foods (bakery goods, such as cookies, cake, sweet rolls), chips and crackers, candy, salad dressings and toppings, specialty coffee drinks, alcoholic beverages, fried foods)
  - Salty foods: salt-shaker use, processed meats, chips and crackers, nuts, convenience foods, restaurant foods
  - Supplement use: nutrient-enhanced food or beverage products, vitamin/mineral supplements, herbal supplements

Physical Activity
- Level of activity at work, school, home
- Frequency, intensity, and duration of planned exercise beyond routine work and leisure activities

Laboratory
- Lipid profile: total serum cholesterol, HDL, LDL, triglycerides
- Glucose: random or fasting glucose, hemoglobin A1c, glucose tolerance test

Medical and Health History
- Obesity severity, extent of physical limitations, impact on activities of daily living
- Potential contributing causes: endocrine, neurological, physical disability, genetics/family history, medications
- Obesity-associated conditions: diabetes, hypertension, cardiovascular diseases, cancer, fatty liver disease, sleep apnea
- Mental health: daily stress level, recent life-changing events (birth, death, marriage, job change or loss, new medical diagnosis), depression, post-traumatic stress disorder, eating disorder (binge-eating, bulimia)
- Active medical diagnoses and medication use

Social History
- Occupation, family composition, caretaking responsibilities
- Economic constraints, food insecurity, food/nutrition program participation, access to health care, coverage for nutrition intervention

Nutrition Knowledge and Attitudes
- Basic understanding about foods and nutrition, guidelines for healthy eating, recommended serving sizes
- Role of nutrition in patient's diseases or conditions; previous diet instruction or lifestyle-management program
- Level of self-care regarding nutrition: experience in meal planning, food purchasing and preparation
- Confidence in ability

Readiness to Change
- Reasons to lose weight at this time, weight-loss goals
- Stage of change: precontemplation, contemplation, preparation, action, maintenance
- Support system

12 to 16 weeks long to build knowledge, modify beliefs and attitudes, and integrate new behaviors through a combination of skills training and analysis of behavior and thought processes. Key features are helping the client recognize and replace automatic and irrational thoughts and beliefs (cognitive restructuring) and increasing awareness and control of cues associated with eating (stimulus control).[24,25]

Components of weight-management programs based on cognitive behavioral therapy:

1. *Realistic goals:* Identify a healthy weight goal and a feasible rate of loss (0.5–1.0 lb/week), and provide the ability to self-monitor progress.

2. *Caloric deficit:* Develop an individualized meal plan with intake adjusted to lose weight gradually.

3. *Meal plan:* Build meal plans around a variety of foods that fit with the patient's lifestyle and budget, can be readily obtained, and can be enjoyed by the entire household.

4. *Skill development:* Provide tools and skills training, including teaching, practicing within sessions, homework, review, and feedback.

5. *Problem-solving techniques:* Assist with development of strategies to anticipate and solve potential weight-management problems.

6. *Self-management:* Provide tools for keeping food and activity records and build confidence in ability to monitor and adjust.

7. *Cognitive restructuring:* Help client examine thought processes and recognize dysfunctional thinking.

8. *Stress management:* Teach strategies other than eating to deal with stressful situations.

9. *Support system:* Encourage having someone to check in with and receive support.

10. *Regular exercise:* Advise initiation and gradual increase of physical activity, tailored to the patient's ability, aiming for 30–60 minutes most days of the week and including aerobic and muscle-strengthening activities.

11. *Maintenance:* Make available support for weight loss and for maintenance of the loss.

12. *Long-term effectiveness:* The weight-management plan is built around learning and practicing behaviors that can be maintained for a lifetime.

Physical Activity for Weight Management

Added physical activity contributes to the energy deficit required for weight loss. Cardiorespiratory fitness and screening for musculoskeletal problems may need to be reviewed before making physical activity recommendations. For obese individuals, exercise should be initiated slowly and the intensity increased gradually. Even 10-minute sessions have been shown to have beneficial effects.[26] Depending on body size, fitness level, and exercise intensity, 30 minutes of moderate physical activity five days a week would burn approximately 1000 calories. Increasing physical activity has the additional benefit of reducing diabetes and cardiovascular risk beyond that produced by weight loss alone through its effect on blood cholesterol, blood pressure, and blood glucose.[26]

Comorbidity The presence of one or more diseases or conditions in addition to the primary disease or disorder.

Physical activity is crucial to the prevention of weight regain. Studies indicate that a high level of daily energy expenditures in the range of 380 to 600 kcal per day may be necessary to sustain energy balance without overly restricting food intake.[20,27]

The Challenge of Weight Maintenance

After six months, the rate of weight loss usually declines and weight plateaus, due in part to a decline in metabolic rate—the body's physiological response to protect against starvation.[28] This metabolic compensation, termed an "energy gap," is about 8 kcal/lb lost/day.[27] Long-term maintenance of weight loss requires lifelong personal adherence to behaviors that balance calories consumed with energy burned in an environment that pushes food and fosters sedentary lifestyles.[21]

A widely held misconception is that most people regain all lost weight and more. Follow-up studies have found that while most regain some weight, up to 50% are below their baseline weight one to five years after completing therapy, and over 20% are successful at maintaining a 10% weight loss for at least one year.[29,30]

Individuals who successfully maintain weight loss use more behavioral strategies to support weight loss and maintenance. These behaviors include consistently controlling caloric intake (through restricted intake of certain types or classes of food, limiting fast food, eating all foods but in limited quantities, counting calories, or limiting percentage of daily energy from fat), exercising more often and more strenuously, tracking weight, and eating breakfast.[27]

National guidelines acknowledge that weight maintenance for many individuals requires ongoing therapy, yet this service is not covered by most health plans unless the patient has a ***comorbidity***. Some health plans employ lifestyle coaches who provide telephone and e-mail coaching to assist patients in sustaining behavioral changes and maintaining the health benefits of reduced risk for diseases. Research backs up the effectiveness of this strategy.[31] See Case Study 17.1 to explore real-life challenges to attaining a healthy weight.

Pharmocotherapy for Weight Loss

In some patients, comorbidities and risk factors warrant addition of weight-loss drugs to the comprehensive treatment/intervention plan. Research studies with overweight adults found that medication combined with lifestyle modification produced a slightly greater weight loss than lifestyle modification and placebo.[32] Mechanisms for

Case Study 17.1

Maintaining a Healthy Weight

Adam is 5'11" tall and weighs 190 pounds. He is a single father with two teenage sons. The commute to and from his software development job takes about 90 minutes. He likes his coworkers and the work environment and is happy that his workplace provides a cafeteria so he doesn't have to bring a lunch. He believes it is important for the family to have a "hot meal" every night, but he mostly relies on frozen entrees or take-out to accomplish this. Frequently, the evening meal is from the drive-through window of a fast-food restaurant on the way to his sons' sport events. Adam is an avid football and basketball fan and spends many hours watching televised games. In his spare time, he is restoring an old car with his sons.

Photodisc

Questions

Assessment

1. Calculate Adam's current BMI. How would you classify his weight status based on the NIH classifications?
2. What would you consider a healthy BMI and "healthy weight" for Adam?
3. What lifestyle and dietary factors are related to Adams weight status?

Diagnosis

4. What is Adam's nutrition diagnosis? What modifiable factors cause or contribute to this problem? What evidence do you have for the problem?

Intervention

5. What dietary prescription would you have for Adam? What goal would you recommend?
   Hint: Calculate a calorie intake level and estimate the number of weeks it would take for Adam to reach a healthy weight.
6. What interventions would you use to help Adam reach the goal?
   What topics and suggestions would you discuss with Adam?

Monitoring and Evaluation

7. What would you measure later to determine if Adam was making progress?

(Answers are located in the Instructor's Manual for the 4th edition of *Nutrition Through the Life Cycle*.)

---

drug action include appetite suppression and inhibiting fat absorption from the gut. Over-the-counter and herbal weight-loss preparations have not been tested for efficacy and safety and are not recommended.

Bariatric Surgery

Weight-loss surgery is reserved for a limited number of patients with clinically extreme obesity who meet criteria established by the National Institutes of Health (BMI ≥ 40 or ≥ 35 with high risk for obesity-related morbidity or mortality, and after other therapy has been tried for 6 or more months without success).[7] In such patients, surgery

is the most effective therapy for weight management and can result in improvement or resolution of obesity-related comorbidities (e.g., lower blood glucose, better lipid profile, lower blood pressure, reduced cardiovascular disease risks, and increased mobility). Seven to 10-year follow-up of bariatric surgery patients found lowered incidence of obesity-related conditions such as cardiovascular disease and cancer.[33,34] Surgical procedures reduce stomach size and restrict intake (stapling) or produce malabsorption by bypassing a section of the small intestine (Roux-en Y procedure). Patients considered for gastric surgery must be highly motivated to adhere to aftercare guidelines to prevent the onset of post operative complications (nausea, vomiting, dehydration, dumping syndrome) and prevent long-term nutritional deficiencies.[35]

Cardiovascular Disease

Cardiovascular diseases (CVD) are diseases related to the heart and blood vessels and are usually associated with atherosclerosis (hardening of the arteries), which is a buildup of plaque in the blood vessel wall. The areas affected include the heart (coronary heart disease or CHD), the brain (cerebral vascular disease or CVD), and blood vessels in the legs (perepherial arterial disease or PAD). Atherosclerotic lesions begin to form in adolescence and may remain silent until a vessel becomes occluded or blocked by the plaque or a blood clot and a heart attack (myocardial infarction) or stroke occurs.

Hyperlipidemia (high blood cholesterol) and hypertension (high blood pressure) are important factors in the progression of CVD. Table 17.5 shows progress made toward the 2010 National Health Objectives for decreasing these factors.[36] In Healthy People 2020 monitoring low density lipoprotein also is emphasized.

Prevalence of CVD

Over 80 million adults have one or more CVD diagnoses. Men develop CVD at a younger age, but women catch up after menopause. The prevalence of CVD in American adults increases with age from 8% and 16%, respectively, for women and men in the 20–39 age group to 38% for both women and men in the 40–59 age group.[37] Many people think of CVD as an "old person's" disease, but hyperlipidemia and hypertension are early precursors. Racial/ethnic differences

Atherosclerosis A disease of the arterial blood vessels (arteries) in which the walls of the blood vessels become thickened and hardened by cholesterol-containing plaque.

in risk factors for CVD are linked with earlier disease onset and higher death rates. Compared to whites, blacks develop hypertension at an earlier age and have higher blood pressure levels, and Mexican American men have the highest cholesterol levels.[38]

CHD is the number-one cause of death for men and women in the United States. However, progress is being made in reducing these deaths. Mortality rates from CHD and stroke fell by 34% and 28%, respectively, from 1995 to 2005, with significant changes in all racial/ethnic groups (see Table 17.6).[37]

Etiology of Atherosclerosis

Atherosclerosis begins when fatty deposits become part of tissues that form over injured arterial wall cells. Fibrous plaques (containing fats, cholesterol, collagen, muscle, and other cells and metabolites) form and gradually become calcified, increasing the extent of atherosclerosis. Chronic inflammation and infection are believed to make arteries susceptible to plaque development.[39] High blood levels of homocysteine, abnormal blood clotting factors, abdominal obesity, elevated blood glucose and insulin levels, and other conditions also influence the development of atherosclerosis.[40–42] The progression of atherosclerosis can be slowed, neutralized, or partially reversed by dietary and lifestyle modifications.[43]

Physiological Effects of Atherosclerosis

The buildup of lesions and plaque inside the blood vessels reduces blood flow. Consequently, the heart has to work harder to pump blood through this narrower space to reach all parts of the body, leading to higher blood pressure levels. Atherosclerosis decreases blood circulation to the heart, resulting in decline in organ function. CHD can present as decreased energy and shortness of

Table 17.5 Healthy People 2010 nutrition-related health objectives to reduce heart disease and stroke among adults

Objective	Percentage of All Adults	
	Baseline	2006 NHANES*
Increase the proportion of adults with high blood pressure whose blood pressure is under control to 50%.	18%	45.4%
Increase the proportion of adults with high blood pressure who are taking action to help control their blood pressure to 95%.	72%	68% taking medication
Reduce the proportion of adults with high total blood cholesterol levels (240 mg/dL or greater) to17%.	21%	16%
	Mean of All Adults	
Reduce the mean total blood cholesterol levels among adults to 199 mg/dl.	206 mg/dL	199 mg/dL

SOURCE: *Healthy People 2010: National Health Promotion and Disease Prevention Objectives.* Washington, D.C.: U.S. Department of Health and Human Services, 2000: Sections 12 and 19.

*Heart Disease and Stroke Statistics—2009 Update: A Report From the American Heart Association Statistics Committee and Stroke Statistics Subcommittee. *Circulation* 2009; 119:e21–e181.

Table 17.6 Deaths from heart disease and stroke compared to national goals to reduce mortality

			Per 100,000 Population		
		Baseline	2005*		
Objective	Target	(1998)	All	Females	Males
Reduce coronary heart disease deaths	166	208			
American Indian or Alaska Native	166	126	96.2		
Asian or Pacific Islander	166	123	81		
Black or African American	166	252		213.9	140.9
Hispanic or Latino	166	145	118		
White	166	206		187.7	110.0
Reduce stroke deaths	48	60			
American Indian or Alaska Native	48	38		31	37
Asian or Pacific Islander	48	51		41.5	36.3
Black or African American	48	80		70.5	60.7
Hispanic or Latino	48	39		38.0	33.5
White	48	58		44.7	44.0

SOURCE: *Healthy People 2010: National Health Promotion and Disease Prevention Objectives.* Washington, D.C.: U.S. Department of Health and Human Services, 2nd ed. 2000: Section 12.

*Heart Disease and Stroke Statistics—2009 Update: A Report from the American Heart Association Statistics Committee and Stroke Statistics Subcommittee. *Circulation* 2009; 119:e21–e181.

Specific indicators and cutoff levels used to identify risk of CHD and other CVDs are listed in Table 17.7. The table includes well-established indicators recommended by the National Cholesterol Education Program, Adult Treatment Panel III, as well as screening criteria for metabolic syndrome and additional indicators that are emerging as important for CVD risk screening but have less consensus.[45]

Nutrition Assessment People identified at high risk should be referred to a registered dietitian for an nutrition assessment (Table 17.4) and individualized intervention.[43,46] Key assessment components include:

- Food and nutrition history to determine usual intake, especially amount and type of fat, fruits and vegetables, bread and grains, meat, fish, and dairy foods; meal and snack patterns; and supplement use
- Nutrition knowledge of healthy-eating recommendations and relationship of food choices to CVD risk, and attitudes about food choices and change
- Physical activity
- Anthropometric measurements of weight, height, BMI, and waist circumference
- Laboratory values for lipid and blood glucose profiles
- Medical and social history to clarify other health and lifestyle factors that impact nutritional status, food choice and access, and motivation and ability to initiate and maintain lifestyle changes

breath after exertion or chest pain (angina) and death from a heart attack. Plaque in carotid and cranial arteries, complicated by hypertension, leads to stroke, with transient or permanent changes in mental and physical functioning. Poor circulation to the extremities (PAD) causes pain and limits physical activity. The progression of these cardiovascular diseases significantly interferes with activities of daily living, reduces quality of life, and can result in early death.

Risk Factors for CVD

The risk factors for the various CVDs have been known and targeted for intervention for many years, and they include *dyslipidemia* (high LDL cholesterol, low HDL cholesterol, and high triglycerides), high blood pressure, and the lifestyle factors of diet, physical activity, and smoking. Genetics, evidenced through a family history of these diseases, gender (women are at lower risk until menopause), and older age are also risk factors.[43] More recently recognized is the interconnectedness of CVD with obesity, diabetes, infection, and inflammation.[39–42]

Screening and Assessment of CVD

A quick screening tool, shown in Illustration 17.2, is used to identify individuals at highest risk for having a heart attack in the next ten years and guide the aggressiveness of intervention. The tool is based on predictive risk factors identified in the Framingham Heart Health Study.[44]

Nutrition Interventions for CVD

Nutrition intervention for CVD begins early in life to prevent or delay the development of atherosclerosis, often through population-oriented messages and dietary guidance, and shifts to the individual level when risk factors develop or CVD is diagnosed.

Primary Prevention

Eating for cardiometabolic health is not on the top of the agenda for many young adults, but eating habits

Dyslipidemia Abnormal blood levels of cholesterol and/or triglycerides resulting from altered lipid metabolism.

Illustration 17.2 Framingham heart health assessment.

Heart Health Assessment

This is a risk assessment tool that uses data from The Framingham Heart Health Study to estimate the 10-year risk of coronary heart disease. This self-assessment is not meant to be a replacement for medical advice.

Men
Estimate of 10-Year Risk for Men

AGE	POINTS
20–34	−9
35–39	−4
40–44	0
45–49	3
50–54	6
55–59	8
60–64	10
65–69	11
70–74	12
75–79	13

Total Cholesterol	Age 20-39	Age 40-49	Age 50-59	Age 60-69	Age 70-79
<160	0	0	0	0	0
160–199	4	3	2	1	0
200–239	7	5	3	1	0
240–279	9	6	4	2	1
≥280	11	8	5	3	1

	Age 20-39	Age 40-49	Age 50-59	Age 60-69	Age 70-79
Nonsmoker	0	0	0	0	0
Smoker	8	5	3	1	1

HDL (mg/dL)	POINTS
≥60	−1
50–59	0
40–49	1
<40	2

Systolic BP (mmHg)	If Untreated	If Treated
<120	0	0
120–129	0	1
130–139	1	2
140–159	1	2
≥160	2	3

Your Points

Age Points

Cholesterol Points

Smoking Points

HDL Points

BP Points

Total Points

< less than
≥ greater than

Women
Estimate of 10-Year Risk for Women

AGE	POINTS
20–34	−7
35–39	−3
40–44	0
45–49	3
50–54	6
55–59	8
60–64	10
65–69	12
70–74	14
75–79	16

Total Cholesterol	Age 20-39	Age 40-49	Age 50-59	Age 60-69	Age 70-79
<160	0	0	0	0	0
160–199	4	3	2	1	1
200–239	8	6	4	2	1
240–279	11	8	5	3	2
≥280	13	10	7	4	2

	Age 20-39	Age 40-49	Age 50-59	Age 60-69	Age 70-79
Nonsmoker	0	0	0	0	0
Smoker	9	7	4	2	1

HDL (mg/dL)	POINTS
≥60	−1
50–59	0
40–49	1
<40	2

Systolic BP (mmHg)	If Untreated	If Treated
<120	0	0
120–129	1	3
130–139	2	4
140–159	3	5
≥160	4	6

Your 10-Year Risk _____ %

Discuss the results of this survey with your physician.

Males

Point Total	<0	0	1	2	3	4	5	6	7	8	9	10	11	12	13	14	15	16	≥17
10-Year Risk %	<1	1	1	1	1	1	2	2	3	4	5	6	8	10	12	16	20	25	≥30

Females

Point Total	<9	9	10	11	12	13	14	15	16	17	18	19	20	21	22	23	24	≥25
10-Year Risk %	<1	1	1	1	1	2	2	3	4	5	6	8	11	14	17	22	27	≥30

Table 17.7 Risk factors and criteria for CVD and CHD

Risk Factor	Criteria for Risk
Major Risk Factors for CHD	
Hyperlipidemia	ATP III classification (lipid profile following 9–12 hour fast)
Low-density lipoprotein (LDL) mg/dL	Optimal: <100 Borderline high: 130–159 High: 160–189 Very high: ≥190
Total cholesterol levels mg/dL	Desirable: <200 Borderline high: 200–240 High: >240
High-density lipoprotein (HDL) cholesterol mg/dL	Low: <40 High: ≥60 (good, high HDL compensates for other risk factor)
Clinical atherosclerotic disease	CHD, PAD, carotid artery disease, abdominal aortic aneurysm
Diabetes	Diagnosis of diabetes or prediabetes
Hypertension	Blood pressure ≥140/90 mmHg or on antihypertensive medication
Cigarette smoking	Current smoker
Family history of premature CHD	Parent or sibling with CHD If male <55 years If female <65 years
Age and gender	Men ≥45 years Women ≥55 years
Risk Factors for Metabolic Syndrome	
Abdominal obesity	Waist circumference Men: ≥40 inches (102 cm) Women: ≥35 inches (88 cm)
Elevated blood triglyceride (TG) mg/dL	Normal: <150 Borderline high: 150–199 High: 200–499 Very high: ≥500
High-density lipoprotein (HDL) cholesterol mg/dL	Men: <40 Women: <50
Blood pressure	Blood pressure ≥130 (systolic) or >85 (diastolic) mmHg
Fasting blood glucose mg/dL	≥110
Overweight Obesity	BMI >25 (overweight) or >30 (obese)
Comorbidities	HIV/AIDS Diabetes, especially if uncontrolled Elevated fasting plasma insulin levels
Lifestyle Factors	
Food intake patterns	Consumption of few vegetables, fruits, and whole grains High saturated fats and *trans*-fatty acid intake Infrequent intake of fish (low omega-3 fatty acid intake) Inadequate folate intake
Sedentary lifestyle Lack of physical activity	Less than 30 minutes of moderate physical activity on most days of the week (<150 minutes a week)
Emotional stress	Unresolved emotional stress Hostility, angry personality

(continued)

Table 17.7 Risk factors and criteria for CVD and CHD (continued)

Risk Factor	Criteria for Risk
Emerging Risk Factors	
Elevated levels of high-sensitivity C-reactive protein (indicator of inflammation) mg/L	Average risk: 1.0 to 3.0 High risk: >3.0
Elevated plasma apolipoprotein B (apo B) (the protein constituent of LDL cholesterol) mg/L	≥1.20 (75th percentile) increased risk for CHD
High plasma homocysteine levels μmol/L (related to folic acid and B-vitamin intake)	≥15

Cardio-Protective Diet A diet that emphasizes plant foods (vegetables, fruits, grains, especially whole grains, and legumes), appropriate fats, and fish, along with smaller amounts of lean meat and dairy.

Pharmacotherapy Treatment of disease through the use of drugs.

Therapeutic Lifestyle Change (TLC) A higher-intensity dietary approach for reducing risk of cardiovascular disease with defined targets for type and amount of fat and dietary fiber, physical activity, and weight reduction. This is considered the first line of treatment.

during those years influence the development of atherosclerosis and risk factors for CHD and stroke. All young and middle-aged adults, with risk factors or not, should follow the principles of a *cardio-protective diet* that emphasizes plant foods (vegetables, fruits, and grains), appropriate fats, fish and lean meat, and dairy.

The American Heart Association (AHA) provides diet and lifestyle goals and recommendations, based on scientific evidence of effect on CVD risk factors, as a public health measure to reduce risk for CVD across the population (see Table 17.8).[47] Recommendations have many commonalities with Dietary Guidelines for Americans, the MyPyramid food guide, and the DASH diet (see Chapter 1). However, AHA goes beyond individual behavior and provides specific recommendations for health care professionals, schools, restaurants and the food industry, and for local policies to prevent CVD. Selected recommendations are shown in Table 17.9.

Medical Nutrition Therapy for CVD

A more restricted plan called Therapeutic Life Changes (TLC), with behavioral counseling and follow-up by health care providers, is recommended for individuals identified at high risk.[43] Diet and lifestyle change is the cornerstone of therapy and is recommended even when *pharmacotherapy* (lipid-lowering medications) is implemented. Higher-intensity intervention is important to support the individual in making lifestyle changes to reduce risk factors and halt or reverse atherosclerotic processes and prevent a coronary event (heart attack) or death. This approach was developed by the third NCEP Expert Panel.

Table 17.8 Diet and lifestyle recommendations for cardiovascular disease reduction[47]

a. Balance calorie intake and physical activity to achieve or maintain a healthy body weight.
b. Consume a diet rich in vegetables and fruits.
c. Choose whole-grain, high-fiber foods.
d. Consume fish, especially oily fish, twice a week.
e. Limit intake of saturated fat to less than 7% of calorie intake, *trans* fat to <1% of calorie intake, and dietary cholesterol to less than 200 mg per day by:
  —choosing lean meats and vegetable alternatives;
  —selecting fat-free (skim), 1%, and low-fat dairy products; and
  —minimizing intake of partially hydrogenated fats.
f. Minimize intake of beverages and food with added sugars.
g. Choose and prepare foods with little or no salt.
h. If you consume alcohol, do so in moderation.
i. When you eat food that is prepared outside the home, follow the AHA Diet and Lifestyle Recommendations.

Therapeutic Lifestyle Changes (TLC) for high-risk individuals:[43]

- Total fat intake: 25–35% of calories
- Saturated fat intake: less than 7% of total calories
- Monounsaturated fat: up to 20% of calories
- Polyunsaturated fat: not more than 10% of calories
- *Trans* fat to <1% of calories
- Dietary cholesterol intake: less than 200 mg per day
- Carbohydrates: 50–60% of total calories
- Dietary fiber intake: 20–30 grams per day, with 5–10 grams from viscous fiber
- Dietary options for additional reduction of LDL

Table 17.9 High-Priority Recommendations to Change Environmental Factors[47]

Practitioners	Advocate for healthy dietary patterns.
	Encourage regular physical activity.
	Calculate BMI and discuss results with patients.
Restaurants	Display calorie content prominently; make calorie and other nutrition information accessible at point of purchase.
	Reduce portion sizes and provide options for smaller portions.
	Allow substitutions of non-fried, low-fat vegetables for fries.
Food industry	Reduce the salt and sugar content of processed foods.
	Increase the proportion of whole-grain food availability.
	Develop packaging for greater stability and palatability of fresh fruits and vegetables.
Schools	Adopt competitive food policies.
	Offer and require daily physical activity taught by qualified teachers at all grade levels.
	Incorporate healthy nutrition and increased physical activity into after-school activities.
Local government	Implement land-use practices that promote non-motorized transportation (walking and biking), such as complete streets and community parks.
	Promote policies that increase the availability of healthy foods (farmers' markets, full-service grocery stores in low-income areas).

- Plant stanols/sterols (2 grams per day) from spreads
- Addition of 5–15 grams of additional viscous fiber
- Expenditure of at least 200 calories per day through physical activity
- Weight reduction if overweight or obese

Avoidance of *trans* fats and encouragement of stanol/sterol and viscous (soluble) fiber intake are newer additions to the guidelines.[46–50] Consumption of foods rich in omega-3 fatty acids or fish oil supplements is also encouraged.[46,47,50] Making therapeutic lifestyle changes requires ongoing intervention using strategies as discussed earlier, with cognitive behavior change. See previous chapters for more information on *trans* fats and omega-3 fatty acids.

Stanols/Sterols Plant stanols and sterols are phytosterols, an essential component of plant cell membranes, that resemble the chemical structure of animal cholesterol. When eaten, they block particles responsible for cholesterol transport, which results in less cholesterol absorption. Regular consumption of 2–3 grams per day is associated with a 7–15% reduction in LDL.[46,47] Stanol and sterol esters are added to food products such as spreads, salad dressings, and yogurt.

Viscous Fiber Viscous fiber is the 'sticky' type of soluble fiber found in oats, barley, and flax, psyllium-enriched cereals, legumes (beans and lentils), some fruits (apples, mangoes, plums, kiwi, pears, berries,

peaches, citrus fruits, and dried apricots, prunes, and figs) and certain vegetables (such as okra and eggplant). Viscous fiber is responsible for the fiber-related physiological effects of decreased LDL cholesterol. Viscous fiber holds water in the gut, forming a thick gel that reduces absorption of cholesterol-rich bile acids. The liver shifts from cholesterol production to replace bile acids lost in stools. In addition, fermentation by colonic microflora inhibits fat absorption and cholesterol transport and synthesis. Eating 5–10 grams of viscous fiber (1½ cups of cooked oatmeal provides 3 grams) a day has been shown to reduce LDL 10–15%.[50,52]

Pharmacotherapy of CVD

Lipid-lowering medications are prescribed for high-risk individuals when LDL cholesterol is >100 mg/dl. Statins are a class of drug used to lower blood cholesterol levels (brand names include Lipitor®, Lescol®, and Crestor®). Statins work by blocking the enzyme (**HMG-CoA**) responsible for making cholesterol in the liver. Lowered blood cholesterol results in reduced formation of new plaques and reduced size of existing plaques lining arterial walls. Statins also stabilize plaques, making them less prone to rupturing and forming clots that can block arteries. They also reduce arterial inflammation, which contributes to atherosclerosis.[43]

HMG-CoA Reductase The primary enzyme in the metabolic pathway that produces cholesterol. Statins lower blood cholesterol because they slow the action of HMG-CoA.

Metabolic Syndrome

Introduction

Metabolic syndrome (also known as syndrome X or the dysmetabolic syndrome) designates a cluster of altered metabolic conditions that come together in a single individual. Metabolic syndrome is associated with *hyperinsulinemia* (high blood insulin level) and places a person at high risk for coronary artery disease, stroke, and type 2 diabetes. The metabolic conditions include abdominal obesity, high blood pressure, elevated fasting glucose, dyslipidemia with elevated LDL ("bad") cholesterol, low HDL ("good") cholesterol, and elevated triglycerides.[45] The diagnosis of metabolic syndrome is made when an individual has three of these conditions. Cutoff levels established by the National Cholesterol Education Program ATP III panel defining metabolic syndrome risk are shown in Table 17.7.[43] The World Health Organization and the International Diabetes Federation include a high insulin level as part of their definitions.[54]

Other factors associated with increased odds for having metabolic syndrome are older age, postmenopausal status, current smoking, low household income, high carbohydrate intake, no alcohol consumption, and physical inactivity.

Prevalence of Metabolic Syndrome

Metabolic syndrome is quite common. Approximately 20%–30% of the population in industrialized countries have metabolic syndrome. In 2010 it is expected to affect 50–75 million people in the United States. Metabolic syndrome is present in about 5% of people with normal body weight, 22% of those who are overweight, and 60% of those who are obese. Adults who continue to gain five or more pounds per year raise their risk of developing metabolic syndrome by up to 45%.[55]

The prevalence of metabolic syndrome varies substantially by ethnicity even after accounting for a person's BMI, age, socioeconomic status, and other factors. American Indian and Alaskan Native populations have a wide range of metabolic syndrome rates that parallel the groups' rates of diabetes.[56] Based on national survey data, Mexican Americans have the highest and blacks have the lowest prevalence of metabolic syndrome for both women and men.[57] However, the low rate of metabolic syndrome in blacks appears to be related to triglyceride concentrations. African Americans with very high BMIs and very high levels of insulin resistance can have very low levels of triglycerides, even though they have a significantly higher prevalence of cardiovascular disease and diabetes compared to whites.[58] The system that relies on triglyceride levels as a marker for insulin resistance leads to under-diagnosis in African Americans. Asians, especially South Asians, can develop metabolic syndrome with only moderate excess in abdominal fat.[14] Thus African Americans and Asians are at risk for metabolic syndrome with only two metabolic risk factors.

Etiology of Metabolic Syndrome

The underlying cause of metabolic syndrome is not entirely clear, but it is thought to result from central obesity and insulin resistance. Insulin resistance refers to the diminished ability of cells to respond to the action of insulin in promoting the transport of glucose from blood into muscles and other tissues. To compensate, the pancreas produces more insulin, resulting in hyperinsulinemia. Several factors contribute to insulin resistance, including sedentary lifestyles, high body fat (especially central obesity), high-calorie diets, high saturated fat intake, inflammation, and existing cancer and HIV.[45]

As is true with many medical conditions, genetics and the environment both play important roles in the development of the metabolic syndrome. Genetic factors influence each individual component of the syndrome, and the syndrome itself. A family history that includes type 2 diabetes, hypertension, and early heart disease greatly increases the chance that an individual will develop the metabolic syndrome. Lifestyle issues such as low activity level, sedentary lifestyle, and progressive weight gain can contribute significantly to the risk of developing the metabolic syndrome. A pro-inflammatory, atherogenic diet high in total and saturated fat and low in whole grains, vegetables, and fruits, can further increase risk for development of this syndrome.[43]

Effects of Metabolic Syndrome

The hyperinsulinemic state related to insulin insensitivity is considered a prediabetic condition, but it is also recognized as a major risk factor for the development of early atherosclerotic cardiovascular disease, the most significant cause of morbidity and mortality among people with type 2 diabetes.[43,45,59] The presence of metabolic syndrome increases the risk of developing type 2 diabetes by 9–30 times, heart disease by 2–4 times, and it nearly doubles the risk of stroke. Metabolic syndrome is also associated with fat accumulation in the liver (fatty liver disease or steatahepitis), chronic kidney disease, obstructive sleep apnea, polycystic ovary syndrome, and cognitive decline and dementia in the elderly.

Metabolic Syndrome   A constellation of metabolic abnormalities that increases the risk of type 2 diabetes and cardiovascular diseases. It is characterized by insulin resistance, abdominal obesity, high blood pressure and triglyceride levels, low HDL cholesterol, and impaired glucose tolerance. Also called Syndrome X, insulin-resistance syndrome, and the dysmetabolic syndrome.

Hyperinsulinemia   A state of excess levels of insulin circulating in the blood. It is common among persons with metabolic syndrome and type 2 diabetes and is caused by the pancreas trying to compensate for insulin resistance of cells.

Case Study 17.2

Managing Metabolic Syndrome in Adults: Dan Goes Dancing

Dan Beek is 59 years old, semiretired, and lives with his wife in a midtown apartment complex. Dan was diagnosed with metabolic syndrome 10 years ago, and he has since gained 15 pounds. He attributes his weight gain to lazy afternoons in front of the television and frequent suppers at a local buffet restaurant. Though he plans to take his wife ballroom dancing on the evening of their wedding anniversary, Dan fears he will be out of shape and uncomfortable in the tight confines of his old suit. His wife suggests that the couple speak with a health professional regarding the management of his metabolic syndrome before attempting to lose weight. The following information is obtained at a recent medical follow-up.

Image Source/Jupiter Images

Height: 5'9"	TCHOL: 218 mg/dl
BMI: 32 kg/m^2	HDL: 33 mg/dl
Waist circumference: 42"	LDL: 154 mg/dl
Weight history (in lb):	TRIG: 155 mg/dl
Current: 225	FBS: 125 mg/dl
Highest: 225	TSH: Normal
Lowest: 200	HgbA1C: 7.1%
Healthy body weight: 155 to 165 lb	Blood pressure: 130/90

Questions

1. From the information gathered at his medical visit, how well do you think Dan is managing his metabolic syndrome? Why?
2. What are the desired goals for metabolic syndrome factors (i.e., anthropometric and laboratory indicators)?
3. List the primary sequelae of poorly managed metabolic syndrome.
4. What sort of lifestyle modifications would you discuss with Dan in order to improve the management of his condition?

(Answers are located in the Instructor's Manual for the 4th edition of *Nutrition Through the Life Cycle*.)

Screening and Assessment

Waist circumference is a simple, low-cost method that can be used to screen for metabolic syndrome in community or clinic settings and identify those who should be referred for laboratory tests.[54] A fasting lipid profile providing LDL cholesterol and HDL cholesterol and triglycerides levels, and fasting blood glucose, along with blood pressure, are necessary for diagnosis and provide baseline measures to track changes over time.[45,45] Screening for metabolic syndrome is recommended beginning at age 45 for asymptomatic adults or earlier for individuals who are overweight and have one additional risk factor.[60] With identification, earlier treatment can be initiated to modify cardiovascular and stroke risk.

Insulin resistance, a hallmark of metabolic syndrome and type 2 diabetes, can present with phenotypic manifestations (physical signs). These are hyperpigmentation of the skin at the back of the neck (acanthosis nigricans),[61] "buffalo hump" (cervical lipmatosis), and double chin.[62] These signs suggest high risk and should signal further assessment. Case Study 17.2 presents a somewhat typical picture of metabolic syndrome.

Nutrition Interventions for Metabolic Syndrome

The goal of clinical management is to reduce the risk of atherosclerotic diseases and progression to diabetes. The first line therapy is directed to the risk factors: for dyslipidemia, achieve an optimal lipid profile; for hypertension, normalize blood pressure; and for elevated glucose, reduce fasting blood glucose and increase insulin sensitivity.[43,45] The preferred treatment is lifestyle change to increase physical activity, adopting healthy eating, and reducing weight, as discussed throughout this chapter. Exercise in itself is an important tool in treating metabolic syndrome because of its beneficial effects on blood pressure, cholesterol levels, and insulin sensitivity, regardless of whether weight loss is achieved.[63] If a period of lifestyle intervention does not reduce risk factors, or the individual is in a very high risk category, medications may be prescribed to treat the dyslipidemia, hypertension, elevated blood glucose, and/or insulin resistance.

Diabetes Mellitus

"What AIDS was in the last 20 years of the twentieth century, diabetes is to be in the first 20 years of this century."

Paul Zimmet, International Diabetes Institute

Diabetes is a chronic disease associated with abnormally high levels of glucose in the blood. Diabetes is due to one of two mechanisms: minimal or no production of the hormone insulin by the pancreas (type 1 diabetes), or insensitivity of cells to the action of insulin (type 2 diabetes). In type 2 diabetes, circulating insulin is high, and, in addition, cholesterol, triglycerides, and blood pressure are commonly elevated. Overweight is also characteristic of type 2 diabetes. Another type of diabetes, gestational diabetes, is discussed in Chapter 5. Of primary concern for adults is type 2 diabetes.

Diabetes is diagnosed when fasting plasma glucose levels exceed 125 mg/dl. People who develop type 2 diabetes often have impaired fasting glucose (IFG) and/or impaired glucose tolerance (IGT) years before type 2 diabetes is diagnosed.[64] Blood glucose levels between 100 and 125 mg/dl indicate IGT, a *prediabetes* state that often converts to diabetes in 5 to 10 years. A landmark study found that an intensive program of weight loss (7% of body weight) and physical activity (150 minutes/week) reduced the conversion from prediabetes to diabetes by 58% and reduced cardiovascular risk factors.[65] The Diabetes Prevention Program (DPP) and similar studies have led to a greater emphasis on screening for prediabetes and intervention before diabetes symptoms occur.

Prevalence of Diabetes

Worldwide, over 80 million people have diabetes and another 35 million have prediabetes. The increase in diabetes correlates with a rise in overweight and obesity. Although most often diagnosed in people over the age of 40, type 2 diabetes is becoming increasingly common in children, adolescents, and younger adults who are overweight.[38] Type 1 diabetes accounts for less than 10% of diabetes cases.

Disparities in the Prevalence of Diabetes

Prevalence of diabetes varies widely across population groups. In the U.S., about 6% of white men and women over age 20 years have physician-diagnosed diabetes; another 3% may have undiagnosed diabetes. African American men and women and Mexican American women have twice the rate of whites, and American Indians/Alaskan Natives have about three times the rate of whites.[37] To address these disparities, many culturally-specific diabetes programs have been developed and promoted by the American Diabetes Association and the Centers for Disease Control and Prevention.

Etiology of Diabetes

Type 1 diabetes is a progressive autoimmune disease in which the beta cells of the pancreas that produce insulin are destroyed by the body's own immune system. It has a relatively quick onset. A genetic predisposition along with environmental factors, such as a childhood viral infection, are involved.

Type 2 diabetes develops over time and is, in part, the result of insulin resistance. There is a strong link between visceral adiposity, insulin resistance, and type 2 diabetes. Insulin resistance affects muscle, fat, and liver cells in different ways. Insulin resistance in fat cells leads to the mobilization of stored lipids in these cells and elevates free fatty acids in the blood plasma. Insulin resistance in muscle cells reduces glucose uptake and interferes with muscle storage of glucose as glycogen. Insulin resistance in liver cells results in impaired glycogen synthesis and a failure to suppress glucose production. These metabolic alterations all contribute to elevated blood glucose levels. High plasma levels of insulin and glucose due to insulin resistance are believed to be the origin of metabolic syndrome and type 2 diabetes, including its complications. Insulin also affects the arterial walls throughout the body and leads to hypertension.[45,59]

Prediabetes   A condition in which blood glucose levels are higher than normal but not high enough for the diagnosis of diabetes. It is characterized by impaired glucose tolerance, or fasting blood glucose levels between 100 and 126 mg/dl.

Type 1 Diabetes   A disease characterized by high blood glucose levels resulting from destruction of the insulin-producing cells of the pancreas. This type of diabetes was called juvenile-onset diabetes and insulin-dependent diabetes in the past.

Type 2 Diabetes   A disease characterized by high blood glucose levels due to the body's inability to use insulin normally, or to produce enough insulin. This type of diabetes was called adult-onset diabetes and non-insulin-dependent diabetes in the past.

Physiological Effects of Diabetes

In the short run, untreated or poorly controlled diabetes produces increased thirst, increased hunger, fatigue, frequent urination, weight loss, blurred vision, increased susceptibility to infection, and delayed wound healing. In the long run, diabetes contributes to heart disease, hypertension, blindness, kidney failure, stroke, and the loss of limbs due to poor circulation.

Micro and macro vascular complications typically develop in people with diabetes. Oxidative stress (see Chapter 3) related to hyperglycemia is believed to be a mechanism underlying vascular changes. Damage to capillaries in the back of the eye results in retinopathy and eventual blindness; nephropathy leads to reduced kidney function and kidney failure requiring renal dialysis; and neuropathy, damage to nerves, causes loss of feeling in hands and feet. Athersclerotic changes in large vessel are accelerated by diabetes resulting in earlier onset of cardiovascular diseases. As a result, heart disease, not diabetes, is the number one cause of death among people with diabetes.

Prevention of Diabetes Complications

Diabetes studies in the 1990s demonstrated that intensive management can prevent or delay the development of complications due to diabetes. The Diabetes Control and Complications Trial (DCCT) found that when blood glucose levels were maintained at near normal levels (A1c <7%), complications of diabetes (retinopathy and nephropathy) could be delayed several years.[65] These intervention programs were intensive in terms of the amount of time dedicated to diabetes self-management education, coaching, and support, and in terms of the targeted level of glucose control. A significant contribution of the DPP and the DCCT studies was verifying the effectiveness of interventions that incorporate cognitive behavioral change strategies, and demonstrating the cost-effectiveness potential of long-term, intensive lifestyle change interventions.[66] The current Look AHEAD study is investigating the effect of greater weight loss (10% of initial weight) and higher physical activity (175 minutes/week) on outcomes of individuals with type 2 diabetes.[67]

Screening and Assessment

Routine screening of asymptomatic adults is not recommended by the U.S. Preventive Services Task Force except when risk factors are present.[68] The risk factors for diabetes and prediabetes include:

- parent or sibling with diabetes,
- history of gestational diabetes or delivery of an infant weighing more than 9 pounds,
- racial or ethnic background associated with an increased risk (African American, Native American,

Asian American, Pacific Islander, or Hispanic American/Latino),
- sedentary lifestyle,
- hypertension,
- low HDL cholesterol, high triglycerides, or CVD.[68]

The American Diabetes Association recommends that adults should be screened for type 2 diabetes or prediabetes if they are 45 years old and overweight and have one or more additional risk factors listed above.[60]

A random capillary blood glucose (requiring a finger prick and a glucometer) is used for screening. An elevated glucose level (100–125 mg/dl for prediabetes or ≥120 mg/dl for diabetes) indicates need for further assessment. The diagnosis of diabetes requires two fasting plasma glucose tests on separate days with elevated levels (≥126 mg/dl).[68] An oral glucose tolerance test is also used for diagnosis.[60]

After diabetes is diagnosed, a hemoglobin A1c (A1c) test is used every 3 to 6 months to monitor glucose control. It indicates average blood glucose concentration over the previous 120 days. The A1c levels can range from below 6% (normal range) to as high as 25% in uncontrolled diabetes.

Nutrition Assessment

Nutrition assessment of the person with diabetes includes many of the same areas as listed in Table 17.4. In diabetes, the following are important for determining the individual's needs and tailoring a diabetes management plan.[60,69]

- Weight status
- Current eating pattern; types and amounts of food typically eaten throughout the day, especially types and amounts of carbohydrate
- Knowledge about diabetes and how food intake and physical activity relate to blood glucose changes
- Usual physical activity and opportunities and interests for increasing physical activity
- Laboratory values (see Table 17.10) and self-monitored blood glucose records
- Medical and social history relevant to management of diabetes; hypoglycemia or hyperglycemia events
- Past education and experience with meal planning, carbohydrate counting, or exchange lists; attitudes about diabetes, expectations for medical management and outcomes, and resources (financial, social, and emotional)

Interventions for Diabetes

The clinical goals of diabetes care are to normalize blood glucose and glucose metabolism and prevent or slow the progression of diabetes complications.

Table 17.10 Laboratory values used for diabetes screening and management in adults

	Normal Range	Treatment Goal[a]
Impaired glucose intolerance and diabetes		
Random capillary glucose, mg/dL	70–150	<120
Fasting plasma glucose, mg/dL	70–100	<126
2 hr 75 gram OGTT[b], mg/dL	<140	—
Hemoglobin A1c, %	4.0 to 5.9%	<7%
Dyslipidemia		
Total cholesterol, mg/dL	120–200	<200
LDL cholesterol, mg/dL	80–120	<100
HDL cholesterol, mg/dL		
Women	40–86	>50
Men	35–80	>40
Triglycerides, mg/dL	70–150	<150

[a]To convert these Conventional Units to SI Units see Appendix B: Conventional Units to SI Units table
[b]OGTT – Oral glucose tolerance test
SOURCE: American Diabetes Association. Standards of Medical Care in Diabetes—2009. *Diabetes Care.* 2009; 32 (Suppl 1):S6–S12.

Diabetes care is provided by a team that includes a physician, nurse, dietitian, and the patient. Additional team members might be a pharmacist, health educator, exercise physiologist, or psychologist. The person with diabetes is the central member of the team because the treatment focus is on empowering him/her to self-manage diabetes and maintain good control of blood glucose levels and other metabolic indicators. The ABCs of preventing diabetes complications are: A1c <7% blood pressure <130/80 mmHg, and LDL cholesterol <100 mg/dL (or <70 mg/dL if in the high-risk category for CVD).

Weight loss is frequently a goal, since modest weight loss (5–10% of body weight) has been repeatedly shown to significantly improve blood glucose control in overweight and obese people with type 2 diabetes and prediabetes.[60]

Medical Nutrition Therapy for Diabetes

Diet and exercise are the cornerstones of diabetes management. Developing the knowledge and skills to self-manage diet, physical activity, and medication and to self-monitor blood glucose requires time and repeated contacts, offered through a diabetes self-management education program or repeated visits with the dietitian and other team members.[69] Diabetes research has established the importance of intensive and ongoing education to make the lifestyle changes necessary to manage the disease and prevent or delay complications.[67] Individuals at very high risk or with diabetes control issues may need additional medical nutrition therapy with the dietitian.[69]

The current philosophy of dietary management of diabetes is diet flexibility within an individualized plan.[60]

a. Diet Plan (nutrition prescription) includes a target calorie level and indicates the percent of calories as carbohydrate, protein, and fat. Carbohydrate is set between 40–60%. The macronutrient mix depends on individual circumstances and is determined by understanding the patient's lifestyle, preferences, diabetes medication, and weight-management goals.

b. The calorie level is based on current weight and weight goal (to maintain current weight or lose weight), amount of physical activity, and other risk factors or concurrent disease.

c. Calories and carbohydrate are then distributed into a meal plan, including snacks, that provides relatively stable levels of carbohydrate throughout the day.

d. A variety of foods are encouraged to meet basic nutrient needs and nutritional recommendations consistent with healthy eating/cardio-protective diet.

A consistent eating pattern (timing of meals and snacks and amount eaten, especially carbohydrate) helps moderate blood glucose levels throughout the day. Strategies for managing carbohydrate intake include food exchange lists, carbohydrate counting, and experience-based estimates. The use of glycemic index and glycemic load (discussed in Chapters 1 and 3) to select foods may provide modest additional benefit.[60]

ADA Exchange Lists

The exchange lists for diabetes are used as a tool to guide meal planning and managing carbohydrate intake throughout the day.[70] Over 700 common foods are grouped into lists. One serving of any food in a list has approximately the same amount of carbohydrate, protein, fat, and calories and has about the same effect on blood glucose as other foods in the list. One serving is called an "exchange." See Table 17.11 for food groups and exchange examples. An individualized meal plan indicates the number of servings or exchanges to select from each list for each meal and snack.

The benefit of the exchange system is that it is easy to learn and allows personal choices and flexibility within a meal plan. A drawback is figuring out exchanges for

Table 17.11 Exchange lists for diabetes: averaged amount of macronutrients(grams) and calories in one serving

Food List	Carbohydrate (grams)	Protein (grams)	Fat (grams)	Calories	Examples
Carbohydrates					
Starches/Grains	15	0–3	0–1	80	½ c pasta, oatmeal ¾ c cold cereal 1 slice bread ½ c corn ⅓ c cooked legumes
Fruits	15	—	—	60	small apple 1 c berries ½ c fruit 4 oz juice
Milk					
Fat-free, low-fat, 1%	12	8	0–3	100	8 oz fat-free milk
Reduced-fat, 2%	12	8	5	120	8 oz kefir
Whole	12	8	8	160	8 oz whole milk yogurt
Sweets, desserts, and other carbohydrates	15	Varies	Varies	Varies	Read the Nutrition Facts label
Nonstarchy vegetables	5	2	—	25	½ c cooked green beans 1 c raw broccoli
Meats and Meat Substitutes					
Lean meat	—	7	0–3	45	1 oz chicken breast
Medium-fat	—	7	4–7	75	1 egg, 1 oz hamburger
High-fat	—	7	8+	100	1 oz cheese
Plant-based proteins	Varies	7	Varies	Varies	1 tbsp peanut butter, 4 oz tofu
Fats			5	45	1 slice bacon 1 tsp olive oil 1 tbsp salad dressing
Alcohol	Varies	—	—	100	8 oz beer 5 oz wine
Free Foods					Lettuce, coffee

Adapted from: American Dietetic Association/American Diabetes Association. *Choose Your Foods: Exchange Lists for Diabetes*, 6th ed, 2007.

combination dishes (such as soups and casseroles), which can be difficult if recipes or package labels are not available.

Carbohydrate Counting

In "carb counting" only food groups that contain carbohydrate are counted and the grams of carbohydrate provided by each serving are totaled. The grain and fruit groups have 15 grams/servings, the milk group has 12 grams/serving, and vegetable group has 5 grams/serving). Adults taking hypoglycemic drugs strive for regular amounts of carbohydrates at meals and snacks throughout the day and from day to day. Adults using insulin can adjust units of insulin to the amount of carbohydrate in a meal. After becoming well versed in the diabetic exchange lists or carbohydrate counting and understanding their blood glucose response, some adults can routinely use experience-based estimates instead of counting exchanges or carbohydrate grams at each meal.

Self-Monitored Blood Glucose

Self-monitoring of blood glucose (SMBG) levels through the use of a glucometer is a part of everyday life for persons with diabetes. Recording the values helps the individual see his or her pattern of blood glucose and how it responds to physical activity and eating various amounts and combinations of food. Information about current blood glucose level can be used to adjust diet and exercise (and to adjust insulin, if type 1) to minimize blood glucose swings. The record is also reviewed at each clinic visit to monitor control and make adjustments in diabetes-management plans.

Physical Activity in Diabetes Management

Physical activity is an integral part of lifestyle interventions to prevent type 2 diabetes and is important in the management of type 1 and type 2 diabetes. The benefits of regular physical activity include aiding weight loss and maintenance and improvement of insulin/glucose profile, as well as reducing lipids and blood pressure.[6,26,71] National physical activity recommendations apply to persons with diabetes.[60,71] Because exercise facilitates the uptake of glucose by muscle cells, individuals with type 1 diabetes must learn to reduce insulin dose and increase carbohydrate supplementation to minimize blood glucose swings during and after exercise. Frequent SMBG is recommended to learn how the body responds to various types, duration, and intensities of physical activity. Consistent daily physical activity that includes aerobic exercise supplemented with resistance exercise is recommended over sporadic, intensive bouts of physical activity.[71]

Pharmacological Therapy of Type 2 Diabetes

Diet alone is successful for about 30% of patients with type 2 diabetes. Insulin, injected intramuscularly, is required in type 1 diabetes. Oral antihyperglycemic drugs are used in type 2 diabetes if glycemic goals are not reached in six weeks of lifestyle interventions.[72] They have three primary mechanisms of action: stimulating the pancreas to produce more insulin (insulin secretagogues like sulfonylureas), increasing the response to insulin at cell receptor sites (e.g., sensitizers such as biguanides), and delaying absorption of glucose in the intestine (e.g., glucosidase inhibitors such as acarbose).

Herbal Remedies and Other Dietary Supplements   Numerous dietary supplements are promoted as being effective for blood glucose control and the prevention of diabetes complications. Many have been tested in studies, and a few are supported to some extent by scientific evidence of safety and effectiveness. For example, bitter melon has a hypoglycemic action, ginseng may decrease carbohydrate absorption and increase glucose transport, and gymnema may stimulate beta cells and increase insulin release.

Chromium supplementation along with biotin may reduce insulin resistance. Dietary supplements are pharmacologically active substances that have side effects and well as potential drug interactions and must be used with caution.[73]

Cancer

Cancer is a group of diseases in which genes malfunction, resulting in unregulated cell growth and tumor formation. *Carcinogenesis*, the process by which normal cells are transformed into cancer cells, is complex and moves through several stages. The stages are activation, initiation (injury or insult to DNA by a carcinogen such as

Carcinogenesis   The process by which normal cells are transformed into cancer cells. It includes activation, initiation, promotion, progression, and invasion and metastasis. Dietary constituents can modify the process at several points along the continuum.

free radicals, toxin, virus, or radiation), promotion (damaged DNA divides during a lag period, potentially over 10 to 30 years), progression (uncontrolled growth of cancer cells), invasion and metastasis (spread to other tissues and organs), and possible remission (successful treatment or reversal). Dietary constituents can modify carcinogenesis at several points along the continuum—some by promoting (e.g., aflatoxins, red meat, alcohol) others by inhibiting (e.g., cruciferous vegetables, phytoestrogens in soy beans).[74] Cancers can originate in any cell, but the majority develop in epithelial tissue, where cells replicate at a high rate, including skin, lungs, prostate, breast, colon and rectum, uterus, pancreas, oral cavity, esophagus, stomach, and urinary tract. Since most cancers take many years to develop, the chance of a cancer diagnosis increases with age.

Cancer is considered a preventable disease because most cancers are caused by modifiable environmental exposures and lifestyle factors.[6] Cancer is also considered a chronic disease, because the majority of those diagnosed have an extended post-treatment survival period where the disease is in remission or "cured." Cancer survivorship is a growing practice area dedicated to improving the length and quality of life of those diagnosed with cancer.[75]

Prevalence of Cancer

In 2008 in the U.S., over 1.4 million people were diagnosed with cancer and 565,650 died with cancer as the cause of death.[76] The American Cancer Society (ACS) estimates that there are 11 million cancer survivors. Cancer types and rates very across populations, and there is a large disparity across racial/ethnic groups. These inequities arise from socioeconomic disparities in work, income, education, housing, and overall standard of living. Economic and social barriers to cancer prevention programs, screening for early detection, and access to high-quality, ongoing treatment services are other factors behind the disparity in cancer morbidity and mortality.[77] Although cancer incidence rates are highest among whites, African Americans have the highest death rate and the shortest cancer survival of any racial and ethnic group. American Indians/Alaska Natives have the lowest incidence of cancer, while Asians/Pacific Islanders have the lowest death rates from cancer.[78]

Physiological Effects of Cancer

Cancer is the first leading cause of death for adults aged 45–65 and the second leading cause of death for adults 24 to 44. Cancer takes a toll on the family, ability to work, disability, work days lost, economics and health care costs.[79] In 2008 the annual cost of cancer was $228.1 billion, including medical costs, lost productivity, and premature death.[76]

Etiology of Cancer

Cancer development is age-associated but not age-dependent. As people age, it becomes more likely that some insult or error will damage RNA or adversely affect the DNA replication process, and ultimately cause cancer. In healthy, resilient individuals, initiation may be repaired and subsequent cancer avoided or delayed. In a person with impaired immunity or suffering from major physiological stress, initiation may proceed through promotion and progression.

Cancer is caused by exogenous (environmental) and endogenous factors. Environmental factors include tobacco use, infectious agents, radiation, chemicals (some of which become concentrated in the food supply), and carcinogenic agents in food or resulting from food preservation or cooking methods.[6,75] Epithelial tissue, where a majority of cancers originate, has the greatest exposure to these environmental carcinogens. Endogenous factors include inherited genes and genetic mutations and accumulated genetic defects as people age, oxidative stress, inflammation response, and hormonal activity. Obesity is recognized as a factor in cancer development. Adipocytes (fat cells) are not passive deposits of excess fat. They comprise an active endocrine organ that secretes hormone-like factors associated with chronic low-grade inflammation and insulin resistance. These factors produce a cellular environment conducive to survival and growth of cancer cells.[80]

ACS estimates that 50% of cancer deaths could be prevented by modifiable lifestyle factors, including diet and exercise/physical activity.[81] Modification of lifestyle behaviors including low-energy diets and increased physical activity to control weight may prevent cancer progression and mortality.[82]

Risk Factors for Cancer

Although smoking is the most recognized contributor to cancer occurrence and death, obesity and insulin resistance; excess alcohol consumption; and low intakes of fruits, vegetables, and calcium are important nutrition-related risks.[6,80,82,83] See Table 17.12 for other nutrition-related factors that increase or decrease cancer risk. The role of specific foods, food components, and eating patterns on cancer is an active area of research. Obesity is associated with cancer occurrence, recurrence, and mortality. High BMI and insulin resistance are predictive of poor cancer prognosis.[83]

Screening and Assessment

Primary Prevention   Screenings for cancer and cancer risk are important public health and clinical measures to reduce cancer development and detect cancer at early stages, when it is most treatable.

Assessment Following Diagnosis and During Treatment   Nutrition screening is used to identify patients who may be at nutritional risk. Nutrition assessment determines the complete nutritional status of the patient

Table 17.12 Nutrition-related factors associated with cancer risk

Increase Cancer Risk*	Decrease Cancer Risk**
Aflatoxins	Allium vegetables
Alcoholic drinks	Dietary fiber
Arsenic in drinking water	Fruits
Cantonese-style salted fish	Garlic
Diets high in calcium	Milk
Mate'	Nonstarchy vegetables
Red meat, processed meat	Foods containing folate, carotenoids, beta-carotene, lycopene, vitamin C, selenium
Salt	
Salted and salty foods	Calcium supplements
Beta-carotene supplements	Selenium supplements
Greater birth weight	Lactation
Body fatness, abdominal fatness, adult weight gain	Physical activity
Increased risk through weight gain, overweight and obesity: energy-dense foods, fast foods, sugary drinks, sedentary living, television viewing	Decreased risk through effect on weight: physical activity, low energy-dense foods, being breastfed

*Probable or convincing evidence for increased risk
**Probable or convincing evidence for decreased risk

SOURCE: Based on World Cancer Research Fund/American Institute for Cancer Research. Food, Nutrition, Physical Activity and the Prevention of Cancer: A Global Perspective. Washington, D.C.: AICR, 2007.

and identifies if and what nutrition therapy is needed. Nutrition assessment areas are outlined in Table 17.4. Areas of focus for cancer include:

- Anthropometrics: usual weight and recent weight loss or gain
- Food and nutrition history: appetite, food tolerance, energy intake, nutrient adequacy or shortfalls, supplement use, knowledge of appropriate strategies to optimize nutritional intake in the context of the specific type of cancer and treatment
- Medical and social history: cancer type and treatment, side effects, support system, and resources to meet nutritional needs

Finding and treating nutrition problems early may help the individual with cancer gain or maintain weight, improve the patient's response to therapy and reduce complications of treatment, and improve the patient's prognosis. Screening and assessment are done before beginning anticancer therapy, and assessment continues throughout treatment.

Nutrition Interventions for Cancer

Nutrition intervention for cancer differs significantly depending on the stage of care:

- **Prevention**
  Healthy diet (see Table 17.13) with caloric intake adjusted to achieve and maintain weight within normal ranges to reduce risk of cancer development[84]
- **Treatment**
  Medical nutrition therapy is a part of care during treatment with chemotherapy, radiation, and surgery and recovery to restore nutrient shortages, maintain nutritional health, and prevent or manage complications. Anticancer medications and radiation treatments are associated with nausea, vomiting, diarrhea or constipation, fatigue, and weight loss. Nutrition modifications help the people with cancer cope with the effects of cancer and its treatment. In addition to those mentioned above, side effects that interfere with eating include taste aversions, anorexia (loss of appetite for food), mouth sores, trouble swallowing, pain, depression, and anxiety. Loss of weight, fat, and muscle is common in cancer and is due to a combination of eating fewer calories, altered absorption, and using more calories. Dietitians individualize nutrition recommendations according to each patient's symptoms, treatment, nutritional status, and tastes for food. Foods and beverages that are high in calories, protein, vitamins, and minerals are usually advised. Some cancer treatments are more effective and better tolerated if the patient is well nourished. Enteral or parenteral nutrition supports are sometimes used. Getting enough calories and protein

Table 17.13 Nutrition and physical activity guidelines for cancer prevention*

Maintain a healthy weight throughout life.
- Balance calorie intake with physical activity.
- Avoid excessive weight gain throughout life.
- Achieve and maintain a healthy weight if currently overweight or obese.

Adopt a physically active lifestyle.
- Adults: engage in at least 30 minutes of moderate to vigorous physical activity, above usual activities, on 5 or more days of the week; 45 to 60 minutes of intentional physical activity are preferable.
- Children and adolescents: Engage in at least 60 minutes per day of moderate to vigorous physical activity at least 5 days per week.

Eat a healthy diet, with an emphasis on plant sources.
- Choose foods and drinks in amounts that help achieve and maintain a healthy weight.
- Eat 5 or more servings of a variety of vegetables and fruits each day.
- Choose whole grains over processed (refined) grains.
- Limit intake of processed and red meats.

If you drink alcoholic beverages, limit your intake.
- No more than 1 drink per day for women or 2 per day for men.

*American Cancer Society Recommendations for Individual Choices (Kushi et al.[84])

is important for healing, fighting infection, and providing energy and maintaining stamina, and being well nourished is linked to better prognosis.[85]

- **Periods of Remission**
  Healthy eating is encouraged as recommended for the general population. Lifestyle interventions can optimize health and nutritional status and help achieve or maintain normal weight. Intervention may include an individualized plan for weight management and physical activity.
- **Nutrition Care During Advanced Stages of the Disease**
  Adjust food and fluid intake, in accordance with patient's wishes, to manage symptoms and improve quality of life.

Alternative Medicine and Cancer Treatment

The hope for remission and cure is a powerful motivator to consider alternative medicine therapies. Many cancer patients or their families seek complementary or alternative treatments, including special regimens and nutritional and herbal supplements. Some herbal products have potentially useful roles in cancer treatment to ameliorate nausea and common symptoms. Examples include ginger capsules before and after chemotherapy treatment to prevent nausea,[86] chamomile or ginger tea for gastrointestinal discomfort, and peppermint tea as a digestive aid. The National Cancer Institute and the National Center for Complementary and Alternative Medicine provide funding for studies information regarding complementary and alternative treatments for cancer.

HIV Disease

The multiple aspects of disease initiated by or surrounding infection with the human immune deficiency virus are referred to as HIV disease. In the early latency stage, the body is able to contain the virus. Without diagnosis or treatment, the virus destroys immune cells and chronic symptoms occur, including weight loss, diarrhea, and cough.

The latency stage can last from several weeks to 20 years. AIDS, acquired immunodeficiency syndrome, is the advanced stage that develops when the body's immune system is severely damaged and unable to contain the virus or defend against opportunistic infections or tumor development. Advances in drug therapy have changed the view of HIV/AIDS from a terminal disease characterized by malnutrition and severe wasting to HIV disease as a chronic condition that can be managed over many years.

Prevalence of HIV

An estimated 40 million people were living with HIV in 2007, 1.1 million of them in the U.S. The U.S. adult prevalence rate for HIV disease is 0.6%. The disease affects nearly seven times more African Americans than whites. Half of new cases occur in gay or bisexual men.[87]

Physiological Effects of HIV

HIV infection raises nutrient requirements. Macro and micro nutrient needs increase with high viral load, decline of immune function, secondary infections, and altered absorption and metabolism. High demand leads to antioxidant depletion, anemias, and protein-energy malnutrition. Nutrient malabsorption due to changes in the gut and gastrointestinal pathogens can further compromise nutrient stores and lead to malnutrition. Loss of lean tissue is present throughout the HIV disease process regardless of weight maintenance. Maintenance of weight and body-protein stores (body cell mass) is associated with a person's ability to survive the HIV disease.[88]

The introduction of highly active antiretroviral therapy (HAART) has been associated with lipodystrophy—a redistribution of body fat stores.[89] HIV-associated lipodystrophy is characterized by loss of body fat from the arms, legs, face, and buttocks and with abnormal buildup of fat in the breasts, on the back of the neck and upper shoulders ("buffalo hump"), deep within the abdomen ("protease paunch"), or in fatty growths known as lipomas. These changes can be very disturbing to patients.

The accumulation of central fat is associated with insulin resistance and hyperlipidemias, and persons with HIV have high rates of metabolic syndrome and are at risk for heart disease and type 2 diabetes.[90] These metabolic changes were initially attributed to HAART, but research has indicated that traditional risk factors play a more significant role in the development of metabolic syndrome than HIV treatment-associated factors.[91]

The medication regimen and drug side effects often require adjustment of meal spacing to minimize gastrointestinal disturbances and food-drug interactions. In later stages, oral lesions, nausea, and diarrhea make eating difficult.

HIV disease imposes significant psychosocial burdens. Many people with HIV disease face social isolation and stigmatization, and some experience economic insecurity and have food access and housing issues. Substance abuse or comorbidities including mental illness and physical disability may complicate the disease and its treatment.[88]

Etiology of HIV

The human immune deficiency virus is able to make its own DNA and replicate itself by using genetic material from the host's cells. It penetrates the body's immune cells and eventually destroys the cells. Oxidative stress related to continued activation of the inflammatory process plays a role in disease progression. Decreased function of the immune system increases the risk that people with HIV will develop infections and cancer. Not all individuals with HIV go on to develop AIDS.

Nutrition status is strongly predictive of survival and functional status of people living with HIV. Nutritional problems can occur at any stage of disease and can contribute to impaired immune response, accelerated disease progression, increased frequency and severity of infections, and impede the effectiveness of medications.

Assessment

Screening and assessment are important for identifying nutritional problems and tailoring nutrition intervention plans. Table 17.4 outlines areas of nutrition assessment. Areas of focus will vary with the stage of HIV disease. Important considerations include:[88]

- Anthropometric: weight status, muscle and fat distribution, wasting
- Food and nutrition history: current food intake, appetite and food intolerances, medication and food timing, use of supplements and herbal remedies
- Physical activity: regular exercise to prevent muscle wasting, barriers to physical activity (fatigue, neuropathy), physical performance (strength/ weakness, stamina)
- Laboratory: viral load, infection, GI function, lipid and glucose profile, anemia
- Medical history: gastrointestinal symptoms, intercurrent health events (including infections and hospitalizations), body-image concerns

- Social history: psychosocial issues, economic constraints, food access, lifestyle and living arrangements, support system
- Knowledge: knowledge of nutritional needs and strategies for managing disease and medication side effects, adherence to past recommendations, personal goals

Nutrition Interventions for HIV

The clinical goal is preventing viral replication and disease progression. Individualized care that integrates medical and social services and is delivered by professionals with HIV expertise is necessary for optimal management.[92]

Maintaining good nutritional status is important for overall health and immune system function. Nutrition care supports optimum nutrition as well as medication therapy, symptom management, improving resistance against infections and complications, and increasing quality of life.

Individualized nutrition plans are an essential feature of medical nutrition therapy for persons with HIV infection and AIDS. Nutrition goals in the early phase of disease include:

- Consume sufficient calories to maintain a healthy weight and prevent rapid weight loss
- Consume adequate protein and other nutrients to maintain lean muscle mass

- Engage in regular weight-bearing or resistance exercise
- Follow a nutrient-rich, heart-healthy diet to maintain nutrient stores and reduce risk of cardiovascular disease and diabetes
- Choose calcium-rich and vitamin D-fortified food and calcium supplements to prevent progressive bone loss
- As a result of greater longevity, managing elevated lipids, insulin resistance, and other metabolic changes become part of the nutrition intervention plan.[90]

Nutrition goals in the later symptomatic stage

- Choose nutrient-rich foods and supplements to achieve adequate nutrient and energy intake and maintain weight and body nutrient stores as long as possible
- Use dietary strategies to manage symptoms—nausea, vomiting, diarrhea, anorexia, pain, chewing/swallowing difficulties, taste changes
- Adjust meal plans and timing to accommodate medication regimens

Currently, even the best nutritional advice and self-care cannot restore immune function and prevent the eventual progression of HIV. However, nutritionally adequate diets can help people with the disease maintain weight and avoid depletion of nutrient stores, and increase their level of control and sense of well-being.

Key Points

1. Overweight, obesity, and abdominal body fat are associated with the major chronic diseases—cardiovascular disease, diabetes, metabolic syndrome, and cancer—in a dose-response manner.

2. Relatively small amounts of weight loss (5–10% of body weight) can reduce risk for insulin resistance, diabetes, hypertension, and cardiovascular disease.

3. Categories of nutrition assessment include anthropometrics, food and nutrition history, physical activity, laboratory tests, medical and health history, social history, nutrition knowledge and attitudes, and readiness to change.

4. Modifiable risk factors for cardiovascular disease are high blood cholesterol, especially high low-density lipoprotein cholesterol, high blood pressure, and the lifestyle factors of diet, physical activity, and smoking.

5. Therapeutic lifestyle changes, alone or with pharmacotherapy, can halt or reverse atherosclerotic process and prevent coronary events.

6. Metabolic syndrome is very common among overweight and obese adults. Therapy is directed to normalizing blood lipids, blood pressure, and blood glucose through diet and physical activity.

7. Studies have shown that among overweight people with prediabetes, progression to type 2 diabetes can be halted or delayed with a weight loss of 7% and 150 minutes of physical activity per week; and that among persons with diabetes, complications can be delayed when blood glucose levels are maintained at near-normal levels.

8. Intensive and ongoing education and counseling are necessary to make lifestyle changes to manage chronic disease and prevent complications.

9. Nutrition intervention in cancer depends on the stage of care: prevention, treatment, remission, or advanced stage.

10. HIV infection raises nutrient requirements; and nutrition intervention is directed toward maintaining weight, lean body mass, and nutrient stores, improving resistance against infections, and increasing quality of life.

Review Questions

The first six questions concern Rebecca, who is 36 years old, weighs 182 pounds, and is 5 feet 4 inches tall. She just found out her LDL cholesterol is 170 and her HDL is 37.

1. True or false: Rebecca is obese.
2. True or false: Rebecca should be screened for metabolic syndrome.
3. True or false: Rebecca needs to lose over 20 pounds to reduce her risk of developing type 2 diabetes.
4. True or false: Eating high-fiber foods a few times a week will help lower her LDL cholesterol.
5. What other information do you need to determine Rebecca's 10-year risk of having a heart attack?
6. Rebecca quit smoking a few years ago. What else can she do to reduce the development of cardio-vascular disease?
7. Identify and discuss at least three themes that are common across all the conditions—overweight and obesity, cardiovascular disease, metabolic syndrome, diabetes, cancer, and HIV disease—presented in this chapter.
8. How would you design a lifestyle-change program based on cognitive behavioral therapy?

Resources

Overweight and Obesity

NHLBI Obesity Education Initiative Expert Panel on the Identification, Evaluation, and Treatment of Overweight and Obesity in Adults. Clinical guidelines on the identification, evaluation, and treatment of overweight and obesity in adults: The evidence report. National Institutes of Health. *Obes Res* 1998; 6(Suppl 2):51S–209S. Available at www.nhlbi.nih.gov.

Physical Activity Guidelines Advisory Committee Report to the Secretary of Health and Human Services, 2008. Available at www.health.gov/paguidelines, accessed 8/31/2009.

Slide show of how obesity has swept across the U.S. Available at: www.cdc.gov/obesity/data/.

Cardiovascular Disease

Third report of the National Cholesterol Education Program (NCEP) Expert Panel on Detection, Evaluation, and Treatment of High Blood Cholesterol in Adults (Adult Treatment Panel III). *J Am Med Assoc* 2001; 285:2486–2497. Available at www.nhlbi.nih.gov/guidelines/, accessed 8/31/2009.

Lichtenstein, A. H. et al. Diet and lifestyle recommendations revision 2006: A scientific statement from the American Heart Association Nutrition Committee. *Circulation* 2006; 114:82–96.

Metabolic Syndrome

Grundy, S. M. et al. Definition of metabolic syndrome: Report of the National Heart, Lung, and Blood Institute/American Heart Association Conference on Scientific Issues Related to Definition. *Circulation* 2004; 109:433–438.

Diabetes Mellitus

ADA Diabetes Type 1 and 2 Evidence-Based Nutrition Practice Guidelines for Adults and Toolkit. American Dietetic Association Evidence Analysis Library Web Site. Available at https://www.adaevidencelibrary.com.

American Diabetes Association. Executive summary: Standards of medical care in diabetes—2010. *Diabetes Care* 2009; 33(Suppl 1):S11—S51.

National Diabetes Education Program and culturally specific initiatives. Available at www.diabetes.org and www.cdc.gov.

A Diabetes Risk Test is available at http://www.diabetes.org/.

Cancer

Kushi, L. H. et al. The American Cancer Society 2006 Nutrition and Physical Activity Guidelines Advisory Committee. American Cancer Society Guidelines on Nutrition and Physical Activity for Cancer Prevention: Reducing the Risk of Cancer with Healthy Food Choices and Physical Activity. *CA Cancer J Clin* 2006; 56:254–281.

The National Cancer Institute website provides up-to-date information regarding complementary and alternative medicine and anticancer remedies. Available at www.cancer.gov.

HIV Disease

Nutrition Intervention in the Care of Persons with Human Immunodeficiency Virus Infection—Position of the American Dietetic Association and Dietitians of Canada. *J Am Diet Assoc* 2004; 104:1425–1441. ADA and Dietitians of Canada Joint Position. Update due in 2009–2010.

Chapter 18

Nutrition and Older Adults

Prepared by **U. Beate Krinke**

Key Nutrition Concepts

1   Enjoying a varied diet contributes to mental and physical well-being.

2   The population of "older adults" is heterogeneous; generalizations about nutrition and health status are unlikely to fit many individuals.

3   Diseases and disabilities are *not* inevitable consequences of aging.

4   Functional status is more indicative of health in older adults than is chronological age.

5   Body-composition changes that occur with aging can alter lifestyle; these changes may modify nutritional needs.

Introduction

> "An ounce of prevention is worth a pound of cure."
>
> Traditional

A positive outlook toward one's own old age is associated with increased *longevity*. In the Ohio Longitudinal Study of Aging and Retirement, "older individuals with more positive self-perceptions of aging . . . lived 7.5 years longer than those with less-positive self-perceptions of aging."[1] Attitude affects how individuals manage the inevitable, irreversible physical changes and increased prevalence of disease associated with normal aging. The leading causes of death for people aged 60 and older are heart disease, cancer, and stroke; these three and diabetes mellitus (the 6th leading cause of death)[2] have nutritional risk factors. Despite the significant increase in disease prevalence that accompanies old age, 71% of adults aged 75 and older consider themselves to be in good, very good, or even excellent health.[3] More than anything, older adults want to remain independent; they certainly do not want to be a burden to others. Older adults feel that good nutrition and exercise are the most important health habits they can maintain in order to avoid losing autonomy and independence.[4]

What constitutes good nutrition in old age? Older adults can meet their decreasing energy requirements by choosing more nutrient-dense foods. They can drink more water to stay hydrated, even when not thirsty. Eating adequate amounts of vegetables, fruits, and whole grains, keeping fats in balance, and drinking alcohol only in moderation will reduce risks of contracting disease. Diet quality is linked to the longevity of older men and women.[5–7] Good health habits help to delay mortality and achieve *compression of morbidity* in older populations.[8–11] This chapter defines aging and provides information about the nutrient requirements, dietary recommendations, and food and nutrition programs designed to support healthy aging.

What Counts as Old Depends On Who Is Counting

Many chronological ages have been used as cut-points to mark the beginning of "old age." While no biological benchmark signals a person's becoming old, there are societal and governmental definitions for *old*. However, these definitions may not reflect the perceptions of older adults. A survey by the Pew Research Center found that the average older adult does not feel old, and that what is considered as *old* depends on one's age.[12] Overall, adult survey respondents said that old age begins at 68. However, adults under 30 considered 60 to be old, while respondents over age 65 believe that old age begins at age 75. The World Health Organization uses age 60 when referring to aging populations. This leaves approximately 20 to 50 years for persons to be "old." The U.S. Census Bureau has separated adults into young old, aged, and oldest old categories.

Despite the lack of biological benchmarks for chronological aging, programs supporting older adults need clear guidelines to define their target populations and services. The Elderly Nutrition Program, first funded under the Older Americans Act in 1972, uses eligibility criteria of 60 years and older or being the younger spouse of a 60-year-old. The Social Security Program identifies people 65 years old as being eligible for **Medicare**. These different age limits for program eligibility reflect society at the time the laws were enacted, and they affect program funding; adjusting them to meet public health needs will happen as science and the public's perceptions evolve. The arbitrarily set retirement age of 65 years to denote an *older adult* is commonly used, and we will use it here. Chronological age is just one of the many factors that affect the nutritional health of older adults.

Chronological age is a simple place to begin to understand someone's nutritional health. However, functional status, a description of how well one can accomplish the desired tasks of daily living, is more indicative of health than chronological age. Rather than ask, "How old are you?" we should ask, "Can you do the things you want and need to do?" and "Can you shop for food?" and "Can you see well enough to know if your food is moldy?" Nonetheless, chronological age commonly serves as a proxy for predicting health status and functional abilities.

Longevity   Length of life, measured in years.

Compression of Morbidity   Shortening the period of illness and decreased functional capabilities at the end of life.

Medicare   Federal health insurance for all people age 65 and older, and for younger individuals with certain conditions.

Food Matters: Nutrition Contributes to a Long and Healthy Life

> "We found that tomatoes were the primary food associated with higher functional status in centenarians ... even when controlled for other factors such as illness, depression, and gastrointestinal problems."

Dr. Mary Ann Johnson, commenting on findings that elderly nuns with higher functional status also had higher blood lycopene levels[13]

In our search for magic bullets, tomatoes will probably not turn out to be nature's perfect food. The cumulative effects of lifelong dietary habits determine nutritional status in old age. Good nutrition throughout life contributes to optimal growth, to appropriate weight, and to nutrient levels in blood and other tissues that boost immunity and provide disease resistance. In trying to assess the contribution good nutrition can make to longer life, the Centers for Disease Control and Prevention (CDC) suggest that longevity depends 19% on genetics, 10% on access to high-quality health care, 20% on environmental factors such as pollution, and 51% on lifestyle factors.

Besides not smoking, diet and exercise are estimated to be the lifestyle factors contributing most to decreased mortality, or longer life. In a longitudinal study including diet monitoring, older women who ate the healthiest diets were 30% less likely to die during a 6-year study period than the women who ate few whole grains, fruits, vegetables, low-fat dairy products, and lean meats.[6]

The role of food and nutrition often changes during aging. Besides reducing risk of disease and delaying death, diet contributes to wellness. Wellness means having the energy and ability to do the things one wants to do and to feel in control of one's life. Being able to choose, purchase or prepare, and eat a satisfying diet every day; enjoying traditional foods at holidays, birthdays, and other special occasions; and having the resources to purchase desired foods on a regular basis all contribute to independence and a higher quality of life. Good nutrition, as defined by dietary guidelines covered later in this chapter, can help "add life to years" as well as add years to life.

Life Expectancy Average number of years of life remaining for persons in a population cohort or group; most commonly reported as life expectancy from birth.

Life Span Maximum number of years someone might live; human life span is projected to range from 110 to 120 years.

A Picture of the Aging Population: Vital Statistics

More and more of us are growing old and older. During Roman times, fewer than 1% of the population reached age 65, in 1900 it was 4%, and today, 13% of the North American population is aged 65 and older. By 2050, 20 percent of the population is expected to be over age 65, and about 5 percent will be 85 and older.[12]

Persons aged 85 years and older are the fastest growing segment of the population. A White House conference on aging called this age wave "a demographic revolution" and predicted that it will change the twenty-first century much as information technologies revolutionized the twentieth century.[14]

Global Population Trends: Life Expectancy and Life Span

Today, *life expectancy* at birth in the United States is 78 years (see Table 18.1 and Illustration 18.1), compared to 47 years for someone born in 1900. At age 65, an individual can expect to live another 19 years, and someone reaching age 85 can expect to live another seven years.[15] Other contributors to life expectancy estimates include rates of childhood mortality, infectious and chronic diseases, and death from violence and accidents.

Since the early 1900s, immunizations and other risk-reduction measures, treatment of disease, decreased infant and childhood mortality rates, and clean water and safe food have increased average life expectancies, which are getting closer to the potential human *life span*. For instance, from 1980 to 2000, life expectancy increased by more than 4 years for men and 2 years for women, largely due to fewer deaths from heart disease, stroke, and accidents among older adults.[16] After China and India, the United States has the highest absolute number of adults aged 60 and over.[17]

Although life expectancy is rising and populations are aging, human life span remains stable at around 110 to 120 years.[18] Jeanne Calment of France was the oldest known person, living to age 122. In the U.S., Sarah DeRemer Knauss lived to age 119.[19] Reaching age 100 has been rare enough that the President of the United States sends personal greetings to individuals for their 100th

Table 18.1 Ten countries leading in longest life expectancy at birth, of 224 ranked, using 2009 data

Rank	Country	Years
1	Macau	84
2	Andorra	83
3	Japan	82
4	Singapore	82
5	San Marino	82
6	Hong Kong	82
7	Australia	82
8	Canada	81
9	France	81
10	Sweden	81
50	United States	78

Countries with the shortest life expectancy are Swaziland, at 32 years; Angola, at 38 years, and Zambia, at 39 years.

SOURCE: Central Intelligence Agency. *The World Factbook.* Available at www.cia.gov/library/publications/the-world-factbook/rankorder/2102rank.html, accessed 6/16/09.

Illustration 18.1 **Life expectancy at birth and at age 65, by race, sex.**

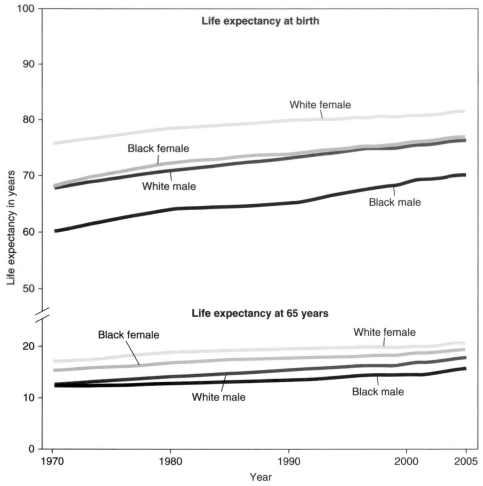

SOURCE: CDC.gov/nchs/data/hus/hus08.pdf, accessed July, 2009.

decreased resistance to disease, and other physical and mental changes that accompany aging.

Biological systems are so complex that no single theory has been robust enough to explain the mechanisms of aging. Genetics are thought to account for up to one third of longevity.[20] Environmental factors influence expression of the genetic code by exacerbating or attenuating certain traits. Nutritional genomics is a field that examines the interplay between genetics and nutrition; it can help to promote healthy aging by tailoring interventions to specific risk factors, such as adding omega-3 fatty acid to one's diet in an attempt to enhance gene-controlled responses to inflammation. Body composition is another area that demonstrates reactions between genes and environment. While height and weight are genetically programmed, individuals have some ability to control diet, activity levels, and other environmental exposures (such as smoking) to determine what their body composition will be in later life.[10,21] Understanding how

birthday. By 2050, when nearly 1% of the population is expected be a *centenarian*, the President may have to reserve this special goodwill gesture for *supercentenarians* (individuals who live to age 110 or more).[12]

Nutrition: A Component of Health Objectives for the Older Adult Population

Behaviors that can enhance the health of an aging population are reflected in national health objectives. Table 18.2 identifies dietary goals related to disease prevention and to health promotion for older adults. New national health goals emphasize overall fitness and the health consequences of obesity. The greatest dietary improvement in the older-adult population would be to eat more vegetables and grains, especially whole-grain products.

Theories of Aging

What triggers aging? Theories explaining aging grow from human desire to understand the biological processes that determine how long and how well we live. Aging theory tries to explain the mechanisms behind loss of physical resilience,

people age can help to separate the effects of age-associated disease from changes naturally caused by aging. Such an understanding can guide prevention efforts, leading to compression of morbidity and mortality.

Theories of aging can be examined from two perspectives: (1) programmed aging theories and (2) "wear-and-tear" theories. Caloric restriction is an intervention that incorporates aspects of several aging theories to manipulate life span and morbidity.

Programmed Aging

Hayflick's Theory of Limited Cell Replication
Hayflick proposed that all cells contain a genetic code that directs them to divide a certain number of times during their life span.[18] After cells divide according to their programmed limit, and barring disease or accident, cells begin to die (some call this replicative *senescence*). For example, if individual cells of a fly have a 3-day life span and replicate 15 times, a fly can live 45 days. Using this

Centenarian Person who reaches age 100 or more.

Super-Centenarian Person who has reached age 110 or more, validated.

Senescence Old age.

Table 18.2 Healthy People 2010, food and nutrition goals for older adults

	Percentage of Population		
	Target	Current, Age 60+ Females	Males
Increase the number who are at a healthy weight	60	37	33
Decrease the number who are obese (BMI 30+)	15	26	21
Increase the proportion of persons who:			
• Eat at least 2 daily servings of fruit	75	35	40
• Eat at least 3 daily servings of vegetables (at least one-third dark green or deep yellow)	50	6	5
• Eat at least 6 daily servings of grain products (at least 3 of those as whole grains)	50	4	11
• Eat less than 10% of calories from saturated fat	75	47	42
• Eat no more than 30% of calories from fat	75	40	34
Baseline information is for age 50 and older			
Consuming 2400 mg of sodium or less	65		
Baseline information is for age 50 and older			
Meet dietary recommendations (1200 mg) for calcium	74	27	35
Baseline information is for age 65+			
Increase the proportion of physician office visits made by patients with a diagnosis of cardiovascular disease, diabetes, or hyperlipidemia that include counseling or education related to diet and nutrition	75	33[a]	33[a]

SOURCE: *Healthy People 2010.*[16]

[a]Adequate data to distinguish levels for males and females is not available.

Telomere A caplike structure that protects the end of chromosomes; it erodes during replication.

theory, Hayflick calculated the potential human life span to be in the range of 110 to 120 years, estimating that human cells replicate from 40 to 60 times. Although most human cells can regenerate (e.g., blood, liver, kidney, and skin cells reproduce themselves), not all cells have that capacity (e.g., spinal cord, nerves, and brain cells). Hayflick's theory is difficult to prove in humans because we die from age-associated chronic disease more often than from old age itself.

Molecular Clock Theory Another theory of programmed aging is that of the molecular clock. *Telomeres* that cap the ends of chromosomes act to mark time, becoming a bit shorter with each cell division. Eventually loss of telomeres stops the ability of chromosomes to replicate. Loss of chromosomal replication may produce signs of aging because new cells cannot be formed, and the function of existing cells declines with time. A major thrust of current research is to identify ways to limit loss of telomeres and thus prolong cell replication.[22]

Wear-and-Tear Theories of Aging

Wear-and-tear theories are built on the concept that things wear out with use. Mistakes in the replication of cells or buildup of damaging by-products from biological processes eventually destroy the organism. Cytotoxicity (poisoning of the cell) results when damaged cell components accumulate and become toxic to healthy cells. Glucose binding to proteins (glycosylation) leads to accumulation of AGEs (Advanced Glycosylation End-products); cross-linking between cells stiffens collagen fibers, including those in tendons, ligaments, blood vessels, and kidneys. According to wear-and-tear theories, the accumulation of damaged cells and waste by-products leads to declining function and aging.

Free-Radical or Oxidative Stress Theory Oxygen is an integral and versatile part of metabolic processes; it can both accept and donate electrons during chemical reactions. One cause of aging is thought to be oxidative stress due to the buildup of reactive (unstable) oxygen compounds. Unstable oxygen, formed normally during metabolism (e.g., hydroxyl radicals), can also initiate reactions that break down cell membranes and damage cells needed to keep the immune system intact. Exposure to oxidizing agents is increased by smoking, ozone, solar radiation, and environmental pollutants. Unstable oxygen compounds are neutralized, however, when they combine with an antioxidant. This prevents them from interfering with normal cell functions. The body produces antioxidant enzymes (such as catalases, glutathione, peroxidase reductases, and superoxide dismutase), but part of our need for antioxidants is met from the diet. Dietary antioxidants include selenium, vitamins E and C, and phytochemicals such as beta-carotene, lycopene, flavonoids, lutein, zeaxanthin, resveratrol, and isoflavones.

Rate-of-Living Theory The rate-of-living theory is similar to the oxidative stress theory in that it suggests that "faster" living results in faster aging. For example, higher metabolic rate and greater energy expenditure leads to greater turnover of all body tissues. Theoretically, fast-paced living shortens life span, whereas living more slowly leads to a longer life. Scientists have not adequately examined old people, including centenarians, to fully understand this theory.

Calorie Restriction to Increase Longevity

Animal studies (e.g., of fruit flies, water fleas, spiders, guppies, mice, rats and other rodents, and primates) show that an energy-restricted diet that meets micronutrient needs can prolong healthy life.[23] For example, laboratory mice and rats fed calorie-restricted diets live longer and have fewer age-associated diseases than their counterparts whose diets are unrestricted. In the 1930s, McCay and colleagues suggested that delays in aging result after food restriction, due to slowed growth and development.[24] But since then, rodent studies have shown that instituting *caloric restrictions* in midlife, after growth and development were completed, results in longer life spans.[23,25]

Study of caloric restriction (CR) in primates is currently underway.[26] Adult rhesus monkeys fed calorie-restricted diets (at 70% of their previous energy maintenance level) had significantly fewer age-related deaths (hazard ratio of 3, meaning the controls were 3 times more likely to die than the CR monkeys during the 20-year study period). Researchers at the Wisconsin National Primate Research Center also found that CR led to delayed onset of age-related illnesses, maintenance of muscle mass at a higher level, and delayed gray matter atrophy. The authors suggest one mechanism driving these results may be that CR leads to changes in cell signaling pathways that induce metabolic reprogramming and subsequent life extension. This study is ongoing; stay tuned.

Could calorie-restricted diets also extend human life? Experimental findings in small animals have led some individuals, such as Dr. Roy Walford of Biosphere 2, to personally adopt very-low-calorie diets. Walford coordinated the calorie-restricted diets of eight normal-weight people living in Biosphere 2.[27] However, such a small study lasting two years is too limited to determine human life-span extension results from caloric restriction, especially in the closed terrarium-like setting of the Biosphere. Luigi Fontana and colleagues tested the theory that caloric reduction also reduces a thyroid hormone (T3) that controls cell respiration and free-radical production.[28] They compared hormone levels of healthy, lean, weight-stable adults eating an 1800 kcal diet for 3–15 years with those of two matched groups eating typical western diets of 2400 (sedentary group) or 2800 kcal (exercise group). Fontana's results suggest that reducing metabolic rate and oxidative stress may slow the rate of aging.

From an ecological view, people in France and Japan have lower caloric intake than do people in the United States, and people in both of those countries also live longer (see Table 18.1). Willcox and colleagues[29] studied a cohort of older Japanese residents of Okinawa whose nutrient-dense diets provided approximately 11% fewer calories than estimated to meet their energy needs. Population records show that this Okinawan cohort experienced caloric restriction at least until middle age, some longer. Lower caloric intake was coupled with reduced mortality from age-associated diseases. The authors concluded that, consistent with animal literature supporting caloric restriction, "an adaptive response to early and mid-life energy restriction in the older cohort of Okinawans may be implicated in their low morbidity and exceptionally long survival." We know that nutrition affects human longevity by moderating risks of developing chronic diseases, ameliorating certain chronic conditions, and aiding in the healing of acute conditions.[16,30] Many physiological relationships seem to function well in a mid-range, where insufficiency and excess are harmful. Body weight is an example. Severe caloric restriction during famine leads to malnutrition and starvation, with poor outcomes in human reproduction, growth, development, immune status, and healing. Severe obesity leads to early death. CR research ensures nutrient density in diets of study subjects. Dietary guidance tools around the world promote nutrient density to decrease chronic disease risk. Researchers seek the perfect balance of energy and nutrient intake to maintain optimal health and long life, and CR is a growing area of aging research.

> **Calorie Restriction** Decreasing the energy level of one's diet by 25 to 30% while meeting protein, vitamin, and mineral needs.
>
> **Resilience** Ability to bounce back, to deal with stress, and recover from injury or illness.
>
> **Lean Body Mass** Sum of fat-free body tissues: muscle, mineral as in bone, and water.
>
> **Fat-Free Mass** Used interchangeably with Lean Body Mass, comprised of bone, muscle, water.

"Everything in moderation. Including moderation."

Joyce Hendley, food writer, citing Julia Child

Physiological Changes

Normal aging is associated with body-composition shifts that most often lead to loss of physical *resilience*. Physiological system changes commonly associated with healthy aging are described in Table 18.3. As scientists come to understand the human aging process, they will learn to sort through age-associated physiological changes to be able to distinguish exactly which changes are due to genetic factors and which are due to poor diets, inactivity, or other lifestyle-related factors. Yet aging is not all loss or decline. Rather, healthy aging is associated with continuing psychosocial, personal, moral, cognitive, and spiritual development. Chronic conditions that affect older persons' health are discussed in Chapter 19.

Body-Composition Changes

Lean Body Mass (LBM) and Fat Individual shifts in body composition are common but neither inevitable nor irreversible.[31] Of all physiologic changes that occur during aging, the biggest effect on nutritional status is due to the shifts in the musculoskeletal system, which loses up to 15% of *fat-free mass* (Table 18.4). On average, there is a decline in lean body mass of 2% to 3% per decade from age 30 to 70, including loss of muscle (**sarcopenia**) beginning around age 40, even when weight is stable.[11,32]

Table 18.3 Age-associated physiological system changes that affect nutritional health*

Cardiovascular System
- Reduced blood vessel elasticity, blood volume, stroke volume output
- Increased arterial stiffening, blood pressure

Endocrine System
- Reduced levels of estrogen, testosterone
- Decreased secretion of growth hormone
- Increase in cortisol (stress)
- Reduced glucose tolerance
- Reduced levels of thyroid gland secretions
- Decreased ability to convert provitamin D to previtamin D in skin

Gastrointestinal System
- Reduced secretion of saliva and of mucus
- Missing or poorly fitting teeth
- Dysphagia or difficulty in swallowing
- Damaged, less-efficient mitochondria produce less ATP, less energy
- Reduced secretion of hydrochloric acid and digestive enzymes
- Slower peristalsis
- Reduced vitamin B_{12} absorption

Musculoskeletal System
- Reduced lean body mass (bone mass, muscle, water)
- Increased fat mass
- Decreased resting metabolic rate
- Reduced work capacity (strength)

Nervous System
- Blunted appetite regulation
- Blunted thirst regulation
- Reduced nerve conduction velocity, affecting sense of smell, taste, touch, cognition
- Changed sleep as the wake cycle becomes shorter

Renal System
- Reduced number of nephrons
- Less blood flow
- Slowed glomerular filtration rate

Respiratory System
- Reduced breathing capacity
- Reduced work capacity (endurance)

*Some of these age-associated changes, such as the increase in blood pressure, are usual but not normal.

During this time, body fat increases, especially in the visceral region. Compared to males in their twenties, males in their seventies have roughly 24 pounds less muscle (a decrease from 53 to 29 lb on average) and 22 pounds more fat (an increase from 33 to 55 lb on average). Think about losing 24 lb of muscle: It is roughly equivalent to the muscle mass that girls or boys gain throughout the puberty growth spurt. Over 50 years, 24 lb of muscle slowly goes

Functional Status Ability to carry out the activities of daily living, including telephoning, grocery shopping, food preparation, and eating.

Table 18.4 Comparison of body composition of a young and an old adult

	20 to 25 Years	70 to 75 Years
Protein/cell solids	19%	12%
Water	61%	53%
Mineral mass	6%	5%
Fat	14%	30%

SOURCE: Data from Chernoff, R., ed. *Geriatric Nutrition: The Health Professional's Handbook.* Sudbury, MA: Jones and Bartlett, 2006: p. 435. Based on Shock, N. W., *Biological Aspects of Aging.* New York, NY: Columbia University Press, 1962.

away, to be replaced by 22 lb of fat. Of course, the extra fat does provide a reserve of energy for periods of low food intake or recovery from illness or surgery, acts as an insulator in cold weather, and cushions falls. After age 70, weight, including fat at all sites, begins to decline.[11] These composition changes are associated with lower levels of physical activity, food intake, and hormonal changes in women.[32] Losses of fat-free mass leave older people with lower mineral, muscle, and water reserves to call upon when needed.

Muscles: Use It or Lose It   Many older people expect decreases in physiologic function with increasing age. However, physical activity contributes to staying strong, no matter the age. For example, the HEalth, RIsk Factors, Training, And GEnetics (HERITAGE) Family Study compared the effects of a 20-week strength-training program on older and younger men and women, finding that the training response differed by gender and by race, but *not* by age.[33] Training exercises led to increases in fat-free mass; decreases in total, subcutaneous, and visceral fat mass; and to weight loss. Weight-bearing and resistance exercise increase lean muscle mass and bone density.[11,34] Because muscle tissue contains more water than fat tissue does, building muscle also results in more stored water. Regular physical activity, including strengthening and flexibility exercise, contributes to maintenance of *functional status.*

> "I would not wish to imagine a world in which there were no games to play and no chance to satisfy the natural human impulse to run, to jump, to throw, to swim, to dance."
>
> Sir Roger Bannister, 1989

Weight Gain   Weight gain, although not inevitable, tends to accompany aging. Large cross-sectional studies show that mean body weight increases gradually during adulthood. Weight and BMI peak between 50 and 59 years, then stabilize and start slowly dropping around age 70.[11,35] Reasons for age-associated gains are uncertain, but longitudinal studies are showing that lack of exercise could be a factor.[31] For example, men in the

Baltimore Longitudinal Study on Aging decreased their energy expenditure by 17 to 24 calories per day after age 55, and gained weight.[36] In the Fels Longitudinal Study, men gained 0.7 pounds per year, and women gained an average of 1.2 pounds as they aged.[32] Subjects were 40 to 66 years old when entering the study and were followed for up to 20 years. Weight gains were concurrent with decreases in lean body mass and increases in body fat. This overall weight and body-composition shift was moderated by physical activity. For example, the two groups with moderate or high physical activity levels (as opposed to the least active group) increased lean body mass and decreased total and percentage of body fat as age increased. Physical activity effects differed by gender. In women, higher levels of physical activity were associated with higher levels of lean body mass. However, lack of estrogen seems to promote fat accumulation, and total weight increased regardless of the group's activity level. Men in the highest physical activity groups in the Fels Longitudinal Study slowed their total body-weight and body-fat gains.

Changing Sensual Awareness: Taste and Smell, Chewing and Swallowing, Appetite and Thirst

Taste and Smell   Although there is some argument about the extent to which aging affects the sense of taste, there is general agreement that taste and smell senses are generally robust until age 60, when they start declining.[37,38] Approximately 60% of individuals over age 80 have some olfactory impairment, compared to 6% of adults in their fifties.[38] Women retain their sense of smell better than men do. The conditions most associated with impaired olfaction—namely, congestion, upper respiratory tract infections, stroke, epilepsy, and current smoking—are not due to aging.

Eating is a sensuous activity, involving not just taste buds but also the olfactory nerves and, many believe, the eyes. A blunted sense of smell can lead to blunted enjoyment of food, which anyone with a stuffy nose can easily affirm. Declining taste and smell also result in decreased ability to detect spoiled or burnt foods.

It is difficult to sort out whether aspects of sensory decline are due to illness or age. The number and structure of taste buds are not significantly altered during aging. In addition, taste perception for sucrose does not decline with age. Bartoshuk[39] argues that taste is so important to biological survival that the body has developed "redundancy in the mediation of taste," meaning that several pathways (nerves and receptors) control taste mechanisms. All of the pathways would have to be damaged before the ability to identify tastes is lost.

Disease, medications, and covering part of the palate with dentures may affect taste more than does age itself. Yet age matters. For example, during illness or with the use of medications, younger individuals maintained

greater ability than older individuals to detect salty, bitter, sour, sweet, and savory tastes.[40]

Oral Health: Chewing and Swallowing   Poor dietary habits are a modifiable risk factor that can contribute to caries and potential tooth loss. Tooth loss is associated with disability and mortality in old age.[41] What and when we eat affects oral health. Oral health affects what and how we eat and, in turn, nutritional status and health.[42] Oral health depends on several organ systems working together: gastrointestinal secretions (saliva), the skeletal system (teeth and jaw), mucous membranes, muscles (tongue, jaw), taste buds, and olfactory nerves for smelling and tasting. Disturbances in oral health and tooth loss are associated with, but not necessarily caused by, aging. Of adults aged 65–74 and 75 and over, 22 and 30%, respectively, no longer have any natural teeth, compared to 7% of adults aged 45–64.[3] A Healthy People 2010 oral health objective for adults aged 65–74 years is to reduce the percentage of individuals who have lost all their teeth from 26% (year 1998) to 20%.[16] Tooth loss has been found to be as high as 42% in Native American elders.[43] It is inversely related to income and education.[3] Reduction of periodontal disease, which may be exacerbated by diabetes, stress, osteopenia, osteoporosis, inflammation, and malnutrition, is another public health goal relating to oral health of older adults.

Healthy teeth are protected by an enamel layer, but bacterial action on the breakdown products of food slowly erodes tooth enamel. For about 15 minutes after we ingest food or drink, oral bacteria feast on the food breakdown products, especially those of sucrose. Foods such as caramels or raisins stick around longer, especially when they lodge between teeth. Frequent eating and drinking of sugary beverages provides a continuous substrate for bacteria. The acid in carbonated beverages adds to the corrosive potential of food.

Saliva, which lubricates the mouth and begins the digestive process (amylase in "spit" begins starch breakdown), also helps to keep tooth enamel clean. However, saliva seems to become thicker and more viscous with age. Lack of saliva slows nutrient absorption. Lack of sufficient and effective saliva, especially in the presence of gingivitis and periodontal disease, also makes the oral cavity more sensitive to temperature extremes and coarse textures, resulting in pain while eating.

Pain and discomfort with chewing foods can result in eating fewer fruits, vegetables, and whole grains. A loss of self-esteem associated with missing teeth and worry regarding how to pay for dental care can affect quality of life. Edentulous older people are less likely to visit the dentist for oral care (denture adjustment, periodontal disease management) than are those individuals having their natural teeth.[44] An oral health assessment includes review of soft tissues, teeth, and other factors that may affect dietary intake. One such factor is functional status, which may affect the ability to brush, floss, and obtain regular dental care, potentially resulting in periodontal disease which can precipitate tooth loss and lead into a vicious cycle.

Appetite and Thirst   Hunger and satiety cues are weaker in older than in younger adults. Roberts and colleagues examined the ability of 17 young men (mean age 24) and 18 older men (mean age 70) to adjust caloric intake after periods of overeating and of undereating.[45] All men were healthy and not taking medications. Food intake and weight were monitored for 10 days, and then men were overfed by roughly 1000 calories, or underfed by about 800 calories, for 21 days. Periods of over- or underfeeding were followed by 46 days of "ad-lib" intake during which all men were free to eat as much or as little as desired. After the periods of over- and underfeeding, young men adjusted their caloric intake to get back to their initial calorie intake level and weight. Older men kept overeating if they had been in the overfed group, and under-eating if they had been in the underfed group. The authors suggest that older adults may need to be more conscious of food intake levels because their appetite-regulating mechanism may be blunted. Whereas healthy young people adjust to cycles of more and less food intake, healthy older people's inability to adapt to these changes may lead to overweight or anorexia.

Elderly people don't seem to notice thirst as clearly as younger people do. A set of papers by Phillips and Rolls demonstrated that the thirst-regulating mechanism of older adults was less effective than that of younger individuals.[46-48] Researchers compared thirst response to fluid deprivation in a group of seven 20- to 31-year-old men, and seven 67- to 75-year-old men. Subjects lost 1.8% to 1.9% of body weight during 24 hours without fluids. Both groups were asked about feeling thirsty, mouth dryness, and how pleasant it would be to drink something. After fluid deprivation, the younger group reported being thirsty and having dry mouth. The older group, however, reported no change in thirst or mouth dryness. Both the older and younger groups thought that it would be pleasant to drink something after fluid deprivation. Blood measures showed that older men lost more blood volume than younger men did, indicated by their plasma concentrations of sodium. Researchers also measured how much water the men drank in the hour after their 24-hour period of fluid deprivation. Older men drank less water than their younger counterparts did. Younger people made up for fluid loss in 24 hours; older people did not drink enough to achieve their prior state of hydration. It appears that dehydration occurs more quickly after fluid deprivation and that rehydration is less effective in older men.

Nutritional Risk Factors

Identifying nutritional risk factors before chronic illness occurs is basic to health promotion. Risk-factor reduction forms the basis of dietary guidance. In adults of all ages, dietary risk factors that increase the likelihood of developing heart disease, cancer, and stroke are consuming a diet that is high in saturated fat; low intake of vegetables, fruits, and whole-grain products; and poor nutritional habits leading to obesity. Healthy People 2010, MyPyramid, and the Dietary Guidelines for Americans all emphasize adopting eating patterns that reduce the risk of the leading killer diseases. Overall dietary patterns only match these guidelines in part.

Another approach to nutritional risk-factor identification is to compare adequacy of current dietary intake to dietary intake recommendations such as the Recommended Dietary Allowances and the Dietary Reference Intakes. Compared to national intake data from earlier surveys, older adults are now consuming enough protein and folic acid, but suboptimal levels of vitamins A, D, E, K, choline, potassium, magnesium, and calcium (see Table 18.11 on page 470).[49]

A third approach is to examine a population and determine how environmental factors combine with dietary factors to predict nutritional health. This approach was used by a consortium of care providers, policy makers, and researchers to develop the Nutrition Screening Initiative (NSI).[50] The American Academy of Family Physicians, the American Dietetic Association (ADA), and the National Council on the Aging, Inc., sponsored the development of this health-promotion campaign. The NSI consortium used literature review, expert discussion, and a consensus process to generate a list of warning signs of poor nutritional health in older adults.[51] See Table 18.5 for a condensed version of the acronym DETERMINE. These warning signs were integrated into a screening tool, the NSI DETERMINE checklist (Illustration 18.2), and pilot-tested for use in community settings. Ten risk factors remained after testing a longer list.[52] Community agencies, educators, and care providers use the NSI tool to screen the aged public for risk of malnutrition.

The nutritional risk factors identified during the NSI process[50] are reflected in the list of risk factors identified in the ADA position on nutrition and aging.[7] Any of the following conditions potentially place older adults at nutritional risk:

- Hunger
- Poverty
- Inadequate food and nutrient intake
- Functional disability
- Social isolation
- Living alone
- Urban and rural demographic areas
- Depression
- Dementia
- Dependency
- Poor dentition and oral health; chewing and swallowing problems
- Presence of diet-related acute or chronic diseases or conditions

Table 18.5 DETERMINE: Warning signs of poor nutritional health

Disease. Any disease, illness, or chronic condition (i.e., confusion, feeling sad or depressed, acute infections) that causes changes in the way you eat, or makes it hard for you to eat, puts your nutritional health at risk.

Eating poorly. Eating too little, too much, or the same foods day after day, or not eating fruits, vegetables, and milk products daily, will cause poor nutritional health.

Tooth loss/mouth pain. It is hard to eat well with missing, loose, or rotten teeth, or dentures that do not fit well or cause mouth sores.

Economic hardship. Having less or choosing to spend less than $41.90 (female) to $46.80 (male) weekly for groceries makes it hard to get the foods needed to stay healthy. [These costs are calculated for individuals living in 4-person households; add 20% to adjust for living alone.]

Reduced social contact. Being with people has a positive effect on morale, well-being, and eating.

Multiple medicines. The more medicines you take, the greater the chance for side effects such as change in taste, increased or decreased appetite and thirst, constipation, weakness, drowsiness, diarrhea, nausea, and others. Vitamins or minerals taken in large doses can act like drugs and can cause harm.

Involuntary weight loss or gain. Losing or gaining a lot of weight when you are not trying to do so is a warning sign to discuss with your health care provider.

Needs assistance in self-care. Older people who have trouble walking, shopping, and buying and cooking food are at risk for malnutrition.

Elder years above 80. As age increases, risk of frailty and health problems also rises.

SOURCE: Warning signs adapted from Nutrition Screening Initiative, a project of the American Academy of Family Physicians, the American Dietetic Association, and the National Council on the Aging, Inc., and funded by a grant from Ross Products Division, Abbott Laboratories, Inc.; dollar amounts under economic hardship inserted by author using the August 2009 USDA low-cost food plan.

- Polypharmacy (use of multiple medications)
- Minority status
- Advanced age

Why are factors such as poverty and minority status included in a list of nutritional risk factors? Economic security contributes to food security, one of the Healthy People 2010 goals (see Table 18.6). Lack of food security in older adults is associated with inadequate food intakes (below 67% of the RDA), especially for energy, magnesium, calcium, zinc, and vitamins E, C, and B$_6$.[53] More recent data show that older African Americans are nearly twice as likely as Caucasian elders to live in low-income households and also more likely to be food-insecure.[54]

Although older people on average are less likely to live in poverty than are children, they are a heterogeneous population, and several groups are at high risk of poverty. Minority status is related to economic status. Black and Hispanic populations aged 65 and over are more likely to be poor or nearly poor (living at or below 199% of the poverty threshold). Furthermore, minority populations are more likely to be in fair or poor health, while non-Hispanic whites are most likely to be in excellent or good health. Health care costs may contribute to food insecurity; for some, that means choosing between drugs and food.

Poverty is one risk factor for malnutrition; polypharmacy is another. The Slone Survey at Boston University has tracked medication use in the United States and found that it increases with age; 4 of 5 older adults reported using prescription drugs in the previous week, compared to 1 of 5 young males and 2 of 5 young females. Increasing use of multiple drugs has occurred in all adults; the population using five or more medications jumped from 23 to 29% in six years.[55] Table 18.7 shows that the most commonly used medications of older adults include anti-inflammatory drugs, cholesterol-lowering agents, diuretics, and anti-hypertensives.

Taken individually, the risk factors identified in the DETERMINE acronym and in the ADA position statement on aging are not unique to older adults. But each is more likely to lead to nutritional problems in a frail, vulnerable population. For instance, functional disability can affect dietary intake at any age, but very old people are more likely to live alone and to have fewer resources to compensate for lost function. Consequently, diet quality may decline. Fewer 65-year-olds live alone than 75- and 85-year-olds, and more women than men live alone. Race and ethnicity affect living situations; Hispanic men and women are least likely to live alone, whereas white women over age 85 are most likely to live alone; 65% of older white women live alone, compared to 47% of African American and 35% of Hispanic women.

A common perception is that older adults living alone eat poorly. Although living alone is a nutritional risk factor, it's not clear whether the issue is eating alone or living alone. On average, meals eaten with other people last longer and supply more calories than do meals eaten alone.[56] The effect of living alone starts at a younger age for men and affects nutrition more extensively than for women.[57] Women aged 75 and older eat less protein and also less sodium than those living with others; men aged 65 and older eat less protein, beta-carotene, vitamin E, phosphorus, calcium, zinc, and fiber. Men and women living alone did not consume greater amounts of any of the 23 nutrients (alcohol was not included) examined than did men and women living with someone.

The purpose of a screening tool is primary prevention. When screening for nutritional risk, it should identify conditions that need further exploration, such as

Illustration 18.2 Determine your nutritional health checklist.

The Warning Signs of poor nutritional health are often overlooked. Use this checklist to find out if you or someone you know is at nutritional risk.

Read the statements below. Circle the number in the yes column for those that apply to you or someone you know. For each yes answer, score the number in the box. Total your nutritional score.

DETERMINE YOUR NUTRITIONAL HEALTH

	Yes
I have an illness or condition that made me change the kind and/or amount of food I eat.	2
I eat fewer than 2 meals per day.	3
I eat few fruits or vegetables, or milk products.	2
I have 3 or more drinks of beer, liquor, or wine almost every day.	2
I have tooth or mouth problems that make it hard for me to eat.	2
I don't always have enough money to buy the food I need.	4
I eat alone most of the time.	1
I take 3 or more different prescribed or over-the-counter drugs a day.	1
Without wanting to, I have lost or gained 10 pounds in the last 6 months.	2
I am not always physically able to shop, cook, and/or feed myself.	2
TOTAL	

Total your nutrition score. It it's . . .

0–2 Good! Recheck your nutritional score in 6 months.

3–5 You are at moderate nutritional risk. See what you can do to improve your eating habits and lifestyle. Your office on aging, senior nutrition program, senior citizens center, or health department can help. Recheck your nutritional score in 3 months.

6 + You are at high nutritional risk. Bring this checklist the next time you see your doctor, dietitian, or other qualified health care or social service professional. Talk with them about any problems you may have. Ask for help to improve your nutritional health.

Remember: warning signs suggest risk, but do not represent diagnosis of any condition.

These materials developed and distributed by the Nutrition Screening Initiative, a project of: AMERICAN ACADEMY OF FAMILY PHYSICIANS, THE AMERICAN DIETETIC ASSOCIATION, and NATIONAL COUNCIL ON AGING, INC.

SOURCE: The Nutrition Screening Iniative, a project of the American Academy of Family Physicians, The American Dietetic Association and the National Council on the Aging, Inc., and funded by a grant from Ross Products Division, Abbott Laboratories, Inc.

detecting significant weight loss. In secondary prevention, such as testing for blood cholesterol or bone density, early symptoms of disease are identified and treated to prevent exacerbation of the condition. The question for the NSI DETERMINE list is "Does this tool categorize older persons' nutritional status for further assessment and intervention?" The results are mixed. The cumulative score resulting from using the NSI tool weakly predicted mortality in an aging population of mostly white, educated adults.[58] The NSI checklist did not consistently identify all those individuals who had poor health or low nutrient

consumption.[52] The search for an easy-to-use, valid, reliable tool for identification of risk of malnutrition has led to the development of the short two-part Mini Nutritional Assessment (MNA) (see Illustration 18.3). The MNA combines six screening questions in part 1 with 12 assessment questions in part 2, and does not require blood work or other biochemical tests, unlike tools based on acute-care use.[59] It has a greater focus on illness than DETERMINE does and has been used in clinical, home-care, and community settings. Note that the MNA uses a BMI of 23 or greater as presenting the least risk of malnutrition in older adults. Evaluation of the NSI and the MNA by comparing results with measured dietary intake, anthropometrics, and blood biochemistries showed that each tool has some limited value.[60] Another tool that is growing in popularity is the brief Malnutrition Universal Screening Tool (MUST), developed for institutional and community use.[61] Implementing MUST is a collaborative effort to improve access to a full spectrum of nutritional care. Assessment of nutritional status is a five-step process that includes 1) height and weight measurement to calculate BMI, 2) scoring unplanned weight loss, 3) assigning an acute disease effect score (for example, no nutritional intake due to illness for more than 5 days rates the maximum risk score of 2), 4) summing scores to determine overall risk of malnutrition, and 5) using the score to plan management.

When height and weight cannot be measured, alternative calculations using the ulna, knee height, and mid-upper-arm circumference are suggested. The focus of scoring the MUST tool is to predict ability to combat disease; MUST uses a BMI of 20 or less to flag potential malnutrition. A BMI of 30 or more is scored as "obese."

Table 18.6 Healthy People 2010, food-security goals for adults aged 65 and older

		Current Goal	
	Target	Males	Females
Increase food security and in so doing, decrease hunger			
Households <130% of poverty, with elderly persons	94	85[a]	85[a]
Households >130% poverty, with elderly persons	94	98[a]	98[a]

SOURCE: *Healthy People 2010*
[a]Adequate data to distinguish levels for males and females is not available.

Table 18.7 Seven most commonly used medications by men and women aged 65 and older, with usage prevalence in percent

Men	Women
Aspirin (46%)	Aspirin (34%)
Lisinopril (16%)	Acetaminophen (19%)
Atorvastatin (16%)	Levothyroxine (17%)
Simvastatin (15%)	Hydrochlorothiazide (17%)
Metoprolol (15%)	Atorvastatin (14%)
Acetaminophen (13%)	Metoprolol (11%)
Furosemide (13%)	Ibuprofen (10%)

SOURCE: Patterns of medication use in the United States, 2006. A report from the Slone Survey. Compiled from Table 1: Thirty Most Commonly Used Prescription and Over-the-Counter Drugs Taken by Adults in 2006, According to Sex and Age. Available at www.bu.edu/slone/SloneSurvey/AnnualRpt/slonesurveyreport2006.pdf, accessed 8/4/2009.

Dietary Recommendations: Pyramids for Older Adults

Sometimes, in all the discussions about nutritional effects on health and disease status of older adults, we forget that old age is not a disease. No matter what the health status of a person is, however, recommendations for specific nutrients change with age, and food consumption patterns need to adjust accordingly.

MyPyramid is a ubiquitous source of dietary advice and has been adapted for many cultures and age groups. Pyramid adaptations for older adults include one by Tufts University researchers (see Illustration 18.4) and one by University of Florida Extension Services (Illustration 18.5). Together, they illustrate special issues concerning nutrient guidance for older adults, which include their greater need for nutrient density and adequate fluid, activities appro-

priate to functional ability, and adaptations of materials in the absence of internet access. The Tufts group pyramid relies almost exclusively on icons and was designed to be useful for print dissemination.[62] Florida's version comes as a two-page guide with text. Both pyramids depict older adults' declining need for calories with a narrower pyramid base. Food product illustrations emphasize whole grains and all forms of fruits and vegetables, not just fresh ones. Adequate fluid intake is emphasized by the row of 8 glasses in Illustration 18.4; in Illustration 18.5, a text paragraph suggests, "Drink water and other beverages that are low in added sugars." Potential need for supplemental calcium and vitamins D and B_{12} is flagged at the top of the Tufts pyramid, while the Florida pyramid adds a sentence about using fortified foods to meet vitamin D and B_{12} requirements. Neither pyramid specifically discusses other nutrients consumed in amounts smaller than the DRI recommendations (vitamins A, K, E, potassium, choline, and magnesium).

Both pyramids show the food groups comprising a healthy diet and offer vegetarian options. The Florida pyramid uses text to describe food groups and corresponding serving sizes for achieving adequate food intake. See Table 18.8 for food groupings matching calorie levels from 1600 to 2400. Each can be modified by choosing more or less nutrient-dense foods from each group. For instance, citrus and berries are among the most nutrient-dense fruits. A baked potato has more vitamins and minerals per bite than french fries do. Among grain-based products, whole-grain breads and cereals provide fiber as well as the usual nutrients found in grain. Consistently choosing nutrient-dense foods minimizes caloric intake. Fortification and enrichment, while enhancing nutrient intake, also complicate nutrient-density calculations. For instance, adequate folic-acid intake was problematic before 1998, when fortification of grain products became mandatory. Now, older adults consume more than enough folic acid and could easily exceed the 1000 mcg Tolerable Upper Intake Level if they take supplements.[63]

The "discretionary calories"   Often, these are snacks. A review of NHANES data showed that 84% of older adults snack and that snacks provide nearly one-fourth of daily energy intake.[64] But snacks contribute more than fat and sugar. In the diets of older adults, snacks make up:

- 14% of protein consumed
- 20% of fat consumed
- 26% of carbohydrate consumed
- 12% of alcohol consumed

Dietary guidance begins with assessment. Information about current food intake helps to answer questions such as "Are there unusual dietary patterns that might reflect some potential problem?" Change can be a good thing, but unexplained changes in diet and nutritional status need to be evaluated, especially weight loss.

Illustration 18.3 MNA nutritional screening and assessment.

Mini Nutritional Assessment
MNA®

Last name:		First name:		
Sex:	Age:	Weight, kg:	Height, cm:	Date:

Complete the screen by filling in the boxes with the appropriate numbers. Add the numbers for the screen. If score is 11 or less, continue with the assessment to gain a Malnutrition Indicator Score.

Screening

A   Has food intake declined over the past 3 months due to loss of appetite, digestive problems, chewing or swallowing difficulties?
0 = severe decrease in food intake
1 = moderate decrease in food intake
2 = no decrease in food intake ☐

B   Weight loss during the last 3 months
0 = weight loss greater than 3kg (6.6lbs)
1 = does not know
2 = weight loss between 1 and 3kg (2.2 and 6.6lbs)
3 = no weight loss ☐

C   Mobility
0 = bed or chair bound
1 = able to get out of bed / chair but does not go out
2 = goes out ☐

D   Has suffered psychological stress or acute disease in the past 3 months?
0 = yes   2 = no ☐

E   Neuropsychological problems
0 = severe dementia or depression
1 = mild dementia
2 = no psychological problems ☐

F   Body Mass Index (BMI) (weight in kg) / (height in m²)
0 = BMI less than 19
1 = BMI 19 to less than 21
2 = BMI 21 to less than 23
3 = BMI 23 or greater ☐

Screening score
(subtotal max. 14 points) ☐☐

12 points or greater: Normal – not at risk – no need to complete assessment

11 points or below: Possible malnutrition – continue assessment

Assessment

G   Lives independently (not in nursing home or hospital)
1 = yes   0 = no ☐

H   Takes more than 3 prescription drugs per day
0 = yes   1 = no ☐

I   Pressure sores or skin ulcers
0 = yes   1 = no ☐

J   How many full meals does the patient eat daily?
0 = 1 meal
1 = 2 meals
2 = 3 meals ☐

K   Selected consumption markers for protein intake
• At least one serving of dairy products (milk, cheese, yoghurt) per day   yes ☐   no ☐
• Two or more servings of legumes or eggs per week   yes ☐   no ☐
• Meat, fish or poultry every day   yes ☐   no ☐
0.0 = if 0 or 1 yes
0.5 = if 2 yes
1.0 = if 3 yes ☐ . ☐

L   Consumes two or more servings of fruit or vegetables per day?
0 = no   1 = yes ☐

M   How much fluid (water, juice, coffee, tea, milk...) is consumed per day?
0.0 = less than 3 cups
0.5 = 3 to 5 cups
1.0 = more than 5 cups ☐ . ☐

N   Mode of feeding
0 = unable to eat without assistance
1 = self-fed with some difficulty
2 = self-fed without any problem ☐

O   Self view of nutritional status
0 = views self as being malnourished
1 = is uncertain of nutritional state
2 = views self as having no nutritional problem ☐

P   In comparison with other people of the same age, how does the patient consider his / her health status?
0.0 = not as good
0.5 = does not know
1.0 = as good
2.0 = better ☐ . ☐

Q   Mid-arm circumference (MAC) in cm
0.0 = MAC less than 21
0.5 = MAC 21 to 22
1.0 = MAC 22 or greater ☐ . ☐

R   Calf circumference (CC) in cm
0 = CC less than 31
1 = CC 31 or greater ☐

Assessment (max. 16 points) ☐☐ . ☐

Screening score ☐☐ . ☐

Total Assessment (max. 30 points) ☐☐ . ☐

Malnutrition Indicator Score

17 to 23.5 points   ☐   at risk of malnutrition

Less than 17 points   ☐   malnourished

Ref.   Vellas B, Villars H, Abellan G, et al. *Overview of MNA® - Its History and Challenges.* J Nut Health Aging 2006; 10: 456-465.
Rubenstein LZ, Harker JO, Salva A, Guigoz Y, Vellas B. Screening for Undernutrition in Geriatric Practice: *Developing the Short-Form Mini Nutritional Assessment (MNA-SF).* J. Geront 2001; 56A: M366-377.
Guigoz Y. The Mini-Nutritional Assessment (MNA®) *Review of the Literature – What does it tell us?* J Nutr Health Aging 2006; 10: 466-487.
® Société des Produits Nestlé, S.A., Vevey, Switzerland, Trademark Owners
© Nestlé, 1994, Revision 2006. N67200 12/99 10M
For more information: www.mna-elderly.com

Illustration 18.4 Tufts University modified food pyramid for 70+ adults.

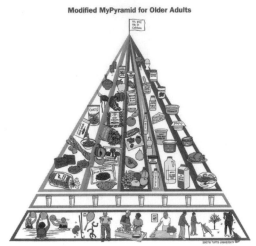

SOURCE: Reprinted by permission from Dr. Robert M. Russell, Tufts University. ©1999 Tufts University.

Nutrient Recommendations

Nutrient recommendations change as scientists learn more about the effect of foods on human function. Nutrient levels specific for population groups above age 51 were first established in 1997 as the Dietary Reference Intakes (DRIs). The complete tables, including Tolerable Upper Intake Levels (UL), are found on the inside covers of this text. This section first addresses the energy nutrient recommendations (see Tables 18.9–11), followed by fluid, vitamin, and mineral recommendations (Table 18.12).

Estimating Energy Needs

The main goal for energy calculations is to maintain a healthy body weight. NHANES data (Table 18.9) shows that on average, older adults eat fewer calories as they age. Women aged 70 and older report the lowest intake and are especially vulnerable to malnutrition, even though approxi-

Illustration 18.5 UF and ENAFS pyramid for nutrition and fitness in aging.

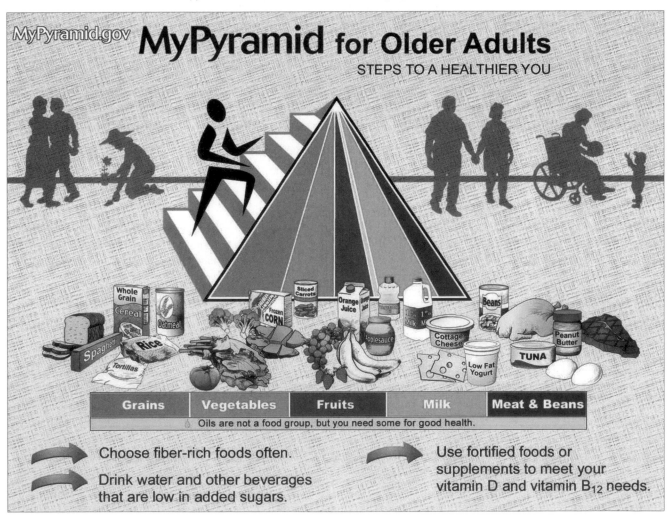

(continued)

Illustration 18.5 (Continued)

GRAINS Make half your grains whole	VEGETABLES Vary your veggies	FRUITS Focus on fruits	MILK Get your calcium-rich foods	MEAT & BEANS Go lean with protein
Eat at least 3 oz. of whole-grain cereals, breads, rice, crackers, or pasta every day. 1 oz. is about 1 slice of bread, 1 cup of cold breakfast cereal, or ½ cup of cooked cereal, rice, or pasta. Eat cereals fortified with vitamin B$_{12}$.	Eat more dark-green veggies, like broccoli, salad greens, and cooked greens. Eat more orange vegetables, such as carrots and sweet potatoes. Eat more dried beans and peas, like pinto, black, or kidney beans and lentils.	Eat a variety of fruit. Choose fresh, frozen, canned, or dried fruit. Eat fruit rather than drinking juice for most of your fruit choices.	Choose low-fat or fat-free milk, yogurt, and other milk products. If you don't or can't consume milk, choose lactose-free products or other calcium sources, such as fortified foods and beverages.	Choose low-fat or lean meats and poultry. Bake, broil, or grill. Vary your protein sources. Include eggs, beans, fish, and nuts/seeds.

For an 1,800-calorie diet, you need the amounts below from each food group. To find the amounts that are right for you, go to MyPyramid.gov.

Eat 6 oz. every day.	Eat 2½ cups every day.	Eat 1½ cups every day.	Eat 3 cups every day.	Eat 5 oz. every day.

Eat Right

- Choose foods rich in fiber to help keep you regular.
- Drink plenty of fluids to stay hydrated.
- Limit sweets to decrease empty calories.
- Get your oils from fish, nuts, and liquid oils such as canola, olive, corn, or soybean oils.
- Choose and prepare foods with less salt or sodium.
- Talk to your doctor or pharmacist about supplements you are taking.

Be Active

- Go for a walk.
- Play with your grandchildren and/or a pet.
- Work in your yard or garden.
- Take an exercise or dance class at a community center or gym.
- Share a fun activity with a friend or family member.
- Remember: all activity adds up! You don't have to do it all at once.

Enjoy Life: Spend time with caring people doing things you enjoy.

 UNIVERSITY *of* FLORIDA
IFAS Extension

MyPyramid for Older Adults was adapted from USDA's MyPyramid by nutrition faculty in the Department of Family, Youth and Community Sciences, IFAS, University of Florida, Gainesville, Florida 32611.
2007

 ENAFS
Elder Nutrition and Food Safety

SOURCE: University of Florida, Department of Family, Youth and Community Sciences, Institute of Food and Agricultural Sciences Extension. Reprinted by permission of Dr. Linda Bobroff and David L. Day.

Table 18.8 Using MyPyramid recommendations to show the daily amount of food needed to meet nutritional needs for older adults at various activity levels (see Chapter 1 for food measure equivalents)

	Female, age 72, sedentary	Female, age 60 and older, active	Male, age 65, moderately active
Calorie level	1600	2000	2400
Fruits	1.5 c	2 c	2 c
Vegetables	2 c	2.5 c	3.5 c
Grains	5 oz equivalent	6 oz-eq	8 oz-eq
Meat and beans	5 oz-eq	5.5 oz-eq	6.5 oz-eq
Milk	3 c	3 c	3 c
Oils	5 tsp	6 tsp	7 tsp
Discretionary calorie allowance	132	267	362

SOURCE: www.mypyramid.gov, accessed 10/29/2006

mately one in five is obese (BMI = 30 or more). Population averages are a rough guide for individual needs. The DRI formulas allow energy calculations to be adapted for physical activity levels of older individuals (see Table 18.10).

From early to late adulthood, a decrease of physical activity and basal metabolic rate leads to an estimated decrease of 7–10 calories in energy needs per year. Over a decade, daily energy needs would decline by 70–100 calories.

Table 18.9 Caloric intake comparison of younger and older adults, by gender, from *What We Eat In America*, NHANES, 2005–2006

	Actual Daily Calorie Intake		Recommended Calorie Intake	
Age	Males	Females	Males	Females
20–29	2821	1959	2900	2200
60–69	2202	1598	2300	1900
age 70+	1984	1495	2300	1900

SOURCE: *What We Eat in America*, NHANES [49]

Table 18.10 Dietary reference intake values for energy by active individuals with sample calculations

Life Stage Group A	EER (kcal/d) B (Representing Active PAL C)	
All adults	Male	Female
>18 y E	3067 D	2403 D (19 y)

SOURCE: Institute of Medicine. *Dietary Reference Intakes for Energy, Carbohydrate, Fiber, Fat, Fatty Acids, Cholesterol, Protein, and Amino Acids*, 2002.

A For healthy, moderately active Americans and Canadians·
B EER = estimated energy requirement
C PAL = Physical Activity Level Factor
D TEE = Total Energy Expenditure, which is the intake that meets the average energy expenditure of individuals at reference height, weight, and age
E Subtract 10 kcals/d for males and 7 kcals/d for females *for each* year above 19 y.

Male Example	Female Example
70 years old	85 years old
3067 kcals	2403 kcals
Subtract	Subtract
10 kcals/d	7 kcals/d
70 y – 19 y = 51 y	85 y – 19 y = 66 y
3067 – 510 = 2557 kcal	2403 – 462 = 1941 kcal

Energy expenditure is primarily determined from basal metabolic rate, which slows with age, diet-induced thermogenesis, and physical-activity energy needs. However, genetics, hormones, and body composition also influence metabolic rates, resulting in broad ranges of energy needs for older populations. Basal metabolic rates (BMR), using doubly labeled water measurements, have ranged from 1004 kcal/d to 2060 kcal/d in older individuals with normal body mass index (BMI).[63] Such a wide range of energy use reflects the heterogeneity of older adult populations and the need for individualized nutrition planning.

Traditionally, the Harris-Benedict equation has been used to estimate individual calorie needs. However,

validation was done with mostly young adults. Newer energy-estimation formulas are being developed to account for greater individual variation. The Mifflin–St. Jeor formula most closely predicts energy requirements of healthy adults in general (see Chapter 16), and validation did include individuals up to age 80, but it was not specifically developed for older adults.

Arcerio and colleagues developed formulas for use with older adults.[65,66] The equation used with older adult females includes a factor for hormonal status (e.g., menopause). Males completed the Minnesota Leisure Time Activity survey as well as a set of three chest skinfold measurements.[67] Using these formulas requires training; they are not in common use.

Energy calculations and weight status are topics that illustrate the dangers of generalizing from populations to individuals, or the reverse. It is difficult to meet vitamin and mineral needs at caloric levels below 1600. Older adults who are active enough to meet the physical-activity recommendations are likely to eat more than 1600 calories and thus must ensure that they consume sufficient calories to meet nutrient needs. Being physically active every day may be easier said than done.

Nutrient Recommendations for Older Adults: Energy Sources

Potentially problematic nutrients for older adults are presented in Tables 18.11 and 18.12. Several of the listed nutrients are consumed in amounts different from the recommended level. The following nutrients are potentially problematic for older adult populations.

Carbohydrate and Fiber Carbohydrate intake between 45% to 65% of calories (the AMDR) is generally not a problem.[49] Following the MyPyramid recommendations ensures that the carbohydrate quantity and quality is adequate and that fiber guidelines are met. For example, men who need 2300 calories/day would need to eat 288 grams of carbohydrate per day to meet 50% of calories from carbohydrate sources; women would need 188 grams. A listing of food that provides at least 50% carbohydrates in a 1500-calorie diet is presented in Table 18.13, with fiber levels.

Depending on caloric intake, a minimum of 21 and 30 grams of dietary fiber daily for females and males aged 51 and older (14 g/1000 cal) is recommended for adults.[63] Females eat 14 grams and males eat 17 grams of fiber daily, which is less than the recommended levels (see Table 18.10).

Table 18.14 shows how common portions of foods eaten by older adults can supply adequate fiber. Dietary fiber reduces the risk of coronary heart disease,[63] but older adults are more often concerned with the role of fiber for gastrointestinal health; this is discussed in Chapter 19.

Table 18.11 Daily energy nutrient and cholesterol intakes of older adults, NHANES 2005–2006,[a] compared to recommendations

Nutrient	Actual Intake[a] Males	Females	Recommended Intake Males/Females
Protein, g	77	57	0.8 grams/kg/day (RDA) 10–35% of calories (AMDR)
Carbohydrate, g	239	192	45–65% of calories (AMDR)
Total sugars, g	109	90	Added sugars, <25% of calories
Dietary fiber, g	17	14	30 gm/day 21gm/day (AI)
Total fat, g	77	56	20–35% of calories (AMDR)
Saturated fat, g	26	19	<10% of calories
Monounsaturated fat, g	28	20	Up to 20% calories
Polyunsaturated fat, g	16	12	Up to 10% calories
Cholesterol, mg	306	205	300 mg or less
Alcohol, g	8	3	Moderation (12–15 g = 1 drink)

[a]Actual intake amounts from *What We Eat in America,* NHANES, 2005–2006, Table 1. Available at www.ars.usda.gov/ba/bhnrc/frsg, accessed 8/04/09. Recommended intake data are based on Dietary Reference Intake (DRI), Institute of Medicine of the National Academies. *Dietary Reference Intakes: Energy, carbohydrate, fiber, fat, fatty acids, cholesterol, protein, and amino acids.* Washington, D.C.: National Academies Press, 2002. Acceptable Macronutrient Distribution Range (AMDR), Institute of Medicine.

Table 18.12 Selected micronutrient intakes of male and female adults aged 70 and older compared with recommendations

Nutrient, Unit of Measure	Daily Intake Males	Females	Dietary Reference Intakes Males/Females
Vitamin A, mcg RAE	814	624	900/700 = RDA
Vitamin D, mcg	5.6	4.5	15 = AI
Vitamin E, mg alpha tocopherol	7.1	5.7	15 = RDA
Vitamin K, mcg	100.3	84.6	120/90 = AI
Lycopene, mcg	4021	3408	
Lutein + zeaxanthin, mcg	1606	1420	
Thiamin, mg	1.69	1.29	1.2/1.1 = RDA
Riboflavin, mg	2.39	1.80	1.3/1.1 = RDA
Niacin, mg	24.0	18.0	16/14 = RDA
Vitamin B_6, mg	2.08	1.59	1.7/1.5 = RDA
Folate, mcg DFE	564	443	400 mcg = RDA
Choline, mg	340	241	550/425 = AI
Vitamin B_{12}, mcg	6.09	4.38	2.4 = RDA
Vitamin C, mg	97.4	81.8	90/75 = RDA
Sodium, mg	3142	2395	1200 = AI
Potassium, mg	2863	2223	4700 = AI
Calcium, mg	881	752	1200 = AI
Phosphorus, mg	1274	993	700 = RDA
Magnesium, mg	289	235	420/320 = RDA
Iron, mg	16.9	12.8	8 = RDA
Zinc, mg	12.0	8.8	11/8 = RDA
Copper, mg	1.4	1.0	0.9 = RDA
Selenium, mcg	102.9	78.3	55 = RDA

SOURCES: Actual consumption from U.S. Department of Agriculture, Agricultural Research Service, 2008. Nutrient Intakes from Food: Mean amounts consumed per individual, one day, 2005–2006. Available at www.ars.usda.gov/ba/bhnrc/fsrg, accessed 8/03/09. Recommended intake data are Dietary Reference Intake (DRI) Institute of Medicine of the National Academies, Otten, J. J., Hellwig, J. P., and Meyers, L. D., eds. Institute of Medicine of the National Academies. DRI Dietary Reference Intakes: The Essential Guide to Nutrient Requirements. Washington, D.C.: The National Academies Press, 2006.

Table 18.13 Using the MyPyramid groups to estimate adequate carbohydrate and fiber for daily intake

Basic Food Groups

Number of Servings/ Grams of Carbohydrate Per Serving	Carbohydrate, Grams	Approximate Total Fiber Content, Grams
6 servings of grain at 15 g each	90	12
2 servings of fruit at 15 g each	30	6
3 servings of milk at 12 g each	36	0
3 servings vegetables at 5 g each	15	9
Total from basic groups	**171**	**27**

Other Carbohydrate-Containing Foods

1 Tbsp sugar for coffee or tea	12	0
2 Fig Newton cookies	20	1
Total, including "other" group	**203**	**28**

NOTE: Mixed dishes such as soups, sandwiches, and salads count as partial servings from their contributing food groups.

Table 18.14 An example of fiber-containing foods that might comprise one day's intake

Food Item	Grams of Fiber
Oatmeal (1/2 cup) with wheat germ (1/4 cup)	8
Banana	2
Peanut butter (2 Tbsp)/whole-wheat bread (2 slices)	6
Orange	3
Baked potato with skin	4
Green beans (1/2 cup)	3
Bran muffin (1 med)	2
Pear	4
Total dietary fiber	**32**

NOTE: Meat, poultry, fish, eggs, milk, sugar, and oils do not contain dietary fiber.

A slow increase of dietary fiber allows the intestinal system to adapt to the additional bacterial substrate.

Protein   Most (94–95%) older adults in North America eat sufficient or even excess amounts of protein.[68] Older adults who are living alone, living in poverty, or have functional limitations are vulnerable to inadequate protein intake; this contributes to muscle wasting (sarcopenia),

weak bones, a weakened immune status, and delayed wound healing. Current protein guidelines for adults do not differentiate among adult age groups, and offer a range of intakes that will meet metabolic needs:

The AMDR is 10–35% of total calories.
The RDA is 0.8 gm per kg body weight/day.

It is not clear what constitutes an optimal amount of dietary protein for older adults. Campbell reported that age alone does not alter protein requirements.[69] Rand's meta-analysis of nitrogen balance studies[70] concluded that daily protein intake should be 0.83 gram of good-quality protein. A more recent review of protein needs of older adults led Morais and associates to propose intakes of 1.0 to 1.3 grams protein per kg per day.[71]

Nitrogen-balance studies used to determine current recommendations of 0.8 gram per kilogram (or 0.36 gram per pound) were done primarily with young adults, who have proportionately more muscle mass than older adults, male or female, and are more efficient at maintaining nitrogen balance. Nitrogen balance is also easier to achieve when an individual is doing resistance training, when protein is ingested with adequate calories, and when the protein eaten is of high quality (such as meat, milk, and eggs; see Table 18.15). People eat mixed diets, so individual protein scores tell only part of the story. For example, using the human protein digestibility scale, the average American and Chinese mixed diets score 96, the rural Mexican diet scores 80, and an Indian rice-and-bean diet scores 78. However, for older adults who are

Table 18.15 Protein sources and protein quality measures: examples of protein scores used by the Institute of Nutrition of Central America and Panama (INCAP)

Protein Source	Chemical Quality Score*	True Digestibility in Humans
Meat, eggs, milk	100	95
Beans	80	78
Soy protein isolate	97	94
Rice	73	88
Oats, oatmeal	63	86
Lentils	60	Not available
Corn	50	85
Wheat	44	Refined, 96; whole wheat, 86

SOURCE: Data from Torun, B., Menchu, M. T., and Elias, L. G. *Recomendaciones dieteticas diarias del INCAP.* Guatemala: INCAP, 1994; and Shils, M. E., Olson, J. A., M. Shike, M., et al. *Modern Nutrition in Health and Disease.* Philadelphia: Lippincott, Williams & Wilkin, 1999.

*Amino acid score relative to amino acid score of egg, the reference protein.

following modified diets, who are cutting back on meat because they can't afford it or lack energy to prepare it, or who may eat too few calories, protein quality can make the difference between a good and a poor diet.

Despite adequate protein-intake levels of older adults in national surveys, individual elders may be at risk for protein malnutrition. The following questions help to determine protein adequacy for older individuals:

● Based on total energy requirements, how much protein will meet the individual need?

● Are enough calories eaten so that protein does not have to be used for energy?

● If marginal amounts of protein are eaten, is the protein of high quality?

● Are there additional needs—wound healing, tissue repair, surgery, fracture, infection?

● Is the individual exercising? It is harder to achieve nitrogen balance while sedentary.

Consuming a low-calorie diet, as many older adults do, leads to a proportionately greater need for protein. Instead of consuming 10% of total calories in protein, needs might be better met near the 35% end of the AMDR.

Decreasing muscle mass does not lead to lower protein requirements. In fact, someone who is losing muscle due to inactivity requires higher protein intake. Consuming protein at levels slightly above the RDA (or at the higher end of the AMDR), and distributed throughout the day, stimulates muscle synthesis and minimizes the risk of sarcopenia in aging adults.[72]

Fats and Cholesterol The role of dietary fat does not seem to change with age; high saturated fat and *trans*-fatty acid intake continues to be a risk factor for chronic disease. Minimizing the amount of saturated fat in the diet and keeping total fat between 20 and 35% of calories is a reasonable goal for older adults to maintain a beneficial blood cholesterol ratio. Cholesterol intake for older adults is near recommended levels (see Table 18.11). Eggs, which have high cholesterol content, are a nutrient-dense, convenient, and safe food for most people (those without lipid disorders, e.g., high triglyceride and high serum cholesterol). Unless their blood cholesterol levels put them at risk for heart disease, older adults can enjoy lean high-cholesterol foods such as shrimp and eggs for their high-quality nutrients without added risk of CVD. Even individuals who are hyper-responders to dietary cholesterol have been shown, albeit in a small study, to maintain their LDL:HDL ratio after high dietary cholesterol intake.[73]

Recommendations for Fluid

The proportion of water in total body weight decreases with age, resulting in a smaller water reservoir and leaving a smaller safety margin for maintaining hydration.

The DRIs for water are constant after age 19 (see inside cover pages). Drinking 6 or more glasses of fluid per day prevents dehydration (and subsequent confusion, weakness, and altered drug metabolism) in individuals whose thirst mechanism has grown less sensitive. Dehydration is discussed further in Chapter 19.

To individualize fluid recommendations, provide 1 milliliter (mL) of fluid per calorie eaten, with a minimum of 1500 mL. For a 2000-calorie diet, that would be 2000 mL or 2 liters of fluid, roughly 8 cups. Foods such as stews, puddings, fruits, and vegetables contribute significant amounts of fluid to the diet but are not counted as part of the fluid allowance in healthy individuals. The Tufts older adults' food guide pyramid (see Illustration 18.4) shows 8 glasses of water, which is adequate for a 2000-calorie diet. Individuals who need additional calories can use milk, juice, shakes, and soups as nutrient-dense fluids.

Age-Associated Changes: Nutrients of Concern

The nutritional health of older adults depends on modifying dietary habits that address age-associated changes in absorption and metabolism. The nutrients discussed in the following sections are of special concern for older adults because of age-associated metabolic changes or low dietary intake.

Vitamin A Diet surveys before the year 2000 reported that older adults were more likely to overdose on vitamin A than to be deficient in the nutrient, but the latest NHANES survey[49] showed dietary intake below the RDAs of 900 mcg and 700 mcg REA (retinol activity equivalents) for males and females, respectively (see Table 18.12). Plasma levels and liver stores of vitamin A increase with age. This may be due to increased absorption but is more likely due to decreased clearance of vitamin A metabolites (retinyl esters) from the blood. Kidney disease further elevates serum vitamin A levels because retinol-binding protein, another vitamin A metabolite, can no longer be cleared from blood. Thus older adults are more vulnerable to vitamin A toxicity and possible liver damage than younger individuals are. The UL for vitamin A is 3000 mcg (3 mg) for adults aged 19 and older, which could be reached if fish liver oils are taken in addition to daily vitamin supplements.

Vitamin A's plant precursor, beta-carotene, will not damage the liver, although supplements used as antioxidants to prevent cardiovascular disease have been linked to higher all-cause mortality.[74] Excess dietary beta-carotene, because it is water-soluble, may give old skin a yellow-orange tint, but it will not lead to hair loss, dry skin, nausea, irritability, blurred vision, or weakness as excess vitamin A would.

Vitamin D, Calciferol Age-associated metabolic changes affect vitamin D status, independent of dietary intake, primarily due to a four-fold decrease in the ability of aged skin to synthesize vitamin D.[75,76] In addition, older adults use more medications. Some of them, such as barbiturates, cholestyramine, phenytoin (Dilantin), and laxatives, interfere with vitamin D metabolism. Declining photochemical production ability may also be compounded by limited sunlight exposure due to wearing more clothes to stay comfortable, institutionalization or being homebound, and sunscreen use. Furthermore, in northern regions (above 42° north, the latitude of Boston and Chicago) between November and February, ultraviolet (UV) light is not powerful enough to synthesize vitamin D in exposed skin.[77] The sun's rays are even weaker in Edmonton, Canada (at 52° N latitude), so that previtamin D_3 is not synthesized between mid-October and mid-April. Dietary recommendations for vitamin D are higher for older adults than for any other population group (see inside cover), so individuals who live in these northern latitudes are at especially high risk for vitamin D deficiency. How far south does one have to go for the winter sun to convert vitamin D precursors? Tests in Los Angeles, at 34° N, showed vitamin D production in skin even in January. The sun's rays in Puerto Rico (latitude of 18° N) were even more effective.[77] Although humans store vitamin D, reaching and maintaining adequate serum 25 hydroxy vitamin D levels of 30 ng/mL (see Chapter 1) is likely to require supplements for people living at northern latitudes. Holick and Chen's review found that 800–1000 IU vitamin D is needed to maintain adequate vitamin D serum levels, and this was especially so for nursing home residents.[76]

NHANES surveys of vitamin D intake[49] report dietary shortfalls (see Table 18.12), which are reflected in lower serum vitamin D levels of older adults, especially older women. Levels of 30 ng/mL (75 nmol/L) or higher are needed for improving muscle strength, for dental health (including reduced periodontal disease), and to reduce risk of colorectal cancer.[78] A vitamin D level of 80 nmol/L maximizes intestinal calcium absorption.[76] Vitamin D's role in the maintenance of blood calcium levels and bone health is addressed in the section on osteoporosis in Chapter 19.

The Tolerable Upper Intake Level (UL) for vitamin D has been set at 50 mcg (2000 IU), although intoxication may not result until reaching higher intakes. Symptoms of toxicity are hypercalcemia (high blood calcium levels), anorexia, nausea, vomiting, general disorientation, muscular weakness, joint pains, bone demineralization, and calcification (calcium deposits) of soft tissues. Toxicity is rare from food sources. Cod liver and other fish oils contain medicinal levels of vitamin D (about 21 mcg, or 840 IU, per teaspoon).

Vitamin E Also known as tocopherol, vitamin E is a potent antioxidant. It is a problematic nutrient because dietary intake is well below the recommended 15 mg or 15 IU alpha-tocopherol equivalents (TE; see Table 18.12). Vitamin E plays a special role in the health of older adults due to its antioxidant functions, such as hindering development of cataracts.[79] Vitamin E is associated with enhanced immune function[80] and cognitive status, although not with reduced cardiovascular disease risk.[74,81] The UL is 1000 mg (or IU, seen on supplement labels) alpha-TE. Even though vitamin E is fat-soluble and is stored in the body, intake seems safe in doses up to 400 IU.[82] At higher doses, vitamin E may increase all-cause mortality and is linked to longer blood-clotting times and increased bleeding. Aspirin, anti-coagulants, and fish-oil supplements also increase blood-clotting time and are incompatible with high vitamin E intake.

Vitamin K Interest in vitamin K is increasing because, in addition to its blood coagulation role, vitamin K-dependent proteins have been found in bone, vascular, and central nervous system tissues.[83] Supplementation has been linked to reduced bone fractures.[84] On average, vitamin K intake is below the AI (see Table 18.12), although it is difficult to diagnose deficiency without blood tests; bacteria in the large intestine synthesize vitamin K. However, for older populations, the concern is to maintain consistent intake levels that do not interfere with warfarin (a vitamin K antagonist) therapy.[85] Additional vitamin K information is provided in Chapter 19, including a table showing vitamin K content of foods.

Vitamin B_{12} Vitamin B_{12} blood levels decrease with age even in healthy adults. Population intakes of vitamin B_{12} are higher than the RDA of 2.4 mcg (see Table 18.12), but many older adults are unable to use B_{12} efficiently. An estimated 30% of older adults suffer from atrophic gastritis and decreased absorption of B_{12}. In atrophic gastritis, a bacterial overgrowth of the stomach leads to inflammation and decreased secretion of hydrochloric acid and pepsin and subsequent inability to split vitamin B_{12} from its food protein carrier (see Chapter 19). It takes years to develop a B_{12} deficiency; but once developed, the neurological symptoms are irreversible. Symptoms include deterioration of mental function, change in personality, and loss of physical coordination.

"Food first" is usually sound advice regarding nutritional needs, but B_{12} is one of the two vitamins better absorbed in synthetic or purified form. Folic acid is the other one. Synthetic, not protein-bound, vitamin B_{12} is found in fortified foods such as cereals and soy products. Protein-bound B_{12} is found in all animal products, although poultry is a surprisingly poor source (0.25 mcg of B_{12} in 3 oz of chicken compared to 2.0 mcg in 3 oz ground beef).

Folate, Folic Acid Fortification of grain products with folic acid has led to intakes well above the RDA (see Table 18.12). Absorption of folate, like vitamin B_{12}, may be impaired by atrophic gastritis. Moreover, alcoholism is associated with folate deficiency and subsequent pernicious anemia. Folic-acid deficiency may mask vitamin B_{12}

deficiency, which is more common in older adults than folic-acid deficiency. For persons with low serum folate levels, dietary increases of folic acid (100 to 400 mcg) can lower serum homocysteine levels and subsequent risk of heart disease.[86] Medications commonly used by older adults—such as antacids, diuretics, phenytoin (Dilantin), sulfonamides, and anti-inflammatory drugs—affect folate metabolism.

Iron   Women's iron needs decrease after menopause, and older men and women eat more iron than the RDA of 8 mg (see Table 18.12). Like vitamin A, iron is stored more readily in the old than in the young. Excess iron contributes to oxidative stress, increasing the need for antioxidants to deal with oxidant overload. Fortunately, vitamin C intake increases iron absorption and also serves as an antioxidant.

Older adults are a heterogeneous population, and not every older person has adequate iron stores. Reasons for inadequate iron status include blood loss from disease or medication (e.g., aspirin), poor absorption due to antacid interference or decreased stomach acid secretion, and low caloric intake.

Calcium   A 1994 consensus conference sponsored by the National Institutes of Health[87] recommended calcium intake for women depending on estrogen status; namely, 1500 mg for women aged 50 to 64 not taking supplementary estrogen and 1000 mg calcium per day for postmenopausal women taking estrogen. It was recommended that all men and women consume 1500 mg calcium per day after age 65. This recommendation of 1500 mg was lowered in 1997, when the National Academy of Sciences set the AI for adults aged 51 and older at 1200 mg daily, independent of gender or hormone status. Recommended levels are met by a small portion of the population; on average, calcium intake of older men and women ranges from 750 to 900 mg per day.

Low calcium intake has been linked to colon cancer, overweight, and hypertension. Study results are strongest for the protective effects of calcium intake in the development of hypertension.[88] For example, Appel and the DASH Collaborative Research Group found that the subgroup of hypertensive individuals with higher calcium intake also had the greatest decrease in blood pressure. On average, participants who adhered to the Dietary Approaches to Stop Hypertension (DASH) diet reduced their blood pressure. The diet includes two or more servings of low-fat dairy products (1265 mg calcium from food in the experimental group); 10 servings of fruits and vegetables; and limited fat, saturated fat, and cholesterol.

The National Academy of Sciences has set the UL for calcium at 2500 mg per day. An explosion of calcium-fortified foods and supplements presents the possibility of adverse effects of excess calcium (reported at 4000 mg per day). These include high blood calcium levels, kidney damage, and calcium deposits in soft tissues and outside the bone matrix, such as bone spurs on the spine. High calcium intake may interfere with zinc, iron, and magnesium absorption, and it may result in elevated urinary excretion of calcium, leading to new kidney stones in individuals with a history of kidney stones.

Magnesium   Adequate magnesium intake (see Table 18.12) is needed for bone and tooth formation, nerve activity, glucose utilization, and synthesis of fat and proteins. An indicator of the wide-ranging functions of magnesium is that it plays a part in over 300 enzyme systems.[89] Age does not seem to affect magnesium metabolism, and the RDA is constant at 420 mg for males and 320 mg for females after age 31. The UL is 350 mg from non-food (supplementary) sources. Older adult intake is below the RDA (see Table 18.2). Magnesium deficiency can result not only from low intake but also from malabsorption due to gastrointestinal disorders, chronic alcoholism, and diabetes. Signs of deficiency include personality changes (irritability, aggressiveness), vertigo, muscle spasms, weakness, and seizures.

Drugs used by older adults, such as magnesium hydroxide or citrate laxatives, may lead to magnesium overdose. Signs of magnesium toxicity are diarrhea, dehydration, and impaired nerve activity. Food sources including milk, yeast breads, coffee, ready-to-eat cereals, beef, and potatoes do not result in toxicity.

Nutrient Supplements: When, Why, Who, What, and How Much?

When to Consider Supplements   Do older adults benefit from taking nutrient supplements? It depends on the nutrient.[90] Recovery from illness and trauma is definitely aided by supplemental formulas, including vitamins, minerals, and energy nutrients such as protein and fatty acids. This section deals with multivitamin and mineral supplements. These can decrease infection,[91,92] but the evidence is weak and conflicting.[93] Here's what an NIH panel concluded after reviewing the role of vitamins and minerals in the prevention of chronic disease:[94]

> "Multivitamin/mineral supplement use may prevent cancer in individuals with poor or suboptimal nutritional status. The heterogeneity in the study populations limits generalization to United States population. Multivitamin/mineral supplements conferred no benefit in preventing cardiovascular disease or cataract, and may prevent advanced age-related macular degeneration only in high-risk individuals. The overall quality and quantity of the literature on the safety of multivitamin/mineral supplements is limited."

Population surveys show that diets of many older persons fall short of meeting recommended nutrient levels (see Table 18.12), and appropriately used supplements fill nutritional gaps.[95] Writing in the Clinician's Corner of the *Journal of the American Medical Association,* Fletcher and Fairfield recommend that all adults take one multivitamin daily.[96]

Discussions about supplements are based on the assumption that whole foods are the ideal source of nutrients, and supplements boost marginal diets. Sometimes it turns out that, as is the case with vitamins A, E, and beta-carotene, the pill form of a nutrient is harmful while the food form promotes health.[90] Furthermore, the interactions among nutrients and the composition of plants and animals making up our food supply are much too complex to replicate in supplements. However, vitamin B_{12} and folic acid are two nutrients better absorbed in a synthetic form than in their protein-bound food form—but this becomes important only when normal metabolic processes fail. Despite the AMA's recommendation[96] of a daily multivitamin/mineral supplement, it is possible for older adults to live well without dietary supplements.

Some age-associated circumstances make an individual vulnerable to malnutrition and more likely to benefit from dietary supplements. Such nutritional risk factors include:

- Lack of appetite resulting from illness, loss of taste or smell, or depression
- Diseases or bacterial overgrowths in the gastrointestinal tract that prevent absorption
- Poor diet due to food insecurity, loss of function, dieting, or disinterest in food
- Avoidance of specific food groups such as meats, milk, or vegetables
- Contact with substances that affect absorption or metabolism: smoke, alcohol, drugs

Who Takes Supplements and Why?   The individuals most likely to take supplements are non-Hispanic white females, females in general, and individuals with more education and with higher incomes.[94] Older adults are motivated by wellness and want to take responsibility for their own health. They use vitamin and mineral supplements to make them feel better, to have more energy, to improve health, and to prevent or treat disease.[94]

What to Take?   The question "What should I take?" may not have a simple answer. An analysis of NHANES data reported in 2000 showed that a multivitamin was the most commonly used vitamin/mineral preparation, and around 30% of elderly men and women use multivitamins.[97] Vitamin E was the second highest vitamin/mineral supplement taken by men (i.e., 14%), whereas calcium was next for women (23%). Finally, the third most often used supplement was vitamin C, with 12% and 19% in men and women, respectively. Most of the nearly 2500

products reported in NHANES III were vitamin–mineral combinations or vitamin–single nutrient combinations.[97] The most common supplement ingredients in this survey were vitamin C, vitamin B_{12}, vitamin B_6, niacin, thiamin, riboflavin, vitamin E, beta-carotene, cholecalciferol or vitamin D, and folic acid. Most of these are consumed in adequate amounts. Research does not support vitamins A, C, E, and antioxidant supplement use for the prevention of CVD or cancer.[98]

Considerations that guide supplement choices are especially important for older adults, who use more medications than younger adults, and these address the following five questions:

1. Does the supplement contain a balance of vitamins and minerals?
2. When all supplements and fortified foods are combined, is the dose still safe?
3. Does the supplement supply the missing nutrients?
4. Does the supplement carry a *USP* (U.S. Pharmacopeia, a mark that indicates that the manufacturer followed recognized standards when making the product) or *NSF* (NSF.org, NSF International tests consumer products, including dietary supplements) code to assure potency and purity?
5. Is the supplement safe? ("Natural" does not mean safe.)

How Much to Take?   In general, multivitamin/mineral supplements should be used in physiologic rather than high-potency doses to maintain nutritional balance. Physiologic dose formulas unique to older adults are available with little or no iron and additional vitamins B_{12} and D.

Generously fortified breakfast cereals, "power" bars, and fortified beverages count as vitamin or mineral supplements. On average, dietary intakes of older adults lack sufficient choline, calcium, magnesium, potassium, and vitamins A, D, E, K, and B_{12}, but are adequate in the rest. Additional vitamins and minerals are superfluous and possibly unsafe (e.g., beta-carotene supplements). Balluz and associates[97] report that adults also use more than 300 non-vitamin and non-mineral products, including some that have toxic effects at normal doses (e.g., blue cohosh, chaparral). Because older adults eat fewer calories and are less resilient physically, they also have a smaller tolerance for mistakes in supplement use.

Table 18.16 is a summary of some of the vitamins, minerals, and other dietary supplements of special interest to older adults.

USP (United States Pharmacopeia)   A nongovernmental, nonprofit organization (since 1820); establishes and maintains standards of identity, strength, quality, purity, processing, and labeling for health care products.

NSF   International, a nongovernmental, nonprofit that also tests dietary supplements

Table 18.16 Dietary supplements potentially used by older adults for selected conditions

Condition or Health Status	Supplement
Poor appetite or dieting, leading to intake below 1200–1600 calories	Multivitamin/mineral
Weight loss, chronic underweight	Add high-calorie/protein foods/fats as oils
Vegetarian or vegan	Vitamins B_{12}, D, calcium, zinc, iron
Arthritis*	Antioxidants potentially useful (vitamins C, E), lactobacillus, fiber
Age-related macular degeneration; eye health and cataracts	AREDS formulation (high levels of C, E, A, Zn, Cu) for AMD; Vitamins E, C, B_2, B_6, selenium to reduce cataract risk
Constipation	Fiber (cellulose, bran, psyllium) Fluid to accompany fiber
Diarrhea	Fluid, multivitamin/mineral
Energy boosters	Evaluate total nutrient intake, adequate calories, iron if blood levels are low
Immune status enhancement	Multivitamin/mineral (DRI dosage)
Memory aids, dementia	Ginkgo leaf extracts as part of comprehensive therapy; (no effect on cognitive decline from B_6 and folic acid supplements)[94,99]
Osteoporosis	Vitamins D, K, calcium, fluoride, magnesium (avoid excess K if on blood thinners)
Sleep aids	Milk and carbohydrate at bedtime; valerian, lavender; avoid guarana, caffeine, alcohol
Stress reduction	Eat well and play in the sunshine (not available in a pill or tonic)

SOURCES: National Institutes of Health, State-of-the-Science conference statement. Multivitamin/mineral supplements and chronic disease prevention. *Ann Intern Med* 2006; 145, Schulz, V. et al. *Rational phytotherapy: A reference guide for physicians and pharmacists*, 5th ed. Berlin: Springer Verlag, 2004, and *Hanninen, K., Rauma, A. L., Nenonen, M., et al. Antioxidants in vegan diet and rheumatic disorders. *Toxicology* 2000; 155:45–53.

Dietary Supplements, Functional Foods, Nutraceuticals, and Older Adults

"Eat leeks in March and wild garlic in May / And all the year after, physicians may play."

Welsh rhyme

The growing availability of functional foods can be a boon for nutrient-deficient older adults. They may benefit from the convenience of nutrient-dense foods such as calcium-fortified breads and orange juice, fortified high-fiber cereals, yogurt–juice beverages, yogurt with live cultures, soy milks, breakfast powders, and various fortified chews, bars, and drinks. However, regular users of functional foods, no matter what age, need to add up all intake in order to avoid potentially toxic nutrient levels (see Table 18.17 for recommended and potentially toxic intake levels). Advice that "you can't overdose on nutrients from foods" does not apply to fortified foods. High nutrient doses act like drugs, and should be treated as such. Just like drugs, nutrient supplements interact with each other (and with medications); side effects may outweigh the desired benefits.

Several other "ingestibles" besides vitamins and minerals affect older adults' health status. Examples include quasi-functional foods, herbs, stimulants, and other non-vitamin or mineral food components used by older adults to promote health and ward off chronic disease. *Herbal therapy* is defined as the use of plants for medicinal purposes rather than for food consumption, and it comes in many forms: capsules, pills, infusions, tea; tinctures or extracts; and oils and salves. Different parts of the plant are used to produce supplements.[99] Herbal therapies used alongside traditional medical prescriptions are called complementary, and those used alone are called alternative.

Here is a guide for botanicals and nutraceuticals that may appeal to older adults:

1. *Herbs and spices.* Rosemary, sage, thyme, cinnamon, clove, and ginger enhance the taste of food and protect against oxidative stress. Doses in research studies tend to be larger than amounts used to flavor foods.

2. *Caffeine and guarana* (Paullinia cupana). Guarana is a popular energy booster in beverages. Guarana seeds contain more caffeine than coffee beans; older adults may be more sensitive to the stimulating effects of caffeine and other methylxanthines.

3. *Black, white, and green teas* (Camellia sinensis). These teas contribute fluid *and* phytochemicals, especially the antioxidant catechins and flavonols.

4. *Garlic.* Garlic contains allicin, which has antibacterial activity, and ajoene, an anticoagulant. Garlic in various forms is heavily promoted

Table 18.17 Vitamin and mineral levels for nutrition labeling (Daily Reference Values or DV) compared with Dietary Reference Intakes (DRI) and Tolerable Upper Intake Levels (UL) for older adults

Mandatory Vitamin and Mineral Components of the Nutrition Label

Nutrient	Daily Values (1993, 1995) All Adults	Dietary Reference Intake (2001) Adults Over 70		Upper Limit (UL)
		Males	Females	
Vitamin A, IU or RE (5 IU = 1 RE = 1 RAE, Retinol Activity Equivalent)	5000 IU	900 RE	700 RE	3000 RE
Vitamin C, mg	60	90	75	2000
Calcium, mg	1000	1200	1200	2500
Iron, mg	18	8	8	45

Voluntary Vitamin and Mineral Components of the Nutrition Label

Nutrient	Daily Values (1993, 1995) All Adults	Males	Females	Upper Limit (UL)
Vitamin D, mcg (1 mcg = 40 IU)	400 IU	15 mcg	15 mcg	50 mcg
Vitamin E, mg (1 mg = 1 TE = 1 IU)	30 IU	15 mg	15 mg	1000 mg
Vitamin K, mcg	80	120	90	–
Thiamin, mg	1.5	1.2	1.1	–
Riboflavin, mg	1.7	1.3	1.1	–
Niacin, mg	20	16	14	35
Vitamin B_6, mg	2.0	1.7	1.5	100
Folic acid, mcg	400	400	400	1000[a]
Vitamin B_{12}, mcg	6.0	2.4	2.4	–
Biotin, mcg	300	30	30	–
Pantothenic acid, mg	10	5.0	5.0	–
Phosphorus, mg	1000	700	700	3000
Iodine, mcg	150	150	150	1100
Magnesium, mg	400	420	320	350[b]
Zinc, mg	15	11	8	40
Selenium, mg	70	55	55	400
Copper, mcg	2000	900	900	10,000
Manganese, mg	2.0	2.3	1.8	11
Chromium, mcg	120	30	20	–
Molybdenum, mcg	75	45	45	2000
Chloride, mg	3400	2300	2300	–
Potassium, mg	3500	4700	4700	–
Choline, mg	–	550	425	3500
Fluoride, mg	–	4.0	3.0	10

SOURCE: Pennington, J. A. and Hubbard, V. S. Derivation of Daily Values Used for Nutrition Labeling. *J Am Diet Assoc* 97 1999; 1407–12. Trumbo, P. et al. Dietary Reference Intakes: Vitamin A, Vitamin K, Arsenic, Boron, Chromium, Copper, Iodine, Iron, Manganese, Molybdenum, Nickel, Silicon, Vanadium, and Zinc. *J Am Diet Assoc* 2001;101:294–301; 2004 DRIs for Macronutrients.

[a]synthetic
[b]nonfood sources

as a nutraceutical and "super food," but eat it for enjoyment, because research does not support claims for cancer prevention, cholesterol reduction, or other health benefits.

5. *Herbal treatments.* Herbal supplements are widely used and often effective, but herbs used as medicinal treatment can also be dangerous. Herbs should help, not hinder, health! Rule #1 in using herbs is also the first rule of medicine: "Do no harm."

6. *Phytochemicals.* These plant-based compounds (such as resveratrol, flavonoids, carotenoids, indoles, isoflavones, lignans, and salicylates) are

of special interest in aging because they have been linked to reduction of chronic disease. Eating fruits, vegetables, and whole grains automatically increases phytochemical intake.

7. *Pre- and probiotics.* Prebiotics are nondigestible food ingredients that feed health-promoting colon bacteria, and probiotics are live, health-promoting bacterial cultures such as lactobacillus acidophilus and bifidobacterium in yogurt. After a course of antibiotic treatment, cultured foods can help to reestablish intestinal bacterial life.

8. *Plant stanols.* These plant products are similar in structure to cholesterol and compete with cholesterol in the small intestine. They are found in corn, wheat, oats, rye, and some other foods, and are also processed from wood. Eating stanol-containing spreads results in decreased levels of LDL, but not HDL, cholesterol.

9. *Hormones.* DHEA (dehydroepiandrosterone) is taken to increase muscle mass and immune function, pregnenolone to enhance memory, and melatonin to enhance sleep. There is no evidence that pregnenolone enhances human memory or improves concentration, but there is limited evidence that the other two can work. Despite equivocal evidence for melatonin, it is popularly used to induce sleep and reduce jet lag.[100] Secretion of this biorhythm regulator normally decreases with age.

Nutrient Recommendations: Using the Food Label

The Nutrition Facts panel on food packages is structured to provide nutrient content information in relation to nutrient needs.

Several of the recommended nutrient levels for older adults (aged 70 years and older) differ from those used as reference values (DV) for the nutrition label (see Table 18.17). For example, older adults need more vitamin C and calcium and less iron and zinc than the reference values listed on the label. Values also differ for vitamins D, E, and B$_{12}$. The UL is listed for convenient comparison; an explosion of functional foods and beverages in the marketplace is making it easier to overdose on vitamins and minerals.

Older adults can still use the DV percentages on food labels for dietary guidance, as long as they adjust them to get more than 100% calcium and vitamins D and C and less than 100% for vitamin A, iron and zinc. The vitamin B$_{12}$ label recommendation is higher than the DRI, but poor absorption in older adults makes unsafe intake from food products unlikely. The Nutrition Facts panel information underscores that "one size does not fit all" in dietary guidance, especially when it comes to older adults!

Cross-Cultural Considerations in Making Dietary Recommendations

Food habits develop in cultural contexts, and we can learn about them in various ways. Travel through North America would allow us to observe cultures that make up our society. Visiting ethnic restaurants, stores, and farmer's markets can be another way to get a glimpse of cultural food diversity. Cookbooks, films, talking with individuals about their food history, and participating in ethnic celebrations are other sources of insight into food patterns of various cultures. Each new immigrant wave adds unique food traditions to the country's mix. Older adults may be stronger advocates for upholding traditional food patterns than young people are. In working with them on food issues, it is useful (and interesting) to determine the cultural history of their food and lifestyle habits.

National food-monitoring programs survey the population with proportionately larger samples drawn from minority groups in order to develop a balanced picture of the whole population. North America is home to rapidly growing Hispanic, Asian, Russian, Middle-Eastern, and African immigrant groups. The U.S. Census tracks minority groups, but it is completed only once every 10 years. The local Area Agencies on Aging and Senior Nutrition Programs also track population trends, and these are likely to yield greater insight about some of the smaller ethnic population groups in their unique communities and regions.

Cultural differences are reflected in approaches to dietary guidance. For example, Chile has separate guidelines for older and younger adults. In New Zealand, older adults are encouraged to socialize at mealtimes to improve appetite. In France, South Korea, and Japan, people of all ages are encouraged to enjoy mealtimes and take pleasure in eating. Other unique guidelines are those of China, which suggest people eat 20 to 25 grams (nearly an ounce) of fish daily. Guatemala uses a bean pot as a nutritional icon. India has developed one set of guidelines for the rich (i.e., overall energy intake should be restricted to levels commensurate with the sedentary occupations of the affluent, so that obesity is avoided; total fat intake is not to exceed 20% of total energy; and use of clarified butter, a prized Indian culinary ingredient, should be restricted to special occasions) and one for the poor (addressing the fact that at least one-third of the households in India are not able to afford even the minimum nutritional requirements, even though they are spending 80% of their income on food, these recommendations identify food combinations that are most likely to meet recommended dietary intakes).

Communicating effectively and avoiding misinterpretation in intercultural settings is probably the most important thing a nutritionist can learn to do when working with older adults from various cultures. Cultural competence requires skill in transferring information, developing and maintaining relationships, and gaining compliance. Developing individual skills takes time, commitment, and practice. On the other hand, nutrition education and guidance tools have been developed in many languages and for diverse cultures, although not typically for elders of ethnic groups. Culturally appropriate resources can be found through local extension services, cross-cultural education centers, diabetes education programs, public health agencies, and some commodity groups.

Food Safety Recommendations

Older adults with a compromised immune status are particularly vulnerable to foodborne illness. No one knows exactly how widespread this problem is because many foodborne illnesses are not reported when individuals think they have "the flu."

Poor food-handling practices leading to microorganism growth are generally to blame for foodborne illnesses. Impaired functioning may result in impaired food handling practices. Signs and symptoms of foodborne illness include gastrointestinal distress, diarrhea, vomiting, and fever and may appear within half an hour of eating a contaminated food or may not develop for up to 3 weeks.

Leading practices that put an older person at risk are as follows:

- Improper holding temperatures of foods
- Poor personal hygiene
- Contaminated food-preparation equipment (cutting boards, knives)
- Inadequate cooking time

The Dietary Guidelines for Americans offer steps to keep food safe:

- Wash hands and surfaces often.
- Separate raw, cooked, and ready-to-eat foods while shopping, preparing, or storing.
- Cook foods—especially raw meat, poultry, fish, and eggs—to a safe temperature.
- Refrigerate or freeze perishable or prepared foods within 2 hours.
- Follow the label for food-safety preparation and storage instructions.
- Serve hot foods hot (140°F or above) and cold foods cold (40°F or below).
- When it doubt, throw it out!

Physical Activity Recommendations

"There is no segment of the population that can benefit more from exercise than the elderly."

William J. Evans[101]

Evans made this statement in 1999, and it is still true today. Exercise is a true fountain of youth. Physical activity builds lean body mass, helps to maintain balance and flexibility, contributes to aerobic capacity and to overall fitness, improves cognitive performance in previously sedentary older adults, and is associated with overall psychological well-being.[11] However, older adults are relatively sedentary, see Table 18.18. For example, only 23% of adults aged 75 or older engaged in regular physical activity, defined as 20 or more minutes at least three times per week, when the Healthy People 2010 goals were published.[16] Low activity levels as well as deteriorating strength, endurance, and sense of balance are associated with, but not caused by, increasing age.

Older people benefit from exercise even more than younger people do because strength training is the only way to maintain and build muscle mass. In addition to strength gains, increased muscle mass increases caloric needs. Higher caloric intake increases the chances of optimal nutrient intake.

Age does not hinder training effects, as shown in Appendix A, from the American College of Sports

Table 18.18 Healthy People 2010 physical activity goals and baseline activity levels comparing older and younger adults

Type of Activity	Baseline for Age 75+	Baseline for Age 18–24	Target for All Adults
Engage in no leisure-time physical activity	65%	31%	20%
Engage in regular physical activity at least 20 min, 3 times/wk	23%	36%	30%
Perform strengthening activities 2 or more days/wk	8%	30%	30%
Perform stretching and flexibility activities	21%	39%	40%

SOURCE: *Healthy People 2010: National Health Promotion and Disease Prevention Objectives.* Washington, D.C.: U.S. Department of Health and Human Services, 2000.

Medicine. This research synopsis and position statement describes the effects of exercise and physical activity on older adults and includes recommended exercise prescriptions made collaboratively with the American Heart Association.[11]

Exercise Guidelines

How can one predict whether physical activity will exacerbate existing medical conditions? Physician screening or assessment by completing a questionnaire like the one in Table 18.19 can identify potential problem areas. For a more detailed assessment, Kligman and colleagues also recommend ways to evaluate cardiovascular fitness, strength, function, balance, flexibility, body composition, bone density, and lipid levels.[102] Evaluation by an individual's physician identifies possible contraindications to exercise. Individuals, including those with physical limitations such as heart disease or inability to stand, can be cleared for participation. They will find that even small increases in exercise add up.[103] Elaine Souza, RD, MPH, has developed the following guide for planning effective exercise sessions for healthy older adults:

- Decide on frequency: 2–3 times per week is effective for strength training, using 8–10 different exercises with 8–12 repetitions each, with the whole routine to be done in 20–30 minutes.
- For general health, exercise for 30 minutes on most days of the week.

- Drink water when exercising (before, during, and after exercise).
- Do warm-up and cool-down activities of 5 to 10 minutes each.

Individuals can take charge of their own healthy aging by developing appropriate and effective exercise habits. Simple (and occasionally challenging) fitness advice is to "eat good food, drink enough water, and play hard!" See Case Study 18.1.

Nutrition Policy and Intervention for Risk Reduction

Nutrition policy promotes health by combining nutrition education for individuals and population interventions. The ultimate goal of nutrition intervention is to improve health outcomes.

Nutrition Education

"The human mind, once stretched to a new idea, never goes back to its original dimension."

Oliver Wendell Holmes

Contrary to some beliefs, older people do learn and change. Someone born in 1930 has seen the invention of microwaves, television and TV dinners, and a whole

---

Table 18.19 Keep moving: fitness after 50 chart

A. Do I get chest pains while at rest and/or during exercise?

B. If the answer to question A is "yes": Is it true that I have not had a physician diagnose these pains yet?

C. Have I ever had a heart attack?

D. If the answer to question C is "yes": Was my heart attack within the last year?

E. Do I have high blood pressure?

F. If you do not know the answer to question E, answer this: Was my last blood pressure reading more than 150/100?

G. Am I short of breath after extremely mild exertion, and sometimes even at rest or at night in bed?

H. Do I have any ulcerated wounds or cuts on my feet that do not seem to heal?

I. Have I lost 10 lb or more in the past 6 months without trying and to my surprise?

J. Do I get pain in my buttocks or the back of my legs—my thighs and calves—when I walk? (This question is an attempt to identify persons who suffer from intermittent claudication. Exercise training may be extremely painful; however, it may also provide relief from pain experienced when performing lower-intensity exercise.)

K. When at rest, do I frequently experience fast irregular heartbeats or, at the other extreme, very slow beats? (Although a low heart rate can be a sign of an efficient and well-conditioned heart, a very low rate can also indicate a nearly complete heart block.)

L. Am I currently being treated for any heart or circulatory condition, such as vascular disease, stroke, angina, hypertension, congestive heart failure, poor circulation to the legs, valvular heart disease, blood clots, or pulmonary disease?

M. As an adult, have I ever had a fracture of the hip, spine, or wrist?

N. Did I have a fall more than twice in the past year (no matter what the reason)? (Many older persons have balance problems and at the initiation of a walking program will have a high chance of falling. These persons may benefit from balance training and resistance exercise before beginning a walking program.)

O. Do I have diabetes?

SOURCE: From Evans, W. J. and Cyr-Campbell, D. Nutrition, Exercise, and Healthy Aging. JADA 1997; 97:632–8. Copyright: American Dietetic Association. Reprinted by permission from the author.

Case Study 18.1

JT—Spiraling Out of Control?

Masterfile

JT, a retired computer company executive, eats out four times a week since his wife died last year. Meals at home consist of microwave dinners or supreme pizza. He belongs to a health club, which he visits three times a week, and where he has many friends. Occasionally, JT and his friends go out for beers after their workout. Shortly after his 69th birthday (2 years ago), he was diagnosed with type 2 diabetes. Last week, he visited the clinic for his annual check-up. He was measured at 5 feet 9 inches (1.75 m) and weighed 235 pounds (106.8 kg). His doctor is worried about JT's family history of heart disease.

Questions to assess, diagnose, intervene, monitor, and evaluate JT's condition

1. What would you ask JT about his food and fitness routine?
2. Calculate JT's BMI. How does it compare to guidelines? He wants to know whether BMI is an accurate measure of his body fat. How might you answer?
3. As his nutritionist, what nutrition remedies would you explore to help JT?
4. What fluid recommendation would you make?
5. What sort of monitoring and evaluation plan could you and JT devise to track his weight management?

(Answers are located in the Instructor's Manual for the 4th edition of *Nutrition Through the Life Cycle*.)

---

host of computer-controlled kitchen appliances. To age is to adapt. Learning new nutrition habits is part of aging. Nutrition education is different from education in general because its goal is changed dietary behaviors. Nutrition education consists of a set of learning experiences to facilitate voluntary adoption of nutrition-related behaviors that are conducive to health and well-being. Several requirements must be met for nutrition education—that is, behavior change—to occur (see Illustration 18.6). Think of them as the 4 C's of nutrition education.[104]

1. *Commitment.* Commitment means being motivated to adopt health-promoting behaviors and intending to adopt and maintain a new food behavior.
2. *Cognitive processing.* Understanding how a food behavior contributes to health and planning how it will fit into your life constitutes cognitive processing.
3. *Capability.* Acquiring the skills to practice new food behaviors is part of nutrition education. An example is learning to identify whole-grain breads or to prepare vegetables when intending to adopt a high-fiber diet.

4. *Confidence.* "Nothing breeds success like success!" The best predictor that someone will practice new dietary habits is their personal confidence in being able to do so.

Educational sessions for older learners are best designed around their potential limitations, such as declines

Illustration 18.6 Four essential elements to achieve and maintain individual dietary behavior change.

SOURCE: Adapted from Krinke, U. B. Effective Nutrition Education Strategies to Reach Older Adults. in Watson, R. R., ed. *Handbook of Nutrition in the Aged.* Boca Raton, FL: CRC Press, 2001.

in visual acuity and hearing loss.[104] Adaptation for written material includes:

- Larger type size
- Serif lettering (Helvetica is a typeface with no serifs; Times Roman is a typeface with serifs)
- Bold type
- High contrast (black on white)
- Non-glossy paper to decrease glare
- Avoidance of blue, green, and violet (paper and print color) due to decreased ability to discriminate among these colors
- Reading level of fifth to eighth grade for general audiences

Educational strategies can result in better diets for individuals, but these alone may not be sufficient to enhance the dietary patterns of populations. Nutrition policies reflect community environments. Cultural environments either support, ignore, or punish desired behavior change. An example of a culturally supportive environment is a peer group that values health and fitness; group members will help each other to achieve and maintain health-promoting behaviors. For instance, members may belong to a biking group, share in farmers' market trips, and serve healthful foods when entertaining. Their environment supports healthy habits.

Food and nutrition policies arise from public values, beliefs, and opinions that define the cultural context in which dietary behaviors exist. Policies can be overt or unspoken. Public policies supporting the health of older adults in the United States are evident in the Social Security program, which provides financial support for post-retirement living, including health care and nutrition counseling.

Community Food and Nutrition Programs

Nutrition Programs Serving Older Adults

Governmental programs for older adults include the USDA's Supplemental Nutrition Assistance Program (SNAP, formerly the Food Stamp Program), Seniors' Farmers Market Nutrition Programs, the Commodity Supplemental Foods, the Child and Adult Care Food Program, and the Ryan White Comprehensive AIDS Resources Emergency Act.[105] The U.S. Department of Health and Human Services (HHS) administers the Older Americans Act programs, including Meals-on-Wheels and other home-delivered meal programs, and nutrition services for tribal and Native older Americans' programs. HHS also administers Social Security programs that provide basic living expenses for older adults and health care coverage through Medicare.

Nongovernmental programs may provide food and nutrition services as part of a broader range of screening and assessment, nursing, and other support services. For instance, home health agencies provide staff who can offer nutritional counseling or aides who will shop for and prepare food, and clean up the kitchen afterward. Home care services allow individuals to receive the necessary support to stay in their homes for as long as they wish. Remaining in one's home indefinitely is sometimes referred to as "aging in place." A broader definition of this concept is found in the Position of the American Dietetic Association.[7] Aging in place does not necessarily mean living in one home for a lifetime. Ideally, it means having the choice to grow old in one's preferred community. Liveable communities offer a spectrum of living options for residents of all ages. Supportive services customized to accommodate those who are fully active and have no impairments, those who require limited assistance, and those with more severe impairments who require care in long-term care facilities are basic to creating liveable communities. Food and nutrition programs help to meet the most basic human needs in a continuum of care.

Other food and nutrition programs that contribute to the continuum of nutrition services include food pantries, soup kitchens, cooperative buying groups such as Fare for All, and screening and referral services.

Store-to-Door: A Nongovernmental Service that Supports Aging in Place

Store-to-Door is an example of a small program that was begun by one person who made a big difference. After "retiring" in the 1980s, Dr. David Berger surveyed his community to see where he might do some good. Many older adults told him that getting to the grocery store was impossible, and that even when they could get to the store, bringing the bags home was difficult. Winter ice and snow made things worse, for people feared falling. The community's grocery delivery service had closed because profits were not meeting shareholder expectations. So Dr. Berger joined forces with his wife, Fran, and with friends, colleagues, and many volunteers to start up the nonprofit Store-to-Door, a home-delivered grocery program for older people and those who are disabled. Volunteers will shop for and deliver your groceries. They buy items that discount grocers offer: food, of course, but also greeting cards, medicines, paper goods, and cleaning supplies, although no alcohol or cigarettes. Customers can get credit for coupons. After starting Store-to-Door in the Midwest, Dr. Berger started other home-delivered grocery programs in Portland, Oregon, and Ventura, California. Does the program work? Yes: the first Store-to-Door celebrated its 25th anniversary in 2009, the Portland group is now 20 years old, and the Shop Ahoy in Ventura is growing. That's no small feat for programs that depend on hundreds of volunteers and funding sources to help older and disabled adults live at home.

Senior Nutrition Program: Promoting Socialization and Improved Nutrition

Congress first appropriated funds under Title VII of the Older Americans Act of 1965 to begin the Senior Nutrition Program, also called the Elderly Nutrition Program (ENP). The Senior Nutrition Program was created to alleviate poor nutritional intake and reduce social isolation among older adults. It was based on evidence that older adults do not eat adequately because of the following:

1. Lack of income limits ability to purchase food.
2. Lack of skills limits ability to select and prepare nourishing meals.
3. Limited mobility affects shopping and meal preparation.
4. Feelings of isolation and loneliness decrease the incentive to eat well.

Senior Nutrition, now Title IIIC of the Older Americans Act, is a community-based nutrition program that provides meals (congregate and home-delivered), increased social contact, nutrition screening and education, and information and linkages to other support programs and services, as well as volunteer opportunities. Anyone who is 60 or more years of age (and spouse regardless of age) is eligible to participate in the congregate dining program; home-delivered meal clients must be homebound and unable to prepare their own meals. Typically, $1.00 of Title III funds spent on congregate services is supplemented by an additional $1.70 from other sources; the average cost of an ENP meal is $5.17, and a home-delivered one is $5.31.[106] Title VI grant programs are similar to Title III funds and were established to help deliver social and nutrition services to older American Indians, Alaskan Natives, and Native Hawaiians. About 25% of participants are minorities, almost twice the national percentage of minority adults over age 60.[106]

Today there is less poverty among younger seniors, many of whom are actively working. But there are greater nutritional and social needs among frail elders and individuals with low incomes, chronic illness, limited mobility, and limited English speaking ability, and among minority and isolated elders. About one-third of Title III congregate meal participants and more than one-half of Title VI meal participants have incomes at or below the Department of Health and Human Services (DHHS) poverty threshold.[107]

Senior dining sites are targeted to neighborhoods where older, frail, impoverished seniors live. Dining sites are located in community centers, senior centers, civic buildings, subsidized housing units, schools, and other accessible locations. Meals are delivered to the homes of individuals who are 60 years of age or older, homebound by reason of illness or disability, and unable to prepare meals. Services are adapted to meet each unique community's setting. For instance, meal vouchers for use at local cafes or diners are available in some small communities.

Other services to meet the needs of frail older seniors include multiple meals, weekend meals, take-home snacks, liquid supplements, nutrition screening and education, and one-to-one nutrition counseling. Also available are diets adapted for medical, religious, and cultural reasons.

Nutrition programs for older adults have successfully brought together millions of people to socialize and enjoy nutritious meals. In 2002, about 250 million congregate and home-delivered meals were served to approximately 2.6 million older adults.[106] The Older Americans Act 2000 Amendment, Section 339 (Nutrition) (H.R. 782), states that nutrition projects shall use a dietician (or person with compatible expertise) to provide meals that comply with the Dietary Guidelines for Americans. Nutrition program meals were found to exceed the one-third RDA standard. Compared to nonparticipants, participants had up to 31% higher intake of recommended nutrients. In other words, the program is working. Surveys show that targeting those who need services most is successful, and a national evaluation of the senior nutrition program found that the program is targeted to those most in need.[107] Increasing socialization was one of the original program goals and continues to be one of the outcomes. Participants have 16–18% more social contacts per month than nonparticipants do. Relative to the general older population, participants are older and more likely to be female, to belong to an ethnic minority, to live alone, and to have incomes well below poverty level.

Grocery shopping assistance is an important service for frail adults. A variety of models are used, including volunteer escorts to the supermarket, bus rides, and grocery delivery to the door.

The Promise of Prevention: Health Promotion

"Grow old along with me, the best is yet to be!"

Robert Browning

Aging baby boomers who expect to enjoy an active adulthood are driving longevity and aging research. Lifelong caloric restriction may not find many converts, but emulating habits of long-lived populations is gaining interest. Dan Buettner has been studying regions of the world inhabited by disproportionately high numbers of centenarians. In describing the lives of these healthy old residents of Okinawa, Sardinia, Costa Rica, and North America in a book called *The Blue Zones*,[108] Buettner compiled a list of nine habits and traits among these population groups:

1. Being active as a regular part of daily life
2. *Hara hachi bu*, which means to stop eating when one is 80% full
3. Eating more beans, whole grains, vegetables, nuts, and fruits while limiting meats and processed foods
4. Drinking red wine, in moderation
5. Having a reason to get up in the morning, a *plan de vida* or *ikigai*
6. Making time to relieve stress
7. Belonging to, and participating in, a spiritual community
8. Making family a priority and upholding rituals and traditions
9. Picking the right tribe, i.e., surrounding yourself with people who share long-life values

Residents of places like Okinawa and Sardinia and Loma Linda are supported by their culture throughout life. Although good nutritional habits make a greater impact when started early in life, sometimes individuals are not motivated to pursue these risk-reduction strategies until later in life or after experiencing a health problem. Successful strategies to reach an older audience address specific population needs and interests. The wave of aging baby boomers will be less accepting of the status quo. Should the myth that a 70-year-old is too old to learn and practice health-promotion strategies still exist when they turn 70, they will certainly disprove it!

Key Points

1. Functional ability (the demonstrated ability to carry out activities of daily living) is more important than chronological age in assessing the health status of older adults.

2. Good nutrition, good health habits, environment, access to health care, and genetics contribute to human life expectancy, which is still significantly shorter than the potential human life span. Theories of aging, such as wear-and-tear theories, help to explore which factors contribute most to a longer, disease-free life.

3. Of all the physiological changes associated with aging, loss of lean body mass and the concomitant gains of body fat may well be the most important in determining functional age.

4. "Use it or lose it" applies to the body and the spirit: keep learning to maintain acute brain function, stay active to build muscle and bone, eat well to maintain and repair tissue, and cultivate a positive approach to aging for enhanced quality of life as well as to live longer.

5. The DETERMINE acronym is a reasonable summary of warning signs associated with poor nutritional health.

6. While adults in general consume more than enough calories and protein, populations of older adults may be lacking in adequate dietary protein and energy.

7. The thirst mechanism of older adults is not as sensitive as that of younger adults, placing them at higher risk of dehydration.

8. Physiological changes that lead to malnutrition in older adults are decreased absorption of vitamins D and B_{12} and increased storage of vitamin A and iron.

9. In general, older adults eat better than do younger adults, but they do not consume enough vitamin A, D, E, K, choline, calcium, magnesium, or potassium to meet recommended intake levels.

10. Vitamin and mineral supplements can be helpful for older adults who have lost their appetite, avoid certain food groups, have poor diets due to food insecurity, loss of function, dieting or depression, or who have gastrointestinal bacterial overgrowth that prevents nutrient absorption.

11. Excellent food safety practices are especially important for older adults, who may be more vulnerable to infection for many reasons, such as a higher prevalence of chronic diseases, sensory and functional losses, and decreased resilience in healing and recovery from illness.

12. Older adults are often more interested in nutrition education and health promotion than are younger adults. The stereotype that older adults will not change is just that: an old stereotype.

Review Questions

1. True or false: Personal attitudes and beliefs contribute to longer life.

2. True or false: Older adults have proportionately more illness than younger adults, and the three leading causes of death are heart disease, cancer, and stroke. The CDC suggests that longevity depends more on lifestyle factors than on genetics.

3. True or false: U.S. life expectancy at birth is the highest in the world.

4. True or false: Fat-free mass or lean body mass tends to shrink in old age. This results in significantly decreased need for calories, vitamins, and minerals.

5. DETERMINE and the MNA are two tools to screen for nutritional risk. List five factors that place older adults at risk of malnutrition.

6. Food guide pyramids for older adults are different from those for the general public. Cite two adjustments in food guidance that are particularly important for older adults.

7. The AMDR for protein is 10–35% of calories, and the RDA is 0.8 gm/kg body weight. If the average older adult meets the RDA, why is protein a nutrient of concern for older adults?

8. Using information from question #7, how would you determine the protein needs of JT in Case Study 18.1? Hint: he weighs 235 pounds (107 kg).

9. True or false: The Senior Nutrition Program of the Older Americans Act serves individuals aged 60 and older. It was established to increase socialization as well as to improve nutritional intakes of older adults.

Resources

AARP

Information about a range of services related to growing old in America, including health, finances, politics, travel, and population trends.
Website: www.aarp.org/

American Dietetics Association

This site offers food and nutrition tips, fact sheets, position papers and help to find a registered dietician to the public. Members can find summaries of carefully done research in an evidence analysis library.
Website: www.eatright.org

Centers for Disease Control and Prevention

Mortality and morbidity data.
Website: www.cdc.gov/mmwr

Florida National Policy and Resource Center on Nutrition and Aging

Aging policy and education center.
Website: www.nutritionandaging.fiu.edu

National Heart, Lung, and Blood Institute

Gateway site for consumers and health professionals; links and informational materials about heart health and more.
Website: www.nhlbi.nih.gov

National Institute of Dental and Craniofacial Research

Educational resources for the public and for health care providers.
Website: www.nidr.nih.gov

Oral Health in America: A Report of the Surgeon General
Website: www.nidr.nih.gov/sgr/sgrohweb/home.htm

Tufts University Center on Aging

Resources at Tufts and a popular gateway to information on aging.
Website: www.hnrc.tufts.edu

Administration on Aging

Fact sheet, news, statistics, and links to other resources.
Websites: **www.aoa.gov, eldercare.gov, and agingstats.gov**

Blue Zones

Descriptions of ongoing research into pockets of long-lived populations and tools to "live younger."
Website: BlueZones.com

Chapter 19

Nutrition and Older Adults:
Conditions and Interventions

Prepared by **U. Beate Krinke**

Key Nutrition Concepts

1 Multiple health problems put older adults at higher nutritional risk.

2 Nutrient thresholds depend on individual circumstances: dietary nutrients that are beneficial in treating a deficiency have little effect when the diet is adequate, and they can be dangerous when consumed in excess of recommendations.

3 Successful nutrition interventions complement other lifestyle choices to stabilize physiological decline and enhance physical and mental resilience.

Introduction: The Importance of Nutrition

Aging adults want to stay *healthy* until death, and between ages 65 and 74, most (63–80%) believe they have *good to excellent* health, according to the respondent-assessed health status question reported in *Older Americans 2008, Key Indicators of Well Being*.[1] However, chronic illnesses are more prevalent in old age. Having a chronic health problem (see Table 19.1) does not prevent someone from having the perception of being healthy. Still, self-perception of good health declines with age. By age 85, self-perceived health status of *good to excellent* decreases to 67% of non-Hispanic whites, 54% of non-Hispanic blacks, and 47% of Hispanics.[1] Good nutrition contributes to *quality of life* and ameliorates effects of illness. Diet quality of adults aged 65 and older scores 68 as measured by the Healthy Eating Index (68 of 100 possible points),[1] compared to an HEI score of 58 for Americans above age 2.[2] But illness often brings with it diet modifications, either as a symptom of disease, such as anorexia in depression, or as part of therapy. *Medical nutrition therapy (MNT)* becomes part of a comprehensive treatment plan that encourages health-promoting food choices once diseases have been diagnosed. Outcome data regarding the effectiveness of MNT by registered dieticians (RD) has resulted in reimbursement provided to Medicare Part B beneficiaries with diabetes mellitus and kidney disease,[3] using nationally recognized nutrition protocols and evidence-based practice guidelines.[4] The most

current Medicare rules can be found at the Center for Medicare and Medicaid Services (www.cms.hhs.gov).

Numerous studies have measured health care utilization of older adults and found that better nutritional status is related to better health outcomes. The Lewin Group researchers estimated that covering medical nutrition therapy for older managed-care clients who had cardiovascular disease, diabetes, or renal disease would recover Medicare costs after three years and would begin to save system dollars by the fourth year.[5] Malnourished older patients have higher postoperative complication rates and longer hospital stays, therefore incurring greater health care costs.[6,7] In free-living older adults, nutritional risk status was found to be the most important predictor of total number of physician visits, visits to physicians in the emergency room, and hospitalization rates.[8] Aging adults use proportionately more health care services and products than younger persons do; therefore, nutrition interventions can play a larger role in their health.

"Mens sana in corpore sano"

Nutrition and Health

Across time, adults of all ages have defined health as "a sound mind in a sound body." Public health professionals monitor health by measuring leading causes of death (mortality) and leading diagnoses of health conditions (morbidity) in order to design interventions that improve the population's health. Leading health problems and disabilities become the focus of nutritional interventions.

Deaths from heart and cerebrovascular diseases have declined by 2% and 1% respectively since the last edition of this text, but they are still among the leading causes of death of older adults. The National Center for Health Statistics reports leading causes of death. For persons aged 65 and older,

Healthy More than the absence of disease, health is a sense of well-being. Even individuals with a chronic condition may properly consider themselves to be healthy. For instance, a person with diabetes mellitus whose blood sugar is under control can be considered healthy.

Quality of Life A measure of life satisfaction that is difficult to define, especially in an aging heterogeneous population. Quality-of-life measures include factors such as social contacts, economic security, and functional status.

Medical Nutrition Therapy (MNT) Comprehensive nutrition services by registered dieticians to treat the nutritional aspects of acute and chronic diseases.

Table 19.1 Percentage of people age 65 and older who reported selected chronic conditions, 2005–2006, by sex

Sex	Heart Disease	Hypertension	Stroke	Any Cancer	Diabetes	Arthritis
Men	37	52	10	24	19	43
Women	26	54	8	19	17	54

SOURCE: Reference 1, Federal Interagency Forum on Aging-Related Statistics. *Older Americans Update 2008: Key Indicators of Well-Being. Federal Interagency Forum on Aging-Related Statistics.* Washington, D.C.: U.S. Government Printing Office, 2008. Accessed Indicator 16, Chronic Health Conditions, 8/18/09.

the leading causes of death (with percentage of mortality) that have nutritional risk factors rank as follows:[9]

1. Heart disease — 29%
2. Cancer — 22%
3. Cerebrovascular disease — 7%
4. Alzheimer's — 4%
5. Diabetes mellitus — 3%

The majority of older Americans have one or more chronic conditions, also called comorbidities. If you sum the percentages in Table 19.1, you'll note that the total is greater than 100. Common comorbid conditions are arthritis, hypertension, heart disease, and, increasingly, obesity. Despite eating patterns that earn better HEI scores than those of younger people, the diets of older adults contribute to the incidence and course of their diseases, especially heart disease, hypertension, and cancer. These diseases can, in turn, impact functional ability. For example, an overweight individual with heart disease may continue to overeat and become obese, further complicating arthritis management. In turn, arthritis can limit functioning. Weight loss of even 1 pound will reduce stress on the knees during daily activities and may lead to enhanced management of heart disease, as well.[10] Old age is not a good reason to forego health-promotion. The Physician's Health Study examined modifiable risk factors contributing to survival and optimal functioning of men at age 90 and beyond. Good health habits (not smoking, moderate alcohol intake, not becoming obese, and regular physical exercise, see Appendix A) contributed to delayed mortality and to higher functional status in old age.[11] An analysis of the dietary habits of postmenopausal women with established heart disease (mean ages of study's tertile groups: 64.1, 66.3, and 66.6 years) assessed how closely food intake met the 2005 Dietary Guidelines for Americans. Using a scoring tool that was weighted for heart health risk showed that women with diets most closely resembling the 2005 guidelines had a slower rate of atherosclerosis progression over the 3.3-year period of follow-up.[12] Although these two examples focus on the contributions of nutrition to heart health, this chapter addresses the role of nutrition as a factor in disease prevention, treatment, and recovery of overall health.

Heart Disease: Coronary Heart Disease, Cerebrovascular Disease, Peripheral Artery Disease

Heart disease (cardiovascular disease or CVD) is the leading cause of death in older adults and, in part, is potentially reversible by adopting a healthy lifestyle. The adult risk factors and course of heart diseases have been discussed in Chapter 17. Specifics for older adults are highlighted in this section, including stroke and hypertension.

Prevalence

Heart disease prevalence varies by race and gender; see Table 19.2, which shows prevalence of heart disease and diabetes in relatively frail older adults.[13] CVD prevalence rises with age. The American Heart Association publishes updated statistics on the prevalence of CVD (they include coronary heart disease, heart failure, stroke, and hypertension) as 73.3% in men and 72.6% in women aged 60 to 79 years.[14] By age 80, these percentages have risen to 79.3% and 85.9%, respectively. Or, if you reverse the numbers, 20.7% of men and 14.1% of women aged 80 and over do not have any heart disease.

Risk Factors

Risk factors for cardiovascular disease in old age remain the same as in younger adults, except that the factors have less predictive value in old age.[15] Of adults aged 65 years and older examined in the NHANES III survey, 86% had one or more modifiable cardiovascular risk factor. These include hypertension (140 mm Hg/90 mm Hg), elevated LDL cholesterol (at least 130 mg/dl), and/or diabetes mellitus (physician diagnosis or fasting plasma glucose greater than 126 mg/dl).[16] Race is associated with risk, and older African Americans are nearly three times more likely to have one of the three cardiovascular risk factors than the average population. Eating a Mediterranean-style diet is associated with lower CVD mortality in 70–90 year olds, as is engaging in physical activity.[14]

Nutritional Remedies for Cardiovascular Diseases

Assertive treatment can modify the course of heart disease at any age, although an older adult is more likely to have comorbid conditions that necessitate balancing multiple goals. Together, the older adult and his or her

Table 19.2 Health disparities in the oldest old: percentage of adults age 85 and older who are in poor or fair health, with heart disease, hypertension, and diabetes, by gender and race

Selected Characteristics	Heart Disease	Hypertension	Diabetes
Men	47.3	49.6	15.6
Women	37.2	56.7	11.7
Non-Hispanic White	41.7	53.0	12.0
Non-Hispanic Black	31.4	65.2	22.4
Non-Hispanic Asian	35.0	62.7	15.4
Hispanic	35.8	58.1	16.5

SOURCE: Reference 12.

health care providers can develop a treatment plan that balances health and quality-of-life goals. Basic questions address an individual's motivation to adopt and maintain heart-healthy routines. The individual may be wondering if potential health gains are worth the necessary changes. Is the individual balancing quality of life with potential life expectancy? For individuals who are new to cooking or have become bored with it, are there classes that demonstrate how to prepare whole grains, cook new vegetables, or find ways to add fish into a meal plan? Will the budget support buying additional fruits and vegetables and eating at restaurants that offer more heart-healthy choices? If function is limited, what are the resources available to build a sustainable exercise routine? Is the person willing to ask for and accept assistance if needed to shop for groceries or to prepare meals? Can the individual maintain heart-healthy habits when eating with family and friends, at home, or in restaurants? The nutrition interventions that support heart health are not inherently different for an older adult than for a younger one. But the day-to-day context for adopting therapeutic lifestyle changes is likely quite different for an 80-year-old

than for a 45-year-old. An individual's view of his or her own physical and emotional status is an integral part of how treatment will be pursued and how heart-healthy habits will be maintained. Nutritional habits alter the progression of atherosclerosis only if an individual adheres to the eating plan.

Nutrition intervention guidelines for CVD in all adults include those from the National Cholesterol Education Program (NCEP) and the American Heart Association.[17] In the updated Adult Treatment Panel III guidelines, Grundy and colleagues suggest that intensive LDL-lowering therapy is appropriate for older adults with established CVD, and that "clinical judgment is required as to when to initiate intensive LDL-lowering therapy in older persons without CVD."[17] Cardiac rehabilitation interventions were described in Chapter 17, and age-related considerations are described in Table 19.3.

Consumption of fish, fish oils, and omega-3 polyunsaturated fatty acids is of special interest to older adults because so many have underlying heart disease. A review by Levie and colleagues[18] concluded that individuals with known coronary heart disease and heart failure

Table 19.3 Treatment factors for older adults with heart disease

Target Area	Adults (≥65 years)
Decrease amount and type of fat	Focus on 1–2 items to decrease saturated fat intake in individual's regular diet rather than change all things
• Use lean meats	Ensure adequate protein intake
• Substitute saturated fatty acids with PUFA & MUFA	Focus on oils currently using and suggest one to change, if appropriate Decrease synthetic *trans*-fatty acids
• Decrease synthetic *trans*-fatty acids	Consider giving a brief description of *trans*-fatty acids and sources: hydrogenated oils and shortening, cookies, pastries, and other processed fats based on mental awareness and readiness to change
Reduce cholesterol intake	Focus on 1–2 food items; research is conflicting on the role of cholesterol in older adults; liver makes less
Increase fiber, fruits, and vegetables	Work with the fruits and vegetables that the individual can chew (e.g., if dentures, do they fit?)
Healthy cooking	May not be controllable if Meals-on-Wheels; overall goal is adequate nutrient intake
Limit salt	Focus on "no added salt," and no salt shaker on the table
Label reading	May be difficult if eyesight is poor; consider financial limits; know bargain strategies
Exercise	Obtain doctor's approval prior to starting; emphasize health benefits that include mobility, agility, and strength; emphasize that walking is exercise
Maintain healthy weight	Strongly influenced by functional status of individual; emphasize adequate nutrient intake
Reduce stress	Exercise, relaxation, and socialization with friends
Quit smoking	Refer to smoking cessation program; continue the no-smoking followed while in hospital; discuss potential for weight gain

SOURCE: Adapted from Gerlach, Anne F. *Principles in a cardiac rehabilitation program.* Guest lecturer for Nutrition for Adults and the Elderly, University of Minnesota, Minneapolis, MN, 2002.

should consume 800–1000 mg/day or more of EPA and DHA combined for cardioprotection. A review of randomized controlled trials on the effects of fish oil supplements on arrhythmias and mortality found that "fish oil supplementation was associated with a significant reduction in deaths from cardiac causes but had no effect on arrhythmias or all cause mortality."[19] The authors added that variation in supplement formulations made it difficult to recommend appropriate dosages, but suggested that the level of 465 mg EPA/386 mg DHA used in the GISSI-Prevenzione trial seemed reasonable.

Stroke

Definition

The American Heart Association and the American Stroke Association describe both stroke and TIAs (transient ischemic attack) as serious conditions involving reduced cerebral blood flow (brain *ischemia*); both are markers for increased risk of disability and death.[20] TIAs, brief episodes of neurological dysfunction such as sudden confusion, trouble speaking or understanding, or sudden dizziness and trouble walking, often precede a stroke. During an ischemic stroke (about 85% of all strokes), an obstruction clogs a blood vessel and prevents oxygen and other nutrients from reaching part of the brain. A hemorrhagic stroke occurs when a weakened blood vessel breaks, such as the rupture of an *aneurysm*. Leaking blood accumulates, putting pressure on the surrounding tissue and eventually destroying brain cells.

Prevalence

Of adults aged 65 and older, 8% of females and 10% of males have had a stroke.[1] The American Heart Association cites statistics from the Framingham study to demonstrate gender differences in the impact of a stroke on survivors.[14] Six months after having a stroke, 34% of women and 16% of men were disabled. At younger ages, incidence of stroke is higher for men than women, but by age 85, incidence is greater for women. Who is at highest risk for having a stroke? Age-adjusted first-ever strokes in 45–84-year-olds occur more often in black males and females (6.6 and 4.9 per 1000 population) than white males and females (3.6 and 2.3 per 1000 population).[14] For individuals having a first stroke at age 70 or older, approximately one-fourth (22–27%) die within a year.

Ischemia Blockage of blood vessel leading to lack of blood supply.

Aneurysm Ballooning of the blood vessel wall.

Thrombus Blood clot.

Cerebral embolism Piece of a blood clot formed elsewhere that travels to the brain.

Carotid Artery Disease The arteries that supply blood to the brain and neck becoming damaged.

Atrial Fibrillation Degeneration of the heart muscle causing irregular contractions.

Transient Ischemic Attacks (TIAs) Temporary and insufficient blood supply to the brain.

Etiology

Factors that can lead to a stroke include blocked arteries (by a *thrombus* or *cerebral embolism*), easily clotting blood cells, and weak heartbeats that are unable to keep blood circulating through the body, allowing pools of blood to form and clot. Hypertension contributes to strokes because the force of blood may break weak vessels.

Effects of Stroke

Strokes deprive the brain of needed oxygen and other nutrients, causing brain and nerve cells to die. As a result, stroke leads to loss of function for parts of the body controlled by the oxygen-deprived cells. For example, stroke victims' bodies may become paralyzed in either the left or the right side, or they may become unable to speak, walk, or swallow. Nutrition is likely to be affected in the stroke aftermath. Quick recognition of stroke results in faster treatment and better recovery. Although dead brain cells cannot be replaced, new nerve pathways can develop in the gray-matter reservoirs of the brain. The ability to develop new neural pathways provides hope for successful rehabilitation therapies. Relearning how to feed oneself, how to chew, and how to swallow may well be a part of the slow, arduous process of rehabilitation.

Risk Factors

Gender is not a risk factor for stroke, although more women die from a stroke than men. The following factors place an individual at higher risk for stroke:

- Long-term high blood pressure (either systolic or diastolic)
- Family history
- African American, Asian, and Hispanic ethnicity
- Physical inactivity
- Cigarette smoking (doubles the risk of ischemic strokes!)
- Comorbid conditions, including diabetes mellitus, *carotid artery disease, atrial fibrillation, transient ischemic attacks (TIAs)*, sickle cell anemia, and depression
- Living in poverty
- Excessive use of alcohol; use of cocaine and illicit intravenous drugs

The role of alcohol is complex and controversial. Moderate amounts of any type of alcohol can be protective against stroke, while excessive amounts increase stroke risk significantly. In Japan, a 10.5-year prospective study with nearly 3000 men aged 40 to 69 years (only 5% of women drank) found the least risk at 42 grams of alcohol per day (a typical drink contains 12–15 grams of alcohol).[21] Men who consumed more than 70 grams of alcohol per day were

2.5 times more likely to have a stroke than the low-risk group. In a Framingham study of alcohol use and ischemic strokes in older adults, up to 2 drinks (or 24–30 grams of alcohol) per day counted as "moderate."[22] Moderate intake was protective. Stroke rates were triple the lowest rate for individuals drinking 7 drinks per day.

Nutritional Remedies

The focus of dietary advice in stroke prevention is to normalize blood pressure,[14,23] which is discussed with hypertension. Individualized medical nutritional therapy is used to promote rehabilitation.

Hypertension

Definition

High blood pressure (HBP) is defined as "untreated systolic pressure of 140 mm Hg or higher, or diastolic pressure of 90 mm Hg or higher or taking antihypertensive medicines."[14] Prehypertension increases risk for CVD and is defined by systolic pressure of 120–139 mm Hg or diastolic pressure of 80–89 mm Hg. Although it has been suggested that older adults can tolerate higher blood pressure and may even benefit from increased blood flow to the brain, old age does not change the diagnosis criteria for high blood pressure. Higher blood pressure puts more force on potential vessel blockages and increases chances of blood vessel breakage. An individual who controls high blood pressure with medication is still considered to have hypertension.

Prevalence

Hypertension (HBP) is the only chronic condition that has higher prevalence in older adults than arthritis (see Table 19.1). In Western societies, prevalence rises with age. Before age 45, a higher proportion of men than women have HBP; then the percentages are similar until age 64, when HBP becomes more common in women.[14] Death rates from high blood pressure are much higher for blacks than for whites. For example, in 2005, mortality rates were 52.1 for black and 15.8 for white males, and 40.3 for black and 15.1 for white females. Rates for Hispanics are similar to or lower than those for whites. Uncontrolled hypertension is a major public health challenge: prevalence is high, the consequences are serious, and HBP is manageable. Seventy percent who have it are aware they have it, yet only 34% of individuals with hypertension have it under control.[23]

Etiology

Family history and ethnic background increase the risk of hypertension; African Americans are most likely to have hypertension. Salt intake can also contribute to hypertension, although not every individual is salt-sensitive.

Researchers with the Intersalt Study calculated that over time, 20% of hypertension in Western societies is attributable to salt intake.

Effects of Hypertension

Prolonged high blood pressure puts extra tension on blood vessels and organs in the body, wearing them out before the natural aging process. Damaged kidneys are a common sign of uncontrolled hypertension.

Risk Factors

Nutritional risk factors are drinking alcohol to excess, high-saturated-fat diets leading to dyslipidemia and atherosclerosis, lifestyles resulting in overweight and obesity, and a diet low in calcium.[23]

Nutritional Remedies

Nutritional strategies to normalize blood pressure include weight management, moderation of alcohol intake for those who drink, and limiting sodium intake while maintaining adequate potassium, magnesium, and calcium intakes. The Dietary Approaches to Stop Hypertension, or DASH, diet (see Illustration 19.1 and Table 19.4) is effective in decreasing blood pressure[24] and risk of stroke in adults under age 65. It has been shown to enhance perceptions of quality of life.[25] Finding beneficial effects of lifestyle modifications on blood pressure led researchers to recommend lifestyle plus the DASH as intervention focus

Illustration 19.1 DASH works to reduce blood pressure.

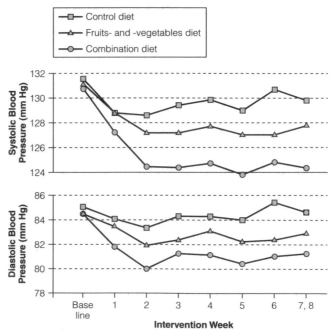

SOURCE: Table taken from Appel LJ et al. A clinical trial of the effects of dietary patterns on blood pressure: DASH collaborative research group. NEngJMed 1997; 336:1117–1124.

Table 19.4 The DASH eating plan for blood-pressure control

	Servings per Day	Serving Sizes of Foods within the Food Group
Grains and Grain Products Especially whole-grain[a]	7–8	Breads: 1 slice or 1 oz Cereal: ½ c cooked or dry Rice, pasta: ½ c cooked
Vegetables Fresh, frozen, no-salt-added canned	4–5	Raw, 1 c; cooked, ½ c
Fruits Fresh, frozen, or canned in juice	4–5	Juice: 6 oz Fresh: 1 med piece Mixed or cut: ½ c Juice: 6 oz Dried: ¼ c
Dairy Foods Skim or 1% milk, fat-free dairy products	2–3	Milk: 8 oz Yogurt: 1 c Cheese: 1½ oz
Meats, Poultry, and Fish	Up to 2	3 oz, cooked
Nuts, Seeds, Dry Beans	4–5 per week	⅓ c or 1½ oz nuts 2 Tbsp or ½ oz seeds ½ c cooked beans (legumes)
Fats and Oils[b] Select olive, canola, corn, and safflower oils	2–3	1 tsp soft margarine, oils, mayonnaise 1 Tbsp low-fat mayonnaise 2 Tbsp light salad dressing
Sweets	Up to 5 per week	1 Tbsp jam, jelly, syrup, or sugar ½ oz fat-free candy, jelly beans, or 12 oz sweetened beverage

SOURCE: Adapted from a 2000-calorie eating plan, the National High Blood Pressure Education Program's HeartFile, Winter 1999, National Heart, Lung, and Blood Institute and the National Dairy Council's "DASH TO THE DIET," 2000.

[a]Whole grain is the entire edible part of wheat, corn, rice, oats, barley, and other grains. Whole-grain bread has the words "whole grain" before the type of flour that is listed; whole-grain breakfast cereals include the word "whole" or "whole-grain" before the grain name (e.g., "whole-grain wheat").
[b]One serving is equivalent to 5 grams of fat.

in individuals over age 50.[26] Researchers found that the over-50 participants benefitted more than their under-50 counterparts. Other non-drug interventions have successfully lowered the blood pressure (e.g., using weight reduction and/or sodium restriction of 1500 to 1800 mg per day over 30 months).[27,28] This sodium level is well below current intakes (see Table 18.12).

In the DASH-sodium study, the greatest overall blood pressure reduction occurred in the subjects with the strictest sodium intake limit (1500 mg a day).[28] Blood pressure reduction occurred whether individuals were normo- or hypertensive. Choosing foods with less processing can help to limit sodium intake, because approximately 75% of dietary sodium is attributable to manufacturing and preservation processes, with salt at the table contributing the rest. For instance, the sodium in a plain potato averages 10–15 mg compared to 150–200 mg sodium in a serving of potato chips.

Effectiveness of the DASH eating plan continues to be tested in various settings. For example, a cardiovascular health-center team, including a dietician, used a modified DASH approach combined with exercise to help patients in an outpatient, office-based counseling program.[29] The team saw patients (aged 55 ± 12 years) for CVD and weight loss. Patients had significant weight loss (5.3% of body weight) that was maintained for the 2.6 years of study follow-up. Diastolic blood pressure reduction was also significant, although systolic pressure reduction was not. Several researchers are developing DASH-adherence scoring systems to help study the effects of following a DASH diet. Scoring systems can enhance the comparability of research using DASH eating patterns as dietary intervention. In contrast to Folsom's study of CVD mortality and hypertension in nearly 21,000 women, which found no statistically significant outcomes using a DASH concordance score,[30] Levitan and colleagues compared four scoring methods and found that following the DASH diet led to decreased heart failure for middle-aged and older women (48–83 years old).[31] The women who most closely following the DASH diet (the top quartile of diet scores) had a 37% lower rate of heart failure than the lowest-scoring quartile in the 7-year-long study.

The DASH diet is one of the meal plan options in the MyPyramid supporting materials because it is considered to be health-promoting for the general public as well as for people with hypertension. Dietary and lifestyle changes that address hypertension are also likely to have a beneficial effect on atherosclerosis.

Diabetes

Special Concerns for Older Adults

In the National Health Interview Survey 2005–2006, nearly 1 in 5 of all adults aged 65 years and older reported having diabetes, primarily type 2 (see Table 19.1). Native American, Latino, African American, Asian American, and Pacific Islander adults face higher risks for diabetes than do Caucasians.[32]

Individuals with diabetes are at greater risk for heart disease and its complications; diabetes itself is an independent risk factor for atherosclerosis. Four of five older people have diabetes as one of several comorbid conditions, and these complicate diabetes management.

Diabetes diagnosis criteria and management goals are the same for older as for younger adults, using individualized treatment plans that include assessment of functional status, cognitive capacities, and motivation.[32] While glycemic goals, such as Hg A_1C, may be relaxed to fit individual situations, hyperglycemia and risk of complications "should be avoided in all patients."[32] For older adults, diabetes may exacerbate declining organ functions, making them less resilient.

Effects of Diabetes

Diabetes leads to a tenfold greater risk of amputations, macular degeneration, visual loss, cataracts, glaucoma, and neuropathies (nerve damage, pain, or tingling) of the hands and feet. Hyperglycemia may lead to sodium depletion and dehydration, trace mineral depletion (zinc, chromium, magnesium), insomnia, nocturia, blurred vision, increased platelet adhesiveness related to atherosclerosis, increased infection and decreased wound healing, and aggravated peripheral vascular disease.

Alcohol and drugs such as salicylates (aspirin) contribute to drops in blood sugar. Hypoglycemia in older adults may lead to weakness, confusion, and possible

falls and fractures. Other reasons for falls include declining vision and nerve function. For example, in the Health, Aging, and Body Composition[33] study of older adults with diabetes, 22% to 31% of participants reported falling in a 12-month span. Mean age at study enrollment was 73.6 years, and the percentage of participants who self-reported falls during each subsequent year of study follow-up was 24% (baseline), 22%, 26%, 31%, and 30%. Authors attributed the increased risk of falling to reduced peripheral nerve function, renal function, and vision; all are diabetes complications.

Nutritional Interventions

One result of increasing life spans is that older adults with diabetes now have more years to maintain desired quality of life and avoid complications. Diabetes self-management training, to the extent that the individual is able to manage his or her own regimen, works in tandem with medical nutrition therapy to achieve glycemic control. Special concerns for older adults include the following:

1. Where nephropathy (chronic kidney disease) is present, limit protein intake to 0.8–1.0 g/kg/day, depending on stage of disease.[32]

2. Assess dietary adequacy and supplement with vitamins and minerals to meet age-appropriate nutrient DRI.

3. Monitor functional status and modify the care plan as appropriate. For example, carbohydrate counting requires vision to read the label and memory to track carbohydrate grams. Glycemic management in older adults may be harder to achieve due to altered senses, decreased mobility, difficulty in buying, preparing, or eating food, social changes, depression, and comorbid conditions. Providers must attend to the psychosocial and the physical needs of an aging individual.

4. Ask about special foods and alternative and complementary therapies. Complementary medicine has long history among Mexican American and Native American populations. Evening primrose oil, milk thistle, fenugreek seeds, and prickly pear cactus leaves (nopales) are foods used as botanical treatments. Safe complementary nutritional remedies can enhance the standard nutritional therapies.

Try to clarify confusing food terms, especially when it comes to pre- and probiotics. For example, the sunchoke or Jerusalem artichoke (a vegetable which can be used like a potato) contains inulin, a fructose polymer that is absorbed more slowly than other starches and serves as a prebiotic because it provides substrate for bacteria in the gut. The *inulin* in sunchokes could be confused with *insulin*. Inulin is fermented by gut bacteria, leading to gas production that may cause distension and discomfort.

5. Sugar alcohols (e.g., xylitol, sorbitol) in candies and gums are much sweeter than sucrose and fructose, and they provide very little energy. High doses of these polyols lead to diarrhea.

6. Carbohydrate and fiber recommendations do not change with age, but older adults who have constipation or diarrhea may also want to know the potential benefits of fiber for glycemic control. Moderate increases in carbohydrate (raising percent of energy from 45 to 55%) have been well tolerated in 42- to 79-year-old adults with diabetes when given as breakfast cereal over a 6-month period.[34] High fiber intake has been significantly associated with reduced inflammation. In Britain, Wannamethee and colleagues tracked nearly 8000 men for 20 years (60–79 years old at follow-up) to study their diet and health outcomes.[35] For 7 of those years, they followed a healthy cohort of 3428 men, assessing dietary fiber intake, inflammation (C-reactive protein and Interleukin-6), hepatic function, and risk of type 2 diabetes. Eating at least 20 grams of fiber (cereal, vegetables, fruit) per day was associated with less hepatic fat, lower levels in markers of inflammation, and decreased risk of developing diabetes.

Glycosylated Hemoglobin   A laboratory test that measures how well the blood sugar level has been maintained over a prolonged period of time; also called Hemoglobin A₁C.

Cost-benefit analyses have shown that tight control of blood sugar (maintaining *glycosylated hemoglobin* below 7%) can lead to better quality of life for older adults with diabetes and can result in fewer long-term complications.[36,37]

Obesity

Definition

The National Heart, Lung, and Blood Institute[38] and the World Health Organization[39] define obesity as a BMI of 30.0 or higher, and extreme obesity as a BMI of 40 or higher. These definitions are intended for adults of all ages. However, the underlying research for BMI cut-points relied primarily on young and middle-aged populations.[40] In older adults, BMI alone is not an adequate indicator of excess body fat associated with morbidity and mortality.

Prevalence

Population mean body weight and BMI tend to peak around age 60.[41] As people pass age 70, obesity rates begin to decline; see Table 19.5. But the girth of the average older adult has grown along with that of the rest of the country. Beginning in the 1980s, obesity rates increased from 13% to 25% in men and from 19% to 24% in women aged 75 and older.[1]

Table 19.5   Obesity rates decline in old age: 2005 NHANES data

Age	Percent of Males Who Are Obese	Percent of Females Who Are Obese
65–74 years	33	37
75 and older	25	24

SOURCE: Reference #1, Indicator 25.

Etiology, Effects, and Risk Factors of Obesity

Decreased functional status in old age may exacerbate obesity, but more importantly, it may interfere with fitness. In 2007, approximately one fourth of adults over age 75 found it difficult or impossible to walk a quarter of a mile or to climb up 10 steps without resting. One of the challenges for an aging and increasingly obese population is to design fitness programs to accommodate adults who have functional limitations as well as those who are fully mobile. Cardiorespiratory fitness contributes to longevity of older adults as described by Sui and colleagues in the Aerobics Center Longitudinal Study.[42] 2603 adults with a mean age 64 years were followed for 12 years to find that body composition and fat distribution did not predict longevity, but that BMI and fitness as measured by treadmill test did. Fit individuals with a BMI of 30–34.9 had greater longevity than lean and normal-weight unfit individuals. This demonstration that fitness contributes more to longevity than does absolute body weight in older adults raises the question: what are healthy weights for older adults?

Analyses from the National Center for Health Statistics at the CDC support the notion that in adults, the body mass index associated with the lowest mortality falls within the combined healthy and overweight range of 18.5–30.0.[43] Flegal's analysis of BMI data and mortality found that "the majority of deaths associated with obesity were associated with BMI 35 and above." Perhaps the associations among health, BMI, and aging are unclear because BMI is not a good measure of fatness. Based on Ernsberger's analysis,[44] morbidity and mortality is not any higher, and sometimes is lower, in older people who are at the high end of the BMI continuum. A study of older community-dwelling Canadians found that increased BMI was associated with lower mortality.[45] Measuring the excess body fat that leads to morbidity and mortality is of increasing interest to researchers and policy makers, in part because of the expanding, increasingly older population. Two large longitudinal studies that followed populations consisting of[46] or including[47] older adults suggested that abdominal obesity is a better measure of premature death than BMI. An exploration of obesity as a risk factor for stroke and transient ischemic attacks found

Illustration 19.2 Markers of obesity.

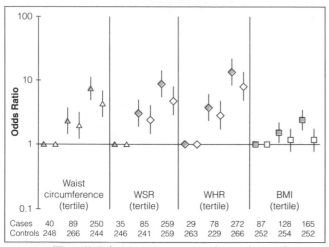

Cases   40   89   250   35   85   259   29   78   272   87   128   165
Controls 248 266 244   246 241 259   263 229 266   252 254 252

SOURCE: Winter, Y., Rohrman, S., Lineisen, J. et al. Contribution of obesity and abdominal fat mass to risk of stroke and transient ischemic attacks. Stroke 2009; 40. Wolters Kluwer Health. Copyright © 2008, American Heart Association, Inc. Available at http://stroke.ahajournals.org, accessed 12/30/08.

NOTE: WSR as Waist-to-Stature Ratio and WHR as Waist-to-Hip Ratio.

that markers of waist circumference, waist-to-stature ratios, and waist-to-hip ratio predicted stroke cases better than BMI (see Illustration 19.2).[48]

For older adults, extra weight during illness episodes, especially hospitalizations, seems to be protective. Materials developed in 1999 by the American Dietetic Association Long Term Care Task Force in conjunction with the Centers for Medicare and Medicaid Services suggest a BMI range of 19–27 as an acceptable and health-promoting weight range for older adults.[49] This range is similar to yet broader than that used by the Nutrition Screening Initiative, where BMI cut-points are 22–27. (See Table 19.6 for a compilation of BMI ranges suggested for use with older adults.) Using BMI to define obesity in older adults comes with so many caveats (difficulty in measuring height, shifts in body composition from youth, hydration status) that assessment will be more meaningful when using more reliable measures of excess body fat. Weight gain or loss as changes in BMI often serve as an initial estimate for health and fitness, awaiting elaboration with a more detailed nutrition assessment. The BMI is a marginal proxy at best because it was designed to measure and compare population thinness and fatness, not to assess the fat or lack of fat in an individual.

Nutritional Remedies

Researchers in the Baltimore Longitudinal Study of Aging (BLSA) wanted to explore the relationships of eating patterns to the gradual weight gain commonly experienced with increasing age.[50] Data from 7-day food records were used to create 41 food groups for a cluster analysis. Five clusters emerged: 1) Healthy, (which was high in fruits, vegetables, reduced-fat dairy, and whole grains, and low in red and processed meat, fast food, and soda), 2) White Bread, 3) Alcohol, 4) Sweets, and 5) Meat & Potatoes. Compared to the Healthy pattern, the mean annual increase of BMI was highest in the Meat & Potatoes cluster, at 0.3 ± 0.06 kg/m², and the annual increase in waist circumference was highest in the White Bread cluster (1.32 ± 0.29 cm), compared to the Healthy pattern. Maintaining lean mass is particularly important for older adults, and the DASH-like pattern of the Healthy cluster in the BLSA ensures adequate nutrient intake.

Table 19.6 BMI values suggested for use with older adults

Screening/Assessment Tool or Researcher	BMI (kg/m²) That Represents Smallest Risk of Malnutrition	Comment
Mini Nutritional Assessment (MNA), see Illustration 18.3	≥23	"Obese" does not generate higher risk scores; tool developed by Nestle Nutrition Institute
Malnutrition Universal Screening Tool (MUST), www.bapen.org.uk	>20 and no recent weight loss	A BMI of 30 or greater does not generate a higher risk score; developed for use in communities and long-term care settings
American Dietetic Association (ADA) Consultant Dietitians in Health Care Facilities	BMI of 19–27 with no change in weight	Tool developed by the ADA in conjunction with the Centers for Medicare and Medicaid Services, tracking weight trends is one of many strategies to assess nutrition risk
Nutrition Screening Initiative (NSI), Physician's Guide to Nutrition in Chronic Disease Management	22–27	Developed by American Academy of Family Physicians, the NSI and ADA; partly funded by Ross Products, Abbott Laboratories
G.M. Price et al.[46]	Assess diet and health status when BMI <23 in men and <22.3 in women	At age 75 or older, use waist-to-hip ratio
K.M. Flegal et al.[43]	18.5 to 30.0	Body mass index associated with lowest mortality

Sarcopenic obesity may complicate attempted weight loss in older adults.[51] Exercise can preserve and build muscle (see Chapter 18) to ensure that the weight lost will not be lean body mass. A healthy eating program is based on sufficient nutrient-dense calories to support a gradual loss of fat without losing bone or muscle tissue. Weight management was discussed in earlier chapters; a balance of servings from basic food groups, such as outlined in the MyPyramid or the DASH eating patterns, can promote health in older as well as younger adults. Comorbidities in older adults may require the help of a registered dietitian to balance various nutrition priorities, such as sodium, protein, and saturated fat restrictions when high blood pressure and kidney disease complicate diabetes in an obese individual.

Sarcopenic Obesity Low lean body mass combined with excessive fat stores.

Physical activity promotes functional independence. The Healthy People 2010 goal was for 40% of the population to perform physical activities that enhance or maintain flexibility. Approximately 22% of adults 65 years or older achieved this goal. Physical activity is the only way to prevent muscle loss. Functional limitations need not be barriers to exercise. Physical therapists and community and senior centers can be helpful resources to develop and promote special routines using chairs, a pool, and adaptive tools.

Osteoporosis

"Watch it when you hug Grandma."

Mary Nelsestuen, afraid that a strong hug might break her frail mother's osteoporotic ribs

Definition

Osteoporosis means "porous bone." Reduced bone mass and disruption of bone architecture can result from an imbalance of available nutrients. Osteoporosis progression depends on the homeostatic mechanism involved. An accelerated phase of bone loss occurs due to estrogen or testosterone loss. Bone mass loss is greater for women, who can lose up to 20% in the 5 to 7 years past menopause.[52] After age 65, rates of loss are typically less than 1% per year.[53] Men develop osteoporosis later than women because they have larger frames, and their testosterone levels fall more slowly, typically over a period between ages 40 to 70. However, men's bone mass losses double after androgen deprivation (a treatment for prostate cancer).

World Health Organization criteria for bone mass density (BMD, based on measures such as dual-energy X-ray absorptiometry or DXA scans) are used worldwide to diagnose osteoporosis. BMD that falls 2.5 or more standard deviations below values for healthy young adults denotes osteoporosis.[54] BMD that is 1–2.5 standard deviations below the adult normal (osteopenia) precedes osteoporosis and may lead to fracture.

Daily dietary patterns affect bone health because bones are constantly being remodeled. But the hormonal shifts associated with aging result in overall loss of bone mass. Up to 50% of trabecular or spongy bone (wrist, vertebrae, and ends of long bones) and up to 35% of cortical or compact bone (shafts of long bones) may be lost during a lifetime.[55]

Prevalence

Osteoporosis is four times more common in women than men (80% compared to 20%), in part due to greater peak and total bone mass in men. Blacks and Hispanics have greater BMD than do whites. In Hispanic populations, 10% of women and 3% of men over 50 are estimated to have osteoporosis.[52] Prevalence rates for osteoporosis are elusive because osteoporosis has no symptoms, such as feeling one's bones becoming weaker. Diagnosis relies on BMD measurement or fracture. Statistics for falls and fractures[56] are presented in Table 19.7. The risks are

Table 19.7 Percentage of persons age 65 and older who self-reported falls and fall-related injuries within the last 3 months, by gender and ethnicity; data from the Behavioral Risk Factor Surveillance System, 2006

Characteristic	Percentage Who Had At Least One Fall During Preceding 3 Months	Among Those Who Fell, Percentage Who Were Injured At Least Once
Women	16.4	35.7
Men	15.2	24.6
Race, ethnicity		
American Indian, Alaska Native	27.8	34.3
Asian, Hawaiian, Pacific Islander	13.0	25.7 (very small data set)
Black	13.0	32.8
Hispanic	17.4	41.0
White	15.8	30.3
Women and men, 80 and older	20.8	32.1

SOURCE: Reprinted from J Prosthet Dent., 73:65–72. Faine, M. P., Dietary factors related to preservation of oral and skeletal bone mass in women. Copyright 1995 with permission from Elsevier.

high: older adults are more likely to suffer disability from a bone fracture than from prostate cancer, rheumatoid arthritis, breast cancer, and hypertension.[54] The chance that an adult over age 50 will have an osteoporosis-related fracture during the remainder of his or her life span is approximately 1 in 2 for women and 1 in 4–5 for men.[52,57]

Etiology

Bone mass is gained primarily during growth periods, with peak bone density reached between ages 18 and 30. Subsequently, bone mass remains stable until about age 40 to 50 for women (menopause) and about age 60 for men. Inadequate building of peak bone mass coupled with significant bone loss leads to a low bone density and increased risk for fractures.

Inadequate Bone Mass   Although osteoporosis is seen most often in the elderly, the risk for developing osteoporosis in later years begins during childhood and adolescence. Development of osteoporosis is delayed when an individual develops bigger, denser bones during youth.[58] For example, an epidemiological study in Yugoslavia showed that higher calcium intake in youth led to higher peak bone mass, independent of exercise and other factors. Higher bone mass is associated with slower decline in later life.[59] Studies in the United States have also shown that getting enough calcium during growth spurts (between ages 11 and 17) reduces the risk of osteoporosis. In an intervention study, girls aged 11 to 12 years who received calcium supplements (500 mg calcium citrate-malate) gained an additional 1.3% bone mineral density per year compared to controls.[60]

Inactivity such as that due to bed rest and sedentary lifestyle leads to bone loss. Weight-bearing or resistance exercises are needed to develop bone mass because bone grows in response to pressure on the bone tissue. The more often and the harder you push on the bone (not enough to break it, of course), the more the body will respond by depositing minerals (calcium, magnesium, phosphorus, fluoride, and boron) into the bone matrix. For instance, tennis players have significantly higher bone mass in their playing arm than do nonathletic controls.[61] However, even the controls had more bone mass in their dominant arm than in the less-used one. Exercise also stimulates growth hormone, which in turn stimulates bone development. "Use it or lose it" applies here. Adults who are frail may fear falling and fracture to the degree that it prevents them from getting much-needed exercise, contributing to a vicious circle.

Increased Bone Loss   The skeleton acts as structural support and as a calcium reservoir for the body. Bone tissue includes jawbones and teeth. Bones and teeth contain about 99% of the calcium in an adult, roughly 2.2 to 3.3 lb.[61] The remaining 1% of calcium is found linked with protein in blood, soft tissues, and extracellular fluids. This reservoir is needed for nerve transmission, muscle contraction, and enzyme systems such as those controlling blood clotting. Maintaining nerve transmissions takes physiologic priority over maintaining bone structure. In order for calcium to be consistently available, it is tightly regulated by hormone systems. PTH levels are increased by low blood 25-hydroxy vitamin D levels, by high phosphate levels, and by low calcium levels. When calcium levels in the blood fall, the body responds by secreting more parathyroid hormone (PTH). PTH acts to raise blood calcium levels by increasing dietary calcium absorption, decreasing urinary excretion, and by releasing calcium from bone. Excess blood calcium stimulates calcitonin secretion. The hormone calcitonin slows release of stored calcium.[61] Bone mineral reserves are dissolved (resorption) and rebuilt continuously, thus maintaining adequate calcium levels for crucial messenger functions.

A consistent dietary supply of bone-building minerals (i.e., calcium, magnesium, phosphorus, fluoride, boron, zinc, copper, and manganese) and vitamins (primarily D and K) coupled with regular weight-bearing exercise helps maintain the skeletal mineral reserves. When a portion of this build-dissolve-rebuild cycle is malfunctioning, the body's first priority is to maintain blood calcium levels for nerve, muscle, and enzyme functions.

Osteoporosis can develop from a shortage of phosphoros during bone mineralization. A varied diet provides a balanced calcium–phosphate ratio and allows both nutrients to be used by the osteoblasts. Lack of sufficient phosphorus promotes release of calcium from the skeleton. Although phosphorus is abundant in the food supply, some antacids bind with phosphorus, making it unusable by the body and delaying bone formation until more phosphate is available. Shortage of vitamin D also delays bone mineralization.

Finally, the process of normal aging results in a slow increase of PTH as well as a decrease in the skin's ability to make vitamin D; both lead to bone loss.

Effects of Osteoporosis

Falls and Fractures   Fractures and the resulting injuries may make it impossible for an older adult to remain independent, but not everyone who falls sustains injury. Table 19.7 shows incidence of falls and subsequent injuries in various population segments. Note the greater risk for persons over age 80 compared to those aged 65 and older. Ten to 20% of older persons who break a hip die within a year.[62] Death is not due to the fracture itself but to complications resulting from the break. One of these complications is impaired mobility, complicating all the activities of daily living (including eating and exercising). If an older adult has also had a stroke, impaired mobility becomes the leading cause of institutionalization in the United States. Furthermore, 50% of older individuals who fracture a hip have permanent functional disabilities.

Shrinking Height, Kyphosis   In contrast to hip fractures, most vertebral fractures (67%) are asymptomatic. Postmenopausal women with compression and/or a bone fracture in the spinal column have a condition known as "shrinking height," leading to dowager's hump (also known as *kyphosis*, meaning a bent upper spine). Shrinking in height is slow and usually not painful. For example, a woman who was 5 feet 6 inches tall at age 30 may measure 5 feet tall at age 83. She may not notice the gradual height loss until someone comments or until she notices that clothes no longer fit.

Risk Factors

A typical osteoporosis patient is a petite elderly white female. Brittle bones develop from a complex array of physiological factors, including poor nutrition and lack of exercise. Major risk factors for osteoporosis are listed in Table 19.8. The WHO has developed an online calculator that generates a 10-year probability of fracture based on several risk factors, such as age, height, weight, smoking, having had a previous fracture, and bone mass density score based on the type of DXA machine used (FRAX™ algorithm available at www.nof.org or www.shef.ac.uk/FRAX).

Table 19.8 Risk factors associated with osteoporosis in older adults

Not Modifiable
Female, multiple pregnancies, length of time between pregnancies
Age at which pregnancy(ies) occurred, breastfeeding
Family history of osteoporosis, maternal history of hip fracture
Caucasian, Asian
Thin, small-boned rather than large-boned
Low body weight, sarcopenia
History of amenorrhea, premature menopause
Hypogonadism, low levels of testosterone
Glucocorticoids, steroid use

Potentially Modifiable
Lack of weight-bearing exercise
Cigarette smoking
Long-term dietary phosphorus deficiency (e.g., use of phosphorus-binding antacid)
Heavy alcohol consumption
Underweight, malnourished
Inadequate dietary calcium (<1200 mg) and vitamin D (<400/600 IU) intake

Still Controversial or Not Yet Clear
Diet high in phosphorus while low in calcium (mixed evidence)
Inadequate fluoride, boron, and magnesium in diet
Consistently high protein and/or low fruit and vegetable intake

Nutritional Remedies

Calcium   On average, older Americans are consuming 300 to 450 mg less calcium than the DRI, which is 1200 mg per day (see Table 18.12), and still less than the 1500 mg calcium recommended by the Canadian osteoporosis management guidelines.[63] The goal is to provide enough available calcium and vitamin D through diet or supplements so that bone loss is minimized despite declining absorption rates. Many foods are good sources of calcium, and a sample meal plan that provides the DRI for calcium is shown in Table 19.9.

Adhering to a calcium supplementation routine may not be any easier than eating a healthful diet every day. In a 5-year, double-blind, placebo-controlled study with nearly 1500 women, mean age 75 years, the authors concluded that supplementation with 1200 mg of calcium carbonate per day "is effective in those patients who are compliant."[64] The problem in recommending calcium supplementation to prevent clinical fractures was "poor long-term compliance."

Balancing Nutrients for Bone Health   While calcium is the best-known nutrient when it comes to bone health, other dietary components contribute to the complex relationships among hormones, muscles, and bones. Depending on level of intake, some nutrients interfere with calcium metabolism (see Table 19.10). High protein levels lead to greater calcium excretion, although losses

Table 19.9 An example of one day's food that provides at least 1200 mg calcium

Food	Amount of Calcium (mg)
Oatmeal made with milk 1 c total	266
Banana, one medium	6
Coffee, 10 oz with 1 oz (2 Tbsp) evaporated milk	87
Turkey sandwich on whole-wheat bread, lettuce, mayonnaise	54
Cheese added to sandwich, 1 oz cheddar	148
Canned fruit cocktail, ½ c	9
Iced tea, plain	0
Orange juice, calcium-fortified, 8 oz	289
Roasted almonds, 2 Tbsp	33
Pasta with chicken, 1½ c	54
Tomato slices, 2	2
Sugar cookie, 1 medium	5
Chocolate milk, 8 oz	287

Table 19.10 Dietary components that contribute to calcium in urine	
	Calcium Lost in Urine
Salt: for 1 gram salt (2300 mg sodium)	~ 26 mg
Protein: for 1 gram dietary protein	~ 1 mg
Protein amount in 1 oz of meat, fish, poultry, or egg	~ 7 mg
Quarter-pound hamburger	~ 25–30 mg
Caffeine: 6 oz of regular coffee	~ 40 mg

are small. In the Nurses Health Study, women (aged 35 to 59, followed for 12 years) who ate more than 95 grams protein per day suffered more osteoporotic forearm (but not hip) fractures.[65] On the other hand, Heaney and Layman reviewed the role of protein in sarcopenia and bone health and concluded that "higher protein diets are actually associated with greater bone mass and fewer fractures when calcium intake is adequate."[66] They found that high-protein diets (30% of energy) did not lead to excess calcium excretion in urine. Instead, low protein intakes were associated with lowered intestinal calcium absorption and increased PTH levels.

To promote the effective daily remodeling of bone, several nutritional habits can improve calcium intake and absorption:

- Consume calcium-rich beverages like milk and kefir or yogurt drinks with a meal: food slows intestinal transit time and allows more calcium to be absorbed from the gut; protein from dairy products becomes substrate for new bone matrix.

- Increase fruit and vegetable intake for their alkalinizing effect; the alkaline environment improves calcium balance by inhibiting bone resorption.[66,67]

- Consume foods that are rich in bone-building vitamins and minerals at recommended levels. This balance will promote bone synthesis of the collagen matrix, into which minerals are deposited during bone mineralization.

- Vitamin D stimulates active calcium transport in the small intestine and colon; for residents of locales with seasonally ineffective sun exposure, supplemental vitamin D can maintain blood levels to suppress PTH levels and increase calcium absorption.

- Greens are especially rich sources of the vitamin K needed for formation of proteins that stimulate *osteoblasts* and *osteoclasts*, active in bone remodeling. (See Table 19.14 for food sources of vitamin K.)

- High sodium intake leads to higher levels of urinary calcium; that is, more calcium is excreted when higher levels of sodium are eaten.

- Caffeine to equal 2 to 3 cups of coffee daily (in postmenopausal women), consumed in conjunction with a low calcium intake, has been associated with bone loss.

- If taking supplements, divide dosage throughout the day for greater absorption. Carbonate is better absorbed with food, while citrate is well absorbed with or without food. Take calcium at a different time than antacids, because stomach acidity enhances absorption. In cases where stomach acidity is decreased, calcium citrate is more soluble than calcium carbonate.

Exercise   Strong bones require exercise and good nutrition, which is why the Canadian and U.S. Clinical Practice Guidelines both include exercise prescriptions.[63,57] Bed rest and immobilization lead to rapid loss of bone mass. Weight training builds bones. Thus, Canadian practice guidelines for nonpharmacologic interventions in osteoporosis recommend physical activity ≥30 minutes per day three times per week.[63]

Other Issues Impacting Nutritional Remedies

Vitamin K Caution   Vitamin K contributes to increased bone mineral content,[68] and some calcium supplements now also include vitamin K. However, vitamin K also plays an important role in the blood clotting process. When older adults with a history of strokes are placed on an anticoagulant like warfarin, nutrition counselors can advise them on how to maintain a stable vitamin K intake.

Two forms of vitamin K are found in foods. Phylloquinone, also known as vitamin K_1, is naturally occurring in plants, and menatetrenone, vitamin K_2, is found in meat, cheese, and fermented products. Big portions of broccoli and greens for a few weeks in summer will decrease the effectiveness of warfarin and thus increase potential for blood clotting and another stroke. In addition, when a vitamin K–containing supplement is added to the diet, intake may quickly become excessive. How much is too much? As yet, no tolerable upper levels are set for vitamin K.

Hormones   Hormones direct the dynamic system of bone remodeling. Estrogen, testosterone, growth hormone, and insulin-like growth factor-1 (IGF-1) increase intestinal calcium absorption. Hormone replacement, with or without additional calcium, can effectively increase bone mineral density in postmenopausal women.

Osteoblasts   Bone cells involved with bone formation; bone-building cells.

Osteoclasts   A bone cell that absorbs and removes unwanted tissue.

However, in 2002, the Women's Health Initiative reported increased rates of breast cancer, coronary heart disease, stroke, and venous thromboembolism in addition to decreased rates of hip fracture and colorectal disease.[69]

Medications   Serotonin reuptake inhibitors (SSRIs) such as Prozac are frequently prescribed for depression; they are associated with bone loss. In trying to understand signaling pathways involving low-density lipoprotein receptors and serotonin, researchers discovered that serotonin produced in the duodenum suppressed osteoblast proliferation. Writing about the deleterious blocking effects of gut-produced serotonin on bone mass in a perspective paper for the *New England Journal of Medicine*, Dr. Clifford Rosen described the intricate links between skeletal tissue and that of the gut and brain.[70] He called his paper *Serotonin rising–the bone, brain, bowel connection*. His paper highlights the complex mechanisms supporting bone health. Calcium intake is one piece of a complex system. It seems easy to accept that blood cells and taste buds are renewed in a matter of days, but it is not so easy to picture bones as a dynamic tissue that is being remodeled continuously, even in old age. The systems of older adults are just not quite as efficient and resilient as those of younger adults.

In summary, the best osteoporosis prevention strategy is exercise and adequate diet in young people when bones are first growing. For older individuals who have brittle bones, a nutrient-dense diet, supplemented with calcium and vitamin D, coupled with appropriate exercise, strengthens bones. The many bone-building medications available for people with osteopenia and osteoporosis should be discussed with a physician.

Oral Health

> "We can't have a healthy mouth, a great smile, or a good conversation without it. Saliva, or 'spit,' lubricates living."
>
> Nelson Rhodus, professor, Division of Oral Medicine and Diagnosis, University of Minnesota

It is possible to be well nourished without a full set of teeth, or even without any teeth at all. However, the ability to bite and chew facilitates fruit and vegetable intake, which is linked to better health. Fortunately, more people are keeping their teeth as a result of better dental care. At the turn of the century, about one-third of all adults had no natural teeth; now that number has dropped to about one-fourth of older adults who have no natural teeth (see Table 19.11, which shows the prevalence of being edentulous by age and poverty).[71] Chewing is more effective with natural

Xerostomia   Dry mouth, or xerostomia; can be a side effect of medications (especially antidepressants), of head and neck cancer treatments, of diabetes, and also a symptom of Sjogren's syndrome, which is an autoimmune disorder for which no cure is known.

Gingiva   Gum tissue.

Table 19.11   Proportion of older people who reported having no natural teeth, by age, race, and poverty, 2007

Age and Poverty Status	No Natural Teeth
Age 65–74	22.4%
Age 75 and older	30.1%
For ages 18 and older, by race and by economic status	
White	7.5%
Black or African American	9.0%
American Indian, AK native	13.9%
Asian	7.0%
Native HI, Pacific Islanders	Insufficient data
Poor	12.5%
Not poor	6.1%

SOURCE: Adapted from Table 12 in Pleis JR, Lucas JW. Summary health statistics for U.S. adults: National Health Interview Survey, 2007. National Center for Health Statistics. Vital Health Stat 10(240), 2009.

teeth; individuals who wear complete dentures have approximately 20% of the chewing ability of those having all natural teeth.[72] Foods can be cut, sliced, chopped, grated, and made into smoothies and soups to ensure a varied diet when chewing ability is impaired. Lack of natural teeth does not have to result in a poor diet.

Older adults seek dental care at similar rates to other age groups. Nearly two-thirds of adults age 65 and over saw a dentist or dental health professional in the last year (41–45% within the last 6 months and another 14–15% within the last year), compared to 57% of all adults over age 18.[71]

Changes in oral health are most likely to be a result of disease, medical treatment, or medications rather than aging itself. Periodontal disease (PD) and *xerostomia* (known as dry mouth) are two conditions that can interfere with food tolerance and with enjoyment of food. PD results from bacterial infections of the *gingiva*, with destruction of the ligaments attaching teeth to the jawbone, and with receding gums. Plaque builds up in the resultant pockets, contributing to further infection and eventual tooth loss. Persons whose overall health and immune system are compromised are at greater risk for periodontal disease. Prevention of PD emphasizes strict oral hygiene to remove plaque, enhancing immune status, and ensuring optimal nutrition. Besides deficiencies of vitamin C, folic acid, and zinc that are associated with PD, correcting potential deficiencies of calcium, vitamin D, and magnesium will help postmenopausal women keep their bones, including the jaw, strong. Ensuring optimal nutrition includes management of diabetes. High blood sugar raises glucose in saliva, leading to caries and accelerated periodontal disease. It can also make the mouth more susceptible to yeast infection (candidiasis). Diabetes control translates into better oral health.

Except for Sjogren's syndrome, where xerostomia is a side effect, medications and other treatments are the likely causes of dry mouth, and thus prevention may be impossible. Amelioration may be likely. For example, diuretic treatment for hypertension leads to decreased salivary secretion; other medications that tend to lead to dry mouth include anti-anxiety drugs, antidepressants, sedatives, and antihistamines. Head and neck cancer treatment can also lead to xerostomia when the salivary glands are involved. Lack of saliva for any reason gives bacteria a better environment to build plaque. Further interfering with enjoyment of food with xerostomia is the tendency to have loss of taste (dysgeusia) and pain of the tongue (glossodynia). The key treatment of xerostomia is good oral hygiene, especially after meals, plus stimulating saliva with sugar-free candy, frozen fruit bits, chewing xylitol-flavored gums, and sipping water liberally. Artificial saliva can also help to keep the oral cavity moist.

Good oral health for older adults also includes caries prevention. Brushing, flossing, and dietary recommendations for older adults match those of younger ones. Tea sippers (black tea, no sugar) may even have an advantage here; the polyphenols in black tea seem to interfere with the bacteria's ability to stick to the plaque. Unfortunately, swishing with tea does not cancel the need to brush and floss to remove plaque.

Gastrointestinal Diseases

The gastrointestinal (GI) system serves so many functions that it should not be surprising to learn that by late adulthood, it occasionally malfunctions. It seems miraculous that we so consistently eat what we like, without much thought, and our body converts that food to energy for daily living. Parts of the GI system most likely to malfunction in old age are:

1. The esophageal-stomach juncture: weakened muscle results in *gastroesophageal reflux disease* (GERD)
2. The stomach: decreased acidity leading to *changes in nutrient absorption* and increased acidity aggravates ulcers
3. The intestines: resulting in *constipation*, diarrhea, and some food intolerance

Often these problems are secondary to other diseases. No matter what the cause, older adults are at higher risk for the GI conditions discussed next, any of which may impair older adults' activities.

Gastroesophageal Reflux Disease (GERD)

Definition   Occasional heartburn that becomes chronic may be due to gastroesophageal reflux disease. GERD results when stomach contents flow back into the esophagus.

Prevalence   Statistics range from 10% of Americans having daily episodes of heartburn to an estimated 25–35% of the U.S. population being affected with GERD.[73]

Etiology and Effects   It is not clear if acid in the esophagus leads to a weakened lower esophageal sphincter (*LES*) or if a weakened sphincter leads to GERD. The main symptoms of GERD are heartburn and acid regurgitation. Stomach contents, which are highly acidic, spill back into the esophagus, resulting in irritation, belching, hoarseness, and substernal pain. Ulceration and swallowing disorders are symptoms of severe cases of GERD. *Helicobacter pylori (H. pylori)*, a cause of peptic ulcer disease, does not cause GERD.[74]

> **LES**   Lower esophageal sphincter, which is the muscle enabling closure of the junction between the esophagus and stomach.

Nutritional Risk Factors   Excess alcohol, obesity, and smoking are consistently linked to GERD episodes. In addition, both regular and decaffeinated coffee are associated with heartburn.

Nutritional Remedies   The primary dietary remedy is to omit foods that are chemically or mechanically irritating. There is little consistency about which foods these are for an individual. However, general guidelines are to choose a low-fat diet, avoid large meals, and to take advantage of gravity by remaining upright for several hours after eating. High-fat meals and alcohol both lower LES pressure. Fermented beverages (wine, beer) and caffeine stimulate gastric acid, so reducing them may give some relief. Protein also stimulates gastric acid, so consuming the necessary protein throughout the day rather than in one or two sittings will help to minimize reflux. Spicy foods, chocolate, peppermint, citrus, and tomato products are among potentially reflux-pain inducing foods.[73]

Stomach Conditions Affect Nutrient Availability: Vitamin B$_{12}$ Malabsorption

Definition and Etiology   Pernicious anemia and atrophic gastritis are two conditions which result in vitamin B$_{12}$ deficiency. Pernicious anemia (macrocytic megaloblastic anemia), which is due to lack of intrinsic factor (IF) being released from the stomach cell wall, is associated with hypochlorhydria and impaired iron absorption, and it is rare in individuals under 35. It is also uncommon even in older adults. The most common cause of vitamin B$_{12}$ deficiency in older adults is related to abnormal stomach function. Prolonged inflammation, followed by atrophied stomach mucosa secreting less acid, leads to atrophic gastritis. Older adults with atrophic gastritis still absorb some dietary vitamin B$_{12}$, but they will become deficient over time. Bacterial overgrowth (usually related to infection with *H. pylori*) will also use vitamin B$_{12}$ and diminish the amount available for absorption. Finally,

Cobalamin Another name for vitamin B_{12}. Important roles of cobalamin are fatty-acid metabolism, synthesis of nucleic acid (i.e., DNA, a complex protein that controls the formation of healthy new cells), and formation of the myelin sheath that protects nerve cells.

Methylmalonic Acid (MMA) An intermediate product that needs vitamin B_{12} as a coenzyme to complete the metabolic pathway for fatty-acid metabolism. Vitamin B_{12} is the only coenzyme in this reaction; when it is absent, the blood concentration of MMA rises.

Krebs Cycle A series of metabolic reactions that produce energy from the proteins, fats, and carbohydrates that constitute food.

Homocysteine Another intermediate product that depends on vitamin B_{12} for complete metabolism. However, both vitamin B_{12} and folate (another B vitamin) are coenzymes in the breakdown of certain protein components in this pathway. Thus, elevated homocysteine levels can result from vitamin B_{12}, folate, or pyridoxine deficiencies.

antacid treatment neutralizes stomach acid and raises stomach pH, preventing vitamin B_{12} from being split from its protein carrier, even though the intrinsic vitamin B_{12} factor is present. Symptoms of vitamin B_{12} deficiency begin to appear after 3–6 years of poor absorption.[75]

Absorption of vitamin B_{12} from foods requires stomach acidity, enzymes (especially pepsin), and intrinsic factor (IF). Stomach acids and enzymes split off and transfer vitamin B_{12} from foods to carrier proteins (mostly secreted by the salivary gland in the mouth). Then the vitamin B_{12}-carrier protein complex moves into the small intestine, where the vitamin B_{12} will again be broken off and

bound to the IF (produced in the stomach and migrated to the small intestine). The vitamin B_{12}–IF linked complex then binds to a specific site in the lining of the small intestine, is transported across, and released into blood serum to be taken to tissue cells. (See Illustration 19.3.)

Prevalence Nilsson-Ehle found prevalence rates for vitamin B_{12} deficiency of up to 80% in the papers he reviewed, but he found that typical estimates were much lower.[75] Protein-bound vitamin B_{12} deficiency is estimated to occur in 20% of people over 69 years old (known and undiagnosed).[76]

Serum *cobalamin* levels are used to detect deficiency, although the cutoff limit may have been set too low for older adults.[75,77] For example, community-dwelling adults aged 61 to 87 entering a 5-year study of brain volume loss all had normal vitamin B_{12} levels (above 185 pmol/L) throughout the study. The groups with the least brain atrophy had significantly higher serum levels of vitamin B_{12} (308–386 pmol/L in the middle tertile and >386 pmol/L in the highest tertile) than the lowest tertile with the greatest brain volume loss (<308 pmol/L).[77] To properly assess vitamin B_{12} status, it may also be useful to measure the blood levels of intermediate breakdown products in pathways where vitamin B_{12} is needed. For example, vitamin B_{12} is needed to change *methylmalonic acid (MMA)* into a component used in the *Krebs cycle*. Blood levels of MMA rise when adequate amounts of vitamin B_{12} are unavailable. A test for levels of MMA is specific to vitamin B_{12} deficiency because vitamin B_{12} is the only coenzyme to catalyze this reaction. Another test is for blood levels of the coenzyme *homocysteine*. If vitamin B_{12} isn't available to complete the pathway that forms DNA, then homocysteine (intermediate metabolite) levels become elevated. Since folate is also needed in this metabolic pathway, lack of either folate or vitamin B_{12} can lead to high levels of homocysteine.

Elevated MMA and homocysteine levels can confirm a vitamin B_{12} deficiency. However, if MMA levels are normal but homocysteine levels are high, folic acid deficiency or another cause of increased homocysteine levels should be investigated.

Effects Vitamin B_{12} deficiency leads to irreversible neurological damage, walking and balance

Illustration 19.3 Overview of vitamin B_{12} absorption.

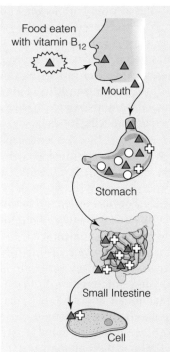

Food eaten with vitamin B_{12}

Mouth — The salivary glands in the mouth secrete enzymes and mucus to begin breaking down foods and provide substrates for binding.

Stomach — The food substrate–enzyme mixture enters the stomach where acid and enzymes (especially pepsin) detach the vitamin B_{12} from protein in foods and reattach it to binder proteins. The acid and enzymes trigger release of an intrinsic factor from the stomach cell wall (known as vitamin B_{12} intrinsic factor, or IF).

Small Intestine — The B_{12} complex moves to the small intestine, where enzymes secreted by the pancreas release vitamin B_{12} and transfer it to the IF that was secreted by the stomach. Next, the vitamin B_{12}–IF complex moves to the ileum where it enters cells through specific receptors. IF is degraded and B_{12} is transported by transcobalamins or through blood to tissues.

Cell

Key

▲ Dietary vitamin B_{12} (extrinsic)

✛ Intrinsic factor vitamin B_{12}

▲✛ Protein-bound vitamin B_{12} complex

○ Protein-binder complex found in stomach at normal acidity

disturbances, and cognitive impairment (including confusion and mood changes). High levels of homocysteine are known to be a risk factor for heart and peripheral vascular disease.[78]

Risk Factors   Risk factors for vitamin B_{12} malabsorption include advanced age, gastrointestinal disorders, genetic family patterns, medications, and (to some extent) inadequate food intake. An example of aging as a risk factor comes from Nilsson-Ehle, who reported that 1.5% of 50-year-olds have achlorhydria and B_{12} deficiency as compared to 18% of 80-year-olds.[75] Gastrointestinal disorders that affect vitamin B_{12} absorption include atrophic gastritis (with higher risks in diabetes mellitus and autoimmune thyroid disorders), partial stomach removal, and *H. pylori* infection. Medications that suppress stomach acid secretion or impair absorption are associated with the risk of deficiency. Examples of these types of medicines are oral biguanides (e.g., metformin used to treat type 2 diabetes mellitus), modified-release potassium preparations, hydrogen-receptor antagonists (e.g., Cimetidine or Tagamet), and proton pump inhibitors (e.g., omeprazole/Prilosec given for gastroesophageal reflux disease). In genetically linked cases, individuals often have antibodies for the intrinsic factor.

Vegetarians not taking a B_{12} supplement are at risk of vitamin B_{12} deficiency. For example, a study of 245 Australian Seventh Day Adventist ministers found that most (70%) had a vitamin B_{12} deficiency due to a low dietary intake.[79] Even with intrinsic vitamin B_{12} factor present, 53% of the group had serum cobalamin levels below the reference level for this study (171–850 picomoles per liter).

Some ethnic and gender differences may exist for developing a vitamin B_{12} deficiency. Elderly white men had the highest prevalence rates, whereas black and Asian women had the lowest rates among community-dwelling seniors (aged 60 or older) in Los Angeles.[80] In contrast, Zeitlin and associates[81] concluded that there were no age- or gender-related differences in the mean levels of serum B_{12} after studying approximately 440 subjects in the Bronx Aging Study (mean age = 79 years).

Nutritional Remedies   The DRI of vitamin B_{12} for men and women over 50 years is 2.4 mcg/day,[82] and the tolerable upper intake level is not determined.[83] However, oral pharmacological doses for vitamin B_{12}-deficient patients are 0.2–1 mg (200–1000 mcg) and are above the DRI because roughly 1–2% is absorbed through passive diffusion in the small intestine.[84] A dose-finding study tested vitamin B_{12} doses from 2.5 mcg to 1000 mcg with healthy, free-living adults over age 70 and found that mild deficiencies can be normalized with doses of 500 to 1000 mcg of cyanocobalamin, depending on the blood indicator measured.[85] Foods fortified with high levels of crystalline or synthetic vitamin B_{12} increase the likelihood of passive

absorption. Foods high in folate and iron are also necessary in cases where folate deficiency and/or iron deficiency has been diagnosed. Fortified flour and breakfast cereals supply almost two-thirds of the total folate in the food supply.[86] Thus, for older adults, vitamin B_{12}- and folic acid-fortified whole-grain cereal is a "power food."

Constipation

Definition   There are many definitions of "normal bowel pattern." Normal patterns of defecation can consist of two or three bowel movements per day to two or three bowel movements per week. To help sort out what constitutes an abnormal pattern, a working group of gastroenterologists developed the Rome II criteria, which are now in use.[87] Functional constipation in adults means that in the past year, an individual had 2 or more of the following for at least 12 weeks (in 25% or more of defecations): straining, lumpy or hard stools, sensation of incomplete evacuation, sensation of blockage in rectum, manual manipulation, and fewer than 3 defecations per week.

Prevalence   When "two or fewer bowel movements per week" is used as the definition of constipation, the prevalence of constipation was 3.8% in older adults aged 60–69 years and 6.3% in older adults aged 80 years and older.[88] These rates are much lower than the percentage of the older population concerned about the regularity of their bowel movements. For example, of people aged 65 years and older living in a Midwestern community, 40% reported having had some type of constipation.[89]

Etiology and Effects   Medications and diseases such as diabetes, cancer, Parkinson's, and irritable bowel and other gastric diseases are associated with constipation. There is also a public perception that "being regular" means having a bowel movement each morning. Media hype around cleansing diets further contributes to perceptions that stool should be cleared from the colon daily. Older adults are not immune to these promotions. Writing about the widely held myths and misconceptions surrounding normal bowel movements, Müller-Lissner and three international gastroenterologist colleagues introduced their paper with an observation that could fit other conditions: "As with any widespread disorder that has been incompletely understood, there are many strongly held beliefs, which are often not evidence-based . . . ."[90] Table 19.12 summarizes the misconceptions they encounter. Stool transit times, weights, and consistencies vary greatly among healthy adults of all ages. This is understandable when considering the main functions of the colon, which are water conservation, serving as a place where bacterial action can digest food remnants to usable nutrients, and acting as a site for waste storage until suitable times and places for a bowel movement. Animal studies indicate that aging intestinal muscles respond less to triggers and

Table 19.12 Misconceptions about chronic constipation

Common Myth	What the Evidence Shows
Stool that remains in the colon for a long time causes "autointoxication"; this theory gave rise to colonic cleansing, enemas.	There is no evidence that toxins are absorbed from the colon.
An elongated colon may cause constipation because too much water is reabsorbed from stool, making it hard, and there is danger of kinking.	Calculated normal colon lengths in human adults ranged from 95 to 180 cm in 6 studies reported; length depends on methodology. No studies link colonic length with transit time.
Hormones lead to constipation.	Some do, in a very small percentage of older adults, hypothyroidism is associated with constipation, hyperthyroidism with diarrhea.
Constipation is due to a diet low in fiber and is best treated with dietary fiber.	Fiber has various functions and is not digested uniformly; this complicates studies. Yes, fiber increases stool bulk. Wheat bran reduces transit time. A low-fiber diet may be a contributor to constipation, but some cases of severe constipation are exacerbated by additional dietary fiber.
Low fluid intake causes hard stool.	Interviews with older adults showed no association between estimated fluid intake and constipation; drinking mineral water may have a slight laxative effect due to magnesium content. Except in cases of dehydration, "available data do not suggest that stools can be manipulated to a clinically relevant extent by modifying fluid ingestion."
A sedentary life style and inactivity leads to constipation.	Bowel-management programs in nursing homes (fiber, fluid, exercise) have reduced laxative consumption. "Intervention programs to increase physical activity as part of a broad rehabilitation program may help."
Laxative intake causes electrolyte disturbances and abdominal complaints.	Abdominal complaints occur with constipation and with laxative use; stimulant laxatives may cause cramping. Laxatives can cause electrolyte disturbances, but these can be managed.
Laxative use induces habituation and tolerance.	Limited studies of laxative tolerance suggest that tolerance is uncommon "in the majority of users." Those with severe cases using stimulant laxatives may need increasingly large doses.

SOURCE: Table content compiled from summary information in Reference 90.

that the brain-muscle transmitters are either inadequate or less responsive. Older adult's thirst mechanisms decline. Lower levels of stomach secretions and potentially less muscle strength affect peristalsis. There is a potential for avoiding fiber-rich foods due to chewing problems; such low-fiber diets can also exacerbate *diverticulitis*. Declining cognitive function may become so severe that the individual may not recognize the urge to defecate; this can also lead to constipation and fecal incontinence.

Diverticulitis Infected "pockets" within the large intestine.

Risk Factors Consistent evidence that exercise relieves constipation in older adults is lacking, although epidemiologic studies have found that higher levels of regular physical activity were associated with a lower risk for constipation.[87] A classic paper by Burkitt in 1984 reported that, compared to developing nations, U.S. populations had lower stool weights (300–500 gm/d and 80–120 gm/d

respectively) and longer average stool transit times (36 hours compared to 72 hours).[91] He also reported that U.S. populations had low fiber intakes compared to the developing countries he studied. Nutritional risk factors for constipation include:

- Dehydration
- Medications, such as opioid and nonsteroidal anti-inflammatory drugs
- High-iron and other mineral supplements and antacids

Nutritional Remedies Traditional remedies are to encourage increased dietary fiber in tandem with increased fluid. In the presence of dehydration, adding fluid improves constipation. There are no studies available to suggest the minimum fluid intake that will maintain the normal bowel routine of an older adult. In healthy adults, fluid intake can vary widely while maintaining regular

Illustration 19.4 High-fiber recipes as alternative to laxatives.

Power Pudding, also known as Behm's Special Recipe
1 cup applesauce
1 cup unprocessed bran, all-bran, or bran buds
½ cup prune juice

Mix all ingredients and refrigerate of freeze.
Serve 2 tablespoons at supper, follow by extra fluid such as 1 cup of water.

You can even find Power Pudding at Cooks.com (accessed 4.21.2010):

1½ cups pitted prunes
1 cup applesauce
½ cup All-Bran
¾ cup prune juice
Put all ingredients into blender and puree until smooth. Serve ¼ cup daily, followed by a glass of water.

bowel movements.[90] Optimizing food and fiber intake will help create stool mass; on average, older adults fall far short of the recommended fiber intake. Insoluble fiber sources include bran and cellulose in fruits and vegetables; these fibers are fermented by the colonic bacteria, producing gas (leading to flatulence). Bacterial type and food residue determine gas production. Sugar alcohols are fermented in the colon, as are stachyose and raffinose, two of the carbohydrates in legumes. Bran in breads and cereals is one way to add to fecal bulk that stimulates peristalsis to move wastes through the colon. The fiber section of Chapter 18 offers suggestions for incorporating dietary fiber (see Tables 18.13 and 18.14). Long-term care nutrition services have devised special recipes to wean residents off laxatives, such as incorporating psyllium fiber, apple sauce, and juice into fruit-flavored gelatin. One particularly popular method is called Behm's Special Recipe or Power Pudding (see Illustration 19.4). There is one caution before increasing fiber: find out if there is potential for fecal impaction due to disease such as colon cancer. Integrating activity and exercise training can improve peristalsis and alleviate constipation; it may also relieve individuals who worry about bloating and flatulence.

Inflammatory Diseases: Osteoarthritis

Definition

Inflammatory diseases include periodontal disease, osteoarthritis, rheumatoid arthritis, celiac disease (*gluten* intolerance), irritable bowel disease, diverticulitis (an infection in the large intestine), atrophic gastritis (typically due to *Helicobacter pylori* infection), and asthma. Of these conditions, arthritis affects the greatest number of older individuals.

Osteoarthritis is the most common form of arthritis, affecting approximately 27 million adults aged 25 and over[92] (of a U.S. population numbering 217.8 million people aged 18 and older in 2008; "25 and older" was not available for comparison) and is the focus of this section. Prevalence of osteoarthritis at the knee is twice as common as at the hip; the knee and hip are the two most common sites for persons aged 30 and older. Prevalence increases with age and peaks between 70 and 79 years. More men than women have osteoarthritis before age 50; after age 50, women are more often affected.

Etiology

Cartilage loss, bone hypertrophy, changes in the synovial membrane, hardening of soft tissues, and inflammation leading to tissue damage constitutes osteoarthritis. Variability in occurrence and progression of knee, hip, and hand osteoarthritis suggests that the disease may be a group of different and unique conditions, but it is primarily a wear-and-tear disease that becomes more prevalent with age. In contrast to osteoarthritis, rheumatoid arthritis is an autoimmune collagen disease characterized by inflammation and increased protein turnover.

Gluten   A protein found in wheat, spelt, kamut, dinkle/dinkel and triticale (genus Triticum), barley, and rye. Spelt, dinkel, and kamut are ancestral forms of today's wheat. Oats appear on some "gluten" lists; oats are inherently gluten-free, but they may be contaminated by gluten-containing grains during processing.

Effects of Osteoarthritis

Osteoarthritis is a degenerative joint disease. Joint movement brings on pain because cartilage that cushions the bone ends has eroded.[93] Osteoarthritis is among the most disabling conditions of older adults. Treatment goals are to control pain,

Case Study 19.1

Photodisc

Bridget Doyle Remembers Laura

Just because she lives at Lenoir Manor, a continuing-care retirement facility, Laura, a petite (4 ft. 8 in., 97 pounds) widow of the local college dean, does not consider herself as old. She is 87. She has had no major nutritional or health problems and her appetite is good. She had been a good cook and entertained graciously, but in the residential care facility, meals are prepared for her. Because she has had slight fluid retention over the past year, she no longer adds salt to her meals. She tells Bridget Doyle, her nutritionist, that yes, occasionally she does not like her meals and misses cooking for herself.

One Monday morning, Laura is found in bed with her left side paralyzed. The diagnosis is a right-sided stroke, resulting in three weeks of hospitalization. Back at the skilled-care wing of Lenoir, Laura needs a nasogastric tube for feeding. She is alert and knows people but is limited in speech. Overnight, Laura's care has changed from an individual needing routine nutritional monitoring to someone with many interrelated problems:

- Inability to communicate her overall medical and nutritional concerns clearly
- Weight loss of 9 pounds during the three-week hospital stay
- Intense dislike of the nasal tube, as demonstrated by repeated attempts to pull it out, leading to restraint of her hands

Questions

Assessment

1. What nutritional parameters should be assessed and monitored now that Laura is back at Lenoir Manor?

Diagnosis

2. What disciplines should be involved in Laura's care plan, and why?
3. The interdisciplinary care team wants to meet Laura's needs in a dignified and respectful manner. How can the care team address both clinical and ethical concerns?
4. What are strategies young adults can adopt to reduce their risk of stroke?

Monitoring and Evaluation

5. How could Bridget ensure that Laura's nutritional needs are met?

(Answers are located in the Instructor's Manual for the 4th edition of *Nutrition Through the Life Cycle*.)

---

improve joint function, maintain a normal body weight, and achieve a healthy lifestyle.

Risk Factors

Obesity, continuous overexposure to oxidants, and possibly low vitamin D levels are risk factors for developing osteoarthritis. Obesity may have two detrimental effects. One is the weight stressing the joint, and the other is due to secretion of cytokines in adipose tissue. Fat is a metabolically active tissue that secretes signaling molecules that can trigger inflammation.[94] Low intakes of vitamins C and D are risk factors for osteoarthritis progression. Research may identify unique risk factors for the various joints involved (knee, hip, hands, lower back, and neck).[93]

Nutritional Remedies

Weight loss is the first remedy advised.[10,95] Felson and colleagues report[95] that women in the Framingham study who lost 11 lb cut their risk for knee osteoarthritis in half. Messier and associates[10] found that weight loss as small as 1 lb reduces load exerted on the knee; the beneficial effect adds up as steps accumulate during the day. Other nutritional approaches to symptom reduction[96] include the following:

Antioxidants   Individuals with the highest levels of vitamin C intake had significantly slower (threefold) disease progression and less knee pain than individuals with the lowest intakes. Results of increased vitamin E and beta-carotene intake were inconsistent, although diets providing modest amounts of carotenoids have been linked with reduced incidence of inflammatory polyarthritis.[93]

Vitamin D   Progression at higher intakes and serum levels is roughly one-third slower than at the lowest levels. The DRI for older adults is 400 to 600 IU, with a tolerable upper intake level of 2000 IU.

Flavonoids   This is a large group of phytochemicals with antioxidant and anti-inflammatory properties that act to maintain cell membranes. Higher levels of antioxidants are required to scavenge oxidized metabolites and free radicals in inflammatory diseases.

Chondroitin and Glucosamine   These two substances naturally occur in the body as substrate for cartilage repair. Clinical trials of their use, separately and together, demonstrate significant pain reduction when knee pain is moderate to severe. Mild pain did not improve significantly more than the placebo effect.[97] A caution: product tests have shown that quality and levels of active compound display surprisingly great variation. Not all products contain the specified levels of active compound. Check with the FDA or go to www.consumerlab.com to find out how specific brands rate.

SAM-e   S-adenosylmethionine may reduce pain and stiffness; in large doses, it may also cause nausea.

Capsaicin   The compound that provides the heat in peppers is used in a topical cream for pain relief.

Other Treatments   Fatty acids and oils are other anti-inflammatory therapies with potential to lessen signs and symptoms in a variety of conditions, including osteo- and rheumatic arthritis. Borage seed and evening primrose oils contain GLA or gamma-linolenic acid, which plays a role in prostaglandin synthesis (GLA competes with other fatty acids, limiting the production of inflammatory omega-6 fatty acids) and can decrease pain of arthritis after several months of use.[96] Flaxseed and purslane also contain a type of omega-3 fatty acid. Traditional Native American medical therapy used echinacea (three varieties) for pain relief, rheumatism, and arthritis; ginseng for asthma and rheumatism; garlic for asthma; and evening primrose oil for obesity.[98] Today, echinacea and ginseng are marketed to boost immune function, and evening primrose oil is marketed as an antioxidant.

Vegetarian Diets   Plants are rich in antioxidants that may play a role in managing inflammatory diseases in general.

Food Allergies   Food allergies have been suggested as a cause for inflammatory diseases such as irritable bowel syndrome, Crohn's disease, and rheumatoid arthritis, but evidence from well-done studies that can point to allergens as a mechanism for osteoarthritis are not available.

Cognitive Disorders: Alzheimer's Disease

"The better the legs, the better the mind."

Dr. Paul Dudley White, cardiologist

Definition

One of the most dreaded aspects of "getting old" is losing one's ability to function independently and becoming dependent on others. *Memory impairment* is a step in the loss of independence. More serious losses of memory and cognitive function are frequently grouped under the term *dementia*. Dementia is a condition of progressive cognitive decline, typically characterized by impaired thinking, memory, decision-making, and linguistic ability. Dementia is not a disease itself, but rather a set of symptoms associated with particular degenerative neurological conditions:[99]

> **Memory Impairment**   Moderate or severe impairment is when 4 or fewer words can be recalled from a list of 20.

Alzheimer's disease
Vascular dementia
Dementia with Lewy bodies
Frontal-temporal lobe dementia
Parkinson's disease
Alcohol-related dementia
AIDS-related dementia

Dementia is not a part of normal aging, but rather the manifestation of various forms of physiological damage.

Prevalence of Dementia

In the United States, approximately 14% of older (≥71 years) adults have dementia and approximately 10% have Alzheimer's disease (AD).[100] Prevalence increases with age, so that more than one-third of individuals age 90 and older have dementia. AD ranks fifth in leading causes of death for adults aged 65 and older in the United States.[101]

Etiology of Cognitive Disorders

Dementia can be caused by a variety of conditions resulting in traumatic physiologic changes (see Table 19.13). The symptoms of dementia may be completely or partially reversible in some conditions, including early treatment of drug abuse, metabolic disorders such as a vitamin B_{12} deficiency and hypoglycemia, and medical interventions such as surgery to remove tumors or change of medications.[102] Nontreatable forms of dementia include those associated with degenerative diseases such as Alzheimer's, Huntington's, and Parkinson's.

Alzheimers is the most common cause of dementia, and searches for a cause (leading to treatment) are ongoing. Aluminum, copper, carnitine, and choline have been examined as possible causes of AD due to their role in neurological function, but they have not been demonstrated to be causal.

Deficiencies of vitamin B_{12} and folate are related to high concentrations of homocysteine, an amino acid associated with the promotion of poor vascular health and cognitive decline.[77,102] In order to prevent the buildup of homocysteine in the blood and neural tissue, vitamin B_{12} and folate are needed to convert it to the amino acid methionine. Methionine contributes to the synthesis of S-adenosylmethionine, which is widely distributed throughout the central nervous system for use in methylation reactions. Primary examples of methylation reactions include vitamin B_{12} in the production of myelin (the insulation cover on nerves) and folate in the DNA cycle (cell replication). Excess homocysteine in brain tissue is thought to contribute to the development of Alzheimer's disease either through vascular mechanisms or as a neurotoxin. High-dose vitamin B supplementation has decreased homocysteine levels but has not slowed progression of cognitive decline in people with AD.[103]

To discover dietary habits that may prevent cognitive decline, and especially AD, researchers are using broader approaches, including review of dietary patterns and physical activity. The Mediterranean diet has shown promise in prolonging longevity and in delay of cardiovascular disease. Researchers in a prospective cohort study in Bordeaux, France, developed an adherence score and assessed the dietary patterns of 1410 adults aged 65 and older.[104] After adjustment for cardiovascular risk factors, higher adherence to the Mediterranean diet was associated with slowed cognitive decline in one of the four cognitive function tests (the Mini-Mental State Exam). In contrast, a prospective cohort study of 1880 older adults in New York showed that the Mediterranean-type diet and physical activity were independently related to lower AD risk.[105] Comparisons of adults with better diet scores and higher levels of physical activity to those with low scores on both showed that a dose-response relationship in risk reduction may exist. This can be an important message for older adults who need some encouragement to start a health and fitness program.

Other promising research into the mechanisms of cognition in aging deals with the theory of caloric restriction and inflammation.[106] A small intervention study in humans (29 healthy women, mean age 60.5 years) showed that a reduced calorie diet (30% below normal intake, not to drop below 1200 kcal) improved insulin sensitivity, inflammatory response (tumor necrosis factor-alpha), and memory performance on the Rey Auditory Verbal Learning Task test. Both obesity and long-term underweight were associated with lower cognitive scores in the larger Whitehall Cohort Study.[107]

Table 19.13 Conditions associated with cognitive disorders and dementia

Condition	Changes Leading to Cognitive Disorders
Vascular dementia	"Mini-strokes" and vascular obstructions deprive brain of oxygen and other nutrients
Degenerative diseases (Alzheimer's and Parkinson's)	Neurological changes eventually affect memory and ability to think
Physical trauma and infection	Brain injuries may impair physical and/or cognitive function
Depression	Change in mood may affect, ability to maintain high-level functioning
Chronic substance, alcohol abuse	Neurological damage may become irreversible
Malnutrition, including B_{12} deficiency and dehydration	Confusion resulting from dehydration is reversible, neurological damage of B_{12} deficiency is permanent, chronic hunger keeps brain from focusing on life beyond acquiring food
Deficiencies of niacin, thiamin, folic acid, biotin, iron, and selenium	Decreased ability to learn and depression are among the effects of these vitamin and mineral deficiencies

The term *cognitive disorders* can be used to describe a large range of mental malfunctions. Older adults may worry less about the classification of cognitive disorders than about how to maintain functional independence. The etiology of AD remains elusive. But as communities face caring for a large numbers of cognitively impaired older adults, any progress into elucidating contributors to cognitive health will find an appreciative audience, both in individuals with the disorder and, perhaps more so, in their caregivers.

Effects of Cognitive Disorders

Confusion, anxiety, agitation, loss of oral muscular control, impairment of hunger and appetite regulation, changes in smell and taste, chewing and swallowing difficulties, and dental problems are all aspects of Alzheimer's disease that make it difficult to maintain good nutritional habits. As the disease advances, individuals with AD will require more and more assistance with meal preparation and eating. In later stages of the disease, wandering and restless movements expend energy and increase caloric need. Behavioral, physical, or neurologic problems may impede adequate food intake. Consequently, individuals with late-stage AD suffer from unintentional weight loss (this is discussed in the "Low Body Weight/Underweight" section).

Nutrition Interventions for Cognitive Disorders

Ensuring food safety and safe use of kitchen tools and equipment are primary considerations for good nutrition; many dangers lurk in the kitchens of cognitively impaired persons who are trying to maintain an independent life at home. When a caretaker is present, concerns turn to ensuring adequate intake rather than preventing foodborne illness or injuries related to food preparation.

Since there is no cure for AD, the dietary focus is to maintain a nutrient-dense diet that is acceptable to the individual, maintains hydration, and supplies needed energy. Additional calories may be needed for increased energy expended by individuals who pace and wander. Persons with dementia can benefit from meals offered in a calm dining environment, free of loud noise and confusion (no highly patterned tableware, nothing extraneous on the table). Additional strategies caregivers might use to promote food and fluid intake are to 1) maintain the focus on eating, 2) provide plenty of time to eat, 3) serve finger foods, and 4) encourage regular drinks between bites. In cases of decreased physical coordination, adaptive eating utensils, such as slip-resistant placemats, mugs or cups without saucers and spoons, and silverware with built-up handles, may promote independent eating.

Medications and Polypharmacy

Prescription medicines are those ordered by a physician or other licensed health care provider, and do not include vitamins, herbal medicines, or over-the-counter (OTC) medicines. OTC medications are any pills, liquids, salves, creams, and supplements that are purchased at a pharmacy, discount, or food store without prescription. Complementary medicines such as botanicals and herbs are typically sold as OTCs. Table 18.7 lists drugs most commonly taken by older adults; OTCs are at the top of the list.

Polypharmacy is common in older adults, and it is associated with adverse drug reactions, hospital admissions and readmissions, and increased mortality. Polypharmacy is one of the 10 risk factors in the Nutrition Screening Initiative DETERMINE tool. Disease progression, drug effects, and functional limitations such as poor eyesight or impaired memory all increase risks for using a drug incorrectly. The potential for error also increases with the number of drugs used. Taking multiple medications becomes inevitable when someone has multiple illnesses.

Effects of Medications   Medications may require dietary restrictions and can interfere with appetite, digestion, metabolism, and general alertness. For example, the blood-thinning drug warfarin requires a stable vitamin K intake (see Table 19.14). The cost of drugs may compete with the food budget. Some medications lead to unintentional weight loss, while others lead to undesired weight gain. Table 19.15 describes nutritional implications associated with medications used to treat diseases that are prevalent in older adults.

Low Body Weight/ Underweight

Definition

There is no consensus or universal definition for underweight in the frail elderly. The most common methods that measure changes in nutrition status are body mass index (BMI) and unplanned weight loss within the past 3 to 6 months, see Case Study 19.1. A BMI falling into the lowest percentile of a reference standard in a comparable population is a starting point. Because of the National Health and Nutrition Examination Survey III, we now

Table 19.14 Amounts of vitamin K in selected fruits and vegetables (DRI for ages 51 and older is 120 mcg/day for males, 90 mcg/day for females)

Less than 5 mcg vitamin K:

1 c servings of corn, mushrooms, onions, baked and navy beans, potatoes, applesauce, cherries, pineapple, strawberries

1 apple, banana, nectarine, peach, pear, orange, tangerine; 5 dried apricots, 5 dates

5 to 10 mcg vitamin K:

1 c canned apricots, fresh raspberries, chickpeas, lima beans, all varieties of squash, stewed tomatoes, sweet potatoes, fruit cocktail, papaya

2 figs

>10 and <50 mcg vitamin K:

1 c yellow and green string beans, blueberries, red cabbage, carrots, cauliflower, celery, cucumber with and without peel, iceberg lettuce, roasted peppers, canned plums, grapes (red or green)

½ cup cooked chard

1 kiwi, 2 sprigs parsley

>50 mcg vitamin K (1 c cooked unless specified):

Beet greens, turnip greens (529–851)

Broccoli (183–220)

Cabbage, any (27–58)

Collards (836–1059)

Kale (1062–1147)

Romaine lettuce (57)

Green leaf lettuce (97)

Mustard greens (419)

Okra (64)

Raw green onion (207)

Prunes or dried plums (65)

Rhubarb (71)

Raw spinach (145)

Cooked spinach (1027)

SOURCE: www.nal.usda.gov/fnic/foodcomp/Data/SR18/nutrlist/sr18a430.pdf, accessed 8/24/06, and NDSR, University of Minnesota, accessed 8/31/09.

have weight percentiles for older adults by decade (beginning at 50 years), gender, and ethnic-racial groups and are able to recognize underweight in older adults (see Table 19.16).[108] Individuals with weights falling at the 5th or lower percentile of this population can be defined as "underweight."

Another approach is to compare an individual's current weight to "usual" body weight. Terms such as sarcopenia (muscle loss), anorexia (no appetite), and cachexia (another way to say "no appetite") are related to undernutrition in older adults and are associated with becoming frail.

The National Heart, Lung, and Blood Institute (NHLBI) defined underweight as a BMI <18.5 kg/meter squared for all adults.[109] The World Health Organization

(WHO) further defines levels of underweight as grades of "thinness."[110]

BMI 17.0–18.49 indicates grade 1 thinness
BMI 16.0–16.99 indicates grade 2 thinness
BMI <16.00 indicates grade 3 thinness

Approximately one-third of older adults are underweight.

Etiology

Underweight is not considered problematic when the individual has had a lifelong low weight. However, weight cycling is problematic. In addition, unintentional weight loss is likely due to disease. A loss of 10% or more of

Table 19.15 Medications associated with chronic conditions in older adults: nutritional implications

Coronary Heart Disease
- Cardiac glycosides (e.g., digitalis) may result in anorexia and/or nausea.
- Statins may result in elevated enzymes.
- High doses of niacin may be associated with flushing, hyperglycemia, hypotension, hypoalbuminemia, upper GI distress, and liver enzyme elevation.

Hypertension
- Diuretics may result in depletion of sodium, calcium, magnesium, and/or potassium.
- Centrally acting anti-hypertensives may result in a decline in food intake due to sedation, confusion, and depression.
- Medications such as beta-blockers may cause constipation and delayed gastric emptying.
- Potential related side effects are dizziness, dry mouth, or mouth pain.

Diabetes Mellitus
- Medications to treat diabetes may cause hypoglycemia, especially if nutritional intake is erratic and/or if there is increased or decreased appetite or diarrhea.
- Oral hypoglycemics (e.g., sulfonylureas such as glipizide) may cause heartburn, nausea, hypoglycemia, decreased appetite; biguanides (such as Glucophage) may cause decreased appetite, diarrhea, and vomiting.

Dementia
- Cholinesterase inhibitors (e.g., rivastigmine or galantamine) may cause nausea, diarrhea, or weight loss.
- Antipsychotic/antidepressants with anticholinergic side effects may lead to dry mouth, delayed stomach emptying, and constipation.
- Antidepressants may enhance appetite in depressed patients, but SSRIs may cause a decrease in appetite.

Congestive Heart Failure
- Diuretics (e.g., Diuril) may lead to electrolyte abnormalities, especially sodium and potassium and/or thiamine deficiency (e.g., furosemide).

Cancer
- Radiation, chemotherapy, and/or surgery can negatively affect nutritional status and metabolism.

Chronic Obstructive Pulmonary Disease (COPD)
- Xanthine derivatives (e.g., theophylline) may result in anorexia and nausea.

Lower Respiratory Infections
- Antibiotics (e.g., amoxicillan) may result in nausea, anorexia, diarrhea, and stomatitis.
- Radiation, chemotherapy, and/or surgery can negatively affect nutritional status and metabolism.

SOURCE: American Academy of Family Physicians, Nutrition Screening Initiative, American Dietetic Association. *A physician's guide to nutrition in chronic disease management for older adults.* Waldorf, MD: 2002. Can be ordered at e-mail address nsi@gmmb.com.

Table 19.16 NHANES III, 1988–1994 reported data of 5th percentile mean weight in pounds by age and gender in older adults residing in the United States

Age	Male	Female
60–69	134.5	109.0
70–79	128.7	100.5
80	114.3	92.0

SOURCE: Third National Health and Nutrition Examination Survey (1988–1994). Available at www.cdc.gov/nchs/about/major/nhanes/Anthropometric%20Measures.htm, accessed 7/9/03.

total body weight in a 6-month period is associated with increased mortality.

For older adults, underweight is much more serious than overweight. Being thin has been related to increased incidence of diseases, but it is impossible to tell from the data whether thin precedes or follows incidence of disease. Overall, the effects of malnutrition impact immune response, muscle and respiratory function, and wound healing; malnutrition is associated with, but not caused by, aging.

Protein–calorie malnutrition leads to underweight. Underlying causes may be illness, poverty, or functional decline. (See Case Study 19.2.)

Case Study 19.2

Ms. Wetter: A Senior Suffering Through a Bad Stretch

Photodisc

Ms. Wetter: About to turn 81, Elizabeth Wetter is 5 feet 6 inches tall and weighs 106 pounds. She has had Parkinson's disease for 5 years, but that is not what concerns her. Her problem is pain from arthritis and lack of energy. She saw an ad on television for a vitamin–mineral supplement with ginseng that promises "more energy." Her son also told her to take a liquid dietary supplement to "feel better." Eighteen months ago, she had successful surgery for colon cancer, which was followed by chemotherapy treatments. She is now free of cancer. After the cancer treatment, she fell and broke a hip. This healed well, but serious leg pains started shortly afterward. There seems to be no cause for the pain, and a cure is unavailable. She is no longer able to take her walks through the neighborhood or tend her prize-winning garden. She would like to weigh 118 pounds again (her "usual") and is seeking nutritional counseling to try to regain some of her energy.

Questions

1. What are some of the nutritional issues faced by Ms. Wetter? (Hint: Calculate her current weight compared to her usual body weight as a percentage.)
2. How would you prioritize these in a nutritional care plan?
3. Calculate her energy needs and suggest strategies she might use to regain some energy.
4. What other information would you want to know in order to counsel Ms. Wetter?

(Answers are located in the Instructor's Manual for the 4th edition of *Nutrition Through the Life Cycle*.)

Nutrition Interventions

Avoiding unintentional weight loss is desirable but not always possible. Weight loss in the elderly is intertwined with disease-related biochemical and physiologic mechanisms, which, in turn, affect functional status and appetite. In a systematic review of studies to evaluate the effect of nutrition therapy in protein–energy malnutrition in connection with multiple disorders in the elderly, 20 studies noted an improvement in anthropometric or biochemical measures, and 10 studies reported an improvement in function. In contrast, there was insufficient evidence to determine how nutrition therapy should be formulated due to inconsistent or uncertain treatment adherence.[111] The inability to separate treatment effects further complicates nutrition monitoring. In a systematic review of 22 studies reported by the Cochrane Library, protein and energy supplementation appears to produce a small but consistent weight gain (2–4%), reduced mortality, and shorter length of hospital stays.[112] Supplementation was associated with nausea, diarrhea, and other gastrointestinal disturbances. Overall, medical nutrition therapy for a frailmalnourished person should occur in consultation with an experienced registered dietician. Refeeding and rehydration are done gradually:

- *Calories:* Eat and exercise to build muscle mass, strength.
- *Protein:* 1 to 1.5 grams of protein per kilogram body weight is adequate; 1.5–2 g/kg/day is recommended for severe depletion by the American Dietetic Association's Consultant Dietitians in Health Care Facilities Practice Group.[113] Exceptions are patients with renal or liver failure, who may need a protein restriction.
- *Water:* Drink 1 mL per kcal; rehydrate slowly (see the "Dehydration" section).

Connection: Fluid

8 ounces = 1 cup = 240 mL (milliliters) = 240 cc (cubic centimeters)

A super-sized soda (32 ounces) equals approximately 4 cups (or 4 × 240 = 960 mL). A 2-liter bottle of soda provides a little more than 8 cups of fluid. Some foods also count as fluid. For example, soup, gelatin products, and sherbet are considered fluids.

Dehydration

Definition

Dehydration is the physiological state in which cell fluid loss interferes with metabolic processes. Normal urination does not cause dehydration. Phillips and colleagues[114] defined dehydration as losing nearly 2% of initial body weight; this can occur after avoiding all fluid and eating only dry foods for 24 hours. Analysts with the American Dietetic Association Evidence Analysis Library concluded that fair evidence indicates that hydration status in young, healthy adults can be evaluated by urine specific gravity, urine osmolality, serum osmolality, and urine color.[115] The same review process found that these clinical and biochemical measures do not work with older populations, although multiple measures using bioelectrical impedance have been able to detect dehydration. The clinical signs of dehydration come from a review of 38 potential indicators of dehydration in older adults (61 to 98 years old, median age 82; *n* = 55; half were free-living; half were admitted to the hospital emergency department from extended care).[116] Seven signs and symptoms were strongly related to dehydration (*p* < 0.01 or *p* < 0.001), although *not* to patient age:

1. Upper-body muscle weakness
2. Speech difficulty
3. Confusion
4. Dry mucous membranes in nose and mouth
5. Longitudinal tongue furrows
6. Dry tongue
7. Sunken appearance of eyes in their sockets

These signs were also confirmed in a systematic review of maintaining oral hydration in older people.[117] In older adults, thirst and skin turgor (pinch a skin fold on the forearm, forehead, or over the breastbone, and observe it fall back) are not good indicators of dehydration.[116] Although regulation of body temperature is one of the functions of the body's water compartment, fever or elevated temperature did not identify dehydration status in Gross's study of individuals admitted to emergency rooms. However, fever or elevated temperature may be an indicator of impending dehydration.

There are three types of dehydration (isotonic, hypotonic, and hypertonic), and they are related to the proportional balance of sodium and water losses. Abnormally high serum sodium levels (>150 milliequivalents per liter) or a high ratio of blood urea nitrogen to creatinine (>25) can also be used to diagnose "significant dehydration."[116] Hypertonic dehydration (i.e., serum sodium is >145 mmol/L) is the type seen in iatrogenic cases (i.e., fluid deprivation with possible neglect). It can also be seen in individuals with fever.

Dehydration can be measured as percentage of body weight lost when normal body weight is known. In a continuum of dehydration shown by Briggs and Calloway[118] (Table 19.17) originally designed for the National Aeronautics Space Administration (NASA), the indicators are somewhat different from the indicators that Gross[116] found. The NASA continuum shows how the human body responds to water losses.

Weight loss of 4% normal weight would be hard to ignore unless the individual had a compromised mental or cognitive status. Flushed skin, nausea, and apathy or lack of energy occur when 4% body weight is lost due to dehydration. The ADA Evidence Library analysts considered a 3% body-weight loss "acute dehydration."[115]

Although the human body is quite resilient in that it can lose up to 10% of its water weight and survive, dehydration in smaller degrees is common for older people. Prevalence data is hard to gather because dehydration is usually temporary, and there is no simple or commonly used standard definition for diagnosing dehydration in the elderly. For example, patients coming to a nursing home from a hospital may have IV tubes, and nursing notes may simply say, "The patient looks well-hydrated."

Etiology

Aging itself does not cause dehydration, even though the percentage of total body water shrinks from infancy to old age. Dehydration occurs more often in the elderly as a result of illness or other problems. Older people are less

Table 19.17 Percent of initial weight lost due to dehydration and physiological signs

Percent Lost	Physiological Signs
1%	Thirst (true for young people, not necessarily in older men or women)
4–6%	Economy of movement, flushed skin, sleepiness, apathy, nausea, tingling in arms, hands, feet, headache, heat exhaustion in fit men, increases in body temperature, pulse rate, respiratory rate
8%	Dizziness, slurred speech, weakness, confusion
12%	Cognitive signs: wakefulness, delirium
20%	Bare survival limit

For someone who normally weighs 160 lb:
1% loss means weight down to 158.4 lb
4% loss means weight down to 153.6 lb
6% loss means weight down to 150.4 lb
20% loss means weight down to 128.0 lb

SOURCE: Adapted from Briggs and Calloway, originally NASA, 1967.[118]

Antidiuretic Hormone Hormone that causes the kidneys to dilute urine by absorbing more water.

sensitive in detecting thirst than are younger people, and they therefore may not think to drink.[114] Once fluids are consumed, aging kidneys may lose the ability to concentrate urine, and *antidiuretic hormone* may become less effective. Swallowing problems, depression, or dementia may cause individuals to avoid food or drink. Decreased mobility impairs older adults' access to water and subsequent ability to reach the bathroom. Fear of incontinence, in general, is another reason leading to decreased fluid intake and subsequent dehydration.

Effects of Dehydration

Dehydration increases the resting heart rate and susceptibility to development of urinary tract infection, pneumonia, and pressure ulcers; it also leads to confusion, disorientation, and dementia.[119,120] Because confusion and delirium are signs of—as well as risk factors for—dehydration, consuming adequate fluids can become a vicious circle for someone at risk for cognitive decline.

Nutritional Interventions

The Institute of Medicine's DRI for water does not change as adults get older:[120] "The AI for total water (drinking water, beverages, and foods) for the elderly is set based on median total water intake of young adults." Some health professionals suggest fluid levels to be 1 mL per calorie eaten, with a minimum of 1500 mL per day (approximately 6 cups). The ultimate beverage is *water*—tap or flavored, *not* sugared. Water is generally accessible, adds no calories to the diet, provides traces of minerals needed for metabolism, and is very low in sodium, even when softened. Pure water does not provide energy, but lacking water to the point of dehydration dramatically reduces an individual's energy.

Many beverages contribute nutrients as well as fluid:

1. Tea, especially green tea, has been reported as "promising, but requiring future studies" in relation to cardiovascular and cancer risk reduction; the flavonoids in tea act as antioxidants.[121]
2. Milk provides calcium, protein, riboflavin, and vitamin D. When it is low in fat, milk enhances hypertension treatment and weight maintenance.
3. Regular use of cranberry juice reduces urinary tract infection in older women.[122]
4. Fruit and vegetable juices count as part of the recommended fruit and vegetable servings.

Rehydrate Slowly

To treat dehydration in older adults, replace fluids slowly. Guidelines are to provide roughly one-fourth to one-third of the overall fluid deficit each day in the form of water or a 5% glucose solution (when the individual's blood values are stable).[117] For individuals with swallowing problems, or dysphagia, thickened liquids count as fluid. When bedridden older adults are offered fluids hourly and also with medication, they achieve higher levels of hydration.[117]

Dehydration at End of Life

Lack of hydration can be an issue for people with a terminal disease or who are near death. Some individuals stop eating and/or drinking hours, days, or even weeks before death. Laboratory values for blood and urine are likely to become abnormal. Treatment for dehydration at the end of life may differ from treatment of dehydration during an acute disease episode. Suggestions for treating dehydration in a dying person are to integrate four C's:[123]

1. *Common sense:* Use approaches that benefit the whole patient; ask, "What does the patient want?"
2. *Communication:* Respect and acknowledge the sadness felt by friends and family.
3. *Collaboration:* With the patient's permission, bring in experienced people such as hospice workers to guide treatment.
4. *Caring:* Listen; respond with love and compassion.

Dehydration at the end of life contributes to an overall slowing down of body systems, including production of body fluids, resulting in less congestion, less edema and ascites, and less gastrointestinal action. A person experiencing dehydration at the end of life may experience slight thirst, although many dying patients are not thirsty. They may experience dry mouth that can be alleviated by sucking on ice chips or using artificial saliva. Decreased urine output can be a benefit because there is less need to go to the toilet. The most commonly reported symptoms in the last week of life of an individual with advanced progressive disease are loss of appetite, asthenia (loss of strength), dry mouth, confusion, and constipation.[124] Dehydration may also lead to increasing levels of drowsiness, which can reduce fear and anxiety related to dying.

Bereavement

Bereavement is the loss felt when someone who is personally significant dies. Losses of friends and family members happen more often in the lives of older persons. Grief, a very powerful emotion, is a natural response to bereavement. The grieving process, with its stages of shock and denial, disorganization, volatile reactions, guilt, loss and loneliness, relief, and reestablishment,[125] diverts attention from normal activities. Shopping and food preparation, eating, and drinking may be ignored in the grieving process. Any loss of long-shared relationships through death, dementia, or moving brings about lack of interest in activities surrounding meal planning, preparation, shopping, and eating. People who are in mourning are vulnerable to malnutrition.

Widowhood has been shown to trigger disorganization and changes in daily routine, especially related to food preparation and eating.[126] Widowed persons who are able to enjoy mealtimes, have good appetites, have higher-quality diets, and receive social support work through the grieving process with fewer health consequences. They demonstrate healthy aging, which is, in the words of Dr. Tamara Harris, chief of Geriatric Epidemiology at the National Institutes of Health, "the ability of the individual to be resilient, to be adaptive, to be flexible, and to mobilize compensatory areas as they face adversities in all areas associated with health, disease, and decline in old age."

Key Points

1. Nearly three of four older adults consider their health to be good, very good, or excellent, despite the high prevalence of chronic disease in older adults.

2. Hypertension, which affects nearly half of older adults, can be moderated with the DASH diet; further blood pressure reductions are achieved by restricting sodium intake in addition to following the DASH eating pattern.

3. Of adults over age 65, nearly one in five reports having diabetes, primarily type 2, suggesting that meals planned for older-adult nutrition programs automatically promote heart health and support diabetes management.

4. Body Mass Index should be combined with other measures such as waist circumference to assess weight status in older adults; a BMI range of 19 to 27 seems reasonable for older adults.

5. Bed rest and inactivity is detrimental to strong bones. Exercise combined with good nutrition—including adequate calcium and vitamins D, C, B_6, and K—helps to maintain bone mass density. Older adults on blood thinners need to monitor vitamin K intake, which is included in some bone-strengthening supplements.

6. Polypharmacy is a risk factor for malnutrition and may contribute to poor oral health; dry mouth and the associated gingival disease can result from drug use (such as antihistamines, antidepressants, sedatives, and anti-anxiety drugs).

7. Despite adequate dietary intake of vitamin B_{12}, older adults may still have low blood levels due to malabsorption. Monitoring B_{12} blood levels in vulnerable populations can help to prevent the irreversible nerve damage caused by B_{12} deficiency. Vitamin B_{12} is one instance when the synthetic version of a vitamin is better absorbed than the food-bound version.

8. Osteoarthritis is the most common inflammatory disease of older adults. Weight loss helps to reduce the load exerted on knees and also reduces risk of developing osteoarthritis (as do adequate intake of vitamins C and D).

9. Dementia associated with old age can be due to vascular and degenerative diseases, physical trauma and infection, depression, and malnutrition. However, most of the dementia in older adults in the United States is associated with Alzheimer's disease. Nutritional efforts focus on treatment of underlying causes and maintaining quality of life.

10. Weight loss in an older adult signals potential illness and malnutrition; an individual who is losing weight without trying needs a thorough assessment.

11. Confusion and muscle weakness related to dehydration is avoidable. Adequate fluid intake prevents dehydration and can also contribute needed nutrients to the diet of an older adult, such as calcium, vitamin D, protein and riboflavin from milk, antioxidants from tea and coffee, vitamins, minerals, and fiber from juices and nectars. Socializing over a beverage may add to quality of life. Some long-term care institutions have implemented "happy hour" programs to encourage fluid intake.

Review Questions

1. What are the three most common chronic conditions affecting men over age 65?

2. What are the three most common chronic conditions affecting women over age 65?

3. True or false: The DASH diet has been shown to reduce HBP; additional sodium restrictions have led to further reductions in blood pressure.

4. BMI is a less effective predictor of morbidity and mortality in older than younger adults. Why is BMI a less reliable tool in older adults?

5. True or false: Even a small loss of body weight can improve osteoarthritis, as can adequate intakes of vitamins C and D.

6. True or false: Vitamin K contributes to bone health, but the levels present in dark green vegetables may interfere with warfarin (anti-coagulant) therapy.

7. True or false: Eating a Mediterranean-style diet is associated with lower CVD mortality in older adults.

8. True or false: Older adults who suffer from sarcopenia may benefit from high-protein diets. High protein intake will also help to keep their bones strong.

9. Vitamin B_{12} deficiency can cause irreversible neurological damage. List at least three reasons why an older adult might have low vitamin B_{12} levels.

10. Dehydration can result in confusion. List three or more strategies that can help to restore hydration and provide nutrients an older adult might be missing.

Resources

Alzheimer's Disease Education and Referral Center
This site provides information about Alzheimer's disease and related disorders. It is a service of the National Institute of Aging (NIA).
Website: www.nia.nih.gov/alzheimers

Alzheimer's Association
This nonprofit organization raises funds for education, caregiver support, and research.
Website: www.alzheimers.org

American Diabetes Association
This site features areas for professionals and for the public and includes links.
Website: www.diabetes.org

American College of Sports Medicine, Exercise is Medicine
ACSM is one of the founders of the Exercise is Medicine campaign. This site offers data and tools to professionals wishing to integrate exercise into their health promotion programs.
Website: www.exerciseismedicine.org

Arthritis Foundations
Arthritis Today is maintained by the Arthritis Foundation; the site includes nutrition information and reviews of treatments for various forms of arthritis.
Website: www.ArthritisToday.org

National Center for Complementary and Alternative Medicine
This site provides fact sheets for complementary medicine associated with dietary supplements, cancer prevention and treatment, and other dietary components. The website is supported by the National Institutes of Health (NIH).
Website: www.nccam.nih.gov

National Institute of Diabetes and Digestive and Kidney Diseases
This site offers health information for diabetes, metabolic illnesses, and kidney disease. Research and clinical trial information is also available.
Website: www.niddk.nih.gov

National Institute of Health, Office of Dietary Supplements
Website: http://ods.od.nih.gov/

National Institute of Mental Health
Major depression is most prevalent in developed nations. This site provides a link for the public with a specific topic discussing depression. A further link relating to older adults provides comprehensive information on definitions of depression and the treatment options available.
Website: www.nimh.nih.gov

National Osteoporosis Foundation
Website: www.nof.org

Partnership for Caring
This is a national hospice and palliative care organization.
Website: www.caringinfo.org/

Answers to Review Questions

Chapter 2

1. False
2. True
3. True
4. True
5. False
6. False
7. False
8. False
9. True
10. Oxidative stress can damage sperm membranes, damage DNA, harm egg and follicular development, and interfere with corpus luteum function and implantation of the egg in the uterine wall.
11. Altering the environment in which eggs and sperm develop and modifying levels of hormones involved in reproductive processes
12. Menstrual cycles in which ovulation does not occur

Chapter 3

1. False
2. True
3. True
4. False
5. True
6. False
7. True
8. False
9. True
10. True
11. False
12. False

Chapter 4

A.

1. True
2. 4 lb, 10 oz
3. False; he was disproportionately small for gestational age.
4. False; he was hospitalized during the postneonatal period.
5. False
6. False
7. True
8. False

B.

9. 52 ng/mL vitamin D x 2.496 = 130 nmol/L
10. 86 ng/mL ferritin x 2.247 = 193 pmol/L
11. 402 ng/mL red cell folate x 2.266 = 911 nmol/L

Chapter 5

1. True
2. True
3. False
4. True
5. False
6. True
7. False
8. 33% of calories come from snacks.
9. Yes. The lunch menu provides 26.8% of total calories, and the recommended range of percent calories from lunch is 20–30%.
10. She is underweight and here's why. There are 2.2 pounds in a kilogram (kg), so 132 pounds/2.2 pounds per kg = 60 kg. The diet provides 2200 calories (or kcal). The kcal per kg body weight would equal 2200 kcal/60 kg, or 37 kcal/kg. It is recommended that underweight women with gestational diabetes consume up to 40 kcal/kg, so she is most likely underweight.
11. True
12. True
13. False
14. True
15. True

Chapter 6

1. d
2. a
3. c
4. b
5. c
6. b
7. c
8. a
9. c
10. Lactogenesis I: the first stage of milk production which extends into the first few postpartum days when suckling is not needed for milk production. Lactose and protein content of milk increases.

    Lactogenesis II: begins 2-5 days postpartum and continues to about 10 days postpartum. This is when milk comes in, the lactose and fat composition are increasing, the protein content decreases, and the volume of milk produced is increasing.

    Lactogenesis III: begins about 10 days postpartum and the milk composition becomes stable.

11. Correct answers include:

    - The balance of nutrients in human milk matches human infant requirements for growth and development closely; no other animal milk or HMS meets infant needs as well.
    - Human milk is isosmotic (of similar ion concentration; in this case human milk and plasma are of similar ion concentration) and therefore meets the requirements for infants without other forms of food or water.
    - The relatively low protein content of breast milk compared to cow's milk meets the infant's needs without overloading the immature kidneys with nitrogen.
    - Whey protein in human milk forms a soft, easily digestible curd.
    - Human milk provides generous amounts of lipids in the form of essential fatty acids, saturated fatty acids, medium-chain-triglycerides, and cholesterol.
    - Long-chain polyunsaturated fatty acids, especially docosahexaenoic acid (DHA), which promotes optimal development of the central nervous system, are present in human milk and are present in only some of the HMS marketed in the United States.
    - Minerals in breast milk are largely protein bound and balanced to enhance their availability and meet infant needs with minimal demand on maternal reserves.

Chapter 7

1. Trauma from poor positioning, poor latch, improper release of suction after a feeding, infection, thrush, incorrect flange size during pumping or too much pressure while pumping and disorganized or dysfuctional suck.
2. d
3. Development after cracked or sore nipples, missing a feeding or mom sleeping through the night, restrictive clothing or a tight bra.
4. True
5. True
6. True
7. Jaundice observed in the first 24 hours of life, blood group incompatibility or haemolytic disease, gestational age 35–36 weeks, previous siblings received phototherapy, significant bruising, exclusive breastfeeding when not going well and/or excessive weight loss, east Asian race.
8. False
9. d
10. d

Chapter 8

1. The correct answer is: Vitamin D supplementation is needed in infants fed exclusively breast milk but not in those fed exclusively formula.
2. The correct answer is: Newborn reflexes promote for the coordination of feeding and breathing.
3. The safe meal is: Bottle of infant formula, cooked carrot slices, cut-up fresh peach and shredded deli ham.
4. The correct answer is: Regarding hunger and satiety cues, parenting skills develop more slowly to read them than infants' skills to give them.
5. The correct answer is: Infants need a higher percent of total calories as dietary fats than recommended for adults.
6. Correct answers include: increase gastrointestinal distress, increase gastroesophageal reflux, stifle infant hunger and

fullness cues, increase spitting up

7. The correct answer is: Infants with normal birth weights reflect a healthy pregnancy.

8. The correct answer is *not* a possible explanation: This shows

that the infant needs an additional nutrition source such as baby cereal.

9. The correct answers include: pureed texture, smooth texture, hypoallergenic type of food.

10. The correct answers include: a medical diagnosis in the infant, a maternal diagnosis such as depression, untreated gastro-esophageal reflux or ear infections, history of intrauterine growth retardation (IUGR).

Chapter 9

1. d
2. a
3. True
4. False
5. False
6. True
7. The correct answers include: cleft lip and palate, diaphragmatic

hernia, DiGeorge syndrome, contenital anomalies.

8. The correct answers include: disorders that impact brain growth, medications and side effects, changes in body composition, limitations in activity.

9. Correct answers include: non-oral feeding, nasogastric feeding,

orogastric feeding, transpyloric feeding, parenteral feeding, gastrostomy feeding.

10. Correct answers include: removal from toxins, such as maternal diabetes or drug abuse, neonatal illness such as sepsis, preterm birth.

Chapter 10

1. c
2. d
3. e
4. a
5. True
6. d
7. • Whole-grain breads and cereals
   • Fruits and vegetables
   • Low-fat dairy products
   • Lean meats, fish, and beans

   • Limited amount of sweet-ened beverages and other sweets
   • Unsaturated fats

8. • Physical activity plays a role in maintaining energy balance.
   • Physical activity helps to build muscle strength.
   • Being physically active is a healthy lifestyle behavior

that is important to establish at a young age.
   • Appropriate activities for children include riding a tricycle or bicycle, walking, skipping, running, and active play.
   • Limiting screen time and removing screens from young children's bedrooms is also recommended.

Chapter 11

1. c
2. d
3. c
4. a
5. True
6. True
7. False
8. Correct answers include: to identify a decreasing rate of head growth, because any change in brain growth can

impact weight and height (or length) growth, conditions such as spastic quadriplegia, cerebral palsy, Rett syndrome, Prater Willi syndrome, or Down syndrome

9. Correct answers include: offering food textures that make the child a successful eater, allowing a monotonous diet if required for limited oral feeding skills, allowing a bottle longer than

expected due to developmental delays requiring it

10. Correct answers include: when a child has to have a restricted diet, such as a result of allergies; when medications cause side effects that lower appetite; when a child has behaviors that limit food choices, such as autism, Prader-Willi syndrome; treatment with a calorie-restricted diet

Chapter 12

Timothy, in Case Study 12.1, returns to his pediatrician 6 months later at 7½ years of age for a weight check. He has gained 10 kg in 6 months and now weighs 90 pounds (41 kg). His height has increased to 51 inches (130 cm).

1. BMI = weight in kg divided by height in meters squared; 41 kg divided by 1.69 = 24.26 kg/m²

2. >99th % for his age and gender

3. e

4. Timothy would enter the Prevention Plus approach to treatment for 3–6 months. If he maintains his weight or his BMI-for-age percentile is deflecting down, he would continue the Prevention Plus. If he gains weight, the next approach would be a structured weight management program. If he has weight loss or his BMI-for-age percentile deflects down, he would continue the structured weight management program. If he continues to gain weight, then he would enter a comprehensive multidisciplinary intervention.

5. d

6. • Find a safe route for walking or biking to school.
   • Encourage at least 60 minutes of active play per day.
   • Work within the community to maintain parks and other safe places for children to play.
   • Child can participate in organized sports that are appropriate for his stage of development and physical abilities.

7. • Decrease "screen time" to less than 2 hours per day.
   • Remove televisions, video games, and computers from children's bedrooms.
   • Be active as a family.

8. c

Chapter 13

1. c (the other conditions are not known to impact growth directly)

2. b (because this child would have a change in body composition)

3. c (as this is the only answer having to do with limited activity)

4. b

5. True

6. True

7. False

8. Examples of support services for families are maternal and child health services, school nutrition programs, and school educational programs.

9. Correct answers include medication side effects and ability to concentrate or attend to meals.

10. An IEP is a school educational plan that could include feeding or eating assistance, access to snacks or meals at set times at school, and selection of foods appropriate for the child's abilities or condition.

Chapter 14

1. a

2. False

3. d

4. False

5. b

6. c

7. answers will vary

8. answers will vary

9. a

Chapter 15

1. True

2. c

3. d

4. c

5. d

6. c

7. While both conditions include eating large amounts of food in a short period of time, a diagnosis of bulimia nervosa requires frequent purging methods to compensate for the calories consumed while the diagnosis of binge-eating disorder does not include a criteria for purging.

8. Energy intake that results from the consumption of alcohol may displace energy intake from food. Because alcohol does not provide micronutrients, the overall dietary quality would be reduced in an individual who consumed alcohol frequently. The use of tobacco increases vitamin C requirements. The use of recreational drugs may result in increased intakes of snack foods and other highly refined simple carbohydrates which are high in sodium and fat and low in micronutrients.

Chapter 16

1. a
2. d
3. b
4. a
5. e
6. A risk factor is an attribute associated with increased occurrence of a condition. Modifiable risk factors are behaviors or conditions that can be improved through behavior changes or environmental modifications to reduce risk.

   Dietary risk factors include, with possible improvements:

   - High total fat, or high saturated (>10% of calories)—Choose lean meats, eat smaller portions, avoid fried and fatty foods.
   - Any *trans* fatty acids—Read the ingredient list and avoid partially hydrogenated oil.
   - Low monosaturated fat—Use olive oil.
   - Low omega-3 fatty acids—Eat fatty fish such as salmon or tuna, and flax-seed oil.
   - Dietary cholesterol >300 mg—Choose lean meats, limit eggs to 1 per week.
   - Low intake of whole grains and fiber—Eat 6 servings of grains per day and make at least 3 of them whole-grain; women should get 25 g and men 38 g of fiber per day.
   - Low fruit and vegetable intake, low antioxidant intake—Eat 2 cups of fruit and 3 cups of vegetables per day.
   - Low folic acid intake—Chose foods rich in folic acid or folate (fortified breakfast cereals and other grains, leafy green vegetables, legumes, orange and grapefruit juice).
   - Intake of nitrosamines, burnt or charred food—Avoid processed meats, and eat medium rather than well-done meat.
   - High meat intake-, low plant-food intake—Fill half of the plate with fruits and vegetables, ¼ with grains, and ¼ with protein source.
   - High sodium intake—Avoid adding salt, minimize use of processed foods, read the label and choose lower-sodium foods.
   - Low potassium intake—Eat more fresh fruit, vegetables, whole grains, and dairy products.
   - Excessive calorie intake—Balance calories eaten with energy burned through basal metabolic rate and physical activity.
   - Unstructured eating—Plan meals and mealtimes.
   - Sugar-sweetened beverages—Drink water as primary beverage; limit soda to 1 serving per day.
   - Energy-dense, low-nutrient food choices—Be careful in use of discretionary calories.

   Students might use other resources to list specific substitutions, such as those noted above, but a preferred answer would be to refer to the role of dietary guidance in reducing risk factors. For example, dietary risk factors could be modified through effective dietary guidance that includes information on what and how much to eat. Advice should be culturally relevant and tailored to the individual. Learning to like a variety of foods should be encouraged.

7. Health disparities are caused by the lack of healthy food, good housing, good education, and safe neighborhoods. Racism and other forms of discrimination are intertwined with these and other social determinants of health. The ten social determinants of health could also be listed.

8. Identified diets should be critiqued using the following: inadequate nutrient supply, severe energy restriction, unusual food restriction, strange food combinations, strict limitations, and gimmicks.

9. At least three of the following should be mentioned in the dietary guidance system:

   - Incorporate components of the Eating Competence Model.
   - Advocate for reduction of disease risk.
   - Ensure adequate intake of specific nutrients, especially those of concern due to low intake in the past.
   - Provide guidance on what and how much to eat.
   - Be specific rather than general.
   - Classify foods into groups, describe what constitutes a serving, and identify number of servings needed.
   - Include guidance on types and amounts of beverages recommended, including alcohol.

Chapter 17

1. True
2. True
3. False
4. False
5. To complete the Heart Health Assessment, we need her total cholesterol, smoking status, and blood pressure.
6. Reduce caloric intake and increase physical activity to lose weight, and select foods that are compatible with therapeutic lifestyle changes, specifically: total fat at 25–35% of calories, saturated fat <7% of calories, dietary cholesterol less than 200 mg per day, *trans* fats a little as possible, 20–30 g of fiber per day with 10 g or more from viscous fiber, and add 2 g per day of plant stanols or sterols. She should get 30 minutes of moderate physical activity most days of the week and work up to longer duration or greater intensity as she is able.
7. Common themes:

   - Maintain a healthy weight—if overweight, this means a weight loss of 5–10% of body weight. For HIV disease, the focus is on maintaining weight and maintaining lean body mass.
   - Avoid or reduce central or abdominal adiposity, because this is associated with insulin resistance and hyperglycemia, hyperlipidemia, metabolic syndrome, cardiovascular disease, and type 2 diabetes.
   - The foundation diet that crosses all chronic diseases is one that keeps fat intake in the 25–35% of calories range, minimizes saturated fat and *trans* fats, and includes more fruits and vegetables and more fiber.
   - Choosing a "healthy diet," being physically active, and maintaining a healthy weight can help prevent the development of chronic disease and delay complications.
   - Regular physical activity at a moderate or vigorous level is important because it aids weight loss and maintenance and improves insulin/glucose profile, as well as reducing lipids and blood pressure.
   - Be screened and "know your numbers" (BMI, waist circumference, blood lipid levels, blood glucose, and blood pressure) to be aware of risk level and perhaps be motivated to take action.
   - Lifestyle changes often require intensive, on going intervention.

8. Essential components of cognitive behavioral change programs include:

   - Long duration—programs are 12–16 weeks long to build knowledge, modify beliefs and attitude, and integrate new behaviors.
   - Combination of skills training and analysis of behavior and thought processes.
   - Two key features are helping the client recognize and replace automatic and irrational thoughts and beliefs (cognitive restructuring) and increasing awareness and control of cues associated with eating (stimulus control).
   - Strategies include setting realistic goals, calorie deficit if needed for weight loss, individualized meal plan that includes variety of enjoyed foods and fits lifestyle and budget, skills development, problem solving, self-management, cognitive restructuring, stress management, having a support system, and regular exercise, followed by maintenance support—all directed to long-term effectiveness.

Chapter 18

1. True
2. True, lifestyle factors are estimated to account for 51% of longevity compared to a 19% contribution from genetics.
3. False
4. False. While energy needs decrease with age, some vitamins (Vitamins B_6, C, D) and minerals (Mg, Ca) increase with age while iron and chromium decrease.
5. Here are six risk factors: 1) food insecurity or low income, 2) eating fewer than two meals per day, 3) illness or a condition that has affected dietary intake, 4) lost or gained more than 10 pounds in the last 6 months, 5) polypharmacy, and 6) physical limitations to shop, cook, and feed self
6. Here are three adjustments of generic food-guide pyramids to make for older adults: Greater risk of dehydration requires fluid assessment and recommendations, 2) lower caloric intake needs to be balanced with

greater nutrient density, especially when activity level is low, and 3) guidance needs to ensure adequate protein intake when caloric intake is marginal.

7. Reasons might include:
   Lack of money may affect ability to buy high quality protein
   Compromised ability to chew, to cut meats once cooked

Food safety skills or ability to hands protein foods safely (vision, smell)
Are there acceptable substitutions for dietary modifications? (eg., how will the individual replace lunch meats or a favorite breakfast sausage?)

8. Calculate caloric needs, then use the AMDR range of 10-35% of calories to calculate a range of

grams of protein and individualize to JT's activity level, usual dietary pattern, and preferences. OR take his weight in kilograms (107) and multiply by 1 gm of protein per kg body weight. Work at translating grams of protein to show JT what this looks like in foods.

9. True

Chapter 19

1. Hypertension, arthritis and heart diseases.

2. Hypertension, arthritis and heart diseases, in other words, older men and women face the same chronic diseases.

3. True, adherence to the DASH diet reduces blood pressure. An additional dose-response effect on blood pressure reduction results from sodium restriction.

4. The BMI is calculated using height and weight. Getting an accurate height may be more difficult in older adults who have compressed vertebrae or kyphosis which has reduced normal height. On average, older adults have lost muscle tissue which has been replaced with fat tissue and BMIs were

developed with measurements of younger adults. Dehydration can introduce further errors into the interpretation of a BMI in an older individual. Thus, less weight may be due to less muscle and body water.

5. True

6. True. Approximately 11% of men above age 65 and 4.5% of women over age 65 take warfarin and must keep their vitamin K levels stable for effective blood-thinning effect.

7. True

8. True

9. B_{12} deficiency may result from malabsorption due to atrophic gastritis, GI infections such as *H. pylori*, lack of animal

products in the diet, family history of pernicious anemia, and medications.

10. Push fluids by structuring reminders to drink (such as water jug in the refrigerator that should be empty at day's end), drinking liberal amounts of water when taking medications, offering nutritious beverages to increase nutrients, observing "happy hour" as a time to socialize and consume fluids, and perhaps most importantly for many older women, deal with fears related to incontinence by fixing underlying structural problems and doing exercise (Kegels).

Summary of Research of Effects of Exercise Activities on Health of Older Adults

Evidence Statements	Evidence Strength: A = Highest, D = Lowest
Section 1: Normal human aging	
Advancing age is associated with physiologic changes that result in reductions in functional capacity and altered body composition.	A
Advancing age is associated with declines in physical activity volume and intensity.	A/B[a]
Advancing age is associated with increased risk for chronic diseases but physical activity significantly reduces this risk.	B
Section 2: Physical activity and the aging process	
Regular physical activity increases average life expectancy through its influence on chronic disease development, through the mitigation of age-related biological changes and their associated effects on health and well-being, and through the preservation of functional capacity.	A
Individuals differ widely in how they age and in how they adapt to an exercise program. It is likely that lifestyle and genetic factors contribute to the wide interindividual variability seen in older adults.	B
Healthy older adults are able to engage in acute aerobic or resistance exercise and experience positive adaptations to exercise training.	A
Regular physical activity can favorably influence a broad range of physiological systems and may be a major lifestyle factor that discriminates between those individuals who have and have not experienced successful aging.	B/C[a]
Regular physical activity reduces the risk of developing a large number of chronic diseases and conditions and is valuable in the treatment of numerous diseases.	A/B[a]
Section 3: Benefits of physical activity and exercise	
Vigorous, long-term participation in AET is associated with elevated cardiovascular reserve and skeletal muscle adaptations, which enable the aerobically trained older individual to sustain a submaximal exercise load with less cardiovascular stress and muscular fatigue than their untrained peers. Prolonged aerobic exercise also seems to slow the age-related accumulation of central body fat and is cardioprotective.	B
Prolonged participation in RET is consistently associated with higher muscle and bone mass and strength, which are not seen as consistently seen with prolonged AET alone.	B
AET programs of sufficient intensity ($\geq$60% of pretraining VO_{2max}), frequency, and length ($\geq$3 dwk^{-1} for $\geq$16 wk) can significantly increase VO_{2max} in healthy middle-aged and older adults.	A
Three or more months of moderate-intensity AET elicits cardiovascular adaptations in healthy middle-aged and older adults, which are evident at rest and in response to acute dynamic exercise.	A/B[a]
In studies involving overweight middle-aged and older adults, moderate-intensity AET has been shown to be effective in reducing total body fat. In contrast, most studies report no significant effect of AET on FFM.	A/B[a]
AET can induce a variety of favorable metabolic adaptations including enhanced glycemic control, augmented clearance of postprandial lipids, and preferential utilization of fat during submaximal exercise.	B
AET may be effective in counteracting age-related declines in BMD in postmenopausal women	B
Older adults can substantially increase their strength after RET.	A
Substantial increases in muscular power have been demonstrated after RET in older adults.	A
Increases in MO are similar between older and younger adults, and these improvements do not seem to be sex-specific.	B

(continued)

Evidence Statements	Evidence Strength: A = Highest, D = Lowest
Improvements in muscular endurance have been reported after RET using moderate- to higher-intensity protocols, whereas lower-intensity RET does not improve muscular endurance.	C
The effect of exercise on physical performance is poorly understood and does not seem to be linear. RET has been shown to favorably impact walking, chair stand, and balance activities, but more information is needed to understand the precise nature of the relationship between exercise and functional performance	C/D[a]
Favorable changes in body composition, including increased FFM and decreased FM have been reported in older adults who participate in moderate or high intensity RET.	B/C
High-intensity RET preserves or improves BMD relative to sedentary controls, with a direct relationship between muscle and bone adaptations.	B
Evidence of the effect of RET on metabolic variables is mixed. There is some evidence that RET can alter the preferred fuel source used under resting conditions, but there is inconsistent evidence regarding the effects of RET on BMR. The effect of RET on a variety of different hormones has been studied increasingly in recent years; however, the exact nature of the relationship is not yet well understood.	B/C
Multimodal exercise, usually including strength and balance exercises, and tai chi have been shown to be effective in reducing the risk of noninjurious and sometimes injurious falls in populations who are at an elevated risk of falling.	C
Few controlled studies have examined the effect of flexibility exercise on ROM in older adults. There is some evidence that flexibility can be increased in the major joints by ROM exercises; however, how much and what types of ROM exercises are most effective have not been established.	D
Regular physical activity is associated with significant improvements in overall psychological well-being. Both physical fitness and AET are associated with a decreased risk for clinical depression or anxiety. Exercise and physical activity have been proposed to impact psychological well-being through their moderating and mediating effects on constructs such as self-concept and self-esteem.	A/B
Epidemiological studies suggest that cardiovascular fitness and higher levels of physical activity reduce the risk of cognitive decline and dementia.	A/B
Experimental studies demonstrate that AET, RET, and especially combined AET and RET can improve cognitive performance in previously sedentary older adults for some measures of cognitive functioning but not others. Exercise and fitness effects are largest for tasks that require complex processing requiring executive control.	
Although physical activity seems to be positively associated with some aspects of QOL, the precise nature of the relationship is poorly understood.	D
There is a strong evidence that high-intensity RET is effective in the treatment of clinical depression. More evidence is needed regarding the intensity and frequency of RET needed to elicit specific improvements in other measures of psychological health and well-being.	A/B

[a]Any review of evidence pertaining to exercise and physical activity in older adult populations will necessarily be interdisciplinary and subject to differences in research design across various subdisciplines within exercise science. Whenever possible, a single SORT rating is provided; however, occasionally, when the strength of evidence varies across studies, a composite rating is provided.

Abbreviations: AET aerobic exercise training; RET resistance exercise training; FFM fat-free mass; FM Fat mass; BMD Bone mineral density; MQ Muscle Quality; ROM Range of Motion; QOL Quality of Life

1. *Evidence Level A.* Overwhelming evidence from RCTs and/or observational studies, which provides a consistent pattern of findings on the basis of substantial data.
2. *Evidence Level B.* Strong evidence from a combination of RCT and/or observational studies but with some studies showing results that are inconsistent with the overall conclusion.
3. *Evidence Level C.* Generally positive or suggestive evidence from a smaller number of observational studies and/or uncontrolled or nonrandomized trials.
4. *Evidence Level D.* Panel consensus judgment that the strength of the evidence is insufficient to place it in categories A through C.

SOURCE: From Medicine & Science in Sports & Exercise:July 2009 - Volume 41 - Issue 7 - pp 1510-1530. SPECIAL COMMUNICATIONS: Position Stand Exercise and Physical Activity for Older Adults Chodzko-Zajko, Wojtek J. Ph.D., FACSM, (Co-Chair); Proctor, David N. Ph.D., FACSM, (Co-Chair); Fiatarone Singh, Maria A. M.D.; Minson, Christopher T. Ph.D., FACSM; Nigg, Claudio R. Ph.D.; Salem, George J. Ph.D., FACSM; Skinner, James S. Ph.D., FACSM. Copyright © 2009 The American College of Sports Medicine, Wolters Kluwer Health.

Measurement Abbreviations and Equivalents

Metric or SI Unit

Unit	Abbreviation
kilogram	kg
gram	g
milligram	mg
microgram	μg, mcg
nanogram	ng
meter	m
centimeter	cm
millimeter	mm
liter	L
deciliters	dL
milliliter	mL
millimole	mmol
micromole	μmol
picomole	pmol

Nonmetric

Unit	Abbreviation
ounce	oz
pound	lb
tablespoon	Tbsp
teaspoon	tsp
cup	c
pint	pt
quart	qt
gallon	gal
inch	in
foot	ft
yard	yd

Equivalents

Weight: Metric

1 kilogram	=	2.2 pounds; 1000 grams
1 gram	=	0.035 ounce; 1000 milligrams
1 milligram	=	1000 micrograms
1 microgram	=	1000 nanograms

Weight: Nonmetric

1 ounce	=	28.35 grams
1 pound	=	0.45 kilograms; 454 grams

Linear

1 millimeter	=	0.039 inch
1 centimeter	=	0.01 meter; 0.39 inch
1 meter	=	100 centimeters; 39.4 inches; 3.28 feet
1 inch	=	2.54 centimeters; 0.025 meter
1 foot	=	30.5 centimeters; 0.31 meter
1 yard	=	3 feet; 0.91 meters

Fluid Volume

1 teaspoon	=	5 mL
1 tablespoon	=	15 mL; 0.5 ounce; 3 teaspoons
1 ounce	=	30 mL; 29.57 grams; 6 teaspoons; 2 tablespoons
1 cup	=	240 mL; 8 ounces; 48 teaspoons; 16 tablespoons
1 pint	=	480 mL; 2 cups; 16 ounces; 1 pound ("A pint is a pound the whole world round.")
1 quart	=	0.95 liter; 2 pints; 4 cups; 32 ounces; 2 pounds
1 liter	=	1.06 quarts; 1000 mL
1 gallon	=	3.79 liters; 4 quarts; 8 pints; 16 cups; 8 pounds ("Two cups in a pint, two pints in a quart, four quarts in a gallon.")

Conventional Units and SI Units[a]

a. To convert conventional units to SI units, multiply by the conversion factor.
b. To convert from SI units to conventional units, divide by the conversion factor.

	Conventional Units	Conversion Factor	SI Unit
Calcium	mg/dL	0.25	mmol/L
Cholesterol	mg/dL	0.0259	mmol/L
HDL cholesterol	mg/dL	0.0259	mmol/L
Ferritin	ng/mL	2.247	nmol/L
Folate	ng/mL	2.266	nmol/L
Glucose	mg/dL	0.0555	mmol/L
Hematocrit	%	0.01	Proportion of 1.0
Hemoglobin	g/dL	10.0	g/L
Homocysteine	mg/L	7.397	μmol/L
Insulin	μIU/mL	6.945	pmol/L
Iron	μg/dL	0.179	μmol/L
LDL cholesterol	mg/dL	0.0259	mmol/L
Lipoprotein (a)	mg/dL	0.0357	μmol/L
Triglycerides	mg/dL	0.0113	mmol/L
Vitamin A (retinol)	μg/dL	0.0349	μmol/L
Vitamin B_6	ng/mL	4.046	nmol/L
Vitamin B_{12}	pg/mL	0.738	nmol/L
Vitamin C	mg/dL	56.78	pmol/L
Vitamin D	ng/mL	2.496	nmol/L

[a] "SI" stands for *le Système International d'unités*. SI values are based on decades of international cooperation in developing a universal system of measurement.

Body Mass Index (BMI)

	18	19	20	21	22	23	24	25	26	27	28	29	30	31	32	33	34	35	36	37	38	39	40
Height												Body Weight (pounds)											
4'10"	86	91	96	100	105	110	115	119	124	129	134	138	143	148	153	158	162	167	172	177	181	186	191
4'11"	89	94	99	104	109	114	119	124	128	133	138	143	148	153	158	163	168	173	178	183	188	193	198
5'0"	92	97	102	107	112	118	123	128	133	138	143	148	153	158	163	168	174	179	184	189	194	199	204
5'1"	95	100	106	111	116	122	127	132	137	143	148	153	158	164	169	174	180	185	190	195	201	206	211
5'2"	98	104	109	115	120	126	131	136	142	147	153	158	164	169	175	180	186	191	196	202	207	213	218
5'3"	102	107	113	118	124	130	135	141	146	152	158	163	169	175	180	186	191	197	203	208	214	220	225
5'4"	105	110	116	122	128	134	140	145	151	157	163	169	174	180	186	192	197	204	209	215	221	227	232
5'5"	108	114	120	126	132	138	144	150	156	162	168	174	180	186	192	198	204	210	216	222	228	234	240
5'6"	112	118	124	130	136	142	148	155	161	167	173	179	186	192	198	204	210	216	223	229	235	241	247
5'7"	115	121	127	134	140	146	153	159	166	172	178	185	191	198	204	211	217	223	230	236	242	249	255
5'8"	118	125	131	138	144	151	158	164	171	177	184	190	197	203	210	216	223	230	236	243	249	256	262
5'9"	122	128	135	142	149	155	162	169	176	182	189	196	203	209	216	223	230	236	243	250	257	263	270
5'10"	126	132	139	146	153	160	167	174	181	188	195	202	209	216	222	229	236	243	250	257	264	271	278
5'11"	129	136	143	150	157	165	172	179	186	193	200	208	215	222	229	236	243	250	257	265	272	279	286
6'0"	132	140	147	154	162	169	177	184	191	199	206	213	221	228	235	242	250	258	265	272	279	287	294
6'1"	136	144	151	159	166	174	182	189	197	204	212	219	227	235	242	250	257	265	272	280	288	295	302
6'2"	141	148	155	163	171	179	186	194	202	210	218	225	233	241	249	256	264	272	280	287	295	303	311
6'3"	144	152	160	168	176	184	192	200	208	216	224	232	240	248	256	264	272	279	287	295	303	311	319
6'4"	148	156	164	172	180	189	197	205	213	221	230	238	246	254	263	271	279	287	295	304	312	320	328
6'5"	151	160	168	176	185	193	202	210	218	227	235	244	252	261	269	277	286	294	303	311	319	328	336
6'6"	155	164	172	181	190	198	207	216	224	233	241	250	259	267	276	284	293	302	310	319	328	336	345

Under-weight	Healthy Weight			Overweight			Obese				
(<18.5)	(18.5–24.9)			(25–29.9)			(≥30)				

Find your height along the left-hand column and look across the row until you find the number that is closest to your weight. The number at the top of that column identifies your BMI. The area shaded in blue-green represents healthy weight ranges. The figure below presents silhouettes of various BMI.

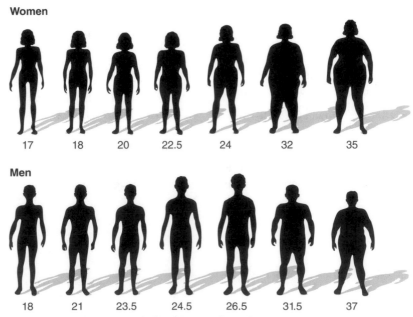

Women

17   18   20   22.5   24   32   35

Men

18   21   23.5   24.5   26.5   31.5   37

SOURCE: Reprinted from material of the Dietitians of Canada.

Carbohydrate Counting for Type 1 Diabetes

Introductory facts

- Only the calories from carbohydrates are considered, approximately 50–55% of total calories needed per day.
- Carbohydrate content of each meal and each snack each day is prescribed.
- One carbohydrate exchange, or "choice," is 15 grams of carbohydrates or approximately 60 calories.
- The number of exchanges is based on age and weight, such as 18 per day.
- The distribution of exchanges across the meals and snacks each day is set by the type of insulin(s) prescribed and its/their method of administration, such as insulin pump or injection.
- Routine blood glucose checks are required from one to four times a day to ensure that planned carbohydrate intake and insulin are well balanced.
- Activity, health status, environmental temperature, and injection site are examples of variables affecting glucose control in addition to carbohydrate intake.

Carbohydrate Counting Lists

Starch (containing little fat)
Fruit
Milk (skim, low-fat, whole)
Other carbohydrates providing variable amounts of protein and fat along with starch (example foods are cake, doughnuts, fruit cobbler, rice milk, sherbet, syrup)
Vegetables

Exchange lists for meat and meat substitutes consider these to have no carbohydrates, contributing just protein, fat, and calories.

Examples

Dividing Carbohydrate Choices into a Meal Plan

A meal plan is devised to provide the total number of carbohydrate (CHO) choices the person can have each day over each meal and snack. The CHO choices are spaced throughout the day, which makes it easier for the insulin to work on the glucose from food.
Example of 18 CHO choices spaced over the day:

Breakfast: 4 choices (or 60 grams CHO)
Snack: 2 choices (or 30 grams CHO)
Lunch: 4 choices (or 60 grams CHO)
Snack: 2 choices (or 30 grams CHO)
Supper: 4 choices (or 60 grams CHO)
Snack: 2 choices (or 30 grams CHO)

This meal plan example is for a young child who eats three meals and three snacks each day and for the child's specific insulin regimen. This meal plan would be adjusted for an older child who also has a total of 18 carbohydrate choices by making three meals larger and reducing to one snack, based on the child's specific insulin regimen.

Example of a Family Education Session About Balancing Diet and Insulin Using an Insulin Pump

Review the prescribed insulin dose and carbohydrate exchanges for the current activity level with the parent and child.
Ask the child to demonstrate how to convert CHO choices into insulin doses before a snack with 2 CHO exchanges. (This step depends on the specific insulin regimen prescribed.) The child should demonstrate that the pump is set for 4 units of rapid-acting insulin if he or she has the usual ratio of 2 units of insulin per carbohydrate choice, and a snack has 2 choices.
Review the food accommodation for increased activity that the family has been prescribed; this may allow an additional CHO choice, or it may be based on insulin dose.

SOURCE: Adapted from American Dietetic Association: Basic Carbohydrate Counting 2003.

References

Chapter 1

1. Wilde, P. E., et al. Individual weight change is associated with household food security status. *J Nutr* 2000; 1395–1400.

2. Summary report on application of the DRIs for dietary assessment. Washington, D.C.: National Academy of Sciences, 2000.

3. Grabitske, H. A. et al. Low-digestible carbohydrates in practice. *J Am Diet Assoc* 2008; 108:1677–81.

4. Position of the American Dietetic Association: Health implications of dietary fiber. *J Am Diet Assoc* 2008; 108:1716–31.

5. Livesey, G. et al. Glycemic response and health—A systematic review and meta-analysis: Relations between dietary glycemic properties and health outcomes. *Am J Clin Nutr* 2008; 87(Suppl):258S–68S.

6. Barclay, A. W. et al. Glycemic index, glycemic load, and chronic disease risk: A meta-analysis of observational studies. *Am J Clin Nutr* 2008; 87:627–37.

7. Dietary Reference Intakes: Energy, carbohydrate, fiber, fat, fatty acids, cholesterol, protein, and amino acids. Washington, D.C.: National Academies Press, 2002.

8. Brand-Miller, J. C. et al. Glycemic index, postprandial glycemia, and the shape of the curve in healthy subjects: Analysis of a database or more than 1000 foods. *Am J Clin Nutr* 2009; 89:97–105.

9. Data extracted from Foster-Powell, K. et al. International table of glycemic index and glycemic load values. *Am J Clin Nutr* 2002; 76:5–56.

10. Gebauer, S. K. et al. n-3 Fatty acid dietary recommendations and food sources to achieve essentiality and cardiovascular benefits. *Am J Clin Nutr* 2006; 83(Suppl):1526S–35S.

11. Arterburn, L. M. et al. Distribution, interconversion, and dose response of n-3 fatty acids in humans. *Am J Clin Nutr* 2006; 83(suppl):1487S–76S.

12. Harris, W. S. et al. Omega-6 fatty acids and risk for cardiovascular disease. *Circulation* 2009; available at http://circ.ahajournals.org, accessed 1/09.

13. Position of the American Dietetic Association and Dietitians of Canada: Dietary fatty acids. *J Am Diet Assoc* 2007; 107:1599–1611.

14. Kris-Etherton, P. M. et al. Adherence to dietary guidelines: Benefits on atherosclerosis progression. *Am J Clin Nutr* 2007; 90:13–4.

15. Ordovas, J. M. Genetic influences on blood lipids and cardiovascular disease risk: Tools for primary prevention. *Am J Clin Nutr* 2009; Suppl:1509S–17S.

16. USDA. What we eat in America. NHANES, 2005–2006. Available at www.are.usda.gov/ba/bhnrc/fsrg.

17. Wolfe, K. L. et al. Cellular antioxidant activity of common fruits. *J Agric Food Chem* 2008; 56:8418–26.

18. Halvorsen, B. L. et al. Content of redox-active compounds in foods consumed in the United States. *Am J Clin Nutr* 2006; 84:95–135.

19. Krauss R. M. et al. Revision 2000: A statement for healthcare providers from the Nutrition Committee of the American Heart Association. *J Nutr* 2001; 131:132–46.

20. Usui, T. Pharmaceutical prospects of phytoestrogens. *Endocr J* 2006; 53:7–20.

21. Dietary reference intakes: Water and electrolytes. Washington, D.C.: National Academies Press, 2004.

22. Grandjean, A. C. et al. Hydration: Issues for the 21st century. *Nutr Rev* 2003; 61:261–71.

23. Noakes, T. D. Too many fluids as bad as too few. *BMJ* 2003; 327:113–14.

24. Kolasa, K. M. Water: The recommendations we give—Are they scientifically accurate? (Conference paper.) Food and Nutrition Conference and Exhibition, Oct. 2002; American Dietetic Association.

25. Corte, C., Johnston, C. S., et al. People with marginal vitamin C status are at high risk of developing vitamin C deficiency. *J Am Diet Assoc* 1994; 99:854–6.

26. Simopoulos, A. P. Genetic variation and nutrition. Experimental Biology Annual Meeting. San Diego, CA, April 12, 2003.

27. Guttmacher, A. E. et al. Genomic medicine—A primer. *N Engl J Med* 2002; 347:1512–20.

28. Yu, S. The life-course approach to health. *Am J Public Health* 2006; 96:768.

29. Marengoni, A. et al. *In utero* and early-life conditions and adult health and disease. *N Engl J Med* 2009; 359:1523.

30. Cornelis, M. C. et al. TCF7L2, dietary carbohydrate, and risk of type 2 diabetes in U.S. women. *Am J Clin Nutr* 2004; 89:1256–62.

31. McBean, L. et al. Allaying fears and fallacies about lactose intolerance. *J Am Diet Assoc* 1998; 98:671–676.

32. Swagerty, D. L. et al. Lactose intolerance. *Am Fam Physician* 2002; 65:1845–50,1855–6.

33. Kittler, P. G. et al. Diet counseling in a multicultural society. *Diabetes Educator* 1989; 16:127–32.

34. Van Horn, L. et al. The evidence for dietary prevention and treatment of cardiovascular disease. *J Am Diet Assoc* 2008; 108:287–331.

35. World Cancer Research Fund/American Institute for Cancer Research. *Food, Nutrition, Physical Activity, and the Prevention of Cancer; A Global Perspective.* Washington, D.C.: AIRC, 2007.

36. Paez, K. More Americans getting multiple chronic illnesses. Retrieved from www.medscape.com/view article/586363, accessed 1/09.

37. Hitz, M. F. et al. Bone mineral density and bone markers in patients with a recent low-energy fracture: Effect of I y of treatment with calcium and vitamin D. *Am J Clin Nutr* 2007; 86:251–9.

38. Adrogue, H. J. et al. Sodium and potassium in the pathogenesis of hypertension. *N Engl J Med* 2007; 356:1966–78.

39. O'Keffe, J. H. et al. Dietary and lifestyle strategies for improving postprandial glucose, lipid profile, markers of inflammation, and cardiovascular health. *Am Coll Cardiol* 2008; 51:249–255.

40. Scarmeas, N. et al. Adherence to the Mediterranean diet and the risk of developing Alzheimer's disease. *Ann Neurol.* Posted online 4/18/06 at www.medscape.com/viewarticle/530121, accessed 4/20/06.

41. Hossain, P. et al. Obesity and diabetes in the developing world: A growing challenge, *N Engl J Med* 2007; 365:213–5.

42. Utsugi, M. T. et al. Fruit and vegetable consumption and the risk of hypertension determined by self-measurement of blood pressure at home: The Ohasama study. *Hypertens Res* 2008; 3:1435–43.

43. Esmaillzadeh, A. et al. Home use of vegetable oils markers of systemic inflammation, and endothelial dysfunction among women. *Am J Clin Nutr* 2008; 88913–21.

44. Freeland-Graves, J. et al. Position of the American Dietetic Association Total Diet Approach to Communicating Food and Nutrition Information. *J Am Diet Assoc* 2002; 102:100–108.

45. Park, Y. K. et al. History of cereal-grain product fortification in the United States. *Nutr Today* 2001; 36:23–36.

46. Position of the American Dietetic Association: Functional foods, 2009; 109:735–46.

47. Hasler, C. M. Functional foods: Benefits, concerns and challenges. *J.Nutr* 2002; 132:377–81.

48. Beale, B. Probiotics: Their tiny worlds are under scrutiny. *The Scientist* 2002; Jul:20–22.

49. Boyle, R. J. et al. Risks of probiotic treatment. *Am J Clin Nutr* 2006; 83:1254–64.

50. Looijer-van Lagen, M. A. C. et al. Probiotics and prebiotics as functional ingredients in inflammatory bowel disease. *Nutr Today* 2008; 43:235–42.

51. Douglas, D. C. et al. Probiotics and prebiotics in dietetics practice. *J Am Diet Assoc* 2008; 108:510–21.

52. Taveras, E. M. et al. Changes in weight status in infancy and later risk of obesity. *Pediatrics* 2009; 123:1177–1183.

53. Blackmore, K. M. et al. Early vitamin D intake and future risk of breast cancer. *Am J Epidemiol* 2008; 168:915–24.

54. Romero-Gwynn, E. et al. Dietary patterns and acculturation among immigrants from El Salvador. *Nutr Today* 2003; 35:233–40.

55. Conti, K. The four winds nutrition model: A new culturally specific food guide and nutrition approach for the Northern Plains Indians. (Seminar presentation.) Minneapolis: University of Minnesota, Oct 28, 2002.

56. Dirks, R. T. et al. African American dietary patterns at the beginning of the 20th century. *J Nutr* 2001; 131:1881–9.

57. McCaffree, J. Dietary restrictions of other religions. *J Am Diet Assoc* 2003; 103:912.

58. Halsted, C. H. Dietary supplements and functional foods: Two sides of a coin? *Am J Clin Nutr* 2003; 77(Suppl):1001S–7S.

59. Dwye, J. Dietary assessment. In Shils, M. E., ed. *Modern nutrition in health and disease*. Philadelphia: Lippincott, Williams & Wilkins: 1998. PP. 937–62.

60. Horner, N. K. et al. Participant characteristics associated with errors in self-reported energy intake from the Women's Health Initiative food-frequency questionnaire. *Am J Clin Nutr* 2004; 76:766–73.

61. Conway, J. M. et al. Accuracy of the dietary recall using the USDA five-step multiple-pass method in men: An observational study. *J Am Diet Assoc* 2004; 104:595–603.

62. Moshfegh, A. J. et al. The U.S. Department of Agriculture Automated Multiple-Pass Method reduces bias in the collection of energy intakes. *Am J Clin Nutr* 2008; 88:324–32.

63. Guenther, P. M. et al. Valuation of the Healthy Eating Index, 2005. *J Am Diet Assoc* 2008; 108:1854–64.

64. Ginde, A. A. et al. Mean serum 25 (OH) vitamin D levels decreasing in all categories of the U.S. population. *Arch Intern Med* 2009; 169:626–32.

65. Selhub, J. et al. The use of blood concentrations of vitamins and their respective functional indicators to define folate and vitamin B_{12} status. *Food Nutr Bull* 2008; 29(2Suppl):S67–73.

66. Kranz, A. et al. Laboratory reference values. *N Engl J Med* 2004; 351:1548–63.

67. Greer, J. P. et al. *Wintrobe's Clinical Hematology*, 11th ed. Philadelphia, PA: Lippincott Williams and Wilkins, 2004.

68. Go, V. L. W. et al. Nutrient-gene interaction: Metabolic genotype-phenotype relationships. *J Nutr* 2003; 135: 3016S–20S.

69. National Nutrition Monitoring System. Available at http://www.cdc.gov/nchs/nhanes.htm, accessed 8/06.

70. Food and Nutrition Assistance. Available at www.ers.usda.gov/Browse/FoodNutritionAssistance, accessed 11/08.

71. Position of the American Dietetic Association: Child and adolescent food and nutrition programs. *J Am Diet Assoc* 2003; 103:887–92.

72. Public health achievements. *Nutrition Today* 1999; May/Jun:2.

73. Nowson, C. A. et al. Blood pressure response to dietary modifications in free-living individuals. *J Nutr* 2004; 134:2322–2329.

Chapter 2

1. Bellver, J. Impact of body weight and lifestyle in IVF outcome. *Expert Rev Obstet Gynceol* 2008; 3:607–25.

2. Faststats A to Z. Available at www.cdc.gov/nchs/fastats/2births.html, accessed 7/09.

3. Wong, W. Y. et al. Male factor subfertility: Possible causes and the impact of nutritional factors. *Fertil Steril* 2000; 73:435–42.

4. Van Voorhis, B. J. In vitro fertilization. *N Engl J Med* 2007; 356:379–86.

5. Kovac, P. Recurrent pregnancy loss. Available at www.medscape.com/viewprogram/5293, accessed 4/06.

6. Kasturi, S. S. et al. The metabolic syndrome and male infertility, *J Androl* 2008; 29:251–9.

7. Health, United States, 2008. Available at www.cdc.gov/fastats, accessed 7/09.

8. Ford, W. C. et al. Increasing paternal age is associated with delayed conception in a large population of fertile couples: Evidence for declining fecundity in older men. The ALSPAC Study Team (Avon Longitudinal Study

of Pregnancy and Childhood). *Hum Reprod* 2000; 15:1703–8.

9. Powers, H. J. Interaction among folate, riboflavin, genotype, and cancer, with reference to colorectal and cervical cancer. *J Nutr* 2005; 135:2960S–6S.

10. Bhasin, S. Approach to the infertile man. *J Clin Endocrinol Metab* 2007; 92:1995–2004.

11. Forti, G. et al. Clinical review 100: Evaluation and treatment of the infertile couple. *J Clin Endocrinol Metab* 1998; 83:4177–88.

12. Homan, G. F. et al. The impact of lifestyle factors on reproductive performance in the general population and those undergoing infertility treatment: A review. *Hum Reprod Update* 2007; 13:209–23.

13. Chavarro, J. E. et al. Diet and lifestyle in the prevention of ovulatory disorder infertility. *Obstet Gynecol* 2007; 110(5):1050–8.

14. Robker, R. et al. Are changes in ovarian environment linked to obesity and infertility? *J Clin Endocrinol Metab* 2009; March 12.

15. van der Spuy, Z. M. Nutrition and reproduction. *Clin Obstet Gynecol* 1985; 12:579–604.

16. DeNoon, D. J. U.S. Ranks in Lowest Third for Infant Mortality, 10/15/08. Available at http://www.medscape.com/viewarticle/582117.

17. Rao, V. G. et al. Nutritional deficiency disorders and high mortality among children of the Great Andamanese tribe. *Natl Med J India* 1998; 11:65–8.

18. Wood, J. W. Maternal nutrition and reproduction: Why demographers and physiologists disagree about a fundamental relationship. *Ann NY Acad Sci* 1994; 709:101–16.

19. Hirschman, C. Why fertility changes. *Annu Rev Sociol* 1994; 20:203–33.

20. Ramakrishnan, U. et al. Early childhood nutrition, education and fertility milestones in Guatemala. *J Nutr* 1999; 129:2196–202.

21. Pena, R. et al. Fertility and infant mortality trends in Nicaragua 1964–1993.

The role of women's education. *J Epidemiol Community Health* 1999; 53:132–7.

22. Kolata, G. B. Kung hunter-gatherers: Feminism, diet, and birth control. *Science* 1974; 185:932–4.

23. Leslie, P. W. et al. Evaluation of reproductive function in Turkana women with enzyme immunoassays of urinary hormones in the field. *Hum Biol* 1996; 68:95–117.

24. Bongaarts, J. Does malnutrition affect fecundity? A summary of evidence. *Science* 1980; 208:564–9.

25. Hofny, E. R. et al. Semen parameters and hormonal profile in obese fertile and infertile males. Fertil Steril, 2009; May 5.

26. The Practice Committee of the American Society for Reproductive Medicine, Obesity and Reproduction: An educational bulletin. *Fertil Steril* 2008; 90:S21–9.

27. De Pergola, G. et al. Abdominal fat accumulation, and not insulin resistance, is associated to oligomenorrhea in non-hyperandrogenic overweight/obese women. *J Endocrinol Invest* 2009; 32:98–101.

28. Chavarro, J. E. et al. Body mass index in relation to semen quality, sperm DNA integrity, and serum reproductive hormone levels among men attending an infertility clinic. *Fertil Steril* 2009; Mar 2.

29. Pasquali, R. Obesity and infertility. *Curr Opin Endocrinol Diabetes Obes* 2007; 14:482–7.

30. Moran, L. J. and Norman, R. J. The obese patient with infertility: A practical approach to diagnosis and treatment. *Nutr Clin Care* 2002; 5:290–7.

31. Zain, M. M. et al. Impact of obesity on female fertility and fertility treatment. *Womens Health* (Lond) 2008; 4:183–94.

32. Reichlin, S. Female fertility and the body fat connection (review). *N Engl J Med* 2003; 348:869–70.

33. Bolúmar, F. et al. Body mass index and delayed conception: A European Multicenter Study on Infertility and Subfecundity. *Am J Epidemiol* 2000; 151:1072–9.

34. Keys, A. et al. *The biology of human starvation*. Minneapolis: University of Minnesota Press, 1950.

35. Williams, N. L. et al. Evidence for a causal role of low energy availability in the induction of menstrual cycle disturbances during strenuous exercise training. *J Clin Endocrinol Metab* 2001; 86:5184–93.

36. Fenichel, R. M. et al. Anorexia, bulimia, and the athletic triad: Evaluation and management. *Curr Osteoporos Rep* 2007; 5:160–4.

37. Loucks, A. B. et al. The female athlete triad: Do female athletes need to take special care to avoid low energy availability? *Med Sci Sports Exerc* 2006; 38:1694–1700.

38. Ruder, E. H. et al. Oxidative stress and antioxidants: Exposure and impact on female fertility. *Hum Reprod Update* 2008; 14:345–57.

39. Mehendale, S. S. et al. Oxidative stress-mediated essential polyunsaturated fatty acid alterations in female infertility. *Hum Fertil* (Camb) 2009; 12:28–33.

40. Agarwal, A. et al. Clinical relevance of oxidative stress in male factor infertility: An update. *Am J Reprod Immunol* 2008; 59:2–11.

41. Tremellen, K. et al. Oxidative stress and male infertility: A clinical perspective. *Hum Reprod Update* 2008; 14:243–58.

42. Agarwal, A. Oxidative stress and its implications in female infertility: A clinician's perspective. *Reprod Biomed Online* 2005; 11:641–50.

43. Mendiola, J. et al. A low intake of antioxidant nutrients is associated with poor semen quality in patients attending fertility clinics. *Fertil Steril* 2009; Jan 13.

44. Eskenazi, B. et al. Antioxidant intake is associated with semen quality in healthy men. *Hum Reprod* 2005; 20:1006–12.

45. Keskes-Ammar, L. et al. Sperm oxidative stress and the effect of an oral vitamin E and selenium supplement on semen quality in infertile men. *Arch Androl* 2003; 49:83–94.

46. Safarinejad, M. R. et al. Efficacy of selenium and/or N-acetyl-cysteine for improving semen parameters in infertile men: A double-blind, placebo controlled, randomized study. *J Urol* 2009; 181:741–51.

47. Hill, P. B. et al. Gonadotropin release and meat consumption in vegetarian women. *Am J Clin Nutr* 1986; 43:37–41.

48. Wyshak, G. et al. Fiber consumption and menstrual regularity in young women. *J Women's Health* 1993; 2:295–9.

49. Reichman, M. E. et al. Effect of dietary fat on length of the follicular phase of the menstrual cycle in a controlled diet setting. *J Clin Endocrinol Metab* 1992; 74:1171–5.

50. Chavarro, J. E. et al. Soy food and isoflavone intake in relation to semen quality parameters among men from an infertility clinic. *Hum Reprod* 2008; 23: 2584–90.

51. Pirke, K. M. Dieting influences the menstrual cycle: Vegetarian versus nonvegetarian diet. *Fertil Steril* 1986; 46:1083–8.

52. Griffith, J. Association between vegetarian diet and menstrual problems in young women: A case presentation and brief review. *J Pediatr Adolesc Gynecol* 2003; 16:319–23.

53. Ebisch, I. M. et al. The importance of folate, zinc and antioxidants in the pathogenesis and prevention of subfertility, *Hum Reprod Update* 2007; 13:163–74.

54. Yuyan, L. et al. Are serum zinc and copper levels related to semen quality? *Fertil Steril* 2008; 89:1008–11.

55. Colagar, A. H. et al. Zinc levels in seminal plasma are associated with sperm quality in fertile and infertile men. *Nutr Res* 2009; 29:82–8.

56. Abbasi, A. A. et al. Experimental zinc deficiency in man. Effect on testicular function. *J Lab Clin Med* 1980; 96:544–50.

57. Omu, A. E. et al. Effect of antioxidant intake on sperm chromatin stability in healthy nonsmoking men. *J Androl* 2005; 26:550–6.

58. Chavarro, J. E. et al. Use of multivitamins, intake of B vitamins, and risk of ovulatory infertility. *Fertil Steril* 2008; 89:668–76.

59. Young, S. S. et al. The association of folate, zinc and antioxidant intake with sperm aneuploidy in healthy non-smoking men. *Hum Reprod* 2008; 23:1014–22.

60. Bolumar, F. et al. Caffeine intake and delayed conception: A European multicenter study on infertility and subfecundity. European study group on infertility subfecundity. *Am J Epidemiol* 1997; 145:324–34.

61. Hatch, E. E. and Bracken, M. B. Association of delayed conception with caffeine consumption. *Am J Epidemiol* 1993; 138:1082–92.

62. Hakim, R. B. et al. Alcohol and caffeine consumption and decreased fertility. *Fertil Steril* 1998; 70:632–7.

63. Chavarro, J. E. et al. Caffeinated and alcoholic beverage intake in relation to ovulatory disorder infertility. *Epidemiology* 2009; 20:374–81.

64. Emanuele, M. A. et al. Alcohol's effects on male reproduction. *Alcohol Health Res World* 1998; 22:195–201.

65. Jensen, T. K. et al. Does moderate alcohol consumption affect fertility? Follow up study among couples planning first pregnancy. *BMJ* 1998; 317:505–10.

66. Tolstrup, J. S. et al. Alcohol use as predictor for infertility in a representative population of Danish women. *Acta Obstet Gynecol Scand* 2003; 82:744–9.

67. Chavarro, J. E. et al. Caffeinated and alcoholic beverage intake in relation to ovulatory disorder infertility. *Epidemiology* 2009; 20:374–81.

68. Maneesh, M. et al. Alcohol abuse-duration dependent decrease in plasma testosterone and antioxidants in males. *Indian J Physiol Pharmacol* 2006; 50:291–6.

69. Hassan, M. A. et al. Negative lifestyle is associated with a significant reduction in fecundity. *Fertil Steril* 2004; 81:384–92.

70. Floyd, R. L. et al. The clinical content of preconception care: Alcohol, tobacco, and illicit drug exposures. *Am J Obstet Gynecol* 2008; 199:S333–9.

71. Benoff, S. et al. Blood lead levels and infertility in males. *Hum Reprod* 2003; 18: 374–83.

72. Lamb, E. J. and Bennett, S. Epidemiologic studies of male factors in infertility. *Ann NY Acad Sci* 1994; 709:165–78.

73. Choy, C. M. et al. Infertility, blood mercury concentrations and dietary seafood consumption: A case-control study. *Brit J Obstet Gynecol* 2002; 109:1121–5.

74. Buck, G. M. et al. Consumption of contaminated sport fish from Lake Ontario and time-to-pregnancy. New York State Angler Cohort. *Am J Epidemiol* 1997; 146:949–54.

75. Meeker, J. D. et al. Cadmium, lead, and other metals in relation to semen quality: Human evidence for molybdenum as a male reproductive toxicant. *Environ Health Perspect* 2008; 116:1473–9.

76. Siu, E. R. et al. Cadmium-induced testicular injury. *Toxicol Appl Pharmacol* 2009; Feb 21.

77. Queiroz, E. K. et al. Occupational exposure and effects on the male reproductive system. *Cad Saude Publica* 2006; 22:485–93.

78. Vandenbroucke, J. P. et al. Oral contraceptives and the risk of venous thrombosis. *N Engl J Med* 2001; 344:1527–35.

79. Summer, A. E. et al. Oral contraceptive use and the risk of cardiovascular disease in African American women. *J Clin Endocrinol Metab* 2008; 93:2097–2103.

80. Beral, V. et al. Long term use of oral contraceptives and the risk of ovarian cancer. The Collaborative Group on Epidemiological Studies on Ovarian Cancer. *Lancet* 2008; 371:275, 277–8, 303–14.

81. Green, T. J. et al. Oral contraceptives did not affect biochemical folate indexes and homocysteine concentrations in adolescent females. *J Am Diet Assoc* 1998; 98:49–55.

82. Lussana, F. et al. Blood levels of homocysteine, folate, vitamin B6 and B12 in women using oral contraceptives compared to non-users. *Thromb Res* 2003; 112:37–41.

83. Sellerberg, U. et al. Women who take the pill should limit consumption of licorice. Available at www.medscape.com, accessed 2/02.

84. Barnhart, K. T. Return to fertility following discontinuation of oral contraceptives. *Fertil Steril* 2009; 91:659–63.

85. Polaneczky, M. et al. Early experience with the contraceptive use of depot medroxyprogesterone acetate in an inner-city clinic population. *Fam Plann Perspect* 1996; 28:174–8.

86. Matson, S. C. et al. Physical findings and symptoms of depot medroxyprogesterone acetate use in adolescent females. *J Pediatr Adolesc Gynecol* 1997; 10:18–23.

87. Pelkman, C. Hormones and weight change. *J Repro Med* 2002; 47(Suppl):791–4.

88. Berenson, A. B. et al. Use of DMPA associated with significant increases in body weight and fat. *Am J Obstet Gynecol* 2009; 200:329.e1–329.e8.

89. Cromer, B. A. et al. DMPA use and loss of bone mineral density in adolescents. *Fertil Steril* 2008; 90:2060–7.

90. Shaarawy, M. et al. Effects of the long-term use of depot medroxyprogesterone acetate as hormonal contraceptive on bone mineral density and biochemical markers of bone remodeling. *Contraception* 2006; 74:297–302.

91. Sivin, I. et al. Prolonged effectiveness of Norplant® capsule implants: A 7-year study. *Contraception* 2000; 61:187–94.

92. Meirik, O., Farley, T. M., and Sivin, I. Safety and efficacy of levonorgestrel implant, intrauterine device, and sterilization. *Obstet Gynecol* 2001; 97:539–47.

93. Kaunitz, A. M. Long-acting hormonal contraception: Assessing impact on bone density, weight, and mood. *Int J Fertil Womens Med* 1999; 44:110–7.

94. Haugen, M. M. et al. Patient satisfaction with a levonorgestrel-releasing contraceptive implant. Reasons for and patterns of removal. *J Reprod Med* 1996; 41:849–54.

95. U.S. Food and Drug Administration. Changes to the Ortho Evra Transdermal Patch label. Available at www.fda.gov/cder/drug/infopage/orthoevra/default.htm, accessed 4/09.

96. Berenson, A. B. et al. Contraceptive outcomes among adolescents prescribed Norplant implants versus oral contraceptives after one year of use. *Am J Obstet Gynecol* 1997; 176:586–92.

97. Amory, J. K. Progress and prospects in male hormonal contraception. *Curr Opin Endocrinol Diabetes Obes* 2008; 15:255–6097.

98. Shulman, L. P. Contraception today: Risks and benefits in scientific, clinical, and sociocultural context. Available at www.medscape.com/viewprogram/6049, accessed 9/06.

99. Institute of Medicine. *Nutrition during pregnancy.* Washington, D.C.: National Academies Press, 1990.

100. Wolff, T. et al. Folic acid supplementation for the prevention of neural tube defects: An update of the evidence for the U.S. Preventive Services Task Force. *Ann Intern Med* 2009; 150:632–9.

101. Poretti, A. et al. Neural tube defects in Switzerland from 2001 to 2007: Are periconceptual folic acid recommendations being followed? *Swiss Med Wkly* 2008; 138:608–13.

102. Ganji, V. et al. Trends in serum folate, RBC folate, and circulating total homocysteine concentrations in the U.S. *J Nutr* 2006; 138:153–8.

103. Kuntz, K. M. et al. Folate intake in U.S. falling short of established targets. *Am J Public Health* 2006; 96. Available at www.medscape.com/ viewartide/545509, accessed 10/06.

104. Shuaibi, A. M. et al. Folate status of young Canadian women after folic acid fortification of grain products. *J Am Diet Assoc* 2008; 108:2090–4.

105. Hubner, R. A. et al. Folate and colorectal cancer prevention. *Br J Cancer* 2009; 100:233–9.

106. Caan, B. et al. Benefits associated with WIC supplemental feeding during the interpregnancy interval. *Am J Clin Nutr* 1987; 45:29–41.

107. Jus'at, I. et al. Reaching young Indonesian women through marriage registries: An innovative approach for anemia control. *J Nutr* 2000; 130:456S–8S.

108. Recommendations to improve preconception health and health care, United States. *MMWR* April 21, 2006.

109. *Summary of the Nutrition Care Process Steps for Nutrition Assessment (Pocket Guide for International Dietetics & Nutrition Terminology), Reference Manual,* 2 ed. Chicago: American Dietetic Association, 2009.

110. Chavarro, et al. Iron consumption and the risk of infertilty. *Obstet Gynecol* 2006; 108:1145–1152.

111. Allen, L. H. Multiple micronutrients in pregnancy and lactation: An overview. *Am J Clin Nutr* 2005; 81:1 206S–1212S.

112. Bates, G. W. Body weight control practice as a cause of infertility. *Clin Obstet Gynecol* 1985; 28:632–44.

Chapter 3

1. King, J. C. Preface. *Am J Clin Nutr* 2000; 71(Suppl):1217S.

2. Pearlstein, T. Premenstrual dysphoric disorder: Burden of illness and treatment update. *J Psychiatry Neurosci* 2008; 33:291–301.

3. Pray, W. S. et al. Menstrual cycle–related discomforts. *US Pharmacist* 2005; 30.

Available at www.medscape.com/ viewarticle/513070, accessed 10/05.

4. Vaitukaitis, J. L. Premenstrual syndrome (editorial). *N Engl J Med* 1984; 311:1371–3.

5. Grady-Weliky, T. Premenstrual dysphoric disorder. *N Engl J Med* 2003; 348:433–8.

6. Daugherty, J. E. Treatment strategies for premenstrual syndrome. *Am Fam Physician* 1998; 58:183–92, 197–8.

7. Chayachinda, C. et al. Premenstrual syndrome in Thai nurses. *J Psychosom Obstet Gynaecol* 2008; 29:199–205.

8. Goodale et al. Alleviation of premenstrual syndrome symptoms with the

relaxation response. *Obstet Gynecol* 1990; 75:649–55.

9. Walker, A. F. et al. Magnesium supplementation alleviates premenstrual symptoms of fluid retention. *J Women's Health* 1998; 7:1157–65.

10. Thys-Jacobs, S. et al. Calcium carbonate and the premenstrual syndrome: Effects on premenstrual and menstrual symptoms. Premenstrual Syndrome Study Group. *Am J Obstet Gynecol* 1998; 179:444–52.

11. Ghanbari, Z. et al. Effects of calcium supplement therapy in women with premenstrual syndrome. Taiwan *J Obstet Gynecol* 2009; 48:124–9.

12. Bertone-Johnson et al. Calcium and vitamin D intake and risk of incident premenstrual syndrome. *Arch Intern Med* 2005; 165:1246–52.

13. Berman, M. K. et al. Vitamin B-6 in premenstrual syndrome. *J Am Diet Assoc* 1990; 90:859–61.

14. Wyatt, K. M. et al. Efficacy of vitamin B_6 in the treatment of premenstrual syndrome: A systematic review. *BMJ* 1999; 318:1275–81.

15. Sharma, P. et al. Role of bromocriptine and pyridoxine in premenstrual tension syndrome. Indian *J Physiol Pharmacol* 2007; 51:368–74.

16. Obesity rates. Available at www.cdc.gov/nchs/data/nhis/earlyrelease/200809_06.pdf, accessed 7/09.

17. Bellver, J. Impact of body weight and lifestyle in IVF outcome. *Expert Rev Obstet Gynceol* 2008; 3:607–25.

18. Kovac, P. Recurrent pregnancy loss. Available at www.medscape.com/viewprogram/5293, accessed 4/06.

19. Robker, R. et al. Are changes in ovarian environment linked to obesity and infertility? *J Clin Endocrinol Metab* 2009; March 12.

20. Michos, E. D. et al. Vitamin D deficiency and the risk of incident Type 2 diabetes. *Future Cardiol* 2009; 5:15–8.

21. Kulkarni, S. Lifestyle management before infertility treatment. Available at www.medscape.com/viewarticle/52224, accessed 1/06.

22. Kasturi, S. S. et al. The metabolic syndrome and male infertility. *J Androl* 2008; 29:251–9.

23. Hofny, E. R. et al. Semen parameters and hormonal profile in obese fertile and infertile males. *Fertil Steril* 2009; May 5.

24. Hu, F. B. Overweight and obesity in women: Health risks and consequences. *J Women's Health* 2003; 12:163–72.

25. Alberti, K. G. et al. Harmonizing the metabolic syndrome. A joint statement of the International Diabetes Federation Task Force on Epidemiology and Prevention; National Heart, Lung, and Blood Institute; American Heart Association; World Heart Federation; International Atherosclerosis Society; and the International Association for the Study of Obesity. *Circulation* 2009; 120:1640–5.

26. Janiszewski, P. M. et al. Themed review: Lifestyle treatment of the metabolic syndrome. *Am J Lifestyle Med* 2009; 2:99–108.

27. Azadbakht, L. et al. Beneficial effects of a Dietary Approaches to Stop Hypertension eating plan on features of the metabolic syndrome. *Diabetes Care* 2005; 28:2823–31.

28. Zaadstra, B. M. et al. Fat and female fecundity: Prospective study of effect of body fat distribution on conception rates. *BMJ* 1993; 306:484–7.

29. Bellver, J. et al. Female obesity impairs in vitro fertilization outcome without affecting embryo quality. *Fertil Steril* 2009; Jan 24.

30. Pasquali, R. et al. Achievement of near-normal body weight as the prerequisite to normalize sex hormone-binding globulin concentrations in massively obese men. *Int J Obes Relat Metab Disord* 1997; 21:1–5.

31. Hammoud, A. O. et al. Impact of male obesity on infertility: A critical review of the current literature. *Fertil Steril* 2008; 90:897–904.

32. Miller, P. B. et al. Effect of short-term diet and exercise on hormone levels and menses in obese, infertile women. *J Reprod Med* 2008; 53:315–9.

33. Hill, J. O. Can a small-changes approach help address the obesity epidemic? A report of the joint task force of the ASN, IFT, and the IFIC. *Am J Clin Nutr* 2009; 89:477–84.

34. Maggard, M. A. et al. Pregnancy and fertility following bariatric surgery: A systematic review. *JAMA* 2008; 300: 2286–96.

35. Ren, C. et al. Gastric banding, weight loss, and the resolution of type 2 diabetes. American Society of Metabolic and Bariatric Surgery 2009. Annual Meeting: Abstract PL-104, presented June 24, 2009. Available at www.medscape.com/viewarticle/705047, accessed 7/09.

36. Guelinckx, I. et al. Reproductive outcome after bariatric surgery: A critical review. *Hum Reprod Update* 2000; 15:189–201.

37. Bariatric surgery in women of reproductive age: Special concerns for pregnancy. Agency for Healthcare Research and Quality, 11/9/08. Available at ww.ahrq.gov/clinic/tp/babarireptp.htm.

38. The Practice Committee of the American Society for Reproductive Medicine, Obesity and Reproduction: An educational bulletin. *Fertil Steril* 2008; 90:S21–9.

39. Ahima, R. S. Body fat, leptin, and hypothalamic amenorrhea. *N Engl J Med* 2004; 352:95962.

40. Meczekalski, B. et al. Functional hypothalamic amenorrhea: Current view on neuroendocrine aberrations. *Gynecol Endocrinol* 2008; 24:4–11.

41. Loucks, A. B. et al. The female athlete triad: Do female athletes need to take special care to avoid low energy availability? *Med Sci Sports Exerc* 2006; 38:1694–1700.

42. Lifshitz, F. Quoted in Welt, C. K. et al. Recombinant human leptin in women with hypothalarnic amenorrhea. *Growth, Genetics & Hormones* 2005; 21(1). Available at www.GGHjoumal.com, accessed 3/05.

43. Joy, E. et al. Physician lack of knowledge if the female athlete triad. American College of Sports Medicine 57th Annual Meeting, Abs. 793, presented 5/28/09.

44. Stafford, D. E. Altered hypothalamic-pituitary-ovarian axis function in young female athletes: Implications and recommendations for management. *Treat Endocrinol* 2005; 4:l47–54.

45. Waldrop, J. Early identification and interventions for female athlete triad. *J Pediatr Health Care* 2005; 19:213–20.

46. Becker, A. E. et al. Eating disorders. *N Engl J Med* 1999; 340:1092–8.

47. Usdan, L. S. et al. The endocrinopathies of anorexia nervosa. *Endocr Pract* 2008; 14:1055–63.

48. Resch, M. et al. Eating disorders from a gynecologic and endocrinologic view: Hormonal changes. *Fertil Steril* 2004; 81:1151–3.

49. Agras, W. S. et al. A multicenter comparison of cognitive behavioral therapy and interpersonal psychotherapy for bulimia nervosa. *Arch Gen Psychiatry* 2000; 57:459–66.

50. Lorber, D. L. Preconception counseling of women with diabetes. *Pract Diabetol* 1995; 14:12–16.

51. Carr, S. R. Effect of maternal hyperglycemia on fetal development. American Dietetic Association Annual Meeting and Exhibition, Kansas City, MO, October 10, 1998.

52. Neff, L. M. Evidence-based dietary recommendations for patients with type 2 diabetes mellitus. *Nutr Clin Care* 2003; 6:51–61.

53. CDC Fastats A to Z. Available at www.cdc.gov/nchs/fastats.htm, accessed 3/09.

54. Schenk, S. et al. Insulin sensitivity: Modulation by nutrients and inflammation. *J Clin Invest* 2008; 118:2992–3002.

55. Romao, I. et al. Genetic and environmental interactions in obesity and type 2 diabetes. *J Am Diet Assoc* 2008; 108:S24–S28.

56. Bantle, J. P. et al. American Diabetes Association updated guidelines for Medical Nutrition Therapy to prevent diabetes, manage existing diabetes, and prevent or slow the rate of development of diabetes complications. *Diabetes Care* 2008; 31(Suppl):S61–78.

57. American Dietetic Association. Evidence-based Nutrition Practice Guidelines. Available at www.adaevidencelibrary.com/topic.cfmf?cat=3731, accessed 2/09.

58. Albright, A. et al. Revised nutrition guidelines for diabetes prevention. *Diabetes Care* 2006; 29:2140–57.

59. Hayes, C. et al. Role of physical activity in diabetes management. *J Am Diet Assoc* 2008; 108:S19–S23.

60. Saaddine, J. B. et al. A diabetes report card for the United States: Quality of care in the 1990s. *Ann Intern Med* 2002; 136:565–74.

61. Ripsin, C. M. et al. Review of blood glucose management for type 2 diabetes. *Am Fam Physician* 2009; 79:29–36.

62. Bedno, S. A. Weight loss in diabetes management. *Nutr Clin Care* 2003; 6:62–72.

63. Bowman, B. A. and Vinicor, F. The knowledge gap in diabetes. *Nutr Clin Care* 2003; 6:49–50.

64. Tabak, A. G. et al. Glucose metabolism changes prior to diabetes diagnosis. American Diabetes Association 69th Scientific Session: Abs. 1050-P, presented 6/8/09.

65. Diabetes Prevention Program Research Group. Reduction in the incidence of type 2 diabetes with lifestyle intervention or metaformin. *N Engl J Med* 2002; 346:393–403.

66. Obesity and polycystic ovary syndrome. *Clin Endocrinol* 2006; 65:137–145.

67. Ehrmann, D. A. Polycystic ovary syndrome. *N Engl J Med* 2005; 352: 1223–36.

68. Cascella, T. et al. Visceral fat is associated with cardiovascular risk in women with polycystic ovary syndrome. *Human Reprod* 2008; 23:153–9.

69. Azziz, R. Diagnostic criteria for polycystic ovary syndrome: A reappraisal. *Fertil Steril* 2004; 83:1343–6.

70. Tolstoi, L. G. and Josimovich, J. B. Weight loss and medication in polycystic ovary syndrome therapy. *Nutr Today* 2002; 37:57–62.

71. Asagami, T. et al. Differential effects of insulin sensitivity on androgens in obese women with polycystic ovary syndrome or normal ovulation. *Metab* 2008; 57:1355–60.

72. Stein, K. Polycystic ovary syndrome: What it is and why registered dietitians need to know. *J Am Diet Assoc* 2006; 106:1738–41.

73. Siassakos, D. et al. Polycystic ovary syndrome: Effects on early and late pregnancy outcomes. *BJOG* 2007; 114:922–32.

74. Pape, J. Polycyctic ovary syndrome: A personal and professional experience. Food and Nutrition Exhibition and Conference, St. Louis, Oct. 23, 2005.

75. McKittrick, M. Diet and polycystic ovary syndrome. *Nutr Today* 2002; 37: 63–69.

76. de Baulny, H. O. et al. Management of phenylketonuria and hyperphenylalaninemia. *J Nutr* 200; 137(6 Suppl 1):1561S–1563S.

77. Margaret Doll. Retrieved from www.savebabies.org/familystories/photos/MargaretDollPKU.jpg, accessed 7/09.

78. Simon, E. et al. Evaluation of quality of life and description of the sociodemographic state in adolescent and young adult patients with phenylketonuria (PKU). *Health Qual Life Outcomes* 2008; 6:25.

79. Committee on Genetics, American Academy of Pediatrics. Maternal Phenylketonuira. *Pediatrics* 2008; 122:445–9.

80. Harding, C. Progress toward cell-directed therapy for phenylketonuria. *Clin Genet* 2008; 74:97–104.

81. Isaacs, J. Metabolic dietician. Personal communication, 12/10/03.

82. Ghazeeri, G. et al. Effect of rosiglitazone on spontaneous and clomiphene citrate-induced ovulation in women with PCOS. *Fertil Steril* 2003; 79:562–6.

83. Nieuwenhuis-Ruifrok, A. E. et al. Insulin-sensitizing drugs for weight loss in women of reproductive age who are overweight or obese: Systematic review and meta-analysis. *Human Reprod Update* 2009; 15:57–68.

84. Moran, L. J. et al. Dietary therapy in polycystic ovary syndrome. *Semin Reprod Med* 2008; 26:85–92.

85. Michals-Matalon, K. et al. Maternal phenylketonuria, low protein intake, and congenital heart defects. *Am J Obstet Gynecol* 2002; 187:221–4.

86. Waisbren, S. E. et al. Outcome at age 4 years in offspring of women with maternal phenylketonuria: The Maternal PKU Collaborative Study. *JAMA* 2000; 283:756–62.

87. Friedman, E. G. et al. The International Collaborative Study on maternal phenylketonuria: Organization, study design and description of the sample. *Eur J Pediatr* 1996; 155(Suppl)1:S158–61.

88. Pellicano, R. et. al. Women and celiac disease: Association with unexplained infertility. *Minerva Med* 2007; 98:217–9.

89. Stazi, A. V. et al. Reproductive aspects of celiac disease. *Ann Ital Med Int* 2005; 20:143–57.

90. Hadziselimovic, F. et al. Celiac disease, pregnancy, small for gestational age: Role of extravillous trophoblast. *Fetal Pediatr Pathol* 2007; 26:125–34.

91. Pope, R. et al. Celiac disease during pregnancy: To screen or not to screen? *Arch Gynecol Obstet* 2009; 279:1–3.

92. Ford, A. C. et al. Relationships between celiac disease and irritable bowel syndrome. *Arch Intern Med* 2009; 169:651–8.

93. Bren, L. Food labels identify allergens more clearly. *FDA Consumer Magazine* Mar–Ap 2006.

94. Catassi, C. et al. Celiac disease. *Curr Opin Gastroenterol* 2008; 24:687–91.

95. Burrowes, J. D. Helping adults with celiac disease to eat well. *Nutr Today* 2008; 43:250–6.

96. Niewinski, M. M. Advances in celiac disease and gluten-free diets. *J Am Diet Assoc* 2008; 108:661–72.

97. Rubio–Tapia, A. et al. Increased Prevalence and Mortality in Undiagnosed Celiac Disease. *Gastroenterol* 2009; 137: 88–93.

98. Rewers, M. et al. Rotavirus infections and the development of celiac disease. *Am J Gastroenterol* 2006; 101:2333–2340.

99. American Dietetic Association's Evidence-Based Library, Celiac. Available at www.adaevidencelibrary.com/topic.cfm?cat=3726, accessed 6/09.

100. Dennehy, C. E. The use of herbs and dietary supplements in gynecology: An evidence-based review. *J Midwifery Womens Health* 2006; 51:402–9.

101. Vitex or chaste tree; it helps balance women's cycles. *Environ Nutr* 1999; 22:8.

102. Collins, A. et al. Essential fatty acids in the treatment of premenstrual syndrome. *Obstet Gynecol* 1993; 81:93–8.

103. Safarinejad, M. R. et al. Efficacy of selenium and/or N-acetyl-cysteine for improving semen parameters in infertile men: A double-blind, placebo controlled, randomized study. *J Urol* 2009; 181:741–51.

104. Balercia, G. et al. Coenzyme Q10 treatment in infertile men with idiopathic asthenozoospermia: A placebo-controlled, double-blind randomized trial. *Fertil Steril* 2009; 91:1785–92.

105. Balercia, G. et al. Coenzyme Q and the treatment of male infertility. *Fertil Steril* 2004; 81:93–8.

106. Balercia, G. et al. Coenzyme Q10 and male infertility, *J Endocrinol Invest* 2009 (May 21; E-pub ahead of print).

107. Wolever, T. M. S. et al. The Canadian Trial of Carbohydrates in Diabetes, a 1-y controlled trial of low-glycemic index dietary carbohydrate in type 2 diabetes. *Am J Clin Nutr* 2008; 87:114–25.

108. Hare-Bruun, H. et al. Should glycemic index and glycemic load be considered in dietary recommendations? *Nutr Rev* 2008; 66:569–90.

109. Hitt, E. Bee propolis may improve infertility associated with mild endometriosis. American Society of Reproductive Medicine Annual Meeting, Abs. 0-84, presented Oct. 13, 2003.

Chapter 4

1. CDC Fastats A to Z. Available at www. cdc.gov/nchs/fastats/births.htm, accessed 7/09.

2. Hamilton, B. E. et al. Births: Preliminary Data for 2007. *National Vital Statistics Report* March 18, 2009; 57(12).

3. Wegman, M. E. Infant mortality in the 20th century, dramatic but uneven progress. *J Nutr* 2001; 131:401S–8S.

4. Krisberg, K. U.S. lagging behind many other nations on infant mortality rates: Healthy behavior, healthier babies. *Nation Health* 2/27/2009; available at www. medscape.com/viewarticle/587840.

5. MacDorman, M. F. et al. NCHS Data Brief: Recent Trends in Infant Mortality in the United States (October 2008). Available at www/medscape.com/viewarticle/582117, accessed 3/09.

6. Leading Causes of Neonatal and Post-neonatal Deaths—United States, 2002. *MMWR Weekly Report* September 30, 2005; 54:966.

7. Health United States, 2000, Vol. 2001; and Deaths: Preliminary Data for 2002. Available at www.cdc.gov/nchs, accessed 2/14/04.

8. Kirkegaard, I. et al. Size at birth and later school performance. *Pediatrics* 2006; 118:1600–1606.

9. Godfrey, K. M. Maternal regulation of fetal development and health in adult life. *Eur J Obstet Gynecol Repro Biol* 1998; 78:141–50.

10. Hytten, F. E. and Leitch, I. *The physiology of human pregnancy*. Oxford: Blackwell Scientific Publications, 1971.

11. Cruikshank, D. P. et al. Maternal physiology in pregnancy. In Gabbe, S. G. et al., eds. *Obstetrics: Normal and problem pregnancies*. New York: Churchill Livingstone, 1996: 91–109.

12. King, J. C. Physiology of pregnancy and nutrient metabolism. *Am J Clin Nutr* 2000; 71:1218S–25S.

13. Rosso, P. *Nutrition and metabolism in pregnancy*. New York: Oxford University Press, 1990: 117–118, 125, 150–151. Winick, M. *Nutrition, pregnancy, and early infancy*. Baltimore: Williams & Wilkins, 1989: 182.

14. King, J. C. Effect of reproduction on the bioavailability of calcium, zinc and selenium. *J Nutr* 2001;131:1355S–8S.

15. Georgieff, M. K. Nutrition and the developing brain: Nutrient priorities and measurement. *Am J Clin Nutr* 2007; 85(Suppl):614S–20S.

16. Fazleabas, A. T. Update and advances in reproductive biology. Experimental Biology Annual Meeting, San Diego, April 11, 2003.

17. Butte, N. F. Carbohydrate and lipid metabolism in pregnancy: Normal compared with gestational diabetes mellitus. *Am J Clin Nutr* 2000; 71:1256S–61S.

18. Naeye, R. L. et al. Effects of maternal acetonuria and low pregnancy weight gain on children's psychomotor development. *Am J Obstet Gynecol* 1981; 139:189–93.

19. Duggleby, S. L. et al. Higher weight at birth is related to decreased amino acid oxidation during pregnancy. *Am J Clin Nutr* 2002; 76:852–7.

20. Saarelainen, H. et. al. Pregnancy-related hyperlipidemia and endothelial function in healthy women. *Circ J* 2006; 70:768–72.

21. van Stiphout, W. A. H. J., Hofman, A., and de Bruijn, A. M. Serum lipids in young women before, during, and after pregnancy. *Am J Epidemiol* 1987; 126:922–8.

22. Martin, U. et al. Is normal pregnancy atherogenic? *Clin Sci* (Colch) 1999; 96:421–5.

23. Ortega, R. M. et al. Influence of maternal serum lipids and maternal diet during the third trimester of pregnancy on umbilical cord blood lipids in two populations of Spanish newborns. *Int J Vitam Nutr Res* 1996; 66:250–7.

24. American College of Obstetricians and Gynecologists. Nutrition during pregnancy. *ACOG Technical Bulletin No. 179* 1993:1–7.

25. Gluckman, P. D. Fetal origins of adult disease: Insulin resistance. *Medscape Ob/Gyn & Women's Health* 2003; 8:4–5.

26. Rosso, P. and Cramoy, C. Nutrition and pregnancy. In Winick, M., ed. *Human nutrition: Pre- and post-natal development*. New York: Plenum Press, 1979:133–228.

27. Trahair, J. F. et al. Ultrastructural anomalies in the fetal small intestine indicate that fetal swallowing is important for normal development: An experimental study. *Virchows Arch A Pathol Anat Histopathol* 1992; 420:305–12.

28. Scholl, T. O., Stein, T. P., et al. Leptin and maternal growth during adolescent pregnancy. *Am J Clin Nutr* 2000; 72:1542–7.

29. Norwitz, E. R. et al. Implantation and the survival of early pregnancy. *N Engl J Med* 2001; 345:1400–8.

30. Rozovski, S. J. et al. Nutrition and cellular growth. In Winick M, ed. *Human nutrition: Pre- and postnatal development*. New York: Plenum Press, 1979: 61–102.

31. Smith, D. W. Growth and its disorders: Basics and standards, approach and classifications, growth deficiency disorders, growth excess disorders, obesity. In Schaeffer, A. J. and Markowitz, M., eds. *Major problems in clinical pediatrics*, Vol. 15. Philadelphia: W. B. Saunders Company, 1979: 134–169.

32. Winick, M. Malnutrition and brain development. *J Pediatr* 1969; 74:667–79.

33. Rosso, P. and Winick, M. Intrauterine growth retardation: A new systematic approach based on the clinical and biochemical characteristics of this condition. *J Perinat Med* 1974; 2:147–60.

34. National Academy of Sciences. *Nutrition during pregnancy. I. Weight gain. II. Nutrient supplements*. Washington, D.C.: National Academy Press, 1990.

35. Godfrey, K. M. et al. Fetal nutrition and adult disease. *Am J Clin Nutr* 2000; 71:1344S–52S.

36. Ziegler, E. E. et al. Body composition of the reference fetus. *Growth* 1976; 40:329–41.

37. Hediger, M. L. et al. Growth of infants and young children born small or large for gestational age: Findings from the Third National Health and Nutrition Examination Survey. *Arch Pediatr Adolesc Med* 1998; 152:1225–31.

38. Lucas, A. et al. Breastfeeding and catch-up growth in infants born small for gestational age. *Acta Paediatr* 1997; 86:564–9.

39. Luo, Z. C. et al. Length and body mass index at birth and target height influences on patterns of postnatal growth in children born small for gestational age. *Pediatrics* 1998; 102:E72.

40. Kovac, P. Recurrent pregnancy loss. Available at www.medscape.com/viewprogram/5293, accessed 4/06.

41. Maconochile, N. et al. Underweight and the risk of miscarriage. Online edition of BJOG: An International Journal of Obstetrics and Gynecology, available at www.medicinenet.com/ script/main/art.asp?articlekey=77914, accessed 12/06.

42. The Spectrum of Gastrointestinal Disorders during Pregnancy. *Nat Clin Pract Gastroenterol Hepatol* 2008. Available at medscape.comCME, www.medscape.com accessed 4/09.

43. Kalk, P. et al. Impact of maternal body mass index on neonatal outcome. *Eur J Med Res* 2009; 14:216–22.

44. Catov, J. M. et al. Inflammation and dyslipidemia related to risk of spontaneous preterm birth. *Am J Epidemiol* 2007; 166:1312–9.

45. Hasan, A. F. et al. Self-reported vitamin supplementation in early pregnancy and risk of miscarriage. *Am J Epidemiol* 2009; 169:1312–1318.

46. Kramer, M. S. et al. The contribution of mild and moderate preterm birth to infant mortality. Fetal and Infant Health Study Group of the Canadian Perinatal Surveillance System. *JAMA* 2000; 284:843–9.

47. Elizabeth, K. E. et al. Umbilical cord blood nutrients in low birth weight babies in relation to birth weight & gestational age. *Indian J Med Res* 2008; 128:128–33.

48. Chery, C. et al. Hyperhomocysteinemia is related to a decreased level of vitamin B_{12} in the second and third trimesters of pregnancy. *Clin Chem Lab Med* 2002; 40:1105–8.

49. Ortega, R. M. et al. Riboflavin levels in maternal milk: The influence of vitamin B_2 status during the third trimester of pregnancy. *J Am Coll Nutr* 1999; 18:324–9.

50. Sommer, A. et al. Assessment and control of vitamin A deficiency: The Annecy Accords. *J Nutr* 2002; 132:2845S–50S.

51. Dietz, P. M. et al; Prepregnancy body mass index and the risk of preterm delivery. *Epidemiol* 2006; 17:170–77.

52. Castro, L. C. et al. Maternal obesity and pregnancy outcomes. *Curr Opin Obstet Gynecol* 2002; 14:601–6.

53. Scholl, T. O. Maternal nutrition before and during pregnancy. *Nestle Nutr Workshop Ser Pediatr Program* 2008; 61:79–89.

54. Vahratian, A. et al. Multivitamin use and the risk of preterm birth. *Am J Epidemiol* 2004; 1(60):886–92.

55. Allen, L. H. Multiple micronutrients in pregnancy and lactation: An overview. *Am J Clin Nutr* 2005; 81:1206S–1212S.

56. Juhl, M. et al. Physical Exercise During Pregnancy and the Risk of Preterm Birth: A Study Within the Danish National Birth Cohort. *Am J Epidemiol* 2008; 167:859–856.

57. Catov, J. M. et al. Early pregnancy lipid concentrations and spontaneous preterm birth. *Am J Obstet Gynecol* 2007; 197:610.e1–7.

58. Marcus, D. M. Do no harm: Avoidance of herbal medicines during pregnancy. *Obstet Gynecol* 2005; 105(5 Pt. 1):1119–22.

59. Chen, X. et al. Association of elevated free fatty acids during late pregnancy with preterm delivery. *Obstet Gynecol* 2008; 112:297–303.

60. Denny, C. H. et al. Pattern of alcohol use during pregnancy since 1991. *MMWR* 2009; 58:529–32.

61. Gluckman, P. D. et al. Effect of in utero and early-life conditions on adult health and disease. *N Engl J Med* 2008; 359:61–73.

62. Simonetti, G. D. et al. Salt sensitivity of children with low birth weight. *Hypertension* 2008; 52625–30.

63. Prentice, A. M. Early influences on human energy regulation: Thrifty genotypes and thrifty phenotypes. *Physiol Behav* 2005; 86:640–5.

64. Wu, G. et al. Maternal nutrition and fetal development. *J Nutr* 2004; 134:2l69–72.

65. Ibanez, L. et al. Visceral adiposity in the absence of overweight in small for gestational age children. *J Clin Endrcinol Metab* 2008; 93:2079–83.

66. Rich-Edwards, J. W. et al. Birth weight and risk of cardiovascular disease in a cohort of women followed up since 1976. *BMJ* 1997; 315:396–400.

67. Bhat, P. V. et al. Role of vitamin A in determining nephron mass and possible relationship to hypertension. *J Nutr* 2008; 138:1407–10.

68. Barnes, S. Nutritional genomics, polyphenols, diets, and their impact on dietetics. *J Am Diet Assoc* 2008; 108:1888–95.

69. Barker, D. J. P. et al. Trajectories of growth among children who have coronary events as adults. *N Engl J Med* 2005; 353:1802–9.

70. Vivian, E. M. Type 2 diabetes in children and adolescents—the next epidemic? *Curr Med Res Opin* 2006; 22:297–306.

71. Robinson, S. M. et al. Combined effects of dietary fat and birth weight on serum cholesterol concentrations: The Hertfordshire Cohort Study. *Am J Clin Nutr* 2006; 84:237–44.

72. Kuzawa, C. W. et al. Lipid profiles in adolescent Filipinos: Relation to birth weight and maternal energy status during pregnancy. *Am J Clin Nutr* 2003; 77:960–6.

73. Villar, J. et al. Nutritional factors associated with low birth weight and short gestational age. *Clin Nutr* 1986; 5:78–85.

74. Eastman, N. J. *Expectant motherhood.* Boston: Little, Brown and Company, 1947: 198.

75. Rassmussen, K. M. et al. Weight gain during pregnancy: Reexamining the guidelines. Institute of Medicine, 2009. Available at www.nap.edu/catalog/12584.html.

76. Parker, J. D. and Abrams, B. Prenatal weight gain advice: An examination of the recent prenatal weight gain recommendations of the Institute of Medicine. *Obstet Gynecol* 1992; 79:664–9.

77. Stotland, N. E. et al. Excess gestational weight gain associated with adverse neonatal outcomes. *Obstet Gynecol* 2006; 108:635–43.

78. Brown, J. E. et al. Variation in newborn size by trimester weight change in pregnancy. *Am J Clin Nutr* 2002; 76:205–9.

79. Carmichael, S. L. and Abrams, B. A critical review of the relationship between gestational weight gain and preterm delivery. *Obstet Gynecol* 1997; 89:865–73.

80. To, W. W. et al. The relationship between weight gain in pregnancy, birthweight and postpartum weight retention. *Aust N Z J Obstet Gynaecol* 1998; 38:176–9.

81. Abrams, B. Prenatal weight gain and postpartum weight retention: A delicate balance (editorial). *Am J Public Health* 1993; 83:1082–4.

82. King, J. C. et al. Leptin targeted in research on obesity, pregnancy. ARS News Service, USDA. Available at www.ars.usda.gov, accessed 10/02.

83. Scholl, T. O. et al. Insulin and the "thrifty" woman: The influence of insulin during pregnancy on gestational weight gain and postpartum weight retention. *Maternal Child Health Journal* 2002; 6:255–61.

84. Keppel, K. G. and Taffel, S. M. Pregnancy-related weight gain and retention: Implications of the 1990 Institute of Medicine guidelines. *Am J Public Health* 1993; 83:1100–3.

85. Brown, J. E. et al. Parity-related weight change in women. *Int J Obes* 1992; 16:627–31.

86. Gigante, D. P. et al. Breast-feeding has a limited long-term effect on anthropometry and body composition of Brazilian mothers. *J Nutr* 2001; 131:78–84.

87. Polley, A. et al. Randomized controlled trial to prevent weight gain in pregnant women. *Int J Obes* 2002; 26:1494–1502.

88. Mussey, R. D. Nutrition and human reproduction: an historical review. *Am J Obstet Gynecol* 1949; 58:1037–48.

89. Stein, A. D. et al. Famine, third-trimester pregnancy weight gain, and intrauterine growth: The Dutch Famine Birth Cohort Study. *Hum Biol* 1995; 67:135–50.

90. Smith, C. A. Effects of maternal undernutrition upon the newborn infant in Holland (1944–1945). *J Pediatr* 1947; 30:229–43.

91. Susser, M. and Stein, Z. Timing in prenatal nutrition: A reprise of the Dutch Famine Study. *Nutr Rev* 1994; 52:84–94.

92. Roseboom, T. J. et al. Plasma lipid profiles in adults after prenatal exposure to the Dutch famine. *Am J Clin Nutr* 2000; 72:1101–6.

93. Antonov, A. N. Children born during the siege of Leningrad in 1942. *J Pediatr* 1947; 30:250–9.

94. Gruenwald, P. et al. Influence of environmental factors on fetal growth in man. *Lancet* 1967; I:1026–8.

95. Guyer, B. et al. Annual summary of vital statistics—1997. *Pediatrics* 1998; 102:1333–49.

96. Simic, S. et al. Nutritional effects of the siege on newborn babies in Sarajevo. *Eur J Clin Nutr* 1995; 49(Suppl., 2):S33–6.

97. Position of the American Dietetic Association: Nutrition and lifestyle for a healthy pregnancy outcome. *J Am Diet Assoc* 2008; 108:553–60.

98. Burke, B. S. Nutritional needs in pregnancy in relation to nutritional intakes as shown by dietary histories. *Obstetr Gynecol Survey* 1948; 3:716–30.

99. Burke, B. S. et al. Nutrition studies during pregnancy. IV. Relation of protein content of mother's diet during pregnancy to birth length, birth weight, and condition of infant at birth. *J Pediatr* 1943; 23:506–15.

100. Butte, N. F. Energy requirements during pregnancy and consequences of deviations from requirement on fetal outcome. *Nestle Nutr Workshop Ser Pediatr Program* 2005; 55:49–67.

101. Brown, J. E. and Kahn, E. S. B. Maternal nutrition and the outcome of pregnancy: A renaissance in research. *Clin Perinatol* 1997; 24:433–49.

102. Dietary reference intakes: energy, carbohydrate, fiber, fat, fatty acids, cholesterol, protein, and amino acids. Washington, DC: National Academies Press; 2002.

103. Anderson, J. W. Health implications of wheat fiber. *Am J Clin Nutr* 1985; 41:1103–12.

104. Deierlein, A. L. et al. Dietary energy density but not glycemic load is associated with gestational weight gain. *Am J Clin Nutr* 2008; 88:693–9.

105. Duffy, V. B. and Anderson, G. H. Position of the American Dietetic Association: Use of nutritive and nonnutritive sweeteners. *J Amer Diet Assoc* 1998; 98:580–7.

106. American Academy of Pediatrics. Committee on Substance Abuse and Committee on Children with Disabilities. Fetal alcohol syndrome and alcohol-related neurodevelopmental disorders. *Pediatrics* 2000; 106:358–61.

107. Duggleby, S. L. et al. Protein, amino acid and nitrogen metabolism during pregnancy: How might the mother meet the needs of her fetus? *Curr Opin Clin Nutr Metab Care* 2002; 5:503–9.

108. Moore, V. M. et al. Dietary composition of pregnant women is related to size of the baby at birth. *J Nutr* 2004; 134:1820–6.

109. Johnston, P. K. Counseling the pregnant vegetarian. *Am J Clin Nutr* 1988; 48:901–5.

110. Position of the American Dietetic Association and Dietitians of Canada: Vegetarian diets. *J Am Diet Assoc* 2003; 103:748–65.

111. Messina, V. et al. A new food guide for North American vegetarians. *J Am Diet Assoc* 2003; 103:771–5.

112. Haddad, E. H. et al. Vegetarian food guide pyramid: a conceptual framework. *Am J Clin Nutr* 1999; 70(Suppl): 615S–19S.

113. Neurologic impairment in children associated with maternal dietary deficiency of cobalamin—Georgia, 2001. *MMWR* 2003; 52(04):61–4.

114. What we eat in America, NHANES, 2005–2006, USDA, 2009. Available at www.ars.usda.gov/ba/bhnrc/fsrg.

115. Muskiet, F. A. et al. Long-chain polyunsaturated fatty acids in maternal and infant nutrition. *Prostaglandins Leukot Essent Fatty Acids* 2006; 75:135–44.

116. Whelan J. (n-6) and (n-3) polyunsaturated fatty acids and the aging brain: Food for thought. *J Nutr* 2008; 2512–22.

117. Simopoulos, A. P. Human requirement for N-3 polyunsaturated fatty acids. *Poult Sci* 2000; 79:961–70.

118. Kolanowski, W. et al. Possibilities of fish oil application for food products enrichment with omega-3 PUFA. *Int J Food Sci Nutr* 1999; 50:39–49.

119. Williams, C. M. et al. Long-chain n-3 PUFA: Plant v. marine sources. Proc Nutr Soc 2006; 65:42–50.

120. Position of the American Dietetic Association and Dietitians of Canada: Dietary fatty acids. *J Am Diet Assoc* 2007; 107:1599–1611.

121. Harris, W. S. et al. Towards establishing dietary reference intakes for EPA and DHA. J Nutr 2009; 139:804S–19S.

122. Makrides, M. Outcomes for mothers and their babies: Do n-3 long-chain polyunsaturated fatty acids make a difference? J Am Diet Assoc 2008; 108:1622–26.

123. Alessandri, J. M. et al. Polyunsaturated fatty acids in the central nervous system: evolution of concepts and nutritional implications throughout life. Reprod Nutr Dev 2004; 44:509–38.

124. Szajewska, H. et al. Effect of n-3 long-chain polyunsaturated fatty acid supplementation of women with low-risk pregnancies on pregnancy outcomes and growth measures at birth: A meta-analysis of randomized controlled trials. Am J Clin Nutr 2006; 83:1337–44.

125. Helland, I. B. et al. Effect of supplementing pregnant and lactating mothers with n-3 very-long-chain fatty acids on children's IQ and body mass index at 7 years of age, Pediatrics 2008; 122:e472–9.

126. Judge, M. P. et al. Maternal consumption of DHA-containing functional food during pregnancy: Benefit for infant performance on problem-solving but not on recognition memory tasks at age 9 mo. Am J Clin Nutr 2007; 85:1572–7.

127. Position of the American Dietetic Association: Vegetarian diets. J Am Diet Assoc 2009; 109:1266–82.

128. United States Department of Agriculture (USDA), Agricultural Research Service (ARS), Nutrient Data Laboratory. Available at www.ars.usda.gov/nutrientdata.

129. What you need to know about the safety of fish and shellfish: 2004 EPA and FDA advice for women who might become pregnant, women who are pregnant, nursing mothers, young children. Available at www.cfsan.fda.gov/~dms/admehg3.html, accessed 1/07.

130. Al-Ardhi, F. M. et al. Maternal fish consumption and prenatal methylmercury exposure: A review. Nutr Health 2008; 19:289–97.

131. Lederman, S. A. et al. Relation between cord blood mercury levels and early child development in a World Trade Center cohort. Environ Health Perspect 2008; 116:1085–91.

132. Dunstan, J. A. et al. Fish-oil-supplements during pregnancy and eye and hand coordination in children. *Arch Dis Child Fetal Neonatal Ed*. Published online December 21, 2006.

133. Ershow, A. G., Brown, L. M., and Cantor, K. P. Intake of tapwater and total water by pregnant and lactating women. *Am J Public Health* 1991; 81:328–34.

134. Lumley, J. et al. Periconceptional supplementation with folate and/or multivitamins for preventing neural tube defects. *Cochrane Database Syst Rev* 2000; 2.

135. Suitor, C. W. and Bailey, L. B. Dietary folate equivalents: Interpretation and application. *J Amer Diet Assoc* 2000; 100:88–94.

136. Bailey, L. B. Evaluation of a new Recommended Dietary Allowance for folate. *J Amer Diet Assoc* 1992; 92:463–71.

137. Fodinger, M. et al. Recent insights into the molecular genetics of the homocysteine metabolism. *Kidney Int* 2001; 59(Suppl., 78):S238–42.

138. Warkany, J. Production of congenital malformations by dietary measures. *JAMA* 1958; 168:2020–3.

139. Smithells, R. W. Availability of folic acid. *Lancet* 1984; 1:508.

140. Eskes, T. K. A. B. From birth to conception, open or closed. *Eur J Obstet Gynecol Reprod Biol* 1998; 78:169–77.

141. Folic acid. *MMWR Morb Mortal Wkly Rep* 2001; 50:185–9.

142. Folate status in women of childbearing age—United States, 1999. *MMWR Morb Mortal Wkly Rep* 2000; 49:962–5.

143. Daly, L. E. et al. Folate levels and neural tube defects. *JAMA* 1995; 274:1698–702.

144. Brown, J. E. et al. Predictors of red cell folate level in women attempting pregnancy. *JAMA* 1997; 277:548–52.

145. Folate Intake in U.S. Falling Short of Established Targets. *MMWR* 2007; 55:1377–80.

146. McDowell, M. A. et al. Blood Folate Levels: The Latest NHANES Results, Number 6, May 2008.

147. Wolff, T. et al. Folic acid supplementation for the prevention of neural tube defects: An update of the evidence for the U.S. Preventive Services Task Force. *Ann Intern Med* 2009; 150:632–9.

148. Shaw, G. M. et al. Periconceptional vitamin use, dietary folate, and the occurrence of neural tube defects. *Epidemiology* 1995; 6:219–26.

149. Smithells, D. Vitamins in early pregnancy. *BMJ* 1996; 313:128–9.

150. Velzing-Aarts, F. V. et al. Plasma choline and betaine and their relation to plasma homocysteine in normal pregnancy. *Am Clin Nutr* 2005; 81:1383–9.

151. Zeisel, S. H. Importance of methyl donors during pregnancy. *Am J Clin Nutr* 2009; 89(Suppl):673S–7S.

152. Zeisel, S. H. Fetal origin of memory: The role of dietary choline. *J Pediatr* 2006; 149(Suppl., 5):S131–6.

153. Signore, C. et al, Choline concentrations in human maternal and cord blood and intelligence at 5 y of age. *Am J Clin Nutr* 2008; 87:896–902.

154. Zile, M. H. Function of vitamin A in vertebrate embryonic development. *J Nutr* 2001; 131:705–8.

155. Lott, I. T. et al. Fetal hydrocephalus and ear abnormalities associated with maternal use of isotretinion. *J Pediatr* 1984; 105:597–600.

156. DeNoon, D. J. US Ranks in Lowest Third for Infant Mortality, 10/15/08. Available at http://www.medscape.com/viewarticle/582117.

157. McCullough, M. L. Vitamin D deficiency in pregnancy: Bringing the issue to light, *J Nutr* 2007; 137:305–6.

158. Schroth, R et al. Influence of maternal vitamin D status on infant oral health, Abs. no. 1646, International Association for Dental Research, July 4, 2008.

159. Specker, B. Vitamin D requirements during pregnancy. *Am J Clin Nutr* 2004; 80(Suppl):1740S–7S.

160. Kovacs, C. S. Vitamin D in pregnancy and lactation: Maternal, fetal, and neonatal outcomes from human and animal studies. *Am J Clin Nutr* 2008; 88(Suppl):520S–8S.

161. Bodnar, L. M. et al. High prevalence of vitamin D insufficiency in black and white women residing in the northern United States and their neonates. *J Nutr* 2007; 137:447–52.

162. Bodnar, M. L. Prepregnancy obesity predicts poor vitamin D status of mothers and their neonates. *J Nutr* 2007; 137:2437–42.

163. O'Brien, K. O. et al. Bone calcium turnover during pregnancy and lactation in women with low calcium diets is associated with calcium intake and circulating insulin-like growth factor 1 concentrations. *Am J Clin Nutr* 2006; 83:317–23.

164. Prentice, A. Maternal calcium metabolism and bone mineral status. *Am J Clin Nutr* 2000; 71:1312S–6S.

165. Olausson, H. et al. Changes in bone mineral status and bone size during pregnancy and the influence of body weight and calcium intake. *Am J Clin Nutr* 2008; 8:1032–9.

166. Stefanie, L. Exploring potential pathways between parity and tooth loss among American women. *Am J Public Health* 2008; 98:1263–1270.

167. Hertz-Picciotto, I. et al. Patterns and determinants of blood lead during pregnancy. *Am J Epidemiol* 2000; 152:829–37.

168. Zhu, M. et al. Lead levels and pregnancy outcomes. American Public Health Association 136th Annual Meeting, Abstract #4105.0, presented October 28, 2008.

169. Cleveland, L. M. et al. Lead hazards for pregnant women and children: Part 1: Immigrants and the poor shoulder most of the burden of lead exposure in this country. *Am J Nurs* 2008; 108:40–9.

170. Gedalia, I. et al. Effect of prenatal and postnatal fluoride on the human deciduous dentition: A literature review. *Adv Dent Res* 1989; 3:168–76.

171. Leverett, D. H. et al. Randomized clinical trial of the effect of prenatal fluoride supplements in preventing dental caries. *Caries Res* 1997; 31:174–9.

172. Fairbanks, V. F. Iron in medicine and nutrition. In Shils, M. E. et al, eds. *Modern Nutrition in Health and Disease*, 9th ed. Philadelphia: Lippincott, Williams & Wilkins, 1999: 193–221.

173. Cogswell, M. E. et al. Iron supplement use among women in the United States: Science, policy, and practice. *J Nutr* 2003; 133:1974S–7S.

174. Allen, L. H. Biological mechanisms that might underlie iron's effects on fetal growth and preterm birth. *J Nutr* 2001; 131:581S–9S.

175. Tamura, T. et al. Cord serum ferritin concentrations and mental and psychomotor development of children at five years of age. *J Pediatrics*, 2002; 140:165–70.

176. Chaparro, C. M. Setting the stage for child health and development: Children in an iodine deficient area. *J Clin Endocrinol Metab* 2008; 93:2466–8, 2616–21.

177. Allen, L. H. Anemia and iron deficiency: Effects on pregnancy outcome. *Am J Clin Nutr* 2000; 71:1280S–4S.

178. Recommendations to prevent and control iron deficiency in the United States. Centers for Disease Control and Prevention. *MMWR Morb Mortal Wkly Rep* 1998; 47:1–29.

179. Zimmerman, M. B. The adverse effects of mild-to-moderate iodine deficiency during pregnancy and childhood: A review. *Thyroid* 2007; 17(9):829–35.

180. Xue-Yi, C. et al. Timing of vulnerability of the brain to iodine deficiency in endemic cretinism. *N Eng J Med* 1994; 331:1739–44.

181. Mahomed, K. and Gulmezoglu, A. M. *Maternal iodine supplements in areas of deficiency* (Cochrane Review). The Cochrane Library. Issue 3. Oxford: Update Software, 1999.

182. Caldwell, K. L. et al. Urinary iodine concentration: United States National Health And Nutrition Examination Survey 2001–2002. *Thyroid*, 2005; 15:692–9.

183. Vermiglio, F. et al. Thyroid function in pregnant women from a mildly iodine deficient area. *J Clin Endocrinol Metab* 2008; 93:2466–8, 2616–21.

184. Becker, D. V. et al. Iodine supplementation for pregnancy and lactation, United States and Canada: Recommendations from the American Thyroid Association. *Thyroid* 2006; 16:959–51.

185. Meydani, M. et al. Diabetes risk: Antioxidants or lifestyle? *Am J Cli Nutr* 2009; 90:253–4.

186. Agarwal, A. et al. Role of oxidative stress in female reproduction. *Reprod Biol Endocrinol* 2005; 14;3:28.

187. Litonjua, A. A. et al. Maternal antioxidant intake in pregnancy and wheezing illnesses in children at 2 y of age. *Am J Clin Nutr* 2006; 84:903–11.

188. Devereux, G. et al. Low intake of vitamin E during pregnancy may increase risk for childhood asthma. *Am J Respir Grit Care Med* 2006; 174:499–507.

189. Delemarre, F. M. et al. Eclampsia despite strict dietary sodium restriction. *Gynecol Obstet Invest* 2001; 51:64–5.

190. Pike, R. L. and Gursky, D. S. Further evidence of deleterious effects produced by sodium restriction during pregnancy. *Am J Clin Nutr* 1970; 23:883–9.

191. Duley, L., Henderson-Smart, D., and Meher, S. Altered dietary salt for preventing pre-eclampsia, and its complications. *Cochrane Database of Syst Rev* 2005; Issue 4, Art. No. CD005548, DOI 10.1002/14651858.CD005548.

192. Nehlig, A. and Debry, G. Consequences on the newborn of chronic maternal consumption of coffee during gestation and lactation: A review. *J Am Coll Nutr* 1994; 13:6–21.

193. Hinds, T. S. et al. The effect of caffeine on pregnancy outcome variables. *Nutr Rev* 1996; 54:203–7.

194. Christian, M. S. et al. Teratogen update: Evaluation of the reproductive and developmental risks of caffeine. *Teratol* 2001; 64:51–78.

195. Jahanfar, S. et al. Effects of restricted caffeine intake by mother on fetal, neonatal and pregnancy outcome. *Cochrane Database Syst Rev* 2009; 15:CD006965.

196. Bech, B. H. et al. Effect of reducing caffeine intake on birth weight and length of gestation: Randomized controlled trial. *BMJ* 2007; 334:409–414.

197. Hook, E. B. Dietary cravings and aversions during pregnancy. *Am J Clin Nutr* 1978; 31:1355–62.

198. Brown, J. E. et al. Taste changes during pregnancy. *Am J Clin Nutr* 1986; 43:412–6.

199. Johns, T. and Duquette, M. Detoxification and mineral supplementation as functions of geophagy. *Am J Clin Nutr* 1991; 53:448–56.

200. Rainville, A. J. Pica practices of pregnant women are associated with lower maternal hemoglobin level at delivery. *J Amer Diet Assoc* 1998; 98:293–6.

201. *Summary of the Nutrition Care Process Steps for Nutrition Assessment* (Pocket Guide for International Dietetics & Nutrition Terminology Reference Manual), 2nd ed. Chicago: American Dietetic Association, 2009.

202. Larsson, A. et al. Reference values for clinical chemistry tests during normal pregnancy. *BJOG* 2008; 115:874-81.

203. Dog, L. T. The use of botanicals during pregnancy and lactation. *Altern Ther Health Med* 2009; 15:54–8.

204. Dickerson, A. et al. Physician and nurses use and recommend dietary supplements: Report of a survey. *Nutr J* 2009; 8:29–31.

205. Position of the American Dietetic Association: Vitamin and mineral supplementation. *J Am Diet Assoc* 1996; 96:73–7.

206. Picciano, M. F. et al. Use of dietary supplements by pregnant and lactating women in North America. *Am J Clin Nutr* 2009; 89(Suppl):663S–7S.

207. FDA statement concerning structure/function rule and pregnancy claims. HHS Statement, U.S. Department of Health and Human Services; 65 FR 1000; January 6, 2000; Docket No. 98N-0044.

208. Zeisel, S. H. Is maternal diet supplementation beneficial? Optimal development of infant depends on mother's diet. *Am J Clin Nutr* 2009; 89:685S–7S.

209. Shah, P. S. et al. Knowledge Synthesis Group on Determinants of Low Birth Weight and Preterm Births. Effects of prenatal multimicronutrient supplementation on pregnancy outcomes: A meta-analysis. *CMAJ* 2009; 180:E99–108.

210. Prevalence of multivitamin use during pregnancy by race and ethnicity: Results from the 2007 Pregnancy Nutrition Surveillance System. Available at www.ndhealth.gov/wic/publications/07%20ND%20PNSS.ppt, accessed 7/09.

211. Meisler, J. G. Toward optimal health: The experts discuss the use of botanicals by women. *J Women's Health* 2003; 12:847–52.

212. Jones, F. A. Herbs: Useful plants. *J R Soc Med* 1996; 89:717–9.

213. Dugoua, J. J. et al. Safety and efficacy of ginkgo (Ginkgo biloba) during pregnancy and lactation. *Can J Clin Pharmacol* 2006; 13:e277–84 (e-pub Nov. 3, 2006).

214. Vutyavanich, T. et al. Ginger for nausea and vomiting in pregnancy: Randomized, double-masked, placebo-controlled trial. *Obstet Gynecol* 2001; 97:577–82.

215. Johnson, K. Pregnancy exercise recommendations growing more liberal. *Medscape Ob/Gyn & Women's Health* 2003; 8:1–2.

216. Artal, R. and Sherman, C. Exercise during pregnancy—Safe and beneficial for most. *Physician and Sports Medicine* 1999; 27:51–56.

217. Dempsey, J. C. et al. No need for a pregnant pause: Physical activity may reduce the occurrence of gestational diabetes mellitus and preeclampsia. *Exerc Sport Sci Rev* 2005; 33:141–9.

218. Physical Activity Guidelines for Americans, 2008, U.S. Department of Health & Human Services. Available at http://www.health.gov/PAGuidelines.

219. Clapp, J. F. et al. Exercise during pregnancy may improve perimenopauseal fitness. *Am J Obstet Gynecol* 2008; 199:489.e1–489.e6.

220. Clapp, J. F. et al. Beginning regular exercise in early pregnancy: Effect on fetoplacental growth. *Am J Obstet Gynecol* 2000; 183:1484–8.

221. Position of the American Dietetics Association: Food Safety. *J Am Diet Assoc* 2003; 103:1203–18.

222. Disson, O. et al. Conjugated action of two species-specific invasion proteins for fetoplacental listeriosis. *Nature* 2008; Sep 17.

223. Smith, J. L. Foodborne illness during pregnancy. *J Food Protection* 1999; 62:818-29.

224. FDA. Food safety for moms to be. Available at www.cfsan.fda.gov/~pregnant/pregnant.html

225. Soto, C. Toxoplasmosis in pregnancy. *Clin Rev* 2002; 12:51–6.

226. EPA fish advisory update. Vol. 2001, 2001.

227. The Spectrum of Gastrointestinal Disorders during Pregnancy. *Nat Clin Pract Gastroenterol Hepatol*, 2008. Available at medscape.comCME, www.medscape.com, accessed 4/09.

228. Badell, M. L. et al. Treatment options for nausea and vomiting during pregnancy. *Pharmacotherapy* 2006; 26:1273–87.

229. Gross, S. et al. Maternal weight loss associated with hyperemesis gravidarum: A predictor of fetal outcome. *Am J Obstet Gynecol* 1989; 160:906–9.

230. Scialli, A. Revived focus on treatment for morning sickness. *Medscape Wire* 2000.

231. Brown, J. E., Kahn, E. S., Hartman, T. J., et al. Profet, profits and proof: Do nausea and vomiting of early pregnancy protect women from "harmful" vegetables? *Am J Obstet Gynecol* 1997; 176:179–81.

232. Jewel, D. and Young, G. Interventions for nausea and vomiting in early pregnancy. *Cochrane Database Syst Rev* 2002; (1):CD000145.

233. Emelianova, S. et al. Prevalence and severity of nausea and vomiting of pregnancy

and effect of vitamin supplementation. *Clin Invest Med* 1999; 22:106–10.

234. Keller, J. et al. The spectrum and treatment of gastrointestinal disorders during pregnancy. *Nat Clin Pract Gastroenterol Hepatol* 2008; 5:430–43.

235. Baron, T. H., Ramirez, B., and Richter, J. E. Gastrointestinal motility disorders during pregnancy. *Ann Intern Med* 1993; 118:366–75.

236. Higgins, A. C. et al. Impact of the Higgins Nutrition Intervention Program on birth weight: A within-mother analysis. *J Amer Diet Assoc* 1989; 89:1097–103.

237. Dubois, S. et al. Ability of the Higgins Nutrition Intervention Program to improve adolescent pregnancy outcome. *J Amer Diet Assoc* 1997; 97:871–8.

238. Duquette, M. P., Director of the Montreal Diet Dispensary. Personal communication, 2000.

239. Seligman, P. A. et al. Measurements of iron absorption from prenatal multivitamin–mineral supplements. *Obstet Gynecol* 1983; 61:356–62.

240. Dawson, E. B. et al. Iron in prenatal multivitamin/multimineral supplements. Bioavailability. *J Reprod Med* 1998; 43:133–40.

241. Singh, K., Fong, Y. F., Kuperan, P., et al. A comparison between intravenous iron polymaltose complex (ferrum hausmann) and oral ferrous fumarate in the treatment of iron deficiency anaemia in pregnancy. *Eur J Haematol* 1998; 60:119–24.

242. Casanueva, E. et al. Iron and oxidative stress in pregnancy. *J Nutr* 2003; 133:1700S–1708S.

243. Casanueva, E. and Viteri, F. E. Iron and oxidative stress in pregnancy. *J Nutr* 2003; 133:1700S–8S.

244. Iron deficiency anemia: Recommended guidelines for prevention, detection, and management among US children and women of childbearing age. Washington, D.C.: National Academy Press, 1993.

245. Cogswell, M. E. et al. Iron supplementation during pregnancy, anemia, and birth weight: A randomized controlled trial. *Am J Clin Nutr* 2003; 78:773–81.

Chapter 5

1. Chu, S. Y. et al. Association between obesity during pregnancy and increased use of health care. *N Engl J Med* 2008; 358:1444–53.

2. Bartha, J. L. et al. Ultrasound evaluation of visceral fat and metabolic risk factors during early pregnancy. *Obesity* (Silver Springs) 1007; 15:2233–9.

3. Wildman, R. P. et al. The obese without cardiometabolic risk factor clustering and the normal weight with cardiometabolic risk

factor clustering: Prevalence and correlates of 2 phenotypes among the US population: (NHANES 1999–2004). *Arch Intern Med* 2008; 168:1617

4. Pi-Sunyer, X. et al. Obesity associated inflammation. Presented at the Experimental Biology annual meetings, Washington, D.C., 4/03/07.

5. Demerath, E. W. et al. Visceral adiposity and its anatomical distribution as predictors

of the metabolic syndrome and cardiometabolic risk factor levels. *Am J Clin Nutr* 2008; 88:1263–71.

6. Forgarty, A. W. et al. A prospective study of weight change and systemic inflammation over 9 y. *Am J Clin* 2008; 87:30–5.

7. Pischon, T. et al. General and abdominal adiposity and risk of death in Europe. *N Engl J Med* 2008; 359:2105–20.

8. Health, United States, 2008. Available at www.cdc.gov/nchs, accessed 8/09.

9. Kominiarek, M. A. et al. Guidelines on managing obesity in pregnancy. American College of Obstetrics and Gynecology. *Obstet Gynecol* 2009; 113:1405–13.

10. Catalano, P. M. et al. Obesity during pregnancy and the risk for metabolic compromise. *Diabetes Care* 2009; 32:1076–80.

11. Puhl, R. M. Weight bias in Healthcare: An important clinical concern. Available at ww.medscape.com/viewarticle/701809, accessed 5/09.

12. Position of the American Dietetic Association and American Society for Nutrition: Obesity, reproduction, and pregnancy outcome. *J Am Diet Assoc* 2009; 109:918–27.

13. Brantsaeter, A. L. et al. A dietary pattern characterized by high intake of vegetables, fruits, and vegetable oils is associated with reduced risk of preeclampsia in nulliparous pregnant Norwegian women. *J Nutr* 2009; 139:1162–8.

14. Karmon, A. et al. Pregnancy after bariatric surgery: A comprehensive review. *Acta Gynecol Obstet* 2008; 277:381–88.

15. Malinowski, S. S. Nutritional and metabolic complications of bariatric surgery. *Am J Med Sci* 2006; 331:219–25.

16. *Bariatric Surgery in Women of Reproductive Age: Special Concerns for Pregnancy*, Structured Abstract. Rockville, MD: Agency for Healthcare Research and Quality, 2008 (Nov.). Available at http://www.ahrq.gov/clinic/tp/barireptp.htm.

17. Maggard, M. A. et al. Pregnancy and Fertility Following Bariatric Surgery: A Systematic Review. *JAMA* 2008; 300:2286–2296.

18. Aasheim, E. T. et al. Vitamin status after bariatric surgery: A randomized study of gastric bypass and duodenal switch. *Am J Clin Nutr* 2009; 90:15–22.

19. Agarwal, A. et al. Role of oxidative stress in female reproduction. *Reprod Biol Endocrinol* 2005; 14:3–28.

20. Makrides, M. Outcomes for mothers and their babies: Do n-3 long-chain polyunsaturated fatty acids make a difference? *J Am Diet Assoc* 2008; 108:1622–26.

21. Adiga, U. et al. Antioxidant activity and lipid peroxidation in preeclampsia. *J Chin Med Assoc* 2007; 70:435–8.

22. Granger, J. P. Hypertension during preeclampsia: A lesson in integrative physiology. Experimental Biology Annual Meeting, San Diego, April 6, 2008.

23. Sibai, B. M. et al. Risk factors for preeclampsia, abruptio placentae, and adverse neonatal outcomes among women with chronic hypertension. National Institute of Child Health and Human Development Network of Maternal-Fetal Medicine Units. *N Engl J Med* 1998; 339:667–71.

24. Haddad, B. and Sibai, B. M. Chronic hypertension in pregnancy. *Ann Med* 1999; 31:246–52.

25. Report of the National High Blood Pressure Education Program Working Group on High Blood Pressure in Pregnancy. *Am J Obstet Gynecol* 2000; 183:S1–S22.

26. Banerjee, M. et al. Pregnancy as the prodrome to vascular dysfunction and cardiovascular risk. *Nat Clin Pract Cardiovasc Med* 2006; 3:596–603.

27. von Versen-Hoeynck, F. M. et al. Maternal-fetal metabolism in normal pregnancy and preeclampsia. *Front Biosci* 2007; 12:2457–70.

28. McDonald, S. D. et al. Cardiovascular sequelae of preeclampsia/eclampsia: A systematic review and meta-analysis. *Am Heart J* 2008; 156:918–30.

29. Thadhani, R. et al. Preclampsia—A glimpse into the future? *N Engl J Med* 2008; 359:858–60.

30. Berends, A. et al. History of preeclampsia and abdominal fat accumulation later in life. *BJOG* 2009; 116:442–51.

31. Roberts, J. M. et al. Nutrient involvement in preeclampsia. *J Nutr* 2003; 133:1684S–92S.

32. Prevent recurrent eclamptic seizures with magnesium sulfate, an unconventional anticonvulsant. *Drug Ther Perspect* 2000; 16:6–8.

33. Signs and symptoms of preeclampsia. Available at www.preeclampsia.org/symptoms.asp, accessed 8/09.

34. Pipkin, F. B. Risk factors for preeclampsia (editorial). *N Engl J Med* 2001; 344:925–6.

35. Hauth, J. C. et al. Pregnancy outcomes in healthy nulliparas who developed hypertension. Calcium for Preeclampsia Prevention Study Group. *Obstet Gynecol* 2000; 95:24–8.

36. Germain, A. M. et al. Endothelial dysfunction: A link among preeclampsia, recurrent pregnancy loss, and future cardiovascular events? *Hypertension* 2007; 43:90–5 (e-pub Nov. 20, 2006).

37. Berkowitz, K. M. Insulin resistance and preeclampsia. *Clin Perinatol* 1998; 25:873–85.

38. Hernandez-Diaz, S. et al. Risk of preeclampsia in first and subsequent pregnancies: Prospective cohort study. *BMJ* 6/26/09. Available at www.medscape.com/viewarticle/704801, accessed 7/09.

39. Wiznitzer, A. et al. Association of lipid levels during gestation with preeclampsia and gestational diabetes mellitus: a population-based study. *Am J Obstet Gynecol* 2009 (e-pub ahead of print, Jul. 23,).

40. Bodnar, L. M. et al. Maternal vitamin D deficiency increases the risk of preeclampsia. *J Clin Endocrinol Metab* 2007; 92:3517–22.

41. Xiong, X. et al. Association of preeclampsia with high birth weight for age. *Am J Obstet Gynecol* 2000; 183:148–55.

42. Klebanoff, M. A. et al. Maternal size at birth and the development of hypertension during pregnancy: A test of the Barker hypothesis. *Arch Intern Med* 1999; 159:1607–12.

43. Rumbold, A. et al. Antioxidants for preventing preeclampsia. *Cochrane Database Syst Rev* 2008; 23(1):CD004227.

44. Haugen, M. et al. Vitamin D Supplementation and Reduced Risk of Preeclampsia in Nulliparous Women. *Epidemiol* 2009 (e-pub ahead of print, May 15).

45. Bodnar, L. M. et al. Periconceptional multivitamin use reduces the risk of preeclampsia. *Am J Epidemiol* 2006; 164:470–7.

46. Catov, J. M. et al. Association of periconceptional multivitamin use with reduced risk of preeclampsia among normal-weight women in the Danish National Birth Cohort. *Am J Epidemiol* 2009; 169:1304–11.

47. Hofmeyr, G. J. et al. Calcium supplementation during pregnancy for preventing hypertensive disorders and related problems. *Cochrane Database Syst Rev* 2006; 3:CD001059 (Jul 19).

48. Qui, C. et al. Dietary fiber intake early in pregnancy and risk of subsequent preeclampsia. *Am J Hypertens* 2008: 21:903–9.

49. Duley, L. et al. Altered dietary salt for preventing pre-eclampsia, and its complications. *Cochrane Database of Syst Rev* 2005; Issue 4. Art. No. CD005548, DOI 10.1002/14651858.CD005548.

50. Rayman, M. P. et al. Iron supplementation in preeclampsia. *Am J Obstet Gynecol* 2002; 187:412–18.

51. Carr, S. R. Effect of maternal hyperglycemia on fetal development. American Dietetic Association Annual Meeting and Exhibition, Kansas City, MO, Oct. 10, 1998.

52. Branchtein, L. et al. Waist circumference and waist-to-hip ratio are related to gestational glucose tolerance. *Diabetes Care* 1997; 20:509–11.

53. Lawrence, J. M. et al. Prevalence of existing and gestational diabetes during pregnancy. *Diabetes Care* 2008 (May). Available at www.medscape.com/viewarticle/573541, accessed 8/08.

54. Butte, N. F. Carbohydrate and lipid metabolism in pregnancy: Normal compared with gestational diabetes mellitus. *Am J Clin Nutr* 2000; 71:1256S–61S.

55. Carducci, A. A. et al. Glucose tolerance and insulin secretion in pregnancy. *Diabetes, Nutrition & Metabolism—Clinical & Experimental* 1999; 12:264–70.

56. Grissa, O. et al. Antioxidant status and circulating lipids are altered in human gestational diabetes and macrosomia. *Transl Res* 2007; 150:164–71.

57. Nielsen, G. et al. Hemogloblin A1c levels and outcomes of pregnancy in women with type 1 diabetes. *Diabetes Care* 2006; 29:2612–16.

58. Landon, M. B. Diabetes mellitus and other endocrine disorders. In Gabbe, S. G. et al., eds. *Obstetrics, normal and problem pregnancies.* New York: Churchill – Livingstone, 1996: 1037–81.

59. Pettitt, D. J. et al. In utero exposure to diabetes and type 2 diabetes risk later in life. *Diabetes Care* 2008; 31:2126–30.

60. Lindsay, R. S. et al. Secular trends in birth weight, BMI, and diabetes in the offspring of diabetic mothers. *Diabetes Care* 2000; 23:1249–54.

61. MacNeill, S. et al. Rates and risk factors for recurrence of gestational diabetes. *Diabetes Care* 2001; 24:659–62.

62. Kjos, S. L. and Buchanan, T. A. Gestational diabetes mellitus. *N Engl J Med* 1999; 341:1749–56.

63. Zhang, C. et al. Dietary fiber intake, dietary glycemic load, and the risk for gestational diabetes mellitus. *Diabetes Care* 2006; 29:2223–30.

64. McMahon, M. J. et al. Gestational diabetes mellitus: Risk factors, obstetric complications and infant outcomes. *J Reprod Med* 1998; 43:372–8.

65. Menato, G. et al. Current management of gestational diabetes mellitus. *Expert Rev of Obstet Gynecol* 2008; 3:73–91.

66. Coleman, T. Patient-centered care of diabetes in general practice. Study failed to measure patient centeredness of GPs' consulting behavior. *BMJ* 1999; 318:1621–2.

67. Adams, K. M. et al. Sequelae of unrecognized gestational diabetes. *Am J Obstet Gynecol* 1998; 178:1321–32.

68. Reader, D. et al. Impact of gestational diabetes mellitus nutrition practice guidelines implemented by registered dietitians on pregnancy outcomes. *J Am Diet Assoc* 2006; 106:1426–33.

69. Pezzarossa, A. et al. Effects of maternal weight variations and gestational diabetes mellitus on neonatal birth weight. *J Diabetes Complications* 1996; 10:78–83.

70. Denice, S. et al. Oral antidiabetic agents in pregnancy and lactation: A paradigm shift? *Ann Pharmacother* 2007; 411174–80.

71. Executive summary of the American Dietetic Association's recommendations for evidence-based practice guidelines for gestational diabetes. Available at www.adaevidencelibrary.com/topic.cfm?cat=3731, accessed 6/09.

72. Fagen, C. et al. Nutrition management in women with gestational diabetes mellitus: a review by ADA's Diabetes Care and Education Dietetic Practice Group. *J Am Diet Assoc* 1995; 95:460–7.

73. Kalkwarf, H. J. et al. Dietary fiber intakes and insulin requirements in pregnant women with type 1 diabetes. *J Am Diet Assoc* 2001; 101:305–10.

74. Moses, R. G. et al. Can a low-glycemic index diet reduce the need for insulin in gestational diabetes mellitus? A randomized trial. *Diabetes Care* 2009; 32:996–1000.

75. Ecker, J. L. et al. Gestational diabetes. *N Engl J Med* 2000; 342:896–7.

76. *Summary of the Nutrition Care Process Steps for Nutrition Assessment* (Pocket Guide for International Dietetics & Nutrition Terminology Reference Manual), 2nd ed. Chicago: American Dietetic Association, 2009.

77. Butte, N. F. Dieting and exercise in overweight, lactating women [editorial]. *N Engl J Med* 2000; 342:502–3.

78. Temple, R. C. et al. Glycemic control during the second trimester of pregnancy and the risk of preeclampsia in women with type 1 diabetes. *BJOG* 2006; 113:1329–32.

79. Reece, E. A. et al. The role of free radicals and membrane lipids in diabetes-induced congenital malformations. *J Soc Gynecol Investig* 1998; 5:178–87.

80. Raychaudhuri, K. and Maresh, M. J. Glycemic control throughout pregnancy and fetal growth in insulin-dependent diabetes. *Obstet Gynecol* 2000; 95:190–4.

81. Mello, G. et al. What degree of maternal metabolic control in women with type 1 diabetes is associated with normal body size and proportions in full-term infants? *Diabetes Care* 2000; 23:1494–8.

82. Martin, J. A. et al. Births: Final data for 2006. *National Vital Statistics Report* Jan. 7, 2009; 57(7).

83. Rebar, R. W. et al. Assisted reproductive technology in the United States. *N Engl J Med* 2004; 350:1603–4.

84. Brown, J. E. and Schloesser, P. T. Prepregnancy weight status, prenatal weight gain, and the outcome of term twin gestations. *Am J Obst Gynecol* 1990; 162:182–6.

85. Guyer, B. et al. Annual summary of vital statistics—1998. *Pediatrics* 1999; 104:1229–46.

86. Hoskins, R. E. Zygosity as a risk factor for complications and outcomes of twin pregnancy. *Acta Genet Med Gemellol* 1995; 44:11–23.

87. Redline, R. W. Non identical twins with a single placenta—disproving dogma in perinatal pathology. *N Engl J Med* 2003; 349:111–14.

88. Imaizumi, Y. A comparative study of twinning and triplet rates in 17 countries, 1971–1996. *Acta Genet Med Gemellol* 1998; 47:101–14.

89. Czeizel, A. E. et al. Higher rate of multiple births after periconceptional vitamin supplementation. *New Engl J Med* 1994; 330:1687–8.

90. Maclennan, A. H. Multiple gestation. Clinical characteristics and management. In Creasy, R. K. and Resnick, R., eds. *Maternal-fetal medicine: Principles and practice.* Philadelphia: W. B. Saunders, 1994: 589–601.

91. Chasen, S. et al. Pregnancy outcomes with a vanishing twin. *Am J Obstet Gynecol* 2006; 195:814–7.

92. Brown, J. E. and Carlson, M. Nutrition and multifetal pregnancy. *J Am Diet Assoc* 2000; 100:343–8. Updated based on MyPyramid, 2/07.

93. Kogan, M. D. et al. Trends in twin birth outcomes and prenatal care utilization in the United States, 1981–1997. *JAMA* 2000; 284:335–41.

94. Ventura, S. J. et al. Births: Final data for 1999. Vol. 2001, available at www.cdc.gov/nchs.

95. Ellings, J. M. et al. Reduction in very low birth weight deliveries and perinatal mortality in a specialized, multidisciplinary twin clinic. *Obstet Gynecol* 1993; 81:387–91.

96. Cohen, S. B. et al. New birth weight nomograms for twin gestation on the basis of accurate gestational age. *Am J Obstet Gynecol* 1997; 177:1101–4.

97. Dubois, S. et al. Twin pregnancy: The impact of the Higgins Nutrition Intervention Program on maternal and neonatal outcomes. *Am J Clin Nutr* 1991; 53:1397–403.

98. Rassmussen, K. M. et al. Weight gain during pregnancy: Reexamining the guidelines. Institute of Medicine, 2009. Available at www.nap.edu/catalog/12584.html.

99. National Academy of Sciences. *Nutrition during pregnancy. I. Weight gain. II. Nutrient supplements.* Washington, D.C.: National Academy Press, 1990.

100. Konwinski, T. et al. Maternal pregestational weight and multiple pregnancy duration. *Acta Genet Med Gemellol* 1973; 22:44–7.

101. Zeijdner, E. E. et al. Essential fatty acid status in plasma phospholipids of mother and neonate after multiple pregnancy. *Prostaglandins Leukotrienes and Essential Fatty Acids* 1997; 56:395–401.

102. HIV infection on the rise worldwide. Press release from the World Health Organization. Available at www.medscape.com/viewarticle/548140, accessed 11/06.

103. Dunham, W. Global AIDS deaths decline slightly. Available at www.medscape.com/viewarticles/578259, accessed 8/09.

104. Rich, K. C. et al. Maternal and infant factors predicting disease progression in human immunodeficiency virus type 1-infected infants. Women and Infants Transmission Study Group. *Pediatrics* 2000; 105:E8.

105. French, M. et al. Highly active antiretroviral therapy. *Lancet* 1998; 351:1056–7.

106. Zulling L. et al. Nutrition for HIV positive women: practical guidelines. Women's Health and Reproductive Nutrition Report. 2005; Winter: 1, 4, 14–16.

107. Fields-Gardner, C. and Ayoob, K. T. Position of the American Dietetic Association and Dietitians of Canada: Nutrition intervention in the care of persons with human immunodeficiency virus infection. *J Am Diet Assoc* 2000; 100:708–17.

108. Becker, A. E. et al. Eating disorders. *N Engl J Med* 1999; 340:1092–8.

109. Morrill, E. S. et al. Bulimia nervosa during pregnancy: A review. *J Am Diet Assoc* 2001; 102:448–54.

110. Little, L. and Lowkes, E. J. Midwifery. Love of women with eating disorders. *Womens Health* 2000; 45:301–7.

111. Mehta, S. et al. Nutritional indicators of adverse pregnancy outcomes and mother-to-child transmission of HIV among HIV-infected women. *Am J Clin Nutr* 2008; 87:1639–49.

112. Marston, B. et al. Multivitamins, nutrition, and antiretroviral therapy for HIV disease in Africa. *N Engl J Med* 2004; 351:78–80.

113. CDC issues guidelines for the Identification of fetal alcohol syndrome. *MMWR* 2005; 54(RR-11):1–15.

114. American Academy of Pediatrics. Committees on Substance Abuse and Children with Disabilities. Fetal alcohol syndrome and alcohol-related neurodevelopmental disorders. *Pediatrics* 2000; 106:358–61.

115. Denny, C. H. et al. Pattern of alcohol use during pregnancy since 1991. *MMWR* 2009; 58:529–32.

116. Shankar, K. et al. Physiologic and genomic analyses of nutrition ethanol interactions during gestation: Implications for fetal ethanol toxicity. *Exp Biol Med* (Maywood). 2006; 231:1379–97.

117. March of Dimes Birth Defect Foundation. Public health education information sheet: Drinking alcohol during pregnancy, 1997.

118. Konovaloov, H. V. et al. Disorders in brain development in the progeny of mothers who used alcohol during pregnancy. *Early Human Dev* 1997; 48:153–66.

119. Aliyu, M. H. et al. Alcohol consumption during pregnancy and the risk of early stillbirth among singletons. *Alcohol* 2008; 42:369–74.

120. Sayal, K. et al. Binge pattern of alcohol consumption during pregnancy and childhood mental health outcomes: Longitudinal population-based study. *Pediatrics* 2009; 123:e289–96.

121. Teen births. Available at www.cdc.gov/nchs/fastats/teenbrth.htm, accessed 1/07.

122. Rosso, P. *Nutrition and metabolism in pregnancy*. New York: Oxford University Press, 1990: 117–118, 125, 150–151.

123. Scholl, T. O. et al. Maternal growth during pregnancy and the competition for nutrients. *Am J Clin Nutr* 1994; 60:183–8.

124. Sukalich, S. et al. Obstetric outcomes in overweight and obese adolescents. *Am J Obstet Gynecol* 2006; 195:851–5.

125. Wallace, J. M. et al. Nutritionally mediated placental growth restriction in the growing adolescent: consequences for the fetus. *Biol Reprod* 2004; 71:1055–62.

126. Scholl, T. O. et al. Leptin and maternal growth during adolescent pregnancy. *Am J Clin Nutr* 2000; 72:1542–7.

127. Baker, P. N. et al. A prospective study of micronutrient status in adolescent pregnancy. *Am J Clin Nutr* 2009; 89:1114–24.

128. Iannotti, L. L. et al. Iron deficiency anemia and depleted body iron reserves are prevalent among pregnant African-American adolescents. *J Nutr* 2005; 135:2572–7.

129. Nielsen, J. N. et al. Interventions to improve diet and weight gain among pregnancy adolescents and recommendations for future research. *J Am Diet Assoc* 2006; 106:1825–40.

130. Lenfant, C. Clinical research to clinical practice—lost in translation. *N Engl J Med.* 2003; 868–74

Chapter 6

1. Institute of Medicine. *Nutrition During Lactation*. Washington, D.C.: National Academy Press, 1991.

2. Ross Products Division, Abbott Laboratories. *Breastfeeding Trends Through 2000*, 2003.

3. U.S. Department of Health and Human Services. Healthy People 2010: Conference Edition—Volumes I and II. Washington, D.C.: U.S. Department of Health and Human Services, Public Health Service, Office of the Assistant Secretary for Health, 2000.

4. Breastfeeding trends and updated national health objectives for exclusive breastfeeding—United States, birth years 2000–2004. *MMWR* 2007; 56(30):760–3.

5. McDowell, M. M., Wang, C. Y., and Kennedy-Stephenson, J. Breastfeeding in the United States: Findings from the National Health and Nutrition Examination Surveys, 1999–2006. NCHS Data Brief. 2008; 5:1–8.

6. Gartner, L. M. Morton, J., Lawrence, R.A., Naylor, A.J., O'Hare, D., Schanler, R.J., and Eidelman, A.I. Breastfeeding and the use of human milk. *Pediatrics* Feb 2005; 115 (2, Feb.):496–506.

7. Centers for Disease Control and Prevention Department of Health and Human

Services. *Breastfeeding: Data and Statistics: Breastfeeding Practices—Results from the 2005 National Immunization Survey.* Atlanta, GA: Centers for Disease Control and Prevention, U.S. Department of Health and Human Services, 2005.

8. Phares, T. M. Phares, T. M., Morrow, B., Lansky, A., Barfield, W. D., Prince, C. B., Marchi, K. S., Braveman, P. A., Williams, L. M. and Kinniburgh, B. Surveillance for disparities in maternal health-related behaviors—selected states. Pregnancy Risk Assessment Monitoring System (PRAMS), 2000–2001. *MMWR Surveill Summ* Jul. 2, 2004; 53(4):1–13.

9. Taylor, J. S., Risica, P. M., Geller, L., Kirtania, U., and Cabral, H.J. Duration of breastfeeding among first-time mothers in the United States: Results of a national survey. *Acat Paediatr* 2006; 95(8):980–984.

10. Ryan, A. S. and Zhou, W. Lower breastfeeding rates persist among the Special Supplemental Nutrition Program for Women, Infants, and Children participants, 1978–2003. *Pediatrics* Apr. 2006; 117(4):1136–46.

11. Hartmann, P. E. Changes in the composition and yield of the mammary secretion

of cows during the initiation of lactation. *J Endocrinol* 1973; 59(2):231–47.

12. Vorherr, H. Human lactation and breastfeeding. In Larson, B. L., ed. *The mammary gland/human lactation/milk synthesis*, Vol. 4. New York, NY: Academic Press, 1978.

13. Lawrence, R. A. and Lawrence, R. M. *Breastfeeding: A guide for the medical professional*, 6th ed. Minneapolis: Mosby, 2005.

14. Cox, D. B. et al. Blood and milk prolactin and the rate of milk synthesis in women. *Exp Physiol* 1996; 81(6):1007–20.

15. Neville, M. C. The physiological basis of milk secretion, Part I:. Basic physiology. *Ann NY Acad Sci* 1990; 5:861–68.

16. Jensen, R. G. *The Lipids of Human Milk.* Boca Raton, FL: CRC Press, 1989.

17. Jensen, R. G. ed. *Handbook of Milk Composition.* New York: Academic Press, 1995.

18. Prentice, A. Constituents of human milk. *Food and Nutrition Bulletin*, Volume 17. The United Nations University Press, 1996.

19. USDA. National Nutrient Database for Standard Reference: United States Department of Health and Human Development, 2009.

20. Williams, H. G. 'And not a drop to drink'—why water is harmful for newborns. *Breastfeeding Review.* 2006;14(2):5–9.

21. Axelson, I., Borulf, S., Righard, L., and Raiha, N. Protein and energy intake during weaning: I. Effects on growth. *Acta Paediatrica Scandinavica* 1987; 76(2):321–27.

22. Butte, N. F., Garza C. Johnson, C.A., Smith, E.O., and Nichols, B.L. Longitudinal changes in milk composition of mothers delivering preterm and term infants. *Early Hum Dev* 1984; 9(2):153–62.

23. Hediger, M. L., Overpeck, M.D. , Ruan, W.J. , and Troendle, J.F. Early infant feeding and growth status of U.S.-born infants and children aged 4–71 mo.: Analyses from the third National Health and Nutrition Examination Survey, 1988–1994. *Am J Clin Nutr* 2000; 72(1):159–67.

24. Connor, W. E., Lowensohn, R., and Hatcher, L. Increased docosahexaenoic acid levels in human newborn infants by administration of sardines and fish oil during pregnancy. *Lipids* 1996; 31(Suppl): S183–87.

25. Insull, W. and Ahrens, E. H. The fatty acids of human milk from mothers on diets taken ad libitum. *Biochem J* 1959; 72:27.

26. Meier, P. P., Engstrom, J. L., Murtaugh, M. A., Vasan, U., Meier, W. A., and Schanler, R. J. Mothers' milk feedings in the neonatal intensive care unit: Accuracy of the creamatocrit technique. *J Perinatol* 2002; 22(8):646–9.

27. Van Goor, S. A., Dijck-Brouwer, D. A., Hadders-Algra, M., Doornbos, B., Erwich, J. J., Schaafsma, A., and Muskiet, F. A. Human milk arachidonic acid and docosahexaenoic acid contents increase following supplementation during pregnancy and lactation. *Prostaglandins Leukot Essent Fatty Acids* 2009; 80(1):65–9.

28. Agostoni, C., Riva, E., Trojan, S., Bellu, R., and Giovannini, M. Docosahexaenoic acid status and developmental quotient of healthy term infants. *Lancet* 1995; 346(8975):638.

29. Kovacs, A., Funke S., Marosvolgyi, T., Burus, I., and Decsi, T. (2005). Fatty acids in early human milk after preterm and fullterm delivery. *J Pediatr Gastroenterol Nutr.* 2005; 41:454–59.

30. Helland, I., Smith, L., and Saarem, K. Maternal supplementation with very-long-chain n-3 fatty acids during pregnancy and lacatation augments children's IQ at 4 years of age. *Pediatr* 2003; 111(1):e39–44.

31. Mosley, E. E., Wright, A.L., McGuire, M.K., and McGuire, M.A. *Trans* fatty acids in milk produced by women in the United States. *Am J Clin Nutr* Dec. 2005; 82(6, Dec.):1292–97.

32. Friesen, R. and Innis, S. M. *Trans* fatty acids in human milk in Canada declined with the introduction of *trans* fat food labeling. *J Nutr* 2006; 136(10): 2558–61.

33. Wong, W. W., Hachey, D.L., Insul, W., Opekun, A.R., and Klein, P.D. Effect of dietary cholesterol on cholesterol synthesis in breast-fed and formula-fed infants. *J Lipid Res* 1993; 34(8):1403–11.

34. Owen, C.G, Whincup, P.H., Odoki, K., Gilg, J.A., and Cook, D.G. Infant feeding and blood cholesterol: a study in adolescents and a systematic review. *Pediatr* 2003; 110(3):597–608.

35. Rosen, J. M., Jones, W.K., Rodgers, J.R., Compton, J.G., Bisbee, C.A., David-Inouye, Y., and Yu-Lee, L.Y. Regulatory sequences involved in the hormonal control of casein gene expression. *Ann NY Acad Sci* 1986; 464:87–99.

36. Phadke, S., Deslouches, B., Hileman, S.E., Montelaro, R.C., Wiesenfield, H.C., and Mietzner, T.A. Antimicrobial peptides in mucosal secretions: the importance of local secretions in mitigating infection. *J Nutr* 2005; 135:1289–93.

37. Velona, T., Abbiati, L., Beretta, B., Gaiaschi, A., Flauto, U., Tagliabue, Pl, Galli, C.L., and Restani, P. Protein profiles in breast milk from mothers delivering term and preterm babies. *Pediatr Res* 1999; 45 (5, Pt. 1):658–63.

38. Lonnerdal, B. and, Atkinson, S. Nitrogenous components of milk. In Jensen, R. G., ed. *Handbook of Milk Composition.* San Diego, CA: Academic Press, 1995: 351–68.

39. Newburg, D. S. and Neubauer, S. H. Carboyhdrates in milks: Analysis, quantities, and signficance. Jensen, R. G. (ed.). (1995). *Handbook of Milk Composition.* San Diego, CA: Academic Press; 1995:. 273–349.

40. Bode, L. Recent advances on structure, metabolism, and function of human milk oligosaccharides. *J Nutr* 2006; 136:2127–30.

41. Morrow, A., Ruiz-Palacios, G.M., Jiang, X., and Newburg, D.S. Human-milk glycans that inhibit pathogen binding protect breast-feeding infants against diarrhea. *J Nutr* 2005; 135, 1304–1207.

42. Food and Nutrition Board, Institute of Medicine. Dietary Reference Intakes for Vitamin A, Vitamin K, Arsenic, Boron, Chromium, Copper, Iodine, Iron, Manganese, Molybdenum, Nickel, Silicon, Vanadium, and Zinc. Washington, D.C.: National Academy Press, 2001.

43. Rothberg, A.D., Pettifor, J.M., Cohen, D.F., Sonnendecker, E.W., and Ross, F.P. Maternal-infant vitamin D relationships during breast-feeding. *J Pediatr* 1982; 101(4):500–503.

44. Ala-Houhala, M., Koskinen, T., Parviainen, M. T., and Visakorpi, J. K. 25-Hydroxyvitamin D and vitamin D in human milk: effects of supplementation and season. *Am J Clin Nutr* 1988; 48(4):1057–60.

45. Dawdou, A. and Wagner, C. L. Mother-child vitamin D deficiency: An international perspective. *Arch Dis Child* 2007; 92(9):737-4046.

46. Lammi-Keefe, C. J. and Jensen, R. G. Fat-soluble vitamins in human milk. *Nutr Rev* 1984; 42(11):365–71.

47. Haug, M., Laubach, C., Burke, M., and Harzer, G. . Vitamin E in human milk from mothers of preterm and term infants. *J Pediatr Gastroenterol Nutr* 1987; 6(4):605–9.

48. Chappell, J. E., Francis, T., and Clandinin, M. T. Vitamin A and E content of human milk at early stages of lactation. *Early Hum Dev* 1985; 11(2):157–67.

49. Andon, M. B., Reynolds, R.D., Moser-Veillon, P.B., and Howard, M.P. Dietary intake of total and glycosylated vitamin B-6 and the vitamin B-6 nutritional status of unsupplemented lactating women and their infants. *Am J Clin Nutr* 1989; 50(5):1050–58.

50. Salmenpera, L., Perheentupa, J., and Simes, M. A. Folate nutrition is optimal in exclusively breast-fed infants but inadequate in some of their mothers and in formula-fed infants. *J Pediatr Gastroenterol Nutr* 1986; 5(2):283–89.

51. Specker, B. L., Black, A., Allen, L, and Morrow, F. Vitamin B-12: Low milk concentrations are related to low serum concentrations in vegetarian women and to methylmalonic aciduria in their infants. *Am J Clin Nutr* 1990; 52(6):1073–76.

52. Butte, N. F., Garza, C., Smith, E.O., Wills, C., and Nichols, B.L. Macro- and trace-mineral intakes of exclusively breast-fed infants. *Am J Clin Nutr* 1987; 45(1):42–48.

53. Fransson, G. B. and Lonnerdal, B. Zinc, copper, calcium, and magnesium in human milk. *J Pediatr* 1982; 101(4):504–8.

54. Atkinson, S. A., Whelan, D., Whyte, R.K., and Lönnderdal, B. Abnormal zinc content in human milk. Risk for development of nutritional zinc deficiency in infants. *Am J Dis Child* 1989; 143(5):608–11.

55. Duncan, B., Schifman, R.B., Corrigan, J.J., Jr., and Schaefer, C. Iron and the exclusively breast-fed infant from birth to six months. *J Pediatr Gastroenterol Nutr* 1985; 4(3):421–25.

56. Pisacane, A., DeVizia, B., Vallente, A., Vaccaro, F., Russo, M., Grillo, G., and Guistardi, A. et al. Iron status in breast-fed infants. *J Pediatr* 1995; 127(3):429–421.

57. Krebs, N. F. and Hambidge, K. M. Zinc requirements and zinc intakes of breast-fed infants. *Am J Clin Nutr* 1986; 43(2):288–92.

58. Sian, L., Krebs, N.F., Westcott, J.E., Fengliang, L., Tong, L. Miller, L.V., Sonko, B., and Hambidge, M. Zinc homeostasis during lactation in a population with a low zinc intake. *Am J Clin Nutr* Jan. 2002; 75(1):99–103.

59. Food and Nutrition Board Institute of Medicine. *Dietary Reference Intakes for Calcium, Phosphorus, Magnesium, Vitamin D, and Fluoride.* Washington, D.C.: National Academy Press, 1999.

60. Palmer, C. and Wolfe, S. H. Position of the American Dietetic Association: The impact of fluoride on health. *J Am Diet Assoc* Oct. 2005; 105(10):1620–28.

61. McDaniel, M. R., Barker, E., and Lederer, C. L. Sensory characterization of human milk. *J Dairy Sci* 1989; 72(5):1149–58.

62. Menella, J. A. Mother's milk: A medium for early flavor experiences. *J Hum Lactation* 1995; 11:39–45.

63. Huasner, H., Bredie, W. L., Molgaard, C., Petersen, M. A., and Moller, P. Differential transfer of dietary flavour compounds into human breast milk. *Physiol Behav* 2008; 95(1–2):118–24.

64. Menella, J. A. and Beauchamp, G. K. Experience with a flavor in mother's milk modifies the infant's acceptance of flavored cereal. *Developmental Psychobiol* 1999; 34:197–203.

65. Gerrish, C. J. and Menella, J. A. Flavor variety enhances food acceptance in formula-fed infants. *Am J Clin Nutr* 2001; 73:1080–85.

66. U.S. Department of Health and Human Services. *Health and Human Services, Blueprint for Action on Breastfeeding.* Washington, D.C.: U.S. Department of Health and Human Services, Office on Women's Health, 2000.

67. Heinig, M. J. and Dewey, K. G. Health effects of breastfeeding for mothers: A critical review. *Nutrition Reviews* 1997; 10:59–73.

68. Kuzela, A. L., Stifter, C. A., and Worobey, J. Breastfeeding and mother-infant interactions. *J Reprod Infant Psychol* 1990; 8:185–94.

69. Narod, S. A. Modifiers of risk of heredity of breast cancer. *Oncogene* 2006; 25:5832–6.

70. Rosenblatt, K. A. and Thomas, D. B. Lactation and the risk of epithelial ovarian cancer. The WHO Collaborative Study of Neoplasia and Steroid Contraceptives. *Int J Epidemiol* 1993; 22(2):192–97.

71. Karlson, E. W. Do breast-feeding and other reproductive factors influence future risk of rheumatoid arthritis? Results from the Nurses' Health Study. *Arthritis Rheum* Nov. 2004; 50(11):3458–67.

72. Lubbock, M. H., Clark, D., and Goldman, A. S. Breastfeeding: Maintaining an irreplaceable immunological resource. *Nature Reviews Immunology* 2004; 4(Jul.):565–572.

73. Keeney, S. E., Schmalstieg, F.C., Palkowetz, K.H., Le, B.M., and Goldman A.S. Activated neutrophils and neutrophil activators in human milk: Increased expression of CD11b and decreased expression of L-selectin. *J Leukoc Biol* Aug. 1993; 54(2):97–104.

74. Brandtzaeg, P. Mucosal immunity: Integration between mother and the breast-fed infant. *Vaccine* Jul. 28 2003; 21(24):3382–88.

75. German, J. B. and Dillard, C. J. Composition, structure and absorption of milk lipids: A source of energy, fat-soluble nutrients and bioactive molecules. *Crit Rev Food Sci Nutr* 2006; 46(1):57–92.

76. Sprong, R. C., Hulstein, M. F., and Van der Meer, R. Bactericidal activities of milk lipids. *Antimicrob Agents Chemother* Apr. 2001; 45(4):1298–1301.

77. Pickering, L. K., Granoff, D. M.Erickson, J. R., Masor, M. L., Cordle, C. T., Schaller, J. P., Winship, T. R., Paule, C. L., and Hilty, M. D. Modulation of the immune system by human milk and infant formula containing nucleotides. *Pediatrics* Feb. 1998; 101(2):242–49.

78. World Health Organization. *Global Strategy for Infant and Young Child Feeding.* Geneva, Switzerland: World Health Organization, 2003.

79. McVea, K. L., Turner, P. D., and Peppler, D. K. The role of breastfeeding in sudden infant death syndrome. *J Hum Lact* 2000; 16(1):13–20.

80. Dewey, K. G., Heinig, M. J., and Nommsen-Rivers, L. A. Differences in morbidity between breast-fed and formula-fed infants. *J Pediatr* 1995; 126(5, Pt. 1):696–702.

81. Kramer, M. S., Guo, T., Platt, R. W., Sevkovskaya, Z., Dzikovich, I., Collet, J. P., Shapiro, S., Chalmers, B., Hodnett, E., Vanilovich, I., Mezen, I., Ducruet, T., Shishko, G., and Bogdanovich, N. Infant growth and health outcomes associated with 3 compared with 6 mo. of exclusive breastfeeding. *Am J Clin Nutr* Aug. 2003; 78(2):291–95.

82. Raisler, J., Alexander, C., and O'Campo, P. Breastfeeding: A dose-response relationship? *Am J Pub Health* 1999; 89:25–30.

83. Ivarsson, A., Persson, L.A., Nystron, L., Ascher, H., Cavell, B., Danielsson, L., Dannaeus, A., Lindberg, T., Lindquist, B., Stenhammar, L., and Hernell, O. Epidemic of coeliac disease in Swedish children. *Acta Paediatr* 2000; 89(2):165–71.

84. Koletzko, S. et al. Role of infant feeding practices in development of Crohn's disease in childhood. *BMJ* 1989; 298(6688):1617–18.

85. Daniels, J. L., Olshan, A.F., Pollock, B.H., Shah, N.R., Stram, D.O. Breastfeeding and neuroblastoma, USA and Canada. *Cancer Causes Control* Jun. 2002; 13(5):401–5.

86. Oddy, W. H., Holt, P.G., Sly, P.D., Read, A.W., Landau, L.I., Stanley, F.J., Kendall, G.E., and Burton, P.R. Association between breastfeeding and asthma in 6-year-old children: Findings of a prospective birth cohort study. *BMJ* 1999; 319(7213):815–19.

87. Scariati, P. D., Grummer-Strawn, L. M., and Fein, S. B. A longitudinal analysis of infant morbidity and the extent of breastfeeding in the United States. *Pediatrics* 1997; 99(6):E5.

88. Dewey, K. G., Heinig, M.J., Nommsen, L.A., Peerson, J.M., and Lonnerdal, B. Growth of breast-fed and formula-fed infants from 0 to 18 months: The DARLING Study. *Pediatrics.* 1992; 89(6, Pt. 1):1035–41.

89. Dewey, K. G. Is breastfeeding protective against child obesity? *J Hum Lact* Feb. 2003; 19(1):9–18.

90. Grummer-Strawn, L. M. and Mei, Z. Does breastfeeding protect against pediatric overweight? Analysis of longitudinal data from the Centers for Disease Control and Prevention Pediatric Nutrition Surveillance System. *Pediatrics* Feb. 2004; 113(2):e81–86.

91. Owen, C. G., Martin, R.M., Whincup, P.H., Davey-Smith, G., Gilman, M.W., and Cook, D.G. The effect of breastfeeding on mean body mass index throughout life: a quantitative review of published and unpublished observational evidence. *Am J Clin Nutr* Dec. 2005; 82(6):1298–1307.

92. Agostoni C. Ghrelin, leptin and the neurometabolic axis of breastfed and formula-fed infants. *Acta Pediatr* 2005; 95(5): 523–5.

93. Savino, F., Fissore, M. F., Liguori, S. A., and Oggero, R. Can hormones contained in mothers' milk account for the beneficial effect of breast-feeding on obesity in children? *Clin Endocrinol* 2009; 1365–2265 (electronic).

94. Rogan, W. J. and Gladen, B. C. Breastfeeding and cognitive development. *Early Hum Dev* 1993; 31(3):181–93.

95. Evenhouse, E. and Reilly, S. Improved estimates of the benefits of breastfeeding using sibling comparisons to reduce selection bias. *Health Serv Res* Dec. 2005; 40(6, Pt. 1): 1781–1802.

96. Rao, M. R., Hediger, M.L., Levine, R.J., Naficy, A.B., and Vik, T. Effect of breastfeeding on cognitive development of infants born small for gestational age. *Acta Paediatr* 2002; 91(3):267–74.

97. Smith, M. M., Durkin, M., Hinton, V.J., Bellinger, D., Kuhn, L. Initiation of breastfeeding among mothers of very low birth weight infants. *Pediatrics* Jun 2003; 111(6, Pt. 1):1337–42.

98. Ors, R., Ozek, E., Baysoy, G., Cebeci, D., Bilgen, H., Turnkuner, M. and Basaran, M. Comparison of sucrose and human milk on pain response in newborns. *Eur J Pediatr* Jan. 1999; 158(1):63–66.

99. Shah, P. S., Aliwalas, L. L. and Shah, V. S. Breastfeeding or breast milk for procedural pain in neonates. *Cochrane Database of Systematic Reviews* 2006(3).

100. Carbajal, R., Veerapen, S., Couderc, S. and Jugie, M. V. Analgesic effect of breastfeeding in term neonates: Randomised controlled trial. *BMJ* 2003; 326(7379):13.

101. Splett, P. L. and Montgomery, D. L. The economic benefits of breastfeeding an infant in the WIC program twelve-month follow-up study. Washington, D.C.: Food and Consumer Service, U.S. Department of Agriculture, 1998.

102. Ball, T. M. and Wright, A. L. Health care costs of formula-feeding in the first year of life. *Pediatrics* 1999; 103(4, Pt. 2): 870–76.

103. Cohen, R., Mrtek, M. B., and Mrtek, R. G. Comparison of maternal absenteeism and infant illness rates among breast-feeding

and formula-feeding women in two corporations. *Am J Health Promot* Nov.–Dec. 1995; 10(2):148–53.

104. Daly, S. E. and Hartmann, P. E. Infant demand and milk supply. Part 1: Infant demand and milk production in lactating women. *J Hum Lact* 1995; 11(1):21–26.

105. Dewey, K. G. and Lonnerdal, B. Infant self-regulation of breast milk intake. *Acta Paediatr Scand* 1986; 75(6):893–98.

106. Daly, S. E., Owens, R. A., and Hartmann, P. E. The short-term synthesis and infant-regulated removal of milk in lactating women. *Exp Physiol* 1993; 78(2):209–20.

107. Saint, L., Maggiore, P., and Hartmann, P. E. Yield and nutrient content of milk in eight women breastfeeding twins and one woman breastfeeding triplets. *Br J Nutr* 1986; 56:49–58.

108. Newton, M. Human lactation. In Kon, S. K., ed. *Milk: The Mammary Gland and Its Secretion*. New York, NY: Academic Press, 1961.

109. Daly, S. E. J. and Hartmann, P. E. Infant demand and milk supply. Part 2: The short-term control of milk synthesis in lactating women. *J Hum Lactation*. 1995; 11:27–37.

110. Hill, P. D., Aldag, J. C., and Chatterton, R. T. J. Breastfeeding experience and milk weight in lactating mothers pumping for preterm infants. *Birth* 1999; 26(4).

111. Neifert, M. and Seacat, J. M. Practical aspects of breastfeeding the premature infant. *Perinatal Neonatol* 1988; 12:24.

112. Souto, G. C., Guigliani, E.R., Guigliani, C., and Schneider, M.A. The impact of breast reduction surgery on breastfeeding performance. *J Hum Lact* Feb. 2003; 19(1):43–49, 66–49, 120.

113. Hurst, N. Breastfeeding after breast augmentation. *J Hum Lact* Feb 2003; 19(1):70–71.

114. Brown, S. L., Todd, J. F., Cope, J. U. and Sachs, H.C. Breast implant surveillance reports to the U.S. Food and Drug Administration: Maternal-child health problems. *J Long Term Eff Med Implants* 2006; 16(4):281–90.

115. Transfer of drugs and other chemicals into human milk. *Pediatrics* 2001; 108(3):776–89.

116. Semple, J. L., Lugowski, S.J., Baines, C.J., Smith, D.C., and McHugh, A. Breast milk contamination and silicone implants: preliminary results using silicon as a proxy measurement for silicone. *Plast Reconstr Surg* Aug. 1998; 102(2):528–33.

117. Ziegler, E. E., Foman, S.J., Nelson, S.E., Rebouche, C.J., Edwards, B.B., Rogers, R.R. and Lehman, L.J. Cow milk feeding in infancy: Further observations on blood loss from the gastrointestinal tract. *J Pediatr* 1990;116:11–18.

118. Inch, S. Breastfeeding problems. Prevention and Management. *Community Practitioner* 2006; 79(5):165–67.

119. Morland-Schultz, K. and Hill, P. D. Prevention of and therapies for nipple pain: A systematic review. *J Obstet Gynecol Neonatal Nurs* Jul.–Aug. 2005; 34(4):428–37.

120. Wooldrige, M. S. and Fischer, C. Colic, "overfeeding" and symptoms of lactose malabsorption in the breast-fed baby. *Lancet* 1988; 2:382–84.

121. Butte, N. F., Wills, C., Jean, C.A., Smith, E.O., and Garza, C. Feeding patterns of exclusively breast-fed infants during the first four months of life. *Early Hum Dev* Dec. 1985; 12(3):291–300.

122. Wagner, C. L. and Greer, F. R. American Academy of Pediatrics Section on Breastfeeding; American Academy of Pediatrics Committee on Nutrition. Prevention of rickets and vitamin D deficiency in infants, children, and adolescents. *Pediatrics* 2008; 122(5):1142–1152.

123. Brams, M. and Maloney, J. "Nursing bottle caries" in breast-fed children. *J Pediatr* 1983; 103:415–16.

124. Palmer, B. The influence of breastfeeding on the development of the oral cavity: A commentary. *J Hum Lact*. 1998; 14(2):93–98.

125. U.S. Department of Health and Human Services, United States Department of Agriculture. Dietary Guidelines for Americans. *Home and Garden Bulletin No. 232*. Washington, D.C: 2005.

126. Food and Nutrition Board, Institute of Medicine. Dietary Reference Intakes for energy, carbohydrate, fiber, fat, fatty acids, cholesterol, protein and amino acids (macronutrients). Washington, D.C.: National Academy Press, 2002.

127. Butte, N. F. and King, J. C. Energy requirements during pregnancy and lactation. *Public Health Nutr* Oct. 2005; 8(7A):1010–27.

128. Goldberg, G. R., Prentice, A.M., Coward, W.A., Davies, H.L., Murgatroyd, P.R., Wensing, C., Black, A.E., Harding, M., and Sawyer, M. Longitudinal assessment of energy expenditure in pregnancy by the doubly labeled water method. *Am J Clin Nutr* 1993; 57:494–505.

129. Prentice, A. M. , Poppitt, S.D., Goldberg, G.R., Murgatroyd, P.R., Black, A.E. and Coward, W.A. Energy balance in pregnancy and lactation. In: Allen, L. H., King, J., and Lonnerda, B. eds. *Nutrient Regulation During Pregnancy, Lactation and Infant Growth*. New York, NY: Plenum Press, 1994.

130. Roberts, S., Cole, T., and Coward, W. Lactational performance in relation to energy intake in the baboon. *Am J Clin Nutr* 1985; 41:1270–76.

131. Butte, N. F. and Hopkinson, J. M. Body composition changes during lactation are highly variable among women. *J Nutr* 1998; 128:381S–385.

132. Strode, M. A., Dewey, K. G., and Lonnerdal, B. Effects of short-term caloric restriction on lactational performance of well-nourished women. *Acta Paediatr Scand* 1986; 75(2):222–29.

133. Dusdieker, L. B., Hemingway, D. L., and Stumbo, P. J. Is milk production impaired by dieting during lactation? *Am J Clin Nutr* 1994; 59(4):833–40.

134. Lovelady, C. A. et al. The effects of dieting on food and nutrient intake of lactating women. *J Am Diet Assoc* Jun. 2006; 106(6):908–12.

135. Mohammad, M. A., Sunehag, A. L., and Haymond, M. W. Effect of dietary macronutrient composition under moderate hypocaloric intake on maternal adaptation during lactation. *Am J Clin Nutr* 2009; 89(6): 1821–7.

136. Lovelady, C. A., Lonnerdal, B., and Dewey, K. G. Lactation performance of exercising women. *Am J Clin Nutr* 1990; 52(1):103–9.

137. Dewey, K. G. and McCrory, M. A. Effects of dieting and physical activity on pregnancy and lactation. *Am J Clin Nutr* 1994; 59(2, Suppl.):446S–452S; discussion, 452S–453S.

138. Lovelady, C. A., Hunter, C. P., and Geigerman, C. Effect of exercise on immunologic factors in breast milk. *Pediatrics* Feb. 2003; 111(2):E148–52.

139. Food and Nutrition Board Institute of Medicine. Dietary reference intakes for thiamin riboflavin, niacin, vitamin B_6, folate, vitamin B_{12}, pantothenic acid, biotin, and choline. Washington, D.C.: National Academy Press, 1998.

140. Food and Nutrition Board, Institute of Medicine. *Dietary Reference Intakes of water, potassium, sodium, chloride, and sulfate*. Washington, D.C.: National Academy Press, 2004.

141. Lust, K. D., Brown, J. E., and Thomas, W. Maternal intake of cruciferous vegetables and other foods and colic symptoms in exclusively breast-fed infants. *J Am Diet Assoc* 1996; 96(1):46–48.

142. Hill, D. J., Hudson, I.L., Sheffield, L.J., Shetlon, M.J., Menahem, S., and Hosking, C.S. A low-allergen diet is a significant intervention in infantile colic: results of a community-based study. *J Allergy Clin Immunol* Dec. 1995; 96(6, Pt. 1):886–92.

143. Oddy, W. H., Holt, P.G., Sly, P.D., Read, A.W., Landau, L.Il, Stanley, F.J., Kendall, G.E., and Burton, P.R. The association of maternal overweight and obesity with breastfeeding duration. *J Pediatr* Aug. 2006; 149(2):185–91.

144. Rasmussen, K. M. and Kjolhede, C. L. Prepregnant overweight and obesity diminish the prolactin response to suckling in the first week postpartum. *Pediatrics* May 2004; 113(5):e465–71.

145. Rasmussen, K. M., Hilson, J. A., and Kjolhede, C. L. Obesity as a risk factor for failure to initiate and sustain lactation. *Adv Exp Med Biol* 2002; 503:217–22.

146. Lazarov, M. and Evans, A. Breastfeeding—encouraging the best for low-income women. *Zero to Three* 2000; (August/September):15–23.

147. Bryant, C., Coreil, J., D'Angelo, S.L., Bailey, D.F.C., and Lazarov, M.A. A strategy for promoting breastfeeding among economically disadvantaged women and adolescents. *NAACOG's Clin Issu Perinat Women's Health Nurs* 1992; 3(4):723–730.

148. Bryant, C. and Roy, M. *Best Start's Three Step Counseling Strategy.* Tampa, FL: Best Start, Inc., 1997.

149. Rinker, B., Veneracion, M., and Walsh, C. P. The effect of breastfeeding on breast aesthetics. *Aesthet Surg J* 2008; 28(5):534-537.

150. Lewin, S. A., Dick, J., Pond, P, Zwarenstein, M., Aja, G., van Wyk, B., Bosch-Capblanch, X., and Patrick, M. Lay health workers in primary and community health care. *Cochrane Database of Systematic Reviews* 2005(1).

151. Bonuck, K. A., Trombley, M., Freeman, K., and McKee, D. Randomized, controlled trial of a prenatal and postnatal lactation consultant intervention on duration and intensity of breastfeeding up to 12 months. *Pediatrics* Dec. 2005; 116(6):1413–26.

152. Shealy, K. R., Li, R., Benton-Davis, S, and Grummer-Strawn, L.M. *The CDC Guide to Breastfeeding Intervention.* Atlanta: U.S. Department of Health and Human Services, Centers for Disease Control and Prevention, 2005.

153. Dyson, L., Mccormick, F., and Renfrew, M. J. Interventions for promoting the initiation of breastfeeding. *Cochrane Database of Systematic Reviews* 2005(2).

154. McCamman, S. and Page-Goetz, S. Breastfeeding success: You can make the difference. *Perinatal Nutr Report* 1998; 4(Winter):2–4.

155. Cadwell, K. Reaching the goals of "Healthy People 2000" regarding breastfeeding. *Clin Perinatol* 1999; 26(2):527–37.

156. Moreland, J. C., Lloyd, L., Braun, S.B., and Heins, J.N. A new teaching model to prolong breastfeeding among Latinos? *J Hum Lact* 2000; 16(4):337–41.

157. Pisacane, A., Continisio, G.I., Aldinucci, M., D'Amora, S., and Continisio, P. A controlled trial of the father's role in breastfeeding promotion. *Pediatrics* Oct. 2005; 116(4):e494–98.

158. Freed, G. L., Fraley, J. K., and Schanler, R. J. Attitudes of expectant fathers regarding breast-feeding. *Pediatrics* 1992; 90(2, Pt. 1): 224–27.

159. United States Department of Agriculture, Food and Nutrition Service. *Fathers Supporting Breastfeeding.* Washington, D.C.: United States Department of Agriculture, 2002.

160. Bentley, M. E., Caulfield, L.E., Gross, S.M., Bronner, Y., Jensen, J., Kessler, L.A. and Paige, D.M. Sources of influence on intention to breastfeed among African-American women at entry to WIC. *J Hum Lact* 1999; 15(1):27–34.

161. Barron, S. P. et al. Factors influencing duration of breastfeeding among low-income women. *J Am Diet Assoc* 1988; 88(12):1557–61.

162. Valaitis, R. K., Sheeshka, J. D., and O'Brien, M. F. Do consumer infant feeding publications and products available in physicians' offices protect, promote, and support breastfeeding? *J Hum Lact* 1997; 13(3):203–8.

163. Howard, C., Howard, F., Lawrence, R., Andresen, E., DeBlieck, E., and Weitzman, M. Office prenatal formula advertising and its effect on breast-feeding patterns. *Obstet Gynecol.* 2000; 95(2):296–303.

164. Heiser, B. and, Walker, M. *Selling Out Mothers and Babies: Marketing of Breast Milk Substitutes in the USA.* Ellicott City, MD: National Alliance for Breastfeeding Promotion, 2003.

165. Avery, A., Zimmermann, K., Underwood, P. W., and Magnus, J. H. Confident commitment is a key factor for sustained breastfeeding. *Birth* 2009; 36(2): 141–8.

166. James, D. C. and Dobson, B. Position of the American Dietetic Association: Promoting and supporting breastfeeding. *J Am Diet Assoc* May 2005; 105(5):810–18.

167. World Health Organisation International Code of Marketing of Breast-milk Substitutes. Geneva, Switzerland: 1981.

168. Perez-Escamilla, R., Pollitt, E., Lonnerdal, B., and Dewey, K.G. Infant feeding policies in maternity wards and their effect on breastfeeding success: An analytical overview. *Am J Public Health* 1994; 84(1):89–97.

169. World Health Organization. Protecting, promoting, and supporting breastfeeding. The special role of maternity services (a joint WHO/UNICEF statement). Geneva, Switzerland: WHO/UNICEF, 1989.

170. Merten, S., Dratva, J., and Ackermann-Liebrich, U. Do baby-friendly hospitals influence breastfeeding duration on a national level? *Pediatrics* Nov. 2005; 116(5):e702–8.

171. UNICEF/WHO. Innocenti Declaration on the Protection, Promotion, and Support of Breast-Feeding. Forence, Italy: UNICEF and WHO, 1990.

172. Ahluwalia, I. B., Morrow, B., and Hsia, J. Why do women stop breastfeeding? Findings from the Pregnancy Risk Assessment and Monitoring System. *Pediatrics* Dec. 2005; 116(6):1408–12.

173. Petrova, A., Ayers, C., Stechna, S., Gerling, J. A., and Mehta, R. Effectiveness of exclusive breastfeeding promotion in low-income mothers: A randomized controlled study. *Breastfeed Med* 2009; 1556–8342 (electronic).

174. Division of Labor Statistics,. *Women in the Labor Force: A Databook.* Washington, D.C.: U.S. Departmeut of Labor, 2004.

175. *Report of the Surgeon General's Workshop on Breastfeeding and Human Lactation.* Washington, D.C.: U.S. Department of Health and Human Services, 1984.

176. Kurinij, N, Shiono, P.H., Ezrine, S.F., and Rhoads, G.G. Does maternal employment affect breast-feeding? *Am J Public Health* 1989; 79:1247–50.

177. Guendelman, S., Kosa, J. L., Pearl, M., Graham, S., Goodman, J., and Kharrazi, M. Juggling work and breastfeeding: Effects of maternity leave and occupational characteristics. *Pediatrics* 2009; 123(1):2e38–46.

178. Meek, J. Y. Breastfeeding in the workplace. *Pediatr Clin North Am* Apr. 2001; 48(2):461–74, xvi.

179. Healthy Mothers Healthy Babies Coalition. *Workplace Models of Excellence 2000: Outstanding programs supporting working women that breastfeed.* Alexandria, VA: National Healthy Mothers, Healthy Babies Coalition, 2000.

180. Meehan, K., Harrison, G. G., Afifi, A. A., Nicket, N., Jenks, E., and Ramirez, A. The association between electric pump loan program and the timing of requests for formula by working mothers in WIC. *J Hum Lact* 2008; 24(2):150–8.

181. Ortiz, J., McGilligan, K., and Kelly, P. Duration of breast milk expression among working mothers enrolled in an employer-sponsored lactation program. *Pediatr Nurs* Mar–Apr 2004; 30(2):111–19.

182. Healthy Mothers Healthy Babies Coalition. *Workplace Models of Excellence 2000: Outstanding programs supporting working women that breastfeed.* Alexandria, VA: National Healthy Mothers, Healthy Babies Coalition, 2000.

183. United States Breastfeeding Committee. *Workplace Breastfeeding Support.* Raleigh, NC: 2002.

184. Ryan, A. S. The resurgence of breast-feeding in the United States. *Pediatrics* 1997; 99(4):e12.

185. U.S. Department of Health and Human Services. *Follow-up Report: The Surgeon General's Workshop on Breastfeeding and Human Lactation.* Washington, D.C.: U.S. Department of Health and Human Services, Public Health Service, Health Resources and Services Administration, 1985.

186. Sharbaugh, C. S. *Call to Action: Better Nutrition for Mothers, Children, and Families.* Washington, D.C.: National Center for Education in Maternal and Child Health; 1990.

187. Spisak, S. G., and Gross, S. S. *Second Followup Report: The Surgeon General's Workshop on Breastfeeding and Human Lactation.* Washington, D.C: National Center for Education in Maternal and Child Health, 1991.

188. Skisser, W. and Thomas, S. Report of the national breastfeeding policy conference: UCLA Center for Healthier Communities, Families and Children, 1999.

189. United States Breastfeeding Committee. *Breastfeeding in the United States: A National Agenda.* Rockville, MD: Department of Health and Human Services, Maternal and Child Health Bureau, 2001.

190. Andreason, A. *Marketing Social Change: Changing Behavior to Promote Health, Social Development, and the Environment.* San Francisco: Jossey-Bass, 1995.

191. Carothers, C. Social Marketing Institute Success Stories: National WIC breastfeeding promotion project. Best Start Social Marketing, Inc., 2001.

192. Khoury, A. *Social Marketing Institute Success Stories: National WIC breastfeeding promotion project.* Mississippi WIC Program; 2001. Available at www.social-marketing.org/success/cs-nationalwic.html.

193. Stremler, J. and Lovera, D. Insight from a breastfeeding peer support pilot program for husbands and fathers of Texas WIC participants. *J Hum Lact.* Nov 2004; 20(4):417–22.

194. Cricco-Lizza, R. The milk of human kindness: Environmental and human interactions in a WIC clinic that influence infant-feeding decisions of black women. *Qual Health Res* Apr. 2005; 15(4):525–38.

Chapter 7

1. From the association: Position of the American Dietetic Association: Promotion and Supporting Breastfeeding. *J Am Diet Assoc* 2005;105(5);810–818.

2. Morland-Schultz, K. and Hill, P. D. Prevention of and therapies for nipple pain: a systematic review. *J Obstet Gynecol Neonatal Nurs* 2005; 34:428–37.

3. Mass, S. Breast pain: engorgement, nipple pain and mastitis. *Clin Obstet Gynecol* 2004; 47:676–82.

4. Inch, S. Breastfeeding problems: prevention and management. *Community Pract* 2006; 79:165–67.

5. Smith, J. W. and Tully, M. R. Midwifery management of breastfeeding: using the evidence. *J Midwifery Womens Health* 2001; 46:423–38.

6. Martin, C. and Krebs, N. F. *The nursing mother's problem solver.* New York: Simon and Schuster, 2000.

7. Walker, M. Conquering Common Breast-feeding Problems. *J Perinat Neonat Nurs* 2008; (22)267–274.

8. Spencer, J. P. Management of Mastitis in Breastfeeding Women. *Am Fam Phys* 2008; (78)6:727–732.

9. Snowden, H. M., Renfrew, M. J., and Woolridge, M. W. Treatments for breast engorgement during lactation. *Cochrane Database Syst Rev* 2003; CD000046.

10. Ayers, J. F. The Use of Alternative Therapies in the Support of Breastfeeding. *J Hum Lact* 2000; (16):1; 52–56.

11. Walker, M. *Breastfeeding Management for the Clinician: Using the Evidence.* MA: Jones and Bartlett Publishers, 2006: 387.

12. Academy of Breastfeeding Medicine. Clinical Protocol #4: Mastitis. Breastfeed Med 2008; (3):3, 177–180.

13. Lau, C. The Effects of Stress on Lactation. *Pediatr Clin North Am* 2001; (48):1; 221–234.

14. Matheson, I. et al. Bacteriological findings and clinical symptoms in relation to clinical outcome in puerperal mastitis. *Acta Obstet Gynecol Scand* 1988; 67:723–26.

15. Academy of Breastfeeding Medicine: Clinical Protocol #9: Use of Galactogogues in initiating or augmenting maternal milk supply, 2004. Available at www.bfmed.org, accessed 5/09.

16. Hale, T. W. and Hartmann, P. E. Hale & *Hartmann's Textbook of Human Lactation,* 1st ed. Hale Publishing, 2007: 479.

17. Import Refusal Report #61–07: Domperidone. Available at www.acessdata.fda.gov/ImportAlerts, accessed 5/09.

18. Lawrence, R. A. *Breastfeeding: a guide for the medical profession,* 6th ed. St Louis: Mosby, Inc., 2005.

19. Hale, T. W. *Medications and Mother's Milk,* 13th ed. Amarillo, TX: Pharmasoft Medical Publishing, 2008.

20. Lagoy, C. T. et al. Medication use during pregnancy and lactation: An urgent call for public health action. *J Womens Health* 2005; 14:104–09.

21. Berlin, C. M. and Briggs, G. G. Drugs and chemicals in human milk. *Semin Fetal Neonatal Med* 2005; 10:149–59.

22. Begg, E. J. et al. Studying drugs in human milk: Time to unify the approach. *J Hum Lact* 2002; 18:323–32.

23. Ito, S. Drug therapy for breast-feeding women. *N Engl J Med* 2000; 343:118–26.

24. Transfer of drugs and other chemicals into human milk. *Pediatrics* 2001; 108:776–89.

25. Committee on Nutrition, American Academy of Pediatrics. *Pediatric Nutrition Handbook,* 1st ed. Elk Grove Village, IL: American Academy of Pediatrics, 2004.

26. Chung, A. M., Reed, M. D., and Blumer, J. L. Antibiotics and breast-feeding: a critical review of the literature. *Paediatr Drugs* 2002; 4:817–37.

27. Guthmann, R. A., Bang, J., and Nashelsky, J. Combined oral contraceptives for mothers who are breastfeeding. *Am Fam Physician* 2005; 72:1303–4.

28. Diaz, S. Contraceptive implants and lactation. *Contraception* 2002; 65:39–46.

29. Nice, F. J., Snyder, J. L., and Kotansky, B. C. Breastfeeding and over-the-counter medications. *J Hum Lact* 2000; 16:319–31.

30. Department of Child and Adolescent Health. Breastfeeding and Maternal Medication: Recommendations for Drugs in the Eleventh WHO Model List of Essential Drugs, 2003. UNICEF World Health Organization, 2003.

31. The complete German Commission E monographs: *Therapeutic guide to herbal medicines.* Boston: Integrative Medicine Communications, 1998.

32. Hardy, M. L. Herbs of special interest to women. *J Am Pharm Assoc* (Wash.) 2000; 40:234–42.

33. Scott, C. R. and Jacobson, H. A. Selection of international nutritional & herbal remedies for breastfeeding concerns. *Midwifery Today* 2005; 38–39.

34. Kopec, K. Herbal medications and breastfeeding. *J Hum Lact* 1999; 15:157–61.

35. Humphrey, S. L. and McKenna, D. J. Herbs and breastfeeding. *Breastfeeding Abstracts* 1997; 17:11–2.

36. Conover, E. and Buehler, B. A. Use of herbal agents by breastfeeding women may affect infants. *Pediatr Ann* 2004; 33:235–40.

37. Woolf, A. D. Herbal remedies and children: do they work? Are they harmful? *Pediatrics* 2003; 112:240–46.

38. Humphrey, S. Sage advice on herbs and breastfeeding. *LEAVEN* 1998; 34:43–47.

39. Leung, A. and Foster, S. *Encyclopedia of common natural ingredients used in foods, drugs, and cosmetics.* New York: John Wiley & Sons, 1996.

40. Cunningham, E. and Hansen, K. Question of the month: Where can I get information on evaluating herbal supplements? *J Am Diet Assoc* 1999; 99:1240.

41. The American Botanical Council. The ABC Clinical Guide to Herbs. American Botanical Council, 2005.

42. Cartwright, M. M. Herbal use during pregnancy and lactation: A need for caution. American Dietetic Association Public Health/Community Nutrition Practice Group. *The Digest (Summer)* 2001; 1–3.

43. Klier, C. M. et al. St. John's wort (*Hypericum perforatum*) and breastfeeding: plasma and breast milk concentrations of hyperforin for 5 mothers and 2 infants. *J Clin Psychiatry* 2006; 67(2):305–9.

44. Lee, A. et al. The safety of St. John's wort (*Hypericum perforatum*) during breastfeeding. *J Clin Psychiatry* 2003; 64(8):966–68.

45. Gabay, M. P. Galactogogues: Medications that induce lactation. *J Hum Lact* 2002; 18:274–79.

46. Betzold, C. M. Galactagogues. *J Midwifery Womens Health* 2004; 49:151–54.

47. American Academy of Pediatrics Policy Statement: Breastfeeding and the use of Human Milk. *Pediatrics* 2005; 115(5:2):496–506.

48. Mennella, J. Alcohol's effect on lactation. *Alcohol Res Health* 2001; 25:230–34.

49. Giglia, R. and Binns, C. Alcohol and lactation: A systematic review. *Nutrition and Dietetics* 2006; 63:103.

50. Ho, E. et al. Alcohol and breast-feeding: Calculation of time to zero level in milk. *Biol Neonate* 2001; 80:219–22.

51. Mennella, J. A., Pepino, M. Y., and Teff, K. L. Acute alcohol consumption disrupts the hormonal milieu of lactating women. *J Clin Endocrinol Metab* 2005; 90:1979–85.

52. Mennella, J. A. Regulation of milk intake after exposure to alcohol in mothers' milk. *Alcohol Clin Exp Res* 2001; 25:590–93.

53. Mennella, J. A. and Gerrish, C. J. Effects of exposure to alcohol in mother's milk on infant sleep. *Pediatrics* 1998; 101:E2.

54. Mennella, J. A. and Garcia Comez, P. L. Sleep disturbances after acute exposure to alcohol in mother's milk. *Alcohol Int Biomed J* 2001; 25:153–58.

55. Gunzerath, L. et al. National Institute on Alcohol Abuse and Alcoholism report on moderate drinking. *Alcohol Clin Exp Res* 2004; 28:829–47.

56. Little, R. E. et al. Maternal alcohol use during breast-feeding and infant mental and motor development at one year. *N Engl J Med* 1989; 321:425–30.

57. Mohrbacher, N. and Stock, J. *The Breastfeeding Answer Book.* Schaumburg, IL: La Leche League International, 2003.

58. Subcommittee on Nutrition During Lactation, Committee on Nutritional Status During Pregnancy and Lactation. *Food and Nutrition Board Institute of Medicine National Academy of Sciences. Nutrition During Lactation: Summary, Conclusions, and Recommendations.* Washington, D.C.: National Academy Press, 1991.

59. Howard, C. R. and Lawrence, R. A. Xenobiotics and breastfeeding. *Pediatr Clin North Am* 2001; 48:485–504.

60. Dempsey, D. A. and Benowitz, N. L. Risks and benefits of nicotine to aid smoking cessation in pregnancy. *Drug Saf* 2001; 24:277–322.

61. Amir, L. H. and Donath, S. M. Does maternal smoking have a negative physiological effect on breastfeeding? The epidemiological evidence. *Birth* 2002; 29:112–23.

62. Ilett, K. F. et al. Use of nicotine patches in breast-feeding mothers: Transfer of nicotine and cotinine into human milk. *Clin Pharmacol Ther* 2003; 74:516–24.

63. Chapman, D. J. Short-term Effects of Smoking on Breastfed Infants. *J Hum Lact* 2008; (24):1, 92–93.

64. Dahlstrom, A. et al. Nicotine and cotinine concentrations in the nursing mother and her infant. *Acta Paediatr Scand* 1990; 79:142–47.

65. Gaffney, K. F. Postpartum smoking relapse and becoming a mother. *J Nurs Scholarsh* 2006; 38:26–30.

66. Nickerson, K. Environmental contaminants in breast milk. *J Midwifery Womens Health* 2006; 51:26–34.

67. Molyneux, A. Nicotine replacement therapy. *British Medical Journal* 2006; 328:454–56.

68. Fernandez-Ruiz, J. et al. Cannabinoids and gene expression during brain development. *Neurotox Res* 2004; 6:389–401.

69. Astley, S. J. and Little, R. E. Maternal marijuana use during lactation and infant development at one year. *Neurotoxicol Teratol* 1990; 12:161–68.

70. Rivera-Calimlim, L. The significance of drugs in breast milk. Pharmacokinetic considerations. *Clin Perinatol* 1987; 14:51–70.

71. Nehlig, A. and Debry, G. Consequences on the newborn of chronic maternal consumption of coffee during gestation and lactation: A review. *J Am Coll Nutr* 1994; 13:6–21.

72. World Health Organization. Safe food: Crucial for child development. Available at www.who.int/ceh/publications/en/poster15new.pdf, accessed 10/06.

73. Massart, F. et al. Human breast milk and xenoestrogen exposure: A possible impact on human health. *J Perinatol* 2005; 25:282–88.

74. Condon, M. Breast is best, but it could be better: What is in breast milk that should not be? *Pediatr Nurs* 2005; 31:333–38.

75. Wang, R. Y. et al. Human milk research for answering questions about human health. *J Toxicol Environ Health* 2005; 68:1771–1801.

76. International Lactation Consultant Association: Position on Breastfeeding, Breastmilk, and Environmantal Contaminants, 2001. Available at www.ilca.org, accessed 5/09.

77. Schreiber, J. S. Parents worried about breast milk contamination. What is best for baby? *Pediatr Clin North Amer* 2001 Oct; 48(5): 11113-27, viii

78. U.S. Department of Health and Human Services. *HHS Blueprint for action on breastfeeding 2000.* Department of Health and Human Services, Office of Women's Health.

79. World Health Organization. Report of the WHO working group on the assessment of health risks for human infants from exposure to PCDDs, PCDFs, PCBs. *Chemosphere* 1998; 57:1627–43.

80. Pronczuk, J. et al. Global perspectives in breast milk contamination: Infectious and toxic hazards. *Environ Health Perspect* 2002; 110:A349–A351.

81. LaKind, J. S., Berlin, C. M., and Naiman, D. Q. Infant exposure to chemicals in breast milk in the United States: What we need to learn from a breast milk monitoring program. *Environ Health Perspect* 2001; 109:75–88.

82. American Academy of Pediatrics Subcommittee on Hyperbilirubinemia. Management of hyperbilirubinemia in the newborn infant 35 or more weeks of gestation. *Pediatrics* 2004; 114:297–316.

83. Cohen, S. M. Jaundice in the full-term newborn *Pediatr Nurs* 2006; 32:202–8.

84. Stokowski, L. A. Early recognition of neonatal jaundice and kernicterus. *Adv Neonatal Care* 2002; 2:101–14.

85. Watchko, J. F. Vigintiphobia revisited. *Pediatrics* 2005; 115:1747–53.

86. Bhutani, V. K. and Johnson, L. Kernicterus in the 21st Century: Frequently asked questions. *J Perinatol* 2009; 29:S20–24.

87. Geiger, A. M., Petitti, D. B, and Yao, J. F. Rehospitalisation for neonatal jaundice: Risk factors and outcomes. *Paediatr Perinat Epidemiol* 2001; 15:352–58.

88. Joint Commission on Accreditation of Healthcare Organizations. Kernicterus threatens healthy newborns. *Sentinel Event Alert* April 2001, Issue 18. Joint Commission on Accreditation of Healthcare Organizations, 2001.

89. Stark, A. R. and Lannon, C. M. System changes to prevent severe hyperbilirubinemia and promote breastfeeding: pilot approaches. *J Perinatol* 2009; 29:S.53–57.

90. Gartner, L. M. and Herschel, M. Jaundice and breastfeeding. *Pediatr Clin North Am* 2001; 48:389–99.

91. Reiser, D. J. Neonatal jaundice: physiologic variation or pathologic process. *Crit Care Nurs Clin North Am* 2004; 16:257–69.

92. Moerschel, S.k., Clanciaruso, L.B., and Tracy, L.R. A Practical Approach to Neonatal Jaundice. *Amer Fam Phys* 2008; (77):9.

93. Reiser, D. J. Hyperbilirubinemia. AWHONN. *Lifelines* 2001; 5:55–61.

94. AAP Subcommittee on Neonatal Pediatrics: Neonatal jaundice and kernicterus. *Pediatrics* 2001; 108:763–65.

95. Gartner, L. M. Breastfeeding and jaundice. *J Perinatol* 2001; 21(Suppl.)1:S25–S29.

96. Gourley, G. R. Breast-feeding, neonatal jaundice and kernicterus. *Semin Neonatol* 2002; 7:135–41.

97. Willis, S. K., Hannon, P. R., and Scrimshaw, S. C. The impact of the maternal experience with a jaundiced newborn on the breastfeeding relationship. *J Fam Pract* 2002; 51:465.

98. Geraghty, S. R. et al. Breast milk feeding rates of mothers of multiples compared to mothers of singletons. *Ambul Pediatr* 2004; 4:226–31.

99. Flidel-Rimon, O. and Shinwell, E. S. Breast-feeding multiples. *Semin Neonatol* 2002; 7:231–39.

100. Gromada, K. K. and Spangler, A. K. Breastfeeding twins and higher-order multiples. *J Obstet Gynecol Neonatal Nurs* 1998; 27:441–49.

101. Gromada, K. K. Breastfeeding more than one: Multiples and tandem breastfeeding. NAACOGS. *Clin Issu Perinat Womens Health Nurs* 1992; 3:656–66.

102. Auer, C. and Gromada, K. K. A case report of breastfeeding quadruplets: Factors perceived as affecting breastfeeding. *J Hum Lact* 1998; 14:135–41.

103. Palmer, G. *The Politics of Breastfeeding*, 2nd ed. London: Pandora Press/Harper Collins, 1993: 32.

104. Saint, L., Maggiore, P., and Hartmann, P. E. Yield and nutrient content of milk in eight women breast-feeding twins and one woman breast-feeding triplets. *Br J Nutr* 1986; 56:49–58.

105. Hattori, R. and Hattori, H. Breastfeeding twins: guidelines for success. *Birth* 1999; 26:37–42.

106. Geraghty, S. R., Khoury, J. C., and Kalkwarf, H. J. Comparison of feeding among multiple-birth infants. *Twin Res* 2004; 7:542–47.

107. Zeiger, R. S. and Friedman, N. J. The relationship of breastfeeding to the development of atopic disorders. *Nestle Nutr Workshop Ser Pediatr Program* 2006; 57:93–105.

108. O'Connell, E. J. Pediatric allergy: A brief review of risk factors associated with developing allergic disease in childhood. *Ann Allergy Asthma Immunol* 2003; 90:53–8.

109. Greer, F. R., Sicherer, S. H., Burks, A. W., and the Committee on Nutrition and Section on Allergy and Immunology. Effects of Early Nutritional Interventions on the Development of Atopic Disease in Infants and Children: The Role of Maternal Dietary Restriction, Breastfeeding, Timing of Introduction of Complementary Foods, and Hydrolyzed Formulas. AAP Clinical Report. *Pediatrics* 2008; (121)1:183–191.

110. Palmer, D. J. and Makrides, M. Diet of lactating women and allergic reactions in their infants. *Curr Opin Clin Nutr Metab Care* 2006; 9:284–88.

111. Mofidi, S. Nutritional management of pediatric food hypersensitivity. *Pediatrics* 2003; 111:1645–53.

112. Hill, D. J. et al. Effect of a low-allergen maternal diet on colic among breastfed infants: A randomized, controlled trial. *Pediatrics* 2005; 116:e709–e715.

113. Mennella, J. A. and Beauchamp, G. K. The effects of repeated exposure to garlic-flavored milk on the nursling's behavior. *Pediatr Res* 1993; 34:805–8.

114. Mennella J. A. and Beauchamp, G. K. Maternal diet alters the sensory qualities of human milk and the nursling's behavior. *Pediatrics* 1991; 88:737–44.

115. Cooper, R. L. and Cooper, M. M. Red pepper-induced dermatitis in breast-fed infants. *Dermatology* 1996; 193:61–2.

116. Zeretzke, K. Allergies and the breastfeeding family. *New Beginnings* 1998; 15(4):100.

117. Davidoff, M. J. et al. Changes in the gestational age distribution among U.S. singleton births: Impact on rages of late preterm birth, 1992–2002. Semin Perinatol 2006; 30(1):8–15.

118. Wight, N. E. Breastfeeding the borderline (near-term) preterm infant. *Pediatr Ann* 2003; 32:329–36.

119. Engle, W. A., Tomashed, K. M., Wallman, C. and the Committee on Fetus and Newborn. "Late-Preterm" Infants: A Population at Risk. AAP Clinical Report. *Pediatrics* 2007; (120):6:1390–1401.

120. American Academy of Pediatrics Work Group on Breastfeeding. Breastfeeding and the use of human milk. *Pediatrics* 1997; 100:1035–39.

121. Shulman, R. J. et al. Early feeding, feeding tolerance, and lactase activity in preterm infants. *J Pediatr* 1998; 133:645–49.

122. Spatz, D. L. State of the science: Use of human milk and breast-feeding for vulnerable infants. *J Perinat Neonatal Nurs* 2006; 20:51–55.

123. Callen, J. and Pinelli, J. A review of the literature examining the benefits and challenges, incidence and duration, and barriers to breastfeeding in preterm infants *Adv Neonatal Care* 2005; 5:72–88.

124. Rodriguez, N. A., Miracle, D. J. and Meier, P. P. Sharing the science on human milk feedings with mothers of very-low-birth-weight infants. *JOCNN* 2005; 34:109–119.

125. Schanler, R. J., Shulman, R. J. and Lau, C. Feeding strategies for premature infants: Beneficial outcomes of feeding fortified human milk versus preterm formula. *Pediatrics* 1999; 103:1150–57.

127. Spatz, D. L. Ten steps for promoting and protecting breastfeeding for vulnerable infants. *J Perinat Neonatal Nurs* 2004; 18:385–96.

128. Sisk, P. M. et al. Lactation counseling for mothers of very low birth weight infants: Effect on maternal anxiety and infant intake of human milk. *Pediatrics* 2006; 117:67–75.

129. Lawrence, R. M. and Lawrence, R. A. Breast milk and infection. *Clin Perinatol* 2004; 31:501–28.

130. Lawrence, R. M. and Lawrence, R. A. Given the benefits of breastfeeding, what contraindications exist? *Pediatr Clin North Am* 2001; 48:235–51.

131. Newell, M. L. Current issues in the prevention of mother-to-child transmission of HIV-1 infection. *Trans R Soc Trop Med Hyg* 2006; 100:1–5.

132. Dorosko, S. M. Vitamin A, mastitis, and mother-to-child transmission of HIV-1 through breast-feeding: Current information and gaps in knowledge. *Nutr Rev* 2005; 63:332–46.

133. Jackson, D. J. et al. HIV and infant feeding: Issues in developed and developing countries. *J Obstet Gynecol Neonatal Nurs* 2003; 32:117–27.

134. American Academy of Pediatrics, Section on Breastfeeding. Breastfeeding and the use of human milk. *Pediatrics* 2005; 115:496–506.

135. Effect of breastfeeding on infant and child mortality due to infectious diseases in less developed countries: A pooled analysis. WHO Collaborative Study Team on the Role of Breastfeeding on the Prevention of Infant Mortality. *Lancet* 2000; 355:451–55.

136. Coutsoudis, A. Breastfeeding and the HIV-positive mother: the debate continues. *Early Hum Dev* 2005; 81:87–93.

137. World Health Organization. HIV and Infant Feeding. *A review of HIV transmission through breastfeeding:* Geneva, Switzerland: World Health Organization, 1998.

138. World Health Organization: HIV and infant feeding. Available at www.who.int/ child.htm, assessed 6/09.

139. International Lactation Consultant Association: Position Paper on HIV and Infant Feeding, 2006. Available at www.ilca. org, accessed 5/09.

140. WHO Technical Consultation on Behalf of the UNFPA/UNICEF/WHO/ UNAIDS Inter-Agency Task Team on Mother-to-Child Transmission of HIV. *New data on the prevention of mother-to-child transmission of HIV and their policy implications: Conclusions and recommendations.* Oct. 13, 2000. Geneva: World Health Organization.

141. Hartmann, S. U., Berlin, C. M., and Howett, M. K. Alternative modified infant-feeding practices to prevent postnatal transmission of human immunodeficiency virus type 1 through breast milk: Past, present, and future. *J Hum Lact* 2006; 22:75–88.

142. Academy of Breastfeeding Medicine. Protocol for the Storage of Human Milk. Available at www.bfmed.org, accessed 11/06.

143. Jones, F. and Tully, M. R. Best Practice for Expressing, Storing and Handling Human Milk in Hospitals, Homes and Child Care Settings, 2nd ed., 2006.

144. Mohrbacher, N. and Kendall-Tackett, K. *Breastfeeding Made Simple: Seven Natural Laws for Nursing Mothers.* Oakland, CA: New Harbinger Publications, Inc., 2005.

145. Lang, S. *Breastfeeding Special Care Babies*. China: Elsevier Health, 2002.

146. Jones, F. History of North American donor milk banking: One hundred years of progress. *J Hum Lact* 2003; 19:313–18.

147. Flatau, G. Milk Banks Keep the Milk Flowing–HMBANA Responds to Increasing Demand. HMBANA Matters June 2008 (5).

148. Paxson, C. I. Survival of human milk leukocytes. *J Pediatr* 1979; 94:61–64.

149. Meier, P. P et al. The Rush Mother's Milk Club: Breastfeeding interventions for mothers with very low birth weight infants. *JOGNN* 2004; 33:164–74.

Chapter 8

1. National Center for Health Statistics. *NCHS growth curves for children 0–19 years*. U.S. Vital and Health Statistics, Health Resources Administration. Washington, D.C.: U.S. Government Printing Office, 2000.

2. Martin, J. A., Hamilton, B. E., Sutton, P. D., Ventura, S. J., et al. Births: Final data for 2006. *Natl Vital Stat Rep* 2009; 57(7).

3. Mathews, T. J. and MacDorman, M. F. Infant mortality statistics from the 2005 period linked birth/infant death data set. *Natl Vital Stat Rep* 2008; 57(2). Available at http://www.cdc.gov/nchs/data/nvsr/nvsr57/nvsr57_02.pdf, accessed 6/09.

4. United States Department of Health and Human Services, Centers for Medicare and Medicaid Services. Retrieved from www.cms.hhs.gov/schip, accessed 6/09.

5. United States Department of Health and Human Services, National Center for Chronic Disease Prevention and Health Promotion. Retrieved from www.cdc.gov/nccdphp/dnpa/pednss.htm, accessed 6/09.

6. Hagan, J. F., Shaw, J. S., and Duncan, P. M., eds. *Bright Futures: Guidelines for Health Supervision of Infants, Children and Adolescents*, 3rd ed. Elk Grove Village, IL: American Academy of Pediatrics, 2008.

7. Engle, W. A. and Boyle, D. W. Newborn care: Perinatal care: Delivery room management and transitional care. In Osborn et al, eds. *Pediatrics*. St. Louis, MO: Mosby, 2005: 1250–38.

8. Kleinman, R. E., ed. *Pediatric nutrition handbook*, 6th ed. Chicago, IL: American Academy of Pediatrics, 2009: 3–28.

9. Kail, R. V. and Cavanaugh, J. C. *Human development: A lifespan view*, 2nd ed. Belmont, CA: Wadsworth/Thomson Learning, 2000: 83–121.

10. Nardella, M. T. and Owens-Kuehner, A. Feeding and eating. In Lucas, B. L. *Children with special health care needs: Nutrition care handbook*. Chicago, IL: American Dietetic Association, 2004: 59–87.

11. Santrock, J. W. *Life-span development*, 7th ed. New York: McGraw Hill, 1999: 124–9.

12. Shelton, T. L., Dobbins, T. R., and Neal, J. M. Principles of child development and developmental assessment. In Osborn et al, eds. *Pediatrics*. St. Louis, MO: Mosby, 2005: 41–57.

13. Institute of Medicine Food and Nutrition Board. Dietary Reference Intakes for energy, carbohydrate, fiber, fat, fatty acids, cholesterol, protein, and amino acids. Washington, D.C.: National Academy Press, 2002.

14. Regalado, M. and Halfon, N. Developmental and behavioral surveillance and promotion of parenting skills. In Osborn et al., eds. *Pediatrics*. St. Louis, MO: Mosby, 2005: 224–33.

15. Batshaw, M. L. Genetics and Developmental Disabilities. In Batshaw, M. L., Pellegrino, L., and Roizen, N. J., eds. *Children with disabilities*, 6th ed. Baltimore, MD: Paul H. Brookes, 2007; 3–21.

16. Hobbie, C., Baker, S., and Bayerl, C. Parental understanding of basic infant nutrition: Misinformed feeding choices. *J Pediatr Health Care* 2000; 14:26–31.

17. Lawrence, R. A. and Lawrence, R. M. *Breastfeeding: A guide for the medical profession*, 6th ed. Philadelphia, PA: Mosby, 2005:105–170.

18. Hendricks, K. M. and Duggan, C. *Manual of Pediatric Nutrition*, 4th ed. Hamilton, ON: BC Decker, 2005:450–468.

19. Trumbo, P. et al. Dietary reference intakes: Vitamin A, vitamin K, arsenic, boron, chromium, copper, iodine, iron, manganese, molybdenum, nickel, silicon, vanadium and zinc. *J Am Diet Assoc* 2001; 101:294–301.

20. Wagner, C. L., Greer, F. R., and the Section on Breastfeeding and Committee on Nutrition of the American Academy of Pediatrics. Prevention of rickets and vitamin D deficiency in infants, children and adolescents. *Pediatrics* 2008; 122:1142–1152.

21. European Commission Health and Consumer Protection Directorate-General. Report of the Scientific Committee on Food on the Revision of Essential Requirements of Infant Formulae and Follow-Up Formulae. Adopted April 4, 2003; Brussels, Belgium.

22. Wright, C. M. and Waterston, A. J. R. Relationships between paediatricians and infant formula milk companies. *Arch Dis Child* 2006: 91;383–85.

23. Clandinin, M. T. et al. Growth and development of preterm infants fed infant formulas containing docosahexaenoic acid and arachidonic acid. *J Pediatr* 2005; 146:461–68.

24. Birch, E. E. et al. A randomized controlled trial of long-chain polyunsaturated fatty acid supplementation of formula in term infants after weaning at 6 wks of age. *Am J Clin Nutr* 2002; 75:570–80.

25. Joanna Briggs Institute. Early childhood pacifier use in relation to breastfeeding, SIDS, infection and dental malocclusion. *Nurs Stand* 2006; 20:52–55.

26. Carver, J. D. Advances in nutritional modifications of infant formulas. *Am J Clin Nutr* 2003; 77(suppl):1550S–4S.

27. Yates, A. A., Schlicker, S. A., and Suitor, C. W. Dietary reference intakes: The new basis for recommendations for calcium and related nutrients, B vitamins, and choline. *J Am Diet Assoc* 1998; 98:699–706.

28. Hampl, J. S., Betts, N. M., and Benes, B. A. The "age + 5" rule: Comparisons of dietary fiber intake among 1–4-year-old children. *J Am Diet Assoc* 1998; 98:1418–23.

29. Infant foods as listed in www.ars.usda.gov/ food search, accessed 6/09.

30. WHO Multicentre Growth Reference Study Group. WHO Child Growth Standards: Length/height-for-age, weight-for-age, weight-for-length, weight-for-height and body mass index-for-age: Methods and development. Geneva: World Health Organization, 2006. Covers infants up to age 5.

31. Isaacs, J. S. and Zand, D. J. Single-Gene Autosomal Recessive Disorders and Prader-Willi Syndrome: An Update for Rood and Nutrition Professionals. J *Am Diet Assoc* 2007; 107:466-478.

32. American Academy of Pediatrics Work Group on Breastfeeding. Breastfeeding and the use of human milk. *Pediatrics* 2005; 115: 496–506.

33. Galson, S. K. Mothers and Children Benefit from Breastfeeding. *J Am Diet Assoc* 2008; 108:1106.

34. Kleinman, R. E., ed. *Pediatric nutrition handbook*, 6th ed. Chicago, IL: American Academy of Pediatrics, 2009: 29–60.

35. Ross products handbook. Available at http://rpdcon40.ross.com/pn/PediatricProducts.NSF/web_Ross.com_XML_PediatricNutrition, accessed 6/09.

36. BPM Products infant formula. Available at http://www.nutritiondata.com/facts/baby-foods/7179/2, accessed 6/09.

37. Nestle Good Start infant formulas. Available at www.verybestbaby.com/GoodStart/Default.asp, accessed 6/09.

38. Amorde-Spalding, K. and Nieman, L., eds. *Pediatric manual of clinical dietetics*. American Dietetic Association, 2008: Appendix 1.

39. Ziegler, E. E. et al. Cow's milk and intestinal blood loss in late infancy. *J Pediatrics* 1999; 135:720–26.

40. Kleinman, R. E., ed. *Pediatric nutrition handbook*, 6th ed. Chicago, IL: American Academy of Pediatrics, 2009: 76.

41. Bhatia, J., Greer, F., and the Committee on Nutrition of the American Academy of Pediatrics. Clinical Report. Use of Soy Protein-Based Formulas in Infant Feeding. *Pediatrics* 2008; 121:1062–1068.

42. American Academy of Pediatrics. The use and misuse of fruit juice in pediatrics. *Pediatrics* 2001; 107:1210–13.

43. Isaacs, J. S. Fluid and bowel problems. In Lucas, B. L., ed. *Children with special health care needs: Nutrition care handbook.* Chicago, IL: American Dietetic Association, 2004: 103–18.

44. Tervo, R. Parent's reports predict their child's developmental problems. *Clin Pediatr* 2005; 44:601–11.

45. Kellogg, N. D. and the Committee on Child Abuse and Neglect. Evaluation of Suspected Child Physical Abuse. *Pediatrics* 2007; 119:1232–1241.

46. Birch, L. L. et al. Infants' consumption of a new food enhances acceptance of similar foods. *Appetite* 1998; 30:283–95.

47. Sandritter, T. Gastroesophageal reflux disease in infants and children. *J Pediatr Health Care* 2003; 17(4):198–205.

48. Kleinman, R. E., ed. *Pediatric nutrition handbook,* 6th ed. Chicago, IL: American Academy of Pediatrics, 2009: 201–224.

49. Mennella, J. A. and Beauchamg, G. K. Maternal diet alters the sensory qualities of human milk and the nurslings' behavior. *Pediatrics* 1991; 88:737–44.

50. Davis, C. M. Self-selection of diet by newly weaned infants: An experimental study. *Am J Dis Child* 1928; 36:651–79.

51. Story, M. and Brown, J. E. Do young children instinctively know what to eat? *N Engl J Med* 1998; 103–6.

52. American Academy of Pediatrics. Policy Statement (RE8132). Infant exercise programs. *Pediatrics* 1988; 82:800. Reaffirmed 11/94.

53. Mangels, A. R. and Messina, V. Considerations in planning vegan diets: Infants. *J Am Diet Assoc* 2001; 101:670–77.

54. Khoshoo, V. and Reifen, R. Use of energy-dense formula for treating infants with nonorganic failure to thrive. *Eur J Clin Nutr* (England) 2002; 56(9):921–24.

55. Goldsmith, E. Nonorganic failure to thrive. *MCN Am J Matern Child Nurs* 2001; 26(4):221.

56. Needell, B. and Barth, R. P. Infants entering foster care compared to other infants using birth status. *Child Abuse Negl* 1998; 22:1179–87.

57. Metcalf, T. et al. Simethicone in the treatment of infant colic: A randomized, placebo-controlled multicenter trial. *Pediatrics* 1994; 94:29–34.

58. Recommendations to prevent and control iron deficiency in the United States. *MNWR* April 3, 1998; 47.

59. Kleinman, R. E., ed. *Pediatric nutrition handbook,* 6th ed. Chicago, IL: American Academy of Pediatrics, 2009: 403–422.

60. Subcommittee on Management of Acute Otitis. Media Diagnosis and Management of Acute Otitis Media. *Pediatrics* 2004; 113: 1451–1465.

61. Wood, R. A. Prospects for the prevention of allergy: A losing battle or a battle still worth fighting? *Arch Pediatr Adolesc Med* 2006; 160:552–4, 502–7.

62. Greer, F. R., Sicherer, S. H., Burks, A. W., and the Committee on Nutrition and Section on Allergy and Immunology. Clinical Report: Effects of early nutritional interventions on the development of atopic disease in infants and children: The role of maternal dietary restriction, breastfeeding, timing of introduction of complementary foods, and hydrolyzed formulas. *Pediatrics* 2008; 121:183–191.

63. Kleinman, R. E., ed. *Pediatric nutrition handbook,* 6th ed. Chicago, IL: American Academy of Pediatrics, 2009: 783–799.

Chapter 9

1. Mathews, T. J. and MacDorman, M. F. Infant mortality statistics from the 2005 period linked birth/infant death data set. *Natl Vital Stat Rep* 2008; 57(2); available at http://www.cdc.gov/nchs/data/nvsr/nvsr57/nvsr57_02.pdf, accessed 06/09.

2. Swamy GK, Ostbye T, and Skjaerven R. Association of preterm birth with long-term survival, reproduction, and next-generation preterm birth. *JAMA* 2008; 299:1429–1436.

3. Cuevas, K. D. et al. The cost of prematurity: Hospital charges at birth and frequency of rehospitalizations and acute care visits over the first year of life: A comparison by gestational age and birth weight. *Am J Nurs* 2005; 105:56–64.

4. Kleinman, R. E. *Pediatric Nutrition Handbook.* Chicago Il: American Academy of Pediatrics 2009; 79–112, 319–324.

5. Committee on Nutrition of the Preterm Infant. European Society of Paediatic Gastroenterology and Nutrition. In *Nutrition and feeding of the preterm infant.* Oxford, UK: Blackwell Scientific Publications, 1987.

6. Rais-Bahrami, K. and Short, B. L. Premature and small-for-dates infants. In Batshaw, M. L., Pellegrino, L., and Roizen, N. J., eds. *Children with disabilities,* 6th ed. Baltimore, MD: Paul H. Brookes, 2007: 107–124.

7. American College of Medical Genetics. *Newborn screening: Toward a uniform screening panel and system.* 2005 Final Report. Available at http://mchb.hrsa.gov/screening, accessed 06/09.

8. Henriksen, C., Haugholt, K., Lindgren, M., Aurvåg, A. K., Rønnestad, A., Grønn, M., et al. Improved Cognitive Development Among Preterm Infants Attributable to Early Supplementation of Human Milk With Docosahexaenoic Acid and Arachidonic Acid. *Pediatrics* 2008; 121:1137–1145.

9. Trumbo, P. et al. Dietary reference intakes: Vitamin A, vitamin K, arsenic, boron, chromium, copper, iodine, iron, manganese, molybdenum, nickel, silicon, vanadium, and zinc. *J Am Diet Assoc* 2001; 101:294–301.

10. Darlow, B. A. and Graham, P. J. Vitamin A supplementation to prevent mortality and short and long-term morbidity in very low birthweight infants. *Cochrane Database of Systematic Reviews* 2007, Issue 4. Art. No. CD000501. DOI 10.1002/14651858. CD000501.pub2.

11. U.S. Vital and Health Statistics, Health Resources Administration. *National Center for Health Statistics: NCHS growth curves for children 0–19 years.* Washington, D.C.: U.S. Government Printing Office, 2000.

12. *Growth references: Third trimester to adulthood.* 2nd ed. Clinton, SC: Greenwood Genetic Center, 1998.

13. The Infant Health and Development Program. Enhancing the outcomes of low-birth-weight, premature infants. *JAMA* 1990; 263(22):3035–42.

14. Fenton, T. R. A new growth chart for preterm babies: Babson and Benda's chart updated with recent data and a new format. Available at http://www.biomedcentral.com/1471–2431/3/13, accessed 06/09.

15. Huysman, W. A. et al. Growth and body composition in preterm infants with bronchopulmonary dysplasia. *Arch Dis Child Fetal Neonatal Ed* 2003; 88:F46–F51.

16. Hintz, S. R. et al. Changes in Neurodevelopmental Outcomes at 18 to 22 Months Corrected Age Among Infants of Less Than 25 Weeks' Gestational Age Born in 1993–1999. *Obste Gynecol Surv* 2005; 60:714–15.

17. Lucas, A., Fewtrell, M. S., and Cole, T. J. Fetal origins of adult disease—the hypothesis revisited. *Br Med J* 1999; 319(7204):245–249.

18. Chyi, L. J., Lee, H. C., Hintz, S. R., Gould, J. B., and Sutcliffe, T. L. School outcomes of later preterm infants: Special needs and challenges for infants born at 32 to 36 weeks gestation *J Pediatric.* 2008; 153:25–31.

19. Franz, A. R., Pohlandt, F., Bode, H., Mihatsch, W. A., Sander, S., Kron, M., and Steinmacher, J. Intrauterine, Early Neonatal, and Postdischarge Growth and Neurodevelopmental Outcome at 5.4 Years in Extremely Preterm Infants After Intensive Neonatal Nutritional Support. *Pediatrics* 2009; 123:e101–e109.

20. Lindström, K., Lindblad, F., and Hjern, A. Psychiatric Morbidity in Adolescents and Young Adults Born

Preterm: A Swedish National Cohort Study. *Pediatrics* 2009; 123:e47–e53

21. Hans, D. M., Pylipow, M., Long, J. D., Thureen, P. J., and Georgieff, M. K. Nutritional Practices in the Neonatal Intensive Care Unit: Analysis of a 2006 Neonatal Nutrition Survey. *Pediatrics* 2009; 123:51–57

22. Weaver, L. Rapid growth in infancy: Balancing the interests of the child. *J Pediatr Gastroenterol Nutr* 2006; 43:428–32.

23. Eicher, P. S. Feeding. In Batshaw, M. L., Pellegrino, L., Roizen, N. J., eds. *Children with disabilities*, 6th ed. Baltimore, MD: Paul H. Brookes, 2007: 479–98.

24. Persson, B. et al. Screening for infants with developmental deficits and/or autism. *J Pediatr Nurs* 2006; 21:313–24.

25. Martin, J. A., Hamilton, B. E., Sutton, P. D., Ventura, S. J., et al. Births: Final data for 2006. *National vital statistics reports* Vol. 57, No 7. Hyattsville, MD: National Center for Health Statistics, 2009.

26. Wood, N. S. et al. Neurological and developmental disability after extremely preterm birth. *N Eng J Med* 2000; 343:378–84.

27. Strauss, R. Adult functional outcome of those born small for gestational age: Twenty-six–year follow-up of the 1970 British birth cohort. *JAMA* 2000; 283:625–32.

28. Matlow, A. et al. Microbial contamination of enteral feed administration sets in a pediatric institution. *Amer J Infect Contr* 2003; 51:49–53.

29. American Dietetic Association. Infant feedings: guidelines for preparation of formula and breast milk in health care facilities, 2003.

30. National Academy of Sciences. *Infant Formula: Evaluating the Safety of New Ingredients*. Washington, D.C.: National Academies Press, 2004.

31. American Academy on Pediatrics. Breastfeeding and the use of human milk. *Pediatrics* 2005; 115:496–506.

32. Lawrence, R. A. and Lawrence, R. M. *Breastfeeding: A guide for the medical profession*, 6th ed. Philadelphia, PA: Mosby, 2005: 479–513.

33. Quigley, M., Henderson, G., Anthony, M. Y., and McGuire, W. Formula milk versus donor breast milk for feeding preterm or low birth weight infants. *Cochrane Database of Systematic Reviews* 2007, Issue 4, Art. No. CD002971. DOI 10.1002/14651858.CD002971.pub2.

34. Mead-Johnson Nutritionals product information website; Available at http://www.meadjohnson.com/professional/

products/enfamillipilwithiron.htm, acccessed 06/09.

35. Gray, J. E. et al. Normal birth weight intensive care unit survivors: Outcome assessment. *Pediatrics* 1996; 97:832–38.

36. Lee, P. J. and Cook, P. Frequency of metabolic disorders: More than one needle in the haystack. *Arch Dis Child* 2006; 91:879–80.

37. Centers for Disease Control and Prevention. Spina bifida and anencephaly before and after folic acid mandate—United States, 1995–1996 and 1999–2000. *MMWR* 2004; 53:362–65.

38. Lee, J. et al. Height and weight achievement in cleft lip and palate. *Arch Dis Child* 1997; 76:70–72.

39. Isaacs, J. S. and Zand, D. J. Single-Gene Autosomal Recessive Disorders and Prader-Willi Syndrome: An Update for Rood and Nutrition Professionals. *J Am Diet Assoc* 2007; 107:466–478.

40. Bassett, A. S. et al. Clinical features of 78 adults with 22q11 deletion syndrome. *Am J Med Genet* 2005; 138A:307–13.

41. General information about disabilities. National Information Center for Children and Youth with Disabilities; available at www.nichcy.org/general.htm, accessed 06/09.

Chapter 10

1. Center on Hunger, Poverty and Nutrition Policy. *Statement on the link between nutrition and cognitive development in children*. Tufts University: School of Nutrition Science and Policy, 1998.

2. The Annie E. Casey Foundation. *2008 kids count data book*. Baltimore, MD: The Annie E. Casey Foundation, 2008.

3. U.S. Department of Health and Human Services, Health Resources and Services Administration, Maternal and Child Health Bureau. *Child Health USA 2007*. Rockville, MD: U.S. Department of Health and Human Services, 2008.

4. U.S. Department of Health and Human Services. *Healthy People 2010* (conference edition, in two volumes). Washington, D.C.: U.S. Department of Health and Human Services, January 2000; Centers for Disease Control. Retrieved from http://wonder.cdc.gov/data2010/focus.htm, accessed 7/09.

5. Kliegman, R. M., Behrman, R. E., Jenson, H. B., and Stanton, B. M. D, eds. *Nelson's Textbook of Pediatrics*, 18th ed. Philadelphia, PA: WB Saunders Co, 2007.

6. Centers for Disease Control and Prevention, National Center for Health Statistics. *CDC growth charts: United States*. Available at www.cdc.gov/ nchs/about/major/nhanes/growthcharts/charts.htm, accessed 05/00.

7. Kuczmarski, R. J. et al. 2000 CDC growth charts for the United States: methods and development. *Vital Health Stat* 11. 2002; 246:1–190.

8. Barlow, S. E. and the Expert Committee. Expert Committee Recommendations regarding the prevention, assessment, and treatment of child and adolescent overweight and obesity: Summary report. *Pediatrics* 2007; 120(S4):S164–S192.

9. WHO Multicentre Growth Reference Study Group. *WHO Child Growth Standards: Length/height-for-age, weight-for-age, weight-for-length, weight-for-height and body mass index-for-age: Methods and development*. Geneva: World Health Organization, 2006. Available at http://www.who.int/childgrowth/standards/technical_report/en/index.html, accessed 06/20/2009.

10. Story, M., Holt, K., and Sofka, D., eds. *Bright Futures in Practice: Nutrition*. Arlington, VA: National Center for Education in Maternal and Child Health, 2000.

11. USDA, MyPyramid for Kids. Available at www.MyPyramid.gov.

12. Satter, E. The feeding relationship: Problems and interventions. *J Pediatr* 1990; 117:181–89.

13. Birch, L. L. Children's food acceptance patterns. *Nutr Today* 1996; 31:234–40.

14. Birch, L. L et al. The variability of young children's energy intake. *N Eng J Med* 1991; 324:232–35.

15. Story, M. and Brown, J. E. Do young children instinctively know what to eat? The studies of Clara Davis revisited. *New Eng J Med* 1987; 316:103–6.

16. Van den Bree, M. B. M. et al. Genetic and environmental influences on eating patterns of twins ages >50 years. *Am J Clin Nutr* 1999; 70:456–65.

17. Chess, S. and Thomas, A. Dynamics of individual behavioral development. In Levine, M. D., Carey, W. B., and Crocker, A. C., eds. *Developmental-behavioral pediatrics*. Philadelphia, PA: WB Saunders Co., 1992: 84–94.

18. Birch L. L. and Fisher, J. O. Development of eating behaviors among children and adolescents. *Pediatrics* 1998; 101: 539–49(S).

19. Birch, L. L. and Fisher, J. A. Appetite and eating behavior in children. *Pediatr Clinic N Amer* 1995; 42:931–53.

20. Fisher, J. O. et al. Parental influences on wwyoung girls' fruit and vegetable, micronutrient, and fat intakes. *J Amer Diet Assoc* 2002; 102:58–64.

21. Fisher, J. O. and Birch, L. L. Restricting access to palatable foods affects children's behavioral response, food selection, and intake. *Am J Clin Nutr* 1999; 69: 1264–72.

22. Connor, S. M. Food-related advertising on preschool television: Building brand recognition in young viewers. *Pediatrics* 2006; 118:1478–85.

23. Birch, L. L. and Fisher, J. O. Food intake regulation in children, fat and sugar substitutes and intake. *Ann NY Acad Sci* 1997; 819:194–220.

24. Rolls, B. J., Engell, D., and Birch, L. L. Serving portion size influences 5-year-old but not 3-year-old children's food intake. *J Amer Diet Assoc* 2000; 100:232–34.

25. Fisher, J. O., Rolls, B. J., and Birch, L. L. Children's bite size and intake of an entrée are greater with large portions than with age-appropriate or self-selected portions. *Am J Clin Nutr* 2003; 77:1164–70.

26. Fisher, J. O. and Birch, L. L. Parents' restrictive feeding practices are associated with young girls' negative self-evaluation of eating. *J Amer Diet Assoc* 2000; 100:1341–46.

27. Davison, K. K. and Birch, L. L. Weight status, parent reaction, and self-concept in five-year-old girls. *Pediatrics* 2001; 107:46–53.

28. Abramovitz, B. A. and Birch, L. L. Five-year-old girls' ideas about dieting are predicted by their mothers' dieting. *J Amer Diet Assoc* 2000; 100:1157–63.

29. Satter, E. Feeding dynamics: Helping children to eat well. *J Pediatr Health Care* 1995; 9:178–84.

30. Davis, M. M. et al. Recommendations for prevention of childhood obesity. *Pediatrics* 2007; 120(S4):S229–53.

31. Institute of Medicine, Food and Nutrition Board. Dietary Reference Intakes for energy, carbohydrate, fiber, fat, protein, and amino acids. Washington, D.C.: National Academy Press, 2002/2005.

32. Institute of Medicine, Food and Nutrition Board. Dietary reference intakes for calcium, phosphorus, magnesium, vitamin D, and fluoride. Washington, D.C.: National Academy Press, 1997.

33. Institute of Medicine, Food and Nutrition Board. Dietary reference intakes for vitamin A, vitamin K, arsenic, boron, chromium, copper, iodine, iron, manganese, molybdenum, nickel, silicon, vanadium, and zinc. Washington, D.C.: National Academy Press, 2001.

34. Centers for Disease Control and Prevention. Recommendations to prevent and control iron deficiency in the United States. *Morbidity and Mortality Weekly Report* April 3, 1998; 47(RR-03):1–29.

35. Centers for Disease Control and Prevention. Iron deficiency—United States, 1999–2000. *Morbidity and Mortality Weekly Report* October 11, 2002; 51(40).

36. American Academy of Pediatric Dentistry. Policy on early childhood caries (ECC): Classifications, consequences, and preventive strategies, 2008; available at http://www.aapd.org/media/policies.asp, accessed 7/1/2009.

37. Casamassimo, P. *Bright futures in practice: oral health.* Arlington, VA: National Center for Education in Maternal and Child Health, 1996.

38. American Academy of Pediatrics, *Committee on Nutrition. Pediatric nutrition handbook,* 6th ed. Kleinman, R. E., ed. Elk Grove Village, IL: American Academy of Pediatrics, 2009.

39. McClung, H. J., Boyne, L., and Heitlinger, L. Constipation and dietary fiber intake in children. *Pediatrics* 1995; 96:999–1001(S).

40. Meyer, P. A. et al. Surveillance for elevated blood lead levels among children—United States, 1997–2001. In Surveillance Summaries, September 12, 2003. *MMWR* 2003; 52(SS-10):1–21.

41. American Academy of Pediatrics, Committee on Environmental Health. Lead exposure in children: Prevention, detection, and management. *Pediatrics* 2005; 116:1036–46.

42. Centers for Disease Control. National Center for Environmental Health. Childhood Lead Poisoning. Available at www.cdc.gov/nceh/lead/tips/population.htm, accessed 7/5/2009.

43. Centers for Disease Control and Prevention. Managing elevated blood lead levels among young children: Recommendations from the advisory committee in childhood lead poisoning prevention. Atlanta, GA: Centers for Disease Control and Prevention, 2002. Available at www.cdc.gov/nceh/lead/CaseManagement/caseManage_main.thm, accessed 7/3/2009.

44. Ballew, C., et al. Blood lead concentration and children's anthropometric dimensions in the Third National Health and Nutrition Examination Survey (NHANES III), 1988–1994. *J Pediatr* 1999; 134:623–30.

45. Binns, H. J. et al. Interpreting and managing blood lead levels of less than 10 mcg/dL in children and reducing childhood exposure to lead: Recommendations of the Centers for Disease Control and Prevention Advisory Committee on Childhood Lead Poisoning Prevention. *Pediatrics* 2007; 120:e1285–98.

46. Centers for Disease Control and Prevention. Screening young children for lead poisoning: Guidance for state and local public health officials. Atlanta, GA: Centers for Disease Control and Prevention, 1997. Available at www.cdc.gov/nceh/programs/lead/guide/1997/guide97.htm.

47. American Academy of Pediatrics. Screening for elevated blood lead levels. *Pediatrics* 1998; 101:1072–78.

48. Food Insecurity in Households with Children. Economic Research Service. Economic Info Bulletin No. (EIB-56), September 2009. Available at www.ers.usda.gov/Publications/EIB56/, accessed 9/30/09.

49. Kleinman, R. E. et al. Hunger in children in the United States: Potential behavioral and emotional correlates. *Pediatrics* [serial online] 1998; 101:E3. Available at www.pediatrics.org/cgi/content/full/101/1/e3, accessed 7/09.

50. Centers for Disease Control and Prevention, National Center for Infectious Diseases, U.S. Department of Health and Human Services. Foodborne Illness. Available at www.cdc.gov/ncidod/diseases/food/index.htm, accessed 7/09.

51. Partnership for Food Safety Education: FightBAC™. Available at www.fightbac.org.

52. U.S. Department of Health and Human Services and U.S. Department of Agriculture. Dietary Guidelines for Americans, 2005, 6th ed. Washington, D.C.: U.S. Government Printing Office, January 2005.

53. Kavey, R. W. et al. American Heart Association guidelines for primary prevention of atherosclerotic cardiovascular disease beginning in childhood. *Circulation* 2003; 107:1562–66.

54. Ogden, C. L. et al. Prevalence of high body mass index in U.S. children and adolescents, 2007-2008. *JAMA* 2010; 303:242–249.

55. Krebs, N. F. et al. Assessment of child and adolescent overweight and obesity. *Pediatrics* 2007; 120(S4):S193–S228.

56. Spear, B. A. et al. Recommendations for treatment of child and adolescent overweight and obesity. *Pediatrics* 2007; 120(S4):S254–S288.

57. Dietz, W. H. and Gortmaker, S. L. Preventing obesity in children and adolescents. *Annu Rev Public Health* 2001; 22:337–53.

58. Whitaker, R. C. et al. Early adiposity rebound and the risk of adult obesity. *Pediatrics* [serial online] 1998; 101:E5. Available at www.pediatrics.org/cgi/content/full/101/3/e5, accessed 6/09.

59. Gidding, S. S. et al. Implementing American Heart Association pediatric and adult nutrition guidelines: A scientific statement from the American Heart Association Nutrition Committee of the Council on Nutrition, Physical Activity and Metabolism, Council on Cardiovascular Disease in the Young, Council on Arteriosclerosis, Thrombosis and Vascular Biology, Council on Cardiovascular Nursing, Council on Epidemiology and Prevention, and Council for High Blood Pressure Research. *Circulation* 2009; 119:1161–75.

60. Gidding, S. S. et al. Dietary recommendations for children and adolescents: A guide for practitioners. Consensus statement from the American Heart Association. *Circulation* 2005; 112:2061–75.

61. American Heart Association, Gidding, S. S., et al. Dietary recommendations for children and adolscents: A guide for practitioners. *Pediatrics* 2006; 117:544–59.

62. Lichtenstein, A. H. et al. Diet and lifestyle recommendations revision 2006: A scientific statement from the American Heart Association Nutrition Committee. *Circulation* 2006; 114:82–96.

63. Rask-Nissila, L. et al. Prospective, randomized, infancy-onset trial of the effects of a low-saturated-fat, low-cholesterol diet on serum lipids and lipoproteins before school age: The special turku coronary risk factor intervention project (STRIP). *Circulation* 2000; 102:1477–83.

64. Daniels, S. R. et al. Lipid screening and cardiovascular health in childhood. *Pediatrics* 2008; 122:198–208.

65. Vital & Health Statistics—Series 11: Data from the National Health Survey. 1999; 244:1–14.

66. Yu, S. M., Kogan, M. D., and Gergen, P. Vitamin-mineral supplement use among preschool children in the United States. *Pediatrics* [serial online] 1997; 100:E4. Available at www.pediatrics.org/ cgi/ content/full/100/5/e4, accessed 7/09.

67. Lohse, B., Stotts, J. L., and Priebe, J. R. Survey of herbal use by Kansas and Wisconsin WIC participants reveals moderate, appropriate use and identifies herbal education needs. *J Amer Diet Assoc* 2006; 106:227–37.

68. Buck, M. L. and Michael, R. S. Talking with families about herbal products. *J Pediatr* 2000; 136:673–78.

69. American Dietetic Association. Position of the American Dietetic Association: Dietary guidance for healthy children aged 2 to 11 years. *J Amer Diet Assoc* 2008; 108:1038–47.

70. A summary of conference recommendations on dietary fiber in childhood: Conference on Dietary Fiber in Childhood, New York, May 24, 1994. *Pediatrics* 1995; 96:1023–28.

71. Hampl, J. S., Betts, N. M., and Benes, B. A. The "age + 5" rule: Comparisons of dietary fiber intake among 4- to 10-year-old children. *J Amer Diet Assoc* 1998; 98:1418–23.

72. Wagner, C. L., Greer, F. R., and the Section on Breastfeeding and Committee on Nutrition. Prevention of rickets and vitamin D deficiency in infants, children, and adolescents. *Pediatrics* 2008; 122:1142–52.

73. Greer, F. R., Krebs, N. F., and Committee on Nutrition. Optimizing bone health and calcium intakes of infants, children, and adolescents. *Pediatrics* 2006; 117:578–85.

74. Wang, Y. C., Bleich, S. N., and Gortmaker, S. L. Increasing caloric contribution from sugar-sweetened beverages and 100% fruit juices among U.S. children and adolescents, 1988–2004. *Pediatrics* 2008; 121:e1604–14.

75. Fox, M. K. et al. Feeding infants and toddlers study: What foods are infants and toddlers eating? *J Am Diet Assoc* 2004; 104:S22–30.

76. Harnack, L., Stang, J., and Story, M. Soft drink consumption among U.S. children and adolescents: Nutritional consequences. *J Amer Diet Assoc* 1999; 99:436–41.

77. American Academy of Pediatrics. Committee on Nutrition. The use and misuse of fruit juices in pediatrics. *Pediatrics* 2001; 107:1210–13.

78. U.S. Department of Agriculture, Agricultural Research Service, 2008. Nutrient Intakes from Food: Mean Amounts and Percentages of Calories from Protein, Carbohydrate, Fat, and Alcohol, One Day, 2005–2006 and Nutrient Intakes from Food: Mean Amounts Consumed per Individual, One Day, 2005–2006. Available at www.ars.usda.gov/ba/bhnrc.fsrg, accessed 6/09.

79. Devaney, B. et al. Nutrient intakes of infants and toddlers. *J Am Diet Assoc* 2004; 104:S15–21.

80. Skinner, J. D. et al. Longitudinal study of nutrient and food intakes of white preschool children aged 24 to 60 months. *J Amer Diet Assoc* 1999; 99:1514–21.

81. Picciano, M. F. et al. Nutritional guidance is needed during dietary transition in early childhood. *Pediatrics* 2000; 106:109–14.

82. Bowman, S. A. et al. Effects of fast-food consumption on energy intake and diet quality among children in a national household survey. *Pediatrics* 2004; 113:112–18.

83. Kranz, S. et al. Adverse effect of high added sugar consumption on dietary intake in American preschoolers. *J Pediatr* 2005; 146:105–11.

84. McConaby, K. L. et al. Food portions are positively related to energy intake and body weight in early childhood. *J Pediatr* 2002; 140:340–47.

85. Sanders, T. A. B. Vegetarian diets and children. *Pediatr Clinic N Amer* 1995; 42:955–65.

86. American Dietetic Association. Position of the American Dietetic Association: Vegetarian diets. *J Amer Diet Assoc* 2009; 109: 1266–82.

87. American Dietetic Association. Position of the American Dietetic Association: Benchmarks for nutrition programs in child-care settings. *J Amer Diet Assoc* 2005; 105:979–86.

88. American Academy of Pediatrics, American Public Health Association, Health Resources and Services Administration, Maternal and Child Health Bureau. *Caring for our children, national health and safety performance standards: Guidelines for out-of-home child care programs*, 2nd ed. Elk Grove Village, IL: American Academy of Pediatrics, 2002.

89. American Academy of Pediatrics, Council on Sports Medicine and Fitness and Council on School Health. Active healthy living: Prevention of childhood obesity through increased physical activity. *Pediatrics* 2006; 117:1834–42.

90. Hagan, J. F., Shaw, J., and Duncan, P. M., eds. *Bright Futures: Guidelines for Health Supervision of Infants, Children, and Adolescents*, 3rd ed. Elk Grove Village, IL: American Academy of Pediatrics, 2008.

91. U.S. Department of Agriculture. Food and Nutrition Services, WIC. Available at www.usda.gov.

92. Rose, D., Habicht, J. P., and Devaney, B. Household participation in the food stamp and WIC programs increases the nutrient intakes of preschool children. *J Nutr* 1998; 128:548–55.

93. U.S. Department of Health and Human Services. Administration for Children and Families. Office of Head Start. Available at www.dhhs.gov.

94. U.S. Department of Agriculture. Supplemental Nutrition Assistance Program. Available at www.fns.usda.gov.

Chapter 11

1. Westbrook, L. E., Silver, E. J., and Stein, R. E. K. Implications for estimates of disability in children: A comparison of definitional components. *Pediatrics* 1998; 101:1025–30.

2. Yeargin-Allsopp, M., Drews-Botsch, C., and Van Naarden, B. Epidemiology of developmental disabilities. In Batshaw, M. L., Pellegrino, L., and Roizen, N. J., eds. *Children with disabilities*, 6th ed. Baltimore, MD: Paul H. Brookes, 2007: 231–243.

3. Staiano, A. Food refusal in toddlers with chronic diseases. *J Ped Gastro Nutr* 2003; 37:225–27.

4. National Information Center for Children and Youth with Disabilities. General information about disabilities. Available at www.nichcy.org/general.htm, accessed 6/09.

5. Hagan, J. F., Shaw, J. S., and Duncan, P. M., eds. *Bright Futures: Guidelines for Health Supervision of Infants, Children and Adolescents*. Elk Grove Village, IL: American Academy of Pediatrics, 2008.

6. Isaacs, J. S. The role of the nutritionist in life care planning. In Riddick-Grisham, S., ed. *Pediatric life care planning and case management*. Boca Raton, FL: CRC Press, 2004: 325–38.

7. Trumbo, P. et al. Dietary reference intakes: Vitamin A, vitamin K, arsenic, boron, chromium, copper, iodine, iron, manganese, molybdenum, nickel, silicon, vanadium, and zinc. *J Amer Diet Assoc* 2001; 101:294–301.

8. Kleinman, R. E., ed. *Pediatric nutrition handbook*, 6th ed. Chicago, IL: American Academy of Pediatrics, 2009: 821–842.

9. National Center for Health Statistics. *NCHS growth curves for children 0–19 years*. U.S. Vital and Health Statistics, Health Resources Administration. Washington, D.C.: U.S. Government Printing Office, 2000.

10. Anneren, G. et al. Normalized growth velocity in children with Down's syndrome during growth hormone therapy. *J Intel Disabil Res* 1993; 371:381–87.

11. Vohr, B. R., Poindexter, B. B., Dusick, A. M., McKinley, L. T., Higgins, R. D., Langer, J. C., et al. Persistent Beneficial Effects of Breast Milk Ingested in the Neonatal Intensive Care Unit on Outcomes of Extremely Low Birth Weight Infants at 30 Months of Age. *Pediatrics* 2007; 120:e953–e959.

12. The Infant Health and Development Program: Enhancing the outcomes of low birth weight, premature infants. *JAMA* 1990; 263(22):3035–42.

13. Nellhaus, G. Composite international and interracial graphs. *Pediatrics* 1968; 41:106–14.

14. Isaacs, J. S. et al. Eating difficulties in girls with Rett syndrome compared with other developmental disabilities. *J Amer Diet Assoc* 2003; 103(2):224–30.

15. Pellegrino, L. Cerebral palsy. In Batshaw, M. L., Pellegrino, L., and Roizen, N. J., eds. *Children with disabilities*, 6th ed. Baltimore, MD: Paul H. Brookes, 2007: 387–408.

16. American Academy of Pediatrics. The use and misuse of fruit juice in pediatrics. *Pediatrics* 2001; 107:1210–13.

17. Black, R. E. et al. Children who avoid drinking cow milk have low dietary calcium intakes and poor bone health. *Am J Clin Nutr* 2002; 76:675–80.

18. Ruottinen, S., Niinikoski, H., Lagström, H., Rönnemaa, T., Hakanen, M., Viikari, J. et al. High sucrose intake is associated with poor quality of diet and growth between 13 months and 9 years of age: The Special Turku Coronary Risk Factor Intervention Project. *Pediatrics* 2008; 121:e1676–e1685.

19. Kellogg, N. D. and the Committee on **Child Abuse** and Neglect. **Evaluation of Suspected Child Physical Abuse.** *Pediatrics* 2007; 119:1232–1241.

20. Berkovitch, M. et al. Copper and zinc blood levels among children with nonorganic failure to thrive. *J Clin Nutr* 2003; 22(2):183–86.

21. Niewinski, M. M. Advances in celiac disease and gluten-free diet. *J Amer Diet Assoc* 2008; 108:661–672.

22. Johnson, C. P., Myers, S. M., and Council on Children with Disabilities. American Academy of Pediatrics Policy Clinical Report: Identification on evaluation of children with autism spectrum disorder. *Pediatrics* 2007; 120:1183–1215.

23. Minns, R. A. Neurological disorders. In Kelnar, C. J. H. et al., eds. *Growth disorders.* Chapman and Hall, 1998: 447–70.

24. Krick, J. et al. Pattern of growth in children with cerebral palsy. *J Amer Diet Assoc* 1996; 96:680–85.

25. Kleinman, R. E., ed. *Pediatric nutrition handbook*, 6th ed. Chicago, IL: American Academy of Pediatrics, 2009: 1001–1020.

26. Centers for Disease Control and Prevention. Asthma prevalence and control characteristics by race/ethnicity—United States, 2002. *MMWR* 2004; 53:145–48.

27. Falci, K. J., Gombas, K. L., and Elliot, E. L. Food allergen awareness: An FDA priority. *Food Safety Magazine* 2001: Feb–March.

28. Council on Children With Disabilities. Role of the Medical Home in Family: Centered Early Intervention Services. *Pediatrics* 2007; 120:1153–1158.

29. Greer, F. R., Sicherer, S. H., Burks, A. W., and the Committee on Nutrition and Section on Allergy and Immunology. Effects of early nutritional interventions on the development of atopic disease in infants and children: The role of maternal dietary restriction, breastfeeding, timing of introduction of complementary foods, and hydrolyzed formulas. *Pediatrics* 2008; 121:183–191.

30. Ball, S. D., Kertesz, D., Laurie, J, and Moyer-Mileur, L. J. Dietary supplement use is prevalent among children with a chronic illness. *J Amer Diet Assoc* 2005; 105:78–84.

31. National Down Syndrome Society. Position statement on vitamin-related therapies, 1997. Available at www.ndss.org/content.cfm, accessed 6/09.

32. American College of Medical Genetics. Statement on nutritional supplements and piracetam for children with Down syndrome, 1996. Available at www.acmg.net/resources/policies/pol-006.asp, accessed 6/09.

33. McGuire, J., Kulkarni, M., and Baden, H. Fatal hypermagnesemia in a child treated with megavitamin/megamineral therapy. *J Pediatr* 2000; 105:318.

34. Maternal and Child Health Bureau, Health and Human Services. Available at http://mchb.hrsa.gov/html/drte.html, accessed 6/09.

Chapter 12

1. Story, M., Holt, K., and Sofka, D., eds. *Bright futures in practice: Nutrition.* Arlington, VA: National Center for Education in Maternal and Child Health, 2002.

2. Meyer, A. F. et al. School breakfast program and school performance. *Am J Dis Child* 1989; 143:1234–9.

3. The Annie E. Casey Foundation. *2008 Kids Count Data Book.* Baltimore, MD: The Annie E. Casey Foundation, 2008.

4. U.S. Department of Health and Human Services, Health Resources and Services Administration, Maternal and Child Health Bureau. *Child Health USA 2007.* Rockville, MD: U.S. Department of Health and Human Services, 2008.

5. Ogden, C. L. et al. Prevalence of high body mass index in US children and adolescents, 2007-2008. *JAMA* 2010; 303:242–249.

6. U.S. Department of Health and Human Services. *Healthy People 2010* (Conference Edition, in two volumes). Washington, D.C.: January 2000.

7. Kliegman, R. M., Behrman, R. E., Jenson, H. B., and Stanton, B. M. D., eds.

Nelson Textbook of Pediatrics, 18th ed. Philadelphia: WB Saunders Co, 2007.

8. Centers for Disease Control and Prevention, National Center for Health Statistics. CDC growth charts: United States. Available at http://www.cdc.gov/nchs/about/major/nhanes/growthcharts/charts.htm, accessed 4/04.

9. Troiano, R. P. and Flegal, K. M. Overweight children and adolescents: Description, epidemiology, and demographics. *Pediatrics* 1998; 101:497–504.

10. Barlow, S. E. Expert committee recommendations regarding the prevention, assessment, and treatment of child and adolescent overweight and obesity: Summary report. *Pediatrics* 2007; 120:S164–S192.

11. Krebs, N. F. et al. Assessment of child and adolescent overweight and obesity. *Pediatrics* 2007; 120:S193–S228.

12. de Oris, M. et al. Development of a WHO growth reference for school-aged children and adolescents. *Bulletin of the World Health Organization* 2007; 85:660–7.

13. Dietz, W. H. Critical periods in childhood for the development of obesity. *Am J Clin Nutr* 1994; 59:955–9.

14. Satter, E. Feeding dynamics: Helping children to eat well. *J Pediatr Health Care* 1995; 9:178–84.

15. Gillman, M. W. et al. Family dinner and diet quality among older children and adolescents. *Arch Fam Med* 2000; 9:235–40.

16. Batada, A. et al. Nine out of 10 food advertisements shown during Saturday morning children's television programming are for foods high in fat, sodium, or added sugars, or low in nutrients. *J Amer Diet Assoc* 2008; 108:673–8.

17. U.S. Department of Agriculture. *MyPyramid.* Available at www.MyPyramid.gov.

18. Moore, E. and the Kaiser Family Foundation. *It's child play: Advergaming and the online marketing of food to children.* Menlo Park, CA: Henry J. Kaiser Family Foundation, 2006.

19. Birch, L. L. and Fisher, J. O. Food intake regulation in children, fat and sugar substitutes and intake. *Ann NY Acad of Sci* 1997; 819:194–220.

20. Birch, L. L. and Fisher, J. A. Appetite and eating behavior in children. *Pediatr Clinic N Amer* 1995; 42:931–53.

21. Birch, L. L. and Fisher, J. O. Mothers' child-feeding practices influence daughters' eating and weight. *Amer J Clin Nutr* 2000; 71:1054–61.

22. Birch, L. L. Psychological influences on the childhood diet. *J Nutr* 1998; 128:407S–10S.

23. Fisher, J. O. and Birch, L. L. Parents' restrictive feeding practices are associated with young girls' negative self evaluation of eating. *J Amer Diet Assoc* 2000; 100:1341–6.

24. Birch, L. L. and Fisher, J. O. Development of eating behaviors among children and adolescents. *Pediatrics* 1998; 101:539–49(s).

25. Cachelin, F. et al. Does ethnicity influence body-size preference? A comparison of body image and body size. *Obesity Research* 2002; 10:158–66.

26. Institute of Medicine, Food and Nutrition Board. *Dietary Reference Intakes for energy, carbohydrate, fiber, fat, protein, and amino acids.* Washington, D.C.: National Academy Press, 2002/2005.

27. U.S. Department of Health and Human Services and U.S. Department of Agriculture. *Dietary Guidelines for Americans, 2005,* 6th ed. Washington, D.C.: U.S. Government Printing Office, January 2005.

28. Institute of Medicine, Food and Nutrition Board. *Dietary Reference Intakes for calcium, phosphorus, magnesium, vitamin D, and fluoride.* Washington, D.C.: National Academy Press, 1997.

29. Institute of Medicine, Food and Nutrition Board. *Dietary Reference Intakes for vitamin A, vitamin K, arsenic, boron, chromium, copper, iodine, iron, manganese, molybdenum, nickel, silicon, vanadium, and zinc.* Washington, D.C.: National Academy Press, 2001.

30. Centers for Disease Control and Prevention. Iron deficiency. United States, 1999–2000. *Morbidity and Mortality Weekly Review* 2002; 51:897–9.

31. Kleinman, R. E., ed. American Academy of Pediatrics, Committee on Nutrition. *Pediatric nutrition handbook,* 6th ed. Elk Grove Village, IL: American Academy of Pediatrics, 2009.

32. Centers for Disease Control and Prevention. Recommendations to prevent and control iron deficiency in the United States. *Morbidity and Mortality Weekly Report* April 3, 1998; 47:RR-03, 1–29.

33. Casamassimo, P. *Bright futures in practice: Oral health.* Arlington, VA: National Center for Education in Maternal and Child Health, 1996.

34. Troiano, R. P. et al. Energy and fat intakes of children and adolescents in the United States: Data from the National Health and Nutrition Examination Surveys. *Am J Clin Nutr* 2000; 72(suppl):1343S–53S.

35. Freedman, D. S. et al. The relation of overweight to cardiovascular risk factors among children and adolescents: The Bogalusa Heart Study. *Pediatrics* 1999; 103:1175–82.

36. Freedman, D. S. et al. Cardiovascular risk factors and excess adiposity among overweight children and adolescents: The Bogalusa Heart Study. *J Pediatr* 2007; 150:12–17.

37. Dietz, W. H. Health consequences of obesity in youth: Childhood predictors of adult disease. *Pediatrics* 1998; 101:518S–25S.

38. American Diabetes Association. Type 2 diabetes in children and adolescents. *Pediatrics* 2000; 105:671–80.

39. Dietz, W. H. Childhood weight affects adult morbidity and mortality. *J Nutr* 1998; 128:411S–14S.

40. Dietz, W. H. and Gortmaker, S. L. Preventing obesity in children and adolescents. *Annu Rev Public Health* 2001; 22:337–53.

41. Strauss, R. S. and Knight, J. Influence of the home environment on the development of obesity in children. *Pediatrics* [serial online] 1999; 103:e85. Available at http://www.aap.publications.org, accessed 7/09.

42. Whitaker, R. C. et al. Predicting obesity in young adulthood from childhood and parental obesity. *New Eng J Med* 1997; 337:869–73.

43. Dowda, M., Ainsworth, B. E., and Addy, C. L. Environmental influences, physical activity, and weight status in 8- to 16-year-olds. *Arch Pediatr Adolesc Med* 2001; 155:711–7.

44. Centers for Disease Control. Retrieved from http://wonder.cdc.gov/data2010/focus.htm, accessed 7/09.

45. Spear, B. A. et al. Recommendations for treatment of child and adolescent overweight and obesity. *Pediatrics* 2007; 120:S254–S288.

46. American Academy of Pediatrics, Committee on Public Education. Children, adolescents, and television. *Pediatrics* 2001; 107:423–6.

47. Dietz, W. H. and Gortmaker, S. L. Do we fatten our children at the television set? Obesity and television viewing in children and adolescents. *Pediatrics* 1985; 75:807–12.

48. Gortmaker, S. L. et al. Television viewing as a cause of increasing obesity among children in the United States, 1986–1990. *Arch Pediatr Adolesc Med* 1996; 150:356–62.

49. Andersen, R. E. et al. Relationship of physical activity and television watching with body weight and level of fatness among children: Results from the Third National Health and Nutrition Examination Survey. *JAMA* 1998; 279:938–42.

50. Crespo, C. J. et al. Television watching, energy intake, and obesity in U.S. children. *Arch Pediatr Adolesc Med* 2001; 155:360–5.

51. Robinson, T. N. Reducing children's television viewing to prevent obesity: A randomized controlled trial. *JAMA* 1999; 282:1561–67.

52. Klesges, R. C., Shelton, M. L., and Klesges, L. M. Effects of television on metabolic rate: Potential implications for childhood obesity. *Pediatrics* 1993; 91:281–6.

53. American Academy of Pediatrics, Committee on Nutrition. Prevention of pediatric overweight and obesity. *Pediatrics* 2003; 112:424–30.

54. American Academy of Pediatrics, Council on Sports Medicine and Fitness and Council on School Health. Active healthy living: Prevention of childhood obesity through increased physical activity. *Pediatrics* 2006; 117:1834–42.

55. Davis, M. M. et al. Recommendations for prevention of childhood obesity. *Pediatrics* 2007; 120:S229–S253.

56. American Dietetic Association. Position of the American Dietetic Association: individual-, family-, school-, and community-based interventions for pediatric overweight. *J Am Diet Assoc* 2006; 106:925–45.

57. Kavey, R. W. et al. American Heart Association guidelines for primary prevention of atherosclerotic cardiovascular disease beginning in childhood. *Circulation* 2003; 107:1562–6.

58. Kavey, R. W. et al. American Heart Association guidelines for primary prevention of atherosclerotic cardiovascular disease beginning in childhood. *J Pediatr* 2003; 142:368–72.

59. Gidding, S. S. et al. Dietary recommendations for children and adolescents: A guide for practitioners. Consensus statement from the American Heart Association. *Circulation* 2005; 112:2061–75.

60. American Heart Association. Gidding, S. S. et al. Dietary recommendations for children and adolescents: A guide for practitioners. *Pediatrics* 2006; 117:544–59.

61. Lichtenstein, A. H. et al. Diet and lifestyle recommendations revision 2006: A scientific statement from the American Heart Association Nutrition Committee. *Circulation* 2006; 114:82–96.

62. Gidding, S. S. et al. Implementing American Heart Association pediatric and adult nutrition guidelines: A scientific statement from the American Heart Association Nutrition Committee of the Council on Nutrition, Physical Activity and Metabolism, Council on Cardiovascular Disease in the Young, Council on Arteriosclerosis, Thrombosis and Vascular Biology, Council on Cardiovascular Nursing, Council on Epidemiology and Prevention, and Council on High Blood Pressure Research. *Circulation* 2009; 119:1161–75.

63. Daniels, S. R. et al. Lipid screening and cardiovascular health in childhood. *Pediatrics* 2008; 122:198–208.

64. Pitetti, R. et al. Complementary and alternative medicine use in children. *Pediatr Emerg Care* 2001; 17:165–9.

65. American Dietetic Association. Position of the American Dietetic Association: Nutrition guidance for healthy children aged 2 to 11 years. *J Amer Diet Assoc* 2008; 108:1038–47.

66. Dwyer, J. T. Dietary fiber for children: How much? *Pediatrics* 1995; 96:1019S–22S.

67. Greer, F. R., Krebs, N. F., and American Academy of Pediatrics, Committee on Nutrition. Optimizing bone health and calcium intakes of infants, children, and adolescents. *Pediatrics* 2006; 117:578–85.

68. Wagner, C. L., Greer, F. R., and American Academy of Pediatrics, Section on Breastfeeding and Committee on Nutrition. Prevention of rickets and vitamin D deficiency in infants, children, and adolescents. *Pediatrics* 2008; 122:1142–52.

69. Heyman, M. B. and American Academy of Pediatrics, Committee on Nutrition. Lactose intolerance in infants, children, and adolescents. *Pediatrics* 2006; 118:1279–86.

70. American College of Sports Medicine, Sawka, M. N., et al. American College of Sports Medicine position statement: Exercise and fluid replacement. *Medicine and Science in Sports and Exercise* 2007; 39:377–90.

71. Wang, V. C. et al. Increasing caloric contribution from sugar-sweetened beverages and 100% fruit juices among U.S. children and adolescents, 1988–2004. *Pediatrics* 2008; 121:e1604–e1614.

72. Ludwig, D. S., Peterson, K. E., and Gortmaker, S. L. Relation between consumption of sugar-sweetened drinks and childhood obesity: A prospective, observational analysis. *Lancet* 2001; 357:505–8.

73. Mrdjenovic, G. and Levitsky, D. A. Nutritional and energetic consequences of sweetened drink consumption in 6- to 13-year-old children. *J Pediatr* 2003; 142:604–10.

74. Rampersaud, G. C. et al. National survey beverage consumption data for children and adolescents indicate the need to encourage a shift toward more nutritive beverages. *J Am Diet Assoc* 2003; 97–100.

75. Bowman, S. A. et al. Effects of fast-food consumption on energy intake and diet quality among children in a national household survey. *Pediatrics* 2004; 113:112–18.

76. Pennington, J. A. T. and Spungen, J. *Bowes and Church's food values of portions commonly used*, 19th ed. Philadelphia: Lippincott Williams Wilkins, 2009.

77. American Dietetic Association. Position of the American Dietetic Association: Use of nutritive and nonnutritive sweeteners. *J Amer Diet Assoc* 2004; 104:255–75.

78. U.S. Department of Agriculture, Agricultural Research Service, 2008. Nutrient Intakes from Food: Mean Amounts and Percentages of Calories from Protein, Carbohydrate, Fat, and Alcohol, 2005–2006 and Nutrient Intakes from Foods; Means Amounts Consumed per Individual, One Day, 2005–2006. Available at www.ars.usda.gov/ba/bhnrc/fsrg, accessed 6/09.

79. Winkleby, M. A. et al. Ethnic variation in cardiovascular disease risk factors among children and young adults: Findings from the Third National Health and Nutrition Examination Survey, 1988–1994. *JAMA* 1999; 281:1006–13.

80. Jahns, L., Siega-Riz, A. M., and Popkins, B. M. The increasing prevalence of snacking among U.S. children from 1977 to 1996. *J Pediatr* 2001; 138:493–8.

81. United States Department of Agriculture, Center for Nutrition Policy and Promotion. The quality of children's diets in 2003–04 as measured by the Health Eating Index, 2005. *Nutrition Insight* 43, April 2009. Available at www. Usda.gov/cnpp, accessed 7/09.

82. Kittler, P. G. and Sucher, K. P. *Food and culture*, 3rd ed. Belmont, CA: Wadsworth/Thomson Learning, 2001.

83. Messina, V., Melina, V., and Mangels, A. R. A new food guide for North American vegetarians. *J Amer Diet Assoc* 2003; 103:771–5.

84. Messina, V. and Mangel, A. R. Considerations in planning vegan diets: Children. *J Amer Diet Assoc* 2001; 101:661–9.

85. American Academy of Pediatrics, Committee on Sports Medicine and Fitness and Committee on School Health. Physical fitness and activity in schools. *Pediatrics* 2000; 105:1156–7.

86. American Academy of Pediatrics, Committee on Environmental Health. The built environment: Designing communities to promote physical activity in children. *Pediatrics* 2009; 123:1591–8.

87. Department of Health and Human Services and Department of Education. Promoting better health for young people through physical activity and sports: A report to the president from the secretary of Health and Human Services and the secretary of Education. Silver Spring, MD: Centers for Disease Control and Prevention, 2000. Available at www.cdc.gov/nccdphp/dash/presphysactrpt.

88. Kohl, H. W. and Hobbs, K. E. Development of physical activity behaviors among children and adolescents. *Pediatrics* 1998; 101:549–54.

89. American Academy of Pediatrics, Committee on Sports Medicine and Fitness and Committee on School Health. Organized sports for children and preadolescents. *Pediatrics* 2001; 107:1459–62.

90. American Dietetic Association. Position of ADA, SNE, and ASFSA: School nutrition services: An essential component of comprehensive school health programs. *J Amer Diet Assoc* 2003; 103:505–14.

91. Lytle, L. Nutrition education for school-aged children. *J Nutr Ed* 1995; 27:298–311.

92. Centers for Disease Control. Guidelines for school health programs to promote lifelong healthy eating. *J Sch Health* 1997; 67:9–26.

93. American Dietetic Association. Position of the American Dietetic Association: Local support for nutrition integrity in schools. *J Amer Diet Assoc* 2006; 106:122–33.

94. Kubic, M. Y. et al. The association of the school food environment with dietary behaviors of young adolescents. *Am J Pub Health* 2003; 93:1168–72.

95. Finkelstein, D. M. et al. School food environments and policies in US public schools. *Pediatrics* 2008; 122:e251–e359.

96. American Academy of Pediatrics, Committee on School Health. Soft drinks in schools. *Pediatrics* 2004; 113:152–4.

97. Centers for Disease Control and Prevention. School Health Index for physical activity, healthy eating and a tobacco-free lifestyle: A self-assessment and planning guide, elementary school version. Atlanta, GA: 2002. Available at www.cdc.gov/nccdphp/dash/SHI, accessed 7/09.

98. Hagan, J. F., Shaw, J. S., and Duncan, P. M., eds. *Bright Futures: Guidelines for Health Supervision of Infants, Children, and Adolescents*, 3rd ed. Elk Grove Village, IL: American Academy of Pediatrics, 2008.

99. Centers for Disease Control and Prevention, National Fruit and Vegetable Program. *Fruits & Veggies–More Matters*. Available at www.fruitsandveggiesmorematters.org, accessed 7/09.

100. Reynolds, K. D. et al. Increasing the fruit and vegetable consumption of fourth-graders: Results from the High 5 project. *Prev Med* 2000; 30:309–19.

101. American Dietetic Association. Position of the American Dietetic Association: Child and adolescent food and nutrition programs. *J Amer Diet Assoc* 2006; 106:1467–75.

102. United States Department of Agriculture, Economic Research Service. Ralston, K. et al. The National School Lunch Program: Background, trends, and issues. *Ecronomic Research Report* Number 61, July 2008. Available at www.ers.usda.gov/Publications/ERR61/ERR61.pdf, accessed 7/09.

103. Martin, J. Overview of federal child nutrition legislation. In Martin, J. and Conklin, M. T., eds. *Managing child nutrition programs leadership for excellence*. Gaithersburg, MD: Aspen Publishers, 1999.

104. U.S. Department of Agriculture, Food and Nutrition Service. A Menu Planner for Healthy School Meals. *FNS-303* 1998, Rev. 2008. Available at www.teamnutrition.usda.gov/Resource/menuplanner, accessed 7/09.

105. U.S. Department of Agriculture, Food and Nutrition Service. *Team Nutrition*. Available at www.fns.usda.gov/tn, accessed 7/09.

Chapter 13

1. General information about disabilities. National Information Center for Children and Youth with Disabilities. Washington, D.C.: U.S. Government Printing Office. Available at www.Nichcy.org/general.htm, accessed 6/09.

2. Perrin, J. M., Bloom, S. R., and Gorkmaker, S. L. The increase of childhood chronic conditions in the United States. *J Am Med Assoc* 2007: 297(24):2755–2759.

3. Krick, J. et al. Pattern of growth in children with cerebral palsy. *J Amer Diet Assoc* 1996; 96:680–85.

4. Isaacs, J. S. and Zand, D. Single-gene autosomal recessive disorders and Prader-Willi syndrome: An update for food and nutrition professionals. *J Amer Diet Assoc* 2007; 107(3):466–478.

5. Institute of Medicine Food and Nutrition Board. Dietary reference intakes for energy, carbohydrate, fiber, fat, fatty acids, cholesterol, protein, and amino acids, 2002.

6. Isaacs, J. S. Non-oral feeding. In Lucas, B., ed. *Children with special health care needs: Nutrition care handbook*. Chicago IL: American Dietetic Association, 2004.

7. Kleinman, R. E., ed. *Pediatric nutrition handbook*, 6th ed. Chicago IL: American Academy of Pediatrics, 2009: 821–842.

8. American Association for Respiratory Care. Metabolic measurement using indirect calorimetry during mechanical ventilation, 2004 revision and update. *Respir Care* 2004; 49(9):1073–9.

9. L'Alleman, D. et al. Cardiovascular risk factors improve during three years of growth hormone therapy in Prader-Willi syndrome. *Eur J Pediatr* 2000; 159:836–42.

10. Minns, R. A. Neurological disorders. In Kelnar, C. J. H. et al., eds. *Growth disorders*. London: Chapman and Hall Medical; Philadelphia, PA, 1998: 447–70.

11. Schertz, M. et al. Predictors of weight loss in children with attention deficit hyperactivity disorder treated with stimulant medication. *Pediatrics* 1996; 98:763–69.

12. Johnson, C. P. and Myers, S. M. Council on Children with Disabilities. American Academy of Pediatrics Policy Clinical Report: Identification on evaluation of children with autism spectrum disorder. *Pediatrics* 2007; 120:1183–1215.

13. Dodge, J. A. and Turck, D. Cystic fibrosis: Nutritional consequences and management. *Best Practice & Research in Clinical Gastroenterology* 2006; 20:531–46.

14. Lefebvre, D. E. et al. Dietary proteins as environmental modifiers of type 1 diabetes mellitus. *Annual Review of Nutrition* 2006; 26:175–202.

15. American Diabetes Association. Standards of medical care in diabetes—2009. *Diabetes Care* 2009; 32:S13–S61.

16. Trumbo, P. et al. Dietary reference intakes: Vitamin A, vitamin K, arsenic, boron, chromium, copper, iodine, iron, manganese, molybdenum, nickel, silicon, vanadium, and zinc. *J Am Diet Assoc* 2001; 101:294–301.

17. Greer, F. R. and Krebs, N. F. Committee on Nutrition. Optimizing bone health and calcium intakes of infants, children, and adolescents. *Pediatrics* 2006; 117:578–585.

18. Ross, E. A., Szabo, N. J., and Tabbett, I. R. Lead content of calcium supplements. *J Am Med Assoc* 2000; 284:1425–29.

19. National Center for Health Statistics. *NCHS growth curves for children 0–19 years*. U.S. Vital and Health Statistics, Health Resources Administration. Washington, D.C.: U.S. Government Printing Office, 2000.

20. *Growth references: Third trimester to adulthood*, 2nd ed. Clinton, SC: Greenwood Genetic Center, 1998.

21. Frisancho, A. R. Triceps skinfold and upper arm muscle size norms for assessment of nutritional status. *Am J Clin Nutr* 1974; 27:1052–157.

22. Fomon, S. et al. Body composition of reference children from birth to age 10 years. *Amer J Clin Nutr* 1982; 35:1169–75.

23. Nassis, G. P. and Sidossis, L. S. Methods for assessing body composition, cardiovascular and metabolic function in children and adolescents: Implications for exercise studies. *Current Opinion in Clinical Nutrition & Metabolic Care* 2006; 5:560–67.

24. Nellhaus, G. Composite international and interracial graphs. *Pediatrics* 1968; 41:106–14.

25. Batshaw, M. L. Chromosomes and heredity. In Batshaw, M. L., Pellegrino, L., and Roizen, N. J., eds. *Children with disabilities*, 6th ed. Baltimore, MD: Paul H. Brookes, 2007: 3–26, 770.

26. Steinkamp, G. and Von der Hardt, H. Improvement of nutritional status and lung function after long-term nocturnal gastrostomy feedings in cystic fibrosis. *J Pediatrics* 1994; 121:244–49.

27. Walkowiak, J. and Przyslawski, J. Five-year prospective analysis of dietary intake and clinical status in malnourished cystic fibrosis patients. *J Hum Nutr Diet* 2003; 16:225–31.

28. Sullivan, P. B. et al. Does gastrostomy tube feeding in children with cerebral palsy increase the risk of respiratory morbidity? *Arch Dis Child* 2006; 91:478–82.

29. Shingadia, D. et al. Gastrostomy tube insertion for improvement of adherence to highly active antiretroviral therapy in pediatric patients with human immunodeficiency virus. *Pediatrics* 2000; 105:1–5. Available at www.pediatrics.org/cgi/content/full/105/6/e80, accessed 6/09.

30. American Diabetes Association. Type 2 diabetes in children and adolescents. *Pediatrics* 2000; 671–80.

31. Crowther, N. J., Cameron, N., Trusler, J., Toman, M., Norris, S. A., and Gray, I. P. Influence of catch-up growth on glucose tolerance and β-cell function in 7-year-old children: Results from the birth to twenty study. *Pediatrics* 2008; 121:e1715–e1722.

32. Kleinman, R. E., ed. *Pediatric nutrition handbook*, 6th ed. Chicago IL: American Academy of Pediatrics, 2009: 1021–1026.

33. Burlina, A. and Blau, N. Effect of BH(4) supplementation on phenylalanine tolerance. *J Inherit Metab Dis* 2009; 32(1):40–5.

34. Pellegrino, L. Cerebral palsy. In Batshaw, M. L., Pellegrino, L., and Roizen, N. J., eds. *Children with disabilities*, 6th ed. Baltimore, MD: Paul H. Brookes, 2007: 387–408.

35. Johnson, R. K. et al. Athetosis increases resting metabolic rate in adults with cerebral palsy. *J Amer Diet Assoc* 1996; 96:145–48.

36. Perrin, J. M., Friedman, R. A., and Knilans, T. K. Black Box Working Groups, Section on Cardiology and Cardiac Surgery. Cardiovascular Monitoring and Stimulant Drugs for Attention-Deficit/Hyperactivity Disorder. *Pediatrics* 2008; 122:451–453.

37. Zachor, D. A. et al. Effects of long-term psychostimulant medication on growth of children with ADHD. *Research in Developmental Disabilities* 2006; 27:162–74.

38. Cala, S., Crismon, M. L., and Baumgartner, J. A survey of herbal use in children with attention-deficit-hyperactivity disorder or depression. *Pharmacotherapy* 2003; 23:222–30.

39. Kleinman, R. E., ed. *Pediatric nutrition handbook*, 6th ed. Chicago IL: American Academy of Pediatrics, 2009: 875–904.

40. Kemper, K. J., Vohra, S., and Walls, R. Task Force on Complementary and Alternative Medicine. The use of complementary and alternative medicine in pediatrics. *Pediatrics* 2008; 122:1374–1386.

41. Ball, S. D., Kertesz, D., and Moyer-Mileur, L. J. Dietary supplement use is prevalent among children with a chronic illness. *J Amer Diet Assoc* 2005; 105(1):78–84.

42. National Down Syndrome Society public position statements. Vitamin-related therapies. Available at www.ndss.org, accessed 6/09.

43. Hagan, J. F., Shaw, J. S., and Duncan, P. M., eds. *Bright futures: Guidelines for health supervision of infants, children and adolescents*, 3rd ed. Elk Grove Village, IL: American Academy of Pediatrics, 2008.

44. Maternal and Child Health Bureau, Health and Human Services. Available at http://mchb.hrsa.gov/html/drte.html, accessed 6/09.

Chapter 14

1. Tanner, J. M. *Growth at adolescence.* Oxford: Blackwell, 1962.

2. Sun, S. S. et al. National estimates of the timing of sexual maturation and racial differences among U.S. children. *Pediatr* 2002; 110:911–19.

3. Barnes, H. V. Physical growth and development during puberty. *Med Clin North Am* 1975; 59:1305–17.

4. Frisch, R. E. Fatness, puberty, and fertility: The effects of nutrition and physical training on menarche and ovulation. In Brooks-Gunn, J. and Peterson, A. C., eds. *Girls at puberty: Biological and psychosocial perspectives.* New York: Plenum Press, 1983: 29–49.

5. Frisch, R. E. and McArthur, J. W. Menstrual cycles: Fatness as a determinant of minimum weight for height necessary for their maintenance or onset. *Science* 1974; 185:949–951.

6. Matkovic, V. et al. Nutrition influences skeletal development from childhood to adulthood: A study of hip, spine and forearm in female adolescents. *J Nutr* 2004; 134:701S–705S.

7. Neumark-Sztainer, D., Story, M., Perry, C., and Casey, M. Factors influencing food choices of adolescents: Findings from focus group discussions with adolescents. *J Am Diet Assn* 1999; 99(8):929–937.

8. Tremblay, L. and Lariviere, M. The influence of puberty onset, body mass index and pressure to be thin on disordered eating behaviors. *Eat Behav* 2009; 10(2):75–83.

9. Bratberg, G. H., Nilesen, T. I., Holmen, T. L., and Vatten, L. J. Perceived pubertal timing, pubertal status and the prevalence of alcohol drinking and cigarette smoking in early and late adolescence: A population-based study of 8950 Norwegian boys and girls. *Acta Paediatr* 2007; 96(2):292–295.

10. Delva, J., O'Malley, P. M., and Johnston, L. D. Racial/ethnic and socioeconomic status difference in overweight and health-related behaviors among American students: National trends 1986–2003. *J Adolesc Health* 2006; Oct(39):536–45.

11. Cusatis, D. C. et al. Longitudinal nutrient intake patterns of U.S. adolescent women: The Penn State Young Women's Health Study. *J Adolesc Health* 2000; 26:194–204.

12. Sebastian, R. S., Cleveland, L. E., and Goldman, J. D. Effect of snacking frequency on adolescents' dietary intakes and meeting national recommendations. *J Adolesc Health* 2008; 42:503–11.

13. Jahns, L., Siega-Riz, A. M., and Popkin, B. M. The increasing prevalence of snacking among US children 1977–1996. *J Pediatr* 2001; 138(4):493–498.

14. Nielsen, S. J., Siega-Riz, A. M., and Popkin, B. M. Trends in food locations and sources among adolescents and young adults. *Prev Med* 2002; 35(2):107–13.

15. Siega-Riz, A. M., Carson, T., and Popkin, B. Three squares or mostly snacks—what do teens really eat? A sociodemographic study of meal patterns. *J Adolesc Health* 1998; 22:29–36.

16. Timlin, M. T., Pereira, M. A., Story, M., and Neumark-Sztainer, D. Breakfast Eating and Weight Change in a 5-year Prospective Analysis of Adolescents: Project EAT. *Pediatrics* 2008; 121:e638–e645.

17. Pearson, N., Biddle, S. J. H., and Gorely, T. Family correlates of breakfast consumption among children and adolescents. A systematic review. *Appetite* 2009; 52:1–7.

18. Woodruff, S. J., Hanning, R. M., Lambraki, I., Storey, K. E., and McCarger, L. Healthy Eating Index-C is compromised among adolescents with body weight concerns, weight loss dieting and meal skipping. *Body Image* 2008; 5:404–408.

19. Larson, N. I., Nelson, M. C., Neumark-Sztainer, D., Story, M., and Hannan, P. J. Making time for meals: Meal structure and associations with dietary intake in young adults. *J Am Diet Assoc* 2009; 109:72–79.

20. Gillman, M. et al. Family dinner and diet quality among older children and adolescents. *Arch Fam Med* 2000; 9:235–40.

21. Taveras, E. et al. Family dinner and adolescent overweight. *Obesity Research* 2005; 13:900–6.

22. Videon, T. and Manning, C. Influences on adolescent eating patterns: The importance of family meals. *Journal of Adolescent Health* 2003; 32:365–73.

23. Kaiser Family Foundation. Generation M: Media in the lives of 8–18-year-olds. March 2005. Available at www.kff.org/entmedia/7251.cfm, accessed 5/06.

24. Robinson-O'Brien, R., Perry, C. L., Wall, M. M., Story, M., and Neumark-Sztainer, D. Adolescent and young adult vegetarianism: Better dietary intake and weight outcomes but increased risk of disordered eating behaviors. *J Am Diet Assoc* 2009; 109(4): 648655.

25. Rosell, M., Appleby, P., and Key, T. Height, age at menarche, body weight and body mass index in life-long vegetarians. *Public Health Nutr* 2005; 8(7):870–875.

26. Winston, C. J. Health effects of vegan diets. *Am J Clin Nutr* 2009; 89(suppl):1627S1633S.

27. Perry, C. L. et al. Characteristics of vegetarian adolescents in a multiethnic urban population. *J Adolesc Health* 2001; 29(6):406–16.

28. Dietary Guidelines Advisory Committee. The Report of the Dietary Guidelines for Americans, 2005. Available at http://www.health.gov/DietaryGuidelines/dga2005/report/default.htm, accessed 6/28/2009.

29. USDA. MyPyramid: Steps to a Healthier You. Available at http://www.mypyramid.gov/, accessed 6/28/2009.

30. Community Nutrition Research Group, Beltsville Human Nutrition Research Center, Agricultural Research Service, USDA. Pyramid serving intakes in the U.S. 1999–2002, 1 day. Available at www.ba.urs.usda.gov/cnrg, accessed 6/8/2009.

31. Stang, J. et al. Relationships between vitamin and mineral supplement use, dietary intake, and dietary adequacy among adolescents. *J Amer Diet Assoc* 2000; 100:905–10.

32. Moshfegh, A., Goldman, J., and Cleveland, L. NHANES 2001–2002: Usual nutrient intakes from food compared to dietary reference intakes. Available at www.ars.usda.gov/foodsurvey, accessed 6/12/2009.

33. Centers for Disease Control and Prevention. Youth risk behavior surveillance—United States 2007. *Morb Mortal Wkly Rep* June 6, 2008; 57(No SS-4).

34. Food and Nutrition Board, Institute of Medicine. Otten, J. J., Hellwig, J. P., and Meyers, L. D., eds. *Dietary Reference Intakes: The essential guide to nutrient requirements.* Washington, D.C.: National Academies Press, 2006.

35. Gidding, S. S., Dennison, B. A., Birch, L. L., Daniels, S. R., Gilman, M. W., Lichtenstein, A. H., Rattay, K. T., Steinberger, J., Stettler, N., and Van Horn, L. Dietary recommendation for children and adolescents: A guide for practitioners: Consensus statement from the American Heart Association. *Circulation* 2005; 112(13):2061–2075.

36. Weaver, C. M. Vitamin D, calcium homeostasis and skeletal accretion in children. *Journal Of Bone And Mineral Research* 2007; 22(2).

37. Whiting, S. J., Vatanparast, H., Baxter-Jones, A., Faulkner, R. A., Mirwald, R., and Bailey, D. A. Factors that affect bone mineral accrual in the adolescent growth spurt. *J Nutr* 2004; 134(3):696S–700S.

38. U.S. Department of Agriculture, Agricultural Research Service, 2008. Nutrient Intakes from Food: Mean Amounts Consumed per Individual, One Day, 2005–2006. Available at www.ars.usda.gov/ba/bnnrc/fsrg, accessed 6/30/2009.

39. Gao, X. et al. Meeting adequate intake for dietary calcium without dairy foods in adolescents aged 9 to 18 years (National Health and Nutrition Examination Survey 2001–2022). *J Amer Diet Assn* 2006; 106:1759–65.

40. Rajeshwari, R., Nicklas, T. A., Yang, S. J., and Berenson, G. S. Longitudinal changes in intake and food sources of calcium from childhood to young adulthood: The Bogalusa Heart Study. *J Am Coll Nutr* 2004; 23:341–350.

41. Reilly, J. K. et al. Acceptability of soymilk as a calcium-rich beverage in elementary school children. *J Am Diet Assoc* 2006; 106:590–93.

42. Hearney, R. P. and Rafferty, K. The settling problem in calcium-fortified soybean drinks. *J Amer Diet Assoc* 2006; 106:1753.

43. Harnack, L., Stang, J., and Story, M. Soft drink consumption among U.S. children and adolescents: Nutritional consequences. *J Am Diet Assoc* 1999; 99:436–41.

44. Forshee, R. A., Anderson, P. A., and Storey, M. L. Changes in calcium intake and association with beverage consumption and demographics: Comparing data from CSFII 1994–1996, 1998 and NHANES 1999–2002. *J Am Coll Nutr* 2006; 25(2):108–116.

45. Lloyd, T. et al. Modifiable determinants of bone status in young women. *Bone* 2002; 30:416–21.

46. Wang, M. C. et al. Diet in midpuberty and sedentary activity in prepuberty predict peak bone mass. *Am J Clin Nutr* 2003; 77:495–503.

47. Centers for Disease Control and Prevention. Recommendations to prevent and control iron deficiency anemia in the United States. *Morb Mortal Wkly Rep* 2002; 51(40):897–99.

48. U.S. Department of Health and Human Services. *Healthy People 2010* (conference edition in two volumes). Washington, D.C.: January 2000.

49. Ginde, A. A., Liu, M. C., and Camargo, C. A. Demographic differences and trends of vitamin D insufficiency in the U.S. population, 1988–2004. *Arch Intern Med* 2009; 169(6):626–632.

50. Wagner, C. L. and Greer, F. R. American Academy of Pediatrics Section on Breastfeeding and Committee on Nutrition. Prevention of rickets and vitamin D deficiency in children. *Pediatrics* 2008; 122:1142–1152.

51. McDowell, M. A., Lacher, D. A., Pfeiffer, C. M., Mulinare, J., Picciano, M. F., Rader, J. L., Yetley, E. A., Kennedy-Stephenson, J., and Johnson, C. L. Blood folate levels: The latest NHANES results. NCHS Data Brief.

2008; 6:1–8. Available at http://www.cdc.gov/nchs/data/databriefs/db06.htm, accessed 6/30/2009.

52. Hagan, J. F., Shaw, J. S., and Duncan, P. M., eds. *Bright Futures: Guidelines for Health Supervision of Infants, Children, and Adolescents*, 3rd ed. Elk Grove Village, IL: American Academy of Pediatrics, 2008.

53. American Medical Association. *Guidelines for adolescent preventive services.* Recommendations Monograph. Chicago, IL: American Medical Association, Department of Adolescent Health, 1997.

54. Olson, A. L., Gaffney, C. A., Hedberg, V. A., and Gladstone, G. R. Use of inexpensive technology to enhance adolescent health screening and counselling. *Arch Pediatr Adolesc Med* 2009; 163(2):172–177.

55. Office of Disease Prevention and Health Promotion, U.S. Department of Health and Human Services, 2008. Physical Activity Guidelines for Americans. Available at www.health.gov/guidelines, accessed 6/30/2009.

56. Centers for Disease Control and Prevention. Guidelines for school and community programs to promote lifelong physical activity among young people. *Morb Mortal Wkly Rep* 1997; 46:1–36.

57. Leverton, R. M. The paradox of teenage nutrition. *J Amer Diet Assoc* 1968; 53:13–16.

58. O'Dougherty, M., Story, M., and Lytle, L. Food choices of young African-American and Latino adolescents: Where do parents fit in? *J Am Diet Assoc* 2006; 106:1846–50.

59. Arcan, C., Neumark-Sztainer, D., Hannan, P., von den Berg, P., Story, M., and Larson, N. Parental eating behaviours, home food environment and adolescent intakes of fruits, vegetables and dairy foods: Longitudinal findings from Project EAT. *Public Health Nutr* 2007; 10(11):1257–1265.

60. Centers for Disease Control and Prevention. Guidelines for school health programs to promote lifelong healthy eating. *Morb Mortal Wkly Rep* 1996; 45:1–37.

61. CDC. *School health index.* Atlanta, GA: U.S. Dept of Health and Human Services, CDC, 2006.

62. O'Toole, T. P., Anderson, S., Miller, C., and Guthrie, J. Nutrition services and foods and beverages available at school: Results from the School Health Policies and Programs Study 2006. *J Sch Health* 2007; 77: 500–521.

63. CDC. Secondary school health education related to nutrition and physical activity—selected sites, United States, 2004. *MMWR* 2006; 55(30):821–24.

64. McGinnis, J. M., Gootman, J. A., and Kraak, V. L., eds. Institute of Medicine Committee on Food Marketing and the Diets of Children and Adolescents. *Food Marketing to Children and Youth: Threat or Opportunity?* National Academies Press: Washington, D.C., 2006.

65. Stallings, V. A. and Yaktine, A. L., eds. Institute of Medicine Committee on Nutrition Standards for Foods in Schools. *Nutrition Standards for Foods in Schools: Leading the Way Toward Healthier Youth.* National Academies Press: Washington, D.C., 2007.

66. United States Department of Agriculture, Food and Nutrition Service. Healthy schools: Local wellness policy. Available at www.fns. usda.gov/tn/Healthy/wellnesspolicy.html, accessed 7/3/2009.

67. School Nutrition Association. A foundation for the future: Analysis of local wellness policies from the largest 100 districts, 2006. Available at www. schoolnutrition.org, accessed 6/30/2009.

68. CANFIT Current Projects. Available at http://www.canfit.org/, accessed 7/5/2009.

Chapter 15

1. Nader, P. R. et al. National Institute of Child Health and Human Development, Early Child Care Research Network. Identifying risk for obesity in early childhood. *Pediatrics* 2006 Sep; 118(3):e594–601.

2. Barlow, S. E. Expert Committee recommendation regarding the prevention, assessment and treatment of child and adolescent overweight and obesity: Summary report. *Pediatrics* 2007; 120:S164–S192.

3. Ogden, C. L., Carroll, M. D., and Flegal, K. M. High body mass index for age among US children and adolescents, 2003–2006. *J Am Med Assn* 2008; 299:2401–2405.

4. Centers for Disease Control and Prevention. Youth risk behavior surveillance—United States 2007. *Morb Mortal Wkly Rep* June 6, 2008; 57(SS-4).

5. Zephier, E., Himes, J. H., Story, M., and Zhou, X. Increasing Prevalences of Overweight and Obesity in Northern Plains American Indian Children. *Arch Pediatr Adolesc Med* 2006; 160:34–39.

6. Deshmukh-Taskar, P. et al. Tracking of overweight status from childhood to young adulthood: The Bogalusa Heart Study. *Eur J Clin Nutr* 2006; 60:48–57.

7. Singh, A. S., Mulder, C., Twisk, J. W. R., van Mechelsen, W., and Chinapaw, M. J. Tracking of childhood overweight into adulthood: A systematic review of the literature. *Obes Rev* 2008; 9:474–488.

8. Cali, A. M. G. and Caprio, S. Obesity in children and adolescents. *J Clin Endocrinol Metab* 2008; 93:S31–C36.

9. Daniels, S. R. et al. Overweight in children and adolescents: Pathophysiology, consequences, prevention, and treatment. *Circulation* 2005; 111(15):1999–2012.

10. Pate, R. R. et al. Cardiorespiratory fitness levels among U.S. youth 12 to 19 years of age: Findings from the 1999–2002 National Health and Nutrition Examination Survey. *Arch Pediatr Adolesc Med* 2006 Oct; 160(10):1005–12.

11. Spear, B. A., Barlow, S. E., Ervin, C., et al. Recommendations for treatment of child and adolescent overweight and obesity. *Pediatrics* 2007; 120:S254–S288.

12. Yanovski, J. A. Intensive therapies for pediatric obesity. *Ped Clinics N America* 2001; 48:1041–53.

13. Inge, T. H. et al. Bariatric surgery for severely overweight adolescents: Concerns and recommendations. *Pediatrics* 2004; 114:217–223.

14. Inge, T. H., Xanthakos, S. A., and Zeller, M. H. Bariatric surgery for pediatric extreme obesity: now or later? *Int J Obesity.* 2007; 31:1–14.

15. Pratt, J. S. A., Lenders, C. M., Dionne, E. A., et al. Best practice updates for pediatric/adolescent weight loss surgery. *Obesity* 2009; 17:901–910.

16. Lawson, M. L. et al. One year outcomes of Roux-en-Y gastric bypass for morbidly obese adolescents: A multicenter study from the Pediatric Bariatric Study Group. *J Ped Surg* 2006; 41:137–43.

17. Shaikh, U., Byrd, R. S., and Auinger, P. Vitamin and mineral supplement use by children and adolescents in the 1999–2004 National Health and Nutrition Examination Survey. *Arch Pediatr Adolesc Med* 2009; 163(2):150–157.

18. Bell, A. et al. A look at nutritional supplement use in adolescents. *J Adol Health* 2004; 34:508–16.

19. Stang, J. et al. Relationships between vitamin and mineral supplement use, dietary intake, and dietary adequacy among adolescents. *J Am Diet Assoc* 2000; 100:905–10.

20. Yussman, S. M., Wilson, K. M., and Klein, J. D. Herbal products and their association with substance use in adolescents. *J Adolescent Health* 2006; 38(4):395–400.

21. O'Dea, J. Consumption of nutritional supplements among adolescents: Usage and perceived benefits. *Health Ed Res* 2003; 18:98–107.

22. Calfee, R., Fadale, P. Popular ergogenic drugs and supplements in young athletes. *Pediatrics* 2006; 117:e577–e589

23. Castillo, E. M. and Comstock, R. D. Pervalence of performance-enhancing substances among United States adolescents. *Pediatr Clin N Am* 2007; 54:663–675.

24. Lattavo, A., Kipperud, A., and Rogers, P. D. Creatine and other supplements. *Pediatr Clin N Am* 2007; 45:735–760.

25. Position of the American Dietetic Association, Dietitians of Canada and the American college of Sports Medicine: Nutrition and athletic performance. *J Am Diet Assoc* 2009; 109:509–527.

26. Pisetsky, E. M., Chao, Y. M., Dierker, L. C., et al. Disordered eating and substance use in high-school students: Results from the Youth Risk Behavior Surveillance System. *Int J Eat Disord* 2008; 41(5):464–70.

27. Hagan, J. F., Shaw, J. S., and Duncan, P., eds. *Bright Futures Guidelines for Health Supervision of Infants, Children, and Adolescents,* 3rd ed. Elk Grove Village, IL: American Academy of Pediatrics, 2008.

28. Centers for Disease Control and Prevention. Recommendations to prevent and control iron-deficiency anemia in the United States. *Morb Mortal Wkly Rep* 2002; 51(40):897–99.

29. National Heart, Lung, and Blood Institute, National High Blood Pressure Education Working Group on Hypertension Control in Children and Adolescents. *4th report on the diagnosis, evaluation and treatment of high blood pressure in children and adolescents.* Bethesda, MD: National Institutes of Health, 2005.

30. Gidding, S. S., Dennison, B. A., Birch, L. L., et al. Dietary recommendations for children and adolescents: A guide for practitioners. *Pediatrics* 2006; 117:544–559.

31. Neumark-Sztainer, D. et al. Obesity, disordered eating, and eating disorders in a longitudinal study of adolescents: How do dieters fare 5 years later? *JADA* 2006; 106(4):559–68.

32. Halvarsson-Edlund, K., Sjödén, P.-O., and Lunner, K. Prediction of disturbed eating attitudes in adolescent girls: A 3-year longitudinal study of eating patterns, self-esteem and coping. *Eating Weight Disord* 2008; 13:87–94.

33. Neumark-Sztainer, D. et al. Does body satisfaction matter? Five-year longitudinal associations between body satisfaction and health behaviors in adolescent females and males. *J Adolesc Health* 2006; 39(2):244–51.

34. Abraham, S., Boyd, C., Lal, M., et al. Time since menarche, weight gain and body image awareness among adolescents girls: Onset of eating disorders? *J Psychoso Obstet Gynaecol* 2009; 30:89–94.

35. Neumark-Sztainer, D. et al. Weight related concerns and behaviors among overweight and nonoverweight adolescents: Implications for preventing weight-related disorders. *Arch Ped Adolesc Med* 2002; 156(2):171–78.

36. American Psychiatric Association. *American Psychiatric Association diagnostic and statistical manual of mental disorders,* 4th ed. (text revision). Washington, D.C.: American Psychiatric Association, 2000.

37. Herpetz-Dahlmann, B. Adolescent eating disorders: Definitions, symptomatology, epidemiology and comorbidity. *Child Adolesc Psychiatr Clin N Am* 2009; 18(1):31-47.

38. Steinhausen, H.-C. Outcome of Eating Disorders. *Child Adolesc Psychiatric Clin N Am* 2008; 18:225–242.

39. Mazzeo, S. E. and Bulik, C. M. Environmental and genetic risk factors for eating disorders: What the clinician needs to know. *Child Adolesc Psychiatr Clin N Am* 2008; 18:67–82.

40. Herpetz-Dahlmann, B. and Slaback-Andrae, H. Overview of treatment modalities in adolescent anorexia nervosa. *Child Adolesc Psychiatr Clin N Am* 2008; 18:131–145.

41. Shaw, H., Stice, E., and Becker, C. B. Preventing eating disorders. *Child Adolesc Psychiatr Clin N Am* 2008; 18:199–207.

Chapter 16

1. Ford, E. S. et al. Healthy living is the best revenge. Findings from the European Prospective Investigation into cancer and nutrition—Potsdam Study. *Arch Intern Med* 2009; 169:1355–1362.

2. Fraser, G. E. et al. Ten years of life: Is it a matter of choice? *Arch Intern Med* 2001; 161:1645–1652.

3. *Healthy People 2010: National health promotion and disease prevention objectives.* Washington, D.C.: U.S. Department of Health and Human Services, 2000. Available at http://healthypeople.gov, accessed 9/09.

4. Slentz, C. A. et al. Exercise, abdominal obesity, sketatal muscle and metabolic risk: evidence of dose response. *Obesity* 2009; 17(suppl):S27–S33.

5. Sipila S. Body composition and muscle performance during menopause and hormonal replacement therapy. *J Endocrinol Invest* 2003; 26:893–901.

6. Goodpaster, B. H. et al. The loss of skeletal muscle strength, mass, and quality in older adults: The health, aging and body composition study. *J Gerontol A Biol Sci Med Sci* 2006; 61(10):1059–64.

7. Canoy D. Distribution of body fat and risk of coronary heart disease in men and women. *Curr Opin Cardio* 2008; 23: 591–298.

8. Ravussin, E. and Bogardus, C. Relationship of genetics, age, and physical fitness to daily energy expenditure and fuel utilization. *Am J Clin Nutri* 1989; 49:968–975.

9. Ravussin, E. et al. Twenty-four hour energy expenditure and resting metabolic rate in obese, moderately obese and control subjects. *Am J Clin Nutr* 1982; 35:566–73.

10. Mifflin, M. C. et al. A new predictive equation for resting energy expenditure in healthy individuals. *Am J Clin Nutr* 1990; 51:241–247.

11. Food and Nutrition Board, Institute of Medicine. *Dietary reference intakes for energy, carbohydrate, fiber, fat, fatty acids, cholesterol, protein, and amino acids.* Washington, D.C.: National Academy Press, 2002.

12. Harris, J. A. and Benedict, F. G. *Standard basal metabolism constants for physiologists and clinicians: A biometric study of basal metabolism in man.* Washington, D.C.: Carnegie Institute of Washington, 1919.

13. Food and Nutrition Board, National Research Council. *Recommended dietary allowances.* Washington, D.C.: National Academy Press, 1989.

14. Harper, E. J. Changing perspectives on aging and energy requirements: Aging and energy intakes in humans, dogs and cats. *J Nutr* 1998; 128:2623S–2626S.

15. McGandy, R. B. et al. Nutrient intakes and energy expenditure of men of different ages. *J Gerontol* 1966; 21:581–587.

16. Farah, H. and Buzby, J. U.S. food consumption up 16 percent since 1970. *Amber Waves*, U.S. Department of Agriculture, Economic Research Service, November 2005.

17. Swinburn, B. A. et al. Estimating change in the energy flux that characterizes the obesity prevalence. *Am J Clin Nutr* 2009; 89:1723–1728.

18. Forman, M. R. et al. Overweight adults in the United States: The behavioral risk factor survey. *Am J Clin Nutr* 1986; 44:410–6.

19. Diaz, V. A. et al. The association between weight fluctuation and mortality: Results from a population-based cohort study. *J Community H* 2005; 30:1532–1565.

20. Nordmann, A. J. et al. Effects of low-carbohydrate vs. low-fat diets on weight loss and cardiovascular risk factors: a meta-analysis of randomized controlled trials. *Arch Intern Med* 2006; 166(3):285–93.

21. World Cancer Research Fund/American Institute for Cancer Research. *Food, nutrition, physical activity, and the prevention of cancer: A global perspective.* Washington, D.C.: AICR, 2007.

22. Arroyave, G. Genetic and biologic variability in human nutrient requirements. *Am J Clin Nutr* 1979; 32:486–500.

23. Leyse-Wallace, R. *Linking nutrition to mental health: A scientific exploration.* Lincoln, NE: iUniverse, Inc., 2008.

24. Health disparities defined. Available at crchd.cancer.gov/disparities/define.html, accessed 8/09.

25. Heart Disease & Stroke Statistics 2009 Update: A report for the American Heart Association Statistics Committee and Stroke Statistics Subcommittee. *Circulation* 2009; 119:e21–e181.

26. *Promoting health equity: A resource to help communities address social determinants of health.* Atlanta, GA: U.S. Department of Health and Human Services, Centers for Disease Control and Prevention, 2008.

27. Commission on Social Determinants of Health. *A conceptual framework for action on the social determinants of health.* Geneva: World Health Organization, April 2007.

28. Wilkinson, R. and Marmot, M. *Social determinants of health: The solid facts.* Copenhagen: WHO Regional Office for Europe, 2006 [Table].

29. Prevention Institute. *Laying the groundwork for a movement to reduce health disparity.* Report II. Oakland, CA: Prevention Institute, April 2007.

30. Krisber K. Healthy People 2020 tackling social determinants of health: input sought from health work force. Nations Health. 2008; 38(10) © 2008 American Public Health Association. Posted: 02/10/2009.

31. Welsh, S. and Saltos, E. Behavioral research and eating competence: USDA-CSREES. *J Nutr Educ Behav* 2007; 39(Suppl.):S142.

32. *Dietary Guidelines for Americans, 2005.* Available at www.healthierus.gov/dietaryguidelines, accessed 8/09.

33. Kushi, L. H. et al. The American Cancer Society 2006 Nutrition and Physical Activity Guidelines Advisory Committee. American Cancer Society guidelines on nutrition and physical activity for cancer prevention: Reducing the risk of cancer with healthy food choices and physical activity. *CA Cancer J Clin* 2006; 56:254–281.

34. Lichtenstein, A. H. et al. Diet and lifestyle recommendations revision 2006: A scientific statement from the American Heart Association Nutrition Committee. *Circulation* 2006; 114:82–96.

35. Stahler, C. How many adults are vegetarian? The Vegetarian Resource Group website. Available at http://www.vrg.org/journal/vj2006issue4/vj2006issue4poll.htm, posted 12/20/06, accessed 9/12/09.

36. Craig, W. J. and Mangels, A. R. Position of the American Dietetic Association: Vegetarian diets. *J Am Diet Assoc* 2009; 109:1266–1282.

37. Popkin, B. et al. A new proposed guidance system for beverage consumption in the Untied States. *Am J Clin Nutr* 2006; 83:529–542.

38. Health behaviors of adults, United States 2002–2004; Chapter 3, "Alcohol use," in *CDC Reports* Series 10, Number 230:8–9, published September 2006.

39. Sensible drinking guidelines. International responsible drinking guidelines. Available at www.drinkingandyou.com, accessed 9/09.

40. Food and Nutrition Board, Institute of Medicine of the National Academy of Sciences. *Dietary reference intakes for water, potassium, sodium, chloride, and sulfate.* Washington, D.C.: The National Academies Press, 2005.

41. Kant, A. K. et al. Intakes of plain water, moisture in foods and beverages, and total water in the adult U.S. population—nutritional, meal pattern, and body weight correlates: National Health and Nutrition Examination Surveys 1999–2006. *Am J Clin Nutr* 2009; 90: 655–663.

42. Grandjean, A. C. et al. The effect of caffeinated, non-caffeinated, caloric and non-caloric beverages on hydration. *J Am Coll Nutr* 2000; 19:591–600.

43. Wilson, J. W. et al. Data tables: Combined results from USDA's 1994 and 1995 Continuing Survey of Food Intakes by Individuals and 1994 and 1995 Diet and Health Knowledge Survey, 1997.

44. Hasler, C. M et al. Position of the American Dietetic Association: Functional foods. *J Am Diet Assoc* 2009; 109:735–746.

45. Sloan, A. E. The top 10 ten functional food trends. *Food Technol* 2008; 62:24–44.

46. Food and Nutrition Board, Institute of Medicine of the National Academy of Sciences. *Dietary reference intakes for energy, carbohydrate, fiber, fat, fatty acids, cholesterol, protein, amino acids (macronutrients).* Washington, D.C.: The National Academies Press, 2005.

47. Pennington, J. A. and Hubbard, V. S. Derivation of daily values used for nutrition labeling. *J Am Diet Assoc* 1997; 97:1407–12.

48. What We Eat in America, NHANES, 2005–2006. U.S. Department of Agriculture, Agriculture Research Service, 2008. Available at www.ars.usda.gov/ba/bhmrc/fsrg, accessed 9/09.

49. Chung, M. et al. Vitamin D and Calcium: Systematic Review of Health Outcomes. AHRQ Publication No. 09-E015, 2009.

50. Lee, J. H. et al. Vitamin D deficiency an important, common, and easily treatable cardiovascular risk factor? *J Am Coll Cardiol* 2008; 52:1949–1956.

51. Vitamin E Fact Sheet, National Insititutes of Health, Office of Dietary Supplements. Available at www.ods.od.nih.gov, accessed 9/09.

52. Miller, E. R. et al. Meta-analysis: High-dosage vitamin E supplementation may increase all-cause mortality. *Ann Intern Med* 2005; 142:37–46.

53. Appel, L. J. et al. Dietary approaches to prevent and treat hypertension: A scientific statement from the American Heart Association. *Hypertension* 2006; 47:296–308. Available at www.hypertensionaha.org, accessed 9/09.

54. Sallis, J. F. and Glanz, K. Physical activity and food environments: Solutions to the obesity epidemic. *The Millbank Quarterly* 2009; 87:123–54.

55. Physical Activity Guidelines Advisory Committee Report to the Secretary of Health and Human Services, 2008. U.S. Department of Health and Human Services website. Available at http://www.halht.gov/PAGuidelines/committeereport.aspx, accessed 9/09.

56. World Health Organization. *Global strategy on diet, physical activity and health.* Geneva: WHO, 2004. Available at www.who.int/dietphysical activity/goals/en/index.html, accessed 9/09.

57. Khan, L. K. et al. Recommended community strategies and measurements to prevent obesity in the United States. *MMWR Recommendation Report* 2009; 58(RR-7):1–26.

58. Rodrigues, N. R. et al. Position paper of the American Dietetic Association and the Canadian Dietetic Association: Nutrition and athletic performance. *J Am Diet Assoc* 2009; 109:509–527.

59. Harnack, L. J. and French, S. A. Effect of point-of-purchase calorie labeling on

restaurant and cafeteria food choices: A review of the literature. *Int J Behav Nutr Phys Act* 2008; 5:51.

60. Wang, M. C. et al. Socioeconomic and food-related physical characteristics of the neighbourhood environment are associated with body mass index. *J Epidemiol Community Health* 2007; 61:491–498.

61. Satter, E. Eating competence: Definition and evidence for the Satter Eating Competence Model. *J Nutr Educ Behav* 2007; 39(Suppl.):S142–153.

62. Satter, E. Eating competence: Nutrition education with the Satter Eating Competence Model. *J Nutr Educ Behav* 2007; 39(Suppl.):S189–194.

63. Sisters Together Program Guide. Weight-control Information Network (WIN), National Institute of Diabetes and Digestive and Kidney Diseases (NIDDK). Available at http://win.niddk.nih.gov/publications/SisPrmGuide2.pdf, accessed 12/09.

64. DeNavas-Walt, C. et al. U.S. Census Bureau, Current Population Reports,

P60-236, *Income, poverty and health insurance coverage in the United States: 2008*. Washington, D.C.: U.S. Census Bureau, 2009. Available at www.census.gov/prod/2008pubs/p60-235.pdf, accessed 12/09.

65. Poverty Definition. Available at http://aspe.hhs.gov/POVERTY/09poverty.shtml, accessed 9/09.

66. Supplemental Food and Nutrition Programs. Available at ers.usda.gov, accessed 9/09.

Chapter 17

1. National Center for Chronic Disease Prevention and Health Promotion. Chronic diseases: The power to prevent, the call to control. Available at www.cdc.gov/nccdphp/publications/AAg/chronic.htm, accessed 8/31/09.

2. *National Vital Statistics Reports* Aug. 2009; 58(1).

3. World Health Organization. Obesity: Preventing and managing the global epidemic. Report of a WHO Consultation. *World Health Organ Tech Rep Ser* 2000; 894:1–253.

4. Overweight and Obesity. Obesity trends: U.S. obesity trends 1985–2008. Available at www.cdc.gov/nccdphp/dnpa/obesity/trend/maps/, accessed 7/09.

5. Differences in prevalence of obesity among black, white, and Hispanic adults: United States, 2006–2008. *MMWR* Jul. 2009; ;58(27):740–4.

6. World Cancer Research Fund/American Institute for Cancer Research. *Food, Nutrition, Physical Activity, and the Prevention of Cancer: A Global Perspective*. Washington, D.C.: AICR, 2007.

7. NHLBI Obesity Education Initiative Expert Panel on the Identification Evaluation, and Treatment of Overweight and Obesity in Adults. Clinical guidelines on the identification, evaluation, and treatment of overweight and obesity in adults: The evidence report. National Institutes of Health. *Obes Res* 1998; 6(Suppl. 2):51S–209S. Available at www.nhlbi.nih.gov.

8. Braveman, P. A health disparities perspective on obesity research. *Prev Chronic Dis* 2009; 6(3). Available at http://www.cdc.gov/pcd/issues/2009/jul/09_0012.htm, accessed 9/20/09.

9. Flegal, K. M. et al. Excess deaths associated with underweight, overweight, and obesity. *JAMA* 2005; 293:1861–67.

10. Berthound, H. R. Mind versus metabolism in the control of food intake and energy balance. *Physiol Behav* 2004; 81:781–793.

11. Clusky, M. and Grobe, D. College weight gain and behavior transitions: Male and female differences. *J Amer Diet Assoc* 2009; 109:325–329.

12. Khan, L. K. et al. Recommended Community Strategies and Measurements to Prevent Obesity in the United States. *MMWR* 2009; 58(RR07):1–26.

13. Razak, F. et al. Defining obesity cut points in a multiethnic population. *Circulation* 2007; 115:2111–2118.

14. Misra, A. Consensus statement for diagnosis of obesity, abdominal obesity and the metabolic syndrome for Asian Indians and recommendations for physical activity, medical and surgical management. *J Assoc Physicians India* Feb. 2009; 57:163–70.

15. Shen, W. et al. Waist circumference correlates with metabolic syndrome indicators better than percentage fat. *Obesity* 2006; 14:727–36.

16. Lohman, T. et al. *Anthropometric Standardization Reference Manual*. Champaign, IL: Human Kinetics, 1991.

17. U.S. Department of Agriculture, U.S. Department of Health and Human Services. *Nutrition and your health: Dietary guidelines for Americans*, 5th ed, 2005.

18. Smith, R. C. *Patient Centered Interviewing*, 2nd ed. Philadelphia: Lippincott Williams & Wilkins, 2002.

19. Wadden, T. A. et al. Randomized trial of lifestyle modifications and pharmacotherapy for obesity. *N Eng J Med* 2005; 353:2111–2120.

20. Adult weight management evidence-based nutrition practice guideline. American Dietetic Association Evidence Analysis Library website. Available at http://www.adaevidencelibrary.comn/topic.cfm?cat=2798, accessed 8/31/ 09.

21. Position of the American Dietetic Association: Weight Management. *J Diet Assoc* 2009; 109:330–346.

22. Shai, I. et al. Dietary Intervention Randomized Controlled Trial (DIRECT) Group. Weight loss with a low-carbohydrate, Mediterranean, or low-fat diet. *N Engl J Med* 2008 Jul 17; 359(3):229–41.

23. Sacks, F. M. et al. Comparison of weight-loss diets with different compositions of fat, protein, and carbohydrates. *N Engl J Med* Feb. 2009; 360(9):859–73.

24. Cooper, Z. et al. *Cognitive-Behavioral Treatment of Obesity: A Clinician's Guide*. New York, NY: The Guilford Press, 2003.

25. Corbalan, M. D. et al. Effectiveness of cognitive-behavioral therapy based on the Mediterranean diet for the treatment of obesity. *Nutrition* 2009; 25:861–869.

26. Physical Activity Guidelines Advisory Committee Report to the Secretary of Health and Human Services, 2008. Available at www.health.gov/paguidelines, accessed 8/31/09.

27. Hill, J. et al. Weight Maintenance: What's Missing? *J Amer Diet Assoc* 2005; 105:63–66.

28. Blundell, J. E. Perspective on the central control of appetite. *Obesity* 2006; 14(Suppl. 4):160S–A63S.

29. Jones, L. R. et al. Lifestyle modification in the treatment of obesity: An educational challenge and opportunity. *Clin Pharmacol Ther* 2007; 81:776–779.

30. Wing, R. R. and Phelan, S. Long-term weight loss maintenance. *Am J Clin Nutr* 2005; 82(Suppl. 1):222S–225S.

31. Digenio, A. G. et al. Comparison of methods for delivering a lifestyle modification program for obese patients: A randomized trial. *Annal Int Med* 2009; 150:255–262.

32. Li, Z. et al. Meta-analysis: Pharmacologic treatment of obesity. *Ann Intern Med* 2005; 142:532–546.

33. Sjostrom, L. et al. Effects of bariatric surgery on mortality in Swedish obese subjects. *N Engl J Med* 2007; 357:741–752.

34. Vogel, J. A. et al. Reduction in predicted coronary heart disease after substantial weight reduction after bariatric surgery. *Am J Cardiol* 2007; 99:222–226.

35. Parkes, E. Nutritional management of patients after bariatric surgery. *Am J Med Sci* 2006; 331:207–213.

36. *Healthy People 2010: National Health Promotion and Disease Prevention Objectives*. Washington, D.C.: U.S. Department of Health and Human Services, 2000.

37. Heart Disease and Stroke Statistics— 2009 Update: A Report from the American Heart Association Statistics Committee and

Stroke Statistics Subcommittee. *Circulation* 2009; 119:e21–e181.

38. Agency for Healthcare Research and Quality. 2007 National Healthcare Disparities Report. Rockville, MD: U.S. Department of Health and Human Services, Agency for Healthcare Research and Quality, February 2008. AHRQ publication No. 08-0041.

39. Willerson, J. T. and Ridker, P. M. Inflammation as a cardiovascular risk factor. *Circulation* 2004; 109:2–10.

40. Burke, G. L. et al. The impact of obesity on cardiovascular disease risk factors and subclinical vascular disease: The Multi-Ethnic Study of Atherosclerosis. *Arch Intern Med* 2008; 168:928–935.

41. Lee, C. D. et al. Abdominal obesity and coronary artery calcification in young adults: The Coronary Artery Risk Development in Young Adults (CARDIA) Study. *Am J Clin Nutr* 2007; 86:48–54.

42. Jeppesen, P. et al. Insulin resistance, the metabolic syndrome, and risk of incident cardiovascular disease: A population-based study. *J Am Coll Cardiol* 2007; 49(21):2112–2129.

43. Third report of the National Cholesterol Education Program (NCEP) Expert Panel on Detection, Evaluation, and Treatment of High Blood Cholesterol in Adults (Adult Treatment Panel III). *J Am Med Assoc* 2001; 285:2486–2497. Available at www.nhlbi.nih.gov/guidelines/cholesterol/atp_iii.htm, accessed 8/31/09.

44. Wilson, P. W. F. et al. Prediction of coronary heart disease using risk factor categories. *Circulation* 1998; 97:1837–1847.

45. Grundy, S. M. et al. Definition of metabolic syndrome: Report of the National Heart, Lung, and Blood Institute/American Heart Association Conference on Scientific Issues Related to Definition. *Circulation* 2004; 109:433–438.

46. *American Dietetic Association evidence-based practice guideline and toolkit: Disorders of lipid metabolism.* Chicago, IL: American Dietetic Association, 2005.

47. Lichtenstein, A. H. et al. Diet and lifestyle recommendations revision 2006: A scientific statement from the American Heart Association Nutrition Committee. *Circulation* 2006; 114:82–96.

48. Riediger, N. D. et al. A systematic review of the roles of n-3 fatty acids in health and disease. *J Amer Diet Assoc* 2009; 109:668–679.

49. American Dietetic Association Evidence Library. Evidence summary: Stanols and sterols and hyperlipidemia. March, 2004. Available at www.adaevidencelibrary.com, accessed 9/4/09.

50. Position of the American Dietetic Association: Health implications of dietary fiber. *J Am Diet Assoc* 2008; 108:1716–1731.

51. Take the *trans* fat out of New York: Cardiovascular disease prevention and control. Available at http://www.nyc.gov/html/doh/cardio/cardio-transfat.shtml, accessed 8/3/09.

52. Shamliyan, T. A. et al. Are your patients with risk of CVD getting the viscous soluble fiber they need? Few patients eat the right amount of fiber known to reduce CVD risks and events. *Journal of Family Practice* Sept. 2006.

53. Leon, H. et al. Effect of fish oil on arrhythmias and mortality: Systematic review. *BMJ* 2009; 338:a2931 doi:10.1136/bmj.a2931.

54. International Diabetes Federation. Worldwide definition of the metabolic syndrome. Available at http://www.idf.org/webdata/docs/MetSyndrome_FINAL.pdf, accessed 8/31/09.

55. Ford, E. S. et al. Prevalence of metabolic syndrome among US adults: Findings from the third National Health and Nutrition Examination Survey. *JAMA* 2002; 287:356–9.

56. Russell, M. et al. The metabolic syndrome in American Indians: The Strong Heart Study. *JCMS* 2007; 2:283–287.

57. Yong-Woo, P. et al. The metabolic syndrome: Prevalence and associated risk factor findings in the US population from the Third National Health and Nutrition Examination Survey, 1988–1994. *Arch Intern Med* 2003; 163:427–436.

58. Sumner, A. and Cowie, C. C. Ethnic differences in the ability of triglyceride levels to identify insulin resistance. *Atherosclerosis* 2008; 196:696–703.

59. Wilson, P. W. et al. Metabolic syndrome as a precursor of cardiovascular disease and type 2 diabetes mellitus. *Circulation* 2005; 112:3066–3072.

60. American Diabetes Association. Executive summary: Standards of medical care in diabetes—2009. *Diabetes Care* 2009; 32(Suppl. 1):S6–S12.

61. Monajemi, H. et al. Inherited lipodystrophies and the metabolic syndrome. *Clin Endocrinol* (Oxf) 2007; 67:479–484.

62. Misra, A. et al. Novel phenotype markers and screening score for the metabolic syndrome in adult Asian Indians. *Diabetes Res Clin Pract* 2008; 79:E1–E5.

63. Phelan, S. et al. Impact of weight loss on the metabolic syndrome. *Int J Obs* (Lond) 2007; 9:1442–1448.

64. American Diabetes Association. Diagnosis and classification of diabetes mellitus. *Diabetes Care* 2006; 29(1):S43–S48.

65. Hamman, R. F. et al. Effect of weight loss with lifestyle intervention on risk of diabetes. *Diabetes Care* 2006; 29: 2102–2107.

66. Urbanski, P. et al. Cost-effectiveness of diabetes education. *J Am Diet Assoc* 2008; 108(Suppl.):S6–S11.

67. Delahanty, L. M. and David, M. Implications of the diabetes Prevention Program and Look AHEAD Clinical Trials for Lifestyle Interventions. *J Am Diet Assoc* 2008; 108(Suppl.):S66–S72.

68. U.S. Preventive Services Task Force. Screening for type 2 diabetes mellitus in adults: U.S. Preventive Services Task Force recommendation statement. *Ann Intern Med* 2008; 148:846–854.

69. Daly, A. et al. Diabetes white paper: Defining the delivery of nutrition services in Medicare medical nutrition therapy vs Medicare diabetes self-management training programs. *J Am Diet Assoc* 2009; 109:528–539.

70. American Dietetic Association/American Diabetes Association. *Choose Your Foods: Exchange Lists for Diabetes*, 6th ed., 2007.

71. Hayes, C. and Kriska, A. Role of physical activity in diabetes management and prevention. *J Am Diet Assoc* 2008; 108(Suppl.):S19–S23.

72. Fonseca, V. A. and Kulkarni, K. D. Management of type 2 diabetes: Oral agents, insulin and injectables. *J Am Diet Assoc* 2008; 108(Suppl.):S29–S33.

73. Geil, P. and Shane-McWhorter, L. Dietary supplements in the management of diabetes: Potential risks and benefits. *J Am Diet Assoc* 2008; 108(Suppl.):S59–S65.

74. World Cancer Research Fund/American Institute for Cancer Research. The cancer process. Chapter 2 in *Nutrition, Physical Activity and the Prevention of Cancer: A Global Perspective*. Washington, D.C.: AICR, 2007.

75. Rowland, J. H. et al. Cancer survivorship: A new challenge in delivering quality cancer care. *J Clin Oncology* 2006; 24:5101–5104.

76. American Cancer Society. *Cancer Facts & Figures 2009*. Atlanta, GA: American Cancer Society, 2009.

77. Krieger, N. Defining and investigating social disparities in cancer: Critical issues. *Cancer and Cancer Causes* 2005; 6:5–14.

78. Ries, L. A. G. et al (eds.) *SEER Cancer Statistics Review, 1975–2004*. Bethesda, MD: National Cancer Institute, 2007. Available at www.seer.cancer.gov/csr/1975_2004/.

79. Jemal, A. et al. Annual Report to the Nation on the Status of Cancer, 1975–2005, Featuring Trends in Lung Cancer, Tobacco Use and Tobacco Control. *Journal of the National Cancer Institute* Dec. 2, 2008; 100(23).

80. Renehan, A. G. et al. Obesity and cancer: Pathophysiological and biological mechanisms. *Arch Physiol Biochem* 2008; 114:71–83.

81. American Cancer Society. Half the cancer deaths could be prevented. *CA Cancer J Clin* 2005; 55:209–210.

82. Doyle, C. et al. The 2006 Nutrition, Physical Activity and Cancer Survivorship Advisory Committee; American Cancer Society. Nutrition and physical activity during and after cancer treatment: An American Cancer Society guide for informed choices. *CA Cancer J Clin* 2006; 56:323–353.

83. Parekh, N. et al. Obesity, insulin resistance, and cancer prognosis: Implications for practice for providing care among cancer survivors. *J Amer Diet Assoc* 2009; 109:346–1353.

84. Kushi, L. H. et al. The American Cancer Society 2006 Nutrition and Physical Activity Guidelines Advisory Committee. American Cancer Society Guidelines on Nutrition and Physical Activity for Cancer Prevention: Reducing the Risk of Cancer with Healthy Food Choices and Physical Activity. *CA Cancer J Clin* 2006; 56:254–281.

85. National Cancer Institute. Nutrition in Cancer Care (PDQ). Available at www.cancer.gov/cancertopics/pdq/supportivecare/nutrition/HealthProfessional/page2, accessed 9/5/09.

86. Ryan, J. L., Heckler, C., Dakhil, S. R. et al. Ginger for chemotherapy-related nausea in cancer patients: A URCC CCOP randomized, double-blind, placebo-controlled clinical trial of 644 cancer patients. Paper presented at the American Society of Clinical Oncology Annual Meeting, May 2009.

87. HIV/AIDS in the United States. Fact Sheet. Centers for Disease Control and Prevention. Division of HIV/AIDS Prevention, National Center for HIV, STD, and TB Prevention. Atlanta, GA. Available at www.cdc.gov/hiv/topics/, accessed 9/09.

88. Position of the American Dietetic Association and the Dietitians of Canada: Nutrition intervention in the care of persons with human immunodeficiency virus infection. *J Am Diet Assoc* 2004; 104:1425–1441.

89. Cofrancesco, J. et al. Treatment options for HIV-associated central fat accumulation. *AIDS Patient Care STDS* 2009; 23:5–18.

90. Samaras, K. et al. Prevalence of metabolic syndrome in HIV-infected patients receiving highly active antiretroviral therapy using International Diabetes Foundation and Adult Treatment Panel III criteria: Associations with insulin resistance, disturbed body fat compartmentalization, elevated C-reactive protein, and hypoadiponectinemia. *Diabetes Care* 2007; 30:113–119.

91. Mondy, K. et al. Metabolic syndrome in HIV-infected patients from an urban, Midwestern U.S. outpatient population. *Clin Infect Dis* 2007; 44:726–734.

92. Coffey, S. (ed.) Clinical manual for management of the HIV-infected adult. Aids Education & Training Centers 2006. Available at www.aids-etc.org/aidsetc?page=cm-00-00, accessed 9/4/09.

Chapter 18

1. Levy, B. R., Slade, M. D., Kunkel, S. R., and Kasl, S. V. Longevity increased by positive self-perceptions of aging. *J Personality and Social Psychology* 2002; 83(2):261–270.

2. National Center for Injury Prevention and Control. *10 Leading Causes of Death, United States, 2006, All Races, Both Sexes, Age 60 to 85+*. Available at http://webappa.cdc.gov/cgi-bin/broker.exe, accessed 7/23/09.

3. Pleis, J. R. and Lucas, J. W. Summary health statistics for U.S. adults: National Health Interview Survey, 2007. National Center for Health Statistics. Vital Health Statistics. *Vital Health Stat* 2009; 10(240).

4. Maloney, S. K., Fallon, B., and Wittenberg, C. K. Executive summary. *Aging and health promotion: Market research for public education*. Washington, D.C.: Office of Disease Prevention and Health Promotion, U.S. Department of Health and Human Services, PHS, 1984.

5. Nube, M. et al. Scoring of prudent dietary habits and its relation to 25-year survival. *J Am Diet Assoc* 1987; 87:171–75.

6. Kant, A. K. et al. A prospective study of diet quality and mortality in women. *JAMA* 2000; 283:2109–15.

7. Position Paper of the American Dietetic Association: Nutrition across the spectrum of aging. *J Am Diet Assoc* 2005; 105:616–33.

8. Vita, A. J. et al. Aging, health risks, and cumulative disability. *N Engl J Med* 1998; 338:1035–41.

9. Trichopoulou, A., Barnia, C., and Trichopoulous, D. Anatomy of health effects of Mediterranean diet: Greek EPIC prospective cohort study. *BMJ* 2009; 338:b2337 doi:10:1136/bmj.b2337.

10. King, D. E., Mainous III, A. G., and Geesey, M. E. Turning back the clock: Adopting a healthy lifestyle in middle age. *The American Journal of Medicine* 2007; 120(7):598–603.

11. American College of Sports Medicine. Exercise and physical activity for older adults. Medicine & Science in Sports and Exercise, Position Statement, 2009. Doi: 10.1249/MSS.0b013e318a0c95c

12. Taylor, P. et al. Pew Research Center. Growing Old in America: Expectations vs. Reality. June 29, 2009. Available at http://pewsocialtrends.org.

13. Johnson, M. A. Achieving 100 candles: The Georgia Centenarian Study lights the way. USDA 2000 Millenium Lecture Symposium, September 28, 2000.

14. 1995 White House Conference on Aging. The road to aging policy for the 21st Century. Washington, D.C.: U.S. National Commission on Libraries, 1996.

15. Center for Disease Control and Prevention. Health Data Interactive. Life expectancy at birth, 65 and 85. Available at http://205.207.175.93/HDI/TableViewer/tableView.aspx, accessed 7/25/09.

16. *Healthy People 2010: National health promotion and disease prevention objectives*. Washington, D.C.: U.S. Dept. of Health and Human Services, 2000.

17. World Health Organization. Aging and life course. Available at http://www.who.int/ageing/en, accessed 6/18/09.

18. Hayflick, L. How and why we age. *Exp Gerontol* 1998; 33:639–53.

19. Gerontology Research Group. Official Tables from the International Committee on Supercentenarians. Available at http://www.grg.org/Adams/B.HTM, accessed 6/18/09.

20. Capri, M., Salvioli, S., Sevini, F., et al. The genetics of human longevity. *Ann NY Acad Scie* 2006; 1067:252–263.

21. Forbes, G. B. *Human body composition: Growth, aging, nutrition, and activity*. New York: Springer-Verlag, 1987: 31.

22. Buys, C. H. Telomeres, telomerase, and cancer. *N Engl J Med* 2000; 342:1282–83.

23. Weindruch, R. and Sohal, R. S. Seminars in medicine of the Beth Israel Deaconess Medical Center. Caloric intake and aging. *N Engl J Med* 1997; 337:986–94.

24. McCay, C., Crowell, M., and Maynard, L. The effect of retarded growth upon the length of life and upon ultimate size. *J Nutr* 1935; 10:63–79.

25. Masoro, E. J. Nutrition and aging in animal models. In Munro, H. N. and Danford, D. E., eds. *Nutrition, aging, and the elderly*. New York: Plenum Press, 1989: 25–41.

26. Colman, R. J. et al. Caloric restriction delays disease onset and mortality in Rhesus monkeys. *Science* 2009; 325:201–204.

27. Walford, R. L. et al. Calorie restriction in biosphere 2: Alterations in physiologic, hematologic, hormonal, and biochemical parameters in humans restricted for a 2-year period. *J Gerontol A Biol Sci Med Sci* 2002; 57:B211–24.

28. Fontana, L. et al. Effect of long-term calorie restriction with adequate protein and micronutrients on thyroid hormones. *J Clin Endocrine Metab* 2006; 91(8):3232–3235.

29. Willcox, B. J., Willcox, D. C., Todoriki, H. et al. Caloric restriction, the traditional Okinawan diet, and healthy aging. The diet of the world's longest-lived people and its potential impact on morbidity and life span. *Ann NY Acad Sci* 2007; 1114:434–455, quotation on page 451.

30. Walker, A. R. and Walker, B. F. Nutritional and non-nutritional factors for 'healthy' longevity. *J R Soc Health* 1993; 113:75–80.

31. Forbes, G. B. Longitudinal changes in adult fat-free mass: Influence of body weight. *Am J Clin Nutr* 1999; 70:1025–31.

32. Guo, S. S. et al. Aging, body composition, and lifestyle: The Fels Longitudinal Study. *Am J Clin Nutr* 1999; 70:405–11.

33. Wilmore, J. H. et al. Alterations in body weight and composition consequent to 20 weeks of endurance training: The HERITAGE Family Study. *Am J Clin Nutr* 1999; 70:346–52.

34. Fiatarone, M. A. et al. High-intensity strength training in nonagenarians: Effects on skeletal muscle. *JAMA* 1990; 263:3029–34.

35. Villareal, D. T. et al. Obesity in older adults: Technical review and position statement of the American Society for Nutrition and NAASO, The Obesity Society. *Obes Res* 2005; 13:1849–63.

36. Elahi, V. K. et al. A longitudinal study of nutritional intake in men. *J Gerontol* 1983; 38:162–80.

37. Kaneda, H. et al. Decline in taste and odor discrimination abilities with age, and relationship between gustation and olfaction. *Chem Senses* 2000; 25:331–37.

38. Murphy, C. et al. Prevalence of olfactory impairment in older adults. *JAMA* 2002; 288:2307–12.

39. Bartoshuk, L. M. Taste: Robust across the age span? *Ann N Y Acad Sci* 1989; 561:65–75.

40. Schiffman, S. Changes in taste and smell: Drug interactions and food preferences. *Nutr Rev* 1994; 52(II):S11–14.

41. Holm-Pedersen, P., Schultz-Larsen, K., Christiansen, N., and Avlund, K. Tooth loss and subsequent disability and mortality in old age. *J Am Geriatr Soc* 2008; 56(3):429–35.

42. American Dietetic Association. Position of the American Dietetic Association: Oral health and nutrition. *J Am Diet Assoc* 2003; 103:615–625.

43. Jones, D. B., Niendorff, W. J., and Broderick, E. B. A review of the oral health of American Indian and Alaska Native elders. *J Public Health Dent* 2000; 60(Suppl. 1):256–60.

44. Vargas, C. M., Kramarow, E. A., and Yellowitz, J. A. *The oral health of older Americans. Aging Trends*, No. 3. Hyattsville, MD: National Center for Health Statistics, 2001.

45. Roberts, S. B. et al. Control of food intake in older men. *JAMA* 1994; 272:1601–6.

46. Rolls, B. J. Regulation of food and fluid intake in the elderly. *Ann N Y Acad Sci* 1989; 561:217–25.

47. Phillips, P. A., Johnston, C. I., and Gray, L. Disturbed fluid and electrolyte homoeostasis following dehydration in elderly people. *Age Ageing* 1993; 22:S26–33.

48. Phillips, P. A. et al. Reduced thirst after water deprivation in healthy elderly men. *N Engl J Med* 1984; 311:753–59.

49. U.S. Department of Agriculture, Agricultural Research Service, 2008. Nutrient Intakes from Food: Mean amounts consumed per individual, one day, 2005–2006. Available at www.ars.usda.gov/ba/bhnrc/fsrg, accessed 8/03/09.

50. White, J. V. et al. Nutrition screening initiative: Development and implementation of the public awareness checklist and screening tools. *J Am Diet Assoc* 1992; 92:163–7.

51. White, J. V. et al. Consensus of the Nutrition Screening Initiative: Risk factors and indicators of poor nutritional status in older Americans. *J Am Diet Assoc* 1991; 91:783–87.

52. Posner, B. M. et al. Nutrition and health risks in the elderly: The nutrition screening initiative. *Am J Public Health* 1993; 83:972–78.

53. Rose, D. and Oliveira, V. Nutrient intakes of individuals from food-insufficient households in the United States. *Am J Public Health* 1997; 87:1956–61.

54. Bowman, S. A. Socioeconomic characteristics, dietary and lifestyle patterns, and health and weight status of older adults in NHANES 1999–2002: A comparison of Caucasians and African Americans. *J Nutr Elderly* 2009; 28(1):30–46.

55. The Slone Survey. Patterns of medication use in the United States 2006. A report from the Slone Survey. Available at www.bu.edu/slone/SloneSurvey/AnnualRpt/slonesurveywebreport2006.pdf, accessed 8/03/09.

56. De Castro, J. M. Social facilitation of food intake: People eat more with other people. *Food Nutr News* 1994; 66:29–30.

57. Gerrior, S. A. et al. How does living alone affect dietary quality? U.S. Department of Agriculture, ARS, Home Economics Research Report No. 51. *Fam Econ Nutr Rev* 1995; 8:44–46.

58. Sahyoun, N. R. et al. Nutrition Screening Initiative checklist may be a better awareness/educational tool than a screening one. *J Am Diet Assoc* 1997; 97:760–64.

59. Cereda, E., Pusani, C., Limonata, D., and Vanotti, A. The ability of the Geriatric Nutritional Risk Index to assess the nutritional status and predict the outcome of home-care resident elderly: A comparison with the Mini Nutritional Assessment. *British J of Nutrition* 2009; 1–8. Doi:10.1017/S0007114509222677.

60. deGroot, L. C. et al. Evaluating the DETERMINE Your Nutritional Health Checklist and the Mini-Nutritional Assessment as tools to identify nutritional problems in elderly Europeans. *Eur J Clin Nutr* 1998; 52(12):877–83.

61. Elia, M., Smith, R. M., and British Association for Parenteral and Enteral Nutrition. *Improving nutritional care and treatment: Perspectives and recommendations from population groups, patients and carers*, 2009. Available at www.bapen.org.uk, accessed 7/20/09.

62. Lichtenstein, A. H., Rasmussen, H., Yu, W. W., et al. Modified MyPyramid for older adults. *J Nutr* 2008; 138:5–11.

63. Otten, J. J., Hellwig, J. P., and Meyers, L. D., eds. Institute of Medicine of the National Academies. DRI Dietary Reference Intakes: The Essential Guide to Nutrient Requirements. Washington, D.C.: The National Academies Press, 2006.

64. Zizza, C. A., Tayie, F. A., and Lino, M. Benefits of snacking in older Americans. *J Am Diet Assoc* 2007; 107:800–806.

65. Arcerio, P. J. et al. A practical equation to predict resting metabolic rate in older men. *Metabolism* 1993; 42(8):950–57.

66. Arcerio, P. J. et al. A practical equation to predict resting metabolic rate in older females. *J Am Geriatric Soc* 1994; 41:389–95.

67. Taylor, H. L. et al. A questionnaire for the assessment of leisure-time physical activities. *J Chronic Dis* 1978; 31:741–55.

68. Fulgoni, V. L. Current protein intake in America: Analysis of the National Health and Nutrition Examination Survey, 2003–2004. *Am J Clin Nutr* 2008; 87(Suppl.);1554S–7S.

69. Campbell, W. W., Johnson, C. A., McCabe, G. P., and Carnell, N. S. Dietary protein requirements of younger and older adults. *Am J Clin Nutri* 2008; 88:1322–9.

70. Rand, W. M., Pellett, P. L., and Young, V. R. Meta-analysis of nitrogen balance studies for estimating protein requirements in healthy adults. *Am J Clin Nutri* 2003; 77:109–27.

71. Morais, J. A., Chevalier, S., and Gougeon, R. Protein turnover and requirements in the healthy and frail elderly. *The Journal of Nutrition, Health, & Aging* 2006; 10(4):272–83.

72. Paddon-Jones, D., Short, K. R., Campbell, W. W., Volpi, E., and Wolfe, R. R. Role of dietary protein in the sarcopenia of aging. *Am J Clin Nutr* 2008; 87(Suppl.):1562S–6S.

73. Heron, K. L. et al. Pre-menopausal women, classified as hypo- or hyper-responders, do not alter their LDL/HDL ratio following a high dietary cholesterol challenge. *J Am Coll Nutr* 2002; 21:250–58.

74. Vivekananthan, D. P. et al. Use of antioxidant vitamins for the prevention of cardiovascular disease: Meta-analysis of randomised trials. *Lancet* 2003; 361:2017–23.

75. Weaver, C. M. and Fleet, J. C. Vitamin D requirements: Current and future. *Am J Clin Nutr* 2004; 80(Suppl.):1735S–9S.

76. Holick, M. F. and Chen, T. C. Vitamin D deficiency: A worldwide problem with health consequences. *Am J Clin Nutri* 2008; 87(Suppl.):1080S–6S.

77. Webb, A. R., Kline, L., and Holick, M. F. Influence of season and latitude on the cutaneous synthesis of vitamin D3: Exposure to winter sunlight in Boston and Edmonton will not promote vitamin D3 synthesis in human skin. *J Clin Endocrinol Metab* 1988; 67:373–78.

78. Bischoff-Ferrari, H. A., Giovannucci, E., Willett, W. C., et al. Estimation of optimal concentrations of 25-hydroxyvitamin D

for multiple health outcomes. *Am J Clin Nutr* 2006; 84:18–28.

79. Jacques, P. F. The potential preventive effects of vitamins for cataract and age-related macular degeneration. *Int J Vitam Nutr Res* 1999; 69:198–205.

80. Meydani, M. Effect of functional food ingredients: Vitamin E modulation of cardiovascular diseases and immune status in the elderly. *Am J Clin Nutr* 2000; 71:1665S–8S; discussion 1674S–5S.

81. Sano, M. et al. A controlled trial of selegiline, alpha-tocopherol, or both as treatment for Alzheimer's disease. The Alzheimer's Disease Cooperative Study. *N Engl J Med* 1997; 336:1216–22.

82. Miller, E. R. et al. Meta-analysis: High-dosage vitamin E supplementation may increase all-cause mortality. *Ann Intern Med* 2005; 142:37–46.

83. Presse, N., Shatenstein, B., Kergoat, M. J., and Ferland, G. Validation of a semi-quantitative food frequency questionnaire measuring dietary vitamin K intake in elderly people. *J Am Diet Assoc* 2009; 109:1251–1255.

84. Cockayne, S., Adamson, J., Lanham-New, S., et al. Vitamin K and the prevention of fractures: Systematic review and meta-analysis of randomized controlled trials. *Arch Intern Med* 2006; 166:1256–1261.

85. Marcason, W. Vitamin K: What are the current dietary recommendations for patients taking Coumadin? *J Am Diet Assoc* 2007; 107:2022.

86. Malinow, M. R. et al. Reduction of plasma homocyst(e)ine levels by breakfast cereal fortified with folic acid in patients with coronary heart disease. *N Engl J Med* 1998; 338:1009–15.

87. Optimal calcium intake. National Institutes of Health Consensus Statement. 1994; 12:1–31.

88. Appel, L. J. et al. A clinical trial of the effects of dietary patterns on blood pressure. DASH Collaborative Research Group. *N Engl J Med* 1997; 336:1117–24.

89. Shils, M. E. Magnesium. In Ziegler, E. E. and Filer, L. J., eds. *Present knowledge in nutrition.* Washington, D.C.: ILSI Press; 1996: 256–64.

90. Bjelakovic, G., Nikolova, D., Gluud, L. L., et al. Antioxidant supplements for prevention of mortality in healthy participants and patients with various diseases. *Coch Dat of System Rev* 2008; Issue 2. Art. No.:CD007176. DOI:10.1002/14651858. CD007176.

91. Prasad, A. S., Beck, F. W. J., Bao, B., et al. Zinc supplementation decreases incidence of infections in the elderly: Effect of zinc on generation of cytokines and oxidative stress. *Am J Clin Nutr* 2007; 85:837–44.

92. Barringer, T. A., Kirk, J. K., Santaniello, A. C. et al. Effect of a multivitamin and mineral supplement on infection and quality of life: A randomized, double-blind, placebo-controlled trial. *Ann Intern Med* 2003; 138:365–371.

93. El-Kadiki, A. and Sutton, A. J. Role of multivitamins and mineral supplements in preventing infection in elderly people: Systematic review and meta-analysis of randomized controlled trials. *BMJ* doi:10.1136/bmj.38399.495648.8F (published 31 March 2005).

94. National Institutes of Health, State-of-the-Science conference statement. Multivitamin/mineral supplements and chronic disease prevention. *Ann Intern Med* 2006; 145.

95. Fairfield, K. M. and Fletcher, R. H. Vitamins for chronic disease prevention in adults: Acientific review. *J Am Med Assn* 2002; 287:3116–26.

96. Fletcher, R. H. and Fairfield, K. M. Vitamins for chronic disease prevention in adults: Clinical applications. *J Am Med Assn* 2002; 287:3127–29.

97. Balluz, L. S. et al. Vitamin and mineral supplement use in the United States. Results from the Third National Health and Nutrition Examination Survey. *Arch Fam Med* 2000; 9:258–62.

98. U.S. Preventive Services Task Force. Routine vitamin supplementation to prevent cancer and cardiovascular disease: Recommendations and rationale. *Ann Intern Med* 2003; 139:51–55. Available at www.preventiveservices.ahrq.gov.

99. Schulz, V. et al. *Rational phytotherapy: A reference guide for physicians and pharmacists*, 5th ed. Berlin: Springer Verlag, 2004.

100. Fragakis, A. S. *The health professional's guide to popular dietary supplements*, 2nd ed. Chicago: American Dietetic Association, 2003.

101. Evans, W. J. Exercise training guidelines for the elderly. *Med Sci Sports Exerc* 1999; 31:12–7.

102. Kligman, E. W., Hewitt, M. J., and Crowell, D. L. Recommending exercise to healthy older adults. *The Physician Sportsmed* 1999; 27(11):1–11.

103. Papadopoulou, S. K. et al. Health status and socioeconomic factors as determinants of physical activity level in the elderly. *Med Sci Monit* 2003; 9(2):79–83.

104. Krinke, U. B. Effective nutrition education strategies to reach older adults. In Watson, R. R., ed. *Handbook of nutrition in the aged*. Boca Raton, FL: CRC Press, 2001: 319–31.

105. Kamp, B., Wellman N.S., and Russell C. Food and Nutrition Programs for Community-Residing Older Adults: Position Statement of the American Dietetic Association, American Society for Nutrition, and Society for NutritionEducation. *J Am Diet Assoc and J Nutr Educ Behav.* 2010 (in press).

106. Wellman, N. S., Rosenzweig, L., and Lloyd, J. L. Thirty years of the Older Americans Nutrition Program. *J Am Diet Assoc* 2002; 102(3):348–50.

107. Ponza, M., Ohls, J., and Millen, B. Serving elders at risk. *The Older Americans Act Nutrition Programs—National Evaluation of the Elderly Nutrition Program, 1993–1995.* Washington, D.C.: Mathematica Policy Research, Inc. and Administration on Aging, 1996.

108. Buettner, D. The Blue Zone Lessons for living longer from the people who've lived the longest. Washington, D.C.: National Geographic Society, 2008.

Chapter 19

1. Federal Interagency Forum on Aging-Related Statistics. Older Americans update 2008: Key indicators of well-being. Washington, D.C.: U.S. Government Printing Office, March 2008. Available at www.agingstats.gov/agingstatsdotnet/main_site/default.aspx, accessed 8/18/09.

2. Guenther, P. M., Juan, W. Y., Reedy, J., et al. Diet Quality of Americans in 1994–96 and 2001–02 as measured by the Healthy Eating Index 2005. *Nutrition Insight* December 2007 (37). United States Department of Agriculture, Center for Nutrition Policy and Promotion.

3. Medicare Program; Revisions to Payment Policies and Five-Year Review of and Adjustments to the Relative Values Units Under the Physician Fee Schedule for Calendar Year 2002: Final Rule. Federal Register, 42 CFR, Part 405; 66(212), November 1, 2001: 55275–81.

4. Lacey, K., and Pritchett, E. Nutrition care process and model: ADA adopts road map to quality care and outcomes management. *J Am Diet Assoc* 2003; 103(8):1061–72.

5. American Dietetic Association Position Paper. Cost effectiveness of medical nutrition therapy. *J Am Diet Assoc* 1995; 95:88–91.

6. Chima, C. S. et al. Relationship of nutritional status to length of stay, hospital costs, and discharge status of patients hospitalized in the medicine service. *J Am Diet Assoc* 1997; 97:975–78; quiz 979–80.

7. Sullivan, D. H., Sun, S., and Walls, R. C. Protein-energy undernutrition among elderly hospitalized patients: A prospective study. *J Am Med Assoc* 1999; 281:2013–19.

8. Wolinsky, F. D. et al. Health services utilization among the noninstitutionalized elderly. *J Health Soc Behav* 1983; 24:325–37.

9. National Center for Health Statistics. Data Warehouse LCWK3.2006. Available at www.cdc.gov/nchs/data/dvs/LCWK3_2006.pdf, accessed 8/20/09.

10. Messier, S. P. et al. Weight loss reduces knee-joint loads in overweight and obese older adults with knee osteoarthritis. *Arthritis & Rheumatism* 2005; 52:2026–32.

11. Yates, L. B., Djousse, L., Kurth, T., et al. Exceptional longevity in men: Modifiable risk factors associated with survival and function to age 90 years. *Arch Intern Med* 2008; 168(3):284–290.

12. Imamura, F., Jacques, P. F., Herrington, D. M. et al. Adherence to 2005 Dietary Guidelines for Americans is associated with a reduced progression of coronary artery atherosclerosis in women with established coronary artery disease. *Am J Clin Nutr* 2009; 90:193–201.

13. Schoenborn, C. A. and Heyman, K. M. Health characteristics of adults aged 55 years and over: United States, 2004–2007. Centers for Disease Control and Prevention, NCHS: National Health Statistics Reports, Number 16, July 8, 2009.

14. American Heart Association. Heart Disease and Stroke Statistics–2009 Update. Dallas, Texas: American Heart Association, 2009.

15. Erlinger, T. P., Pollack, H., and Appel, L. J. Nutrition-related cardiovascular risk factors in older people: Results from the Third National Health and Nutrition Examination Survey. *J Am Geriatr Soc* 2000; 48:1486–89.

16. American Diabetes Association. Diagnosis and classification of diabetes mellitus. *Diabetes Care* 2006; 29(1):S43–S48.

17. Grundy, S. M. et al. Implications of recent clinical trials for the National Cholesterol Education Program Adult Treatment Panel III Guidelines. *Circulation* 2004; 110:227–39.

18. Lavie, C. J., Milani, R. V., Mehra, M. R., et al. Omega-3 polyunsaturated fatty acids and cardiovascular diseases. *J Am Coll Cardiol* 2009; 54:585–94.

19. Leon, H., Shibata, M. C., Sivakumaran, S. et al. Effect of fish oil on arrhythmias and mortality: Systematic review. *BMJ* 2009; 338:a2931 doi:10.1136/bmj.a2931.

20. Easton, D., Saver, J. L., Albers, G. W., et al. Definition and evaluation of transient ischemic attack. *Stroke* 2009; 40:2276–2293. Available at http://stroke.ahajournals.org, accessed 9/1/09

21. Hiroyasu et al. Alcohol intake and the risk of cardiovascular disease in middle-aged Japanese men. *Stroke* 1995; 26(5):767–73.

22. Sacco, R. L. et al. The protective effects of moderate alcohol consumption on ischemic stroke. *JAMA* 1999; 281:53–60.

23. The seventh report of the Joint National Committee on Prevention, Detection, Evaluation, and Treatment of High Blood Pressure (JNC 7). NIH Pub No. 04–5230.

August 2004. Available at www.nhlbi.nih.gov/guidelines/, accessed 8/28/09.

24. Appel, L. J. et al. Dietary approaches to prevent and treat hypertension. A scientific statement from the American Heart Association. *Hypertension* 2006; 47:296–308.

25. Plaisted, C. S. et al. The effects of dietary patterns on quality of life: A substudy of the Dietary Approaches to Stop Hypertension trial. *J Am Diet Assoc* 1999; 99:S84–89S.

26. Svetkey, L. P. et al. Effect of lifestyle modifications on blood pressure by race, sex, hypertension status, and age. *Journal of Human Hypertension* 2005; 19:21–31.

27. Whelton, P. K. et al. Sodium reduction and weight loss in the treatment of hypertension in older persons: A randomized controlled trial of nonpharmacologic interventions in the elderly (TONE). TONE Collaborative Research Group. *J Am Med Assoc* 1998; 279:839–46.

28. Sacks, F. M. et al. Effects on blood pressure of reduced dietary sodium and the Dietary Approaches to Stop Hypertension (DASH) diet. DASH-Sodium Collaborative Research Group. *N Engl J Med* 2001; 344:3–10.

29. Welty, F. K., Nasca, M. M., Lew, N. S. et al. Effect of onsite dietitian counseling on weight loss and lipid levels in an outpatient physician office. *Am J Cardiol* 2007; 100:73–75.

30. Folsom, A. R., Parker, E. D., and Harnack, L. J. Degree of concordance with DASH diet guidelines and incidence of hypertension and fatal cardiovascular disease. Am J Hypertens 2007 (Mar); 20(3):225–32.

31. Levitan, E. B., Wolk, A., and Mittleman, M. A. Consistency with the DASH diet and incidence of heart failure. *Arch Intern Med* 2009; 169(9):851–857.

32. American Diabetes Association. Executive Summary: Standards of medical care in diabetes–2009. *Diabetes Care* 2009; 32(1):S6–S12. DOI:10.2337/dc09-S006.

33. Barclay, L., and Lie, D. Risk factors for falls indentified in older adults with diabetes. Diabetes Care 2008; 31:391–396. Available at http://cme.medscape.com/viewarticle/570911?src=cmemp, accessed 3/23/09.

34. Tsihlias, E. B. et al. Comparison of high- and low-glycemic-index breakfast cereals with monounsaturated fat in the long-term dietary management of type 2 diabetes. *Am J Clin Nutr* 2000; 72:439–49.

35. Wannamethee, S. G., Whincup, P. H., Thomas, M. et al. Associations between dietary fiber and inflammation, hepatic function and risk of type 2 diabetes in older men: Potential mechanisms for benefits of fiber on diabetes risk. *Diabetes Care* (published ahead of print, online 7/23/09.

36. Effect of intensive blood-glucose control with metformin on complications in overweight patients with type 2 diabetes (UKPDS 34). UK Prospective Diabetes Study (UKPDS) Group. *Lancet* 1998; 352:854–65.

37. Intensive blood glucose control with sulphonylureas or insulin compared with conventional treatment and risk of complications in patients with type 2 diabetes (UKPDS 33). UK Prospective Diabetes Study (UKPDS) Group. *Lancet* 1998; 352:837–53.

38. National Heart, Lung, and Blood Institute Expert Panel on the Identification, Evaluation, and Treatment of Overweight and Obesity in Adults. Clinical guidelines on the identification, evaluation, and treatment of overweight and obesity in adults: The evidence report. *Obes Res* 1998; 6(Suppl. 2):51S–209S.

39. World Health Organization Consultation on Obesity. *Obesity: Preventing and managing the global epidemic.* Geneva, Switzerland: World Health Organization, 2000;WHO Technical Report Series 894.

40. Heiat, A., Vaccarino, V., and Krumholz, H. M. An evidence-based assessment of federal guidelines for overweight and obesity as they apply to elderly persons. *Arch Intern Med* 2001; 161:1194–1203.

41. Villareal, D. T., Apovian, C. M., Kushner, R. F. et al. Obesity in older adults: Technical review and position statement of the American Society for Nutrition and NAASO, the Obesity Society. *Obesity Research* 2005; 13(1):1849–1863.

42. Sui, X., LaMonte, M. J., Laditka, J. N. et al. Cardiorespiratory fitness and adiposity as mortality predictors in older adults. JAMA 2007; 298(21):2507–2516.

43. Flegal, K. M. et al. Excess deaths associated with underweight, overweight, and obesity. JAMA 2005; 293:1861–67.

44. Ernsberger, P. and Koletsky, R. J. Biomedical rationale for a wellness approach to obesity: An alternative to a focus on weight loss. *J Soc Issues* 1999; 55:221–59.

45. Ostbye, T. et al. Predictors of five-year mortality in older Canadians: The Canadian Study of Health and Aging. *J Am Geriatr Soc* 1999; 47:1249–54.

46. Price, G. M., Uauy, R., Breeze, E., et al. Weight, shape, and mortality risk in older persons: Elevated waist-hip ratio, not high body mass index, is associated with a greater risk of death. *Am J Clin Nutr* 2006; 84:449–60.

47. Pischon, T., Boeing, H., Hoffman, K., et al. General and abdominal adiposity and risk of death in Europe. *N Eng J Med* 2008; 359:2105–20.

48. Winter, Y., Rohrman, S., Lineisen, J. et al. Contribution of obesity and abdominal fat mass to risk of stroke and transient ischemic attacks. *Stroke* 2009; 40. Available at http://stroke.ahajournals.org, accessed 12/30/08

49. The American Dietetic Association Long Term Care Task Force. *Nutrition risk assessment form, guides, strategies and interventions.* Chicago, IL: American Dietetic Association, 1999.

50. Lewis, C. E., McTigue, K. M., Burke, L. E. et al. Mortality, health outcomes, and

body mass index in the overweight range: A science advisory from the American Heart Association. *Circulation* 2009; 119:3263–3271.

51. Newby, P. K., Muller, D., Hallfrisch, J. et al. Dietary patterns and changes in body mass index and waist circumference in adults. *Am J Clin Nutr* 2003; 77:1417–25.

52. National Osteoporosis Foundation. *Fast Facts, 2008.* Available at nof.org, accessed 9/4/09.

53. Office of Dietary Supplements. Dietary Supplement Fact Sheet: Calcium. Available at http://ods.od.nih.gov/factsheets/Calcium_pf.asp, posted 7/8/09; accessed 9/6/09.

54. World Health Organization Scientific Group on the assessment of osteoporosis at primary health care level. Summary meeting report, Brussels, Belgium, 5–7 May 2004. Available at http://who.org, accessed 9/4/09.

55. Faine, M. P. Dietary factors related to preservation of oral and skeletal bone mass in women. *J Prosthet Dent* 1995; 73:65–72.

56. Centers for Disease Control and Prevention. Self-reported falls and fall-related injuries among persons aged ≥ 65 years–United States, 2006. *MMWR* 2008; 57:225–229.

57. National Osteoporosis Foundation. Clinician's guide to prevention and treatment of osteoporosis, 2008. Available at www.nofstore.org, Health Professional Resources; accessed 9/4/09.

58. Rubin, L. A. et al. Determinants of peak bone mass: Clinical and genetic analyses in a young female Canadian cohort. *J Bone Miner Res* 1999; 14:633–43.

59. Matkovic, V. et al. Bone status and fracture rates in two regions of Yugoslavia. *Am J Clin Nutr* 1979; 32:540–49.

60. Lloyd, T. et al. Calcium supplementation and bone mineral density in adolescent girls. *J Am Med Assoc* 1993; 270:841–44.

61. Heaney, R. P. Bone biology in health and disease. In Shils, M. E. et al., eds. *Modern nutrition in health and disease.* Philadelphia: Lippincott, Williams & Wilkins, 1999: 1327–38.

62. Bischoff-Ferrari, H. A. et al. Fracture prevention with vitamin D supplementation: A meta-analysis of randomized controlled trials. *JAMA* 2005; 293:2257–64.

63. Brown, J. P. and Josse, R. G. for the Scientific Advisory Council of the Osteoporosis Society of Canada. 2002 clinical practice guidelines for the diagnosis and management of osteoporosis in Canada. *CMAJ* 2002; 167(10; Suppl.):S1–S34.

64. Prince, R. L. et al. Effects of calcium supplementation on clinical fracture and bone structure. *Arch Intern Med* 2006; 166:869–75.

65. Feskanich, D. et al. Protein consumption and bone fractures in women. *Am J Epidemiol* 1996; 143:472–79.

66. Heaney, R. P. and Layman, D. K. Amount and type of protein influences bone health. *Am J Clin Nutr* 2008; 87(Suppl.):1567S–70S.

67. Burckhardt, P. The effect of the alkali load of mineral water on bone metabolism: Interventional studies. *J Nutr* 2008; 138:435S–437S.

68. Bolton-Smith, C., McMurdo, M. E. T., Paterson, C. R. et al. Two-year randomized controlled trial of vitamin K_1 (Phylloquinone) and vitamin D_3 plus calcium on the bone health of older women. *J Bone Miner Res* 2007; 22:509–519.

69. Writing Group for the Women's Health Initiative Investigators. Risks and benefits of estrogen plus progestin in healthy postmenopausal women. Principal results from the Women's Health Initiative randomized controlled trial. *J Am Med Assoc* 2002; 288:321–33.

70. Rosen, C. J. Serotonin rising–the bone, brain, bowel connection. *N Engl J Med* 2009; 360(March;10):957–9.

71. Pleis, J. R. and Lucas, J. W. Summary health statistics for U.S. adults: National Health Interview Survey, 2007. National Center for Health Statistics. *Vital Health Stat* 2009; 10(240).

72. Position of the American Dietetic Association: Oral health and nutrition. *J Am Diet Assoc* 2007; 107:1418–1428.

73. Scott, M. and Gehot, A. R. Gastroesophageal reflux disease: Diagnosis and management. *American Family Physician* 1999. Available at www.aafp.org/afp, accessed 9/8/09.

74. Ables, A. Z., Simon, I., and Melton, E. R. Update on *Helicobacter pylori* treatment, 2007. *American Family Physician.* Available at www.aafp.org/afp, accessed 9/8/09.

75. Nilsson-Ehle, H. Age-related changes in cobalamin (vitamin B_{12}) handling. Implications for therapy. *Drugs Aging* 1998; 12:277–92.

76. Ho, C., Kauwell, G. P., and Bailey, L. Practitioner's guide to meeting the vitamin B-12 Recommended Dietary Allowance for people aged 51 years and older. *J Am Diet Assoc* 1999; 99:725–27.

77. Vogiatzoglou, A., Refsum, H., and Johnston C. Vitamin B_{12} status and rate of brain volume loss in community-dwelling elderly. *Neurology* 2008; 71:826–832.

78. Lindenbaum, J. et al. Prevalence of cobalamin deficiency in the Framingham elderly population. *Am J Clin Nutr* 1994; 60:2–11.

79. Hokin, B. D. and Butler, T. Cyanocobalamin (vitamin B-12) status in Seventh-Day Adventist ministers in Australia. *Am J Clin Nutr* 1999; 70:576S–78S.

80. Carmel, R. et al. Serum cobalamin, homocysteine, and methylmalonic acid concentrations in a multiethnic elderly population: Ethnic and sex differences in cobalamin and metabolite abnormalities. *Am J Clin Nutr* 1999; 70:904–10.

81. Zeitlin, A., Frishman, W. H., and Chang, C. J. The association of vitamin B_{12} and folate blood levels with mortality and cardiovascular morbidity incidence in the old: The Bronx aging study. *Am J Ther* 1997; 4:275–81.

82. Yates, A. A., Schlicker, S. A., and Suitor, C. W. Dietary Reference Intakes: the new basis for recommendations for calcium and related nutrients, B vitamins, and choline. *J Am Diet Assoc* 1998; 98:699–706.

83. Otten, J. J., Hellwig, J. P., and Meyers, L. D., eds. Institute of Medicine of the National Academies. DRI Dietary Reference Intakes: The Essential Guide to Nutrient Requirements. Washington, D.C.: The National Academies Press, 2006.

84. Norberg, B. Turn of tide for oral vitamin B_{12} treatment. *J Intern Med* 1999; 246:237–38.

85. Eussen, S. J. P. M. et al. Oral Cyanocobalamin supplementation in older people with vitamin B_{12} deficiency a dose-finding trial. *Arch Intern Med* 2005; 165:1167–72.

86. Bente, L. B. and Gerrior, S. A. Selected food and nutrient highlights of the 20th century: U.S. food supply series. *Family Econ Nutr Rev* 2002; 14(1):43–51.

87. Hsieh, C. Treatment of constipation in older adults. *Am Fam Physician* 2005; 72:2277–84.

88. Harari, D. et al. Bowel habit in relation to age and gender. Findings from the National Health Interview Survey and clinical implications. *Arch Intern Med* 1996; 156:315–20.

89. Talley, N. J. et al. Constipation in an elderly community: A study of prevalence and potential risk factors. *Am J Gastroenterol* 1996; 91:19–25.

90. Müller-Lissner, S. A., Kamm, M. A., Scarpignato, C. et al. Myths and misconceptions about chronic constipation. *Am J Gastroenterol* 2005; 100:232–242.

91. Burkitt, D. Fiber as protective against gastrointestinal diseases. *Am J Gastroenterol* 1984; 79(4):249–252.

92. National Institute of Arthritis and Musculoskeletal and Skin Diseases: Osteoarthritis, handout on health. July 2002, revised May 2006. Available at www.niams.nih.gov/health_info/osteoarthritis, accessed 9/9/09.

93. Arthritis Foundation. Osteoarthritis fact sheet. Available at www.arthritis.org, accessed 9/9/09.

94. Arthritis Foundation. Research update: Clarifying the role of fat in osteoarthritis. Available at www.arthritis.org, accessed 9/9/09.

95. Felson, D. T. et al. Osteoarthritis: New insights. Part 1: The disease and its risk factors. *Ann Intern Med* 2000; 133:635–46.

96. Felson, D. T. et al. Osteoarthritis: New insights. Part 2: Treatment approaches. *Ann Intern Med* 2000; 133:726–37.

97. Clegg, D. O., Reda, D. J., Harris, C. L., et al. Glucosamine, chondroitin sulphate, and the two in combination for painful knee osteoarthritis. *N Engl J Med* 2006; 354:795–808.

98. Borchers, A. T. et al. Inflammation and Native American medicine: The role of botanicals. *Am J Clin Nutr* 2000; 72:339–47.

99. American Psychiatric Association. *Diagnostic and statistical manual of mental disorders*, 4th edition, text revision (DSM-IV-TR). Washington, D.C.: American Psychiatric Press, 2000.

100. Plassman, B. L., Langa, K. M., Fisher, G. G. et al. Prevalence of dementia in the United States: The aging, demographics, and memory study. *Neuroepidemiology* 2007; 29:125–132. DOI: 10.1159/000109998.

101. Alzheimers Association. 2009 Alzheimers disease facts and figures. Available at www.alz.org/national/documents/summary_alzfactsfigures2009.pdf, accessed 9/10/09.

102. Cleveland Clinic Health System. 2004, for your health. Types of dementia. www.cchs.net/health/main.asp, accessed 9/06.

103. Aisen, P. S., Schneider, L. S., Sano, M. et al. High-dose B vitamin supplementation and cognitive decline in Alzheimer disease: A randomized controlled trial. *JAMA* 2008; 300(15):1774–1783.

104. Feart, C., Samieri, C., Rondeau, V. et al. Adherence to a Mediterranean diet, cognitive decline, and risk of dementia. *JAMA* 2009; 302(6):638–648.

105. Scarmeas, N., Luchsinger, J. A., Schupf, N. et al. Physical activity, diet, and risk of Alzheimer Disease. *JAMA* 2009; 302(6):627–637.

106. Witte, A. V., Fobker, M., Gellner, R. et al. Caloric restriction improves memory in elderly humans. Neuroscience, National Academy of Sciences of the USA. *PNAS* January 27, 2009; 106(4)1255–1260. Available at www.pnas.org/cgi/doi/10.1073/pnas.0808587106, requested from author February 2009.

107. Sabia, S., Kivimaki, M., Shipley, M., et al. Body mass index over the adult life course and cognition in late midlife: The Whitehall II Cohort study. *Am J Clin Nutr* 2009; 89:601–7.

108. Kuczmarski, M. F., Kuczmarski, R. J., and Najjar, M. Descriptive anthropometric reference data for older Americans. *J Am Diet Assoc* 2000; 100:59–66.

109. National Institutes of Health. Clinical guidelines on the identification, evaluation, and treatment of overweight and obesity in adults—the evidence report. *Obes Res* 1998; 6(Suppl 2):51S–209S.

110. The World Health Organization/Food and Agricultural Organization/United Nations University. *Physical status: The use and interpretation of anthropometry*. Report of a WHO Expert Committee. Geneva: World Health Organization, 1995.

111. Akner, G. and Cederholm, T. Treatment of protein-energy malnutrition in chronic nonmalignant disorders. *Am J Clin Nutr* 2001; 74:6–24.

112. Milne, A. C., Potter, J., and Avenell, A. Protein and energy supplementation in elderly people at risk from malnutrition. *The Cochrane Database of Systematic Reviews* 2002; Issue 4:1–71. Online access via Ovid, 11/14/02.

113. Niedert, K. and Dorner, B. Consultant Dietitians in Health Care Facilities. In *Nutrition care of the older adult*, 2nd edition. Chicago: The American Dietetic Association, 2004.

114. Phillips, P. A. et al. Reduced thirst after water deprivation in healthy elderly men. *N Eng J Med* 1984; 311:753–59.

115. American Dietetic Association Evidence Analysis Library. Assessing Hydration Status. Available at www.adaevidencelibrary.com, accessed 8/20/09.

116. Gross, C. R. et al. Clinical indicators of dehydration severity in elderly patients. *J Emerg Med* 1992; 10:267–74.

117. Hodgkinson, B., Evans, D., and Wood, J. Maintaining oral hydration in older people. The Joanna Briggs Institute for Evidence Based Nursing and Midwifery. *Systematic Review* 2001; 12:1–66.

118. Briggs, G. M. and Calloway, D. H. *Bogert's nutrition and physical fitness.* New York: Holt, Rinehart and Winston, 1984.

119. Chidester, J. C. and Spangler, A. A. Fluid intake in the institutionalized elderly. *J Am Diet Assoc* 1997; 97:23–38; quiz 29–30.

120. Institute of Medicine of the National Academies. Dietary Reference Intakes for water, potassium, sodium, chloride, and sulfate. Washington, D.C.: The National Academies Press, 2005.

121. Cabrera, C., Artacho, R., ad Gimenez, R. Beneficial effects of green tea—a review. *J Am Coll Nutri* 2006; 25:79–99.

122. Fleet, J. C. New support for a folk remedy: Cranberry juice reduces bacteriuria and pyuria in elderly women. *Nutr Rev* 1994; 52:168–78.

123. Fordyce, M. Dehydration near the end of life. *Ann Long-Term Care* 2000; 8. Available at www.mmhc.com/nhm/articles/NHM0005/fordyce.html, accessed 5/00.

124. Conill, C. et al. Symptom prevalence in the last week of life. *J Pain Symptom Manage* 1997; 14:328–31.

125. Leming, M. R. and Dickinson, G. E. *Understanding dying, death, and bereavement.* Fort Worth, TX: Holt, Rinehart, and Winston, 1990.

126. Rosenbloom, C. A. and Whittington, F. J. The effects of bereavement on eating behaviors and nutrient intakes in elderly widowed persons. *J Gerontol* 1993; 48:S223–29.

Glossary

Adiposity or BMI Rebound A normal increase in body mass index that occurs after BMI declines and reaches its lowest point at 4 to 6 years of age.

AIDS Acquired Immunodeficiency Syndrome.

Allergy Hypersensitivity to a physical or chemical agent.

Alveoli Rounded or oblong cavities present in the breast (singular = *alveolus*).

Alzheimer's Disease A brain disease that represents the most common form of dementia. It is characterized by memory loss of recent events that expands to more distant memories over the course of five to ten years. It eventually produces profound intellectual decline characterized by dementia and personal helplessness.

Amenorrhea Absence of menstrual cycle.

Amino Acids The "building blocks" of protein. Unlike carbohydrates and fats, amino acids contain nitrogen.

Amniotic Fluid The fluid contained in the amniotic sac that surrounds the fetus in the uterus.

Amylophagia Compulsive consumption of laundry starch or cornstarch.

Anaphylaxis Sudden onset of a reaction with mild to severe symptoms, including a decrease in ability to breathe, which may be severe enough to cause a coma.

Androgens Types of steroid hormones produced in the testes, ovaries, and adrenal cortex from cholesterol. Some androgens (testosterone, dihydrotestosterone) stimulate development and functioning of male sex organs.

Anemia A reduction below normal in the number of red blood cells per cubic mm in the quantity of hemoglobin, or in the volume of packed red cells per 100 mL of blood (hematocrit). This reduction occurs when the balance between blood loss and blood production is disturbed.

Anencephaly Condition initiated early in gestation of the central nervous system in which the brain is not formed correctly, resulting in neonatal death.

Angiogenesis Inhibitor Angiogenesis is the formation of new blood vessels. An angiogenesis inhibitor slows or stops vessel formation. Tumors cannot grow or expand without additional blood vessels to carry oxygen and other nutrients.

Anorexia Nervosa A disorder characterized by extreme underweight, malnutrition, amenorrhea, low bone density, irrational fear of weight gain, restricted food intake, hyperactivity, and disturbances in body image.

Anovulatory Cycles Menstrual cycles in which ovulation does not occur.

Anthropometry The science of measuring the human body and its parts.

Antidiuretic Hormone Hormone that causes the kidneys to dilute urine by absorbing more water.

Antioxidants Chemical substances that prevent or repair damage to cells caused by exposure to oxidizing agents such as oxygen, ozone, and smoke and to other oxidizing agents normally produced in the body. Many different antioxidants are found in foods; some are made by the body.

Appropriate for Gestational Age (AGA) Weight, length, and head circumference are between the 10th and 90th percentiles for gestational age.

Arteries Blood vessels that carry oxygenated blood to cells.

Arteriosclerosis Age-related thickening and hardening of the artery walls, much like an old rubber hose that becomes brittle or hard.

Assisted Reproductive Technology (ART) An umbrella term for fertility treatments such as *in vitro* fertilization (IVF, a technique in which egg cells are fertilized by sperm outside the woman's body), artificial insemination, and hormone treatments.

Asthma Condition in which the lungs are unable to exchange air due to lack of expansion of air sacs. It can result in a chronic illness and sometimes unconsciousness and death if not treated.

Atherosclerosis A type of hardening of the arteries in which cholesterol is deposited in the arteries. These deposits narrow the coronary arteries and may reduce blood flow to the heart.

Athetosis Uncontrolled movements of the large muscle groups as a result of damage to the central nervous system.

Atrial Fibrillation Degeneration of the heart muscle, causing irregular contractions.

Attention Deficit Hyperactivity Disorder (ADHD) Condition characterized by low impulse control and short attention span, with and without a high level of overall activity.

Autoimmune Disease A disease related to the destruction of the body's own cells by substances produced by the immune system that mistakenly recognize certain cell components as harmful.

Autism Condition of deficits in communication and social interaction with onset generally before age 3 years, in which mealtime behavior and eating problems occur along with other behavioral and sensory problems.

B-lymphocytes White blood cells that are responsible for producing immunoglobulins.

Baby-Bottle Tooth Decay Dental caries in young children caused by being put to bed with a bottle or allowed to suck from a bottle for extended periods of time. Also called baby- or nursing-bottle dental caries.

Basal Metabolic Rate (BMR) Measured energy expenditure in an individual who has been awake less than 30 minutes and is still at absolute rest, has fasted for 10 hours or more, and is in a quiet room with normal, comfortable temperatures.

Bioactive Food Components Constituents in foods or dietary supplements, other than those needed to meet basic human nutritional needs, that are responsible for changes in health status.

Binge-Eating Disorder (BED) An eating disorder characterized by periodic binge eating, which normally is not followed by vomiting or the use of laxatives. People must experience eating binges twice a week on average for over 6 months to qualify for this diagnosis.

Body Mass Index An index that correlates with total body fat content or percent body fat, and is an acceptable measure of adiposity or body fatness in children and adults. It is calculated by dividing weight in kilograms by the square of height in meters (kg/m^2).

Bone Age Bone maturation; correlates well with stage of pubertal development.

Bronchopulmonary Dysplasia (BPD) Condition in which the underdeveloped lungs in a preterm infant are damaged so that breathing requires extra effort.

Bulimia Nervosa A disorder characterized by repeated bouts of uncontrolled, rapid ingestion of large quantities of food (binge eating) followed by self-induced vomiting, laxatives or diuretic use, fasting, or vigorous exercise in order to prevent weight gain. Binge eating is often followed by feelings of disgust and guilt. Menstrual cycle abnormalities may accompany this disorder.

Calorie A unit of measure of the amount of energy supplied by food. Also known as the "kilocalorie," or the "large Calorie."

Carotenemia A condition, caused by ingestion of high amounts of carotenoids (or carotenes) from plant foods, in which the skin turns yellowish orange.

Carotid Artery Disease Condition in which the arteries that supply blood to the brain and neck become damaged.

Catch-Up Growth Period of time shortly after a slow growth period when the rate of weight and height gains is likely to be faster than expected for age and gender.

Celiac Disease An autoimmune disease characterized by inflammation of the small-intestine lining resulting from a genetically based intolerance to a component of gluten. The inflammation produces diarrhea, fatty stools, weight loss, and vitamin and mineral deficiencies. Also called tropical sprue and gluten-sensitive enteropathy.

Cerebral Palsy A group of disorders characterized by impaired muscle activity and coordination present at birth or developed during early childhood.

Cerebral Spinal Atrophy Condition in which muscle control declines over time as a result of nerve loss, causing death in childhood.

Children with Special Health Care Needs A federal category of services for infants, children, and adolescents with or at risk for physical or developmental disability, or with a chronic medical condition caused by or associated with genetic/metabolic disorders, birth defects, prematurity, trauma, infection, or perinatal exposure to drugs.

Cholesterol A fat-soluble, colorless liquid primarily found in animals. Cholesterol is used by the body to form hormones such as testosterone and estrogen and is a component of cell membranes.

Chronic Condition Disorder of health or development that is the usual state for an individual and unlikely to change, although secondary conditions may result over time.

Chronic Disease Slow-developing, long-lasting diseases that are not contagious (e.g., heart disease, cancer, diabetes). They can be treated but not always cured.

Chronic Inflammation Low-grade inflammation that lasts weeks, months, or years. Inflammation is the first response of the body's immune system to infection or irritation. Inflammation triggers the release of biologically active substances that promote oxidation and other potentially harmful reactions in the body.

Cleft Lip and Palate Condition in which the upper lip and roof of the mouth are not formed completely and are surgically corrected, resulting in feeding, speaking, and hearing difficulties in childhood.

Cobalamin Another name for vitamin B_{12}. Important roles of cobalamin are fatty acid metabolism, synthesis of nucleic acid (i.e., DNA, a complex protein that controls the formation of healthy new cells), and formation of the myelin sheath that protects nerve cells.

Coenzymes Chemical substances that activate enzymes.

Cognitive Function The process of thinking.

Colic A condition marked by a sudden onset of irritability, fussiness, or crying in a young infant between 2 weeks and 3 months of age who is otherwise growing and healthy.

Colostrum The milk produced in the first 2–3 days after the baby is born. Colostrum is higher in protein and lower in lactose than milk produced after the milk supply is established.

Commodity Program A USDA program in which food products are sent to schools for use in the child nutrition programs. Commodities are usually acquired for farm price support and surplus removal reasons.

Competitive Foods Foods sold to children, in food service areas during mealtimes, that compete with the federal meal programs.

Congenital Abnormality A structural, functional, or metabolic abnormality present at birth. Also called congenital anomalies. These may be caused by environmental or genetic factors, or by a combination of the two. Structural abnormalities are generally referred to as congenital malformations, and metabolic abnormalities as inborn errors of metabolism.

Congenital Anomaly Condition evident in a newborn that is diagnosed at or near birth, usually as a genetic or chronic condition, such as spina bifida or cleft lip and palate.

Corpus Luteum (*corpus* = body, *luteum* = yellow) A tissue about 12 mm in diameter formed from the follicle that contained the ovum prior to its release. It produces estrogen and progesterone. The "yellow body" derivation comes from the accumulation of lipid precursors of these hormones in the corpus luteum.

Critical Periods Preprogrammed time periods during embryonic and fetal development when specific cells, organs, and tissues are formed and integrated, or functional levels established. Also called sensitive periods.

Cystic Fibrosis Condition in which a genetically changed chromosome 7 interferes with all the exocrine functions in the body, but particularly pulmonary complications, causing chronic illness.

Daily Values (DVs) Scientifically agreed-upon standards for daily intakes of nutrients from the diet developed for use on nutrition labels.

Development Progression of the physical and mental capabilities of an organism through growth and differentiation of organs and tissues, and integration of functions.

Developmental Delay Conditions represented by at least a 25% delay by standard evaluation in one or more areas of development, such as gross or fine motor, cognitive, communication, social, or emotional development.

Developmental Disabilities General term used to group specific diagnoses together that limit daily living and functioning and occur before age 21.

Developmental Plasticity The concept that development can be modified by particular environmental conditions experienced by a fetus or infant.

Diaphragmatic Hernia Displacement of the intestines up into the lung area due to incomplete formation of the diaphragm *in utero*.

Dietary Fiber Complex carbohydrates and lignins naturally occurring and found mainly in the plant cell wall. Dietary fiber cannot be broken down by human digestive enzymes.

Dietary Reference Intakes (DRIs) Quantitative estimates of nutrient intakes, used as reference values for assessing the diets of healthy people. DRIs include Recommended Dietary Allowances (RDAs), Adequate Intakes (AI), Tolerable Upper Intake Level (UL), and Estimated Average Requirement (EAR).

Dietary Supplements Any product intended to supplement the diet, including vitamin and mineral supplements, proteins, enzymes, amino acids, fish oils, fatty acids, hormones and hormone precursors, and herbs and other plant extracts. In the United States, such products must be labeled "Dietary Supplement."

Differentiation   Cellular acquisition of one or more characteristics or functions different from that of the original cells.

DiGeorge Syndrome   Condition in which chromosome 22 has a small deletion, resulting in a wide range of heart, speech, and learning difficulties.

Diplegia   Condition in which the part of the brain controlling movement of the legs is damaged, interfering with muscle control and ambulation.

Disproportionately Small for Gestational Age (dSGA)   Newborn weight is ≤10th percentile of weight for gestational age; length and head circumference are normal. Also called asymmetrical SGA.

Diverticulitis   Infected "pockets" within the large intestine.

Doula   An individual who gives psychological encouragement and physical assistance to a mother during pregnancy, birth, and lactation; the doula may be a relative, friend, or neighbor and is usually but not necessarily female.

Down Syndrome   Condition in which three copies of chromosome 21 occur, resulting in lower muscle strength, lower intelligence, and greater risk for overweight.

Dumping Syndrome   A condition characterized by weakness, dizziness, flushing and warmth, nausea, and palpitation immediately or shortly after eating and produced by abnormally rapid emptying of the stomach, especially in individuals who have had part of the stomach removed.

Dysmenorrhea   Painful menstruation due to abdominal cramps, back pain, headache, and/or other symptoms.

Early Intervention Services   Federally mandated evaluation and therapy services for children in the age range from birth to 3 years under the Individuals with Disabilities Education Act.

Eicosanoids   Molecules synthesized from essential fatty acids. They exert complex control over many bodily systems, mainly in inflammation and immunity, and act as messengers in the central nervous system.

Edema   Swelling (usually of the legs and feet, but can also extend throughout the body) due to an accumulation of extracellular fluid.

Embryo   The developing organism from conception through 8 weeks.

Empty-Calorie Foods   Foods that provide an excess of calories relative to their nutrient content.

Endocrine   A system of ductless glands, such as the thyroid, adrenal glands, ovaries, and testes, that produces secretions that affect body functions.

Endometriosis   A disease characterized by the presence of endometrial tissue in abnormal locations, such as deep within the uterine wall, in the ovary, or in other sites within the body. The condition is quite painful and is associated with abnormal menstrual cycles and infertility in 30–40% of affected women.

Endothelium   The layer of cells lining the inside of blood vessels.

Enrichment   The replacement of thiamin, riboflavin, niacin, and iron that are lost when grains are refined.

Enteral Feeding   Fluid or food being delivered directly into the gastrointestinal system. The delivery can be by mouth or through a tube that is placed into the stomach or intestines.

Epigenetics   (*epi* = over, above) Biological mechanisms that change gene function without changing the structure of DNA. Epigenetic mechanisms are affected by environmental factors.

Epididymis   Tissues on top of the testes that store sperm.

Epithelial Cells   Cells that line the surface of the body.

EPSDT   The Early Periodic Screening, Detection, and Treatment Program is a part of Medicaid and provides routine checkups for low-income families.

Essential Amino Acids   Amino acids that cannot be synthesized in adequate amounts by humans and therefore must be obtained from the diet. Also, called "indispensible amino acids."

Essential Fatty Acids   Components of fat that are a required part of the diet (i.e., linoleic and alpha-linolenic acids). Both contain unsaturated fatty acids.

Essential Nutrients   Substances required for growth and health that cannot be produced, or produced in sufficient amounts, by the body. They must be obtained from the diet.

Exposure Index   The average infant milk intake per kilogram body weight per day × (the milk to plasma ratio divided by the rate of drug clearance) × 100. It is indicative of the amount of the drug in the breast milk that the infant ingests and is expressed as a percentage of the therapeutic (or equivalent) dose for the infant.

Extremely Low-Birth-Weight Infant (ELBW)   An infant weighing under 1000 g, or 2 lb 3 oz, at birth.

Failure to Thrive (FTT)   Condition of inadequate weight or height gain thought to result from a caloric deficit, whether or not the cause can be identified as a health problem.

Familial Hyperlipidemia   A condition that runs in families and results in high levels of serum cholesterol and other lipids.

Fatty Acids   The fat-soluble components of fats in foods.

Fecundity   Biological ability to bear children.

Fertility   Actual production of children. The word best applies to specific vital statistic rates, but is commonly taken to mean the ability to bear children.

Fetal-Origins Hypothesis   The theory that exposures to adverse nutritional and other conditions during critical or sensitive periods of growth and development can permanently affect body structures and functions. Such changes may predispose individuals to cardiovascular diseases, type 2 diabetes, hypertension, and other disorders later in life. Also called *metabolic programming* and the *Barker Hypothesis*.

Fetus   The developing organism from 8 weeks after conception to the moment of birth.

Fine Motor Skills   Development and use of smaller muscle groups demonstrated by stacking objects, scribbling, and copying a circle or square.

Fluorosis   Permanent white or brownish staining of the enamel of teeth caused by excessive ingestion of fluoride before teeth have erupted.

Food Allergy (Hypersensitivity)   Abnormal or exaggerated immunologic response, usually immunoglobulin E (IgE) mediated, to a specific food protein.

Food Insecurity   Limited or uncertain availability of safe, nutritious foods, or the inability to acquire them in socially acceptable ways.

Food Intolerance   An adverse reaction involving digestion or metabolism but not the immune system.

Food Security   Access at all times to a sufficient supply of safe, nutritious foods.

Fortification   The addition of one or more vitamins or minerals to a food product.

Free Radicals   Chemical substances (often oxygen-based) that are missing electrons. The absence of electrons makes the chemical substance reactive and prone to oxidizing nearby molecules by stealing electrons from them. Free radicals can damage lipids, cell membranes, DNA, and tissues by altering their chemical structure and functions. They also form as a normal part of metabolism. Over time, oxidative stress causes damage to lipids, cell membranes, DNA, cells, and tissues.

Full-Term Infants   Infants born between 37 and 42 weeks of gestation.

Functional Fiber   Nondigestible carbohydrates, including plant, animal, or commercially produced sources, that have beneficial effects in humans.

Functional Foods   Generally taken to mean food, fortified foods, and enhanced food products that may have health benefits beyond the effects of essential nutrients they contain.

Functional Status   Ability to carry out the activities of daily living, including telephoning, grocery shopping, food handling and preparation, and eating.

Galactosemia   A rare genetic condition of carbohydrate metabolism in which a blocked or inactive enzyme does not allow breakdown of galactose. It can cause serious illness if not identified and treated soon after birth.

Gastroesophageal Reflux (GER)   Movement of the stomach contents backward into the esophagus, due to stomach muscle contractions. The condition may require treatment depending on its duration and degree. Also known as *gastroesophageal reflux disease (GERD)*.

Gastrostomy Feeding   Form of enteral nutrition support for delivering nutrition by tube placement directly into the stomach, bypassing the mouth through a surgical procedure that creates an opening through the abdominal wall and stomach.

Geophagia   Compulsive consumption of clay or dirt.

Gestational Diabetes   Carbohydrate intolerance with onset or first recognition in pregnancy.

Glucogenic Amino Acids   Amino acids such as alanine and glutamate that can be converted to glucose.

Gluten   A protein found in wheat, oats, barley, rye, and triticale (all in the genus *Triticum*); gliadin is the toxic fraction of gluten.

Gluten-Free   A food labeling term that indicates a product does not contain any species of wheat, rye, barley, or their hybrids, or ingredients that contain these grains, or 20 or more parts per million (ppm) gluten (about 6 mg per serving). (FDA-proposed definition.)

Glycemic Index (GI)   A measure of the extent to which blood glucose levels are raised by a specific amount of carbohydrate-containing food compared to the same amount of glucose or white bread.

Glycemic Load (GL)   A measure of the extent to which blood glucose levels are raised by a specific amount of carbohydrate-containing food. It is calculated by multiplying the carbohydrate content of an amount of food consumed by the glycemic index of the food, and dividing the result by 100.

Glycerol   A component of fats that is soluble in water. It is converted to glucose in the body.

Glycosylated Hemoglobin   A laboratory test that measures how well the blood sugar level has been maintained over a prolonged period of time; also called Hemoglobin A_1C.

Gravida   Number of pregnancies a woman has experienced.

Gross Motor Skills   Development and use of large muscle groups as exhibited by walking alone, running, walking up stairs, riding a tricycle, hopping, and skipping.

Growth   Increase in an organism's size through cell multiplication (hyperplasia) and enlargement of cell size (hypertrophy).

Growth Velocity   The rate of growth over time.

Gynecological Age   Defined as chronological age minus age at menarch. For example, a female with the chronological age of 14 years minus age at first menstrual cycle of 12 equals a gynecological age of 2.

Health   More than the absence of disease, health is a sense of well-being. Even individuals with a chronic condition may properly consider themselves to be healthy. For instance, a person with diabetes mellitus whose blood sugar is under control can be considered healthy.

Heart Disease   The leading cause of death and a common cause of illness and disability in the United States. Coronary heart disease, the principal form of heart disease, is caused by buildup of cholesterol deposits in the coronary arteries that feed the heart.

Hematocrit   An indicator of the proportion of whole blood occupied by red blood cells. A decrease in hematocrit is a late indicator of iron deficiency.

Heme Iron   Iron contained within a protein portion of hemoglobin that is in the ferrous state.

Hemoglobin   A protein that is the oxygen-carrying component of red blood cells. A decrease in hemoglobin concentration in red blood cells is a late indicator of iron deficiency.

Hemolytic Anemia   Anemia caused by shortened survival of mature red blood cells and inability of the bone marrow to compensate for the decreased life span.

Hemolytic Uremic Syndrome (HUS)   A serious, sometimes fatal complication associated with illness caused by *E. coli* O157:H7, which occurs primarily in children under the age of 10 years. HUS is characterized by renal failure, hemolytic anemia, and a severe decrease in platelet count.

High-Poverty Neighborhoods   Neighborhoods where 40% or more of the people are living in poverty.

HIV   Human immunodeficiency virus.

Homeostasis   Constancy of the internal environment. The balance of fluids, nutrients, gases, temperature, and other conditions needed to ensure ongoing, proper functioning of cells and, therefore, all parts of the body.

Homocysteine   Another intermediate product that depends on vitamin B_{12} for complete metabolism. However, both vitamin B_{12} and folate (another B vitamin) are coenzymes in the breakdown of certain protein components in this pathway. Thus, elevated homocysteine levels can result from vitamin B_{12}, folate, or pyridoxine deficiencies.

Hydrolyzed Protein Formula   Formula that contains enzymatically digested protein, or single amino acids, rather than protein as it naturally occurs in foods.

Hyperbilirubinemia   Elevated blood levels of bilirubin, a yellow pigment that is a by-product of the breakdown of fetal hemoglobin.

Hypertension   High blood pressure. It is defined as blood pressure exerted inside blood vessel walls that typically exceed 140/90 mmHg (millimeters of mercury).

Hypertonia   Condition characterized by high muscle tone, stiffness, or spasticity.

Hypoallergenic   Foods or products that have a low risk of promoting food or other allergies.

Hypocalcemia   Condition in which body pools of calcium are unbalanced, and low levels are measured in blood as a part of a generalized reaction to illnesses.

Hypogonadism   Atrophy or reduced development of testes or ovaries. Results in immature development of secondary sexual characteristics.

Hypothyroidism   A condition characterized by growth impairment and mental retardation and deafness when caused by inadequate maternal intake of iodine during pregnancy. Used to be called cretinism.

Hypotonia   Condition characterized by low muscle tone, floppiness, or muscle weakness.

Iatrogenic   Used in reference to disease, it is a condition induced by a medical treatment.

Immunoglobulin   A specific protein that is produced by blood cells to fight infection.

Immunological   Having to do with the immune system and its functions in protecting the body from bacterial, viral, fungal, or other infections and from foreign proteins (i.e., those proteins that differ from proteins normally found in the body).

Indirect Calorimetry   Measurement of energy requirements based on oxygen consumption and carbon dioxide production.

Infant Health and Development Program (IHDP)   Growth charts with percentiles for VLBW (<1500g birthweight) and LBW (<2500g birthweight).

Infant Mortality   Death that occurs within the first year of life.

Infant Mortality Attributable to Birth Defects (IMBD)   Category used in tracking infant deaths in which specific diagnoses have a high mortality.

Infecundity   Biological inability to bear children after 1 year of unprotected intercourse.

Infertility   Commonly used to mean a biological inability to bear children.

Innocenti Declaration   The Innocenti Declaration on the Protection, Promotion, and Support of Breastfeeding was produced and adopted by participants at the WHO/UNICEF policy makers' meeting on "Breastfeeding in the 1990s: A Global Initiative," held at the Spedale degli Innocenti, in Florence, Italy, on August 1, 1990. The Declaration established exclusive breastfeeding from birth to 4–6 months of age as a global goal for optimal maternal and child health.

Insulin   Hormone usually produced in the pancreas to regulate movement of glucose from the bloodstream into cells within organs and muscles.

Insulin Resistance   A condition in which cells "resist" the action of insulin in facilitating the passage of glucose into cells.

Intrauterine Growth Retardation (IUGR)   Fetal undergrowth from any cause, resulting in a disproportionality in weight, length, or weight-for-length percentiles for gestational age. Sometimes called intrauterine growth restriction.

Iron Deficiency   A condition marked by depleted iron stores. It is characterized by weakness, fatigue, short attention span, poor appetite, increased susceptibility to infection, and irritability.

Iron-Deficiency Anemia   A condition often marked by low hemoglobin level. It is characterized by the signs of iron deficiency plus paleness, exhaustion, and a rapid heart rate.

Jejunostomy Feeding   Form of enteral nutrition support for delivering nutrition by tube placement directly into the upper part of the small intestine.

Juvenile Rheumatoid Arthritis   Condition in which joints become enlarged and painful as a result of the immune system; generally occurs in children or teens.

Kernicterus or Bilirubin Encephalopathy   The end result of very high untreated bilirubin levels. Excessive bilirubin in the system is deposited in the brain, causing toxicity to the basal ganglia and various brain-stem nuclei.

Ketogenic Diet   High-fat, low-carbohydrate meal plan in which ketones are made from metabolic pathways used in converting fat as a source of energy.

Ketones   Metabolic by-products of the breakdown of fatty acids in energy formation. b-hydroxybutyric acid, acetoacetic acid, and acetone are the major ketones, or "ketone bodies."

Klinefelter's Syndrome   A congenital abnormality in which testes are small and firm, legs abnormally long, and intelligence generally subnormal.

Krebs Cycle   A series of metabolic reactions that produce energy from the proteins, fats, and carbohydrates that constitute food.

Kwashiorkor   A severe form of protein-energy malnutrition in young children. It is characterized by swelling, fatty liver, susceptibility to infection, profound apathy, and poor appetite. The cause of kwashiorkor is unclear.

L. Monocytogenes, or Listeria   A foodborne bacterial infection that can lead to preterm delivery and stillbirth in pregnant women. Listeria infection is commonly associated with the ingestion of soft cheeses, unpasteurized milk, ready-to-eat deli meats, and hot dogs.

Lactation Consultant   A health care professional who provides education and management to prevent and solve breastfeeding problems and to encourage a social environment that effectively supports the breastfeeding mother-infant dyad. Those who successfully complete the International Board of Lactation Consultant Examiners (IBLCE) certification process are entitled to use IBCLC (International Board Certified Lactation Consultant) after their names (www.iblce.org/).

Lactiferous Sinuses   Larger ducts for storage of milk behind the nipple.

Lactogenesis   Another term for human milk production.

Lactose   A form of sugar or carbohydrate composed of galactose and glucose.

Large for Gestational Age (LGA)   Weight for gestational age exceeds the 90th percentile for gestational age. Also defined as birthweight greater than 4500 g (≥10 lb) and referred to as excessively sized for gestational age, or macrosomic.

LDL Cholesterol   Low-density lipoprotein cholesterol, the lipid most associated with atherosclerotic disease. Diets high in saturated fat, trans fatty acids, and dietary cholesterol have been shown to increase LDL-cholesterol levels.

Le Leche League   An international, nonprofit, nonsectarian organization dedicated to providing education, information, support, and encouragement to women who want to breastfeed. It was founded in 1956 by seven women who had learned about successful breastfeeding while nursing their own babies. (www.lelecheleague.org).

Lean Body Mass   Sum of fat-free body tissues: muscle, mineral as in bone, and water.

Leptin   A protein secreted by fat cells that, by binding to specific receptor sites in the hypothalamus, decreases appetite, increases energy expenditure, and stimulates gonadotropin secretion. Leptin levels are elevated by high, and reduced by low, levels of body fat.

Life Expectancy   Average number of years of life remaining for persons in a population cohort or group; most commonly reported as life expectancy from birth.

Life Span   Maximum number of years someone might live; human life span is projected to range from 110 to 120 years.

Lignin   Noncarbohydrate polymer that contributes to dietary fiber.

Linseed   From the flax plant, linum; linseed is another name for flaxseed. Linseed oil is used in paints, varnishes, and inks but is also produced in food form for its rich nutrient content.

Liveborn Infant   The World Health Organization developed a standard definition of *liveborn* to be used by all countries when assessing an infant's status at birth. By this definition, a liveborn infant is the outcome of delivery when a completely expelled or extracted fetus breathes, or shows any sign of life such as beating of the heart, pulsation of the umbilical cord, or definite movement of voluntary muscles, whether or not the cord has been cut or the placenta is still attached.

Lobes   Rounded structures of the mammary gland.

Long-Chain Fats   Carbon molecules that provide fatty acids with 12 or more carbons, which are commonly found in foods.

Longevity   Length of life; it is a measure of life's duration in years.

Low-Birth-Weight Infant (LBW)   An infant weighing <2500 g or <5 lb 8 oz at birth.

Lower Esophageal Sphincter (LES)   The muscle enabling closure of the junction between the esophagus and stomach.

Macrobiotic Diet   This diet falls between semivegetarian and vegan diets and includes foods such as brown rice and other grains, vegetables, fish, dried beans, spices, and fruits.

Macrocephaly   Large head size for age and gender as measured by centimeters (or inches) of head circumference.

Macrophages   A white blood cell that acts mainly through phagocytosis.

Malnutrition   Poor nutrition resulting from an excess or lack of calories or nutrients.

Mammary Gland   The source of milk for offspring, also commonly called the breast. The presence of mammary glands is a characteristic of mammals.

Maple Syrup Urine Disease   Rare genetic condition of protein metabolism in which breakdown by-products build up in blood and urine, causing coma and death if untreated.

MCT Oil   A liquid form of dietary fat used to boost calories; composed of medium-chain triglycerides.

Meconium   Dark green mucilaginous material in the intestine of the full-term fetus.

Medical Neglect   Failure of parent or caretaker to seek, obtain, and follow through with a complete diagnostic study or medical, dental, or mental health treatment for a health problem, symptom, or condition that, if untreated, could become severe enough to present a danger to the child.

Medical Nutrition Therapy (MNT)   Comprehensive nutrition services by registered dieticians to treat the nutritional aspects of acute and chronic diseases.

Medicinal Herbs   Plants used to prevent or remedy illness.

Medium-Chain Fats   Carbon molecules that provide fatty acids with 6–10 carbons, again not typically found in foods.

Memory Impairment   Moderate or severe impairment is present when four or fewer words can be recalled from a list of 20.

Menarche   The occurrence of the first menstrual cycle.

Meningitis   Viral or bacterial infection in the central nervous system that is likely to cause a range of long-term consequences in infancy, such as mental retardation, blindness, and hearing loss.

Menopause   Cessation of the menstrual cycle and reproductive capacity in females.

Menses   The process of menstruation.

Menstrual Cycle   An approximately 4-week interval in which hormones direct a buildup of blood and nutrient stores within the wall of the uterus and ovum maturation and release. If the ovum is fertilized by a sperm, the stored blood and nutrients are used to support the growth of the fertilized ovum. If fertilization does not occur, they are released from the uterine wall over a period of 3 to 7 days. The period of blood flow is called the menses, or the menstrual period.

Mental Retardation   Substantially below-average intelligence and problems in adapting to the environment, which emerge before age 18 years.

Metabolic Syndrome   A constellation of metabolic abnormalities that increase the risk of type 2 diabetes, heart disease, and other disorders. It is characterized by insulin resistance, abdominal obesity, high blood pressure and triglyceride levels, low levels of HDL cholesterol, and impaired glucose tolerance. Also called Syndrome X and insulin-resistance syndrome.

Metabolism   The chemical changes that take place in the body. The conversion of glucose to energy or body fat is an example of a metabolic process.

Methylmalonic Acid (MMA)   An intermediate product that needs vitamin B_{12} as a coenzyme to complete the metabolic pathway for fatty acid metabolism. Vitamin B_{12} is the only coenzyme in this reaction; when it is absent, the blood concentration of MMA rises.

Microcephaly   Small head size for age and gender as measured by centimeters (or inches) of head circumference.

Middle Childhood   Children between the ages of 5 and 10 years; also referred to as school-age.

Milk/Plasma Drug Concentration Ratio (M/P Ratio)   The ratio of the concentration of drug in milk to the concentration of drug in maternal plasma. Since the ratio varies over time, a time-averaged ratio provides more meaningful information than data obtained at a single time point. It is helpful in understanding the mechanisms of drug transfer and should not be viewed as a predictor of risk to the infant, as it is the concentration of the drug in milk, and not the M/P ratio, that is critical to the calculation of infant dose and assessment of risk.

Miscarriage   Generally defined as the loss of a conceptus in the first 20 weeks of pregnancy. Also called spontaneous abortion.

Mitochondria   Intracellular unit in which fatty-acid breakdown takes place and many enzyme systems for energy production inside cells are regulated.

Monounsaturated Fats   Fats in which only one pair of adjacent carbons in one or more of its fatty acids is linked by a double bond (e.g., –C–C=C–C–).

Monovalent Ion   An atom with an electrical charge of +1 or –1.

Morbidity Rate   The rate of illnesses in a population.

Mortality Rate   The rate of death.

Myoepithelial Cells   Specialized cells that line the alveoli and can contract to cause milk to be secreted into the duct.

Necrotizing Enterocolitis (NEC)   Condition with inflammation or damage to a section of the intestine, with a grading from mild to severe.

Neonatal Death   Death that occurs in the period from the day of birth through the first 28 days of life.

Neural Tube Defects (NTDs)   Spina bifida and other malformations of the neural tube. Defects result from incomplete formation of the neural tube during the first month after conception.

Neurobehavioral   Pertains to control of behavior by the nervous system.

Neuromuscular   Term pertaining to the central nervous system's control of muscle coordination and movement.

Neuromuscular Disorders   Conditions of the nervous system characterized by difficulty with voluntary or involuntary control of muscle movement.

Neutrophils   A class of white blood cells that are involved in protecting against infection.

NF (National Formulary)   A uniformity standard for herbs and botanicals.

Nonessential Nutrients   Nutrients required for growth and health that can be produced by the body from other components of the diet.

Nonheme Iron   Iron contained within a protein of hemoglobin that is in the ferric state.

Nonorganic Failure to Thrive   Inadequate weight or height gain without an identifiable biological cause, so that an environmental cause is suspected.

Nutrient-Dense Foods   Foods that contain relatively high amounts of nutrients compared to their caloric value.

Nutrients   Chemical substances in foods that are used by the body for growth and health.

Nutrigenomics   The science of gene-nutrient interactions.

Nutrition Programming   The process by which exposure of the fetus to certain levels of energy and nutrients modify the function of genes in ways that affect metabolism and the development of diseases later in life.

Nutrition Support   Provision of nutrients by methods other than eating regular foods or drinking regular beverages, such as directly accessing the stomach by tube or placing nutrients into the bloodstream.

Obesity   BMI-for-age greater than the 95th percentile with excess fat stores as evidenced by increased triceps skinfold measurements above the 85th percentile.

Oral-Gastric (OG) Feeding   A form of enteral nutrition support for delivering nutrition by tube placement from the mouth to the stomach.

Organic Failure to Thrive   Inadequate weight or height gain resulting from a health problem, such as iron-deficiency anemia or a cardiac or genetic disease.

Osmolarity   Measure of the number of particles in a solution, which predicts the tendency of the particles to move from high to low concentration. Osmolarity is a factor in many systems, such as in fluid and electrolyte balance.

Osteoblasts   Bone cells involved with bone formation; bone-building cells.

Osteoclasts   Bone cells that absorb and remove unwanted tissue.

Osteoporosis   Condition in which low bone density or weak bone structure leads to an increased risk of bone fracture.

Ova   Eggs of the female produced and stored within the ovaries (singular = *ovum*).

Overweight   Body mass index at or above the 95th percentile.

Oxidative Stress   A condition that occurs when cells are exposed to more oxidizing molecules (such as free radicals) than to antioxidant molecules that neutralize them. Over time oxidative stress causes damage to lipids, DNA, cells, and tissues. It increases the risk of heart disease, type 2 diabetes, cancer, and other diseases.

Oxytocin   A hormone produced during letdown that causes milk to be ejected into the ducts.

Pagophagia   Compulsive consumption of ice or freezer frost.

Parenteral Feeding   Delivery of nutrients directly to the bloodstream.

Palpebral Fissure   The space between the top and bottom eyelid when the eye is open. This opening is small in children with FAS.

Parity   The number of previous deliveries experienced by a woman; *nulliparous* = no previous deliveries, *primiparous* = one previous delivery, *multiparous* = two or more previous deliveries. Women who have delivered infants are considered to be "parous."

Pediatric AIDS   Acquired immuno-deficiency syndrome, in which infection-fighting abilities of the body are destroyed by a virus.

Pelvic Inflammatory Disease (PID)   A general term applied to infections of the cervix, uterus, fallopian tubes, or ovaries. Occurs predominantly in young women and is generally caused by infection with a sexually transmitted disease, such as gonorrhea or chlamydia, or with intrauterine device (IUD) use.

Periconceptional Period   Around the time of conception, generally defined as the month before and the month after conception.

Perinatal Death   Death occurring at or after 20 weeks of gestation and through the first 28 days of life.

Philtrum   The vertical groove between the bottom of the nose and the upper lip. The philtrum is smooth, or flat, when there is no groove.

Phytochemicals   (*phyto* = plants) Chemical substances in plants, some of which affect body processes in humans that may benefit health.

Phytoestrogen   A hormone-like substance found in plants, about 1/1000 to 1/2000 as potent as the human hormone, but strong enough to bind with estrogen receptors and mimic estrogen and anti-estrogen effects.

PICA   An eating disorder characterized by the compulsion to eat substances that are not food.

PKU (Phenylketonuria)   An inherited error in phenylalanine metabolism most commonly caused by a deficiency of phenyl-alanine hydroxylase, which converts the essential amino acid phenylalanine to the nonessential amino acid tyrosine.

Placenta   A disk-shaped organ of nutrient and gas interchange between mother and fetus. At term, the placenta weighs about 15% of the weight of the fetus.

Platelets   A component of the blood that plays an important role in blood coagulation.

Polycystic Ovary Syndrome (PCOS)   (*polycysts* = many cysts; i.e., abnormal sacs with membranous linings) A condition in females characterized by insulin resistance, high blood insulin and testosterone levels, obesity, menstrual dysfunction, amenorrhea, infertility, hirsutism (excess body hair), and acne.

Polyunsaturated Fats   Fats in which more than one pair of adjacent carbons in one or more of its fatty acids are linked by two or more double bonds (e.g., –C–C=C–C=C–).

Postictal State   Time after a seizure of altered consciousness; appears like a deep sleep.

Pouring Rights   Contracts between schools and soft-drink companies whereby the schools receive a percentage of the profits of soft-drink sales in exchange for the school offering only that soft-drink company's products on the school campus.

Prader-Willi Syndrome   Condition in which partial deletion of chromosome 15 interferes with control of appetite, muscle development, and cognition.

Preadolescence   The stage of development immediately preceding adolescence; 9 to 11 years of age for girls and 10 to 12 years of age for boys.

Prebiotics   Certain fiberlike forms of indigestible carbohydrates that support the growth of beneficial bacteria in the lower intestine. Nicknamed "intestinal fertilizer."

Preeclampsia   A pregnancy-specific condition that usually occurs after 20 weeks of pregnancy (but may occur earlier). It is characterized by increased blood pressure and protein in the urine and is associated with decreased blood flow to maternal organs and through the placenta.

Prediabetes A condition in which blood glucose levels are higher than normal but not high enough for the diagnosis of diabetes. It is characterized by impaired glucose tolerance, or fasting blood glucose levels between 110 and 126 mg/dl.

Preloads Beverages or foods such as yogurt in which the energy/macronutrient content has been varied by the use of various carbohydrate and fat sources. The preload is given before a meal or snack and subsequent intake is monitored. This study design has been employed by Birch and Fisher in their studies of appetite, satiety, and food preferences in young children.

Premenstrual Syndrome (*premenstrual* = the period of time preceding menstrual bleeding; *syndrome* = a constellation of symptoms) A condition occurring among women of reproductive age that includes a group of physical, psychological, and behavioral symptoms with onset in the luteal phase and subsiding with menstrual bleeding. Also called premenstrual dysphoric disorder (PMDD).

Preschool-Age Children Children between the ages of 3 and 5 years, who are not yet attending kindergarten.

Preterm Infants Infants born at or before 37 weeks of gestation.

Primary Malnutrition Malnutrition that results directly from inadequate or excessive dietary intake of energy or nutrients.

Probiotics Strains of Lactobacillus and bifidobacteria that have beneficial effects on the body. Also called "friendly bacteria."

Programming The process by which exposure to adverse nutritional or other conditions during sensitive periods of growth and development produces long-term effects on body structures, functions, and disease risk.

Prolactin A hormone that stimulates milk production.

Proportionately Small for Gestational Age (pSGA) Newborn weight, length, and head circumference are ≤10th percentile for gestational age. Also called symmetrical SGA.

Prostacyclins Biologically active substances produced by blood vessel walls that inhibit platelet aggregation (and therefore blood clotting), dilate blood vessels, and reduce blood pressure.

Prostaglandins A group of physiologically active substances derived from the essential fatty acids. They are present in many tissues and perform such functions as the constriction or dilation of blood vessels, and stimulation of smooth muscles and the uterus.

Psychostimulant Classification of medication that acts on the brain to improve mental or emotional behavior.

Puberty The time frame during which the body matures from that of a child to that of a young adult, and becomes biologically capable of reproduction.

Pulmonary Related to the lungs and their movement of air for exchange of carbon dioxide and oxygen.

Quality of Life A measure of life satisfaction that is difficult to define, especially in heterogeneous aging population. Quality of life measures include factors such as social contacts, economic security, and functional status.

Recommended Dietary Allowances (RDAs) The average daily dietary intake levels sufficient to meet the nutrient requirements of nearly all (97% to 98%) healthy individuals in a population group. RDAs serve as goals for individuals.

Recumbent Length Measurement of length while the child is lying down. Recumbent length is used to measure toddlers less than 24 months of age, and those between 24 and 36 months who are unable to stand unassisted.

Reflex An automatic (unlearned) response that is triggered by a specific stimulus.

Registered Dietitian An individual who has acquired knowledge and skills necessary to pass a national registration examination and who participates in continuing professional education.

Resilience Ability to bounce back, to deal with stress and recover from injury or illness.

Resting Energy Expenditure The amount of energy needed by the body in a state of rest.

Resting Metabolic Rate (RMR) Measuring energy expenditure in an individual who has fasted, had no vigorous physical activity prior to the test, has been given time to relax (e.g., rest) for 30 minutes before starting measurement, and is in a quiet, private room with privacy and normal, comfortable temperatures.

Rett Syndrome Condition in which a genetic change on the X chromosome results in severe neurological delays, causing children to be short, thin appearing, and unable to talk.

Rooting Reflex Action that occurs if one cheek is touched, resulting in the infant's head turning toward that cheek and the infant opening his mouth.

Saturated Fats Fats in which adjacent carbons in the fatty acid component are linked by single bonds only (e.g., –C–C–C–C–).

Scoliosis Condition in which the vertebral bones in the back show a side-to-side curve, resulting in a shorter stature than expected if the back were straight.

Scrotum A muscular sac containing the testes.

Secondary Condition Common consequence of a condition, which may or may not be preventable over time.

Secondary Malnutrition Malnutrition that results from a condition (e.g., disease, surgical procedure, medication use) rather than primarily from dietary intake.

Secondary Sexual Characteristics Physiological changes that signal puberty, including enlargement of the testes, penis, and breasts and the development of pubic and facial hair.

Secretory Cells Cells in the acinus (milk gland) that are responsible for secreting milk components into the ducts.

Secretory Immunoglobulin A One of the proteins found in secretions that protect the body's mucosal surfaces from infections. It may act by reducing the binding of a microorganism with cells lining the digestive tract. It is present in human colostrum but not transferred across the placenta.

Seizures Condition in which electrical nerve transmission in the brain is disrupted, resulting in periods of loss of function that vary in severity.

Self-Efficacy The ability to make effective decisions and to take responsible action based upon one's own needs and desires.

Semen The penile ejaculate containing a mixture of sperm and secretions from the testes, prostate, and other glands. It is rich in zinc, fructose, and other nutrients. Also called seminal fluid.

Senescent Old to the point of nonfunctional.

Sensorimotor An early learning system in which the infant's senses and motor skills provide input to the central nervous system.

Serum Iron, Plasma Ferritin, and Transferring Saturation Measures of iron status obtained from blood plasma or serum samples.

Sex Hormone-Binding Globulin (SHBG) A protein that binds with the sex hormones testosterone and estrogen. Also called steroid hormone binding globulin, because testosterone and estrogen are produced from cholesterol and are thus considered to be steroid hormones. These hormones are inactive when bound to SHBG, but are available for use when needed. Low levels of SHBG are related to increased availability of testosterone and estrogen in the body.

Short-Chain Fats   Carbon molecules that provide fatty acids less than 6 carbons long, as products of energy generation from fat breakdown inside cells. Short-chain fatty acids are not usually found in foods.

Shoulder Dystocia   Blockage or difficulty of delivery due to obstruction of the birth canal by the infant's shoulders.

Small for Gestational Age (SGA)   Newborn weight is ≤10th percentile for gestational age. Also called small for date (SFD).

Social Marketing   A marketing effort that combines the principles of commercial marketing with health education to promote a socially beneficial idea, practice, or product.

Spastic Quadriplegia   A form of cerebral palsy in which brain damage interferes with voluntary muscle control in both arms and legs.

Stature   Standing height.

Steroid Hormones   Hormones such as progesterone, estrogen, and testosterone produced primarily from cholesterol.

Stroke   The event that occurs when a blood vessel in the brain becomes occluded due to a clot or ruptures cutting off blood supply to a portion of the brain. Also called a cerebral vascular accident.

Subfertility   Reduced level of fertility characterized by unusually long time to conception (over 12 months) or repeated, early pregnancy losses.

Subscapular Skinfold Thickness   A skinfold measurement that can be used with other skinfold measurements to estimate percent body fat; the measurement is taken with skinfold calipers just below the inner angle of the scapula, or shoulder blade.

Suckle   A reflexive movement of the tongue moving forward and backward; earliest feeding skill.

T-lymphocytes   White blood cells that are active in fighting infection may also be called T-cells; the T in T-cell stands for thymus). These cells coordinate the immune system by secreting hormones that act on other cells.

T. Gondii, **or Loxoplasmosis**   A parasitic infection that can impair fetal brain development. The source of the infection is often hands contaminated with soil or the contents of a cat litter box; or raw or partially cooked pork, lamb, or venison.

Telomere   A cap-like structure that protects the end of chromosomes; it erodes during replication.

Teratogenic   Exposures that produce malformations in embryos or fetuses.

Testes   Male reproductive glands located in the scrotum. Also called testicles.

Thromboxanes   Biologically active substances produced in platelets that increase platelet aggregation (and therefore promote blood clotting), constrict blood vessels, and increase blood pressure.

Toddlers   Children between the ages of 1 and 3 years.

Tolerable Upper Intake Levels   Highest level of daily nutrient intake that is likely to pose no risk of adverse health effects to almost all individuals in the general population; gives levels of intake that may result in adverse effects if exceeded on a regular basis.

Total Fiber   Sum of dietary fiber and functional fiber.

Tracheoesophageal Atresia   Incomplete connection between the esophagus and the stomach in utero, resulting in a shortened esophagus.

Trans Fatty Acids   Fatty acids that have unusual shapes resulting from the hydrogenation of polyunsaturated fatty acids. Trans fatty acids also occur naturally in small amounts in foods such as dairy products and beef.

Transient Ischemic Attack (TIA)   Temporary and insufficient blood supply to the brain.

Transpyloric Feeding (TP)   Form of enteral nutrition support for delivering nutrition by tube placement from the nose or mouth into the upper part of the small intestine.

Triceps Skinfold   A measurement of a double layer of skin and fat tissue on the back of the upper arm. It is an index of body fatness and measured by skinfold calipers. The measurement is taken on the back of the arm midway between the shoulder and the elbow.

Type I Diabetes   A disease characterized by high blood glucose levels resulting from destruction of the insulin-producing cells of the pancreas. In the past, this type of diabetes was called juvenile-onset diabetes and insulin-dependent diabetes.

Type 2 Diabetes   A disease characterized by high blood glucose levels due to the body's inability to use insulin normally, or to produce enough insulin. In the past this type of diabetes was called adult-onset diabetes and non-insulin-dependent diabetes.

Unsaturated Fats   Fats in which adjacent carbons in one or more fatty acids are linked by one or more double bonds (e.g., –C–C=C–C=C–).

USP (United States Pharmacopeia)   A nongovernmental, nonprofit organization (since 1820); establishes and maintains standards of identity, strength, quality, purity, processing, and labeling for health care products.

Vegan Diet   The most restrictive of vegetarian diets, allowing only plant foods.

Venous Thromboembolism   A blood clot in a vein.

Vermillion Border   The exposed pink or reddish margin of a lip. A thin vermillion border in FAS denotes a thin upper lip.

Very-Low-Birth-Weight Infant (VLBW)   An infant weighing, <1500 g or, <3 lb 5 oz at birth.

Waist-to-hip Ratio   The ratio of the waist circumference, measured at its narrowest, and the hip circumference, measured where it is widest. This ratio is an easy way to measure body fat distribution, with a higher ratio indicative of an abdominal fat pattern. A high waist-to-hip ratio is associated with a high risk of chronic disease.

Weaning   Discontinuation of breastfeeding or bottle-feeding and substitution of food for breast milk or infant formula.

Work of Breathing (WOB)   A common term used to express extra respiratory effort in a variety of pulmonary conditions.

Working-Poor Families   Families where at least one parent worked 50 or more weeks a year and the family income was below the poverty level.

Xerostomia   Dry mouth, or xerostomia, can be a side effect of medications (especially antidepressants), of head and neck cancer treatments, and of diabetes, and is also a symptom of Sjogren's syndrome, which is an autoimmune disorder for which no cure is known.

Index

A

Academy of Breastfeeding Medicine, 183
Acquired immune deficiency syndrome (AIDS). *see* HIV/AIDS
Acute undernutrition, 57–58
Adequate Intakes (AIs), 4
ADHD, 300–301, 350
Adiposity rebound, 282–284, 313–314, 319–320, 431
Adolescent nutrition
    for athletes, 391–392
    cardiovascular disease, 393–395
    chronic health conditions, 403
    eating disorders and, 395–403
    energy/nutrient needs, 365–372
    environmental factors and, 361–362, 386
    ergogenic supplement use and, 390–391
    fluids and, 391
    health/eating-related behaviors and, 361–365
    iron deficiency and, 392–393
    media influence in, 362
    normal growth/development and, 358–360
    normal psychosocial development and, 360–365
    nutritional needs of, 357–358
    nutrition screening/assessment/ intervention, 372–378
    overweight/obesity and, 386–389
    physical activity/sports, 378–380, 391–392
    pregnancy nutrition and, 156–167
    promoting healthy behaviors, 380–383
    resources, 384, 404
    risk indicators, 373–377
    substance abuse and, 392
    vitamin/mineral supplement use and, 389–391
Adolescent pregnancies, 156–167
Adult nutrition. *see also* Older adult nutrition
    alcohol and, 415–416
    beverage intake, 415–417
    cancer, 448–451
    cardiovascular disease and, 436–441
    diabetes and, 444–448
    dietary recommendations for, 413–417
    diets, 410
    energy and, 408–410
    exercise and, 421–422
    health disparities among adults, 412–413
    healthy objectives for, 406–407
    HIV/AIDS, 451–452
    interventions for risk reduction, 422–426
    metabolic syndrome and, 442–444
    nutrient recommendations, 417–421
    overweight/obesity and, 429–441
    physiological changes of adulthood, 407–413
    resources, 425
    vegetarian diets, 414–415

Adult-onset diabetes. *see* Type 2 diabetes
Aging adults. *see* Older adult nutrition
Aging theories, 457–459
AIDS. *see* HIV/AIDS
Alcohol. *see also* Fetal alcohol syndrome
    adolescent nutrition and, 392
    adult nutrition and, 415–416
    cardiovascular disease and, 490–491
    fertility and, 60
    fetal growth and, 64
    lactation and, 203–205
    male factors of fertility and, 60
    older adult nutrition and, 473–474
    pregnancy outcome and, 110, 155–156
Alcohol (ethanol), 4
Allergic diseases, 212
Allergies
    infant nutrition and, 212–214, 242–243
    toddlers/preschoolers nutrition and, 306–307
Alpha-linolenic acid, 3, 9, 322–323
Alternative medicine. *see* Herbal remedies/ supplements
Alveoli, 161
Alzheimer's Disease, 33, 507–509
Amenorrhea, 58–59, 74–75
American Academy of Pediatric Dentistry, 280
American Academy of Pediatrics (AAP), 199, 211, 218–219, 231–233, 280
American College of Obstetricians and Gynecologists (ACOG), 200
American Dental Association, 280
American Diabetes Association, 76, 78
American Dietetic Association, 193
American Heart Association (AHA), 285, 440
Amino acids, 3, 6
Amniotic fluid, 97
Amylophagia, 123
Anabolic phase of pregnancy, 92
Anaphylaxis, 306
Androgens, 56
Androstenedione, 390
Anemia. *see also* Iron
    adolescent nutrition and, 392–393
    cow's milk and, 233
    hemolytic, 281
    infant nutrition and, 241–242
    pregnancy and, 119
    toddlers/preschoolers nutrition and, 278–279
Anencephaly, 115, 259
Aneurysm, 490
Anorexia nervosa, 75–76, 395, 397–398. *see also* Eating disorders
Anovulatory cycles, 58
Anthropometric assessments, 41
Anthropometry, 41
Antibiotics, 37
Antidiuretic hormone, 514
Antioxidants, 13, 22, 60, 507
Appetite/satiety, 276–277, 461–462

Appropriate for gestational age (AGA), 101
Artificial sweeteners and pregnancy, 110
Assessments. *see also* Dietary assessments
    for adolescents, 372–378
    of adults with overweight or obesity diseases, 433
    anthropometric assessment, 41
    biochemical assessment, 41–42
    cancer, 449–450
    cardiovascular disease, 437
    clinical/physical assessment, 39
    community-level assessment, 39
    diabetes, 445
    dietary history, 39–40
    food frequency questionnaires, 40
    Framingham heart health assessment, 438
    HIV/AIDS, 451–452
    individual-level assessment, 39
    for metabolic syndrome, 443
    MNA nutritional screening and assessment, 466
    obesity, 431
    pregnancy, 124–125
    USDA Automated Multiple-Pass Method, 40
    Web dietary assessment resources, 40
Assisted reproductive technology (ART), 148
Asthma, 306
Atherogenic diet, 407
Atherosclerosis, 285, 436–437
Athetosis, 348
Atrial fibrillation, 490
At-risk infants and nutrition, 244, 248–249
Attention deficit hyperactivity disorder (ADHD), 300–301, 350
Autism, 255, 300–301, 304
Autoimmune disease, 32, 77
Autoimmune response, 84

B

Babinski reflex, 225
Baby-bottle caries, 242. *see also* Caries
Baby-bottle tooth decay, 279–280. *see also* Caries
Baby foods, commercial, 237–238. *see also* Infant nutrition
Baby Friendly Hospital Initiative, 185
Bannister, Roger, 460
Bariatric surgery, 74, 136–137, 389, 435–436
Basal metabolic rate (BMR)
    adolescent nutrition and, 367–368
    adult nutrition and, 409
Behavioral Risk Factor Surveillance System, 42
Bereavement, 514–515
Berger, David, 482–483
Bilirubin metabolism, 208–209
Binge-eating disorder, 399–400. *see also* Eating disorders
Bioactive food components, 122
Bioavailability of minerals in breastmilk, 168

Biochemical assessments, 41–42
Biotin
    function/deficiency of, 14
    overdose consequences/food source of, 15
Birthweight. see also Weight
    as measure of health, 223
    of single/multifetal newborns, 149
Blessed thistle, 203
Blink reflex, 225
Blood pressure control. see DASH Eating
    Plan
Blue cohosh, 127
The Blue Zones (Buettner), 483–484
B-lymphocytes, 169
BMI. see Body mass index (BMI)
BMI rebound, 282–284, 313–314,
    319–320, 431
Body image/dieting
    adolescents and, 395–396
    children/preadolescents and, 317
Body mass index (BMI). see also Adiposity
    rebound; Weight
    adolescent nutrition and, 386
    children/preadolescents and, 313–315
    defined, 58
    obesity and, 431, 494–495
    toddlers/preschoolers nutrition and,
        270, 282
Body water changes during pregnancy,
    92–93
Bogalusa Heart Study, 319
Bone age, 319
Bone loss, 497
Bone mass density (BMD), 496–497
Bottle caries, 242. see also Caries
Breast augmentation after lactation, 172
Breastfeeding: A Guide for the Medical
    Professional, 164
Breastfeeding and Human Lactation Study
    Center, 199
Breastfeeding Coalition, 183
Breastfeeding Promotion in Physicians'
    Office Practices (BPPOP), 219
Breast-milk/breastfeeding. see also Lactation
    nutrition
    benefits for infants, 169–171
    benefits for mothers, 168–169
    collection/storage of, 218–219
    community support for, 187
    contamination of, 198–200
    energy and, 165–166
    exercise and, 179–180
    feeding frequency, 175
    flavor of, 168
    formula vs., 231–233
    Health Care System support for, 182
    HIV/AIDS and, 216–217
    hospitals and birthing centers support for,
        183–185
    human milk composition, 164–168
    identifying hunger and satiety, 175
    identifying malnutrition, 175–176
    maternal energy balance and milk
        composition, 179–181
    mechanics of, 174–175
    multiples, 212
    optimal duration for, 172–173
    prenatal education for, 182–183
    preparing the breast for, 173
    presenting to infant, 173–174

preterm delivery and, 102
preterm infants and, 215–216, 256–257
rates in birth cohorts, 161
resources, 191–192, 221
supply and demand, 171–172
support after discharge, 185–186
tooth decay, 176–177
workplace support for, 186–187
Breast milk jaundice syndrome, 211
Breast-nonfeeding jaundice, 209–211
Breast pump, 172
"Bright Futures Guidelines for Health
    Supervision of Infants, Children, and
    Adolescents, 291–292, 330
"Bright Futures in Practice: Nutrition",
    291–292
Bronchopulmonary dysplasia (BPD), 253,
    298, 306
Browning, Robert, 483
Buettner, Dan, 483–484
Bulbourethral gland, 54
Bulimia nervosa, 75–76, 398. see also
    Eating disorders

C
Cabbage leaves, 196
Caffeine
    child/preadolescents nutrition and, 326
    fertility and, 60
    food and beverage content of, 62
    lactation and, 206
    older adult nutrition and, 476
    PMS and, 71
    during pregnancy, 122
    water need and, 416–417
Calcium
    adolescent nutrition and, 369–370, 380
    adult nutrition and, 419–420
    child/preadolescents nutrition and, 325
    food sources for, 28
    function/deficiency of, 24
    hypocalcemia in infants, 251
    older adult nutrition and, 474
    osteoporosis and, 498
    overdose consequences/food source of, 25
    PMS and, 72
    preeclampsia and, 140
    during pregnancy, 118
    toddlers/preschoolers nutrition and, 289
California Adolescent Nutrition and Fitness
    (CANFIT) Program, 383
Calorie. see also Energy
    defined, 2, 3
    infant nutrition and, 228
    intake during pregnancy, 109–110
Calorie restriction, 459
Calorimetry, 409
Campylobacter, 281
Cancer
    adult nutrition and, 448–451
    medications and, 511
    risk factors for, 407
CANFIT, 383
Capsaicin, 507
Carbohydrate counting, 447
Carbohydrate loading, 391
Carbohydrates
    adolescent nutrition and, 368–369
    breastmilk and, 167
    description of, 4–6

DRIs for, 6
as essential nutrients, 3
food sources for, 6, 7
galactosemia, 244
gestational diabetes and, 145
intake during pregnancy, 110
maternal metabolism of, 94
older adult nutrition and, 469–471
Carcinogenesis, 448
Carcinogenic diet, 407
Cardio-protective diet, 440
Cardiovascular disease
    alcohol and, 490–491
    green tea and, 514
    medications and, 511
    older adult nutrition and, 488–490
    prevalence among adults, 436–441
    in school-age children, 322–323
    in teenagers, 393–395
    in toddlers, 284–285
    treatment factors for, 489
Caries
    bottle, 242
    children/preadolescents and, 318
    infant nutrition and, 242
    toddlers/preschoolers and, 279–280
Carotid artery disease, 490
Casein, 166
Catabolic phase of pregnancy, 92
Catch-up growth, 251
CDC. see Centers for Disease Control and
    Prevention (CDC)
The CDC Guide to Breastfeeding
    Interventions, 183
Celiac disease, 32, 79, 81–82, 84, 86, 303
Centenarian, 457
Centers for Disease Control and Prevention
    (CDC)
    growth charts, 313, 341
    iron deficiency and, 279
    lead poisoning and, 281
    longevity and, 456
    preconception nutrition and, 66
Cerebral palsy, 102, 304–305, 339,
    347–348
Cerebral spinal atrophy, 341
Cerebrovascular disease. see Cardiovascular
    disease
Chaste tree berry extract, 83–84
Child and Adult Care Food Program
    (CACFP), 43
Child Nutrition Program, 351
Child/preadolescent nutrition
    body image/dieting and, 317
    caffeine and, 326
    cardiovascular disease, 322–323
    and development of feeding skills,
        315–317
    dietary recommendations for, 324–327
    energy/nutrient needs, 317–318
    exercise recommendations and, 327–328
    herbal remedies, 351
    media influence on, 316
    normal growth/development and, 313,
        341–342
    nutrition problems and, 318
    nutrition recommendations for, 342–345
    nutrition services and, 351–353
    physiological/cognitive development and,
        313–317

prevention disorders related to, 319–324
public food/nutrition programs, 331–334
resources, 335–337, 355
risk reduction programs in, 328–331
special health care needs and, 339–341, 345–351
tracking of, 311–313
Children with special health care needs. *see* Special health care needs
Chloride
 function/deficiency of, 26
 overdose consequences/food source of, 27
Cholesterol
 breastmilk and, 166
 child/preadolescents nutrition and, 326
 defined, 10
 HMG-CoA reductase, 441
 infant nutrition and, 228
 miscarriages and, 102
 older adult nutrition and, 472
 toddlers/preschoolers nutrition and, 285
Cholestin, 418
Choline
 adult nutrition and, 419
 food sources for, 20
 function/deficiency of, 16
 overdose consequences/food source of, 17
 during pregnancy, 116
Chondroitin. *see* Glucosamine and chondroitin
Chromium
 function/deficiency of, 26
 overdose consequences/food source of, 27
Chronic conditions
 adolescent nutrition and, 403
 toddlers/preschoolers nutrition and, 297–299
Chronic diseases
 breastfeeding and, 170
 defined, 33, 406
 risk factors for, 407
Chronic hypertension, 138
Chronic inflammation
 defined, 33
 obesity and, 73
 osteoarthritis, 505–507
Chronic obstructive pulmonary disease (COPD), 511
Chronic undernutrition, 57
Chronological aging, 455–456
Cleft lip/palate, 259
Climacteric change, 408
Clitoris, 54
Cobalamin, 502
Coenzyme Q$_{10}$ and fertility, 84–85
Coenzymes, 13
Cognitive development
 average age of maturation, 357
 in infants, 225–226
 in older adults, 507–509
 in school-age children, 314–315
 in toddler/preschool-age children, 271, 274
Cognitive function, breastfeeding and, 170–171
Colic, 180–181, 241
Colostrum, 164–165
Commodity program, 332
Comorbidity, 434, 488
Competitive foods, 329

Complex carbohydrates, 5. *see also* Carbohydrates
Compression of morbidity, 455
Congenital abnormality
 defined, 76
 infant nutrition and, 257–261
Congenital anomalies
 defined, 259
 gestational diabetes and, 141–142
 infant nutrition and, 257–261
Congestive heart failure, 511
Constipation
 fiber and, 504–505
 infant nutrition and, 242
 misconceptions about, 504
 older adult nutrition and, 503–505
 pregnancy and, 130
 toddlers/preschoolers nutrition and, 280
Continuum of nutritional health, 410–412
Contraceptives, 62–63
Contraindications to breastfeeding, 216–217
Copper
 function/deficiency of, 26
 overdose consequences/food source of, 27
Coronary heart disease. *see* Cardiovascular disease
Corpus luteum, 53
Counseling and adolescent nutrition, 373–378
Cow's milk, 165, 233, 307
Craig, F.W., 123
Creatine, 390–391
Critical periods in embryonic growth, 97
Critical periods in oral feeding development, 225
Cross-sectional study, 179
Cultural considerations
 child/preadolescents nutrition and, 327
 infant nutrition and, 243
 older adult nutrition and, 478–479
 pregnancy diets and, 124
 toddlers/preschoolers nutrition and, 290
Cup drinking in infants, 236
Curie, Marie, 296
Cystic fibrosis
 children/preadolescents and, 339
 defined, 298
 eating/feeding problems in children with, 346

D

Daily Values (DVs), 4
DASH Eating Plan, 45–46, 48, 440, 491–493
Death, leading causes of, 429
Deficiencies, 23, 31. *see also specific nutrient, i.e.,* Selenium
Dehydration, 513–514
Dehydroepiandrosterone (DHEA), 390
Dementia, 507–509, 511. *see also* Alzheimer's Disease
Dental caries
 children/preadolescents and, 318
 infant nutrition and, 242
 toddlers/preschoolers and, 279–280
Depo-Provera, 63, 200
Depot medroxyprogesterone acetate, 63, 200

DETERMINE, 462–464
Development, embryonic/fetal. *see* Embryonic/fetal growth and development
Developmental delay, 254–255, 306
Developmental disabilities, 240
Developmental plasticity, 103
DHEA, 390, 418
Diabetes. *see also* Gestational diabetes
 adult nutrition and, 444–448
 children/preadolescents and, 319, 339
 discussion, 77
 eating/feeding problems in children with, 346–347
 exchange lists for, 446–447
 exercise and, 78, 447–448
 fetal growth and, 64
 management plan for, 445–446
 medications and, 511
 older adult nutrition and, 493–494
 prediabetes, 78, 444–445
 pregnancy nutrition and (*see* Gestational diabetes)
 prevalence among adults, 444
 resources, 86
 risk factors for, 407
 Type 1, 76, 146–147, 444
 Type 2, 5, 76–78, 144, 319, 444, 448
Diabetes Control and Complications Trial (DCCT), 445
Diabetes Mellitus, 76–78
Diabetogenic effect of pregnancy, 94
Diagnostic and Statistical Manual of Mental Disorders (DSM IV), 71
Diaphragmatic hernia, 259
Diarrhea
 infant nutrition and, 242
 toddlers/preschoolers nutrition and, 303
Dietary Approaches to Stop Hypertension (DASH) Eating Plan. *see* DASH Eating Plan
Dietary assessments. *see also* Assessments
 adolescents and, 372–378
 during pregnancy, 124–125
Dietary fiber, 324. *see also* Fiber
Dietary folate equivalents (DFE), 116
Dietary guidance, 412
Dietary guidance system, 413
Dietary Guideline on Weight Management, 284
Dietary Guidelines for Americans, 44–45
 pediatric obesity and, 284
Dietary history for assessment, 39–40
Dietary intake
 adolescent nutrition and, 365
 in twin pregnancy, 150–151
Dietary recommendations
 for adults, 413–417
 for children/preadolescents, 324–327
 for older adults, 465
 for toddler/preschoolers, 286–291
Dietary Reference Intakes (DRIs)
 adolescent nutrition and, 365
 child/preadolescents nutrition and, 340–341
 children/preadolescents and, 317–318
 defined, 4
 toddlers/preschoolers nutrition and, 277, 285

Dietary supplements, 35, 417, 448, 474–478. *see also* Herbal remedies/supplements
Diets. *see also* Eating disorders
  adolescents and, 395–396
  for adults, 410
  for breastfeeding women, 177–178, 180
  children/preadolescents and, 317
  fads, 410
  for gestational diabetes, 145
  Mediterranean diet, 46
  obesogenic diet, 407
  pregnancy nutrition and, 122–125, 177–178
  resources, 86
  weight cycling, 410
  for weight loss, 73–74
  yo-yo dieting, 410
Differentiation of cells, 97
DiGeorge Syndrome, 261
Digestive system development in infants, 226–227
Diplegia, 298
Disabilities Education Act (IDEA), 339, 353
Disordered eating disorders, 397
Disproportionately small for gestational age (dSGA), 100–101
Diverticulitis, 504
Dizygotic (DZ) twins, 148–149
Docosahexaenoic acid (DHA)
  breastmilk and, 166
  pregnancy outcome and, 113–114
Domperidone, 198
Double Talk, 212
Doula, 182
Dowager's hump, 498
Down syndrome, 249, 255, 342
Drinking from cup, 236
DRIs. *see* Dietary Reference Intakes (DRIs)
Drug abuse. *see* Substance abuse
Drugs
  adolescent nutrition and, 392
  human milk contamination and, 198–200
Drug use. *see specific substance*
Dryden, John, 310
Dumping syndrome, 136
Dutch Hunger Winter (1943-1944), 108
Dyslipidemia, 437

E
Ear infections in infants, 242
Early Childhood Caries (ECC), 279
Early Head Start program, 244, 263, 293, 307
Early Intervention Program, 258
Early intervention services for toddlers, 297–298
Early Periodic Screening, Detection, and Treatment Program (EPSDT), 224
Eating Competence Model, 423–424
Eating disorders
  anorexia nervosa, 75–76, 395, 397–398
  binge eating, 399–400
  body dissatisfaction, 396
  bulimia nervosa, 75–76, 398
  dieting behaviors, 395–396
  discussion, 75
  disordered eating behaviors, 397
  etiology of, 400–401

fertility and, 75–76
media influence and, 400
pregnancy and, 154
prevalence among adolescents, 397
prevention of, 402–403
treatment of, 401–402
Eating-related behaviors in adolescents, 361–365
Echinacea, 201, 418
E. coli, 281
Edema, 93
Education
  adolescent nutrition and, 373–378
  older adult nutrition and, 480–482
  prevention of weight-related disorders, 402–403
Eicosapentaenoic acid (EPA), 113–114
Ejaculatory duct, 54
Elderly. *see* Older adult nutrition
Elderly Nutrition Program (ENP), 483
Embryo, 64
Embryonic/fetal growth and development
  critical periods of, 97–99
  normal, 98
  nutrition, miscarriages and preterm delivery, 102
  variations in, 99–101
Empty-calorie foods, 34
Encephalocele, 115
Endocrine, 52
Endometriosis, 56
Endometrium, 54
Endothelium, 137
Energy
  adolescent nutrition and, 365–372
  adult nutrition and, 408–410
  breastmilk and, 165–166
  child/preadolescents nutrition and, 317–318, 340
  children's ability to control intake of, 274
  infant nutrition and, 228, 249–251
  lactation and, 178
  older adult nutrition and, 467–469
  requirements in pregnancy, 109–110
  toddlers/preschoolers nutrition and, 277–278
Energy balance, 408
Engorgement, 195–196
Enrichment, defined, 36
Enrichment/fortification labeling, 36
Enteral feeding, 255
Environmental factors
  adolescent nutrition and, 361–362, 386
  breastfeeding and, 217
  to eating disorders, 400–401
  exposure during lactation of, 206–207
  high-priority recommendations to change, 441
  lead exposure and, 229–230
  obesity and, 431
  older adult nutrition and, 463
Ephedra, 391, 418
Epididymis, 54, 56
Epigenetics, 103
Epithelial cells, 169
EPSDT, 224
Ergogenic supplement use, 390–391
Essential amino acids, 3
Essential fatty acids, 8
Essential nutrients, 3

Estimated Average Requirements (EARs), 4
Estrogen, 53–56, 94
Ethnicity and dietary needs, 38
European Society for Gastroenterology and Nutrition, 250
Evans, William, 479
Exercise
  adolescent nutrition and, 378–380, 391–392
  adult nutrition and, 421–422
  breastfeeding and, 179–180
  children/preadolescents and, 327–328
  diabetes and, 78, 447–448
  fertility and, 59–60
  gestational diabetes and, 144
  infants and, 239
  infertility and, 75
  older adult nutrition and, 479–480
  osteoporosis and, 497, 499
  PMS and, 71–72
  pregnancy and, 128
  resources, 86
  for toddler/preschoolers, 291
  weight management and, 434
Exposure index, 198
Extremely low-birth-weight infant (ELBW), 248–250, 255–256, 261
Extreme obesity, 431

F
Fad diets, 410
Failure to thrive (FTT), 240–241, 261, 302–303
Familial hyperlipidemia, 285
Famines
  pregnancy outcome and, 107–109
  undernutrition and, 58
Farmers' Market Nutrition Program, 293
Fast-food restaurants, 363
Fat-free mass, 459
Fats (lipids)
  adolescent nutrition and, 369
  body fat changes in pregnancy, 107
  breastmilk and, 166
  child/preadolescents nutrition and, 325
  description of, 8–13
  DRIs for, 10
  food sources for, 10–13
  gestational diabetes and, 145
  infant nutrition and, 228–229, 250–251
  intake during pregnancy, 112–114
  maternal metabolism of, 95
  older adult nutrition and, 472
  toddlers/preschoolers nutrition and, 286–287, 289
Fatty acids, 8, 507
Fecundity, 52
Feeding Infants and Toddlers Study (FITS), 289–290
Feeding skills development/behaviors
  in school-age children, 315–317
  in toddlers, 271–273
Female athlete triad, 75. *see also* Exercise
Fenugreek, 203
Fertility
  alcohol and, 60
  amenorrhea and, 58–59
  antioxidant nutrition and, 60
  breastfeeding and, 168
  caffeine and, 60

defined, 52
diet and, 58–59
disruptions in, 56
eating disorders and, 75–76
effects of obesity on, 58, 72–74
exercise and, 59–60
factors related to altered, 57
female athletes and, 75
folate and, 60
heavy metal exposure and, 60–62
hypothalamic amenorrhea and, 74–75
iron deficiency and, 60, 66
male factors of, 60, 72
oxidative stress and, 60
plant foods and, 61
premenstrual syndrome (PMS) and, 71–72
undernutrition and, 57–58
weight loss and, 58–59
zinc status and, 60
Fertility Blend, 84
Fetal alcohol spectrum, 155–156
Fetal alcohol syndrome, 155–156
Fetal growth and alcohol, 64
Fetal-origins hypothesis of later disease risk, 102–104
Fetus, 64
Fiber. see also Carbohydrates
adolescent nutrition and, 366, 369
child/preadolescents nutrition and, 324
constipation and, 504–505
infant nutrition and, 229
older adult nutrition and, 469–471
toddlers/preschoolers nutrition and, 288
viscous fiber, 441
FightBAC, 282
Fine motor skills, 267, 272
First foods for infants, 237–238
Fish
contamination of, 207
mercury in, 114, 129, 287
5 A Day for Better Health, 330
504 Accommodation, 353
Flat nipples, 194–195
Flavonoids, 507
Florida pyramid, 465
Fluids. see also Water
adolescent nutrition and, 391
adult nutrition and, 415–417
caffeine content in, 62
child/preadolescents nutrition and, 325
connection, 512
older adult nutrition and, 472, 513–514
special health care needs and, 345
toddlers/preschoolers nutrition and, 289, 301–302
Fluoride
function/deficiency of, 24
infant nutrition and, 229, 239
overdose consequences/food source of, 25
during pregnancy, 118–119
toddlers/preschoolers nutrition and, 280
Fluorosis, 280
Focus group interviews, 188
Folate
adolescent nutrition and, 371–372
adult nutrition and, 419
breastmilk and, 167
dementia and, 508
dietary sources of, 116

fertility and, 60
folic-acid deficiency, 473–474
food sources for, 20
function/deficiency of, 14
neural tube defects and, 64–65, 115–116
older adult nutrition and, 473–474
overdose consequences/food source of, 15
during pregnancy, 114–116
recommended intake, 116
status in women, 115–116
Folic acid. see Folate
Follicle-stimulating hormone (FSH), 53
Follicular phase, 53, 55
Fontana, Luigi, 459
Food allergy (hypersensitivity)
infant nutrition and, 242–243
inflammatory diseases and, 507
toddlers/preschoolers nutrition and, 306–307
Food and Drug Administration (FDA), 35
Food frequency questionnaires for assessment, 40
Food Guide Pyramid. see MyPyramid Food Guide
Food guides, 414
Food insecurity, defined, 2
Food intolerance in infants, 213–214
Food preferences
children/preadolescents and, 316–317
infant nutrition and, 239
toddlers/preschoolers and, 275–276
Food safety/security
defined, 2
older adult nutrition and, 479
pregnancy outcome and, 128–129
toddlers/preschoolers nutrition and, 281–282, 302
Food shortages in Japan (WWII), 108–109
Food Stamp Program, 43, 293, 425
Food texture and infants, 236–237
Ford, Earl S., 406
Formula. see also Infant formulas
breast-milk vs., 231–233
children/preadolescents and, 344–345
Fortification, 36
Framingham heart health assessment, 438
Free radicals and fertility, 60
Free-radical theories of aging, 458
Friendly bacteria, 37
Fruits & Veggies--More Matters, 330
Full-term infants, 223
Functional fiber, 324. see also Fiber
Functional foods, 36–37, 417
Functional status, 460

G
Galactogogue, 203
Galactosemia, 244, 345
Garlic, 418, 476–477
Gastric bypass, 389, 435–436
Gastroesophageal reflux (GER)
infant nutrition and, 227
older adult nutrition and, 501
Gastrointestinal diseases, 501–505
Gastrostomy feeding, 256, 302, 344–345
Genes and nutrigenomics, 32–33
Genetic disorders in infants, 260–261
Geophagia, 123

Gestational age, 91
Gestational diabetes
case study, 144
consequences of, 141–142
defined, 140
diagnosis of, 142–143
diet plan, 145
exercise and, 144
nutritional management of, 145
overview, 140–141
postpartum follow-up, 146
risk factors for, 142
treatment of, 143–144
Gestational hypertension, 138
Ginger, 128, 130, 418
Gingiva, 500
Ginkgo biloba, 418
Ginseng root, 127, 201, 418
Gleason, Jackie, 429
Glucogen amino acids, 95
Glucosamine and chondroitin, 418, 507
Glucose, 447. see also Diabetes; Gestational diabetes
Gluten, 505
Gluten-free, 81–82
Gluten-Free Certification Organization, 83
Glycemic Index (GI) of carbohydrates, 5–6, 76–77, 145–146
Glycerol, 8
Glycosylated hemoglobin, 494
Goat's-Rue, 203
Goethe, 486
Gonadotropin-releasing hormone (GnRH), 53, 56
Gravida, 105
Green tea, 33, 476, 514
Gross motor skills, 267, 271–272
Growth. see also Embryonic/fetal growth and development
2000 CDC Growth Charts (children), 269–270
adolescent psychosocial, 360–365
adolescents and, 358–360
catch-up, 251
children/preadolescents, 313, 341–342
intrauterine growth as predictor, 252–253
normal toddler/preschooler, 267–271
physiological changes of adulthood, 407–413
preterm infants and, 252
specialty charts for, 342
Growth assessment
of infants, 230–231
toddlers/preschoolers nutrition and, 299–300
Growth charts, 252, 269–270, 313, 341, 342
Growth hormone (GH), 390
Growth velocity, 267
Guarana, 476
Guidelines for Adolescent Preventive Services, 372
Guidelines for School and Community Programs to Promote Lifelong Physical Activity Among Young People, 378, 381
Guidelines for School Health Programs to Promote Lifelong Healthy Eating, 381

H

Hale, Thomas, 199
The Handbook of Milk Composition, 164
Harris, Tamara, 515
Hayflick's theory of limited cell replication, 457–458
Head Start Program, 43, 293, 307
Health disparity, 412–413
Health/eating-related behaviors in adolescents, 361–365
Health Objectives for the Nation, 406–407
Healthy, defined, 487
Healthy Eating Index. *see* HEI (Healthy Eating Index)
Healthy objectives for adult nutrition, 406–407
Healthy People 2010 objectives
  for adults, 408, 429–430
  breastfeeding goals, 160–161
  oral health goals, 461
  for school-age children, 312–313
  toddlers/preschoolers and, 267, 268
Healthy weight, 406
Heartburn and pregnancy, 130
Heart disease
  as cause of death, 284–285
  defined, 284
  risk factors for, 407
Heart health assessment, 438
Heavy metal exposure and male fertility, 60–62
Hedonic system, 431
HEI (Healthy Eating Index), 40, 327
Hematocrit, 279
Heme iron, 370
Hemochromatosis, 32
Hemoglobin, 279
Hemoglobin A1c, 141
Hemolytic anemia, 281
Hemolytic Uremic Syndrome (HUS), 281
Hendley, Joyce, 459
Herbal remedies/supplements. *see also* Medicinal herbs; Vitamins
  adolescent nutrition and, 390
  adult nutrition and, 418
  chaste tree berry extract, 83–84
  child/preadolescents nutrition and, 351
  coenzyme Q$_{10}$ and, 84–85
  defined, 36
  diabetes and, 448
  for infertility, 82–85
  lactation and, 200–203
  n-acetylcysteine and, 84
  older adult nutrition and, 476–478
  pregnancy outcome and, 127–128
  selenium and, 84
  toddlers/preschoolers nutrition and, 286, 307
Herbal teas, 201, 202
Herbal therapy. *see* Herbal remedies/ supplements
High 5 Alabama, 330–331
High blood pressure. *see* Hypertension
Highly active antiretroviral therapy (HAART), 451
Hippocrates, 35, 429
HIV/AIDS
  in adults, 451–452
  breastfeeding and, 216–217
  defined, 216

Innocenti Declaration, 185
  mastitis and, 196
  pediatric AIDS, 298, 350–351
  during pregnancy, 151–154
  St. John's Wort and, 203
  treatment of, 152
HMG-CoA reductase, 441
Holmes, Oliver Wendell, 480
Homeostasis, 23
Homocysteine, 502, 508
Hormones
  breast development/lactation and, 162–163
  female reproductive system and, 53–55
  male reproductive system and, 55–56
  older adult nutrition and, 478
  osteoporosis and, 499–500
  physiological changes in adults, 408
  during pregnancy, 93–94
  that affect reproduction, 56
Hospitals and breastfeeding, 183–185
Human chorionic gpnadotropin (hCG), 94
Human chorionic somatotropin (hCS), 94
Human milk. *see* Breast-milk/breastfeeding
Human Milk Banking Association of North America, 218, 219
Human nutrition principles, 2
Hydration, 391
Hydrogen and trans fat, 9–10
Hydrolyzed protein formula, 243
Hyperactive letdown, 195
Hyperbilirubinemia, 208–211
Hyperemesis gravidarum, 129
Hyperglycemia, 493. *see also* Diabetes
Hyperinsulinemia, 442
Hyperlactation, 195
Hyperlipidemia, 394, 436
Hyperplasia, 98–99
Hypertension
  adolescent nutrition and, 393–394
  adult nutrition and, 436
  chronic, 138
  defined, 33
  gestational, 138
  medications and, 511
  milk and, 514
  older adult nutrition and, 491–493
  risk factors for, 407
Hypertensive disorders of pregnancy, 137–140
Hypertonia, 302, 342
Hypertrophy, 99
Hypoallergenic, 237
Hypocalcemia, 251
Hypoglycemia, 493. *see also* Diabetes
Hypogonadism, 81
Hypothalamic amenorrhea, 74–75
Hypothyroidism, 121, 244
Hypotonia, 302, 342

I

Ibn Ezra, Abraham, 247
Ibuprofen, 196
Illicit drugs. *see* Drugs
Immunoglobin, 169
Immunological, 52
Inappropriate/unsafe foods, 238
Indirect calorimetry, 249, 409
Individuals with Disabilities Education Act (IDEA), 307, 339, 353

Infant allergies, 212–214
Infant feeding skills development, 233–239
Infant formulas, 231–233, 234, 256–257
Infant mortality
  attributable to birth defects, 259
  birth weight/preterm delivery and, 90
  defined, 223
  in developing countries, 170
  rates by race, 223–224
  reduction of, 90
  as reflection of general population's health, 89–90
Infant nutrition
  allergies and, 212–214, 242–243
  at-risk infants and, 248–249
  breastfeeding and, 172–177 (*see also* Lactation nutrition)
  breast-milk/breastfeeding benefits for, 169–171
  colic and, 180–181
  common problems, 240–243
  congenital anomalies/chronic illness and, 257–261
  cross-cultural considerations, 243
  drinking from cup, 236
  in early infancy, 231–233
  energy needs, 249–251
  energy/nutrient needs, 228–230
  exercise and, 239
  feeding problems and, 261
  first foods, 237–238
  growth and, 251–254
  inappropriate/unsafe foods, 238
  infant development and, 224–230
  infant feeding position, 236
  infant feeding skills development, 233–239
  interventions to, 261–263
  introduction of solid food, 234–236
  learning food preferences, 239
  newborn health assessment and, 223–224
  nutrient needs, 249–251
  nutrition guide for, 239–240
  nutrition intervention for risk reduction, 244
  nutrition services, 263–264
  physical growth assessment and, 230–231
  quantity of food for, 238–239
  resources, 246, 265
  severe preterm birth and, 255–257
  special health care needs, 254–255
  vegetarian diets, 243–244
  water and, 238
Infectious diseases and breastfeeding, 217
Infecundity, 52
Infertility, 52, 75, 82–85. *see also* Fertility
Inflammatory diseases and older adults. *see also* Chronic inflammation
Innocenti Declaration, 185
Innocenti Declaration on the Protection Promotion and Support of Breastfeeding, 189
Institute of Medicine Committee on Nutrition Standards for Foods in Schools, 382
Institute of Nutrition of Central America and Panama (INCAP), 471
Insulin, 346

Insulin-dependent diabetes. *see* Type 1 diabetes
Insulin resistance, 5, 73, 78, 138, 430. *see also* Type 2 diabetes
Intestinal fertilizer, 37
Intrauterine growth, 252–253
Intrauterine Growth Retardation (IUGR), 224
Inverted nipples, 194–195
Iodine
  function/deficiency of, 24
  overdose consequences/food source of, 25
  during pregnancy, 121
Iron
  adolescent nutrition and, 370, 392–393
  adult nutrition and, 417–420
  child/preadolescents nutrition and, 324
  children/preadolescents and, 318
  deficiency, 119, 241–242, 279, 392–393
  feeding in early infancy, 233
  fertility and, 60, 66
  food sources for, 29
  function/deficiency of, 24
  infant nutrition and, 240
  older adult nutrition and, 474
  overdose consequences/food source of, 25
  during pregnancy, 119–121
  supplements, 120–121
  toddlers/preschoolers nutrition and, 278–279, 287–288
Iron-deficiency anemia. *see* Anemia
Ischemia, 490

J
Jaundice/kernicterus, 208–212
Jejunostomy feeding, 256
Johnson, Mary Ann, 456
Johnston, Patricia, 111
Juvenile-onset diabetes. *see* Type 1 diabetes
Juvenile rheumatoid arthritis, 342

K
Kernicterus, 208–212
Ketogenic diet, 345
Ketones, 95, 146, 347
King, J.C., 70
Kogan, M.D., 134
Krebs, N.E., 195
Krebs cycle, 502
Kwashiorkor, 8
Kyphosis, 498

L
Labeling
  dietary supplements and, 36
  enrichment/fortification and, 36
  ingredients labeling, 35–36
  nutrition facts panel and, 35
  vitamin/mineral levels for nutrition labeling, 477
Labels and Daily Values (DVs), 4
Labium major, 54
Labium minor, 54
Lactation consultant, 186
Lactation nutrition
  alcohol/drug exposures and, 203–207
  breastfeeding benefits for mother/infants, 168–171

breastfeeding infants, 172–177
breastfeeding multiples, 212
breastfeeding promotion/support, 181–187
breast milk supply and demand, 171–172
caffeine and, 206
common breastfeeding conditions, 194–198
factors of breastfeeding initiation/duration, 181
feeding in early infancy, 231–233
herbal remedies, 200–203
human/cow milk comparison, 165
human milk and preterm infants, 215–216
human milk collection/storage, 218–219
human milk composition, 164–168
infant allergies, 212–214
late-preterm infants, 214–215
maternal diet and, 177–178
maternal energy balance and milk composition, 179–181
maternal medications and, 198–200
medical contraindications to breastfeeding, 216–217
model breastfeeding promotion programs, 188–189, 219–220
neonatal jaundice/hernicterus of, 208–212
physiology and, 161–164
public food/nutrition programs, 187–188
resources, 191–192, 221
United States goals for, 160–161
Lactation support. *see* Breast-milk/breastfeeding
Lactiferous sinuses, 163
Lactogenesis, 162
Lacto-ovo vegetarians. *see* Vegetarian diets
Lactose, defined, 243
Lactose intolerance
  in infants, 243
  in school-age children, 325
  single-gene defects and, 32
La Leche League, 186
Lamb, Charles, 222
Language skills in toddlers, 271
Large for gestational age (LGA), 101
Late-preterm infants, 214–215
Lawrence, R.A., 198, 210
Laxatives, 504–505
LDL cholesterol
  defined, 284
  in toddlers, 284–285
Lead
  infant nutrition and, 229–230
  toddlers/preschoolers nutrition and, 280–281
Lean body mass, 407–408, 459–460
Lenfant, Claude, 157
Leptin, 58, 94
LES, 501
Letdown failure, 195
Letdown reflex of milk, 164
Liability/legal issues, accuracy of *Physician's Desk Reference*, 199
Licorice, 63
Life expectancy/span, 456–457

Lifshitz, Fima, 75
Lignins, 324
Linoleic acid
  adolescent nutrition and, 364
  children/preadolescents and, 322–323
  as essential nutrient, 3, 8–9
  pregnancy nutrition and, 112–113
Listeria, 128
Listeria monocytogenes, 128
Liveborn infant, 90
L. Monocytogenes, 128
Lobes of mammary glands, 161
Long-chain fats, 228
Longevity, 455–456, 459
Look AHEAD, 445
Loss of appetite control, 431
Loving Support Makes Breastfeeding Work, 188–189
Low-birth-weight infant (LBW), 248
Low body weight, 509–512
Lower esophageal sphincter, 501
Lower respiratory infections, 511
Luteal phase, 53, 55
Luteinizing hormone (LH), 53, 56

M
Macrobiotic diet, 290
Macrocephaly, 253
Macrophages, 169
Magnesium
  adult nutrition and, 420
  food sources for, 28
  function/deficiency of, 24
  older adult nutrition and, 474
  overdose consequences/food source of, 25
  PMS and, 72
Male fertility. *see* Fertility
Malnutrition, 31–32, 175–176, 464
Malnutrition Universal Screening Tool (MUST), 464
Mammary gland, 161–162
Manganese
  function/deficiency of, 26
  overdose consequences/food source of, 27
Maple Syrup Urine Disease, 260
Marijuana and lactation, 205–206
Martin, C., 195
Maslow's hierarchy of needs, 3
Mastitis, 196–197
Maternal and Child Health Bureau (MCHB), 263, 351–353
Maternal medications, 198–200
Maternal physiology, 91–92
Maternal PKU, 80–81
McKittrick, Martha, 79
MCT oil, 251
Meals-on-Wheels, 482
Meconium, 211
Media influence
  eating disorders and, 400
  in school-age children, 316
  in teenagers, 362
  in toddlers/preschoolers, 276
Medical foods, 419
Medical neglect, 302
Medical nutrition therapy (MNT), 487
Medicare, 455

Medications
  cardiovascular disease and, 511
  diabetes and, 511
  human milk contamination and, 198–200
  older adult nutrition and, 509
  osteoporosis and, 500
"Medications and Mother's Milk" (Hale), 199, 200
Medicinal herbs, 201–203. *see also* Herbal remedies/supplements
Mediterranean diet, 46
Mediterranean diet pyramid, 49
Medium-chain fats, 228
Memory impairment, 507
Menarche, 359
Meningitis, 300
Menopause, 53, 408
Menses, 358–359
Menstrual age, 91
Menstrual cycle
  in athletes/sedentary women, 59–60
  defined, 53
Mental retardation, 306
Mercury contamination, 114, 129, 287
Metabolic disorders, 79–82, 217
Metabolic syndrome, 73, 442–444
Metabolism
  adult exercise and, 421–422
  bilirubin metabolism, 208–209
  defined, 13
  inborn errors of, 348–350
  infant nutrition and, 229
  during pregnancy, 93–95
Meter, Margaret, 428, 454
Methylmalonic acid (MMA), 502
Microcephaly, 253
Middle childhood, 311
Mifflin-St. Jeor Energy Estimation Formula, 409
Milk. *see* Lactation nutrition
Milk banking, 218–219
Milk secretion, 163–164
Milk thistle, 203
Milk to plasma drug concentration ratio (M/P ratio), 198
Minerals. *see also* Vitamins
  adolescent nutrition and, 365–367, 389–391
  adult nutrition and, 417–420
  in breastmilk, 167–168
  children/preadolescents and, 318, 324, 345
  description of, 22
  as essential nutrients, 3
  food sources for, 28–30
  lactating women and, 180
  maternal metabolism of, 95
  multifetal pregnancies and, 151
  older adult nutrition and, 472–474
  during pregnancy, 118–121
  toddlers/preschoolers nutrition and, 278, 285–286
  types of, 24–27
Miscarriage, 52, 102
Mitochondria, 229
MNA nutritional screening and assessment, 466
Modified foods, 419
Molecular clock theory, 458

Molybdenum
  function/deficiency of, 26
  overdose consequences/food source of, 27
Monosaccharides, 4–5
Monounsaturated fats, defined, 9
Monovalent ion, 167
Monozygotic (MZ) twins, 148–149
Montreal Diet Dispensary (MDD), 130–131
Morbidity, 170
Moro reflex, 225
Mortality, 170, 248
Motor development
  infants and, 225, 226
  toddlers/preschoolers and, 267, 272
Multifetal pregnancy nutrition, 147–152
Multivitamins. *see* Vitamins
Muscle coordination problems in toddlers, 304–305
Myoepithelial cells, 161
Myometrium, 54
MyPyramid Food Guide
  for adolescents, 380
  for adults, 414
  aging and, 465, 467, 471
  cardiovascular disease and, 440
  for children/preadolescents, 317–318
  description of, 46
  food measure equivalents, 48
  maternal diet, 177–178
  menu, 47
  for preconceptional women, 65
  for pregnant women, 124
  for toddler/preschoolers, 287

N
N-acetylcysteine, 84
Natality statistics, 88–90
National Breastfeeding Policy, 187–188
National Center for Chronic Disease Prevention and Health Promotion, 231
National Health and Examination Survey (NHES), 313
National Health and Nutrition Examination Survey (NHANES), 42, 285, 313, 363, 365
National School Lunch Program, 330–331, 332, 381–382
Nationwide Food Consumption Survey (NFCS), 42
Nausea and vomiting, 128–130, 139
Near-term infant breastfeeding, 215
Nelson's Textbook of Pediatrics, 271
Neocrotizing enterocolitis (NEC), 256
Neonatal death, 248
Neonatal jaundice/kernicterus, 208–212
Neural tube defects (NTDs), 64–65, 115–116
Neurobehavioral condition, 350
Neuromuscular disorders
  children/preadolescents and, 342
  toddlers/preschoolers feeding problems and, 302
Neutraceuticals. *see* Functional foods
Neutrophils, 169
Niacin
  food sources for, 19
  function/deficiency of, 14
  overdose consequences/food source of, 15
Nicotine, 205
Nipples, 194–195
Nitrogen, 471

Nonessential nutrients, 3
Nonheme iron, 370
Non-insulin-dependent diabetes. *see* Type 2 diabetes
Nonorganic failure to thrive, 240
Nonprotein nitrogen, 166–167
Norplant, 63
NSF International, 475
NTDs. *see* Neural tube defects (NTDs)
Nurses' Health Study, 315, 499
Nutrient-dense foods, 34
Nutrients
  adolescent nutrition and, 365–372
  adult nutrition and, 417–421
  categories of, 3
  children/preadolescents and, 317–318
  deficiencies, 23, 31
  defined, 2, 3
  inadequate/excessive, 23, 31
  infant nutrition and, 228–230, 249–251
  older adult nutrition and, 469–472, 478
  placenta transport of, 96–97
  preeclampsia and, 140
  of term/preterm formulas, 257
  toddlers/preschoolers nutrition and, 277–278
Nutrigenomics, 32–33
Nutritional assessments. *see* Assessments
Nutritional risk of aging, 462–465
Nutrition basics
  for adolescent pregnancies, 156–157
  anthropometric assessment, 41
  biochemical assessment, 41–42
  chronic diseases and, 33–34
  clinical/physical assessment, 39
  community-level assessment, 39
  contemporary prenatal research results, 109
  dietary assessment, 39–40
  dietary history for assessment, 39–40
  dietary needs based on ethnicity, 38
  dietary needs based on religion, 38–39
  food frequency questionnaires for assessment, 40
  good/bad foods, 35
  importance of adequacy/balance, 33–34
  individual-level assessment, 39
  key concepts, 2
  monitoring U.S., 42
  priorities for improvement, 43–44
  public food/nutrition programs, 42–43
  resources, 49–50
  USDA Automated Multiple-Pass Method, 40
  Web dietary assessment resources, 40
Nutrition biomarkers during pregnancy, 126
Nutrition Care Process (NCP), 66–68
  for celiac disease, 83
Nutrition during Lactation, 180
Nutrition Facts panel, 35, 478
Nutrition labeling. *see* Labeling
Nutrition Labeling and Education Act, 35
Nutrition monitoring, 42
Nutrition programming, 103–104
Nutrition screening for adolescents, 372
Nutrition Screening Initiative (NSI), 462–464
Nutrition support, 255
Nutrition surveillance, 42

O

Obesity
adolescent nutrition and, 386–389
adolescent pregnancies and, 156
adult nutrition and, 429–441
assessment for, 431, 433
breastfeeding and, 170, 181
defined, 282
effects of, 430
etiology of, 430–431
fertility and, 58, 72–74
gestational diabetes and, 142–143
infant outcomes and, 136
infertility and, 72–74
metabolic syndrome and, 73
older adult nutrition and, 494–496
oral contraceptives and, 63
pediatric, 282–284
pregnancy and, 135–137
prevalence among adults, 429–430
resources, 86
risk factors for, 407
in school-age children, 319–322
in toddlers, 282–284
treatment for school-age children, 321–322
U.S. trends, 430
Obesogenic diet, 407
Ohio Longitudinal Study of Aging and Retirement, 455
Older adult nutrition
aging theories, 457–459
alcohol and, 473–474
Alzheimer's Disease and, 507–509
appetite/satiety and, 461–462
bereavement and, 514–515
cardiovascular disease and, 488–490
chronological age and, 455–456
community food/nutrition programs for, 482–484
constipation and, 503–505
dehydration and, 513–514
diabetes and, 493–494
dietary recommendations for, 465
dietary supplements and, 474–478
energy needs for, 467–469
exercise and, 479–480
fluids and, 472, 513–514
food safety, 479
gastrointestinal diseases, 501–505
health and, 487–488
hypertension and, 491–493
importance of, 487
low body weight/underweight, 509–512
medications and, 509
nutrient recommendations for, 469–472, 478
nutritional risk factors and, 462–465
obesity and, 494–496
oral health and, 500–501
osteoarthritis, 505–507
osteoporosis and, 496–500
physiological changes and, 459–462
resources, 485, 516
risk reduction programs in, 480–482
stroke and, 490–491
vital statistics of aging population, 456–457
Vitamin B$_{12}$ deficiency and, 473, 501–503
vitamin/mineral supplement use and, 472–474

Older Americans Act, 483
Oligosaccharides, 167
Omega-3 fatty acid
for cardiovascular disease, 489–490
intake of, 9
pregnancy outcome and, 113–114
Oral contraceptives, 63, 200
Oral-gastric (OG) feeding, 256
Oral glucose tolerance test (OGTT), 143
Oral health in older adults, 500–501
Organic failure to thrive, 240
Osmolality, 167
Osmolarity, 227
Osteoarthritis, 505–507
Osteoblasts, 499
Osteoclasts, 499
Osteoporosis, 33, 345, 496–500
Ova, 53
Ovary, 54, 55
Over-the-counter drugs and breastfeeding, 217
Overweight. *see also* Obesity; Weight
defined, 282–284
school-age children, 319–322
teenagers, 386–389
toddlers, 282–284
Oviduct, 54
Oxidative stress, 33, 58, 60, 137–138, 458
Oxytocin, 161, 163

P

Pager, Stephen, 135
Pagophagia, 123
Palmar, reflex, 225
Palpebral fissure, 156
Pantothenic acid
function/deficiency of, 16
overdose consequences/food source of, 17
Paracelsus, 417
Parathyroid hormone (PTH), 371, 497
Parenteral feeding, 255
Parity, 105
Pathologic jaundice, 209
Pediatric AIDS, 298, 350–351. *see also* HIV/AIDS
Peers, influence among adolescents, 361
Pelvic inflammatory disease (PID), 56
Penis, 54
Peppermint, 418
Periconceptional period, 71
Perinatal death, 248
Periodontal disease (PD), 500
Peripheral artery disease. *see* Cardiovascular disease
Persistent organic pollutants (POPs), 206
Pharmacotherapy, 440
Phenylketonuria (PKU). *see* PKU (phenylketonuria)
Philtrum, 156
Phosphorus
food sources for, 30
function/deficiency of, 24
overdose consequences/food source of, 25
Phototherapy, 211
Physical activity. *see* Exercise
Physical Activity Guidelines for Americans, 378
Physician's Desk Reference (PDR), 199
Physiologic jaundice, 209

Physiology
of adults, 407–413
lactation, 161–164
of older adults, 459–462
of pregnancy, 91–95
during puberty, 359
reproductive, 53–56
of school-age children, 313–317
of toddlers, 271–273, 274–275
Phytochemicals, 13, 477–478
Phytosterols, 441
Pica, 123–124
PID. *see* Pelvic inflammatory disease (PID)
Pima Indians and gestational diabetes, 142
Pituitary gland, 53
PKU (phenylketonuria)
children/preadolescents and, 339–340, 348–350
in infants, 244
ketogenic diet and, 345
preconception nutrition and, 79–81
single-gene defects and, 32
Placenta, 92, 96–97
Plant foods and fertility, 61
Plant sterols, 441, 478
Plasma ferritin, 370
Platelets, 281
Plugged duct, 196
PMS, 71–72
Polycystic ovary syndrome (PCOS), 73, 78–79
Polypharmacy and older adults, 509
Polysaccharides, 5
Polyunsaturated fats, defined, 9
Poor nutrition, 31–32
Postictal state, 347
Postpartum weight retention, 107
Potassium
adult nutrition and, 420
food sources for, 30
function/deficiency of, 26
overdose consequences/food source of, 27
Pouring rights, 329
Poverty and older adult nutrition, 463
Prader-Willi Syndrome, 298, 340
Preadolescence, defined, 311. *see also* Child/preadolescent nutrition
Preadolescent nutrition. *see* Child/preadolescent nutrition
Prebiotics, 37, 478
Preconception nutrition
2010 objectives for, 52–53
contraceptives and, 62–63
female infertility factors, 56
female reproductive system and, 53–55
fetal growth and, 64
folic acid and, 64–65
male infertility factors, 60
male reproductive system and, 55–56
multivitamin supplements and, 60
overview, 52–53
PKU (phenylketonuria) and, 79–81
premenstrual syndrome (PMS) and, 71–72
programs for, 65–68
RDIs for, 65
Prediabetes, 78, 444, 445
Preeclampsia, 114–115, 138–141

Pregnancy nutrition. *see also* Lactation
  nutrition; Preconception nutrition
adolescents and, 156–157
alcohol and, 110, 155–156
caffeine and, 122
catabolic phase, 92
common pregnancy problems, 129–130
course/outcome of pregnancy and,
  107–109
diabetes and (*see* Gestational diabetes)
eating disorders and, 154
exercise and, 128
fetal alcohol syndrome and, 155–156
and fetal-origins hypothesis of later
  disease risk, 102–104
food safety and, 128–129
healthy diets for, 122–125
HIV/AIDS and, 151–154
and hypertensive disorders of pregnancy,
  137–140
and multifetal pregnancies, 147–152
pregnancy weight gain and, 104–107
RDIs for, 123
reduction of infant mortality/morbidity
  and, 90
resources, 132–133, 158
risk reduction programs in, 130–131
status of outcomes, 88–90
Preloads, 276
Premenstrual dysphoric disorder
  (PMDD), 71
Premenstrual syndrome (PMS), 71–72
Prenatal supplements, 113–114, 126
Prentice, A., 130
Preschool-age children, 267
Preschooler nutrition. *see* Toddler/
  preschooler nutrition
Preterm infants
  breastmilk and, 215–216, 256–257
  defined, 223
  delivery, 102
  feeding and, 257
  feeding problems and, 261
  growth and, 252
  infant formulas and, 256–257
  lactation nutrition and, 214–215
  nutrition and severe VLBW, 255–257
Primary malnutrition, 32
Primordial follicles. *see* Ova
Probiotics, 37, 478
Produce for Better Health, 330
Progesterone, 53–55, 56, 94
Programmed aging, 457–458
Prolactin, 162
Promotion of healthy behaviors among
  teenagers, 380–383
Proportionately small for gestational age
  (pSGA), 100, 101
Prostacyclins, 8, 138
Prostaglandins, 8, 55
Prostate gland, 54
Protein
  adolescent nutrition and, 368
  breastmilk and, 166–167
  child/preadolescents nutrition and,
    317–318, 340
  description of, 6–8
  DRIs for, 8
  food sources for, 8
  gestational diabetes and, 145

hydrolyzed protein formula, 243
infant nutrition and, 228, 250
intake during pregnancy, 110–111
maternal metabolism of, 95
older adult nutrition and, 471–472
osteoporosis and, 498–499
toddlers/preschoolers nutrition and, 278
Psychosocial growth for adolescents,
  360–365
Psychosocial maturation, 357
Psychostimulant medication, 350
Puberty, 53, 357–360. *see also* Adolescent
  nutrition
Public food programs, 42–43, 187–188,
  331–334, 425
Pulmonary problems
  defined, 297
  toddlers/preschoolers nutrition and, 306
Pumping/expressing breastmilk, 172

Q
Quality of life, 487
Quantity of food for infant nutrition,
  238–239

R
Rate-of-living theory, 458
Recommended Dietary Allowances (RDAs)
  defined, 4
  preconception nutrition and, 65
  toddlers/preschoolers nutrition and, 278
Recommended Dietary Intakes (RDIs)
  for lactating women, 178
  multifetal pregnancies and, 150–151
Recumbent length, 269
Reflex, 224
Registered dietitian, 39
Religion, dietary needs based on, 38–39
Remission, 450
Reproductive physiology, 53–56
Resilience, 459
Resources
  adolescent nutrition, 384, 404
  adult nutrition, 425
  breastfeeding, 191–192, 221
  child/preadolescents nutrition,
    335–337, 355
  diabetes, 86
  diets, 86
  exercise, 86
  fish advisories, 132
  food safety, 133
  gestational diabetes, 158
  human milk contamination, 199
  infant nutrition, 246, 265
  lactation, 191–192, 221
  natality statistics, 133
  nutrition basics, 49–50
  older adult nutrition, 485, 516
  for preconception nutrition, 86
  pregnancy, 132–133, 158
  toddlers/preschoolers nutrition,
    294–295, 309
  WIC, 133
Resting energy expenditure (REE),
  321, 409
Retinoic acid, 116–118
Retinoic acid syndrome, 117
Retinol, 116–118
Rett Syndrome, 299–300

Rhodus, Nelson, 500
Riboflavin
  food sources for, 18
  function/deficiency of, 14
  overdose consequences/food source of, 15
The "Ripple Effect", 31
Risk groups for malnutrition, 33
Risk reduction programs
  for adults, 422–426
  for children/preadolescents, 328–331
  in older adult nutrition, 480–482
  in pregnancy nutrition, 130–131
  for toddler/preschoolers, 291–292
Rooting reflex, 224, 225
Rosen, Clifford, 500
Rotavirus, 84
Rush Mother's Milk Club, 219–220

S
Salmonella, 281
SAMe, 418, 507
Sandwich generation, 406
Sarcopenia, 459
Sarcopenic obesity, 496
Saturated fats, 9
Saw palmetto, 418
School-age child nutrition. *see*
  Child/preadolescent nutrition
School Breakfast and Lunch Programs, 43,
  333, 381–382
School Health Index (SHI), 330, 331
School Health Policies and Practices Survey
  (SHPPS), 381
School health programs, 380–382
School lunch programs, 330–332, 381–382
School nutrition programs, 329–330
Science of nutrition, principles of, 2
Scoliosis, 341
Screening and assessment. *see* Assessments
Screen time and obesity, 320–321
Secondary condition, 341
Secondary malnutrition, 32
Secondary sexual characteristics, 358
Secretory cells, 161
Secretory immunoglobin A, 169
Seizures, 249, 341, 347
Selenium
  fertility and, 84
  food sources for, 28
  function/deficiency of, 26
  overdose consequences/food source of, 27
Self-efficacy, 361–362
Self-monitoring of blood glucose
  (SMGB), 447
Semen, 56
Seminal vesicle, 54
Senescence, 457
Senior Nutrition Program, 483
Senses and older adults, 461–462
Sensorimotor development in infants,
  225–227
Serum iron, 370
Sex hormone binding globulin (SHBG), 72
Sexuality, maturation ratings and, 358
Sexual Maturation Rating (SMR), 358,
  378–379
Shakespeare, William, 226
Short-chain fats, 228
Shoulder dystocia, 101
Siege of Leningrad (1942), 108

Silicone breast implants, 172
Simonetti, Peter D., 102
Simple sugars. *see* Carbohydrates
Single-gene defects. *see* Celiac disease; PKU (phenylketonuria)
Sisters Together: Move More, Eat Better, 424
Skinfold fat measurements, 342
Small for gestational age (SGA), 100
Smoking. *see* Tobacco
Social determinants of health, 413
Social marketing, 188
Social Security Program, 455
Socioeconomics of breastfeeding, 181
Sodium
    food sources for, 28–29
    function/deficiency of, 26
    infant nutrition and, 229
    overdose consequences/food source of, 27
    preeclampsia and, 140
    during pregnancy, 121
Soft drinks
    adolescent nutrition and, 363
    child/preadolescents nutrition and, 325–326
Solid food introduction to infants, 234–236
Sore nipples, 194
Spastic quadriplegia, 304
Special health care needs
    child/preadolescent nutrition and, 339–341, 345–351
    fluids and, 345
    infant nutrition, 254–255
    toddler/preschooler nutrition, 297–298
Special Supplemental Nutrition Program for Women, Infants, and Children (WIC). *see* WIC program for nutrition
Spermarche, 359
Spina bifida, 115, 342
Sports. *see also* Exercise
    adolescent nutrition and, 378–380, 391–392
    children/preadolescents and, 328
Stang, Jamie, 356, 385
Stanols, 441
Stature, 269
Stepping reflex, 225
Steptococcus mutans, 280
Steroid hormones, 93
Steroid use in teenagers, 390
Sterols, 441
St. John's Wort, 201–203, 418
Storage of human milk, 218–219
Store-to-Door, 482–483
Stroke
    defined, 33
    older adult nutrition and, 490–491
    risk factors for, 407
Structured Weight Management (SWM), 283
Subfertility, 52
Substance abuse. *see also specific substance*
    infant seizures and, 249
    lactation and, 203–206
    in teenagers, 392
Sucking reflex, 225
Suckle, 224
Summer Food Service Program, 43, 333–334
Super-centenarian, 457
Supplemental Nutrition Assistance Program (SNAP), 425

Swift, Jonathan, 236
Syndrome X, 73, 442–444

T
Tanner Stages, 358
Tea, 33, 476, 514
Team Nutrition, 334
Teenagers. *see* Adolescent nutrition
Television and obesity, 320–321
Telomeres, 458
Temperament of toddlers, 271, 275
Teratogenic, 76
Testes, 55, 359
Testis, 54
Testosterone, 56, 408
T. Gondii, 128
Therapeutic Life Changes (TLC), 440–441
Thermal effect of food (TEF), 408–409
Thiamin
    food sources for, 18
    function/deficiency of, 14
    overdose consequences/food source of, 15
Thirst and aging, 462
Thromboxanes, 8, 138
Thrombus, 490
Title V Maternal and Child Health programs, 187–188
T-lymphocytes, 169
Tobacco, 392
Toddler/preschooler nutrition
    allergies and, 306–307
    appetite/satiety and, 276–277
    atherosclerosis and, 285
    autism and, 300–301, 304
    chronic conditions and, 298–299
    common problems, 278–282
    constipation and, 280
    developmental delay, 306
    diarrhea/celiac disease and, 303
    dietary recommendations for, 286–291
    energy/nutrient needs, 277–278
    excessive fluid intake, 301–302
    exercise recommendations and, 291
    failure to thrive (FTT) and, 302–303
    feeding problems and, 300–302
    food allergies/intolerance and, 306–307
    food preferences and, 275–276
    growth assessment and, 299–300
    herbal remedies, 286, 307
    interventions for risk reduction, 291–292
    media influence on, 276
    muscle coordination problems, 304–305
    neuromuscular disorders, 302
    normal growth/development and, 267–271
    nutrition services and, 307
    obesity and, 282–284
    palsy and, 304–305
    physiological/cognitive development and, 271–277
    prevention disorders related to, 282–286
    public food/nutrition programs, 292–293
    pulmonary problems and, 306
    resources, 294–295, 309
    special health care needs, 297–298
    temperament differences and, 275
    tracking of, 267
Toddlers, defined, 267

Tolerable Upper Intake Levels (ULs), 4
    in toddlers, 286
Tooth decay, 176–177
Total Diet Study, 42
Total fiber, 324. *see also* Fiber
Toxicity, 23, 31. *see also special nutrient, i.e.* Selenium
TOXNET LactMed database, 199
Toxoplasma gondii, 128
Toxoplasmosis, 128
Trace minerals in breastmilk, 168
Tracheoesophageal atresia, 259
Trans fat, 9–10
Trans fatty acids
    breastmilk and, 166
    toddlers/preschoolers nutrition and, 285
Transferrin saturation, 370
Transient ischemic attacks (TIAs), 490
Transpyloric feeding (TP), 256
Triglyceride, structure of, 9
Triplets. *see* Multifetal pregnancy nutrition
Tufts pyramid, 465, 467
22q11 microdeletion, 261
24-hour dietary recalls/records, 39
Twins. *see* Multifetal pregnancy nutrition
Type 1 diabetes, 76, 146–147, 444
Type 2 diabetes
    adult nutrition and, 444
    children/preadolescents and, 319
    defined, 5
    management of, 76–78
    pharmacological therapy of, 448
    treatment of, 144

U
Undernutrition and fertility, 57–58. *see also* Weight
Underweight, 509–512
United States Pharmacopeia, 475
Unsaturated fats, 9
Urethra, 54
Urinary ketone testing, 146
U.S. Breast Milk Monitoring Program, 207
USDA Automated Multiple-Pass Method, 40
USDA Child Nutrition Program, 353
U.S. Department of Health & Human Services, 159, 187–188
USP (United States Pharmacopeia), 475
Uterus, 54, 55

V
Vagina, 54
Vanishing twin phenomenon, 149
VCFS, 261
Vegan diet
    adolescent nutrition and, 364
    adult nutrition and, 415
    linoleic acid and, 364
    toddlers/preschoolers nutrition and, 290
Vegetarian diets
    adolescent nutrition and, 363–365
    adult nutrition and, 414–415
    child/preadolescents nutrition and, 318, 327
    infant nutrition and, 243–244
    inflammatory diseases and, 507
    during pregnancy, 111–112
    toddlers/preschoolers nutrition and, 290–291
Vermillion border, 156

Very low-birth-weight infant (VLBW), 248–250, 255–256, 261
Visceral fat, 135
Viscous fiber, 441. *see also* Fiber
Vital statistics, 456–457
Vitamin A
  adult nutrition and, 417
  breastmilk and, 167
  food sources for, 21
  function/deficiency of, 16
  older adult nutrition and, 472
  overdose consequences/food source of, 17
  during pregnancy, 116–117
Vitamin B$_1$. *see* Thiamin
Vitamin B$_2$. *see* Riboflavin
Vitamin B$_3$. *see* Niacin
Vitamin B$_6$
  food sources for, 19–20
  function/deficiency of, 14
  nausea/vomiting during pregnancy and, 130
  overdose consequences/food source of, 15
  PMS and, 72
Vitamin B$_{12}$
  breastmilk and, 167
  deficiency during pregnancy, 111
  dementia and, 508
  food sources for, 20
  function/deficiency of, 14
  infant nutrition and, 240
  older adult nutrition and, 473, 501–503
  overdose consequences/food source of, 15
  toddlers/preschoolers nutrition and, 290
Vitamin C
  adolescent nutrition and, 372
  food sources for, 21
  function/deficiency of, 16
  overdose consequences/food source of, 17
Vitamin D
  adolescent nutrition and, 370–371
  adult nutrition and, 417–419
  breastmilk and, 167
  child/preadolescents nutrition and, 325
  food sources for, 22
  function/deficiency of, 18
  infant nutrition and, 229, 240
  inflammatory diseases and, 507
  older adult nutrition and, 472–473
  overdose consequences/food source of, 19
  PMS and, 72
  during pregnancy, 117–118
  toddlers/preschoolers nutrition and, 290
Vitamin E
  adult nutrition and, 419
  breastmilk and, 167
  food sources for, 21
  function/deficiency of, 16
  older adult nutrition and, 473
  overdose consequences/food source of, 17

Vitamin K
  breastmilk and, 167
  function/deficiency of, 18
  older adult nutrition and, 473
  osteoporosis and, 499
  overdose consequences/food source of, 19
Vitamin K-deficiency bleeding (VKDB), 175
Vitamins. *see also* Herbal remedies/supplements; Minerals
  adolescent nutrition and, 365–367, 370–372, 389–391
  adult nutrition and, 417–420
  breastmilk and, 167, 215
  children/preadolescents and, 318, 324, 345
  description of, 13
  as essential nutrients, 3
  fertility and, 61
  food sources for, 18–22
  infant nutrition and, 229, 239–240, 251
  lactating women and, 180
  multifetal pregnancies and, 151
  older adult nutrition and, 472–474
  during pregnancy, 114–118, 126–127
  preterm delivery and, 102
  toddlers/preschoolers nutrition and, 278, 285–286, 290
  types of, 14–19
Vomiting. *see* Nausea and vomiting

W

Walford, Roy, 459
Water. *see also* Fluids
  breastmilk and, 165
  dehydration and, 513–514
  effects of caffeine on, 416–417
  as essential nutrient, 3, 22–23
  infant nutrition and, 238
  older adult nutrition and, 513–514
  during pregnancy, 114
Weaning, 236
Wear-and-tear theories of aging, 458
Web dietary assessment resources, 40
Weight. *see also* Diets; Obesity
  adolescent nutrition and, 359–360, 386–389
  adult nutrition and, 409–410
  adult status by race/ethnicity, 430
  aging and, 460–461
  children/preadolescents and, 313, 319–322
  diabetes and, 78
  eating disorders and, 395–403
  fertility and, 58–59, 73–74
  fetal growth and, 64
  infant growth and, 230–231
  loss during breastfeeding, 179
  milk and, 514
  multifetal pregnancy gain of, 150

  older adult nutrition and, 509–512
  pregnancy gain and, 104–107
Weight-control Information Network (WIN), 424
Weight cycling/yo-yo dieting, 410
Weight loss, 432, 434–435
Weight management
  cognitive behavioral therapy, 432–434
  exercise and, 434
  medical nutrition therapy, 432
  nutrition intervention for, 432
Wellness, 456
Wellstart International, 189
"What We Eat in America" (WWEIA), 289–290
What We Eat in America 2001-2002, 366
Whey proteins, 166
WIC program for nutrition
  adolescent nutrition and, 382
  breastfeeding and, 161
  defined, 42–43
  infant nutrition and, 263
  lactation and, 188
  preconception benefits of, 66
  pregnancy outcome and, 131
  resources, 133
  toddlers/preschoolers nutrition and, 292–293, 307
Widowhood, 514–515
Withdrawal reflex, 225
Women, Infants, and Children (WIC) program. *see* WIC program for nutrition
Work of breathing (WOB), 306
Workplace and breastfeeding support, 186–187
World Health Organization (WHO)
  on AIDS and breastfeeding, 216–217
  bone mass density and, 496–497
  breast milk substitutes and, 183
  growth standards, 271
  infant growth and, 231
  oral contraceptives and, 200
  and U.S. life expectancy, 44

X

Xerostomia, 500–501

Y

Yo-yo dieting, 410

Z

Zimmer, Paul, 444
Zinc
  breastmilk and, 168
  fertility and, 60
  food sources for, 29–30
  function/deficiency of, 24
  overdose consequences/food source of, 25

Dietary Reference Intakes (DRIs): Tolerable Upper Intake Levels (UL^a), Vitamins
Food and Nutrition Board, Institute of Medicine, National Academies

Life Stage Group	Vitamin A (μg/d)^b	Vitamin C (mg/d)	Vitamin D (μg/d)	Vitamin E (mg/d)^{c,d}	Vitamin K	Thiamin	Riboflavin	Niacin (mg/d)^d	Vitamin B$_6$ (mg/d)	Folate (μg/d)^d	Vitamin B$_{12}$	Pantothenic Acid	Biotin	Choline (g/d)	Carotenoids^e
Infants															
0–6 mo	600	ND^f	25	ND	ND	ND	ND	ND	ND	ND	ND	ND	ND	ND	ND
7–12 mo	600	ND	25	ND	ND	ND	ND	ND	ND	ND	ND	ND	ND	ND	ND
Children															
1–3 y	600	400	50	200	ND	ND	ND	10	30	300	ND	ND	ND	1.0	ND
4–8 y	900	650	50	300	ND	ND	ND	15	40	400	ND	ND	ND	1.0	ND
Males, Females															
9–13 y	1700	1200	50	600	ND	ND	ND	20	60	600	ND	ND	ND	2.0	ND
14–18 y	2800	1800	50	800	ND	ND	ND	30	80	800	ND	ND	ND	3.0	ND
19–70 y	3000	2000	50	1000	ND	ND	ND	35	100	1000	ND	ND	ND	3.5	ND
>70 y	3000	2000	50	1000	ND	ND	ND	35	100	1000	ND	ND	ND	3.5	ND
Pregnancy															
14–18 y	2800	1800	50	800	ND	ND	ND	30	80	800	ND	ND	ND	3.0	ND
19–50 y	3000	2000	50	1000	ND	ND	ND	35	100	1000	ND	ND	ND	3.5	ND
Lactation															
14–18 y	2800	1800	50	800	ND	ND	ND	30	80	800	ND	ND	ND	3.0	ND
19–50 y	3000	2000	50	1000	ND	ND	ND	35	100	1000	ND	ND	ND	3.5	ND

^{a}UL = The maximum level of daily nutrient intake that is likely to pose no risk of adverse effects. Unless otherwise specified, the UL represents total intake from food, water, and supplements. Due to lack of suitable data, ULs could not be established for vitamin K, thiamin, riboflavin, vitamin B$_{12}$, pantothenic acid, biotin, carotenoids. In the absence of ULs, extra caution may be warranted in consuming levels above recommended intakes.

^{b}As preformed vitamin A only.

^{c}As α-tocopherol; applies to any form of supplemental α-tocopherol.

^{d}The ULs for vitamin B, niacin, and folate apply to synthetic forms obtained from supplements, fortified foods, or a combination of the two.

^{e}β-Carotene supplements are advised only to serve as a provitamin A source for individuals at risk of vitamin A deficiency.

^{f}ND = Not determinable due to lack of data of adverse effects in this age group and concern with regard to lack of ability to handle excess amounts. Source of intake should be from food only to prevent high levels of intake.

SOURCES: *Dietary Reference Intakes for Calcium, Phosphorous, Magnesium. Vitamin D, and Fluoride* (1997); *Dietary Reference Intakes for Thiamin, Riboflavin, Niacin, Vitamin B$_6$, Folate, Vitamin B$_{12}$, Pantothenic Acid, Biotin, and Choline* (1998); *Dietary Reference Intakes for Vitamin C, Vitamin E, Selenium, and Carotenoids* (2000); and *Dietary Reference Intakes for Vitamin A, Vitamin K, Arsenic, Boron, Chromium, Copper, Iodine, Iron, Manganese, Molybdenum, Nickel, Silicon, Vanadium, and Zinc* (2001). These reports may be accessed via http://www.nap.edu.

Dietary Reference Intakes (DRIs): Tolerable Upper Intake Levels (UL[a]), Elements
Food and Nutrition Board, Institute of Medicine, National Academies

Life Stage Group	Arsenic[b]	Boron (mg/d)	Calcium (g/d)	Chromium	Copper (µg/d)	Fluoride (mg/d)	Iodine (µg/d)	Iron (mg/d)	Magnesium (mg/d)	Manganese (mg/d)	Molybdenum (µg/d)	Nickel (mg/d)	Phosphorus (g/d)	Potassium	Selenium (µg/d)	Silicon[c]	Sulfate[d]	Vanadium (mg/d)[e]	Zinc (mg/d)	Sodium (g/d)	Chloride (g/d)
Infants																					
0–6 mo	ND[f]	ND	ND	ND	ND	0.7	ND	40	ND	ND	ND	ND	ND	ND	45	ND	ND	ND	4	ND	ND
7–12 mo	ND	ND	ND	ND	ND	0.9	ND	40	ND	ND	ND	ND	ND	ND	60	ND	ND	ND	5	ND	ND
Children																					
1–3 y	ND	3	2.5	ND	1000	1.3	200	40	65	2	300	0.2	3	ND	90	ND	ND	ND	7	1.5	2.3
4–8 y	ND	6	2.5	ND	3000	2.2	300	40	110	3	600	0.3	3	ND	150	ND	ND	ND	12	1.9	2.9
Males, Females																					
9–13 y	ND	11	2.5	ND	5000	10	600	40	350	6	1100	0.6	4	ND	280	ND	ND	ND	23	2.2	3.4
14–18 y	ND	17	2.5	ND	8000	10	900	45	350	9	1700	1.0	4	ND	400	ND	ND	ND	34	2.3	3.6
19–70 y	ND	20	2.5	ND	10000	10	1100	45	350	11	2000	1.0	4	ND	400	ND	ND	1.8	40	2.3	3.6
>70 y	ND	20	2.5	ND	10000	10	1100	45	350	11	2000	1.0	3	ND	400	ND	ND	1.8	40	2.3	3.6
Pregnancy																					
14–18 y	ND	17	2.5	ND	8000	10	900	45	350	9	1700	1.0	3.5	ND	400	ND	ND	ND	34	2.3	3.6
19–50 y	ND	20	2.5	ND	10000	10	1100	45	350	11	2000	1.0	3.5	ND	400	ND	ND	ND	40	2.3	3.6
Lactation																					
14–18 y	ND	17	2.5	ND	8000	10	900	45	350	9	1700	1.0	4	ND	400	ND	ND	ND	34	2.3	3.6
19–50 y	ND	20	2.5	ND	10000	10	1100	45	350	11	2000	1.0	4	ND	400	ND	ND	ND	40	2.3	3.6

[a] UL = The maximum level of daily nutrient intake that is likely to pose no risk of adverse effects. Unless otherwise specified, the UL represents total intake from food, water, and supplements. Due to lack of suitable data, ULs could not be established for arsenic, chromium, silicon, potassium, and sulfate. In the absence of ULs, extra caution may be warranted in consuming levels above recommended intakes.

[b] Although the UL was not determined for arsenic, there is no justification for adding arsenic to food or supplements.

[c] The ULs for magnesium represent intake from a pharmacological agent only and do not include intake from food and water.

[d] Although silicon has not been shown to cause adverse effects in humans, there is no justification for adding silicon to supplements.

[e] Although vanadium in food has not been shown to cause adverse effects in humans, this data could be used to set a UL for adults but not children and adolescents.

[f] ND = Not determinable due to lack of data of adverse effects in this age group and concern with regard to lack of ability to handle excess amounts. Source of intake should be from food only to prevent high levels of intake.

SOURCES: Dietary Reference Intakes for Calcium, Phosphorous, Magnesium, Vitamin D, and Fluoride (1997); Dietary Reference Intakes for Thiamin, Riboflavin, Niacin, Vitamin B$_6$, Folate, Vitamin B$_{12}$, Pantothenic Acid, Biotin, and Choline (1998); Dietary Reference Intakes for Vitamin C, Vitamin E, Selenium, and Carotenoids (2000); Dietary Reference Intakes for Vitamin A, Vitamin K, Arsenic, Boron, Chromium, Copper, Iodine, Iron, Manganese, Molybdenum, Nickel, Silicon, Vanadium, and Zinc (2001); and Dietary Reference Makes for Water, Potassium, Sodium, Chloride, and Sulfate (2004). These reports may be accessed via hllp://www.nap.edu.